FOURTH EDITION

Introductory Statistics for
MANAGEMENT AND
ECONOMICS

James L. Kenkel
University of Pittsburgh

Duxbury Press
An Imprint of Wadsworth Publishing Company
An International Thomson Publishing Company

I(T)P®

Belmont • Albany • Bonn • Boston • Cincinnati • Detroit • London • Madrid • Melbourne
Mexico City • New York • Paris • San Francisco • Singapore • Tokyo • Toronto • Washington

This book is dedicated to Terri and Mom.

Editor: *Curt Hinrichs*
Editorial Assistant: *Cynthia Mazow*
Production Editor: *Susan L. Reiland*
Print Buyer: *Karen Hunt*
Permissions Editor: *Peggy Meehan*
Designer: *Susan M. C. Caffey; Brian Betsill for TECHarts*
Cover Designer: *Ellen Pettengell*
Copy Editor: *Christine M. Levesque*
Technical Illustrator: *Lori Heckelman*
Compositor: *G & S Typesetters, Inc.*
Printer: *R. R. Donnelley, Crawfordsville*
Cover: *William Conger,* New City II, *oil on wood, 12" × 12", 1994.*
    Courtesy of Roy Boyd Gallery, Chicago. Photo: Michael Tropea.

This book is printed on acid-free recycled paper.

COPYRIGHT © 1996
by Wadsworth Publishing Company
A Division of International Thomson Publishing Inc.
   I(T)P The ITP logo is a registered trademark under license.

Printed in the United States of America
1 2 3 4 5 6 7 8 9 10

For more information, contact:

Wadsworth Publishing
  Company
10 Davis Drive
Belmont, California 94002

International Thomson
  Publishing
Berkshire House 168-173
High Holborn
London, WC1V7AA England

Thomas Nelson Australia
102 Dodds Street
South Melbourne 3205
Victoria, Australia

Nelson Canada
1120 Birchmount Road
Scarborough, Ontario
Canada M1K 5G4

International Thomson
  Publishing GmbH
Königwinterer Strasse 418
53227 Bonn, Germany

International Thomson
  Publishing Asia
221 Henderson Road
#05-10
Singapore 0315

International Thomson
  Publishing — Japan
Hirakawacho-cho Kyowa
  Building, 3F
2-2-1 Hirakawacho-cho
Chiyoda-ku, 102 Tokyo
Japan

Library of Congress Cataloging-in-Publication Data

Kenkel, James L.
   Introductory statistics for management and economics / James L.
Kenkel.—4th ed.
    p.   cm.
   Includes bibliographical references and index.
   ISBN 0-534-20370-1 (casebound)
   1. Social sciences—Statistical methods. 2. Economics—
Statistical methods. 3. Commercial statistics. 4. Statistics.
I. Title.
HA29.K428 1995
519.5—dc20
                             95-34991

Preface

This text is written for students in business, economics, management, and related fields for use in an introductory statistics course. The entire book can be covered in two semesters, or selected chapters can be used in a one-semester course. The book evolved from over 25 years of teaching undergraduate and graduate students in economics, statistics, and econometrics courses at the University of Pittsburgh. The purpose of this book is to give students a background in statistics that will enable them to apply statistical techniques when solving real-world problems and making decisions. Emphasis is placed on explaining statistical concepts in an applied setting. I have made a special effort to use interesting examples that are relevant to students pursuing careers in business and economics. The text assumes a background in college algebra; calculus is not required.

The text emphasizes model building and applications in a real-world economic and business setting. It contains more material on building, testing, and interpreting regression models than the competition because I believe these subjects are of greatest importance.

Changes in the Fourth Edition

Approximately 30% of the material in the fourth edition is new, and parts of every chapter have been rewritten with a greater emphasis on applications. Some of the major changes in the fourth edition are as follows:

1. There is a reduced emphasis on probability theory and an increased emphasis on applications. Certain topics such as quality control have received much greater emphasis.

2. The exercises for each section have been changed extensively. Each set of exercises is separated into three sections called Statistical Concepts, Statistical Drills, and Statistical Applications. The exercises on Statistical Concepts ask theoretical questions to test a student's knowledge of the basic concepts covered in the section. The answers to most of these questions can be found directly in the text. The purpose of the Statistical Concepts questions is to force the student to think carefully about the statistical concepts introduced in each section and, if necessary, to influence the student to reread the material a second time (and more carefully with greater attention to details). A typical Statistical Concepts question would be:

Can R^2 ever decrease when an additional explanatory variable is added to a regression model?

The correct answer is "No." The answer is simple *if and only if* the student understands the material; an incorrect answer shows that the student does not fully understand the material. The Statistical Drills exercises provide questions that require the students to perform the basic mathematical and statistical calculations presented in each section. The purpose of these exercises is to require the students to actually use the statistical formulas and to train them to be careful and precise in performing their work. Incorrect answers here could result from incorrect understanding of the material, arithmetic mistakes, or sloppiness. Finally, the exercises on Statistical Applications require the students to apply and interpret the material covered in each section while solving real-world problems, many with the aid of a computer software package.

3. Most of the exercises in the Statistical Concepts and Statistical Drills sections are entirely new. In addition, numerous new Statistical Applications exercises have been added, many of which involve larger data sets and place an emphasis on quality control applications in business situations. Many of these problems are designed to be solved by using statistical packages on a computer.

4. Nearly every chapter contains a section discussing computer applications of the material covered in the chapter. Data disks have been prepared that contain the data used in the examples and exercises so that students can reproduce the examples and solve the exercises by using various statistical packages on the computer.

5. New Case studies have been added to nearly every chapter. The case studies discuss real-world statistical problems and applications that I have analyzed in more than 20 years as an economic and statistical consultant for corporations, government entities, and small business, and in legal cases involving forensic economics.

6. In the chapters that discuss descriptive statistics, greater emphasis has been placed on understanding the statistical concepts and applications and less emphasis has been placed on performing numerical calculations.

7. In response to suggestions by reviewers, the coverage of the formal theory of probability has been reduced and emphasis on applications has been increased.

8. New material involving quality control has been added to nearly every chapter.

9. In Chapter 9, the discussion of sampling distributions has been entirely rewritten. In this edition, the notion of a sampling distribution is introduced by relying on a business context rather than by relying on pure statistical theory. This approach will give students a better understanding of what is meant by a sampling distribution.

10. Much of the material in Chapter 10 on confidence intervals and Chapter 11 on hypothesis testing has been rewritten. As in Chapter 9, the explanations have become less theoretical and more oriented toward business applications. In each of these chapters, the basic concepts are explained via examples in a business context.

11. In Chapter 17, the section discussing dummy variables has been rewritten.

12. Chapter 19 concerning time series analysis has been rewritten and expanded using new data.

13. Two-parameter exponential smoothing has been added to Chapter 22.

14. Chapter 23 is entirely new and introduces the basic concepts, methods, and applications of statistical quality control.

Organization of the Book

The material in Chapters 1–11 is best covered in the order presented. Chapter 1 provides a brief introduction to what statistics is all about and provides a real-world example of how I used statistics in a case in which I served as a consultant. Hopefully, the material in Chapter 1 will whet the student's interest to learn more about statistics by showing how statistical techniques can be used to solve real-world business and economics problems.

Chapter 2 discusses data collection and sampling theory and explains why we frequently have to rely on samples to get information about the characteristics of some population. The chapter discusses many of the biases that can occur if we do not take random samples. As examples, Chapter 2 contains a discussion of two legal cases involving millions of dollars in potential damages where the issues discussed in the chapter were actually applied.

Chapters 3 and 4 provide the basics of descriptive statistics with an emphasis on determining the shape of a frequency distribution, its center of gravity, and some measure of its dispersion. Some new exercises contain large data sets that should be analyzed by using a computer.

Chapters 5 and 6 cover elementary probability theory and discrete probability distributions. For instructors who do not have the inclination (or time) to cover probability theory in detail, some of the material in these chapters can be omitted without interrupting the flow of ideas. For example, some instructors may choose to skip the material on Bayes' theorem, or on permutations and combinations. In Chapter 6, the expected value and standard deviation of a discrete random variable are emphasized. Expected value is illustrated by referring to common economic notions such as the average rate of return on an investment in the investment industry, the average rate of return from a client in the insurance industry, and the average rate of profit in a business enterprise. The standard deviation is illustrated as a measure of risk in an investment situation.

Chapter 7 covers two important discrete probability distributions—the binomial distribution and the Poisson distribution. Numerous examples illustrate how these distributions can be used when analyzing business problems.

Chapter 8 provides an introduction to continuous probability distributions and covers the uniform and normal distributions. The normal distribution, particularly the standard normal distribution, receives special emphasis because of its central role in statistical theory. All instructors should cover Chapter 8 in detail.

Chapters 9 and 10 cover sampling distributions, estimation theory, and the construction of confidence intervals for means, proportions, and variances. Much of the material in these chapters has been rewritten to make clearer the concepts of a sampling distribution and a confidence interval. These are two of the most important concepts in elementary statistics, so special care was taken to develop economic examples that illustrate the concepts in a clear and concise manner. Instructors will find the new discussions of sampling distributions and confidence intervals superior to other approaches. The Student t distribution is introduced in Chapter 10 and is used in the construction of a confidence interval for the population mean. New exercises involving large data sets have been added; they should be analyzed using a computer.

Chapter 11 provides a comprehensive discussion of hypothesis testing concerning a population mean, a population proportion, and a population variance. Greater emphasis has been placed on the interpretation of p-values and test results. This text contains a thorough discussion of Type II errors and the calculation of the power of a test. The chi-square distribution is introduced in Chapter 11 and is used to test hypotheses and construct confidence intervals concerning a population variance. The material concerning the power of the test can be omitted without loss of continuity.

Chapter 12 is a logical extension of Chapter 11 and covers two tests involving two means or two proportions. The F distribution is introduced in Chapter 12 and is used to test hypotheses concerning the equality of two population variances and to construct confidence intervals concerning the ratio of two population variances.

Chapter 13 covers chi-square goodness-of-fit tests and tests of independence. I view Chapter 13 as a logical extension of Chapter 12, since the chi-square tests discussed in Chapter 13 can be viewed as tests of hypotheses involving more than two proportions. This chapter can be omitted without loss of continuity, and some instructors may prefer to cover this material in conjunction with Chapter 21, which covers some nonparametric tests.

Chapter 14 covers one-way analysis of variance and is another logical extension of Chapter 12. In one-way ANOVA, we are testing hypotheses involving more than two population means.

Chapters 15–18 follow a logical sequence in discussing regression and correlation analysis in progressively more detail. Chapter 15 presents a discussion of the simple linear regression model and shows how to obtain the least squares estimates of the slope and intercept of the simple linear regression model. Calculation and interpretation of R^2 as a measure of goodness of fit is emphasized.

Chapter 16 is a logical extension of Chapter 15 and treats the multiple regression model with an emphasis on the analysis and interpretation of computer results. Chapter 17 covers various special topics in regression analysis, including polynomial regression models, the use and interpretation of dummy variables, logarithmic regression models, and stepwise regression. Emphasis is placed on analyzing and interpreting computer results.

Chapter 18 provides an introduction to the elementary econometric techniques for testing the adequacy of assumptions underlying the classical regression model. Analysis of the residuals in the regression model and interpretation of computer results are emphasized.

Chapters 19 and 20 treat model building using time series data. Much of Chapter 19 has been rewritten and emphasizes estimation of the trend component in time series models using polynomial regression and logarithmic regression techniques. Chapter 20 covers estimation of the seasonal component in time series models. New material has been added concerning the exponential smoothing model, forecasting, and measurement of mean forecast error.

Chapters 21–23 are relatively independent of one another and can be treated in any order. Chapter 21 covers various nonparametric tests.

Chapter 22 covers basic decision theory and emphasizes different criteria that can be used to determine an optimal decision.

Chapter 23 is entirely new and covers quality control in the manufacturing and business environment. The chapter emphasizes the idea of a process as an ongoing procedure that produces output over time. Process capability and improvement are discussed.

Supplements

A complete set of supplementary materials is available to support the text.

For the instructor, there is an Instructor's Resource Book containing a **complete solutions manual** and a set of **transparency masters** derived from important art in the text. In addition, the publisher offers to adopters of the text a **test bank file** available in either printed form or in a computer format.

The student edition of JMP (SAS Institute) and StataQuest for both Windows and Macintosh environments are available from the publisher. A data disk containing all the data appearing in the examples and exercises is available with new books upon request. A student's solutions manual, which contains complete solutions to all exercises with answers in the appendix of the text, along with additional odd-numbered exercises (including computer applications), is available for sale to students.

Acknowledgments

I am indebted to the thousands of students who class-tested my material in its earlier versions. I especially want to thank Dr. Terri Gollinger, who reviewed much of the document in its draft stages and made numerous helpful suggestions.

I appreciate the assistance of the many reviewers who provided useful suggestions for this revision. They include Mohamed H. Askalani, Mankato State University; Kamvar Farahbod, Arkansas State University; Ron Hook, University of Northern Iowa; Winston T. Lin, State University of New York, Buffalo; Ata Nahouraii, Indiana University of Pennsylvania; Thomas J. Page, Michigan State University; Roxy Peck, California Polytechnic State University, San Luis Obispo; Mohammed A. Quasem, Howard University; and Toni M. Somers, Wayne State University.

Special thanks to Susan Reiland, who managed the production of the book and provided hundreds of useful suggestions, and to Carol Reitz, whose proofreading skills are extraordinary.

I also want to thank the editorial staff of Duxbury Press, including my editor Curt Hinrichs, project development editor Jennifer Burger, and editorial assistant Cynthia Mazow.

Finally, I wish to thank my wife, Terri, and my kids, Julie and Tim, for their continued assistance and support.

—**James L. Kenkel**

Contents

1   *What Statistics Is All About*   *1*

   1.1   The Role of Statistics   1
           Case Study: A Study Using Descriptive Statistics   4
   1.2   A Word About Computers   6

2   *Data Collection and Sampling Theory*   *7*

   2.1   Populations, Samples, and Types of Data   7
   2.2   Reasons for Sampling and Sources of Bias   13
   2.3   Types of Samples   19
   2.4   Computer Analysis of Multivariate Data Sets   22
           Case Study: Firesafing Problems at the Mellon Bank Building   37
           Case Study: Internal Revenue Service vs. the Pittsburgh Press Club   38
           Supplementary Exercises   39

3   *Summarizing Data in Tables and Graphs*   *42*

   3.1   Frequency Distributions and Histograms   43
   3.2   Common Shapes of Distributions   59
   3.3   Lying with Statistics (Optional)   64
   3.4   The Stem-and-Leaf Diagram (Optional)   65
   3.5   Computer Applications   68
           Supplementary Exercises   73

4   *Summary Statistics: Measures of Location and Dispersion*   *77*

   4.1   Summation Notation   78
   4.2   The Mean, the Median, and the Mode   81
   4.3   Quartiles and Percentiles   91

4.4   Measures of Dispersion   **95**

4.5   The Empirical Rule and Standardized Scores   **108**

4.6   Construction of Box Plots (Optional)   **115**

4.7   Computer Applications   **119**

Case Study: Paint Problems on the Mellon Bank Building   **122**

Supplementary Exercises   **123**

5   *Introduction to Probability*   *133*

5.1   Experiments, Outcomes, Events, and Sample Spaces   **133**

5.2   Assigning Probabilities to Events   **136**

5.3   Some Basic Rules of Probability   **142**

5.4   Probabilities of Compound Events   **145**

5.5   Conditional Probability   **153**

5.6   Independence   **161**

5.7   Bayes' Theorem (Optional)   **167**

5.8   Counting Techniques   **172**

5.9   Computer Applications   **179**

Supplementary Exercises   **182**

6   *Discrete Probability Distributions*   *192*

6.1   Random Variables   **192**

6.2   Properties of Discrete Probability Distributions   **195**

6.3   Cumulative Distribution Function   **199**

6.4   Expected Value of a Discrete Random Variable   **202**

6.5   Variance of a Discrete Random Variable   **208**

6.6   Computer Applications   **213**

Supplementary Exercises   **214**

7   *Some Important Discrete Distributions*   *220*

7.1   The Binomial Distribution   **220**

7.2   The Poisson Distribution   **231**

Supplementary Exercises   **239**

8   *Some Useful Continuous Probability Distributions*   *248*

8.1   Properties of Continuous Probability Distributions   **248**

8.2   The Uniform Distribution   **254**

8.3   The Normal Distribution   **258**

8.4   Calculating Areas Under Any Normal Curve   **266**

8.5   The Normal Distribution As an Approximation to the Binomial Distribution   **272**

Case Study: Examining Loss of Intelligence As a Result of an Injury   **278**

Supplementary Exercises   **279**

9   *Sampling Theory and Some Important Sampling Distributions*   **288**

9.1   Introduction to Sampling Distributions   **289**

9.2   Sampling Distribution of the Sample Mean   **299**

9.3   The Central Limit Theorem   **310**

9.4   Sampling Distribution of the Difference Between Two Sample Means   **316**

9.5   Sampling Distribution of the Sample Proportion   **321**

9.6   Sampling Distribution of the Difference Between Sample Proportions   **327**

9.7   Computer Applications   **330**

Case Study: Examining the Sulfur Content of Coal   **331**

Supplementary Exercises   **332**

10   *Estimating and Constructing Confidence Intervals*   **339**

10.1   Interval Estimation   **339**

10.2   Confidence Intervals for the Mean with Known Population Variance   **342**

10.3   Student's *t* Distribution   **352**

10.4   Confidence Intervals for the Mean with Unknown Population Variance   **357**

10.5   Confidence Intervals for Proportions (Large Samples)   **366**

10.6   Determining the Sample Size   **372**

10.7   Confidence Intervals for the Difference of Two Means   **377**

10.8   Confidence Intervals for the Difference of Two Population Proportions   **381**

10.9   Computer Applications   **383**

Supplementary Exercises   **384**

11   *Hypothesis Testing*   **391**

11.1   Concepts of Hypothesis Testing   **392**

11.2   Testing Hypotheses About a Population Mean When Variance Is Known   **405**

11.3   *p*-Values: Interpretation and Use   **419**

11.4   Testing Hypotheses About a Population Mean with Large Sample Sizes   **425**

11.5   Testing Hypotheses About the Mean of a Normal Population with Unknown Variance   **428**

11.6   Tests of the Population Proportion   **433**

11.7   Measuring the Power of a Test (Optional)   **438**

11.8   The Chi-Square Distribution   **461**

11.9   Testing Hypotheses and Constructing Confidence Intervals
About a Population Variance   **464**

11.10   Computer Applications   **469**

Supplementary Exercises   **470**

12   *Tests of Hypotheses Involving Two Populations*   **476**

12.1   Tests for the Differences of Means   **476**

12.2   Tests of Differences of Means Using Small Samples from Normal Populations
When the Population Variances Are Equal but Unknown   **480**

12.3   Tests for Differences of Means of Paired Samples   **486**

12.4   Tests Concerning Differences of Proportions   **490**

12.5   The *F* Distribution   **493**

12.6   Testing Hypotheses and Constructing Confidence Intervals
for Two Population Variances   **497**

12.7   Computer Applications   **501**

Case Study: Paint Problems on the Mellon Bank Building   **502**

Supplementary Exercises   **503**

13   *Chi-Square Tests*   **509**

13.1   The Chi-Square Goodness-of-Fit Test   **509**

13.2   Tests of Independence and Contingency Tables   **520**

13.3   Computer Applications   **530**

Case Study: Testing for Age Discrimination   **534**

Supplementary Exercises   **537**

14   *Analysis of Variance*   **546**

14.1   The One-Factor ANOVA Model   **546**

14.2   The Statistical Model for One-Way ANOVA   **565**

14.3   Computer Applications   **574**

Supplementary Exercises   **577**

15   *Regression and Correlation*   **584**

15.1   Stochastic Relationships and Scatter Diagrams   **584**

15.2   The Simple Linear Regression Model   **593**

15.3   Method of Least Squares   **597**

15.4   Explanatory Power of a Linear Regression Equation   **609**

15.5   Estimating Variances of Estimated Coefficients in the Regression Model   **615**

15.6   Hypothesis Testing in the Linear Regression Model   **621**

15.7   Confidence Intervals for the Regression Coefficients   **629**

15.8   Prediction Using the Regression Model   **634**

15.9   The Correlation Coefficient   **641**

15.10   Computer Applications   **655**

Case Study: Paint Problems on the Mellon Bank Building   **658**

Supplementary Exercises   **659**

16   *Multiple Regression Models*   *667*

16.1   Models with Two Explanatory Variables   **667**

16.2   The General Multiple Regression Model   **676**

16.3   Measuring Goodness of Fit   **683**

16.4   Confidence Intervals and Tests of Hypotheses Concerning the Regression Coefficients   **687**

16.5   The Multicollinearity Problem   **696**

16.6   Computer Applications   **701**

Case Study: Estimating the Value of Coal Used by a Nationwide Steel Company   **704**

Supplementary Exercises   **705**

17   *Special Topics in Multiple Regression Analysis*   *709*

17.1   Models Involving Polynomials   **709**

17.2   Dummy Variables in Regression Models   **714**

17.3   Estimating Equations in Logarithmic Form   **734**

17.4   Stepwise Regression   **738**

17.5   The Correlation Matrix   **745**

17.6   Computer Applications   **751**

Supplementary Exercises   **752**

18   *Residual Analysis and Violations of the Basic Assumptions*   *757*

18.1   Searching for Model Inadequacies   **757**

18.2   Violations of Basic Assumption 1: Nonnormal Errors   **761**

18.3   Violations of Basic Assumption 2: Nonzero Mean   **762**

18.4   Violations of Basic Assumption 3: Heteroscedasticity   **772**

18.5   Violations of Basic Assumption 4: Serial Correlation   **784**

18.6   Violations of Basic Assumption 5: Correlation Between
Errors and Explanatory Variables   **801**

18.7   Interpreting the Results of a Regression Model   **803**

18.8   Computer Applications   **805**
Supplementary Exercises   **807**

19   *Time Series Analysis I: Estimation of the Trend Component*   *813*

19.1   Components of a Time Series   **813**

19.2   The Linear Trend Model   **825**

19.3   The Polynomial Trend Model   **836**

19.4   The Exponential Trend Model   **841**

19.5   Autoregressive Forecasting Models   **846**

19.6   Measuring Forecast Accuracy   **851**

19.7   Computer Applications   **858**
Case Study: Estimating Lost Profits for a Regional Electrical Contractor   **860**
Supplementary Exercises   **862**

20   *Time Series Analysis II: Estimation of the Seasonal Component*   *867*

20.1   Seasonal Adjustment by Using Moving Averages   **867**

20.2   Seasonal Adjustment by Using Dummy Variables   **877**

20.3   Seasonal Adjustment by Using the Ratio-to-Trend Method   **883**

20.4   Exponential Smoothing   **888**
Supplementary Exercises   **898**

21   *Some Nonparametric Tests*   *901*

21.1   The Sign Test   **902**

21.2   The Mann–Whitney Test   **906**

21.3   The Wilcoxon Signed-Rank Test   **912**

21.4   The Kruskal–Wallis Test   **918**

21.5   The Runs Test   **922**

21.6   Rank Correlation   **928**

21.7   Computer Applications   **934**
Supplementary Exercises   **939**

22   *Introduction to Statistical Decision Theory*   *945*

22.1   Payoff Tables and Opportunity Loss Tables   **945**

22.2   Criteria for Making Decisions   **952**

22.3   Utility Theory and the Expected Utility Criterion   **962**

22.4   Decision Tree Analysis   **971**

Supplementary Exercises   **975**

23   *Quality Control*   *978*

23.1   Introduction   **978**

23.2   Some Patterns That Reveal Assignable Causes of Variation   **984**

23.3 $\bar{x}$-Chart to Monitor the Process Mean   **990**

23.4 R-Chart to Monitor Process Variation   **1011**

23.5 p-Chart to Monitor Proportion of Defectives   **1019**

Supplementary Exercises   **1032**

Appendix A   *Statistical Tables*   *1036*

Appendix B   *Answers to Selected Exercises*   *1065*

Index   *1091*

1 What Statistics Is All About

1.1 ## The Role of Statistics

The word *statistics* has two different meanings. More commonly, *statistics* means a collection of numerical facts or data, such as stock prices, profits of firms, annual incomes of college graduates, and so forth. In its second meaning, *statistics* refers to an academic discipline, just as physics, biology, and computer science are academic disciplines.

DEFINITION **Statistics**

Statistics is a branch of mathematics that consists of a set of analytical techniques that can be applied to data to help us make judgments and decisions in problems involving uncertainty.

Statistics is a scientific discipline consisting of procedures for collecting, describing, analyzing, and interpreting numerical data. One of the main objectives of statistics is to provide a set of procedures that enables us to make inferences, predictions, and decisions about the characteristics of a population of data based on the information obtained from only a part of the population.

The field of statistics can be divided into two parts, descriptive statistics and inferential statistics. Years ago, the study of statistics consisted mainly of the study of methods for summarizing and describing numerical data. This study has become known as *descriptive statistics* because it primarily describes the characteristics of large masses of data. In many such statistical studies, data are presented in a convenient, easy-to-interpret form, usually as a table or graph. Such studies usually describe certain characteristics of the data, such as the center and the spread of the data.

DEFINITION **Descriptive Statistics**

Descriptive statistics consists of procedures for (1) tabulating or graphing the general characteristics of a set of data and (2) describing some characteristics of this set, such as measures of central tendency or measures of dispersion.

The methods used to collect statistical data often require knowledge of sampling theory and experimental design; this topic is discussed in Chapter 2. Once the data have been collected, descriptive statistics can be used to summarize the raw data in a simple, meaningful way. Often this is done by grouping the data into classes, thus forming what are called **frequency distributions.** When data are grouped into classes or categories, it becomes easy to see where most of the values are concentrated, but some information or detail may be lost due to the grouping. Methods of summarizing data by using tables and graphs are discussed in Chapter 3. In Chapter 4, we discuss ways of summarizing the characteristics of data sets by using measures of central tendency and measures of dispersion.

The second branch of statistics is called *inferential statistics.* Whereas descriptive statistics describes characteristics of the observed data, inferential statistics provides methods for making generalizations about the population based on the sample of observed data.

DEFINITION   **Inferential Statistics**

Inferential statistics consists of a set of procedures that helps us make inferences and predictions about a whole population based on information from a sample of the population.

For example, many manufacturers have established quality control programs designed to monitor the uniformity of output. Successive products will never be exactly identical; they will differ slightly because of various random factors. For example, no two bottles of medicine contain exactly the same amount of ingredients, no two components of a fine watch have exactly the same weight, and no two car doors have exactly the same dimensions. For many products, small deviations from the characteristics specified in the product design may be tolerable, but large variations in product uniformity are unacceptable. For drugs and medicines, tiny deviations in the quantities of various chemical components from those specified in the product design may not hinder the effectiveness of the product, but large deviations from the intended quantities may make the medicine useless or even harmful. Every dimension of some commodity, such as a car door, or every ingredient of a medical product has an average value over time, as well as a typical amount of variation over time. The mean level of the variable being analyzed can change as a result of many factors: Perhaps the machinery is becoming worn, so the dimensions of the product are slowly becoming larger; perhaps a valve is sticking, so the amount of a certain chemical being introduced into a medical product is slowly increasing. Also, the typical amount of variability in the process can change: Perhaps the labor force is becoming tired and bored and, as a result, product uniformity decreases; perhaps a supplier begins to send lower-quality raw materials, which increases the variability of product output. These and numerous other factors can affect the average and the variability of any output variable.

Quality control engineers are interested in detecting these possible changes in the average level of output and in the typical variability of output. Early detection of process changes can help reduce, or totally eliminate, the production of defective output. Improvements in the production process can help prevent loss of consumer goodwill and lead to lower production costs, lower repair costs, and increased profits.

To monitor any process, quality control engineers take samples of output at regular time intervals and collect data on the relevant variables. This requires some knowledge of sampling theory. Various decisions have to be made about how the samples should be collected: How large should the samples be? When should they be obtained? Next, the methods of descriptive statistics are applied to describe the important characteristics of the sample data. For each sample, the average value (called the sample mean) is calculated along with various measures of product variability. (Two measures of variability are the standard deviation and the range of the variable being examined.) Frequency distributions show the distribution of the output variable: Within what range do the values fall? How many values fall outside the minimum and maximum values indicated in the product specifications?

The set of data for each sample is then examined to see whether the process is deteriorating over time. If it is, the source of the problem should be found and eliminated. The quality control engineer uses the methods of inferential statistics to make inferences about the characteristics of the entire population of process output and to test hypotheses about potential changes in the mean level of some output variable or in the variability of process output. Does it appear that the mean level of some variable is changing over time? Is the variability of the process changing over time? Answering these questions requires some knowledge of probability theory and statistical theory, topics that are covered in detail later in this book.

We illustrate the application of statistical techniques to real-world business problems with another example. Before building a large shopping center in a certain location, a real estate developer wants the following information about the local community: the income and age distribution of residents in the area; their annual expenditures on items such as jewelry, clothing, sporting goods, and books; the proportion of residents who dine in expensive restaurants; and so forth.

To get this information, the developer would probably rely on a sample of data obtained by interviewing some residents. The developer would then organize the data so that general characteristics would become evident. Using the methods of descriptive statistics, the developer would construct tables that show how many of the observations lie in a specific interval or category. For each relevant variable, an average would be calculated—for example, the average annual income in the community, the average age, the average annual expenditure on sporting goods, and so forth.

By relying on this sample information, the developer might make inferences or predictions about the characteristics of the entire population of people in the community. In general, however, the characteristics of the sample of residents will not exactly match the characteristics of all the residents in the community. Thus, it is always possible to make an error when ascribing properties to an entire population based on a sample of observations. The application of inferential statistics can help us estimate the magnitude of any estimation error that might be made. Inferential statistics depends on probability theory.

Examples of Studies Using Descriptive Statistics

In the business world, a wide variety of interesting and important problems arise that require the application of statistical techniques presented in this book. For more than 20 years, I have served as an economic and statistical consultant for numerous government agencies and for more than 100 different local, national, and international corporations. In this work, I have encountered statistical applications arising from legal disputes over breaches of contract, sometimes involving hundreds of millions of dollars. Examples of legal disputes include alleged age, race, or gender discrimination in employment situations involving hiring, promotions, layoffs, pay raises, and so forth. Throughout this text, we will use some of these real cases to show how statistics can be applied in the business world.

STATISTICS IN ACTION: CASE STUDY

A Study Using Descriptive Statistics

In the early 1980s, Pioneer Supply Company, a firm in Venango County, Pennsylvania, applied for a government patent on a newly designed pump jack with an articulating beam. (A pump jack is the mechanical device that sits above an oil well and lifts the oil from the underground reservoir. When operating, the pump jack looks roughly like a playground seesaw.) This pump jack had a unique articulating balance beam that greatly reduced the amount of electrical power required to operate the jack, thus decreasing the costs of recovering oil and increasing the profits derived from the oil well.

Pioneer intended to produce different models of pump jacks for wells of varying depths. In general, as the depth of the well increases, a larger (and more expensive) pump jack is required. Pioneer's jacks were designed primarily for use on wells under 5,000 feet deep. The management of Pioneer purchased motors to drive the pump jacks from an English engineering firm. The English firm guaranteed Pioneer that the motors would be strong enough to operate the pump jacks and that the motors would have a long life. Pioneer inserted these motors in its pump jacks and, for approximately 6 months, sales of the pump jacks skyrocketed. The articulating beam pump jacks appeared to be far superior to any competing pump jacks. Suddenly, however, many of the pump motors began to break. Pioneer claimed that the motors failed because they were too small and were not designed correctly. Furthermore, Pioneer produced evidence showing that

an engineer at the English firm that manufactured the motors had alerted his bosses about the potential problem. Pioneer alleged that the management of the English firm disregarded this information and produced the motors anyway. As the motors began to fail in the field, Pioneer was inundated with numerous complaints and encountered extensive repair costs. The company's reputation was destroyed and, as a result of the fiasco, it soon went bankrupt and closed.

When management of Pioneer filed a lawsuit against the English firm, they needed an estimate of the economic damages resulting from the English firm's failure to produce high-quality pump jack motors. The economic losses suffered by Pioneer are its potential lost profits, both past and future, reduced to present value. Potential lost profits are the difference between potential sales revenue and potential costs. Potential sales revenue in any year depends on the potential selling price of the pump jack and the quantity of pump jacks that could be sold.

One aspect of the problem concerned estimating potential sales of different sizes of pump jacks. To make these estimates, data had to be gathered showing the depths of oil wells in many counties in the Appalachian Basin where the pump jack was likely to be sold. Government records showed every drilling permit filed in the previous few years, including the location and depth of each well. Various records showed the depth of every well drilled in

Venango County. (This set of data is called a population because it includes every observation of interest in Venango County.) In addition, some of the wells drilled in other counties adjacent to Venango County were selected randomly and examined. (This set of data is called a sample because it includes only a portion of all possible observations.)

A list was constructed showing the depths of several thousand oil wells in and around Venango County. (These observations are called raw data because no grouping techniques or statistical procedures have been applied to them.) Because there were so many observations, it was necessary to summarize the data in some way. This is the purpose of descriptive statistics.

Eventually, a report containing many tables and graphs was written. A table was created for each county. For example, one of the tables contained information only on the wells drilled in Venango County. The table showed how many wells were under 500 feet deep, how many were between 500 and 1,000 feet, how many were between 1,000 and 1,500 feet, and so forth. (This table is an example of a frequency distribution.) In addition, a table showing the proportion of wells in Venango County drilled in each depth category was constructed. (This table is called a relative frequency distribution because it shows the relative frequency, or proportion, of observations in each interval.) Similar frequency distributions and relative frequency distributions were created for each of the counties in the survey.

To compare the depths of wells in different counties, the average depths of the wells drilled in each county were calculated. (These averages are called sample means when they are based on a sample of data and population means when they are based on a population of data. They provide a measurement of the middle or center of the data.) By examining the mean, or average, depth of wells in a county, it was possible to determine where Pioneer's pump jacks might sell best.

In some counties the depths of the wells varied considerably, whereas in other counties they varied only slightly. Because of this, estimates were calculated indicating the amount of variability in well depths in each county. One number used to measure the amount of variation, or dispersion, present in a set of data is called the standard deviation. If the standard deviation is large, then the observations tend to be widely dispersed; if it is small, the observations tend to be highly concentrated about a central value.

The mean and standard deviation are two examples of what are called summary statistics. A summary statistic describes some characteristic of the data as a single number. For example, there are various ways of measuring the center of a set of data and the amount of dispersion in the set. Methods of describing these and other general characteristics of a set of observations are part of the study of descriptive statistics, discussed in Chapters 3 and 4.

After months of litigation, Pioneer and the English firm reached an out-of-court settlement in which the owners of Pioneer received several million dollars. The amount of the settlement was far less than the total amount of damages projected. From Pioneer's point of view, it probably was reasonable to accept the settlement. By doing so, Pioneer was avoiding the risk that it may have received nothing if the case went to trial and Pioneer was unable to prove its case.

Deductive and Inductive Statistics

The field of statistics can also be divided into deductive statistics and inductive statistics. In **deductive statistics,** we deduce the properties of a sample of observations from the known characteristics of the population. In **inductive statistics,** we reverse the procedure—we start with a sample of observations and try to draw general conclusions about the properties of the population based on the characteristics of the sample.

DEFINITIONS   **Deductive Statistics and Inductive Statistics** ——————————————————

In **deductive statistics,** we try to deduce properties of a given sample based on known characteristics of the population. In **inductive statistics,** we infer properties of a population based on observations of a selected sample of the population.

———

The field of deductive statistics is often referred to as *probability theory.* As an example of a problem involving probability theory, suppose that an insurance agent has data showing that about 1% of the population will be involved in an accident this year. Thus, the agent knows something about the entire population. Now suppose that the agent insures a sample of 50 people. The agent might want to know the probability that no people in this particular sample will be involved in a serious accident. In this example, the agent is trying to use probability theory to deduce what is likely to be observed in a sample when given information about the characteristics of the entire population.

In Chapters 5 and 6, we will study probability theory or deductive statistics. In Chapters 7 and 8, we will study some discrete and continuous probability distributions and lay the foundation for the study of inferential statistics. In the remaining chapters of the book, we will reverse the direction of inquiry and study inductive statistics.

1.2   A Word About Computers

The widespread availability of computers has dramatically changed how statistics is taught in universities and how statistical methods are used in business. With the use of a statistical software package, it is relatively easy to store and manipulate large data sets. Statistical calculations that were extremely tedious and time-consuming a few years ago can now be performed in seconds by inserting a few lines into a computer program. Throughout this textbook, various applications of statistical methods will be illustrated for three widely used statistical packages: SPSSX (Statistical Package for the Social Sciences), SAS (Statistical Analysis System), and Minitab (a statistical software package designed specifically for students). At the end of nearly every chapter, a brief section on Computer Applications shows how these three computer packages can be used to perform many of the calculations discussed in the text.

References

American Statistical Association and Institute of Mathematical Statistics. *Careers in Statistics.* Washington, D.C.: 1974.

Owen, Donald B. *Handbook of Statistical Tables.* Reading, Mass.: Addison-Wesley, 1962.

Tanur, Judith M., Frederick Mosteller, William H. Kruskal, Richard F. Link, Richard S. Pieters, and Gerald R. Rising, eds. *Statistics: A Guide to the Unknown.* 2d ed. San Francisco: Holden-Day, 1978. Prepared by a joint committee of the American Statistical Association and the National Council of Teachers of Mathematics.

U.S. Department of Labor, Bureau of Labor Statistics. *BLS Handbook of Methods,* 2 vols. Bulletins 2134-1 and 2134-2. Washington, D.C.: U.S. Government Printing Office, December 1982. Explains how the BLS obtains and prepares its economic data. Volume 1 contains this information for all BLS programs except the Consumer Price Index, which is described in Volume 2.

2 Data Collection and Sampling Theory

Before summarizing the characteristics of a set of data, we must collect the data. In this chapter, we will discuss different types of data as well as some of the purposes of collecting data.

2.1 Populations, Samples, and Types of Data

Population and Sample

A statistical **population** is the set or collection of all possible observations of some specific characteristic. A **sample** is a portion of a population.

One of the main objectives of statistics is to make inferences about a population based on information contained in a sample of that population. The method used to select a sample from the population is called the *sampling design*. The main objective of the sampling design is to provide guidelines for selecting a sample that will provide a specific amount of information about the population at a minimum cost.

Sampling Design

The **sampling design** is the set of decisions that must be made before the data are collected.

If the elements in the population are relatively uniform, then almost any small sample will provide acceptable results. For example, doctors need to test only a few drops of blood to determine a person's hemoglobin count because, for a given individual, one drop of blood generally provides the same information as any other drop of blood.

When the elements in the population are not relatively uniform, we must be careful in determining how to obtain a sample of data. We want to obtain a sample that is representative of *all* the elements in the population, not just some subsector of the population. For example, when trying to determine people's attitudes about an issue such as drug use, it is important to represent the opinions of individuals of different genders, ages, religious beliefs, ethnic backgrounds, incomes, and so forth.

One way of gathering information about a population is to examine every single element in the population. We conduct a **census** of a population when we record every element in the population. Because of time and cost constraints as well as other factors to be discussed later, it is often preferable to obtain a sample rather than to conduct a complete census.

We introduce some terms commonly used in survey sampling.

DEFINITIONS **Elementary Units and Frame**

An **elementary unit** is a person or object on which a measurement is taken. A listing of all elementary units in a given problem is called the **frame.**

The statistical population is not the set of elementary units but the set of *observations* or *measurements* of some characteristic of interest associated with the elementary units. The frame is a listing of all the individuals or objects—that is, all the elementary units—on which a measurement is taken.

For example, suppose we want to study the incomes of all 500 employees at a corporation. To do this, we decide to select a sample of the employees. Each employee is an *elementary unit.* The *frame* is a list containing the names of all 500 employees. The *sample observations* are the incomes of the selected employees, and the *population* is the set of incomes of all 500 employees.

Quantitative and Qualitative Variables

There are two basic kinds of populations, **quantitative populations** and **qualitative populations.** When the characteristic or variable being studied is nonnumerical, the variable is called a **qualitative variable,** and the population of all observations is called a qualitative population. When the variable being studied can be expressed numerically, the variable is called a **quantitative variable,** and the population of all observations is called a quantitative population. Examples of qualitative variables are gender, marital status, race, religion, and occupation. An individual's eye color, model of car, ethnic background, state of residence, and academic major are other qualitative variables. Examples of quantitative variables are age, income, years of education, height, and weight. The diameter of a pipe, the lifetime of a test tube, the amount of money in a checking account, the number of cars owned by a family, the number of customers at a restaurant, and so forth are other quantitative variables.

Qualitative variables can be **dichotomous** or **multinomial.** Observations of a *dichotomous qualitative variable* have only two categories, such as male or female, correct or incorrect, defective or satisfactory, employed or unemployed. Observations about a *multinomial qualitative variable* can have more than two categories, such as a person's religion, occupation, or state of residence. Qualitative data can be stored in a computer by assigning a numerical code to each possible value of the variable. For example, for a dichotomous variable like gender, we can record 0 for male and 1 for female. For a multinomial variable like the manufacturer of an automobile, we can record 1 for Ford, 2 for General Motors, 3 for Chrysler, and 4 for other. Assigning a numerical code to qualitative data is just a convenience, and the values assigned to the various categories carry no special significance. For example, for car manufacturer, we could just as easily have recorded 1 for General Motors, 2 for Chrysler, and so on.

For dichotomous data, it usually is best to assign the values 0 and 1 to the two categories rather than 1 and 2, because the average of a set of 0s and 1s represents the proportion of values that are 1. For example, suppose we have a sample of 20 people, 8 of whom are males, coded as 0, and 12 of whom are females, coded as 1. The sum of the 20 coded values is 12, and the average is $^{12}/_{20}$, or .60. This indicates that 60% of the people in the sample are females. No special meaning can be attached to the average of a multinomial variable. For example, if we code car manufacturers as previously described, the average value for a set of observations might be 1.78. This value has no special meaning.

When the variable being studied is qualitative, we usually want to know how many observations and what proportion of the observations fall in each category. For example, we may be interested in what proportion of adults are married or what proportion of employees at a company have college degrees.

Quantitative variables can be classified as **continuous** or **discrete.** Observations of a *discrete quantitative variable* can assume values only at specific points on a scale of values, with gaps between them. Examples of discrete quantitative variables are number of children in a family, number of rooms in a house, number of cars owned by a family, number of students in a class, and so forth. Observations of a *continuous quantitative variable* can assume all values within an interval. Examples of continuous quantitative variables include the volume of fluid in a bottle, the weight of a shipment of cotton, and the length of time required to produce a product.

When the population is quantitative, we usually are interested in determining how the observations are distributed among all possible values and in determining the average value of the numerical observations. Often we also want to determine the amount of variability in the data.

Scales of Measurement

In statistics we deal with various types of measurements, which represent our raw data. It is useful to define the different kinds of measurement scales that can be used to describe data. There is a hierarchy of scales of measurement, going from weakest to strongest. The weaker the scale of measurement, the less we are able to assume about the relations among elements on the scale. There are four generally used measurement scales, listed here from weakest to strongest: **nominal, ordinal, interval,** and **ratio.**

Nominal Scale

DEFINITION   **Nominal Data** ————————————————————————————————————

In the **nominal scale** of measurement, values are used merely to represent the class or category to which an observation belongs. Nominal data are labels for groups or classes. Nominal data may be verbal or may be recorded as numerical codes.

——

Examples of nominal data are as follows:

1. A survey of 50 households indicates that 20 households own a Ford car, 25 own a Chevrolet, and 5 own a Toyota. For computer entry, we may designate Ford as 1,

Chevrolet as 2, and Toyota as 3. In this case, the numbers 1, 2, and 3 are simply codes representing types of cars.

2. In a set of 40 employees, 15 are male and 25 are female. The names Male and Female are nominal values. Once again, for computer entry, it would be useful to introduce numerical codes, such as 0 = male and 1 = female.

3. Occupations, such as teacher, carpenter, truck driver, and baker, represent nominal data.

If the nominal data are converted to numerical values, the numerical values are only codes. It does not make sense to compare the code values. For example, if 1 = Ford and 2 = Chevrolet, it does not make sense to say that Chevrolet is preferred to Ford because 2 is greater than 1.

Ordinal Scale

DEFINITION   **Ordinal Data** ————————————————————————————

In the **ordinal scale** of measurement, the values or labels may be ranked or ordered in some meaningful way, for example, from worst to best. Ordinal data may be verbal or may be recorded by using numerical codes.

The following are examples of ordinal data:

1. A potential car buyer is asked to rank four automobiles as 1, 2, 3, or 4, where 4 is the best and 1 is the worst. In this scale of measurement, a car with a ranking of 4 is preferred to a car with a ranking of 2, but we cannot say that the preference for the car with the ranking of 4 is twice as strong as the preference for the car with the ranking of 2. In addition, we cannot say that the difference in preference between a car ranked 4 and a car ranked 3 is the same as the difference in preference between a car ranked 2 and a car ranked 1.

2. Many companies require managers to complete employee evaluation forms on which they evaluate an employee's job performance based on a scale from 1 to 10, where 1 is the lowest and 10 is the highest.

3. Business data relating to consumer preferences or the results of taste tests are examples of ordinal data.

Like nominal data, ordinal data enable us to put observations into specific categories. Unlike nominal data, ordinal data can be ranked. That is, we can put them in some specific order, for example, from smallest to largest or from least preferred to most preferred. Thus, ordinal data represent a higher level of data measurement.

Ordinal data may be verbal or numerical. For example, a job candidate could be classified as poor, fair, good, superior, or excellent. Alternatively, we might use the numerical scale 1 = poor, 2 = fair, 3 = good, 4 = superior, and 5 = excellent. With these ordinal data, good is a higher classification than fair, so 3 is preferred to 2.

When ordinal data and nominal data are coded numerically, differences between the numerical values are not meaningful indicators. For example, suppose coffee is rated as 1 = poor, 2 = good, and 3 = excellent. Coffee with a rating of 3 is not necessarily three times as good as coffee with a rating of 1. Also, the difference between poor coffee and good coffee is not necessarily the same as the difference between good coffee and excellent coffee.

Interval Scale

DEFINITION   **Interval Data**
In the **interval scale** of measurement, we can assign a meaning to distances between any two observations, but the ratio of two different measurements is not a meaningful indicator. Interval data are always numerical.

Examples of interval data follow:

1. The temperatures at noon in five different cities represent interval values.
2. The IQs of a sample of five children represent interval values. A child with an IQ of 130 is not necessarily 30% smarter than a child with an IQ of 100.

Interval data are a higher level of data than ordinal data because, in addition to being ranked, interval data indicate the differences between the units being measured, and these absolute differences are meaningful. For example, the difference between 20° Fahrenheit and 40° Fahrenheit is the same as the difference between 10° Fahrenheit and 30° Fahrenheit.

Interval data are characterized by not having a meaningful zero value. As a consequence, although differences between different values on an interval scale are meaningful, the ratios of different values are not. For example, 40° Fahrenheit is not twice as hot as 20° Fahrenheit.

Ratio Scale

DEFINITION   **Ratio Data**
The **ratio scale** is the strongest scale of measurement. With ratio data, distances between pairs of observations, as well as ratios of values, are meaningful. Ratio data are always numerical.

The following are examples of ratio data:

1. Salaries of employees represent ratio data. For example, a salary of $50,000 is twice as large as a salary of $25,000. Such a comparison is not possible with temperature data, which are on an interval scale.
2. The numbers of hours worked in a week by a sample of employees represent ratio data. An employee who works 40 hours per week works twice as much as an employee who works 20 hours per week.
3. The numbers of miles driven by truck drivers during a given month represent ratio data.
4. The numbers of workers employed in different divisions of a corporation represent ratio data.

The ratio scale contains a meaningful zero. That is, working zero hours or having zero employees is a meaningful value. Because of this, ratios of different values are meaningful. In addition, differences between different values are meaningful. For example, the difference between working 10 hours per week and working 20 hours per week is the same as the difference between working 40 hours per week and working 50 hours per week.

Ratio data are the strongest, or most powerful, form of measurement. With ratio data, we can put data into categories, we can rank the values, we can determine meaningful distances between values, and we can determine ratios of values.

Business data relating to prices, revenues, units sold, numbers of employees, volumes, areas, lengths, heights, percentages, and so forth are all ratio data.

Exercises

Statistical Concepts

2.1.1 A bar owner lists the types of beer he sells and records the number of bottles of each brand he sold last week. Suppose the four brands of beer are: Budweiser, Stroh's, Miller, and Coor's. Explain how to create a numerical code to represent the brands of beer. Does this code have any special meaning or could the code values be assigned randomly?

2.1.2 In a taste test, 20 individuals are asked to taste four different diet colas and to rate each cola as poor, fair, good, or excellent. Are the results of this taste test nominal data, ordinal data, interval data, or ratio data? Explain.

2.1.3 Suppose we use numerical codes to record ordinal data. *True or false:* The ratio of two different numerical codes is not meaningful. Explain.

2.1.4 Every week during the college football season the Associated Press publishes a ranking of the top 25 college teams. *True or false:* These rankings represent interval data. Explain.

2.1.5 Categorize each of the following variables as nominal, ordinal, interval, or ratio:
  a. For computer data entry purposes, a garden supply store classifies flowers as follows: 1 = roses, 2 = tulips, 3 = marigolds, 4 = gardenias.
  b. A customer classifies her preferences in flowers from worst to best as follows: 1 = roses, 2 = tulips, 3 = marigolds, 4 = gardenias.
  c. A garden supply store lists its total sales of flowers as follows: 100 roses, 25 tulips, 37 marigolds, 49 gardenias.

2.1.6 Categorize each of the following variables as nominal, ordinal, interval, or ratio. Explain your choice.
  a. Twelve teenagers list the last movie they saw.
  b. Twelve teenagers are asked to rate a newly released movie from 1 to 10.
  c. Twelve teenagers are asked to report how many hours they spent on the telephone during the last month.

2.1.7 Classify the following data as discrete or continuous. Explain your choice.
  a. The number of pumps at a gas station
  b. The SAT score of a randomly selected student
  c. The annual income of a bank president
  d. The number of elevators in a hotel lobby
  e. The number of courses taken by a college freshman

2.1.8 Classify the following data as qualitative or quantitative. Explain your choice.
  a. The state of birth of the President of the United States

    b. The marital status of a corporation president
    c. The price of a new textbook
    d. The number of cars that enter a parking lot during a given day

2.1.9 Suppose that an auditor wants to obtain a sample of 50 checking account balances from the 1,000 customers at a small bank.
    a. What is the frame?
    b. What is an elementary unit?
    c. What is the population?
    d. What is a sample observation?

2.1.10 An office manager wants to obtain a sample showing the length of phone calls for 50 randomly selected phone calls originating in her department last month.
    a. What is the frame?
    b. What is an elementary unit?
    c. What is the population?
    d. What is a sample observation?

2.1.11 Explain how to use a coding system to record a qualitative variable as a number.

2.1.12 Explain why it is preferable to code a dichotomous qualitative variable as 0 and 1 rather than as 1 and 2.

2.1.13 Explain whether the following variables are discrete or continuous:
    a. The length of time for a long-distance phone call
    b. The volume of gasoline remaining in a car's tank
    c. The number of customers waiting in line at a cash register
    d. The number of courses taken by a college student

2.2   Reasons for Sampling and Sources of Bias

Since a sample is only a part of a population, any inferences we make about population characteristics based on the sample may be erroneous. Despite this possibility, there are various reasons for taking a sample rather than a census of the entire population. The most important reasons for sampling are as follows:

    1. *Expense:* It is less expensive to obtain a sample than to survey the entire population.

    2. *Speed of response:* Information is often needed quickly because important decisions have to be made.

    3. *Infinite number of observations:* Sometimes populations of observations are generated as a result of a continuing or recurring process. In these cases, it is not possible to observe every element because new observations are continually being made.

    4. *Destructive sampling:* Testing a product may require its destruction, as is the case when determining the lifelength of light bulbs or the range of ammunition. In such cases, studying more than a sample is not practical.

    5. *Accuracy:* At times a sample can be more accurate than a census of the entire population. This is especially true when the study involves large amounts of repetitive, tedious work, thus increasing the chance of making mistakes because of boredom, fatigue, or the use of unskilled people. A sample obtained by well-trained, skilled people can be more accurate than a census obtained by unskilled people.

A sample is expected to represent the population from which it was selected, but there is no guarantee that a sample will exactly reproduce the characteristics of the population. A sample may not be representative of the population for several reasons. The first reason is just chance or bad luck; we may just happen to obtain a sample containing a large number of atypical elements, thus leading to inaccurate estimates of the population parameters. (A *parameter* is a number that describes some characteristic of a population, such as a mean.) The second reason is the existence of *sampling bias.*

DEFINITION   **Sampling Bias**

A sample is **biased** if it is obtained by a method that favors the selection of elementary units having particular characteristics.

One form of sampling bias is called **selection bias.** Selection bias is a systematic exclusion of certain groups from the sample.

EXAMPLE 2.1   **A Biased Sample—The *Literary Digest* Poll**

Public opinion polls are probably the most difficult random sample to construct. For example, in an election, the statistical population consists of every person who votes. Unfortunately, when a sample is taken before an election, the set of people who will actually vote is unknown. In addition, many potential voters are very difficult to contact, and many people tell pollsters that they do not know who they will vote for. These problems make it difficult to predict the outcome of elections. Probably the most famous example of an inaccurate election prediction is the 1936 *Literary Digest* prediction of a 57% to 43% landslide victory for Alf Landon over Franklin Roosevelt. The *Digest* polled 2.4 million people, the largest political poll ever conducted, and claimed that its prediction would be "within a fraction of 1 percent" of the actual vote. The *Digest* mailed questionnaires to 10 million people, nearly a quarter of the voting population. In sampling, however, quality is much more important than quantity. A sample of 2,000 randomly sampled voters is much better than a sample of millions of unrepresentative voters. The *Literary Digest* got most of the 10 million names by selecting names from every single telephone book in the United States.

In 1936, telephone service was still a luxury item. There were only 11 million residential phones, and for the most part they were in the homes of the well-to-do, who tended to favor the Republican candidate Landon. Those without phones tended to have lower incomes and were overwhelmingly Democratic. Thus, the *Literary Digest* was soliciting information from a biased sample that was unrepresentative of the population. The magazine also got names from car registrations, club memberships, and its own subscription lists. These sources were even more biased than the telephone directories.

The best way to minimize errors resulting from chance or bad luck is to take a sufficiently large random sample. (By a random sample, we mean that every sample of size n has the same probability of being selected.) Increasing the sample size increases the probability that the sample will be representative of the population. In fact, much of statistics is concerned with determining how large a random sample should be to minimize errors due to chance.

Errors caused by sampling bias are extremely important. Ensuring that we will not obtain a biased sample is very difficult, and it frequently is very difficult to tell whether a sample is biased. The best way to minimize errors caused by sampling bias is to choose a proper sampling procedure.

In some cases, extreme accuracy is required; in other cases, rough estimates may be sufficient. Consequently, the nature of any sampling procedure and the amount of time and money spent on conducting the procedure should depend on the accuracy required as well as on the potential benefits derived from obtaining an accurate estimate of the characteristic being studied.

The conclusions drawn from a particular set of data depend on how and from whom the data were collected.

EXAMPLE 2.2   **Selecting the Source of the Data**

Suppose you want to obtain a sample of observations to estimate the mean income of adult males. If you examine an individual's income tax records, you probably will get different information than if you conduct a personal interview and ask the person how much he or she earned last year. This demonstrates that the method used to obtain the data can affect the results of the study.

EXAMPLE 2.3   **Selection of the Sample**

When you estimate the average income of graduates from a certain college, the sample of graduates from whom the data are collected is important. Suppose the names of the subjects are obtained from a list of graduates from the college who have contributed to the alumni association. The people who contribute to the alumni association probably earn more income on the average than people who do not contribute. Because of this, an estimate of average income based on a sample selected from alumni association contributors is likely to be higher than the true average income of all graduates of the college.

Data Acquisition

There are various methods for gathering data in statistical studies. Two of the most frequently used methods are the personal interview and self-enumeration. These methods and their advantages and disadvantages are as follows:

1. Personal interview: In a personal interview, an interviewer asks questions that are printed on a questionnaire and records the respondents' answers on a prepared form.

Advantages:

a. People tend to respond if they are approached directly; thus a personal interview survey has a high response rate.

b. Direct contact with the respondent enables the interviewer to clear up any problems the respondent might have in interpreting the questions.

Disadvantages:

a. The interviewer may not follow the directions for selecting respondents. For example, the interviewer may be told to contact people living at certain addresses but may select other addresses that are more easily found.

b. The interviewer may influence the respondent by the manner or tone of voice in which the questions are asked.

c. The interviewer may record the respondents' answers incorrectly.

2. **Self-enumeration:**   With self-enumeration, the respondent is given or mailed a questionnaire to complete along with a set of instructions, if necessary.

Advantage:

a. Self-enumeration eliminates the errors introduced by the interviewer in a personal interview.

Disadvantages:

a. When a questionnaire is mailed to a household, there is no control over which person answers the questions.

b. Typically, response rates to a questionnaire mailed to a household are lower than response rates to a personal interview. This can cause a serious bias in survey results, because there is no guarantee that the people who do respond to the survey are representative of all the people who received the survey. When the nonresponse rate is high, it is useful to perform some sort of follow-up survey of the nonrespondents.

c. With a mailed questionnaire, it is difficult to clear up any problems that the respondents might have in interpreting questions.

Some Causes of Errors of Estimation

Suppose we wish to estimate some population parameter θ. We obtain a sample of data and calculate the sample estimate $\hat{\theta}$. In general, $\hat{\theta}$ will not be equal to θ.

DEFINITION   **Sampling Error or Error of Estimation** ────────────────────────────

Let $\hat{\theta}$ be a sample estimate of some population parameter θ. The **sampling error,** or **error of estimation,** is equal to $|\hat{\theta} - \theta|$.

──

Because of chance or bad luck, the possibility of obtaining a large error of estimation always exists. By choosing a sampling procedure wisely and selecting a large enough sample, we can control the probability of obtaining a sampling error of any given size. To reduce the probability of getting any given error of estimation, we have to increase the sample size. We should realize, however, that at some point the costs of gathering more data will eventually exceed the benefits obtained from a more reliable estimate.

Errors of estimation caused by a biased sample generally cannot be reduced by selecting a larger sample, and such errors will systematically lead to overestimates or underestimates of the population parameter θ.

Some of the conditions that contribute to bias follow:

1. *Sensitive questions* may yield inaccurate or dishonest answers. For example, questions involving criminal activity, drinking habits, drug use, and sexual habits are likely to elicit dishonest, exaggerated, or incorrect answers.

2. The *bias of nonresponse* can occur when some individuals are systematically omitted from the sample. For example, daytime telephone surveys will tend to underrepresent adult males and overrepresent adult females because men are more likely than women to be working away from the home during the day.

3. The *bias of self-selection* can occur when the individuals being studied rather than the statistician determine which units shall be included in the sample. Self-selection bias is systematic refusal of some groups to respond to a poll. For example, radio talk shows frequently ask listeners to call and cast a vote on some controversial issue. The listener then decides whether to call and vote. In general, those people who call and vote may have different (and stronger) opinions than those who do not call. The same argument holds for conclusions based on mail received by members of Congress. People who write letters have selected themselves to be in the sample, and letter writers, in general, tend to be different (if only more opinionated) from nonwriters. Conclusions based on a sample of observations obtained by mail are likely to be subject to bias because not everyone will mail in a response. Often those people who do not respond to voluntary opinion polls are more willing to leave things as they are, whereas those who do respond tend to prefer a change.

4. *Interviewer bias* occurs when the interviewer has some effect on the answer. The personal characteristics of the interviewer, such as age, gender, race, or occupation, may influence the response. For example, when blacks in the army were asked about how they felt the army treated them, their responses were vastly different depending on whether the interviewer was black or white. The manner in which the interviewer poses the question may elicit an incorrect or skewed response. For example, "The federal deficit is too large, isn't it?" is more likely to elicit a positive response than "Do you think the federal deficit is too large?" or "What do you think about the size of the federal deficit?"

5. Bias can be caused by recording answers incorrectly or making errors in processing the data. For example, an inaccurate radar gun might be used to estimate the speed of a car, or an inaccurate watch might be used to clock the speed of an athlete. Probably the most common error in recording data occurs when an individual hits the wrong key while inputting data into a computer.

After the 1936 *Literary Digest* fiasco, political pollsters turned from mailed questionnaires to quota sampling as an inexpensive way of obtaining a representative sample. In **quota sampling,** a sample is constructed by filling quotas of certain characteristics that are thought to reflect the population as a whole. For example, the interviewer may be told to solicit responses from a certain number of males, females, whites, blacks, young people, old people, rich people, poor people, and so forth. The goal is to guarantee that the sample has certain characteristics that match characteristics believed to exist in the population.

However, for the reasons just discussed, quota sampling is biased, too. Whenever the interviewer is given freedom to choose who will or will not be included in the sample, there is the danger of selection bias. Quota sampling leaves too much discretion in the hands of the interviewer. In addition, most interviewers will tend to meet their quotas in the easiest possible way, which means that people in risky, unfamiliar, or inconvenient places will be ignored.

At the time of the *Literary Digest* poll, this selection bias tended to favor Republicans. This would lead one to suspect that if a poll used quota sampling to predict the outcomes of presidential elections, there would be a systematic tendency to overestimate the proportion of Republican votes.

The Gallup Poll, for example, was just starting at the time of the 1936 presidential election. Using quota sampling, the Gallup Poll did better than the *Literary Digest,* but it still overestimated the vote for the Republican candidate in each of the first three elections it covered. However, the selection bias was not large enough to cause an erroneous prediction of a Republican victory. In the 1948 election, though, the vote was close, and Gallup was one of several pollsters who incorrectly predicted that the Republican candidate Thomas Dewey would be elected over Harry Truman.

Pollsters sought a way to make their samples more random without incurring the extremely high expenses associated with a simple random sample. (In a simple random sample, every sample of size n has the same probability.) Eventually they settled upon a **sequential cluster sampling** procedure. First a random number table is used to select the cities that will be sampled; the chance of a city being selected is proportional to its population. Random numbers are then used to select voting precincts within the city and to select voters from the voter registration lists in those precincts. The sequential nature of this process makes it unnecessary to list every single voter in the country, and every voter has an equal chance of being included in the sample. In addition, clustering observations at the precinct level makes it a much more economical sampling procedure. However, political polls still contain some selection bias. In general, pollsters do not consider very remote, difficult-to-reach households, and many interviewers still have some discretion over whom to interview.

At the present time, polling is a combination of random and quota sampling with some interviewer discretion. This is an economic compromise that greatly reduces the extreme expense of a simple random sample. Apparently this procedure works fairly well. The accompanying data show the results of the Gallup Poll since 1936. Since 1952, when new sampling procedures were instituted, the size of the sampling error has been greatly reduced. In addition, the tendency to overestimate the Republican vote has largely been eliminated. Now the Republican proportion is underestimated about as often as it is overestimated. The amazing statistical fact is that the pre-1950 polls were based on samples of 50,000 registered voters, whereas current polls yield better predictions using samples of less than 4,000 people.

Year	Predicted Republican vote (percent)	Actual Republican vote (percent)	Difference (percent)
1936	44	38	6
1940	48	45	3
1944	48	46	2
1948	50	45	5
1952	51	55.4	−4.4
1956	59.5	57.8	1.7
1960	49	49.9	−.9
1964	36	38.7	−2.7
1968	43	43.4	−.4
1972	62	61.7	.3
1976	49	48.0	1.0
1980	47	50.7	−3.7
1984	59	58.8	.2

Exercises

Statistical Concepts

2.2.1   Define destructive sampling and give an example.

2.2.2   Define selection bias and give an example.

2.2.3   Explain the self-enumeration method of data acquisition and give an example.

2.2.4   Explain the personal interview method of data acquisition and give an example.

2.2.5   Define quota sampling and give an example. Give some possible advantages and disadvantages of this form of sampling.

2.2.6   Give an example of a situation where nonresponse bias is likely to occur, and explain your answer.

2.2.7   Give an example of a situation where self-selection bias is likely to occur, and explain your answer.

2.2.8   Give an example of a situation where interviewer bias is likely to occur, and explain your answer.

2.2.9   Give an example of a sensitive question that is likely to elicit an inaccurate or dishonest answer, and explain your answer.

2.2.10   List three examples where we rely on sample information rather than a census because the product is destroyed in the process of making an observation.

2.2.11   List three examples where we rely on sample information rather than a census because a census would be too expensive.

2.2.12   List three examples where we rely on sample information rather than a census because a census would be too time-consuming.

2.2.13   List three examples where we rely on sample information rather than a census because the population is generated by a recurring process.

2.2.14   Explain why inferential statistics is not required if we have a census of data.

2.2.15   List some advantages of obtaining data from a personal interview. List some disadvantages.

2.2.16   List some advantages of obtaining data from a mailed questionnaire. List some disadvantages.

2.2.17   List five causes of errors of estimation that result from bias.

2.3   Types of Samples

The two basic types of samples are **probability samples** and **nonprobability samples.** In a probability sample, the selected items are determined according to some randomization or chance procedure. For this reason, probability samples are often called *random samples.* One of the objectives in taking a probability sample is to guarantee that each member of the population has an equal chance of being selected for the sample. We will describe how to obtain a probability sample after a brief discussion of nonprobability sampling.

The two primary types of nonprobability samples are *convenience* and *judgment samples.*

DEFINITION   **Convenience Sample**

A **convenience sample** is obtained by selecting elementary units that can be acquired simply and conveniently.

If we stand on a street corner and select the next *n* people who pass by or if we obtain observations by interviewing our neighbors or our family members, we will produce a convenience sample. Letters received by a member of Congress or calls made to a radio talk show also represent convenience samples.

DEFINITION   **Judgment Sample**

A **judgment sample** is obtained by selecting elementary units according to the judgment, intuition, and discretion of an expert or someone familiar with the relevant characteristics of the population.

Both the convenience sample and the judgment sample are likely to yield biased results, although the adequacy of a judgment sample depends to a great extent on the wisdom of the investigator.

DEFINITION   **Simple Random Sample**

Suppose a sample of *n* measurements is selected from a *finite* population of *N* measurements. A **simple random sample** is one in which (1) every possible set of *n* elementary units has the same probability of being selected and (2) the selection of any one elementary unit in no way affects the chance of selecting any other elementary unit.

The great advantage of using a random sample is that there is no tendency to favor or select elementary units possessing certain characteristics. No particular item is systematically excluded from the study, and no particular item is more likely to be included than any other. Because each unit is selected independently, including one particular unit does not affect the chance of including another. Because of this property, the random sample is free from sampling bias. Furthermore, the theorems of probability and statistics can be used to assess the reliability of estimates obtained from random samples. This is not the case with convenience and judgment samples.

Simple Random Sampling

Suppose that we have a frame containing a finite number, *N*, of elements. To select a simple random sample of size *n*, you could proceed as follows. Number each of the elementary units in the frame from 1 to *N*, and write the numbers 1 to *N* on small pieces of paper. Put the pieces of paper in a bowl, mix them thoroughly, and select a sample of size *n* of the papers. To obtain a simple random sample, select those elementary units whose numbers match the numbers pulled from the bowl. This procedure is tedious if the population is large, but it illustrates the idea that every combination of *n* observations is equally likely.

Another way of selecting a random sample is to number each elementary unit in the frame and then choose *n* numbers from a **random number table.** A *random number table* contains a set of digits generated by a process that gives each digit 0, 1, . . . , 9 an equal chance of being next anywhere in the series. Random numbers can be generated by physical processes, mathematical processes, or a combination of the two. Physical processes include the following: (1) drawing a numbered capsule from a bowl, recording the number, replacing the drawn capsule, remixing the capsules in the bowl, and drawing again; and (2) flashing a beam of light at irregular intervals onto a rotating disk and reading off numbers from the disk. Most mathematical processes entail programming a digital computer with a recurrence relation such that the next random number is always derived from one or more prior numbers.

Tables of random numbers generated by computer actually contain pseudorandom numbers, because if the same computer program is executed twice using the same start-up value, or "seed," the same set of random numbers will be generated. A set of random numbers was first published in tabular form in 1927. In 1955, the Rand Corporation published *A Million Random Digits,* a portion of which is reprinted in Table A.1 in the Appendix.

It does not matter how a random number table is read (right to left, left to right, top to bottom, or bottom to top). If the numbers in the table are larger than required, then leading or trailing digits can be ignored. If a random number having no counterpart in the population is obtained, skip it and go on to the next random number. For example, if a population contains 8,000 elementary units and the random number table tells you to select the elementary unit numbered 9061, go to the next random number. Example 2.4 illustrates the use of a random number table.

EXAMPLE 2.4   **Using a Random Number Table** ———————————————————

Suppose a corporation has 600 employees, who are listed in alphabetical order and numbered sequentially from 1 to 600. To estimate the average number of days missed because of illness last year, an auditor wants to select a simple random sample of 40 of the employees. We have to determine which 40 of the 600 employees to examine.

S O L U T I O N   We can use the random numbers given in Table A.1 in the Appendix. Because the table contains five-digit numbers, we delete the first two digits of each number to get numbers from 000 to 999, and we ignore any number greater than 600. We can start at any position in the table; let's choose row 6, column 1. If we decide to read the numbers across the rows, we will record the next 40 appropriate numbers from the table and contact the appropriate employees.

Row 6, column 1 contains the number 74146; we record 146. Row 6, column 2 contains the number 47887; we skip this number because 887 is greater than 600. Row 6, column 3 contains the number 62463; we record 463. Proceeding in this fashion, we would select the 40 employees having the following numbers for our simple random sample:

146	463	45	490	597	12	410	179	75	51
385	378	360	547	78	238	540	219	563	61
492	447	568	333	201	185	154	529	116	42
536	157	172	473	123	533	437	256	596	416

The numbers 540 and 201 each appeared twice in the random number table, and so the second occurrence of each value was omitted.

Exercises

Statistical Concepts

2.3.1 Define simple random sampling and give an example.

2.3.2 Explain how to use a random number table to obtain 50 random numbers between 0 and 500.

 2.3.3 The President has given a speech that has been broadcast on nationwide television. Listeners are asked to call the White House to state whether they agree or disagree with the President's position. People who call are said to be "self-selected."
  a. Does the sample of people who call represent a simple random sample?
  b. Discuss potential sources of bias caused by using a self-selected sample.

 2.3.4 A TV sportscaster asks listeners to call the station and cast a vote for the college football team they think should be ranked Number 1.
  a. Do the resulting calls represent a simple random sample?
  b. List some potential sources of bias resulting from such a sampling procedure.

2.3.5 Explain why statisticians favor random samples to judgment samples or convenience samples.

2.3.6 Explain how to obtain a simple random sample of 50 observations from a population containing 250 observations. Describe how to use the random number table to determine which elementary units to select.

2.4   Computer Analysis of Multivariate Data Sets

When performing a statistical investigation, we gather samples or populations of data concerning the characteristics possessed by the elementary units. Suppose you want to examine the characteristics of a sample of employees at the Computech Corporation. A **data value** is a single measurement, such as an employee's income, an employee's age, or an employee's gender. The information about each employee—name, gender, age, income, and so forth—makes up one observation. An **observation** is a set of data values for the same experimental unit. Any characteristic being observed is called a **variable.** Possible variables include the ages of all the employees at Computech, the incomes of all the employees at Computech, and so forth.

Each individual piece of information is a data value, and all data values for the same person or object form an observation. For each observation, you have one data value for each variable. Any collection of data values about one or more variables is called a **data set,** or **file.** A data set is **univariate, bivariate,** or **multivariate** depending on whether it contains information on one variable only, two variables only, or more than two variables, respectively.

Many multivariate data sets for employees at a corporation contain hundreds or thousands of observations on numerous variables such as the employee's name, gender, race, age, monthly salary, education status, occupation, and so forth. When the number of observations is large, a computer frequently is used to analyze the data. The computer can perform calculations in a few seconds that might take hours or days to perform by hand. In the remainder of this section, we will show how to use the computer to analyze a multivariate data set for a sample of employees at the Computech Corporation.

Selected Characteristics of Employees
at the Computech Corporation

Table 2.1 (pages 24–26) is a multivariate data set for a sample of 120 employees at Computech. Each row contains information on a particular employee, and each column shows the data values of a particular variable for all 120 employees.

◼ Column (1) of Table 2.1 shows the last names followed by the first names of the 120 employees.

◼ Column (2) of Table 2.1 shows the gender of each employee. Because SEX is a qualitative variable that can take only two values, it is a dichotomous qualitative variable. In Table 2.1, SEX is coded as {0 = female, 1 = male}. Be sure to include the code with any statistical output so that the reader will be able to decode and understand the data.

◼ Column (3) of Table 2.1 shows the race of each employee. In Table 2.1, RACE is a dichotomous variable that is coded {0 = black, 1 = white}. Sometimes a third value {2 = other} is included.

◼ Column (4) of Table 2.1 shows the variable DEGREE, which indicates the highest academic degree attained by each employee. DEGREE is a multinomial variable that is coded as follows: {1 = high school degree, 2 = college degree, 3 = postgraduate degree}.

◼ Column (5) of Table 2.1 shows the age of each employee. Theoretically AGE is a continuous variable, but in Table 2.1 it has been rounded down to the largest integer value.

◼ Column (6) shows the variable DIV (short for "division"), which indicates the division of the company where the employee works. DIV is a multinomial variable that is coded as follows: {1 = office, 2 = manufacturing, 3 = sales}.

◼ Column (7) shows the monthly salary of each employee. SALARY can be treated as a continuous variable.

◼ Column (8) indicates the years of seniority at Computech for each employee. SENIOR is a continuous variable that has been rounded down to the largest integer value.

◼ Column (9) indicates the number of tax exemptions claimed by each employee. Computech needs this information for deducting federal income taxes from each employee's paycheck. EXEMPT is a discrete quantitative variable that can take on whole-number values.

With this data set, we can determine general characteristics of the employees, such as the following:

1. The proportion of employees that are male
2. The proportion of employees that are white
3. The proportion of employees that have high school, college, or postgraduate degrees
4. The proportion of employees that work in the office, in manufacturing, or in sales
5. The proportion of employees that have claimed one exemption, two exemptions, and so forth

TABLE 2.1   *A multivariate data set: Selected characteristics of Computech employees*

(1) NAME	(2) SEX	(3) RACE	(4) DEGREE	(5) AGE	(6) DIV	(7) SALARY	(8) SENIOR	(9) EXEMPT
Abbott, Marie	0	1	1	35	1	1,575	16	5
Allen, Joseph	1	1	1	31	2	1,980	10	3
Anthony, Michael	1	1	1	51	2	2,480	28	2
Ard, Debra	0	1	1	57	1	1,925	30	2
Bardol, Stacey	0	1	2	28	3	1,900	6	2
Bates, Robert	1	0	3	64	3	3,050	40	2
Bollman, Phil	1	0	1	30	2	1,920	7	3
Booth, James	1	1	1	35	2	1,970	10	3
Brantly, Cynthia	0	1	1	26	2	1,900	7	3
Brice, Derek	1	0	1	39	2	2,250	18	4
Brown, Dave	1	1	2	25	3	2,300	3	1
Brown, Gretchen	0	1	1	29	1	1,425	10	2
Burns, Wesley	1	0	1	47	2	2,200	20	1
Caste, Felix	1	1	1	22	2	1,840	3	1
Cohen, Richard	1	1	1	24	2	1,880	5	2
Coleman, Joyce	0	1	2	41	3	1,840	12	3
Cooper, Jerry	1	1	1	58	2	2,600	35	2
Cottrell, Bruce	1	1	3	45	3	5,000	20	3
Cusick, Barb	0	1	1	52	1	1,600	17	2
Dillon, Deborah	0	0	1	33	1	1,300	6	1
Dixon, Katie	0	1	2	46	1	1,850	25	5
Doane, David	1	0	1	49	2	2,180	20	1
Doherty, Terry	1	1	1	21	2	1,820	2	1
Donnelly, Mark	1	1	2	43	2	2,580	24	5
Egan, Bryan	1	1	2	35	3	3,000	14	4
Emerson, Michael	1	1	1	33	2	2,080	15	3
Finnegan, John	1	1	1	64	2	2,780	42	2
Folino, John	1	0	1	55	2	2,680	36	2
Fowler, Eileen	0	1	1	33	1	1,475	12	4
Gavett, Bob	1	0	1	55	2	2,600	32	2
Ghent, Barb	0	1	1	42	1	1,400	9	5
Grant, Linda	0	1	1	30	1	1,275	4	3
Gray, Renee	0	1	3	44	3	2,400	20	4
Green, Edward	1	1	1	23	2	1,850	3	1
Gross, Bryan	1	1	2	35	3	3,020	14	3
Hart, Julie	0	1	1	57	1	1,725	22	2
Hicks, Mark	1	1	2	46	2	2,580	23	5
Hill, Betsy	0	0	1	35	1	1,450	11	3
Holtz, Al	1	1	2	36	2	2,000	14	4
Horner, Susan	0	0	1	26	1	1,860	4	1
Hunt, Penny	0	0	1	25	1	1,880	5	1
Jackson, Tom	1	1	3	45	3	3,200	21	5
Johnson, Sue	0	0	1	33	1	1,450	11	4
Jones, Carol	0	0	1	33	1	1,320	6	2
Katila, Ron	1	1	1	19	2	1,800	1	1

TABLE 2.1 *Continued*

(1)	(2)	(3)	(4)	(5)	(6)	(7)	(8)	(9)
NAME	SEX	RACE	DEGREE	AGE	DIV	SALARY	SENIOR	EXEMPT
Katz, Marie	0	1	1	55	1	1,600	17	2
Kennedy, Sue	0	1	2	53	1	1,975	32	2
Kesel, Laura	0	1	1	64	1	1,925	30	2
King, Boyd	1	1	1	38	3	3,400	18	5
Knight, Sharon	0	1	1	44	1	1,275	4	6
Laird, William	1	1	1	21	2	1,840	3	1
Lane, Katie	0	1	2	46	1	1,850	25	5
Lerach, Dick	1	1	1	46	2	2,340	22	4
Lerner, Kathryn	0	1	1	61	1	2,050	35	2
Levering, Jeff	1	0	1	36	2	2,010	11	2
Levine, Harvey	1	1	1	53	2	2,560	33	2
Lewis, Carolyn	0	1	2	24	3	1,980	1	1
Linn, Ron	1	1	1	23	2	1,840	1	1
Lund, Dick	1	1	2	30	3	2,200	9	1
Manke, Terry	1	1	1	37	2	2,200	16	4
Martini, Gregory	1	1	1	20	2	1,820	2	1
Masters, Bryan	1	1	1	54	2	2,520	31	2
Meyer, Jane	0	1	1	41	1	1,420	9	3
Miller, Wayne	1	1	1	30	2	1,980	10	3
Moffitt, Brenda	0	1	1	31	1	1,820	2	4
Molnar, Charles	1	1	3	55	3	4,100	30	2
Morgan, Carl	1	0	2	52	3	2,500	28	2
Morris, George	1	0	1	22	2	1,800	4	1
Neufarth, Ray	1	1	1	41	2	2,320	21	6
Nichols, William	1	1	2	23	3	2,000	2	2
Nixon, Eileen	0	1	1	32	1	1,465	12	2
Noll, Betty	0	1	1	40	1	1,890	5	4
Nunn, Alex	1	1	2	49	3	3,500	27	4
O'Connor, Tom	1	1	1	44	2	2,400	25	5
Otto, Laura	0	1	1	54	1	1,915	25	2
Park, Sandy	0	1	1	41	1	1,375	8	3
Pasurka, Carl	1	1	1	32	2	2,040	13	5
Peplin, Robert	1	1	1	26	2	1,920	7	1
Perkins, Marcia	0	0	1	28	1	1,425	10	2
Phillips, Laura	0	1	1	35	1	1,200	1	4
Pollack, James	1	1	1	55	2	2,560	33	2
Quinn, Denise	0	1	2	48	3	2,800	17	3
Racine, Mary	0	1	1	30	2	1,880	5	3
Ramsey, Henry	1	1	3	57	3	3,300	22	2
Reed, Keith	1	1	1	48	2	2,560	28	2
Reynolds, James	1	1	1	42	2	2,330	22	6
Rosen, Donald	1	1	1	47	2	2,420	23	4
Russell, Mark	1	1	2	21	3	2,000	1	1
Schmidt, Mildred	0	1	1	39	1	1,425	10	3
Schmidt, Peter	1	1	1	20	2	1,820	2	1

(continued)

TABLE 2.1   *Continued*

(1) NAME	(2) SEX	(3) RACE	(4) DEGREE	(5) AGE	(6) DIV	(7) SALARY	(8) SENIOR	(9) EXEMPT
Scott, Donald	1	1	1	45	2	2,400	25	4
Segal, Carl	1	1	1	55	1	1,625	18	2
Seibert, Joan	0	0	1	32	1	1,400	9	3
Shields, Terry	1	1	2	60	3	4,800	38	2
Stacey, Leon	1	0	1	54	2	2,580	34	3
Steen, Mark	1	1	1	28	2	1,900	6	2
Stein, Janet	0	0	1	32	1	1,410	9	3
Stockton, Matt	1	1	1	43	2	2,360	23	3
Sullivan, Mark	1	0	1	26	2	1,880	5	3
Sweeney, Tom	1	1	1	47	2	2,500	30	4
Tinsley, Steven	1	1	2	32	2	1,940	8	3
Tomlin, Shirley	0	1	3	34	3	2,000	6	3
Troy, Guy	1	1	1	51	2	2,480	29	2
Tunis, Kathie	0	0	1	28	1	1,400	9	2
Volk, Sandy	0	1	1	28	1	1,425	10	2
Voss, Jerry	1	1	2	64	3	5,000	35	2
Walker, Bill	1	1	3	33	3	2,400	5	3
Ward, Robert	1	1	1	41	2	2,320	21	6
Warren, Jim	1	1	2	25	2	1,860	4	1
West, Greg	1	1	1	20	2	1,820	2	1
Wetzel, Sharon	0	1	1	44	1	1,255	3	2
Wicks, Robert	1	0	1	27	2	1,900	6	2
Williams, Stephen	1	1	1	46	2	2,320	20	3
Wiltman, Robert	1	1	1	49	2	2,400	25	4
Wollen, Fred	1	1	1	31	2	1,960	9	5
Wood, Brenda	0	1	1	42	1	1,350	7	5
Wooley, Jackie	0	0	2	29	3	2,400	3	2
Yates, Michael	1	1	1	36	2	2,000	11	3
Zielke, William	1	1	1	49	2	2,460	28	4
Zimmer, Ernie	1	1	1	55	1	1,675	20	2

Codes: SEX: (0 = female, 1 = male); RACE: (0 = black, 1 = white); DEGREE: (1 = high school degree, 2 = college degree, 3 = postgraduate degree); DIV: (1 = office, 2 = manufacturing, 3 = sales).

Using a computer, we can construct a frequency distribution. This topic will be discussed in more detail in the next chapter.

An SPSSX Computer Program

Figure 2.1 shows the SPSSX computer program that was used to construct the frequency distributions for the variables SEX, RACE, DEGREE, DIV, and EXEMPT. Each line in Figure 2.1 represents a specific command, as follows:

■ The TITLE command gives a name to the SPSSX run.
■ The DATA LIST command gives each variable a name and describes the data as

```
TITLE 'EMPLOYEE DATA'
DATA LIST FREE/ LNAME (A8), FNAME (A8), SEX, RACE, DEGREE, AGE, DIV, SALARY, SENIOR, EXEMPT
VALUE LABELS
    SEX 0 'FEMALE' 1 'MALE'
    RACE 0 'BLACK' 1 'WHITE'
    DEGREE 1 'HIGH SCHOOL' 2 'COLLEGE' 3 'POST-GRAD'
    DIV 1 'OFFICE' 2 'MANUFACTURING' 3 'SALES'
BEGIN DATA
ABBOTT, MARIE    0  1  1  35  1  1575  16  5
ALLEN, JOSEPH    1  1  1  31  2  1980  10  3
 .
 .
 .
ZIMMER, ERNIE    1  1  1  55  1  1675  20  2
END DATA
FREQUENCIES VARIABLES = SEX, RACE, DEGREE, DIV, EXEMPT
FINISH
```

F I G U R E 2 . 1 *SPSSX program to obtain frequency distributions for data in Table 2.1*

being in FREE format. (In the SPSSX computer program, the employees' last and first names are denoted by the variable names LNAME and FNAME. The (A8) indicates that the data have been entered as letters rather than numbers. Each variable must be given a name, which cannot exceed eight characters in length. This feature is common to most statistical software packages. Consequently, the names of variables often must be abbreviated.)

■ The VALUE LABELS command tells SPSSX to print out how the variables are coded.

■ The BEGIN DATA command indicates to SPSSX that the data immediately follow.

■ The next 120 lines contain the data.

■ The END DATA command indicates the end of the data set.

■ The FREQUENCIES command requests SPSSX to generate a frequency distribution for each of the variables listed.

■ The FINISH command indicates the end of an SPSSX program.

To generate additional output from other commands, just insert the additional commands anywhere between the END DATA command and the FINISH command. (Any SPSSX command that is mentioned in the rest of the book can be inserted anywhere after the END DATA command and before the FINISH command.) The SPSSX program executes the commands in the order in which they appear in the program.

Figure 2.2 (pages 28–29) shows the computer output generated by the SPSSX program in Figure 2.1. The output indicates that there are 45 female and 75 male employees. Thus, 37.5% of the employees are female and 62.5% are male. For the variable DEGREE, the value 1 occurs 89 times, 2 occurs 23 times, and 3 occurs 8 times. This means that

```
SEX


                                                      Valid     Cum
Value Label                   Value  Frequency  Percent  Percent  Percent

FEMALE                          .00         45     37.5     37.5     37.5
MALE                           1.00         75     62.5     62.5    100.0
                                        -------  -------  -------
                              Total        120    100.0    100.0

Valid cases     120    Missing cases      0
```

- -

```
RACE


                                                      Valid     Cum
Value Label                   Value  Frequency  Percent  Percent  Percent

BLACK                           .00         24     20.0     20.0     20.0
WHITE                          1.00         96     80.0     80.0    100.0
                                        -------  -------  -------
                              Total        120    100.0    100.0

Valid cases     120    Missing cases      0
```

- -

```
DEGREE


                                                      Valid     Cum
Value Label                   Value  Frequency  Percent  Percent  Percent

HIGH SCHOOL                    1.00         89     74.2     74.2     74.2
COLLEGE                        2.00         23     19.2     19.2     93.3
POST-GRAD                      3.00          8      6.7      6.7    100.0
                                        -------  -------  -------
                              Total        120    100.0    100.0

Valid cases     120    Missing cases      0          (continued)
```

FIGURE 2.2 *SPSSX output for the program in Figure 2.1*

74.2% of the employees have high school degrees, 19.2% have college degrees, and 6.7% have postgraduate degrees. The other three variables can be similarly evaluated.

A SAS Computer Program

The SAS statistical software package is another widely used program that can produce frequency distributions. The SAS program in Figure 2.3 was used to request frequency distributions for the variables in Table 2.1. These distributions are shown in Figure 2.4 (page 30). The program in Figure 2.3 also generates a vertical bar chart for the variable DEGREE, which is shown in Figure 2.5 (page 31). In SAS, every command ends with a

DIV

Value Label	Value	Frequency	Percent	Valid Percent	Cum Percent
OFFICE	1.00	38	31.7	31.7	31.7
MANUFACTURING	2.00	58	48.3	48.3	80.0
SALES	3.00	24	20.0	20.0	100.0
	Total	120	100.0	100.0	

Valid cases   120   Missing cases   0

- -

EXEMPT

Value Label	Value	Frequency	Percent	Valid Percent	Cum Percent
	1.00	21	17.5	17.5	17.5
	2.00	40	33.3	33.3	50.8
	3.00	26	21.7	21.7	72.5
	4.00	17	14.2	14.2	86.7
	5.00	12	10.0	10.0	96.7
	6.00	4	3.3	3.3	100.0
	Total	120	100.0	100.0	

Valid cases   120   Missing cases   0

F I G U R E 2.2   *Continued*

```
DATA
INPUT LNAME$, FNAME$, SEX, RACE, DEGREE, AGE, DIV, SALARY, SENIOR, EXEMPT
CARDS:
ABBOTT, MARIE    0  1  1  35  1  1575  16  5
ALLEN, JOSEPH    1  1  1  31  2  1980  10  3
  .
  .
  .
ZIMMER, ERNIE    1  1  1  55  1  1675  20  2
;
PROC FREQ:
   TABLES  SEX  RACE  DEGREE  DIV  EXEMPT;
PROC CHART
   VBAR DEGREE;
```

F I G U R E 2.3   *SAS program to obtain frequency distributions for data in Table 2.1*

```
                        The SAS System

                                Cumulative  Cumulative
      SEX    Frequency    Percent  Frequency    Percent
      ------------------------------------------------------
       0         45        37.5        45        37.5
       1         75        62.5       120       100.0
```

```
                                Cumulative  Cumulative
      RACE   Frequency    Percent  Frequency    Percent
      ------------------------------------------------------
       0         24        20.0        24        20.0
       1         96        80.0       120       100.0
```

```
                                Cumulative  Cumulative
     DEGREE  Frequency    Percent  Frequency    Percent
      ------------------------------------------------------
       1         89        74.2        89        74.2
       2         23        19.2       112        93.3
       3          8         6.7       120       100.0
```

```
                                Cumulative  Cumulative
      DIV    Frequency    Percent  Frequency    Percent
      ------------------------------------------------------
       1         38        31.7        38        31.7
       2         58        48.3        96        80.0
       3         24        20.0       120       100.0
```

```
                                Cumulative  Cumulative
     EXEMPT  Frequency    Percent  Frequency    Percent
      ------------------------------------------------------
       1         21        17.5        21        17.5
       2         40        33.3        61        50.8
       3         26        21.7        87        72.5
       4         17        14.2       104        86.7
       5         12        10.0       116        96.7
       6          4         3.3       120       100.0
```

FIGURE 2.4   *SAS output for the program in Figure 2.3*

semicolon. To generate additional results, new commands are added at the end of the program. The SAS program works as follows:

■ The program starts with a DATA command.
■ The INPUT command gives names to the variables. (A dollar sign following a variable name indicates that the data are letters rather than numbers.)
■ The CARDS command indicates that the data follow.
■ The data are listed.
■ The semicolon (;) indicates the end of the data.
■ The PROC FREQ command requests a frequency distribution.

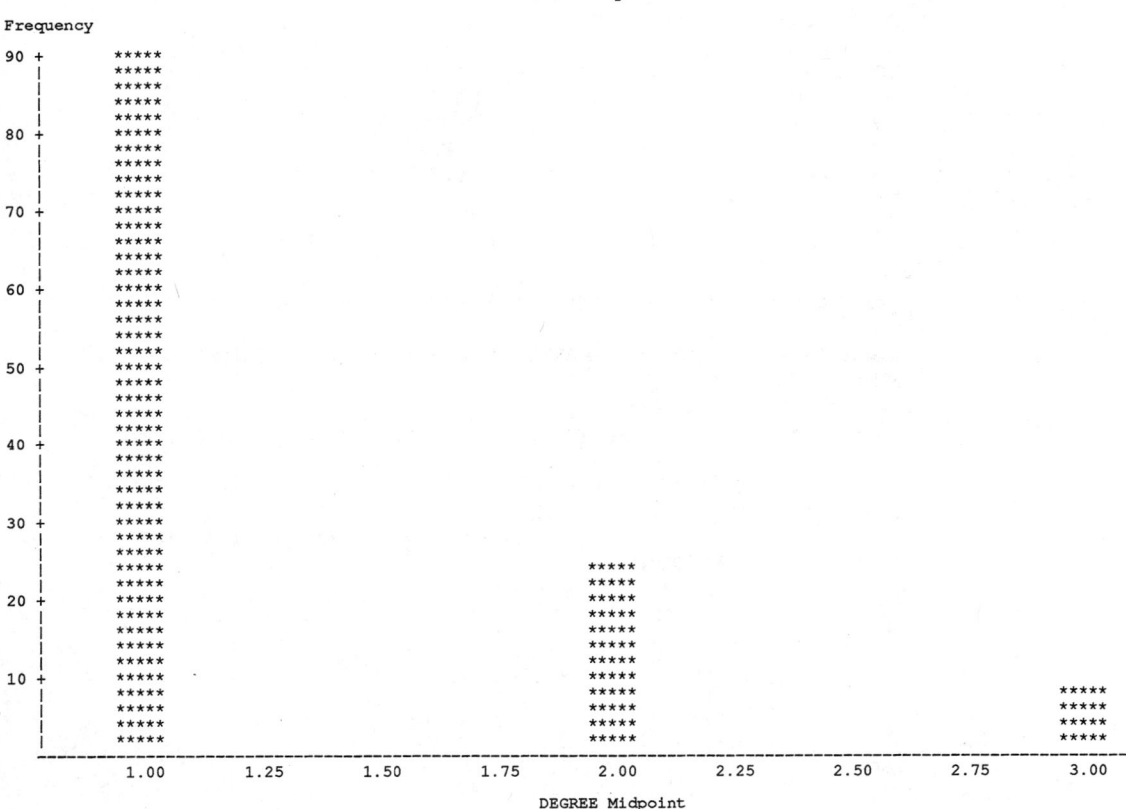

FIGURE 2.5   *SAS bar chart for the variable DEGREE in Table 2.1*

■ The TABLES command requests a frequency distribution for each of the variables specified.

■ The PROC CHART command requests a bar chart.

■ The VBAR command generates a vertical bar chart for the specified variable. The bar chart produced in Figure 2.5 shows how many employees have each type of academic degree.

Suppose the manager of Computech wanted to examine the variables AGE, SENIOR, and SALARY. For example, what is the average salary of the sample of 120 employees? Does salary vary with years of seniority? In later chapters, we will show how to use the SPSSX and SAS programs to answer these and other questions. Occasionally we will also show output from the Minitab statistical package. For information on Minitab, consult the study guide for this book. Figure 2.6 (page 32) shows a Minitab program and the corresponding output of frequency distributions for the variables SEX, RACE, DEGREE, DIV, and EXEMPT.

```
MTB ⟩
READ 'COMPUTECH.DAT' INTO C1-C8
    120 ROWS READ
ROW    C1    C2    C3    C4    C5    C6     C7    C8
 1      0     1     1    35     1    1575   16     5
 2      1     1     1    31     2    1980   10     3
 3      1     1     1    51     2    2480   28     2
 4      0     1     1    57     1    1925   30     2
 .
 .
 .
MTB ⟩
NAME C1 = 'SEX', C2 = 'RACE', C3 = 'DEGREE', C4 = 'AGE'
MTB ⟩
NAME C5 = 'DIV', C6 = 'SALARY', C7 = 'SENIOR', C8 = 'EXEMPT'
MTB ⟩
TALLY C1, C2, C3, C5, C8
        SEX COUNT      RACE COUNT      DEGREE COUNT      DIV COUNT
          0   45          0   25           1   89          1   38
          1   75          1   96           2   23          2   58
        N=  120         N=  120            3    8          3   24
                                         N=  120         N=  120

        EXEMPT COUNT
          1    21
          2    40
          3    26
          4    17
          5    12
          6     4
        N=   120
MTB ⟩
STOP
```

FIGURE 2.6   *Minitab program and output showing frequency distributions*
for data in Table 2.1

Throughout the book, we will use the data in Table 2.1 to demonstrate new analyses and relationships.

Exercises

2.4.1   The data in Table 2.2 are selected characteristics of a sample of 50 students in a statistics class. The qualitative variables are coded at the end of the table. Use the computer to do the following:

a. Generate a frequency distribution and determine what proportion of students are male and female.

b. What proportion of students are freshmen? Sophomores? Juniors? Seniors?

c. What proportion of students are in-state students? Out-of-state?

d. What proportion of students are economics majors? Math majors?

2.4.2   Table 2.3 (pages 34–35) gives information on the 50 states and the District of Columbia from *The Statistical Abstract of the United States.* The variables are coded at the end of the table; in your data set, name the six variables PCTUNION, SALARY, REVENUE, EXPEND, INCOME, and WAGE.

TABLE 2.2 *Selected characteristics of statistics students*

Name	Gender	Class	Resi-dence	Major	SAT score	GPA
Abel	0	2	1	1	1350	3.80
Barth	1	3	1	2	1240	3.25
Bell	0	3	0	0	1200	3.06
Bond	1	1	1	0	1200	3.32
Cohen	0	3	0	2	1230	3.23
Cross	0	4	1	3	1140	2.94
Cruz	1	4	1	2	900	2.43
Daley	1	2	1	1	1120	2.70
Davis	0	3	0	4	990	2.98
Dunne	1	4	0	3	1020	2.60
Ebel	0	4	1	1	990	2.65
Fong	0	3	1	1	1000	2.78
Ghent	1	4	0	4	910	2.37
Grant	0	4	1	3	970	2.59
Hale	0	1	1	0	1300	3.45
Hand	1	2	1	2	1000	2.55
Hart	1	2	1	4	1120	3.24
Hill	0	3	1	0	1140	3.02
James	1	4	0	1	1200	3.11
Jones	0	3	0	3	990	2.87
Kane	1	2	1	3	1040	2.69
Keim	0	2	1	1	1460	3.50
King	1	4	0	2	880	2.29
Klem	1	4	0	4	920	2.54
Land	0	3	1	3	1220	3.14
Lane	1	4	0	0	950	2.44
Levy	1	2	1	2	1140	3.12
Link	0	2	1	3	1230	3.22
Lott	0	3	0	3	1240	3.20
Luddy	0	4	0	4	930	2.45
Lyle	1	4	0	2	1000	2.61
Mann	1	3	1	3	1030	2.82
Mills	0	1	0	2	970	2.35
Mott	0	3	0	3	1140	2.94
Otis	0	3	0	1	1220	3.22
Ott	1	2	1	2	1130	2.69
Rand	1	1	1	1	1140	2.73
Reams	0	2	0	4	1330	3.72
Riley	0	1	0	1	1220	2.94
Rose	0	3	0	0	1310	3.51
Sand	1	2	1	4	1360	3.48
Smith	1	4	1	1	960	2.43
Thomas	0	3	1	3	1000	2.82
Tsiao	1	4	0	4	950	2.51

(continued)

TABLE 2.2   *Continued*

Name	Gender	Class	Resi-dence	Major	SAT score	GPA
West	0	2	1	1	1490	3.68
Wiley	1	3	1	2	1300	3.00
Wolfe	0	3	0	3	1190	2.87
Wulf	0	3	0	2	1240	2.94
Yates	1	4	1	2	1160	2.71
Zorn	1	4	1	4	960	2.56

Codes: Gender: (0 = male, 1 = female); Class: (1 = freshman, 2 = sophomore, 3 = junior, 4 = senior); Residence: (0 = in-state, 1 = out-of-state); Major: (0 = undecided, 1 = economics, 2 = business, 3 = math, 4 = other).

TABLE 2.3   *Selected characteristics of 50 states and Washington, D.C.*

	X_1	X_2	X_3	X_4	X_5	X_6
Alabama	18.2	22,934	2,729	1,931	10,673	8.48
Alaska	30.4	41,480	8,349	11,886	18,187	12.19
Arizona	12.8	24,680	2,829	2,082	12,795	9.47
Arkansas	13.2	19,538	2,642	1,619	10,476	7.57
California	25.4	29,132	3,608	2,524	16,065	10.12
Colorado	18.0	25,892	4,042	2,333	14,182	9.52
Connecticut	18.9	26,610	4,888	2,426	18,189	9.57
Delaware	20.3	24,624	4,517	2,735	14,272	9.84
D.C.	33.4	33,990	5,020	4,722	18,168	10.48
Florida	9.6	22,250	3,731	1,836	13,742	7.86
Georgia	12.7	22,080	2,980	2,010	12,543	8.02
Hawaii	31.5	25,845	3,766	2,535	13,814	8.65
Idaho	16.1	20,969	2,509	1,772	11,120	9.41
Illinois	27.5	27,190	3,621	2,232	14,738	10.37
Indiana	25.1	24,274	3,159	1,876	12,446	10.71
Iowa	20.5	21,690	3,390	2,141	12,594	10.32
Kansas	12.0	22,644	3,476	2,187	13,775	9.46
Kentucky	20.4	20,940	2,853	1,768	10,824	9.53
Louisiana	13.8	20,460	3,124	2,210	11,274	10.43
Maine	18.5	19,583	3,346	2,042	11,887	8.40
Maryland	18.6	27,186	4,349	2,414	15,864	9.73
Massachusetts	19.7	26,800	4,642	2,435	16,380	9.00
Michigan	33.7	30,168	3,782	2,630	13,608	12.64
Minnesota	24.5	27,360	3,982	2,926	14,087	10.05
Mississippi	9.3	18,443	2,305	1,761	9,187	7.22
Missouri	26.6	21,974	3,155	1,726	13,244	9.56
Montana	21.7	22,482	3,947	2,600	10,974	10.97
Nebraska	16.3	20,939	3,285	2,202	13,281	9.03
Nevada	22.1	25,610	2,932	2,351	14,488	9.15
New Hampshire	22.3	20,263	3,114	1,827	14,964	8.39
New Jersey	19.9	27,170	5,536	2,518	17,211	9.90

(*continued*)

TABLE 2.3 *Continued*

	X_1	X_2	X_3	X_4	X_5	X_6
New Mexico	12.8	22,644	3,402	3,044	10,914	8.42
New York	35.8	30,678	5,710	3,296	16,050	9.67
North Carolina	8.9	22,795	3,366	1,750	11,617	7.29
North Dakota	14.2	20,816	3,059	2,694	12,052	8.06
Ohio	27.4	24,500	3,547	2,073	13,226	11.38
Oklahoma	12.9	21,419	2,752	2,021	12,232	9.86
Oregon	27.5	25,788	4,123	2,467	12,622	10.49
Pennsylvania	27.0	25,853	4,168	2,115	13,437	9.57
Rhode Island	19.4	29,470	4,669	2,470	13,906	7.59
South Carolina	5.8	21,570	2,920	1,743	10,586	7.61
South Dakota	10.3	18,095	2,967	2,221	11,161	7.44
Tennessee	17.3	21,800	2,533	1,666	11,243	8.29
Texas	12.5	25,160	3,429	1,923	13,483	9.41
Utah	16.8	22,341	2,297	2,275	10,493	9.40
Vermont	11.9	20,325	3,554	2,304	12,117	8.41
Virginia	10.9	23,382	3,594	1,952	14,542	8.52
Washington	32.9	26,015	3,705	2,369	13,876	11.70
West Virginia	28.9	20,627	2,821	1,930	10,193	10.42
Wisconsin	24.5	26,525	4,247	2,474	13,154	10.26
Wyoming	15.6	27,244	5,440	4,929	13,223	9.93

Code: X_1 = Percentage of employees who belong to a union
X_2 = Average annual salary of teachers in the state
X_3 = Average tax revenue per capita in the state
X_4 = Average government expenditures per pupil in the state
X_5 = Average personal income per capita
X_6 = Average hourly earnings in manufacturing

a. Create an SPSSX program to read these data. After the END DATA command, insert a LIST command, which tells the computer to print out the data.
b. Insert the following command into your program:

SCATTERGRAM SALARY WITH INCOME

This command will plot the salary and income data for each state. Examine this plot. Do states with the highest incomes per capita seem to pay the highest teacher salaries?
c. Are any of these variables discrete?

2.4.3 The data in Table 2.4 (page 36) show annual values from 1960 through 1986 for the following variables: the Gross National Product (GNP), federal government spending (TOTALGOV), federal government defense spending (DEFENSE), the national unemployment rate (URATE), the Consumer Price Index (CPI), the 6-month Treasury bill rate (TBILL), and the Dow Jones Industrial Average (DOWJONES). All data were taken from the *1987 Economic Report of the President*.
a. Create an SPSSX program that will read the data in Table 2.4. Use the variable names given in the table.
b. For each variable, create a time plot in which the values of the variable are plotted against time

TABLE 2.4   *Various economic variables*

DATE	GNP	TOTALGOV	DEFENSE	URATE	CPI	TBILL	DOWJONES
1960	515	54.5	45.3	5.4	88.7	3.247	618.04
1961	534	58.2	47.9	6.5	89.6	2.065	691.55
1962	575	64.6	52.1	5.4	90.6	2.908	639.76
1963	607	65.7	51.5	5.5	91.7	3.253	714.81
1964	650	66.4	50.4	5.0	92.9	3.686	834.05
1965	705	68.7	51.0	4.4	94.5	4.055	910.88
1966	772	80.4	62.0	3.7	97.2	5.082	873.60
1967	816	92.7	73.4	3.7	100.0	4.630	879.12
1968	873	100.1	79.1	3.5	104.2	5.470	906.00
1969	964	100.0	78.9	3.4	109.8	6.853	876.72
1970	1,016	98.8	76.8	4.8	116.3	6.562	753.19
1971	1,103	99.8	74.1	5.8	121.3	4.511	884.76
1972	1,213	105.8	77.4	5.5	125.3	4.466	950.71
1973	1,359	106.4	77.5	4.8	133.1	7.178	923.88
1974	1,473	116.2	82.6	5.5	147.7	7.926	759.37
1975	1,598	129.2	89.6	8.3	161.2	6.122	802.49
1976	1,783	136.3	93.4	7.6	170.5	5.266	974.92
1977	1,991	151.1	100.9	6.9	181.5	5.510	894.63
1978	2,250	161.8	108.9	6.0	195.4	7.572	820.23
1979	2,508	178.0	121.9	5.8	217.4	10.017	844.40
1980	2,732	208.1	142.7	7.0	246.8	11.374	891.41
1981	3,053	242.2	167.5	7.5	272.4	13.776	932.92
1982	3,166	272.2	193.8	9.5	289.4	11.084	884.36
1983	3,406	283.5	214.4	9.5	298.4	8.750	1,190.34
1984	3,756	311.3	235.0	7.4	311.1	9.800	1,178.48
1985	3,998	354.1	259.4	7.1	322.2	7.660	1,328.23
1986	4,209	367.2	278.4	6.9	328.4	6.032	1,792.76

on a scatter diagram, and connect the points. For example, to create a scatter diagram of GNP versus time, issue the command

SCATTERGRAM GNP WITH DATE

2.4.4   A survey was conducted to examine voter preferences. The following data were obtained from a sample of five people who were interviewed:

Name	Gender	Age	Voter registration	Employment status
Rogers	Male	38	Yes	Looking
Jones	Male	52	No	Employed
Kelly	Female	31	Yes	Housewife
Harmon	Female	21	Don't know	Student
Berry	Male	66	No	Retired

a. Discuss how to code the responses for each variable.
b. Using your coding scheme, code each case in the table.

STATISTICS IN ACTION: CASE STUDY 1

Firesafing Problems at the Mellon Bank Building

The Mellon Bank Building in downtown Pittsburgh is a 58-story building that serves as the corporate headquarters for Mellon Bank Corporation. In 1985, during a remodeling project, it was discovered that some of the firesafing on one of the floors was defective: A gap was found in the foam insulation between the inner and outer walls of the building. Such gaps in the firesafing violate the building code and must be repaired because a fire in the office on the floor below could spread through this gap to the floor above.

Management of Mellon Bank filed a lawsuit against the contractors who built the building. In hopes of settling the case out of court, management needed an estimate of the magnitude of the problem; that is, they needed an estimate of the number of gaps in firesafing around the perimeter of the building.

To detect gaps, a small hole is drilled in the wall, and a tube containing a light and a camera is inserted. A photo is taken of approximately a 2-foot segment of the firesafing between the walls. The photo will reveal whether any gap exists in that section. The experts wanted to examine a random sample of points on the perimeter of the building to estimate the proportion of the perimeter that was in violation of the building code. The experts wanted to be 95% certain that the estimate was within 3 percentage points of the true proportion.

The blueprints of the building were obtained and the outer perimeter of each floor was plotted. The sum of the perimeters of the 58 floors was 234,000 feet. The outer perimeter of the building was coded into 117,000 2-foot segments. For example, the first floor has a perimeter of 5,000 feet. By starting at the northeast corner of the first floor and proceeding in a clockwise direction, the perimeter of the first floor was coded into 2,500 2-foot segments numbered 1 to 2,500. Similarly, the perimeter of the second floor was coded into 2,500 2-foot segments numbered 2,501 to 5,000.

Two problems needed to be solved. First, the experts needed to know how many sample points should be examined. By using the techniques discussed in Chapter 10, it was determined that a random sample of at least 1,068 points was necessary.

Second, the experts needed to know how to select the locations of the 1,068 points that were to be examined. A computer program was used to generate a random sample of 1,068 random numbers between 1 and 117,000. Each of these 1,068 random numbers represents one of the 117,000 2-foot segments on the outer perimeter of the building.

Based on the sample of 1,068 random numbers, the firesafing experts drilled the appropriate holes and obtained a photo at each of the 1,068 randomly selected locations. The information obtained in this study helped the management of Mellon Bank reach a settlement with the contractors. The faulty firesafing was just one of several alleged construction deficiencies in the Mellon Bank Building, so the size of the eventual settlement award was only partly a result of the firesafing problems. On September 25, 1992, the *Allegheny Bulletin* reported that the case was settled out of court for $13.1 million.

This case study provides a classic example of how to select a random sample. The following case study shows that life isn't always so easy!

STATISTICS IN ACTION: CASE STUDY 2

Internal Revenue Service vs. the Pittsburgh Press Club

The Pittsburgh Press Club (the Club) is a non-profit organization whose members work in newspaper, television, and radio industries or are public relations managers of various companies. One benefit that members receive is the right to rent the Club's dining and meeting room facilities for various functions. In the early 1980s, the Internal Revenue Service (IRS) filed a lawsuit against the Club claiming that the Club had violated the guidelines for nonprofit organizations and, as a result, would lose its tax-exempt status and would owe a substantial amount of income tax.

To maintain its tax-exempt status, the Club had to establish that, over a period of years, less than 5% of its gross revenues were generated by "nonexempt" activities. An activity is "exempt" if a club member rents Club facilities for a legitimate business purpose, such as holding a press conference. An example of a nonexempt activity is renting a meeting room to hold a banquet for your bowling league.

The Club had daily records showing the name of each club member who rented a room and the amount that was paid in rental fees and food costs, but the Club did not have information showing the specific purpose of each activity. Thus, the Club could establish, for example, that John Smith rented a banquet room on September 3, 1985, and paid a bill of $4,500, but the Club could not establish whether the room was rented to fulfill a legitimate business purpose or to host a party for Mr. Smith's championship softball team.

The Club wanted to conduct a survey to determine the percentage of Club business that was generated by exempt (business-related) activities. A letter was mailed to each sponsor of the several thousand activities that were held at the Club during the early 1980s. The letter asked the recipient to describe the nature of the

event, so the event could be classified as exempt or nonexempt. Based on the information obtained from the sample of responses, the Court ruled that the Club had not exceeded the 5% limit for nonexempt activities.

The IRS appealed this decision and the case eventually reached the Supreme Court of the United States. The Supreme Court ruled that the case should be retried because the data did not represent a true random sample of replies.

The Court's ruling raises the following question: How is it possible to obtain a random sample of responses when people are free to choose whether to reply to a request? Another problem was that the Supreme Court thought that the letter had been sent to only a sample of the Club members who had sponsored activities at the Club. In fact, every sponsor of every event was sent a letter. The problem was that only a sample of the population replied, and the members who replied did not represent a random sample of the population of event sponsors.

The sample of replies is subject to potential *self-selection bias.* That is, the members themselves decided whether to reply to the survey. Of course, everyone recognized these problems before undertaking the survey. Can this problem be avoided? Bringing each event sponsor to court to testify under oath was too expensive and time-consuming. In this situation, the self-selection bias problem could not be avoided.

This case shows that, at times, it may be impossible to obtain a random sample of the population. This case contrasts with the previous case study, in which a computer generated random numbers that were used to select a random sample of points.

Perhaps you're wondering how the case unfolded. The case was never retried. A few days after the Supreme Court issued its decision demanding a new trial, the IRS agreed that the evidence tended to support the Club's position (even with the admitted self-selection bias). The IRS dropped the case, and the Pittsburgh Press Club still operates today as a tax-exempt entity.

Chapter 2 Supplementary Exercises

2.S.1 Hotels often leave questionnaires concerning the quality of their service in the hotel rooms and ask the guests to fill them out. Do the results from such a procedure constitute a random sample? Why or why not?

2.S.2 Suppose that a firm wants to determine the mean income of the families in a particular city. The firm divides the city into six areas, each containing approximately the same number of families. The firm selects two of these areas randomly and then chooses a random sample of 100 families in each of these two areas.
  a. What are the advantages of this sample design over a simple random sample of 200 families drawn from the entire city?
  b. Is the result of this sample design likely to be as precise as the result of a random sample of 200 families drawn from the entire city?

2.S.3 Classify the following variables as qualitative or quantitative. Then classify the qualitative variables as dichotomous or multinomial and the quantitative variables as discrete or continuous.
  a. The number of telephone calls received in a day by a secretary
  b. The employment/unemployment status of an individual
  c. The religion of an individual
  d. The language spoken by an individual
  e. The speed of a car on a racetrack

2.S.4 Consider the following situations and determine whether a census or a sample would be more appropriate.
  a. A doctor wants to study the side effects of a drug that appears to be effective against some types of cancer.
  b. The Republican Party wants to determine what proportion of the electorate favors increased defense spending.
  c. NASA wants to check all the components of a space shuttle fuel system.

2.S.5 Describe the type of sample that was taken in each of the following situations. Discuss whether you would expect to obtain biased results from each sample.
  a. A newspaper reporter learns people's views on a presidential speech by interviewing people as they exit from a movie theater.
  b. A reporter analyzes the stock market by examining the movements in the Dow Jones Industrial Average, a weighted average of 30 stock prices.
  c. An inspector uses a random number table to select a value from 1 to 20. Then the inspector examines that numbered item for defects and examines every 20th item thereafter. The goal is to find the proportion of defective items in a day's production run.
  d. A congressman sends a questionnaire to all registered voters in the district. Replies are received from 20% of those receiving the questionnaire.
  e. The dean's office wants to estimate the average annual income of the parents of students at a college. The dean tells an office worker to put the files in alphabetical order and to examine every 50th file.
  f. The dean's office wants to estimate the average annual income of the parents of students at the

college. The dean tells an office worker to sort the files according to the gender of the students. The dean tells the office worker to select every 50th file from the males and every 50th file from the females.

2.S.6   Describe several ways of obtaining a sample of 10% of the employees of a firm containing 200 employees.

2.S.7   Describe what types of errors or biases, if any, are likely in the following situations:

a. A typist hits the number 2 rather than the number 3 when recording the number of children in families.

b. An interviewer is told to contact 50 randomly selected families as part of a large survey to determine the average income in a community. The interviewer is unable to contact one of the families.

c. An interviewer is told to contact 50 randomly selected families as part of a large survey to determine the average income in a community. The interviewer contacts the families that live in the nice section of town and reports that the other families could not be contacted.

d. A radio announcer asks listeners to call the show to give their opinion about a controversial subject.

2.S.8   Which of the following methods is more likely to elicit correct information concerning the age of individuals: asking individuals to state their age or asking them to state their date of birth?

2.S.9   Describe a situation where it might be useful to take a census.

2.S.10   Describe several situations where biased results are obtained because of self-selection bias.

2.S.11   Suppose a firm has 1,000 employees. Describe how to take a simple random sample of 50 of the employees using a random number table.

 2.S.12   You want to study the employees of the Vates Department Store. You have data like the following for the employees:

Name	Gender	Age	Education	Previous employment	Religion
Jones	Male	45	High school	Accountant	Catholic
Smith	Female	22	College	None	Not reported
Clark	Male	?	High school	Clerk	Protestant
Hall	Female	35	No reply	Secretary	None

a. Discuss how to code the variable Gender so that it can be entered into a computer and analyzed.

b. For the variable Age, what will you do about Mr. Clark?

c. How will you code the variable Education so that it can be analyzed by the computer? What about Ms. Hall?

d. How will you code the variable Previous employment so that it can be analyzed by the computer? What will you do about Ms. Smith?

e. How will you code the variable Religion so that it can be analyzed by the computer? What about Ms. Hall and Ms. Smith?

2.S.13   Obtain a sample of 50 students in order to estimate the proportion of students at your college who are left-handed. Tell how you selected these students and explain why you think this is a reasonable sampling procedure.

2.S.14   Estimate the proportion of students at your college who are from out of state by obtaining a sample of 50 students. Tell how you selected these 50 students, and explain why you think this is a reasonable sampling procedure. Can you think of any reason why more care should be exercised in selecting this sample than in selecting a sample to estimate the proportion of left-handed people?

2.S.15   A television station includes a poll as part of its nightly news show. A provocative "question of the day" flashes on the screen and viewers are asked to call in their yes or no votes. Why might these poll results not be representative of community opinion?

2.S.16   *Consumer Reports'* annual automobile issue includes frequency-of-repair records based on letters received from its readers. Is this a random sample? If not, what biases might appear?

References

Cochran, William G. *Sampling Techniques.* 3d ed. New York: Wiley, 1977.

Cochran, William G., and G. M. Cox. *Experimental Designs.* 2d ed. New York: Wiley, 1957.

Huff, Darrell. *How to Take a Chance.* New York: Norton, 1959.

Huff, Darrell, and Irving Geis. *How to Lie with Statistics.* New York: Norton, 1954.

Morgenstern, Oskar. *On the Accuracy of Economic Observations.* 2d ed. Princeton: Princeton University Press, 1963.

Nie, Norman, C. Hadlai Hull, Jean G. Jenkins, Karin Steinbrenner, and Dale H. Bent. *SPSS Statistical Package for the Social Sciences.* 2d ed. New York: McGraw-Hill, 1975.

Norusis, Marija J. *SPSSX Introductory Statistics Guide.* New York: McGraw-Hill, 1990.

Norusis, Marija J. *SPSSX Advanced Statistics Guide.* Chicago: SPSS, 1990.

Norusis, Marija J. *The SPSS Guide to Data Analysis.* Chicago: SPSS, 1986.

Ryan, Thomas A., Brian L. Joiner, and Barbara F. Ryan. *Minitab Reference Manual.* University Park, Penn.: Minitab Project, 1985.

Ryan, Thomas A., Brian L. Joiner, and Barbara F. Ryan. *Minitab Handbook.* 2d ed. Boston: PWS-Kent, 1985.

SAS Introductory Guide. 3d ed. Cary, N.C.: SAS Institute, 1985.

SAS Procedures Guide for Personal Computers. Version 6 ed. Cary, N.C.: SAS Institute, 1986.

SAS Statistics Guide for Personal Computers. Version 6 ed. Cary, N.C.: SAS Institute, 1986.

SAS User's Guide: Basics. Version 5 ed. Cary, N.C.: SAS Institute, 1985.

SAS User's Guide: Statistics. Version 5 ed. Cary, N.C.: SAS Institute, 1985.

SPSSX User's Guide. Chicago: SPSS, 1988.

Tanur, Judith M., Frederick Mosteller, William H. Kruskal, Richard F. Link, Richard S. Pieters, and Gerald R. Rising., eds. *Statistics: A Guide to the Unknown.* 2d ed. San Francisco: Holden-Day, 1978.

Wald, Abraham. *Sequential Analysis.* New York: Wiley, 1947.

Wald, Abraham. *Statistical Decision Functions.* New York: Wiley, 1950.

3 Summarizing Data in Tables and Graphs

As mentioned in Chapter 1, *descriptive statistics* is a branch of statistics consisting of procedures used to describe the general characteristics of a set of data. Basically, descriptive statistics entails grouping, graphing, and summarizing data so that an investigator can get insights or information that may not be apparent from the raw data.

Any set of data worth studying possesses one quite obvious characteristic: The observations are not all the same; that is, there is a spread, or *distribution,* of observations. The purpose of descriptive statistics is to describe this distribution in a concise manner.

For example, suppose we have a sample of observations showing the annual incomes of pilots for Delta Airlines. Suppose we draw a horizontal number line and place a mark on the line at the value of each annual income. It would be natural to see how the observations are distributed over different portions of the line and where most of the observations are concentrated, to look for the center of the data, and to notice the amount of variation present. The center of the data would give us an estimate of the average income of the pilots. By observing the amount of variation, we could determine whether the incomes of all the pilots are approximately equal to one another or whether they differ substantially.

Three commonly used methods of describing the characteristics of a set of data follow:

1. *Tables:* The data can be presented in **tables,** which show how the data are dispersed or distributed. We will discuss four types of tables in Section 3.1: *frequency distributions, relative frequency distributions, cumulative frequency distributions,* and *cumulative relative frequency distributions.*
2. *Graphs:* Each of the four types of frequency distributions can be presented as a **graph.** The graph of a frequency distribution is called a *histogram.* Histograms provide a visual display of how the data are distributed.
3. *Summary statistics:* The general characteristics of the data can be described by means of **summary statistics,** which describe the center of a data set, the amount of variation present, and other interesting characteristics of the data. The *mean,* the *median,* and the *mode* are different measures of the center of data. These summary statistics are called **measures of central tendency.** The *range,* the *standard deviation,* the *variance,* and the *mean absolute deviation* are different measures of the amount of variation present in data and are called **measures of dispersion.**

In this chapter, we discuss tables and graphs. In Chapter 4, we discuss how to calculate and interpret summary statistics. All of the statistical procedures discussed in this chapter can be performed quite easily using computer programs that are readily available. Occasionally computer output generated by one of three popular computer programs—SPSSX (Statistical Package for the Social Sciences, Version X), SAS, or Minitab—will be used as illustrations. SPSSX and SAS are probably the best and most widely used computer programs for performing statistical analyses. There are numerous other statistical packages available, however, that generate essentially the same type of output. Minitab is a more user-friendly but less powerful program. Once the student learns how to read and understand the output from one statistical package, it is usually a simple task to learn how to read the output from other statistical packages.

3.1   Frequency Distributions and Histograms

As a first step in almost every statistical study, it is useful to describe the data in terms of a *frequency distribution.* The task of interpreting most data sets is usually made somewhat easier by reducing the amount of information that must be absorbed by grouping the observations in intervals and counting how many observations fall in each interval. The intervals are called **classes,** and the number of observations falling in any class is called the **frequency** of that class.

DEFINITION   **Frequency Distribution**

A **frequency distribution** is a table that shows the number of observations falling in each class. The number of observations in the *i*th class is called the **class frequency** and is denoted by the symbol f_i.

Frequency distributions can be constructed for both populations and samples of data. Frequency distributions are especially useful when the number of observations is so large that it is difficult to get a good feel for the data by examining the observations individually.

Constructing a Frequency Distribution

The first step in constructing a frequency distribution is to define the class intervals. The second step is to count and record how many observations fall in each class.

Usually there are many different ways of choosing the classes. For example, the Bureau of the Census annually publishes a document called the *Current Population Report,* which contains frequency distributions of the annual incomes earned by full-time employees in the United States. One frequency distribution is a table that shows how many employees earn from $0 to less than $10,000 per year, how many earn from $10,000 to less than $20,000, and so forth. The bureau could have used different classes in the tables, such as $0 to $15,000, $15,000 to $30,000, and so forth. The person constructing the frequency distribution determines the number of classes and their location, but there are some general guidelines, which will be discussed later in this section.

Except for what are called *open-ended classes,* each class for a quantitative variable must have a **lower limit** and an **upper limit,** which together are called **class boundaries.** For example, the frequency distribution for incomes described above by the Bureau of the Census uses $0 and $10,000 for the lower and upper limits, respectively, of the first class.

A class is **open-ended** if it has no lower limit or no upper limit. Open-ended intervals occur frequently when the data have a large amount of variation and most of the observations are concentrated in a fairly narrow range. This is common with income data, where an open-ended class might consist of all individuals who earn, say, $80,000 per year or more.

The **width** of any class is the upper limit minus the lower limit. For example, if a class consists of all annual incomes from $20,000 to less than $30,000, then the width of the class is $10,000. (Note that an open-ended class does not have a width because one of the limits is not defined.)

The **class mark,** also called the **class midpoint,** is the arithmetic average of the two class limits. For example, if the class limits are $20,000 and $30,000, then the class mark is $25,000. (An open-ended class does not have a class mark because one of the limits is not defined.)

Once the number of classes and the class limits have been chosen, the final step in constructing a frequency distribution is to count and record how many observations fall in each class.

Two Rules for Selecting Classes

Any system of classification used for a frequency distribution must satisfy the following two requirements:

1. The classes in any frequency distribution must be *mutually exclusive.* The classes must be defined so that no observation can fall in more than one class. Frequently it is useful to choose classes so that the upper limit is excluded from that class. For example, when exact incomes are reported, the classes could be, say, $0 to less than $10,000, $10,000 to less than $20,000, and so forth. This notation makes it clear that an individual whose income is exactly $20,000 would be placed in the third class. Sometimes it can be useful to choose as class boundaries values that do not or cannot appear in the raw data. For example, suppose we wish to construct a frequency distribution showing the years of education of individuals. Here it might be useful to use as classes 5.5 to 7.5 years, 7.5 to 9.5 years, and so forth.

2. The classes should be *exhaustive;* that is, every observation must fall in some class. It may be necessary to construct an open-ended class to satisfy this requirement.

Helpful Guidelines

Frequently the construction of a satisfactory frequency distribution requires experimentation with the number of classes and the class boundaries. There are no rules available that apply to all situations. There are, however, some general guidelines that are helpful in constructing a frequency distribution.

1. *Number of classes:* Usually the number of classes should be between 5 and 20. As the number of observations increases, more classes are appropriate. If there are too many classes, the graph will be less effective because the frequency distribution will tend to have a choppy appearance. If there are too few classes, the distribution will give little insight into the underlying pattern of how the data are distributed, and much of the variation in the data will be obscured. A large number of classes provides more detail, but too many classes produce meaningless oscillations. It is common to construct several frequency distributions using different numbers of classes and different class limits to see whether the shape of the distribution changes much when the classes are changed. Select the distribution that appears to give the most logical explanation of the underlying population pattern.

2. *Width of classes:* The choice of the class width depends on the range of the data (that is, the difference between the largest and the smallest observations), the number of observations, and the number of classes desired. For ease of interpretation, it is preferable that all classes have equal widths.

Occasionally class intervals must have unequal widths. This is often the case with income data and other data sets where the difference between the largest value and the smallest value is very large and where most of the observations are concentrated in a relatively narrow range. If most of the observations fall into a relatively narrow interval whereas others are widely dispersed, it may be desirable to have narrow classes where the bulk of the observations lie and wider classes elsewhere. In addition, it may be necessary to use open-ended classes.

3. *Choice of class limits:* Try to select class limits so that the class midpoint, or class mark, of each class is close to the average of the observations included in each class. The reason for this is that when calculating averages and other summary measures from frequency distributions, we assume that the class midpoint is representative of all the values in the class. (This is discussed further in Chapter 4.) Often this can be accomplished without special concern about the class limits. Sometimes, however, the values tend to be bunched at regular intervals throughout the data. For example, many small items cost a few cents less than an even dollar, say, $1.99 or $2.99, whereas many large consumer items tend to be a few dollars less than $100, $200, $300, and so forth. In this latter case, it would probably be best to define class intervals as $50 to under $150, $150 to under $250, and so forth, rather than as $0 to under $100, $100 to under $200, and so forth. Another reason for requiring that class midpoints be representative of the observations falling in a class is that a graph of the distribution will present a more reliable picture of the true location of the data.

Relative Frequency Distributions

A frequency distribution shows the number of observations falling in each class. Usually it is more useful to indicate the *relative frequency,* or *proportion,* of observations that fall in each specific class.

DEFINITIONS **Relative Frequency and Relative Frequency Distribution** ⎯⎯⎯⎯⎯⎯⎯⎯

The **relative frequency** of the ith class is the proportion of observations falling in that class. If the ith class contains f_i observations, then the relative frequency of the ith class is f_i/n, where n denotes the total number of observations in the sample. If we are working with a population containing N observations, then the relative frequency of the ith class is f_i/N. A **relative frequency distribution** is a table that shows the proportion of observations that fall in each class.

Every relative frequency must be a nonnegative number greater than or equal to 0 and less than or equal to 1, and the sum of all the relative frequencies must equal 1.

EXAMPLE 3.1   **Frequency and Relative Frequency Distributions** ——————————————

The raw data in Table 3.1 show the monthly salaries of all the employees at Lang's Trucking Company, a small trucking firm with 80 employees. To get a feel for the salary structure of the company, Mr. Lang needs to have the data simplified and condensed. Construct a frequency distribution and a relative frequency distribution to obtain information on the distribution of salaries.

SOLUTION   First, it is necessary to choose the number of classes and the class limits. The smallest observation is \$1,300 and the largest observation is \$2,025, so the range is \$2,025 $-$ \$1,300, or \$725. Because it is convenient to use classes with a width of \$100, let's form eight classes using the class limits \$1,300, \$1,400, . . . , \$2,100. The first class contains all observations from \$1,300 to under \$1,400, the second class contains all observations from \$1,400 to under \$1,500, and so on up to the eighth class, which contains all observations from \$2,000 to under \$2,100.

After selecting the classes, the only task remaining is to count and record the number of observations belonging to each class. An easy way to do this is to prepare a *tally sheet* like the one shown in Table 3.2. On the tally sheet, each slash indicates one observation. The number of slashes for each class is the class frequency. In addition, the tally sheet provides a visual display of how the data are distributed.

The data in the column titled Frequency show the number of observations falling in each class. The sum of the frequencies must equal the total number of observations. The data in the column headed Relative frequency show the proportion, or relative frequency, of observations falling in each class. The relative frequencies are obtained by dividing each class frequency by 80, the total number of observations. For example, the frequency of the first class is 8, so the relative frequency for the first class is $^8/_{80}$, or .1000. The relative frequencies sum to 1.

TABLE 3.1   *Monthly salaries (in dollars) of employees at Lang's Trucking Company*

1,550	1,310	1,575	1,675	1,585	1,590	1,580	1,475	1,300	1,650
1,380	1,730	1,640	2,000	1,400	1,325	1,900	1,600	1,600	1,555
1,565	1,320	1,750	1,725	1,650	1,740	1,650	1,875	1,620	1,550
1,590	1,570	2,015	1,620	1,860	1,625	2,000	1,850	1,640	1,900
1,700	1,380	1,620	1,650	2,000	1,455	1,625	1,340	1,530	1,410
1,450	1,815	1,440	1,420	1,550	1,550	1,660	1,760	1,550	1,650
1,500	1,620	1,600	1,580	1,705	1,780	1,400	1,550	1,390	1,600
1,775	2,025	1,450	1,425	1,820	1,900	1,700	1,900	1,475	1,850

TABLE 3.2 *Tally sheet and frequency distribution of data in Table 3.1*

Monthly salary (dollars)	Tally	Frequency	Relative frequency
1,300 to under 1,400	///// ///	8	.1000
1,400 to under 1,500	///// ////// /	11	.1375
1,500 to under 1,600	///// ///// ///// //	17	.2125
1,600 to under 1,700	///// ///// ///// ////	19	.2375
1,700 to under 1,800	///// /////	10	.1250
1,800 to under 1,900	////// /	6	.0750
1,900 to under 2,000	////	4	.0500
2,000 to under 2,100	/////	5	.0625
Total		80	1.0000

The frequencies in Table 3.2 indicate that more monthly salaries fall in the $1,600–$1,700 interval than any other. The relative frequencies show that only 6.25% of the employees earn more than $2,000 per month, and only 11.25% of the employees earn more than $1,900 per month. On the other hand, 71.25% of the employees earn from $1,400 to $1,800. A glance at the distribution of tallies indicates that the center of the data appears to be between $1,600 and $1,700. The tally sheet also shows that the distribution of salaries is slightly bell shaped, with most salaries concentrated in a middle range and fewer salaries located in the tails of the distribution.

Graphing Frequency Distributions

Along with a frequency distribution, it usually is useful to prepare a graph showing how the values are distributed.

DEFINITION **Frequency Histogram**

A **frequency histogram** is a graphical presentation of a frequency distribution.

A histogram shows the general shape of the distribution and gives a quick visual impression of where most of the observations are concentrated. We can see whether the distribution has long tails in either direction and whether there are any extreme or unusual values. We can also see where the data are centered and whether the center of the data is a good indicator of a typical observation. In addition, we can see how tightly the observations are clustered around a central value.

The histogram for the data in Table 3.2 is shown in Figure 3.1 (page 48). In a histogram, the class boundaries are marked on the horizontal scale as in Figure 3.1, and the frequency of each class is measured on the vertical axis. Above each class interval, a rectangle is constructed whose area is proportional to the frequency of the class. If the classes all have equal widths, the height of each rectangle can be used to represent the frequency of the class.

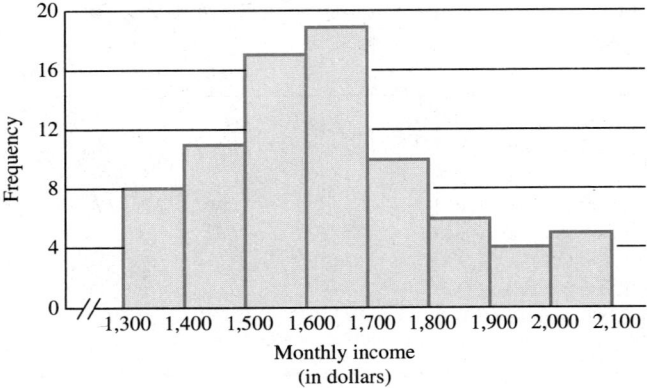

FIGURE 3.1   *Histogram for data in Table 3.2*

Relative Frequency Histograms

A *relative frequency histogram* is similar to a frequency histogram except that the areas of the rectangles above each class interval indicate the relative frequency of the class rather than the actual frequency of the class, although the overall shapes of the frequency histogram and the relative frequency histogram are identical. The relative frequency distribution for the data in Table 3.2 is shown in Figure 3.2.

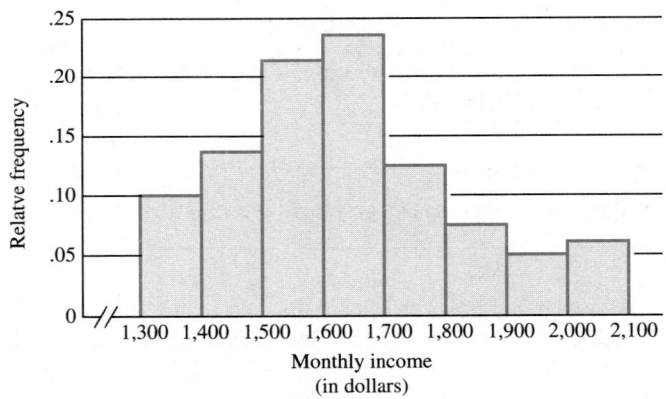

FIGURE 3.2   *Relative frequency histogram for data in Table 3.2*

Computer Output

Figure 3.3 shows a histogram for the salary data in Table 3.1, generated by the SPSSX command FREQUENCIES. (Many other computer programs will generate output comparable to the SPSSX output.) The actual command that generated this output was the following:

FREQUENCIES VARIABLES = SALARY/
    HISTOGRAM MIN(1300) MAX(2100) INCREMENT(100)

This command tells the computer to use the procedure called FREQUENCIES to generate a frequency distribution and a histogram for the variable SALARY. The histogram should start at the minimum value 1,300 and end at the maximum value 2,100, with a class width of 100.

In Figure 3.3, the numbers in the first column entitled Count show the class frequencies. The second column, labeled Midpoint, gives the class mark, or midpoint, for each class. For example, the midpoint for the first class is $1,350. Each row of asterisks represents the number of observations that fall in that class interval. For example, the second

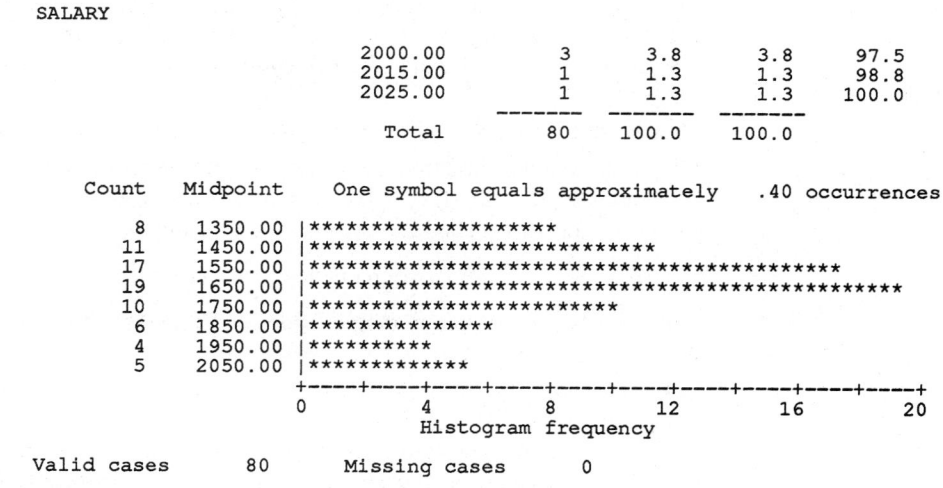

FIGURE 3.3   *SPSSX-generated histogram of data in Table 3.1*

row of the histogram has 28 asterisks, and each asterisk in the graph equals approximately .40 occurrence. Thus, the 28 asterisks represent approximately 28 × .40 = 11.2 occurrences. This indicates that 11 employees have monthly salaries between $1,400 and $1,500. The midpoint of this interval is $1,450.

Interpretation of Relative Frequencies As Probabilities

The relative frequencies shown in Table 3.2 and depicted in Figure 3.2 can be interpreted as population proportions or probabilities. That is, a relative frequency distribution is analogous to a probability distribution, which will be studied in detail in later chapters. For example, the relative frequency for the first class in Table 3.2 indicates that 10% of the employees at Lang's Trucking Company have monthly salaries from $1,300 to less than $1,400. If 1 of the 80 employees was selected randomly, the probability would be 10% that the employee had a monthly salary from $1,300 to less than $1,400.

Using Relative Frequencies to Compare Different Variables

Relative frequencies are more useful than absolute frequencies for comparing different samples of data, especially when the samples are of different sizes. For example, suppose that 20 women in a sample of 200 women have a college degree and that 40 men in a sample of 400 men have a college degree. A person who examined just the absolute frequencies might get the mistaken impression that men are more likely than women to have college degrees, although 10% of each group have college degrees.

Cumulative Frequency Distributions

Frequently we want to consider not only the number of observations in a particular class but also the total number of observations that fall in a particular class and all previous classes. This number is called the *cumulative frequency* of the class.

DEFINITIONS **Cumulative Frequency and Cumulative Frequency Distribution** ———————————

Let f_i represent the class frequency of class i in a frequency distribution. The **cumulative frequency** of the ith class is the sum of the first i class frequencies. The cumulative frequency for the ith class can be obtained by using the formula

$$\text{Cumulative frequency for class } i = f_1 + f_2 + \cdots + f_i$$

The **cumulative frequency distribution** is a table that shows the cumulative frequency for each class in the frequency distribution.

———————————

Finally, we may want to consider the *proportion* of observations in a particular class and all previous classes. These proportions are called *cumulative relative frequencies*. A cumulative relative frequency shows the proportion of observations that are less than or equal to any specific value.

DEFINITIONS **Cumulative Relative Frequency and Cumulative Relative Frequency Distribution**

The **cumulative relative frequency** of the ith class shows the proportion of observations falling in the ith class and all previous classes. It is the sum of the first i relative frequencies. A **cumulative relative frequency distribution** is a table that shows the cumulative relative frequency for each class in a frequency distribution.

———————————

EXAMPLE 3.2 **A Cumulative Frequency Distribution** ———————————

Table 3.3 shows the cumulative frequency and cumulative relative frequency distributions obtained from the frequency distribution in Table 3.2. The second and third columns of Table 3.3 repeat the class frequencies and class relative frequencies of Table 3.2. The last

TABLE 3.3 *Cumulative frequency distribution of data in Table 3.2*

Monthly salary (dollars)	Frequency	Relative frequency	Cumulative frequency	Cumulative relative frequency
1,300 to under 1,400	8	.1000	8	.1000
1,400 to under 1,500	11	.1375	19	.2375
1,500 to under 1,600	17	.2125	36	.4500
1,600 to under 1,700	19	.2375	55	.6875
1,700 to under 1,800	10	.1250	65	.8125
1,800 to under 1,900	6	.0750	71	.8875
1,900 to under 2,000	4	.0500	75	.9375
2,000 to under 2,100	5	.0625	80	1.0000
Total	80	1.0000		

two columns of Table 3.3 show the cumulative frequency distribution and the cumulative relative frequency distribution.

For example, under the column headed Cumulative frequency, the third entry is the value 36. This indicates that 36 employees have a monthly income less than $1,600. This cumulative frequency equals the sum of the first three class frequencies, 8 + 11 + 17.

The cumulative relative frequency for the third class is .4500, which indicates that 45% of the employees have monthly salaries less than or equal to $1,600. The cumulative relative frequency for the third class is the sum of the class relative frequencies for the first three classes, .1000 + .1375 + .2125. Alternatively, the cumulative relative frequency for the third class is the class cumulative frequency divided by the number of observations; that is, $36/80 = .45$.

The information in a cumulative frequency distribution and in a cumulative relative frequency distribution can also be presented graphically. Figure 3.4 (page 52) shows the cumulative relative frequency histogram for the data in Table 3.3. In the figure, the area above each class interval is proportional to the cumulative relative frequency of the class.

Frequency Distributions Using Integer Data

Some variables take only integer values. Examples are the number of children in a family and the number of rooms in a house. In these cases, we may want to form classes that contain both their upper and lower limits (such as 0 to 2, 3 to 5, etc.), that contain only a single point, or that have noninteger class limits (such as 3.5 to 5.5, 5.5 to 7.5, etc.).

As an example, suppose a real estate agent has a computer printout showing all houses for sale in a given community. The agent might classify houses as having 1 to 5 rooms, 6 to 10 rooms, and 11 to 15 rooms. In this case, the classes could be 1 to 5, 6 to 10, and 11 to 15, and the class midpoints would be 3, 8, and 13, respectively.

In this example, it is not appropriate to denote the first class as 1 to under 6, since the class mark would be 3.5 and would not accurately represent the data contained in the

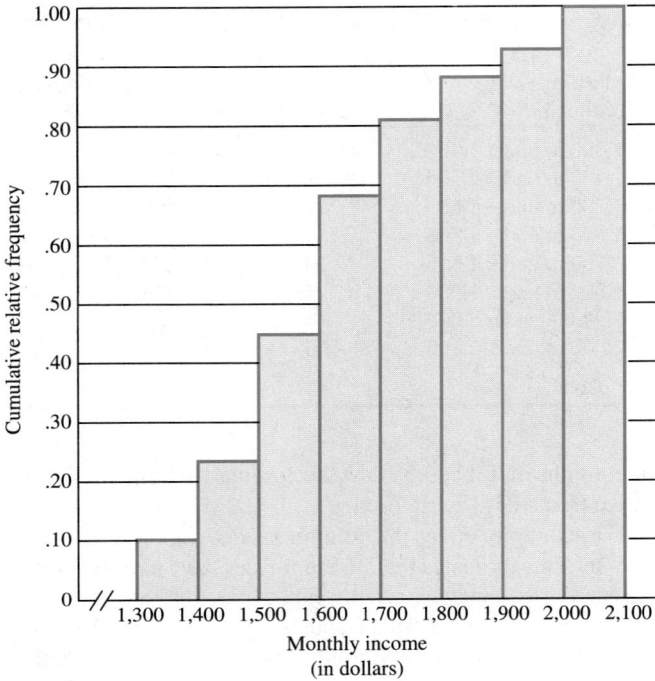

FIGURE 3.4   *Cumulative relative frequency distribution of data in Table 3.3*

class. As an alternative, the number of rooms in the house could be represented by the classes .5 to 5.5, 5.5 to 10.5, and 10.5 to 15.5; in this case, the class marks would again be 3, 8, and 13.

EXAMPLE 3.3   **A Frequency Distribution Based on Integer Data**

The following data show the number of cars owned by a sample of 120 families who live in Detroit, Michigan:

```
2  1  3  1  2  1  2  3  4  2  2  0  5  2  1  1  2  2  0  2
2  3  1  1  0  3  2  4  2  1  2  3  1  1  2  2  1  2  4  3
2  0  1  1  2  3  1  2  0  1  2  3  2  5  2  1  1  2  3  0
0  2  1  1  3  1  2  1  3  0  1  2  2  1  1  2  3  2  1  2
2  3  0  2  2  1  1  2  2  3  2  1  2  2  1  4  2  2  1  2
2  1  2  1  2  1  4  5  2  1  1  2  1  1  2  3  1  2  1  4
```

Table 3.4 shows the frequency, relative frequency, cumulative frequency, and cumulative relative frequency distributions for these data. In the table, each class contains just a single point.

Observe that the relative frequency of the value 0 is .075, which means that 7.5% of

TABLE 3.4 *A frequency distribution based on integer data*

Number of cars	Frequency	Relative frequency	Cumulative frequency	Cumulative relative frequency
0	9	.075	9	.075
1	39	.325	48	.400
2	48	.400	96	.800
3	15	.125	111	.925
4	6	.050	117	.975
5	3	.025	120	1.000
Total	120	1.000		

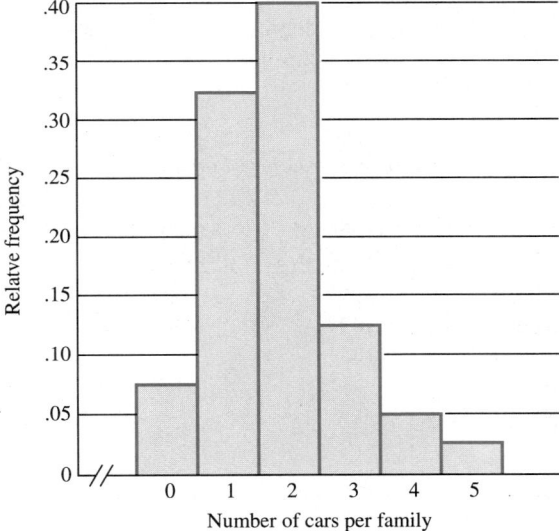

FIGURE 3.5 *Relative frequency histogram for data in Table 3.4*

the families in the sample do not own a car. Conversely, this means that 92.5% of the families in the sample own at least one car. The most frequent number of cars per family is 2. In the sample, 40% of the families fall in this category.

Figure 3.5 shows the relative frequency histogram for the data in Table 3.4. In the histogram, the proportion of families with no car is represented by the area of the rectangle above the interval extending from −.5 to .5; the proportion of families who own one car is represented by the area of the rectangle above the interval extending from .5 to 1.5; and so forth. Each class interval has a width of 1 unit and is centered at the appropriate integer. Because each rectangle has a width of 1 unit, the height of each rectangle represents the relative frequency of the class.

Analysis of Qualitative Variables

When the variable being studied is qualitative, we want to determine how many observations in a sample or population possess each specific value of the variable. Each value can be used to denote a class, or several values can be combined to form a single class. A table showing how many observations fall into each class is a frequency distribution of the qualitative variable.

EXAMPLE 3.4 **A Frequency Distribution Based on Qualitative Data** ——————————————

When studying the productivity of employees at a company, it is customary for economists to examine the educational background of the employees. The data in Table 3.5 show the highest academic degree earned by the 550 employees of the Cottrell Corporation. The data in the column headed Frequency show how many employees have the specified degree as their highest degree; these frequencies form a qualitative frequency distribution. The data in the column headed Relative frequency show the proportion of employees falling in each class.

T A B L E 3.5 *Frequency distribution based on qualitative data*

Highest degree	Frequency (number of employees)	Relative frequency
Grade school	15	.027
High school	200	.364
Bachelor's	185	.336
Master's	55	.100
Doctorate	70	.127
Other	25	.045
Total	550	1.000

Bar Charts and Pie Charts

Qualitative frequency distributions and relative frequency distributions can be illustrated using **bar charts** or **pie charts.** In a bar chart, the height or length of each bar represents the frequency or relative frequency of a particular class. In a pie chart, the area of each slice of the pie represents the relative frequency of the particular class.

Figure 3.6 is a bar chart that depicts the data presented in the frequency distribution in Table 3.5. For example, in Table 3.5, 200 of the employees listed high school as their highest degree. Thus, in the bar chart, the height of the bar above the title "High school" is 200 units. Figure 3.7 is a pie chart depicting the relative frequencies in Table 3.5. For example, in Table 3.5, 36.4% of the employees listed high school as their highest degree. Thus, in the pie chart, the area of the slice titled "High school" represents 36.4% of the entire pie.

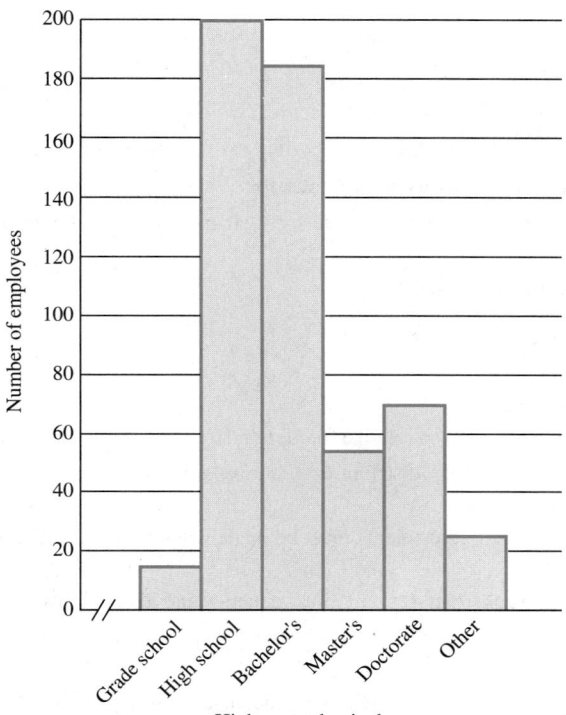

FIGURE 3.6 *Bar chart for the data in Table 3.5*

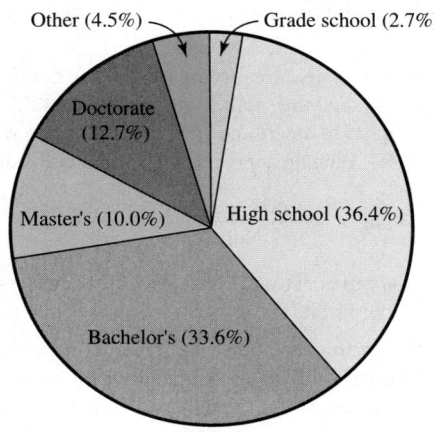

FIGURE 3.7 *Pie chart for the data in Table 3.5*

Exercises

Statistical Concepts

3.1.1 Explain what is meant by descriptive statistics.

3.1.2 Explain three common methods of describing the characteristics of a data set.

3.1.3 Explain what is meant by a summary statistic.

3.1.4 Explain what is meant by a frequency distribution. Describe how to construct a frequency distribution.

3.1.5 Define the following terms:
a. Class boundaries
b. Open-ended class
c. Class width
d. Class mark

3.1.6 Explain what is meant by a relative frequency. How is a relative frequency calculated?

3.1.7 Explain why a frequency distribution and a relative frequency distribution must have the same shape.

3.1.8 Explain what special adjustments must be made when drawing a histogram when the classes have unequal widths.

3.1.9 Describe how to construct a cumulative frequency and a cumulative relative frequency distribution.

3.1.10 *True or false:* A cumulative relative frequency shows the proportion of values falling in a specific class. Explain.

3.1.11 *True or false:* The sum of all relative frequencies must equal 1.00. Explain.

3.1.12 What is meant by integer data? Give an example of an integer variable. Describe how to construct a frequency distribution for integer data.

3.1.13 *True or false:* In a pie chart, the area of each slice of the pie represents the frequency of the particular class.

3.1.14 *True or false:* A pie chart is useful for showing the relative frequencies for different values of a qualitative variable. Explain.

3.1.15 Suppose you want to construct a relative frequency distribution containing k class intervals. Suppose the largest measurement is L and the smallest measurement is S.
a. Describe how to use k, L, and S to determine an appropriate class width W.
b. Let $k = 8$, $L = 13.8$, $S = 5.7$. Find an appropriate class interval using the formula

$$\text{Width} = \frac{\text{Largest value } - \text{ Smallest value}}{\text{Number of intervals}}$$

c. Let $k = 8$. Suppose we want to construct a frequency distribution that begins at $S = 240$ and ends at $L = 400$. Find an appropriate class width.

3.1.16 Suppose you want to construct a frequency distribution. A reasonable way to begin would be to select a reasonable value W for the width of each interval that is some convenient number, such as 5, 10, 25, or 100. Next, find the range R of the data, which is the largest measurement minus the smallest measurement. Divide R by W to get an approximate value for k, the number of classes. If k is not an integer, round up to the next larger integer.
a. Let the range of a set of data be $730. Suppose you choose $100 as an appropriate value for W. How many classes should you use?

    b. Let the range of a set of data be 33. Suppose you choose 5 as an appropriate value for *W.* How many classes should you use?

3.1.17 Suppose we want to construct a pie chart showing the proportions of workers employed in each division of a company.
    a. Suppose the manufacturing division employs 10% of the total work force. How large is the central angle of the slice of pie representing the manufacturing division?
    b. Suppose the accounting division employs one-third of the total work force. How large is the central angle of the slice of pie representing the accounting division?
    c. Explain a procedure for determining the central angle of a slice of pie to represent a particular proportion of the work force.

Statistical Drills

3.1.18 According to the *Statistical Abstract of the United States, 1992* (p. 39), there were 66.1 million families in the United States in 1991. The following data show the number of families (in millions) by income in 1990:

Income	Number of families
Less than $5,000	2.2
$5,000 to $9,999	3.9
$10,000 to $14,999	5.0
$15,000 to $24,999	10.8
$25,000 to $34,999	10.7
$35,000 to $49,999	13.3
$50,000 or more	20.2

    a. Find the relative frequency for each income category.
    b. Draw a histogram showing the relative frequencies. Adjust the areas to account for the unequal class widths. For the last class, use $200,000 as the upper limit.
    c. Construct the cumulative relative frequency distribution and graph it.
    d. What type of variable is Income: nominal, ordinal, interval, or ratio? Explain.

3.1.19 According to the *Statistical Abstract of the United States, 1992* (p. 350), in 1990, there were 39.2 million people in the United States who were receiving Social Security benefits. The following data show the proportions of beneficiaries by type of benefit:

Type of benefit	Proportion
Retired workers	.62
Survivors of deceased workers	.18
Disabled workers	.08
Other	.12

    a. Find the absolute frequency for each type of benefit.
    b. Construct a bar chart showing the absolute frequencies.
    c. Construct a pie chart showing the relative frequencies.
    d. What type of variable is Type of benefit: nominal, ordinal, interval, or ratio? Explain.

3.1.20 According to the *Statistical Abstract of the United States, 1992* (p. 345), there were 469,000 enlistments and reenlistments in the United States military in 1990. The following data show the numbers of military accession (in thousands) by branch of service:

Branch of service	Number
Army	182
Navy	135
Marine Corps	48
Air Force	104

a. Find the relative frequency for each branch of service.
b. Construct a bar chart showing the absolute frequencies.
c. Construct a pie chart showing the relative frequencies.
d. What type of variable is Branch of service: nominal, ordinal, interval, or ratio? Explain.

3.1.21 The following data show the years of service for 25 employees at a manufacturing firm:

    3.2   14.3   13.1   5.7   19.6   22.0   13.3   12.2   17.9   14.5
    14.2 22.8   16.2   8.2   5.3   11.2   6.0   10.1   2.9   12.5
    6.2   8.5   5.3   2.8   .2

a. Construct a frequency distribution and draw a graph of the distribution. Use the intervals 0 to under 5.0, 5.0 to under 10.0, and so forth.
b. Construct a relative frequency distribution and draw a graph of the distribution.

3.1.22 The price-to-earnings ratio of a stock shows the selling price of the stock as a multiple of its annual earnings during the previous year. If investors think that a firm has a bright future and expect its future earnings to increase, they will bid up the price of the stock, resulting in a high price-to-earnings ratio (the PE ratio). For example, suppose a stock has a price of $50 per share and had earnings of $5.00; then its PE ratio is 10. Suppose another stock has a price of $100 per share and had earnings of $5.00 per share; its PE ratio is 20. The second stock costs twice as much as the first stock because purchasers evidently believe that the future prospects of the second stock are much better than future prospects of the first stock. Thus, a high PE ratio is an indicator that investors have high expectations about a stock's future performance. The following data show the PE ratios for a sample of 40 firms as reported in the *Wall Street Journal:*

    12   15   9   7   12   10   8   19   24   32
    16   25   13   8   21   13   37   22   18   33
    42   12   49   31   22   8   17   29   14   23
    21   11   19   31   15   14   11   29   14   12

a. Construct a tally sheet and a frequency distribution, and draw a graph of the distribution. As intervals, use 0 to under 5.0, 5.0 to under 10.0, and so forth.
b. Construct the relative frequency distribution and draw a graph of the distribution.

Statistical Applications

3.1.23 The following data list the miles (in thousands) driven during the past 4 months by 50 sales representatives of a food distributor:

34	23	31	25	4	32	17	19	42	30
8	19	26	35	36	24	47	22	27	29
12	5	26	16	7	46	35	34	27	38
15	27	38	32	10	9	12	24	27	45
15	26	27	25	24	44	18	27	23	29

a. Construct a frequency distribution. As classes, use 0 to under 10, 10 to under 20, and so forth.

b. Construct the corresponding relative frequency distribution and cumulative frequency distribution.

c. Graph the histogram.

3.1.24 Repeat Exercise 23 but use the following classes: 0 to under 5, 5 to under 10, 10 to under 15, and so forth.

a. Comment on how the shape of the distribution changed and why.

b. Does the histogram in Exercise 23 or in this exercise better represent the data?

3.1.25 The following data show the number of cars owned by a sample of 50 households:

1	2	0	2	1	1	3	1	1	2
1	3	1	2	2	0	4	1	1	2
2	3	1	1	2	1	1	2	2	1
0	1	1	2	0	1	1	2	2	1
2	2	1	3	1	1	1	2	4	1

a. Construct a frequency distribution.

b. Construct the corresponding relative frequency distribution.

c. Construct the corresponding cumulative frequency distribution.

d. Construct the corresponding cumulative relative frequency distribution.

e. Graph each of the distributions above.

3.1.26 Draw a pie chart to illustrate the following market shares by company in the U.S. soft drink industry: Coca-Cola, 36%; Pepsi-Cola, 25%; Dr Pepper, 7%; Seven-Up, 6%; Royal Crown, 4%; others, 22%.

3.2   Common Shapes of Distributions

If we have the entire population of data, then we can construct histograms and relative frequency distributions for the population. Usually, however, only a sample of data is available, and our relative frequency graph only approximates the curve for the population relative frequency. Frequency distributions and relative frequency distributions come in all shapes and sizes, but many special distributions and shapes of distributions occur regularly in statistical studies. Some of these distributions have been given special names, and a few will be described now.

Skewness and Symmetry

A distribution is said to be **symmetric** if the left half of the graph of the distribution is the mirror image of the right half. Figure 3.8 (page 60) shows a symmetric distribution. A distribution is said to be **skewed** if it is not symmetric. A distribution is *skewed to the right* if the right-hand tail of the distribution is longer than the left tail and most of the

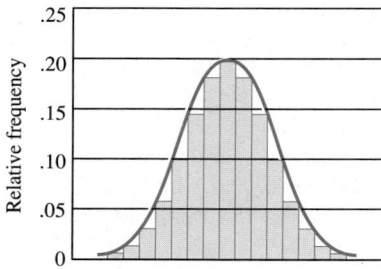

FIGURE 3.8   *A symmetric distribution*

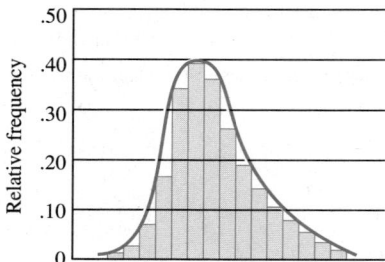

FIGURE 3.9   *A distribution that is skewed to the right*

observations are concentrated in the left side of the distribution. Figure 3.9 shows such a distribution.

The distribution of annual incomes earned by adults and the distribution of the wealth of individuals are obvious and important examples of distributions in economics that are skewed to the right. Most people have relatively modest incomes, but the incomes of the highest-income earners extend over a very wide range. Thus, most incomes are concentrated at the left side of the distribution, and the distribution has an extremely long right tail. The same pattern holds for the distribution of family wealth. Many other economic and demographic variables have distributions that are skewed to the right, including the annual sales of business firms in the United States, the population of cities and towns in the United States, and the number of cars owned by families.

A distribution is *skewed to the left* if the left-hand tail of the distribution is longer than the right-hand tail and most of the observations are concentrated in the right side of the distribution. Figure 3.10 shows a distribution that is skewed to the left. The age at death of U.S. citizens is an example of a variable whose distribution is skewed to the left. Most people die between the ages of 65 and 80, so most observations are concentrated toward the right side of the distribution. A long tail extends to the left, however, reflecting the frequencies of people who die between the ages of 0 and 60.

For many variables, most of the observations are concentrated near the middle of the distribution. As the distance from the middle increases, the frequencies and relative frequencies decrease. Such distributions are often described as "bell shaped." One example of such a distribution is called the **normal distribution.** A number of variables in nature

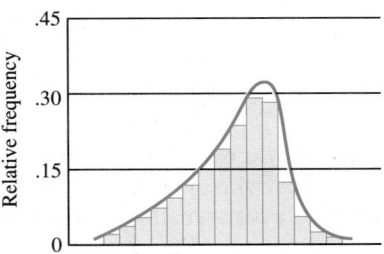

FIGURE 3.10 *A distribution that is skewed to the left*

and in society are approximately normally distributed, such as the height, IQ, and blood pressure of adults. Figure 3.8 is an example of a normal distribution. The normal distribution, by far the most important distribution in statistics, is discussed in detail in Chapter 8.

Other Important Distributions

Figure 3.11 represents a distribution that peaks at the origin and then tails off to the right. This distribution, a type of **exponential distribution,** is skewed to the right and is useful in studying waiting-line or queuing problems. For example, the time between the arrivals of people entering a line at a grocery store or at a toll booth generally has an exponential distribution. The exponential distribution is studied in Chapter 8.

Figure 3.12 shows a graph of a **uniform distribution.** For variables with a uniform distribution, no particular interval of values has a higher relative frequency than any other interval of equal width. The uniform distribution is symmetric about its central value. Thus, if a variable takes some value between a lower limit a and an upper limit b and all

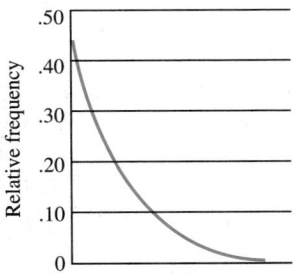

FIGURE 3.11 *An exponential distribution*

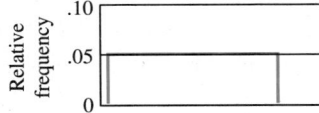

FIGURE 3.12 *A uniform distribution*

values in this interval are equally likely, then the variable follows a uniform distribution. The uniform distribution is studied in Chapter 8.

Unimodal and Bimodal Distributions

The location at which a population relative frequency distribution peaks is called the **mode** of the distribution. Distributions with only one peak are called *unimodal distributions*; distributions with two distinct peaks are called *bimodal distributions*. For example, the normal distribution shown in Figure 3.8 is a unimodal distribution, whereas Figure 3.13 is an example of a bimodal distribution.

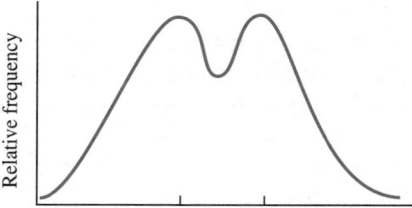

FIGURE 3.13 *A bimodal distribution*

A bimodal distribution is common in a population containing two nonhomogeneous sectors. For example, a relative frequency distribution of the heights of adults will have two distinct peaks, one at a value of about 64 inches and one at about 70 inches. The heights of males cluster at about 70 inches, and those of females at 64 inches. If we separate the data into two subpopulations, one containing just males and one containing just females, we get two unimodal distributions. Similarly, a bimodal distribution results if we combine income data for college graduates with income data for high school graduates.

Populations Versus Samples

Keep in mind that the relative frequency distribution of a sample is different from the relative frequency distribution of the population from which the sample was drawn; although you may not have all the data needed to construct a population relative frequency distribution, such a distribution exists or can be imagined. Try to visualize this population relative frequency distribution, because the questions that statisticians try to answer concern this distribution.

Although we are interested in describing the characteristics of a sample of observations, we are more interested in describing the characteristics of the population from which the sample was selected. The most important feature of the graph of the sample relative frequency distribution is the information it provides about the population relative frequency distribution. The sample relative frequency distribution has approximately the same shape as the population relative frequency distribution, and the degree of similarity increases as the sample size increases.

Consider two relative frequency histograms, one based on a small sample of data and one based on a large sample. As the sample size increases, we obtain a better description of the data by decreasing the width of the class intervals. When the class intervals become

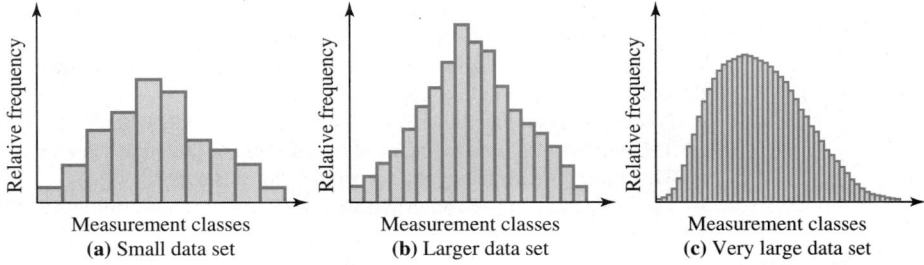

FIGURE 3.14   *The relative frequency distribution approaches a smooth curve as sample size increases*

small enough, a relative frequency histogram begins to look like a smooth curve. In the limit, this smooth curve represents the relative frequency histogram of the population. Figure 3.14 shows how the relative frequency distribution approaches a smooth curve as the number of observations increases and the width of the classes decreases.

Exercises

Statistical Concepts

3.2.1   *True or false:*   The total area under any relative frequency distribution must sum to 1.00. Explain.

3.2.2   *True or false:*   Suppose a relative frequency distribution is symmetric about the value $X = 50$. The area to the right of $X = 60$ is the same as the area to the left of $X = 40$. Explain.

3.2.3   *True or false:*   Suppose a distribution is symmetric about the value $X = 100$. The area between $X = 60$ and $X = 80$ is the same as the area between $X = 80$ and $X = 100$. Explain.

3.2.4   Explain what is meant by a distribution that is skewed to the right. Draw a graph of such a distribution. Give an example of a variable that is skewed to the right.

3.2.5   Suppose a variable has a uniform relative frequency distribution and takes values from 10 to 20. *True or false:*   The area to the right of $X = 16$ is the same as the area between $X = 10$ and $X = 14$. Explain.

3.2.6   Explain what is meant by the mode of a distribution.

3.2.7   *True or false:*   Suppose a relative frequency distribution is unimodal with mode at $X = 25$. Then the area to the right of $X = 25$ is .5. Explain.

3.2.8   *True or false:*   Suppose a distribution is bimodal with modes at 25 and 35. Then the area to the left of $X = 25$ is the same as the area to the right of $X = 35$. Explain.

3.2.9   Suppose a variable has a symmetric normal distribution, with a mode of $X = 50$. Suppose the total area under the distribution is 1.00.
a. Find the area to the right of $X = 50$.
b. *True or false:*   The area to the right of $X = 60$ is the same as the area to the left of $X = 40$.
c. *True or false:*   The area between $X = 50$ and $X = 60$ is the same as the area between $X = 50$ and $X = 40$.

Statistical Drills

3.2.10   For each of the following variables, describe the likely shape of a histogram. Discuss whether the distribution is likely to be symmetric, bell shaped, skewed to the right, skewed to the left, exponential, or uniform.

a. Random numbers selected from the integers 0 to 9
b. Ages of children at an elementary school
c. Annual incomes of adult male college graduates in the United States
d. Daily sales revenues at a small pizza shop
e. Last digits of all telephone numbers in the Miami phone book
f. The number of accidents incurred by a sample of 200 high school students last year
g. The mileages traveled by all cars in the United States last year
h. The amount deducted for charitable contributions when families file their income tax returns

3.3   Lying with Statistics (Optional)

In their famous book *How to Lie with Statistics,* Huff and Geis (New York: Norton, 1954) list numerous examples in which unscrupulous statisticians distorted data to mislead readers. At times it is possible to present a distorted picture of data even when trying to be fair and honest. In this section, we describe some common mistakes that people make when analyzing statistical data.

1. *Improper sizes of graphs:*   Artists frequently draw a picture of some variable (such as a picture of a car to represent the annual dollar sales of cars) and indicate the magnitude of the variable by the size of the picture. For example, a person might be interested in describing the annual dollar sales of new cars between 1980 and 1995. Suppose that the dollar volume of car sales doubled during that time. To denote sales in 1980, the artist might draw a car 2 inches long and 1 inch high. To denote sales in 1995, the artist might then draw a car approximately 4 inches long and 2 inches high. The problem with this scale is that people usually compare areas of objects, not lengths. Because the area of the 4-by-2-inch car is four times the area of the 2-by-1-inch drawing, the unwary reader might think that the dollar value of car sales had quadrupled between 1980 and 1995, rather than doubled.

2. *Emotional presentation of facts:*   Occasionally a statistician can try to influence the opinion of the reader by using suggestive terms. For example, a statistician might report, "The monthly government income supplement given to permanently disabled war veterans is only $450 per month." By using the word *only* in this statement, the statistician would convey the impression that the monthly income supplement was too low. Whether the supplement was too low would be subject to debate, but the statistician should not impose his or her views on the reader.

3. *Incomplete or inadequate numerical presentations:*   In a recent legal case in Pennsylvania, a woman claimed that her salary was too low and that she was the victim of gender discrimination. To support her claim, the woman produced a report comparing the salaries of males and females in five different departments of the company. The report stated that in three of the departments, every woman had a lower monthly salary than every man. On further analysis, it was found that there were only two women and three men in each of the three departments. In each department, each of the men had a master's degree in engineering, had been with the company for more than 10 years, and was responsible for doing complicated engineering analyses. In contrast, each of the females was a secretary without a college degree and with only a few years of experience. By hiding this information, the woman was trying to lead people to believe that women were underpaid. In the other two departments of the company, there were approximately 50 males and 50 females of varying educational backgrounds and with differing amounts of seniority, performing a myriad of different jobs. After taking into account their educational backgrounds, their seniority, and their duties, it was determined that neither men nor women were being paid inordinately high or low salaries.

Thus, before comparing incomes of individuals, we should take into account other factors, such as education, experience, length of service, and so forth. In the U.S. economy, males tend to earn more than females, and whites tend to earn more than nonwhites. At the same time, males tend to have more years of education than females and whites tend to have more years of education than nonwhites. Only after we have controlled for education (and other relevant variables) can we make conclusions concerning discrimination.

4. *Reading too much into a coincidence:*   For 13 of the first 14 years in which the Super Bowl was played, the stock market went down when a team from the American Conference of the National Football League won, and went up when a team from the National Conference won. Suppose an American Conference team wins the next Super Bowl. Does this indicate that the stock market will drop during the following year? Even though the rule worked in 13 years out of 14, it is unreasonable to believe that the outcome of a football game could influence the stock market.

Given the large number of variables for which data have been collected, it should not be surprising that occasionally, just out of luck, some enterprising person will come across a few variables that are completely unrelated to one another but which tend to move together for a short period of time. Admittedly, the correlation between the behavior of the stock market and the outcome of the Super Bowl is unusual and surprising, but in this case, common sense should outweigh statistical evidence. Without a doubt, most statisticians would say that the correlation between the outcome of the Super Bowl and the behavior of the stock market is due to random chance. This is not unlike a person tossing a coin 14 times and calling heads or tails correctly on 13 of the tosses. If enough people try this feat, sooner or later someone will do it. This does not prove that the person can foretell the outcome of the next coin toss.

5. *Making general statements based on small samples:*   It is very dangerous to make general statements about the properties of a population on the basis of a very small sample. Usually small samples contain too little information to make a conclusion with a high level of certainty. As we shall see when we discuss estimation and the construction of confidence intervals, generalization from a sample to a population becomes more reliable as the amount of sample evidence increases. When the sample is very small, one or two unusual observations can have a big effect on the outcome of the study. This is especially the case if the observations are selected from a population with a large amount of variation.

3.4   The Stem-and-Leaf Diagram (Optional)

When we are presented with a set of numerical observations, it is natural for us to want to condense or simplify the data in some way so that we can recognize general characteristics of the data. For example, most people would like to know the following things:

1. How are the data distributed? What is the general shape of the distribution? Where are most of the observations concentrated?
2. Where is the center of the data?
3. How spread out, or dispersed, are the data about this central value?
4. Are there any very large or very small values that appear to be quite different from the other observations?

The most frequently used technique for answering these questions is to construct a frequency or relative frequency distribution and to calculate some summary statistics, which are discussed in the next chapter. Another closely related way of answering these questions is to construct a **stem-and-leaf diagram,** a technique proposed by John Tukey.* The stem-and-leaf diagram is partly tabular and partly graphical in nature. The purposes of the stem-and-leaf diagram are as follows:

1. To show the range of the data, that is, to make it easy to determine the difference between the largest observation and the smallest observation
2. To show how the data are distributed
3. To show in a general way where the center of the data is
4. To show in a general way how spread out the data are
5. To show whether there are any extremely high or low values, called **outliers**

EXAMPLE 3.5   **Construction of a Stem-and-Leaf Diagram** ———————————————

In 1994, a female employee who was laid off sued her employer alleging that she had been released because of age and gender discrimination. A statistician was asked to examine the plaintiff's allegations. In the course of the study, it became necessary to compare the ages of the employees who were laid off with the ages of the employees who were not. The accompanying data show the ages of 60 employees who were laid off in 1994:

43	34	21	26	57	64	51	30	38	60
19	26	41	47	58	50	42	32	25	18
36	27	38	37	24	49	56	61	20	39
32	51	60	52	45	44	33	25	29	27
31	35	39	19	20	30	46	47	32	24
29	22	20	37	34	29	60	55	30	41

In a stem-and-leaf diagram, the first or leading digit serves as the **stem** and the trailing digit serves as the **leaf.** The leading digit determines the row in which an observation is placed; the trailing digit is written in this row to the right of a vertical bar that separates the stems from the leaves. In this example, the stems are the first digits for each person's age; thus the stems are the digits 1 through 6. For example, for a person aged 53, the stem is 5 and the leaf is 3. To improve the appearance of the diagram, it is useful to arrange the leaves in each row from lowest to highest. The stem-and-leaf diagram for the age data follows.

———————
* J. Tukey, *Exploratory Data Analysis* (Reading, Mass.: Addison-Wesley, 1977).

1	8	9	9														
2	0	0	0	1	2	4	4	5	5	6	6	7	7	9	9	9	
3	0	0	0	1	2	2	2	3	4	4	5	6	7	7	8	8	9	9
4	1	1	2	3	4	5	6	7	7	9							
5	0	1	1	2	5	6	7	8									
6	0	0	0	1	4												

The stem-and-leaf diagram allows us to pick out several important characteristics of the data. In this example, the smallest observation is 18 and the largest is 64. There are 18 leaves following the 3 stem and 16 leaves following the 2 stem, indicating that 18 people laid off were in their 30's and 16 employees were in their 20's. These are the largest classes. Few people under 20 or over 60 were laid off.

Each stem in the diagram defines a class of ages, just as a class interval defines a class of ages in a frequency distribution. The number of leaves next to each stem represents the class frequency associated with that stem. If the stem-and-leaf diagram is rotated 90^6 counterclockwise, so that the leaves rise vertically rather than horizontally, the stem-and-leaf diagram looks very similar to a histogram, the height of the leaves in each column visually representing the frequency of that class.

The stem-and-leaf diagram has two advantages over the histogram: (1) it allows you to reconstruct the original data set, and (2) it lists the observations in order of magnitude. The stem-and-leaf diagram has some disadvantages also. It is suitable only for relatively small data sets; no one would want to list hundreds of observations in a stem-and-leaf diagram. Second, there is little flexibility in the choice of stems and thus in the number of classes.

With a bit of care, the stem-and-leaf diagram can be used with any data set. If the observations consist of numbers with, say, six digits, we might use the first digit as the stem and the second digit as the leaf and ignore the remaining digits. For the number 453,210, the stem could be 4 and the leaf 5, or the stem might be 45 and the leaf 3.

If narrower classes are desired, a stem can be used more than once. For example, each stem could be used twice in Example 3.5. The first time the stem 3 is recorded, it would be associated with the values 30, 31, 32, 33, and 34; the second time it would be associated with the values 35, 36, 37, 38, and 39. The same applies to the other stems.

The stem-and-leaf diagram has a few advantages over the frequency distribution. Unlike in the frequency distribution, no information on the value of each observation is lost. In addition, the stem-and-leaf diagram conveniently orders the data from lowest to highest, which makes it easy to pick out the middle value.

Exercises

Statistical Concepts

3.4.1 List some of the reasons for constructing a stem-and-leaf diagram.

3.4.2 *True or false:* In a stem-and-leaf diagram, it is easy to determine the smallest value and the largest value. Explain.

3.4.3 When data are arranged in chronological order, we obtain a time series of data. Plotting data in chronological order can help us detect patterns over time. *True or false:* In a stem-and-leaf diagram, it is possible to determine the chronological order in which the observations occurred. Explain.

Statistical Applications

3.4.4 The following data show the number of customers at a General Motors service center on a sample of 30 days:

67	76	58	82	59	51	63	69	70	75
67	43	58	61	40	58	46	57	72	71
65	73	58	45	48	49	50	64	53	64

Construct a stem-and-leaf diagram. Based on the diagram, guess the average number of customers per day.

3.4.5 The following data show the ages of 40 truck drivers employed by the Eastern Trucking Company:

26	34	22	45	37	30	19	26	34	20
32	43	51	50	40	44	45	36	35	42
49	38	26	28	37	42	40	38	32	43
41	23	18	19	26	52	43	37	47	50

Construct a stem-and-leaf diagram. Based on the diagram, guess the average age.

3.4.6 The following data show the number of years of seniority for the employees at the Hammermill Paper Company.

20	8	13	24	16	12	10	5	3	12	26	24	19	12	24
2	12	9	6	4	11	10	17	18	2	1	15	22	13	18
3	14	16	19	22	25	26	23	14	17	16	13	19	7	5

Construct a stem-and-leaf diagram. Guess the average years of seniority of employees at the firm.

3.5   Computer Applications

You can use the SPSSX FREQUENCIES command to make a frequency table, bar chart, or histogram, or to calculate various descriptive statistics. For example, refer to the data in Table 2.1 in Chapter 2. To generate a frequency table showing how many employees at Computech work in each division, issue the following command:

```
FREQUENCIES VARIABLES = DIV
```

To generate a bar chart showing the same information, issue the command

```
FREQUENCIES VARIABLES = DIV/BARCHART
```

The bar chart showing the frequency distribution for the variable DIV is shown in Figure 3.15.

```
DIV


                                                    Valid    Cum
Value Label             Value  Frequency  Percent  Percent  Percent

OFFICE                   1.00      38      31.7     31.7     31.7
MANUFACTURING            2.00      58      48.3     48.3     80.0
SALES                    3.00      24      20.0     20.0    100.0
                                 -------  -------  -------  -------
                        Total     120     100.0    100.0


                  |
         OFFICE   |********************************* 38
                  |
  MANUFACTURING   |************************************************** 58
                  |
          SALES   |******************** 24
                  |
                  +---------+---------+---------+---------+---------+
                  0        12        24        36        48        60
                                      Frequency

Valid cases      120      Missing cases       0
```

FIGURE 3.15 **SPSSX-generated bar chart for the variable DIV in Table 2.1**

To generate a histogram showing the ages of employees at Computech, issue the following command:

FREQUENCIES VARIABLES = AGE
   /HISTOGRAM MINIMUM (18) MAXIMUM (68) INCREMENT (5)

The histogram will group employees into classes 5 units wide, beginning with age 18. The histogram generated by this FREQUENCIES command is shown in Figure 3.16.

In the histogram, the first line of stars represents employees from age 18 to 23, the second represents employees from age 23 to 28, and so forth. For each 5-year group of ages, the computer output shows a row of asterisks (∗) whose length indicates the number

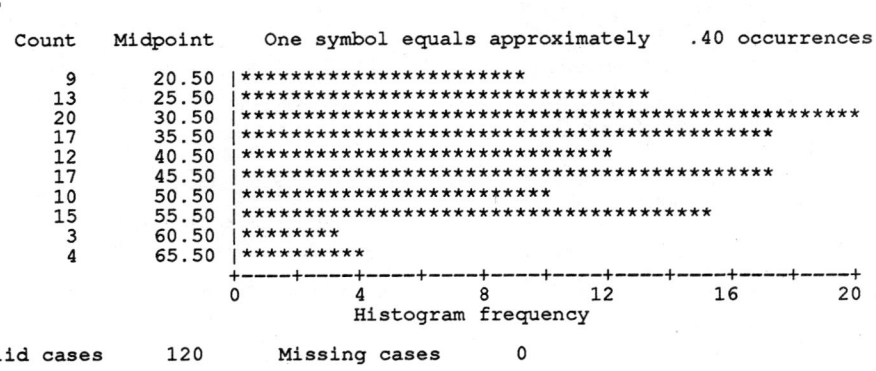

```
AGE
    Count   Midpoint    One symbol equals approximately   .40 occurrences

      9      20.50  |**********************
     13      25.50  |********************************
     20      30.50  |**************************************************
     17      35.50  |******************************************
     12      40.50  |*****************************
     17      45.50  |******************************************
     10      50.50  |************************
     15      55.50  |************************************
      3      60.50  |********
      4      65.50  |**********
                    +----+----+----+----+----+----+----+----+----+----+
                    0         4         8        12        16        20
                                Histogram frequency

Valid cases      120      Missing cases       0
```

FIGURE 3.16 **SPSSX-generated histogram for the variable AGE in Table 2.1**

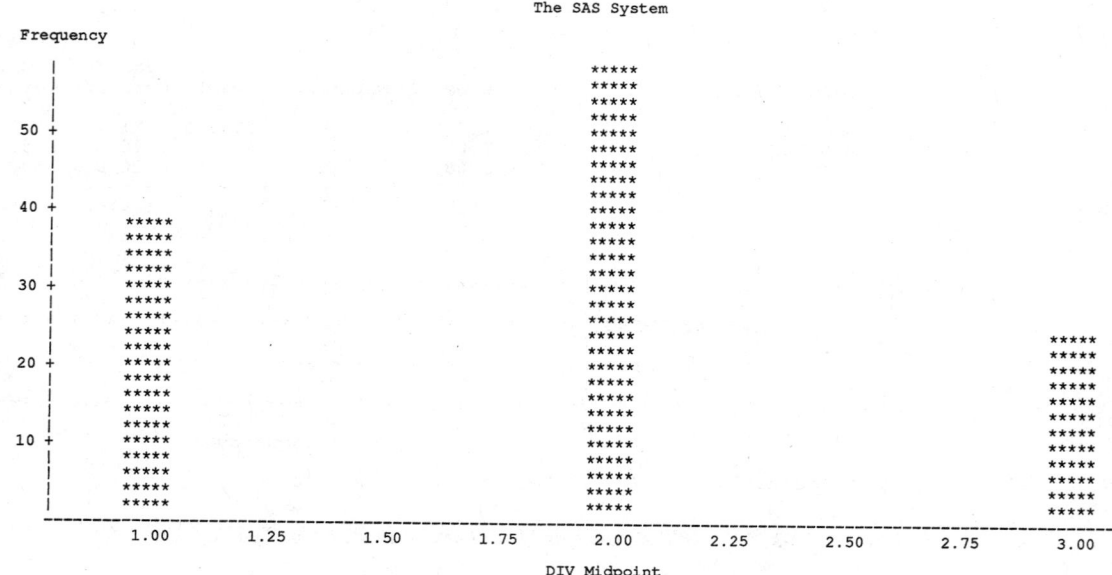

FIGURE 3.17   *SAS-generated bar chart for the variable DIV in Table 2.1*

of observations in that group. The actual number of cases appears in the column labeled Count. The middle value of each interval is printed in the column labeled Midpoint. For example, the midpoint age for employees between age 18 and age 23 is 20.5.

The SAS program can be used to produce bar charts and pie charts for discrete variables and histograms for continuous variables. Suppose we wish to generate a bar chart showing the number of employees in each division at the Computech Corporation according to Table 2.1. We would insert the following commands at the end of the SAS program shown in Figure 2.3 in Chapter 2:

PROC CHART;
   VBAR DIV;

The command PROC CHART tells SAS to draw a chart, and the command VBAR DIV tells SAS to construct a vertical bar chart for the variable DIV. The output is in Figure 3.17.

Generating a Bivariate Frequency Table

SPSSX and SAS programs can be used to generate bivariate frequency tables. Suppose we wish to examine the data in Table 2.1 to see whether there is any relationship between academic degree and division in which an individual is employed. In SPSSX, the command

CROSSTABS TABLES = DEGREE BY DIV

```
DEGREE   by   DIV

                            DIV                       Page 1 of 1
                  Count  |
                         |OFFICE   MANUFACT SALES
                         |          URING
                         |    1.00|     2.00|    3.00| Row
DEGREE          ---------+--------+--------+--------+  Total
           1.00 |    35   |    53   |    1   |      89
    HIGH SCHOOL |         |         |        |      74.2
                +--------+--------+--------+
           2.00 |     3   |     5   |   15   |      23
    COLLEGE     |         |         |        |      19.2
                +--------+--------+--------+
           3.00 |         |         |    8   |       8
    POST-GRAD   |         |         |        |      6.7
                +--------+--------+--------+
         Column      38        58       24       120
          Total      31.7      48.3     20.0     100.0

Number of Missing Observations:   0
```

FIGURE 3.18 *SPSSX output obtained by using the CROSSTABS command on the variables DEGREE and DIV in Table 2.1*

will generate a bivariate frequency distribution for the variables DEGREE and DIV. If the additional command

---

OPTIONS 3, 4, 5

---

is inserted into the SPSSX program, the computer will also print out various relative frequencies along with the bivariate frequency distribution. These two commands can be inserted anywhere between the END DATA command and the FINISH command. Figure 3.18 shows the output generated by these two commands using the data in Table 2.1.

In Figure 3.18, of the 89 employees with a high school degree, only 1 is employed in sales, whereas only 3 of the 23 employees with college degrees are employed in the office. All 8 of the employees with postgraduate degrees are employed in sales. Of the 38 employees in the office, 35 have a high school degree and none have a postgraduate degree. Finally, of the 58 employees in manufacturing, 53 have high school degrees and 5 have college degrees. Thus, the data in Figure 3.18 show that there is a strong relationship between academic degree and division of employment.

Exercises

3.5.1 Refer to the data in Table 2.1.
   a. Use the SPSSX program and a CROSSTABS command to generate a bivariate frequency table showing the relationship between SEX and DIV.
   b. Verify that your result in part **a** is correct by constructing the table by hand.
   c. Does there appear to be a relationship between these two variables?

3.5.2 Refer to the data in Table 2.1.
   a. Use the SPSSX program and a CROSSTABS command to generate a bivariate frequency table showing the relationship between SEX and DEGREE.

   b. Does there appear to be a relationship between these two variables?

3.5.3   Refer to the data in Table 2.1.
   a. Use the SPSSX program and a CROSSTABS command to generate a bivariate frequency table showing the relationship between race and division.
   b. Does it appear that there is a relationship between race and division of employment?

3.5.4   Refer to the data in Table 2.1.
   a. Use the SPSSX program and a CROSSTABS command to generate a bivariate frequency table showing the relationship between race and academic degree.
   b. Does it appear that there is a relationship between race and academic degree?

3.5.5   Refer to the data in Table 2.2.
   a. Generate a frequency distribution and determine what proportion of students are male and what proportion are female.
   b. What proportion of students are freshmen? Sophomores? Juniors? Seniors?
   c. What proportion of students are in-state students? Out-of-state?
   d. What proportion of students are economics majors? Math majors?

3.5.6   Refer to the data in Table 2.2.
   a. Generate a table showing Gender by Class.
   b. Generate a table showing Gender by Residence.
   c. Generate a table showing Gender by Major.
   d. Generate a table showing Class by Major.
   e. Generate a table showing Class by Residence.
   f. Generate a table showing Residence by Major.

3.5.7   Use the computer to generate frequency distributions for the data in Table 2.1.
   a. What proportion of the employees in the sample are male?
   b. What proportion of the employees in the sample are white?
   c. What proportion of the employees in the sample are office workers?
   d. What proportion of the employees in the sample are employed in sales?
   e. Construct a pie chart showing the distribution of employees by division.
   f. Construct a bar chart showing the distribution of employees by degree.

3.5.8   Refer to the data in Table 2.1.
   a. Construct a histogram showing the age of the employees. Start at age 18 and use increments of 5 years.
   b. Construct a histogram as in part **a** but use increments of 15 years. Compare this histogram with the histogram produced in the text and with the one produced in part **a.**
   c. Construct a histogram showing the years of seniority of the employees. Start at 0 and use 5-year increments.
   d. Repeat part **c** using 10-year increments. Which histogram better represents the data?

3.5.9   Refer to the data in Table 2.1. Construct a bar chart showing the number of exemptions claimed by the employees. Examine this bar chart and guess the average number of exemptions claimed.

3.5.10   Refer to the data in Table 2.2. Use the computer to generate various frequency distributions.
   a. Construct a histogram showing the SAT scores of the students. Based on this histogram, guess the average SAT score in the sample.
   b. Construct a histogram showing the GPAs of the students. Guess the average GPA in the sample.

Chapter 3 Supplementary Exercises

3.S.1 Hospital records show the following number of days of hospitalization for 40 patients:

15	19	7	15	8	9	6	14
23	17	22	10	12	15	17	8
14	13	8	18	5	2	10	12
24	32	8	8	7	15	19	18
39	26	6	10	10	15	19	3

    a. Put the data in order from lowest to highest.
    b. Construct a frequency and relative frequency distribution for the data. As classes, use 0 to under 5, 5 to under 10, and so forth.
    c. Construct a cumulative relative frequency distribution for the data.
    d. Draw the histogram.

3.S.2 The duration of 40 long-distance telephone calls is shown here (time recorded in minutes).

13.5	6.0	18.3	26.8	4.2	12.1	11.0	2.0	14.2	2.2
10.2	17.5	8.5	14.2	18.2	3.2	14.6	8.8	14.5	12.6
16.6	13.4	2.1	3.4	5.6	14.5	14.0	5.5	15.5	8.9
23.3	21.0	3.7	7.6	2.3	3.4	.9	23.3	2.7	8.6

    a. Put the data in order from lowest to highest.
    b. Construct a frequency distribution and a relative frequency distribution for the data. As classes, use 0 to under 5, 5 to under 10, and so forth.
    c. Repeat part **b** but use as classes 0 to under 3, 3 to under 6, and so forth.
    d. Draw the histograms in part **b** and part **c,** and compare them to see how the shapes differ when the classes are changed.

3.S.3 Construct a stem-and-leaf diagram for the data in Exercise 2.

3.S.4 The Stop 'n' Go convenience store sold the following numbers of gallons of milk during a sample of 30 weeks:

36	39	49	45	27	26	34	41	61	53
28	46	54	50	31	40	48	42	35	39
44	52	41	47	35	55	60	30	46	38

    a. Put the data in order from lowest to highest.
    b. Construct a frequency distribution using class intervals 24.5 to 29.5, 29.5 to 34.5, and so forth.
    c. Graph the histogram.
    d. Construct the relative frequency distribution.
    e. Construct the cumulative frequency distribution.
    f. Construct the cumulative relative frequency distribution and graph it.

3.S.5 Construct a stem-and-leaf diagram for the data in Exercise 4.

3.S.6 A food distributor has 40 salespeople. The data (in thousands of dollars) at the top of page 74 show their annual salaries:

16.8	19.2	24.5	23.6	19.0	23.2	17.7	24.3	25.4	29.3
15.4	18.9	26.5	24.7	23.4	22.1	26.5	18.6	19.3	20.0
17.6	16.8	26.9	25.4	24.7	20.2	25.3	18.2	16.8	22.3
25.1	24.0	28.2	28.6	24.6	24.3	22.1	27.4	26.8	21.9

a. Put the data in an ordered array.

b. Construct a frequency distribution using class intervals 15 to under 18, 18 to under 21, and so forth.

c. Graph the histogram.

d. Construct the relative frequency distribution.

e. Construct the cumulative frequency distribution.

f. Construct the cumulative relative frequency distribution and graph it.

g. Suppose someone asked you to guess the average salary of the salespeople. By looking at the histogram, make an educated guess.

3.S.7   Construct a stem-and-leaf diagram for the data in Exercise 6.

3.S.8   Refer to the data in Table 2.1 of Chapter 2. Construct a stem-and-leaf diagram for the variable AGE. Based on this diagram, guess the average age of the employees at Computech.

3.S.9   At a small college, there are 40 assistant professors, 60 associate professors, and 20 full professors.

a. Construct a pie chart to represent the data.

b. Construct a bar chart to represent the data.

3.S.10   Suppose that in Exercise 9 an artist drew a square having 1-inch sides to represent the full professors, a square having 2-inch sides to represent the assistant professors, and a square having 3-inch sides to represent the associate professors.

a. Explain what is wrong with this procedure.

b. Describe the dimensions of the square that accurately reflect the number of assistant professors given that the full professors are represented by a 1-inch square.

3.S.11   Suppose that we have data showing the annual salaries of all the players on the New York Yankees baseball team. Give an example where these data could be viewed as the following:

a. A population

b. A sample

3.S.12   Describe some of the biases that are likely to be involved in the following sampling situations:

a. A White House spokesperson reports that 85% of the phone calls received at the White House following a presidential speech were favorable.

b. A newspaper asks for readers' opinions about gun control laws. The paper reports that 400 readers voluntarily replied, and that 70% of the respondents were against gun control laws.

3.S.13   Describe the shape of each of the following distributions. Do you think it is symmetric, skewed to the left or right, uniform, normal, or exponential?

a. Lifetimes of copier machines at accounting firms

b. Actual diameters of 10-inch water pipes

c. Gross profits of corporations in the United States

d. Ages of students at a local high school

3.S.14   Describe the shape of each of the following distributions. Do you think it is symmetric, skewed to the left or right, uniform, normal, or exponential?

a. Household income in the United States

b. Contributions to charity by American households

c. Ages of assistant professors at a university

d. Ages of full professors at a university

3.S.15 Draw a pie chart illustrating the following market shares in a certain town: Budweiser, 50%; Stroh's, 25%; Miller, 15%; others, 10%. What are the sizes of the central angles?

3.S.16 Draw a time plot illustrating the number of employees in a company from 1986 to 1995. Does this time series appear to have a trend?

> 12   17   23   29   44   56   78   101   95   117

3.S.17 Draw a time plot illustrating a consultant's annual income (in thousands of dollars) from 1986 to 1995. Does this time series appear to have a trend?

> 23   27   13   49   24   56   18   21   15   27

3.S.18 The following 40 readings, listed in chronological order, represent a pilot run on the production of a temperature control device. They give the "on" temperature at which a thermostatically controlled switch began to operate.

1	67.2	11	67.0	21	67.7	31	67.7
2	67.1	12	67.3	22	67.4	32	67.6
3	67.4	13	67.5	23	67.3	33	67.8
4	67.5	14	67.5	24	67.6	34	67.8
5	67.3	15	67.6	25	67.5	35	67.6
6	67.3	16	67.5	26	67.8	36	67.6
7	67.4	17	67.1	27	67.7	37	67.6
8	67.5	18	67.3	28	67.9	38	67.8
9	67.2	19	67.4	29	67.5	39	67.7
10	67.3	20	67.6	30	67.5	40	67.9

a. Construct a tally sheet and a frequency distribution. Use 67.0, 67.1, and so forth as the classes.
b. Construct the relative frequency distribution and graph it.
c. Plot the 40 values in chronological order. Does the time plot make it appear that the "on" temperatures of the thermostats may be increasing? This may be an indication that the manufacturing process for producing the thermostats may have gone out of control.
d. Explain why it is important to plot time series data in chronological order. What can be learned by plotting time series data that is not learned by plotting a relative frequency distribution?

References

Fisher, Ronald A. *The Design of Experiments*. 9th ed. New York: Hafner, 1971.

Neter, John, William Wasserman, and G. A. Whitmore. *Applied Statistics*. 3d ed. Boston: Allyn and Bacon, 1988.

Nie, Norman E., C. Hadlai Hull, Jean G. Jenkins, Karin Steinbrenner, and Dale H. Bent. *SPSS Statistical Package for the Social Sciences*. 2d ed. New York: McGraw-Hill, 1975.

Norusis, Marija J. *SPSSX Introductory Statistics Guide*. New York: McGraw-Hill, 1990.

Norusis, Marija J. *SPSSX Advanced Statistics Guide*. Chicago: SPSS, 1990.

Norusis, Marija J. *The SPSS Guide to Data Analysis*. Chicago: SPSS, 1986.

Ryan, Thomas A., Brian L. Joiner, and Barbara F. Ryan. *Minitab Handbook*. 2d ed. Boston: PWS-Kent, 1985.

Ryan, Thomas A., Brian L. Joiner, and Barbara F. Ryan. *Minitab Reference Manual*. University Park, Penn.: Minitab Project, 1985.

SAS Introductory Guide. 3d ed. Cary, N.C.: SAS Institute, 1985.

SAS Procedures Guide for Personal Computers. Version 6 ed. Cary, N.C.: SAS Institute, 1986.

SAS Statistics Guide for Personal Computers. Version 6 ed. Cary, N.C.: SAS Institute, 1986.

SAS User's Guide: Basics. Version 5 ed. Cary, N.C.: SAS Institute, 1985.

SAS User's Guide: Statistics. Version 5 ed. Cary, N.C.: SAS Institute, 1985.

SPSSX User's Guide. Chicago: SPSS, 1988.

Wainer, Howard. "How to Display Data Badly." *American Statistician* 38 (1984): 137–147.

4 Summary Statistics: Measures of Location and Dispersion

In Chapter 3, we discussed how to use tables and graphs to describe and summarize a set of data. These tables and graphs enable us to visualize the shape of the distribution. In addition, we can visualize the amount of spread in the observations and the tendency of the data to cluster about a central value. In this chapter, we discuss how to characterize the center and the amount of spread in the data by computing numerical measures.

When confronted with a large sample of data, the statistician typically wants to determine (1) the general shape or distribution of the data, (2) the average or typical value in the data set, and (3) the amount of dispersion or variation present in the data. For each of these three problems, several solutions are possible.

1. *What is the general shape or distribution of the data?* This problem can be solved by constructing frequency and relative frequency distributions, by drawing a histogram or stem-and-leaf diagram, or by constructing a box plot (discussed later in this chapter). The most popular solution is to construct a histogram.

2. *Where is the center of the data, or what is the average value of the data?* This problem can be solved in several ways. One solution is to calculate the mean, or arithmetic average. Another solution is to calculate the median. At times, a third possible solution is to calculate the mode of the data. The most popular solution to this problem is to calculate the mean, although for certain problems the median or mode is more appropriate.

3. *How dispersed, or spread out, are the data?* There are many ways of measuring the amount of variation in a set of data. Some of these measures are the range, the interquartile range, the mean absolute deviation, the variance, and the standard deviation. The most frequently used measure of variation is the standard deviation.

In Chapter 3, we discussed how to solve the first problem of determining the general shape or distribution of the data. In Chapter 4, we discuss how to locate the center of the data and how to measure the amount of dispersion. If we have all the observations in a population, then we can actually calculate the values for various population characteristics.

DEFINITION   **Parameters**

Numbers that describe population characteristics are called **parameters.**

Examples of population parameters are the population mean, the population variance, the population standard deviation, and the population proportion. Much of the field of statistics is devoted to drawing inferences from a sample concerning the value of a population parameter. If we have only a sample of data, then we calculate *estimates* of the population parameters.

DEFINITIONS   **Estimate and Estimator** ────────────────────────────────

Values calculated from a sample of data that are used to estimate population parameters are called **estimates.** The formula used to calculate an estimate is called an **estimator.**

───

An *estimator* is a function, whereas an *estimate* is a specific value. Examples of estimators are the formulas for determining the sample mean, the sample variance, and the sample standard deviation; the results of these operations are the estimates.

The mean, the median, and the mode are three different numerical measures used to describe the center of a data set. The range, the variance, the standard deviation, and the mean absolute deviation are four different quantities used to measure the amount of spread, dispersion, or variability in a set of data. All of these numerical measures are population parameters if they refer to calculations made from a population of data and sample estimates if they are calculated from a sample of data.

In this text, variables will be denoted by uppercase letters and specific values of the variable by lowercase letters.

4.1   Summation Notation

Before describing these numerical measures, we introduce summation notation to simplify the formulas. The Greek letter Σ (uppercase sigma) is called the *summation sign* and indicates that we should find the sum of the values following the Σ.

DEFINITION   **Summation Sign** ──────────────────────────────────────

The expression $\sum_{i=1}^{n} x_i$ is used to represent the sum of the values $x_1, x_2, \ldots, x_n$ and is defined by the equation

$$\sum_{i=1}^{n} x_i = x_1 + x_2 + \cdots + x_n$$

───

The symbols above and below the summation sign Σ define the limits of the summation. Thus, the subscript $i = 1$ and the superscript n tell us to calculate the sum of all values from x_1 to x_n. When these subscripts and superscripts are omitted, as is usually the case, the symbol Σ means that we should obtain the sum of all the available values of the variable. For example, suppose we have seven observations concerning some variable X, denoted by $x_1, x_2, \ldots, x_7$. The sum of these seven numbers is represented by the symbol $\Sigma\, x_i$, or by the more detailed symbol

$$\sum_{i=1}^{7} x_i$$

EXAMPLE 4.1 **Sum of a Set of Values**

On a sample of 7 days, the numbers of customers at a small restaurant were

$$x_1 = 92, \quad x_2 = 84, \quad x_3 = 70, \quad x_4 = 76, \quad x_5 = 66, \quad x_6 = 80, \quad x_7 = 71$$

Calculate $\Sigma\, x_i$.

S O L U T I O N $\Sigma\, x_i$ represents the total number of customers during the week. The total number of customers is thus

$$\Sigma\, x_i = x_1 + x_2 + \cdots + x_7 = 539$$

EXAMPLE 4.2 **Salary Payroll at Lang's Trucking Company**

Refer to the data in Table 3.1, the monthly salaries earned by the 80 employees at Lang's Trucking Company. Let x_i denote the monthly salary of the ith employee. Calculate $\Sigma\, x_i$.

S O L U T I O N $\Sigma\, x_i$ represents the sum of the monthly salaries for all 80 employees; that is, $\Sigma\, x_i$ represents the monthly payroll. From the data in Table 3.1, we obtain $\Sigma\, x_i = \$130,135$.

EXAMPLE 4.3 **Total Number of Cars Owned**

Refer to the data in Example 3.3, the number of cars owned by each family in a sample of 120 families. Let x_i denote the number of cars owned by the ith family. Calculate $\Sigma\, x_i$.

S O L U T I O N $\Sigma\, x_i$ denotes the total number of cars owned by all 120 families. From the data in Example 3.3, we obtain $\Sigma\, x_i = 219$.

Formulas Involving Summation Notation

Several formulas involving summation signs will be used throughout this book. They are summarized as follows:

1. $\Sigma\, cx_i = c\, \Sigma\, x_i$, where c is any constant
2. $\Sigma(x_i - c) = \Sigma\, x_i - nc$
3. $\Sigma\, x_i^2 = x_1^2 + x_2^2 + \cdots + x_n^2$
4. $\Sigma\, cx_i^2 = c\, \Sigma\, x_i^2$
5. $\Sigma(x_i - c)^2 = \Sigma\, x_i^2 - 2c\, \Sigma\, x_i + nc^2$

The following examples illustrate the use of summation notation.

EXAMPLE 4.4   **Total Parking Revenue** ————————————————————————

Use the data in Example 4.1. Suppose that each customer at the restaurant pays $3.00 for valet parking. Calculate the total parking revenue during the week.

S O L U T I O N   We obtain

$$\Sigma\, 3x_i = 3x_1 + 3x_2 + \cdots + 3x_7$$
$$= 276 + 252 + \cdots + 213 = 1{,}617$$

Alternatively, by using formula 1 above, we obtain

$$\Sigma\, 3x_i = 3\, \Sigma\, x_i = 3 \times 539 = 1{,}617$$

EXAMPLE 4.5   **Sum of Squares** ————————————————————————

Use the data in Example 4.1 and calculate $\Sigma\, x_i^2$.

S O L U T I O N   We obtain

$$\Sigma\, x_i^2 = 92^2 + 84^2 + \cdots + 71^2 = 41{,}993$$

It is tedious to calculate sums by hand, and it is easy to make mistakes. When performing such calculations by hand, you should always do the work twice to check your result. Just as computer programs are useful for constructing frequency distributions, they are also useful for calculating sums of variables. For example, the FREQUENCIES procedure in the SPSSX computer program will calculate the sum of any set of numerical observations. For example, to get the sum of all values of the variable SALARY, issue the commands

```
FREQUENCIES VARIABLES = SALARY/
        STATISTICS SUM
```

Exercises

Statistical Concepts

4.1.1   Prove that $\Sigma\, cx_i = c\, \Sigma\, x_i$. Verify your result by using the following five values of x and let $c = 4$: 5, 3, 2, 8, 7.

4.1.2   Prove that $\Sigma\, (x_i - c) = \Sigma\, x_i - nc$. Verify your result by using the following five values of x and let $c = 4$: 5, 3, 2, 8, 7.

4.1.3   Provide an example to show that, in general, $\Sigma\, (x_i)^2 \neq (\Sigma\, x_i)^2$. Verify your result by using the following five values of x: 5, 3, 2, 8, 7.

4.1.4   Denote the weekly salary of the ith employee as x_i. Suppose an accountant calculates the sum of the weekly salaries of 40 employees of a company. Use summation notation and represent the total payroll. Suppose the total payroll amounts to $10,000. Suppose the company deducts 7.65% of each salary as the employee's contribution to the Social Security fund. Use summation notation to denote the total amount actually received by the employees after Social Security taxes are deducted. Calculate this sum.

4.1.5   *True or false:*

$$\Sigma \left(\frac{x_i}{n} \right) = \frac{(\Sigma x_i)}{n}$$

Prove your answer algebraically. Check your answer using the following $n = 5$ values: 7, 3, 2, 5, 8.

Statistical Drills

4.1.6   Use the following $n = 10$ values of a variable X: 7, 5, 1, 4, 8, 10, 4, 7, 5, 6.
   a. Calculate Σx_i.
   b. Calculate Σx_i^2. Calculate $(\Sigma x_i)^2$. Are these two values equal? Why or why not?
   c. Calculate $\Sigma (x_i - 4)$.
   d. Calculate $\Sigma (x_i/n)$. This value is called the *mean* of the 10 values and is denoted $\bar{x}$.
   e. For each value x_i, calculate $(x_i - \bar{x})$. These values are called the *deviations from the mean.*
   f. Calculate the sum of the 10 deviations from the mean.
   g. Prove algebraically that $\Sigma (x_i - \bar{x}) = 0$.

4.1.7   Use the following $n = 5$ values of a variable X: 14, 26, 30, 20, 10. Use the following $n = 5$ values of a variable Y: 4, 6, 100, 8, 3.
   a. Calculate Σx_i and Σy_i.
   b. Calculate $\Sigma x_i y_i$.
   c. *True or false:* $\Sigma x_i y_i = (\Sigma x_i) (\Sigma y_i)$. Explain.

4.1.8   Use the following $n = 5$ values of a variable X: 14, 26, 30, 20, 10. Use the following $n = 5$ values of a variable Y: 4, 6, 100, 8, 3.
   a. Calculate Σx_i^2 and Σy_i^2.
   b. Calculate $\Sigma (x_i + y_i)^2$.
   c. *True or false:* $\Sigma (x_i + y_i)^2 = \Sigma x_i^2 + \Sigma y_i^2$. Explain.

Statistical Applications

4.1.9   A man has five sons. Let x_i denote the weekly income of the ith son. During a given week, their incomes were as follows: $542, $433, $289, $400, and $456. Each son had $100 per week deducted for taxes.
   a. Calculate the total weekly income of the five sons; that is, find the sum Σx_i.
   b. Calculate the sons' total take-home pay after taxes were deducted. That is, calculate $\Sigma (x_i - 100)$.

4.1.10   Write the following values in summation notation:
   a. $x_1 + x_2 + \cdots + x_{15}$
   b. $5x_1 + 5x_2 + \cdots + 5x_{22}$
   c. $x_1 + y_1 + x_2 + y_2 + \cdots + x_9 + y_9$

4.1.11   Let $x_1 = 3$, $x_2 = 2$, $x_3 = 4$, and $x_4 = 1$. Calculate the following:
   a. $\Sigma x_i - 3$
   b. $\Sigma (x_i - 3)$
   c. $\Sigma (x_i - 3)^2$
   d. $(\Sigma x_i - 3)^2$

4.2   The Mean, the Median, and the Mode

By far the most important and most frequently used measure of central tendency of a data set is the *arithmetic mean*, or simply the *mean*. The mean is frequently called the *average* of a set of data.

If we calculate the mean from a population of N values, then we obtain the *population mean.*

DEFINITION   **Population Mean**

The **population mean** is denoted by the symbol μ (Greek mu) and is calculated by using the following formula, where N is the number of observations of the population:

$$\mu = \frac{x_1 + x_2 + \cdots + x_N}{N} = \frac{\Sigma x_i}{N}$$

When the data set consists of a sample of n observations, we calculate the *sample mean* to estimate the population mean. The sample mean is denoted by the symbol $\bar{x}$ (called "x bar").

DEFINITION   **Sample Mean**

Given the sample of n observations $x_1, x_2, \ldots, x_n$, the **sample mean** $\bar{x}$ is

$$\bar{x} = \frac{x_1 + x_2 + \cdots + x_n}{n} = \frac{\Sigma x_i}{n}$$

EXAMPLE 4.6   **Calculating the Sample Mean**

The data in Example 4.1 showed that, on a sample of 7 days, the numbers of customers at a small restaurant were

$$x_1 = 92, \quad x_2 = 84, \quad x_3 = 70, \quad x_4 = 76, \quad x_5 = 66, \quad x_6 = 80, \quad x_7 = 71$$

Calculate the sample mean.

SOLUTION   The sample mean is

$$\bar{x} = \frac{\Sigma x_i}{7} = \frac{539}{7} = 77$$

EXAMPLE 4.7   **Mean Salary at Lang's Trucking Company**

Table 3.1 shows the monthly salaries of all 80 employees at Lang's Trucking Company. Example 4.2 showed that the sum of all 80 salaries is $130,135. Calculate the population mean salary.

SOLUTION   The population mean is

$$\mu = 130{,}135/80 = 1{,}626.6875$$

This means that the average monthly salary for the 80 employees is approximately $1,626.69.

EXAMPLE 4.8 **Average Number of Cars Owned** —————————————————

Example 3.3 shows the number of cars owned by 120 different families. Example 4.3 showed that the total number of cars owned is 219. Calculate the sample mean.

SOLUTION   The sample mean is

$$\bar{x} = 219/120 = 1.825$$

The Median

Another important measure of the center or middle of a set of data is called the *median.*

DEFINITION   **Median** —————————————————————————————

The **median** is the middle value of data ordered from lowest to highest. If the sequence contains an odd number of observations, the median is the *middle value* in the ordered sequence. If the sequence contains an even number of observations, the median is the *arithmetic average* of the two central values.

When a population of data is available, we calculate the *population median.* The median calculated from a sample of data is called the *sample median* and is an estimate of the population median. Only rarely will the sample median be exactly equal to the population median, but as the sample size becomes larger, the chances increase that the sample median will be close to the population median.

EXAMPLE 4.9   **Median for an Odd Number of Observations** —————————————

Calculate the median for the data on restaurant customers in Example 4.1.

SOLUTION   To calculate the median, the observations must be ordered from lowest to highest. The ordered sequence is

66   70   71   76   80   84   92

The number of observations is odd, so the median is the middle, or fourth, value in the ordered sequence:

Sample median = 76

Recall from Example 4.6 that the sample mean was 77. In this example, the median is slightly less than the mean. This shows that the mean and median are not necessarily equal to one another.

EXAMPLE 4.10   **Median for an Even Number of Observations** —————————————

Calculate the median salary for the monthly salary data presented in Table 3.1.

SOLUTION   To calculate the median salary, the 80 values must be placed in order from lowest to highest. The ordered values are shown in Table 4.1 (page 84). Because the

TABLE 4.1 *Ordered monthly salaries from Table 3.1*

1,300	1,310	1,320	1,325	1,340	1,380	1,380	1,390	1,400	1,400
1,410	1,420	1,425	1,440	1,450	1,450	1,455	1,475	1,475	1,500
1,530	1,550	1,550	1,550	1,550	1,550	1,550	1,555	1,565	1,570
1,575	1,580	1,580	1,585	1,590	1,590	1,600	1,600	1,600	1,600
1,620	1,620	1,620	1,620	1,625	1,625	1,640	1,640	1,650	1,650
1,650	1,650	1,650	1,660	1,675	1,700	1,700	1,705	1,725	1,730
1,740	1,750	1,760	1,775	1,780	1,815	1,820	1,850	1,850	1,860
1,875	1,900	1,900	1,900	1,900	2,000	2,000	2,000	2,015	2,025

number of observations is even, the median is the average of the 40th and 41st values. The 40th value is 1,600; the 41st value is 1,620. The median monthly salary is the arithmetic average of these two values:

$$\text{Population median} = (1,600 + 1,620)/2 = 1,610$$

Recall from Example 4.7 that the mean monthly salary for the 80 employees is approximately \$1,626.69. In this example, the median is slightly less than the mean.

EXAMPLE 4.11   **Median Number of Cars Owned**

Calculate the median number of cars owned per family for the car ownership data in Example 3.3.

SOLUTION   In Table 3.4, the 120 values are in order from lowest to highest. The number of observations (120) is even, so the median is the average of the 60th and 61st values. After the data are ordered, you can see that the first 9 families own no cars, the next 39 families each own one car, the next 48 families own two cars each, and so forth. Thus, the 60th family owns two cars and the 61st family also owns two cars. The median number of cars owned is

$$\text{Median} = \frac{2 + 2}{2} = 2$$

Recall from Example 4.8 that the mean number of cars owned is 1.825. In this example, the median is slightly greater than the mean.

The Mode

A third measure of central tendency for a population or sample of observations is called the *mode*.

DEFINITION   **Mode**

The **mode** of a set of observations is the value that occurs with the greatest frequency. The mode is not necessarily unique.

EXAMPLE 4.12   **Mode of Number of Restaurant Customers** ─────────────

Calculate the mode for the number of customers at the restaurant in Example 4.1.

S O L U T I O N   Each of the seven values occurs exactly once. Because each value has the same frequency, each value could be called a mode. In this problem, the mode conveys no useful information.

EXAMPLE 4.13   **Mode of Monthly Salaries** ─────────────

Calculate the mode for the data in Table 3.1.

S O L U T I O N   The salary $1,550 occurs six times. Because no other value occurs this often, the mode is $1,550.

This value is not necessarily a good measure of the center of the salary data. For example, the value $1,650 occurs five times; the value $1,600 occurs four times; $1,620, four times; and $1,900, four times. Thus there is no reason to expect the most frequently occurring salary to be located near the center of the data. With just a few minor changes in salaries, the mode could easily be $1,900, and this value certainly does not fall near the center of the data. For the data in Table 3.1, the mode is not a useful measure of central tendency because it varies considerably from the mean salary of $1,626.69 and the median salary of $1,610.

EXAMPLE 4.14   **Mode of Number of Cars Owned** ─────────────

Calculate the mode for the data in Example 3.3.

S O L U T I O N   The most common value in Example 3.3 is the value 2, which has a frequency of 48. The value 1 is the second most common, with a frequency of 39. In this case, the mode is a useful measure of location. In some studies, it could be useful to know that more families have two cars than any other number. Note from Example 4.11 that the median for this data set equals the mode.

EXAMPLE 4.15   **Mode for Highest Academic Degree** ─────────────

Calculate the mode for the highest academic degree earned by employees at the Cottrell Corporation. Use the data in Table 3.5.

S O L U T I O N   The modal class is the class having the highest frequency. For this set of data, it is the group having a high school education, which has a frequency of 200. In this case, the mode conveys useful information about the distribution of degrees. Note that since the data are qualitative, there is no mean or median.

Comparison of the Mean, Median, and Mode

To get a visual interpretation of the mean, examine the histogram in Figure 4.1 (page 86). Suppose we cut a piece of plywood in the exact shape of the histogram. Now stand the piece of plywood upright on its lower edge and try to balance the piece of plywood on a

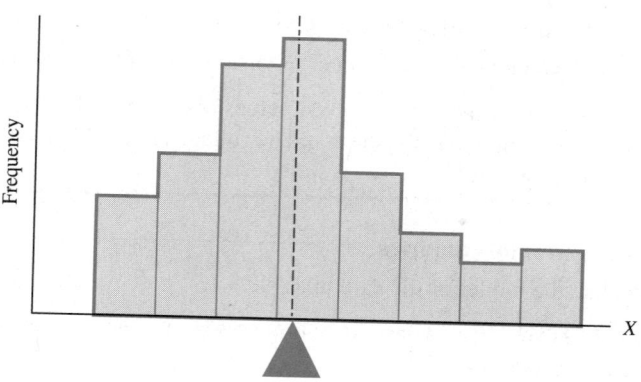

FIGURE 4.1   *The mean of a distribution can be interpreted as the distribution's center of gravity*

fulcrum. There will be exactly one point on the lower edge where we can place the fulcrum so that the plywood histogram will balance. That point is the mean of the distribution. The mean can be thought of as the *center of gravity* of the distribution.

If we think of the mean in this way, it is easy to see how a few extreme values can have a big influence on the value of the mean. If the extreme values are much larger than the rest of the data, then the fulcrum will have to be shifted far to the right to keep the plywood histogram in balance.

When the relative frequency distribution is represented by a smooth curve or histogram, the median is the value such that 50% of the area under the curve lies to its left and 50% of the area lies to its right. When the relative frequency distribution is represented by a smooth curve, the mode is the value lying beneath the highest point on the curve. When the relative frequency distribution is represented by a histogram, the *modal class* is the class with the greatest frequency.

The mean is affected by the value of every observation; if there are a few extreme values, then the mean may not be the best way to describe the center of the set of data. This is often the case with data such as the incomes of individuals, the profits or sales of corporations, and so forth. In these cases the median may be a better measure of the typical value, because it is not so strongly affected by extreme values. In general, the mean does not equal the median, nor do they have to be close to one another. The mean and the median are quite different ways of measuring the center of a set of data.

Both the mean and the median are sometimes called the *average* of a set of data, but most people use the term *average* to indicate the mean. Some serious problems can arise when it is not clear which average is being used.

EXAMPLE 4.16   Comparing the Mean and the Median

Consider a sample of eight individuals whose incomes are as follows:

$$x_1 = \$15,000 \qquad x_2 = \$16,000 \qquad x_3 = \$19,000 \qquad x_4 = \$21,000$$
$$x_5 = \$24,000 \qquad x_6 = \$25,000 \qquad x_7 = \$26,000 \qquad x_8 = \$210,000$$

Calculate the mean and median.

SOLUTION   The mean income for these eight individuals is $44,500, and the median income is ($21,000 + $24,000)/2 = $22,500. In this example, the mean income is almost twice as large as the median income, because it is strongly influenced by the extreme value $210,000.

A Graphical Comparison of the Mean, Median, and Mode

When a distribution is unimodal and symmetric like the bell-shaped normal distribution, the mean, median, and mode all coincide. If the distribution is skewed to the right, the mean is larger than the mode because it is affected by the large values in the tail, whereas the mode is influenced only by the most frequently occurring values, which are small. The median is less influenced by the values in the right-hand tail than is the mean, so the median will be less than the mean. Furthermore, the peak of a right-skewed distribution will be to the left of the value that divides the area under the curve into two equal parts, and so the median will exceed the mode.

Relationships among the mean, median, and mode

> If a distribution is skewed to the right, the following relationships hold among the mean, median, and mode:
>
> *Skewed to the right:*   mode < median < mean
>
> If a distribution is skewed to the left, then the relationships are reversed:
>
> *Skewed to the left:*   mean < median < mode

The relationships among the mean, median, and mode are illustrated in Figure 4.2 (page 88).

When the Median Is Useful

There are two situations in which the median is preferable to the mean as a measure of central tendency.

1. *Open-ended class intervals:*   If you want to calculate an average from a frequency distribution with an open-ended class, there may be no alternative but to use the median, because calculating the mean requires knowledge of the midpoint of the open-ended class or of the sum of the measurements in the open-ended class.
2. *Outliers:*   The mean is far more influenced by extreme values than is the median. Consequently, the median provides a better measure of central tendency than the mean when there are some extremely large or small values.

When the Mode Is Useful

There are three situations in which the mode is preferable as a measure of central tendency.

1. *Qualitative data:*   Frequently the mode is a useful measure of the typical value of a qualitative variable.

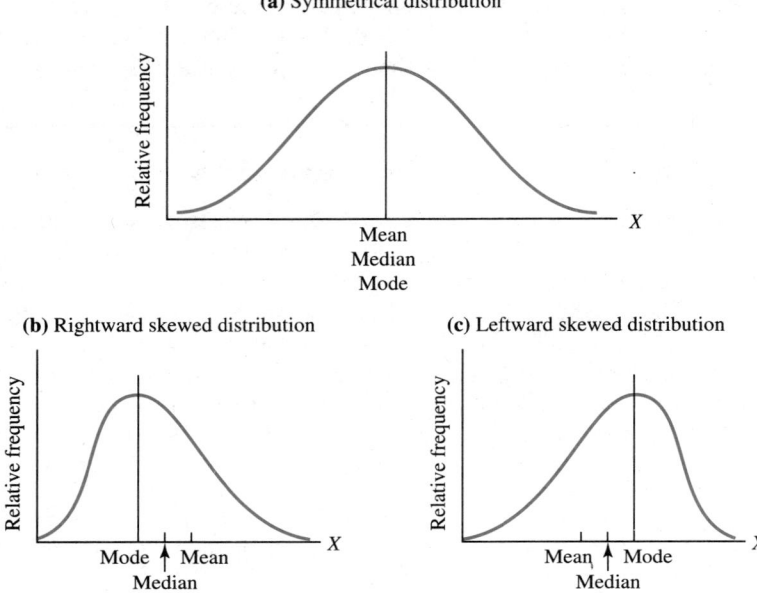

FIGURE 4.2   *Relationships among the mean, median, and mode*

2. *Sizes of products:*   The mode can be useful for data on the sizes of products. For example, managers of clothing stores are interested in the most frequently purchased sizes of dresses and shirts, and managers of supermarkets are interested in the most frequently purchased size of a box of laundry detergent.

3. *Grouped data:*   The mode is usually a more meaningful measure of central tendency of numerical data if it is calculated from a frequency distribution or a relative frequency distribution. When data are grouped into classes, as in a frequency distribution or a relative frequency distribution, the *modal class* is the class having the greatest frequency or relative frequency per unit width.

Exercises

Statistical Concepts

4.2.1   *True or false:*   Suppose a distribution is skewed to the right. Then the mode exceeds the mean. Explain.

4.2.2   *True or false:*   Suppose a distribution is symmetric with mean μ. Then the median must equal μ. Explain.

4.2.3   *True or false:*   Suppose a distribution is symmetric with mean μ. Then the mode must equal μ. Explain.

4.2.4   *True or false:*   Suppose a unimodal distribution is symmetric with mean μ. Then the mean, median, and mode are all equal. Explain.

4.2.5   *True or false:*   The median of a distribution represents the center of gravity of the distribution. Explain.

4.2.6 *True or false:* The mode is more useful than the mean or median when describing nominal data. Explain.

4.2.7 *True or false:* The mode is more useful than the mean or median when describing ordinal data. Explain.

4.2.8 *True or false:* The mode is more useful than the mean or median when describing interval or ratio data. Explain.

4.2.9 To plan for Father's Day, a department store manager wants to know which shirt size is most popular among adult males. Which measure of central tendency would be most useful: the mean, the median, or the mode?

4.2.10 Suppose a financial planner lists the annual incomes of all her clients. Suppose the mean, median, and mode of these values are listed as μ, Md, and Mo. Suppose each income is scaled by 1,000 (that is, an income of $22,800 is listed as 22.8). What are the scaled values of the mean, median, and mode?

4.2.11 Suppose we have ordinal data showing the ratings given to the taste of a certain brand of coffee, where we use the code 1 = poor, 2 = fair, and 3 = good. Suppose we have three ratings of poor, five ratings of fair, and six ratings of good.

a. Explain why it is inappropriate to use the mean as a measure of central tendency in this case.

b. Suppose that you rate the coffee as poor and I rate the coffee as good. Does this mean that, on the average, the coffee is fair?

c. Suppose we change the code to 0 = poor, 1 = fair, and 100 = good. What happens to the mean? Suppose that you rate the coffee as poor and I rate the coffee as good. Does this mean that, on the average, the coffee is a 50 (or almost good)?

d. Explain why it is inappropriate to use the mean as a measure of central tendency when referring to nominal or ordinal data.

4.2.12 Explain how to calculate the median when the number of observations is odd and when it is even.

4.2.13 Suppose we add an additional observation to a data set and that this observation is larger than all the previous observations.

a. *True or false:* The new observation always causes the median to increase.

b. *True or false:* The new observation sometimes causes the median to increase.

c. *True or false:* The new observation always causes the mean to increase.

Statistical Drills

4.2.14 The following five values represent the annual incomes earned by the secretaries in a small office:

$22,000 $25,500 $18,200 $31,700 $24,600

a. Find the mean and median.

b. Divide each salary by $1,000 and record the scaled values. Find the mean and median for the scaled values.

4.2.15 The following five values of x_i represent the annual incomes earned by the secretaries in a small office:

$24,200 $25,500 $17,100 $26,700 $24,600

a. Find the mean μ and median Md.

b. Calculate each deviation from the mean, $(x_i - \mu)$. Find the sum of the deviations from the mean. That is, find $\Sigma(x_i - \mu)$.

c. Calculate each deviation from the median, $(x_i - \text{Md})$. Find the sum of the deviations from the median. That is, find $\Sigma(x_i - \text{Md})$.

    d. *True or false:*   The deviations from the mean must always sum to zero. Prove your result.

    e. *True or false:*   The deviations from the median will never sum to zero. Prove your result.

4.2.16   Consider the following 40 values:

27	17	32	24	18	13	21	16	19	20
15	19	17	25	28	18	24	12	31	27
12	14	21	16	17	15	23	16	17	20
17	21	12	13	16	23	18	13	14	22

    a. Calculate the mean and median.

    b. Construct a frequency distribution using the classes 10 to 14.99, 15.00 to 19.99, and so forth. Find the class with the greatest frequency. Denote the midpoint of this class as the mode of the distribution.

    c. Use the values of the mode, median, and mean to guess whether the data are skewed.

    d. Draw a graph of the distribution.

4.2.17   Suppose a population consists of the following five values: 24, 32, 56, 72, 60.

    a. Find the mean and median.

    b. Replace the value 72 by the value 200. Find the new mean and the new median.

    c. In the original set of data, replace the value 72 by the value 2,000. Find the new mean and median.

    d. Explain how an extreme value influences the mean and median.

4.2.18   **a.** There are 10 people in a room. Their mean weight is 160 lb. An 11th person who weighs 200 lb enters the room. Find the mean weight of all 11 people.

    b. There are 20 people in a room. Their mean weight is 160 lb. A 21st person who weighs 200 lb enters the room. Find the mean weight of all 21 people.

    c. Compare your answers in part **a** and part **b**. Explain how the effect of a single observation is diminished as the sample size increases.

4.2.19   On a test, 20% of the class received a grade of 70, 30% of the class received a grade of 80, and 50% of the class received a grade of 90.

    a. Suppose there were 10 people in the class. Find the mean grade.

    b. Suppose there were 100 people in the class. Find the mean grade.

    c. Can you determine the mean grade without knowing the number of people in the class?

4.2.20   Suppose we take a survey of full-time students at universities in the United States. Which is larger, the mean age or the median age? Explain.

4.2.21   Suppose you have entered the annual incomes of 100 people into a computer file. The largest income was $96,000. By mistake, this income was typed as $960,000.

    a. Suppose the sample mean based on the incorrect income of $960,000 is $46,000. What is the mean using the correct income of $96,000?

    b. Suppose the sample median based on the incorrect income of $960,000 is $32,000. What is the median using the correct income of $96,000?

Statistical Applications

4.2.22   The data that follow show the monthly rental payments (in dollars) for a random sample of 50 apartment dwellers in a certain city:

200	310	150	225	140	135	110	190	130	240
145	95	165	185	150	310	100	175	165	130
165	110	160	150	130	210	210	315	125	245
220	140	210	165	90	190	200	185	150	220
290	125	220	370	75	185	215	160	180	170

    a. Calculate the sample mean.

    b. Calculate the sample median.

    c. Form a frequency distribution using as classes 0 to under 50, 50 to under 100, and so forth.

    d. Find the modal class in part **c.**

4.2.23 Let $x_1, x_2, \ldots, x_N$ denote the N observations in a population with mean μ. Let K be any number. Show that

$$\Sigma(x_i - K)^2 = \Sigma(x_i - \mu)^2 + N(K - \mu)^2$$

Hence, deduce that the value of K for which $\Sigma(x_i - K)^2$ is smallest is $K = \mu$.

[*Hint:* $\Sigma(x_i - K)^2 = \Sigma[(x_i - \mu) + (\mu - K)]^2$.]

4.2.24 For each of the following populations, discuss the shape of the distribution and indicate whether the mean, median, or mode would be the best measure of central tendency:

    a. Annual incomes of adult males

    b. Ages of cars in the United States

    c. Collar sizes of men's shirts sold at a department store

    d. Amounts in savings accounts at a large bank

    e. Number of bus passengers on weekdays during a given year on a particular bus route at 8 A.M.

 4.2.25 The accompanying data show a sample of home mortgage loan amounts (in dollars) handled by a particular loan officer at a savings and loan association.

20,000	38,500	33,000	27,500	34,000
12,500	25,900	43,200	37,500	36,200
25,200	30,900	23,800	28,400	13,000
31,000	35,500	25,400	33,500	20,200
39,000	38,100	30,500	45,500	30,500
52,000	40,500	51,600	42,500	44,800

Find the sample mean, median, and mode.

4.2.26 According to *Family Economics Review, 1991*, in 1989, the mean contribution to charity was $492 per household. The median contribution to charity was $48 per household. Can you explain why these values are so different?

▌4.3 Quartiles and Percentiles

The mean, median, and mode are three different measures of the center or the most representative value of a set of data. There are other summary statistics that can be used to locate specific values in a distribution. Sometimes we want to know the position of a particular observation relative to the others. For example, if you received a score of 1100 on the SAT, you might want to know the percentage of participants who scored lower than 1100. Such a *measure of relative standing* of an observation within a data set is called a *percentile*.

DEFINITION **Percentiles**

Suppose the observations $x_1, x_2, \ldots, x_n$ have been arranged in ascending order. The pth **percentile** is the value x_p such that $p\%$ of the observations are less than or equal to x_p and $(100 - p)\%$ of the observations are greater than or equal to x_p.

Consider the eight observations 4, 6, 8, 10, 12, 14, 16, and 18. Twenty-five percent of the observations are less than 6.5 and 75% are greater. Similarly, 25% are less than 7 and 75% are greater. Thus, both 6.5 and 7.0 could be called the 25th percentile; in fact, any value between 6 and 8 could be called the 25th percentile. Some statisticians recommend interpolating to find percentiles. In most problems of practical interest, however, there are a large number of observations, and interpolation changes the value obtained only slightly. We shall adopt the rules in the accompanying box for calculating a percentile.

Procedure for calculating percentiles

> 1. Arrange the n observations in ascending order. The smallest observation is given rank 1, the second smallest observation is given rank 2, and so forth.
> 2. Calculate the index $i = (p \times n)/100$, where p is the percentile of interest and n is the sample size.
> 3. If i is an integer, the pth percentile is the arithmetic average of the values having ranks i and $(i + 1)$.
> 4. If i is not an integer, the next integer value greater than i denotes the rank of the value that is the pth percentile.

EXAMPLE 4.17 **Calculating Percentiles**

Calculate the 70th and 80th percentiles for the following set of 15 values:

18   14   45   32   65   43   25   41   83   51   26   36   40   55   20

SOLUTION First, arrange the 15 values in ascending order:

14   18   20   25   26   32   36   40   41   43   45   51   55   65   83

For the 70th percentile, the index is

$$i = (70 \times 15)/100 = 10.5$$

Because i is not an integer, the position of the 70th percentile is the next integer greater than 10.5, that is, the 11th position. The 70th percentile corresponds to the 11th data value in the ordered ranking, 45.

For the 80th percentile, the index is

$$i = (80 \times 15)/100 = 12.0$$

Because i is an integer, the 80th percentile is the average of the data values in positions 12 and 13. Thus the 80th percentile is

$$(51 + 55)/2 = 53$$

For a continuous distribution, the pth percentile is the value such that the area to its left is p and the area to its right is $(1 - p)$. Figure 4.3 shows the 30th percentile for a continuous distribution. The median is the 50th percentile.

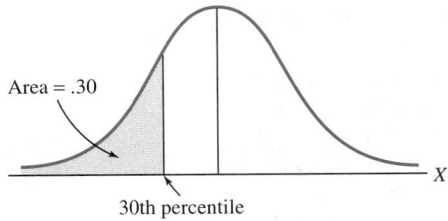

FIGURE 4.3 *The 30th percentile for a continuous distribution*

Other than the median, the most commonly used percentiles are *quartiles*. Quartiles divide the set of observations into four groups, each containing 25% of the data.

DEFINITION **Quartiles**

The **lower quartile** of a set of observations, also called the **first quartile**, is the 25th percentile of the data. That is, 25% of the data lie below the first quartile and thus 75% lie above it. The first quartile is denoted Q_1. The **upper quartile**, also called the **third quartile**, is denoted by the symbol Q_3 and is the 75th percentile of the data. That is, 75% of the observations lie below the third quartile and 25% lie above it. The median is the 50th percentile (or second quartile), because 50% of the data lie below the median and 50% lie above it.

EXAMPLE 4.18 **Calculation of Quartiles**

The following 24 observations have been ordered from lowest to highest:

> 6   8   12   17   20   22   25   26   26   28   30   32
> 34   37   40   50   61   63   65   67   69   80   82   86

Find the first and third quartiles.

SOLUTION   For the lower quartile, calculate the index with $p = 25$ and $n = 24$. Using the formula $i = (p \times n)/100$, we get $i = 6.0$. Because i is an integer, Q_1 is the average of observations having ranks 6 and 7. Because the 6th observation is 22 and the 7th observation is 25, the lower quartile is $Q_1 = (22 + 25)/2 = 23.5$.

For the upper quartile, calculate the index $i = (p \times n)/100$, where $p = 75$ and $n = 24$. The index is $i = 18.0$. Because i is an integer, Q_3 is the average of observations having ranks 18 and 19. The 18th observation is 63 and the 19th observation is 65. The upper quartile Q_3 is $(63 + 65)/2 = 64.0$.

EXAMPLE 4.19 **Quartiles for Salaries at Lang's Trucking Company**

Calculate Q_1 and Q_3 for the data in Table 4.1.

SOLUTION   To find the lower quartile, first calculate the index, $i = (25 \times 80)/100 = 20$. Because this is an integer, the lower quartile is the average of the values whose ranks are 20 and 21. These values are 1,500 and 1,530, and therefore the lower quartile is (1,500 + 1,530)/2 = 1,515. This tells us that 25% of the employees at the company earn less than or equal to $1,515 per month.

To find the upper quartile, we calculate the index $i = (75 \times 80)/100 = 60$. Because this is an integer, the upper quartile is the average of the values whose ranks are 60 and 61. Because these values are 1,730 and 1,740, the upper quartile is (1,730 + 1,740)/2 = 1,735.

Figure 4.4 shows the lower and upper quartiles for a continuous distribution. The area to the left of Q_1 is .25, and the area to the right of Q_3 is .25.

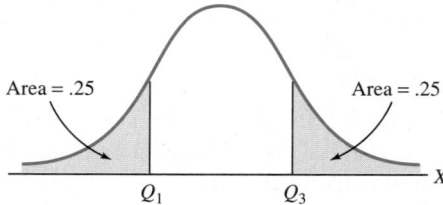

FIGURE 4.4 *The lower and upper quartiles for a continuous distribution*

Exercises

Statistical Concepts

4.3.1   State the procedure for calculating the pth percentile of a set of observations.

4.3.2   *True or false:*   Suppose Ralph Smith's SAT score falls in the 75th percentile. Then 75% of the students earned SAT scores higher than Ralph. Explain.

4.3.3   *True or false:*   The median is another name for the third quartile. Explain.

4.3.4   *True or false:*   If a distribution is symmetric, the distance between Q_1 and Q_3 is twice the distance between Q_1 and the median. Explain.

Statistical Drills

4.3.5   An industrial firm has 45 employees. The following data show the number of weeks of vacation credited to these 45 employees.

Number of weeks	Number of employees
2	19
3	14
4	8
8	4

a. Calculate the first quartile for number of weeks of vacation.
b. Find the median.
c. Find the third quartile.

Statistical Applications

4.3.6 In each of the following circumstances, select the value that you think best represents Q_1:
   a. The mean price of a gallon of gasoline at a sample of 10 service stations in a certain neighborhood is $1.20. Is the first quartile $.10, $1.08, or $10.00?
   b. The mean price of a can of three tennis balls at a sample of 10 sporting goods stores is $4.00. Is the first quartile $1.40, $3.35, or $40.00?
   c. The mean weight of a sample of 100 male college seniors is 170 pounds. Is the first quartile 120, 158, or 250 pounds?
   d. The mean height of a sample of 100 female college seniors is 66 inches. Is the first quartile 40, 62, or 86 inches?

4.3.7 The following data show the number of tennis players who used a racquet club's facilities during a random sample of 40 weekdays:

46	87	69	50	64	35	37	76	84	46
53	66	90	43	27	47	86	74	45	65
36	48	76	78	79	63	47	60	50	52
49	69	82	58	46	58	67	64	35	58

   a. Calculate the sample mean.
   b. Calculate the sample median.
   c. Calculate the first and third quartiles.
   d. Calculate the 30th percentile.
   e. Construct a frequency distribution and plot it.

4.3.8 The following data show the number of personal computers in use at 8 P.M. on a sample of 50 weekday evenings at a university computer center:

27	34	36	38	46	52	35	46	12	25
46	37	62	52	53	36	42	28	56	43
37	39	40	50	52	42	47	49	51	36
39	28	30	38	45	47	48	45	36	48
49	45	43	42	46	48	49	52	36	45

   a. Calculate the sample mean.
   b. Calculate the sample median.
   c. Calculate the first and third quartiles.
   d. Calculate the 30th percentile.
   e. Construct a frequency distribution and plot it.

4.4   Measures of Dispersion

A measure of central tendency does not completely characterize a set of observations because it tells us nothing about variability in the data. Summary statistics that measure the amount of variation in a data set are called **measures of dispersion**.

For example, the notion of variability in statistical analysis is important in production management, where one major concern is reducing variations in the quality of manufactured products or in some important characteristics of a product, such as the diameter of a bearing or the hardness of steel. In many financial problems, risk is measured by the amount of variability in the potential returns from an investment.

In addition to a measure of central tendency, some measure of dispersion is needed to tell whether a particular observation ranks near the top or bottom of a set of observations. Suppose you receive a grade of 70 on a mathematics test and a grade of 70 on a chemistry test. To guess what letter grade you will receive on the tests, you would probably want some information concerning the typical score of all the students who took the test and the amount of variation in the grades for the two tests. Suppose the math scores have a mean of 60 and are highly concentrated about the mean so that all the scores fall between 50 and 72. In this case, a score of 70 would place you near the top of the class, and you could expect an A. Assume that the average score on the chemistry test is also 60 but that the scores are much more widely dispersed than the math scores. Some scores are in the 30s and 40s, and some are in the 80s and 90s. A score of 70 would put you closer to the middle of the class than to the top, and you might expect a B or a C.

We discuss five different summary measures of dispersion: the range, the interquartile range, the mean absolute deviation, the variance, and the standard deviation. Of these, the standard deviation is the measure used most frequently by statisticians.

The Range

DEFINITION   **Range**

The **range** of a set of observations is the difference between the largest value and the smallest value.

EXAMPLE 4.20   **Calculating a Range**

A sporting goods store has seven employees whose ages are 22, 33, 26, 21, 44, 63, and 58. A paint store has eight employees whose ages are 32, 30, 27, 30, 27, 34, 40, and 35. Calculate the ranges of the ages of employees at these stores.

SOLUTION   At the sporting goods store, the oldest employee is 63 and the youngest is 21. The range is thus $63 - 21 = 42$ years. At the paint store, where the oldest employee is 40 and the youngest is 27, the range is $40 - 27 = 13$ years. The range at the sporting goods store is more than 3 times as large as that at the paint store.

EXAMPLE 4.21   **Range of Salaries at Lang's Trucking Company**

Calculate the range for the data in Table 4.1.

SOLUTION   The smallest salary is $1,300 per month and the largest salary is $2,025. Therefore, the range of salaries is $2,025 - $1,300, or $725.

The range is very easy to calculate and interpret. Because of this, it is frequently used by the general public to describe the amount of variation in a set of data. It is less frequently used in scientific circles, however, because its value is easily influenced by the number of observations. In general, the range of a large sample of observations taken from a population is likely to be larger than the range of a small sample taken from the same population.

The range can be an unsatisfactory measure of dispersion because it is influenced too much by a single very high or very low observation. Because it depends only on the largest and smallest observations, the range is susceptible to considerable distortion if there are any extreme values. One way around this difficulty is to arrange the observations in ascending order, discard a few of the highest and lowest observations, and find the range of the remaining observations. Such a measurement is called a *trimmed range*. If we discard the lowest and the highest 25% of the observations, the trimmed range will measure the range of the middle 50% of the data. This measure is called the *interquartile range*.

DEFINITION **Interquartile Range**

For any population or sample of values, the **interquartile range** is $(Q_3 - Q_1)$, where Q_3 represents the upper quartile (or 75th percentile) and Q_1 represents the lower quartile (or 25th percentile).

The **semi-interquartile range** (half the interquartile range) can be thought of as an average distance from the center of the distribution. If the distribution is symmetric, the semi-interquartile range is the distance between Q_1 and the median (or Q_3 and the median).

One advantage of the interquartile range over the range is that the interquartile range can usually be calculated from a frequency distribution having open-ended classes, whereas the range cannot.

Deviations from the Mean

To measure the amount of variation in a set of data, we could calculate how far each observation is from the mean and then calculate the average (or mean) of these deviations. If the population mean is known, then we calculate deviations from the population mean; otherwise, we calculate deviations from the sample mean.

DEFINITION **Deviation from the Mean**

For any value x, the **deviation from the mean** is the difference $(x - \mu)$ if the population mean is known or $(x - \bar{x})$ if the sample mean is used.

If x exceeds the mean, the deviation is positive; if x is less than the mean, the deviation is negative. Deviations from the mean convey important information about the amount of variation in a set of data. If they tend to be large in absolute value, then the data are very spread out and are highly variable. If the deviations are small in absolute value, then the data are tightly clustered about the mean and do not exhibit much variability.

Averaging all the deviations from the mean might seem a useful way to measure dispersion. However, the average of all the deviations from the mean is 0 for any set of data, as shown in the following:

$$\Sigma\,(x_i - \bar{x}) = (x_1 - \bar{x}) + (x_2 - \bar{x}) + \cdots + (x_n - \bar{x})$$
$$= (x_1 + x_2 + \cdots + x_n) - n\bar{x}$$
$$= \Sigma\,x_i - n\frac{\Sigma\,x_i}{n} = 0$$

EXAMPLE 4.22   **Deviations from the Mean** ───────────────────────────

Calculate the sum of the deviations from the mean for the data in Example 4.6.

SOLUTION   In Example 4.6, the mean was shown to be 77. The seven deviations from the mean are as follows:

$$92 - 77 = 15 \qquad 84 - 77 = 7 \qquad 70 - 77 = -7 \qquad 76 - 77 = -1$$
$$66 - 77 = -11 \qquad 80 - 77 = 3 \qquad 71 - 77 = -6$$

The sum of the seven deviations is:

$$\Sigma\,(x_i - \bar{x}) = 15 + 7 - 7 - 1 - 11 + 3 - 6 = 0$$

To avoid the problem of deviations from the mean summing to 0, we have two alternatives. We can take the absolute values of the deviations before we average them, or we can square the deviations before we average them.

The Mean Absolute Deviation

The *mean absolute deviation* is the arithmetic average of the absolute deviations from the mean.

DEFINITION   **Mean Absolute Deviation** ───────────────────────────

The **mean absolute deviation** (M.A.D.) is calculated using the formula

$$\text{M.A.D.} = \frac{|x_1 - \bar{x}| + |x_2 - \bar{x}| + \cdots + |x_n - \bar{x}|}{n} = \frac{\Sigma\,|x_i - \bar{x}|}{n}$$

EXAMPLE 4.23   **Calculating the Mean Absolute Deviation** ───────────────────────────

Refer to the data in Example 4.22, which showed the numbers of customers at a restaurant on a sample of 7 days. Calculate the mean absolute deviation.

SOLUTION   Example 4.22 showed the deviations from the mean for the number of customers at a restaurant on 7 days. The absolute values of the deviations are:

$$|92 - 77| = 15 \qquad |84 - 77| = 7 \qquad |70 - 77| = 7 \qquad |76 - 77| = 1$$
$$|66 - 77| = 11 \qquad |80 - 77| = 3 \qquad |71 - 77| = 6$$

The mean absolute deviation is

$$\text{M.A.D.} = (15 + 7 + 7 + 1 + 11 + 3 + 6)/7 = 7.143$$

The M.A.D. is 7.143, which indicates that on any given day the number of customers deviates from the mean by about 7. This deviation is relatively small and indicates that the number of customers at the restaurant does not vary much.

EXAMPLE 4.24 **Calculating Another Mean Absolute Deviation**

The following data show the adjusted gross incomes reported on a sample of five income tax returns:

$$\$6,800 \quad \$10,000 \quad \$33,000 \quad \$14,500 \quad \$27,700$$

Calculate the mean absolute deviation.

SOLUTION The sample mean is $\bar{x} = \$18,400$. The absolute deviations from the mean are as follows:

$$|\$6,800 - \$18,400| = \$11,600 \qquad |\$10,000 - \$18,400| = \$8,400$$
$$|\$33,000 - \$18,400| = \$14,600 \qquad |\$14,500 - \$18,400| = \$3,900$$
$$|\$27,700 - \$18,400| = \$9,300$$

The mean absolute deviation is

$$\text{M.A.D.} = (\$11,600 + \$8,400 + \$14,600 + \$3,900 + \$9,300)/5 = \$9,560$$

In this example, the data are much more spread out than in Example 4.23, and consequently the mean absolute deviation is much larger.

The mean absolute deviation is not used very frequently because absolute values are difficult to work with mathematically.

The Variance and Standard Deviation

By far the most important and most frequently used summary statistics for measuring variation in a set of data are the *variance* and the *standard deviation*. Because the standard deviation is the square root of the variance, the two statistics convey exactly the same amount of information. Like the mean absolute deviation, these statistics are based on the deviations from the mean. To calculate the variance, we square the deviations and average the results; to get the standard deviation, we then take the square root.

How we calculate the variance depends on whether we have a population or sample of data. If we are using the entire population of data, then we can calculate the *population variance*, denoted by the symbol σ^2 (read "sigma squared").

DEFINITION **Population Variance**

The **population variance** is calculated using the formula

$$\sigma^2 = \frac{\Sigma(x_i - \mu)^2}{N}$$

where N is the population size.

To calculate the population variance, we square each deviation from the population mean, sum these squared values, and divide the sum by N. Thus, the population variance measures the average of the squared deviations from the population mean.

The variance is an inconvenient measure of spread because it measures the average of the *squared* deviations from the mean. Thus, the variance is not expressed in the same units as the original observations. For example, if the original data are in dollars, then the deviations from the mean are measured in dollars and the variance is measured in squared dollars. In addition, the variance is frequently an extremely large number relative to the original observations. For example, if an individual's annual income is $5,000 above the mean, then the squared deviation is 25,000,000 squared dollars. When we calculate the variance of income data where the deviations from the mean exceed $1,000, the variance will usually exceed 1,000,000. To overcome these difficulties, we work with the square root of the variance rather than the variance itself. This leads us to the concept of the *standard deviation,* the most important and most frequently used measure of variation present in a data set.

DEFINITION   **Population Standard Deviation** ⸻⸻⸻⸻⸻⸻⸻⸻

The **population standard deviation** is the positive square root of the population variance and is denoted by the symbol σ.

⸻⸻⸻⸻⸻⸻⸻⸻⸻⸻⸻⸻⸻⸻⸻⸻⸻⸻⸻

If we have a sample of data, we calculate the *sample variance* and the *sample standard deviation.* The sample variance is denoted by the symbol s^2, and the sample standard deviation by the symbol s.

DEFINITIONS   **Sample Variance and Sample Standard Deviation** ⸻⸻⸻⸻⸻

To calculate the **sample variance,** we use the formula

$$s^2 = \frac{\Sigma\ (x_i - \bar{x})^2}{n - 1}$$

where n is the sample size. The **sample standard deviation** is the positive square root of the sample variance and is denoted by the symbol s.

⸻⸻⸻⸻⸻⸻⸻⸻⸻⸻⸻⸻⸻⸻⸻⸻⸻⸻⸻

The formula for calculating the sample variance differs from the formula for calculating the population variance in two important ways:

1. The denominator of the population variance is N, whereas the denominator of the sample variance is $(n - 1)$.
2. The population variance is based on deviations from the population mean μ, whereas the sample variance is based on deviations from the sample mean $\bar{x}$.

The use of $(n - 1)$ rather than n in the denominator of s^2 should be explained. Recall that $\bar{x}$ and s^2 are calculated from sample data and are estimates of the population parameters μ and σ^2. In general, the sample will not exactly reproduce the characteristics of the population, and it is unlikely that $\bar{x}$ will be exactly equal to μ. When estimating a population parameter, we want to use a procedure that has no tendency to overestimate or underestimate the population value on the average. If an estimator equals the population value on the average (i.e., overestimates and underestimates balance out), then we have an *unbiased* estimator.

We use $(n - 1)$ rather than n in the denominator of s^2 so that it will be an unbiased estimator of σ^2. If n were used as the denominator, s^2 would be smaller than σ^2, on the average, and thus would be a biased estimator. The reason for this bias is that, on the average, the sample squared deviations $(x_i - \bar{x})^2$ are slightly smaller than the population squared deviations $(x_i - \mu)^2$. The use of $(n - 1)$ in the denominator corrects this bias.

The number $(n - 1)$ is sometimes called the *degrees of freedom*. The phrase "degree of freedom" arises from the fact that to calculate the variance, we need to find the n deviations from the mean; if $(n - 1)$ of these deviations are known, the last deviation can be calculated. That is, if we know the first $(n - 1)$ deviations from the mean, the last deviation can be deduced from the fact that $\Sigma(x_i - \bar{x}) = 0$.

EXAMPLE 4.25 **Sample Standard Deviation for Number of Customers**

Calculate the sample standard deviation for the data in Example 4.1.

SOLUTION From Example 4.6, we know the mean is $\bar{x} = 77$. The subsequent calculations are shown in Table 4.2. The second column of the table shows the deviations from the mean, and the third column shows the squares of these deviations. The sum of these squared deviations is $\Sigma(x_i - \bar{x})^2 = 490$. We divide this sum by $n - 1 = 6$ to obtain the sample variance

$$s^2 = 490/6 = 81.667$$

The sample standard deviation is simply

$$s = \sqrt{81.667} = 9.037$$

In Example 4.23, we showed that the mean absolute deviation for the same sample of data is M.A.D. $= 7.143$. In this example, the standard deviation is slightly larger than the mean absolute deviation.

TABLE 4.2 *Calculating the sample standard deviation*

x	$x - \bar{x}$		$(x - \bar{x})^2$
92	$92 - 77 =$	15	225
84	$84 - 77 =$	7	49
70	$70 - 77 =$	-7	49
76	$76 - 77 =$	-1	1
66	$66 - 77 =$	-11	121
80	$80 - 77 =$	3	9
71	$71 - 77 =$	-6	36
Total 539		0	490

Alternative Formula for Calculating the Variance

The following formula provides another way of calculating the sample variance. This formula is frequently more convenient when performing the calculations by hand.

Alternative formula for sample variance

$$s^2 = \frac{\sum x_i^2 - n\bar{x}^2}{n - 1}$$

EXAMPLE 4.26   **Calculating the Sample Standard Deviation with the Alternative Formula** ———

Calculate the sample standard deviation for the data in Example 4.1 using the alternative formula.

SOLUTION   The squared values of the data are

8,464   7,056   4,900   5,776   4,356   6,400   5,041

We sum these values to obtain $\sum x_i^2 = 41{,}993$. Since we know the sample mean is $\bar{x} = 77$, the alternative formula for the sample variance yields

$$s^2 = \frac{41{,}993 - 7(77)^2}{7 - 1} = \frac{490}{6} = 81.667$$

The sample standard deviation, $s = \sqrt{81.667} = 9.037$, is identical to the value calculated in Example 4.25.

EXAMPLE 4.27   **Sample Standard Deviation for Data in Example 4.24** ———————

Calculate the sample standard deviation for the data in Example 4.24, the adjusted gross incomes for a sample of five individuals.

SOLUTION   From Example 4.24, we know that the sample mean is $\bar{x} = \$18{,}400$. Because the mean is an integer, it is convenient to use the deviations from the mean to calculate the variance and standard deviation. The deviations from the mean are

−11,600   −8,400   14,600   −3,900   9,300

Squaring these deviations, we get

134,560,000   70,560,000   213,160,000   15,210,000   86,490,000

Finally, the sum of these squared deviations is $\sum(x_i - \bar{x})^2 = 519{,}980{,}000$. Dividing by $(n - 1)$, we get the sample variance of

$$s^2 = 519{,}980{,}000/4 = 129{,}995{,}000$$

The sample standard deviation is thus

$$s = \sqrt{129{,}995{,}000} = 11{,}401.53$$

Recall from Example 4.24 that the mean absolute deviation is 9,560 for these data.

Example 4.27 shows that calculating the variance and standard deviation can be very tedious when the observations consist of large values. Fortunately, numerous computer programs, such as SPSSX, SAS, and Minitab, can perform the calculations for us.

EXAMPLE 4.28   **Standard Deviation of Data in Table 3.1** ────────────────────

Calculate the population standard deviation for the data in Table 3.1.

S O L U T I O N   The 80 observations represent the entire population of data, so we calculate the variance and standard deviation by using $N = 80$ in the denominator. It is left to the reader to show that the variance is $\sigma^2 = 32{,}441.21$ and the standard deviation is $\sigma = 180.11$. Examine the histogram plotted in Figure 3.1 (page 48). The center of gravity of the histogram is the mean value of $1{,}626.69$, and the standard deviation shows that the typical deviation from this central value is approximately $180.00.

If the data are widely scattered about the mean, deviations from the mean will be large, and the variance and the standard deviation will also be large. If the observations are concentrated near the mean, then the variance and standard deviation will be small. This idea is illustrated in Figure 4.5.

Figure 4.5 shows the distributions of two different hypothetical variables, X and Y. Because the distributions are centered at the same value, both X and Y have the same mean μ. However, the distribution of X is more spread out than the distribution of Y, so the variance of X (denoted σ_x^2) is larger than the variance of Y (σ_y^2). In Figure 4.5, consider any interval of values centered about the mean μ, such as the interval $(\mu - c, \mu + c)$ for any positive value c. A larger proportion of the Y values than the X values will fall in the interval. In Figure 4.5, about 95% of the Y values fall between $(\mu - c)$ and $(\mu + c)$, whereas only about 75% of the X values fall in this interval.

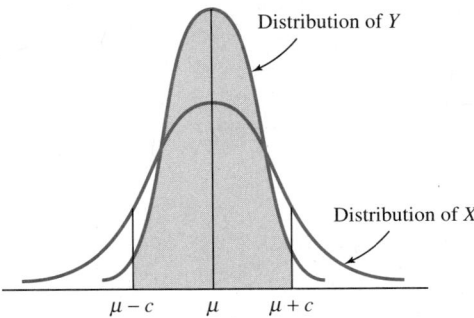

F I G U R E   4.5   *Comparing the variances for two distributions*

EXAMPLE 4.29   **Comparing Two Standard Deviations** ———————————————

To see how the sample standard deviation measures the spread in a set of data, calculate the sample standard deviation for the following two samples of data. The first data set shows the weights of a random sample of 50 7-year-old boys, and the second the weights of a random sample of 50 adult men.

Weights of boys (in pounds)

53	48	45	47	52	61	44	52	48	51
68	57	53	56	55	46	47	47	41	53
63	48	59	57	54	44	66	64	66	67
51	52	59	62	51	65	48	46	49	52
52	53	47	49	56	63	67	71	61	73

Weights of men (in pounds)

126	154	171	204	165	174	178	184	167	172
139	146	128	147	159	157	163	174	181	188
162	154	203	148	171	198	186	162	171	184
142	138	166	159	178	196	174	183	188	191
172	177	196	212	168	171	229	217	187	166

SOLUTION   The sample means, sample variances, and sample standard deviations for the boys' weights and the men's weights are as follows:

$$\bar{x}_{boys} = 54.78 \qquad \bar{x}_{men} = 172.52$$
$$s^2_{boys} = 62.91 \qquad s^2_{men} = 475.48$$
$$s_{boys} = 7.93 \qquad s_{men} = 21.81$$

The sample standard deviation for the men is much larger than that for the boys, indicating that the distribution of the men's weights is much more spread out than the distribution

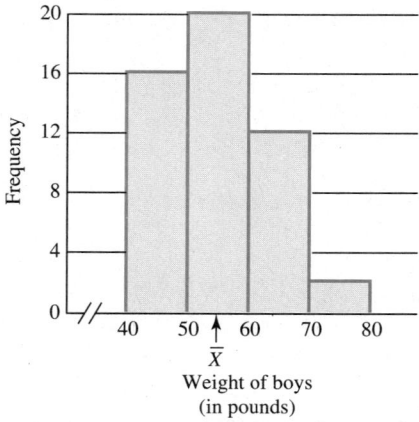

FIGURE 4.6   *Distribution of weights for a sample of boys*

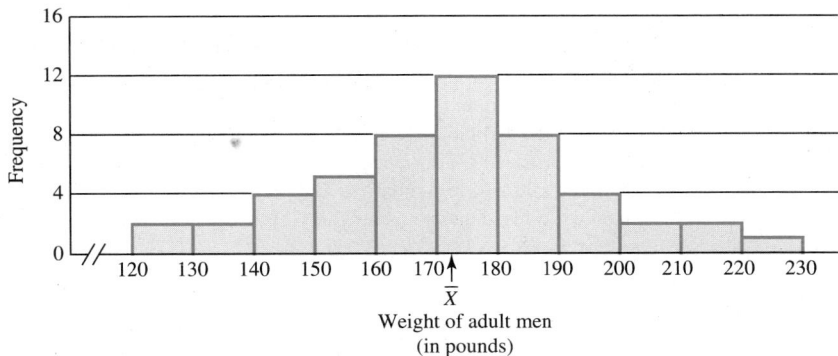

FIGURE 4.7 *Distribution of weights for a sample of men*

of the boys' weights. For the boys, the sample standard deviation is only 7.93. For the men, the standard deviation is 21.81 pounds, or nearly 3 times as large. The histograms in Figures 4.6 and 4.7 show the differences in the spread of the two distributions.

Exercises

Statistical Concepts

4.4.1 State several reasons why the range can be an unsatisfactory measure of dispersion.

4.4.2 *True or false:* Suppose we obtain a sample of data from some population. The range calculated from the sample can never exceed the population range, but it can be less than the population range. Explain.

4.4.3 *True or false:* The interquartile range is twice the range. Explain.

4.4.4 *True or false:* The range is four times the distance between Q_1 and the median. Explain.

4.4.5 *True or false:* The range is the same as the largest absolute deviation from the mean. Explain.

4.4.6 Explain the difference between σ^2 and s^2. When do we calculate σ^2 and when do we calculate s^2?

4.4.7 Explain why it is more useful to use the standard deviation rather than the variance when describing the variability of a data set.

4.4.8 State two different, but equivalent, formulas for calculating the sample variance.

4.4.9 *True or false:* Suppose the variance of population 1 is 4 times the variance of population 2. Then it is reasonable to state that the variability of population 1 is about twice the variability of population 2. Explain.

4.4.10 Can the standard deviation ever be negative? Why or why not?

4.4.11 *True or false:* Suppose you have a population of 100 observations.
  a. If you add 10 to each observation, the mean increases by 10.
  b. If you add 10 to each observation, the standard deviation increases by 10.
  c. If you multiply each observation by 3, the mean is multiplied by 3.
  d. If you multiply each observation by 3, the standard deviation is multiplied by 3.
  e. If you change the sign of each observation, the sign of the mean is changed.
  f. If you change the sign of each observation, the value of the standard deviation is unchanged.

Statistical Drills

4.4.12   Use the following sample of $n = 5$ measurements: 73, 66, 45, 28, 59.
  a. Calculate the sample mean.
  b. Calculate the range.
  c. Calculate the sample mean absolute deviation.
  d. Calculate the sample variance.
  e. Calculate the sample standard deviation.
  f. Change the five data values to 730, 660, 450, 280, and 590. Repeat parts **a–e** and compare with your previous answers.

4.4.13   Suppose a sample of income data yields the following $n = 5$ incomes:

  $23,000 $36,500 $47,200 $20,200 $61,300

  a. Calculate the sample mean.
  b. Calculate the range.
  c. Calculate the sample mean absolute deviation.
  d. Calculate the sample variance.
  e. Calculate the sample standard deviation.
  f. Scale the data values by a factor of 1,000. That is, change the five data values to 23.0, 36.5, 47.2, 20.2, and 61.3. Repeat parts **a–e** and compare with your previous answers.

4.4.14   Each of the following samples of data has a mean of 50.

Sample 1:	0,	30,	40,	50,	60,	70,	100
Sample 2:	0,	46,	48,	50,	52,	54,	100
Sample 3:	0,	2,	4,	50,	96,	98,	100

  a. Determine the range for each sample.
  b. Determine by inspection which sample has the largest standard deviation and which has the smallest. Explain your reasoning.

Statistical Applications

4.4.15   In each of the following circumstances, select the value that you think best represents the standard deviation.
  a. We obtain data showing the price of a gallon of gasoline at a sample of 10 service stations in a certain neighborhood. The mean price is $1.00. Which is a reasonable guess for the standard deviation of prices: $.01, $1.00, or $10.00?
  b. The mean starting salary is $21,000 for 100 recent college graduates. Which is a reasonable guess of the standard deviation: $1.00, $10.00, or $1,000?
  c. The mean weight of a sample of 100 male college seniors is 170 pounds. Is the standard deviation 12, 120, or 250 pounds?
  d. The mean height of a sample of 100 female college seniors is 66 inches. Is the standard deviation 2, 12, or 86 inches?

4.4.16   Select the value that you think best represents the variance.
  a. The mean price of a gallon of gasoline at a sample of 10 service stations in a certain neighborhood is $1.00. Is a reasonable guess concerning the variance of prices .0001, 1.04, or 10.00?
  b. The mean starting salary is $21,000 for 100 recent college graduates. Which is a reasonable guess of the variance: 1.00, 100.00, or 1,000,000?

   c. The mean weight of a sample of 100 male college seniors is 170 pounds. Is the variance 12, 144, or 1,250?

   d. The mean height of a sample of 100 female college seniors is 66 inches. Is the variance .04, 4, or 400?

4.4.17 A sample of depths (in feet) of recent oil drilling is as follows:

    1,500   1,200   1,600   1,700   1,500   2,000

   a. Compute the sample mean, median, and mode.
   b. Calculate the range, the sample variance, and sample standard deviation.
   c. Calculate the deviations from the mean and verify that they sum to 0.
   d. Calculate the mean absolute deviation and compare it to the sample standard deviation.

4.4.18 Refer to the data in Exercise 17. Multiply each depth by 10.
   a. Compute the sample mean, median, and mode.
   b. Calculate the range, the sample variance, and sample standard deviation.
   c. Calculate the deviations from the mean and verify that they sum to 0.
   d. Calculate the mean absolute deviation and compare it to the sample standard deviation.
   e. Calculate the range.

4.4.19 The Walker Printing Company has 12 employees. Last year those employees missed the following number of days of work because of illness:

    8 6 1 13 20 9 8 6 7 14 2 3

   a. Calculate the population mean and median.
   b. Calculate the variance. These observations represent the entire population, so the denominator in calculating σ^2 is 12, not 11.
   c. Calculate the standard deviation.
   d. Calculate the mean absolute deviation and compare it to the standard deviation.
   e. Calculate the range.

4.4.20 The following data represent the population of test scores received by 50 applicants on a civil service exam:

    69 45 78 98 87 73 46 68 69 91
    83 87 78 89 84 56 78 92 47 62
    74 81 80 76 59 62 93 88 77 65
    56 67 78 79 81 80 64 67 78 83
    68 69 80 94 93 76 79 89 92 59

   a. Find the range of the data.
   b. Find the mean.
   c. Find the population variance.
   d. Find the population standard deviation.
   e. Find the median.
   f. Find the first and third quartiles.
   g. Find the interquartile range.
   h. Find the mean absolute deviation.
   i. Plot the data on a histogram and locate the mean, the first quartile, and the third quartile.

| 4.5 | # The Empirical Rule and Standardized Scores |

The standard deviation is an important characteristic of any distribution. The proportion of observations expected to fall between any two specific values depends, at least partly, on the standard deviation of the distribution.

An important theorem, called Chebyshëv's theorem after the Russian mathematician Pafnuti Lvovich Chebyshëv (1821–1894) who proved it, illustrates the importance of the role of the standard deviation in statistical theory.

DEFINITION   **Chebyshëv's Theorem** ────────────────────────────────

Let c be any number greater than 1. For any sample or population of data, the proportion of observations that lie *fewer than* c standard deviations from the mean is at least $1 - 1/c^2$.

───

Chebyshëv's theorem applies to both populations and samples. For a population, the interval $(\mu - c\sigma, \mu + c\sigma)$ contains all values fewer than c standard deviations from the mean. When $c = 2$, we have $(1 - 1/c^2) = (1 - 1/2^2) = .75$. This implies that at least 75% of the observations in any population or sample will lie within 2 standard deviations of the mean. When $c = 3$, we obtain $(1 - 1/c^2) = (1 - 1/3^2) = .89$. Thus Chebyshëv's theorem tells us that at least 89% of the observations in any population fall in the interval $(\mu - 3\sigma, \mu + 3\sigma)$, and at least 89% of the observations in any sample fall in the interval $(\bar{x} - 3s, \bar{x} + 3s)$.

EXAMPLE 4.30   **Application of Chebyshëv's Theorem** ───────────────────────────

Use Chebyshëv's theorem to find an interval that covers at least 75% of the data in Table 4.1.

SOLUTION   In Example 4.7, we found that the mean monthly salary is

$$\mu = \$1,626.69$$

In Example 4.28, we found that the standard deviation is

$$\sigma = 180.11$$

From Chebyshëv's theorem, we can deduce, using $c = 2$, that at least 75% of the employees have incomes between $(1,626.69 - 2 \times 180.11)$ and $(1,626.69 + 2 \times 180.11)$, i.e., at least 75% of the monthly salaries must fall in the interval $(1,266.47, 1,986.91)$. By actually counting the number of salaries, we find that 75 of the 80 salaries, or 93.75%, fall in this interval.

───

Example 4.30 shows that Chebyshëv's theorem gives a conservative estimate of the proportion of observations in an interval about the mean. When the exact form of the distribution is known, the proportion of values falling in any interval can be determined more precisely. Chebyshëv's theorem is useful because it highlights the importance of the standard deviation in statistical theory and because it places a lower limit on the propor-

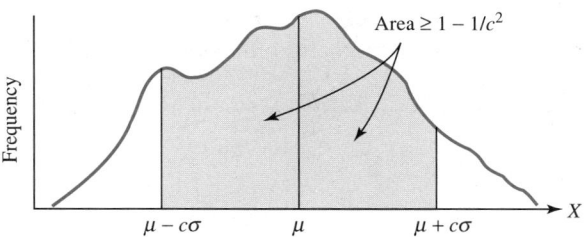

FIGURE 4.8 *Illustration of Chebyshëv's theorem*

tion of values falling in any interval around the mean and an upper limit on the proportion of values falling in the tails of the distribution. Figure 4.8 illustrates Chebyshëv's theorem.

Application of Chebyshëv's theorem

For any sample or population of data,

1. At least 75% of the observations lie within 2 standard deviations of the mean.
2. At least 89% of the observations lie within 3 standard deviations of the mean.
3. At least 93% of the observations lie within 4 standard deviations of the mean.

The Empirical Rule

When a distribution is bell shaped, the proportion of values that fall in the interval $(\mu - c\sigma, \mu + c\sigma)$ will exceed the values given by Chebyshëv's theorem. When the distribution is approximately bell shaped, the following rule, called the *Empirical Rule,* holds at least approximately.

DEFINITION **Empirical Rule**

When the distribution of a population or sample of data is approximately bell shaped, approximately 68% of the values will fall within 1 standard deviation of the mean, approximately 95% of the values will fall within 2 standard deviations of the mean, and approximately 99.7% of the values will fall within 3 standard deviations of the mean.

EXAMPLE 4.31 **Application of the Empirical Rule**

The distribution of IQs of adults in the United States is known to have a mean of 100, a standard deviation of 15, and a bell-shaped distribution. Suppose that a job applicant has an IQ of 150. Is it reasonable to say that this job applicant is very intelligent?

SOLUTION An IQ of 150 is 50 units, or 3.33 standard deviations, above the mean of 100. The Empirical Rule tells us that approximately 99.7% of all IQs should be between 55 and 145, that is, within 3 standard deviations of the mean. The job applicant's IQ lies outside the interval (55, 145); thus the applicant ranks in the upper .3% of the population.

Standardized Scores

Often it is desirable to describe the *relative* position of an observation within a distribution. For example, knowing that an individual finished a downhill ski race in 125 seconds conveys little information about performance. Judgment of the performance would depend on whether 125 seconds is a slow, medium, or fast time.

One way of describing the location of an observation in a distribution is to calculate its *standardized score*. This score, sometimes called the *Z score,* indicates how many standard deviations an observation lies above or below the mean.

DEFINITION   **Standardized Score**

If the population mean and the population standard deviation are known, the **standardized score** for an observation x is

$$z = \frac{x - \mu}{\sigma}$$

When the sample mean and sample standard deviation are used, the standardized score is

$$z = \frac{x - \bar{x}}{s}$$

The standardized score is sometimes called the **Z score.**

The Z score is the signed distance from the mean measured in units of standard deviation. For any population or sample of data, the mean of all the Z scores is 0, and the standard deviation of all the Z scores is 1.

If an observation is above the mean, then the Z score will be positive. For example, if $z = 1$, then an observation falls exactly 1 standard deviation above the mean. If $z = -2$, then the observation is exactly 2 standard deviations below the mean.

Chebyshëv's theorem shows how the standardized scores can be of use. The theorem tells us, for example, that at least 75% of the data lie within 2 standard deviations of the mean. Any value within 2 standard deviations of the mean has a standardized score between -2 and 2. Thus, at least 75% of the standardized scores must be between -2 and 2. The Empirical Rule tells us that for a bell-shaped distribution, approximately 95% of the observations have Z scores between -2 and 2.

EXAMPLE 4.32   **Calculation of Standardized Scores**

Suppose three students' scores on a test were 675, 600, and 625. The mean score was $\mu = 625$ and the standard deviation was $\sigma = 25$. Calculate the standardized scores for these students.

SOLUTION   The first student's standardized score is

$$z = \frac{675 - 625}{25} = 2$$

The second student's standardized score is

$$z = \frac{600 - 625}{25} = -1$$

The third student's standardized score is

$$z = \frac{625 - 625}{25} = 0$$

Use of standardized scores can help us compare observations from different distributions. For example, suppose that in a ski race a competitor records a time of 125 seconds on his first run and 135 seconds on his second run. To judge the competitor, we have to compare his time to the times of all other competitors. The competitor's time on the first run could translate to a standardized score of 2, whereas the time on the second run could translate to a standardized score of -2. By examining the standardized scores, we could determine that the competitor was one of the slowest racers on the first run and one of the fastest racers on the second.

Detecting Outliers

An **outlier** is an observation that falls far out in the tail of a distribution. Sometimes an outlier can be an indication of a faulty or incorrectly recorded observation. Because of this possibility, outliers should be examined carefully. A Z score can be helpful in detecting an outlier. From Chebyshëv's theorem, we know that for any distribution of data, fewer than 11% of the observations lie more than 3 standard deviations from the mean. If the data are approximately normally distributed, then fewer than .3% of the observations have Z scores less than -3 or greater than $+3$.

Suppose we observe a measurement that has a Z score of 4.5. This means that the observation lies 4.5 standard deviations above the mean. The Empirical Rule tells us that a Z score this large is very improbable and could indicate a faulty observation.

Coefficient of Variation

In some situations, we may be more interested in a *relative* measure of the amount of variation present in a data set than in the absolute measure provided by the standard deviation or the variance. For example, the price of a share of stock in company A varies over time in such a way that the mean price during a given year is $100 and the standard deviation is $5. Suppose that the stock of company B has a mean price of $10 per share and a standard deviation of $4. Comparing the variability of the prices of the two stocks is difficult because when the means of the variables differ greatly, we do not get an accurate picture of the relative variability in the two data sets by comparing the standard deviations. A measure of variability that overcomes these difficulties is the *coefficient of variation,* denoted as CV. The coefficient of variation, also called the *relative standard deviation,* expresses the standard deviation as a percentage of the mean.

DEFINITION   **Coefficient of Variation** ───────────────────────────────────

The **coefficient of variation** is calculated as follows:

$$\text{Coefficient of variation} = (\text{Standard deviation/Mean}) \times 100\%$$

For a population of data, the coefficient of variation is

$$CV = (\sigma/\mu) \times 100\%$$

For a sample of data, the coefficient of variation is

$$CV = (s/\bar{x}) \times 100\%$$

───

The coefficient of variation is useful for comparing the relative variation in data sets. For example, for the high-priced stock of company A, the coefficient of variation is $\$5/\$100 \times 100\% = 5\%$. For the low-priced stock of company B, the coefficient of variation is $\$4/\$10 \times 100\% = 40\%$. In relative terms, the price of the company B stock is much more variable than the price of the company A stock.

As another example, suppose the standard deviation of the weights of steel bars is 1 oz. How meaningful is the information conveyed by this measure? If the mean weight of the steel bars is, say, 4 oz, then there appears to be a very large amount of variation in the weights of the bars. On the other hand, if the mean weight of the bars is, say, 700 lb (or 11,200 oz), then the bars would be considered to be virtually identical in weight. Thus, in some instances, a standard deviation of 1 oz might be considered very large and in other cases, it would be considered very small. The coefficient of variation is designed especially to be a relative measure of dispersion. The CV allows us to consider the dispersion as a proportion of the mean, that is, the dispersion in proportion to the average magnitude of the data.

Graphical Methods Versus Summary Measures

Frequently the important fe. ures of a set of data can be described quite concisely by a few summary statistics. In general, however, summary measures provide much less information than the entire frequency distribution. Actually, it is quite surprising that data analysts do not emphasize tables and graphs of frequency distributions more. Whenever possible, you should present the entire frequency distribution in either tabular or graphical form when analyzing a set of data.

One reason that summary measures of central tendency and dispersion are so popular in statistics is that they are the primary tools used by statisticians for testing hypotheses. Just as we use the sample histogram to make inferences about the shape and location of the population frequency distribution, we use the sample mean, the sample standard deviation, and the sample proportion to test inferences about the comparable population measures.

Exercises

Statistical Concepts

4.5.1 Draw a graph of a frequency distribution and indicate the mean of the distribution. Show (approximately) the interval $(\mu - 2\sigma, \mu + 2\sigma)$. At least what proportion of the data must fall inside this interval if we use Chebyshëv's theorem?

4.5.2 *True or false:* Chebyshëv's theorem tells us that, in any data set, approximately 25% of the measurements will fall more than 2 standard deviations from the mean. Explain.

4.5.3 Draw a graph of a bell-shaped frequency distribution and indicate the mean of the distribution. Show (approximately) the interval $(\mu - 2\sigma, \mu + 2\sigma)$. At least what proportion of the data must fall inside this interval if we use the Empirical Rule?

4.5.4 *True or false:* Suppose a distribution is approximately bell shaped. Suppose we have a random sample of 1,000 observations from this distribution. The Empirical Rule tells us that approximately 997 of the measurements will fall less than 3 standard deviations from the mean. Explain.

4.5.5 *True or false:* Suppose a population has mean $\mu = 50$ and standard deviation $\sigma = 10$. All observations less than $x = 40$ have standardized scores less than -1. Explain.

4.5.6 *True or false:* The coefficient of variation is a relative measure of the amount of variation present in a data set, whereas the standard deviation is an absolute measure of the amount of variation present in a data set. Explain.

4.5.7 *True or false:* Most observations will lie within 2 standard deviations of the mean and very few will be more than 3 standard deviations from the mean. Explain.

4.5.8 *True or false:* If a set of numerical observations is roughly bell shaped, approximately 32% of the observations fall more than 1 standard deviation from the mean. Explain.

Statistical Drills

4.5.9 A set of SAT scores has mean $\mu = 890$ and standard deviation $\sigma = 120$. Assume the data are bell shaped.
  a. Calculate the coefficient of variation for this data set.
  b. Joan's SAT score is $x = 1130$. Calculate her standardized score.
  c. Based on the Empirical Rule, approximately what proportion of students received a score higher than Joan's?
  d. Mike's SAT score is $x = 770$. Calculate his standardized score.
  e. Based on the Empirical Rule, approximately what proportion of students received a score lower than Mike's?
  f. Art's SAT score is $x = 1250$. Calculate his standardized score.
  g. Based on the Empirical Rule, approximately what proportion of students received a score higher than Art's?
  h. Approximately what proportion of students received a score between 770 and 1010?
  i. Approximately what proportion of students received a score between 650 and 1130?

Statistical Applications

4.5.10 Suppose that on a statistics test the mean grade in a class is 75 and the population standard deviation is 8.
  a. Find the standardized score for a student whose grade was 83.
  b. Find the standardized score for a student whose grade was 59.
  c. Find the standardized score for a student whose grade was 75.
  d. Based on Chebyshëv's theorem, at least what proportion of students received grades between 59 and 91?

e. Assume that the distribution of grades is approximately bell shaped. Approximately what proportion of students received grades between 59 and 91?

4.5.11   The following data represent a sample of test scores:

88   76   67   90   98   68   75   86   82   90

   a. Calculate the sample mean.
   b. Calculate the sample standard deviation.
   c. Calculate the standardized score for a student whose grade was 90.
   d. Find the interval that contains all values within 2 standard deviations of the mean. What proportion of the sample values fall within this interval?
   e. Calculate the coefficient of variation.

4.5.12   On a placement exam, an applicant was told that his score was 660 and that his standardized score was 2.5. Another applicant was told that her score was 690 and her standardized score was 3.0.
   a. Find the mean and standard deviation for the population.
   b. Find the standardized score for an applicant whose score was 600.
   c. Find the coefficient of variation.

4.5.13   On a certain test the mean score was 70 and the standard deviation was 12.
   a. Ann's score was 1 standard deviation above the mean. What was Ann's score, and what was her standardized score?
   b. Jim's score was 2 standard deviations below the mean. What was Jim's score, and what was his standardized score?
   c. Al's score was A standard deviations above the mean. Calculate Al's standardized score.

4.5.14   Prove that the mean and standard deviation of a set of standardized scores are equal to 0 and 1, respectively.

4.5.15   A public utility claims that at a large manufacturing plant the daily demand for electricity has an average of 25 kilowatts and a standard deviation of 4 kilowatts. Assume that daily energy usage can be described by a bell-shaped distribution. Use the Empirical Rule to find the following:
   a. The proportion of days that demand exceeds 17.5 kilowatts
   b. The proportion of days that demand is less than 20.0 kilowatts

4.5.16   For distribution *A,* the mean is 20 and the standard deviation is 7. Distribution *B* has a mean of 23 and a standard deviation of 2. In which distribution will a raw score of 27 have a higher standing?

4.5.17   Quality control techniques are used to monitor the quality of output in production processes. The goal is to guarantee uniformity and consistency in product output. A quality control engineer for a food distributor examined the production process at a time when the process was known to be in control. The engineer selected a random sample of boxes of cereal and weighed their contents. He found the average weight to be 16.2 oz and the standard deviation to be .1 oz.
   a. The manufacturer wants to virtually guarantee that all the boxes of cereal contain between 15.9 and 16.5 oz of cereal. The values 15.9 oz and 16.5 oz are called the lower and upper specification limits (LSL and USL). Use Chebyshëv's theorem and discuss the proportion of boxes of cereal that will fall inside the specification limits.
   b. Suppose the weights follow approximately a normal distribution. Use the Empirical Rule and discuss the proportion of boxes of cereal that will fall inside the specification limits.

4.5.18   To help detect calculation errors, compare the value of the sample standard deviation s with the value of the range R. According to the Empirical Rule, if the data are approximately bell shaped, approximately 95% of measurements should lie within 2 standard deviations of the mean and approximately 99% of the data should lie within 3 standard deviations of the mean. Thus, the range of the data should be approximately four to six times greater than the standard deviation. This leads to the approximation that the standard deviation should fall approximately between the values $(R/6)$ and $(R/4)$.

a. Suppose an examination of a sample of data reveals that the range is $R = 96$. Make an approximate interval guess of the sample standard deviation.

b. Suppose an individual computed the sample standard deviation and obtained the value $s = 8$. Would you accept this value? Explain.

c. Suppose an individual computed the sample standard deviation and obtained the value $s = 40$. Would you accept this value? Explain.

d. Suppose an individual computed the sample standard deviation and obtained the value $s = 20$. Would you accept this value? Explain.

4.5.19 Suppose a stock analyst is following two stocks, A and B. Suppose a random sample of daily closing prices of the stocks shows that their sample standard deviations are $s_A = \$.60$ and $s_B = \$6.00$. Because the standard deviation for stock B is much greater than the standard deviation for stock A, we might conclude that stock B is much more volatile and, therefore, more risky than stock A. Suppose that further analysis shows that the mean prices of the stocks were $\bar{x}_A = \$2.40$ and $\bar{x}_B = \$120.00$.

a. Calculate the coefficient of variation for stock A.

b. Calculate the coefficient of variation for stock B.

c. Which stock is more volatile?

4.6   Construction of Box Plots (Optional)

We have presented several graphical techniques, including histograms and stem-and-leaf diagrams, for displaying frequency distributions. The **box plot** is another such technique. The box plot is designed to indicate the median of the data and to highlight the behavior of the data at the ends of the distribution. The box plot also highlights outliers, the observations that lie very far from the center of the distribution. By graphing two box plots on a single page, it is easy to visualize differences in the center and spread of two data sets and to detect outliers.

A Z score identifies outliers by computing their distance from the mean in standard deviation units. In a box plot, we measure the distance between the outlier and the upper or lower quartile using the interquartile range (abbreviated IQR) as the unit of measurement. The advantage of this method is that the sample quartiles, unlike the sample mean and sample standard deviation, do not depend on the values of the extreme observations.

To construct a box plot, first locate the median of the data and the lower and upper quartiles Q_1 and Q_3. Draw a horizontal line and mark the positions of the median, Q_1, and Q_3. Draw a rectangle, or *box*, extending from Q_1 to Q_3. Approximately half the observations should fall within the box.

The next step is to decide how far away from the median a measurement must lie before it is classified as an outlier. The decision is made by constructing two sets of *fences*. The *inner fences* are located 1.5 IQR below Q_1 and 1.5 IQR above Q_3. The inner fences are defined as follows:

$$\text{Lower inner fence} = Q_1 - 1.5(\text{IQR})$$
$$\text{Upper inner fence} = Q_3 + 1.5(\text{IQR})$$

The second set of fences, the *outer fences,* are located 3 IQR below Q_1 and 3 IQR above Q_3:

$$\text{Lower outer fence} = Q_1 - 3(\text{IQR})$$
$$\text{Upper outer fence} = Q_3 + 3(\text{IQR})$$

Values lying between an inner fence and the corresponding outer fence are considered to be "mild" outliers; values outside the outer fences are considered to be "extreme" outliers. An example of a box plot is given in Figure 4.10.

EXAMPLE 4.33   **Constructing a Box Plot** ———————————————————————————

In the early 1980s, a large corporation confronted with financial losses laid off a substantial portion of its work force. One of the male employees who was laid off sued the corporation, alleging that he was the victim of age discrimination. To see whether the employee had been discriminated against, the characteristics of all the employees who were laid off were compared with the characteristics of the employees who were not laid off. Every year the corporation rated every employee from 0 to 5 in each of 13 job categories. Thus, in a given year, an employee received a total rating between 0 and 65. The corporation's employment manual indicated that promotions, raises, and layoffs would be determined mainly by these ratings. The data in Table 4.3 show the employee ratings for a sample of 50 employees.

The stem-and-leaf diagram for the data in Table 4.3 is shown in Figure 4.9. In the stem-and-leaf diagram, each stem is used twice. The first time it is associated with the trailing digits 0, 1, 2, 3, and 4; the second time, with the trailing digits 5, 6, 7, 8, and 9.

For the data in Table 4.3, the median is the average of the 25th and 26th observations, after the observations have been arranged in ascending order. The median is thus 39. The lower quartile is the 13th observation, which is 37, and the upper quartile is the 38th observation, which is 42. The interquartile range is

$$\text{IQR} = 42 - 37 = 5$$

TABLE 4.3 *Employee ratings*

39	37	32	40	40	37	39	35	42	43
40	41	31	38	38	37	24	34	42	47
40	51	33	37	36	29	39	44	49	37
34	37	39	38	35	36	43	45	38	42
41	41	59	40	42	48	37	42	39	41

```
2 | 4
2 | 9
3 | 1 2 3 4 4
3 | 5 5 6 6 7 7 7 7 7 7 7 8 8 8 8 9 9 9 9 9
4 | 0 0 0 0 0 1 1 1 1 2 2 2 2 2 3 3 4
4 | 5 7 8 9
5 | 1
5 | 9
```

FIGURE 4.9 *Stem-and-leaf diagram for data in Table 4.3*

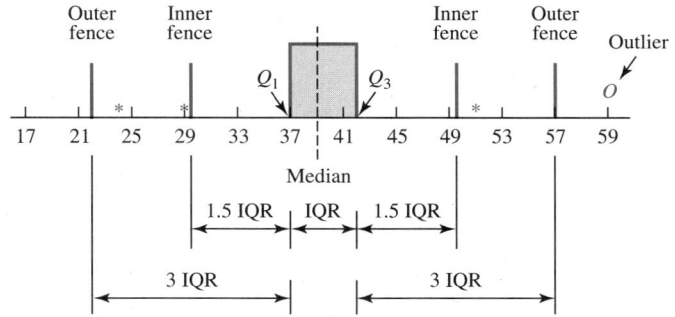

FIGURE 4.10 *Box plot for data in Table 4.3*

The inner and outer fences are calculated as follows:

$$\text{Lower inner fence} = Q_1 - 1.5(\text{IQR}) = 37 - 7.5 = 29.5$$
$$\text{Upper inner fence} = Q_3 + 1.5(\text{IQR}) = 42 + 7.5 = 49.5$$
$$\text{Lower outer fence} = Q_1 - 3.0(\text{IQR}) = 37 - 15 = 22.0$$
$$\text{Upper outer fence} = Q_3 + 3.0(\text{IQR}) = 42 + 15 = 57.0$$

The box plot for these data is shown in Figure 4.10. In the box plot, a rectangle is drawn whose ends correspond to the values of the lower and upper quartiles, 37 and 42. Next, a line is drawn through the rectangle at the value of the median, which is 39. Mild outliers are values between the inner and outer fences. The mild outliers are 24, 29, and 51, which are denoted on the plot by asterisks. Extreme outliers lie outside the outer fences. In this example, the value 59 is the only extreme outlier and is denoted by an uppercase *O* on the box plot.

The employee who filed the suit was the employee with a rating of 59. He claimed that, based on his job performance, he should not have been terminated. He won in court and was awarded more than $200,000 in back pay and other damages.

Both *Z* scores and box plots can be used to detect outliers. A disadvantage of the *Z* score is that it is based on the mean and standard deviation, both of which are sensitive to extreme values. The box plot is less sensitive to extreme values, since it is based on the interquartile range, which uses only the middle 50% of the data and ignores the largest and smallest values.

Table 4.4 (page 118) provides a summary of many important measures of centrality and dispersion mentioned in this chapter.

Exercises

Statistical Concepts

4.6.1 Explain how to construct a box plot.

4.6.2 *True or false:* In a box plot, the lower inner fence is placed at the value $Q_1 - 3(\text{IQR})$. Explain.

4.6.3 *True or false:* Both *Z* scores and box plots are useful for detecting outliers, but the box plot is less sensitive to extreme values. Explain.

TABLE 4.4 *Summary of some measures of centrality and dispersion*

Name	Formula	Comments
Sample mean	$$\bar{x} = \frac{\sum\limits_{i=1}^{n} x_i}{n}$$	1. $\bar{x}$ is the most frequently used measure of the center of a distribution. 2. We have a sample of n observations $x_1, x_2, \ldots, x_n$. 3. $\bar{x}$ represents the center of gravity of the data. 4. $\bar{x}$ can be adversely affected by outliers.
Sample median	Middle observation after putting the data in order from lowest to highest	1. Half the observations are larger than the median, and half are smaller. 2. The median is not affected much by outliers.
Range	Range = Largest value − Smallest value	
Sample variance	$$s^2 = \frac{\sum\limits_{i=1}^{n} (x_i - \bar{x})^2}{n-1}$$ or $$s^2 = \frac{\sum\limits_{i=1}^{n} x_i^2 - n\bar{x}^2}{n-1}$$	1. $n-1$ is called the degrees of freedom of s^2. 2. s^2 is an unbiased estimator of the population variance σ^2. 3. $s^2 \geqslant 0$. $s^2 = 0$ if and only if all observations are identical.
Sample standard deviation	$s = +\sqrt{s^2}$	1. s is the most frequently used measure of the spread of a distribution.
Mean absolute deviation	$$\text{M.A.D.} = \frac{\sum\limits_{i=1}^{n} |x_i - \bar{x}|}{n}$$	1. Another measure of the spread of a distribution
Standardized score	$$z = \frac{x - \bar{x}}{s}$$	1. Average value of Z is 0. 2. Variance of Z is 1. 3. Standardized score adjusts data for mean and variance so that different data sets can be meaningfully compared.

Statistical Applications

4.6.4 The following data show the yields of a sample of mutual funds:

10.61	11.42	12.28	14.45	13.32
12.42	13.34	12.75	14.18	13.44
12.92	12.78	13.05	11.31	13.24
11.47	10.24	13.63	11.48	13.26
12.10	12.70	13.64	12.39	13.30
11.41	13.31	13.15	13.57	13.58
10.51	12.59	12.34	13.26	13.37
11.04	13.34	13.57	13.59	13.60
12.31	14.06	13.34	13.73	13.54
14.23	13.33	13.09	13.40	13.36

a. Put the data in order.
b. Represent these data with a histogram.
c. Find the mean and median for these data.
d. Draw a box plot for these data.

4.6.5 Refer to the civil service exam scores in Exercise 20 of Section 4.4.
a. Represent these data with a histogram.
b. Draw a box plot for these data.

4.7 Computer Applications

The SPSSX, SAS, and Minitab computer programs can be used to find means, standard deviations, and other descriptive statistics. For example, to request descriptive statistics in SPSSX, we use the DESCRIPTIVES command. The command

DESCRIPTIVES AGE, SALARY, SENIOR

tells SPSSX to find the mean and standard deviation for each of the variables AGE, SALARY, and SENIOR. The computer output using the data in Table 2.1 for the sample of 120 employees at the Computech Corporation is shown in Figure 4.11. In addition to the mean and standard deviation, the DESCRIPTIVES command also gives the lowest and highest values, as well as the number of observations used in the calculation.

```
Number of valid observations (listwise) =        120.00

                                                      Valid
   Variable      Mean    Std Dev   Minimum    Maximum    N   Label

   AGE          39.24     12.02     19.00      64.00    120
   SALARY     2133.00    687.43   1200.00    5000.00    120
   SENIOR       15.36     10.87      1.00      42.00    120
```

FIGURE 4.11 *SPSSX-generated descriptive statistics for data in Table 2.1*

To get summary statistics for all the numerical variables in the data set in SAS, issue the command

```
PROC MEANS;
```

Place this command anywhere after the semicolon that ends the data set in the SAS program. If you want the summary statistics for just a specific set of variables, say, AGE, SALARY, SENIOR, and EXEMPT, issue the following two commands:

```
PROC MEANS;
    VAR AGE SALARY SENIOR EXEMPT;
```

Figure 4.12 shows the SAS computer output generated by these two commands using the data in Table 2.1.

```
                              The SAS System

Variable    N        Mean        Std Dev      Minimum      Maximum
-----------------------------------------------------------------------
AGE        120    39.2416667    12.0216862   19.0000000   64.0000000
SALARY     120     2133.00     687.4332359    1200.00      5000.00
SENIOR     120    15.3583333    10.8711289    1.0000000   42.0000000
EXEMPT     120     2.7583333     1.3534632    1.0000000    6.0000000
-----------------------------------------------------------------------
```

FIGURE 4.12   *SAS-generated descriptive statistics for data in Table 2.1*

Figure 4.13 is a Minitab output for the data in Table 2.1. The Minitab output shows the number of observations (N) and the sample mean, median, standard deviation, minimum value, maximum value, lower quartile, and upper quartile for each variable. (The notation TRMEAN denotes the *trimmed mean,* which is the sample mean after deleting the lowest 5% and the highest 5% of the values. The notation SEMEAN denotes the standard deviation of the sample mean and equals $STDEV/\sqrt{N}$.)

Exercises

4.7.1   Refer to the data in Table 2.1 of Chapter 2. Use the SPSSX program and a DESCRIPTIVES command to calculate the mean number of exemptions for the employees in the table. Verify that this sample mean is correct by actually calculating the mean using the output from the FREQUENCIES command.

4.7.2   Refer to the data in Table 2.1. Use the SAS program and a PROC MEANS; command to find the mean number of exemptions.

4.7.3   Refer to the data in Table 2.2. Use the SPSSX program and a DESCRIPTIVES command to find the mean GPA and the mean SAT score for the students in the sample. Find the sample standard deviation for GPA and for SAT scores.

4.7.4   Refer to the data in Table 2.2. Use the SAS program and a PROC MEANS; command to find the mean GPA and the mean SAT score for the students. What are the sample standard deviations?

4.7.5   Refer to the data in Table 2.3.

```
MTB ⟩
READ 'COMPUTEC.DAT' INTO C1-C8
    120 ROWS READ
ROW    C1    C2    C3    C4    C5    C6     C7    C8
 1      0     1     1    35     1   1575    16     5
 2      1     1     1    31     2   1980    10     3
 3      1     1     1    51     2   2480    28     2
 4      0     1     1    57     1   1925    30     2

MTB ⟩
NAME C1 = 'SEX', C2 = 'RACE', C3 = 'DEGREE', C4 = 'AGE'
MTB ⟩
NAME C5 = 'DIV', C6 = 'SALARY', C7 = 'SENIOR', C8 = 'EXEMPT'
MTB ⟩
DESCRIBE C4 C6 C7 C8
               N      MEAN    MEDIAN   TRMEAN    STDEV   SEMEAN
AGE           120    39.24    38.50    38.99     12.02    1.10
SALARY        120    2133.0   1972.5   2060.5    687.4    62.8
SENIOR        120    15.358   12.000   14.907    10.871   0.992
EXEMPT        120    2.758    2.000    2.694     1.353    0.124

               MIN     MAX      Q1       Q3
AGE           19.00   64.00    30.00    48.75
SALARY        1200.0  5000.0   1820.0   2400.0
SENIOR        1.000   42.000   6.000    23.750
EXEMPT        1.000   6.000    2.000    4.000

MTB ⟩
STOP
```

FIGURE 4.13 *Minitab-generated descriptive statistics for data in Table 2.1*

a. Use the SPSSX program and a DESCRIPTIVES command to find the mean of teachers' salaries in the United States.

b. Find the mean and standard deviation of tax revenue per capita.

c. Find the mean government expenditure per pupil.

d. Find the mean income per capita.

e. Find the mean wage of employees in manufacturing.

4.7.6 A sample of 50 individuals at a medical clinic consists of 25 patients with a mild case of a disease (coded 1), 10 patients with a moderately severe case (coded 2), and 15 patients with an extremely severe case (coded 3). Does it make sense to determine the following statistics? If so, compute them.

a. Modal severity

b. Median severity

c. Mean severity

4.7.7 The "average" gender for a sample where gender is coded {female = 1, male = 0} is .72. What, if anything, does this average mean?

4.7.8   The sample mean and standard deviation on a history test are 70 and 12. Calculate the standardized scores for the following students:

Student	Test score
1	70
2	58
3	94

STATISTICS IN ACTION: CASE STUDY

Paint Problems on the Mellon Bank Building

The case study at the end of Chapter 2 described part of a lawsuit that the management of Mellon Bank filed against 20 local and national companies. The lawsuit concerned flaws in the construction of the Mellon Bank Building. The case study in Chapter 2 considered problems in the firesafing. Another problem involved paint that had been peeling on the exterior of the building. The outer walls of the Mellon Bank Building consist of more than 2,000 metal panels, each measuring 6 feet by 10 feet. Each panel was painted with a thin coat of primer and a thicker topcoat. After a few years, the paint began to chip and crack in numerous places on the building.

A lawsuit arose concerning who, if anyone, was responsible for the defective paint job. Various theories were proposed concerning why the paint was peeling. One theory was that the humidity was too high when the panels were painted. Another theory was that the warehouse temperature was not correct when the panels were painted. Yet another theory was that the total thickness of the primer coat and the topcoat was too great. In this case study we examine this last theory.

A company that specializes in the chemical analysis of paint on buildings and bridges was hired to collect the data. Data were obtained from a sample of 30 points where the paint had cracked and peeled (the "failure points") and a sample of 247 points where the paint had not cracked or peeled (the "nonfailure points"). The following data refer to the total thickness of the paint at these 30 failure points and 247 nonfailure points. All values are in millimeters.

	Failure points	Nonfailure points
Sample size	30	247
Mean total thickness	.1355	.0805
Standard deviation	.0643	.0272
Range	.3000	.1600

Total thickness	Relative frequency at failure points	Relative frequency at nonfailure points
.0300–.0599	0.0%	26.7%
.0600–.0999	16.7%	43.4%
.1000–.1299	50.0%	25.9%
.1300–.1699	10.0%	4.0%
.1700–.3600	23.3%	0.0%

The data indicated that the mean thickness of the paint at the failure points was 68% greater than the mean thickness at the nonfailure points. This information supported the hypothesis that the thickness of the paint was an important determinant of paint failure. The relative frequency distributions also provided evidence supporting the hypothesis that the thickness of the paint was an important determinant. Only 16.7% of the failure points had a thickness of .0999 millimeter or less, whereas 70.1% of the nonfailure points had a thickness of .0999 or less. Also, 23.3% of the failure points had a thickness of .1700 or more, whereas none of the nonfailure points had a thickness of .1700 or more.

Many additional tests were performed on the paint to rule out various other potential causes of paint failure. The evidence strongly supported the hypothesis that excessive paint thickness was a major determinant of paint failure. The large standard deviation of paint thickness indicated that the paint thickness varied considerably from one spot to another. This provided evidence that there was a lack of quality control in the painting process. According to a story published in the *Allegheny Bulletin* newspaper, the case was settled out of court for $13.1 million.

This case study provides a classic example of how elementary descriptive statistics can help us make important decisions about data sets.

Chapter 4 Supplementary Exercises

4.S.1 At a grocery store, customers purchased 90 bags of grapes. Each bag was weighed to the closest half pound. The data are shown in the accompanying table.

Weight of bag (pounds)	Frequency (number sold)
.5	25
1.0	15
1.5	20
2.0	12
2.5	8
3.0	6
3.5	4

a. Plot the histogram.
b. Calculate the sample mean.
c. Calculate the sample variance.
d. Calculate the median.

 4.S.2 The following data represent a sample of weekly wages (in dollars) earned by part-time employees at a department store:

80	98	75	69	81	88	78	96
70	88	85	88	75	58	97	67
61	52	76	81	83	70	98	83
85	90	64	95	63	82	108	109
100	96	92	100	73	94	105	78

a. Put the data in order.

b. Construct a frequency distribution using the classes $50 to under $60, $60 to under $70, and so forth.

c. Graph the histogram.

d. Construct and graph the relative frequency distribution.

e. Explain the difference between the height and area of a bar in the histogram.

f. Construct the cumulative frequency distribution.

g. Calculate the sample mean, sample variance, and sample standard deviation from the raw data.

h. What is the median?

i. Find the first and third quartiles.

4.S.3   The following data represent a sample of test scores for a group of 50 students:

```
44  13  47  27  55  41  58  35  58  48
37  45  55  32  45  48  54  78  66  58
66  57  30  72  57  81  33  63  54  79
45  82  36  45  51  24  79  26  33  60
53  35  22  18  58  47  35  64  68  42
```

a. Put the data in order.

b. Construct a frequency distribution using the classes 10 to under 20, 20 to under 30, and so forth.

c. Graph the histogram.

d. Construct and graph the relative frequency distribution.

e. Construct the cumulative frequency distribution.

f. Calculate the sample mean, sample variance, and sample standard deviation from the raw data.

g. Find the median and the first and third quartiles.

4.S.4   Can the mean absolute deviation ever be negative? Why or why not?

4.S.5   Consider the sample of observations, $x_1, x_2, \ldots, x_n$, which has sample mean $\bar{x}$ and sample variance s^2. Suppose that each observation is multiplied by a constant k so that the new observations are $kx_1, kx_2, \ldots, kx_n$. Find the mean and variance of the new observations.

4.S.6   Suppose a sample consists of n observations that are all identical and equal to C.
a. Calculate the sample mean, median, and mode.
b. Calculate the range, mean absolute deviation, variance, and standard deviation.

4.S.7   Is it possible for a sample variance to be 0 or negative? Why or why not?

4.S.8   A transportation agency is interested in how far people travel to see pro football games. At a certain football game, a random sample of 10 people were asked how many miles each person traveled to get to the game. The data were as follows:

```
16  5  7  42  10  6  10  3  4  7
```

Find the standardized score for a person who traveled 5 miles.

4.S.9   Given a sample of observations $x_1, x_2, \ldots, x_n$, prove that $\Sigma(x_i - \bar{x}) = 0$.

4.S.10   Given a sample of observations $x_1, x_2, \ldots, x_n$, prove that $\Sigma(x_i - \bar{x})^2 = \Sigma x_i^2 - n\bar{x}^2$.

4.S.11   Consider a sample of data $x_1, x_2, \ldots, x_n$ having sample mean $\bar{x}$ and sample variance s^2. Suppose that a constant d is subtracted from each observation. Show that the sample mean and sample variance of the new data are $\bar{x} - d$ and s^2, respectively.

4.S.12   During a 5-week period, a salesperson's weekly incomes were $400, $250, $175, $300, and $375.
a. Calculate the sample mean.

    b. Calculate the deviations from the mean and verify that they sum to 0.

    c. Calculate the mean absolute deviation.

    d. Calculate the sample variance and sample standard deviation.

4.S.13 A sample consists of 50 patients at a hospital. In the sample, 28 patients have a mild case of the flu (coded 1), 10 have a moderate case of the flu (coded 2), and 12 have a severe case of the flu (coded 3). Does it make sense to determine the following statistics? If so, compute them.

    a. Modal severity

    b. Median severity

    c. Mean severity

4.S.14 If a sample has 155 observations ranked so that the first observation is the largest, the second is next largest, and so on, which observation is the median?

4.S.15 A researcher transformed data into standardized scores and obtained a mean standardized score of 1.247. Does this constitute grounds for rechecking the calculations? Why or why not?

4.S.16 Which of the following statements are true?

    a. Unlike the variance, the range is not greatly affected by extreme values.

    b. Distributions with similar means, medians, and modes also tend to have similar variances.

    c. The variance measures the spread of the observations, with a larger variance indicating greater spread.

4.S.17 In a certain corporation, a very small group of employees have extremely high salaries, whereas the majority of employees receive much lower salaries. If you were the bargaining agent for the union, which statistic would you calculate to illustrate the low pay level and why? If you were the employer, which statistic would you use to demonstrate a higher pay level and why?

4.S.18 A sample contains 12 students who graduated from Michigan State (coded 1), 15 students who graduated from Boston College (coded 2), 8 students who graduated from the University of Michigan (coded 3), and 24 students who graduated from Northwestern (coded 4). Calculate the listed statistic if you think it is meaningful.

    a. Modal college

    b. Median college

    c. Mean college

    d. Range

4.S.19 In a sample of 60 families, 10 own no car, 25 own one car, 15 own two cars, and 10 own three cars. Calculate the listed statistic if you think it is meaningful.

    a. Mode

    b. Median

    c. Mean

4.S.20 A sample contains 321 observations ranked from highest to lowest.

    a. Which observation is the median?

    b. Suppose the observations are ranked from lowest to highest. Which observation is the median?

4.S.21 Suppose a variable has two categories, such as "acceptable" and "defective." Suppose you have coded the variable so that acceptable products are coded as 0 and defectives are coded as 1.

    a. What, if anything, does the mean of the 0s and 1s tell you?

    b. Suppose the sample mean is .20. What does this mean?

4.S.22 At a university, a small group of faculty members have extremely high salaries, whereas the majority of faculty members receive much lower salaries.

    a. If you were the bargaining agent for the faculty, which statistic would you use to illustrate the low pay level, the mean or the median?

    b. If you were the bargaining agent for the university, which statistic would you use to illustrate the high pay level, the mean or the median? Explain.

4.S.23 Based on Chebyshëv's theorem, what can you say about the proportion of observations that fall outside the limits $\mu \pm 2.5\sigma$ in a frequency distribution?

4.S.24 In quality control testing, output is inspected at regular time intervals to see whether the production process has changed in any way. Inspections typically involve small samples of approximately five items per time period. For each sample, we might calculate the sample mean, the sample standard deviation, and the sample range. Changes over time in the sample mean indicate that the process has gone out of control because the average product characteristic is getting larger or smaller. Changes over time in the sample standard deviation or in the range indicate that the process has gone out of control because of increased variability in output. These ideas will be discussed in Chapter 24.

In the production of an electrical device operated by a thermostatic control, five control switches were tested each hour to determine the "on" temperature at which the thermostat actually operated under a given setting. Results of the test during a 25-hour production period were as follows:

Subgroup number	"ON" TEMPERATURE AT WHICH THERMOSTATIC SWITCH OPERATES				
	A	B	C	D	E
1	54	56	56	56	55
2	51	52	54	56	49
3	54	52	50	57	55
4	56	55	56	53	50
5	53	54	57	56	52
6	53	47	58	55	54
7	52	55	54	55	56
8	56	53	53	54	55
9	55	52	53	52	53
10	50	54	53	55	55
11	57	54	53	52	53
12	52	52	54	53	55
13	54	53	55	52	52
14	59	55	54	51	55
15	58	53	57	56	51
16	58	57	56	51	59
17	55	55	55	56	53
18	54	57	54	55	54
19	54	53	56	53	55
20	53	53	57	54	53
21	58	57	57	56	59
22	59	57	56	59	59
23	56	58	58	58	58
24	57	55	57	58	59
25	58	56	59	59	57

a. Make a tally sheet of these measurements and arrange them in a frequency distribution.

b. Construct and plot the frequency distribution showing the class boundaries, class midpoints, and observed frequencies.

c. Calculate the mean of all the observations. Calculate the standard deviation of all the measurements.

  d. Construct and plot the cumulative frequency distribution and cumulative relative frequency distribution.
  e. What proportion of observations fall within 2 standard deviations of the overall mean?
  f. For each sample of five measurements, calculate the sample mean. Plot these sample means versus time and look for any obvious sudden changes or upward or downward trends in the sample mean. Does it appear that the process has gone out of control at any point?
  g. For each sample of five measurements, calculate the sample range. Plot these sample ranges versus time and look for any obvious changes or upward or downward trends in the range. Does it appear that the process has gone out of control at any point?
  h. For each sample of five measurements, calculate the sample standard deviation. Plot these sample standard deviations versus time and look for any obvious changes or upward or downward trends in the sample standard deviation. Does it appear that the process has gone out of control at any point?

4.S.25 A quality control engineer wants to examine the dimensions of various electrical components in a manufacturing device. Every hour, a sample of five items was measured. The following data show the distance from the end of an electrical source to its contact point expressed in units of .0001 in. in excess of .14 in.

Sample number	Measurement on each item of five items per hour				
1	140	143	137	134	135
2	138	143	143	145	146
3	139	133	147	148	139
4	143	141	137	138	140
5	142	142	145	135	136
6	136	144	143	136	137
7	142	147	137	142	138
8	143	137	145	137	138
9	141	142	147	140	140
10	142	137	145	140	132
11	137	147	142	137	135
12	137	146	142	142	140
13	142	142	139	141	142
14	137	145	144	137	140
15	144	142	143	135	144
16	140	132	144	145	141
17	137	137	142	143	141
18	137	142	142	145	143
19	142	142	143	140	135
20	136	142	140	139	137
21	142	144	140	138	143
22	139	146	143	140	139
23	140	145	142	139	137
24	134	147	143	141	142
25	138	145	141	137	141

  a. Make a tally sheet of these measurements and arrange them in a frequency distribution.
  b. Construct and plot the frequency distribution showing the class boundaries, class midpoints, and observed frequencies.

c. Calculate the mean of all the observations. Calculate the standard deviation of all the measurements.

d. Construct and plot the cumulative frequency distribution and cumulative relative frequency distribution.

e. What proportion of observations fall within 2 standard deviations of the overall mean?

f. For each sample of five measurements, calculate the sample mean. Plot these sample means versus time and look for any obvious sudden changes or upward or downward trends in the sample mean. Does it appear that the process has gone out of control at any point?

g. For each sample of five measurements, calculate the sample range. Plot these sample ranges versus time and look for any obvious changes or upward or downward trends in the range. Does it appear that the process has gone out of control at any point?

h. For each sample of five measurements, calculate the sample standard deviation. Plot these sample standard deviations versus time and look for any obvious changes or upward or downward trends in the sample standard deviation. Does it appear that the process has gone out of control at any point?

 4.S.26 A quality control engineer wants to examine the pitch of threads on aircraft fittings. At 1-hour intervals, a sample of five items was measured. The following data show the pitch of threads expressed in units of .0001 in. in excess of .4000 in.

Sample number	Measurement on each item of five items per hour				
1	36	35	34	33	32
2	31	31	34	32	30
3	30	30	32	30	32
4	32	33	33	32	35
5	32	34	37	37	35
6	32	32	31	33	33
7	33	33	36	32	31
8	23	33	36	35	36
9	43	36	35	24	31
10	36	35	36	41	41
11	34	38	35	34	38
12	36	38	39	39	40
13	36	40	35	36	33
14	36	35	37	34	44
15	30	37	33	34	35
16	28	31	33	33	33
17	33	30	34	33	35
18	27	28	29	27	30
19	25	26	29	27	32
20	23	25	25	29	26

a. Make a tally sheet of these measurements and arrange them in a frequency distribution.

b. Construct and plot the frequency distribution showing the class boundaries, class midpoints, and observed frequencies.

c. Calculate the mean of all the observations. Calculate the standard deviation of all the measurements.

d. Construct and plot the cumulative frequency distribution and cumulative relative frequency distribution.

e. What proportion of observations fall within 2 standard deviations of the overall mean?
f. For each sample of five measurements, calculate the sample mean. Plot these sample means versus time, and look for any obvious sudden changes or upward or downward trends in the sample mean. Does it appear that the process has gone out of control at any point?
g. For each sample of five measurements, calculate the sample range. Plot these sample ranges versus time, and look for any obvious changes or upward or downward trends in the range. Does it appear that the process has gone out of control at any point?
h. For each sample of five measurements, calculate the sample standard deviation. Plot these sample standard deviations versus time, and look for any obvious changes or upward or downward trends in the sample standard deviation. Does it appear that the process has gone out of control at any point?

4.S.27 Food distributors use a system of quality control inspections to ensure that the contents of food containers meet various standards. To monitor the weight of tomatoes in cans of output, a sample of five cans of output is examined every hour. The following data show the drained weights in ounces of the contents of cans of standard grade tomatoes.

Sample number	Drained weight of tomatoes (oz)				
1	22.0	22.5	22.5	24.0	23.5
2	20.5	22.5	22.5	23.0	21.5
3	20.0	20.5	23.0	22.0	21.5
4	21.0	22.0	22.0	23.0	22.0
5	22.5	19.5	22.5	22.0	21.0
6	23.0	23.5	21.0	22.0	20.0
7	19.0	20.0	22.0	20.5	22.5
8	21.5	20.5	19.0	19.5	19.5
9	21.0	22.5	20.0	22.0	22.0
10	21.5	23.0	22.0	23.0	18.5
11	20.0	19.5	21.0	20.0	20.5
12	19.0	21.0	21.0	21.0	20.5
13	19.5	20.5	21.0	20.5	21.0
14	20.0	21.5	24.0	23.0	20.0
15	22.5	19.5	21.0	21.5	21.0
16	21.5	20.5	22.0	21.5	23.5
17	19.0	21.5	23.0	21.0	23.5
18	21.0	20.5	19.5	22.0	21.0
19	20.0	23.5	24.0	20.5	21.5
20	22.0	20.5	21.0	22.5	20.0
21	19.0	20.5	21.0	20.5	22.5
22	21.5	25.0	21.0	19.0	21.0
23	22.5	22.0	23.0	22.0	23.5
24	22.5	22.0	22.0	19.5	20.5
25	18.5	22.0	22.5	21.0	21.5
26	21.5	20.5	20.5	16.5	21.5
27	24.0	22.0	17.5	21.0	22.5
28	19.5	22.5	15.5	20.0	22.5
29	22.0	17.5	21.0	22.0	23.5
30	22.0	20.0	20.5	24.0	21.5

a. Make a tally sheet of these measurements and arrange them in a frequency distribution.

b. Construct and plot the frequency distribution showing the class boundaries, class midpoints, and observed frequencies.

c. Calculate the mean of all the observations. Calculate the standard deviation of all the measurements.

d. Construct and plot the cumulative frequency distribution and cumulative relative frequency distribution.

e. What proportion of observations fall within 2 standard deviations of the overall mean?

f. For each sample of five measurements, calculate the sample mean. Plot these sample means versus time, and look for any obvious sudden changes or upward or downward trends in the sample mean. Does it appear that the process has gone out of control at any point?

g. For each sample of five measurements, calculate the sample range. Plot these sample ranges versus time, and look for any obvious changes or upward or downward trends in the range. Does it appear that the process has gone out of control at any point?

h. For each sample of five measurements, calculate the sample standard deviation. Plot these sample standard deviations versus time, and look for any obvious changes or upward or downward trends in the sample standard deviation. Does it appear that the process has gone out of control at any point?

4.S.28   In Chapter 3, a distribution was said to be skewed to the right (left) if the distribution had a longer tail extending to the right (left). Suppose we have a sample of observations $x_1, \ldots, x_n$. Let $\bar{x}$ denote the sample mean, and let s denote the sample standard deviation. A measure of *skewness* is defined as follows:

$$\text{Skewness} = \text{Sk} = \frac{m_3}{(m_2)^{3/2}}$$

where

$$m_2 = \frac{\Sigma(x_i - \bar{x})^2}{n} \quad \text{and} \quad m_3 = \frac{\Sigma(x_i - \bar{x})^3}{n}$$

a. Explain why Sk will be negative if the distribution is skewed to the left and positive if the distribution is skewed to the right.

b. Construct and plot the frequency distribution for the following set of measurements:

13	17	19	23	21	25	26	34	56	78
15	24	26	37	69	21	19	42	51	87
9	11	21	32	27	25	6	5	12	17
34	42	51	62	71	23	36	22	18	59
21	15	8	5	42	51	13	26	31	25

c. Does the distribution appear to be skewed to the right or to the left, or does it appear to be symmetric?

d. Calculate Sk.

4.S.29   **a.** Construct and plot the frequency distribution for the following set of measurements:

13	56	67	78	82	76	57	21	67	89
77	65	43	57	68	79	82	80	73	54
9	11	67	85	54	31	78	82	81	73
34	42	51	62	71	83	84	76	71	79
21	15	8	65	67	76	82	56	71	75

b. Does the distribution appear to be skewed to the right or to the left, or does it appear to be symmetric?

c. Calculate the skewness of the data set where

$$\text{Skewness} = \text{Sk} = \frac{m_3}{(m_2)^{3/2}}$$

where

$$m_2 = \frac{\Sigma(x_i - \bar{x})^2}{n} \quad \text{and} \quad m_3 = \frac{\Sigma(x_i - \bar{x})^3}{n}$$

4.S.30 Suppose the measurement of some dimension of an industrial product must fall between two specification limits, LSL, the lower specification limit, and USL, the upper specification limit, to be considered nondefective. All possible situations may be grouped into three general classes, as follows:

1. The range of output is appreciably less than the difference between the specification limits (USL − LSL).

2. The range of output is approximately equal to the difference between the specification limits (USL − LSL).

3. The range of output is appreciably greater than the difference between the specification limits (USL − LSL).

In the first situation, all the products manufactured will meet specifications as long as the process stays in control. This is the ideal manufacturing situation. In the second situation, an occasional product might fail to meet specifications. In the third situation, even when the process is in control some defective product will be made. In this case, fundamental changes must be made in the production process to reduce the variability of output, thereby reducing the proportion of defective output.

Suppose a company produces precision parts for aircraft engines. A precision screw is considered acceptable provided its length falls between LSL = 1.20 in. and USL = 1.22 in.; otherwise it is defective. Suppose the lengths of the screws follow approximately a bell-shaped distribution. Suppose the producer can easily adjust the mean length of the screws by adjusting a few machine settings. The standard deviation cannot be changed because it indicates the inherent random variability present in the manufacturing process.

In each of the following situations, draw a graph representing the frequency distribution of output. On each graph draw vertical lines indicating LSL and USL. In each situation, use the Empirical Rule to determine the approximate proportion of output that will be defective. In each situation, indicate whether the mean can be adjusted to reduce the proportion of defective output.

a. $\mu = 1.205, \sigma = .05$

b. $\mu = 1.205, \sigma = .025$

c. $\mu = 1.205, \sigma = .005$

d. $\mu = 1.2075, \sigma = .05$

e. $\mu = 1.2075, \sigma = .025$

f. $\mu = 1.2075, \sigma = .005$

g. $\mu = 1.21, \sigma = .05$

h. $\mu = 1.215, \sigma = .01$

4.S.31 **Process capability** refers to the capability of a process to stay within its specification limits, LSL and USL. In the previous exercise, graphs were used to indicate process capability by simply looking at how much of the frequency distribution fell between the upper and lower specification limits.

Quality control engineers have devised an index that is used to measure process capability. First, assume that the measurements generated by the process approximately follow a bell-shaped distribution with mean μ and standard deviation σ. Then approximately 99.7% of the measurements should fall between $\mu - 3\sigma$ and $\mu + 3\sigma$. In other words, almost all of the measurements should fall within a range of 6σ. The quantity 6σ is referred to as the **actual process spread;** the quantity (USL − LSL), which measures the distance between the specification limits, is called the **allowable process spread.** The process capability index, denoted C_p, is defined as follows:

$$C_p = \frac{\text{Allowable process spread}}{\text{Actual process spread}} = \frac{\text{USL} - \text{LSL}}{6\sigma}$$

If μ and σ are unknown, they can be replaced by the sample mean and sample standard deviation. The index C_p is interpreted as follows:

If $C_p = 1.0$, then the process is just barely capable of meeting its specification limits. If $C_p > 1.0$, then the process is more capable of meeting its specification limits. If $C_p < 1.0$, then the process is not capable of meeting its specification limits. Larger values of C_p are preferable because the probability is higher that the measurements will be able to stay within the specification limits. Values of C_p exceeding 1.33 are often set as a goal.

Suppose a company produces precision parts for aircraft engines. A precision screw is considered acceptable provided its length falls between LSL = 1.20 in. and USL = 1.22 in.; otherwise it is defective. Suppose the lengths of the screws follow approximately a bell-shaped distribution. Assume the mean length of screws is 1.205 in. In each of the following situations, draw a graph indicating the distribution of output. On each graph, locate USL and LSL. In each situation, determine the actual process spread and the allowable process spread; calculate C_p. Indicate whether the process is not capable, just barely capable, or capable of meeting the specification limits.

a. $\sigma = .05$

b. $\sigma = .01$

c. $\sigma = .025$

d. $\sigma = .5$

References

Neter, John, William Wasserman, and G. A. Whitmore. *Applied Statistics.* 3d ed. Boston: Allyn and Bacon, 1988.

Nie, Norman E., C. Hadlai Hull, Jean G. Jenkins, Karin Steinbrenner, and Dale H. Bent. *SPSS Statistical Package for the Social Sciences.* 2d ed. New York: McGraw-Hill, 1975.

Norusis, Marija J. *SPSSX Introductory Statistics Guide.* New York: McGraw-Hill, 1990.

Norusis, Marija J. *SPSSX Advanced Statistics Guide.* Chicago: SPSS, 1990.

Norusis, Marija J. *The SPSS Guide to Data Analysis.* Chicago: SPSS, 1990.

Ryan, Thomas A., Brian L. Joiner, and Barbara F. Ryan. *Minitab Reference Manual.* University Park, Penn.: Minitab Project, 1985.

Ryan, Thomas A., Brian L. Joiner, and Barbara F. Ryan. *Minitab Handbook.* 2d ed. Boston: PWS-Kent, 1985.

SAS Introductory Guide. 3d ed. Cary, N.C.: SAS Institute, 1985.

SAS Procedures Guide for Personal Computers. Version 6 ed. Cary, N.C.: SAS Institute, 1986.

SAS Statistics Guide for Personal Computers. Version 6 ed. Cary, N.C.: SAS Institute, 1986.

SAS User's Guide: Basics. Version 5 ed. Cary, N.C.: SAS Institute, 1985.

SAS User's Guide: Statistics. Version 5 ed. Cary, N.C.: SAS Institute, 1985.

SPSSX User's Guide. Chicago: SPSS, 1990.

5 Introduction to Probability

The previous chapters showed how to use raw data to describe the general characteristics of any population or sample. In this chapter, we assume that the characteristics of the population are known, and, using probability theory, we describe what we can expect to observe if we select an observation from the population.

5.1 Experiments, Outcomes, Events, and Sample Spaces

In general, when some activity is performed, several different outcomes are possible. In this chapter, we are interested in determining the probability or likelihood that any particular outcome occurs. The probability of an outcome is a numerical measure of the chance of an outcome's occurrence. Probability is measured on a scale from 0 to 1. A probability near 0 indicates that the outcome is very unlikely to occur, whereas a probability near 1 indicates that the event is almost certain to occur.

To analyze a probability problem, we must develop a *model* that describes the problem. The model consists of a listing (perhaps theoretical) of all the possible basic outcomes of the activity being studied and the probability associated with each outcome. In statistics, the activities being studied are given the formal name *experiments*. Do not confuse the statistical meaning of the word *experiment* with the notion of mixing chemicals in a laboratory.

DEFINITION **Experiment**

An **experiment** is any activity from which an outcome, measurement, or result is obtained. When the outcomes cannot be predicted with certainty, the experiment is a **random experiment.**

Some examples of experiments are as follows:

1. Measuring the lifetime (time to failure) of a given product
2. Inspecting an item to determine whether it is defective
3. Recording the income of a bank employee
4. Recording the balance in an individual's checking account

Before an experiment is performed, several different outcomes are possible. The set of all these possible outcomes is called the *sample space* of the experiment. At the conclusion of an experiment, there will be exactly one outcome.

DEFINITIONS   **Basic Outcomes and Sample Space** ————————————————

Each possible outcome of a random experiment is called a **basic outcome.** The set of all possible basic outcomes for a given experiment is called the **sample space**.

———————————————————————————————

It is customary to denote each basic outcome by the letter o followed by a subscript. Thus, if an experiment has n possible outcomes, then the basic outcomes will be denoted $o_1, o_2, \ldots, o_n$, where o_i denotes the ith possible outcome and $i = 1, 2, \ldots, n$. The sample space will be denoted by the symbol S and will be represented as $S = \{o_1, o_2, \ldots, o_n\}$.

EXAMPLE 5.1   **Sample Space and Basic Outcomes** ————————————————

Suppose that a corporation has offices in six cities, San Diego, Los Angeles, San Francisco, Denver, Paris, and London. Suppose a new employee will be assigned to work in one of these six offices. Determine the sample space for this experiment.

SOLUTION   The possible assignments are as follows:

$o_1 =$ San Diego $o_2 =$ Los Angeles $o_3 =$ San Francisco
$o_4 =$ Denver $o_5 =$ Paris $o_6 =$ London

The sample space is denoted $S = \{o_1, o_2, o_3, o_4, o_5, o_6\}$. If the employee is assigned to work in Denver, then we say that the basic outcome o_4 has occurred.

Venn Diagrams

In many cases, it is useful to represent the sample space graphically. This can be done by using a **Venn diagram,** named after the English logician John Venn (1834–1923), who popularized their use. In a Venn diagram, the basic outcomes of the experiment are represented as points on a graph. The basic outcomes in Example 5.1 are represented by a Venn diagram in Figure 5.1.

At times, we form sets or collections of basic outcomes, which are called *events.*

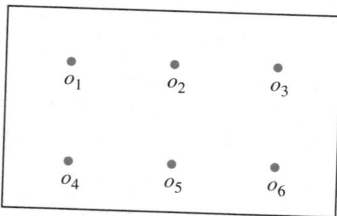

FIGURE 5.1   *A Venn diagram showing possible outcomes in Example 5.1*

DEFINITION   **Event** ————————————————————————————

An **event** is a specific collection of basic outcomes, that is, a set containing one or more of the basic outcomes from the sample space. We say that an event occurs if any one of the basic outcomes in the event occurs.

———————————————————————————————

An event can be thought of as any collection of basic outcomes or as any subset of the sample space. Usually events are denoted by an uppercase letter.

EXAMPLE 5.2 **Events**

Refer to Example 5.1, in which an employee was assigned to work in one of six different cities. Suppose we want to know whether the employee was assigned to work in California. Describe this event.

SOLUTION Let symbol A denote the event {Employee assigned to work in California}. We obtain

$$A = \{o_1, o_2, o_3\}$$

Thus we say that event A has occurred if the employee is assigned to work in San Diego, Los Angeles, or San Francisco.

Let symbol B denote the event that the employee is assigned to work in Europe. We obtain

$$B = \{o_5, o_6\}$$

Event B occurs if the employee is assigned to work in Paris or London.

Keep in mind that an experiment will have one and only one basic outcome, and a particular event occurs if any basic outcome in that event occurs. An event can be represented on a Venn diagram by encircling its basic outcomes in that event. Events A and B in Example 5.2 are shown in Figure 5.2.

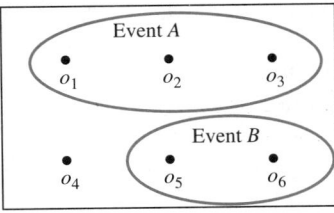

FIGURE 5.2 *A Venn diagram showing events A and B in Example 5.2*

Exercises

Statistical Concepts

5.1.1 Give an example of an experiment whose sample space has five basic outcomes.

5.1.2 Describe an experiment where the sample space contains an infinite number of possible outcomes.

5.1.3 Suppose a sample space contains three basic outcomes. How many different events are there? That is, how many different subsets can you form? *Note:* The empty set, which contains no elements, is a subset of every set.

Statistical Drills

5.1.4   Suppose a fair die is tossed, and we record the value that appears. Construct the sample space and represent it by a Venn diagram.

5.1.5   Suppose a pair of dice are tossed and we record the sum of the values that appear. Construct the sample space and represent it using a Venn diagram.

5.2   Assigning Probabilities to Events

There are two types of random experiments, those that can be repeated over and over again under essentially identical conditions and those that are unique and cannot be repeated. Corresponding to these two types of experiments are two types of probability numbers. A numerical measure that indicates the likelihood of a specific outcome in a repeatable random experiment is called an **objective probability,** whereas the probability associated with a specific outcome of a unique and nonrepeatable random experiment is called a **subjective probability.** Objective probabilities can be determined in one of two ways: empirically, by actually conducting the experiment a large number of times, and theoretically, by studying the basic underlying symmetry in the problem. Objective probabilities that can be found theoretically often involve experiments where different outcomes can be considered equally likely.

In this section, we discuss three different approaches to assigning probabilities to basic outcomes, as follows:

1. The relative frequency approach
2. The equally likely approach
3. The subjective approach

The Relative Frequency Approach

Consider an experiment that can be repeated many times. Assume that after each experiment, it is possible to return to the initial state and repeat the same experiment so that the resulting outcome is unaffected by the previous outcomes. When the experiment is repeated many times, a certain proportion of experiments will result in basic outcome o_1, a certain proportion will result in basic outcome o_2, and so forth. In practical applications, it is customary to think of the proportion of experiments resulting in basic outcome o_i as the *probability of* o_i.

If an experiment has been repeated n times and the basic outcome o_i occurs f_i times, the proportion of times that o_i occurred is f_i/n. The fraction f_i/n is the relative frequency with which basic outcome o_i occurs (or, more simply, the relative frequency of o_i). Similarly, if an event A occurs f_A times in n replications of the experiment, then the relative frequency of event A is f_A/n. As n increases, the proportion of occurrences of A will eventually stabilize and approach a constant value. This idea underlies the *relative frequency* concept of probability.

Examples include predicting whether a fair coin that is tossed will come up heads, whether oil exists at a site based on existing geological conditions, and whether it will rain based on existing meteorological conditions. In assessing the probability of finding oil, a geologist would cite the frequency with which oil had been found in the past at sites with essentially the same conditions.

Relative frequency concept of probability

> Let f_A be the number of occurrences, or frequency of occurrence, of event A in n repeated identical trials. The probability that A occurs is the limit of the ratio f_A/n as the number of trials n becomes infinitely large.

Under the relative frequency approach, the probability of an event A can be thought of as the relative frequency of event A in an infinite number of repeated trials of the experiment. The probability of a basic outcome o_i is denoted $P(o_i)$, and the probability of event A is denoted $P(A)$.

The relative frequency approach has several difficulties. Because we can never replicate an experiment an infinite number of times, it is impossible to determine the limit of the ratio f_A/n as n approaches infinity. Furthermore, we can never be sure that we have repeated an experiment under identical conditions. When we use the relative frequency approach, we use the observed ratio f_A/n to approximate the theoretical probability that event A occurs. That is, we assume that $P(A) \approx f_A/n$ when n is sufficiently large.

EXAMPLE 5.3 **Relative Frequency Approach to Probability** ——————————————

A marketing research agency surveyed 1,000 randomly selected residents in a community. Each resident was asked whether he or she favored expansion of the local shopping mall. Of those questioned, 300 favored the expansion. Let's use the relative frequency approach to estimate the probability that a randomly selected individual will favor expansion.

S O L U T I O N The relative frequency of favoring expansion is $^{300}/_{1,000} = .3$. Thus we approximate the population proportion of all residents who favor expansion by the proportion .3 and estimate that the probability is .3 that a randomly selected resident will favor expansion.

The Equally Likely Approach

In many cases, the relative frequency with which some event occurs in a large number of replications of an experiment can be determined theoretically by relying on a basic underlying symmetry in the problem. This situation occurs when all basic outcomes are considered to be equally likely. In this case, the idea of repeated experimentation is a purely conceptual notion, and we do not have to carry out the experiments repeatedly.

DEFINITION **Probability of Equally Likely Outcomes** ——————————————

Suppose that an experiment must result in one of n equally likely outcomes. Then each possible basic outcome is considered to have probability $1/n$ of occurring on any replication of the experiment.

That the basic outcomes of some experiment are equally likely cannot be proved; it is an assumption based on some underlying symmetry in the experiment. In Example 5.3, randomly selected residents were asked whether they favored expanding a shopping mall.

Although there are only two possible answers (basic outcomes), yes or no, these outcomes are *not* equally likely.

EXAMPLE 5.4   **The Draft Lottery**

In the early 1970s, men were selected to be drafted into the armed services according to the day of the year on which they were born. The 366 dates of the year (including February 29) were written on balls placed in a box, mixed thoroughly, and withdrawn one at a time. Each ball had an equal chance of being selected first. Thus, the probability that your birthday would have been the first one selected was $\frac{1}{366}$.

EXAMPLE 5.5   **Criticism of the Draft Lottery**

Some people criticized the results of the draft lottery described in Example 5.4, claiming that all birthdays did not have an equal chance of being selected first. They argued that the balls were dropped into the box in such a way that balls representing dates early in the year tended to be near the bottom of the box and the balls representing dates late in the year tended to be near the top. The critics further argued that the balls were not mixed thoroughly and that the person selecting the balls tended to choose balls from the top of the pile. If these critics were correct, then dates in November and December were more likely to be selected first than dates in January and February, and thus all dates were not equally likely to be selected first.

EXAMPLE 5.6   **Some Common Equally Likely Models**

Some of the most frequently used examples of equally likely events involve coin tossing, die tossing, selecting cards, and lotteries.

■ *Coin tossing:*   Toss a fair coin. The sample space S consists of two possible basic outcomes, {Heads, Tails}. If the coin being tossed is fair, we can assume that these two outcomes are equally likely. Thus, based on the equally likely approach, the probability of each basic outcome is $\frac{1}{2}$.

■ *Die tossing:*   Toss a symmetrically balanced fair die. It is reasonable to assume that each of the numbers 1 through 6 is equally likely. The sample space S is thus {1, 2, 3, 4, 5, 6}. Based on the equally likely model, it is assumed that each basic outcome has probability $\frac{1}{6}$.

■ *Card selection:*   Select a card randomly from a 52-card poker deck that has been thoroughly shuffled. Based on the assumption that each basic outcome is equally likely, each card has a probability of $\frac{1}{52}$ of being selected.

■ *Lottery gambling:*   Many states conduct a lottery called something like the Daily Number Game. To play the game, you buy a ticket that is imprinted with a number from 000 to 999. Then a number from 000 to 999 is selected randomly. You win a prize if the number selected matches the number on your ticket. It is reasonable to assume that each of the 1,000 numbers from 000 to 999 has an equal chance of being selected. Thus, the probability that the number selected matches the number on your ticket is $\frac{1}{1,000}$, or .001.

In many problems of interest, there is no reason a priori to believe that all possible outcomes of an experiment are equally likely. In many of these instances, we use the relative frequency method to estimate the probability of any particular occurrence.

EXAMPLE 5.7 **Inapplicability of Equally Likely Model**

In a marketing study, a woman is chosen randomly and asked which of three brands of iced tea she prefers. What is the probability that the woman selects brand A?

SOLUTION There is no reason to assume that all brands are equally likely to be selected, so we need to know the relative frequency with which brand A is selected. Suppose that in a random sample of $n = 1,000$ women, brand A was selected with frequency $f_A = 200$. Thus, the relative frequency of selecting brand A is $f_A/n = {}^{200}/_{1,000} = .2$. We estimate that the probability is about .2 that a randomly selected woman prefers brand A.

DEFINITION **Objective Probability**

A probability obtained by using a relative frequency approach or an equally likely approach is called an **objective probability.**

The probabilities discussed so far are termed *objective* because they are based on actual sample data or on theoretical considerations concerning the underlying symmetry in a problem. If several independent researchers were given the same information and asked to calculate the probabilities of different outcomes, each researcher should obtain exactly the same answer. That is, these probabilities are not based on personal beliefs or individual opinions.

The Subjective Approach

In many problems, the equally likely model does not apply because it is unreasonable to assume that the basic outcomes are all equally likely. In addition, at times the relative frequency approach may be difficult or impossible to apply because we may not be able to perform an experiment a large number of times. In these cases, we may have to use our *subjective judgment* to assign a probability to some specific event.

For example, suppose a company is planning to produce an entirely new and untested product. An executive asks, "What is the probability that the company will earn a profit from marketing this product?" How can such a probability be assigned? There are two possible outcomes, earning a profit or not earning a profit, but there is no reason to believe that these outcomes are equally likely. Thus, the equally likely model cannot be applied. In addition, the relative frequency approach is not applicable because the experiment cannot be replicated.

As another example, suppose a stockbroker states that there is a 40% chance that two companies will merge. This is a one-time event that will not be replicated. In situations like this, the probability is the stockbroker's own subjective assessment of the likelihood of a merger. There is no objective way of testing this assessment, and there is no well-defined technique for enabling us to duplicate the stockbroker's thought processes in arriving at the assessment.

Economists speculate about the likelihood of a recession next year; lawyers speculate about the likelihood of winning a particular case; executives speculate about the likelihood of a new product's success. All of these situations involve subjective probabilities.

DEFINITION   **Subjective Probability**

A **subjective probability** is a number in the interval [0, 1] that reflects a person's degree of belief that an event will occur.

Subjective probabilities are usually applied to experiments that occur only once. It may be useful to think of a subjective probability as representing the opinion of a so-called expert. If you do not think that the person is an expert, then it is reasonable to ignore the subjective opinion when formulating your own opinion about the event's probability.

Assigning Probabilities to Basic Outcomes

We have seen that there are three different ways to assign probabilities to basic outcomes of an experiment: the relative frequency approach, the equally likely approach, and the subjective approach. The nature of the problem determines which approach is best. Problems with an underlying symmetry, such as coin, dice, and card problems, are especially suited to the equally likely approach. Problems for which we have large samples of data based on many replications of an experiment are especially suited to the relative frequency approach. Problems that occur only once, such as a sporting event, are especially suited to the subjective approach. The major criticism of the subjective approach is that the conclusions depend on the subjectively assigned probabilities, and these, in turn, depend on the insight and wisdom of the analyst. However, in many problems, the subjective approach is the only way of assigning probabilities.

Odds

Occasionally people express their opinions about the probability of some event occurring in terms of **odds** rather than probabilities. Suppose, for example, that a person claims that the odds in favor of some event occurring are 4 to 1. This means that the person thinks the probability that the event will occur is .8.

Converting odds to a probability

If the odds in favor of event A occurring are a to b, then

$$P(A) = \frac{a}{a + b}$$

EXAMPLE 5.8   **Converting Odds to Probabilities**

A meteorologist states that the odds are 5 to 2 that it will rain tomorrow. Convert this statement to a probability statement.

SOLUTION We have $a = 5$ and $b = 2$. We obtain

$$P(\text{Rain}) = 5/(5 + 2) = .714$$

Thus, the meteorologist thinks the probability is .714 that it will rain tomorrow.

Exercises

Statistical Concepts

5.2.1 Explain the difference between the equally likely approach and the relative frequency approach to assigning a probability to an event.

5.2.2 Can we ever determine the "true" probability of an event by using the relative frequency approach? Comment.

5.2.3 List three examples of situations where you would use the equally likely approach to assign a probability.

5.2.4 List three examples of situations where you would use the relative frequency approach to assign a probability.

5.2.5 List three examples of situations where you would use the subjective approach to assign a probability.

Statistical Drills

5.2.6 Suppose we toss a pair of coins and record how each lands, heads or tails. List the sample space. How many basic outcomes are there? Assign a probability to each basic outcome.

5.2.7 Six horses run in a race. A handicapper assigns probabilities for each horse to win the race. What method do you think was used to assign the probabilities?

5.2.8 *True or false:* An objective probability can be determined in one of two ways: theoretically, by conducting repeated mental experiments, or empirically, by conducting repeated actual experiments. Explain.

Statistical Applications

5.2.9 A family has three children. There are four possible outcomes: no boys, one boy, two boys, or three boys. Are these outcomes equally likely?

5.2.10 In a certain gambling game at a carnival, a box contains six red and six blue envelopes. Each of the six red envelopes contains one of the following: $1, $2, ... , $6; the same is true for each of the six blue envelopes. A contestant randomly selects one red and one blue envelope and keeps the contents. For example, if the contestant selects the red envelope containing $5 and the blue envelope containing $4, then the contestant gets to keep $9. The contestant will win from $2 to $12.
a. Are all these sums equally likely?
b. In how many ways can the contestant win $7?
c. What is the probability of winning $8?

5.2.11 A certain town has 500 voters. A poll is taken by choosing a random sample of 50 voters. All 500 names are placed in a box and thoroughly mixed; then the sample of 50 names is selected. Are all samples equally likely?

5.2.12 The credit department at a department store had the following history for 850 receivables: 120 were paid early, 340 were settled on time, 220 were paid late, and 170 were uncollectible. Assuming that this experience is representative of the future, estimate the probabilities that a particular account receivable will fall into each of the four categories.

5.2.13   Suppose there are four roads (A, B, C, and D) from town X to town Y, and three roads (E, F, and G) from town Y to town Z. Suppose a criminal travels from town X to town Y to town Z, and a detective is asked which route the criminal took from town X to town Z. The detective does not know which route was taken and considers each possible route equally likely. Eventually the detective is forced to guess and hypothesizes that the criminal went via road A between town X and town Y and via road F from town Y to town Z. What is the probability that the detective is correct?

5.2.14   At the racetrack, a spectator hears a gambler claim that the odds are 2 to 5 that a particular horse will win. What is the gambler's subjective assessment of the probability that the horse will win?

5.2.15   The relative frequency approach to assigning probabilities relies on performing a "large number" of experiments. What is a "large number" of experiments? Is it one thousand? One million? Or does the answer depend on the situation?

5.3   Some Basic Rules of Probability

This section covers some basic rules that must be satisfied by any logical system of assigning probabilities to basic outcomes. The probability of any basic outcome o_i can be thought of as the relative frequency with which the basic outcome o_i occurs in a large number of trials. Every relative frequency is a number in the interval [0, 1], so the probability of any basic outcome o_i must be a value from 0 to 1.

Probability of a basic outcome

> For each basic outcome o_i, $0 \leq P(o_i) \leq 1$.

Suppose that event A contains the k basic outcomes $o_1, o_2, \ldots, o_k$. The probability of event A is the relative frequency of occurrence of the basic outcomes $o_1, o_2, \ldots, o_k$. We obtain the following rule:

Probability of an event

> Let event $A = \{o_1, o_2, \ldots, o_k\}$, where $o_1, o_2, \ldots, o_k$ are k different basic outcomes. The probability of any event A is the sum of the probabilities of the basic outcomes in A. That is,
>
> $$P(A) = P(o_1) + P(o_2) + \cdots + P(o_k) = \Sigma_A P(o_i)$$
>
> where Σ_A means to obtain the sum over all basic outcomes in event A.

The probability of any event represents the relative frequency of occurrence of all the basic outcomes in A. Thus, the probability of any event must fall in the interval [0, 1]; no event can have a probability that is negative or that exceeds 1.

Probability of an event

> For any event A, $0 \leq P(A) \leq 1$.

The sample space S contains all possible outcomes of an experiment, one of which must occur. It follows that exactly one of the basic outcomes in S must occur, and the probability that event S occurs must be 1. The probabilities assigned to all the basic outcomes in an experiment must sum to 1.

Probability of a sample space

Let $S = \{o_1, o_2, \ldots, o_n\}$ represent the sample space of an experiment. The probability of S is

$$P(S) = \Sigma_S P(o_i) = 1$$

EXAMPLE 5.9   **Probability of a Sample Space** ——————————————————————————

The quality control division of an electronics firm tests a product for defects. The basic outcomes are $o_1 = \{\text{Good}\}$ and $o_2 = \{\text{Defective}\}$, and so the sample space is $S = \{o_1, o_2\} = \{\text{Good, Defective}\}$. Suppose that past experience indicates that about 90% of the products produced are good and 10% of the products are defective. Show that the sample space has a probability equal to 1.

SOLUTION   Based on the relative frequency approach, we assign the probabilities $P(o_1) = .9$ and $P(o_2) = .1$. Then we obtain $P(S) = P(o_1) + P(o_2) = 1$. Thus, the probabilities assigned to o_1 and o_2 satisfy all of the rules. That is, $0 \leq P(o_i) \leq 1$ for $i = 1$ and 2, and $P(S) = \Sigma P(o_i) = 1$.

Probabilities for Infinite Sample Spaces

At times we want to consider problems in which the sample space contains an infinite number of equally likely outcomes. For example, suppose a tiny leak has developed in a pipe that is 12 feet long. Let us assume that all points on the pipe are equally likely to be the source of the leak. How can we assign probabilities to the infinite number of points on the pipe and still satisfy the basic rules of probability? We cannot assign positive probabilities to the infinite number of individual points, because the sum of all the probabilities would exceed 1. In this case, the probability assigned to each individual point is 0. Some point on the pipe has to be the source of the leak, but chances are that any particular point is not the source.

Now consider a slightly different problem. What is the probability that the source of the leak occurs within 3 feet from the left end of the pipe? That is, what is the probability that the source of the leak falls in the interval $(0, 3)$? By assumption, the source of the leak must fall in the interval $(0, 12)$, and all points in the interval are equally likely. A reasonable way to assign a probability to the event that the source of the leak falls in the interval $(0, 3)$ is to assume that the probability of the interval $(0, 3)$ is the ratio of the length of the interval of interest, $(0, 3)$, to the length of the entire interval, $(0, 12)$. That is, the probability that the point falls in the interval $(0, 3)$ is $\frac{3}{12} = .25$. Similarly, the probability that the source of the leak falls in the interval $(3, 9)$ is $\frac{6}{12} = .50$.

In problems in which all points on a line interval are assumed to be equally likely, the probability that a randomly selected point falls in a specific subinterval is equal to the ratio of the length of the subinterval of interest to the length of the entire interval. Thus, in equally likely problems involving all points in an interval, we measure probabilities by the ratios of lengths of lines. In other equally likely problems involving an infinite number of points, we measure probabilities by ratios of areas or of volumes rather than by ratios of lengths.

EXAMPLE 5.10   **Probabilities of Infinite Sample Spaces** ————————————————

Suppose that an airplane crashed in a rectangular wilderness area A, which is 10 miles wide and 20 miles long. Assume that all points in area A are equally likely to represent the location of the plane. Searchers will thoroughly cover a rectangular area B, which is 4 miles wide and 5 miles long inside A. Find the probability that the plane is found.

S O L U T I O N   Since the area of A is 200 square miles and that of B is 20 square miles, we calculate the ratio of the areas:

$$P(\text{Plane is in area B}) = \frac{\text{Area of B}}{\text{Area of A}} = \frac{20}{200} = .1$$

Exercises

Statistical Concepts

5.3.1   Explain why the probability of every basic event must be nonnegative.

5.3.2   Explain why the probability of a basic event cannot exceed 1.

5.3.3   Explain why the probability of the sample space must equal 1.

Statistical Drills

5.3.4   The first number on a license plate is any digit from 0 to 9. All digits are equally likely. What is the probability the first digit is a 3? What is the probability the first digit is less than 4?

5.3.5   An individual purchases a ticket to the opera and is given a seat in row M. Row M contains 12 seats, with an aisle at each end of the row. The computer picks the ticket randomly, so we can assume that all seats are equally likely.
a. What is the probability that the individual gets a seat next to an aisle?
b. What is the probability that the individual does not get an aisle seat?

Statistical Applications

5.3.6   The research department of the Karscig Chemical Company has four chemists and three engineers. The vice president of the firm randomly selects two of these individuals and asks their opinions about the chances of developing a certain product. Suppose that none of the engineers thinks that the product can be developed and that all the chemists think that it can.
a. List all possible ways of selecting two individuals to be questioned.
b. How many outcomes are possible?
c. Are all possible outcomes equally likely?
d. How many of the possible outcomes contain two chemists?
e. What is the probability that both individuals being questioned are chemists?

5.3.7 In Exercise 6, what is the probability that at least one person being questioned is a chemist?

5.3.8 In Exercise 6, what is the probability that neither person being questioned is a chemist?

5.3.9 An oil company that has purchased a square tract of land in Alaska 20 miles on a side is going to pick a site on the tract at random and drill a well. Assume that oil exists in two rectangular pools, each having dimensions 2 miles by 3 miles. If one well is drilled, what is the probability of striking oil?

5.3.10 The public works department must repair a section of sewer pipe that is cracked. The pipe starts under the center of the street and connects to another pipe under the front wall of a warehouse. The wall is 150 feet from the center of the street. The plumbers think that the crack could be at any point on the pipe with equal probability. Find the probability that the crack is located as follows:
a. Within 30 feet of the wall
b. More than 40 feet from the wall
c. Between 80 and 100 feet from the wall

5.3.11 The U.S. Coast Guard must search for a fishing vessel that sank in a lake during a fierce storm. The lake has a circular shape with a radius of 10 miles. The commanding officer wants to concentrate the search efforts in the areas where the vessel most likely sank.
a. Suppose that all points are considered equally likely. Find the probability that the vessel sank less than 2 miles from shore.
b. Suppose we know that the vessel sank in the eastern half of the lake. Find the probability that the vessel sank more than 1 mile from shore.
c. Suppose we know that the vessel sank in the northeast quarter of the lake. Find the probability that the ship sank between 3 and 4 miles from shore.

5.3.12 In the daily lottery, a number from 000 to 999 is picked randomly. Yesterday the number 463 won the lottery.
a. What is the probability that 463 wins today?
b. What is the probability that 463 does not win today?
c. Pick any integer from 000 to 999. What is the probability that your number will win the lottery today?
d. During the last 200 days, the number 463 won the lottery twice. What is the probability that 463 will be the winning number tomorrow?

5.4 Probabilities of Compound Events

In many probability problems, it is necessary to combine events to define new events. In this section, we will discuss some rules that are helpful in determining the probabilities of these new events.

DEFINITION **Complement of an Event**

Let A denote some event in the sample space S. The **complement** of A, denoted by $\overline{A}$, represents the event composed of all basic outcomes in S that do not belong to A.

Exactly one of the basic outcomes in the sample space S must occur. If the basic outcome that occurs is not in A, then one of the basic outcomes of $\overline{A}$ must occur. It follows that the probability of $\overline{A}$ is 1 minus the probability of A.

Probability of the complement of an event

Let $\bar{A}$ denote the complement of A. Then $P(\bar{A}) = 1 - P(A)$.

For a sample space containing n basic outcomes, it is easy to prove this rule. Consider the sample space $S = \{o_1, o_2, \ldots, o_n\}$. Let $A = \{o_1, o_2, \ldots, o_k\}$. Then $\bar{A} = \{o_{k+1}, o_{k+2}, \ldots, o_n\}$.

We have

$$P(S) = \sum_{i=1}^{k} P(o_i) + \sum_{i=k+1}^{n} P(o_i) = 1$$

Thus

$$P(S) = P(A) + P(\bar{A}) = 1$$

or

$$P(\bar{A}) = 1 - P(A)$$

Although this proof is based on the assumption that the sample space contains a finite number of basic outcomes, the basic theorem holds for all sample spaces.

The complement of an event can be represented graphically in a Venn diagram. In Figure 5.3, the rectangle represents all basic outcomes in the sample space, the unshaded area in the circle represents all basic outcomes in event A, and the shaded area represents $\bar{A}$.

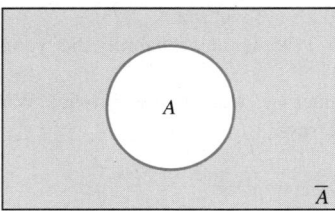

FIGURE 5.3   *Venn diagram showing the complement of an event*

EXAMPLE 5.11   **Probability of the Complement of an Event**

Suppose that the computer center has 10 personal computers. Seven of the computers are in perfect condition and three have keyboards with a key that sticks occasionally. A student randomly selects one computer. What is the probability that the student gets a defective computer?

SOLUTION   By using the equally likely approach, we obtain

$$P(A) = P(\text{Perfect computer}) = \frac{7}{10} = .7$$

For the complement, we obtain

$$P(\bar{A}) = P(\text{Defective computer}) = 1 - P(A) = 1 - .7 = .3$$

EXAMPLE 5.12   **Probability of the Complement of Another Event** ─────────────────

An underground pipe 12 miles long connecting two storage facilities for fuel has a tiny hole that is leaking. Suppose that all points on the pipe are equally likely to be the source of the leak. Investigators begin by examining the 5-mile stretch of pipe that is above ground. Find the probability that the leak is not found above ground.

SOLUTION   By using the equally likely model for events containing an infinite number of possible outcomes, we obtain

$$P(A) = P(\text{Leak is above ground}) = \frac{\text{Length of pipe examined}}{\text{Total length of pipe}} = \frac{5}{12}$$

$$P(\bar{A}) = P(\text{Leak is not above ground}) = 1 - P(A) = \frac{7}{12}$$

Unions and Intersections of Two Events

Let S denote the sample space for an experiment, and let A and B be two events in S. At times we may wish to determine the probability that the basic outcome of an experiment is a member of A or B or both. All the basic outcomes that belong to at least one of the events A or B belong to the event called the *union* of A and B. At other times we may wish to determine the probability that the basic outcome is a member of *both* events A and B. All basic outcomes that belong to both A and B belong to the event called the *intersection* of A and B.

DEFINITION   **Union of Two Events** ─────────────────────────────

Let A and B be two events in the sample space S. Their **union,** denoted $A \cup B$, is the event composed of all basic outcomes in S that belong to *at least one* of the two events A or B. Hence, the union $A \cup B$ occurs if either A or B (or both) occurs.

DEFINITION   **Intersection of Two Events** ─────────────────────────

Let A and B be two events in the sample space S. The **intersection** of A and B, denoted $A \cap B$, is the event composed of all basic outcomes in S that belong to both A and B. Hence, the intersection $A \cap B$ occurs if *both* A and B occur.

The notions of union and intersection can be illustrated by Venn diagrams. In Figures 5.4 and 5.5 (page 148), the rectangle represents the sample space, and the circles inside the rectangle represent events A and B. In Figure 5.4, the shaded area represents the union of A and B, and in Figure 5.5, the shaded area represents the intersection of A and B. The intersection of two events appears in a Venn diagram as the overlapping area between A and B.

Additive Law of Probability

The probability that the union of events A and B occurs is the sum of the probabilities of all the basic outcomes in A and B. If A and B have any basic outcomes in common, then

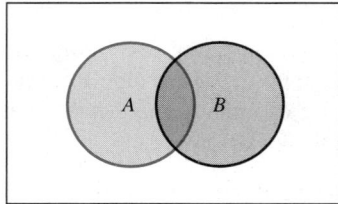

FIGURE 5.4 *Venn diagram showing the union of two sets*

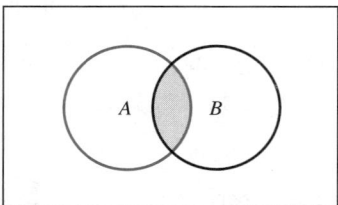

FIGURE 5.5 *Venn diagram showing the intersection of two sets*

the sum of $P(A)$ and $P(B)$ counts the probabilities of the common basic outcomes twice. Thus to obtain the probability of their union, we must subtract from the sum of $P(A)$ and $P(B)$ the probability of all the basic outcomes common to both events, which is the probability of their intersection. This leads to the theorem called the **additive law of probability.**

Additive law of probability

Let A and B be events in the sample space S. The probability of the union of A and B is

$$P(A \cup B) = P(A) + P(B) - P(A \cap B)$$

This theorem is illustrated by the Venn diagram in Figure 5.6. The rectangle S contains all the basic outcomes o_i of the sample space. Portion A contains all the basic outcomes in A, and portion B contains all the basic outcomes in B. The shaded area contains

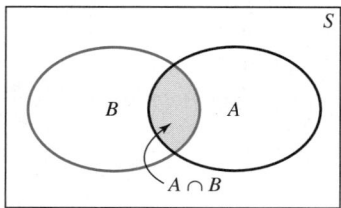

FIGURE 5.6 *Venn diagram illustrating the additive law of probability*

all the basic outcomes common to both A and B. From Figure 5.6, we see that $P(A) +$ $P(B)$ includes the probability of the shaded portion twice.

EXAMPLE 5.13 **Additive Law of Probability** ————————————————

At a certain university, all first-year students must take a chemistry course and a math course. Suppose that 15% of the freshmen fail chemistry, 12% fail math, and 5% fail both. Suppose a first-year student is picked at random. Find the probability that the student failed at least one of the courses.

SOLUTION Let A = {Person failed chemistry} and let B = {Person failed math}. Then $(A \cap B)$ = {Person failed both math and chemistry} and $(A \cup B)$ = {Person failed math, chemistry, or both}.

We obtain

$$P(A \cup B) = P(A) + P(B) - P(A \cap B)$$
$$= .15 + .12 - .05 = .22$$

This result is illustrated in the Venn diagram in Figure 5.7.

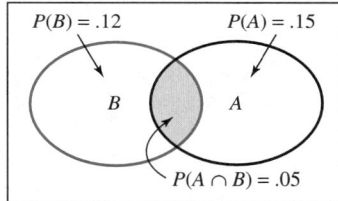

FIGURE 5.7 *Venn diagram for Example 5.13*

DEFINITION **Mutually Exclusive Events** ——————————————————

Let A and B be two events in a sample space S. If A and B have no basic outcomes in common, then they are said to be **mutually exclusive.** If A and B are mutually exclusive events, we write $(A \cap B) = \varnothing$, where $\varnothing$ denotes the empty set.

Two events are mutually exclusive if they cannot both occur in a single trial of an experiment. If events A and B are mutually exclusive, then $P(A \cap B) = 0$. In a Venn diagram, mutually exclusive events have no overlapping area.

EXAMPLE 5.14 **Mutually Exclusive Events** ——————————————————

At the Gromyko Printing Company, 10% of the employees have college degrees. Suppose that 50% of the employees are male, 50% are female, and all of the college graduates are male. Find the probability that a randomly selected person is either a college graduate or a female or both.

SOLUTION Let A = {Person is a college graduate} and B = {Person is female}. There are no female college graduates. Thus, events A and B have no common basic outcomes. Because being a college graduate and being female are mutually exclusive events at the Gromyko Printing Company, we have $(A \cap B) = \varnothing$ and $P(A \cap B) = 0$. We thus obtain

$$P(A \cup B) = P(A) + P(B) - P(A \cap B)$$
$$= .10 + .50 - .00 = .60$$

The probability that neither A nor B occurs is denoted $P(\bar{A} \cap \bar{B})$ and is computed as

$$P(\bar{A} \cap \bar{B}) = 1 - P(A \cup B)$$

By drawing a Venn diagram, you should be able to show that this statement is true. In addition, you should be able to show that the following statement is true by drawing a Venn diagram:

$$P(\bar{A} \cup \bar{B}) = 1 - P(A \cap B)$$

Tips for Problem Solving

1. When solving a probability problem involving compound events, always list the information presented in the problem.
2. Whenever possible, draw a Venn diagram. Label events A, B, $\bar{A}$, $\bar{B}$, $A \cap B$, and $A \cup B$, and label any probabilities stated in the problem.
3. The following relationships are helpful:

$$P(A \cup B) = P(A) + P(B) - P(A \cap B)$$
$$P(\bar{A}) = 1 - P(A)$$
$$P(A) = P(A \cap B) + P(A \cap \bar{B})$$
$$P(\bar{A} \cap \bar{B}) = 1 - P(A \cup B)$$
$$P(\bar{A} \cup \bar{B}) = 1 - P(A \cap B)$$

Exercises

Statistical Concepts

5.4.1 Explain the difference between the union and the intersection of two events. Illustrate this on a Venn diagram.

5.4.2 Explain what we mean when we say two events are mutually exclusive. Give an example of two such events. Illustrate this on a Venn diagram.

5.4.3 State the additive law of probability. Suppose that events A and B are mutually exclusive. Now state the additive law of probability. Illustrate this on a Venn diagram.

5.4.4 *True or false:* $P(A)$ can never exceed $P(A \cup B)$. Explain.

5.4.5 *True or false:* $P(A)$ can never exceed $P(A \cap B)$. Explain.

5.4.6 *True or false:* $P(A$ and $B)$ can never exceed $P(A \cup B)$. Explain.

Statistical Drills

5.4.7 What is wrong with the following assignment of probabilities?

$$P(A) = .6 \qquad P(B) = .3 \qquad P(A \cap B) = .7$$

5.4.8 What is wrong with the following assignment of probabilities?

$$P(A) = .6 \qquad P(B) = .3 \qquad P(A \cup B) = .2$$

5.4.9 We are given $P(A) = .4$, $P(B) = .5$, and $P(A \cap B) = .1$. Find $P(A \cup B)$.

Statistical Applications

5.4.10 The energy commission of a state consists of four people, two of whom are in favor of nuclear power and two of whom are opposed. The governor must select at random two members from the commission to represent the state at a conference.

a. List the sample points contained in the following events:

$$S = \{\text{Sample space}\}$$
$$A = \{\text{Exactly one pronuclear member chosen}\}$$
$$B = \{\text{Exactly two pronuclear members chosen}\}$$
$$C = \{\text{Both antinuclear members chosen}\}$$

b. Find the probabilities of events A, B, $(A \cap B)$, $(A \cap C)$, and $(A \cup B)$.

5.4.11 A large department store claims that 60% of shoplifters in the store will be detected by the store's closed-circuit TV system, 40% by the store's security officers, and 20% by both. If the claims are true, what is the probability that a shoplifter will be detected?

5.4.12 A personnel manager constructed the table shown here to define nine combinations of talent and motivation. The numbers in the table are the manager's estimates of the probabilities that a managerial prospect will be classified in the respective categories.

| | TALENT | | |
Motivation	High	Medium	Low
High	.05	.16	.05
Medium	.19	.32	.05
Low	.11	.05	.02

Suppose the personnel manager has decided to hire a new manager. Define the following events:

$$A = \{\text{Prospect places high in motivation}\}$$
$$B = \{\text{Prospect places high in talent}\}$$
$$C = \{\text{Prospect places medium or better in both categories}\}$$
$$D = \{\text{Prospect places low in at least one category}\}$$
$$E = \{\text{Prospect places high in both categories}\}$$

a. Does the sum of the probabilities in the table equal 1?
b. Find the probability of each event defined above.
c. Find $P(A \cup B)$, $P(A \cap B)$, and $P(A \cup C)$.

5.4.13 The following data show the percentages of employees at a company based on gender and level of education.

| | HIGHEST DEGREE OBTAINED | | | |
	High school	Bachelor's	Master's	Doctorate
Males	5	20	12	11
Females	18	15	14	5

Suppose an employee is selected at random from the firm's 5,000 employees and the following events are defined:

$$A = \{\text{Employee is male}\}$$
$$B = \{\text{Employee is female}\}$$
$$C = \{\text{Employee has doctorate}\}$$
$$D = \{\text{Employee has master's}\}$$
$$E = \{\text{Employee has bachelor's}\}$$
$$F = \{\text{Employee has high school diploma}\}$$

Find the probabilities of the following events:

a. A, B, C, D, E, F **b.** $A \cup B$ **c.** $B \cap C$ **d.** $A \cap F$ **e.** $A \cap B$ **f.** $C \cap D$

5.4.14 A small bank has 1,000 customers having both checking and savings accounts. The customers are categorized according to their average monthly account balances as follows:

Checking account	SAVINGS ACCOUNT Under $1,000	$1,000 or more
Under $500	300	400
$500 or more	200	100

Let A denote the set of people with savings of $1,000 or more, and let B denote the set of people with checking accounts of $500 or more. If a person is selected at random, find the following:

a. $P(A)$ **b.** $P(B)$ **c.** $P(A \cup B)$ **d.** $P(\bar{A} \cup \bar{B})$ **e.** $P(A \cap B)$

5.4.15 Twenty percent of the population watched a TV football game on Saturday, 30% watched a football game on Sunday, and 15% watched both games. An automobile firm purchased advertising on both telecasts and wants to know the probability that a person saw at least one of the games.

a. What proportion of the population saw at least one of the games?
b. What proportion of the population saw neither of the games?

5.4.16 In the Department of Labor, 70% of the employees are male, 30% are female, and 35% of the employees have business degrees. Of those having business degrees, 40% are female. Calculate the following probabilities:

a. A randomly selected employee is a female with a business degree.
b. A randomly selected male has a business degree.

5.4.17 At the Middletown Appliance Company, 10% of the employees are in accounting, 40% are in sales, 20% are in manufacturing, and the rest are in other departments. On any given day, the absentee rates for these departments are 2%, 6%, 3%, and 5%, respectively.

a. What proportion of absentees are from the accounting department?
b. What proportion of absentees are from the sales department?

5.4.18 A roulette wheel has 38 numbered slots; 18 are black, 18 are red, and 2 are green. Half of the black slots are even numbers and half are odd; the same is true for the red slots. The green slots are numbered 0 and 00.

a. You bet $1 that a black slot will occur. Find the probability that you win the bet.
b. On a single trial, you bet $1 that a black value will occur and $1 that an odd value will occur. Find the probability that you will lose both bets.
c. On a single trial, you bet $1 that a black value will occur and another $1 that an odd value will occur. Find the probability that you will win both bets.

d. On a single trial, you bet $1 that a black value will occur and another $1 that an odd value will occur. Find the probability that you will win one bet and lose the other.

5.4.19 State the complement of the following events:
  a. The inflation rate will be less than 4% next year.
  b. A family with two children has two boys.
  c. A basketball player who shoots two free throws will make at least one of the two shots.

5.4.20 Suppose that 65% of the students in a class are male and 40% are seniors. The teacher picks one of the students randomly. Indicate whether the following statements are true, false, or indeterminate.
  a. The probability that the student is a male or a senior (or both) is 1.05.
  b. The probability that the student is a male or a senior (or both) is .7.
  c. The probability that the student is a male or a senior (or both) is .6.
  d. The probability that the student is both a male and a senior is .6.
  e. The probability that the student is both a male and a senior is .7.
  f. The probability that the student is a male or a senior (or both) is 1.00.
  g. The probability that the student is a female is .35.

5.5 Conditional Probability

In many applications, we may be interested in determining the probability that some event A occurs given that some other event B has already occurred. This probability is called a **conditional probability.** If the probability of one event varies depending on whether a second event has occurred, the two events are said to be **dependent.** Let A denote the event that a person has a college degree and B the event that a person earns more than $50,000 per year. Events A and B are related because the probability that a randomly selected person earns more than $50,000 per year, $P(B)$, is not the same as the probability that the person earns more than $50,000 per year given that the person is a college graduate. For example, across the United States, less than 20% of all adults earn more than $50,000 per year, but among college graduates more than 20% of adults earn more than $50,000 per year.

The probability $P(B)$ is an unconditional probability; it shows the proportion of all adults who earn more than $50,000 per year. Now examine just the group of adults who are college graduates. Of this subpopulation, what proportion earns more than $50,000 per year? This latter proportion is the *conditional probability of event B given A* and is denoted $P(B|A)$.

Joint Probability Tables

Before discussing conditional probability, it will be helpful to introduce the concepts of *joint probability* and *marginal probability.* Suppose that a male student who was rejected for admission into graduate school presented the data in Table 5.1 (page 154) in support of his claim that the university was discriminating against male applicants. The student argued that of the 12,500 applicants, 4,700 (or 37.6%) were males who were rejected. On the other hand, only 2,400 of the students (or 19.2%) were females who were rejected. Should we accept the student's claim that the university seems to be discriminating against males?

Table 5.1 shows how all students who applied for admission to the graduate school were distributed according to the two variables Gender and Admission status. Table 5.1

TABLE 5.1   *Joint frequency table for applicants to graduate school*

| Gender | ADMISSION STATUS | | Total |
	Admitted	Rejected	
Male	3,800	4,700	8,500
Female	1,600	2,400	4,000
Total	5,400	7,100	12,500

TABLE 5.2   *Joint probability table for applicants to graduate school*

| Gender | ADMISSION STATUS | | Marginal probability |
	Admitted	Rejected	
Male	.304	.376	.680
Female	.128	.192	.320
Marginal probability	.432	.568	1.000

is an example of a *joint frequency table,* or a *joint frequency distribution.* It is a *bivariate* table because only two variables are being considered. If three variables (such as Race, Gender, and Admission status) had been considered, we would have had a *trivariate* table. Joint frequency tables show the frequency, or number, of observations possessing several characteristics simultaneously.

If we divide each entry in Table 5.1 by the total number of applicants (12,500), we obtain a set of relative frequencies, which can be used to represent probabilities. Table 5.2 shows these relative frequencies or probabilities. Table 5.2 is an example of a *joint probability distribution,* or a *joint relative frequency distribution.*

DEFINITION   **Joint Probability**

A **joint probability** shows the probability that an observation will possess two (or more) characteristics simultaneously. Every joint probability must be a number in the closed interval [0, 1], and the sum of all joint probabilities must be 1.

EXAMPLE 5.15   **Calculating Joint Probabilities**

Use the frequency data in Table 5.1 to find the probability that an individual was both male and rejected. Also find the probability that a student was female and was rejected.

SOLUTION   Table 5.1 indicates that there were 12,500 applicants to graduate school, of whom 4,700 were males who were rejected. The relative frequency 4,700/12,500 = .376 is the joint probability that an individual possesses both characteristics, namely, being male and being rejected. This probability is shown at the intersection of the row Male and the column Rejected in Table 5.2. We obtain

$$P(\text{Male} \cap \text{Rejected}) = \frac{4,700}{12,500} = .376$$

Similarly, the probability that a student was female and rejected is

$$P(\text{Female} \cap \text{Rejected}) = .192$$

Marginal Probability

Whereas a joint probability shows the proportion of observations possessing two characteristics simultaneously, a *marginal probability* shows the proportion of observations that possess any single specific characteristic. The numbers at the bottom and on the right side of Table 5.2 (bottom and right margins) are *marginal probabilities.*

Rule for marginal probabilities

> To obtain the marginal probability of any particular value of a variable, sum the appropriate joint probabilities over all values of the other variable. Every marginal probability must be a number in the closed interval [0, 1]. For any variable, the sum of all marginal probabilities must be 1.

EXAMPLE 5.16 **Marginal Probability**

In Table 5.2, the variable Admission status has two categories, Admitted and Rejected. Find the marginal probabilities for these categories.

SOLUTION   The marginal probability that a randomly selected applicant was admitted is the sum of the appropriate joint probabilities summed over the variable Gender. We obtain

$$P(\text{Admitted}) = P(\text{Male} \cap \text{Admitted}) + P(\text{Female} \cap \text{Admitted})$$
$$= .304 + .128 = .432$$

This result can also be obtained from Table 5.1, which indicates that 5,400 of the 12,500 applicants were admitted. Thus, $P(\text{Admitted}) = 5,400/12,500 = .432$.

The marginal probability that an applicant was rejected (.568) can be obtained as follows:

$$P(\text{Rejected}) = P(\text{Male} \cap \text{Rejected}) + P(\text{Female} \cap \text{Rejected})$$
$$= .376 + .192 = .568$$

We have $P(\text{Admitted}) = .432$ and $P(\text{Rejected}) = .568$. These two marginal probabilities sum to 1.

Now suppose that we want to compare the rejection rate for males with the rejection rate for females. To do this, we have to calculate the conditional probabilities.

To calculate $P(A|B)$, we restrict ourselves to the subpopulation of experiments that resulted in event B. The conditional probability $P(A|B)$ measures the fraction of those occurrences of B that also resulted in event A. Figure 5.8 (page 156) illustrates the notion of conditional probability. The rectangle S denotes the sample space. The two portions labeled A and B represent the basic outcomes in events A and B, respectively. The shaded portion represents the basic outcomes common to both A and B, that is, the intersection

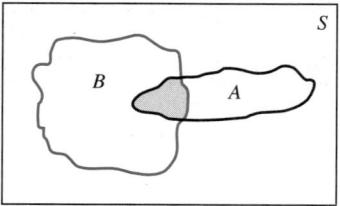

FIGURE 5.8   *Conditional probability*

of *A* and *B*. Let us use the areas of *A*, *B*, and their intersection to represent their probabilities. For any sample space *S*, $P(S) = 1$, so the area of the rectangle is 1.

   We want to know the probability that a basic outcome in *A* has occurred given that a basic outcome in *B* has occurred. This requires that the basic outcome be in the intersection $(A \cap B)$. The conditional probability $P(A|B)$ equals the ratio of the area of the intersection of *A* and *B* to the area of *B*, where these areas measure the probabilities of the events.

   We obtain the following rule for calculating conditional probabilities:

Formula for conditional probability

> The conditional probability that event *A* occurs given that event *B* has occurred is
>
> $$P(A|B) = \frac{P(A \cap B)}{P(B)} \quad \text{provided } P(B) \neq 0$$

   If *A* and *B* have no basic outcomes in common, then $(A \cap B) = \varnothing$ and $P(A \cap B) = 0$. Thus, it follows that if *A* and *B* have no outcomes in common, then $P(A|B) = 0$. That is, if we know *B* has occurred and *A* has no basic outcomes in common with *B*, then it follows that *A* has not occurred.

EXAMPLE 5.17   Conditional Probability

Refer again to Tables 5.1 and 5.2. To determine whether the university has been discriminating against male applicants, we must compare the two conditional probabilities P(Applicant was rejected given that applicant was male) and P(Applicant was rejected given that applicant was female). We will denote these two probabilities by P(Rejected|Male) and P(Rejected|Female), respectively. Calculate these two conditional probabilities.

SOLUTION   From the joint and marginal probabilities in Table 5.2, we obtain

$$P(\text{Rejected}|\text{Male}) = \frac{P(\text{Rejected} \cap \text{Male})}{P(\text{Male})} = \frac{.376}{.680} = .553$$

We can obtain this same result by using the data in Table 5.1. Table 5.1 indicates that there were 8,500 male applicants, of whom 4,700 were rejected. The proportion of males who were rejected is given by

$$P(\text{Rejected}|\text{Male}) = \frac{4,700}{8,500} = .553$$

Similarly, by using the data in Table 5.2, we obtain the conditional probability

$$P(\text{Rejected}|\text{Female}) = \frac{P(\text{Rejected} \cap \text{Female})}{P(\text{Female})} = \frac{.192}{.320} = .600$$

By using the data in Table 5.1, we obtain the same result

$$P(\text{Rejected}|\text{Female}) = \frac{2,400}{4,000} = .600$$

Does the evidence in Tables 5.1 and 5.2 support the claim that the university is discriminating against male applicants? The joint probabilities in Table 5.2 show that 37.6% of all applicants were rejected males, whereas only 19.2% of the applicants were rejected females. This shows that the proportion of applicants who were both male and rejected is approximately twice as high as the proportion of applicants who were both female and rejected. Does this indicate that the university is discriminating against male applicants? Not necessarily. You have to examine conditional probabilities, not joint probabilities, to determine whether there has been discrimination.

The appropriate conditional probabilities are $P(\text{Rejected}|\text{Male}) = .553$ and $P(\text{Rejected}|\text{Female}) = .600$. These probabilities indicate that 60% of the female applicants and 55.3% of the male applicants were rejected. Therefore, the data indicate that the rejection rate for females was actually slightly higher than that for males. These values are close enough, however, to support the argument that the university was not discriminating against either gender.

Can you find the flaw in the original claim that males were being discriminated against? The reason the joint probabilities are larger for males than for females is that there were far more male applicants than female. If males and females are treated exactly alike and there are far more male applicants than female applicants, the majority of the applicants who are rejected will be male. (It also follows that the majority of applicants who will be accepted will be male.)

Multiplicative Law of Probability

The joint probability $P(A \cap B)$ shows the probability that *both* events A and B occur. By manipulating the rule for conditional probability,

$$P(A|B) = \frac{P(A \cap B)}{P(B)}$$

we can obtain a useful formula for calculating $P(A \cap B)$, called the *multiplicative law of probability*.

Multiplicative law of probability

The probability that the two events A and B both occur is given by

$$P(A \cap B) = P(B)P(A|B) \quad \text{and} \quad P(A \cap B) = P(A)P(B|A)$$

EXAMPLE 5.18   **Multiplicative Law of Probability**

Suppose that Valencia's Discount Store has received a shipment of 60 television sets, 5 of which are defective. On the following day, two televisions are sold. Find the probability that both of the televisions are defective.

SOLUTION   Let A denote the event {First television is defective} and B the event {Second television is defective}. Then the intersection of A and B denotes the event {Both televisions are defective}. By using the equally likely model, the probability that the first television is defective is $P(A) = 5/60$. Also, we know that $P(B|A) = 4/59$, because after the first defective television is sold, 59 televisions remain, of which 4 are defective. Now we use the multiplicative law of probability to obtain

$$P(A \cap B) = P(A)P(B|A) = \left(\frac{5}{60}\right)\left(\frac{4}{59}\right) = \frac{1}{177} = .0056$$

EXAMPLE 5.19   **Multiplicative Law of Probability**

Refer to Example 5.18. Find the probability that the first television sold is defective and that the second is not defective.

SOLUTION   Let C denote the event that the second television is not defective. We obtain $P(C|A) = 55/59$, because if the first television is defective, then 59 televisions remain, 55 of which are not defective. We obtain

$$P(A \cap C) = P(A)P(C|A) = \left(\frac{5}{60}\right)\left(\frac{55}{59}\right) = .0777$$

Tips for Problem Solving

1. Before calculating any compound probability, be sure to record all the information listed in the problem.

2. When possible, draw a Venn diagram and label all the relevant events, including A, B, $A \cap B$, $A \cup B$, $A \cap \bar{B}$, $\bar{A} \cap B$, $\bar{A}$, and $\bar{B}$. Also record the corresponding probabilities.

3. When calculating a conditional probability, use the following formula:

$$P(A|B) = \frac{P(A \cap B)}{P(B)}$$

4. When calculating a joint probability, use the multiplicative law of probability:

$$P(A \cap B) = P(A)P(B|A) \quad \text{or} \quad P(A \cap B) = P(B)P(A|B)$$

5. At times, it may be helpful to use the following formula:

$$P(A) = P(A \cap B) + P(A \cap \bar{B})$$

or

$$P(A \cap B) = P(A) - P(A \cap \bar{B})$$

Exercises

Statistical Concepts

5.5.1 Suppose we have constructed a 2-by-2 joint probability table. Explain the difference between a joint probability and a marginal probability.

5.5.2 Explain how to calculate marginal probabilities if we know all the joint probabilities.

5.5.3 *True or false:* The marginal probability $P(A)$ can never exceed the joint probability $P(A \cap B)$.

5.5.4 *True or false:* $P(A)$ can never exceed $P(A|B)$. Explain.

5.5.5 *True or false:* $P(A)$ can never equal $P(A|B)$. Explain.

5.5.6 *True or false:* $P(A \cap B)$ can never be less than $P(A|B)$. Explain.

5.5.7 *True or false:* $P(A \cap B)$ can never exceed $P(B)$. Explain.

Statistical Drills

5.5.8 Is anything wrong with the following assignment of probabilities?

$$P(A) = .6 \qquad P(B) = .3 \qquad P(A|B) = .7$$

5.5.9 Is anything wrong with the following assignment of probabilities?

$$P(A) = .6 \qquad P(B) = .5 \qquad P(A|B) = 0$$

5.5.10 What is wrong with the following assignment of probabilities?

$$P(A \cap B) = .6 \qquad P(A|B) = .3$$

5.5.11 We are given $P(A|B) = 0$. Are A and B mutually exclusive?

Statistical Applications

5.5.12 A quality control engineer has been asked to examine a complex electronic system that is out of control. The examination will involve testing five mechanisms within the system to identify the one that is faulty. Assume that one and only one mechanism is faulty. If the mechanisms are randomly chosen for examination by the engineer but excluded from further consideration if found satisfactory, what is the probability that the faulty switching mechanism will be discovered at the following times:
   a. During the second examination
   b. During the third examination
   c. Before the third examination

5.5.13 An inspector for a food processing firm has accepted 98% of all good shipments and has incorrectly rejected 2% of good shipments. In addition, the inspector accepts 94% of all shipments, and it is known that 5% of all shipments are of inferior quality.
   a. Find the probability that a shipment is rejected.
   b. Find the probability that a shipment is good.
   c. Find the probability that a shipment is good and that it is accepted.
   d. Find the probability that a shipment is of inferior quality and that it is accepted.
   e. Find the probability that a shipment is accepted given that it is of inferior quality.
   f. Find the probability that a shipment is rejected given that it is good.

5.5.14 To see whether a trainee is doing an acceptable job as a product inspector, the trainee is told to inspect a shipment of 1,000 products. It is known that 800 of the products are good (G) and 200 are defective (D). The trainee rejects 40 of the good products and accepts 6 of the defective products.

Let R denote the set of rejected products and A the set of accepted products. Find the following probabilities:

a. $P(G)$ **b.** $P(D)$ **c.** $P(R|G)$ **d.** $P(R|D)$
e. $P(G|R)$ **f.** $P(G|A)$ **g.** $P(A)$ **h.** $P(R)$

5.5.15 For a nationwide oil company, 15% of its credit card customers have annual incomes of $10,000 or less, 55% have annual incomes between $10,000 and $30,000, and the rest have annual incomes of $30,000 or more. In the lowest income category, 20% of the accounts are delinquent; in the middle category, 10% are delinquent; in the highest category, 5% are delinquent.

a. If an account is selected at random, what is the probability that it is delinquent?
b. An account is selected and found to be delinquent. What is the probability that it is in the lowest income category?
c. If the company has 10,000 accounts, how many would you expect to be delinquent and in the middle category?

5.5.16 A research study found that 60% of the employees at a financial institution read the *Wall Street Journal*, 45% read *Time*, and 30% read both.

a. What is the probability that an employee selected at random does not read either publication?
b. What is the probability that a person selected at random reads at least one of the publications?
c. Given that a person reads *Time*, what is the probability that the person also reads the *Wall Street Journal*?

5.5.17 The dean of the college of arts and sciences is undertaking a study to see whether grades are being inflated in various academic departments. The dean obtains the following data on the grades of 200 students in three courses:

Grade	History	Economics	Math	Total
A	25	25	30	80
B	10	60	20	90
C or less	5	15	10	30
Total	40	100	60	200

Suppose one of the 200 students is chosen randomly. Find the following probabilities:

a. P(Student got an A)   **b.** P(Student got B or less)
c. P(Student is in economics)   **d.** $P(A|\text{Economics})$
e. $P(\text{Math}|B)$ **f.** $P(\text{History}|B \text{ or less})$

5.5.18 Three different production lines produce valves for the O'Toole Corporation. They make large and small valves, and their output (in thousands) for a given week was as follows:

Type of valve	Line 1	Line 2	Line 3	Total
Large	300	200	100	600
Small	200	100	100	400
Total	500	300	200	1,000

A customer comes to Mr. O'Toole with a large defective valve. Mr. O'Toole wants to determine which line made this defective valve. Find the following probabilities:

a. $P(\text{Line 1}|\text{Large})$
b. $P(\text{Line 2}|\text{Large})$
c. $P(\text{Line 3}|\text{Large})$

5.5.19 Mr. Wilson, the owner of a used car agency, meets about 45% of the agency's potential customers and makes a sale to about 60% of those he meets. The other salespeople meet the other 55% of the potential customers and make a sale to about 50% of these individuals.
  a. What proportion of potential customers eventually buy a car?
  b. What proportion of sales can be attributed to Mr. Wilson?

5.6   Independence

If the conditional probability $P(A|B)$ differs from the unconditional probability $P(A)$, then the probability that A occurs depends on whether B has occurred. If $P(A|B)$ equals $P(A)$, then the probability of occurrence of A does not depend on the occurrence of B, and we say that A and B are independent. If A and B are **independent,** then events A and B have no influence on one another; the probability of A occurring is unaffected by whether B has occurred.

If A is independent of B, then the probability that A occurs given that B has occurred should be the same as the probability of A given no information about B. Recall that it is always the case that

$$P(A|B) = \frac{P(A \cap B)}{P(B)}$$

If A is independent of B, we also have

$$P(A|B) = P(A)$$

If A and B are independent, we can equate the two expressions for $P(A|B)$ and obtain the following rule: If A and B are independent, then

$$\frac{P(A \cap B)}{P(B)} = P(A)$$

After multiplying both sides by $P(B)$, we obtain the important *formula for independent events.*

Formula for independent events

> Events A and B are **independent** if and only if
>
> $$P(A \cap B) = P(A)P(B)$$

Thus, when two events are independent, the probability that they both occur is obtained by multiplying their respective individual probabilities. This rule can be generalized to apply to any number of independent events. Thus, if events A, B, and C are independent of one another, then the probability that all three events will occur is equal to the product of their individual probabilities.

Note that this theorem works in two directions. That is, if we know a priori that events A and B are independent, then we also know that $P(A \cap B) = P(A)P(B)$. On the other hand, if we can show that $P(A \cap B) = P(A)P(B)$, then we know that events A and B are independent.

The definition of statistical independence can be explained intuitively as follows. Suppose that the probability that event A will occur is $P(A)$. Suppose now that additional information is provided that event B has occurred. If this information does not change our opinion about the likelihood of the occurrence of event A, then our conditional probability assessment $P(A|B)$ will be the same as $P(A)$. We will have concluded that knowing the occurrence of event B is of no use in determining whether A will occur.

In some circumstances, independence can be deduced, or at least reasonably inferred, from the nature of the random experiment. In these cases, the probability of the joint occurrence of events A and B can be calculated as the product of their individual probabilities. This idea is particularly useful for evaluating experiments repeated many times under identical conditions. Because successive outcomes are independent of one another, the probability of any specific set of basic outcomes equals the product of the individual probabilities.

Consider the relationship between mutually exclusive events and independent events. If events A and B are mutually exclusive, then A and B cannot occur together. Are they independent? That is, does the assumption that B has occurred alter our assessment of the probability that A will occur? It certainly does, because if we assume that B has occurred, then we know that A will not or has not occurred. Thus, mutually exclusive events are *dependent* events.

EXAMPLE 5.20   Probability of Independent Events

Approximately 30% of the sales representatives hired by a firm quit in less than 1 year. Suppose that two sales representatives are hired and assume that the first sales representative's behavior is independent of the second sales representative's behavior.
(a) What is the probability that both quit within a year?
(b) Find the probability that exactly one representative quits.

SOLUTION   **a.** Use the theorem for independent events. Let $A = \{$First sales representative quits$\}$ and $B = \{$Second sales representative quits$\}$. Then

$$P(\text{Both quit}) = P(A \cap B) = P(A)P(B) = (.3)(.3) = .09$$

b. Let $\bar{A} = \{$First sales representative stays$\}$ and $\bar{B} = \{$Second sales representative stays$\}$. Applying the rule about complementary events yields the probabilities $P(\bar{A}) = 1 - P(A) = .7$, and similarly $P(\bar{B}) = .7$.
We obtain

$$
\begin{aligned}
P(\text{Exactly one quits}) &= P(\text{First quits and second stays}) \\
&\quad + P(\text{First stays and second quits}) \\
&= P(A \cap \bar{B}) + P(\bar{A} \cap B) \\
&= P(A)P(\bar{B}) + P(\bar{A})P(B) \\
&= (.3)(.7) + (.7)(.3) = .42
\end{aligned}
$$

Tree Diagrams

When events are independent, a **tree diagram** can be a useful device for calculating probabilities. In a tree diagram, each branch represents a possible outcome. We find the

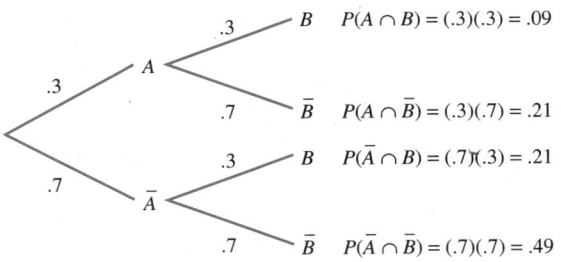

$$P(A \cap B) = (.3)(.3) = .09$$
$$P(A \cap \bar{B}) = (.3)(.7) = .21$$
$$P(\bar{A} \cap B) = (.7)(.3) = .21$$
$$P(\bar{A} \cap \bar{B}) = (.7)(.7) = .49$$

FIGURE 5.9 *Tree diagram for Example 5.20*

joint probability of a sequence of outcomes by multiplying the corresponding probabilities. This is illustrated in Figure 5.9, which shows the tree diagram for Example 5.20.

EXAMPLE 5.21   **Probability of Independent Events**

A husband and wife, each 20 years old, are debating whether to set up a retirement program for themselves. Benefits are paid to the man or the woman at the age of 70. If both have died before reaching age 70, no benefits are paid. Assume that the probability that a man aged 20 lives to age 70 is approximately .6 and the probability that a woman aged 20 lives to age 70 is approximately .7. If the husband and wife join the program, what is the probability that either the man or the woman will collect benefits? Assume that the chances of the man or woman dying are independent of each other.

SOLUTION   Benefits are paid if either the man or the woman or both live to age 70. The appropriate probabilities can be found by applying the rule for probability of independent events. Let $M = \{$Man lives to age 70$\}$; $\bar{M} = \{$Man dies before age 70$\}$; $W = \{$Woman lives to age 70$\}$; and $\bar{W} = \{$Woman dies before age 70$\}$. The probability of collecting benefits is the probability that both live to age 70 plus the probability that just the woman lives to age 70 plus the probability that just the man lives to age 70. The possible outcomes and their probabilities are as follows:

$$P(M \cap W) = (.6)(.7) = .42$$
$$P(\bar{M} \cap W) = (.4)(.7) = .28$$
$$P(M \cap \bar{W}) = (.6)(.3) = .18$$
$$P(\bar{M} \cap \bar{W}) = (.4)(.3) = .12$$
$$P(\text{At least one lives}) = P(M \cap W) + P(\bar{M} \cap W) + P(M \cap \bar{W})$$
$$= .42 + .28 + .18 = .88$$

Alternatively, by using the rule for complementary events, we have

$$P(\text{At least one lives}) = 1 - P(\text{Both die}) = 1 - (.4)(.3) = .88$$

Figure 5.10 (page 164) shows the tree diagram for Example 5.21.

Sampling With and Without Replacement

There is a broad class of problems in which successive outcomes can be considered independent that involve *sampling with replacement.* In contrast, when we *sample without replacement* from a finite population, the successive outcomes will not be independent.

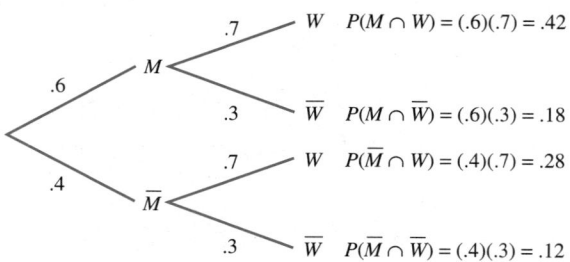

W $P(M \cap W) = (.6)(.7) = .42$

$\overline{W}$ $P(M \cap \overline{W}) = (.6)(.3) = .18$

W $P(\overline{M} \cap W) = (.4)(.7) = .28$

$\overline{W}$ $P(\overline{M} \cap \overline{W}) = (.4)(.3) = .12$

FIGURE 5.10   *Tree diagram for Example 5.21*

Selecting a random sample can be viewed as a process in which we sequentially obtain one observation after another. At each step of the process, we select one observation from the remaining population of observations. When we sample with replacement, we put each unit back into the original population before selecting the next observation. When we sample without replacement, we do not put a unit back before selecting the next observation.

When we sample with replacement, successive outcomes are independent because the probability of obtaining any particular observation on any selection does not depend on which units were previously selected. When we sample without replacement, successive outcomes are not independent because the composition of the remaining population changes after each selection. Therefore, at each selection the probability of a specific outcome depends on the previous outcomes. When we sample without replacement, we find the joint probability $P(A \cap B)$ by using the general multiplicative law of probability

$$P(A \cap B) = P(A)P(B|A) \quad \text{or} \quad P(A \cap B) = P(B)P(A|B)$$

When we sample with replacement, we can use the law for independent events, which is a special case of the multiplicative law:

$$P(A \cap B) = P(A)P(B)$$

EXAMPLE 5.22   **Sampling Without Replacement** —————————————————————————

Suppose that a car rental agency has 10 seemingly identical cars. Unknown to anyone, two of these cars contain defective brakes because of abuse by former drivers. Two customers arrive and rent cars. Find the probability that the first customer gets a good car and the second gets a car with defective brakes.

SOLUTION   Define $A = \{$First customer gets a good car$\}$ and $B = \{$Second customer gets a defective car$\}$. This problem involves sampling without replacement, so we use the multiplicative law $P(A \cap B) = P(A)P(B|A)$. Initially, 10 cars are available, of which 8 are good. By using the equally likely model, we obtain $P(A) = 8/10$. After 1 good car is rented to the first customer, 9 cars are available for the second customer, of which 2 are defective. We have $P(B|A) = 2/9$. We obtain the desired probability

$$P(A \cap B) = P(A)P(B|A) = \left(\frac{8}{10}\right)\left(\frac{2}{9}\right) = .178$$

EXAMPLE 5.23   **Sampling With Replacement**

The Internal Revenue Service (IRS) wants to investigate whether gambling casinos have been making illegal deductions on their income tax returns. The IRS makes a list with the names of 10 potential offenders, 3 of which have reported no illegal deductions. The IRS selects 1 of the 10 casinos randomly and conducts a full-scale audit. During the following year, to keep all the casinos honest, the IRS again selects 1 of the 10 casinos randomly and conducts a full-scale audit. In this problem, the casino audited in the first year could be audited again in the second; thus, the sampling is done with replacement. Find the probability that the casino audited first reported no illegal deductions and that the casino audited second reported illegal deductions.

SOLUTION   Because this problem involves sampling with replacement, successive events are independent. Define A = {First casino made no illegal deductions} and B = {Second casino made illegal deductions}. We apply the equally likely model and obtain $P(A) = .3$ and $P(B) = .7$. We obtain the joint probability by using the law for independent events:

$$P(A \cap B) = P(A)P(B) = (.3)(.7) = .21$$

Exercises

Statistical Concepts

5.6.1   Explain in your own words what it means for two events to be independent.

5.6.2   *True or false:*   Event A and its complement are independent. Explain.

5.6.3   *True or false:*   If A and B are independent, then they are mutually exclusive. Explain.

Statistical Drills

5.6.4   A jar contains one red, one blue, and one yellow ball. Suppose you select three balls from the jar *with replacement* after each selection of a ball. Draw a tree diagram showing all the possible outcomes. How many different outcomes are there?

5.6.5   A jar contains one red, one blue, one green, and one yellow ball. Suppose you select four balls from the jar *without replacement.* Draw a tree diagram showing all the possible outcomes. How many different outcomes are there?

5.6.6   Suppose $P(A) = .4$, $P(B) = .5$, and A and B are mutually exclusive; find $P(A \cap B)$.

5.6.7   Suppose $P(A) = .4$, $P(B) = .5$, and A and B are independent; find $P(A \cap B)$.

5.6.8   *True or false:*   If A and B are independent, then $P(A \cap B)$ cannot exceed $P(A)$. Explain.

5.6.9   We toss a fair coin twice. Find the probability of two heads.

5.6.10   We toss a fair coin 10 times. Find the probability of 10 consecutive heads.

5.6.11   At an industrial plant, the median width of pieces of steel is .382 inch. We randomly select three pieces of steel. Find the probability that all three pieces of steel exceed the median width.

5.6.12   Let A and B denote any two events. Draw a Venn diagram to verify that $A = (A \cap B) \cup (A \cap \bar{B})$.

5.6.13   Let $P(B) = .85$, $P(A|\bar{B}) = .6$, and $P(A|B) = .25$; find $P(B|A)$.

Statistical Applications

5.6.14   A real estate agent is showing houses to two potential buyers, Mr. Smith and Mr. Jones. The agent assigns .5 as the subjective probability that Smith will buy a house and .5 as the subjective probability that Jones will buy a house. Let *S* indicate that Smith buys a house, and let *J* denote that Jones buys a house. Assume that the decisions of Smith and Jones are independent of one another.

   a. Construct a tree diagram showing all possible outcomes denoting whether Smith and/or Jones bought a house. List the sample space.

   b. Find the probability of each element in the sample space.

   c. Let *A* denote the event {Agent has at least one sale}. List the elements of *A* and find $P(A)$.

   d. List the elements of $\bar{A}$ and find $P(\bar{A})$.

5.6.15   Two teams A and B are of equal ability, so each has a probability of .5 of defeating the other. Assume that the outcome of any game is independent of the outcome of any other game. What is the probability that team A wins four games in a row?

5.6.16   The CPA examination has four parts, one each on accounting problems, auditing, business law, and the theory of accounting. To be certified, an applicant must pass all four parts. Past history shows that on the first attempt, 30% pass the accounting part, 35% pass auditing, 30% pass law, and 20% pass theory.

   a. If success on one part of the examination is independent of success on any other part, what is the probability that an applicant becomes certified on the first attempt to pass the CPA examination?

   b. Do you believe that this probability is a reasonable approximation if you know that more than 10% of applicants pass all parts of the examination on their first attempt? Comment on the assumption of independence among parts of the test.

5.6.17   In handling a customer's order, the order department will fill it incorrectly with probability .03. Also, the order will be delivered to the wrong address with probability .01. Assume that the two variables Filling of order and Delivery of order are statistically independent.

   a. What is the probability that a customer's order is filled incorrectly and delivered to the wrong address?

   b. What is the probability that an order is filled correctly and delivered to the right address?

5.6.18   At the McVay Advertising Agency, 80% of newly hired employees quit within 1 year. A man and a woman are hired on the same day. Assume that their decisions to stay or leave are independent of one another. What is the probability of each of the following:

   a. Both will work there more than a year.

   b. At least one will work there more than a year.

   c. Neither works there for a year.

5.6.19   The Gavett Management Company wants to purchase stock in the Wiltman Manufacturing Company. Mr. Gavett believes that if inflation decreases, the probability is .9 that the investment will be profitable. If inflation remains constant or increases, the probability is .6 that the investment will be profitable. Assume that the probability that the inflation rate declines is .3. What is the probability that the investment will be profitable?

5.6.20   A bank has two emergency sources of power for its computers. There is a 95% chance that source 1 will operate during a total power failure, and an 80% chance that source 2 will operate. Find the probability that in a total power failure, neither emergency power source will operate. Assume the energy sources are independent.

5.6.21   A doctor claims that the probability of a patient surviving a particular type of open heart surgery is .7. If the doctor performs three such operations, find the probability for each of the following:

   a. All three patients survive.

   b. None of the patients survives.

   c. At least one patient survives.

5.6.22   About 5% of the population becomes ill during whitewater raft tours. Suppose a raft contains five randomly selected tourists, and assume that all illnesses are independent of one another. What is the probability of each of the following?

a. None of the five tourists becomes ill.

b. At least one becomes ill.

 5.6.23 A group of individuals concerned about environmental problems claims that 30% of the adults in a certain town have been adversely affected by a new nuclear power plant that pollutes the air and causes lung damage. To test their claim, you randomly select four adult residents of the town.

a. If the environmental group is correct, what is the probability that all four people have been adversely affected?

b. What is the probability that at least one of the four individuals has been adversely affected?

5.6.24 In a southern California county, the probability of a forest fire in any year is .2. An investor has just bought a home near the edge of the forest and intends to keep the home for 4 years before selling it. The investor decides not to buy fire insurance.

a. If events are independent of one another, what is the probability that no fire occurs during the next 4 years?

b. What is the probability that at least one fire occurs during the next 4 years?

5.7 Bayes' Theorem (Optional)

An interesting application of probability theory involves estimating or revising probabilities when new information on a random experiment is obtained. An English philosopher, the Reverend Thomas Bayes (1702–1761), was one of the first to work with rules for revising probabilities in light of sample information. Bayes' contribution, published in 1763, consists of a method for calculating conditional probabilities.

To see how Bayes' theorem works, suppose the sample space of an experiment is partitioned into k mutually exclusive and exhaustive events $E_1, E_2, \ldots, E_k$. Assume the probabilities of these events are equal to $P(E_1), P(E_2), \ldots, P(E_k)$, respectively. These probabilities are called **prior probabilities**, because they are determined before any new information is taken into account.

A probability that has been revised based on new information is called a **posterior probability**, because it represents a probability calculated after new information is taken into account. By using Bayes' theorem, a prior (unconditional) probability that an event E_i will occur or has occurred is revised to a new probability, a posterior (conditional) probability.

Let E_i represent some event whose prior probability is $P(E_i)$. Let A represent the new information. For any conditional probability, we have the rule

$$P(E_i|A) = \frac{P(E_i \cap A)}{P(A)}$$

In addition, from the multiplicative rule of probability, we have

$$P(A \cap E_i) = P(E_i)P(A|E_i)$$

If we substitute this result into the preceding equation, we obtain the formula for calculating the posterior probability that E_i occurred given that A occurred:

$$P(E_i|A) = \frac{P(E_i)P(A|E_i)}{P(A)}$$

Bayes' theorem

Let E_i be an event having a prior probability $P(E_i)$. Let A be some new event or information. Then the posterior probability of E_i is given by

$$P(E_i|A) = \frac{P(E_i)P(A|E_i)}{P(A)}$$

Let $E_1, E_2, \ldots, E_n$ be a set of mutually exclusive and exhaustive events so that $E_i \cap E_j = \varnothing$ and $(E_1 \cup E_2 \cup \cdots \cup E_n)$ is the entire sample space. Frequently, when applying Bayes' theorem, $P(A)$ will not be directly given, but it can be calculated by using the formula

$$P(A) = P(A \cap E_1) + P(A \cap E_2) + \cdots + P(A \cap E_n)$$

Recall that $P(A \cap E_i) = P(E_i)P(A|E_i)$. Therefore, to calculate $P(A)$, we can use the equivalent formula

$$P(A) = P(E_1)P(A|E_1) + P(E_2)P(A|E_2) + \cdots + P(E_n)P(A|E_n)$$

In this case, Bayes' theorem becomes

$$P(E_i|A) = \frac{P(E_i)P(A|E_i)}{\Sigma \ P(E_i)P(A|E_i)}$$

When $n = 2$, we obtain

$$P(E_i|A) = \frac{P(E_i)P(A|E_i)}{P(E_1)P(A|E_1) + P(E_2)P(A|E_2)}$$

The following examples illustrate applications of Bayes' theorem.

EXAMPLE 5.24   **An Application of Bayes' Theorem**

A hospital has developed a test to discover whether newborn babies have a certain type of mental disorder. From past records, the doctors know that the probability that a baby has this disorder is .003. If the baby does have the disorder, the test will be positive 98% of the time and negative 2% of the time. If the baby does not have the disorder, the test will be negative 99% of the time and positive 1% of the time. A test administered on a newborn baby is positive. What is the probability that the baby actually has the disorder?

SOLUTION 1   Before applying Bayes' theorem, let us solve the problem intuitively. Suppose the test was administered to 100,000 randomly selected babies. Based on the prior information, we would expect 300 of the 100,000 babies (or .3%) to have the disorder and 99,700 to be healthy. Of the 300 who have the disorder, we would expect 294 (or 98%) to test positive and 6 to test negative. Of the 99,700 who do not have the disorder, we would expect 98,703 (or 99%) to test negative and 997 to test positive. Thus, of 100,000 babies, we would expect

$$294 + 997 = 1,291$$

to test positive and

$$6 + 98,703 = 98,709$$

to test negative. All of these expected frequencies are shown in Table 5.3.

TABLE 5.3 *Joint frequency table for testing 100,000 babies*

	TEST RESULT		
Health status	Positive	Negative	Total
Healthy	997	98,703	99,700
Mental disorder	294	6	300
Total	1,291	98,709	100,000

Of the 1,291 babies who yield a positive test result, we would expect 294 to actually have the disorder. Thus, $P(\text{Disorder}|\text{Positive test}) = {}^{294}\!/_{1,291} = .2277$.

SOLUTION 2   Now let us apply Bayes' theorem to solve this problem. Define the following events: $P_1 = \{\text{Test is positive}\}$, $P_2 = \{\text{Test is negative}\}$, $D_1 = \{\text{Baby has mental disorder}\}$, and $D_2 = \{\text{Baby is healthy}\}$. The following information is known:

$$P(D_1) = .003 \qquad P(P_1|D_1) = .98 \qquad P(P_2|D_2) = .99$$

We want to find $P(D_1|P_1)$. From Bayes' theorem, we have

$$P(D_1|P_1) = \frac{P(D_1)P(P_1|D_1)}{P(P_1)}$$

To solve the problem, we must find $P(P_1)$. We obtain

$$\begin{aligned} P(P_1) &= P(P_1 \cap D_1) + P(P_1 \cap D_2) \\ &= P(D_1)P(P_1|D_1) + P(D_2)P(P_1|D_2) \\ &= (.003)(.98) + (.997)(.01) = .01291 \end{aligned}$$

Thus,

$$P(D_1|P_1) = \frac{(.003)(.98)}{.01291} = \frac{.00294}{.01291} = .2277$$

This is the same answer we obtained in solution 1. This result is illustrated by the tree diagram in Figure 5.11. When applying Bayes' theorem, you should first construct a tree diagram like that in Figure 5.11 or a table like Table 5.3 to help you visualize the necessary calculations.

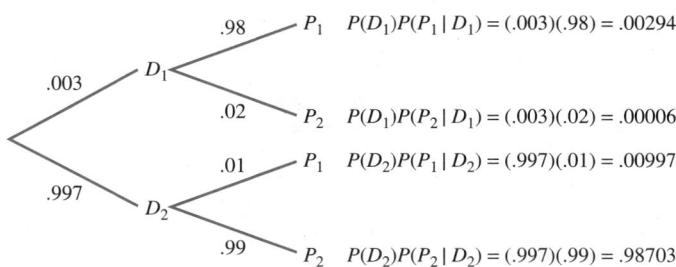

FIGURE 5.11 *Tree diagram for Example 5.24*

What did we learn by applying Bayes' theorem in Example 5.24? Without testing, the probability that a baby has the disorder is .003. Given a positive test result, many people mistakenly think that the probability is .98 that the baby has the mental disorder. This is not correct. The probability .98 represents $P(P_1|D_1)$, that is, the probability that the test is positive, given that the disorder is present. What we are seeking is $P(D_1|P_1)$— the probability that the disorder is present, given that the test is positive. After learning that the test is positive, the probability that the baby has the disorder is only .2277.

In the tree diagram illustrated in Figure 5.11, the probabilities assigned to the branches labeled P_1 and P_2 emanating from the branch labeled D_1 differ from the probabilities emanating from the branch labeled D_2. This is because the occurrence of event P_1 is not independent of the occurrence of event D_1 or event D_2. When D_1 occurs, we have $P(P_1|D_1) = .98$. This conditional probability is listed above the branch connecting D_1 with P_1. When D_2 occurs, we have $P(P_1|D_2) = .01$. This conditional probability is listed above the branch connecting D_2 with P_1. Similarly, the conditional probability listed above the branch connecting D_1 with P_2 is $P(P_2|D_1) = .02$, and the conditional probability listed below the branch connecting D_2 with P_2 is $P(P_2|D_2) = .99$.

EXAMPLE 5.25   **Another Application of Bayes' Theorem** ————————————————

A corporation operates two factories that make engines for automobiles. Let E_1 denote an engine made at plant 1 and E_2 an engine made at plant 2. Let A denote the event that an engine is defective. From past records, it is known that 2% of the engines at plant 1 are defective, and 3% of the engines at plant 2 are defective. That is, $P(A|E_1) = .02$ and $P(A|E_2) = .03$. Assume that plant 1 makes 40% of the engines and plant 2 makes the rest. That is, $P(E_1) = .4$ and $P(E_2) = .6$. An engine selected at random is defective. Find the probability that it was made at plant 1.

S O L U T I O N   We want to find $P(E_1|A)$. We first find $P(A)$ as follows:

$$\begin{aligned} P(A) &= P(A \cap E_1) + P(A \cap E_2) \\ &= P(E_1)P(A|E_1) + P(E_2)P(A|E_2) \\ &= (.4)(.02) + (.6)(.03) = .026 \end{aligned}$$

From Bayes' theorem we have

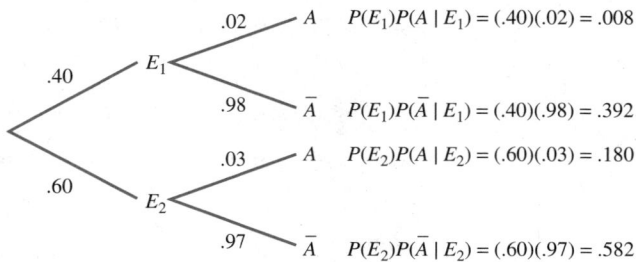

FIGURE 5.12   *Tree diagram for Example 5.25*

$$P(E_1|A) = \frac{P(E_1)P(A|E_1)}{P(A)} = \frac{(.4)(.02)}{.026} = \frac{.008}{.026} = .308$$

This result is illustrated in Figure 5.12.

Exercises

Statistical Concepts

5.7.1 Draw a Venn diagram and show that $P(A) = P(A \cap B) + P(A \cap \bar{B})$.

5.7.2 Let A_1, A_2, and A_3 be mutually exclusive and exhaustive events. Draw a Venn diagram to prove that $P(B) = P(B \cap A_1) + P(B \cap A_2) + P(B \cap A_3)$.

Statistical Drills

5.7.3 Let $P(B) = .7$, $P(A|B) = .2$, and $P(A|\bar{B}) = .3$. Find $P(A \cap B)$.

5.7.4 Let $P(B) = .7$, $P(A|B) = .2$, and $P(A|\bar{B}) = .3$. Find $P(A)$.

5.7.5 Let $P(B) = .7$, $P(A|B) = .2$, and $P(A|\bar{B}) = .3$. Find $P(A \cap \bar{B})$.

5.7.6 Let $P(B) = .7$, $P(A|B) = .2$, and $P(A|\bar{B}) = .3$. Find $P(B|A)$.

Statistical Applications

5.7.7 At a particular factory, every product is examined by two inspectors. The first inspector catches 80% of the defectives and sends them back for repairs. All remaining items are sent to a second inspector, who misses about 40% of the defectives that get past the first inspector.
a. What proportion of the defectives will get by both inspectors?
b. Given that a product was found to be defective, what is the probability that it was found by the first inspector?

5.7.8 A trucking firm receives a message that one of its trucks has broken down on the highway. The firm has three types of trucks and does not know which type has broken down. Because the firm wants to send a replacement truck of the same type, it wants to calculate which type is most likely. Of the firm's trucks, 20% are type T_1, 30% are type T_2, and 50% are type T_3. Let B denote that a truck broke down. From past experience, the firm has the following data:

$$P(B|T_1) = .1 \qquad P(B|T_2) = .1 \qquad P(B|T_3) = .05$$

Calculate the following probabilities:
a. $P(T_1|B)$ **b.** $P(T_2|B)$ **c.** $P(T_3|B)$ **d.** $P(B)$

5.7.9 The producers of a new movie think that the movie has a prior probability for success of .3. An influential film critic, who has liked 70% of all successful films and disliked 80% of all unsuccessful films that she has reviewed, is going to review the film. Find the posterior probability that the movie will be a success given the following:
a. The critic likes it.
b. The critic dislikes it.

5.7.10 An oil wildcatter has assigned a probability of .5 to striking oil on a certain plot of property. The wildcatter orders a seismic survey that has proven to be 90% reliable in the past. That is, when oil is present, the survey predicts favorably 90% of the time; when no oil is present, it predicts no oil 90% of the time.
a. Given a favorable seismic result, what is the probability for oil?
b. Given an unfavorable seismic result, what is the probability for oil?

5.8 ## Counting Techniques

In many probability problems, it is necessary to determine the number of ways in which some sequence of events can occur. If a sequence of activities must be performed, the total number of possible outcomes can be determined by using the **fundamental rule of counting.**

Fundamental rule of counting

> If event A can occur in n_1 different ways and event B can occur in n_2 different ways, then A and B can occur in sequence in $(n_1 \times n_2)$ different ways. Similarly, if A can occur n_1 ways, B can occur n_2 ways, and C can occur n_3 ways, then A, B, and C can occur together in $n_1 \times n_2 \times n_3$ different ways.

EXAMPLE 5.26   **The Fundamental Rule of Counting**

Let event A consist of choosing a car from the set {Ford, Chevrolet, Buick}; thus, $n_1 = 3$. Let event B consist of selecting an automatic or manual transmission, so $n_2 = 2$. Let event C consist of purchasing tires, either blackwalls or whitewalls; thus, $n_3 = 2$. Let event D consist of getting a red, brown, or blue car; thus, $n_4 = 3$.

There are

$$n_1 \times n_2 \times n_3 \times n_4 = 3 \times 2 \times 2 \times 3 = 36$$

different ways to choose A, B, C, and D together. These 36 possibilities are listed in the tree diagram of Figure 5.13.

Factorial Notation

At times, it is necessary to calculate the product of all the positive integers from 1 to N. A shorthand way to denote this product is by the symbol $N!$, which is read "N factorial."

DEFINITION   **Factorial**

Let N be a positive integer. The product of all integers from 1 to N is called N **factorial** and is denoted $N!$:

$$N! = N(N - 1)(N - 2) \cdots (3)(2)(1)$$

We define 0! to be equal to 1.

EXAMPLE 5.27   **Factorial Notation**

$$5! = 5 \times 4 \times 3 \times 2 \times 1 = 120$$
$$8! = 8 \times 7 \times 6 \times 5 \times 4 \times 3 \times 2 \times 1 = 40{,}320$$
$$1! = 1$$

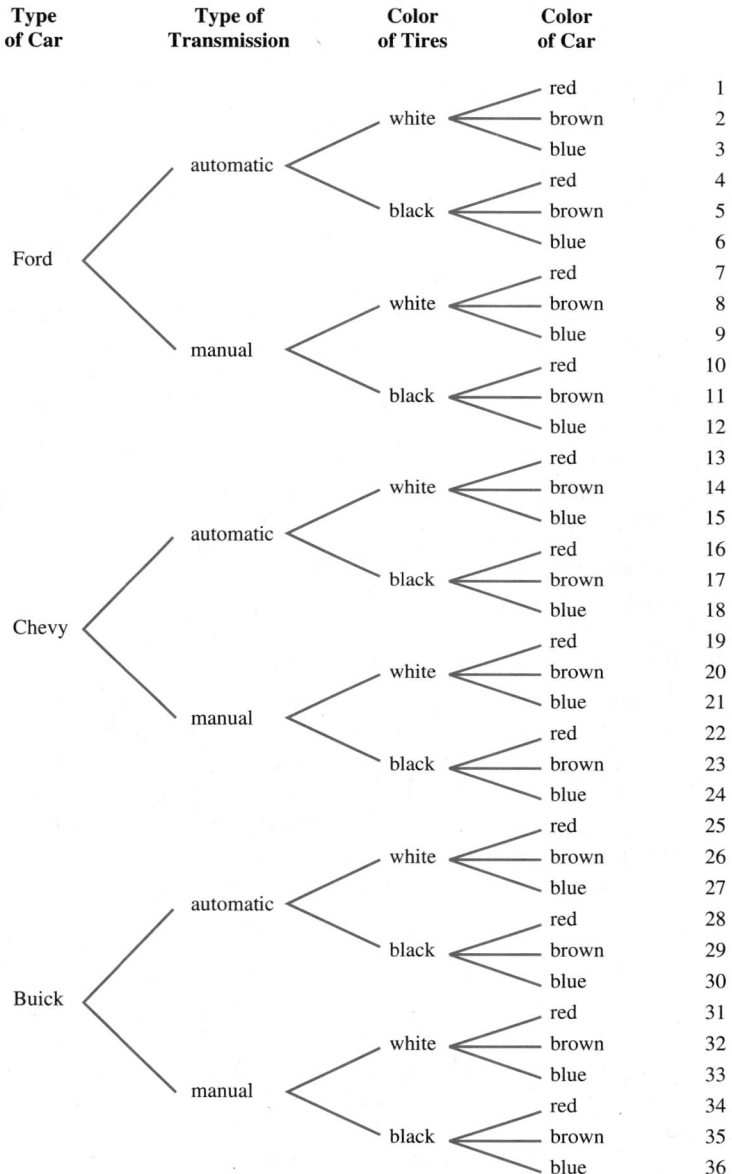

FIGURE 5.13 *Tree diagram for Example 5.26*

Factorial notation is very useful when we try to count the number of ways that N things can be arranged. Suppose we want to rank four students. How many possible rankings are there? We have four choices for the first student, three choices for the next, and so forth. Thus, there are $4 \times 3 \times 2 \times 1 = 4!$ ways of arranging the four students.

Permutations

At times we are interested in determining how many different ways we can select R items from N items and then arrange them in some specific order. For example, suppose there are $N = 8$ planes requesting clearance to land at the Buffalo airport. Because of a snowstorm, only $R = 5$ of the planes can land, and the remaining 3 planes must be rerouted to Cleveland. The flight control attendant must select 5 of the 8 planes for landing in Buffalo and then put them in order (i.e., the pilots must be told who will land first, second, and so on). This arrangement of the 5 planes in a specific order is an example of a *permutation* of 8 things taken 5 at a time.

DEFINITION   **Permutation**

A **permutation** of N different things taken R at a time, denoted $_NP_R$, is an arrangement in a specific order of any R of the N things.

EXAMPLE 5.28   **Counting the Number of Permutations**

The board of directors of a corporation is planning the annual meeting. Five different proposals are under consideration for the meeting. Because each proposal is sure to be subjected to a long and emotional debate at the meeting, the board believes that there will be enough time to discuss only two of the proposals; thus it must decide which two to discuss and in which order. The listing of all possibilities is as follows (each listing represents a different permutation):

AB	AC	AD	AE
BA	BC	BD	BE
CA	CB	CD	CE
DA	DB	DC	DE
EA	EB	EC	ED

In this example, $N = 5$ and $R = 2$, and there are 20 different possible permutations.

Determining the number of possible permutations of N things taken R at a time can be quite time-consuming if we actually list all the possibilities. We have N choices for the object selected first, $(N - 1)$ choices for the object selected second, and so on down to $(N - R + 1)$ choices for the object selected on the Rth choice. The number of permutations of N things taken R at a time, denoted by the symbol $_NP_R$, can be computed using the formula

$$_NP_R = N(N - 1)(N - 2) \cdots (N - R + 1)$$

This formula can be simplified as shown in the next box.

Formula for number of permutations

> Suppose we select R objects from N distinct objects and put the R objects in order. This is a permutation of N things taken R at a time. The total number of different permutations is
>
> $$_N P_R = \frac{N!}{(N - R)!}$$
>
> As a special case, when $R = N$, we obtain $_N P_N = N!$.

When we think of a permutation, it might be useful to think of a batting lineup in a baseball game. There are $N = 24$ players on a professional team. The manager must select $R = 9$ to play the game and then set up the batting order for these 9 players. Each different lineup is a different permutation.

EXAMPLE 5.29   **Calculating the Number of Permutations**

The personnel manager at a corporation has interviewed 10 job applicants for three jobs at three different salaries. The manager must select the three best applicants and rank them. In how many ways can three applicants be selected from 10 and ranked?

SOLUTION   We need to calculate $_{10} P_3$. We obtain

$$_{10} P_3 = \frac{10!}{(10 - 3)!} = \frac{10!}{7!} = 10 \times 9 \times 8 = 720$$

EXAMPLE 5.30   **Calculating the Number of Permutations**

Every year the American Economic Association holds an annual meeting. The directors want to choose the sites for the next four meetings. If the four sites will be New York, Chicago, Miami, and Dallas, in how many ways can the locations be ordered?

SOLUTION   There are four choices for the first location, three choices for the second location, two for the third, and one for the fourth. From the fundamental rule of counting, we obtain

$$_4 P_4 = 4 \times 3 \times 2 \times 1 = 24$$

Alternatively, if we use the formula for permutations, we have $N = 4$, $R = 4$, and

$$_4 P_4 = \frac{4!}{(4 - 4)!} = \frac{4!}{0!} = 24$$

Combinations

DEFINITION   **Combination**

A **combination** of N things taken R at a time, denoted $_N C_R$, is an arrangement of any R of these things without regard to order.

As an example of a combination, think of the starting lineup for a basketball team. The team contains a total of, say, $N = 12$ players, and the coach selects $R = 5$ of the players to play the game. Each group of 5 players represents a different combination. Unlike a permutation, however, the order in which the objects are picked does not matter.

The notion of a combination is important in statistical theory because a combination can be thought of as a particular sample of R items selected from a population containing N items. When we obtain a sample, the order in which the observations appear usually does not matter. In this case, the sample $\{A, B, C\}$ is the same as the samples $\{B, A, C\}$, $\{C, B, A\}$, and so forth.

EXAMPLE 5.31   **Calculating the Number of Combinations** ——————————————————

Suppose an event at the Olympics has four participants, denoted by A, B, C, and D. After the event, two of the participants are selected randomly to be tested for drugs. The selection of two participants from the set of four participants represents a combination of four things taken two at a time. The order in which the participants are selected does not matter. List all possible combinations of participants to be selected for drug testing.

SOLUTION   There are six combinations of four participants taken two at a time, namely, AB, AC, AD, BC, BD, and CD. The combination AB is identical to the combination BA because the order in which the participants are selected does not matter. Each of the six combinations could form two separate permutations. Thus, there are 12 permutations of four things taken two at a time.

EXAMPLE 5.32   **Calculating the Number of Combinations** ——————————————————

In Example 5.31, suppose three participants were selected for drug testing rather than two. List all possible combinations of participants to be tested.

SOLUTION   There are four different ways to select three participants to be tested, namely, ABC, ACD, ABD, and BCD.

For every combination containing three items, there are 3! permutations of those same items. That is, the combination ABC yields the 3! = 6 permutations ABC, ACB, BAC, BCA, CAB, and CBA. Similarly, the combination BCD yields the six permutations BCD, BDC, CBD, CDB, DBC, and DCB.

We want to derive a formula to determine how many combinations exist of N things taken R at a time. To do this, we start with the formula for the number of permutations of N things taken R at a time. Consider a listing of all the possible permutations of N things taken R at a time. Many of these permutations are just rearrangements of the same R objects. In fact, because R objects can be ordered in $R!$ ways, each combination of R objects will yield $R!$ different permutations. Thus, if we divide the number of permutations by $R!$, we will obtain the number of combinations. That is,

$$_N C_R = \frac{_N P_R}{R!}$$

Now recall that

$$_NP_R = \frac{N!}{(N - R)!}$$

After substitution, we obtain the important formula for the number of combinations of N things taken R at a time.

Formula for number of combinations

> Suppose we select R objects from N distinct objects. The number of different combinations of these R objects is
>
> $$_NC_R = \frac{N!}{(N - R)!R!}$$
>
> Note that $_NC_R = {}_NC_{N-R}$.

EXAMPLE 5.33 **Calculating the Number of Combinations**

Refer to Example 5.31. Use the formula to calculate the number of combinations of four things taken two at a time.

SOLUTION

$$_4C_2 = \frac{4!}{(4 - 2)!\,2!} = \frac{4 \times 3 \times 2 \times 1}{2 \times 1 \times 2 \times 1} = 6$$

which agrees with the answer calculated in Example 5.31.

EXAMPLE 5.34 **Calculating the Number of Combinations**

Refer to Example 5.32. Calculate the number of combinations of four things taken three at a time.

SOLUTION

$$_4C_3 = \frac{4!}{(4 - 3)!\,3!} = \frac{4 \times 3 \times 2 \times 1}{1 \times 3 \times 2 \times 1} = 4$$

which agrees with the answer calculated in Example 5.32.

EXAMPLE 5.35 **Calculating the Number of Combinations**

A hospital laboratory employs 11 technicians. A sample of three must be selected for training on a new computer. How many different samples are possible?

SOLUTION

$$_{11}C_3 = \frac{11!}{(11 - 3)!\,3!} = \frac{11!}{8!\,3!} = \frac{11 \times 10 \times 9}{3 \times 2 \times 1} = 165$$

Tips for Problem Solving

> **1.** To visualize the fundamental rule of counting, it often helps to draw a tree diagram. When the number of possible outcomes in each activity is small, you can actually list all the possible sequences of outcomes and count them. If the numbers in the problem are large, construct a miniature version of the problem and visualize how to count the number of possible outcomes.
>
> **2.** Remember that combination and permutation problems always refer to problems where we are selecting R items from N items without replacement. After each selection, the number of remaining possible choices is reduced by 1. Problems involving the fundamental rule of counting do not necessarily share this feature.
>
> **3.** Be sure to note whether the problem involves combinations or permutations. If ordering matters, use the permutation formula. If ordering does not matter, use the combination formula.
>
> **4.** Formulas for calculating the number of possible combinations or permutations involve calculation of factorials:
>
> $$n! = n(n - 1)(n - 2) \cdots (3)(2)(1) \qquad 0! = 1$$
>
> **5.** $\displaystyle {}_NC_R = \frac{N!}{(N - R)!R!}$
>
> **6.** $\displaystyle {}_NP_R = \frac{N!}{(N - R)!}$

Exercises

Statistical Concepts

5.8.1   Explain the fundamental rule of counting in your own words.

5.8.2   Prove the fundamental rule of counting by drawing a tree diagram.

5.8.3   What is a permutation? State the rule for the number of permutations of N things taken R at a time. Prove that ${}_NP_N = N!$.

5.8.4   What is a combination? State the rule for the number of combinations of N things taken R at a time. Prove that ${}_NP_R = R!\,{}_NC_R$.

5.8.5   Explain the difference between a permutation and a combination. Prove that ${}_NC_R = {}_NC_{N-R}$. Explain what this relationship means.

Statistical Drills

5.8.6   Calculate $6!$, $5!$, $7!/5!$, and $10!/8!$.

5.8.7   Evaluate ${}_5P_4$, ${}_6P_4$, ${}_8P_8$, ${}_7P_2$, and ${}_NP_N$.

5.8.8   Evaluate ${}_8C_3$, ${}_9C_1$, ${}_{11}C_1$, and ${}_NC_{N-2}$.

5.8.9   Evaluate ${}_5C_2$, ${}_5C_3$, ${}_9C_2$, and ${}_9C_7$.

Statistical Applications

5.8.10   A clothing manufacturer has a line of coordinated children's wear consisting of six different pairs of pants, five shirts, and three vests. How many different three-piece ensembles are available to the consumer?

5.8.11 A ballot contains the names of five candidates. Because it is supposedly an advantage to have your name near the top of the ballot, different ballots are printed listing the candidates' names in all possible ways. How many different ballots are possible?

5.8.12 A publishing company publishes five different how-to books for the handy person, and it has a special offer of three books for $10. A woman has decided to buy three books and give them to her husband one at a time to entice him to make three desired home repairs. In how many ways can three books be selected and ordered from the list of five books?

5.8.13 A construction firm builds homes using four basic floor plans. The purchaser has to choose one of these. In addition, the purchaser must select one of four colors of carpet, one of three colors of wallpaper, and one of five types of kitchen cabinet. How many variations of the firm's homes are there?

5.8.14 The menu at a cafeteria offers a choice of four drinks, three salads, five entrees, four vegetables, two kinds of potatoes, and five desserts. If a meal consists of one of each item, how many different meals are possible?

5.8.15 An Olympic gymnastic event has 10 competitors. At the end of the event, four competitors are selected randomly to be tested for the presence of certain illegal drugs.
  a. How many different samples are possible?
  b. Suppose that three of the competitors are from Japan. How many of the samples contain no Japanese competitors?
  c. Find the probability that the sample contains no Japanese competitors.
  d. Find the probability that the sample contains at least one Japanese competitor.

5.8.16 An agent for the Internal Revenue Service is presented with 12 income tax returns. The agent has time to carefully audit four of the returns. The remaining returns go unaudited. The agent randomly selects a sample of four returns.
  a. How many different samples are possible?
  b. Suppose that 2 of the 12 returns contain major errors and the remaining 10 are correct. How many of the samples contain no returns with errors?
  c. Find the probability that the sample contains no returns with errors.
  d. Find the probability that the sample contains at least one return with errors.

5.9 Computer Applications

As discussed in Section 5.2, the probabilities of many events are estimated by calculating the relative frequency of occurrence of the event in a large number of repeated trials. When we have a large data set, it is convenient to calculate relative frequencies with a computer. In the SPSSX program, the FREQUENCIES command requests the computer to construct frequency distributions and relative frequency distributions for discrete and for qualitative variables.

Refer to the data in Table 2.1 of Chapter 2. To obtain a relative frequency distribution to estimate the proportions of employees at the Computech Corporation who are employed in the various divisions, insert the command

```
FREQUENCIES VARIABLES = DIV
```

into the SPSSX program.

Figure 5.14 (page 180) shows the computer output obtained by executing the FREQUENCIES command. The output shows that 38 of the 120 employees (or 31.7%) are

```
- - - - - - - - - - - - - - - - - - - - - - - - - - - - - - - - - - - - -

DIV

                                                          Valid      Cum
  Value Label                 Value   Frequency  Percent  Percent  Percent

  OFFICE                       1.00       38       31.7     31.7     31.7
  MANUFACTURING                2.00       58       48.3     48.3     80.0
  SALES                        3.00       24       20.0     20.0    100.0
                                        -------  -------  -------  -------
                             Total       120      100.0    100.0

  Hi-Res Chart   # 4:Bar chart of div

  Valid cases      120     Missing cases      0
```

FIGURE 5.14   *SPSSX output from executing FREQUENCIES command on data in Table 2.1*

employed in the office. The value 31.7% is shown in the Percent column on the computer output. Similarly, we see that 48.3% of the employees are in manufacturing and 20.0% are in sales.

The SPSSX program also can be used to generate bivariate frequency and relative frequency distributions. For example, suppose we wish to determine what proportion of employees have a specific academic degree and are employed in a particular division. By inserting the commands

---

CROSSTABS TABLES = DEGREE BY DIV
OPTIONS 3, 4, 5

---

into the SPSSX program, we can generate a bivariate table that shows how many employees have a particular degree and are employed in a particular division. In addition, the output will show the relative frequency for each pair of values along with the row and column relative frequencies. Figure 5.15 shows the computer output generated by the CROSSTABS statement.

The computer output shows that 35 of the 120 employees, or 29.2%, have a high school degree and are employed in the office. The value 29.2% is located in the row labeled Tot Pct in Figure 5.15 (in the upper left-hand corner of the table). Similarly, 53 of the 120 employees, or 44.2%, have a high school degree and are employed in the manufacturing division.

The computer output also shows that 89 of the 120 employees, or 74.2%, have a high school degree. Similarly, 19.2% of the employees have a college degree and 6.7% have a postgraduate degree. The data also show that 31.7% of the employees work in the office, 48.3% work in manufacturing, and 20% work in sales. These percentages, when divided by 100, represent marginal relative frequencies and are located in the margins of the table.

The output in Figure 5.15 can be used to find conditional relative frequencies, which are used to estimate conditional probabilities. For example, we see that of the 89 employees who have a high school degree, 35 are employed in the office. This means that the conditional relative frequency of being in the office, given that the employee has a high school degree, is $^{35}\!/_{89}$, or 39.3%. This conditional probability is shown in the row titled

```
                          DIV                    Page 1 of 1
                Count  |
                Row Pct |OFFICE   MANUFACT SALES
                Col Pct |         URING                 Row
                Tot Pct |   1.00|    2.00|    3.00| Total
     DEGREE    --------+--------+--------+--------+
        1.00   |    35 |   53 . |     1 |      89
 HIGH  SCHOOL  |  39.3 |  59.6  |   1.1 |    74.2
               |  92.1 |  91.4  |   4.2 |
               |  29.2 |  44.2  |    .8 |
               +--------+--------+--------+
        2.00   |     3 |     5 |    15 |      23
    COLLEGE    |  13.0 |  21.7 |  65.2 |    19.2
               |   7.9 |   8.6 |  62.5 |
               |   2.5 |   4.2 |  12.5 |
               +--------+--------+--------+
        3.00   |       |       |     8 |       8
   POST-GRAD   |       |       | 100.0 |     6.7
               |       |       |  33.3 |
               |       |       |   6.7 |
               +--------+--------+--------+
       Column      38      58      24     120
        Total    31.7    48.3    20.0   100.0

     Number of Missing Observations:   0
```

FIGURE 5.15 *SPSSX output from executing CROSSTABS command on data in Table 2.1 to show relationship between DEGREE and DIV*

Row Pct in Figure 5.15. Similarly, we see that 59.6% of the employees with a high school degree are employed in the manufacturing division, and so forth. Observe that 100% of the employees having a postgraduate degree are employed in the sales division.

In Figure 5.15, the values in the row titled Col Pct show the conditional probabilities of having any specific academic degree, given that the employee is employed in any specific division. For example, of the 58 employees in the manufacturing division, 53 have a high school degree and 5 have a college degree. This indicates that the conditional probability of having a high school degree, given that the person is employed in the manufacturing division, is $^{53}/_{58}$ or 91.4%. The value 91.4% is reported in Figure 5.15 in the Col Pct row of the cell for manufacturing and high school. When the percentages labeled Row Pct or Col Pct are divided by 100, we obtain conditional relative frequencies.

Exercises

5.9.1 Refer to the data in Table 2.1. Generate a joint frequency distribution for DIV and SEX.
  a. Find the proportion of employees in the sample who are male and work in the office, who are male and work in manufacturing, and who are male and work in sales.
  b. Repeat part **a** for females.
  c. Estimate the conditional probability that an employee is in sales given that the employee is a male and then given that the employee is a female. Does the probability that a person is in sales seem to depend on the gender of the individual?
  d. Estimate the conditional probability that an employee is in the office given that the employee is a male and then given that the employee is a female. Does the probability that a person is in the office seem to depend on the gender of the individual?
  e. Estimate the conditional probability that an employee is in the manufacturing division given that the employee is a male and then given that the employee is a female. Does the probability that an employee is in the office seem to depend on the gender of the individual?

5.9.2   Refer to the data in Table 2.2. Generate a joint frequency distribution for the variables Gender and Major.

  a. Find the proportion of students in the sample who are male and are undecided about their major, who are male and are economics majors, who are male and are math majors, who are male and are business majors, and who are male and have other majors.

  b. Repeat part **a** for females.

  c. Estimate the conditional probability that a student is majoring in economics given that the student is a male and then given that the student is a female. Does the probability that a student is majoring in economics seem to depend on the gender of the individual?

  d. Estimate the conditional probability that a student is majoring in an unspecified major (Other) given that the student is a male and then given that the student is a female. Does the probability that a person has an Other major seem to depend on the gender of the individual?

  e. Estimate the conditional probability that a student is majoring in math given that the student is a male and then given that the student is a female. Does the probability that a student is majoring in math seem to depend on the gender of the individual?

5.9.3   Refer to the data in Table 2.1.

  a. Use the SPSSX program and a CROSSTABS command to determine the relative frequency of employees who are male and have a high school degree.

  b. Find the proportions of all employees who are male and have a college degree and who are male and have a postgraduate degree.

  c. Find the proportions of all employees who are female and have a high school degree, a college degree, and a postgraduate degree.

  d. Find the proportion of all employees with a high school degree who are male and the proportion with a high school degree who are female.

  e. Find the proportion of all employees with a college degree who are male and the proportion with a college degree who are female.

  f. Find the proportion of all employees with a postgraduate degree who are male and the proportion with a postgraduate degree who are female.

  g. Given that an employee is male, find the conditional probabilities that the employee has a high school degree, a college degree, and a postgraduate degree.

  h. Given that an employee has a college degree, find the conditional probabilities that the employee is male or female.

  i. Based on the conditional probabilities in parts **g** and **h,** does it appear that SEX and DEGREE are independent of one another?

Chapter 5 Supplementary Exercises

5.S.1   A company found that only 20% of its products introduced over the last few years have become profitable. When two new products were introduced during the same year, both products became profitable only 5% of the time. Suppose the company plans to introduce two new products, A and B, next year. What is the probability of the following?

  a. Product A will become profitable.

  b. Product B will not become profitable.

  c. At least one of the two products will become profitable.

  d. Neither of the two products will become profitable.

  e. Either product A or product B (but not both) will become profitable.

  f. Product A becomes profitable, given that product B is profitable.

  g. Given that at least one of the products is profitable, product A is profitable.

5.S.2 Two reviewers for a publishing house independently screen unsolicited manuscripts, giving a grade of "good," "fair," or "poor" to each manuscript.
  a. Show the sample space describing the possible outcomes of the joint review of a manuscript. Is this sample space univariate or bivariate?
  b. Let E_1 denote the event that at least one reviewer assigns a grade of "good." What basic outcomes constitute E_1?
  c. Let E_2 denote the event that the reviewers give different grades. What basic outcomes constitute E_2?
  d. Are E_1 and E_2 mutually exclusive here?

5.S.3 Suppose we define the following events: A_1 = {Family owns a home}, A_2 = {Family does not own a home}, B_1 = {Family income is under $20,000}, B_2 = {Family income is $20,000 to $50,000}, and B_3 = {Family income is $50,000 or more}. Suppose we also know the following probabilities:

$$P(A_2) = .52 \qquad P(B_1) = .50 \qquad P(B_3) = .10$$
$$P(A_1|B_1) = .20 \qquad P(A_1|B_3) = .80$$

  a. Find $P(A_1 \cap B_3)$.
  b. Find $P(A_1 \cup B_3)$.
  c. Find $P(B_3|A_1)$.

5.S.4 For each of the following probability statements, state whether it is always true, always false, or neither for any pair of events E and F of a sample space. Justify each answer.
  a. Both $P(E|F) < P(E)$ and $P(E|\bar{F}) < P(E)$
  b. $P(\bar{E}|F) = 1 - P(E|F)$

5.S.5 A woman is selling her house. She believes there is a .3 chance that a person who inspects the house will purchase it. What is the probability that more than two people will have to inspect the house before the woman finds a buyer? (Assume that the decisions of the people inspecting the house are independent.)

5.S.6 The probability that a launch of a particular spacecraft will occur on time is .4. Whether a launch occurs on time is independent of whether previous launches occurred on time. If three such launches take place, compute the probability that the first two will occur on time but the third will not.

5.S.7 The Kenmont Carbide Company has two machines that produce a certain product. Machine 1 produces 40% of the product, and machine 2 produces 60%. Machine 1 produces 3% defective products, and machine 2 produces 5% defective products. What is the probability of each of the following?
  a. A randomly selected product is defective.
  b. A product that is found to be defective was made on machine 1.
  c. A product that is found to be good was made on machine 2.

5.S.8 The probability that a candidate passes the bar exam on the first try is .8. Of those who fail on the first try, 70% pass on the second try, and of those who fail on the second try, 88% pass on the third. Calculate the following probabilities:
  a. A candidate does not pass the bar exam after two tries.
  b. A candidate takes three tries to pass the bar exam.

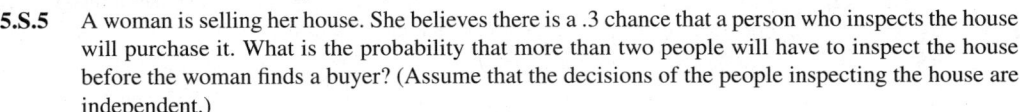

5.S.9 A company has the following data on the age and marital status of 140 employees:

Age	Single	Married
Under 30	77	14
30 or over	28	21

  a. What proportion of employees are single and under the age of 30?
  b. If an employee is under 30, what is the probability that he or she is single?

5.S.10   The following are data from a sample of 80 families in a midwestern city on the record of college attendance by fathers and their oldest sons.

| | SON | |
Father	College	No college
College	18	7
No college	22	33

a. What is the probability that a son attended college given that his father did not attend college?
b. Is attending college by the son independent of whether his father attended college? Explain using probability values.

5.S.11   Assume that we have two mutually exclusive events A and B. Assume further that it is known that $P(A) = .3$ and $P(B) = .4$.
a. What is $P(A \cap B)$?
b. What is $P(A|B)$?
c. A student argues that the concepts of mutually exclusive events and independent events are really the same and that if events are mutually exclusive they must be independent. Do you agree with this statement? Use the probability information in this problem to justify your answer.
d. What general conclusion would you make about mutually exclusive and independent events given the results of this problem?

5.S.12   A market survey of 800 people found the following facts about the ability to recall a television commercial for a product and the actual purchase of the product:

	Recalled	Did not recall	Total
Purchased	150	90	240
Did not purchase	250	310	560
Total	400	400	800

Let R denote the event that a person could recall the television commercial, and let P denote the event that a person purchased the product.
a. Find $P(R)$, $P(P)$, and $P(R \cap P)$.
b. What is the probability that a person who recalled seeing the television commercial actually purchased the product?
c. Are R and P independent events? Use probability values to explain.
d. Does the commercial seem to help sell the product?

5.S.13   A research study investigating the relationship between smoking and heart disease in a sample of 1,000 men over 50 years of age provided the following data:

	Smoker	Nonsmoker	Total
Heart disease	100	80	180
No heart disease	200	620	820
Total	300	700	1,000

a. For the entire population, estimate the probability that a man over 50 years of age is a smoker and has a record of heart disease.

    b. Given that a man over 50 years of age is a smoker, what is the estimated probability that he has heart disease?

    c. Given that a man over 50 years of age is a nonsmoker, what is the estimated probability that he has heart disease?

    d. Does the research indicate that heart disease and smoking are independent events? Explain.

5.S.14 In the setup of a manufacturing process, the probability that a machine is correctly adjusted is .98. When correctly adjusted, the machine operates with a 4% defective rate. However, if it is incorrectly adjusted, the defective rate is 75%.

    a. After the machine starts a production run, what is the probability that a defect is observed when one part is tested?

    b. Suppose that the part selected by an inspector is found to be defective. What is the probability that the machine is incorrectly adjusted?

5.S.15 Sixty percent of the graduates of a driver training school pass their driver's license test on the first attempt and the other 40% fail. The school gives a pretest to graduates before they take the official test. Of the graduates who pass the official test on the first attempt, 85% passed the pretest. Of the graduates who fail the official test on the first attempt, 10% pass the pretest. Let A_1 denote that a graduate passed the official test on the first attempt and A_2 that the graduate failed, and let B denote that the graduate passed the pretest.

    a. Find the following probabilities: $P(A_1)$, $P(A_2)$, $P(B|A_1)$, and $P(B|A_2)$.

    b. Suppose that a student has passed the pretest. Obtain the posterior probabilities $P(A_1|B)$ and $P(A_2|B)$ and interpret them. Does the information provided by the pretest lead to a substantial modification of the prior probabilities? Discuss.

5.S.16 The probability of an accident on any day at a large manufacturing plant is .05.

    a. What is the probability that on 5 successive days no accidents occur?

    b. What is the probability that an accident occurs on at least 1 of 3 days?

5.S.17 An investment broker has provided all clients with a list of several possible independent investments, each having a probability of .7 of yielding a substantial profit. A client, Mr. McManus, decides to purchase four of the investments. What is the probability that all four yield a profit?

5.S.18 In Exercise 17, what is the probability that Mr. McManus's first four investments are not profitable?

5.S.19 A bank vault has a computerized lock that opens when three numbers are selected in sequence. The same number can be repeated three times as a possible solution, and each of the three numbers is an integer from 1 to 25. The bank manager has forgotten the correct sequence and decides to try all the possible combinations (actually, they are permutations!) until the vault opens. How many possibilities are there?

5.S.20 The Bollman Brewing Company has a new brand of beer and wishes to begin an intense advertising campaign. It has produced eight different commercials and purchased four advertising spots during a televised football game. The company must now decide which four commercials to use and in which order. How many arrangements of four commercials are possible?

5.S.21 The secretaries at the Metro Life Insurance Company file insurance policies by classifying a person according to age, gender, and marital status. Age is classified as under 25 years, 25 to 50 years, and over 50. Marital status is single or married. What is the total number of classifications?

5.S.22 In a particular city, 45% of all households have a color television and 22% have a microwave oven. If 15% of all households have both items, what is the probability that a randomly chosen household will have at least one of them?

5.S.23 Candidates for employment at a large corporation must go through two initial screening procedures, a written aptitude test and an oral interview. Fifty percent of the candidates are unsuccessful on the written test, 30% are unsuccessful in the interview, and 11% are unsuccessful in both. The corporation gives further consideration only to candidates who are successful in both procedures.

a. What is the probability that a randomly chosen candidate will receive further consideration for employment?

b. What is the probability that a candidate who is unsuccessful in the written test will be successful in the interview?

c. What is the probability that a candidate who is successful in the interview will also be successful on the written test?

d. Are successes in the two screening procedures independent of one another?

5.S.24   A bank classifies borrowers as "high risk" or "low risk," and 15% of its loans are made to those in the "high risk" category. Of all its loans, 5% are in default, and 40% of those in default are to high-risk borrowers. What is the probability that a high-risk borrower will default?

5.S.25   Households in a certain city were surveyed to determine whether they would subscribe to a cable television service. The households were classified as "low," "middle," or "high" income. The table gives the proportions of households in each of the six joint classifications.

Income	Will subscribe	Will not subscribe
Low	.03	.17
Middle	.14	.46
High	.10	.10

a. Find the probability that a randomly chosen household would subscribe.

b. Are income and willingness to subscribe independent?

c. Given that a high-income family has been contacted, find the probability that the household will subscribe.

5.S.26   There are many consulting organizations that predict whether increases in stock prices will be unusually low, unusually high, or normal. Before deciding whether to continue purchasing these forecasts, a stockbroker compares past predictions with actual outcomes. The accompanying table shows the past performance of a certain consultant.

	PREDICTION		
Outcome	Unusually high	Normal	Unusually low
Unusually high	.20	.14	.04
Normal	.07	.22	.06
Unusually low	.02	.05	.20

a. What proportion of predictions were for unusually high increases?

b. What proportion of outcomes were for unusually high increases?

c. When the consultant predicted unusually high increases in stock prices, what was the probability that he was correct?

d. Given that there were unusually high increases, what was the probability that the consultant had predicted unusually low increases?

5.S.27   A publisher sends advertising material for an economics text to 75% of all professors teaching the appropriate economics course. Thirty percent of the professors who received this material adopted the book, and 8% of the professors who did not receive the advertising material adopted the book.

a. What proportion of all professors adopted the book?

b. What is the probability that a professor who adopts the book has received the advertising material?

5.S.28 A professor finds that she awards a final grade of A to 22% of her students, and of these students, 72% got an A on the midterm examination. Also, 12% of students who did not get a final grade of A earned an A on the midterm exam. What is the probability that a student with an A on the midterm examination will obtain a final grade of A?

5.S.29 A company based in Alaska has found that 30% of the employees who come to Alaska from California have difficulty adjusting to the Alaskan climate and quit within 2 years of moving. Four California residents have recently been hired.
a. Find the probability that all four quit within 2 years. Assume that all decisions to stay or quit are independent of one another.
b. What is the probability that none of the four employees quits within 2 years?
c. What is the probability that at least one of the four employees quits within 2 years?

5.S.30 Four machines, labeled A, B, C, and D, are used to produce golf balls. Of the total output, machine A produces 10%, B produces 20%, C produces 30%, and D produces 40%. Of the golf balls produced by machine A, 1% are defective; by machine B, 2%; by machine C, 3%; and by machine D, 4%.
a. What proportion of golf balls are defective?
b. Given that a ball is defective, what is the probability that machine D made it?
c. Given that a ball is defective, what is the probability that machine A did not make it?

5.S.31 Suppose that when certain geological conditions exist, there is a 20% chance of striking oil. A drilling company finds five independent locations where these geological conditions exist and drills one well at each location. If five wells are drilled, what is the probability that at least one of them will strike oil?

5.S.32 On an assembly line, an inspector checks every article and destroys those that are defective, passing the remaining articles to a second inspector. The first inspector catches 70% of the defectives and destroys them. The second inspector catches and destroys 80% of the defectives that he or she inspects. What percentage of the defectives will get by both inspectors?

5.S.33 A sales representative has a 30% probability of making a sale to any customer. Assume all customers are independent of one another.
a. What is the probability that the first sale is made to the third customer?
b. What is the probability that the first four customers do not purchase the product?

5.S.34 The sales representative at a garden center claims that about 40% of all newly planted apple trees eventually bear fruit. How many trees should be planted so that there is at least a 90% probability of having at least one apple tree that bears fruit?

5.S.35 A basketball promoter has decided to hold a holiday tournament at the University of Kentucky involving eight highly ranked college teams. The promoter hopes that Kentucky and Louisville eventually play each other in the final game so that television rights can be sold for a handsome profit. Because Louisville and Kentucky are in opposite halves of the draw, they cannot play each other before the final game. To reach the finals, a team must win its two preceding games. The probability that Louisville will win its opening game is .7; given that it wins its first game, the probability that Louisville wins its second game is .4. The probability that Kentucky wins its opening game is .8; given that it wins its first game, the probability that Kentucky wins its second game is .5. What is the probability of each of the following?
a. Louisville reaches the finals.
b. Kentucky reaches the finals.
c. Kentucky plays Louisville in the finals.
d. Neither Kentucky nor Louisville plays in the finals.

5.S.36 The Peterson Lumber Company specializes in the production of wood paneling for game rooms. Mr. Peterson knows from past experience that about 10% of the panels contain minor flaws, 5% contain major flaws, and 2% contain both types of flaws. A customer orders three panels selected randomly. Find the probability of the following:
  a. The first panel is nondefective.
  b. None of the panels contain a defect.
  c. At least one of the panels contains a major defect.

5.S.37 The Gruver Glass Company produces thermal pane glass windows in three states. The Ohio plant produces 50% of the output, the Kentucky plant produces 30%, and the Missouri plant produces 20%. Of the windows produced in Ohio, 1% are defective; in Kentucky, 3%; and in Missouri, 2%.
  a. What proportion of all windows are defective?
  b. Given that a window is defective, find the probability that it was produced at the Kentucky plant.

5.S.38 Approximately 80% of all new cars sold in the United States have air conditioning and 40% have power windows.
  a. Determine the maximum possible percentage of these cars that have both air conditioning and power windows.
  b. Determine the minimum possible percentage of these cars that have both air conditioning and power windows.
  c. Determine the maximum possible percentage of these cars that have neither air conditioning nor power windows.

5.S.39 Last year an investment advisor picked four stocks as good investments. As it turned out, although 75% of the stocks on the New York Stock Exchange increased in price that year, all four of this advisor's stocks went down. If a monkey had thrown four darts at the financial pages at the beginning of the year, what is the probability that all four picks would have gone down in price?

5.S.40 A team of experts is examining the reliability of lie detector tests. To determine the probability that a lie detector test is correct, a lie detector test is given to a group of students. Unknown to the experts, 80% of the students are told to be truthful and 20% are told to lie occasionally. At the conclusion of the test, the experts judged 85% of the truthful students as truthful and 60% of the liars as liars.
  a. If the lie detector test tells the expert that the person is truthful, what is the probability that the person is a liar?
  b. If the test indicates that the person is a liar, what is the probability that the person is truthful?

5.S.41 In a laboratory experiment, pigeons were trained to push one of two buttons to receive food. The experiment was set up so that button A would give a reward 70% of the time and no reward 30% of the time. Button B was constructed to give a reward 30% and no reward 70% of the time. The pigeons eventually began choosing button A 70% of the time and button B 30% of the time. When a group of rats was offered the same sort of options, the rats soon began choosing button A every time. Did the pigeons or rats act more intelligently? Prove your answer.

5.S.42 In marketing research and other survey fields, overcoming nonresponse to survey questions and questionnaires is an important problem. With telephone surveys or home interviews, the problem arises from absence from home at the time of the call or visit. With mail surveys, refusal to respond to the survey is a major problem. Suppose a telephone survey is conducted and a market researcher believes that a respondent who is home at the time of the call will answer all questions with probability .90; that is, about 10% of the respondents refuse to answer any questions. Further, it is believed that the probability that a given person will be found at home is .55. Given this information, what proportion of the interviews will be successfully completed?

5.S.43 The calculation of "exposure probabilities," which reflect the proportion of the population exposed to a specific advertisement, is important in the evaluation of advertising efforts. Suppose the probability that a consumer will be exposed to an advertisement for a certain product by seeing a commercial on television is .23. Suppose the probability that the consumer will be exposed to the product by hearing an advertisement on the radio is .04. Assume that the two events, Being exposed to the TV commercial and Being exposed to the radio commercial, are independent.

a. What is the probability that the consumer will be exposed to both ads?

b. What is the probability that he or she will be exposed to at least one of the ads?

5.S.44 Suppose an electronic device contains four independent components, each having a reliability of .95. The device works only if all four components are functional. What is the probability that the device will work?

5.S.45 Suppose a company sells magazine subscriptions by telephone. From experience, it is known that the probability of making a sale during any call is approximately .03. Suppose a salesperson makes 15 calls during an hour. What is the probability that at least one subscription will be sold? What assumption did you make to solve the problem?

5.S.46 In statistical quality control, one way of looking for problems in the production process is by searching for unusual patterns in the sequence of output. Suppose that the diameters of machine screws are being monitored. Supposed eight machine screws are selected as they come off the production line. If the process is in control, successive diameters should be independent of one another. If the process is in control, what is the probability that all eight diameters will come from the top quartile of the population distribution? If this event actually occurred, would you assume that the process had gone out of control?

5.S.47 Refer to the previous exercise. What is the probability that at least one of the eight diameters will come from the bottom quartile of the distribution?

5.S.48 In the Massachusetts lottery, six numbers are selected without replacement from the 36 numbers 1 to 36. To win, a person must have the correct six numbers drawn, regardless of their order. The jackpot depends on the number of players and is usually worth several million dollars. What is the probability of winning the jackpot?

5.S.49 A construction firm has a .75 probability of winning a contract to build a power plant in India if a major competitor does not bid for the contract, and a .25 probability of concluding the deal if the competitor does bid for it. It is estimated that the probability that the competitor will submit a bid for the contract is .45. What is the probability of winning the contract?

5.S.50 While searching for oil in Australia, an oil explorer orders seismic tests to determine whether oil is likely to be found in a certain drilling area. The following data summarize the past results concerning test reliability: When oil does exist in the testing area, the test will indicate so 85% of the time; when oil does not exist in the testing area, the probability is .03 that the test will erroneously indicate that oil does exist. Preliminary exploration by geologists indicates that the probability of existence of an oil deposit in the test area is .45. Suppose a seismic test is conducted and indicates the presence of oil. What is the probability that an oil deposit really exists?

5.S.51 In statistical quality control, it is important to be on the lookout for changes in the production process that could lead to damaged or defective output. When computer components are produced, each unit of output must be virtually identical to every other unit. One way of detecting changes in uniformity of output is to look for trends in the sequence of output. Recall that the median is that number such that half the observations lie above it and half the observations lie below it. Suppose a sequence of eight items is selected from some production process and some key variable is measured. What is the probability that all eight items lie above the median? If you observed such an event, would you suspect that the process had gone out of control?

5.S.52 The military frequently buys large quantities of goods from a single supplier. Sampling inspection schemes are efficient and inexpensive methods of judging the quality of a shipment of goods based on checking only a sample of the shipment. The supply of goods is accepted or rejected on the basis of examining a few sample items. From past experience, we know that a certain government inspector has correctly accepted 99% of all good products that were tested. The inspector has incorrectly accepted 2% of the defective products that were inspected. Suppose it is known that, on the average, about 3% of all products are defective.

a. Find the probability that a given product is rejected.
b. Find the probability that a given product is good and that it is accepted.
c. Find the probability that a product is accepted, given that it is of inferior quality.
d. Suppose a given product is rejected. Find the probability that the product was not defective.

5.S.53   A company that produces pump jacks for oil wells uses motors from two sources. Supplier A provides 68% of the motors, and supplier B provides the remaining 32% of the motors. Past experience indicates that about 1% of the motors provided by supplier A are defective, and 3% of the motors provided by supplier B are defective. Suppose a pump jack fails in the field because of a defective motor. What is the probability that this motor was provided by supplier B?

5.S.54   A businesswoman drives to work on Route 33 or Route 47, with probabilities .2 and .8, respectively. When she takes Route 33, she is late on 4% of the days. When she takes Route 47, she is late on 10% of the days. What proportion of days will the businesswoman be late?

 5.S.55   Assume that stock prices are efficient in the sense that the current price of a stock fully reflects all available information concerning the stock and its potential future returns. If the market is efficient, then, in the absence of new information, the price change today for a stock should be independent of yesterday's price change. That is, the prices should follow a "random walk." The accompanying data show the price change today (up or down) and the price change yesterday (up or down) for a sample of 100 stocks.

Today's Price Change	YESTERDAY'S PRICE CHANGE		
	Up	**Down**	**Total**
Up	44	24	68
Down	18	14	32
Total	62	38	100

a. Find the probability that the price of a randomly selected stock was up today, i.e., find P(Up today).
b. Find P(Up yesterday).
c. Find the joint probability P(Up today and Up yesterday).
d. Find the conditional probability P(Up today|Up yesterday).
e. Determine whether the events Up today and Up yesterday are independent of one another.
f. If you knew that a stock's price was up yesterday, what would you predict for today?

 5.S.56   Five percent of the home buyers who get their mortgages through First Fidelity National default on their loans. Of those who default, 60% received a "good credit risk" rating from a loan officer. Of those who did not default, 90% received a "good credit risk" rating.

a. Given that a randomly selected home buyer has a "good credit risk" rating, find the probability of default.
b. Does the credit rating help in determining the home buyers who will default? Explain.

5.S.57   The following is a summary of an actual event that occurred in 1964. A woman was mugged in an alley. Shortly thereafter, a witness saw a blond woman with a ponytail run away and jump into a yellow car driven by a bearded black male with a mustache. Later, the police arrested Janet and

Malcolm Collins, a married couple who fit the description given by the witness and who owned a yellow Lincoln. At the trial, a mathematician used the following probabilities:

$P(\text{Blond}) = .33$ $P(\text{Yellow car}) = .10$

$P(\text{Ponytail}) = .10$ $P(\text{Mustache}) = .25$

$P(\text{Interracial couple}) = .001$ $P(\text{Bearded black male}) = .10$

a. The mathematician assumed that the six events were independent of one another. If we make this assumption, find the probability of all six events occurring simultaneously.

b. Based on the result in part **a,** the jury convicted the Collinses. Later, the California Supreme Court overturned the verdict. It was learned that the mathematician had assumed the individual probabilities, and there was no evidence at trial that any of the individual probabilities were even approximately correct. Also, there was no evidence at trial that the events were independent. For example, the facts that the black male had a beard and the woman had blond hair were not independent of the fact that they were an interracial couple. This court case is very famous and is frequently cited to show how people misused probability theory. Do you think the evidence was strong enough to support a conviction? What is your opinion concerning the overturning of the verdict?

5.S.58 Communities along the Mississippi River frequently suffer major damage when the river floods. Suppose a farmer believes that he can earn a profit as long as there is not a flood during the next 5 years. Historically, there has been a flood about once every 6 years. Assume that the occurrences of floods during different years are independent events. Find the probability that no flood occurs during the next 5 years.

5.S.59 Experts have claimed that the probability that a tornado will strike a particular community during any given year is .01.

a. Find the probability that no tornado occurs during the next 15 years.

b. Find the probability that at least one tornado strikes during the next 15 years.

References

Feller, William. *An Introduction to Probability Theory and Its Applications.* Vol. 1. 3d ed. New York: Wiley, 1968.

Heron House, eds. *The Odds on Virtually Everything.* New York: Putnam's, 1980.

Huff, Darrell. *How to Take a Chance.* New York: Norton, 1959.

Huff, Darrell, and Irving Geis. *How to Lie with Statistics.* New York: Norton, 1954.

Neter, J., W. Wasserman, and G. A. Whitmore. *Applied Statistics.* 3d ed. Boston: Allyn & Bacon, 1987.

Summers, G. W., W. S. Peters, and C. P. Armstrong. *Basic Statistics in Business and Economics.* 4th ed. Belmont, California: Wadsworth, 1985.

6 Discrete Probability Distributions

In this chapter we discuss the properties of discrete random variables, including how to construct a discrete probability distribution and how to calculate the mean, variance, and standard deviation of a discrete random variable. A probability distribution is the theoretical or analytical counterpart of the relative frequency distribution discussed in Chapter 3. As described in Chapter 4, the mean is a measure of the center of a distribution, and the variance and standard deviation are measures of a distribution's spread.

6.1 Random Variables

As we saw in Chapter 5, the sample space of an experiment may or may not consist of a set of real numbers. When the variable studied is a qualitative variable, the sample space consists of the possible values or attributes that the variable can have. In this chapter, however, we are especially interested in problems in which the variable studied is a quantitative variable, which takes on values that are real numbers.

Sometimes a qualitative variable is converted into a quantitative variable by assigning a numerical value or code to each qualitative category. For example, suppose that an adult male is selected randomly and classified as employed or unemployed; the employment status of the individual male is thus a qualitative variable. This variable can be transformed to a quantitative variable by using 0 to represent Unemployed and 1 to represent Employed. Thus, even when the outcomes of an experiment are not real numbers, we can think of a process (or function) that associates some real number with each possible outcome.

DEFINITION   **Random Variable** ————————————————————————————

A variable X is a **random variable** if the value that X assumes at the conclusion of an experiment is a chance or random occurrence that cannot be predicted with certainty in advance.

Suppose we select a random sample of 30 adults and ask each individual whether he or she is employed. The number of individuals in the sample who are employed is a random variable X that can take any of the integer values from 0 to 30. Each of these values corresponds to a particular experimental outcome, where the experiment consists of selecting a sample of 30 adults and noting how many are employed. The variable X is a random variable because the value that X assumes cannot be predicted with certainty before performing the experiment.

The set of possible values $\{x_1, x_2, \ldots, x_k\}$ assumed by the random variable X is called the *sample space of the random variable X*. In the definition, it has been assumed that the sample space contains a finite number of elements, but this is not always the case. The sample space may contain an infinite but countable number of outcomes, or the sample space may contain all values in some interval.

DEFINITIONS **Discrete and Continuous Random Variables**

A random variable X is **discrete** if X can assume only a finite or countably infinite number of different values. A random variable X is **continuous** if it can assume all the values in some interval.

By *countably infinite*, we mean that the possible values of X can be put in one-to-one correspondence with the positive integers. *Countable* means that you can count the possible values that the variable can take. The total number can be infinite because we can conceive of the count as proceeding forever without end.

EXAMPLE 6.1 **Discrete Random Variables**

a. X = Number of days of rainfall in New York City in May (possible values of X are 0, 1, 2, ..., 31)
b. X = Number of defective parts in a sample of 1,000 parts (possible values of X are 0, 1, 2, ..., 1,000)
c. X = Number of highway deaths on the Fourth of July (possible values of X are 0, 1, 2, ...)

EXAMPLE 6.2 **Continuous Random Variables**

a. X = Height of an individual
b. X = Time required for an airplane to travel from Chicago to New York
c. X = Distance traveled by a truck driver in a given month
d. X = Diameter of a ball bearing
e. X = Volume of gasoline in a storage tank

Measuring a continuous random variable is always problematic because, at least conceptually, a continuous variable can take all values in an interval. Thus we cannot measure, say, the diameter of a ball bearing exactly; similarly, we cannot exactly measure heights, weights, times, distances, volumes, and so forth. Consequently, there is always a certain amount of approximation and round-off error in measuring a continuous variable.

In this chapter, we discuss discrete random variables and their properties. In Chapter 7, we discuss some special discrete probability distributions, and in Chapter 8, we discuss continuous random variables.

It is common to use uppercase letters (e.g., X, Y, and Z) to denote random variables and lowercase letters (e.g., x, y, and z) or lowercase subscripted letters (e.g., x_1, x_2, and

x_3) to denote particular values assumed by random variables. The expression $P(X = x_i)$ denotes the probability that a random variable X assumes the value x_i. Similarly, the expression $P(X = x)$ denotes the probability that the random variable X assumes the specific value x. Frequently the notation $P(X = x)$ will be abbreviated to $P(x)$, where it is understood that x is a specific value of the random variable X.

Exercises

Statistical Concepts

6.1.1   Define random variable.

6.1.2   Describe a random variable for which the sample space contains an infinite number of possible values.

6.1.3   Describe the difference between a discrete and a continuous random variable.

6.1.4   Identify the following as discrete or continuous random variables:
   a. The number of loan applications received by a savings and loan during a particular month
   b. The length of time for an employee to fill out a business report
   c. The number of defects on a roll of coated film

6.1.5   Which of the following describe continuous random variables, and which describe discrete random variables?
   a. Number of houses sold by a real estate developer during a given week
   b. Quantity of natural gas used per month for heating an apartment building
   c. Volume of milk in a quart container
   d. Number of accidents per week at a manufacturing plant
   e. Number of people per day who report for work at a manufacturing plant
   f. Number of errors found in an audit of a company's financial records
   g. Length of time a customer waits for service at a supermarket checkout counter
   h. Number of automobiles recalled by General Motors next year

6.1.6   Which of the following random variables are discrete and which are continuous?
   a. X = Income of a secretary
   b. X = Number of heads obtained in 10 tosses of a coin
   c. X = Weight of a child at birth
   d. X = Time required to run 100 yards
   e. X = Weight of fluid in a bottle
   f. X = Number of Democrats in a certain precinct

Statistical Applications

6.1.7   A company drills three oil wells at an expense of $10,000 each. Any well that strikes oil will produce $25,000 in revenue, resulting in a profit of $15,000. Let X denote the total profit or loss achieved by the company drilling the wells. Describe the sample space and list all possible values of X.

6.1.8   A student is registered for courses in economics and in math. A new math book costs $18 and a used one costs $10; a new economics book costs $20, a good used one costs $15, and a worn one costs $8. The student will buy the cheapest math and economics texts that are still available at the bookstore. Let X denote the total cost of the two books. Describe the sample space and list all possible values of X.

6.2   Properties of Discrete Probability Distributions

When we discuss any discrete random variable X, we usually want to know all possible values of X and the probability with which each possible value will occur. This set of facts is called a *discrete probability distribution* or a *probability function*.

DEFINITION   **Discrete Probability Distribution**

A **discrete probability distribution** is a table, graph, or rule that associates a probability $P(X = x_i)$ with each possible value x_i that the discrete random variable X can assume.

Every discrete probability distribution must satisfy the following rules:

1. No probability can be negative. That is, for any value x,

$$P(X = x) \geqslant 0$$

2. The sum of the probabilities of all the possible values of the random variable X must equal 1. That is,

$$\Sigma \, P(X = x) = 1$$

where the summation is over all possible values of X.

EXAMPLE 6.3   **A Discrete Probability Distribution**

Suppose that a car salesman is trying to sell a new car to a customer who is interested in buying one of three different cars. The selling prices of the cars are $10,000, $15,000, and $20,000, and the salesman gets a 5% commission on every sale. Suppose the salesman believes that there is a 30% chance that the customer will purchase the $10,000 car, a 20% chance that the customer will purchase the $15,000 car, a 10% chance that the customer will purchase the $20,000 car, and a 40% chance that the customer will not buy a car. Let the random variable X denote the salesman's potential commission. Construct the discrete probability distribution of the random variable X.

SOLUTION   Based on a 5% commission, the salesman will receive $0 if the customer does not buy a car, $500 if the $10,000 car is sold, $750 if the $15,000 car is sold, and $1,000 if the $20,000 car is sold. If it is assumed that the customer will not buy two cars, the possible outcomes are $x_1 = \$0$, $x_2 = \$500$, $x_3 = \$750$, and $x_4 = \$1,000$. The sample space of the random variable X is $S = \{\$0, \$500, \$750, \$1,000\}$. The corresponding probabilities are .4, .3, .2, and .1. The probability distribution of the discrete random variable X is shown in Table 6.1 (page 196). This probability distribution can also be drawn as a graph, as illustrated in Figure 6.1.

A discrete probability distribution shows the relative frequency distribution of the population. For a discrete random variable, a relative frequency distribution shows the proportion of times each value occurred in a large number of random experiments. When

TABLE 6.1   *Discrete probability distribution for Example 6.3*

x	P(x)
$ 0	.4
500	.3
750	.2
1,000	.1
	1.0

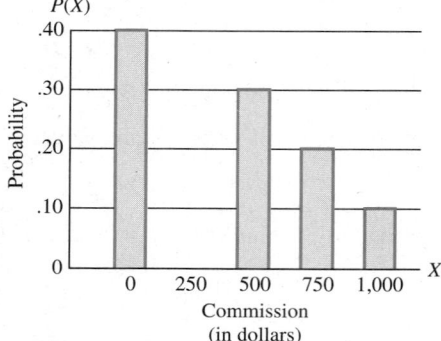

FIGURE 6.1   *Graph of the discrete probability distribution for Example 6.3*

we assign various probabilities to the basic outcomes of an experiment, we are trying to approximate population relative frequencies.

The data in Table 6.1 indicate that 10% of the time the salesman can expect to earn a $1,000 commission, 20% of the time he can expect to earn $750, and so forth. That is, the probabilities in Table 6.1 can be interpreted as predictions of the relative frequencies of occurrence of each value of X in a large number of repeated trials of the experiment, provided all potential customers have the same tendencies.

A discrete probability function like that in Table 6.1 is similar to the relative frequency distributions presented in Chapter 3. The probability function tells us the limiting relative frequency of occurrence of each possible value of X when the random experiment is repeated many times.

Suppose that a random experiment can be repeated a large number of times and that each occurrence of the experiment generates a value of the random variable X. If the discrete random variable X is observed a very large number of times and the values that occur are arranged in a relative frequency distribution, then this relative frequency distribution will be indistinguishable from the probability distribution of the random variable X. Thus, the probability distribution of a random variable is a theoretical model for the relative frequency distribution of a population.

The probability distribution contains all the information about the probability properties of the random variable, and a graph of the probability distribution will also reveal the general characteristics of the probability distribution.

Tips for Problem Solving

1. When constructing a discrete probability distribution, list all possible values of the random variable X and all the associated probabilities.
2. Verify that every probability assumes a value in the closed interval [0, 1].
3. Verify that the probabilities sum to 1.
4. Construct a bar graph showing each value of X and its associated probability.

Exercises

Statistical Concepts

6.2.1 Every discrete probability distribution must satisfy two rules. State them and explain what they mean.

6.2.2 Explain why it is not feasible to describe a continuous probability distribution by constructing a probability table.

6.2.3 Let the random variable X denote the amount that an investor gains or loses from a certain investment. The investor states that her subjective probability distribution is as follows: $P(X = -200) = .40$; $P(X = 400) = .60$. A person claims that this subjective probability distribution is invalid because, in a discrete probability distribution, X can never take a negative value. Is this correct? Explain.

Statistical Drills

6.2.4 Are the following valid probability distributions? If not, explain why.

x	P(X = x)	x	P(X = x)	x	P(X = x)
2	.2	3	−.2	−2	.3
4	.3	5	.5	2	.4
6	.4	6	.6	4	.1
8	.2	9	.1	5	.1

6.2.5 A random variable X has the following probability distribution:

x	P(x)
2	.10
4	.20
5	.15
8	.45
9	.10

a. Verify that this is a valid probability distribution.
b. Find $P(X > 4)$, $P(X \leq 4)$, and $P(X < 4)$.
c. Find $P(X > 5)$, $P(X \geq 5)$, and $P(X \leq 5)$.
d. Graph the probability distribution for X.

Statistical Applications

6.2.6   Suppose 50% of the government employees at a large agency work more than 40 hours per week. A sample of two of these employees is taken. Let X denote the number of employees in the sample who work more than 40 hours per week. Construct and graph the probability distribution of X.

6.2.7   The information that follows shows the distribution of the number of days of sick leave taken by 200 employees during a year:

Days	0	1	2	3	4	5	6	7
Number of employees	20	40	40	30	20	10	10	30

Let X denote the number of days of sick leave taken by an employee.
a. Construct and graph the probability distribution of X.
b. Find $P(X \geq 4)$.
c. Find $P(X \leq 3)$.
d. Find $P(X \geq 3)$.
e. Find $P(3 \leq X \leq 6)$.

6.2.8   A new math book costs $20 and a used one costs $10. A new chemistry book costs $18, a good used one costs $14, and a worn one costs $9. A student wants to buy the cheapest math book and chemistry book available in the bookstore. The probability of getting a used math book is .4, of getting a worn chemistry book is .3, and of getting a good used chemistry book is .2. Let X denote the cost of the two books the student purchases. Assume that the purchases are independent of one another.
a. Construct and graph the probability distribution of X.
b. Find $P(X > \$15)$.
c. Find $P(X > \$25)$.
d. Find $P(\$27 \leq X \leq \$35)$.

6.2.9   An investor makes two independent investments and believes the following: Investment 1 will yield a profit of $1,000 with probability .6 or a loss of $400 with probability .4; investment 2 will yield a profit of $2,000 with probability .1, a profit of $700 with probability .4, and a loss of $500 with probability .5.
a. Construct and graph the probability distribution of X, where X represents the total profit or loss yielded on the two investments.
b. Find $P(X < \$1,600)$.

6.2.10   The Chesler Car Rental Agency claims that 70% of its customers rent compact cars. Suppose four customers rent cars. (Assume the rentals are independent of one another.) What is the probability of the following?
a. All four cars are compacts.
b. At least one car is a compact.

6.2.11   The manager of a retail television outlet thinks that it is equally likely that it will sell zero, one, two, or three televisions in a day. (Suppose that the store has never sold more than three televisions on any day.) Let X denote the number of televisions sold on any given day.
a. Construct a table showing the probability distribution of X.
b. Plot the probability distribution of X.
c. What is the probability that the number of televisions sold on a given day will be less than two?
d. Suppose the outlet has only one salesman, whose income depends on the number of televisions he sells per day. Suppose he receives $10 on the first television sold per day, a $20 commission on the second, and a $45 commission on the third. (Thus, if he sells three televisions in a given day, his commissions will total $75.) Let Y denote the salesman's daily commission. Construct a table showing the probability distribution of Y.

e. What is the probability that the salesman's commission in part **d** on a particular day will exceed $10?

6.2.12 Suppose that a random variable X has the following discrete probability distribution:

x	-2	-1	0	1	2
$P(x)$	.10	.15	.40	.20	.15

a. Find $P(X \leq 0)$.   **b.** Find $P(X > 0)$.
c. Find $P(-1 \leq X \leq 1)$.   **d.** Find $P(X < 2)$.
e. Find $P(-1 < X < 2)$.   **f.** Find $P(X < 1)$.

6.3 Cumulative Distribution Function

For a random variable X, the cumulative distribution function shows the probability that X assumes a value less than or equal to any specific value. The cumulative distribution function of X is denoted by $F(x)$, where x is any specified value.

DEFINITION **Cumulative Distribution Function**

Let X be a discrete or continuous random variable and let x be any real number. The **cumulative distribution function** (CDF) of X is the function

$$F(x) = P(X \leq x)$$

Every cumulative distribution function must satisfy the following rules:

1. For any value x, $0 \leq F(x) \leq 1$.
2. If $x_1 < x_2$, then $F(x_1) \leq F(x_2)$.
3. $F(-\infty) = 0$ and $F(\infty) = 1$.

Tips for Problem Solving

1. To construct a cumulative distribution function for a discrete random variable, construct a table with three columns.
2. In column 1, list the possible values of X in ascending order.
3. In column 2, list the corresponding probabilities.
4. Verify that all the probabilities fall in the closed interval $[0, 1]$.
5. Verify that the probabilities sum to 1.
6. In column 3, list the cumulative probabilities for each value of X. The cumulative probability $F(X = x)$ is the sum of the probabilities for all values of X less than or equal to x. Thus,

$$F(x) = P(X \leq x)$$

Let X be a discrete random variable that can assume the values $x_1, x_2, \ldots, x_n$, where $x_1 < x_2 < \cdots < x_n$. Then $F(x_r)$ denotes the probability that X assumes a value that is less than or equal to x_r and is given by

$$F(x_r) = P(X \leq x_r) = \sum_{i=1}^{r} P(X = x_i)$$
$$= P(X = x_1) + P(X = x_2) + \cdots + P(X = x_r)$$

The CDF for a discrete random variable is a step function. A step (or jump) occurs at each point x_i where $P(X = x_i)$ is positive, and the height of the step is $P(X = x_i)$. If X can assume only the n values $x_1, x_2, \ldots, x_n$, then there are n steps in the CDF. The cumulative distribution function will change values only at those points x_i that can occur with positive probability. For any values a and b such that $b \geq a$, we have

$$P(a < X \leq b) = F(b) - F(a)$$

EXAMPLE 6.4   **A Cumulative Distribution Function** ——————————————

The cumulative distribution function for Example 6.3 is as follows:

$$
\begin{aligned}
F(0) &= P(X \leq 0) &&= .4 \\
F(500) &= P(X \leq 500) &&= .4 + .3 = .7 \\
F(750) &= P(X \leq 750) &&= .4 + .3 + .2 = .9 \\
F(1{,}000) &= P(X \leq 1{,}000) &&= .4 + .3 + .2 + .1 = 1.0
\end{aligned}
$$

Figure 6.2 shows the graph of the cumulative distribution. The vertical axis shows the cumulative probabilities and the horizontal axis shows the values of the random variable X. To interpret the graph, examine the bar above the value $X = 500$. For $X = 500$, we have the cumulative probability $F(X = 500) = P(X \leq 500) = .70$. This implies that in repeated trials under identical conditions, the random variable X would assume a value less than or equal to 500 approximately 70% of the time.

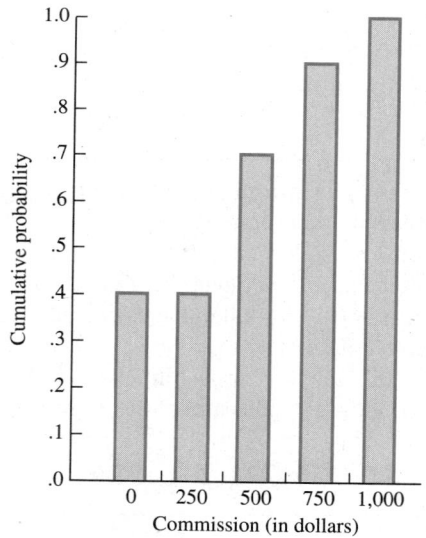

FIGURE 6.2 *Cumulative distribution for data in Example 6.4*

Interpreting Cumulative Probabilities

The *Statistical Abstract of the United States, 1993* (p. 463) contains data concerning the cumulative distribution of family income in the United States. The accompanying table shows the maximum family income received by families at different positions of the income distribution. The incomes are reported as follows:

Position in Distribution	Maximum Income
Lowest fifth	$17,000
Second fifth	$29,111
Third fifth	$43,000
Fourth fifth	$62,991
Lower limit of top 5%	$102,824

We have the following cumulative probabilities:

$$P(X \leqslant \$17,000) = .20$$
$$P(X \leqslant \$29,111) = .40$$
$$P(X \leqslant \$43,000) = .60$$
$$P(X \leqslant \$62,991) = .80$$
$$P(X \leqslant \$102,824) = .95$$

The last cumulative probability is obtained by noting that, if 5% of families have incomes exceeding $102,824, then 95% of families have incomes less than or equal to $102,824.

Exercises

Statistical Concepts

6.3.1   Every cumulative probability distribution must satisfy three rules. State them and explain what they mean.

6.3.2   Explain why it is not feasible to describe a continuous probability distribution by constructing a cumulative probability table.

6.3.3   Show how to express the cumulative probability distribution of a discrete random variable as a graph. Explain why the values of $F(x)$ begin at 0 and rise to a maximum of 1.

6.3.4   *True or false:* $P(X \leqslant x) \geqslant P(X < x)$. Explain.

6.3.5   *True or false:* $P(X = x) = P(X \leqslant x) - P(X < x)$. Explain.

6.3.6   *True or false:* $P(X \leqslant x) - P(X < x) = 0$. Explain.

Statistical Drills

6.3.7   For each of the following discrete probability distributions, construct the cumulative probability distribution and graph the distribution:

x	P(X = x)	x	P(X = x)	x	P(X = x)
2	.2	3	.2	−2	.3
4	.3	5	.5	2	.4
6	.4	6	.2	4	.2
8	.1	9	.1	5	.1

6.3.8   For each of the distributions in Exercise 7, find $P(X \leqslant 5)$.

6.3.9   For each of the distributions in Exercise 7, find $P(X < 5)$.

6.3.10   For each of the distributions in Exercise 7, find $P(X > 5)$.

6.3.11   Let X have the following probability distribution:

x	2.0	4.0	6.0	8.0	10.0
$P(x)$	.05	.20	.35	.30	.10

a. Find $P(X \leqslant 6.0)$.
b. Graph the cumulative probability distribution.
c. Find $P(X > 6.0)$.

6.3.12   A random variable X has the following probability distribution:

x	10	20	30	40	50
$P(x)$	.1	.3	.4	.1	.1

a. Find $P(X \leqslant 30)$.
b. Graph the cumulative probability distribution.
c. Find $P(X > 20)$.

Statistical Applications

6.3.13   A credit card company examines its records to determine the length of time (in weeks) between the time a bill is sent and payment is received. The following table shows the results:

Number of weeks	1	2	3	4	5
Proportion	.35	.20	.20		.05

a. Find the missing proportion.
b. Graph the probability distribution.
c. Calculate and graph the cumulative distribution function.
d. For a randomly chosen bill, what is the probability that payment will be received within three weeks from the time the bill is sent out?

6.3.14   In a study to investigate traffic patterns, the numbers of occupants in cars on an interstate highway were counted. The following table shows the proportions for all cars:

Number of occupants	1	2	3	4	5	6
Proportion	.40	.36	.10	.11	.02	.01

a. Find the missing proportion.
b. Graph the probability distribution.
c. Calculate and graph the cumulative distribution function.
d. What is the probability that a randomly chosen car will have at least four occupants?

6.4   Expected Value of a Discrete Random Variable

In Chapter 4, we saw that the population mean and the population variance are useful population parameters that help us describe and compare different quantitative populations. In the same manner, the mean and variance of a random variable X help us describe the characteristics of its distribution and compare it with other random variables.

The *expected value,* or *mean,* of a random variable is used as a measure of the center of the probability distribution. The expected value of a discrete random variable is the weighted average of the possible values of the random variable where the weight assigned to x_i is the probability $P(X = x_i)$.

Throughout the remainder of this text, we will occasionally use the notation $P(x)$ instead of the more detailed notation $P(X = x)$. That is, $P(x)$ denotes the probability that the random variable X assumes the specific value x.

DEFINITION **Expected Value, or Mean, of a Discrete Random Variable** ———————

The **expected value**, or **mean**, of a discrete random variable X is denoted by the symbol $E(X)$ or the symbol μ_X. The expected value of X is

$$E(X) = \mu_X = \Sigma \, xP(x)$$

where the sum is taken over all possible values of X.

When there is no danger of confusion, we will occasionally use the symbol μ instead of μ_X to denote the expected value of X. The mean is called the *expected value* because it denotes the average value that we would expect to occur if the experiment were repeated a large number of times.

EXAMPLE 6.5 **Expected Value of a Random Variable** ———————————

Find the expected value for the salesman's commission in Example 6.3.

SOLUTION   Suppose the situation described for the salesman occurred day after day. Thus, on 40% of the days the salesman would receive no commission, on 30% of the days the salesman would receive $500, on 20% of the days the salesman would receive $750, and on 10% of the days the salesman would receive $1,000. The calculations to determine the salesman's expected commission are shown in Table 6.2.

TABLE 6.2 *Calculation of expected commission in Example 6.3*

x	$P(x)$	$xP(x)$
$ 0	.4	$ 0
500	.3	150
750	.2	150
1,000	.1	100
	1.0	$400

In the third column, we calculate the weighted value $xP(x)$ for each value of X. The sum of these values is the mean, or the expected value, of X. We obtain

$$
\begin{aligned}
E(X) &= \Sigma \, xP(x) \\
&= (0)(.4) + (500)(.3) + (750)(.2) + (1,000)(.1) \\
&= \$400
\end{aligned}
$$

This result means that, over a long period of time, the salesman would average about $400 per day in commissions.

Some Applications of Expected Value

In many problems in business and economics, the random variable X measures the possible profit or loss of an investment or gamble. Then $E(X)$ measures the amount that we can expect to win or lose by undertaking the investment or gamble. A negative $E(X)$ implies that, in the long run, we can expect to lose money by participating in similar investments or gambles.

EXAMPLE 6.6

Expected Profit from an Investment

The L. M. Corporation purchases old, run-down buildings, remodels them, and then sells them. The corporation has the opportunity to purchase a building for $100,000, which will cost $60,000 to remodel. The manager thinks that there is a 50% chance that the remodeled building will sell for $120,000, yielding a $40,000 loss. (We are ignoring interest costs to keep the problem simple.) However, there is a 20% chance that the building will sell for $180,000, yielding a $20,000 profit, and a 30% chance that the building will sell for $230,000, yielding a $70,000 profit. Calculate the expected profit or loss from buying, remodeling, and selling the building.

S O L U T I O N   The data are shown in Table 6.3, where the random variable X denotes the profit (loss is shown as negative profit). The expected profit is $E(X) = \$5,000$. Because the expected profit is positive, the corporation could expect to earn about $5,000 per project on the average by undertaking a large number of similar projects.

T A B L E 6.3   *Calculation of expected return on investment in Example 6.6*

x	$P(x)$	$xP(x)$
− $40,000	.5	− $20,000
20,000	.2	4,000
70,000	.3	21,000
	1.0	$5,000

In almost all gambling situations, the gambler's expected profit is negative. This means that the gambler will lose money if he or she plays the game long enough. If the gambler's expected profit is $0, then the game is a fair game and the gambler can expect to break even in the long run. Generally, there are no gambling situations at casinos where the gambler's expected profit is positive. If such a gamble existed, the casino would be subsidizing the gambler!

EXAMPLE 6.7

Expected Gain or Loss from Playing Roulette

Roulette is a gambling game that is popular in Las Vegas and Atlantic City. A roulette wheel contains 38 slots, numbered 00, 0, 1, 2, . . . , 36. Eighteen of the slots are red, 18 are black, and 2 (0 and 00) are green. A ball is dropped onto the wheel while it is spinning. Suppose you bet $1 that the ball will stop in a red slot. If it does, you win $1; if it stops in a black slot, you lose $1; if it stops in a green slot, you lose $.50. Calculate the expected gain or loss from betting $1 on red in roulette.

S O L U T I O N   Let X denote the payoff from betting $1 on red. There are 38 equally likely possibilities for the slot where the ball will stop. Thus, each slot has probability $\frac{1}{38}$. There are 18 red slots, so the probability that the ball stops in a red slot is $\frac{18}{38}$. The probability that the ball stops in a black slot is $\frac{18}{38}$, and the probability that the ball stops in a green slot is $\frac{2}{38}$. The calculation of the expected gain is shown in Table 6.4.

TABLE 6.4   *Calculation of expected return for roulette problem in Example 6.7*

x	P(x)	xP(x)
$ 1.00	18/38	$ .474
− .50	2/38	− .026
− 1.00	18/38	− .474
	1.00	− .026

The expected gain is

$$E(X) = -.026$$

This means that, in the long run, playing roulette will result in a loss of 2.6 cents per $1 bet. If you play the game 1,000 times and bet $1 on red each time, you can expect to be $26.00 behind.

Expected Values of Functions of X

Let X be a random variable and let $g(X)$ denote some function of X, such as X^2, $5X$, $(X - 5)^2$, and so on. At times we may be interested in finding the expected value of $g(X)$. We do this by using the rule on the expected value of a function of X.

DEFINITION   **Expected Value of a Function of X** ─────────────────────────

Let X be a discrete random variable, and let Y be any function of X such that $Y = g(X)$. Then the **expected value** of Y, or the expected value of $g(X)$, is

$$E(Y) = E[g(X)] = \Sigma\ g(x)P(x)$$

where the sum is over all values of X.

EXAMPLE 6.8   **Expected Value of a Function of X** ─────────────────────────

Suppose a wealthy gambler bets $100 on the red in roulette. Find the gambler's expected gain or loss. Refer to Example 6.7.

SOLUTION   Define the random variable $Y = g(X) = 100X$, where X is the random variable described in Example 6.7. To calculate the expected value of Y, we weight each possible value of Y by its probability of occurrence. The expected value of Y, calculated in Table 6.5, is $E(Y) = -\$2.63$. The expected gain is negative, which means the gambler can expect to lose $2.63 per $100 bet.

TABLE 6.5   *Calculation of expected return from roulette problem in Example 6.8*

x	y = g(x) = 100x	P(x)	g(x)P(x)
$ 1.00	$ 100	18/38	$ 47.37
− .50	− 50	2/38	− 2.63
− 1.00	− 100	18/38	− 47.37
		1.00	− $2.63

Tips for Problem Solving

To find the expected value of a discrete random variable X, construct a table that has three columns. The first column shows all the values of X, the second shows all the probabilities $P(X = x)$, and the third shows the products $xP(X = x)$. The sum of the values in column 3 is the expected value of X.

Exercises

Statistical Concepts

6.4.1   *True or false:*   Since probabilities must be nonnegative, the expected value of a discrete random variable must be nonnegative. Explain.

6.4.2   *True or false:*   The expected value of a discrete random variable can never be 0. Explain.

6.4.3   *True or false:*   The expected value of a discrete random variable can be thought of as the weighted average of the possible values of X, using the appropriate probabilities as weights. Explain.

6.4.4   *True or false:*   Suppose the discrete random variable X takes values such that $5 \leqslant X \leqslant 12$. Then, $5 \leqslant E(X) \leqslant 12$. Prove your answer mathematically and by drawing a graph.

Statistical Drills

6.4.5   For each of the following discrete probability distributions, find $E(X)$ and locate $E(X)$ on the graph of the probability distribution.

x	P(X = x)	x	P(X = x)	x	P(X = x)
2	.2	3	.2	−2	.3
4	.3	5	.5	2	.4
6	.4	6	.2	4	.2
8	.1	9	.1	5	.1

6.4.6   For each of the distributions in Exercise 5, let $Y = 10X$. Graph the probability distribution of Y. Find $E(Y)$ and locate it on the graph. Explain what happened to the expected value of Y relative to the expected value of X.

6.4.7   For each of the distributions in Exercise 5, let $Y = 10X + 10$. Graph the probability distribution of Y. Find $E(Y)$ and locate it on the graph.

Statistical Applications

6.4.8   Consider the following probability distribution for the random variable X:

x	−2	−1	0	1	2
P(x)	.08	.12	.40	.30	.10

a. Find $E(X)$.
b. Graph $P(x)$ and locate $E(X)$ on the graph.

6.4.9   A roulette wheel at a Las Vegas casino has 18 red numbers, 18 black numbers, and 2 green numbers. Assume that a \$5 bet is placed on the black numbers. If a black number comes up, the player wins \$5; if a green number comes up, the player loses \$2.50; otherwise, the player loses \$5.
a. Let X be a random variable indicating the player's net winnings on one bet. Find the probability distribution for X.

   b. What are the expected winnings? Interpret this value.
   c. If a player makes 100 bets of $5 each, what is the expected profit or loss? Can you explain why casinos like a high volume of betting?

6.4.10   A grocer has shelf space for three units of a highly perishable item that must be disposed of at the end of the day if it is not sold. Each unit costs $2.80 and sells for $5.50. Demand probabilities are as follows: $P(\text{Demand} = 0) = .35$, $P(\text{Demand} = 1) = .40$, and $P(\text{Demand} = 2) = .25$. Let X be a random variable indicating daily profit if a retailer stocks two units each day. Let Y be a random variable indicating daily profit if a retailer stocks one unit each day.
   a. Show the probability distributions for X and Y.
   b. Using the expected values of X and Y, determine whether the retailer would be better off stocking one or two units per day.

6.4.11   A manufacturing representative is considering the purchase of an insurance policy to cover possible losses from marketing a new product. If the product is a complete failure, the company will lose $88,000; if it is only moderately successful, the company will lose $23,000. The probabilities that the product will be a failure or only moderately successful are .04 and .16, respectively. Assuming that the manufacturing representative would be willing to ignore all other possible losses, what premium should the insurance company charge for the policy to break even?

6.4.12   A basketball player who makes 75% of his free throws comes to the line to shoot a one-and-one. (That is, if the first shot is successful, the player is given a second shot; if not, he gets no second shot. One point is scored for each successful shot.) Assume that the outcome of the second shot, if there is one, is independent of that of the first.
   a. Find the expected number of points resulting from the one-and-one.
   b. Compare this with the expected number of points from a two-shot foul, where a second shot is always given.

6.4.13   A person owns a new car valued at $5,000. Past records indicate that the probability is .005 that a car like this will be totally destroyed in an accident in any year. The car may also suffer damages of $1,000 or $3,000 with probabilities .1 and .01, respectively. If we ignore all other partial losses, what premium should the insurance company charge for a yearly policy to break even?

6.4.14   An individual makes three independent investments. The possible profits or losses from each investment are listed in the accompanying table along with their probabilities. Let X denote the total profit or loss yielded by these investments.

INVESTMENT 1		INVESTMENT 2		INVESTMENT 3	
Profit	**Probability**	**Profit**	**Probability**	**Profit**	**Probability**
$700	.6	$2,000	.3	$900	.5
− 300	.4	1,000	.5	400	.5
		− 600	.2		

300   980   650

   a. Construct and graph the probability distribution of X.
   b. Calculate the expected value of X

6.4.15   Financial analysts classify investments according to their expected profitability and according to their riskiness, where risk is measured by the variance of the potential profits. Investments that yield a negative expected profit are rejected: investments that yield a positive expected profit are compared with one another in terms of risk. For example, suppose a real estate investor plans to invest $10 million in a new office building in downtown Detroit. Analysts believe that the success of the venture depends on local business conditions. After a detailed analysis of the economy, the analysts concluded that the following probability distribution represented their beliefs about the probability distribution of the potential profitability of the project:

X = Potential gain (or loss) from investment (in \$millions)

x	-8	-4	0	5	10	25
$P(x)$	.1	.2	.2	.1	.3	.1

Find the expected profit for investing in the venture.

6.4.16 Farming is a very risky business because of the possibility that bad weather might cause financial ruin. Farmers usually insure their crops each year against possible losses resulting from bad weather. Determine the annual premium for an insurance policy to cover a farmer's \$200,000 wheat crop. Insurance representatives believe that the probability is .03 that the crop will be totally destroyed by adverse weather conditions, and the probability is .05 that there will be a loss of \$100,000.

6.5   Variance of a Discrete Random Variable

Suppose that the discrete random variable X can take only the values $x_1, x_2, \ldots, x_k$ with probabilities $P(X = x_1), P(X = x_2), \ldots, P(X = x_k)$. Thus, we would observe the squared deviation from the mean $(x_1 - \mu)^2$ with probability $P(X = x_1)$, the squared deviation from the mean $(x_2 - \mu)^2$ with probability $P(X = x_2)$, and so forth.

The squared deviations $(x_i - \mu)^2$ can be used to provide a measure of how spread out a probability distribution is. To obtain the *variance* of a discrete probability distribution, we calculate the expected value of all possible squared deviations from the mean $(x_1 - \mu)^2, (x_2 - \mu)^2, \ldots, (x_k - \mu)^2$. To obtain the expected value of these squared deviations, we weight each squared deviation by its probability of occurrence.

The variance of the random variable X, denoted as Var(X) or by the symbol σ^2 (Greek sigma and read "sigma squared"), is defined as follows.

DEFINITIONS   **Variance and Standard Deviation of a Discrete Random Variable** ——————

The **variance** of a discrete random variable X is given by the formula

$$\sigma^2 = (x_1 - \mu)^2 P(X = x_1) + (x_2 - \mu)^2 P(X = x_2) \\ + \cdots + (x_k - \mu)^2 P(X = x_k)$$

Using summation notation, we obtain

$$\sigma^2 = \Sigma(x - \mu)^2 P(X = x)$$

where the summation is over all values of X.

The **standard deviation**, denoted by the symbol σ, is the square root of the variance.

The variance represents the average or expected value of the squared deviations from the mean and is often denoted by $E[(X - \mu)^2]$. By using the rule $E[g(X)] = \Sigma g(x)P(x)$ and letting $g(x) = (x - \mu)^2$, we obtain

$$\sigma^2 = E[(X - \mu)^2] = \Sigma(x - \mu)^2 P(x)$$

The formula for calculating the variance of a random variable X can be expressed in a different but equivalent way, as indicated in the accompanying box. In some cases, especially when the mean is not an integer, it may be easier to calculate the variance by using the alternative formula rather than the original definition.

Alternative formula for the variance

$$\text{Var}(X) \;=\; \Sigma \, x^2 P(x) \;-\; \mu^2$$

That is, $\text{Var}(X) = E(X^2) - [E(X)]^2$.

Because $(x - \mu)^2$ is always nonnegative and $P(x)$ is always nonnegative, it follows that $\Sigma(x - \mu)^2 P(x)$ is always nonnegative. Thus, every variance is nonnegative, and $\text{Var}(X) \geqslant 0$. If $\text{Var}(X) = 0$, then it must be that $(x - \mu)^2 = 0$ for all possible values of X; that is, X must take the value μ with probability 1. Thus if $\text{Var}(X) = 0$, then X is a constant.

Application: Portfolio Selection

Large investors, such as corporations, banks, and insurance companies, do not hold just a single financial asset; rather, they hold *portfolios* of financial assets. These investors are less concerned with the rate of return achieved by any single particular asset than with the overall rate of return earned by the entire portfolio. Because the future rate of return of the portfolio is uncertain, a probability distribution is used to characterize the portfolio's possible rates of return.

Investors frequently use the mean and variance of a portfolio's future rate of return to judge its quality. The standard deviation of a portfolio's rate of return is frequently used as a measure of the *risk* associated with the portfolio. The greater the standard deviation, the greater the risk associated with the portfolio—that is, the greater the uncertainty of the portfolio's rate of return.

In forming a portfolio, an investor can select from many individual assets. Because of this, the investor has numerous different portfolios to choose from. Which portfolio should the investor select? This problem was addressed by Harry M. Markowitz in his famous book *Portfolio Selection* (New Haven: Yale University Press, 1959). He argued that the mean return and the standard deviation of the possible return should be used to characterize all portfolios.

Markowitz argued that any portfolio that is not an "efficient portfolio" should not be considered for investment. He defined an efficient portfolio as any portfolio that provides the highest possible mean rate of return for any *given* degree of risk (i.e., any given standard deviation) or that provides the lowest possible degree of risk for any given mean rate of return. From the set of efficient portfolios, the investor should choose the portfolio that best suits his or her needs.

EXAMPLE 6.9   **Portfolio Selection** ———————————————————————————————————————

Suppose that an investor is contemplating investing $10,000 in one of two different portfolios of stocks, portfolio A or portfolio B. The potential profits from investing in the portfolios are assumed to be random variables whose probability distributions are shown in Table 6.6 (page 210). The random variables X and Y are used to denote the potential profits from portfolios A and B, respectively. Determine which of the portfolios is an efficient portfolio.

TABLE 6.6   *Probability distributions for portfolio returns in Example 6.9*

PORTFOLIO A		PORTFOLIO B	
Profit x	Probability $P(x)$	Profit y	Probability $P(y)$
$ 600	.40	$600	.10
800	.30	700	.40
1,000	.30	880	.50

TABLE 6.7   *Calculation of mean and variance of portfolio returns in Example 6.9*

PORTFOLIO A

x	$P(x)$	$xP(x)$	$x - \mu$	$(x - \mu)^2$	$(x - \mu)^2P(x)$
$ 600	.40	240	-180	32,400	12,960
800	.30	240	20	400	120
1,000	.30	300	220	48,400	14,520
	1.00	780		Variance = 27,600	

PORTFOLIO B

y	$P(y)$	$yP(y)$	$y - \mu$	$(y - \mu)^2$	$(y - \mu)^2P(y)$
$600	.10	60	-180	32,400	3,240
700	.40	280	-80	6,400	2,560
880	.50	440	100	10,000	5,000
	1.00	780		Variance = 10,800	

SOLUTION   We need to calculate the mean return and variance for each portfolio; these calculations are shown in Table 6.7. For portfolio A, the expected return is $E(X) = \$780$, the variance is $\sigma_X^2 = 27,600$, and the standard deviation is $\sigma_X = \sqrt{27,600} = \$166.13$. For portfolio B, the expected return is $E(Y) = \$780$, the variance is $\sigma_Y^2 = 10,800$, and the standard deviation is $\sigma_Y = \sqrt{10,800} = \$103.92$.

Because both portfolios have the same expected return of $780 but the variance for portfolio B is less than that for portfolio A, portfolio A is not an efficient portfolio.

Tips for Problem Solving

To find the mean and variance of a discrete random variable X, construct a table like Table 6.7 containing six columns. The first column shows all the values of the random variable X, the second shows all the probabilities $P(X = x)$, and the third shows the products $xP(x)$. The sum of the values in column 3 is the mean or expected value of X. Column 4 shows the deviations from the mean $(x - \mu)$, column 5 shows the squared deviations from the mean $(x - \mu)^2$, and column 6 shows the products $(x - \mu)^2 P(x)$. The sum of the values in column 6 is the variance σ^2.

Exercises

Statistical Concepts

6.5.1 *True or false:*   Since probabilities must sum to 1, the variance of a discrete random variable cannot exceed 1. Explain.

6.5.2 *True or false:*   The variance of a discrete random variable must be nonnegative. Explain.

6.5.3 For the value $X = x_i$, the **deviation from the expected value**, also called the **deviation from the mean**, is $(x_i - E(X))$. Interpret these deviations. Can they ever be negative? Explain.

6.5.4 *True or false:*   Let $Y = X - E(X)$. Then $E(Y) = 0$. Explain. Can you prove your answer?

6.5.5 The **standard deviation** of a discrete random variable is the positive square root of the variance. The standard deviation is denoted $\sigma(X)$, or simply σ when there is no danger of confusion. The standard deviation can be interpreted as the typical deviation from the expected value. Explain.

6.5.6 *True or false:*   Suppose the discrete random variable X takes values such that $7 \leqslant X \leqslant 12$. Then, $7 \leqslant \sigma(X) \leqslant 12$. Prove your answer mathematically and by drawing a graph.

6.5.7 *True or false*:   Suppose the discrete random variable X takes values such that $7 \leqslant X \leqslant 12$. Then, $\text{Var}(X) \leqslant 25$. Prove your answer mathematically and by drawing a graph.

Statistical Drills

6.5.8 For each of the following discrete probability distributions, find $\text{Var}(X)$ and $\sigma(X)$.

x	P(X = x)	x	P(X = x)	x	P(X = x)
2	.2	3	.2	−2	.3
4	.3	5	.5	2	.4
6	.4	6	.2	4	.2
8	.1	9	.1	5	.1

6.5.9 For each of the distributions in Exercise 8, let $Y = 10X$. Find $\text{Var}(Y)$ and $\sigma(Y)$. Compare these values with $\text{Var}(X)$ and $\sigma(X)$. Can you explain the relationships between these variances and standard deviations?

6.5.10 For each of the distributions in Exercise 8, let $Y = 10X + 10$. Find $\text{Var}(Y)$ and $\sigma(Y)$. Compare these values with $\text{Var}(X)$ and $\sigma(X)$. Can you explain the relationships between these variances and standard deviations?

6.5.11 An alternative method of calculating the variance of a discrete random variable is by using the following formula:

$$\sigma^2 = E(X^2) - \mu^2$$

This formula is less subject to rounding errors because it does not involve calculation of the deviations from the mean $(x - \mu)$. To find $E(X^2)$, follow the procedure described in the Tips for Problem Solving, except ignore column 4. In column 5, enter the values of x^2 rather than $(x - \mu)^2$. In column 6, calculate the products $x^2 P(x)$. The sum of the elements in column 6 will give $E(X^2)$. Calculate σ^2 for each of the probability distributions listed in Exercise 8 using this alternative method.

6.5.12 The probability distribution for the random variable X is described as follows:

x	.5	1.0	1.5	2.0	2.5
P(x)	.04	.46	.30	.12	.08

    a. Calculate $E(X)$.

    b. Calculate σ^2 as $E(X - \mu)^2$.

    c. Calculate σ^2 as $E(X)^2 - \mu^2$.

6.5.13 The random variable X has the following probability distribution:

x	0	1	2	3
$P(x)$	.3	.4	.2	.1

    a. Find the expected value and variance of X.

    b. Construct a graph of the probability distribution.

    c. Find the probability that X exceeds $(\mu + 2\sigma)$.

    d. What is the probability that exceeds $(\mu + \sigma)$?

Statistical Applications

6.5.14 You have been given a ticket for a sweepstakes with five possible mutually exclusive outcomes. You could win a grand prize valued at $8,000 with probability .00005, a major prize of value $4,000 with probability .00005, a prize valued at $150 with probability .0002, and a prize of value $3 with probability .003. Finally, you could win nothing (valued at $0). Find the mean and standard deviation of the value of your ticket.

6.5.15 A publisher finds that the probability distribution for the number of errors per page of text depends on whether the material is mathematical or nonmathematical. The two probability distributions are shown in the accompanying table. Suppose that a mathematics book contains 300 pages and a nonmathematics book, 200 pages. Assume that the number of errors on any one page is independent of the number on any other. Find the mean and standard deviation of the number of errors per page in each book.

Number of errors per page: x	0	1	2	3	4	5	6
Mathematical text: $P(x)$	.45	.15	.10	.10	.10	.05	.05
Nonmathematical text: $P(x)$	.75	.13	.10	.01	.01	.00	.00

6.5.16 In Wildwood, N.J., the police give tickets to individuals who park in designated no-parking zones. The amount of the fine depends on the time of day the ticket is given. Past data indicate that 50% of the tickets are for $2, 20% are for $4, 20% are for $6, and 10% are for $8. Let X denote the amount of the fine.

    a. List the probability distribution of X.

    b. Find the mean value of X.

    c. Find the variance of X.

6.5.17 In New York City, the fines described in Exercise 16 cost 10 times as much as in Wildwood. That is, the fines are $20, $40, $60, and $80.

    a. List the probability distribution of X.

    b. Find the mean value of X.

    c. Find the variance of X.

6.5.18 Let X be a discrete random variable with the following probability distribution:

x	1	2	3	4	5
$P(x)$	.2	.2	.3	.1	.2

Find the mean and variance of X.

Computer Applications

The SPSSX computer program can be used to find relative frequencies on a discrete random variable based on a sample of observations, as well as the mean and standard deviation of the variable. For example, suppose we use the sample of data in Table 2.1 to estimate the discrete probability distribution for the number of exemptions claimed by all employees at the Computech Corporation. We make this estimate by determining the relative frequency distribution for the sample of employees. To obtain the frequency, relative frequency, and cumulative relative frequency distributions for the variable EXEMPT, issue the command

FREQUENCIES VARIABLES = EXEMPT

To obtain the sample mean and sample standard deviation of the variable EXEMPT, insert the command

STATISTICS = ALL

on the line following the FREQUENCIES command. Thus to get the estimated discrete probability distribution as well as the estimated sample mean and sample standard deviation, issue the commands

FREQUENCIES VARIABLES = EXEMPT
STATISTICS = ALL

Figure 6.3 shows the computer output generated by these two commands. For example, 17.5% of the 120 employees claimed one exemption, 33.3% of the employees

Value Label		Value	Frequency	Percent	Valid Percent	Cum Percent
		1.00	21	17.5	17.5	17.5
		2.00	40	33.3	33.3	50.8
		3.00	26	21.7	21.7	72.5
		4.00	17	14.2	14.2	86.7
		5.00	12	10.0	10.0	96.7
		6.00	4	3.3	3.3	100.0
		Total	120	100.0	100.0	

Mean	2.758	Std err	.124	Median	2.000
Mode	2.000	Std dev	1.353	Variance	1.832
Kurtosis	−.446	S E Kurt	.438	Skewness	.595
S E Skew	.221	Range	5.000	Minimum	1.000
Maximum	6.000	Sum	331.000		

Valid cases	120	Missing cases	0

FIGURE 6.3 *SPSSX-generated output showing the discrete probability distribution, mean, and standard deviation for the variable EXEMPT in Table 2.1*

claimed two exemptions, and so forth. The data in the column titled Cum Percent show cumulative relative frequencies expressed as percentages. For example, 72.5% of the employees claimed three or fewer exemptions; 86.7% of the employees claimed four or fewer exemptions.

The computer output shows that the mean of the variable EXEMPT is 2.758. This indicates that, on the average, an employee has 2.758 exemptions. The sample standard deviation is 1.353.

Exercises

6.6.1   Refer to the sample of data in Table 2.2 in Chapter 2. The variable Class indicates whether the student is a freshman, sophomore, junior, or senior.
a. Use the computer to estimate the discrete probability distribution for the variable Class.
b. Find the mean of the variable Class by using the computer. What does this mean indicate?
c. Verify the answer in part **b** by using a hand calculator.

6.6.2   Refer again to Table 2.2. The variable Major indicates the academic major of each student.
a. Use the computer to estimate the discrete probability distribution for the variable Major.
b. Does it make sense to find the mean of the variable Major?

Chapter 6 Supplementary Exercises

6.S.1   The Watson Construction Company is bidding on two contracts, one in New York and one in Boston. The probability is .6 of winning the New York contract and .3 of winning the Boston contract. Suppose the contracts are independent of one another. If Watson gets the New York contract, the net profit will be $20,000; if they do not get the contract, the loss will be $4,000 in costs for preparing the bids. If Watson gets the Boston contract, the net profit will be $54,000; if they do not get the Boston contract, the loss will be $9,000 in costs. Let X denote the net profit or loss for the Watson Construction Company after making the two bids.
a. Determine the probability distribution of X.
b. Find the mean value of X.

 6.S.2   A personnel officer suspects that job applicants who have had many previous jobs are more likely to leave the company relatively quickly after employment than those with few previous jobs. A review of the records for all employees who stayed 4 years or less yielded the joint probability distribution shown in the table.

Years before leaving (Y)	NUMBER OF PREVIOUS JOBS (X)				
	1	2	3	4	
1	.03	.06	.08	.12	.29
2	.05	.07	.08	.07	.27
3	.05	.08	.06	.02	.21
4	.07	.09	.05	.02	.23
	.2	.3	.27	.23	

a. Find the marginal probability distribution of X.
b. Find the mean number of previous jobs.
c. Find the marginal probability distribution of Y.
d. Find the mean number of years before leaving.

e. Find the conditional probability distribution of Y given $X = 4$.

f. Are the number of previous jobs and the years before leaving independent of one another?

6.S.3 A potential customer for a $60,000 fire insurance policy has a home that may sustain a total loss in a given year with a probability of .0005 and a 50% loss with a probability of .001. Ignoring all other partial losses, what premium should the insurance company charge for a yearly policy to break even?

6.S.4 The probability distribution of X, the number of passengers on a daily helicopter shuttle run from airport A to airport B, follows:

x	0	1	2	3	4
$P(x)$	.1	.4	.3	.1	.1

a. Find the probability that fewer than three customers will appear on a given day.

b. Find the expected number of customers on a given day.

6.S.5 Suppose the probability that a $50 million oil drilling platform will be totally lost at sea next year is .006. How much should the platform's owners expect to pay for complete insurance against this contingency in the next year (excluding administrative and other costs and profits of the insurance company)?

6.S.6 A farmer's crop will receive a grade of A, B, or C. If the crop is graded A, the farmer will receive $5.20 per bushel. For grades B and C, the prices per bushel will be $4.00 and $3.20, respectively. The probabilities for the grade of the harvest when done at the normal time are $P(A) = .35$, $P(B) = .55$, and $P(C) = .10$. The yield of the harvest at the normal time is expected to be 66,000 bushels. If the harvest is early, the probabilities are $P(A) = .50$, $P(B) = .50$, and $P(C) = 0$, but the yield is reduced to 58,000 bushels. Should the farmer harvest early or at the normal time if he wishes to maximize expected revenue?

6.S.7 Consider the following probability distribution for the random variable X:

x	10	20	30	40	50	60
$P(x)$	.05	.25	.35	.20	.10	.05

a. Find the expected value and the variance of X.

b. Graph the distribution.

c. Locate μ and the interval $(\mu - 2\sigma, \mu + 2\sigma)$ on your graph. What is the probability that X will fall within this interval? Does this probability satisfy Chebyshëv's theorem?

6.S.8 A company sells packages of chocolate chip cookies. The numbers of cookies per package vary as follows:

Number of cookies	20	21	22	23
Proportion of packages	.10	.40	.30	

a. Find the missing proportion.

b. Find the mean number of cookies per package.

c. Find the standard deviation of the number of cookies per package.

d. What proportion of packages contain at least 21 cookies?

e. What proportion of packages contain fewer than 22 cookies?

f. The cost (in cents) of producing a package of cookies is $3 + 2.2X$, where X is the number of cookies in the package. That is, packaging costs $.03 and cookies cost $.022 each. The revenue from selling a package, regardless of how many cookies it contains, is $.70. If profit is defined as the difference between revenue and cost, find the expected profit per package.

6.S.9   A company is drilling for oil. When it strikes oil, it expects a profit of $100,000. The cost of drilling is $5,000. If the probability of striking oil is .1, what is the expected profit for the company from this project?

6.S.10   The following information shows the amount of time in minutes by which a mechanic on an assembly line misses the design completion time when working on a certain job. Negative values indicate early completion.

Minutes (x)	-2	-1	0	1	2
$P(x)$	.1	.1	.2	.35	.25

a. Graph the probability distribution of X.
b. Find the mean and variance of X.
c. Find $P(X < 0)$.

6.S.11   An individual has $10,000 to invest. If he or she invests in project A, the estimated profit is $40,000, $20,000, or $-$10,000$ with probabilities .2, .5, and .3, respectively. If he or she invests in project B, the profit is estimated to be $20,000, $15,000, or $0 with probabilities .3, .4, and .3, respectively. Let the random variables X and Y denote the return from projects A and B, respectively. Assume that the investments are independent of one another.
a. Graph the probability distributions of X and Y.
b. Find $E(X)$ and $E(Y)$.
c. In your opinion, which project is the better investment? Note that $E(X)$ exceeds $E(Y)$, but in project A it is possible to lose $10,000, whereas in project B it is not possible to lose money.

6.S.12   A soap manufacturing company is planning to introduce two new laundry detergents. The potential profits or losses are partly dependent on whether the leading competitor is already planning to market similar products. The first product will yield a profit of $100,000 if the competitor has not developed a competing product but a loss of $50,000 otherwise. The second product will yield a profit of $80,000 without competition but a loss of $30,000 otherwise. The probability that the competitor has developed a rival is .6 for the first product and .7 for the second. Let X denote the total profit or loss from developing the two products. Assume that the success or failure of the first product is independent of the success or failure of the second.
a. Find the probability distribution of X and graph it.
b. Find the mean value of X.

6.S.13   A computer sales representative for T.L.G. Industries gets a certain commission on all sales made. Several months were spent trying to get two large banks to purchase equipment from T.L.G. The sales representative thinks there is a 70% chance that bank 1 will purchase a small computer, a 20% chance that bank 1 will purchase a large computer, and a 10% chance that bank 1 will purchase nothing. Bank 2 will see what bank 1 does before making its decision. If bank 1 buys a small computer, there is a 60% chance that bank 2 will buy a small computer, a 25% chance that it will buy a large computer, and a 15% chance that it will buy nothing. If bank 1 buys a large computer, there is a 10% chance that bank 2 will buy a small computer, an 85% chance that it will buy a large computer, and a 5% chance that it will buy nothing. If bank 1 buys nothing, there is a 5% chance that bank 2 will buy a small computer and a 95% chance that it will buy nothing. The sales representative's commission is $14,000 on a small computer and $32,000 on a large computer. Let x denote the total commission earned.
a. Find the probability distribution of X.
b. Find the sales representative's expected commission.

6.S.14 Suppose that a race car driver will drive in four races and always has a probability of ⅓ of winning any specific race. The driver signs a contract with an advertiser. The driver's bonus is based on how many races he or she wins. For one win and three losses, the bonus is $5,000; for two wins and two losses, the bonus is $10,000; for three wins and one loss, the bonus is $20,000. If the driver wins all four races, the bonus is $50,000, and if the driver loses all four races, $5,000 must be paid back from his or her salary; that is, the bonus is − $5,000. Let X denote the driver's bonus.
  a. Find the probability distribution of X.
  b. Find the mean value of X.

6.S.15 When it is sunny, a man walks to work at a cost of $.00. When it is cloudy, he takes the bus, which costs $1.00. When it is raining, he drives his car, which costs $2.00. It is sunny 50% of the time, cloudy 30% of the time, and raining 20% of the time. Find the average or expected cost of a trip to work.

6.S.16 Julie is shopping for an automobile insurance policy. During the following year, there is a .01 probability that her car will be totally destroyed in an accident and she will lose $5,000. There is a .05 probability that she will have an accident in which she will lose $2,500. There is a .20 probability that an accident will cause $300 in damage. Suppose she pays $300 for her insurance policy. What is the insurance company's expected profit on her policy? That is, find the difference between the price of the policy and the company's expected payment to Julie.

6.S.17 Ed Conner contacts people and tries to convince them to purchase a 1-, 2-, or 3-year subscription to *Sports Illustrated* magazine. Of those people contacted, 10% buy a 1-year subscription, 5% buy a 2-year subscription, and 4% buy a 3-year subscription. The remaining 81% do not buy anything. Ed's commission is $2 for each 1-year subscription, $5 for each 2-year subscription, and $10 for each 3-year subscription.
  a. On his next contact, what is Ed's expected commission?
  b. If Ed contacts 100 people, how much money can he expect to earn?

6.S.18 A store owner stocks an out-of-town newspaper that a small number of customers sometimes request. Each copy of this newspaper costs the owner 22¢ and is sold for 35¢. Any copies left over at the end of the day have no value and are destroyed. The probability distribution of the number of requests for the newspaper in a day is as follows:

Number of requests	0	1	2	3	4	5
Probability	.10	.15	.20	.30	.15	.10

If the store owner defines total daily profit as total revenue from newspaper sales less total cost of newspapers ordered, how many copies per day should he order to maximize expected profit?

6.S.19 You have two friends who want to bet with you on an upcoming basketball game. One wants to bet you that the Boston Celtics will beat the Los Angeles Lakers, whereas the second wants to bet you that the Lakers will beat the Celtics. The first will put up $1 for every $1 that you bet. The second will put up $1.30 for every $.70 that you wager.
  a. If you bet $50 with each friend, find your net profit if the Celtics win. Find your net profit if the Lakers win.
  b. If you bet $40 with the first friend and $60 with the second, find your net profit if the Celtics win. Find your net profit if the Lakers win.
  c. Assume the probability that the Celtics win is .5. If you have $100 available, how should you divide your wagers between these two gamblers to maximize your expected net profit?

6.S.20 An automobile dealership records the number of cars sold each day. The data at the top of page 218 show the probability distribution of daily sales.

x	P(x)
0	.2
1	.3
2	.2
3	.1
4	.1
5	.1

a. Verify that this is a valid probability distribution.
b. Find the probability that the number of cars sold tomorrow will exceed 2.
c. Find the cumulative distribution function of the number of cars sold per day.
d. Find the expected number of cars sold per day.
e. Find Var(X).

\$\$ 6.S.21 Financial analysts have been presented with two investment opportunities, investment X and investment Y. The accompanying data record the analysts' beliefs about the possible gains or losses from each investment and their associated probabilities. All numbers are in millions of dollars.

x	P(x)	y	P(y)
−50	.03	−60	.12
−30	.05	−40	.06
−10	.14	−20	.08
0	.18	0	.20
10	.23	20	.16
30	.15	40	.13
50	.14	60	.11
70	.06	80	.04
90	.02	100	.10

a. Verify that each is a valid probability distribution.
b. Plot each distribution.
c. Calculate $E(X)$ and $E(Y)$ and show these values on the graph in part **b.** Which investment is preferable in terms of expected return?
d. Find Var(X) and Var(Y). Which investment is preferable in terms of minimizing variance?
e. Find the cumulative distribution for each distribution.
f. Find $P(20 \leqslant X \leqslant 60)$.
g. Find $P(20 \leqslant Y \leqslant 60)$.
h. What is the probability that X will fall in the interval $(\mu_X - 2\sigma_X, \mu_X + 2\sigma_X)$?
i. What is the probability that Y will fall in the interval $(\mu_Y - 2\sigma_Y, \mu_Y + 2\sigma_Y)$?

6.S.22 An individual investor has purchased two stocks whose price movements are assumed to be independent of one another. For each stock, potential profit from stock A is a random variable X_A having a certain mean and variance. For the following year, the investor believes the expected profit (dividends plus capital gains) for stock A is $E(X_A) = \$10$. For stock B, the expected profit is $E(X_B) = \$12$. For stock A, the standard deviation of potential profits is assumed to be $\sigma_A = \$3$. For stock B, the standard deviation of potential profits is assumed to be $\sigma_B = \$4$. The portfolio profit P is the sum of X_A and X_B: $P = X_A + X_B$.
a. Find the expected profit from the portfolio that contains both stock A and stock B.
b. Find the variance of each stock profit.

c. Find the variance of each portfolio profit.

d. Find the standard deviation of the portfolio profit.

References

Canavos, George C., and Don M. Miller. *Modern Business Statistics*. Belmont, CA: Duxbury Press, 1995.

Cryer, Jonathan D., and Robert B. Miller. *Statistics for Business: Data Analysis and Modelling*. Boston: PWS-Kent, 1991.

Feller, William. *An Introduction to Probability Theory and Its Applications*. 2 vols. New York: Wiley, 1950–1966.

Freund, John E., and R. E. Walpole. *Mathematical Statistics*. 4th ed. Englewood Cliffs, N.J.: Prentice-Hall, 1987.

Kohler, Heinz. *Statistics for Business and Economics*. Glenview, IL: Scott, Foresman, 1988.

McClave, James T., and Frank H. Dietrich. *Statistics*. 6th ed. New York: Macmillan, 1994.

Mendenhall, William, James E. Reinmuth, and Robert Beaver. *Statistics for Management and Economics*. 6th ed. Boston: PWS-Kent, 1989.

Mood, Alexander M., and Franklin A. Graybill. *Introduction to the Theory of Statistics*. 3d ed. New York: McGraw-Hill, 1973.

Neter, J., W. Wasserman, and G. A. Whitmore. *Applied Statistics*. 3rd ed. Boston: Allyn & Bacon, 1987.

Shiffler, Ronald E., and Arthur J. Adams. *Introductory Business Statistics with Computer Applications*. Belmont, CA: Wadsworth, 1995.

Watson, Collin J., Patrick Billingsley, D. James Croft, and David V. Huntsberger. *Statistics for Management and Economics*. 4th ed. Needham Heights, MA: Allyn & Bacon, 1990.

7 Some Important Discrete Distributions

In Chapter 6, we discussed how to construct a discrete probability distribution as well as how to calculate the expected value and variance of any discrete random variable. Recall that for any probability distribution we are interested in several important characteristics:

1. *Formula:* Is there a general mathematical formula that we can use to determine the probability that certain events will occur?
2. *Shape:* What is the general shape of the distribution? Is it symmetric? Is it skewed to the right or left? Over what interval is most of the probability concentrated? Are there long tails in either direction?
3. *Mean:* Where is the center of gravity of the distribution? That is, what is the expected value, or mean, of the random variable?
4. *Standard deviation and variance:* How spread out is the distribution? What is the variance or standard deviation of the distribution?

Several different discrete probability distributions occur over and over again in business, economics, and numerous other fields. In this chapter, we discuss the characteristics of the most important discrete probability distributions: the binomial and the Poisson distributions.

7.1 The Binomial Distribution

An important distribution occurs when a random experiment with two possible outcomes (denoted "success" and "failure") is repeated several times. This type of experiment leads to the concept of the **binomial distribution.**

In a binomial experiment, the random variable X denotes the number of successes obtained in n repeated independent trials of the experiment. The probability of obtaining a success on any trial is denoted by p. Because the trials are independent of one another, this probability remains constant for all trials. If p denotes the probability of a success, then $q = (1 - p)$ is the probability of a failure on any trial. Numerous problems fit this model, at least approximately.

EXAMPLE 7.1 **Examples of Binomial Experiments**

a. A manufacturing firm claims that about 2% of its products are defective. Suppose we randomly select 10 products and test them. If the products are randomly selected, the outcome of any test is independent of any other test. Thus, we have performed the experiment 10 independent times. Let X denote the number of defective products obtained in the sample of size $n = 10$. The random variable X is a binomial random variable where $p = .02$.

b. In Alabama 10% of the labor force is unemployed. Select a random sample of $n = 100$ individuals. Let X denote the number of individuals in the sample who are unemployed. Then X is a binomial random variable with $p = .1$.

c. A brokerage firm sends an application form to a random sample of $n = 300$ individuals asking them to invest in the firm's new mutual fund. Industry data indicate that approximately 5% of the people who receive such an application eventually reply. Let X denote the number of replies received in this sample of $n = 300$ individuals. Then X is a binomial random variable with $p = .05$.

d. Industry data show that 25% of the people who purchase contact lenses experience difficulty and stop wearing the lenses within 1 year. We take a random sample of $n = 60$ individuals who have just purchased contact lenses. Let X denote the number of these individuals who stop wearing the lenses within 1 year. Then X is a binomial random variable with $p = .25$.

If a binomial experiment has been performed n independent times, the possible number of successes in n trials must be one of the values $0, 1, 2, \ldots, n$. Thus, the sample space for the binomial random variable X is $S = \{0, 1, 2, \ldots, n\}$. Now we want to derive a general mathematical formula, or probability function, that will enable us to calculate the probability of obtaining a specific number of successes in independent trials of an experiment.

EXAMPLE 7.2   **Derivation of the Binomial Probability Function** ————————————————

Based on past data, approximately 30% of the oil wells drilled in areas having a certain favorable geological formation have struck oil. A company has identified $n = 5$ locations that possess this formation. Because the locations are widely separated, it is safe to assume that the chance of striking oil at any location is independent of the chance of striking oil at any of the others. The company decides to drill a well on each of the five sites. Calculate the probability that exactly two of the five wells strike oil.

SOLUTION   The number of independent trials is $n = 5$. The probability of a success (outcome s) on any trial is $p = .3$. Thus, the probability of a failure (outcome f) is $q = (1 - p) = .7$. Let the random variable X denote the number of successes in five attempts. We want to know the probability that $X = 2$. Let A denote the event $A = \{$Exactly 2 successes in 5 attempts$\}$. We obtain the basic outcomes $A = \{ ssfff, sfsff, sffsf, sfffs, fssff, fsfsf, fsffs, ffssf, ffsfs, fffss \}$. Since we obtain 2 successes in 5 attempts if any 1 of the 10 basic outcomes of A occurs, we have

$$P(A) = P(X = 2) = P(ssfff) + P(sfsff) + \cdots + P(fffss)$$
$$= (.3)^2(.7)^3 + (.3)^2(.7)^3 + \cdots + (.3)^2(.7)^3$$
$$= 10(.3)^2(.7)^3 = .3087$$

In this equation, the number 10 represents the number of basic outcomes in event A; that is, it represents the number of ways of arranging two successes and three failures among five attempts. The number of ways of arranging two successes and three failures is given by the combination formula

$$_5C_2 = \frac{5!}{2!(5 - 2)!} = 10$$

The number $(.3)^2(.7)^3$ represents the probability of any single basic outcome in event A, such as $\{sfsff\}$.

The method used in Example 7.2 can be used to determine any binomial probability.

General Formula for the Binomial Distribution

Formula for the binomial distribution

Let the random variable X denote the number of successes obtained in n independent trials of an experiment where the probability of a success on any trial is p. Let $q = (1 - p)$. Let x denote a certain value of the random variable X. The probability of x successes in n trials is given by the formula

$$P(X = x) = {_nC_x}\, p^x q^{n-x} \qquad x = 0, 1, 2, \ldots , n$$

where the symbol $_nC_x$ denotes the combination formula

$$_nC_x = \frac{n!}{x!(n - x)!}$$

EXAMPLE 7.3   **Using the Binomial Formula**

In Example 7.2, find the complete probability distribution of X, where X denotes the number of successes when five independent wells are drilled.

SOLUTION   The complete probability distribution of X is shown in Table 7.1 and illustrated in Figure 7.1. The probability $P(X = 0)$ is represented by the area of the rectangle extending from $-.5$ to $.5$ that is centered at $X = 0$; the probability $P(X = 1)$ is represented by the area of the rectangle extending from $.5$ to 1.5 that is centered at $X =$

TABLE 7.1   *Distribution of oil strikes in Example 7.3*

x	Probability	$P(X = x)$
0	$_5C_0(.3)^0(.7)^5$	.16807
1	$_5C_1(.3)^1(.7)^4$	.36015
2	$_5C_2(.3)^2(.7)^3$	.30870
3	$_5C_3(.3)^3(.7)^2$	.13230
4	$_5C_4(.3)^4(.7)^1$	.02835
5	$_5C_5(.3)^5(.7)^0$	.00243
		1.00000

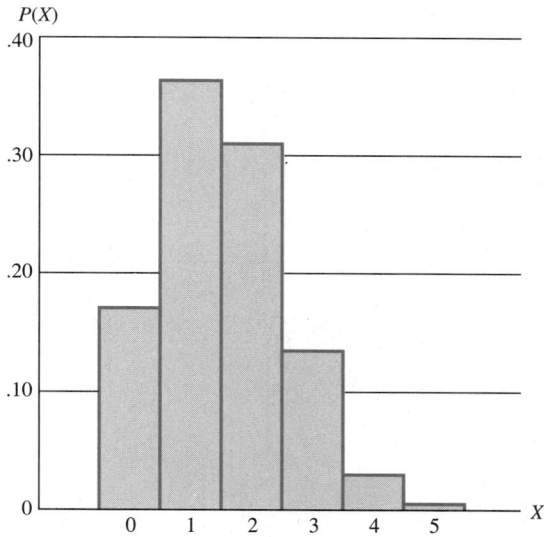

FIGURE 7.1   *Binomial distribution for Example 7.3 where n = 5 and p = .3*

1; and so forth. The areas of all the rectangles sum to 1. Because the width of each rectangle is 1 unit, the height of each rectangle is the same as the area of the rectangle.

Shape of the Binomial Distribution

Actually, the binomial distribution is a family of distributions. There is a different binomial distribution for each set of values of the parameters n and p, and the shape of the binomial distribution varies with these values. For $p = .5$, the graph of the binomial distribution is symmetric. That is, the probability of obtaining 0 successes and n failures in n trials will be the same as the probability of obtaining n successes and 0 failures in n trials; the probability of 1 success and $(n - 1)$ failures will be the same as the probability of 1 failure and $(n - 1)$ successes; and so forth.

If p is greater than .5, successes are more likely than failures, and the graph of the binomial distribution is not symmetric. The probabilities on the right side of the graph will be larger than those on the left side. If p is less than .5, the argument is just reversed, as is evident in Example 7.3.

An interesting characteristic of the graph of the binomial distribution for p close to .5 is its bell shape. It can be shown that when the number of trials n becomes larger and larger, the graph of the binomial distribution always becomes bell shaped. If p is very close to .5, this bell-shaped appearance is evident even when n is fairly small. In general, a bell-shaped curve can be used to approximate the probabilities of the binomial distribution provided n and p are such that $np \geqslant 5$ and $nq \geqslant 5$. This will be discussed in more detail in Chapter 8.

Figure 7.2 (page 224) shows the graphs of several binomial distributions for various

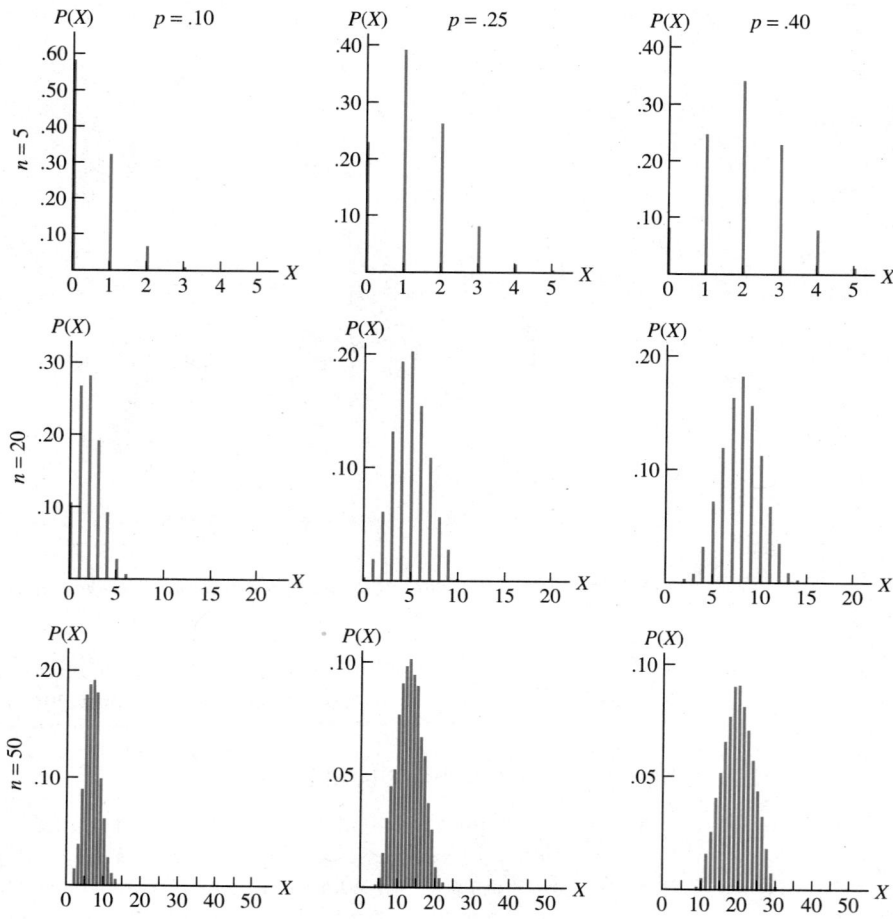

FIGURE 7.2   *Graphs of binomial distributions for various values of n and p*

values of n and p. The probability shifts to the right as p increases. Also, the distribution becomes more symmetric as p approaches .5.

Mean and Variance of the Binomial Distribution

Mean, variance, and standard deviation of the binomial distribution

> Let X denote the number of successes obtained in n independent trials of an experiment where the probability of success on any trial is p. Let $q = 1 - p$. The mean, variance, and standard deviation of X are as follows:
>
> $$\mu = E(X) = np$$
> $$\sigma^2 = \text{Var}(X) = npq$$
> $$\sigma = \sqrt{npq}$$

The mean value of X represents the average number of successes that would be obtained in n independent trials.

EXAMPLE 7.4   **Mean and Variance of the Binomial Distribution**

a. Find the mean, variance, and standard deviation of the random variable X in Example 7.3, where $n = 5$ and $p = .3$.

b. Verify that the mean is 1.5 by actual computation.

SOLUTION   **a.** We obtain

$$E(X) = np = 5(.3) = 1.5$$
$$\text{Var}(X) = npq = 5(.3)(.7) = 1.05$$
$$\sigma = \sqrt{npq} = \sqrt{1.05} = 1.025$$

b. For any discrete random variable, the mean is obtained by using the formula

$$E(X) = \Sigma\, xP(X = x)$$

By using the probabilities given in Example 7.3, we obtain

$$
\begin{aligned}
E(X) &= 0P(0) + 1P(1) + \cdots + 5P(5) \\
&= 0(.16807) + 1(.36015) + 2(.30870) \\
&\quad + 3(.13230) + 4(.02835) + 5(.00243) \\
&= 1.5
\end{aligned}
$$

Using the Binomial Table

In many practical applications, we want to calculate the probability that a binomial variable X assumes any one of the values 0, 1, 2, . . . , c, where c is some integer between 0 and n. That is, we want to determine the *cumulative probability* $P(X \leq c)$. The cumulative binomial distribution function is given in Table A.2 of the Appendix, which shows $P(X \leq c)$ for selected values of n, p, and c. More extensive tables are available, but the tables rarely go beyond $n = 50$. As will be shown in Chapter 8, the normal distribution often provides an excellent approximation to the binomial distribution, thus reducing the need for tables for $n > 50$.

EXAMPLE 7.5   **Use of the Binomial Table**

According to the Internal Revenue Service, approximately 20% of all income tax returns contain mathematical errors.

a. Find the probability that 3 or fewer returns out of a sample of 10 contain mathematical errors.

b. Find the probability that fewer than three of the returns contain errors.

c. Find the probability that exactly three of the returns contain errors.

d. Find the probability that three or more of the returns contain errors.

SOLUTION   **a.** Let X denote the number of returns containing a mathematical error. Then X follows the binomial distribution with $n = 10$ and $p = .20$. We seek

$$P(X \leq 3) = P(X = 0) + P(X = 1) + P(X = 2) + P(X = 3)$$

Using $n = 10$, $p = .20$, and $c = 3$ in Table A.2 in the Appendix, we obtain

$$P(X \leqslant 3) = .879$$

(The values in Table A.2 have been rounded to three decimal places.)

b. We seek

$$P(X < 3) = P(X = 0) + P(X = 1) + P(X = 2)$$

From Table A.2, using $n = 10$, $p = .20$, and $c = 2$, we obtain

$$P(X < 3) = P(X \leqslant 2) = .678$$

c. We seek

$$P(X = 3) = P(X \leqslant 3) - P(X \leqslant 2) = .879 - .678 = .201$$

d. We seek

$$P(X \geqslant 3) = P(X = 3) + P(X = 4) + \cdots + P(X = 10)$$
$$= 1 - P(X \leqslant 2) = 1 - .678 = .322$$

Sampling With and Without Replacement

Suppose that a proportion p of the items in a certain population possess a certain characteristic. There are two ways to obtain a random sample of size n from the population—with or without replacement. If we sample with replacement, then the probability is always p that the next item chosen will possess the characteristic. If we sample without replacement from a finite population, then the probability that the second item selected has the characteristic will be greater than or less than p, depending on whether the first item possessed the characteristic. If sampling is done without replacement from a finite population, then successive selections are not independent; the probabilities change after each selection. If the population is large relative to the sample size, however, then the probabilities change very little on successive drawings, and we can proceed as if successive drawings were independent of one another. For this reason, the binomial distribution is often used despite the fact that a sample has been selected without replacement.

Sample Proportion of Successes

Statisticians frequently are more interested in the proportion of successes in a sample than in the number of successes. If we obtain X successes in n trials, then the sample proportion $\hat{p}$ is calculated using the formula $\hat{p} = X/n$.

The random variable X must take one of the discrete values $0, 1, 2, \ldots, n$; similarly, the random variable $\hat{p}$ must take one of the values $0/n, 1/n, 2/n, \ldots, 1$. For example, when $X = 3$, $\hat{p} = 3/n$. Also, the probability that $X = 3$ is the same as the probability that $\hat{p} = 3/n$. For any specific value x, we obtain

$$P(X = x) = P(\hat{p} = x/n)$$

This shows that the probability distribution for the random variable $\hat{p}$ can be derived from the probability distribution of X.

EXAMPLE 7.6 **Probabilities for the Sample Proportion**

An experiment is tried 10 independent times, and the probability of a success on any trial is $p = .3$. Find the probability of obtaining a success on 40% of the trials.

SOLUTION Let X denote the number of successes in 10 trials. Then X follows the binomial distribution with $n = 10$ and $p = .3$. We have

$$P(\hat{p} = .4) = P\left(\hat{p} = \frac{4}{10}\right) = P(X = 4)$$

$$= {}_{10}C_4(.3)^4(.7)^6 = .2001$$

Mean and Variance of $\hat{p}$

Let $\hat{p}$ denote the sample proportion of successes obtained in n independent trials of an experiment. The expected value of $\hat{p}$ is

$$E(\hat{p}) = p$$

This result means that the sample proportion $\hat{p}$ is an unbiased estimator of the population proportion p.

In addition, it is possible to show that the variance of $\hat{p}$ is

$$\text{Var}(\hat{p}) = \frac{pq}{n}$$

Of course, the standard deviation of the random variable $\hat{p}$ is then $\sqrt{pq/n}$.

Knowledge of the mean and variance of $\hat{p}$ can help us test hypotheses about the population parameter p.

EXAMPLE 7.7 **Testing a Claim about the Population Proportion**

A councilman claims that at least 30% of the voters of a large city are in favor of increasing taxes on alcoholic beverages. To test this claim, a polling agent obtains a random sample of 500 voters. Suppose that $X = 100$ voters in the sample say they favor the tax. Thus, the sample proportion is $\hat{p} = {}^{100}\!/_{500} = .2$. Is it reasonable to reject the claim that, in the population, p is at least .3?

SOLUTION If the claim is true, then the sample proportion $\hat{p}$ has expected value $E(\hat{p}) = p = .3$. The variance of $\hat{p}$ is given by

$$\text{Var}(\hat{p}) = \frac{pq}{n} = \frac{(.3)(.7)}{500} = .00042$$

and the standard deviation of $\hat{p}$ is $\sqrt{.00042} \approx .02$.

Because the distribution of X is bell shaped, so is the distribution of $\hat{p}$. By applying the Empirical Rule, we can state that in repeated sampling more than 99.7% of the values of $\hat{p}$ should fall within 3 standard deviations of the mean. This interval is

$$(.3 - .06, .3 + .06) = (.24, .36)$$

Since our observed value $\hat{p} = .2$ lies outside this interval, we have strong evidence that, in the population, $\hat{p}$ does not equal .3. If p were .3, it would be quite unusual to observe a value as extreme as $\hat{p} = .2$ in a sample size of $n = 500$.

Tips for Problem Solving

1. You can recognize a binomial random variable by the fact that the same two-outcome experiment is performed independently n times.

2. To calculate any binomial probability, you must know the probability p of a success on any single trial. Thus, this probability has to be stated somewhere in the problem. Record the values of n and p.

3. Next record the value of X, the desired number of successes.

4. If n is small, you can calculate a binomial probability on a hand calculator or you can use the tables in the Appendix.

5. To calculate a binomial probability, use the formula

$$P(X = x) = \frac{n!}{x!(n - x)!}\, p^x(1 - p)^{n-x}$$

6. If n is large, you will want to calculate approximate binomial probabilities by using the Poisson distribution, discussed in Section 7.2, or the normal distribution, discussed in Section 8.5.

Exercises

Statistical Concepts

7.1.1　*True or false:*　Let X be a binomial random variable with $n = 10$ and $p = .6$. Then $P(X < 5) = P(X \leq 4)$. Explain.

7.1.2　*True or false:*　Let X be a binomial random variable with $n = 10$ and $p = .6$. Then $P(X < 5) = 1 - P(X < 5)$. Explain.

7.1.3　Suppose X is a binomial random variable with $n = 10$ trials. Describe how the shape of the probability function changes as p varies from .1 to .9.

7.1.4　Suppose X is a binomial random variable with $p = .2$. Describe how the shape of the probability function changes as n varies from 3 to 300.

7.1.5　*True or false:*　Let X be a binomial random variable with $n = 10$ and $p = .6$. Then $P(X = 4) = P(X \leq 4) - P(X < 4)$. Explain.

7.1.6　*True or false:*　Let X be a binomial random variable with $n = 10$ and $p = .6$. Then $P(X = 4) = P(X < 5) - P(X < 4)$. Explain.

Statistical Drills

7.1.7　Let X be a binomial random variable with $n = 4$ and $p = .5$. Calculate the binomial probabilities $P(X = x)$ for $x = 0, 1, 2, 3, 4$.

7.1.8　Let X be a binomial random variable with $n = 4$ and $p = .5$. Calculate $E(X)$ by doing the arithmetic calculations. Verify that you get the same answer by using the formula $E(X) = np$.

7.1.9　Let X be a binomial random variable with $n = 4$ and $p = .5$. Calculate $\text{Var}(X)$ by doing the arithmetic calculations. Verify that you get the same answer by using the formula $\text{Var}(X) = npq$.

7.1.10 Let X be a binomial random variable with $n = 3$ and $p = .2$. Calculate the binomial probabilities $P(X = x)$ for $x = 0, 1, 2, 3$. Verify that these probabilities sum to 1. Graph the distribution.

7.1.11 Let X be a binomial random variable with $n = 3$ and $p = .7$. Calculate $E(X)$ by doing the arithmetic calculations. Verify that you get the same answer by using the formula $E(X) = np$.

7.1.12 Let X be a binomial random variable with $n = 3$ and $p = .7$. Calculate $\text{Var}(X)$ by doing the arithmetic calculations. Verify that you get the same answer by using the formula $\text{Var}(X) = npq$.

7.1.13 Let X be a binomial random variable with $n = 8$ and $p = .2$. Use Table A.2 in the Appendix to find $P(X \leqslant 4)$.

7.1.14 Let X be a binomial random variable with $n = 10$ and $p = .6$. Use Table A.2 in the Appendix to find $P(X > 6)$.

7.1.15 Let X be a binomial random variable with $n = 8$ and $p = .3$. Use Table A.2 in the Appendix to find $P(X \geqslant 4)$.

7.1.16 Use Table A.2 in the Appendix to evaluate the following binomial probabilities when $n = 10$ and $p = .4$.
 a. $P(X \leqslant 3)$ **b.** $P(X > 3)$ **c.** $P(X \geqslant 3)$
 d. $P(X = 3)$ **e.** $P(X < 3)$

7.1.17 Use Table A.2 in the Appendix to evaluate the following binomial probabilities when $n = 20$ and $p = .2$.
 a. $P(X \leqslant 3)$ **b.** $P(X > 2)$ **c.** $P(X \geqslant 4)$
 d. $P(X = 3)$ **e.** $P(X < 5)$

Statistical Applications

7.1.18 An advertising agency seeks comments on an advertisement that appeared during television coverage of the Super Bowl. About 40% of the adult population saw the Super Bowl on TV. If the agency takes a random sample of four adults, find the probability that exactly two of them saw the Super Bowl.

7.1.19 On any given day, about 45% of the stocks on the American Stock Exchange increase in value.
 a. If you randomly select four independent stocks, what is the probability that exactly three of them increase in value?
 b. What is the probability that at least three of the four stocks increase in value?

7.1.20 Approximately 20% of the cars produced by a certain manufacturer have a defect. A sales representative receives a shipment of five randomly selected cars.
 a. What is the probability that one car has a defect?
 b. What is the probability that at least one car has a defect?
 c. Let X denote the number of defective cars. List the possible values of X and their probabilities.
 d. Calculate the mean value of X using part **c.**

7.1.21 About 20% of all professional football players are injured during a given season. A team has four star players. What is the probability that at least one of the star players gets injured?

7.1.22 A manufacturer of digital watches claims that the probability of its watch running more than 1 minute fast or slow after a year of use is .05. A consumer protection agency has purchased three of the manufacturer's watches with the intention of testing the claim.
 a. If the manufacturer's claim is correct, what is the probability that all three of the watches are accurate within a minute?
 b. If the manufacturer's claim is correct, what is the probability that exactly two of the three watches fail to meet the claim?

7.1.23 A plumber installs six hot water heaters in a particular housing development. The probability that any individual heater will last more than 10 years is .7, and their lifelengths are independent. Let X denote the number of water heaters that last more than 10 years.

a. Construct the probability distribution of X.

b. Find the probability that more than three of the water heaters will last more than 10 years.

c. Find the mean and variance of the random variable X.

7.1.24 A critical component of a machine operates successfully only 75% of the time. To increase the reliability of the system, four of the components will be installed so that the system will operate successfully if at least one of the components is functioning correctly. What is the probability that the system will fail? Assume that the components operate independently.

7.1.25 A doctor estimates that 5% of her patients have no real physical ailment. We randomly sample the records of the doctor and find that 3 of 15 patients had no ailment.

a. What is the probability of observing 3 or more patients who have no ailment in a sample of 15 patients if the population proportion is actually .05?

b. Why might your answer to part **a** make you believe that p is larger than .05?

7.1.26 Approximately 10% of all computers are returned for repair while their guarantee is still in effect. If a firm purchased nine computers, what is the probability that three or more will need repairs while their guarantees are still in effect?

7.1.27 An apartment manager knows that 15% of new washing machines purchased require maintenance during the first year of operation. The manager purchases four new machines. Assume the performances of the machines are independent of one another.

a. What is the probability that all four machines will require maintenance during the first year of operation?

b. What is the probability that none of them will require maintenance during the first year of operation?

c. What is the probability that at least two of them will require maintenance during the first year of operation?

7.1.28 Approximately 50% of a bank's customers would be willing to pay a fee to have pay-by-phone privileges. A sample is taken of six of the bank's customers. Let X denote the number who would pay for pay-by-phone privileges.

a. Construct the probability distribution of X.

b. Graph the probability distribution of X.

c. Calculate $E(X)$.

7.1.29 A doctor knows that about 20% of the population will have an adverse reaction to a certain type of medicine. The doctor prescribes the medicine for 20 patients. Find the probability that the number of patients who suffer an adverse reaction is the following:

a. Less than five

b. More than two

c. Exactly three

7.1.30 Suppose that X is a random variable having the binomial distribution. Given the following data, graph the distribution of X using rectangles.

a. $n = 5$, $p = .1$

b. $n = 5$, $p = .9$

c. $n = 5$, $p = .5$

7.1.31 Let X be a binomial variable with $p = .5$ and $n = 10$. Which of the following statements are true? Can you prove your answers?

a. $P(X = 0) = P(X = 1)$

b. $P(X = 0) = P(X = n)$

c. $P(X = 1) = P(X = n)$

d. $P(X = 1) = P(X = n - 1)$

e. $P(X = 2) = P(X = n - 2)$

7.1.32 A survey indicates that the probability of any student purchasing the school yearbook is .7. The school has 1,000 students and the yearbook costs $12.
    a. What is the expected number of sales?
    b. What is the expected revenue from yearbook sales?

 7.1.33 The National Record Store has taken surveys that indicate that 25% of the people who enter the store make a purchase. A survey of the cash register tapes indicates that the average revenue from a sale is $5.50. Next year are expected to enter the 100,000 people store.
    a. What is the expected number of people who will make a purchase?
    b. What is the expected revenue for the year?

7.2   The Poisson Distribution

The Poisson distribution is named after the French mathematician Simeon Denis Poisson (1781–1840), who published an article in 1837 discussing the distribution. The Poisson distribution has many applications in problems concerning the number of occurrences of some event during some continuous time interval or in some continuous region of space, including many applications in waiting-time problems and in queuing theories. The Poisson random variable X is a discrete random variable and indicates the number of occurrences of some event in a certain amount of time or space. The set of possible values of X is the set containing all the nonnegative integers $S = \{0, 1, 2, \ldots\}$.

EXAMPLE 7.8 **Examples of Poisson Random Variables** ─────────────────────────

In waiting-time problems, some variables that may have the Poisson distribution are the following:

    1. Number of telephone calls to a switchboard in a given minute (This assumes that incoming calls are independent of one another, an assumption that could be violated if there has been some type of emergency that leads to a rush of phone calls.)
    2. Number of customers entering a checkout line at a supermarket in a given minute
    3. Number of airplanes arriving at an airport in an hour
    4. Number of accidents at a factory in a day
    5. Number of customers purchasing a product during a day
    6. Number of cars crossing a bridge during a 5-second interval

The Poisson distribution also relates to problems concerning the number of occurrences of an event in a certain interval of space, such as the following:

    7. Number of misprints on a page of newsprint
    8. Number of flaws in a sheet of glass
    9. Number of paint scratches on a new car
    10. Number of bacteria in an ounce of fluid
    11. Number of flaws in a bolt of fabric

Characteristics of the Poisson Process

Variables that follow the Poisson distribution are said to be generated by a **Poisson process,** which has the following characteristics:

1. The occurrences of events are independent; that is, they have no effect on the probability of another occurrence of the event in the same or any other interval of time or space.
2. The probability of occurrence is approximately proportional to the length of the time or space interval.
3. The probability of more than one occurrence in any infinitesimally small interval is negligible.

In addition, the Poisson distribution can be used to approximate the binomial distribution in cases where the number of trials of the binomial experiment is large (say, $n \geqslant 50$) and where p is close to 0 so that $np < 5$. When p is close to 0 in a binomial experiment, we are studying what is called a *rare event*. The Poisson distribution is frequently used to determine probabilities in situations involving rare events.

The following formula gives the probability function for a Poisson random variable:

Formula for the Poisson distribution

$$P(X = x) = \frac{\mu^x e^{-\mu}}{x!} \quad \text{for } x = 0, 1, 2, \dots$$

In the formula, e is the number $2.71828 \dots$, and μ is a nonnegative number representing the mean of the Poisson distribution. The value μ represents the average number of occurrences of some rare event over a given interval of time or space. The values of $P(X = x)$ can be computed with a hand calculator that can calculate powers of e. If the mean has only one decimal place, Poisson probabilities can be obtained directly from Table A.4 in the Appendix.

EXAMPLE 7.9 **Calculating Poisson Probabilities** ─────────────────────────

From noon to 4 P.M., the number of airplanes taking off from a certain runway during any 1-minute interval has a Poisson distribution with mean $\mu = 1.5$.
a. Find the probability that during a specific 1-minute interval, no planes take off.
b. Find the probability that exactly two planes take off during a specific 1-minute interval.
c. Find the probability that during a 1-minute interval at least one plane takes off.

SOLUTION   **a.** We obtain

$$P(X = 0) = \frac{(1.5)^0 e^{-1.5}}{0!} = .223$$

b. We obtain

$$P(X = 2) = \frac{(1.5)^2 e^{-1.5}}{2!} = .251$$

c. We obtain

$$P(X \geqslant 1) = 1 - P(X = 0) = 1 - .223 = .777$$

EXAMPLE 7.10   **Calculating Poisson Probabilities** ————————————————

Suppose the number of misprints on a newspaper page has a Poisson distribution with mean $\mu = 2.2$. Construct the probability distribution of X.

SOLUTION
$$P(X = 0) = \frac{(2.2)^0 e^{-2.2}}{0!} = e^{-2.2} \approx .111$$

$$P(X = 1) = \frac{(2.2)^1 e^{-2.2}}{1!} \approx .244$$

Similarly, we obtain the results shown in Table 7.2. Figure 7.3 shows the Poisson distribution for the case when $\mu = 2.2$.

TABLE 7.2   *Poisson distribution for $\mu = 2.2$ (Example 7.10)*

x	Probability
0	.111
1	.244
2	.268
3	.197
4	.108
5	.048
6	.017
7	.005
8	.002
9 or more	.000
	1.000

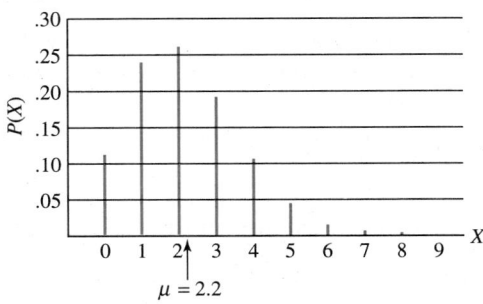

FIGURE 7.3   *Poisson distribution for $\mu = 2.2$ (Example 7.10)*

EXAMPLE 7.11   **Mean of the Poisson Random Variable** ————————————————

Use the formula for the mean of a discrete random variable and verify that the mean of the Poisson random variable in Example 7.10 is $\mu = 2.2$.

SOLUTION   For any discrete random variable, the mean is given by

$$E(X) = \Sigma \, xP(X = x)$$

We obtain

$$
\begin{aligned}
\mu = {}& 0(.111) + 1(.244) + 2(.268) + 3(.197) + 4(.108) \\
& + 5(.048) + 6(.017) + 7(.005) + 8(.002) + 9(.000) \\
= {}& 2.2
\end{aligned}
$$

(When X is 10 or more, the additional terms $xP(x)$ are approximately 0 because the values of $P(x)$ get progressively smaller as x increases.)

Use of the Poisson Probability Table

The Poisson distribution depends on the single parameter μ, and both the mean and variance of the Poisson random variable are equal to μ. Each value of μ determines a different distribution. Thus, like the binomial distribution, the Poisson distribution is really a family of distributions, where each member is determined by the specific value of μ.

If the distribution is going to be used extensively, it is convenient to use a table of Poisson probabilities. Poisson probabilities for selected values of μ are given in Table A.4 of the Appendix.

EXAMPLE 7.12   Using a Table to Calculate Poisson Probabilities ————————————

Assume that X has a Poisson distribution with mean $\mu = 1.7$. Calculate $P(X \leq 4)$.

SOLUTION   From Table A.4, we obtain for $\mu = 1.7$

$$
\begin{aligned}
P(X \leq 4) &= P(X = 0) + P(X = 1) + P(X = 2) + P(X = 3) + P(X = 4) \\
&= .1827 + .3106 + .2640 + .1496 + .0636 \\
&= .9705
\end{aligned}
$$

Shape of the Poisson Distribution

The shape of the Poisson distribution varies with the value of the mean μ. Figure 7.4 shows Poisson probability distributions for different values of μ. As μ increases, the probability shifts to the right and the distribution becomes more bell shaped. When μ is no more than 1, the distribution is extremely skewed to the right with almost all of the probability located at $P(0)$ and $P(1)$.

Poisson Distribution As Approximation to the Binomial Distribution

The Poisson distribution is used to approximate the binomial distribution when the number of trials n is large and the probability of a success on any trial p is small. The rule for this application of the Poisson distribution is given in the accompanying box.

Poisson approximation to the binomial distribution

> The Poisson distribution provides a good approximation to the binomial distribution provided $n \geq 50$ and $np < 5$. Recall that the mean of the binomial distribution is given by $\mu = np$. When using the Poisson distribution to approximate the binomial distribution, set the mean of the Poisson distribution equal to np.

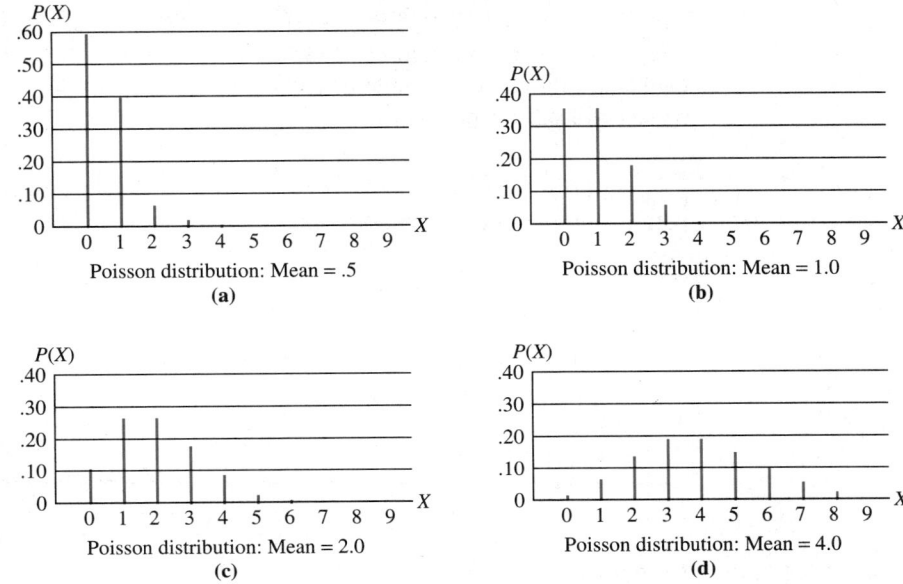

FIGURE 7.4 *Poisson distributions for various values of μ*

EXAMPLE 7.13 **Poisson Approximation to the Binomial**

A company sells insurance policies to a random sample of $n = 1,000$ men who are 35 years of age. The probability that a 35-year-old man dies within a year is approximately .002 (from *Vital Statistics of the United States, Life Tables*). What is the probability that the insurance company will have to pay claims on two or more policies in the next year?

SOLUTION Let X denote the number of men in the sample who die within 1 year. Then X is a binomial random variable with $n = 1,000$ and $p = .002$. Because $n > 50$ and $np < 5$, we use the Poisson distribution to approximate the binomial where $\mu = np = (1,000)(.002) = 2$. We obtain

$$
\begin{aligned}
P(X \geqslant 2) &= 1 - P(X = 0) - P(X = 1) \\
&\approx 1 - \frac{2^0 e^{-2}}{0!} - \frac{2^1 e^{-2}}{1!} \\
&= 1 - .1353 - .2707 \\
&= .5940
\end{aligned}
$$

The exact binomial probability is

$$
P(X \geqslant 2) = 1 - .13506 - .27067 = .59427
$$

The Poisson approximation to the binomial is extremely accurate in this case.

EXAMPLE 7.14   **Another Poisson Approximation to the Binomial** ——————————

About 1 person in 1,000 suffers a bad reaction to a certain medicine. If a doctor gives this medicine to a random sample of 2,600 patients, what is the probability that exactly 1 person will suffer a bad reaction?

SOLUTION   We have $n = 2,600$, $p = .001$, and $np = 2.6$. We use the Poisson distribution with $\mu = 2.6$ and obtain

$$P(X = 1) \approx \frac{2.6^1 e^{-2.6}}{1!}$$
$$= .1931$$

The exact binomial probability is $P(X = 1) = .19305$, which is virtually identical to the Poisson approximation.

EXAMPLE 7.15   **Comparing Poisson and Binomial Probabilities** ——————————

To show that the Poisson distribution provides a good approximation to the binomial distribution, let us calculate some exact binomial probabilities when $n = 100$ and $p = .02$ and the corresponding Poisson probabilities using $\mu = 2$.

SOLUTION   Table 7.3 shows corresponding binomial and Poisson probabilities. The Poisson probabilities were obtained from Table A.4. Note the similarities in the probabilities and in the graphs of the probability distributions, shown in Figures 7.5 and 7.6. Figure 7.6 shows the typical shape of a Poisson random variable, with most of the probability concentrated near 0 and a long tail extending to the right.

Tips for Problem Solving

1. The Poisson distribution provides a good approximation to the binomial distribution when we are dealing with a **rare event.**
2. The number of trials should be large. Verify that $n \geqslant 50$.
3. The expected number of occurrences should be small. Verify that $np < 5$.
4. To approximate a binomial probability by a Poisson probability, use $\mu = np$ in the Poisson formula.
5. Use the formula

$$P(X = x) = \frac{\mu^x e^{-\mu}}{x!}$$

where $\mu = np$ and x denotes the number of successes.

TABLE 7.3 *Comparison of probabilities for binomial ($n = 100, p = .02$)*
and Poisson ($\mu = 2$) distributions, $P(X = x)$

	VALUE OF x							
	0	1	2	3	4	5	6	7
Binomial $P(X = x)$	.1326	.2707	.2734	.1823	.0902	.0353	.0114	.0031
Poisson $P(X = x)$	.1353	.2707	.2707	.1804	.0902	.0361	.0120	.0034

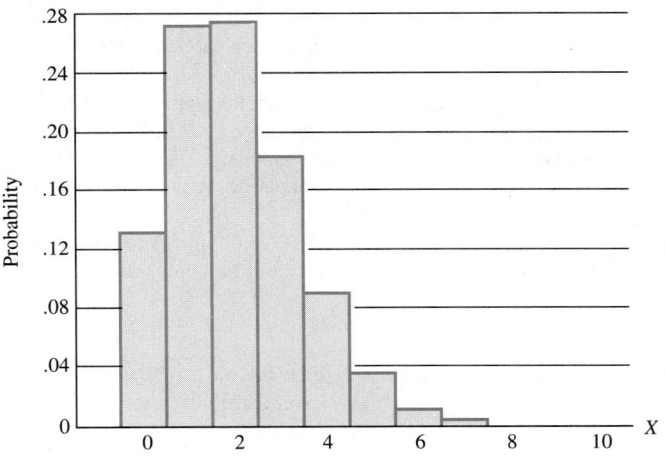

FIGURE 7.5 *Binomial probability distribution for $n = 100$ and $p = .02$*

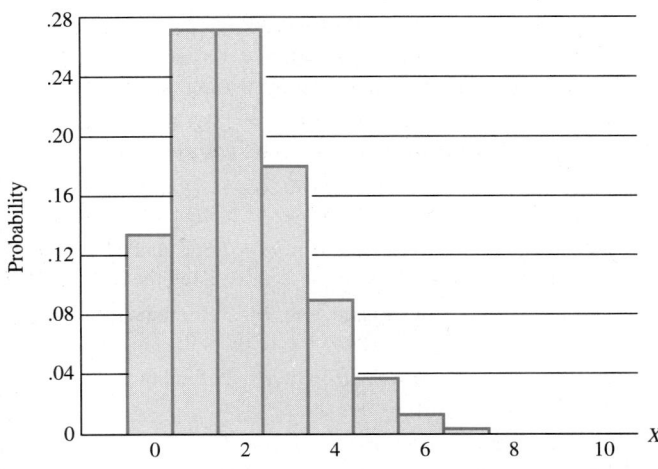

FIGURE 7.6 *Poisson probability distribution for $\mu = 2$*

Exercises

Statistical Concepts

7.2.1   State the formula for calculating a Poisson probability.

7.2.2   State the formulas for the mean, variance, and standard deviation of a Poisson random variable.

7.2.3   Suppose X is a Poisson random variable. Describe how the shape of the probability function changes as μ varies from .1 to 3.0.

7.2.4   State the conditions under which the Poisson distribution provides a good approximation to the binomial distribution.

7.2.5   *True or false:*   Let X be a Poisson random variable with $\mu = 2$. Then $P(X = 4) = P(X \leq 4) - P(X \leq 3)$. Explain.

7.2.6   Can any of the following probabilities be calculated using the Poisson distribution? If not, explain why.
    a. The probability that more than three patients arrive at a hospital emergency room during a given hour. On the average, arrivals occur randomly at the rate of one every 15 minutes.
    b. The probability that an emergency ambulance service receives more than five calls during a 4-hour period. Records show that calls occur randomly and independently at the rate of one per hour.
    c. The probability that at least 5 of 15 football players recruited by a college coach will eventually agree to play for the coach. Past records show that, on the average, the coach signs 20% of the players he recruits.

Statistical Drills

7.2.7   Show that the Poisson probability distribution provides a good approximation to the binomial probability distribution. Use the Poisson distribution to calculate $P(X = 1)$ and $P(X = 2)$ when $n = 20$ and $p = .02$. Calculate the exact probabilities using the binomial distribution.

7.2.8   Let X be a Poisson random variable with $\mu = 1$. Calculate the Poisson probabilities $P(X = x)$ for $x = 0, 1, 2, 3, 4, \ldots$. Verify that these probabilities sum to 1. Graph the distribution.

7.2.9   Let X be a Poisson random variable with $\mu = 1$. Calculate $E(X)$ by doing the arithmetic calculations. Verify that you get the same answer by using the formula $E(X) = \mu$.

7.2.10   Let X be a Poisson random variable with $\mu = 1$. Calculate $\text{Var}(X)$ by doing the arithmetic calculations. Verify that you get the same answer by using the formula $\text{Var}(X) = \mu$.

7.2.11   Let X be a Poisson random variable with $\mu = 1.5$. Calculate the Poisson probabilities $P(X = x)$ for $x = 0, 1, 2, 3, \ldots$. Verify that these probabilities sum to 1. Graph the distribution.

7.2.12   Let X be a Poisson random variable with $\mu = 1.5$. Calculate $E(X)$ by doing the arithmetic calculations. Verify that you get the same answer by using the formula $E(X) = \mu$.

7.2.13   Let X be a Poisson random variable with $\mu = 1.5$. Calculate $\text{Var}(X)$ by doing the arithmetic calculations. Verify that you get the same answer by using the formula $\text{Var}(X) = \mu$.

7.2.14   Let X be a Poisson random variable with $\mu = 3$. Find the following:
    a. $P(X = 0)$ **b.** $P(X = 1)$ **c.** $P(X = 2)$ **d.** $P(X \geq 3)$

7.2.15   Let X be a Poisson random variable with $\mu = 2$. Find the following:
    a. $P(X \leq 2)$ **b.** $P(X > 2)$

7.2.16   Let X have the Poisson distribution with $\mu = .5$.
    a. Calculate the probability distribution of X and graph it.
    b. Calculate the mean and verify that the mean is .5.

7.2.17   Let X have the Poisson distribution with $\mu = .5$.
    a. Calculate $P(X = 2)$ by performing the computations.

b. Find $P(X = 2)$ by using the Poisson distribution table (Table A.4) in the Appendix.

c. Verify that the answers in parts **a** and **b** are the same.

d. Repeat parts **a**–**c** for $P(X = 3)$ and $P(X = 4)$.

Statistical Applications

7.2.18 The number of flaws in a sheet of glass follows the Poisson distribution. The average number of flaws in a sheet of glass measuring 10 square feet is 1. Find the probability of the following:

    a. A 2-by-5-foot sheet contains no flaws.

    b. A 2-by-5-foot sheet contains two flaws.

    c. A 3-by-10-foot sheet contains no flaws.

7.2.19 During any 1-minute interval, the number of cars arriving at a tollbooth has a Poisson distribution. The average number of arrivals in any 1-minute period is 3. Find the probability of the following:

    a. Two cars arrive between 3 P.M. and 3:01 P.M.

    b. More than two cars arrive between 3 P.M. and 3:01 P.M.

7.2.20 On Saturday nights, the number of crimes reported to police headquarters follows a Poisson process. During any 10-minute interval, the average number of reports is 2. Find the probability that no crimes are reported during intervals of the following lengths:

    a. 10 minutes   **b.** 30 minutes

7.2.21 Assume that the number of mufflers purchased at a service station in a week follows the Poisson distribution with $\mu = 2$. Deliveries are made every Monday, so every Monday the mechanic has to decide how many mufflers to stock. How large a stock should the mechanic have on hand on Monday to be 95% certain that the week's demand can be satisfied?

7.2.22 The number of people entering a bank during any 2-minute interval follows the Poisson distribution with mean $\mu = 2$. Find the probability of the following during a 5-minute interval:

    a. No one enters the bank.   **b.** At least two people enter the bank.

7.2.23 Suppose that the number of accidents per week over a certain stretch of highway has a Poisson distribution with mean $\mu = 2$. Find the probability that on that stretch there will be the following:

    a. No accidents next week   **b.** Two accidents next week

    c. More than three accidents next week

7.2.24 At a certain university, computer breakdowns occur randomly at the rate of one every 10 days. Assume that the number of breakdowns in any 10-day period has a Poisson distribution with mean $\mu = 1$. Find the probability of the following:

    a. No breakdowns during the next 10 days

    b. More than one breakdown during the next 20 days

7.2.25 An insurance office has 400 typewriters. The probability that any one of them will require repair on a given day is .01. Find the probability that fewer than four of the typewriters require repair on a particular day. Use the Poisson approximation to the binomial distribution.

7.2.26 The Internal Revenue Service reports that 2.5% of all taxpayers make arithmetic mistakes on their income tax returns. A random sample of 100 returns will be checked.

    a. What is the probability that none of them will contain errors?

    b. What is the probability that at least three of them will contain errors?

Chapter 7 Supplementary Exercises

7.S.1 The probability that a sales representative makes a sale to any customer is .5.

    a. Find the probability of making exactly two sales to six customers. Assume that events are independent.

    b. Find the probability of getting more than one sale in six tries.

7.S.2   A stockbroker has suggested three industrial stocks to a client. The broker claims that each of these stocks has a probability of .6 of increasing in price within a week. The client decides to buy the stocks. Assume that movements in the prices of these three stocks are independent of one another. What is the probability of the following?

a. At least one of these stocks does not increase in value within a week.

b. All three increase in price.

7.S.3   In the past, about 30% of the new toys manufactured by the Martel Toy Company have yielded a profit. The profits from the successful products have been large enough to cover the losses from marketing the unsuccessful ones. For the next Christmas season, the company is planning to introduce four independent new toys. What is the probability of the following?

a. At least one of the toys yields a profit.

b. None of the toys yields a profit.

c. All four toys yield a profit.

7.S.4   It is known that about .3% of the population have an allergic reaction to the chemicals used at the Moore Chemical Company. Suppose that the company hires 500 new employees. What is the probability of the following?

a. None of the new employees has an allergic reaction.

b. More than one have an allergic reaction.

7.S.5   A sales representative sells life insurance policies to five men, each 30 years of age. Suppose that the probability is .8 that any 30-year-old man will live for 20 more years. Let X denote the number of these men alive after 20 years.

a. Construct and graph the probability distribution of X.

b. Calculate the mean and variance of X. Use the general formulas for mean and variance, and check your results using the formulas $\mu = np$ and $\sigma^2 = npq$.

7.S.6   A certain doctor has a great reputation for performing successful heart transplants. She has performed four heart transplants, and two of the patients survived for more than 5 years. The doctor thinks that she is not better than other doctors in the field and that she has been very lucky. She thinks that the probability of a heart transplant patient surviving for 5 years or more is .2. What is the probability of having two patients out of four survive for 5 years if the probability of survival is .2?

7.S.7   Approximately 80% of the people in a community would pay to have cable TV. A salesperson visits a random sample of nine homes. Find the probability that fewer than seven of the homes want cable TV.

7.S.8   A copying machine has an average rate of one defective copy per 100 pages. Suppose you need to copy 200 pages.

a. Find the probability of no defective pages.

b. Find the probability of exactly one defective page.

c. Find the probability of at least two defective pages.

7.S.9   About 70% of the driving population would use self-service gas stations if the price of gas were lower than at full-service gas stations. What is the probability that, in a random sample of eight drivers, from four to six inclusive will say that they would use self-service gas stations?

7.S.10   The actuary of a life insurance company has found that the probability of a person having a certain type of fatal accident in a given year is .0002. If the company holds 8,000 life insurance policies, what is the probability that the company will have to pay three or more claims next year?

7.S.11   A gasoline station sells gas at a 5¢-per-gallon discount if the customer uses self-service. Past data indicate that 70% of all customers choose the self-service system. During a given time period, 10 customers enter the station. What is the probability of the following?

    a. More than five customers use self-service.
    b. Fewer than eight customers use self-service.

7.S.12 Surveys show that 10% of the population is left-handed. A classroom contains 10 students and 10 chairs, all of which are designed for right-handed people.
    a. What is the expected number of left-handed students in the class?
    b. What is the probability that all of the students are right-handed?

7.S.13 A sales representative for the Zavos Air Conditioning Company makes 10 house calls a day. The probability of making a sale at a randomly selected house is .1. Find the probability that the sales representative makes the following:
    a. No sales in a day
    b. One or more sales in a day
    c. Exactly two sales in a day

7.S.14 A major oil company has drilled six independent test wells in Alaska. If the probability of any well producing oil is .4, find the probability of the following:
    a. Two of the wells are productive.
    b. Five of the wells are productive.
    c. None of the wells is productive.

7.S.15 At an urban medical clinic, 80% of the incoming patients are covered by some type of medical insurance. Find the probability of the following among the next 10 incoming patients:
    a. More than four are covered.
    b. Fewer than eight are covered.

7.S.16 Thirty percent of the employees at the Conroy Insurance Company travel to work on the bus and will be hours late whenever the bus system fails. A sudden snowstorm has shut down all bus service in the city. Suppose that the accounting department contains 10 people. Assume independence. Find the probability of the following in the accounting department:
    a. At least two people are late.
    b. Fewer than five people are late.

7.S.17 On the average, an airline loses about 1 piece of luggage out of 1,000. Suppose that the airline handles 3,000 pieces of luggage per day. If the airline promises its employees a bonus for every day on which no luggage is lost, what is the probability that the employees get a bonus on any given day?

7.S.18 In a suburban area, it is thought that 20% of the households will purchase an automatic garage door opener. If a salesperson randomly contacts six different families, what is the probability that the salesperson makes the following:
    a. No sales
    b. One sale
    c. More than two sales

7.S.19 A columnist claims that 50% of all cars on the state highways have something wrong with the brake system. If the claim is correct, what is the probability that, in a random sample of six cars, none is defective?

7.S.20 Let $\mu = .7$. Construct the probability distribution of X, where X has the Poisson distribution.

7.S.21 Construct and graph the probability distribution of the Poisson variable X having mean $\mu = 1$.

7.S.22 The probability is .005 that a person will buy a product that is brought to his or her attention through a phone call by a sales representative. If a salesman makes 500 independent phone calls, what is the probability that he will sell at least two products? That is, find $P(X \geq 2)$.

7.S.23   The number of customers who enter a service station to buy gas in any 5-minute period has the Poisson distribution with mean .3. Find the probability that during a 5-minute period, four customers enter the station.

7.S.24   A car manufacturer places a 1-year guarantee on the steering mechanism of its cars, and about 1 car in 5,000 has a faulty steering mechanism on the average. If 20,000 cars are sold, what is the probability that more than three cars will have faulty steering mechanisms? Assume independence.

7.S.25   On the average, 1 person in 1,000 is allergic to a certain type of food. If the food is served to 500 schoolchildren, what is the probability that no one is allergic to the food? Assume independence.

7.S.26   The number of telephone calls passing through a switchboard has a Poisson distribution with mean equal to $3t$, where t is the time in minutes. Find the probability of the following:
   a. Two calls in any one minute
   b. Four calls in 2 minutes
   c. At least two calls in 2 minutes
   d. At least one call in 1 minute

7.S.27   At a particular computer center, a keypunch operator must have an error rate of .1% or less. That is, on the average there can be no more than 1 incorrect keystroke for every 1,000. A job applicant punches 2,000 keystrokes and makes 5 errors.
   a. If the applicant really has a .1% error rate, what is the probability that more than four errors will be made on 2,000 keystrokes?
   b. What is the probability that fewer than three errors will be made on 2,000 keystrokes?
   c. What is the probability that no errors will be made on 2,000 keystrokes?

7.S.28   Assume that on the average 1 person out of 25 who make plane reservations fails to show. An airline takes 100 independent reservations and has 96 seats. What is the probability that every passenger will have a seat?

7.S.29   On the average, about 1 person out of 100 dies between the ages of 20 and 30. An insurance company insures 350 twenty-year-old people. What is the probability that fewer than three of these people die before they are 30?

7.S.30   Assume that about 1 prospective homeowner in 1,000 defaults on a home mortgage. A local savings and loan has issued 500 mortgages, and three have defaulted. The Federal Home Loan Bank Board is accusing the savings and loan of shoddy lending practices. If 1 person in 1,000 defaults, what is the probability of getting two or fewer defaults from 500 customers?

7.S.31   A company claims that, on the average, a certain rocket motor will fail to start only 2 times out of 1,000 tries. In a test, the motor failed to start 3 times out of 500 tries. What is the probability of this occurring if the company's claim is true?

7.S.32   A political science professor has argued that the probability of a nuclear war occurring during any given year is .02. Assume that annual events are independent of one another. Using the binomial distribution, find the probability of the following:
   a. There is no nuclear war in the next 10 years.
   b. There is no nuclear war in the next 50 years.

7.S.33   The Koch Electric Company makes electric shavers. If the probability that an electric shaver is defective is .01, what is the probability of the following in a shipment of 500 electric shavers?
   a. None is defective.
   b. One is defective.
   c. More than three are defective.

7.S.34   Assume that at the college library, books can be checked out for 1 month. Assume that the number of requests per month for *The General Theory* by John Maynard Keynes follows a Poisson distribution with mean $\mu = 2$. How many copies should the library have so that in a given month it can fulfill all requests with a probability of .95?

7.S.35 A company has determined that the probability that a newly appointed apprentice will remain with them for at least 1 year is .7. Four apprentices are appointed.
 a. Find the probability distribution for the number of apprentices who remain with the company after a year's employment if their decisions to remain are made independently.
 b. Find the probability that at least three of these apprentices will remain with the company for a year or more.

7.S.36 A delivery service claims that 90% of its parcels are delivered within 48 hours. If the claim is true, find the probability of the following in a random sample of 10 parcels:
 a. Exactly seven are delivered in less than 48 hours.
 b. Fewer than six are delivered in less than 48 hours.
 c. Assume that fewer than six of the parcels were delivered in less than 48 hours. Would this make you doubt the shipper's claim? Explain.

7.S.37 When the mailroom clerk of a publishing house is in a hurry to go home, the probability that he or she will make a mistake in an address is .25. If five shipments have to be sent out just before closing time, find the probability that a mistake will be made in addressing at least two of these shipments. Assume errors are independent of one another.

7.S.38 A store sells an average of three jars of pizza sauce a week and brings its inventory up to five jars every Saturday when the delivery truck stops by. Assume that the Poisson distribution applies. What is the probability that more than five jars will be sold in a week's time?

7.S.39 A bookie offers gamblers the following bet. The gambler must guess the gender of the next 10 babies born at the city hospital. There is a $2 fee, but the gambler wins $1,000 if the gender of each child is guessed correctly. Assume that boys and girls are equally likely.
 a. What is a gambler's probability of winning the bet?
 b. What is the gambler's expected return?

7.S.40 There are 40,000 people and one hospital in a town. On any given day, there is a .0001 probability that a person will require a hospital bed. How many beds must the hospital have to be 99% sure that there will be enough beds for everyone who requires one on any given day?

7.S.41 A con artist mails gambling advice concerning an evenly matched football game to 320 wealthy gamblers. Half of the letters predict team A will win and half predict team A will lose. Suppose team A wins. The con artist crosses out the 160 gamblers who were given bad advice and mails another prediction concerning another evenly matched football game to the 160 who were given correct advice. Half of these 160 people are told that team A will win the game and half are told that it will lose. At the end of this game, 80 gamblers have been given two correct predictions in a row. These people are given a third prediction; again, half are told that one team will win and the other half are told that the team will lose. By the end of five games, there are 10 gamblers who have received five correct predictions in a row, and they should be eager to subscribe to the con artist's expensive advice.
 a. What is the probability of five correct predictions in a row?
 b. If you had received these five correct predictions, would you be impressed?
 c. If you knew how these predictions were made, would you be impressed by five correct predictions?

7.S.42 A computer magazine claims that about 1 out of 100 floppy disks is defective. Suppose an academic department buys a random sample of 300 floppy disks. Find the probability that fewer than two are defective.

7.S.43 The economics department has mailed fellowship offers to five prospects for graduate school. In the past, 32% of all offers have been accepted. Find the probability that exactly three of the five offers will be accepted.

7.S.44   The probability that a college female will live to age 60 or more is .99. In a random sample of 200 college women, what is the probability that more than three of these women die before reaching age 60?

7.S.45   An office has five computer terminals, and each terminal is broken 10% of the time.
   a. Find the probability that at least two terminals are working.
   b. Find the probability that at least one terminal is broken.

7.S.46   In a certain county in Mississippi, 70% of the homes are believed to be insured against flood damage. Four homeowners are chosen at random from the entire population of homeowners in the county. What is the probability that at least two of the four have insurance?

7.S.47   Which of the following problems can be solved by using the binomial distribution? If the use of the binomial distribution is inappropriate, explain why.
   a. Calculate the probability that a customer files a repair claim for a product. Past records indicate that 12% of all buyers file claims.
   b. Calculate the probability that a company receives one defective computer in a shipment of four computers. The four computers were purchased from a company that had 5 defective and 80 nondefective computers in its warehouse.
   c. Calculate the probability that at least 2 out of 25 people respond to the mailing of an advertisement. The typical response rate is 10%.

7.S.48   Maintenance records indicate that 1% of all new camcorders require repairs during the first year of use. A camera store has sold 10 camcorders during the Christmas season.
   a. Find the probability that none of these camcorders requires repairs during the first year of use.
   b. Find the probability that fewer than two of these camcorders require repairs during the first year of use.

7.S.49   Suppose the number of people entering the emergency room at a particular hospital in any hour possesses a Poisson probability distribution with mean equal to three persons per hour.
   a. What is the probability that the number of people entering the emergency room during a particular hour will exceed five?
   b. Suppose the emergency room can handle 15 patients per hour. Is it likely that the number of arrivals during any hour will exceed 15? Explain your answer.

7.S.50   Mail questionnaires usually have low response rates, but they tend to be less expensive than personal interviews or telephone interviews. The public relations director for the National Rifle Association (NRA) has sent a questionnaire to 8,000 NRA members asking for an opinion concerning a new gun control law being discussed in Congress. Past questionnaires on similar issues had a 20% return rate. Let the random variable X denote the number of replies. If questionnaires were sent to 8,000 members, find the expected number of replies $E(X)$ and the variance of the number of replies. Based on the Empirical Rule, within what limits would X be expected to fall? [*Hint:*   Use the mean and variance and find the interval $(\mu - 2\sigma, \mu + 2\sigma)$.]

7.S.51   Brand preference studies for beverages are often conducted by providing complimentary drinks of the two beverages to a selected group of consumers. The consumer is blindfolded or the drinks are served in identical containers, so the consumer does not know which drink is brand A and which is brand B. After tasting the drinks, each consumer states his or her brand preference. To test the hypothesis that one brand is preferred to the other, we start with the assumption that there is no preference between brands; that is, it is assumed that $p = .5$, where p is the probability any specific consumer favors brand A. Suppose a brand preference study of two brands is conducted among 20 consumers and assume that there is actually no difference in the quality of the brands.

    a. If the hypothesis that $p = .5$ is true, find the probability that 16 or more consumers would state a preference for brand A.

    b. Suppose 17 of the consumers state that they prefer brand A. Would this convince you that, in fact, $p > .5$? Explain your reasoning.

    c. Suppose 12 of the consumers state that they prefer brand A. Would this convince you that, in fact, $p > .5$? Explain your reasoning.

    d. Suppose four of the consumers state that they prefer brand A. Would this convince you that, in fact, $p < .5$? Explain your reasoning.

7.S.52 A quality control engineer wants to monitor the production of curved glass windows that are used in solar rooms. It is known that about 10% of the windows will need to be reworked because they do not have the correct degree of curvature. Suppose a random sample of 1,000 windows is inspected. Let the random variable X denote the number of windows that must be reworked. Find the expected value and variance of X. Based on the Empirical Rule, within what limits would X be expected to fall? [*Hint:* Use the mean and variance and find the interval $(\mu - 2\sigma, \mu + 2\sigma)$.] Suppose that there were 130 defective windows in the sample. Would this raise your suspicion that the population proportion had changed and was greater than .1? Explain your reasoning.

7.S.53 The mayor of a city claims that 60% of the taxpayers in the city favor raising taxes to increase the size of the police force. To test this claim, a random sample of 20 taxpayers is questioned. Suppose that 17 of the taxpayers favor the proposed tax increase.

    a. What is the probability of observing at least 17 taxpayers who favor the increase if, in fact, $p = .6$?

    b. Suppose that only six of the taxpayers favor the tax increase. What is the probability of observing at most six taxpayers who favor the increase if, in fact, $p = .6$? Would this occurrence raise your suspicion that, in fact, $p < .6$? Explain.

7.S.54 A manufacturer of 3.5-in. computer diskettes claims that less than 5% of its diskettes are defective. You purchase a set of 20 diskettes and discover that 5 are defective.

    a. Suppose the manufacturer's claim is true. Let the random variable X denote the number of defective diskettes in a random sample of 20 diskettes. Find $E(X)$ and the variance of X.

    b. Suppose the manufacturer's claim is true. What is the probability of getting exactly 5 defective diskettes in a random sample of 20 diskettes?

    c. Suppose the manufacturer's claim is true. What is the probability of getting 5 or more defective diskettes in a random sample of 20 diskettes?

    d. Suppose your sample of 20 diskettes contained 5 defectives. Would this raise your suspicion that, in fact, $p > .05$? Explain.

7.S.55 Quality control engineers monitor production processes by keeping track of the proportion of defective items being produced. Suppose that a certain process is considered to be in control if the proportion of defective items manufactured by the process is at most 5%. To check whether the process is in control, 15 items are randomly selected from the output. Let the random variable X denote the number of defective items obtained in the random sample of $n = 15$ items. Suppose the quality control engineers adopt the following decision rule: Reject the hypothesis that the process is in control if three or more defectives are found.

    a. Suppose that, in fact, the process is in control and that $p = .05$. Find the probability that the engineers will incorrectly conclude that the process is out of control.

    b. Suppose that, in fact, the process is out of control and that $p = .10$. Find the probability that the engineers will incorrectly conclude that the process is in control.

7.S.56 A retailer receives a large shipment of lawn tractors. To protect the company against a bad shipment, the department manager will carefully inspect 10 tractors and accept the entire lot if she observes 0 or 1 defective. Suppose that, in fact, 5% of all the lawn tractors are defective. What is the probability that she accepts the entire shipment?

7.S.57 A manufacturer of disposable cameras claims that at most 1% of the cameras manufactured and sold by his firm contain defects. A department store purchases 400 of these cameras for sale during the holiday season. Let the random variable X denote the number of defective cameras received by the department store.

a. Find the probability that none of these cameras is defective.

b. Find the probability that at most five are defective.

c. Find the probability that at least two are defective.

d. Find the expected value and variance of X.

e. Based on the Empirical Rule, within what limits would X be expected to fall? [*Hint:* Use the mean and variance and find the interval $(\mu - 2\sigma, \mu + 2\sigma)$.]

f. Suppose that the 400 cameras are sold and consumers return 8 defective cameras. Would this make you suspect that, in fact, $p > .01$? Explain.

7.S.58 Quality control engineers monitor production processes to detect instances when the process might be changing because of assignable causes of variation, that is, causes that can be readily identified and eliminated. These engineers have adopted various rules, called Rules for Pattern Analysis, to help them assess whether a process is out of control. Some important characteristic, such as the diameter of bearings made at an aircraft manufacturing plant, is measured. One rule states that a process may be out of control if we obtain nine consecutive observations that are above the median. Suppose that, in fact, the process is in control; that is, for every bearing, the probability is .5 that its diameter will exceed the median. Suppose the engineers inspect a sample of nine bearings. Find the probability that all nine diameters will exceed the median and thus provide an incorrect indication that the process is out of control. This mistake is called a Type I error—the engineers have incorrectly rejected the hypothesis that the process is in control when, in fact, the hypothesis is true.

7.S.59 Quality control engineers consider a process to be out of control if at least two out of three consecutive observations are more than 2 standard deviations above the mean. Suppose the probability that any specific observation will be more than 2 standard deviations above the mean is approximately .0228. Suppose the process is in control. Find the probability of making a Type I error and incorrectly rejecting the hypothesis that the process is in control based on observing at least two out of three consecutive observations more than 2 standard deviations above the mean.

7.S.60 Quality control engineers also consider a process to be out of control if at least four out of five consecutive observations are more than 1 standard deviation above the mean. Suppose the probability that any specific observation will be more than 1 standard deviation above the mean is approximately .1587. If the process is in control, find the probability of making a Type I error and considering the process to be out of control based on observing at least four out of five consecutive observations more than 1 standard deviation above the mean.

References

Canavos, George C., and Don M. Miller. *Modern Business Statistics.* Belmont, CA: Duxbury Press, 1995.

Cryer, Jonathan D., and Robert B. Miller. *Statistics for Business: Data Analysis and Modelling.* Boston: PWS-Kent, 1991.

Feller, William. *An Introduction to Probability Theory and Its Applications.* 2 vols. New York: Wiley, 1950–1966.

Freund, John E., and R. E. Walpole. *Mathematical Statistics.* 4th ed. Englewood Cliffs, N.J.: Prentice-Hall, 1987.

Handbook of Tables for Probability and Statistics. 2d ed. Cleveland: Chemical Rubber Company, 1968.

Kohler, Heinz. *Statistics for Business and Economics.* Glenview, IL: Scott, Foresman, 1988.

McClave, James T., and Frank H. Dietrich. *Statistics.* 6th ed. New York: Macmillan, 1994.

Mendenhall, William, James E. Reinmuth, and Robert Beaver. *Statistics for Management and Economics.* 6th ed. Boston: PWS-Kent, 1989.

National Bureau of Standards. *Tables of the Binomial Probability Distribution.* Washington, D.C.: Government Printing Office, 1949.

Neter, J., W. Wasserman, and G. A. Whitmore. *Applied Statistics.* 3d ed. Boston: Allyn & Bacon, 1987.

Neter, John, William Wasserman, and G. A. Whitmore. *Fundamental Statistics for Business and Economics.* 4th ed. Boston: Allyn & Bacon, 1973.

Shiffler, Ronald E., and Arthur J. Adams. *Introductory Business Statistics with Computer Applications.* Belmont, CA: Wadsworth, 1995.

Watson, Collin J., Patrick Billingsley, D. James Croft, and David V. Huntsberger. *Statistics for Management and Economics.* 4th ed. Needham Heights, MA: Allyn & Bacon, 1990.

8 Some Useful Continuous Probability Distributions

Properties of Continuous Probability Distributions

As we discussed in Chapter 6, a *continuous random variable X* is a random variable that can take all values in an interval. In this chapter, we study two of the most important continuous probability distributions. The simplest one, the *uniform distribution,* is discussed first. Next, the *normal distribution* is treated in detail. The normal curve is a smooth, bell-shaped curve that can be used to approximate many different probability distributions. The normal distribution is by far the most important distribution in statistics.

Strictly speaking, every measured variable is discrete because no continuous variable can be measured exactly. Any measurement has to be rounded off after a finite number of significant digits. Nevertheless, it is conceptually advantageous to view many variables as strictly continuous. In fact, some discrete variables are treated as though they were continuous, either because it would be too tedious to list all the values of X and their associated probabilities or because the differences between successive values of the random variable X are insignificant. It is convenient to regard as continuous those essentially discrete random variables that are measured on such a fine grid that the probability of occurrence of any specific value is extremely small. Examples of discrete variables that usually are treated as continuous variables are family income, college grade point average, taxes paid by a corporation, and the sales revenue of a company. Any monetary figure can be expressed only to the nearest unit of currency, so every monetary amount is inherently discrete, but in most cases we treat monetary amounts as continuous variables.

Representing Probabilities by Areas

Because the number of points in an interval is infinite, we cannot assign a positive probability to every point and still have the probabilities sum to 1. Thus we need to use a different approach when constructing and interpreting a continuous probability distribution. The approach relies on the relative frequency histogram and the notion that the probability of some event can be approximated by the relative frequency of occurrence of that event in a large number of repeated experiments.

EXAMPLE 8.1   **Approximating a Continuous Distribution**

We select a random sample of 100 individuals and record the time required for each individual to perform a certain task. Table 8.1 shows a hypothetical frequency distribution and relative frequency distribution for 100 observations, where each class interval has a width of 4 seconds. Figure 8.1 shows the relative frequency histogram for the data in Table 8.1.

TABLE 8.1 *Hypothetical distribution for a continuous random variable (Example 8.1)*

Time (seconds)	Frequency	Relative frequency
16 to under 20	6	.06
20 to under 24	9	.09
24 to under 28	22	.22
28 to under 32	27	.27
32 to under 36	21	.21
36 to under 40	12	.12
40 to under 44	3	.03
Total	100	1.00

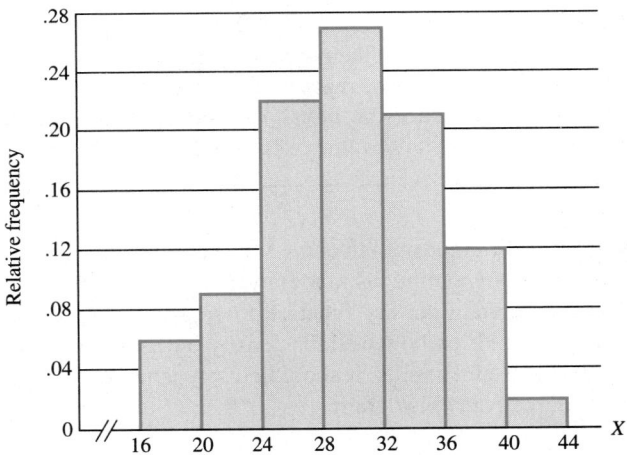

FIGURE 8.1 *Relative frequency distribution for data in Table 8.1*

Now suppose we repeat the experiment using a sample of 100,000 individuals and then construct a relative frequency distribution with a very large number of very narrow classes, say, .01 or .001 unit wide. The graph of the resulting relative frequency distribution will contain a large number of very narrow rectangles.

As the number of observations becomes very large and the class intervals become very narrow, the shape of the relative frequency histogram approaches a smooth curve, as shown in Figure 8.2 (page 250). This smooth curve, which is approximately the same as the population probability distribution or population relative frequency distribution, is called a *density function*.

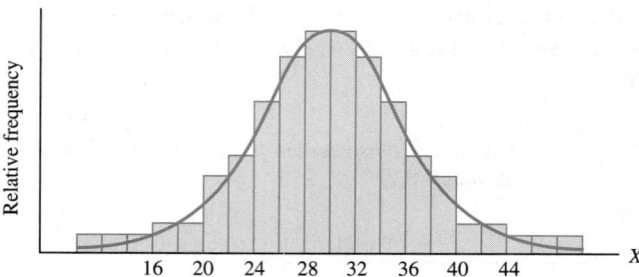

FIGURE 8.2   *Relative frequency distribution*
approaching a continuous distribution

For continuous variables, probabilities are measured by areas under the density func-tion. The probability that the random variable X falls in any particular interval (a, b) is the area under the curve between the points a and b.

DEFINITION   **Density Function**

Let a smooth curve represent the probability distribution of a continuous random variable X, and let the smooth curve be represented in mathematical notation by the function $f(x)$. The function $f(x)$ is called the **density function** of the continuous random variable X, where x represents a specific value of the random variable X.

Numerous different smooth curves have been used to represent the probability distri-butions of different continuous random variables. Whether or not we obtain good results by selecting a certain density function to approximate a population relative frequency distribution depends on how well the density function reflects reality. For a density func-tion to provide useful results, it should closely approximate the true underlying popula-tion relative frequency histogram.

Characteristics of density functions

All density functions must satisfy the following two requirements:

1. The curve must never fall below the horizontal axis. That is,

$$f(x) \geq 0 \quad \text{for all } x$$

2. The total area between the curve and the horizontal axis must be 1. In calcu-lus this is expressed as

$$\int f(x)\,dx = 1$$

A density function must be nonnegative. If $f(x)$ could be negative, it would be possible to obtain negative areas between the curve and the horizontal axis, and negative probabilities are meaningless. The total area under the curve must equal 1 to guarantee that the probability is 1 that X takes some value between $-\infty$ and $+\infty$. If the area exceeded 1, we could find some interval having a probability greater than 1.

Because the area above any single point is 0, every specific single value of X has a probability of 0. That is, if X is a continuous random variable, then for any specific value x, $P(X = x)$ is equal to 0.

Figure 8.3 shows the hypothetical probability distribution of a continuous random variable X. The probability that X falls between two values a and b is denoted $P(a < X < b)$ and is equal to the area under the curve between the points $X = a$ and $X = b$. The shaded portion of the graph represents this area.

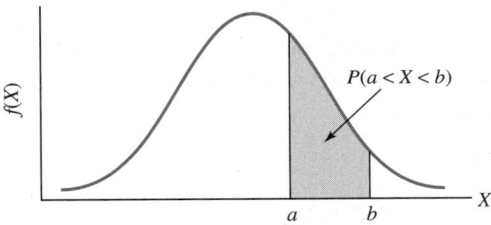

FIGURE 8.3 *For a continuous distribution, the area between a and b represents P(a < X < b).*

For the specific points $X = a$ and $X = b$, we have $P(X = a) = 0$ and $P(X = b) = 0$. Thus we have

$$P(a \leq X \leq b) = P(a < X < b)$$

That is, if X is a continuous random variable, including or excluding the endpoints of the interval will not change the probability that X falls within the interval.

A particular density function $f(x)$ provides a mathematical model that approximates the population relative frequency distribution that exists in reality. Presented graphically, the density function represents an approximation to the population relative frequency histogram. In actual practice, however, there will almost always be some disparity between the theoretical model $f(x)$ and the actual relative frequency distribution generated when an experiment is repeated an extremely large number of times.

Just as with discrete probability distributions, we are interested in several characteristics of the probability distribution of any random variable. First, we would like to know the *general shape* of the graph representing the relative frequency distribution—whether it is symmetric or skewed, whether it has long tails in either direction, whether it is relatively flat or bell shaped, and so forth. Second, we would like to determine the *center of gravity* of the distribution. This is the *mean,* or *expected value,* of the random variable X. Third, we would like some measure of the *amount of spread,* or dispersion, present in the distribution. As before, this is measured by the *variance* of the random variable X and is denoted by the symbol σ_X^2. Again, the square root of the variance is the *standard deviation,* denoted by the symbol σ_X. When there is no danger of confusion, the subscript X can be deleted.

Mean and Variance of a Continuous Distribution

Just as with discrete probability distributions, the mean μ and the variance σ^2 are important summary statistics that reveal characteristics of a continuous probability distribution. Except in simple cases, integral calculus is required to calculate the mean and variance of a continuous random variable. Because this is beyond the scope of this book, in most cases we will state the mean and variance of various continuous distributions without providing an analytical proof.

Just as with discrete variables, the mean μ measures the center of gravity of the distribution. If we cut a piece of plywood in the exact shape of the density function, it would just balance if we placed a fulcrum at one particular point perpendicular to its horizontal axis. This point is the mean μ of the variable X (see Figure 8.4). As with discrete random variables, the standard deviation σ and the variance σ^2 measure the spread of the distribution.

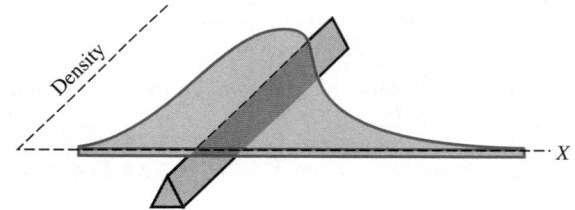

FIGURE 8.4   *Mean as the center of gravity*
of a continuous probability distribution

Tips for Problem Solving

1. Before calculating a probability for a continuous random variable, always draw a graph to represent the appropriate density function.
2. On the horizontal axis, label the variable being analyzed, typically the letter X but sometimes Z, Y, or some other letter.
3. Always indicate where the center of gravity of the distribution is and label this point μ. Also note whether the desired distribution is symmetric.
4. Shade the area that represents the desired probability.
5. After calculating the desired probability, compare your answer with the shaded area to guarantee that your answer makes sense. A frequent mistake is to report a probability as, say, .95, when the correct probability is .05. Such a mistake would indicate that you reported the area of the unshaded portion under the density function rather than the shaded portion.

Exercises

Statistical Concepts

8.1.1 Explain what is meant by a continuous random variable.

8.1.2 Describe four different continuous random variables.

8.1.3 Describe the two requirements that must be satisfied by all density functions. Explain what these requirements mean and why they are needed.

8.1.4 Suppose X is a continuous random variable. Explain why $P(a \leq X \leq b) = P(a < X < b)$.

8.1.5 *True or false:* Let X be a discrete random variable. It is possible that $P(a \leq X \leq b) > P(a < X < b)$. Explain.

8.1.6 *True or false:* Let X be a discrete random variable. It is possible that $P(a \leq X \leq b) = P(a < X < b)$. Explain.

Statistical Drills

8.1.7 Let X be a continuous random variable. Suppose $P(X < 3) = .2$ and $P(X < 4) = .5$. Find $P(3 \leq X \leq 4)$. Draw a picture and verify your answer.

8.1.8 Let X be a continuous random variable. Suppose $P(X \leq 1) = .3$. Find $P(X \geq 1)$. Draw a picture and verify your answer.

8.1.9 Let X be a continuous random variable having a density function symmetric about $X = 2$. Find $P(X > 2)$. Draw a picture and verify your answer.

8.1.10 Let X be a continuous random variable having a density function symmetric about $X = 0$. Suppose $P(X \leq 3) = .8$. Find $P(-3 \leq X \leq 3)$. Draw a picture and verify your answer.

Statistical Applications

8.1.11 Which of the following variables are continuous and which are discrete?
   a. Number of heads in 10 tosses of a coin
   b. Length of a telephone call
   c. Temperature of liquid in a test tube
   d. Number of books on a library shelf
   e. Miles per gallon averaged by a car

8.1.12 A distribution is symmetric about its mean. What can we say about the median of the distribution?

8.1.13 Is $P(X \leq \mu - 1)$ equal to $P(X \geq \mu + 1)$ if X is symmetric about μ?

8.1.14 For a certain random variable, $P(X \geq \mu + 2)$ equals $P(X \leq \mu - 2)$. Does this imply that the distribution of X is symmetric about μ?

8.1.15 At a certain airport, the amount of time people must wait to get their luggage is a random variable that is evenly distributed between 10 minutes and 20 minutes. (This is an example of a uniform distribution, which will be discussed in the next section.)
   a. Find the probability that the luggage will arrive in less than 13 minutes.
   b. Can you determine the expected waiting time?

8.1.16 A certain random variable X has the following density function:

$$f(x) = \begin{cases} 2 - 2x & \text{if } 0 \leq x \leq 1 \\ 0 & \text{otherwise} \end{cases}$$

Graph this density function and determine the following probabilities:
   a. $P(0 \leq X \leq 1)$ **b.** $P(X \geq .5)$ **c.** $P(X \leq .25)$

8.1.17   A certain random variable X has the following density function:

$$f(x) = \begin{cases} 1 + x & \text{if } -1 \leq x \leq 0 \\ 1 - x & \text{if } 0 \leq x \leq 1 \\ 0 & \text{otherwise} \end{cases}$$

Graph this probability distribution and determine the following probabilities:

a. $P(-1 \leq X \leq 1)$ **b.** $P(-1 \leq X \leq 0)$ **c.** $P(0 \leq X \leq .5)$

8.2   The Uniform Distribution

In this section, we discuss the simplest of all continuous probability distributions, the *uniform distribution*. The uniform distribution is the continuous counterpart of the equally likely probability model for discrete random variables, which we discussed in Chapter 6. A continuous random variable X follows the uniform distribution over the interval $[a, b]$ if X can take any value in the closed interval $[a, b]$ and if the density function of X is constant (or flat) over this interval. Figure 8.5 shows a uniform density function.

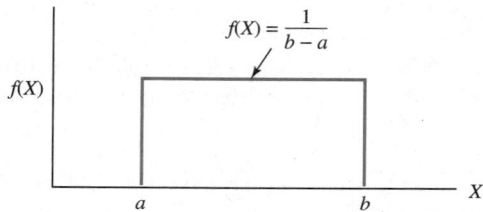

FIGURE 8.5   *A uniform distribution*

DEFINITION   **Uniform Density Function** ─────────────────────────────

Let the continuous random variable X follow the uniform distribution over the interval $[a, b]$. The **uniform density function** of X is

$$f(x) = \frac{1}{b - a} \quad \text{for } a \leq x \leq b$$

Characteristics of the Uniform Distribution

The uniform distribution has the simplest shape of all probability distributions—a rectangle that extends from a to b that has a height of $1/(b - a)$. The distribution is symmetric about its mean value, which is $(b + a)/2$. The characteristics of the uniform distribution are described in the accompanying box.

Characteristics of the uniform distribution

Let X be a uniform random variable defined over the interval $[a, b]$.

1. The mean value of X, or the expected value of X, is

$$\mu = E(X) = \frac{a + b}{2}$$

2. The median is

$$\text{Median} = \frac{a + b}{2}$$

3. The variance is

$$\sigma^2 = \text{Var}(X) = \frac{(b - a)^2}{12}$$

4. The standard deviation is

$$\sigma = \frac{b - a}{\sqrt{12}}$$

Calculating Uniform Probabilities

Let c and d be any two numbers such that $a \leq c \leq d \leq b$. The probability $P(c \leq X \leq d)$ is the area under the density function between c and d. The width of this rectangle is $(d - c)$, and the height is $1/(b - a)$; therefore, the area of the rectangle is $(d - c) \div (b - a)$.

Calculating probabilities with the uniform distribution

Let X follow the uniform distribution over the interval $[a, b]$. The probability that X takes a value between c and d is

$$P(c \leq X \leq d) = \frac{d - c}{b - a} \quad \text{where } a \leq c \leq d \leq b$$

EXAMPLE 8.2   **The Uniform Distribution**

An investor in a bank's trust department thinks that in 100 days the price of gold will be between $420 and $460 an ounce and that all prices in the interval [$420, $460] are equally likely. Let X denote the price of gold in 100 days. Find the probability that X is between $425 and $435.

SOLUTION X can be treated as a continuous random variable having a uniform distribution. The value of X varies from $a = 420$ to $b = 460$. The probability that X takes a value between $c = 425$ and $d = 435$ is given by

$$P(425 \leqslant X \leqslant 435) = \frac{435 - 425}{460 - 420} = .25$$

This probability is represented by the shaded area in Figure 8.6.

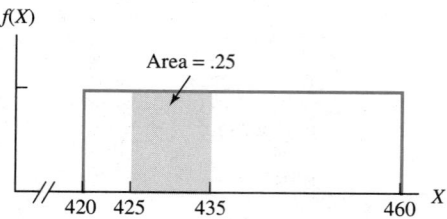

FIGURE 8.6 *Uniform distribution associated with Example 8.2*

Tips for Problem Solving

1. Before calculating a probability for a uniform random variable, always draw a graph to represent the appropriate density function.
2. Suppose the variable X takes values in the interval $[a, b]$. On the graph, place labels at the values a and b. Sometimes it is helpful to place a label at the center of the distribution to indicate the mean value of X. Label this point μ, where $\mu = (a + b)/2$.
3. Suppose you want to find $P(c \leqslant X \leqslant d)$. On the graph, shade the desired area between the values c and d.
4. To calculate the probability $P(c \leqslant X \leqslant d)$, find the area between c and d by using the formula

$$P(c \leqslant X \leqslant d) = \frac{d - c}{b - a}$$

5. Compare your answer with the shaded area to make sure that your answer makes sense.

Exercises

Statistical Concepts

8.2.1   State the formula for calculating a uniform probability.

8.2.2   State the formulas for the mean, variance, and standard deviation of a uniform random variable.

8.2.3   Suppose X is a uniform random variable, where X takes values in the interval $[3, 7]$. Draw a graph of the appropriate density function. What is the expected value of X?

8.2.4   *True or false:* Let X be a uniform random variable defined on the interval $[0, 10]$. The mean of X is 5 and the variance is 10. Explain.

8.2.5   *True or false:* Let X be a uniform random variable defined on the interval $[2, 6]$. Then $P(X = 4) = P(X \leqslant 4) - P(X \leqslant 3)$. Explain.

8.2.6 *True or false:* Let X be a uniform random variable defined on the interval [2, 6]. Then $P(X = 4) = 0$. Explain.

8.2.7 *True or false:* Let X be a uniform random variable defined on the interval [2, 6]. Then $P(X > 4) = 1 - P(X \leq 4)$. Explain.

Statistical Drills

8.2.8 Let X be a uniform random variable defined on the interval [0, 8].
  a. State the formula for the density function and graph it.
  b. Calculate $P(0 \leq X \leq 8)$ and verify that the requirements of a density function are satisfied.
  c. Find the mean and variance of X.
  d. Calculate $P(2 \leq X \leq 6)$.
  e. Calculate $P(X \leq 6)$.
  f. Calculate $P(X \geq 6)$.
  g. Calculate $P(X = 6)$.

8.2.9 Let X be a uniform random variable defined on the interval [10, 20].
  a. State the formula for the density function and graph it.
  b. Calculate $P(10 \leq X \leq 20)$ and verify that the requirements to be a density function are satisfied.
  c. Find the mean and variance of X.
  d. Calculate $P(12 \leq X \leq 16)$.
  e. Calculate $P(X \leq 16)$.
  f. Calculate $P(X \geq 16)$.
  g. Calculate $P(X = 16)$.

Statistical Applications

8.2.10 If X is uniformly distributed between 0 and 1, find the following:
  a. $P(X \leq .5)$ **b.** $P(0 \leq X \leq .4)$ **c.** $P(0 < X < .4)$ **d.** $P(X > .8)$

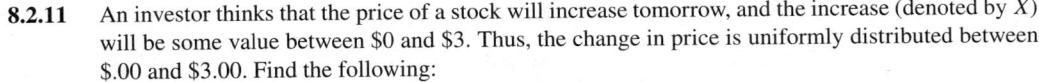

8.2.11 An investor thinks that the price of a stock will increase tomorrow, and the increase (denoted by X) will be some value between \$0 and \$3. Thus, the change in price is uniformly distributed between \$.00 and \$3.00. Find the following:
  a. $P(.50 \leq X \leq 1.00)$ **b.** $P(X \leq 1.20)$ **c.** $P(X \leq 1.00)$
  d. $P(.25 \leq X \leq .75)$ **e.** Mean of X

8.2.12 An economist claims that next year the rate of change of sales at a corporation will fall in the interval $[-5\%, 5\%]$. If the rate of change is a random variable X having the uniform distribution, find the following:
  a. $P(X < 0)$ **b.** $P(X < 2)$ **c.** $P(X < -2)$
  d. $P(-2 \leq X \leq 2)$ **e.** Mean of X

8.2.13 A car breaks down on an interstate highway. The driver has noticed the frequency of police patrols along the highway and believes that the time before a patrol car arrives on the scene is a uniformly distributed random variable between 0 and 30 minutes.
  a. Graph the probability density function.
  b. Find the probability that a patrol car arrives within 20 minutes of the breakdown.
  c. Find the probability that a patrol car does not arrive within 10 minutes of the breakdown.
  d. Find the probability that a patrol car arrives between 15 and 20 minutes after the breakdown.

8.2.14 In St. Louis, the rescue team of the city fire department is responsible for approximately a 7-mile stretch of river. The distance between the northernmost point and the location of an emergency is a uniformly distributed random variable over the interval [0, 7].
  a. Graph the probability density function.
  b. Find the probability that a given emergency arises within 1 mile of the northernmost point of this stretch of river.

c. Find the probability that a given emergency arises between 2 and 4 miles from the northernmost point of the stretch.

d. Suppose that the rescue team has its headquarters at the midpoint of this stretch of river. Find the probability that a given emergency arises more than 2 miles from the team's headquarters.

8.2.15 The travel time for a truck traveling from Fairmount to Wheeling, West Virginia, is uniformly distributed between 100 and 140 minutes.

a. Give a mathematical expression for the probability density function.

b. Compute the probability that the truck will make the trip in 125 minutes or less.

c. Compute the probability that the trip will take longer than 125 minutes.

d. Find the expected value and standard deviation of the travel time.

e. What is the probability that the trip will take exactly 115 minutes?

8.3   The Normal Distribution

The normal distribution, first used in 1733 by Antoine de Moivre (1667–1745), is the most important statistical distribution. It is sometimes called the *Gaussian distribution* in honor of Karl F. Gauss (1777–1855), a famous German mathematician who did extensive work with it.

The normal curve is represented by the density function

$$f(x) = \frac{1}{\sqrt{2\pi\sigma^2}}\, e^{-(x-\mu)^2/(2\sigma^2)}$$

where $\pi = 3.14159\ldots$ and $e = 2.71828\ldots$ and where x is a specified value of the random variable X. The numbers μ and σ^2 in the formula represent the mean and variance, respectively, of the distribution. Consequently, different values of μ and σ^2 give us different members of the family of normal distributions.

Characteristics of the normal distribution

Some of the most important characteristics of the density function of the normal distribution are as follows:

1. The curve is bell shaped and symmetric about the value $X = \mu$.
2. The curve extends from $-\infty$ to $+\infty$.
3. The total area under the curve is 1. (This is required of all density functions.)
4. The curve is always above the X-axis. (Thus, $f(x) \geq 0$ for all x.)
5. The mean, median, and mode are all equal to the parameter μ.

The three normal distributions graphed in Figure 8.7 all have the same variance σ^2 (and thus the same standard deviation σ), but they have different means. Increasing the mean shifts the curve to the right, but each curve remains symmetric about its mean. In contrast to Figure 8.7, the three normal curves in Figure 8.8 all have the same mean but different variances. Curve 1 has the smallest variance and thus is least spread out; curve 3 has the largest variance and is most spread out.

The value of μ determines the center of the normal curve, and the value of σ^2 determines the spread. Together μ and σ^2 completely determine the shape of the curve. It is

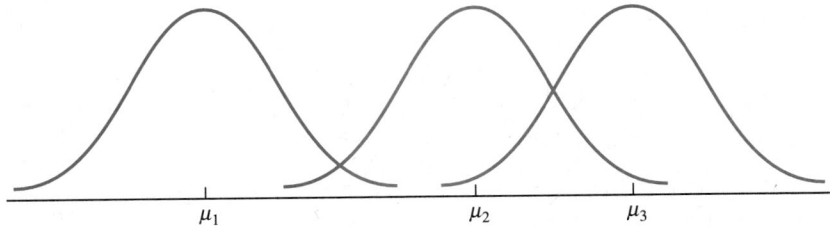

FIGURE 8.7 *Three normal distributions*
with different means but equal variances

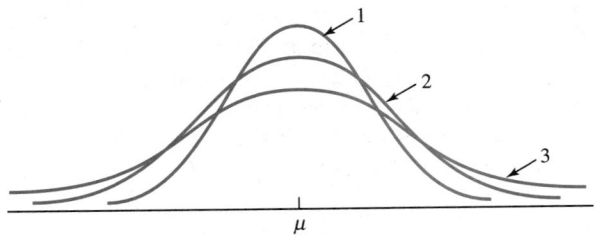

FIGURE 8.8 *Three normal distributions*
with equal means but different variances

customary to denote the normal distribution that has mean μ and variance σ^2 by the notation $N(\mu, \sigma^2)$. Thus, if we write that X is distributed as $N(8, 68)$, we mean that the random variable X is distributed normally with a mean of 8 and a variance of 68.

The Standard Normal Distribution

The normal distribution is actually a family of distributions made up of a different normal distribution for every unique combination of mean and variance. A special case of the normal distribution, called the *standard normal distribution,* exists when the mean is 0 and the variance is 1.

DEFINITION **Standard Normal Distribution**

A random variable is said to have the **standard normal distribution** if it has the normal distribution with mean $\mu = 0$ and variance $\sigma^2 = 1$, denoted as $N(0, 1)$. It is common to denote the standard normal random variable by the letter Z rather than the letter X.

Figure 8.9 (page 260) shows the standard normal distribution. In it the area between $z = 0$ and $z = 1$ is .341, which indicates that the probability is 34.1% that the standard normal variable Z is between 0 and 1. By symmetry, the probability is 34.1% that Z falls between 0 and -1. By adding these two probabilities, we determine that the probability is .682 that Z falls between -1 and $+1$. More than 95% of the area under the curve lies between -2 and $+2$, and more than 99% of the area lies between -3 and $+3$. This is the reasoning behind the Empirical Rule, discussed in Chapter 4.

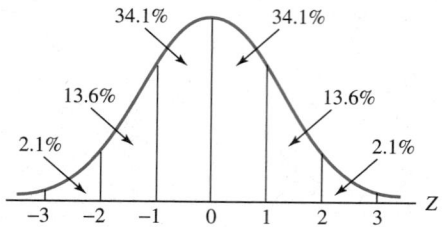

FIGURE 8.9   *The standard normal distribution and its area*

To calculate probabilities in problems involving the normal distribution, we must know how to calculate areas under the normal curve $N(\mu, \sigma^2)$ between any two points a and b. The standard normal distribution is so important because any area under any normal curve can be found by calculating an appropriate related area under the standard normal curve $N(0, 1)$. To determine any area under any normal curve, simply convert the $N(\mu, \sigma^2)$ curve of interest to the $N(0, 1)$ curve and then calculate the area under the standard normal, or $N(0, 1)$, curve.

Calculating Areas Under the Standard Normal Curve

To calculate areas under the standard normal curve, we use Table A.5 in the Appendix. The numbers in the body of the table show the area under the standard normal curve between 0 and a positive value of Z, say, z_0. That is, the values in the table show $P(0 \leq Z \leq z_0)$. Values of z_0 to the nearest tenth are shown in the left column of the table, and the second decimal place, corresponding to hundredths, is listed across the top row of the table.

EXAMPLE 8.3   **Area Between the Mean and a Positive Value Under the Standard Normal Curve**

Find the area under the standard normal curve between 0 and 2.34. That is, find $P(0 \leq Z \leq 2.34)$.

SOLUTION   To find the shaded area in Figure 8.10, first move down the left side of Table A.5 to the row value 2.3. Now move across to the hundredths column headed by the digit 4. The number found at the intersection of this row and column is .4904, which represents the area between 0 and 2.34. Thus we write

$$P(0 \leq Z \leq 2.34) = .4904$$

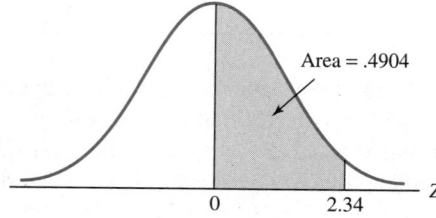

FIGURE 8.10   *Area between 0 and 2.34 under the standard normal curve*

EXAMPLE 8.4 **Area Between the Mean and a Negative Value Under the Standard Normal Curve**

Find the area under the standard normal curve between 0 and -1.67.

SOLUTION   We want to find the shaded area in Figure 8.11. Because the normal distribution is symmetric, the area between -1.67 and 0 is the same as the area between 0 and 1.67. From Table A.5, we see that the area between 0 and 1.67 is .4525. We thus obtain

$$P(-1.67 \leqslant Z \leqslant 0) = .4525$$

To find areas under the standard normal distribution that are to the left of 0, calculate the corresponding area to the right of 0.

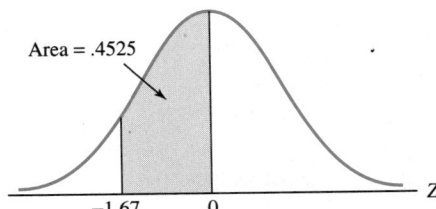

Area = .4525

-1.67 0 Z

FIGURE 8.11   *Area between -1.67 and 0 under the standard normal curve*

EXAMPLE 8.5 **Area Between a Negative and a Positive Value Under the Standard Normal Curve**

Let Z be distributed as $N(0, 1)$. Calculate the probability that Z is between -1.21 and 2.15; that is, evaluate $P(-1.21 \leqslant Z \leqslant 2.15)$.

SOLUTION   We want to find the area under the curve between -1.21 and 2.15 in Figure 8.12. Calculating this area requires two steps. First, find these areas using Table A.5: the area between 0 and 2.15 is .4842, and the area between 0 and -1.21 is .3869. Next, simply sum these two areas to find the area between -1.21 and 2.15:

$$P(-1.21 \leqslant Z \leqslant 2.15) = .4842 + .3869 = .8711$$

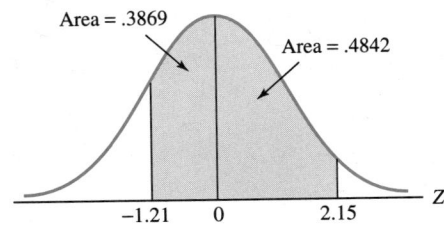

Area = .3869

Area = .4842

-1.21 0   2.15 Z

FIGURE 8.12   *Area between -1.21 and 2.15 under the standard normal curve*

EXAMPLE 8.6 **Area in the Right Tail of the Standard Normal Curve**

If Z is a standard normal variable, find the probability that Z exceeds 1.64; that is, find $P(Z > 1.64)$.

SOLUTION   We must evaluate the shaded area in Figure 8.13. First, note that $P(Z > 1.64) = P(Z \geqslant 1.64)$. That is, the probability does not change if we include or exclude the single point $Z = 1.64$, because $P(Z = 1.64) = 0$. By symmetry, the total area under the curve to the right of $Z = 0$ is .5000. Since from Table A.5 we know that the area between 0 and 1.64 is .4495, the area to the right of 1.64 must be .5000 − .4495. We thus obtain

$$P(Z > 1.64) = .5000 - .4495 = .0505$$

To calculate the area in a tail of the distribution beyond some value z_0, first calculate the area between 0 and z_0 and subtract this amount from .5000.

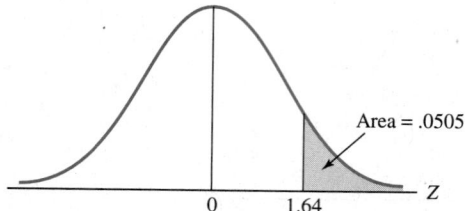

FIGURE 8.13   *Area to the right of 1.64 under the standard normal curve*

Whenever you are required to use the normal distribution to calculate a probability, it is a good practice to draw a graph and shade the appropriate area. Students sometimes report a probability as 45%, say, when the correct answer is 5% because they record the area between 0 and some positive value z_0 rather than the area in the tail of the distribution to the right of z_0.

EXAMPLE 8.7   **Area in the Left Tail of the Standard Normal Curve**

Find $P(Z \leqslant -2.02)$.

SOLUTION   We want to evaluate the shaded area in Figure 8.14. From Table A.5, the area between 0 and 2.02, and thus between −2.02 and 0, is .4783. Because the area to the left of 0 is .5000, we obtain

$$P(Z \leqslant -2.02) = .5000 - .4783 = .0217$$

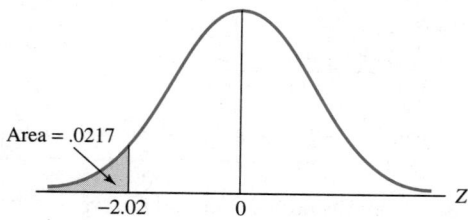

FIGURE 8.14   *Area to the left of −2.02 under the standard normal curve*

EXAMPLE 8.8   **Areas in the Extreme Tail of the Standard Normal Curve**
Find $P(Z > 4.63)$.

SOLUTION   Note that Table A.5 does not contain Z values of 4 or above. To four decimal places, the area between 0 and 4 is .5000, so the area to the right of $Z = 4$ is 0. Refer to Figure 8.15. Because the normal curve extends from $-\infty$ to $+\infty$, there is always some area between any two values of Z. In this case, however, the area is so small that we would say that the area is 0 if we measure it to only four decimal places. Thus, to an accuracy of four decimal places, we have

$$P(Z \leqslant 4.63) = 1 \quad \text{and} \quad P(Z \geqslant 4.63) = 0$$

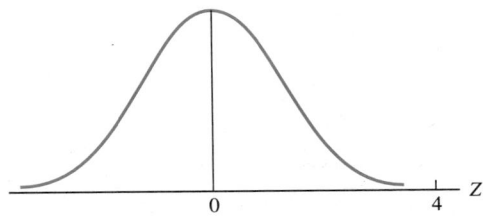

FIGURE 8.15   *Area to the right of z = 4.00 under the standard normal curve*

EXAMPLE 8.9   **Finding a Z Score Associated with a Specific Area**
Find the values z_1 and z_2 such that the area to the right of z_2 is .025 and the area to the left of z_1 is .025.

SOLUTION   Refer to Figure 8.16. If the area to the right of z_2 is .025, then the area between 0 and z_2 is .475. From Table A.5 we see that when $z_2 = 1.96$, the area between 0 and z_2 is .4750. Similarly, the area between 0 and -1.96 is also .4750. Thus, we obtain

$$P(-1.96 \leqslant Z \leqslant 1.96) = .95$$

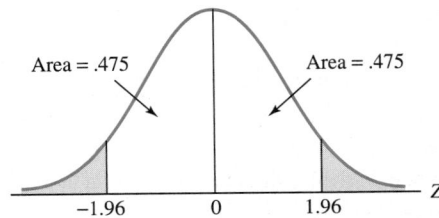

FIGURE 8.16   *Area between − 1.96 and 1.96 under the standard normal curve*

In Example 8.9, we showed that the area under the standard normal distribution between $Z = 0$ and $Z = 1.96$ is .4750. Because the area to the left of $Z = 0$ is .5000, we obtain $P(Z < 1.96) = .9750$. This is a *cumulative probability*. Some other points on the cumulative distribution are shown in Table 8.2.

TABLE 8.2   *Some cumulative probabilities for the standard normal distribution*

z	$P(Z < z)$
-2.58	.005
-2.33	.01
-1.96	.025
-1.645	.05
-1.28	.10
$-.67$	.25
.00	.50
.67	.75
1.28	.90
1.645	.95
1.96	.975
2.33	.99
2.58	.995

Tips for Problem Solving

1. Before calculating a probability for a standard normal random variable, always draw a graph to represent the appropriate density function.
2. Draw a vertical line to indicate the mean of the distribution and label this point 0.
3. To visualize the probability $P(a \le Z \le b)$, shade the area between the points a and b. Sometimes you want to find $P(X \le a)$ or $P(X \ge b)$. Be sure to shade the appropriate area. Shading the wrong area under the density function is a frequent mistake.
4. Find the desired probability by referring to Table A.5.
5. Compare your answer with the shaded area to make sure that your answer makes sense.

Exercises

Statistical Concepts

8.3.1   List the important characteristics of the normal distribution.

8.3.2   Suppose X is a normal random variable with mean μ and standard deviation σ. Describe how the density function changes as μ increases.

8.3.3   Suppose X is a normal random variable with mean μ and standard deviation σ. Describe how the density function changes as σ increases.

8.3.4   *True or false:*   Let X be a standard normal random variable. Then $P(0 \le Z \le 3) = P(-3 \le Z \le 0)$. Explain.

8.3.5   *True or false:*   Let X be a standard normal random variable. Then $P(Z \le -2) = P(Z \le 2)$. Explain.

8.3.6   *True or false:*   Let X be a standard normal random variable. Then $P(Z \le 2) = P(Z < 2)$. Explain.

8.3.7   *True or false:*   Let X be a standard normal random variable. Then $P(X = 1) = P(X \leq 1) - P(X \leq 0)$. Explain.

8.3.8   *True or false:*   Let X be a standard normal random variable. Then $P(X > -2) = .5 + P(0 \leq X \leq 2)$. Explain.

8.3.9   *True or false:*   Let X be a standard normal random variable. Then $P(X < -2) = 1.0 - P(0 \leq X \leq 2)$. Explain.

8.3.10   Sometimes you may want to calculate a probability based on Z values with greater than two-decimal accuracy. To find such a probability, we use a linear interpolation between two probabilities obtained from Table A.5. For example, $P(0 \leq Z \leq 1.645)$ is found as the midpoint between the two probabilities $P(0 \leq Z \leq 1.64)$ and $P(0 \leq Z \leq 1.65)$. Show that $P(0 \leq Z \leq 1.645) = .45$.

8.3.11   Use the interpolation method and find the value z_0 such that $P(-z_0 \leq Z \leq z_0) = .99$. Verify that $z_0 = 2.575$. Explain.

8.3.12   In many situations, instead of finding the probability that a standard normal random variable will be within a given interval, we may be interested in finding an interval that encompasses a specified probability. For example, suppose you need to find the value z_0 such that $P(0 \leq Z \leq z_0) = .40$. To find z_0, look inside Table A.5 and locate the area closest to .40. Verify that the area closest to .40 is .3997. The Z score corresponding to the area .3997 is approximately the desired value z_0. Verify that the approximate value z_0 is 1.28. Find z_0 such that $P(0 \leq Z \leq z_0) = .475$.

8.3.13   *True or false:*   If z_0 is the value such that $P(0 \leq Z \leq z_0) = .40$, then $P(Z \leq z_0) = .90$ and $P(Z \leq -z_0) = .10$. Explain.

Statistical Drills

In Exercises 14–25, let Z be distributed as $N(0, 1)$:

8.3.14   Find $P(0 \leq Z \leq 1.5)$.

8.3.15   Find $P(Z \leq 1.82)$.

8.3.16   Find $P(Z \geq -1.64)$.

8.3.17   Find $P(-1.61 \leq Z \leq 2.34)$.

8.3.18   Find $P(Z \leq 4.20)$.

8.3.19   Find $P(-2 \leq Z \leq 2)$.

8.3.20   Find $P(Z > 5.10)$.

8.3.21   Find $P(1.26 \leq Z \leq 2.45)$.

8.3.22   Find $P(-2.12 \leq Z \leq -.74)$.

8.3.23   Find z_1 such that $P(Z > z_1) = .2296$.

8.3.24   Find z_1 such that $P(Z < z_1) = .9495$.

8.3.25   Find z_1 such that $P(|Z| < z_1) = .9544$.

Statistical Applications

8.3.26   When monitoring a production process, quality control engineers search for clues that a production process may have gone out of control. One indication that a process may have gone out of control is the occurrence of one or more unusual observations. These unusual observations lie far out in the tail of the probability distribution and should occur very infrequently if the process is in control. Quality control engineers consider a process to be out of control when an observation falls more than 3 standard deviations from the mean. Suppose a random variable follows the standard normal distribution.

    a. Find the probability that an observation will fall more than 3 standard deviations above the mean.

    b. Find the probability that an observation will fall more than 3 standard deviations below the mean.

    c. Find the probability that an observation will fall more than 3 standard deviations above or below the mean.

    d. Find the probability that an observation will fall within 3 standard deviations of the mean.

8.3.27 In quality control analysis, there are many other rules that can be adopted as indicating that a process may have gone out of control. Some of the rules are based on sequences of observations that fall more than 2 standard deviations from the mean. Suppose a random variable follows the standard normal distribution.

    a. Find the probability that an observation will fall more than 2 standard deviations above the mean.

    b. Find the probability that an observation will fall more than 2 standard deviations below the mean.

    c. Find the probability that an observation will fall more than 2 standard deviations above or below the mean.

    d. Find the probability that an observation will fall within 2 standard deviations of the mean.

8.3.28 When applying statistical quality control methods to monitor a process, it is customary to reject the hypothesis that the process is in control if we get an observation that falls far from the mean. Suppose a random variable follows the standard normal distribution.

    a. Find the value z_0 such that $1 - P(-z_0 \leq Z \leq z_0) = .10$.

    b. Find the value z_0 such that $1 - P(-z_0 \leq Z \leq z_0) = .05$.

    c. Find the value z_0 such that $1 - P(-z_0 \leq Z \leq z_0) = .01$.

8.4　Calculating Areas Under Any Normal Curve

Now that we know how to find areas under the standard normal curve, we can proceed to the problem of finding areas under any normal curve $N(\mu, \sigma^2)$. By using Table A.5, we can find any area under the normal curve $N(\mu, \sigma^2)$ by exploiting the fact that any normal variable X having mean μ and variance σ^2 can be transformed into the standard normal variable Z by using what is called the *standardizing transformation*.

DEFINITION　**Standardizing Transformation**

Suppose X is distributed as $N(\mu, \sigma^2)$; that is, X has a normal distribution with mean μ and variance σ^2. The transformation to the standard normal distribution is accomplished by means of the formula

$$Z = \frac{X - \mu}{\sigma}$$

This formula is called the **standardizing transformation,** and the random variable Z is called the **standardized score,** or the **Z score.**

The formula

$$Z = \frac{X - \mu}{\sigma}$$

transforms each value of X from the $N(\mu, \sigma^2)$ distribution into a value of Z from the standard normal distribution $N(0, 1)$.

EXAMPLE 8.10   **Area Under the Normal Curve $N(\mu, \sigma^2)$** ————————————————

Suppose the random variable X has a normal distribution with mean 10 and variance 25, as shown in Figure 8.17. Find the area under the curve between $x_1 = 12$ and $x_2 = 16$.

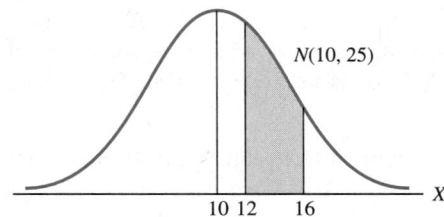

FIGURE 8.17 *Area under the normal distribution $N(10, 25)$*
between $x_1 = 12$ and $x_2 = 16$

SOLUTION   In Figure 8.17, the variable X is distributed as $N(10, 25)$; that is, $\mu = 10$ and $\sigma = 5$. To find the area under the curve between the values $x_1 = 12$ and $x_2 = 16$, we first have to find the standardized scores associated with 12 and 16. Using the formula $Z = (X - \mu)/\sigma$, we obtain

$$z_1 = \frac{x_1 - \mu}{\sigma} = \frac{12 - 10}{5} = .4$$

$$z_2 = \frac{x_2 - \mu}{\sigma} = \frac{16 - 10}{5} = 1.2$$

The area under the curve $N(10, 25)$ between $x_1 = 12$ and $x_2 = 16$ is thus equal to the area under the curve $N(0, 1)$ between $z_1 = .4$ and $z_2 = 1.2$. This is illustrated in Figure 8.18.

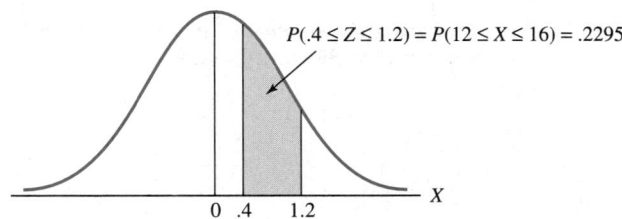

FIGURE 8.18 *Area under the standard normal distribution*
between $z_1 = .4$ and $z_2 = 1.2$

From Table A.5, the area between 0 and 1.2 is .3849 and the area between 0 and .4 is .1554. Therefore the area between .4 and 1.2 under the standard normal curve is

$$P(.4 \leq Z \leq 1.2) = .3849 - .1554 = .2295$$

Because this area is the same as the area between 12 and 16 under the $N(10, 25)$ curve, we obtain $P(12 \leqslant X \leqslant 16) = .2295$.

EXAMPLE 8.11   ### Application of the Standardizing Transformation

A company has been hired to dig a tunnel as part of a new subway system. The amount of tunnel that can be constructed during any week is a random variable that depends mainly on geological conditions. Let the random variable X denote the number of feet of tunnel completed during any week. Assume that X is approximately normally distributed with mean 100 and variance 400. Find the probability that during the next week the length of tunnel completed is between $x_1 = 80$ feet and $x_2 = 120$ feet. That is, find $P(80 \leqslant X \leqslant 120)$.

S O L U T I O N   The problem as stated and as transformed to standardized form is shown in Figures 8.19 and 8.20, respectively. We use the standardizing transformation and obtain

$$z_1 = \frac{x_1 - \mu}{\sigma} = \frac{80 - 100}{20} = -1$$

$$z_2 = \frac{x_2 - \mu}{\sigma} = \frac{120 - 100}{20} = 1$$

From Table A.5, we obtain

$$P(80 \leqslant X \leqslant 120) = P(-1 \leqslant Z \leqslant 1) = .3413 + .3413 = .6826$$

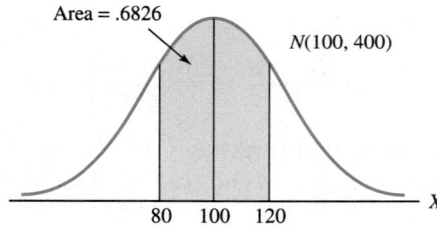

Area = .6826

$N(100, 400)$

80 100 120

X

FIGURE 8.19   *Area under the normal distribution N(100, 400)*
between $x_1 = 80$ and $x_2 = 120$

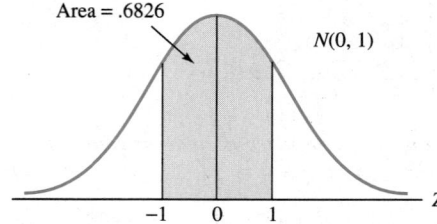

Area = .6826

$N(0, 1)$

−1   0   1

Z

FIGURE 8.20   *Area under the standard normal distribution*
between $z_1 = -1.0$ and $z_2 = 1.0$

EXAMPLE 8.12   **Verification of the Empirical Rule** —————————————————

In Chapter 4, we introduced the Empirical Rule, which stated that if a variable has an approximately bell-shaped distribution, about 95% of the values fall within 2 standard deviations of the mean and approximately 99.7% of the observations fall within 3 standard deviations. The Empirical Rule is nothing more than a statement about areas under the normal distribution. For example, suppose that a random variable X is distributed as $N(\mu, \sigma^2)$. Let us find the probability that X is within 2 standard deviations of the mean. According to the Empirical Rule, this probability should be approximately .95.

SOLUTION   If X is within 2 standard deviations of the mean, then X falls between $(\mu - 2\sigma)$ and $(\mu + 2\sigma)$. Let x_1 equal $(\mu - 2\sigma)$ and x_2 equal $(\mu + 2\sigma)$.

After making the standardizing transformation, we obtain the Z scores as follows:

$$z_1 = \frac{x_1 - \mu}{\sigma} = \frac{(\mu - 2\sigma) - \mu}{\sigma} = -2$$

$$z_2 = \frac{x_2 - \mu}{\sigma} = \frac{(\mu + 2\sigma) - \mu}{\sigma} = +2$$

Thus,

$$
\begin{aligned}
P(\mu - 2\sigma \le X \le \mu + 2\sigma) &= P(-2 \le Z \le 2) \\
&= .4772 + .4772 \\
&= .9544
\end{aligned}
$$

If a variable follows the normal distribution exactly, then approximately 95% of the values fall within 2 standard deviations of the mean.

———

EXAMPLE 8.13   **Finding Values Associated with Specified Tail Areas** —————————————

Scores on a civil service exam follow a normal distribution with mean 250 and standard deviation 11.5. To rank in the highest 2.5%, how high must a person's score be?

SOLUTION   Let the random variable X follow a normal distribution with mean 250 and standard deviation 11.5. We want to find the value x_0 such that $P(X \ge x_0) = .025$. Equivalently, we need to find the value x_0 such that $P(X \le x_0) = .975$.

First, we use the standardizing transformation

$$z_0 = \frac{x_0 - \mu}{\sigma} = \frac{x_0 - 250}{11.5}$$

The probability that X is less than x_0 is the same as the area to the left of z_0 for the standard normal distribution. That is,

$$P(Z \le z_0) = .975$$

For the standard normal distribution, the area to the left of 0 is .50. Thus, the area to the right of 0 between 0 and z_0 must be .475. From Table A.5, we find that $z_0 = 1.96$.

Substituting $z_0 = 1.96$ into the equation for z_0 yields

$$1.96 = \frac{x_0 - 250}{11.5}$$

Solving for x_0, we obtain

$$x_0 = (1.96)(11.5) + 250 = 272.54$$

To rank in the highest 2.5%, the applicant must score 272.54 or higher.

Tips for Problem Solving

1. Before calculating a probability for a normal random variable, always draw a graph to represent the appropriate density function.
2. Draw a vertical line to indicate the mean of the distribution and label this point μ.
3. To visualize the probability $P(a \leq X \leq b)$, shade the area between the points a and b. Sometimes you want to find $P(X \leq a)$ or $P(X \geq b)$. Be sure to shade the appropriate area. Shading the wrong area under the density function is a frequent mistake.
4. Find the corresponding area under the standard normal curve by using the standardizing transformation

$$z = \frac{X - \mu}{\sigma}$$

5. Find the desired probability by referring to Table A.5.
6. Compare your answer with the shaded area to make sure that your answer makes sense.

Exercises

Statistical Concepts

8.4.1   *True or false:*   Let X be a normal random variable with mean 10 and variance 16. Then $P(0 \leq X \leq 3) = P(-3 \leq X \leq 0)$. Explain.

8.4.2   *True or false:*   Let X be a normal random variable with mean 10 and variance 16. Then $P(10 \leq X \leq 13) = P(7 \leq X \leq 10)$. Explain.

8.4.3   *True or false:*   Let X be a normal random variable with mean 10 and variance 16. Then $P(X > 13) = P(X \leq 7)$. Explain.

8.4.4   *True or false:*   Let X be a normal random variable with mean 10 and variance 16. Then $P(X \leq 2) = P(X < 2)$. Explain.

8.4.5   *True or false:*   Let X be a normal random variable with mean 10 and variance 16. Then $P(X > 12) = 1 - P(X \leq 12)$. Explain.

8.4.6   *True or false:*   Let X be a normal random variable with mean 10 and variance 16. Then $P(X < 12) = .5 + P(10 \leq X \leq 12)$. Explain.

8.4.7   *True or false:*   The location and shape of the normal distribution are determined by the population mean μ and the population standard deviation σ. Large values of σ reduce the height of the curve and increase the spread; small values of σ increase the height and reduce the spread of the curve. Explain.

8.4.8   *True or false:*   The probability model for a continuous random variable differs greatly from the model for a discrete random variable when we consider the probability that X equals some particular value, say, *a*. For a continuous variable, $P(X = a) = 0$. This statement is not necessarily true for a discrete random variable because $P(X = a)$ may not equal 0. Explain.

Statistical Drills

8.4.9   Let *X* be a normal random variable with mean 8 and variance 4. Find the following:
a. $P(X \leq 6)$ **b.** $P(X > 12)$ **c.** $P(X \leq 12)$
d. $P(X \geq 6)$ **e.** $P(6 \leq X \leq 12)$

8.4.10   Let *X* be a normal random variable with mean 20 and variance 25. Find the following:
a. $P(X \leq 5)$ **b.** $P(X > 10)$ **c.** $P(X \leq 15)$
d. $P(X \geq 25)$ **e.** $P(10 \leq X \leq 25)$

Statistical Applications

8.4.11   Let *X* be a random variable denoting the percentage change in the cost of living in the next year in a South American country. If *X* is distributed as $N(8, 36)$, find $P(6 \leq X \leq 12)$.

8.4.12   A company has recently raised the price of its product and simultaneously instituted a large advertising campaign. In the past, the company had about 70 customers per day. Let *X* denote the increase in customers per day after the price increase and advertising campaign. If *X* is distributed as $N(10, 100)$, find $P(X \leq 22)$.

8.4.13   A clerk at a grocery store tries to break apart bunches of grapes into 1-pound bunches. After breaking them apart, the clerk weighs each bunch. Let *X* denote the weight of a bunch of grapes. If *X* is distributed as $N(1, .09)$, find the following:
a. $P(1.01 \leq X \leq 1.04)$ **b.** $P(.7 \leq X \leq 1.06)$

8.4.14   At Halloween, a grocery store brings in a large supply of pumpkins. Let *X* be a random variable denoting a pumpkin's weight in pounds. If *X* is distributed as $N(22, 16)$, find the following:
a. $P(21 \leq X \leq 25)$ **b.** $P(18 \leq X \leq 20)$ **c.** $P(X \leq 18)$
d. $P(X \geq 17)$ **e.** $P(X > 30)$

8.4.15   If *X* is distributed as $N(50, 25)$, find the following:
a. *c* such that $P(X \leq c) = .95$
b. *d* such that $P(X \leq d) = .05$
c. *e* such that $P(X \leq e) = .99$
d. *f* such that $P(X \leq f) = .50$
e. *a* and *b,* which are equidistant from 50, such that $P(a \leq X \leq b) = .95$

8.4.16   The lives of light bulbs follow a normal distribution. If 90% of the bulbs have lives exceeding 2,000 hours and 3% have lives exceeding 6,000 hours, what are the mean and variance of the lives of light bulbs?

8.4.17   A flashlight battery is guaranteed to last for 40 hours. Tests indicate that the length of life of these batteries is normally distributed with mean 50 and variance 16. What percentage of the batteries will fail to meet the guarantee?

8.4.18   Scores on a personnel examination at the Anderson Electronics Company are normally distributed with mean $\mu = 73$ and standard deviation $\sigma = 5$. Mary Jonathon scored 80 on the test. What proportion of people taking the test got a score higher than Mary's?

8.4.19   The number of patients entering the St. Joseph Hospital emergency room during any given weekday is approximately normally distributed with mean $\mu = 46$ and standard deviation $\sigma = 10$. Find the probability that during a given weekday, the number of patients entering the emergency room is the following:
a. At least 30 **b.** More than 50 **c.** Less than 60 **d.** Between 40 and 50

8.4.20   On a civil service exam, the grades are distributed as $N(70, 100)$. The police department will hire the applicants whose grades are among the top 10% of the population. What is the minimum grade required to be hired?

8.4.21 X is a normal random variable with mean $\mu = 500$. Find the probability $P(490 \leqslant X \leqslant 510)$ for $\sigma = 10$, 16, and 25.

8.4.22 A review of a plumber's records indicates that the time taken for a service call can be represented by a normal distribution with mean 50 minutes and standard deviation 10 minutes.
  a. What proportion of the service calls take more than 30 minutes?
  b. What proportion of the service calls take less than an hour?
  c. A plumber is scheduled to make three calls in a morning. The times taken for each of these calls are independent of one another. What is the probability that at least one of them will take more than an hour?

8.4.23 Seniors at a public school who take a placement test have scores that are normally distributed with a mean of 280 and a standard deviation of 40. Seniors at a private school who take the same test have scores that are normally distributed with a mean of 310 and a standard deviation of 60. A student qualifies for a state scholarship if his or her score exceeds 380.
  a. For a randomly chosen senior from the public school, what is the probability his or her score on the test will qualify the student for the scholarship?
  b. For a randomly selected senior from the private school, what is the probability his or her score will qualify the student for the scholarship?

8.4.24 The length of time required for a crew of mechanics to complete a routine pit stop during an automobile race can be represented by a normal random variable with mean 15 seconds and standard deviation 2.0 seconds.
  a. The probability is .95 that a pit stop will take at least how many seconds?
  b. The probability is .70 that a pit stop will take at most how many seconds?

8.4.25 Tom drives to work every day and has meticulously recorded his driving time, which has a mean of 15 minutes and is normally distributed with a standard deviation of 4 minutes. If he leaves home at 9:05 A.M., what is the probability that he will be late for work if he has to punch in by 9:30 A.M.?

8.4.26 The scores for a statistics test have a normal $N(80, 8)$ distribution. Tom's score was 84, Mary's score was at the 60th percentile, and John's score corresponded to a Z score of $+.75$. If these students were listed from highest score to lowest score, what would be the correct ordering?

8.5 The Normal Distribution As an Approximation to the Binomial Distribution

The normal distribution can be used to solve problems involving the binomial distribution when the number of trials n is relatively large and the binomial distribution has a bell-shaped appearance closely resembling the normal curve. Solving these problems is very difficult and time-consuming when we must calculate a large number of binomial probabilities. However, these probabilities can be approximated by an area under the appropriate normal curve. In Chapter 7, we indicated that the binomial distribution is approximately bell shaped (i.e., it resembles a normal distribution) when $np \geqslant 5$ and $nq \geqslant 5$, where $q = (1 - p)$.

Normal approximation to the binomial

> Use of the normal curve to approximate binomial probabilities is appropriate provided $np \geqslant 5$ and $nq \geqslant 5$. When approximating the binomial distribution, use the normal distribution having mean $\mu = np$ and variance $\sigma^2 = npq$.

Recall from Chapter 7 that the mean of the binomial distribution is $\mu = np$ and the variance is $\sigma^2 = npq$. To use the normal curve to approximate the binomial distribution, use the normal curve having the same mean and variance as the binomial distribution it is to approximate. After verifying that the two conditions $np \geqslant 5$ and $nq \geqslant 5$ both hold, draw a graph of a normal distribution and shade the appropriate area.

A *continuity correction* is needed when we use the normal distribution to approximate the binomial distribution. When we graph the relative frequency histogram of a binomial variable X, we represent the probability that $X = x_0$ by the area of a rectangle extending from $(x_0 - .5)$ to $(x_0 + .5)$. When we use the normal distribution to approximate the binomial distribution, we approximate the area of this rectangle in the binomial histogram by the area under the normal curve between $(x_0 - .5)$ and $(x_0 + .5)$.

For example, if X follows the binomial distribution, the probability that X is, say, 6, 7, 8, or 9 is represented by the areas of the four rectangles that extend from 5.5 to 9.5. The areas of these rectangles can be approximated by the area under the normal curve between $X = 5.5$ and $X = 9.5$.

DEFINITION   **Continuity Correction** ─────────────────────────────────

The **continuity correction** is made when we approximate a discrete probability $P(X = x)$ by the continuous probability $P(x - .5 \leqslant X \leqslant x + .5)$.

───

EXAMPLE 8.14   **Normal Approximation to the Binomial Distribution** ──────────────

The discrete random variable X follows the binomial distribution with $n = 10$ and $p = .5$. Let us find the probability that X takes one of the values 3, 4, 5, or 6. We will calculate the exact binomial probability and then use the normal distribution to approximate it.

SOLUTION   For the exact binomial probabilities, we obtain $P(X = 3) = .1172$, $P(X = 4) = .2051$, $P(X = 5) = .2461$, and $P(X = 6) = .2051$. The exact binomial probability of getting 3, 4, 5, or 6 successes in 10 trials, when $p = .5$, is thus

$$P(3 \leqslant X \leqslant 6) = .1172 + .2051 + .2461 + .2051 = .7735$$

Figure 8.21 (page 274) shows the relative frequency histogram of the binomial distribution for the case when $n = 10$ and $p = .5$. The probability that X is 3, 4, 5, or 6 is represented by the area of the shaded rectangles extending from 2.5 to 6.5.

Our rule indicates that areas under the normal curve should yield good approximations to binomial probabilities provided $np \geqslant 5$ and $nq \geqslant 5$. In this case, $np = 5$ and $nq = 5$, so our conditions are just satisfied.

To approximate the binomial distribution, we use the normal distribution having mean $\mu = np = 5$ and variance $\sigma^2 = npq = 2.5$. Figure 8.22 (page 274) shows the normal curve with mean $\mu = np = 5$ and variance $\sigma^2 = npq = 2.5$ (the standard deviation is $\sigma = 1.58$). The shaded area in Figure 8.22 is used to approximate the shaded area in Figure 8.21.

If we use the normal curve to approximate the binomial, we should make the continuity correction. To approximate the binomial probability $P(3 \leqslant X \leqslant 6)$, we find the area under the normal curve between $x_1 = 2.5$ and $x_2 = 6.5$.

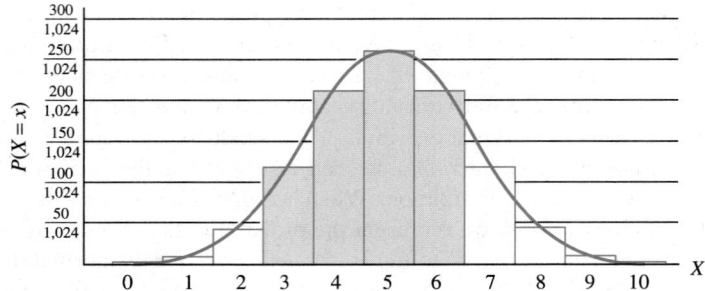

FIGURE 8.21 $P(3 \leq X \leq 6)$ *using the binomial distribution*

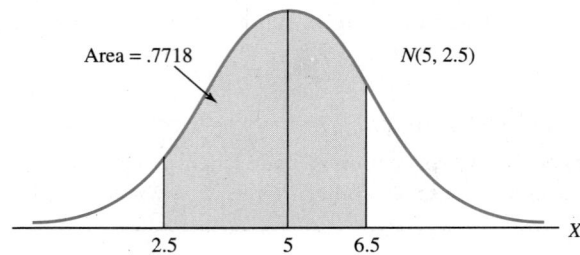

FIGURE 8.22   *Area between 2.5 and 6.5 using the normal distribution* $N(5, 2.5)$

To find the area under the normal distribution, make the standardizing transformation and obtain the Z scores associated with the values $x_1 = 2.5$ and $x_2 = 6.5$. We obtain

$$z_1 = \frac{x_1 - \mu}{\sigma} = \frac{2.5 - 5}{1.58} = \frac{-2.5}{1.58} = -1.58$$

$$z_2 = \frac{x_2 - \mu}{\sigma} = \frac{6.5 - 5}{1.58} = \frac{1.5}{1.58} = .95$$

From Table A.5, we obtain

$$P(-1.58 \leq Z \leq .95) = \text{(Area between 0 and .95)}$$
$$+ \text{(Area between } -1.58 \text{ and 0)}$$
$$= .3289 + .4429 = .7718$$

The approximate answer .7718 obtained from the normal distribution agrees quite well with the exact binomial value .7735. The normal approximation to the binomial improves as the value of n gets larger and p gets closer to .5.

EXAMPLE 8.15   **Another Normal Approximation to the Binomial** ―――――――――――

About 50% of the individuals who purchase a personal computer from a certain supplier also order a hard disk drive. The supplier is placing an order for 100 PCs and has to decide how many to order with hard disk drives. Of the next 100 purchasers of a PC from this supplier, let X denote the number who want a hard disk drive. Find $P(X > 60)$.

S O L U T I O N   If we assume that the purchasers are independent of one another, then X follows the binomial distribution with $n = 100$ and $p = .5$. To see whether the normal approximation to the binomial is appropriate, we evaluate np and nq. We obtain $np = 50$ and $nq = 50$. Since both exceed 5, our rule indicates that the normal distribution should give a good approximation to the correct answer. Thus we use the normal distribution having mean $\mu = np = 50$ and variance $\sigma^2 = npq = 100(.5)(.5) = 25$.

The binomial probability that X exceeds 60 is the probability that X is 61, 62, 63, and so on, which is represented by the area under the normal curve to the right of $x_1 = 60.5$. Refer to Figure 8.23.

Using the standardizing transformation, we obtain the Z score as follows:

$$z_1 = \frac{x_1 - \mu}{\sigma} = \frac{60.5 - 50}{5} = 2.10$$

The desired probability is the area to the right of $z = 2.10$. (Refer to Figure 8.24.) The area between $z = 0$ and $z = 2.10$ is .4821, so the required area in the right tail is $(.5000 - .4821) = .0179$. This answer shows that the probability is very small that more than 60 of the next 100 customers will order a PC with a hard disk drive.

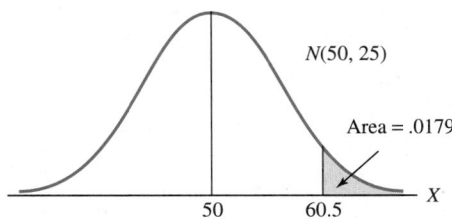

FIGURE 8.23 *Area to the right of X = 60.5 under the normal distribution N(50, 25)*

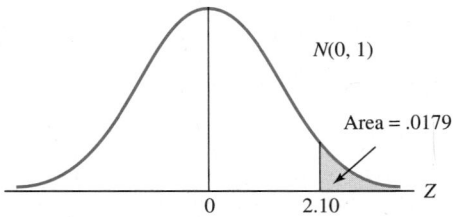

FIGURE 8.24 *Area to the right of 2.10 under the standard normal distribution*

Tips for Problem Solving

1. Before using the normal distribution to approximate the binomial distribution, verify that $np \geqslant 5$ and $n(1 - p) \geqslant 5$.
2. Draw a graph showing the normal curve that is used to approximate the binomial distribution. Record the mean and variance of the normal curve, $\mu = np$ and $\sigma^2 = np(1 - p)$, respectively. Draw a vertical line at the mean of the distribution.
3. If necessary, sketch a few rectangles to show how the normal curve approximates the binomial distribution.
4. Remember to make the continuity correction when using the normal distribution to approximate the binomial distribution.
5. When using the continuity correction, use the following rules:
   a. Approximate the discrete probability $P(X < a)$ by the continuous probability $P(X < a - .5)$.
   b. Approximate the discrete probability $P(X \leqslant a)$ by the continuous probability $P(X \leqslant a + .5)$.
   c. Approximate the discrete probability $P(X > a)$ by the continuous probability $P(X > a + .5)$.
   d. Approximate the discrete probability $P(X \geqslant a)$ by the continuous probability $P(X \geqslant a - .5)$.

Exercises

Statistical Concepts

8.5.1   State the conditions under which the normal distribution provides a good approximation to the binomial.

8.5.2   State the formulas for the mean, variance, and standard deviation of the normal random variable that is used to approximate the binomial.

8.5.3   Explain what the continuity correction is and how it is implemented.

8.5.4   Is it appropriate to use the normal distribution to approximate the binomial where $n = 20$ and $p = .01$? Explain.

8.5.5   Is it appropriate to use the normal distribution to approximate the binomial where $n = 20$ and $p = .4$? Explain.

8.5.6   *True or false:*   Suppose we want to use the normal distribution to approximate the binomial where $n = 20$ and $p = .5$. Then the binomial probability $P(X = 12)$ is approximately equal to the normal probability $P(11.5 \leqslant X \leqslant 12.5)$. Explain.

8.5.7   *True or false:*   Suppose we want to use the normal distribution to approximate the binomial where $n = 20$ and $p = .6$. The appropriate mean is 12 and the appropriate variance is 4.8. Explain.

8.5.8   *True or false:*   Suppose we want to use the normal distribution to approximate the binomial where $n = 200$ and $p = .6$. Then, the binomial probability $P(180 \leqslant X < 194)$ is approximately equal to the normal probability $P(179.5 \leqslant X \leqslant 193.5)$. Explain.

8.5.9   Suppose the random variable X follows the binomial distribution with $n = 25$ and $p = .5$. The exact probability that $X = 6, 7,$ or 8 is equal to the area of the three rectangles lying over $X = 6, 7,$

and 8. *True or false:* This probability can be approximated by the area under the normal curve from $X = 6.5$ to $X = 8.5$. Explain.

Statistical Drills

8.5.10 Use the normal distribution to approximate the binomial where $n = 200$ and $p = .5$. Calculate $P(X \le 190)$.

8.5.11 Use the normal distribution to approximate the binomial where $n = 144$ and $p = .8$. Calculate $P(140 < X)$.

8.5.12 Use the normal distribution to approximate the binomial where $n = 20$ and $p = .4$. Calculate $P(4 \le X \le 9)$. Find the exact probability using the binomial distribution (see Table A.2). Compare your answers.

8.5.13 Use the normal distribution to approximate the binomial where $n = 15$ and $p = .5$. Calculate $P(5 < X < 9)$. Find the exact probability using the binomial distribution (see Table A.2). Compare your answers.

8.5.14 Use the normal distribution to approximate the binomial where $n = 25$ and $p = .4$. Calculate $P(5 \le X \le 9)$. Find the exact probability using the binomial distribution (see Table A.2). Compare your answers.

8.5.15 Use the normal distribution to approximate the binomial where $n = 25$ and $p = .4$. Calculate $P(5 < X < 9)$. Find the exact probability using the binomial distribution (see Table A.2). Compare your answers.

Statistical Applications

8.5.16 Of the students who enter law school, 80% eventually graduate. If we take a sample of 100 new law students, what is the probability that fewer than 75 eventually graduate?

8.5.17 A doctor claims that a certain drug will cure 20% of the patients suffering from a certain illness within 1 week. The drug is given to a random sample of 100 patients.
  a. If the doctor's claim is correct, what is the probability that fewer than 15 patients are cured?
  b. What is the probability of fewer than 12 of the patients being cured if the doctor's claim is correct?
  c. If fewer than 12 patients are cured, would you doubt the doctor's claim? Explain.

8.5.18 An airline knows that about 20% of the people who buy tickets for a certain flight cancel their reservations. The airline sells 100 tickets for a flight that contains only 90 seats. What is the probability that there will be enough seats for all the passengers?

8.5.19 A true–false test containing 100 questions is given to a student who is totally ignorant of the subject matter. What is the probability that the student gets 65 or more correct? Use the continuity correction.

8.5.20 On any given day, about 30% of the stocks listed on the New York Stock Exchange increase in value, 40% stay the same, and 30% decrease in value. If you select a random sample of 100 stocks, what is the probability that more than 40 of them increase in value?

8.5.21 In a certain community, 60% of the residents are in favor of building a new library. A random sample of 200 residents is taken. What is the probability that fewer than 100 of these residents favor the construction of the library?

8.5.22 A civil service exam consists of 100 multiple choice problems. Each problem has five possible answers, only one of which is correct. What is the probability that a person with no knowledge of the test material could guess more than 35 correct answers on the exam?

8.5.23 A large shipment of a product is received at a manufacturing plant in boxes containing 100 items. Suppose that 5% of all products are defective. We randomly select 10 items from a box. If two or more of the items are defective, the entire box is sent back. What is the probability of the following?

   a. A box will be sent back.

   b. All 10 items will be good.

   c. Exactly 1 of the 10 items is defective.

8.5.24 National data indicate that 80% of those who purchase a certain brand of color television will make no claims covered by the guarantee. A dealer sells 30 of these color TVs to different customers. What is the probability of the following?

   a. At least three of these customers will have claims against the guarantee.

   b. Fewer than five customers will have claims.

8.5.25 In many cities, downtown parking lots operate on an honor system after 9 P.M. A survey indicates that the probability is 25% that a person will not pay for his or her parking under the system. On a given night, a parking lot has 40 cars parked in it. What is the probability of the following?

   a. More than 10 of the drivers do not pay.

   b. More than 35 of the drivers do pay.

8.5.26 A census report states that 15% of the country's households have incomes below the poverty line. Suppose that 1,000 households are randomly selected. Find the probability that fewer than 200 of these households are below the poverty line.

8.5.27 A credit card company claims that 80% of all clothing purchases in excess of $60 are made with credit cards. A random sample of 90 clothing purchases in excess of $60 is obtained. If, in fact, 80% of all clothing purchases in excess of $60 are made with credit cards and if X is the number in a sample of 90 that make credit card purchases, find values for the following:

   a. $P(X \leq 75)$ **b.** $P(75 \leq X \leq 85)$

8.5.28 In the last election, 60% of all voters in a certain county voted for a tax increase to pay for a better school system. Suppose we select a random sample of 400 of all those who voted.

   a. What is the probability that more than 260 of the sample members voted for the tax increase?

   b. What is the probability that at least 200 of the sample members voted for the tax increase?

8.5.29 A survey research organization has found that answers are received from 25% of people sent a particular mail questionnaire. If this questionnaire is sent to 1,000 people, what is the probability that at least 300 answers will be received?

STATISTICS IN ACTION: CASE STUDY

Examining Loss of Intelligence As a Result of an Injury

A recent court case involved the personal injury of a 6-year-old boy who had suffered serious head injuries in a car accident. His parents sued for the loss of his potential earning power.

   Before he entered kindergarten at age 5, the boy took an IQ test on which he scored 105. For the general public, IQs are approximately normally distributed with mean $\mu = 100$ and standard deviation $\sigma = 15$. The boy's preinjury test score of 105 exceeded the population mean of 100. Thus, before the injury, the boy was slightly above average in intelligence. To further support this conclusion, his parents and neighbors testified that the boy was a typical child with approximately average intelligence.

   A year after the accident, the boy was given a second intelligence test, which indicated that his IQ had fallen to 55. This value is 45 points below the population mean of 100. Given that the standard deviation of IQs is 15, the boy's IQ is 3 standard deviations below the mean. By referring to Table A.5 showing areas under the standard normal distribution, we can determine that 99.87% of the general public has an IQ above 55. (Can you verify this?) Prior to the accident, the boy ranked approximately in the middle of the general population in intelligence. After the accident, the boy ranked in the lowest .13% of the general public—that is, in the lowest 13 out of 10,000.

This information helped convince the jury that the boy had suffered a severe loss of intelligence.

The jury awarded the boy over $1 million to cover his medical expenses and to compensate him for lost potential income and fringe benefits, for his pain and suffering, and for the loss of some of the joys of life.

Chapter 8 Supplementary Exercises

8.S.1 Let Z be distributed as $N(0, 1)$. Find the following probabilities:

   a. $P(0 \leq Z \leq 1.53)$ **b.** $P(-1.67 \leq Z \leq 0)$ **c.** $P(-2.25 < Z < 1.96)$

   d. $P(Z < -2.44)$ **e.** $P(Z > -2.57)$ **f.** $P(Z > 2.72)$

   g. $P(Z \leq 1.90)$ **h.** $P(Z < 4.61)$ **i.** $P(Z \leq -4.44)$

   j. $P(Z > -6.21)$ **k.** $P(Z \geq 6.24)$ **l.** $P(1.31 \leq Z \leq 2.42)$

   m. $P(-2.61 \leq Z \leq -1.44)$

8.S.2 Let X be distributed as $N(7, 100)$. Find the following probabilities:

   a. $P(0 < X < 8)$ **b.** $P(-2 < X < 5)$ **c.** $P(8.5 < X < 10.7)$

   d. $P(2.7 < X < 6.6)$ **e.** $P(X < 11.8)$ **f.** $P(X \geq 18)$

   g. $P(X < 0)$ **h.** $P(1.4 < X < 27.2)$

8.S.3 Let X be distributed as $N(\mu, \sigma^2)$. Find the following probabilities:

   a. $P(X < \mu)$ **b.** $P(X \geq \mu)$ **c.** $P(\mu - \sigma < X < \mu + \sigma)$

   d. $P(\mu - 2\sigma \leq X \leq \mu + 2\sigma)$ **e.** $P(\mu - 3\sigma \leq X \leq \mu + 3\sigma)$

   f. $P(\mu - 1.96\sigma \leq X \leq \mu + 1.96\sigma)$

8.S.4 Find the probability that a student can guess more than 65 answers correctly on a true–false test with 100 questions.

 8.S.5 Candidate A is favored by 64% of the registered voters. You select a random sample of 100 registered voters. What is the probability that fewer than 50 of these people favor candidate A?

8.S.6 Students' grade point averages at a school are normally distributed with mean 2.9 and variance .16. What average must a student attain to be in the upper 10% of the school?

8.S.7 On the average, 20% of the individuals who reserve a table at a restaurant will not appear. If the restaurant has 90 tables and takes 100 reservations, what is the probability that it will have enough tables to accommodate everyone? Assume all reservations are independent of one another.

8.S.8 Scores on a typing exam are normally distributed with mean 245 and variance 100. To be hired, an individual must score in the top 30% of the population. What is the minimum score required to get hired?

8.S.9 On a civil service exam, the scores are normally distributed with a mean of 70 and a standard deviation of 10. The highest 10% get A's, the next 20% get B's, the next 40% get C's, the next 20% get D's, and the lowest 10% get F's. Determine the scores that separate A from B, B from C, C from D, and D from F.

8.S.10 The weight of cereal in a box is normally distributed with mean 16.5 ounces and standard deviation .1 ounce. What percentage of the boxes contain less than 16 ounces of cereal?

 8.S.11 The diameters of steel rods manufactured by a company are normally distributed with mean 2 inches and standard deviation .02 inch. A rod is defective if its diameter is less than 1.95 inches. What percentage of the rods are defective?

8.S.12   The amount of liquid placed in a soft drink cup is normally distributed with mean 10.5 ounces and standard deviation .5 ounce. If we purchase two drinks, what is the probability that each cup contains less than 10 ounces of liquid?

8.S.13   A college basketball coach is seeking tall recruits who are smart enough to be eligible for college. The recruit must be at least 74 inches tall and have an IQ of 115 or above. Height and IQ are independent of one another. IQ is normally distributed with mean 100 and standard deviation 12, and height is normally distributed with mean 70 and standard deviation 2 inches. What percentage of the population satisfies the coach's requirements?

8.S.14   According to actuarial tables, the probability is .03 that a person aged 70 will die within 1 year. A company insures 6,400 people of age 70. What is the probability that 200 or more of these people will die within 1 year?

8.S.15   During the last year at Sandy's Hamburger Shop, the amount of time customers took to complete lunch after being seated and served was normally distributed with mean 24 minutes and standard deviation 4. A customer is randomly selected. When eating lunch, what proportion of customers require the following amounts of time?
   a. Less than 20 minutes
   b. More than 29 minutes
   c. Less than 32 minutes
   d. More than 14 minutes
   e. Between 16 and 18 minutes

8.S.16   The Acme Computer Company has two sales representatives, Mr. Smith and Ms. Brown, who sell computer equipment. Smith sells to banks and Brown to the insurance companies. Sales to banks are normally distributed with mean $\mu = \$200,000$ per month and standard deviation $\sigma = \$30,000$, and sales to insurance companies are normally distributed with mean $\mu = \$260,000$ per month and standard deviation $\sigma = \$40,000$.
   a. During a given month, Smith sells $260,000 worth of computer equipment. To do an equally good job, how much equipment must Brown sell to insurance companies?
   b. If Smith sold $160,000 worth of computer equipment, how much equipment must Brown sell to do an equally good job?

8.S.17   The mayor of Oakmont wants to know what proportion of the population favor raising taxes to build a new community swimming pool. Suppose 40% of the population favor a new pool. The mayor's office contacts a random sample of 100 Oakmont citizens. Find the probability of the following:
   a. More than 50 citizens favor raising taxes.
   b. Fewer than 35 favor raising taxes.

8.S.18   In a recent survey, it was found that 15% of the residents of Bedford listened to radio station WAMX between 5 P.M. and 6 P.M. To see whether the ratings have changed, a random sample of 400 residents are contacted. If p is still .15, find the probability that the number of residents in the sample who listen to WAMX during the prescribed hour is as follows:
   a. Fewer than 40
   b. More than 50
   c. More than 100
   d. If more than 100 people in the sample listen to WAMX, would this convince you that the population proportion exceeds .15? Explain.

8.S.19   The Polychrome Camera Company produces flashbulbs for inexpensive cameras, and about 5% of the flashbulbs are defective. If a random sample of 2,000 flashbulbs is shipped to a certain retail outlet, find the probability that the number of defective bulbs is as follows:

    a. More than 120
    b. Fewer than 75
    c. Between 80 and 90

8.S.20 The Cuneo Cosmetics Company has decided to sponsor a daytime television program because an industry report stated that 70% of the program's viewers are female. If this assumption is correct, find the probability that in a random sample of 150 viewers, the number of females is as follows:
    a. Fewer than 100
    b. Fewer than 90
    c. Suppose Cuneo took a random sample of 150 viewers and fewer than 90 were female. Would this convince you that the population proportion is less than .70? Explain.

8.S.21 Gavett's Appliance Store has a rule that any credit purchase exceeding $200 has to be approved by a department supervisor. It is known that credit purchases are normally distributed with mean $\mu = \$130$ and standard deviation $\sigma = \$50$.
    a. What proportion of credit purchases must be approved by department supervisors?
    b. Three credit sales are made. What is the probability that approval is needed at least once?

8.S.22 Past experience indicates that 30% of the spectators at major league baseball games buy a hot dog. The attendance at a baseball game is 25,000, and the team has a supply of 15,000 hot dogs.
    a. How many spectators do you expect to buy a hot dog?
    b. Assume that no one buys more than one hot dog. What is the probability that the supply will run out?

8.S.23 The manager of Walton's Discount Shoe Store knows that daily revenue from shoe sales is normally distributed with mean $4,000 and standard deviation $700. Find the probability of the following:
    a. On 3 successive days, revenue never exceeds $3,000.
    b. On 2 consecutive days, revenue exceeds $5,000.

8.S.24 The weight of a randomly selected piece of luggage on international flights has an average of 22.6 pounds and a standard deviation of 5.4 pounds.
    a. If the weights are normally distributed, what proportion of the luggage weighs more than 30 pounds?
    b. In a shipment of 200 pieces of luggage, how many can be expected to weigh less than 20 pounds?

8.S.25 The probability that a critical component in a space shuttle will fail in its first 100 hours of use is .06. Suppose that 1,400 of these components are randomly selected. Find the probability of the following:
    a. More than 100 of these components fail in their first 100 hours of use.
    b. Fewer than 60 of these components fail in their first 100 hours of use.
    c. How many of the 1,400 components would you expect to fail in their first 100 hours of use?

8.S.26 Traffic studies indicate that the number of cars using a certain bridge on weekday afternoons between 4 P.M. and 5 P.M. is normally distributed with mean $\mu = 2,000$ and standard deviation $\sigma = 200$. Severe traffic jams occur if more than 2,500 cars try to use the bridge in any 1-hour period.
    a. Find the probability that on any given weekday, a severe traffic jam occurs between 4 P.M. and 5 P.M.
    b. During a period covering 200 weekdays, how many traffic jams would you expect to occur?

8.S.27 In a midwestern state, the age of members of the state legislature is normally distributed with mean $\mu = 50$ years and standard deviation $\sigma = 5$. What proportion of the legislators have the following ages?
    a. Over 60
    b. 55 or younger

8.S.28   The Cooper Auto Repair Center has just received a shipment of 1,000 spark plugs from a supplier. In past shipments, the gap in 6% of the spark plugs obtained from the supplier has been set incorrectly.

    a. In the new supply, what is the expected number of misgapped plugs?

    b. What is the probability that there will be 50 or fewer misgapped plugs?

    c. What is the probability that there will be 74 or more misgapped plugs?

8.S.29   The Montgomery Employment Agency claims that it finds jobs for 75% of its clients.

    a. What is the probability that in a sample of 160 job applicants the agency finds employment for fewer than 100?

    b. If fewer than 100 applicants find jobs, would this convince you that Montgomery's claim is false? Explain.

8.S.30   An airline analyst predicts an average monthly demand of 3,900 passengers and a standard deviation of 400 passengers on a particular plane route. Assume the monthly demand is approximately normally distributed.

    a. Find the probability that in a given month the demand for the flight is between 3,500 and 4,300.

    b. Next month the actual demand is 2,500. Find the probability that the demand for this route would be 2,500 or less if the analyst's predictions are correct.

    c. If the actual demand next month turns out to be 2,500 passengers, would you be convinced that the analyst's predictions are wrong? Explain.

8.S.31   At a fast-food restaurant, the daily demand for ground beef is normally distributed with a mean of 680 pounds and a standard deviation of 80 pounds. The manager would like to ensure that sufficient ground beef is available in the restaurant each day so that the probability is no greater than 2% that the day's supply is exhausted. How many pounds of ground beef should the manager have available for use each day?

8.S.32   GMAT (Graduate Management Admissions Test) scores are normally distributed with a mean of 500 and a standard deviation of 100.

    a. What proportion of scores fall in the interval from 350 to 450?

    b. What proportion of scores fall in the interval from 550 to 650?

    c. A certain graduate business school automatically accepts all applicants whose GMAT scores exceed 620. Approximately what proportion of all those taking the GMAT would qualify for admission to this school?

8.S.33   The duration of a flight between two cities is normally distributed with mean $\mu = 4.6$ hours and standard deviation $\sigma = .30$ hour. Let X denote the duration of a randomly selected flight.

    a. Find $P(X \le 4.0)$.

    b. Find $P(4.2 \le X \le 5.0)$.

    c. Find the 80th percentile of the probability distribution.

    d. During 90% of the flights, within what interval centered about μ will the flight duration fall?

8.S.34   In restoring an office building, the required number of carpentry man-hours (X) is assumed to be a random variable that is distributed $N(100, 144)$. The required number of painting man-hours (Y) is assessed to be $N(25, 25)$. Both X and Y are assumed to be independent random variables.

    a. Let $T = X + Y$. What does T represent?

    b. If X and Y are normally distributed, then the probability distribution of T is normal. What is the expected value of T?

    c. The variance of T is the sum of the variances of X and Y. What is the variance of T?

    d. Find the probability that T exceeds 150.

    e. The labor cost of either type of work is $10 per hour. A sum of $1,600 has been budgeted for all labor. What is the probability that the total labor cost will not exceed the budget?

8.S.35 If $p = .20$ and $n = 100$, should you use the normal distribution or the Poisson distribution to approximate the binomial distribution?

8.S.36 Given $p = .01$ and $n = 300$, should you use the normal distribution or the Poisson distribution to approximate the binomial distribution?

8.S.37 The sulfur content of a shipment of coal can be assumed to be uniformly distributed between 1% and 3%. Let X denote the sulfur content of a particular shipment of coal.
  a. Draw the probability density function for the random variable X.
  b. Find the probability that a randomly chosen shipment has sulfur content over 1.6%.

8.S.38 The time of arrival of the early bus is uniformly distributed between 7:00 A.M. and 7:15 A.M.
  a. Determine the probability that the bus will arrive before 7:05 A.M.
  b. If you get to the bus stop at exactly 7:00 A.M., determine the expected waiting time.

8.S.39 Use the table for the standard normal distribution and find the values of Z such that the area in the right tail of the standard normal distribution is as follows:
  a. 10%   **b.** 5%   **c.** 2.5%   **d.** 1%   **e.** .5%

8.S.40 A certain college has found that 40% of the applicants accepted eventually decide to attend school elsewhere. How many students should be admitted if the admissions committee wants the expected size of the freshman class to be 500?

8.S.41 Suppose the school in Exercise 40 can accommodate a maximum of 550 new students. How many students should be accepted if the committee wants the probability of more than 550 new students to be less than .10?

8.S.42 If too few students show up at the school in Exercise 40, then a severe financial strain is put on the college. How many students should be accepted if the committee wants the probability of fewer than 450 students coming in to be less than .10?

8.S.43 Suppose the time it takes an international telephone operator to place an overseas call is normally distributed with mean 45 seconds and standard deviation 10 seconds.
  a. Find the probability that your call will be placed in less than 60 seconds.
  b. Find the probability that your call will be placed in less than 30 seconds.
  c. What is the probability that you will have to wait more than 65 seconds for your call to be placed?

8.S.44 Suppose that employee records obtained from the railroad industry indicate that the probability that a railroad worker will be involved in an injury accident during any given year is approximately .02. In a random sample of $n = 500$ railroad workers in Texas, 22 were involved in injury accidents within the past year. Do these data suggest that railroad workers in Texas have a higher injury accident rate than railroad workers in general?

8.S.45 Most colleges require applicants to submit a score on the SAT (Scholastic Aptitude Test). Suppose that the mean SAT score is 900 and the standard deviation is 125. Past evidence indicates that the distribution of SAT scores is approximately normal.
  a. What fraction of scores would you expect to find in the interval from 1000 to 1100?
  b. What fraction of scores would you expect to find in the interval from 850 to 1050?
  c. A certain state university automatically accepts all applicants whose SAT scores exceed 750. Approximately what fraction of all those taking the SAT would qualify for admission to this state university?

8.S.46 A manufacturing company produces bearings for use on military vehicles. The diameters of the bearings are normally distributed, with a mean of .3750 in. and a standard deviation of .0025 in. In the language of statistical quality control, the minimum and maximum allowable values for some measurement are called the *lower (LSL) and upper specification limits (USL)*. Suppose military specifications require that the bearing diameter be .3750 inch ± .0045 inch.

a. Determine what fraction of the output will be unacceptable because it fails to meet specifications.

b. Almost all output should fall within 3 standard deviations above and below the process mean. Thus, almost all output should fall in the interval $(\mu - 3\sigma, \mu + 3\sigma)$. In quality control language, the distance 6σ is called the *actual process spread*. Find the actual process spread.

c. The distance between the upper and lower specification limits (USL − LSL) is called the *allowable process spread*. Calculate the allowable process spread.

d. The *process capability index* is defined by

$$C = \frac{\text{USL} - \text{LSL}}{6\sigma}$$

Values of C less than 1.0 indicate that the process is not capable of meeting its specification limits because the allowable process spread is less than the actual process spread. That is, it is inevitable that eventually some defective output will be produced. If $C \geq 1.0$, the process is said to be *capable*. Calculate C and determine whether the process is capable. Explain.

8.S.47 The popularity of various television programs is determined by the weekly Nielsen ratings. Suppose that a random sample of 1,600 television viewers is obtained to see whether the popularity of a particular program has declined. Historically, the Nielsen ratings showed that 30% of all television viewers watch a particular program. In the random sample of 1,600 viewers, 390 viewers watch the program. Do these data suggest that the program's popularity has declined?

8.S.48 The buyer for the toy department of a chain of department stores has to decide how many dolls to purchase for potential sale during the Christmas season. The buyer believes that, at a specific price, the quantity of dolls demanded by the public is a random variable X that follows the normal distribution with mean 800 and standard deviation 100.

a. Find the probability that the quantity demanded is 650 or less. [Technically, X is a discrete random variable. Thus, it might be useful to make a slight continuity adjustment to take into account the fact that $P(X = 650) > 0$. To take this into account, represent $P(X = 650)$ by the area of a rectangle extending from 649.5 to 650.5. Thus, the discrete probability $P(X \leq 650)$ is represented by the continuous probability $P(X \leq 650.5)$.]

b. Find $P(X \leq 650)$ and compare with $P(X \leq 650.5)$. Does the adjustment for continuity make much difference? Explain why or why not.

c. Find the probability that the quantity demanded is at least 930. Ignore the continuity adjustment.

d. Find the probability that the quantity demanded is from 700 to 900. Ignore the continuity adjustment.

8.S.49 In the previous problem, the buyer knows that if too many dolls are stocked, the leftover dolls will have to be sold at a loss. How many dolls should be stocked if the buyer wants no more than a 25% chance that there are unsold dolls left at the end of the Christmas season? [This type of problem has numerous possible applications in the field of retail inventory management. One obvious difficulty with applying this type of problem is getting information about the probability distribution of the random variable X. It is difficult for the buyer to get accurate information about the mean and standard deviation of X. Sometimes, market surveys are used for this purpose.]

8.S.50 A common practice in advertising is to make various claims praising the quality of the product being advertised. For example, suppose a company manufactures a product that supposedly can help people quit smoking. Suppose the company claims that 40% of the people that use its product will be able to give up smoking permanently. To test this claim, a consumer group contacts a random sample of 400 people who have used the product. In the sample, 110 people succeeded in giving up smoking. On the basis of the results of this experiment, what conclusions would you make concerning the company's claim about the effectiveness of its product? Explain.

8.S.51 People who sustain head injuries in car accidents frequently experience a diminished capacity to perform tasks that require fine manual dexterity. Suppose that, for unimpaired adults, the average length of time required to complete a series of tasks requiring manual dexterity is 250 seconds and the standard deviation is 25 seconds. Accident victims are tested at the hospital; patients who take significantly longer than the average amount of time are placed in a rehabilitation program. Let the random variable X denote the time required to complete the series of tasks. What should be the threshold time if the rehabilitation center can treat only those patients whose test times fall in the highest 10% of the population? That is, find x_0 such that $P(X \geqslant x_0) = .10$. Assume that the time required to complete the task is normally distributed.

8.S.52 Every retail store must solve a difficult inventory management problem when ordering from its suppliers. If too little of a product is ordered, potential sales are lost and potential profits are diminished. If too much product is ordered, the store may be left with unwanted inventory. The inventory problem is especially important if there is a lengthy lag time between the date that an order is placed and the date of delivery. Let the random variable X denote the number of units of output that could potentially be sold. The solution of the inventory problem requires the store manager to know (or assume) something about the probability distribution of X. Suppose a manager believes that, at a certain price, the quantity demanded of a new book has a probability distribution that is approximately normal, with mean $\mu = 140$ units and standard deviation $\sigma = 20$ units. Suppose the manager wants to avoid excess inventory after the book loses its initial appeal. How many books should be ordered if the manager wants the probability to be at most 15% that there will be unsold items left over?

8.S.53 The owner of a bar regularly serves free chicken wings to patrons to attract a crowd at the end of the business day. The owner believes that the daily demand for chicken wings is approximately normally distributed with a mean of 180 lb and a standard deviation of 25 lb. Storage facilities are very limited, so new supplies are delivered every day. The manager wants to keep excess supply to a minimum, but also wants to have enough on hand so there is only a small probability of running out. How many pounds of chicken wings should the manager have available each day so that the probability is no greater than 5% that the day's supply is exhausted?

8.S.54 A manufacturer of electrical products purchases many parts from outside suppliers. A lot of 20,000 of a certain small component is received from a new vendor. The receiving inspection department for the manufacturer has taken a random sample of 200 components from this lot and measured the resistance of each component. These resistances (in ohms) have been arranged in the following frequency distribution:

Cell boundaries (ohms)	Frequency
66.5–68.5	2
68.5–70.5	7
70.5–72.5	13
72.5–74.5	22
74.5–76.5	25
76.5–78.5	44
78.5–80.5	40
80.5–82.5	24
82.5–84.5	16
84.5–86.5	5
86.5–88.5	2

a. Compute the sample mean and sample standard deviation of this frequency distribution.

b. What percentage of a normal distribution with μ and σ as computed in part **a** would fall outside the specification limits 75 ± 10 ohms?

c. If you make the arbitrary assumption that resistances are distributed uniformly throughout each cell, approximately what percentage of the actual distribution fell outside these limits?

d. What conclusion, if any, can you reach about whether the vendor was maintaining good statistical control of this quality characteristic? Explain your answer.

8.S.55 Tests of the stiffness of a number of aluminum-alloy channels gave the following frequency distribution. Stiffness was measured in "effective EI in psi."

Stiffness	Frequency	Stiffness	Frequency	Stiffness	Frequency
2,160	1	2,360	35	2,520	11
2,200	3	2,400	41	2,560	7
2,240	5	2,440	33	2,600	2
2,280	14	2,480	25	2,640	1
2,320	22				

a. Compute the sample mean and sample standard deviation.

b. If a normal distribution had this mean and standard deviation, what percentage of the distribution would fall below 2,250?

8.S.56 Tests have indicated that the tensile strengths of certain aluminum-alloy castings have an average of 22,300 psi and a standard deviation of 2,700 psi. If the distribution is normal, what percentage of the castings will have tensile strengths less than 17,000 psi?

8.S.57 During recent years, there has been a decline in Scholastic Aptitude Test (SAT) scores in mathematics. In 1967, the mean SAT math score was 492. Assume that in 1993 the mean math score was 475. Assume the standard deviation was 100 during both periods. Assume that test scores approximately followed a normal distribution during both periods.

a. In 1967, what proportion of students achieved a math score of 700 or higher?

b. In 1993, what proportion of students achieved a math score of 700 or higher?

8.S.58 A flight is scheduled to land in New York at 2:35 P.M., but it is equally likely to arrive at any time between 2:20 P.M. and 3:15 P.M.

a. Find the probability that the plane is late.

b. Find the probability that the plane arrives before 2:45 P.M.

 8.S.59 During weekends in June and July, the daily revenue at a toll booth on the Pennsylvania Turnpike is a random variable following the normal distribution with mean $14,000 and standard deviation $1,500.

a. Find the probability that on a given day revenue exceeds $15,300.

b. Find the probability that on a given day revenue exceeds $11,000.

References

Canavos, George C., and Don M. Miller. *Modern Business Statistics*. Belmont, CA: Duxbury Press, 1995.

Cryer, Jonathan D., and Robert B. Miller. *Statistics for Business: Data Analysis and Modelling.* Boston: PWS-Kent, 1991.

Freund, John E., and R. E. Walpole. *Mathematical Statistics*. 4th ed. Englewood Cliffs, N.J.: Prentice-Hall, 1987.

Handbook of Tables for Probability and Statistics. 2d ed. Cleveland: Chemical Rubber Company, 1968.

Hoel, Paul G. *Elementary Statistics.* 4th ed. New York: Wiley, 1976.

Kohler, Heinz. *Statistics for Business and Economics.* Glenview, IL: Scott, Foresman, 1988.

McClave, James T., and Frank H. Dietrich. *Statistics.* 6th ed. New York: Macmillan, 1994.

Mendenhall, William, James E. Reinmuth, and Robert Beaver. *Statistics for Management and Economics.* 6th ed. Boston: PWS-Kent, 1989.

National Bureau of Standards. *Tables of the Binomial Probability Distribution.* Washington, D.C.: Government Printing Office, 1949.

Neter, J., W. Wasserman, and G. A. Whitmore. *Applied Statistics.* 3d ed. Boston: Allyn & Bacon, 1987.

Neter, John, William Wasserman, and G. A. Whitmore. *Fundamental Statistics for Business and Economics.* 4th ed. Boston: Allyn & Bacon, 1973.

Shiffler, Ronald E., and Arthur J. Adams. *Introductory Business Statistics with Computer Applications.* Belmont, CA: Wadsworth, 1995.

Watson, Collin J., Patrick Billingsley, D. James Croft, and David V. Huntsberger. *Statistics for Management and Economics.* 4th ed. Needham Heights, MA: Allyn & Bacon, 1990.

9 Sampling Theory and Some Important Sampling Distributions

Many important statistical problems involve estimating or testing hypotheses about population parameters. For example, we may be interested in estimating the mean starting salary of college graduates, the mean number of miles traveled per year by rental cars, the proportion of adults who earn more than $100,000 per year, or the proportion of the labor force that is unemployed. In each of these problems, we are interested in finding the mean of some population or the proportion of individuals in a population that possess some characteristic. In other problems, we may seek information about a population variance or a population standard deviation. In each case, we seek information about some numerical value that describes a characteristic of the population.

DEFINITION **Parameter**

A **parameter** is a numerical quantity that describes some characteristic of a population.

The population mean μ, the population standard deviation σ, the population median, the population range, and the population variance are all examples of parameters. For categorical populations, the proportion (p) of the population that possesses a certain characteristic is a population parameter.

In many cases, it is unreasonable to examine the entire population to get information about a population parameter. To estimate a population parameter, we use sample data. Given the sample data, we calculate *sample statistics* that help us estimate the population parameters and test hypotheses about them.

DEFINITION **Sample Statistic**

A **sample statistic** is a numerical quantity that describes some characteristic of a sample. A sample characteristic is a function of the observations in the sample.

The sample mean, $\bar{X}$, the sample variance, S^2, the sample standard deviation, S, and the sample proportion, $\hat{p}$, are all examples of sample statistics.

The value taken by a sample statistic depends on the sample observations and will vary from sample to sample. For this reason, each sample statistic is a random variable that follows some probability distribution. We discuss these probability distributions in detail in this chapter.

Introduction to Sampling Distributions

A sample statistic such as the sample mean $\bar{X}$ is a function of the sample observations. For example, to find the sample mean, we sum the n sample values and divide by n; thus, $\bar{X}$ depends solely on the n sample observations $x_1, x_2, \ldots, x_n$. Sample statistics such as the sample mean and sample variance provide information about population parameters such as the population mean μ and the population variance σ^2. Whereas the population mean μ depends on all the values in the population, the sample mean $\bar{X}$ depends only on the observations in the sample.

Sometimes it is useful to make a fine distinction between a statistic such as $\bar{X}$ (denoted by an uppercase letter), which is a random variable, and the specific value $\bar{x}$ (denoted by a lowercase letter), which we get after substituting the values of the n sample observations into the formula. The distinction that is made between a sample statistic $\bar{X}$ and a specific value $\bar{x}$ is exactly the same as the distinction that is made between a random variable X and a specific value x of the random variable.

When a statistic such as $\bar{X}$ is used to estimate a population parameter such as μ, the statistic is called an *estimator* and the specific value $\bar{x}$ that is obtained after taking the random sample is called an *estimate*. Thus, an estimate is a specific value of an estimator.

In this book, we will use uppercase symbols, such as $\bar{X}$, S^2, and S, to denote sample statistics, and we will use lowercase symbols, such as $\bar{x}$, s^2, and s, to denote specific values of the sample statistics.

DEFINITIONS **Estimator and Estimate** ────────────────────────────────

Suppose we have a random sample $x_1, x_2, \ldots, x_n$ of observations from some population. An **estimator** of a population parameter is a *sample statistic*. It is a *rule* that tells us how to use the values $x_1, x_2, \ldots, x_n$ to calculate a single number that can be used as an estimate of the parameter. An **estimate** is the *value* obtained after the observations $x_1, x_2, \ldots, x_n$ have been substituted into the formula.

Suppose we take a random sample of n observations from the population and calculate the sample statistic $\bar{X}$. For any given population, numerous different samples of size n can be selected. Each sample of n observations $x_1, x_2, \ldots, x_n$ yields one specific value $\bar{x}$; this value varies from sample to sample and depends on chance. Thus, the estimator $\bar{X}$ is a random variable. Before the sample is selected, we cannot know with certainty what the value of $\bar{X}$ will be.

The probability distribution of any sample statistic is called its *sampling distribution*. For example, the sampling distribution of the random variable $\bar{X}$ shows how the different possible values of $\bar{X}$ are distributed when different samples of size n are selected.

DEFINITION   **Sampling Distribution** ─────────────────────────────

A sample statistic is a random variable whose possible values vary from sample to sample. Thus, the sample statistic follows a probability distribution. This probability distribution is called the **sampling distribution** of the sample statistic.

Desirable Properties of Estimators

Many different sample statistics can be used to estimate any population parameter. For example, to estimate the population mean μ, we could use the sample mean $\bar{X}$ or the sample median M. For some samples, $\bar{X}$ might provide a better estimate of μ; for other samples, M might provide a better estimate of μ. Thus, the sample mean does not always provide a better estimate of μ, nor does M. When deciding which of two different estimators to use to estimate a population parameter, therefore, we do not base our decision on the specific values that we obtain in a specific sample. We need a set of criteria for evaluating estimators that recognizes the fact that the sample statistic is a random variable. What properties does a good estimator have? In general, a good estimator produces estimates that are close to the unknown population parameter being estimated.

DEFINITION   **Sampling Error** ─────────────────────────────────

The distance between an estimate and the estimated parameter is called the **sampling error.**

Just as the value assumed by an estimator varies from sample to sample, the sampling error also will vary from sample to sample. For some samples, $\bar{X}$ will be close to μ and the sampling error will be small; for other samples, the sampling error might be quite large. A good estimator is one in which the typical sampling error is small. The typical sampling error will be small if most samples result in values of $\bar{X}$ that are close to μ. That is, the typical sampling error will be small if the probability distribution of the estimator is highly concentrated about the parameter being estimated.

A Word About Random Sampling

There are various types of random samples that can be selected from a population. Suppose a population contains a finite number N of elements. A **simple random sample** of size n is a collection of n of the N elements where the n elements are chosen (without replacement) in such a way that each of the possible samples of size n has the same probability of being selected. There are $_NC_n = N!/[n!(N-n)!]$ different ways of selecting n items from N items. Each possible simple random sample would have a selection probability of $1/(_NC_n)$.

Derivation of a Sampling Distribution

The following example shows how to derive the sampling distribution of a sample statistic in a simple case where the population size is small ($N = 5$) and the sample size is small ($n = 3$).

EXAMPLE 9.1 **Derivation of a Sampling Distribution** ———————————————

Suppose a population contains $N = 5$ elements whose values are 1, 3, 4, 8, and 9. Suppose we select a simple random sample of size $n = 3$ from this population to estimate the mean μ. [The population mean is $\mu = (1 + 3 + 4 + 8 + 9)/5 = 5$.] For each sample, let us calculate both the sample mean $\bar{X}$ and the sample median M. By examining the sampling distribution for each estimator, we can decide which estimator in general provides a better estimate of the population mean. In this example, we derive the sampling distributions of both $\bar{X}$ and M.

SOLUTION There are $_5C_3 = 10$ different ways of selecting three items from a population containing five items. Thus, there are 10 different possible samples, each of which occurs with probability $\frac{1}{10}$. For each possible sample, Table 9.1 shows the sample observations, the sample mean, the sample median, and the appropriate sample probability.

TABLE 9.1 *Values of the sample mean and sample median using simple random sampling*

Sample	Sample values	Sample mean	Sample median	Probability
1	1, 3, 4	2.67	3.00	.1
2	1, 3, 8	4.00	3.00	.1
3	1, 3, 9	4.33	3.00	.1
4	1, 4, 8	4.33	4.00	.1
5	1, 4, 9	4.67	4.00	.1
6	1, 8, 9	6.00	8.00	.1
7	3, 4, 8	5.00	4.00	.1
8	3, 4, 9	5.33	4.00	.1
9	3, 8, 9	6.67	8.00	.1
10	4, 8, 9	7.00	8.00	.1

Examine sample 1, which contains the values 1, 3, and 4. The sample mean is 2.67, and the sampling error is $|2.67 - 5.00| = 2.33$. For sample 1, the sample median is 3.00, and the sampling error is $|3.00 - 5.00| = 2.00$. In sample 1, the sampling error of the sample median is smaller than the sampling error of the sample mean. In sample 1, the sample median provides a better estimate of μ than does the sample mean.

Now, examine sample 2, which contains the values 1, 3, and 8. The sample mean is 4.00, and the sampling error is $|4.00 - 5.00| = 1.00$. The sample median is 3.00, and the sampling error is $|3.00 - 5.00| = 2.00$. In sample 2, the sampling error of the sample mean is smaller than the sampling error of the sample median. In sample 2, the sample mean provides a better estimate of μ than does the sample median.

This illustrates that, in general, neither the sample mean nor the sample median will always fall closer to the population mean μ. To decide which sample statistic to use to estimate μ, we should compare the properties of the sampling distributions of the two estimators, not the values taken in any individual sample.

Table 9.2 (page 292) shows the sampling distribution of the sample mean $\bar{X}$ and the sampling distribution of the sample median M using the data in Table 9.1. For example, there are two samples that yield a sample mean of 4.33. Thus, the probability of obtaining

$\bar{x} = 4.33$ is .2. Similarly, there are four samples that yield a sample median of 4.00. Thus, the probability of obtaining $m = 4.00$ is .4.

TABLE 9.2　*Sampling distribution of the sample mean and the sample median*

Sample mean, $\bar{x}$	Probability	Sample median, m	Probability
2.67	.1	3.00	.3
4.00	.1	4.00	.4
4.33	.2	8.00	.3
4.67	.1		
5.00	.1		
5.33	.1		
6.00	.1		
6.67	.1		
7.00	.1		

When the population size N and the sample size n are both small, the sampling distribution of a sample statistic can be derived directly by finding every possible value of the sample statistic and calculating its associated probability. Another way of finding the sampling distribution of a sample statistic at least approximately is to repeatedly select random samples of size n from the population and calculate the value of the sample statistic for each sample. If many random samples are selected, the relative frequency distribution of the observed values of the sample statistic will approximate the true sampling distribution of the sample statistic.

In many cases, mathematicians have been able to derive the exact sampling distribution of a sample statistic. The theory and level of mathematics needed to do this are beyond the scope of this text. Later, we will show that in many cases, the sampling distributions of various sample statistics can be closely approximated by the normal distribution.

In general, we are interested in three important characteristics of the sampling distribution of an estimator: (1) What is the shape of this sampling distribution? Is it normal? Is it symmetric? (2) What is the mean of the distribution? (3) What is the variance or standard deviation of the sampling distribution? Later, we show that many estimators have sampling distributions that are approximately normal, at least when the sample size is sufficiently large. In addition, in many cases, the mean of the sampling distribution of an estimator equals the mean of the original population.

For example, consider the data in Table 9.2. The sample mean underestimates the population mean μ for some samples and for other samples it overestimates μ. What can we say about the average value of $\bar{X}$ in all possible samples? Does the sample mean tend to consistently overestimate or underestimate μ, or is it correct "on the average"?

An estimator is an *unbiased* estimator of a population parameter if the mean of its sampling distribution is equal to the estimated population parameter. The sampling distribution of an unbiased estimator is centered at the population parameter. If the sampling distribution is not centered at the value of the estimated parameter, then the estimator is said to be *biased.* An estimator is *biased* if there is a systematic tendency either to over-estimate or to underestimate the population parameter being estimated.

Let A be any sample statistic. The mean of the sampling distribution of A is called the *expected value* of the sample statistic. The mean value of A is denoted $E(A)$ or as μ_A. The standard deviation of the sampling distribution of A is denoted σ_A.

DEFINITION

Unbiased Estimator

Let A be an estimator of a population parameter θ. A is an **unbiased estimator** of θ if

$$E(A) = \theta$$

Otherwise, A is said to be a **biased** estimator of θ. The *bias* of the estimator A is

$$\text{Bias} = E(A) - \theta$$

The mean of the sampling distribution of $\overline{X}$ is denoted as $E(\overline{X})$, or $\mu_{\overline{X}}$. Similarly, the mean of the sampling distribution of the sample median is denoted $E(M)$, or μ_M. Let us show how to find $E(\overline{X})$ and $E(M)$ based on the data in Table 9.2. Are $\overline{X}$ and M both unbiased estimators of the population mean μ?

In Table 9.2, both $\overline{X}$ and M are discrete random variables that assume certain discrete values with specific probabilities. Recall from Chapter 6 that the mean value of a discrete random variable is obtained by summing each possible value of the random variable weighted by its associated probability. (For a continuous random variable, the mean is the center of gravity of the probability distribution.)

EXAMPLE 9.2

Calculating $E(\overline{X})$ and $E(M)$

Use the data in Table 9.2 to determine the mean value of the sample mean, $E(\overline{X})$, and the mean value of the sample median, $E(M)$. To obtain $E(\overline{X})$ or $E(M)$, we must weigh each possible value $\overline{x}$ or m by its corresponding probability and find the sum of the products. The appropriate calculations are shown in Table 9.3.

TABLE 9.3 *Calculation of $E(\overline{X})$ and $E(M)$ for the data in Table 9.2*

$\overline{x}$	$P(\overline{x})$	$\overline{x}P(\overline{x})$	m	$P(m)$	$mP(m)$
2.67	.1	.267	3.00	.3	.90
4.00	.1	.400	4.00	.4	1.60
4.33	.2	.866	8.00	.3	2.40
4.67	.1	.467		$E(M) = 4.90$	
5.00	.1	.500			
5.33	.1	.533			
6.00	.1	.600			
6.67	.1	.667			
7.00	.1	.700			
	$E(\overline{X}) = 5.000$				

SOLUTION To calculate $E(\overline{X})$, we obtain

$$E(\overline{X}) = 2.67(.1) + \cdots + 7.00(.1) = 5.00$$

Note that the mean of the random variable $\bar{X}$ is the same as the population mean μ. That is, $E(\bar{X}) = \mu = 5.0$. This shows that $\bar{X}$ is an unbiased estimator of μ.

To calculate $E(M)$, we obtain

$$E(M) = 3.00(.3) + 4.00(.4) + 8.00(.3) = 4.90$$

Note that the expected value of M is *not* the same as the population mean μ. That is, $E(M) \neq \mu$. This means that, in this instance, M is a biased estimator of μ.

The sample mean $\bar{X}$, the sample proportion $\hat{p}$, and the sample variance S^2 are unbiased estimators of the corresponding population parameters μ, p, and σ^2. In general, however, the sample standard deviation S is *not* an unbiased estimator of the population standard deviation σ. In Example 9.2, we showed that the sample median is not necessarily an unbiased estimator of μ. If the population is symmetric, then the sample median is an unbiased estimator of μ.

Figure 9.1 shows the sampling distributions of two hypothetical estimators A and B used to estimate some population parameter θ. Observe that the sampling distribution of A is centered at θ, so $E(A) = \theta$ and A is an unbiased estimator of θ. In contrast, the sampling distribution of B is centered at some value higher than θ, which means that B is a biased estimator of θ. Since $E(B) > \theta$, if we use B to estimate θ, the specific estimate will usually be too high. Other things being equal, we prefer an unbiased estimator to a biased estimator.

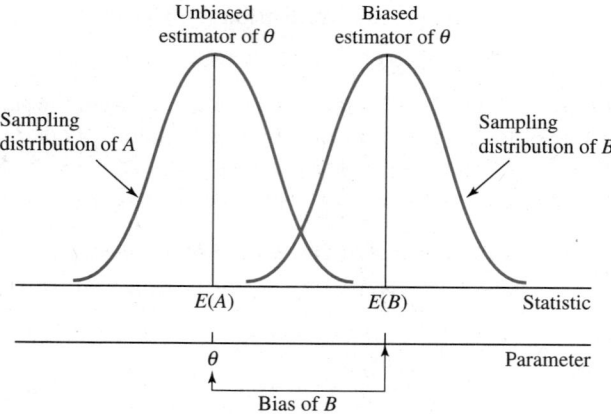

FIGURE 9.1   *Biased and unbiased estimators*

The variance and standard deviation of a sampling distribution convey information about the amount of variation that exists in values of the sample statistic obtained from different samples. If the standard deviation is small, the possible values of the sample statistic tend to be concentrated close to one another. If the standard deviation is large, the possible values of the sample statistic can vary greatly from one sample to the next.

Now suppose we have an estimator A that is an unbiased estimator of some population parameter θ. The fact that the sampling distribution of A is centered at θ does not imply that any specific value of A will be close to θ. Even though the sampling distribution of A is centered at θ, it may not be highly concentrated about θ. A desirable estimator would be one whose sampling distribution is centered at θ *and* whose sampling distribution has a small standard deviation. In this case, the sampling distribution of A will be highly concentrated about θ.

For any population parameter θ, several different unbiased estimators can usually be obtained. For example, if a random variable has a symmetric probability distribution, the sample mean and the sample median are both unbiased estimators of the population mean. Naturally, we prefer the estimator whose distribution is more highly concentrated about the population parameter being estimated.

EXAMPLE 9.3 **Comparing Two Unbiased Estimators**

Let the random variable X follow the normal distribution with mean μ and variance σ^2. To estimate the mean, take a random sample of size n from the population. To estimate μ, we can use either the sample mean $\overline{X}$ or the sample median M. Which estimator is preferable?

SOLUTION   Since the population is normal, both $\overline{X}$ and M are unbiased estimators of μ. For any specific sample, either $\overline{x}$ or m could be closer to μ. That is, for any specific sample, we cannot tell which sampling error will be smaller, $|\overline{x} - \mu|$ or $|m - \mu|$.

When deciding which estimator to use, we should consider which estimator would yield smaller sampling errors if we could repeat the sampling procedure many times. That is, we want to use the estimator whose sampling distribution is most highly concentrated about μ. Since both $\overline{X}$ and M are unbiased estimators of μ, we want to use the estimator whose sampling distribution has the smaller standard deviation.

The standard deviation of the population is σ. It can be shown that the standard deviation of the sampling distribution of $\overline{X}$ is $\sigma_{\overline{x}} = (\sigma/\sqrt{n})$, where n is the sample size. The standard deviation of M is approximately equal to $\sigma_M = 1.25(\sigma/\sqrt{n})$.

It follows that the typical sampling error $|\overline{X} - \mu|$ is smaller than the typical sampling error $|M - \mu|$. Consequently, we prefer the sample mean $\overline{X}$ over the sample median M when estimating μ.

When comparing two unbiased estimators, select the estimator whose sampling distribution has the smaller variance (or the smaller standard deviation). This leads to the concept of the *relative efficiency* of an estimator.

DEFINITION   **Relative Efficiency**

Let A and B be two unbiased estimators of some population parameter. The **relative efficiency** of A with respect to B is the ratio of their variances; that is,

$$\text{Relative efficiency} = \frac{\text{Var}(B)}{\text{Var}(A)} = \frac{\sigma_B^2}{\sigma_A^2}$$

The estimator A is said to be *more efficient* than B if

$$\text{Var}(A) < \text{Var}(B)$$

EXAMPLE 9.4   **Relative Efficiency of the Sample Mean to the Sample Median** ——————

In Example 9.3, we explained that, when sampling from a normal population, the sample mean $\bar{X}$ is a more efficient estimator of the population mean than is the sample median M. Figure 9.2 shows that the sampling distribution of the sample mean $\bar{X}$ is more highly concentrated about μ than is the sampling distribution of the sample median M. Let us find the relative efficiency of the sample mean $\bar{X}$ to the sample median M.

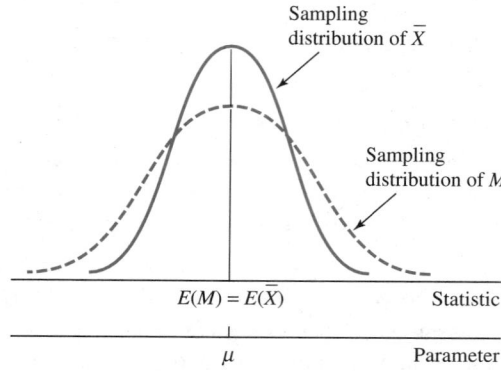

Sampling distribution of $\bar{X}$

Sampling distribution of M

$E(M) = E(\bar{X})$ Statistic

μ Parameter

FIGURE 9.2   *Sampling distributions of the sample mean and the sample median*

SOLUTION   The variance of the sample mean is σ^2/n. When sampling is from a normal distribution, the variance of the sample median is approximately $1.57(\sigma^2/n)$. The relative efficiency of $\bar{X}$ to M is

$$\text{Relative efficiency} = \frac{\text{Var}(M)}{\text{Var}(\bar{X})} = 1.57$$

Thus, the variance of the sample median is 57% larger than the variance of the sample mean. Figure 9.2 shows that $\bar{X}$ has a higher probability than M of falling within any specific interval about the population mean.

————

In a few cases, it is possible to find the most efficient of all unbiased estimators of a population parameter. This leads to the concept of the *minimum variance unbiased estimator,* also called the *most efficient unbiased estimator.*

DEFINITION   **Minimum Variance Unbiased Estimator** ——————

An estimator A is a **minimum variance unbiased estimator** of θ if A is an unbiased estimator of θ and if no other unbiased estimator has a smaller variance.

————

It is not always possible to find minimum variance unbiased estimators; however, some do exist. The sample mean when sampling is from a normal distribution, the sample variance when sampling is from a normal distribution, and the sample proportion are minimum variance unbiased estimators of μ, σ^2, and p, respectively.

Exercises

Statistical Concepts

9.1.1 Describe what is meant by a sample statistic and a sampling distribution.

9.1.2 Define sampling error.

9.1.3 What is the definition of an unbiased estimator?

9.1.4 *True or false:* The sample mean is always an unbiased estimator of the population median. Explain.

9.1.5 *True or false:* Sometimes the sample median is an unbiased estimator of the population mean. Explain.

9.1.6 *True or false:* A desirable estimator is one that is both unbiased and has a small variance. Explain.

9.1.7 *True or false:* For estimation of the population mean of a normal population, the sample mean is preferred over the sample median because the sample mean is an unbiased estimator. Comment.

9.1.8 *True or false:* For estimation of the population mean, the sample mean is preferred over the sample median because the sample mean has a smaller variance. Comment.

9.1.9 Let A and B be two unbiased estimators of some population parameter. Define the relative efficiency of A with respect to B. Explain how the relative efficiency can be helpful in determining which of the two estimators is preferred.

9.1.10 Define minimum variance unbiased estimator. Explain why this is a good property for an estimator to possess.

9.1.11 Consider two sample statistics A and B, which are both unbiased estimators of some population parameter. Explain how knowledge of the sampling distributions of A and B can help us decide which is the better estimator.

9.1.12 Explain the difference between a parameter and a sample statistic.

9.1.13 Explain how to obtain a random sample of n items from a population containing N items. Would the following procedure be acceptable? Make a list of all the elements in the population of interest. Assign a number from 1 to N to each of the items. Using a computer or a random number table, generate n integers from 1 to N. If any integer falls outside the range 1 to N or if any number is repeated, omit it and select another integer. Does this procedure give every set of n items in the population an equal chance of being included in the sample?

9.1.14 You need to select a random sample of $n = 10$ items from a population containing 200 items. Use the random number table to generate a random sample of 10 values from 1 to 200. Explain how to use the random number table to accomplish your task.

9.1.15 Let A and B be two estimators of some population parameter θ. Suppose A is an unbiased estimator and B is a biased estimator of θ. Suppose the two estimators have equal variances. Draw a picture illustrating why A would be preferable to B.

9.1.16 Let A and B be two biased estimators of some population parameter θ. Suppose the two estimators have equal variances. Draw a picture illustrating a situation where A would be preferable to B.

9.1.17 Let A and B be two estimators of some population parameter. Suppose A is an unbiased estimator and B is a biased estimator of θ. Suppose the two estimators have unequal variances. Draw a picture illustrating a situation where A would be preferable to B. Draw a picture illustrating a situation where B would be preferable to A.

Statistical Drills

9.1.18 Suppose a population contains $n = 5$ elements: 1, 4, 8, 9, 11.
  a. Calculate μ.
  b. Suppose we select three elements without replacement. How many different samples are possible? List them. [*Note:*   The order in which the elements are selected does not matter.]
  c. For each sample, calculate the sample mean and the sample median.
  d. Find the sampling distribution of the sample mean.
  e. Find $E(\bar{X})$.
  f. Find the sampling distribution of the sample median.
  g. Find $E(M)$.

9.1.19 Suppose a population contains $n = 4$ values: 2, 5, 6, 10.
  a. Calculate μ.
  b. Suppose we select two elements without replacement. How many different samples are possible? List them. [*Note:*   The order in which the elements are selected does not matter.]
  c. For each sample, calculate the sample mean $\bar{x}$.
  d. Find the sampling distribution of $\bar{X}$.
  e. Calculate the expected value of $\bar{X}$.
  f. Suppose we select two elements with replacement. How many different samples are possible? List them. [*Note:*   The order in which the elements are selected does not matter.]
  g. For each sample, calculate the sample mean $\bar{x}$.
  h. Find the sampling distribution of $\bar{X}$.
  i. Calculate the expected value of $\bar{X}$. Denote this as $\mu_{\bar{x}}$ Verify that $\mu_{\bar{x}} = \mu$.
  j. *True or false:*   Regardless of whether we sample with or without replacement, the sample mean is an unbiased estimator of the population mean.

9.1.20 Suppose a population contains $n = 3$ values: 0, 3, 9.
  a. Calculate the population mean μ.
  b. Suppose we select three elements with replacement. How many different samples are possible? List them.
  c. For each sample, calculate the sample mean $\bar{x}$.
  d. Find the sampling distribution of $\bar{X}$.
  e. Calculate the expected value of $\bar{X}$. Denote this as $\mu_{\bar{x}}$. Verify that $\mu_{\bar{x}} = \mu$.

Statistical Applications

9.1.21 Rotor bearings are produced with mean weight $\mu = 1.64$ grams and standard deviation $\sigma = .03$ gram. What is the mean of the sampling distribution of $\bar{X}$ when $n = 10$? When $n = 50$?

9.1.22 In Exercise 21, what is the standard deviation of the sampling distribution of $\bar{X}$ when $n = 10$? When $n = 50$?

9.1.23 In Exercise 21, for which sample size ($n = 10$ or $n = 50$) is it more likely that the sample mean will be within .02 gram of μ? Why?

9.1.24 Suppose the random variable X has a normal distribution with mean μ and standard deviation σ. In repeated sampling, approximately what proportion of values of X should fall in the interval $(\mu - 1.96\sigma, \mu + 1.96\sigma)$? Draw a graph to illustrate your finding. Assume that the sampling distribution of the sample mean is normal with mean μ and standard deviation $\sigma/\sqrt{n}$. In repeated sampling, approximately what proportion of sample means should fall in the interval $(\mu - 1.96\sigma/\sqrt{n}, \mu + 1.96\sigma/\sqrt{n})$? Draw a graph to illustrate your finding.

9.1.25 Suppose the random variable X has a normal distribution with mean μ and standard deviation σ. Suppose we use the sample mean to estimate μ. Assume that the sampling distribution of the sample mean is normal with mean μ and standard deviation $\sigma/\sqrt{n}$. Find the probability that the sampling error will be less than $1.645\sigma/\sqrt{n}$. Draw a graph to illustrate your finding.

9.1.26   Let E_1 and E_2 be two unbiased estimators of some parameter θ. Suppose $\mathrm{Var}(E_1) = 12$ and $\mathrm{Var}(E_2) = 16$. Calculate the relative efficiency of E_1 with respect to E_2. Which estimator is more efficient? Illustrate by drawing a graph.

9.2   Sampling Distribution of the Sample Mean

It is important to distinguish between the distribution of the population of values X and the sampling distribution of the sample mean $\bar{X}$. The following example utilizes sampling without replacement and illustrates how to obtain the approximate sampling distribution of $\bar{X}$ by taking repeated samples of size n from the population.

EXAMPLE 9.5   **Finding the Sampling Distribution of $\bar{X}$** ————————————————————

The H & R Block Company is one of the nation's largest tax consulting firms. Suppose the manager of a regional office is planning an advertising campaign to generate more business. Before instituting the campaign, the manager wants to estimate the mean adjusted gross income (AGI) claimed on the income tax returns of past clients. Suppose the office has had $N = 1,000$ clients during the past year. It is too time-consuming to review each client's file to find the person's AGI, so the manager decides to select a random sample of $n = 20$ files and calculate the mean AGI in the sample to estimate the population mean μ.

The data in Table 9.4 (page 300) show the population of 1,000 values of AGI rounded to the nearest $1,000. First, we will calculate the population mean μ and the population standard deviation σ, and we will construct a histogram to show the distribution of AGI values. Next, we will take various samples of size $n = 20$ from the population, and we will calculate the sample mean and sample standard deviation for each sample. Finally, we will construct a histogram to show the sampling distribution of the sample means.

SOLUTION   Table 9.5 (page 301) shows the frequency distribution and relative frequency distribution of the 1,000 values. Figure 9.3 (page 301) is a graph of the relative frequency distribution. The relative frequency distribution has a bell-shaped appearance, but the distribution is not symmetric: It is skewed to the right.

The mean of the 1,000 values in Table 9.4 is $\mu = 30.47$ and the population standard deviation is $\sigma = 16.54$.

To show how the sample mean varies from sample to sample, let us take various samples of size $n = 20$. There are $_{1,000}C_{20} = 3.39 \times 10^{41}$ different samples of size 20 that could be selected. To show the exact sampling distribution of $\bar{X}$, it would be necessary to calculate the sample mean for each of these samples. Obviously, taking such a large number of samples is not feasible. To approximate the sampling distribution, let us take just a few random samples of size $n = 20$, and calculate the sample mean for each sample.

Suppose we use the 20 values listed across the first row of Table 9.4 and calculate the sample mean. For row 1, the sample mean is $\bar{x}_1 = 32.95$ and the sample standard deviation is $s_1 = 20.48$. If we use the 20 values in row 2, we obtain the sample mean $\bar{x}_2 = 29.45$ and the sample standard deviation $s_2 = 16.67$. If we use the 20 values in row 50, we obtain the sample mean $\bar{x}_{50} = 30.20$ and the sample standard deviation $s_{50} = 14.80$.

Suppose we calculate the sample mean and sample standard deviation for each of the

TABLE 9.4   *Adjusted gross incomes of 1,000 clients (in $1,000s)*

																				Sample mean
13	23	60	4	13	35	15	63	26	6	19	31	53	72	36	49	48	55	17	21	32.95
3	30	68	14	51	20	50	22	17	24	9	28	37	39	11	25	32	17	47	45	29.45
25	60	12	5	10	42	3	18	29	46	63	60	29	35	28	19	17	13	28	36	28.90
42	12	8	25	18	29	27	34	41	52	10	29	57	9	10	24	13	61	31	11	27.15
45	26	17	7	30	25	16	81	43	36	38	25	15	17	33	47	38	29	17	32	30.85
42	11	9	22	14	51	50	19	27	14	3	47	19	22	20	43	22	20	43	72	28.50
12	30	29	11	42	32	67	44	22	31	15	29	25	46	21	20	31	24	32	10	28.65
2	17	43	13	5	54	29	42	29	57	69	43	10	17	27	28	17	40	30	23	29.75
34	15	55	37	4	56	25	29	22	25	18	29	25	44	27	18	43	46	18	16	29.30
22	20	26	5	18	22	20	44	46	24	78	39	35	31	25	46	17	26	54	47	32.25
13	33	25	42	45	42	12	2	34	22	25	60	58	12	25	13	55	20	51	30	30.95
23	30	53	12	26	11	30	17	15	20	42	12	34	5	8	35	37	26	7	25	23.40
60	58	12	8	17	9	29	43	55	26	45	26	31	10	29	54	34	25	19	16	30.30
4	34	5	25	7	22	11	13	37	5	42	31	20	42	27	29	56	18	27	81	26.80
13	31	10	18	30	14	42	5	4	18	12	30	50	23	34	42	45	52	24	43	27.00
35	20	42	29	25	51	32	54	56	22	2	17	22	18	41	29	29	20	3	36	29.15
15	50	3	27	16	50	67	29	25	20	34	15	17	29	52	57	22	44	47	38	32.85
63	22	18	34	81	19	44	42	29	44	22	20	24	46	10	69	25	46	19	25	35.10
26	17	29	41	43	27	22	29	22	46	24	42	29	63	19	63	8	14	2	14	29.00
6	24	46	52	36	14	31	57	25	24	31	33	28	60	57	10	29	78	20	17	33.90
19	29	63	10	38	3	15	69	18	78	25	42	37	29	9	17	25	39	43	33	32.05
31	28	60	19	25	47	29	43	29	39	19	24	39	55	10	17	54	15	12	57	32.60
53	37	29	47	15	19	20	10	25	35	10	17	11	8	24	18	27	31	20	18	23.70
72	39	35	9	17	22	46	17	44	31	41	31	25	19	13	17	18	25	43	29	29.65
36	11	28	20	33	20	21	27	27	25	23	27	32	17	51	40	43	46	72	17	30.80
49	25	19	24	47	43	20	28	18	46	72	13	67	13	21	30	46	17	9	32	31.95
48	32	17	13	38	22	31	17	43	17	17	19	47	28	21	23	78	56	42	11	31.00
55	27	13	51	29	20	24	40	46	26	15	25	35	36	22	53	16	54	14	45	32.30
17	47	28	31	17	43	32	30	18	54	26	28	23	14	29	21	65	37	61	21	32.10
21	35	36	21	32	72	10	23	16	47	13	34	17	18	35	39	27	30	60	7	29.65
45	26	17	7	30	25	16	81	43	36	38	17	9	29	43	55	26	25	60	12	32.00
42	11	9	22	14	51	50	19	27	14	3	7	22	11	13	37	65	42	82	18	27.95
12	30	29	11	42	32	67	44	22	31	15	30	14	42	5	94	18	45	26	17	31.30
2	17	43	13	5	54	29	42	29	57	69	25	51	32	54	56	22	42	11	9	33.10
34	15	55	37	4	56	25	29	22	25	18	16	50	67	29	25	20	12	30	29	29.90
22	20	26	5	18	22	20	44	46	24	78	81	19	44	42	29	44	12	17	33	32.30
25	42	45	42	12	2	34	22	24	31	25	43	27	22	29	22	46	34	5	65	29.85
60	12	26	31	30	17	15	20	42	33	42	36	14	31	57	25	44	2	20	26	29.15
58	34	31	20	50	22	17	24	29	28	37	38	3	15	69	18	78	25	42	45	34.15
12	5	10	42	23	18	29	46	63	60	29	25	47	29	43	29	39	60	12	26	32.35
18	34	81	19	44	42	29	44	2	17	28	23	47	35	23	17	9	22	14	51	29.95
29	41	43	27	22	29	22	46	34	15	36	51	28	36	4	28	9	11	62	32	30.25
46	52	36	14	31	57	25	24	22	20	19	44	21	22	29	35	43	13	5	54	30.60
63	10	38	3	15	69	18	78	25	42	45	36	23	53	21	19	55	37	24	36	35.50
60	19	25	47	29	43	29	39	60	12	34	29	18	16	45	27	26	5	18	22	30.15
29	57	15	19	25	10	25	35	58	34	37	17	26	54	47	30	25	42	45	42	33.60
35	9	17	22	46	17	44	31	12	5	29	28	22	14	51	50	60	12	26	11	27.05
28	20	33	20	21	27	27	25	25	28	56	31	11	45	21	37	12	58	17	9	27.55
19	24	47	43	20	28	18	46	13	35	51	42	22	24	31	25	43	27	72	57	34.35
17	13	38	22	31	17	43	17	55	37	9	29	20	42	33	42	36	14	65	24	30.20

Population mean   30.47

TABLE 9.5 *Frequency distribution of adjusted gross incomes for data in Table 9.4*

Adjusted gross income (in $1,000s)	Frequency	Relative frequency
0– 9.99	66	.066
10.00–19.99	220	.220
20.00–29.99	285	.285
30.00–39.99	147	.147
40.00–49.99	143	.143
50.00–59.99	75	.075
60.00–69.99	42	.042
70.00–79.99	14	.014
80.00–89.99	7	.007
90.00–99.99	1	.001
Sum	1,000	1.000

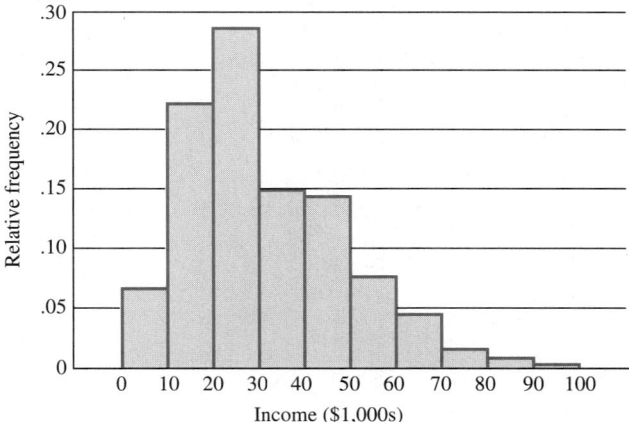

FIGURE 9.3 *Relative frequency distribution of data in Table 9.4*

50 rows of data in Table 9.4. The 50 sample means are shown on the right side of Table 9.4. Because the sample mean varies from sample to sample, it is a random variable that has a sampling distribution. This sampling distribution is approximated by the relative frequency distribution shown in Table 9.6 (page 302). Table 9.6 also shows the frequency distribution of the 50 sample means. Figure 9.4 (page 302) shows the relative frequency histogram of the 50 sample means.

It is interesting to compare Figure 9.3, which shows the relative frequency distribution of the raw data, with Figure 9.4, which shows the relative frequency distribution of the sample means. First, examine the shapes of the two distributions. In Figure 9.3, the population of 1,000 values has a bell-shaped appearance and is slightly skewed to the right. In Figure 9.4, the distribution for the 50 sample means is bell shaped and is closely approximated by a normal distribution.

TABLE 9.6   *Distribution of 50 sample means in Table 9.4*

Value of $\bar{X}$	Frequency	Relative frequency
23.00–24.99	2	.04
25.00–26.99	1	.02
27.00–28.99	8	.16
29.00–30.99	19	.38
31.00–32.99	13	.26
33.00–34.99	5	.10
35.00–36.99	2	.04
Sum	50	1.00

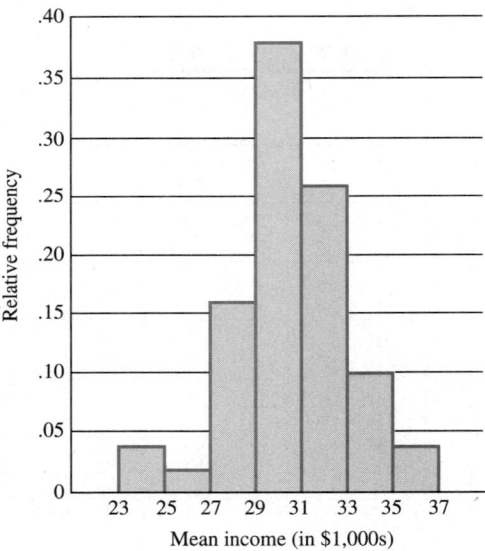

FIGURE 9.4   *Relative frequency histogram of 50 sample means in Table 9.4*

Second, examine the centers of gravity of the two distributions. The average of the 50 sample means is 30.47, which is the same as the population mean. This result occurs because we used each population value exactly once when constructing the 50 samples. It can be proved that if we took every possible sample of size $n = 20$ and calculated each sample mean, the average of all these sample means would be identical to the population mean μ. Thus, the sampling distribution of $\bar{X}$ is centered at the population mean $\mu = 30.47$. The mean of the sampling distribution of $\bar{X}$ is denoted as $E(\bar{X})$ or $\mu_{\bar{x}}$. Thus, we have $\mu_{\bar{x}} = \mu$.

Third, examine the standard deviations of the two distributions. The population standard deviation for the 1,000 values of adjusted gross income is $\sigma = 16.54$. The sample standard deviation computed from the 50 sample means is $s_{\bar{x}} = 2.573$. A comparison of

Figures 9.3 and 9.4 shows that the sample means are much more highly concentrated about their central value than are the individual values of adjusted gross income.

In summary, we have taken 50 samples of size $n = 20$ and have learned the following about the sampling distribution of $\overline{X}$:

1. The *shape* of the sampling distribution of $\overline{X}$ is a bell and is approximately normal. It can be proved that this approximation to the normal distribution would improve if the size of the sample (n) were increased. That is, if we repeated this example by using samples of size, say, $n = 40$, rather than $n = 20$, the sampling distribution would appear to be even closer to a normal curve. This result holds even though the population does not follow a normal distribution.

2. The *mean* of the sampling distribution of $\overline{X}$ is the same as the mean of the population, i.e., $\mu_{\overline{x}} = \mu_x$.

3. The *standard deviation* of the sampling distribution of $\overline{X}$ is much smaller than the standard deviation for the population of X values. Thus, the sample means are much more highly concentrated about their central value than are the individual values of $\overline{X}$. (Later we will explain that the standard deviation of the sampling distribution of $\overline{X}$ is approximately $\sigma_{\overline{x}} = \sigma/\sqrt{n}$, where σ is the population standard deviation.)

4. Observe that none of the 50 sample means is exactly equal to the population mean, but most of the 50 sample means are relatively close to the population mean. Thus, most of the sampling errors are relatively small.

In Example 9.5, the sampling distribution of the sample mean appeared to have approximately a normal distribution even though the population of adjusted gross incomes was skewed to the right. You might think that the approximate normality of the sampling distribution arises from the fact that the population of adjusted gross incomes from which the samples were selected was approximately bell shaped.

The following example presents a case where the population of values follows approximately a uniform distribution. It is interesting to note that, even when the samples are selected from a population that is not bell shaped, the sampling distribution of the sample mean tends to have a bell-shaped appearance.

EXAMPLE 9.6 **The Sampling Distribution of $\overline{X}$** ———————————————

A real estate developer ran a promotion in which each client who visited a new housing development was given nine coupons for reduced-price dinners at local restaurants. The developer gave the coupons to 200 clients. To determine whether the promotion was effective, the developer wanted to get information concerning the average number of coupons redeemed per client. The data in Table 9.7 (page 304) are the number of restaurant coupons redeemed by each of the 200 clients. Consider these X values to be a population. The population mean for these values is $\mu = 4.62$. Table 9.8 (page 304) shows the frequency and relative frequency distributions for the data in Table 9.7. Figure 9.5 is a relative frequency histogram of the data.

Suppose that the real estate developer thinks that it is too expensive and too time-consuming to contact all 200 clients to determine how many coupons they redeemed.

TABLE 9.7   *Number of restaurant coupons redeemed*

1	2	6	5	1	6	1	6	4	6	1	1	7	6	9	7	5	1	0	9
8	6	9	9	6	9	7	6	6	9	2	5	7	5	7	3	2	5	3	5
0	7	1	2	2	7	6	7	6	3	8	1	7	6	9	7	4	4	3	6
0	2	6	3	0	7	2	3	1	0	4	5	0	4	9	1	8	0	2	9
0	7	4	6	9	4	2	3	4	1	9	8	1	7	3	7	9	2	6	0
3	6	7	3	7	9	8	8	6	3	7	7	2	4	0	7	6	2	5	1
0	0	6	5	4	6	4	6	8	8	0	9	3	4	3	7	0	2	7	8
6	7	3	3	1	9	8	7	2	9	8	2	8	6	1	5	4	3	7	1
7	6	6	1	0	9	4	9	3	4	7	2	7	4	8	4	4	1	2	4
0	5	6	1	0	5	3	7	5	0	9	5	9	3	8	0	1	4	8	5

TABLE 9.8   *Distribution of data in Table 9.7*

X	0	1	2	3	4	5	6	7	8	9
Frequency	19	20	17	19	20	15	27	28	15	20
Relative frequency	.095	.100	.085	.095	.100	.075	.135	.140	.075	.100

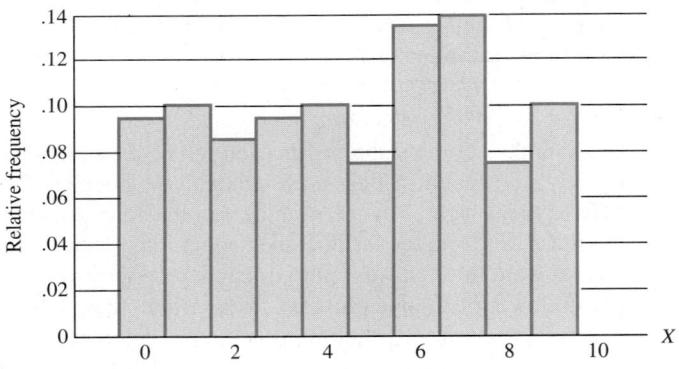

FIGURE 9.5   *Relative frequency distribution of data in Table 9.7*

Instead, the developer decides to rely on a small sample of $n = 10$ clients. Before taking the sample, it would be reasonable to ask whether a small sample of only 10 observations could be expected to provide reliable information about the population mean. To answer this question, we need to know something about the sampling distribution of $\bar{X}$. As we showed in Example 9.5, one way of getting information about the sampling distribution of $\bar{X}$ is to take a series of samples from the population and examine the resulting relative frequency distribution of sample means.

   To get information about the sampling distribution of the sample mean, we took 80 different samples of size $n = 10$ and calculated the sample mean for each sample. The 80 values of $\bar{X}$ that were obtained are shown in Table 9.9. We have chosen 80 samples to illustrate that $\bar{X}$ is a random variable with a value that varies from sample to sample. Table 9.10 shows the relative frequency distribution of the 80 sample means. Figure 9.6 is the relative frequency histogram of these 80 sample means.

TABLE 9.9 *Eighty sample means using samples of size 10 for data in Table 9.7*

3.8	4.6	7.4	4.4	4.1	5.5	2.4	4.2	4.0	5.2
6.0	4.1	4.7	4.3	5.5	4.5	4.9	4.3	3.2	5.2
2.5	4.8	5.4	3.8	3.0	7.1	4.5	6.2	4.5	4.3
5.5	4.5	5.1	4.9	5.7	4.8	4.3	2.4	4.3	4.8
3.3	5.1	5.1	4.3	4.0	4.4	5.0	6.2	4.0	4.3
4.2	4.1	5.0	6.4	4.8	5.6	5.0	4.3	2.7	4.8
3.5	5.7	4.6	3.2	4.7	5.1	3.5	4.8	4.8	4.9
4.9	6.2	5.0	3.4	4.5	5.0	5.5	5.2	4.4	3.5

TABLE 9.10 *Distribution of 80 sample means in Table 9.9*

$\bar{X}$	Frequency	Relative frequency
2.0 to under 2.5	2	.0250
2.5 to under 3.0	2	.0250
3.0 to under 3.5	5	.0625
3.5 to under 4.0	5	.0625
4.0 to under 4.5	19	.2375
4.5 to under 5.0	20	.2500
5.0 to under 5.5	13	.1625
5.5 to under 6.0	7	.0875
6.0 to under 6.5	5	.0625
6.5 to under 7.0	0	.0000
7.0 to under 7.5	2	.0250
Total	80	1.0000

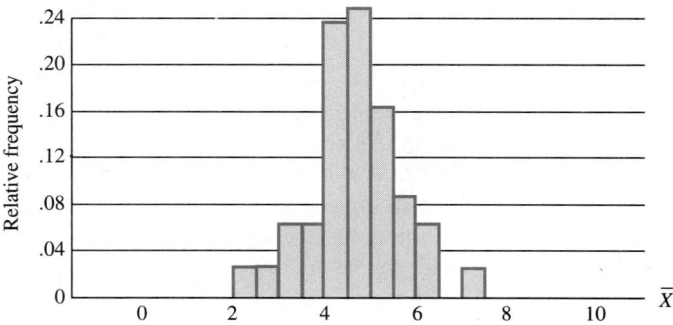

FIGURE 9.6 *Relative frequency histogram of 80 sample means in Table 9.9*

Figure 9.5 shows that the relative frequency distribution of the population of data is approximately a uniform distribution. This result indicates that approximately an equal number of clients redeemed 0, 1, 2, . . . , 9 restaurant coupons.

Figure 9.6 shows the distribution of $\bar{X}$ values. This distribution is approximately bell shaped, and most of the values of $\bar{X}$ are clustered around 4.62, which is the mean of the original population of data. Even though the population is approximately uniformly distributed, the sampling distribution of $\bar{X}$ is approximately normal. It can be shown that this approximation to the normal distribution improves as the sample size increases. For example, if we had taken samples of size $n = 30$ rather than $n = 10$, the approximation to normality would have been even better.

Mean and Variance of the Random Variable $\bar{X}$

The mean and variance of the sampling distribution of $\bar{X}$ depend on the mean and variance of the population and on the sample size n. The formulas for the mean and variance of the sampling distribution of $\bar{X}$ are given in the accompanying box.

Formulas for the mean and variance of $\bar{X}$

Let $x_1, x_2, \ldots, x_n$ denote a random sample of size n selected from a population having mean μ_X (or μ) and variance σ_X^2 (or σ^2). Let $\bar{X}$ denote the sample mean.

1. The mean value, or expected value, of $\bar{X}$ is

$$E(\bar{X}) = \mu_X = \mu$$

2. If the population is infinitely large or if sampling is done with replacement, then the variance of $\bar{X}$ is

$$\text{Var}(\bar{X}) = \frac{\sigma_X^2}{n}$$

and the standard deviation of $\bar{X}$ is given by $\sigma_X/\sqrt{n}$.

3. If sampling is done *without replacement from a finite population* containing N elements, then the variance of $\bar{X}$ is

$$\text{Var}(\bar{X}) = \frac{\sigma^2}{n}\left(\frac{N - n}{N - 1}\right)$$

The value $[(N - n)/(N - 1)]$ in the formula for calculating the variance when sampling is done without replacement is called the *finite population correction factor*, where N is the population size and n is the sample size. If the population size N is much larger than the sample size n, this correction factor is close to 1; in such cases, it makes little difference whether we sample with or without replacement. Unless explicitly noted, we will assume henceforth that sampling is done from an extremely large population so that this correction factor can be ignored.

According to the formulas, the standard deviation of the $\bar{X}$ values is $\sigma_{\bar{X}} = \sigma/\sqrt{n}$. This shows that $\sigma_{\bar{X}}$ will be small if σ is small or if n is large. If the population standard deviation σ is small, there is little variation in the population, and we would expect sample means obtained from different samples to be highly concentrated. If the sample size is large, we would expect the different estimates $\bar{X}$ obtained from different samples to be highly concentrated about the mean μ.

EXAMPLE 9.7   **Effect of a Small Population Standard Deviation** ———————————

An employment agency wants to estimate the mean hourly wage earned by high school students who work as cashiers at supermarkets. The agency hires several pollsters and asks each one to obtain a random sample of $n = 25$ students. Because most students earn between \$4.50 and \$5.00 per hour, the population standard deviation is relatively small, say, $\sigma = \$1.00$. Because the students' wages are highly concentrated and do not exhibit much variation, the estimates obtained by the different pollsters should not exhibit much variation either. That is, we would expect the values of $\overline{X}$ obtained by each pollster to be quite close to one another. Thus, if the population has a small standard deviation, the sampling distribution of sample means will also be highly concentrated.

EXAMPLE 9.8 **Effect of a Large Population Standard Deviation** ————————————

The employment agency in Example 9.7 is also interested in estimating the mean annual income earned by lawyers. Again pollsters are hired, and each is told to obtain a random sample of $n = 25$ lawyers. The population mean is, say, \$70,000 per year. However, the standard deviation of this population is very large, say, $\sigma = \$20,000$, because the annual incomes are widely dispersed about the mean. As a result, the sample mean obtained by one pollster could be quite different from the sample mean obtained by another. We would not necessarily expect the values of $\overline{X}$ obtained by the pollsters to be quite so close to one another. If the population has a large standard deviation, the sampling distribution of the sample mean can also have a large amount of dispersion.

The following example illustrates the point that the variance of the sampling distribution of $\overline{X}$ depends on the sample size n. If the sample size is large, then the sampling distribution of $\overline{X}$ will be highly concentrated about the population mean.

EXAMPLE 9.9 **Effect of Sample Size on the Variance of $\overline{X}$** ————————————

Several economists are interested in estimating the mean annual income of accountants. Suppose the unknown population mean is $\mu = \$50,000$ and the population standard deviation is $\sigma = \$10,000$. Now suppose one economist takes a random sample of $n = 4$ employees. Because this sample size is very small, the estimate obtained by the economist could be quite different from the true population mean. For example, the standard deviation of the sampling distribution of $\overline{X}$ is $\sigma/\sqrt{n} = 10,000/\sqrt{4} = \$5,000$. From the Empirical Rule, we know that approximately 68% of the data lie within 1 standard deviation of the mean if the distribution is approximately bell shaped, so only about 68% of the potential estimates of the population mean will be within \$5,000 of the true mean. Since the remaining 32% of the estimates will be more than \$5,000 from the mean, the probability is approximately .32 that the sampling error will be more than \$5,000. An estimate of the mean income that is in error by more than \$5,000 could be useless in many economic studies. Because the sample size is very small, the sampling distribution of $\overline{X}$ is widely dispersed about the population mean.

Now suppose another economist takes a random sample of $n = 1,600$ employees. Because this sample size is quite large, we would expect to obtain a very good estimate of the population mean. In this case, the standard deviation of the sampling distribution of $\overline{X}$ is $\sigma/\sqrt{n} = \$10,000/\sqrt{1,600} = \$250$. From the Empirical Rule, we know that the probability is approximately .95 that the sampling error $|\overline{X} - \mu|$ will be less than 2 standard deviations, namely, \$500, if the distribution is approximately bell shaped. Thus, in approximately 95% of the potential samples of size 1,600, the estimated sample mean will be within $500 of the true population mean. An estimate of the mean income that is in error by less than $500 would probably be quite useful in most economic studies. When the sample size is large, the sampling distribution of the sample mean is highly concentrated about the population mean.

Graphical Representation of the Distribution of $\bar{X}$

The graphs in Figure 9.7 show that the sampling distribution of $\bar{X}$ is centered at the population mean and becomes more concentrated as the sample size increases. Also as the

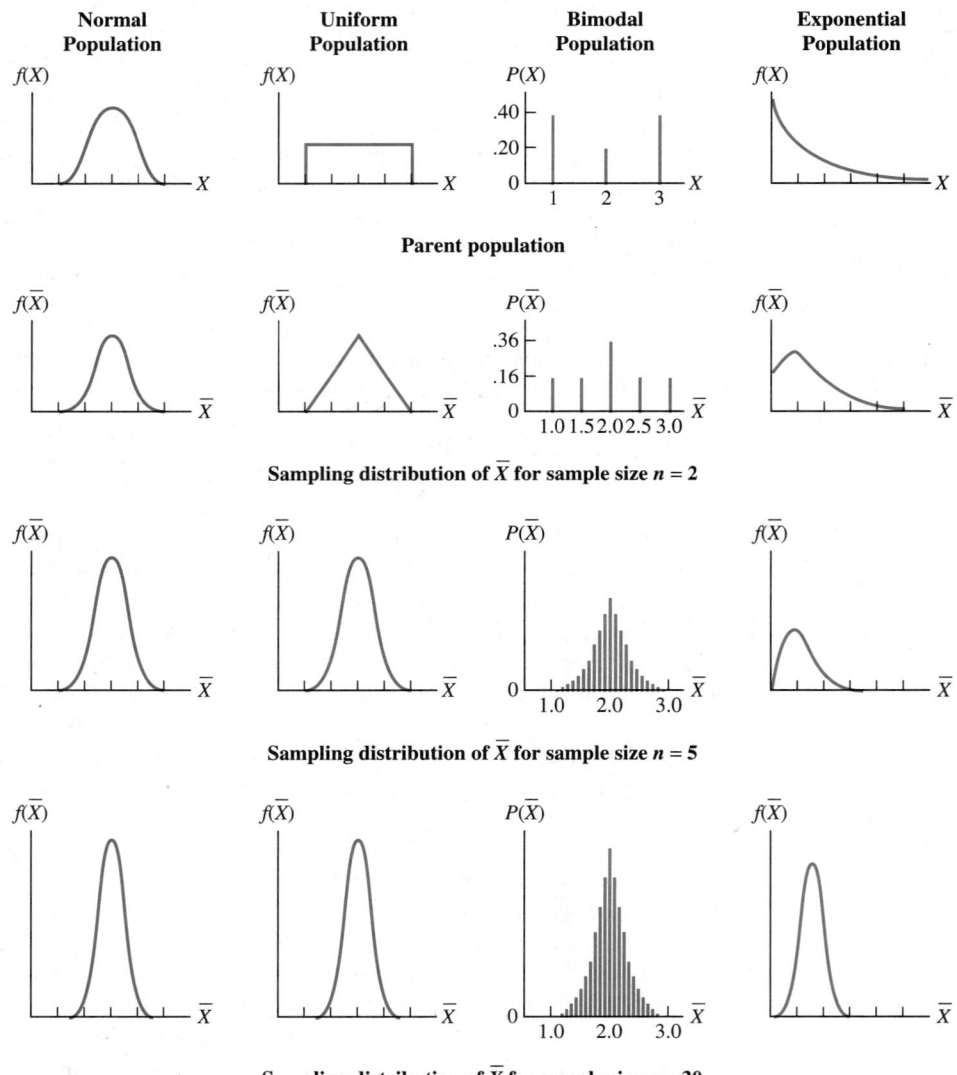

FIGURE 9.7 *Sampling distribution of $\bar{X}$ for n = 2, 5, and 30*

sample size gets larger, the distribution of $\bar{X}$ begins to approach normality regardless of the shape of the original distribution.

Exercises

Statistical Concepts

9.2.1   *True or false:*   It is possible that no single value of the sample mean equals μ, but we still obtain $E(\bar{X}) = \mu$. Comment.

9.2.2   *True or false:*   The population standard deviation σ decreases as the population size N increases. Comment.

9.2.3   *True or false:*   The standard deviation of $\bar{X}$ decreases as the population size N increases. Comment.

9.2.4   *True or false:*   The standard deviation of $\bar{X}$ decreases as the sample size n increases. Comment.

9.2.5   *True or false:*   Let X be a normal variable with mean 100 and standard deviation 5. Suppose we take a random sample of $n = 25$ items and calculate the sample mean. Then $P(90 < \bar{X} < 110) > P(90 < X < 110)$. Comment.

9.2.6   *True or false:*   Let X be a normal variable with mean 100 and standard deviation 5. Suppose we take a random sample of $n = 25$ items and calculate the sample mean. Then $P(98 < \bar{X} < 102) = P(-2 < Z < 2)$. Comment.

9.2.7   The standard deviation of a sampling distribution is also called the *standard error* of the sampling distribution. For the sample mean, the standard error is equal to $\sigma/\sqrt{n}$, where σ is the standard deviation of the population from which the sample was taken. Explain what happens to the sampling distribution of $\bar{X}$ as the sample size is increased.

Statistical Drills

9.2.8   Suppose a random sample of $n = 25$ items is selected from a population having mean $\mu = 220$ and standard deviation $\sigma = 44$. Calculate the mean and standard deviation of the sampling distribution of $\bar{X}$. Ignore the finite population correction factor.

9.2.9   Suppose random sample of $n = 100$ items is selected from a population having mean $\mu = 220$ and standard deviation $\sigma = 44$. Calculate the mean and standard deviation of the sampling distribution of $\bar{X}$. Ignore the finite population correction factor. Compare your answers with the answers in the previous problem and comment.

9.2.10   Suppose a random sample of $n = 100$ items is selected from a population containing 250 items having mean $\mu = 220$ and standard deviation $\sigma = 44$. Calculate the mean and standard deviation of the sampling distribution of $\bar{X}$. Use the finite population correction factor. Compare your answers with the answers in the previous problem and comment.

9.2.11   Calculate the finite population correction factor if $N = 100$ and $n = 20$.

9.2.12   Calculate the finite population correction factor if $N = 1,000$ and $n = 50$.

9.2.13   Use the following information and determine the mean and standard deviation of the sampling distribution of the sample mean, $\bar{X}$, for a random sample of size n drawn from a normal population with mean μ and standard deviation σ:

a. $n = 25, \mu = 25, \sigma = 10$ **b.** $n = 25, \mu = 25, \sigma = 20$
c. $n = 100, \mu = 50, \sigma = 10$ **d.** $n = 100, \mu = 50, \sigma = 20$

Statistical Applications

9.2.14   A city council is studying the tourist trade. During a certain year, the mean expenditure per tourist was $\mu = \$500$ with variance 8,100. The city takes a random sample of 10 tourists and calculates their average expenditure $\overline{X}$. What are the mean, variance, and standard deviation of this distribution?

9.2.15   In Exercise 14, the city takes a random sample of 1,000 tourists and calculates the mean expenditure. Now what are the mean, variance, and standard deviation of the distribution of $\overline{X}$? In this case, would you be very confident that your value of $\overline{X}$ is very close to μ?

9.2.16   A person takes a random sample of 10 college seniors to obtain their IQs and gets the values $x_1 = 110$, $x_2 = 145$, $x_3 = 150$, $x_4 = 130$, $x_5 = 120$, $x_6 = 105$, $x_7 = 130$, $x_8 = 105$, $x_9 = 125$, and $x_{10} = 120$. Use the formulas in Chapter 4 to calculate the sample mean and sample variance. Use the formula S^2/n to estimate the variance of $\overline{X}$.

9.2.17   If the population is approximately normal, the variance of the sample median equals $(\pi/2)\,(\sigma^2/n)$. If a sample median is to have a standard deviation equal to that of a sample mean (from the same population), how much bigger must the sample size be?

9.2.18   Suppose we want to estimate the mean number of courses taken per semester by students at a certain university. Suppose we select a random sample of 25 observations from a population where, in fact, 10% of the students take one course, 20% take two courses, 20% take three courses, 30% take four courses, and the remaining 20% take five courses.
a. Let the random variable X denote the number of courses a student takes. Find μ and σ.
b. Find $\mu_{\overline{x}}$ and $\sigma_{\overline{x}}$.

9.2.19   Suppose we select a random sample of n observations from a large population having mean $\mu = 50$ and standard deviation $\sigma = 25$. Find the standard deviation of $\overline{X}$ for $n = 5, 10, 25, 50$, and 100. Plot these values versus the sample size n to show how $\sigma_{\overline{x}}$ varies with n.

9.2.20   Suppose you take a random sample of n items from a population and calculate the sample mean. Suppose the population mean is $\mu = 125$ and the population standard deviation is $\sigma = 50$.
a. If $n = 25$, find $\sigma_{\overline{x}}$.
b. Suppose you want to reduce the standard deviation in part **a** by 50%. How large does the sample size have to be?
c. Suppose you want $\sigma_{\overline{x}} = 2$. How large does the sample size have to be?
d. Suppose you want $\sigma_{\overline{x}} = 1$. How large does the sample size have to be?

9.2.21   Does the use of the finite population correction factor cause the standard deviation of $\overline{X}$ to get larger or smaller? Explain why this is reasonable.

9.3   The Central Limit Theorem

In the previous section, we showed that $\overline{X}$ is an unbiased estimator of the population mean μ $(\mu_{\overline{x}} = \mu)$ and that the standard deviation of $\overline{X}$ is inversely related to the sample size n $(\sigma_{\overline{x}} = \sigma/\sqrt{n})$. We also showed two examples where the shape of the sampling distribution of $\overline{X}$ could be approximated by a bell-shaped relative frequency distribution that closely resembled a normal distribution. When making inferences about the possible value of the population mean μ, we rely on the sample mean $\overline{X}$. It is important to know, at least approximately, the shape of the sampling distribution of $\overline{X}$. There are two important theorems that describe the sampling distribution of $\overline{X}$ in different situations. The first theorem applies when the sample is selected from a population that has a normal distribution.

THEOREM   **Sampling Distribution of $\overline{X}$ When the Population Is Normal** ——————————

Suppose we select a random sample of n observations from a population with a normal distribution having mean μ and standard deviation σ. The sampling distribution of $\overline{X}$ will be a normal distribution with mean μ and standard deviation $\sigma/\sqrt{n}$. This result holds regardless of the sample size.

This result is of limited applicability because there are many populations that do not have a normal distribution. Fortunately, there is another theorem, called the Central Limit Theorem, which tells us that, if the sample size is sufficiently large, the sampling distribution of $\overline{X}$ is approximately normal even when the population is not normal.

The Central Limit Theorem is one of the most important theorems in statistics and is the reason that the normal distribution is the most important of all probability distributions. One version of the Central Limit Theorem is given in the accompanying box.

THEOREM **Central Limit Theorem**

Suppose we select a random sample of n observations from any population having mean μ and standard deviation σ. If n is sufficiently large, the sampling distribution of $\overline{X}$ will be approximately a normal distribution with mean μ and standard deviation $\sigma/\sqrt{n}$. The approximation improves as the sample size increases.

Unfortunately, the Central Limit Theorem does not state how large the sample size n must be before the approximation is accurate. The reason for this silence is that the accuracy of the approximation depends on the shape of the relative frequency distribution of the population. If the relative frequency distribution of the population is symmetric and bell shaped or mound shaped, the approximation may be accurate for sample sizes as small as $n = 5$. If the relative frequency distribution is highly skewed or if it is not mound shaped, larger sample sizes may be needed to get an accurate approximation. Recall that, in Example 9.6, the relative frequency distribution of the population was relatively flat like a uniform distribution. In Example 9.6, we showed that the relative frequency distribution of the 80 sample means was approximately a normal distribution when the sample size was as small as $n = 10$.

The Central Limit Theorem would be of little practical value if the approximation was applicable only when n is extremely large, but in fact it works in most cases, even when n is fairly small. When monitoring production processes, quality control engineers frequently rely on the Central Limit Theorem to make inferences about a population mean based on samples containing as few as $n = 5$ observations. A common rule of thumb is that the sampling distribution of $\overline{X}$ can reliably be approximated by the normal distribution when n is 20 or 30. We shall assume that the normal distribution provides an accurate approximation to the sampling distribution of $\overline{X}$ when n is at least 30.

Recall that if the original population follows the normal distribution, then the distribution of $\overline{X}$ is *exactly* a normal distribution, regardless of the sample size. The Central Limit Theorem is important because it states that *even if the original population is not normal,* the distribution of $\overline{X}$ will be approximately normal provided the sample size is sufficiently large; we assume that $n \geqslant 30$ is sufficient.

EXAMPLE 9.10 **Application of the Central Limit Theorem**

Suppose the Bureau of Labor Statistics wants to estimate the mean starting salary of newly graduated chemical engineers. Suppose the population mean is actually $\mu =$

$25,000 and the population standard deviation is $\sigma = \$2,000$. To estimate μ, the Bureau obtains a random sample of $n = 100$ recently graduated chemical engineers. Find the probability that the sample mean will be within $400 of the true mean.

SOLUTION   We have $n = 100$, $\mu = \$25,000$, and $\sigma = \$2,000$. Because the sample size greatly exceeds 30, we are confident in assuming that the Central Limit Theorem applies and that the sampling distribution of $\bar{X}$ can be accurately approximated by a normal distribution with mean $\mu = \$25,000$ and standard deviation

$$\sigma_{\bar{x}} = \frac{\sigma}{\sqrt{n}} = \frac{\$2,000}{\sqrt{100}} = \$200$$

If $\bar{X}$ is within $400 of the population mean, then $\bar{X}$ will be between $24,600 and $25,400. Thus the required probability is represented by $P(24,600 \leq \bar{X} \leq 25,400)$. Refer to Figure 9.8.

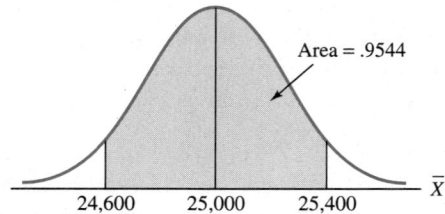

Area = .9544

24,600   25,000   25,400 $\bar{X}$

FIGURE 9.8   *Normal distribution for Example 9.10*

We obtain the standardized Z scores

$$z_1 = \frac{24,600 - 25,000}{200} = -2.00$$

$$z_2 = \frac{25,400 - 25,000}{200} = 2.00$$

From Table A.5, we obtain

$$P(24,600 \leq \bar{X} \leq 25,400) = P(-2.00 \leq Z \leq 2.00)$$
$$= .4772 + .4772 = .9544$$

Thus, if the Bureau of Labor Statistics takes a random sample of 100 chemical engineers, the probability is .9544 that the sample mean will be within $400 of the true population mean.

___

Tips for Problem Solving

To calculate the probability that the sample mean $\bar{X}$ falls in any interval, follow these steps:

1. List the sample size n, the population mean μ, and the population standard deviation σ. Verify that the normal distribution provides an accurate approximation to the sampling distribution of $\bar{X}$.

2. If the population has a normal distribution, then the sampling distribution of $\bar{X}$ is exactly normal with mean μ and standard deviation $\sigma_{\bar{x}} = \sigma/\sqrt{n}$.

3. If the population is not normal, but the sample size is sufficiently large, the sampling distribution of $\bar{X}$ is approximately normal with mean μ and standard deviation $\sigma_{\bar{x}} = \sigma/\sqrt{n}$. (For most cases, $n \geq 30$ is more than sufficient.)

4. Sketch the appropriate normal distribution and shade the appropriate area.

5. Calculate the Z scores corresponding to the values of interest in the problem. Sketch a standard normal curve and find the appropriate area.

6. The accuracy of your answers will depend on the number of significant digits used in the calculations. When calculating a value of Z, try not to round off the value $\sqrt{n}$ or the value $\sigma/\sqrt{n}$. After calculating the Z value, round off to the nearest hundredth and find the required probability from Table A.5.

7. Examine your sketch and verify that the calculated answer agrees with the shaded area.

8. *Warning:* A common mistake is to use σ, rather than $\sigma/\sqrt{n}$, for the standard deviation of $\bar{X}$. If you make this mistake, you will be finding probabilities using the probability distribution of the population rather than the probability distribution of $\bar{X}$.

Exercises

Statistical Concepts

9.3.1 Explain the difference between the probability distribution of the random variable X and the probability distribution of $\bar{X}$. Suppose X follows a normal distribution with mean μ and standard deviation σ. What can be said about the probability distribution of $\bar{X}$?

9.3.2 State the Central Limit Theorem. Why is it so important?

9.3.3 *True or false:* One difficulty with implementing the Central Limit Theorem is that we need some idea of how large the sample size n must be for the approximation to be useful. The appropriate value for n depends on the shape of the population probability distribution. Comment.

9.3.4 *True or false:* When the probability distribution of X is symmetric about its mean μ, the Central Limit Theorem might apply fairly well even with small sample sizes such as $n = 10$. Comment.

9.3.5 *True or false:* Suppose we take a sample of $n = 16$ observations from a standard normal distribution. Then the sampling distribution of $\bar{X}$ will also follow the standard normal distribution. Explain.

9.3.6 Explain how the shape, the mean, and the standard deviation of the sampling distribution of $\bar{X}$ change as n increases.

9.3.7 Explain how the shape, the mean, and the standard deviation of the sampling distribution of $\bar{X}$ change as σ increases.

9.3.8 *True or false:* Suppose we take a random sample of n items from a normal population. The sampling distribution of $\bar{X}$ is normal so there is no benefit in taking a large sample rather than a small sample.

9.3.9 Suppose we have two different unbiased estimators of the mean. The sampling distribution of estimator A has a variance σ^2/n, whereas the sampling distribution of estimator B has variance $\sigma^2/(.8n)$. Which estimator would be preferred? Explain.

9.3.10 *True or false:* Suppose we take a random sample of n items from a normal population having unknown mean μ and unknown standard deviation σ. The sampling distribution of $\bar{X}$ is no longer normal because the value of σ is unknown. Explain.

Statistical Drills

9.3.11 Suppose a random sample of 30 observations is selected from a population having mean $\mu = 28$ and standard deviation $\sigma = 15$. Suppose the population contains 500,000 observations.
  a. Find the probability that the value of $\bar{X}$ will exceed 29.
  b. Does your answer in part **a** change if the population contains 5 million observations?
  c. Does your answer in part **a** change if the sample size is $n = 100$?
  d. Repeat part **a** with a sample size $n = 400$.

9.3.12 A sample of size $n = 16$ is randomly selected from a normal population with mean $\mu = 87$ and standard deviation $\sigma = 20$.
  a. Find $P(\bar{X} \geqslant 93)$.
  b. Find $P(80 < \bar{X} < 94)$.
  c. Find $P(\bar{X} < 77)$.

9.3.13 A sample of size $n = 100$ is randomly selected from a nonnormal population with mean $\mu = 240$ and standard deviation $\sigma = 40$.
  a. Find $P(\bar{X} \geqslant 230)$.
  b. Find $P(228 < \bar{X} < 252)$.
  c. Find $P(\bar{X} < 226)$.

9.3.14 A sample of size $n = 400$ is randomly selected from a nonnormal population with mean $\mu = 240$ and standard deviation $\sigma = 40$.
  a. Find $P(\bar{X} \geqslant 235)$.
  b. Find $P(236 < \bar{X} < 244)$.
  c. Find $P(\bar{X} < 243)$.

9.3.15 A random sample of $n = 100$ observations is drawn from a population with mean $\mu = 274$ and standard deviation $\sigma = 38$.
  a. State the mean and standard deviation of the sampling distribution of $\bar{X}$.
  b. Describe the shape of the sampling distribution of $\bar{X}$. Did you invoke the Central Limit Theorem to get your answer?
  c. Calculate $P(267.4 \leqslant \bar{X} \leqslant 277.8)$.

Statistical Applications

9.3.16 A consumer research agency takes a random sample of 100 people to estimate the average number of hours per month that adults watch TV. Suppose the population mean is $\mu = 100$ hours and the standard deviation is $\sigma = 20$. Find the probability that the sample mean $\bar{X}$ is between 100 and 105.

9.3.17 A certain random variable X has mean 50 and variance 144.
  a. We obtain a random sample of size 81. What is the distribution of the sample mean $\bar{X}$, at least approximately?
  b. If X is normally distributed and the sample size is 10, what is the distribution of the sample mean $\bar{X}$?

9.3.18 On weekdays at noon, the number of people in a downtown cafeteria has mean $\mu = 140$ and standard deviation $\sigma = 50$. The manager counts the number of noontime customers on 100 randomly selected days and calculates the sample mean $\bar{X}$. What is the probability that $\bar{X}$ is between 130 and 150?

9.3.19 In Exercise 18, what is the probability that $\bar{X}$ is between 135 and 145? Do the results from Exercise 18 and this exercise convince you that it is possible to get a fairly good estimate of a population mean by taking a relatively small sample?

9.3.20   At a certain bank, the mean checking account balance is $\mu = \$150$ and the population standard deviation is $\sigma = \$60$. What is the probability that the sample mean of 400 randomly chosen checking account balances is as follows:
a. Between $135 and $165
b. Between $144 and $156

9.3.21   The IRS wants to determine the mean amount of money that married couples contribute to charity. Assume the standard deviation of contributions is $360. What is the probability that in a random sample of 100 married couples, the sample mean contribution is within $10 of the population mean?

9.3.22   A study is being undertaken to determine whether the new cash registers installed at Foodland Supermarkets have sped up the checkout process. In the past, the mean time spent in a checkout line was $\mu = 11$ minutes and the standard deviation was $\sigma = 3$. A random sample of 36 customers is observed. If μ is in fact still 11, find the probability of the following results in the sample:
a. $\bar{X}$ exceeds 12.
b. $\bar{X}$ exceeds 15.
c. $\bar{X}$ is less than 10.5.

9.3.23   A teacher gives a test to a class containing several hundred students. It is known that the standard deviation of the scores is about 12 points. A random sample of 36 scores is obtained. What is the probability that the sample mean will differ from the population mean by more than 6 points?

9.3.24   A local bank reported to the federal government that its savings accounts have a mean balance of $1,890 and a standard deviation of $264. Government auditors randomly sample 144 of the bank's accounts to assess the reliability of the mean balance reported by the bank. The auditors will certify the bank's report only if the sample mean balance is within $50 of the reported mean balance. What is the probability that the auditors will not certify the bank's report, even if the mean balance really is $1,890? (Assume the standard deviation reported by the bank is accurate.)

9.3.25   Suppose a company sells canned vegetables, such as peas, corn, and so forth. Producers closely monitor the weight of vegetables being placed into containers to be sure that the canning process is in control. Suppose a random sample of $n = 16$ containers is selected, and the weight of the vegetables in each can is recorded. Prior experience has shown that this weight has a normal distribution with $\mu = 14.36$ ounces and $\sigma = .58$ ounce. The process is considered to be out of control and the process is stopped for inspection and possible adjustment if the sample mean is less than 14.24 ounces.
a. If the process is in control, find the probability that the inspector will mistakenly think that the process is out of control.
b. Suppose, in fact, the process is out of control with $\mu = 14.22$ and $\sigma = .66$. Find the probability that the inspector will mistakenly think that the process is in control.
c. Suppose the inspector increases the sample size to $n = 100$. If the process is in control, find the probability that the inspector will mistakenly think that the process is out of control.
d. Suppose, in fact, the process is out of control with $\mu = 14.22$ and $\sigma = .66$. If the inspector increases the sample size to $n = 100$, find the probability that the inspector will mistakenly think that the process is in control.
e. Suppose the inspector increases the sample size to $n = 400$. If the process is in control, find the probability that the inspector will mistakenly think that the process is out of control.
f. Suppose, in fact, the process is out of control with $\mu = 14.22$ and $\sigma = .66$. If the inspector increases the sample size to $n = 400$, find the probability that the inspector will mistakenly think that the process is in control.

9.3.26   Suppose a chemical manufacturer produces various products that are marketed in plastic bottles. The material is toxic, so the bottles must be tightly sealed. Thus, the bottles and the caps must be produced within very tight specification limits. Suppose that bottle caps having a diameter between .397 and .403 inch are acceptable to the chemical manufacturer. The manufacturer of the bottle caps has instituted a tight quality control program to prevent the production of defective output. As part of its quality control program, the manufacturer measures the diameters of a random sample of $n = 16$ bottle caps each hour and computes the sample mean diameter. The manufacturer knows that when the process is in control the mean diameter is $\mu = .400$ inch and the population standard deviation is .001.

a. If the process is in control, what proportion of bottle caps would have diameters outside the buyer's specification limits?

b. The manufacturer has a rule that the process will be stopped and inspected any time the sample mean falls below .399 inch or above .401 inch. If the process is in control, find the probability that the process will be stopped during any given inspection period.

c. Suppose the process has changed; the mean diameter is now $\mu = .4005$ inch and the standard deviation is $\sigma = .001$. Find the probability that the manufacturer will halt the production process.

 9.3.27 In a typical quality control analysis of a production process, a random sample of output is obtained and some variable X, such as the weight or length of each item, is recorded. Suppose that, when a process is in control, X follows a normal distribution with $\mu = 6.34$ and $\sigma = .10$. To check whether the process has gone out of control, the quality control engineers take random samples of n items at, say, k different intervals and record the sequence of k sample means. The engineers construct what is called a control chart to monitor the production process. The control chart is a graph showing the sequence of sample means. On the control chart, two boundaries are drawn, called the upper and lower control limits. Any time a sample mean falls outside the control limits, the production process is stopped, inspected, and possibly adjusted. Typically, the control limits are placed at a distance of $3\sigma_{\bar{x}} = 3(\sigma/\sqrt{n})$ units above and below the mean.

a. Suppose the engineers take a random sample of $n = 25$ items from a process having $\mu = 6.34$ and $\sigma = .10$. Calculate the upper and lower control limits.

b. Suppose the process is in control. Find the probability that the next sample of items will indicate that the process is out of control.

c. Suppose the engineers take a sample of $n = 25$ items during each of the next 10 hours and suppose the process is in control. Find the probability that at least one of the sample means falls outside the control limits.

9.4 Sampling Distribution of the Difference Between Two Sample Means

Frequently we are interested in determining whether the mean of one population is equal to the mean of another. For example, we might want to answer questions of the following type:

1. Is the mean credit card balance for males the same as the mean credit card balance for females?

2. Is the mean starting salary for college graduates who majored in economics the same as the mean starting salary for college graduates who majored in English?

3. Is the mean number of cavities for children who use brand A toothpaste the same as the mean number of cavities for children who use brand B?

In each case we have two different populations, which we shall call population 1 and population 2. Population 1 has mean μ_1 and variance σ_1^2, and population 2 has mean μ_2 and variance σ_2^2.

To make such comparisons, we take independent random samples of n_1 observations from population 1 and n_2 observations from population 2 and then calculate the respective sample means, denoted $\bar{x}_1$ and $\bar{x}_2$. Frequently, the two sample sizes are equal ($n_1 = n_2$), but this does not have to be the case. To obtain information about the difference between two population means ($\mu_1 - \mu_2$), we rely on the sample estimate ($\bar{x}_1 - \bar{x}_2$).

Recall that $\bar{X}_1$ and $\bar{X}_2$ are random variables with values that vary from sample to sample. Similarly, ($\bar{X}_1 - \bar{X}_2$) is a random variable whose value varies from sample to sample. Before examining the distribution of ($\bar{X}_1 - \bar{X}_2$), we should note that at this point we are actually concerned with five different probability distributions. First, we have the two distributions of the two populations, which have means μ_1 and μ_2 and variances σ_1^2 and σ_2^2. Second, we have the two sampling distributions of $\bar{X}_1$ and of $\bar{X}_2$. According to the Central Limit Theorem, if n_1 is sufficiently large, then $\bar{X}_1$ has approximately a normal distribution with mean μ_1 and variance σ_1^2/n_1, and similarly for $\bar{X}_2$. Finally, we have the sampling distribution of ($\bar{X}_1 - \bar{X}_2$). If both samples are sufficiently large (say, $n_1 \geq 30$ and $n_2 \geq 30$), then the distribution of ($\bar{X}_1 - \bar{X}_2$) will be approximately normal no matter how the original populations are distributed. If the original populations are normally distributed, then ($\bar{X}_1 - \bar{X}_2$) is exactly normally distributed for any values of n_1 and n_2.

Figures 9.9 and 9.10 (page 318) show the five distributions described previously for two different cases, where both populations are approximately normal and where both

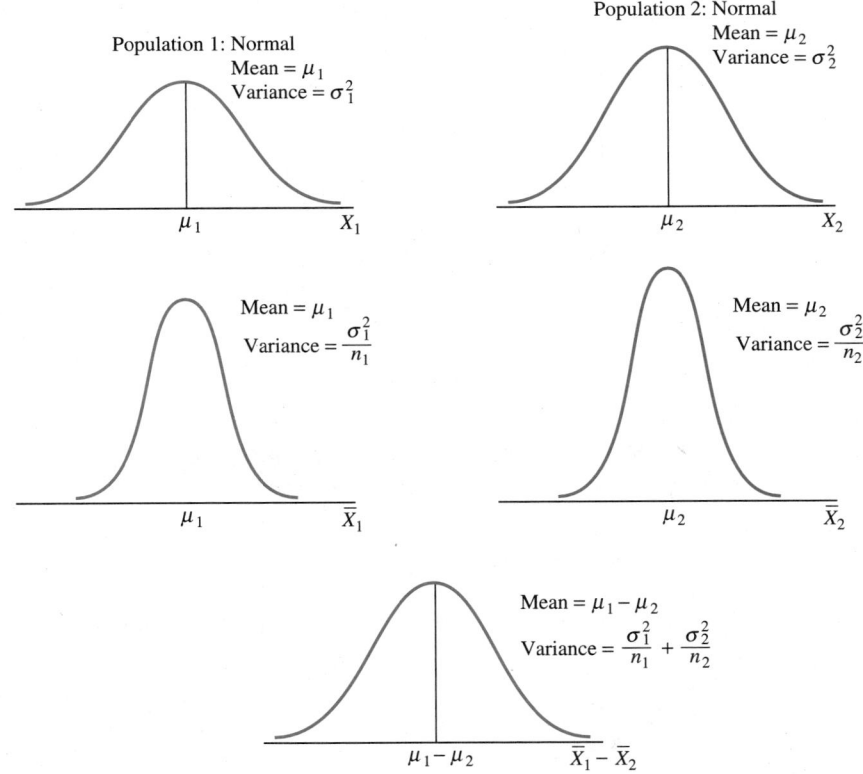

FIGURE 9.9 *Population and sampling distributions of two normal populations*

populations are not normal. In Figure 9.9, the distributions of $\bar{X}_1$ and $\bar{X}_2$ are normal because the populations are normal (i.e., we do not need the Central Limit Theorem). In Figure 9.10, the distributions of $\bar{X}_1$ and $\bar{X}_2$ are approximately normal because the sample sizes are assumed to be large (i.e., we *do* need the Central Limit Theorem here). Once again, the sampling distribution of $(\bar{X}_1 - \bar{X}_2)$ will be approximately normal.

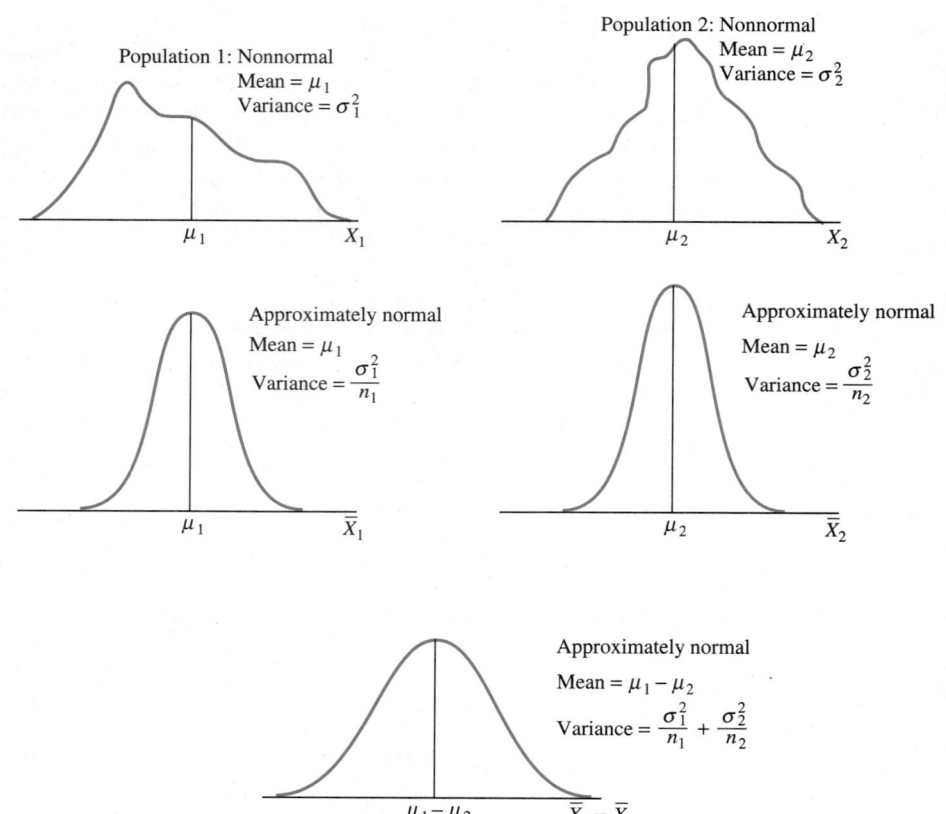

FIGURE 9.10   *Population and sampling distributions of two nonnormal populations*

Mean and variance of $(\bar{X}_1 - \bar{X}_2)$

The expected value of the random variable $(\bar{X}_1 - \bar{X}_2)$ is

$$E(\bar{X}_1 - \bar{X}_2) = \mu_1 - \mu_2$$

This implies that the random variable $(\bar{X}_1 - \bar{X}_2)$ is an unbiased estimator of $(\mu_1 - \mu_2)$.

The variance of $(\bar{X}_1 - \bar{X}_2)$ depends on the four numbers σ_1^2, σ_2^2, n_1, and n_2.

> If the two random samples are independent of one another, then the variance of $(\bar{X}_1 - \bar{X}_2)$ is given by the formula
>
> $$\text{Var}(\bar{X}_1 - \bar{X}_2) = \frac{\sigma_1^2}{n_1} + \frac{\sigma_2^2}{n_2}$$
>
> where σ_1^2 and σ_2^2 represent the variances of the two populations, and n_1 and n_2 are the sample sizes.

EXAMPLE 9.11   **Distribution of $(\bar{X}_1 - \bar{X}_2)$**
_____

A financial loan officer claims that the mean monthly payment for credit cards is $80 and the variance is 1,400 for females; for males, the mean is $80 and the variance is 1,320. You take a random sample of 100 females (population 1) and an independent random sample of 120 males (population 2). What is the probability that the sample mean for females will be at least $5 higher than the sample mean for males?

SOLUTION   We have

$$\mu_1 = 80, \quad \sigma_1^2 = 1,400, \quad n_1 = 100$$
$$\mu_2 = 80, \quad \sigma_2^2 = 1,320, \quad n_2 = 120$$

The random variable $(\bar{X}_1 - \bar{X}_2)$ is approximately normally distributed with mean $(\mu_1 - \mu_2) = 0$ and variance

$$\frac{\sigma_1^2}{n_1} + \frac{\sigma_2^2}{n_2} = \frac{1,400}{100} + \frac{1,320}{120} = 14 + 11 = 25$$

We seek the probability that $\bar{X}_1$ is at least $5 more than $\bar{X}_2$. That is, we seek $P(\bar{X}_1 - \bar{X}_2 \geq 5)$. (See Figure 9.11.) We obtain the Z score

$$Z = \frac{5 - 0}{\sqrt{25}} = 1$$

Thus,

$$P(\bar{X}_1 - \bar{X}_2 \geq 5) = P(Z \geq 1) = .5000 - .3413 = .1587$$

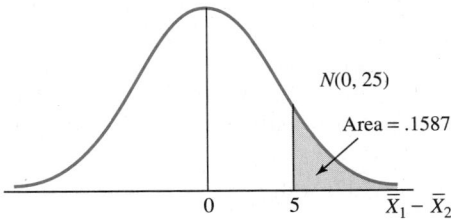

$N(0, 25)$

Area $= .1587$

FIGURE 9.11   *Area under the normal distribution for Example 9.11*

Exercises

Statistical Concepts

9.4.1   Suppose that population 1 has mean μ_1 and standard deviation σ_1. Suppose that population 2 has mean μ_2 and standard deviation σ_2. Suppose we take independent random samples of n items from each population. State the mean and variance of the random variable $(\bar{X}_1 - \bar{X}_2)$. What is the standard deviation of $(\bar{X}_1 - \bar{X}_2)$?

9.4.2   Suppose that population 1 has mean μ_1 and standard deviation σ_1. Suppose that population 2 has mean μ_2 and standard deviation σ_2. Suppose we take independent random samples of n items from each population. Under what conditions will the random variable $(\bar{X}_1 - \bar{X}_2)$ have exactly a normal distribution? Under what conditions will the random variable $(\bar{X}_1 - \bar{X}_2)$ have approximately a normal distribution?

9.4.3   *True or false:*   Let X_1 be a normal random variable with mean 10 and variance 16. Let X_2 be a normal random variable with mean 10 and variance 16. Suppose we take independent random samples of 4 items from each population. The random variable $(\bar{X}_1 - \bar{X}_2)$ has a normal distribution. Explain.

9.4.4   *True or false:*   Let X_1 be a normal random variable with mean 10 and variance 16. Let X_2 be a normal random variable with mean 10 and variance 16. Suppose we take independent random samples of 4 items from each population. The variance of the random variable $(\bar{X}_1 - \bar{X}_2)$ is 8. Explain.

9.4.5   *True or false:*   Let X_1 be a normal random variable with mean 12 and variance 25. Let X_2 be a normal random variable with mean 14 and variance 20. Suppose we take independent random samples of 50 items from population 1 and 25 items from population 2. The mean of the random variable $(\bar{X}_1 - \bar{X}_2)$ is 2. Explain.

9.4.6   *True or false:*   Let X_1 be a normal random variable with mean 12 and variance 25. Let X_2 be a normal random variable with mean 14 and variance 20. Suppose we take independent random samples of 50 items from population 1 and 25 items from population 2. The variance of the random variable $(\bar{X}_1 - \bar{X}_2)$ is $^{45}/_5$. Explain.

9.4.7   *True or false:*   Let X_1 be a normal random variable with mean 12 and variance 25. Let X_2 be a normal random variable with mean 14 and variance 20. Suppose we take independent random samples of 50 items from population 1 and 25 items from population 2. $P(\bar{X}_1 - \bar{X}_2 < -2) = .5$. Explain.

Statistical Drills

9.4.8   Let X_1 be a normal random variable with mean 18 and variance 16. Let X_2 be a normal random variable with mean 18 and variance 16. Suppose we take independent random samples of 25 items from population 1 and 25 items from population 2. Find the following:
a. $P(\bar{X}_1 - \bar{X}_2 \leqslant 0)$ **b.** $P(\bar{X}_1 - \bar{X}_2 > 1)$ **c.** $P(\bar{X}_1 - \bar{X}_2 \leqslant 2)$

9.4.9   Let X_1 be a normal random variable with mean 134 and variance 64. Let X_2 be a normal random variable with mean 147 and variance 81. Suppose we take independent random samples of 25 items from population 1 and 25 items from population 2. Find the following:
a. $P(\bar{X}_1 - \bar{X}_2 \leqslant -7)$ **b.** $P(\bar{X}_1 - \bar{X}_2 > 8)$ **c.** $P(-11 \leqslant \bar{X}_1 - \bar{X}_2 \leqslant 2)$

9.4.10   Let X_1 be a normal random variable with mean 34 and standard deviation 7. Let X_2 be a normal random variable with mean 34 and standard deviation 9. Suppose we take independent random samples of 30 items from population 1 and 30 items from population 2. Find the following:
a. $P(-2 \leqslant \bar{X}_1 - \bar{X}_2 \leqslant 2)$ **b.** $P(\bar{X}_1 - \bar{X}_2 \geqslant 3)$ **c.** $P(-1 \leqslant \bar{X}_1 - \bar{X}_2 \leqslant 1)$

9.4.11   Let X_1 be a normal random variable with mean 1,134 and standard deviation 65. Let X_2 be a normal random variable with mean 1,096 and standard deviation 56. Suppose we take independent random samples of 64 items from population 1 and 81 items from population 2. Find the following:

a. $P(0 \leqslant \bar{X}_1 - \bar{X}_2 \leqslant 29)$ **b.** $P(\bar{X}_1 - \bar{X}_2 \geqslant 32)$ **c.** $P(-11 \leqslant \bar{X}_1 - \bar{X}_2 \leqslant 21)$

Statistical Applications

9.4.12 The mean annual income of union carpenters is $1,000 higher than the mean annual income of nonunion carpenters. For each population, the standard deviation is $3,000. We take independent random samples of $n_1 = 100$ union members and $n_2 = 200$ nonunion members. What is the probability that in the two samples the mean annual incomes differ by more than $1,500?

9.4.13 Hertz claims that the mean annual repair bill for its rental cars is $290 and the standard deviation is $50. Avis also claims its mean annual repair bill is $290 and the standard deviation is $50. If independent random samples of 100 cars from each company are obtained, what is the probability that $|\bar{X}_1 - \bar{X}_2|$ exceeds $5?

9.4.14 A company manufactures two varieties of golf balls, the Pro ($2.25 each) and the Maxi ($2.00 each). When struck with the same force, the two types of balls travel the same distance on the average. However, the Pro ball has a better cover, which is more difficult to cut. Suppose that a mechanical device hits $n_1 = 100$ randomly selected Pro balls and $n_2 = 100$ randomly selected Maxi balls, and the distance each ball travels is measured. Assume the population variances are $\sigma_1^2 = 400$ and $\sigma_2^2 = 400$. If the population means are equal, find the probability that $(\bar{X}_1 - \bar{X}_2)$ is as follows:

a. Between -5 and $+5$ yards   **b.** Between -3 and $+3$ yards

---

9.5

Sampling Distribution of the Sample Proportion

In numerous problems in economics and business, we want to know the proportion of items in a population that possess a certain characteristic. For example, a TV producer might want to know what proportion of viewers of a certain program have incomes exceeding $40,000 per year, a quality control engineer might want to know what proportion of products off an assembly line are defective, or a labor economist might want to know what proportion of the labor force is unemployed.

In each case, we have to estimate the parameter p, the proportion of the population having the characteristic of interest. Let $q = (1 - p)$. To estimate the population proportion p, we take a random sample of n observations from the population. Let X denote the number of observations in the sample that possess the characteristic of interest. We use the sample proportion $\hat{p} = X/n$ to estimate the population proportion p. In Chapter 8, we showed that, when $np \geqslant 5$ and $nq \geqslant 5$, the distribution of X is approximately normal with mean np and variance npq; under the same constraints, the sampling distribution of $\hat{p}$ is approximately normal.

Sampling distribution of the sample proportion

Let p denote the proportion of a population possessing some characteristic of interest. Take a random sample of n observations from the population. Let X denote the number of items in the sample possessing the characteristic. We estimate the population proportion p by the sample proportion $\hat{p} = X/n$. If $np \geqslant 5$ and $nq \geqslant 5$, the random variable $\hat{p}$ has approximately a normal distribution with mean

$$E(\hat{p}) = p$$

and variance

$$\text{Var}(\hat{p}) = \frac{pq}{n}$$

The standard deviation of $\hat{p}$ is

$$\sigma_{\hat{p}} = \sqrt{pq/n}$$

The standard deviation of $\hat{p}$ is called the **standard error of $\hat{p}$.**

Notice that for fixed p, the standard error of the sample proportion decreases as the sample size increases. This implies that as the sample size increases, the distribution of $\hat{p}$ becomes more concentrated about its mean, as illustrated in Figure 9.12.

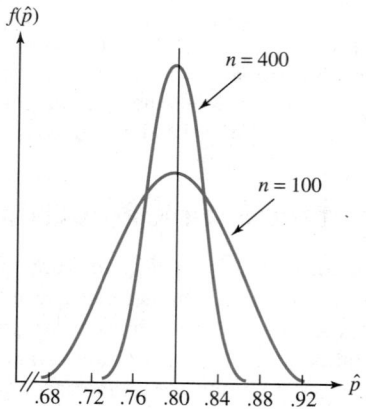

FIGURE 9.12   *Sampling distribution of the sample proportion when $p = .8$*

If the distribution of $\hat{p}$ is approximately normal, then the random variable

$$Z = \frac{\hat{p} - p}{\sqrt{pq/n}}$$

is approximately distributed as standard normal.

EXAMPLE 9.12   **Distribution of the Sample Proportion** ─────────────────────────

In an election, 55% of the registered voters favor a certain candidate. If we take a random sample of 400 voters, what is the probability that, based on the sample proportion, we will predict the wrong winner? That is, what is the probability that $\hat{p}$ will be less than .5?

SOLUTION   We have $n = 400$, $p = .55$, and $q = (1 - p) = .45$. First, we check to see whether $np \geq 5$ and $nq \geq 5$, so that we can use the normal distribution. We have $np = 400(.55) = 220 \geq 5$ and $nq = 400(.45) = 180 \geq 5$. Thus, $\hat{p}$ is approximately normally distributed with mean $p = .55$ and variance $pq/n = (.55)(.45)/400 = .00062$. We seek $P(\hat{p} < .50)$.

We obtain the Z score

$$Z = \frac{.50 - .55}{\sqrt{.00062}} \approx \frac{-.05}{.025} = -2$$

Thus,

$$P(\hat{p} < .5) = P(Z < -2)$$
$$= .5000 - .4772 = .0228$$

Therefore, the probability is only .0228 that we will predict the wrong winner if we take a sample of size 400.

A Note About the Continuity Correction

When we used the normal distribution to approximate binomial probabilities, we used a continuity correction of $\pm.5$ to the endpoints of the interval when finding areas under a normal curve. To be precise, a continuity correction should also be applied when using the normal distribution to approximate the sampling distribution of the sample proportion. Let X be a binomial random variable where the sample size is n and the population proportion is p. For any integers x_1 and x_2,

$$P(x_1 \leq X \leq x_2) = P\left(\frac{x_1}{n} \leq \hat{p} \leq \frac{x_2}{n}\right)$$

because $\hat{p} = X/n$.

The discrete probability, $P(x_1 \leq X \leq x_2)$, is approximated by the area under a normal curve between $(x_1 - .5)$ and $(x_2 + .5)$. That is, we find the normal probability $P(x_1 - .5 \leq X \leq x_2 + .5)$. But

$$P(x_1 - .5 \leq X \leq x_2 + .5) = P\left(\frac{x_1 - .5}{n} \leq \hat{p} \leq \frac{x_2 + .5}{n}\right)$$

This is equivalent to the probability

$$P\left(p_1 - \frac{.5}{n} \leq \hat{p} \leq p_2 + \frac{.5}{n}\right)$$

where $p_1 = x_1/n$ and $p_2 = x_2/n$. The amount $\pm.5/n$, or equivalently $\pm 1/(2n)$, is the required continuity correction when calculating probabilities involving sample proportions.

In the remainder of this book, we will ignore the continuity correction when finding probabilities involving sample proportions. For large sample sizes, there is very little difference between

$$P(p_1 \leq \hat{p} \leq p_2) \quad \text{and} \quad P\left(p_1 - \frac{.5}{n} \leq \hat{p} \leq p_2 + \frac{.5}{n}\right)$$

Thus, when n is large, the effect of making the continuity correction is negligible. For small values of n, it might be useful to take the continuity correction into account if a precise probability is required. Recall, however, that if n is small, the sampling distribution of the sample proportion is not necessarily closely approximated by a normal distribution. Throughout this book, ignore the continuity correction unless you are explicitly instructed to use it.

A Note About Deciding Whether the Normal Approximation Is Appropriate

We have indicated that the normal distribution provides a good approximation to the sampling distribution of the sample proportion provided $np \geq 5$ and $n(1 - p) \geq 5$. What do we do when p is unknown? When p is unknown, many statisticians have adopted the following guideline: Assume that the normal distribution provides a good approximation to the sampling distribution of the sample proportion provided the interval $(\hat{p} \pm 3\sqrt{\hat{p}(1-\hat{p})/n})$ does not include either 0 or 1. This guideline is almost equivalent to the former rule except here we rely on the sample proportion $\hat{p}$ rather than the population proportion p.

Tips for Problem Solving

To calculate the probability that the random variable $\hat{p}$ falls in some interval, follow these steps:

1. List the sample size n, the population proportion p, and $q = (1 - p)$.
2. The sampling distribution of $\hat{p}$ is (approximately) normal if $np \geq 5$ and $nq \geq 5$. Verify that these conditions hold.
3. List the variance of the random variable $\hat{p}$, which is pq/n.
4. List the standard deviation of $\hat{p}$, which is $\sqrt{pq/n}$.
5. Sketch the appropriate normal distribution and shade the appropriate area. The mean of the distribution is p.
6. Calculate the Z scores corresponding to the values of interest in the problem. Sketch a standard normal curve and find the appropriate areas.
7. Examine your sketch and verify that the calculated answer agrees with the shaded area.

Exercises

Statistical Concepts

9.5.1   State the conditions under which the normal distribution provides a good approximation to the sampling distribution of the sample proportion.

9.5.2   The guidelines given in this chapter indicate that the sampling distribution of the sample proportion is approximately normal if $np \geq 5$ and $n(1 - p) \geq 5$. These guidelines are useless because we need to know p to implement them, and if we knew p, there would be no need to take a sample to estimate p. Comment.

9.5.3   Suppose we want to use the normal distribution to approximate the sampling distribution of the sample proportion where $n = 20$ and $p = .01$. Is it appropriate to use the approximation? Explain.

9.5.4   Suppose we want to use the normal distribution to approximate the sampling distribution of the sample proportion where $n = 20$ and $p = .4$. Is it appropriate to use the approximation? Explain.

9.5.5 *True or false:* The sample proportion is an unbiased estimator of the population proportion. Explain.

9.5.6 Suppose you take a random sample of 225 adults to estimate the proportion of adults who have been to an opera during the past year. In your sample, the sample proportion was .10. Do you think that the sample size is large enough so that it is reasonable to assume that the sampling distribution of the sample proportion is normal? Comment.

9.5.7 *True or false:* Suppose we want to use the normal distribution to approximate the sampling distribution of the sample proportion when $n = 20$ and $p = .5$. The binomial probability $P(6 \le X \le 14)$ is the same as the probability $P(.3 \le \hat{p} \le .7)$. Comment.

9.5.8 *True or false:* Suppose we want to use the normal distribution to approximate the sampling distribution of the sample proportion when $n = 20$ and $p = .5$. If we use the continuity correction, the binomial probability $P(6 \le X \le 14)$ is approximately equal to the normal probability $P(5.5 \le X \le 14.5)$. Comment.

9.5.9 Suppose the random variable X follows the binomial distribution with $n = 25$ and $p = .5$. The exact probability that $X = 6, 7,$ or 8 is equal to the area of the three rectangles lying over $X = 6, 7,$ and 8.
  a. *True or false:* This probability can be approximated by the area under the normal curve from $x_1 = 5.5$ to $x_2 = 9.5$. Explain.
  b. *True or false:* Since $\hat{p} = X/n$, $P(5.5 \le X \le 9.5) = P(5.5/25 \le \hat{p} \le 9.5/25) = P(.22 \le \hat{p} \le .38)$. Explain.

Statistical Drills

9.5.10 Let $\hat{p}$ denote the proportion of successes obtained in a random sample of $n = 64$ items selected from a population where the population proportion is $p = .4$.
  a. Is it reasonable to use the normal distribution to approximate the sampling distribution of $\hat{p}$? Verify that $np \ge 5$ and $n(1 - p) \ge 5$.
  b. State the mean and variance of the random variable $\hat{p}$.
  c. Find $P(.36 \le \hat{p} \le .44)$. Do not use the continuity correction.
  d. Find $P(\hat{p} > .44)$. Do not use the continuity correction.

9.5.11 Let $\hat{p}$ denote the proportion of successes obtained in a random sample of $n = 400$ items selected from a population where the population proportion is $p = .4$.
  a. Is it reasonable to use the normal distribution to approximate the sampling distribution of $\hat{p}$? Verify that $np \ge 5$ and $n(1 - p) \ge 5$.
  b. State the mean and variance of the random variable $\hat{p}$.
  c. Find $P(.36 \le \hat{p} \le .44)$. Do not use the continuity correction.
  d. Find $P(\hat{p} > .44)$. Do not use the continuity correction.

9.5.12 Let $\hat{p}$ denote the proportion of successes obtained in a random sample of $n = 100$ items selected from a population where the population proportion is $p = .2$.
  a. Is it reasonable to use the normal distribution to approximate the sampling distribution of $\hat{p}$? Verify that $np \ge 5$ and $n(1 - p) \ge 5$.
  b. State the mean and variance of the random variable $\hat{p}$.
  c. Find $P(|\hat{p} - p| \le .04)$.
  d. Find $P(|\hat{p} - p| \ge .03)$.
  e. Find the value p_1 such that $P(\hat{p} \ge p_1) = .05$.

9.5.13 Let $\hat{p}$ denote the proportion of successes obtained in a random sample of $n = 400$ items selected from a population where the population proportion is $p = .8$.
  a. Is it reasonable to use the normal distribution to approximate the sampling distribution of $\hat{p}$? Verify that $np \ge 5$ and $n(1 - p) \ge 5$.

b. State the mean and variance of the random variable $\hat{p}$.

c. Find $P(|\hat{p} - p| > .02)$.

d. Find the value p_1 such that $P(\hat{p} \geq p_1) = .025$.

e. Find the values p_1 and p_2 such that $P(p_1 \leq \hat{p} \leq p_2) = .05$. Let p_1 and p_2 be equidistant from $p = .8$.

9.5.14   Let $\hat{p}$ denote the proportion of successes obtained in a random sample of $n = 400$ items selected from a population where the population proportion is $p = .8$.

a. Find $P(.77 \leq \hat{p} \leq .83)$. Do not use the continuity correction.

b. Find $P(.77 \leq \hat{p} \leq .83)$. Use the continuity correction. Compare your answer with part **a.**

c. Find $P(.74 \leq \hat{p} \leq .86)$. Do not use the continuity correction.

d. Find $P(.74 \leq \hat{p} \leq .86)$. Use the continuity correction. Compare your answer with part **c.**

e. Explain why the answers to parts **c** and **d** are more similar than the answers to parts **a** and **b.**

f. When $n = 400$, does it appear worthwhile to make the continuity correction, or is its effect negligible?

9.5.15   Let $\hat{p}$ denote the proportion of successes obtained in a random sample of $n = 25$ items selected from a population where the population proportion is $p = .5$.

a. Find $P(.46 \leq \hat{p} \leq .54)$. Do not use the continuity correction.

b. Find $P(.46 \leq \hat{p} \leq .54)$. Use the continuity correction. Compare your answer with part **a.**

c. Find $P(.40 \leq \hat{p} \leq .60)$. Do not use the continuity correction.

d. Find $P(.40 \leq \hat{p} \leq .60)$. Use the continuity correction. Compare your answer with part **c.**

e. Explain why the answers to parts **c** and **d** are more similar than the answers to parts **a** and **b.**

f. *True or false:*   When the sample size is as small as 25, the continuity correction has a sizeable effect on probabilities. Discuss.

Statistical Applications

9.5.16   A newspaper claims that 50% of high school students have used drugs at some time during their senior year. A random sample of 400 high school seniors is obtained. If the claim is correct, find the probability that the sample proportion is between .47 and .53.

9.5.17   Repeat Exercise 16 using a sample size of 900. Find the probability that $\hat{p}$ is between .47 and .53.

9.5.18   Just prior to an election, 54% of the electorate favor candidate A. If we take a random sample of 1,500 voters, what is the probability that $\hat{p}$ exceeds .5? That is, what is the probability that we will correctly predict the winner of the election?

9.5.19   Government data show that 10% of males under age 25 are unemployed. A random sample is taken of 400 males who are in the labor force and under age 25. Find the probability that the sample unemployment rate is .12 or more.

9.5.20   Solve Exercise 19 if the sample size is 900.

9.5.21   The commissioners of Lane County are trying to decide whether to pass a law that will loosen restrictions on environmental pollution in an effort to attract manufacturing firms to the county. To learn how the county's residents feel about the law, a random sample of 100 residents is questioned. Suppose that 70% of the county's residents favor the new law. Find the probability that in the sample the proportion who favor the law is as follows:

a. Between .60 and .70

b. Between .60 and .65

c. Less than .50

9.5.22   The Chrysler Corporation claims that 80% of its cars meet the tough new standards of the Environmental Protection Agency (EPA). The EPA tests a random sample of 400 Chrysler cars. Find the probability that the percentage of cars that pass the test is as follows:

a. Less than .75

b. Between .70 and .78

c. More than .82

9.5.23 A medical journal states that the probability of surviving open heart surgery is .9. To check this claim, you examine the records of a sample of 200 patients who had open heart surgery. In the sample, 77% survived.

a. If the journal's claim is correct, find the probability that the proportion of survivors is 77% or less.

b. Does the result in part **a** make you doubt the journal's claim? Explain.

9.5.24 The Internal Revenue Service claims that 65% of all tax returns lead to a refund. A random sample of 100 tax returns is taken. What is the probability that the sample proportion exceeds .72?

9.5.25 A record store owner claims that 25% of customers entering the store make a purchase. One morning 120 people, who can be regarded as a random sample of all customers, enter the store. What is the probability that the sample proportion is less than .2?

9.5.26 The Nielsen Company takes random samples of viewers to estimate the proportion of the public that watches particular TV shows. Suppose a particular program was watched by 30% of all adults and that Nielsen takes a random sample of 1,500 adults.

a. What is the probability that the sample proportion is between .27 and .33?

b. What is the probability that the sample proportion is between .25 and .35?

9.6 Sampling Distribution of the Difference Between Sample Proportions

We frequently are interested in determining whether the proportion of items in one population that possess a certain characteristic is the same as the proportion possessing the characteristic in another population. For example, a doctor who gives one type of medicine to some patients and another type of medicine to others may want to determine whether the percentage of people cured by the first medicine is the same as the percentage of people cured by the second.

This and many other types of problems concern the difference between two proportions. Let p_1 and p_2, respectively, denote the proportions of populations 1 and 2 possessing a certain characteristic. If we don't know these proportions, we can take independent random samples of size n_1 and n_2 from the two populations to obtain estimates of p_1 and p_2. Let X_1 and X_2 denote the number of items in each sample possessing the characteristic of interest. The two sample proportions are then

$$\hat{p}_1 = \frac{X_1}{n_1} \quad \text{and} \quad \hat{p}_2 = \frac{X_2}{n_2}$$

Because inferences about the difference between two population proportions are based on the random variable $(\hat{p}_1 - \hat{p}_2)$, it is necessary to study the sampling distribution of the random variable $(\hat{p}_1 - \hat{p}_2)$.

Sampling distribution of $(\hat{p}_1 - \hat{p}_2)$

Suppose we take independent samples of size n_1 and n_2 from two populations. Let p_1 and p_2 be the proportion of items in each population that possess a certain characteristic, and let $q_1 = (1 - p_1)$ and $q_2 = (1 - p_2)$. If $n_1 p_1 \geq 5$, $n_1 q_1 \geq 5$,

$n_2 p_2 \geqslant 5$, and $n_2 q_2 \geqslant 5$, then the random variable $(\hat{p}_1 - \hat{p}_2)$ is approximately normally distributed with mean

$$E(\hat{p}_1 - \hat{p}_2) = p_1 - p_2$$

and variance

$$\text{Var}(\hat{p}_1 - \hat{p}_2) = \frac{p_1 q_1}{n_1} + \frac{p_2 q_2}{n_2}$$

EXAMPLE 9.13   **Distribution of the Difference of Sample Proportions** ————————

A marketing agency is interested in determining whether a certain TV program appeals equally to upper-income and lower-income people. Suppose that 40% of the upper-income people and 50% of the lower-income people in the population like the show. The agency takes a random sample of 100 upper-income people (population 1) and an independent sample of 200 lower-income people (population 2). Let $\hat{p}_1$ and $\hat{p}_2$ denote the proportions in each sample that like the program. Find the probability that the two sample proportions differ by less than .05.

SOLUTION   We need to calculate the probability

$$P(-.05 \leqslant \hat{p}_1 - \hat{p}_2 \leqslant .05)$$

We have $p_1 = .4$, $q_1 = .6$, $n_1 = 100$, $p_2 = .5$, $q_2 = .5$, and $n_2 = 200$. The two samples are independent. The distribution of $(\hat{p}_1 - \hat{p}_2)$ is approximately normal because $n_1 p_1 \geqslant 5$, $n_1 q_1 \geqslant 5$, $n_2 p_2 \geqslant 5$, and $n_2 q_2 \geqslant 5$. The mean is $(p_1 - p_2) = .40 - .50 = -.10$ and the variance is

$$\frac{p_1 q_1}{n_1} + \frac{p_2 q_2}{n_2} = \frac{(.4)(.6)}{100} + \frac{(.5)(.5)}{200} = .00365$$

We obtain the Z scores

$$z_1 = \frac{-.05 - (-.1)}{\sqrt{.00365}} = \frac{.05}{\sqrt{.00365}} = .83$$

$$z_2 = \frac{.05 - (-.1)}{\sqrt{.00365}} = \frac{.15}{\sqrt{.00365}} = 2.48$$

Thus,

$$P(-.05 \leqslant \hat{p}_1 - \hat{p}_2 \leqslant .05) = P(.83 \leqslant Z \leqslant 2.48)$$
$$= .4934 - .2967 = .1967$$

Exercises

Statistical Concepts

9.6.1   State the conditions under which the normal distribution provides a good approximation to the sampling distribution of the difference between two sample proportions.

9.6.2 State the formulas for the mean, the variance, and the standard deviation of the difference between two sample proportions.

9.6.3 Suppose we want to use the normal distribution to approximate the sampling distribution of the difference between two sample proportions where, for each population, $n = 20$ and $p = .01$. Is it appropriate to use the approximation? Explain.

9.6.4 Suppose we want to use the normal distribution to approximate the sampling distribution of the difference between two sample proportions where, for each population, $n = 20$ and $p = .4$. Is it appropriate to use the approximation? Explain.

9.6.5 *True or false:* If the samples are independent, the difference between two sample proportions is an unbiased estimator of the difference between the two population proportions. Explain.

Statistical Drills

9.6.6 In each of the following situations, determine whether the sample sizes are large enough to justify use of the normal distribution to represent the approximate sampling distribution of the random variable $(\hat{p}_1 - \hat{p}_2)$.
  a. $n_1 = 16$, $p_1 = .5$, $n_2 = 10$, $p_2 = .3$
  b. $n_1 = 25$, $p_1 = .3$, $n_2 = 30$, $p_2 = .3$
  c. $n_1 = 80$, $p_1 = .2$, $n_2 = 90$, $p_2 = .1$
  d. $n_1 = 6$, $p_1 = .5$, $n_2 = 10$, $p_2 = .3$

9.6.7 Independent random samples of size $n_1 = 100$ and $n_2 = 100$ are taken from two binomial populations where $p_1 = .6$ and $p_2 = .5$.
  a. Find $P(0 \le \hat{p}_1 - \hat{p}_2 \le .2)$.
  b. Find $P(.15 \le \hat{p}_1 - \hat{p}_2 \le .25)$.
  c. Find $P(|\hat{p}_1 - \hat{p}_2| > .16)$.

9.6.8 Independent random samples of size $n_1 = 25$ and $n_2 = 50$ are taken from two binomial populations where $p_1 = .4$ and $p_2 = .4$.
  a. Find $P(-.03 \le \hat{p}_1 - \hat{p}_2 \le .03)$.
  b. Find $P(-.05 \le \hat{p}_1 - \hat{p}_2 \le .05)$.
  c. Find $P(|\hat{p}_1 - \hat{p}_2| > .04)$.
  d. Find the value a such that $P(-a \le \hat{p}_1 - \hat{p}_2 \le a) = .95$.

Statistical Applications

9.6.9 A company is interested in determining whether a new product appeals equally to men and women. Independent random samples are taken of 200 men and 400 women. In the populations, $p_m = .7$ and $p_w = .6$. Find

$$P(-.04 \le \hat{p}_m - \hat{p}_w \le .04)$$

9.6.10 On a Sunday afternoon, a random sample of 400 people is taken to estimate p_1, the proportion of the population that watched a hockey game on TV. On the following Sunday, an independent random sample of 400 people is taken to estimate p_2, the proportion of the population who watched a basketball game on TV. If $p_1 = .3$ and $p_2 = .4$, find the probability that $\hat{p}_2 < \hat{p}_1$ in our samples. That is, find $P(\hat{p}_1 - \hat{p}_2 > 0)$. If this event actually occurred, the producer would mistakenly think that more people watched hockey than basketball.

9.6.11 At a certain university, there is a movement to form a faculty union. Approximately 40% of the entire faculty favor unionizing. A prounion professor takes a random sample of 100 faculty members. Let $\hat{p}_1$ denote the proportion in this sample who favor a union. An antiunion professor takes an independent random sample of 100 professors. Let $\hat{p}_2$ denote the proportion in this sample who favor a union. Calculate the probability that $\hat{p}_1$ exceeds $\hat{p}_2$ by .1 or more.

9.7　**Computer Applications**

The SPSSX computer program can be used to find the sample mean and sample standard deviation for any sample of data. To find the mean and standard deviation, use the DES-CRIPTIVES command followed by the names of the variables to be analyzed. For example, to find the mean and standard deviation for the variables AGE, SALARY, and SENIOR in Table 2.1 of Chapter 2, use the command

DESCRIPTIVES AGE, SALARY, SENIOR

Figure 9.13 shows the resulting computer output. The mean age of the sample of employees is 39.24 and the standard deviation is 12.02. The respective values for SALARY are $2,133.00 and $687.43; for SENIOR, 15.36 and 10.87.

```
Number of valid observations (listwise) =       120.00

                                                 Valid
Variable      Mean     Std Dev   Minimum   Maximum   N    Label

AGE          39.24      12.02     19.00     64.00    120
SALARY     2133.00     687.43   1200.00   5000.00    120
SENIOR       15.36      10.87      1.00     42.00    120
```

FIGURE 9.13　*SPSSX-generated output showing descriptive statistics for variables AGE, SALARY, and SENIOR for data in Table 2.1*

By issuing a BREAKDOWN command, we can request the SPSSX program to split the data into subsamples and compute statistics for each subsample. Procedure BREAK-DOWN prints means and standard deviations of a variable within subgroups defined by another variable. For example, it will provide descriptive statistics on the variable SALARY broken down by SEX, or RACE, or academic DEGREE, and so forth. For example, the command

BREAKDOWN TABLES = SALARY BY SEX

tells SPSSX to calculate the mean and standard deviation of the variable SALARY for all individuals whose gender is classified as 0 (female) and as 1 (male). The output from the BREAKDOWN command is shown in Figure 9.14. The mean salary for females is

```
Summaries of      SALARY
By levels of      SEX

Variable      Value  Label                    Mean      Std Dev    Cases

For Entire Population                       2133.0000   687.4332     120

SEX            .00   FEMALE                 1676.8889   347.3721      45
SEX           1.00   MALE                   2406.6667   696.9733      75

   Total Cases = 120
```

FIGURE 9.14　*SPSSX-generated output showing descriptive statistics for variable SALARY broken down according to SEX for data in Table 2.1*

$1,676.89 and the standard deviation is $347.372; for males, the comparable figures are $2,406.67 and $696.973.

Exercises

In Exercises 1–4, use the SPSSX program and the BREAKDOWN command on the data in Table 2.1 of Chapter 2.

9.7.1 Find the sample mean and sample standard deviation for the variables AGE and SENIOR broken down according to the variable SEX.

9.7.2 Find the sample mean and sample standard deviation for the variables AGE, SALARY, and SENIOR broken down according to the variable RACE.

9.7.3 Find the sample mean and sample standard deviation for the variables AGE, SALARY, and SENIOR broken down according to the variable DEGREE.

9.7.4 Find the sample mean and sample standard deviation for the variables AGE, SALARY, and SENIOR broken down according to the variable DIV.

9.7.5 Use the SPSSX program and the DESCRIPTIVES command on the data in Table 2.2 of Chapter 2 to find the sample mean and sample standard deviation for SAT scores and GPAs.

In Exercises 6–9, use the SPSSX program and the BREAKDOWN command on the data in Table 2.2 of Chapter 2.

9.7.6 Find the sample mean and sample standard deviation for SAT scores and GPAs broken down according to the variable SEX.

9.7.7 Find the sample mean and sample standard deviation for SAT scores and GPAs broken down according to the variable CLASS.

9.7.8 Find the sample mean and sample standard deviation for SAT scores and GPAs broken down according to the student's residence.

9.7.9 Find the sample mean and sample standard deviation for SAT scores and GPAs broken down according to the student's major.

STATISTICS IN ACTION: CASE STUDY

Examining the Sulfur Content of Coal

A power company purchased millions of tons of coal per year from various suppliers. The price of the coal varied from barge load to barge load and was based on the quality of the coal, which depended on its BTU (British Thermal Units) content and its sulfur, ash, and moisture contents. The higher the BTU content, the higher the price; the higher the sulfur, ash, or moisture content, the lower the price. The power company had a chemist take a sample of coal from each barge to estimate the quality of the coal. The purchase price was based on the results obtained in the sample.

Eventually, the management of the power company began to suspect that one of the chemists was reporting incorrect results because the power generated by the coal did not correspond with the estimated quality of the coal. The company suspected that the chemist was being paid by the coal supplier to misrepresent the quality of the coal, thereby generating a higher purchase price. The power company filed criminal charges against its employee and filed a lawsuit with its insurance company to recover the money allegedly stolen by the chemist and the coal suppliers.

Several questions had to be answered. First, could it be proved that the chemist was overestimating the value of the coal? Second, if the insurance company had to compensate the power company for its losses, what was the total amount of the losses; that is, what was the true value of the coal? We address only the first issue here.

Before filing criminal charges, the power company secretly had a second chemist inspect a sample of coal from each barge for a period of several months. The results obtained by the second chemist were compared with the results obtained by the first chemist who was suspected of altering the data.

A typical barge contains more than 100 tons of coal, so it is reasonable to expect that the two chemists would not exactly agree on the qualities of the coal on any specific barge. It would not be reasonable, however, to find that one chemist consistently found higher BTU contents and lower sulfur, ash, and moisture contents. Also, if the number of barges sampled was relatively large, it would not be reasonable to find that the average result obtained by one chemist would differ substantially from the average result obtained by the other chemist.

A sample of coal was inspected from $n = 50$ barges. Suppose the average sulfur content reported by the suspected chemist was $\bar{x}_1 = .020$, whereas the average sulfur content reported by the second chemist was $\bar{x}_2 = .033$. Thus the difference between the two sample means reported by the two chemists was

$$\bar{x}_1 - \bar{x}_2 = .020 - .033 = -.013$$

The chemists were sampling from the same population, so we know that $\mu_1 = \mu_2$ and $\sigma_1 = \sigma_2$. Based on past experience, the power company believed the population standard deviations to be equal to .005 (i.e, $\sigma_1 = \sigma_2 = .005$). What we want to find is the probability of obtaining a result as extreme as or even more extreme than that which actually occurred. You should be able to show that this probability is less than .000001. That is, if $\mu_1 - \mu_2 = 0$, $\sigma_1 = \sigma_2 = .005$, and $n_1 = n_2 = 50$, then $P(\bar{X}_1 - \bar{X}_2 \leq -.013) < .000001$, or less than 1 in 1 million.

Based on this information, the chemist and one of the coal suppliers who was paying the chemist to misrepresent the quality of the coal were indicted and convicted. The insurance company eventually paid over $1 million in damages to the power company to compensate them for the theft perpetrated by their employee and the coal supplier.

Chapter 9 Supplementary Exercises

9.S.1   Suppose light bulbs have a lifelength with an average of 4,000 hours and a variance of 40,000. If we take a sample of 100 light bulbs and calculate $\bar{X}$, what is the probability that $\bar{X}$ will be between 3,950 and 4,050?

9.S.2   We want to determine the mean income of college professors. We take a random sample of 144 professors, and the population standard deviation is assumed to be $12,000. What is the probability that the value of $\bar{X}$ is within $2,000 of the population mean μ?

9.S.3   Four hundred ball bearings are selected from a population having a mean weight of 5 ounces and a variance of .09. What is the probability that the mean weight of bearings in our sample is between 4.95 and 5.05 ounces?

9.S.4   The probability is approximately .1 that a person aged 80 will die within 1 year. An insurance company insures 900 eighty-year-olds. What is the probability that more than 12% of these people will die in the following year?

9.S.5 A government agency wants to estimate the mean amount of money spent per week for gasoline by truck drivers. A sample of 400 drivers is taken. If the population standard deviation is $40, what is the probability that $\bar{X}$ will be within $6 of the true mean μ?

9.S.6 A student commission wants to know the mean amount of money spent by college students for textbooks in the senior year. Assume the variance of the population is 400. If a sample of 625 students is taken, what is the probability that $\bar{X}$ will be within $3 of the population mean?

9.S.7 In a certain district, 55% of the voters want Congress to decrease expenditures on defense. A member of Congress takes a random sample of 2,475 voters. What is the probability that the sample proportion who want defense expenditures cut is less than .5?

9.S.8 We want to estimate the mean grade point average of college seniors. We take a random sample of nine seniors and obtain the following values:

$$x_1 = 3.0 \quad x_2 = 3.3 \quad x_3 = 2.4$$
$$x_4 = 2.7 \quad x_5 = 2.4 \quad x_6 = 3.6$$
$$x_7 = 3.3 \quad x_8 = 2.4 \quad x_9 = 2.7$$

a. Calculate $\bar{X}$ to estimate μ.
b. Calculate S^2 to estimate σ^2.
c. Calculate S^2/n to estimate the variance of $\bar{X}$.

9.S.9 In a small town, 30% of the residents use the library regularly. The city council is debating a bill that would increase the library's budget. Before increasing the budget, the council wants to estimate the proportion of residents who use the library regularly. Suppose a sample of 400 residents is taken. What is the probability that the sample proportion will be within .03 of the population proportion?

9.S.10 A company claims that method 2 of making TV tubes is superior to the old method 1. In a random sample of 1,000 tubes produced by method 1, the mean lifetime was $\bar{x}_1 = 4,000$ hours. Assume that the variance in population 1 is 40,000. In a random sample of 1,000 tubes made by method 2, the mean lifetime was 4,025 hours and the variance was 41,000. If, in fact, $(\mu_1 - \mu_2)$ is 0, find the probability that $(\bar{X}_1 - \bar{X}_2)$ will be between -25 and $+25$.

9.S.11 The United States Golf Association wants to determine whether golf balls of type 1 are livelier than balls of type 2. A mechanical device hits a sample of 400 balls of type 1 and 300 balls of type 2. The distance (in yards) traveled by each ball is measured and the sample means are obtained. If $\sigma_1^2 = 2,000$ and $\sigma_2^2 = 1,200$, and if $(\mu_1 - \mu_2) = 0$, find $P(-10 \leq \bar{X}_1 - \bar{X}_2 \leq 10)$.

9.S.12 In a congressional election, 55% of the electorate prefers candidate A. If you take a random sample of 900 voters, what is the probability that candidate A will get less than half the vote? That is, what is the probability that you will project the wrong candidate as the winner?

9.S.13 Suppose that during the last year, 10% of the adults in a city have been victims of some unreported crime. The police contact a random sample of 900 adults to estimate the proportion of people who were victims of an unreported crime. What is the probability that the sample proportion will be between .07 and .13?

9.S.14 In Exercise 13, let the sample size be 2,500. What is the probability that $\hat{p}$ will be between .09 and .11?

9.S.15 A rehabilitation counselor wants to estimate the proportion of high school dropouts who have used drugs. Assume that 80% of all dropouts have used drugs. If the counselor obtains a random sample of 240 dropouts, what is the probability that fewer than 70% have used drugs?

9.S.16 Suppose that about 5% of the population has a serious accident or illness during any given year. If an insurance company insures 47,500 people, what is the probability that more than 6% of these people will have a serious accident or illness next year?

9.S.17   A random variable X has mean μ and variance σ^2. We take a random sample of size n and calculate $\overline{X}$. Find the number z such that

$$P[\mu - z(\sigma/\sqrt{n}) \leqslant \overline{X} \leqslant \mu + z(\sigma/\sqrt{n})] = .90$$

9.S.18   In Exercise 17, find z such that

$$P[\overline{X} \geqslant \mu + z(\sigma/\sqrt{n})] = .05$$

9.S.19   Ball bearings have a mean weight of .5 ounce and a variance of .01. The bearings are sold in boxes of 400. If we randomly choose two boxes of bearings, what is the probability that the mean weights of the two boxes of bearings differ by more than .0125 ounce?

9.S.20   A federal judge ruled that the selection of juries in a particular county in Alabama was discriminatory. The judge observed that blacks made up 30% of the jury-age population but only 12% of the jury rolls, and that women made up 54% of the jury-age population but only 16% of the jury rolls. If there are 100 people on a jury roll, what are the probabilities that random selection yields the following:
   a. No more than 12% blacks
   b. No more than 16% females
   c. Do these data lend evidence for or against the hypothesis that jury rolls are chosen without regard to race or gender?

9.S.21   If you take a random sample of size n from a population that is normally distributed with mean μ and variance σ^2, the sample variance S^2 is a random variable having mean σ^2 and variance $2\sigma^4/(n - 1)$. As the sample size increases, the distribution of S^2 approaches a normal distribution. Suppose you take a sample of size 50 from a normal population having $\sigma^2 = 196$.
   a. What is the approximate probability that S^2 will exceed 202?
   b. Find $P(190 \leqslant S^2 \leqslant 202)$.
   c. Find $P(S^2 \leqslant 180)$.

9.S.22   An environmental agency wants to estimate the number of fish in a lake. They catch 10,000 fish, tag them, and put them back in the lake, distributing them evenly. A short while later they catch another 10,000 fish and find that 500 have been tagged.
   a. Based on this sample, what is your estimate of the proportion of fish in the lake that are tagged?
   b. What is your estimate of the total number of fish in the lake?

9.S.23   The mean daily output at a small West Virginia coal mine is 38 tons of coal. The log book showing the tonnage mined each day indicates that the standard deviation of the daily output is $\sigma = 5$ tons. What is the probability of the following?
   a. During a random sample of 40 days, the sample mean output exceeds 42 tons.
   b. During a random sample of 60 days, the mean output is less than 35 tons.

9.S.24   The mean tread life of all Jensen radial tires is $\mu = 44,500$ miles with standard deviation $\sigma = 1,900$ miles. If the population is normally distributed, what is the probability of the following?
   a. The mean life of six randomly selected tires exceeds 43,000 miles.
   b. The mean life of 100 randomly selected tires exceeds 43,000 miles.

9.S.25   At the Caste Automobile Company, the mean profit from selling new cars is $400 per car and the standard deviation is $\sigma = \$80$. If Mr. Levine is a typical salesperson, find the probability that in a random sample of 60 sales, he shows a mean profit $\overline{X}$ of less than $320.

9.S.26   The customers who rent compact cars from the Rankin Car Rental Agency travel 400 miles on the average. The population standard deviation is $\sigma = 70$. A random sample of 64 customers is checked during a certain time period. Find the probability of the following:
   a. $\overline{X}$ exceeds 420 miles.
   b. $\overline{X}$ is less than 390 miles.

9.S.27 The light bulbs used in the U.S. space shuttle have an average life expectancy of $\mu = 100$ hours. The distribution of life expectancies is normal with $\sigma = 10$ hours. For a trip that is expected to last 380 hours, the crew takes along four new bulbs (one in place and three spares). If each bulb is replaced immediately when it burns out, what is the probability that the four bulbs will be sufficient for the entire 380-hour trip? [*Note:* For the four bulbs to total 380 hours, they must average 95 hours per bulb.]

9.S.28 Suppose the life of a tire is normally distributed with a mean of 30,000 miles and a standard deviation of 6,000 miles. What is the probability of the following:
a. A tire will last more than 40,000 miles.
b. The average life of four tires will be more than 40,000 miles.

9.S.29 The gasoline mileage for a brand of compact car is normally distributed with a mean of 31 miles per gallon (mpg) and a standard deviation of 4 mpg. If an impartial group measures the mileages for a sample of these cars, what is the probability that their average will be greater than 30 mpg if only one car is tested? If 25 cars are tested? If 100 cars are tested?

9.S.30 When monitoring a production process, quality control engineers search for clues that a production process may have gone out of control. One indication that a process may have gone out of control is the occurrence of one or more unusual observations. These unusual observations lie far out in the tail of the probability distribution and should occur very infrequently if the process is in control. One rule frequently used by quality control engineers is to consider a process to be out of control if a sample mean falls more than 3 standard deviations ($3\sigma_{\bar{x}} = 3\sigma/\sqrt{n}$) from the mean. Suppose a random variable follows a normal distribution with mean μ and standard deviation σ.
a. Find the probability that a sample mean will fall more than $3\sigma_{\bar{x}}$ above the mean.
b. Find the probability that a sample mean will fall more than $3\sigma_{\bar{x}}$ below the mean.
c. Find the probability that a sample mean will fall more than $3\sigma_{\bar{x}}$ above or below the mean.
d. Find the probability that a sample mean will fall less than $3\sigma_{\bar{x}}$ above or below the mean.

9.S.31 In quality control analysis, there are many other rules that indicate that a process may have gone out of control. Some of the rules are based on observing sequences of sample means that fall more than 2 standard deviations ($2\sigma_{\bar{x}} = 2\sigma/\sqrt{n}$) from the mean. Suppose a random variable follows the standard normal distribution.
a. Find the probability that a sample mean will fall more than $2\sigma_{\bar{x}}$ above the mean.
b. Find the probability that a sample mean will fall more than $2\sigma_{\bar{x}}$ below the mean.
c. Find the probability that a sample mean will fall more than $2\sigma_{\bar{x}}$ above or below the mean.
d. Find the probability that a sample mean will fall less than $2\sigma_{\bar{x}}$ above or below the mean.

9.S.32 A producer of oil drums wants to estimate the proportion of damaged oil drums in a large warehouse. The producer selects a random sample of 500 drums and determines the proportion of damaged drums. If the actual proportion in the entire warehouse is .10, what is the probability that the sample proportion will deviate from the population proportion by less than .03?

9.S.33 A government official wants to get information concerning the average cost of employee health insurance across the United States. Suppose the mean and standard deviation of the yearly cost of employee medical insurance are approximately $3,400 and $450, respectively. Suppose an official from the Bureau of Labor Statistics obtains a random sample of 400 employees from around the country.
a. What is the probability that the sample mean medical insurance cost will deviate from the population mean by less than $50?
b. What is the probability that the sample mean will deviate from the population mean by more than $60?

9.S.34 On February 16, 1993, the *Wall Street Journal* reported that General Motors announced a recall of more than 500,000 Quad 4 engines—a recall that could cost as much as $250 million—because of problems with the head gasket allowing engine coolant to leak into the cylinders. To determine whether a machine that produces head gaskets is properly adjusted, every hour a random sample of 25 gaskets is collected and the thickness of each is measured. Suppose $\sigma = .035$ inch.

a. If the machine is properly adjusted, what is the probability that the mean thickness of the head gaskets in the sample will lie within .01 inch of the population mean?

b. Assume the desired mean thickness of the gaskets is $\mu = .25$ inch. The machine will be considered out of control if the mean of the sample of $n = 25$ thicknesses is less than .237 inch or larger than .263 inch. Suppose the process has gone out of control and needs to be adjusted because the true mean thickness of the gaskets produced by the machine is .255 inch with $\sigma = .041$. What is the probability that the test will detect that the process is out of control?

c. Suppose the process is in control so that $\mu = .25$ and $\sigma = .035$. Find the probability that the test will incorrectly indicate that the process is out of control.

d. Repeat part **b** but assume that the sample size is increased to $n = 100$.

e. Repeat part **c** but assume that $n = 100$.

9.S.35 A manufacturer of copper wire claims that it produces 100-foot rolls of wire that have a mean length of $\mu = 100.20$ feet with $\sigma = .06$ foot. Suppose the length of a roll of wire is a random variable having a normal distribution.

a. What proportion of the rolls would have a length less than 100 feet?

b. To determine whether the process is in control, a quality control engineer selects a random sample of $n = 25$ rolls of wire and finds the sample mean. If the company claims are true, what is the probability that the sample mean will be less than 100.15 feet?

c. Suppose the sample mean shifts to $\mu = 100.04$ feet. What is the probability that the sample mean will be less than 100 feet?

9.S.36 *True or false:* The standard deviation of $\bar{X}$ is $\sigma/\sqrt{n}$, so $\sigma_{\bar{x}}$ is proportional to $1/\sqrt{n}$. Thus, to reduce the standard deviation of the sampling distribution by 50%, you need to make the sample size 4 times as large. Illustrate by drawing a graph.

 9.S.37 The credit manager of a department store has reported that customers have 9,863 charge accounts outstanding having a mean balance of $347 and a standard deviation of $116. Auditors have selected a random sample of 70 of the charge accounts to assess the reliability of the mean balance reported by the store. What is the probability that the sample mean obtained by the auditors is within $30 of the population mean?

9.S.38 The distribution of annual salaries of full-time carpenters has a mean equal to $24,000 and a standard deviation equal to $2,500. The distribution of annual salaries of full-time welders has a mean equal to $25,000 and a standard deviation equal to $3,000. Suppose we take random samples of 50 carpenters and 50 welders.

a. What is the probability that, in the samples, the mean salary for the welders exceeds the mean salary for the carpenters?

b. What is the probability that, in the samples, the mean salaries for the welders and carpenters are within $500 of one another?

9.S.39 An official for a fast-food restaurant is monitoring the length of time customers must wait in line before receiving their food. The official claims that the mean waiting time is $\mu = 64$ seconds. The official obtains the following random sample of 40 waiting times in seconds:

35	48	123	69	80	64	14	47	156	116
46	58	69	87	73	99	23	59	184	176
45	65	176	53	24	16	34	78	131	67
36	51	110	25	36	91	30	82	114	83

a. Calculate the sample mean and sample standard deviation.
b. Assume that the sample standard deviation is a good estimate of the population standard deviation. Based on the sample data, is it reasonable to reject the claim that the mean is $\mu = 64$? Explain.

9.S.40 A real estate developer claims that the mean annual income of households in a certain town is $\mu =$ \$72,000. To check this claim, a random sample of 50 households is obtained and their incomes (in \$1,000s) are as follows:

25	45	133	73	85	34	64	27	106	106
36	27	79	97	63	79	53	39	84	76
44	63	136	63	23	26	44	48	121	57
45	54	104	35	46	81	40	82	134	73
53	44	93	47	66	92	51	98	111	63

a. Calculate the sample mean and sample standard deviation.
b. Assume that the sample standard deviation is a good estimate of the population standard deviation. Based on the sample data, is it reasonable to reject the claim that the mean is $\mu = 72$? Explain.

9.S.41 *True or false:* A desirable property of an estimator is that the standard deviation of its sampling distribution is small. We would prefer that the variation in potential estimates generated from different samples be as small as possible. Comment and explain by drawing graphs of various sampling distributions.

9.S.42 *True or false:* For many estimators, the standard deviation of the sampling distribution is controllable. That is, we can make the standard deviation of the sampling distribution as small as we wish by increasing the sample size. Comment.

9.S.43 *True or false:* We prefer estimators that are unbiased and have small standard deviations. If the size of the standard deviation is unacceptably large, nothing can be done about it. Explain.

9.S.44 Suppose the sampling distribution of an estimator A is approximately normal with mean μ_A and standard deviation σ_A.
a. Find the probability that the sampling error for A will be less than $2\sigma_A$.
b. Find the probability that the sampling error for A will be less than $1.96\sigma_A$.

References

Canavos, George C., and Don M. Miller. *Modern Business Statistics.* Belmont, CA: Duxbury Press, 1995.

Cochran, William G. *Sampling Techniques.* 3d ed. New York: Wiley, 1977.

Cochran, William G., and G. M. Cox. *Experimental Designs.* 2d ed. New York: Wiley, 1957.

Cryer, Jonathan D., and Robert B. Miller. *Statistics for Business: Data Analysis and Modelling.* Boston: PWS-Kent, 1991.

Freund, John E., and R. E. Walpole. *Mathematical Statistics.* 4th ed. Englewood Cliffs, N.J.: Prentice-Hall, 1987.

Handbook of Tables for Probability and Statistics. 2d ed. Cleveland: Chemical Rubber Company, 1968.

Hoel, Paul G. *Elementary Statistics.* 4th ed. New York: Wiley, 1976.

Kohler, Heinz. *Statistics for Business and Economics.* Glenview, IL: Scott, Foresman, 1988.

McClave, James T., and Frank H. Dietrich. *Statistics.* 6th ed. New York: Macmillan, 1994.

Mendenhall, William, James E. Reinmuth, and Robert Beaver. *Statistics for Management and Economics*. 6th ed. Boston: PWS-Kent, 1989.

Neter, J., W. Wasserman, and G. A. Whitmore. *Applied Statistics*. 3rd ed. Boston: Allyn & Bacon, 1977.

Shiffler, Ronald E., and Arthur J. Adams. *Introductory Business Statistics with Computer Applications*. Belmont, CA: Wadsworth, 1995.

Watson, Collin J., Patrick Billingsley, D. James Croft, and David V. Huntsberger. *Statistics for Management and Economics*. 4th ed. Needham Heights, MA: Allyn & Bacon, 1990.

10 Estimating and Constructing Confidence Intervals

Interval Estimation

In this chapter, we discuss how to construct an interval estimate for the population mean μ, the population proportion p, the difference between means ($\mu_1 - \mu_2$), and the difference between proportions ($p_1 - p_2$). For most problems of interest, a point estimate is inadequate without some additional information concerning its reliability. A sample estimate is a random variable and will not exactly equal the population parameter being estimated. After obtaining a point estimate, it is reasonable to ask, "How close is the estimate to the true value of the population parameter being estimated?"

When we try to evaluate the goodness, or reliability, of an estimator $\hat{\theta}$, we are trying, in general, to put some bound on the possible error of estimation $|\hat{\theta} - \theta|$, where θ represents the true value of the parameter being estimated. The error of estimation $|\hat{\theta} - \theta|$, called the *sampling error,* measures the distance between the estimated value and the true value of the population parameter.

A systematic method of indicating the precision of an estimator $\hat{\theta}$ exists, provided we know the form of the sampling distribution of $\hat{\theta}$. We indicate the precision of our estimator by constructing **confidence intervals** for θ, where we use the estimate $\hat{\theta}$ to determine two values $\hat{\theta}_1$ and $\hat{\theta}_2$ such that the interval ($\hat{\theta}_1, \hat{\theta}_2$) contains the value θ with a specified probability. The probability is usually denoted as ($1 - \alpha$), and the percentage $100(1 - \alpha)\%$ is called the **confidence coefficient** or **level of confidence** of the confidence interval ($\hat{\theta}_1, \hat{\theta}_2$).

In any estimation problem, we would like our estimators to have two important properties:

1. We want an *unbiased* estimator, so there is no systematic tendency to overestimate or underestimate the true value of the population parameter.
2. We want an estimator with a sampling distribution that is highly concentrated about the value of the population parameter θ. That is, we want the variance of the random variable $\hat{\theta}$ to be small.

Now consider $\bar{X}$ and $\hat{p}$ as estimates of μ and p, respectively. Recall that the variance of the sample mean $\bar{X}$ is given by $\text{Var}(\bar{X}) = \sigma_{\bar{X}}^2/n$. Thus, the variance of the sampling distribution of $\bar{X}$ depends on n and is controllable. That is, we can make $\text{Var}(\bar{X})$ as small as we desire by increasing the sample size. Similarly, the variance of the sample proportion $\hat{p}$, given by $\text{Var}(\hat{p}) = pq/n$, can also be controlled. Since the sampling distributions of $\bar{X}$ and $\hat{p}$ are centered at μ and p, respectively, we can virtually guarantee that our estimate $\bar{x}$ or $\hat{p}$ will be close to the desired value μ or p by choosing a sufficiently large sample.

Before discussing the idea of a confidence interval, it is useful to introduce some new notation that will help us identify areas in the tail of the standard normal distribution. The value z_α is the value of the standard normal variable such that the area to its right is α. The following examples explain this notation.

Value of z_α

Let Z be a standard normal random variable and let α be any number such that $0 < \alpha < 1$. Then z_α denotes the number for which

$$P(Z \geqslant z_\alpha) = \alpha$$

EXAMPLE 10.1   **Finding a Value of z_α**

Suppose we select $\alpha = .025$. Find the value z_α.

SOLUTION   If 2.5% of the area falls in the right tail of the distribution beyond z_α, then the area between 0 and z_α is .475. From Table A.5 in the Appendix, the area between 0 and 1.96 is .475. Thus, $P(Z > 1.96) = .025$ and $z_{.025} = 1.96$. This value is found by locating .475 in the body of Table A.5 and finding the corresponding value of Z. By symmetry, we know that 2.5% of the area lies to the left of the value $-z_{.025} = -1.96$. Thus we have

$$P(Z > 1.96) = .025 \quad \text{and} \quad P(Z < -1.96) = .025$$

This idea is illustrated in Figure 10.1.

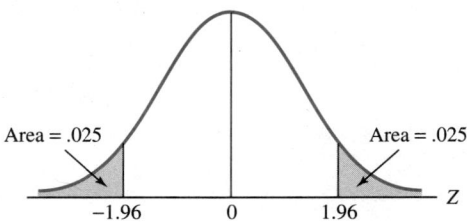

Area = .025   Area = .025

-1.96 0   1.96 Z

FIGURE 10.1 z_α *for* $\alpha = .025$

EXAMPLE 10.2   **Finding More Values of z_α**

Find the values $z_{.05}$ and $z_{.005}$ such that $P(Z > z_{.05}) = .05$ and $P(Z > z_{.005}) = .005$.

SOLUTION   If 5% of the area falls in the right tail of the distribution, then $P(Z > z_{.05}) = .05$ and $P(0 \leqslant Z \leqslant z_{.05}) = .45$. From Table A.5 in the Appendix, we obtain $z_{.05} \approx 1.645$. This is illustrated in Figure 10.2. Similarly, if .5% of the area under the standard normal distribution falls to the right of $z_{.005}$, then 49.5% of the area falls between 0 and $z_{.005}$. We obtain $P(Z > z_{.005}) = .005$ and $P(0 \leqslant Z \leqslant z_{.005}) = .495$. From Table A.5, we obtain $z_{.005} \approx 2.58$.

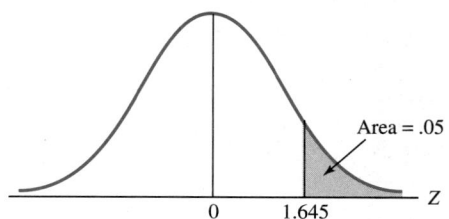

FIGURE 10.2 z_α *for* $\alpha = .05$

The symbol $z_{\alpha/2}$ denotes the value of the standard normal variate Z such that the area in the right tail of the distribution is $\alpha/2$. We have

$$P(Z > z_{\alpha/2}) = \alpha/2 \quad \text{and} \quad P(Z < -z_{\alpha/2}) = \alpha/2$$

We obtain

$$P(-z_{\alpha/2} \leq Z \leq z_{\alpha/2}) = 1 - \alpha$$

Thus, the area under the standard normal distribution between $-z_{\alpha/2}$ and $z_{\alpha/2}$ is $(1 - \alpha)$. The area in each tail of the distribution is $\alpha/2$, and the total area in the two tails of the distribution is α. (See Figure 10.3.)

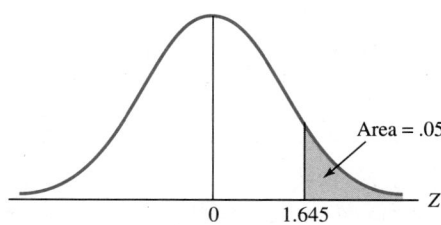

FIGURE 10.3 $P(-z_{\alpha/2} \leq Z \leq z_{\alpha/2}) = 1 - \alpha$

Exercises

Statistical Concepts

10.1.1 *True or false:* The value of z_α increases as α increases. Explain.

10.1.2 *True or false:* If $z_\alpha = 1.96$, then $\alpha = .05$. Explain.

10.1.3 *True or false:* If $z_\alpha = 1.645$, then $\alpha = .05$. Explain.

10.1.4 *True or false:* If $z_{\alpha/2} = 1.96$, then $\alpha = .05$. Explain.

Statistical Drills

10.1.5 Find z_α for the following values of α: .005, .01, .025, .05, .10.

10.1.6 Find $z_{\alpha/2}$ for the following values of α: .005, .01, .025, .05, .10.

10.1.7 Find the value $z_{\alpha/2}$ such that $P(-z_{\alpha/2} \leq Z \leq z_{\alpha/2}) = K$ for the following values of K: .80, .90, .95, .99.

Confidence Intervals for the Mean with Known Population Variance

Suppose that we take a random sample of size n from a normal population having mean μ and variance σ^2. The random variable $\bar{X}$ will have exactly a normal distribution with mean μ and variance $\text{Var}(\bar{X}) = \sigma^2/n$. In addition, the standardized random variable

$$Z = \frac{\bar{X} - \mu}{\sigma/\sqrt{n}}$$

will follow the standard normal distribution.

Our previous discussion showed that, for the standard normal distribution, the area to the right of $z_{\alpha/2}$ is $\alpha/2$ and the area to the left of $-z_{\alpha/2}$ is $\alpha/2$. Thus the area between $-z_{\alpha/2}$ and $z_{\alpha/2}$ is $1 - \alpha$. We obtain

$$1 - \alpha = P(-z_{\alpha/2} < Z < z_{\alpha/2})$$

After substituting for Z, we obtain

$$1 - \alpha = P\left(-z_{\alpha/2} < \frac{\bar{X} - \mu}{\sigma/\sqrt{n}} < z_{\alpha/2}\right)$$
$$= P(-z_{\alpha/2}\sigma/\sqrt{n} < \bar{X} - \mu < z_{\alpha/2}\sigma/\sqrt{n})$$
$$= P(\bar{X} - z_{\alpha/2}\sigma/\sqrt{n} < \mu < \bar{X} + z_{\alpha/2}\sigma/\sqrt{n})$$

This last probability statement tells us that the probability is $(1 - \alpha)$ that the random interval $(\bar{X} - z_{\alpha/2}\sigma/\sqrt{n}, \bar{X} + z_{\alpha/2}\sigma/\sqrt{n})$ will contain the true population mean μ.

DEFINITION   **Level of Confidence** ―――――――――――――――――――――――――――――

The **level of confidence** of a confidence interval measures the probability that a population parameter will be contained in an interval calculated after a random sample has been selected from a population. The level of confidence is denoted by the symbol $(1 - \alpha)$.

――

The value α indicates the proportion of times that we will be incorrect in assuming that an interval contains the population parameter. If $\alpha = .05$, then the level of confidence of the interval is $(1 - \alpha) = .95$, or 95%; if $\alpha = .10$, then the level of confidence of the interval is $(1 - \alpha) = .90$, or 90%; and so forth. Thus, it is possible to construct a confidence interval having any desired level of confidence. To do this, we must find the Z scores from Table A.5 such that the combined areas in the two tails of the distribution total α.

DEFINITION   **Confidence Interval for the Mean of a Normal Population with Known Population Variance** ――――――――――――――――――――――――――

Suppose we take a random sample of n observations from a normal population with mean μ and variance σ^2. If σ^2 is known and the observed sample mean is $\bar{x}$, then the **confidence**

interval for the mean with a level of confidence $100(1 - \alpha)\%$ is given by

$$(\bar{x} - z_{\alpha/2}\sigma/\sqrt{n}, \ \bar{x} + z_{\alpha/2}\sigma/\sqrt{n})$$

where $z_{\alpha/2}$ is the number for which

$$P(Z > z_{\alpha/2}) = \alpha/2$$

and where the random variable Z has a standard normal distribution.

EXAMPLE 10.3 **Constructing a 95% Confidence Interval for the Mean** ——————

A student advisor wants to estimate the mean annual income of all college students who graduated last year. It is believed that the incomes follow a normal distribution with population standard deviation $2,000. Based on a random sample of 25 college graduates, the advisor obtains $\bar{x} = \$19,500$. Let us construct a 95% confidence interval for the unknown population mean μ.

SOLUTION   We have $\sigma = \$2,000$, $n = 25$, $\bar{x} = \$19,500$, and $(1 - \alpha) = .95$. Thus, we have $\alpha = .05$ and $\alpha/2 = .025$. From Table A.5 we see that 2.5% of the area under the standard normal curve lies to the right of 1.96. We obtain $z_{\alpha/2} = 1.96$. The desired confidence interval is

$$\left(\bar{x} - z_{\alpha/2}\frac{\sigma}{\sqrt{n}}, \ \bar{x} + z_{\alpha/2}\frac{\sigma}{\sqrt{n}}\right)$$

$$\left(19,500 - 1.96\frac{2,000}{\sqrt{25}}, \ 19,500 + 1.96\frac{2,000}{\sqrt{25}}\right)$$

$$(18,716, \ 20,284)$$

Thus, we are 95% confident that the population mean is between $18,716 and $20,284. We cannot be sure that this interval contains the population mean, but if we repeated this process a large number of times, 95% of the confidence intervals obtained would contain the population mean.

The confidence interval obtained in Example 10.3 depends on the specific value $\bar{x} = \$19,500$ obtained from the sample of n observations. Suppose we take another random sample of n observations, calculate a new value of $\bar{x}$, and obtain the 95% confidence interval $(\bar{x} - 1.96\sigma/\sqrt{n}, \ \bar{x} + 1.96\sigma/\sqrt{n})$. Like the first confidence interval, this one may or may not contain the population mean μ. If we repeated this process, say, 1,000 times, we would have 1,000 different sample means and 1,000 different confidence intervals. A 95% level of confidence means that approximately 95% (or 950) of these confidence intervals would contain μ and 5% (or 50) would not.

The idea that 95% of the samples of size n produce intervals $(\bar{x} - 1.96\sigma/\sqrt{n}, \ \bar{x} + 1.96\sigma/\sqrt{n})$ containing μ is illustrated in Figure 10.4 (page 344). This figure shows 10 confidence intervals that could be obtained by repeated sampling. In the figure, 9 of the 10 intervals contain μ. If many repeated samples were taken from the same population, the proportion of intervals containing μ would be approximately .95.

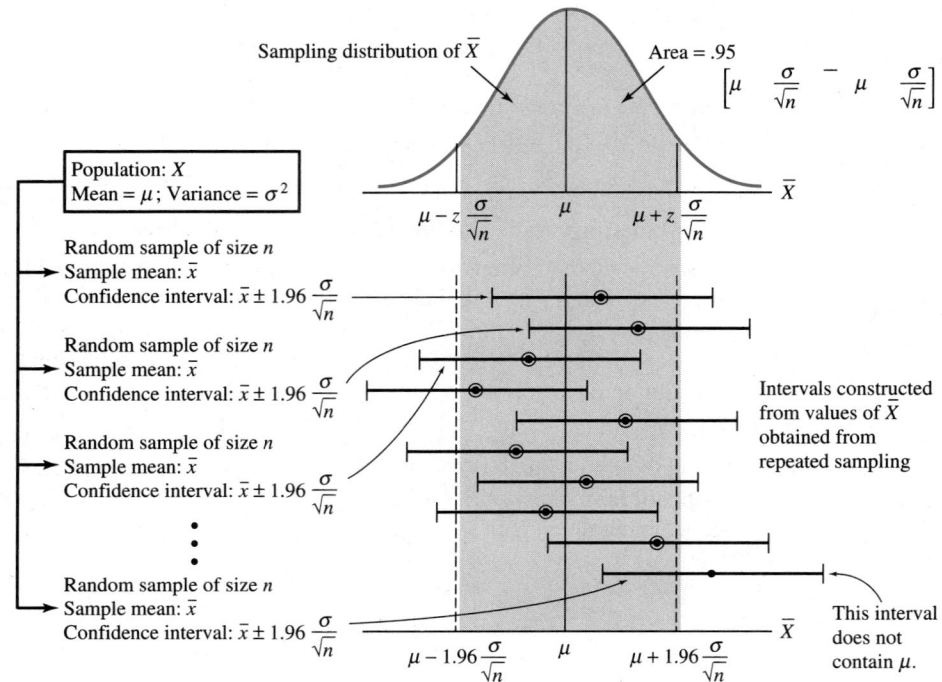

FIGURE 10.4   Ten possible confidence intervals for the mean

It is important that the probability statement

$$P(\bar{X} - z_{\alpha/2}\sigma/\sqrt{n} < \mu < \bar{X} + z_{\alpha/2}\sigma/\sqrt{n}) = 1 - \alpha$$

be interpreted correctly. In this statement, the parameter μ is *not* a random variable and does not vary from sample to sample; rather the mean μ is an unknown population parameter. On the other hand, $\bar{X}$ is a random variable and varies from sample to sample. If we take many samples of size n from the population, we get a different value of $\bar{X}$ for each sample. For a particular estimate $\bar{x}$, we can calculate the endpoints of the interval $(\bar{x} - z_{\alpha/2}\sigma/\sqrt{n}, \bar{x} + z_{\alpha/2}\sigma/\sqrt{n})$; these endpoints vary from sample to sample. The probability statement says that $100(1 - \alpha)\%$ of these random intervals contain the value μ. We say that we are $100(1 - \alpha)\%$ confident that our interval contains μ because, essentially, our interval is just one of many possible intervals.

Before we select the random sample, the probability is $(1 - \alpha)$ that the confidence interval will contain the population mean. After we have taken the sample, we say that we are $100(1 - \alpha)\%$ confident that our interval contains the mean, because if we performed the same experiment a large number of times, then in $100(1 - \alpha)\%$ of the cases, the interval would contain the mean.

Formulas for commonly constructed confidence intervals

The most frequently constructed confidence intervals use levels of confidence 90%, 95%, and 99%. The corresponding confidence intervals are as follows:

LEVEL OF CONFIDENCE				CONFIDENCE INTERVAL
$(1 - \alpha)$	α	$\alpha/2$	$z_{\alpha/2}$	$(\bar{x} - z_{\alpha/2}\sigma/\sqrt{n},\ \bar{x} + z_{\alpha/2}\sigma/\sqrt{n})$
.90	.10	.05	1.645	$(\bar{x} - 1.645\sigma/\sqrt{n},\ \bar{x} + 1.645\sigma/\sqrt{n})$
.95	.05	.025	1.96	$(\bar{x} - 1.96\sigma/\sqrt{n},\ \bar{x} + 1.96\sigma/\sqrt{n})$
.99	.01	.005	2.58	$(\bar{x} - 2.58\sigma/\sqrt{n},\ \bar{x} + 2.58\sigma/\sqrt{n})$

Desirable Properties of Confidence Intervals

Two properties are desirable in a confidence interval:

1. The interval should have a high level of confidence $(1 - \alpha)$.
2. The confidence interval should have a narrow width.

In most cases, we would like the probability that our confidence interval contains the mean to be very high, say, 90% or more. We would also like the confidence interval to be very narrow so that our estimate is very precise.

DEFINITION   **Width of a Confidence Interval for μ** ———————————————————

The **width W of a confidence interval** for the population mean is

$$W = 2z_{\alpha/2}\frac{\sigma}{\sqrt{n}}$$

The width of a confidence interval for the mean depends on three factors:

1. The level of confidence of the confidence interval, $(1 - \alpha)$
2. The standard deviation of the population, σ
3. The sample size, n

The following properties hold for all confidence intervals:

1. The larger the level of confidence $(1 - \alpha)$, the larger will be $z_{\alpha/2}$ and the wider will be the confidence interval. That is, if we hold σ and n constant, an increase in $(1 - \alpha)$ causes an increase in W. This means that if we want to increase the probability that our interval will contain the true mean, we have to make the confidence interval wider. For example, all other things being equal, a 99% confidence interval will be wider than a 95% confidence interval.
2. The smaller the population standard deviation σ, the narrower the confidence interval. If the population is highly concentrated, our estimate of the mean is very reliable. Because the estimate is very reliable, only a narrow confidence interval is required.

3. As the sample size increases, the width of the confidence interval decreases. As we obtain more information, our estimate should become better, as reflected by a narrower confidence interval. Note that to cut the width of a confidence interval in half, it is necessary to multiply the sample size by a factor of 4.

EXAMPLE 10.4   **Comparing Widths of Confidence Intervals**

Suppose we take a random sample of size n from a population having known variance σ^2. Construct 99%, 95%, and 90% confidence intervals for the population mean and compare their widths.

SOLUTION   For 99%, 95%, and 90% confidence intervals, the respective Z scores are $z_{.005} = 2.58$, $z_{.025} = 1.96$, and $z_{.05} = 1.645$. The widths of these three confidence intervals are $W_1 = 2(2.58)\sigma/\sqrt{n}$, $W_2 = 2(1.96)\sigma/\sqrt{n}$, and $W_3 = 2(1.645)\sigma/\sqrt{n}$. The fact that $W_1/W_2 = 1.32$ means that a 99% confidence interval is 32% wider than a 95% confidence interval. We obtain $W_2/W_3 = 1.19$, which indicates that a 95% confidence interval is 19% wider than a 90% confidence interval. This idea is illustrated in Figure 10.5. In the figure, note how rapidly the width of the confidence interval increases as the level of confidence gets near 100%.

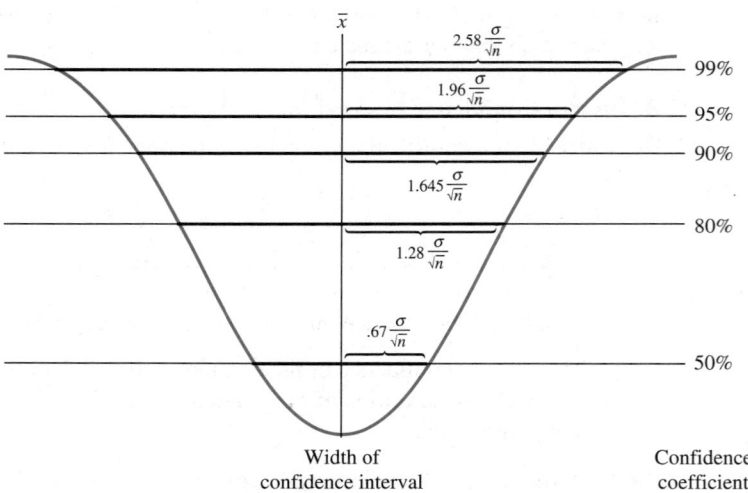

FIGURE 10.5   *Width of several confidence intervals*

To decrease the width of the confidence interval, we must either use a smaller level of confidence $(1 - \alpha)$, which decreases $z_{\alpha/2}$, or increase the sample size n. By making the sample size larger and larger, one can make the confidence interval (for any value of α) as narrow as desired, but at an increased cost of sampling.

EXAMPLE 10.5   **99% Confidence Interval for the Mean**

Construct a 99% confidence interval for the mean in Example 10.3, and compare with the 95% confidence interval calculated there.

SOLUTION   We have $\sigma = 2,000$, $n = 25$, $\bar{x} = \$19,500$, and $(1 - \alpha) = .99$. Thus, $\alpha = .01$ and $\alpha/2 = .005$. From Table A.5, we obtain $z_{\alpha/2} = 2.58$. The desired confidence interval is thus

$$\left(\bar{x} - z_{\alpha/2} \frac{\sigma}{\sqrt{n}}, \; \bar{x} + z_{\alpha/2} \frac{\sigma}{\sqrt{n}} \right)$$

$$\left(19,500 - 2.58 \frac{2,000}{\sqrt{25}}, \; 19,500 + 2.58 \frac{2,000}{\sqrt{25}} \right)$$

$$(18,468, \; 20,532)$$

The width of this confidence interval is $W_1 = \$2,064$. The 95% confidence interval is $(18,716, 20,284)$ with a width of $\$1,568$. We obtain $W_1/W_2 = 2,064/1,568 = 1.32$, which shows that the 99% confidence interval is 32% wider than the 95% confidence interval. This coincides with the result obtained in Example 10.4.

Confidence Intervals for Large Samples

The development of confidence intervals for the mean presented here is limited by two requirements:

1. The population distribution must be normal.
2. The population variance or standard deviation must be known.

When the sample size is large, neither requirement is too important. Because the sampling distribution of $\bar{X}$ will be approximately normal because of the Central Limit Theorem for large n, requirement 1 is eliminated. In addition, the sample standard deviation should provide a fairly accurate estimate of the population standard deviation for large n, thus eliminating requirement 2.

Thus, when the sample size is large, say, 30 or more, we can find confidence intervals that are approximately correct by using the observed sample standard deviation s as an estimate of σ and following the procedures developed in this section. It should not be inferred, however, that the confidence intervals calculated as described in this section will be excellent approximations to the exact confidence intervals when $n \geq 30$ and terrible approximations when $n < 30$. The use of $n = 30$ as the breaking point at which the approximation is good is just a rule of thumb that most statisticians recommend.

The quality of the approximation gradually improves as n increases because the sampling distribution of $\bar{X}$ gradually approaches the normal distribution; thus as n increases, our estimate of σ gradually improves.

EXAMPLE 10.6   **Confidence Interval for the Mean, Large Sample Size**

Suppose a health care official for the federal government wants to estimate the mean number of sick days per year taken by mail carriers. The official obtains a random sample of $n = 100$ observations. The population distribution is unknown and the population standard deviation is unknown. Suppose the sample mean is $\bar{x} = 8.2$ days and the sample standard deviation is $s = 2.7$ days. Construct a 95% confidence interval for the population mean μ.

SOLUTION In this problem, the population may not be normal and the population standard deviation is unknown. Thus, it might appear that the methods discussed in this section do not apply. Because the sample size is large, however ($n \geqslant 30$), the methods discussed in this section apply at least approximately. To construct the confidence interval, we substitute the sample standard deviation s for the unknown population standard deviation σ. We have $n = 100$, $\bar{x} = 8.2$, $s = 2.7$, and $(1 - \alpha) = .95$. Thus, $\alpha = .05$ and $\alpha/2 = .025$. From Table A.5, we obtain $z_{\alpha/2} = 1.96$. The desired (approximate) confidence interval is

$$\left(\bar{x} - z_{\alpha/2} \frac{s}{\sqrt{n}}, \ \bar{x} + z_{\alpha/2} \frac{s}{\sqrt{n}} \right)$$

$$\left(8.2 - 1.96 \frac{2.7}{\sqrt{100}}, \ 8.2 + 1.96 \frac{2.7}{\sqrt{100}} \right)$$

$$(7.6708, \ 8.7292)$$

One-Sided Confidence Intervals for the Mean

Occasionally, a one-sided confidence interval that has either no upper or no lower limit is required. For example, suppose we take a random sample of n observations from some normal population having unknown mean μ and known standard deviation σ. Suppose we wish to find the lower confidence limit LCL such that the probability is $(1 - \alpha)$ that μ exceeds LCL. The one-sided interval (LCL, ∞) is a *left-sided confidence interval*. The lower confidence limit is given by LCL $= \bar{x} - z_{\alpha}\sigma/\sqrt{n}$. Suppose we wish to find the upper confidence limit UCL such that the probability is $(1 - \alpha)$ that μ is less than UCL. The one-sided interval $(-\infty, \text{UCL})$ is a *right-sided confidence interval*. The upper confidence limit is given by UCL $= \bar{x} + z_{\alpha}\sigma/\sqrt{n}$.

If the population standard deviation is unknown but the sample size is large, say, $n \geqslant 30$, we can substitute the sample standard deviation s for the unknown population standard deviation σ, and the resulting one-sided confidence interval will be approximately correct. The approximation holds even if the original population is not normal. The approximation improves as the sample size increases and as the population distribution approaches normality.

A left-sided confidence interval can be interpreted as follows: Suppose we take repeated samples of size n from some population. For each sample, calculate the left-sided confidence interval (LCL, ∞). In repeated sampling, the proportion of these intervals that will contain the population mean μ is $(1 - \alpha)$. Alternatively, for each sample, calculate the right-sided confidence interval $(-\infty, \text{UCL})$. In repeated sampling, the proportion of these intervals that will contain the population mean μ is $(1 - \alpha)$.

EXAMPLE 10.7 **One-Sided Confidence Interval for the Mean, Large Sample Size**

Refer to Example 10.6. Suppose the health care official wants to construct a one-sided confidence interval for the mean number of sick days per year taken by mail carriers. Suppose the official obtains a random sample of $n = 100$ observations. The population distribution is unknown and the population standard deviation is unknown. Suppose the sample mean is $\bar{x} = 8.2$ days and the sample standard deviation is $s = 2.7$ days. Construct a 95% left-sided confidence interval for the population mean μ.

SOLUTION   In this problem, the population may not be normal and the population standard deviation is unknown. Because the sample size is large, however ($n \geq 30$), the left-sided confidence interval will be at least approximately correct. To construct the confidence interval, we substitute the sample standard deviation for the unknown population standard deviation σ. We have $n = 100$, $\bar{x} = 8.2$, $s = 2.7$, and $(1 - \alpha) = .95$. Thus, $\alpha = .05$. From Table A.5, we obtain $z_\alpha = 1.645$. The desired (approximate) lower confidence limit is

$$\text{LCL} = \bar{x} - z_\alpha \frac{s}{\sqrt{n}} = 8.2 - 1.645 \frac{2.7}{\sqrt{100}} = 7.75585$$

The 95% left-sided confidence interval for μ is $(7.75585, \infty)$. Thus, we can be 95% confident that the population mean μ exceeds 7.75585.

Tips for Problem Solving

To construct a confidence interval for the population mean μ of a normal population having known standard deviation σ, follow these steps.

1. Determine the sample mean $\bar{x}$.
2. Record the sample size n, the level of confidence $(1 - \alpha)$, and the population standard deviation σ. Record the standard deviation of the sampling distribution of $\bar{X}$. This is $\sigma_{\bar{x}} = \sigma/\sqrt{n}$, not σ.
3. Find the values $\pm z_{\alpha/2}$ from Table A.5.
4. Calculate the lower confidence limit

$$\text{LCL} = \bar{x} - z_{\alpha/2} \frac{\sigma}{\sqrt{n}}$$

and the upper confidence limit

$$\text{UCL} = \bar{x} + z_{\alpha/2} \frac{\sigma}{\sqrt{n}}$$

5. The two-sided confidence interval is

$$\left(\bar{x} - z_{\alpha/2} \frac{\sigma}{\sqrt{n}}, \bar{x} + z_{\alpha/2} \frac{\sigma}{\sqrt{n}} \right)$$

6. If the population standard deviation is unknown but the sample size is large, the sample standard deviation s can be used in place of σ. The resulting confidence interval will be approximately correct even when the population is only approximately normal.
7. The large-sample approximate confidence interval is

$$\left(\bar{x} - z_{\alpha/2} \frac{s}{\sqrt{n}}, \bar{x} + z_{\alpha/2} \frac{s}{\sqrt{n}} \right)$$

8. As a general rule, the large-sample confidence interval is approximately correct when $n \geq 30$.

Exercises

Statistical Concepts

10.2.1   *True or false:*   For a certain level of confidence, we would prefer a wide confidence interval over a narrow confidence interval because the wide confidence interval is more likely to contain the true value of the population parameter. Explain.

10.2.2   To construct a confidence interval as described in this section, what assumption is made concerning the distribution of the population?

10.2.3   Suppose X is not necessarily a normal random variable, but assume the sample size is large ($n \geqslant 30$). Explain how to construct a confidence interval for the population mean. Do we need any assumptions concerning σ, the standard deviation of the population?

10.2.4   *True or false:*   Let LCL and UCL denote the lower and upper confidence limits of a confidence interval for the population mean. The width of the confidence interval is $W = \text{UCL} - \text{LCL}$. Explain.

10.2.5   *True or false:*   Suppose we construct a confidence interval for the mean of a normal population having a known standard deviation σ. The width of a confidence interval for the population mean is $W = 2\sigma/\sqrt{n}$. Explain.

10.2.6   *True or false:*   Let LCL and UCL denote the lower and upper confidence limits of a confidence interval for the population mean. Then the sample mean is $(\text{LCL} + \text{UCL})/2$. Explain.

10.2.7   List the two desirable properties of a confidence interval.

10.2.8   The width of a confidence interval depends on three factors. What are they, and how do they affect the width of a confidence interval?

10.2.9   Suppose a confidence interval having level of confidence $(1 - \alpha)$ has been constructed for the mean of a normal population. Suppose the width of the confidence interval is $W = 20$ and the sample size is $n = 100$. How large must the sample size be to decrease the width of the confidence interval to $W = 10$?

10.2.10   *True or false:*   A good confidence interval has a large width and a high level of confidence. Explain.

10.2.11   Explain how to construct a left-sided confidence interval for the population mean.

10.2.12   *True or false:*   Suppose we construct a two-sided and a left-sided confidence interval for the population mean. In each case, suppose the level of confidence is $(1 - \alpha)$. The lower confidence limit for the two-sided interval exceeds the lower confidence limit for the left-sided interval. Explain.

Statistical Drills

10.2.13   Suppose we take a random sample of n observations from a normal population with known standard deviation $\sigma = 40$. We obtain the sample mean $\bar{x} = 136$. Construct a 95% confidence interval for the population mean using the following sample sizes:
a. 20   **b.** 40   **c.** 80   **d.** 160   **e.** 320
f. Plot these confidence intervals on a graph to show how the width of a confidence interval is affected by the sample size.

10.2.14   Suppose we take a random sample of $n = 25$ observations from a normal population. We obtain the sample mean $\bar{x} = 136$. Construct a 95% confidence interval for the population mean based on the following population standard deviations:
a. 10   **b.** 20   **c.** 40   **d.** 80
e. Plot these confidence intervals on a graph to show how the width of a confidence interval is affected by the standard deviation.

10.2.15   Let X be a normal random variable with known standard deviation $\sigma = 40$. Suppose we take a random sample of $n = 25$ observations and obtain the sample mean $\bar{x} = 136$. Construct a confidence interval for the population mean having the following levels of confidence:

a. 99%   **b.** 95%   **c.** 90%   **d.** 80%   **e.** 50%

f. Plot these confidence intervals on a graph to show how the width of a confidence interval is affected by the level of confidence.

10.2.16 Let X be a normal random variable with known standard deviation $\sigma = 50$. Suppose we take a random sample of $n = 25$ observations and obtain the sample mean $\bar{x} = 89$. Construct a left-sided confidence interval for the population mean having the following levels of confidence:

a. 99%   **b.** 95%   **c.** 90%

10.2.17 Let X be a normal random variable with known standard deviation $\sigma = 15$. Suppose we take a random sample of $n = 20$ observations and obtain the sample mean $\bar{x} = 73$. Construct a right-sided confidence interval for the population mean having the following levels of confidence:

a. 99%   **b.** 95%   **c.** 90%

10.2.18 A random sample of 100 observations from a population produced the following data:

$$\Sigma x = 1{,}640 \qquad \Sigma(x_i - \bar{x})^2 = 6{,}336$$

a. Calculate the sample mean and sample standard deviation.

b. Are the methods discussed in this section appropriate for constructing a confidence interval for μ? Explain.

c. Construct a 95% confidence interval for the population mean.

10.2.19 A random sample of 100 observations from a normally distributed population produced a sample mean of 231.7 and a sample standard deviation of 9.3.

a. Are the methods discussed in this section appropriate for constructing a confidence interval for μ? Explain.

b. Construct a 95% confidence interval for μ.

c. Construct a 99% confidence interval for μ.

d. Repeat parts **b** and **c** assuming the sample size is 400. Explain what happens to the width of a confidence interval as the sample size is held fixed and the level of confidence is increased.

e. Do the confidence intervals of parts **b** and **c** rely on the assumption that the population is normal? Explain.

Statistical Applications

10.2.20 To estimate the mean age of subscribers to *Sports Illustrated* magazine, a random sample of 100 subscribers is taken. The sample mean is $\bar{x} = 31$, and the sample variance is 144. Calculate the following confidence intervals for μ:

a. 95%   **b.** 99%

10.2.21 The Summerhill Trucking Company owns a large fleet of rental trucks. Many of the trucks need substantial repairs from time to time. The company president takes a random sample of 64 trucks and finds that the sample mean annual repair bill is $1,245 and the sample standard deviation is $s = \$288$. Construct the following confidence intervals for μ:

a. 95%   **b.** 99%

10.2.22 A statistical report states that a 95% confidence interval for the mean salary of new Ph.D.'s in economics is ($24,000, $26,000) with a population standard deviation σ of $2,000. How large was the sample used to find the confidence interval?

10.2.23 An executive for the Rankin Car Rental Agency wants to know the average mileage driven by customers who rent cars from Rankin. A random sample of 200 customers is obtained. The sample mean mileage is $\bar{x} = 325$ miles, and the sample standard deviation is $s = 60$. Construct the following confidence intervals for μ:

a. 90%   **b.** 95%

10.2.24   For winter months the standard deviation for monthly heating bills for residential homes in a certain area is believed to be $100. A random sample of 16 homes in a particular subdivision will be used to estimate the mean monthly heating bills for all homes in this type of subdivision.

a. What is the standard error of the mean?

b. Find the sampling distribution for the sample mean heating bill.

c. Find the 95% confidence interval for the mean monthly heating bill if the sample mean is $195.00.

10.2.25   A research report stated that the mean return on invested capital was between 6.1% and 10.8% per year with a confidence coefficient of 95%. One person interpreted this as meaning that 95% of investors had investment returns between 6.1% and 10.8%. Another person interpreted the statement to mean that if many random samples were taken, 95% would have sample means between 6.1% and 10.8%. Why are these interpretations incorrect?

10.3   Student's *t* Distribution

In Section 10.2, we showed how to use the *Z* score to construct confidence intervals for the population mean in the following cases:

■   *Case 1*:   The population is normal and the population variance is known.
■   *Case 2*:   The sample size is large, say, $n \geqslant 30$. In this case, the population may or may not be normal, and the variance may or may not be known.

How do we construct a confidence interval for the mean when the sample size is small and the population variance is unknown? There are two different cases to consider:

■   *Case 3*:   The population is approximately normal, the variance is unknown, and the sample size is small.
■   *Case 4*:   The population is nonnormal and the sample size is small.

In Section 10.4, we discuss how to construct confidence intervals for the population mean μ for case 3. Under these circumstances, confidence intervals can be obtained by using a *t* score and the *t* distribution, which we will describe in this section.

In case 4, where the population is nonnormal and the sample is small, the Z score is not appropriate because the population is nonnormal, the variance is unknown, and the Central Limit Theorem does not apply. In addition, the use of the *t* distribution is not theoretically correct because the population is nonnormal. Nevertheless, many people use the *t* distribution in this situation and then qualify their confidence intervals by stating that they are only approximate.

The use of the *t* distribution is based on the following argument. If the population is normally distributed with mean μ and variance σ^2, the sample mean $\bar{X}$ is distributed as $N(\mu, \sigma^2/n)$ and the standardized Z score

$$Z = \frac{\bar{X} - \mu}{\sigma/\sqrt{n}}$$

is a standard normal variable. If σ is unknown and replaced by the sample standard deviation S in the Z score formula, the Z score no longer follows the standard normal distribution (although the approximation is good when $n \geqslant 30$). If we replace the true standard deviation σ by its estimate S, we obtain what is called the *t score*, where

$$t = \frac{\overline{X} - \mu}{S/\sqrt{n}}$$

It can be shown that when the population is normally distributed, the t score follows what is called the *Student t distribution,* or more simply the *t distribution.* The t distribution was first studied and used by William Sealy Gosset (1876–1937) in a 1908 paper, "The Probable Error of the Mean." Because Gosset was employed by Arthur Guinness and Son, a Dublin brewery, and not permitted to publish company research, he published under the pseudonym "Student." Thus the t distribution is frequently referred to as Student's t distribution; the distribution is also called the t distribution because Gosset used the letter t to denote the random variable whose distribution was being studied.

Gosset showed that the random variable t

$$t = \frac{\overline{X} - \mu}{S/\sqrt{n}}$$

follows the t distribution provided the population is normal, where S is the sample standard deviation and n is the sample size.

The statistic t might be thought of as an estimated standardized normal variable because the estimated standard deviation S, rather than the true standard deviation σ, is used in its calculation. As the sample size gets large, the estimated standard deviation S approaches the true standard deviation σ. Thus, as the sample size becomes large, the statistic $t = (\overline{X} - \mu)/(S/\sqrt{n})$ approaches $Z = (\overline{X} - \mu)/(\sigma/\sqrt{n})$. Because the random variable Z is a standard normal variable, the t distribution approaches the standard normal distribution as n gets large.

Characteristics of the *t* Distribution

1. The t distribution is symmetric about 0 and ranges from $-\infty$ to ∞.
2. The t distribution is bell shaped and has approximately the same appearance as the standard normal distribution.
3. The mean of the t distribution is 0.
4. The t distribution depends on a parameter ν (Greek nu), called the **degrees of freedom** of the distribution. When constructing confidence intervals for the population mean, the appropriate number of degrees of freedom is $\nu = (n - 1)$, where n is the sample size.
5. The variance of the t distribution is $\nu/(\nu - 2)$ for $\nu > 2$.
6. The variance of the t distribution always exceeds 1.
7. As ν increases, the variance of the t distribution approaches 1 and the shape approaches that of the standard normal distribution.
8. Because the variance of t exceeds 1 whereas the variance of the standard normal variable Z equals 1, the t distribution is slightly flatter in the middle than the standard normal distribution and has thicker tails.
9. The t distribution is actually a family of distributions with a different density function corresponding to each different value of the parameter ν.

Figure 10.6 (page 354) compares the shapes of the t distribution and the standard normal distribution.

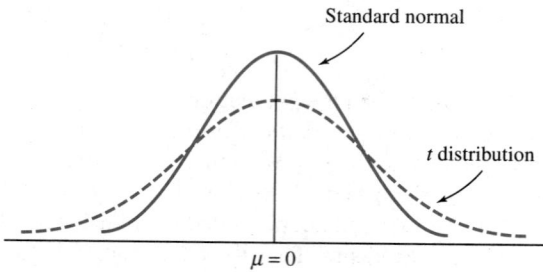

FIGURE 10.6   *Student's t distribution and the standard normal distribution*

Because each value of ν yields a different density function, it is not feasible to present tables showing areas under each possible t distribution. However, Table A.6 in the Appendix shows the values of t such that the area in one tail of the distribution is 10%, 5%, 2.5%, 1%, and .5% for various values of ν. To use the table, we must know the degrees of freedom of the distribution and how much area we want in the right tail of the curve.

Value of $t_{\alpha,\nu}$

> The symbol $t_{\alpha,\nu}$ denotes the value of t such that the area to its right is α and t has ν degrees of freedom. The value $t_{\alpha,\nu}$ satisfies the equation
>
> $$P(t > t_{\alpha,\nu}) = \alpha$$
>
> where the random variable t has the t distribution with ν degrees of freedom.

EXAMPLE 10.8   **Area Under the Right Tail of the t Distribution** ———————————

Consider the t distribution having $\nu = 9$ degrees of freedom. Find the value $t_{.05,9}$, such that the area in the right tail of the t distribution is .05.

SOLUTION   To find $t_{.05,9}$ in Table A.6, we search down the column entitled Degrees of Freedom until we reach the row $\nu = 9$. Move across this row to the column headed by the value .05. The number at the intersection of this row and column is $t_{.05,9} = 1.833$. Thus, if the random variable t follows the t distribution with 9 degrees of freedom, then $P(t > 1.833) = .05$. This is illustrated in Figure 10.7.

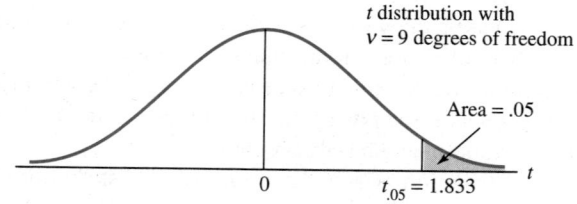

FIGURE 10.7   *t distribution having 9 degrees of freedom and right-tail area of 5%*

EXAMPLE 10.9   **Area Under the Left Tail of the *t* Distribution**

Consider the *t* distribution having $\nu = 9$ degrees of freedom. Find the value of *t* such that the area in the left tail of the distribution is 5%. Refer to Figure 10.8.

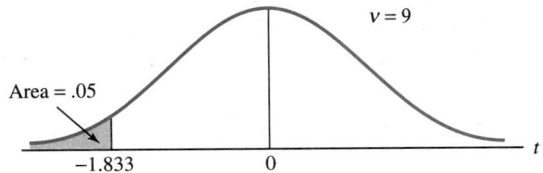

Area = .05

−1.833   0   *t*

FIGURE 10.8   *t distribution having 9 degrees of freedom and left-tail area of 5%*

SOLUTION   In Example 10.8, we showed that 5% of the area under the distribution lies to the right of $t_{.05,9} = 1.833$. Because the *t* distribution is symmetric, 5% of the area under the distribution lies to the left of -1.833. We obtain $P(t < -1.833) = .05$.

EXAMPLE 10.10   **Area Under the *t* Distribution**

Consider the *t* distribution with 9 degrees of freedom. Find the values $t_{.025,9}$ and $-t_{.025,9}$ such that each tail of the distribution contains area .025.

SOLUTION   From Table A.6, we obtain $t_{.025,9} = 2.262$. By symmetry, the value of *t* such that the left tail contains an area of .025 is -2.262. We have $P(t > 2.262) = .025$ and $P(t < -2.262) = .025$. It follows that $P(-2.262 \le t \le 2.262) = .95$. Refer to Figure 10.9.

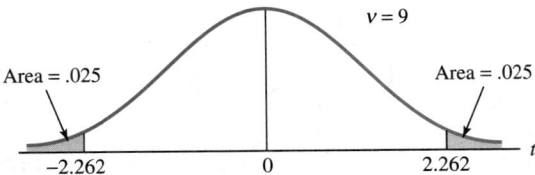

Area = .025   Area = .025

−2.262   0   2.262   *t*

FIGURE 10.9   *t distribution having 9 degrees of freedom and area of .025 in each tail*

Example 10.10 shows that for $\nu = 9$ degrees of freedom, 95% of the area under the *t* distribution lies between -2.262 and 2.262. For the standard normal distribution, 95% of the area under the curve lies between -1.96 and 1.96. This shows that the tails of the *t* distribution are fatter than the tails of the standard normal distribution.

EXAMPLE 10.11   **Determining $t_{.025,20}$**

Consider the *t* distribution with 20 degrees of freedom. Find the value of $t_{.025,20}$ such that the right tail of the distribution contains area .025.

SOLUTION   From Table A.6, the desired value of t associated with $\nu = 20$ and $\alpha = .025$ is $t_{.025,20} = 2.086$. Thus, $P(t \leqslant 2.086) = .975$ and $P(t > 2.086) = .025$. See Figure 10.10.

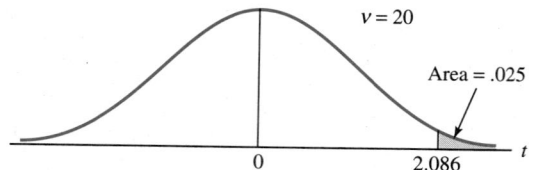

FIGURE 10.10   *t distribution having 20 degrees of freedom and right-tail area of .025*

Exercises

Statistical Concepts

10.3.1   List the important characteristics of the Student t distribution.

10.3.2   Suppose t is a random variable having the t distribution. Describe how the density function changes as the number of degrees of freedom increases from 10 to 20.

10.3.3   *True or false:*   Let the random variable t follow the t distribution. Then $P(0 \leqslant t \leqslant 3) = P(-3 \leqslant t \leqslant 0)$. Explain.

10.3.4   *True or false:*   Let the random variable t follow the t distribution. Then $P(t \leqslant -2) = P(t \geqslant 2)$. Explain.

10.3.5   *True or false:*   Let the random variable t follow the t distribution. Then $P(1 \leqslant t \leqslant 2) = P(2 \leqslant t \leqslant 3)$. Explain.

10.3.6   *True or false:*   Let the random variable t follow the t distribution. Then $P(t \leqslant 2) = P(t < 2)$. Explain.

10.3.7   *True or false:*   Let the random variable t follow the t distribution. Then $P(t > 2) = .5 + P(0 \leqslant t \leqslant 2)$. Explain.

10.3.8   *True or false:*   Let the random variable t follow the t distribution. Then $P(t > -2) = .5 + P(0 \leqslant t \leqslant 2)$. Explain.

10.3.9   *True or false:*   Let the random variable t follow the t distribution. Then $P(t < -1) = .5 - P(0 \leqslant t \leqslant 1)$. Explain.

10.3.10   Use an interpolation method to find the value t_0 such that $P(-t_0 \leqslant t \leqslant t_0) = .99$ if t is a random variable having 50 degrees of freedom. Verify that $t_0 \approx 2.682$. Explain.

10.3.11   In many situations, we may be interested in finding an interval that encompasses a specified probability. For example, suppose you need to find the value t_0 such that $P(0 \leqslant t \leqslant t_0) = .40$, or equivalently $P(t > t_0) = .10$, where the random variable t follows the t distribution with 20 degrees of freedom. To find t_0, look at Table A.6 and locate the column headed by $t_{.10}$.
a. Verify that the appropriate value t_0 is 1.325.
b. Find t_0 such that $P(0 \leqslant t \leqslant t_0) = .475$.

Statistical Drills

10.3.12   Find the value of t such that the area in the right tail of the t distribution is .05 (that is, find $t_{.05}$) if the number of degrees of freedom is as follows:
a. 2   **b.** 5   **c.** 10   **d.** 20   **e.** 25

10.3.13 Recall that for the standard normal distribution, 95% of the area lies between $-z_{.025} = -1.96$ and $z_{.025} = 1.96$. Find the corresponding values of t (that is, find $t_{.025}$ and $-t_{.025}$) if the number of degrees of freedom is as follows:
   a. 2   **b.** 5   **c.** 10   **d.** 20   **e.** 25

10.3.14 We want 1% of the area to lie in the left tail of the t distribution. Find the critical value of t if the number of degrees of freedom is as follows:
   a. 2   **b.** 5   **c.** 10   **d.** 20   **e.** 25
   f. What is the corresponding critical value for a standard normal variable?

10.3.15 Compare the t distribution to the standard normal distribution when the degrees of freedom of the t distribution is infinity. For example, for the random variable t and the random variable Z, show that the probability is the same that each will exceed the following values:
   a. 1.645   **b.** 1.96   **c.** 2.58

10.3.16 Suppose the random variable t follows the t distribution with 17 degrees of freedom.
   a. Find the value t_0 such that $P(-t_0 \leq t \leq t_0) = .90$.
   b. Find the value t_0 such that $P(-t_0 \leq t \leq t_0) = .95$.
   c. Find the value t_0 such that $P(-t_0 \leq t \leq t_0) = .99$.

10.3.17 Suppose the random variable t follows the t distribution with 10 degrees of freedom.
   a. Find the value t_0 such that $P(t \leq t_0) = .99$.
   b. Find the value t_0 such that $P(t \geq -t_0) = .95$.
   c. Find the value t_0 such that $P(-t_0 \leq t \leq t_0) = .95$.

10.3.18 Let t_0 be a particular value of t. Use Table A.6 to find t_0 values such that the following statements are true:
   a. $P(t \geq t_0) = .025$ where degrees of freedom (df) $= 12$
   b. $P(t \geq t_0) = .01$ where df $= 15$
   c. $P(t \leq t_0) = .01$ where df $= 11$
   d. $P(t \leq -t_0 \text{ or } t \geq t_0) = .05$ where df $= 15$
   e. $P(t \leq -t_0 \text{ or } t \geq t_0) = .10$ where df $= 15$

10.4 Confidence Intervals for the Mean with Unknown Population Variance

We know that if a population is normal with mean μ and variance σ^2, then the statistic

$$Z = \frac{\bar{X} - \mu}{\sigma/\sqrt{n}}$$

has exactly a normal distribution with mean 0 and variance 1. The statement also holds, at least approximately, when the population is not normal provided the sample size is large, even when the sample standard deviation has been used to estimate the population standard deviation.

Usually, however, the population variance σ^2 is unknown and must be estimated from the sample data. When the population standard deviation σ is replaced by the sample standard deviation S, we obtain the t score

$$t = \frac{\bar{X} - \mu}{S/\sqrt{n}}$$

DEFINITION   *t* **Statistic or** *t* **Score**
_____

Suppose we take a random sample of size n from some population having mean μ and calculate the sample mean $\bar{X}$ and the sample variance S^2. If the population is normal, then the *t* **statistic**

$$t = \frac{\bar{X} - \mu}{S/\sqrt{n}}$$

has the t distribution with $\nu = (n - 1)$ degrees of freedom. The t statistic is also called the *t* **score.**

_____

Constructing Confidence Intervals Using the *t* Distribution

The area to the right of $t_{\alpha/2,\nu}$ is $\alpha/2$ for the t distribution having ν degrees of freedom. Similarly, the area to the left of $-t_{\alpha/2,\nu}$ is $\alpha/2$. Thus, we obtain

$$P(-t_{\alpha/2,\nu} < t < t_{\alpha/2,\nu}) = 1 - \alpha$$

If we replace t by

$$t = \frac{\bar{X} - \mu}{S/\sqrt{n}}$$

we obtain

$$P\left(-t_{\alpha/2,\nu} \leq \frac{\bar{X} - \mu}{S/\sqrt{n}} \leq t_{\alpha/2,\nu}\right) = 1 - \alpha$$

Thus, the probability is $(1 - \alpha)$ that the t score will fall between the values $-t_{\alpha/2,\nu}$ and $t_{\alpha/2,\nu}$. Multiplying through by the denominator of the t score yields the following:

$$P(-t_{\alpha/2,\nu} S/\sqrt{n} \leq \bar{X} - \mu \leq t_{\alpha/2,\nu} S/\sqrt{n}) = 1 - \alpha$$

A bit of algebra yields

$$P(\bar{X} - t_{\alpha/2,\nu} S/\sqrt{n} \leq \mu \leq \bar{X} + t_{\alpha/2,\nu} S/\sqrt{n}) = 1 - \alpha$$

This expression indicates that when sampling is from a normal population, the probability is $(1 - \alpha)$ that the interval

$$(\bar{X} - t_{\alpha/2,\nu} S/\sqrt{n}, \; \bar{X} + t_{\alpha/2,\nu} S/\sqrt{n})$$

will contain the population mean.

DEFINITION   **Confidence Interval for the Mean of a Normal Population with Unknown Population Variance**
_____

Suppose we take a random sample of n observations from a normal population with mean μ and unknown variance σ^2. If the observed sample mean is $\bar{x}$ and the observed sample standard deviation is s, then the **confidence interval for the mean** having level of confidence $100(1 - \alpha)\%$ is given by

$$\left(\bar{x} - t_{\alpha/2,\nu} \frac{s}{\sqrt{n}}, \ \bar{x} + t_{\alpha/2,\nu} \frac{s}{\sqrt{n}} \right)$$

where $t_{\alpha/2,\nu}$ is the number such that

$$P(t > t_{\alpha/2,\nu}) = \alpha/2$$

and the random variable t has a t distribution with $\nu = (n - 1)$ degrees of freedom.

EXAMPLE 10.12   **Constructing a 95% Confidence Interval for the Mean** ————————

In Table 9.4 in Chapter 9, we presented a population of 1,000 values of adjusted gross income (AGI) for clients at an H & R Block tax office. The data are repeated in Table 10.1 (page 360). The population mean is $\mu = 30.47$ and the population standard deviation is $\sigma = 16.54$. In Chapter 9, we constructed a relative frequency distribution of the 1,000 incomes and showed that the distribution is approximately bell shaped, although it is slightly skewed to the right. Suppose we take a sample of $n = 20$ observations from this population, estimate the sample mean and sample standard deviation, and use the t distribution to construct a 95% confidence interval for μ. Since the population is not exactly normal, the 95% confidence interval will be only approximately correct. Nevertheless, this is the problem that statisticians are typically confronted with. That is, when we use the t distribution to construct a confidence interval, we typically have to assume that the population is normal—we do not know for sure that the population is normal.

Table 10.1 contains 50 rows of data, with 20 observations in each row. To illustrate the concept of a confidence interval for the population mean, let us obtain 50 different samples containing 20 observations each and construct 50 different 95% confidence intervals for μ. If the confidence intervals are correct, then approximately 95% of these confidence intervals should contain μ.

SOLUTION   Suppose we use the 20 observations in the first row as our sample. For this sample, the sample mean is $\bar{x}_1 = 32.95$ and the sample standard deviation is $s_1 = 20.48$. We have $n = 20$ and $(1 - \alpha) = .95$. Thus, we have $\alpha = .05$ and $\alpha/2 = .025$. The appropriate number of degrees of freedom is $(n - 1) = 19$. From Table A.6, we find $t_{.025,19} = 2.093$. The desired confidence interval is

$$\left(\bar{x} - t_{\alpha/2} \frac{s}{\sqrt{n}}, \ \bar{x} + t_{\alpha/2} \frac{s}{\sqrt{n}} \right)$$

$$\left(32.95 - 2.093 \frac{20.48}{\sqrt{20}}, \ 32.95 + 2.093 \frac{20.48}{\sqrt{20}} \right)$$

$$(23.36, \ 42.54)$$

Thus, we are 95% confident that the mean adjusted gross income of the population is between 23.36 and 42.54 (or $23,360 to $42,540). Recall that the population mean is $\mu = 30.47$, so this confidence interval does contain μ.

TABLE 10.1 *Data for Example 10.12:*
50 Different 95% Confidence Intervals for μ

																				Sample mean	Sample std. dev.	Lower conf. limit	Lower conf. limit
13	23	60	4	13	35	15	63	26	6	19	31	53	72	36	49	48	55	17	21	32.95	20.48	23.36	42.54
3	30	68	14	51	20	50	22	17	24	9	28	37	39	11	25	32	17	47	45	29.45	16.67	21.65	37.25
25	60	12	5	10	42	3	18	29	46	63	60	29	35	28	19	17	13	28	36	28.90	18.01	20.47	37.33
42	12	8	25	18	29	27	34	41	52	10	29	57	9	10	24	13	61	31	11	27.15	16.56	19.40	34.90
45	26	17	7	30	25	16	81	43	36	38	25	15	17	33	47	38	29	17	32	30.85	16.17	23.28	38.42
42	11	9	22	14	51	50	19	27	14	3	47	19	22	20	43	22	20	43	72	28.50	17.80	20.17	36.83
12	30	29	11	42	32	67	44	22	31	15	29	25	46	21	20	31	24	32	10	28.65	13.68	22.25	35.05
2	17	43	13	5	54	29	42	29	57	69	43	10	17	27	28	17	40	30	23	29.75	17.88	21.38	38.12
34	15	55	37	4	56	25	29	22	25	18	29	25	44	27	18	43	46	18	16	29.30	13.87	22.81	35.79
22	20	26	5	18	22	20	44	46	24	78	39	35	31	25	46	17	26	54	47	32.25	16.61	24.48	40.02
13	33	25	42	45	42	12	2	34	22	25	60	58	12	25	13	55	20	51	30	30.95	17.04	22.97	38.93
23	30	53	12	26	11	30	17	15	20	42	12	34	5	8	35	37	26	7	25	23.40	12.85	17.38	29.42
60	58	12	8	17	9	29	43	55	26	45	26	31	10	29	54	34	25	19	16	30.30	17.00	22.35	38.25
4	34	5	25	7	22	11	13	37	5	42	31	20	42	27	29	56	18	27	81	26.80	19.03	17.89	35.71
13	31	10	18	30	14	42	5	4	18	12	30	50	23	34	42	45	52	24	43	27.00	15.12	19.92	34.08
35	20	42	29	25	51	32	54	56	22	2	17	22	18	41	29	29	20	3	36	29.15	14.85	22.20	36.10
15	50	3	27	16	50	67	29	25	20	34	15	17	29	52	57	22	44	47	38	32.85	17.06	24.87	40.83
63	22	18	34	81	19	44	42	29	44	22	20	24	46	10	69	25	46	19	25	35.10	19.00	26.21	43.99
26	17	29	41	43	27	22	29	22	46	24	42	29	63	19	63	8	14	2	14	29.00	16.40	21.32	36.68
6	24	46	52	36	14	31	57	25	24	31	33	28	60	57	10	29	78	20	17	33.90	18.85	25.08	42.72
19	29	63	10	38	3	15	69	18	78	25	42	37	29	9	17	25	39	43	33	32.05	20.04	22.67	41.43
31	28	60	19	25	47	29	43	29	39	1	24	39	55	10	17	54	15	12	57	32.60	15.80	25.21	39.99
53	37	29	47	15	19	20	10	25	35	10	17	11	8	24	18	27	31	20	18	23.70	12.18	18.00	29.40
72	39	35	9	17	22	46	17	44	31	41	31	25	19	13	17	18	25	43	29	29.65	14.92	22.67	36.63
36	11	28	20	33	20	21	27	27	25	23	27	32	17	51	40	43	46	72	17	30.80	14.24	24.14	37.46
49	25	19	24	47	43	20	28	18	46	72	13	67	13	21	30	46	17	9	32	31.95	17.90	23.57	40.33
48	32	17	13	38	22	31	17	43	17	17	19	47	28	21	23	78	56	42	11	31.00	17.17	22.97	39.03
55	27	13	51	29	20	24	40	46	26	15	25	35	36	22	53	16	54	14	45	32.30	14.37	25.57	39.03
17	47	28	31	17	43	32	30	18	54	26	28	23	14	29	21	65	37	61	21	32.10	14.80	25.18	39.02
21	35	36	21	32	72	10	23	16	47	13	34	17	18	35	39	27	30	60	7	29.65	16.36	21.99	37.31
45	26	17	7	30	25	16	81	43	36	38	17	9	29	43	55	26	25	60	12	32.00	18.67	23.26	40.74
42	11	9	22	14	51	50	19	27	14	3	7	22	11	13	37	65	42	82	18	27.95	21.33	17.97	37.93
12	30	29	11	42	32	67	44	22	31	15	30	14	42	5	94	18	45	26	17	31.30	20.94	21.50	41.10
2	17	43	13	5	54	29	42	29	57	69	25	51	32	54	56	22	42	11	9	33.10	19.93	23.77	42.43
34	15	55	37	4	56	25	29	22	25	18	16	50	67	29	25	20	12	30	29	29.90	16.13	22.35	37.45
22	20	26	5	18	22	20	44	46	24	78	81	19	44	42	29	44	12	17	33	32.30	19.95	22.96	41.64
25	42	45	42	12	2	34	22	24	31	25	43	27	22	29	22	46	34	5	65	29.85	14.90	22.88	36.82
60	12	26	31	30	17	15	20	42	33	42	36	14	31	57	25	44	2	20	26	29.15	14.80	22.22	36.08
58	34	31	20	50	22	17	24	29	28	37	38	3	15	69	18	78	25	42	45	34.15	18.72	25.39	42.91
12	5	10	42	23	18	29	46	63	60	29	25	47	29	43	29	39	60	12	26	32.35	17.21	24.29	40.41
18	34	81	19	44	42	29	44	2	17	28	23	47	35	23	17	9	22	14	51	29.95	18.00	21.52	38.38
29	41	43	27	22	29	22	46	34	15	36	51	28	36	4	28	9	11	62	32	30.25	14.42	23.50	37.00
46	52	36	14	31	57	25	24	22	20	19	44	21	22	29	35	43	13	5	54	30.60	14.74	23.70	37.50
63	10	38	3	15	69	18	78	25	42	45	36	23	53	21	19	55	37	24	36	35.50	20.30	26.00	45.00
60	19	25	47	29	43	29	39	60	12	34	29	18	16	45	27	26	5	18	22	30.15	14.96	23.15	37.15
29	57	15	19	25	10	25	35	58	34	37	17	26	54	47	30	25	42	45	42	33.60	14.01	27.04	40.16
35	9	17	22	46	17	44	31	12	5	29	28	22	14	51	50	60	12	26	11	27.05	16.00	19.56	34.54
28	20	33	20	21	27	27	25	25	28	56	31	11	45	21	37	12	58	17	9	27.55	13.29	21.33	33.77
19	24	47	43	20	28	18	46	13	35	51	42	22	24	31	25	43	27	72	57	34.35	15.28	27.20	41.50
17	13	38	22	31	17	43	17	55	37	9	29	20	42	33	42	36	14	65	24	30.20	14.80	23.28	37.12

Pop. mean 30.47   16.54 Pop. st. dev.

Rather than using the 20 observations in row 1 of Table 10.1 as our sample of data, suppose we use the 20 observations in row 2 and construct another 95% confidence interval for μ. In this case, the sample mean would be $\bar{x}_2 = 29.45$ and the sample standard deviation would be $s_2 = 16.67$. The desired confidence interval would be

$$\left(\bar{x} - t_{\alpha/2} \frac{s}{\sqrt{n}}, \ \bar{x} + t_{\alpha/2} \frac{s}{\sqrt{n}} \right)$$

$$\left(29.45 - 2.093 \frac{16.67}{\sqrt{20}}, \ 29.45 + 2.093 \frac{16.67}{\sqrt{20}} \right)$$

$$(21.65, \ 37.25)$$

Thus, we are 95% confident that the mean adjusted gross income of the population is between 21.65 and 37.25 (or $21,650 to $37,250). Recall that the population mean is $\mu = 30.47$, so this confidence interval also contains μ.

We could repeat this process over and over using different samples of 20 observations from Table 10.1. For each sample, we could calculate the sample mean and sample standard deviation and then construct a 95% confidence interval. Approximately 95% of these confidence intervals would contain μ and approximately 5% would not contain μ. To illustrate this, we have constructed 50 of these confidence intervals based on using the 20 observations in each row of Table 10.1 as the appropriate sample. On the right side of Table 10.1, we show the sample mean, the sample standard deviation, the lower confidence limit, and the upper confidence limit for each of the 50 samples. Note that 48 of the 50 confidence intervals (or 96%) contain $\mu = 30.47$.

Two of the 50 confidence intervals shown in Table 10.1 do not contain $\mu = 30.47$. Sample 12 contains the 20 observations in row 12 of Table 10.1. The confidence interval based on this sample of data is (17.38, 29.42). Sample 23 contains the 20 observations in row 23 of Table 10.1. The confidence interval based on this sample of data is (18.00, 29.40).

Approximately 95% of the confidence intervals do contain μ. If the population was exactly normal and if we constructed every possible confidence interval using every possible sample of 20 observations, exactly 95% of the confidence intervals would contain μ.

The confidence interval based on the t score is always wider than the corresponding confidence interval based on the Z score because the latter confidence interval uses more information (the population variance is known). One way of gauging the greater dispersion of the t distribution is to see how wide an interval must be to encompass 95% of the observations. The probability is .95 that Z will be between -1.96 and $+1.96$, whereas with degrees of freedom $\nu = 9$, for example, there is a .95 probability that t will be between -2.262 and $+2.262$.

As the sample size increases, the width of the confidence interval decreases for two reasons: First, $\sqrt{n}$ is in the denominator, and second, the t value decreases as the number of degrees of freedom increases. The following confidence intervals show how the width of the confidence interval decreases as the sample size increases:

Sample size n	Degrees of freedom $n - 1$	95% confidence interval
5	4	$\bar{x} \pm 2.776(s/\sqrt{n})$
10	9	$\bar{x} \pm 2.262(s/\sqrt{n})$
20	19	$\bar{x} \pm 2.093(s/\sqrt{n})$
30	29	$\bar{x} \pm 2.045(s/\sqrt{n})$
∞	∞	$\bar{x} \pm 1.96(s/\sqrt{n})$

When the number of degrees of freedom exceeds 30, the confidence interval computed using the t distribution is approximately the same as the confidence interval obtained using the standard normal distribution. That is, for $\nu = 29$, we have $t_{.025} = 2.045$, whereas the corresponding Z score is $z_{.025} = 1.96$. Thus, if you have the 30 or more observations recommended for invoking the Central Limit Theorem, you also have a large enough sample to ignore the distinction between the normal distribution and Student's t distribution. As a rule of thumb, the t distribution rather than the standard normal distribution should be used when the sample size is small, say, 29 or less.

EXAMPLE 10.13   **99% Confidence Interval for the Mean** ───────────────

As part of a traffic control program, the city planner needs to know the number of cars that pass through a certain intersection on weekday mornings. On a sample of eight Monday mornings, the numbers of cars passing through the intersection between 7 A.M. and 9 A.M. are counted. Assume the population is normal. Construct a 99% confidence interval for the population mean if the sample mean is $\bar{x} = 1,500$ and the sample standard deviation is 300.

SOLUTION   There are $\nu = (n - 1) = 7$ degrees of freedom. For a 99% confidence interval, we have $(1 - \alpha) = .99$. Thus we obtain $\alpha = .01$ and $\alpha/2 = .005$. We obtain $t_{\alpha/2,\nu} = t_{.005,7} = 3.499$. The 99% confidence interval is given by

$$\left(\bar{x} - t_{\alpha/2,\nu} \frac{s}{\sqrt{n}}, \ \bar{x} + t_{\alpha/2,\nu} \frac{s}{\sqrt{n}} \right)$$

$$\left[1,500 - 3.499 \left(\frac{300}{\sqrt{8}} \right), \ 1,500 + 3.499 \left(\frac{300}{\sqrt{8}} \right) \right]$$

$$(1,500 - 371.12, \ 1,500 + 396.75)$$

The 99% confidence interval is

$$(1,128.88, \ 1,871.12)$$

EXAMPLE 10.14   **Confidence Intervals Using Large Samples** ───────────────

The purpose of this example is to show that when the sample size is large, confidence intervals calculated using the standard normal distribution will be approximately the same as those obtained using the t distribution. Suppose we have taken a random sample of $n = 121$ observations and obtain the estimates $\bar{x} = \$20,000$ and $s = \$4,000$; also, assume that the population is normal so that the use of the t distribution is theoretically correct.

In addition, because the sample size is large, the use of the standard normal distribution is justified even though the population variance is unknown. Let us construct two 95% confidence intervals for the mean, one using the t distribution and one using the standard normal distribution.

SOLUTION There are $\nu = (n - 1) = 120$ degrees of freedom. For a 95% confidence interval, we have $\alpha = .05$ and $\alpha/2 = .025$. We obtain $t_{.025,120} = 1.984$ and $z_{.025} = 1.96$.

If we use the t distribution, a 95% confidence interval for the mean is given by

$$\left(\bar{x} - t_{\alpha/2,\nu} \frac{s}{\sqrt{n}}, \bar{x} + t_{\alpha/2,\nu} \frac{s}{\sqrt{n}} \right)$$

$$\left[20,000 - 1.984 \left(\frac{4,000}{\sqrt{121}} \right), 20,000 + 1.984 \left(\frac{4,000}{\sqrt{121}} \right) \right]$$

Thus, the 95% confidence interval for the mean based on the t distribution is

$$(\$19,278.55, \$20,721.45)$$

If we use the standard normal distribution, a 95% confidence interval for the mean is given by

$$\left(\bar{x} - z_{\alpha/2} \frac{s}{\sqrt{n}}, \bar{x} + z_{\alpha/2} \frac{s}{\sqrt{n}} \right)$$

$$\left[20,000 - 1.96 \left(\frac{4,000}{\sqrt{121}} \right), 20,000 + 1.96 \left(\frac{4,000}{\sqrt{121}} \right) \right]$$

The 95% confidence interval for the mean based on the standard normal distribution is

$$(\$19,287.27, \$20,712.73)$$

This is very similar to the interval obtained using the t distribution. The confidence interval based on the standard normal distribution is centered at $20,000 and has width $1,425.46, whereas that based on the t distribution is centered at $20,000 and has width $1,442.90. Thus the confidence interval based on the t distribution is approximately 1.2% wider than the confidence interval based on the standard normal distribution.

Consider the difference between a confidence interval based on the standard normal value $z_{\alpha/2}$ and a confidence interval based on the value $t_{\alpha/2,\nu}$. As the number of degrees of freedom increases, the t distribution approaches the standard normal distribution. If the sample size is large, then $t_{\alpha/2,\nu} \approx z_{\alpha/2}$, so it will make little practical difference whether the t score or the Z score is used in constructing the confidence interval. For this reason, statisticians emphasize that the use of the t distribution is especially important when the sample size is small. When the sample size is large, the difference between using $t_{\alpha/2,\nu}$ and $z_{\alpha/2}$ is relatively minor.

EXAMPLE 10.15 **Comparing Confidence Intervals** ———————————————

Suppose a sample of size $n = 10$ is obtained and we want to construct a confidence interval for μ having level of confidence $(1 - \alpha) = .95$. If the population is normal with known variance, then we should use $z_{\alpha/2} = 1.96$ to construct the interval. If the variance is unknown and has to be estimated, we use $t_{\alpha/2,9} = 2.262$. We have $2.262/1.96 = 1.15$. In this case there is a substantial difference of 15% between the Z value and the t statistic. Thus the confidence interval based on the t value will be 15% wider than that based on the Z value.

The use of the t distribution is based on the assumption that the sample of observations has been selected from a normal population. In practical situations, it may be difficult to determine the exact distribution of the population, especially if the sample size is small. Fortunately, the t distribution is relatively "robust"; that is, the confidence intervals obtained by using the t distribution are still approximately correct provided the population does not differ significantly from a normal distribution and provided the population distribution is approximately symmetric.

One-Sided Confidence Intervals for the Mean

Suppose we need a one-sided confidence interval that has either no upper or no lower limit. For example, suppose we take a random sample of n observations from some normal population having unknown mean μ and unknown standard deviation σ. Suppose we wish to find the lower confidence limit LCL such that the probability is $(1 - \alpha)$ that μ exceeds LCL. The one-sided interval (LCL, ∞) is a *left-sided confidence interval*. The lower confidence limit is given by LCL $= \bar{x} - t_\alpha \, s/\sqrt{n}$, where $\bar{x}$ is the sample mean and s is the sample standard deviation. Suppose we wish to find the upper confidence limit UCL such that the probability is $(1 - \alpha)$ that μ is less than UCL. The one-sided interval $(-\infty, \text{UCL})$ is a *right-sided confidence interval*. The upper confidence limit is given by UCL $= \bar{x} + t_\alpha \, s/\sqrt{n}$.

EXAMPLE 10.16   **One-Sided Confidence Interval for the Mean, Small Sample Size**

Suppose an official wants to use a one-sided confidence interval to estimate the mean of a normal population. Suppose we obtain a random sample of $n = 10$ observations. The population standard deviation is unknown. Suppose the sample mean is $\bar{x} = 14.5$ and the sample standard deviation is $s = 2.5$. Construct a 95% left-sided confidence interval for the population mean μ.

SOLUTION   In this problem, the population is assumed to be normal, and the population standard deviation is unknown. Because the sample size is small ($n \leq 30$), the left-sided confidence interval is based on the t distribution. We have $n = 10$, $\bar{x} = 14.5$, $s = 2.5$, and $(1 - \alpha) = .95$. Thus, $\alpha = .05$. From Table A.6, using 9 degrees of freedom, we obtain $t_{.05,9} = 1.833$. The desired lower confidence limit is

$$\text{LCL} = \bar{x} - t_\alpha \frac{s}{\sqrt{n}} = 14.5 - 1.833 \frac{2.5}{\sqrt{10}} = 13.051$$

The 95% left-sided confidence interval for μ is $(13.051, \infty)$. This means that we can be 95% confident that the population mean μ exceeds 13.051.

Tips for Problem Solving

To construct a confidence interval for the population mean, follow these steps:

1. Record the sample size n, the level of confidence $100(1 - \alpha)$, and the sample mean $\bar{x}$.

2. Record the population standard deviation σ (if it is known). Otherwise, record the sample standard deviation s.

3. If σ is known, find the Z scores $\pm z_{\alpha/2}$. If the population is approximately normal or if the sample size is at least 30, calculate the confidence interval

$$\left(\bar{x} \pm z_{\alpha/2} \frac{\sigma}{\sqrt{n}} \right)$$

4. If σ is unknown, estimate it using the sample standard deviation s. If the population is approximately normal, calculate the confidence interval

$$\left(\bar{x} \pm t_{\alpha/2} \frac{s}{\sqrt{n}} \right)$$

5. If the sample size is large and σ is unknown, the two confidence intervals

$$\left(\bar{x} \pm t_{\alpha/2} \frac{s}{\sqrt{n}} \right) \quad \text{and} \quad \left(\bar{x} \pm z_{\alpha/2} \frac{s}{\sqrt{n}} \right)$$

yield approximately the same result, so either can be used.

Exercises

Statistical Concepts

10.4.1   *True or false:*   Although the t distribution is the correct distribution to use whenever σ is not known (assuming the population is normal), when the number of degrees of freedom is large, we may use the standard normal distribution as an adequate approximation to the t distribution. For example, instead of using 1.984 in a 95% confidence interval based on 120 degrees of freedom, we can use the Z value 1.96. Explain.

10.4.2   State the formula for calculating a confidence interval for the population mean of a normal population having unknown standard deviation based on a small sample of observations.

10.4.3   To construct a confidence interval based on the t distribution, is any assumption made concerning the standard deviation of the population?

10.4.4   Suppose a confidence interval having level of confidence $(1 - \alpha)$ has been constructed for the mean of a normal population. Suppose the population standard deviation is unknown and has been estimated to be $s = 14$; the sample size is $n = 61$. Find the width of the confidence interval based on using the t distribution. Find the width of the confidence interval based on using the standard normal distribution.

10.4.5   Explain how the width of a confidence interval changes as the number of degrees of freedom increases, given a fixed level of confidence.

Statistical Drills

10.4.6   Let X be a normal random variable with unknown standard deviation. Suppose we take a random sample of n observations and obtain the sample mean $\bar{x} = 212$. Suppose the sample standard deviation is $s = 12.4$. Construct a 95% confidence interval for the population mean using the following sample sizes:

   a. 10   **b.** 20   **c.** 30

10.4.7   Let X be a normal random variable with unknown standard deviation. Suppose we take a random sample of n observations and obtain the sample mean $\bar{x} = 212$. Suppose the sample standard deviation is $s = 12.4$. Construct a 99% confidence interval for the population mean using the following sample sizes:

   a. 10   **b.** 20   **c.** 30

10.4.8   Let X be a normal random variable. Suppose a random sample of 21 observations is used to estimate the mean and standard deviation. What is the confidence level associated with each of the following confidence intervals for μ?

   a. $\bar{x} \pm 2.086 \dfrac{s}{\sqrt{n}}$ **b.** $\bar{x} \pm 1.725 \dfrac{s}{\sqrt{n}}$ **c.** $\bar{x} \pm 2.845 \dfrac{s}{\sqrt{n}}$

10.4.9   Let X be a normal random variable. Suppose we take a random sample of $n = 25$ observations and obtain the sample mean $\bar{x} = 136$ and sample standard deviation $s = 9.8$. Construct a confidence interval for the population mean having the following levels of confidence:

   a. 99%   **b.** 95%   **c.** 90%   **d.** 80%

   e. Plot these confidence intervals on a graph to show how the width of a confidence interval is affected by the level of confidence.

10.4.10   Let X be a normal random variable. Suppose we take a random sample of $n = 25$ observations and obtain the sample mean $\bar{x} = 89$ and sample standard deviation $s = 11.3$. Construct a left-sided confidence interval for the population mean having the following levels of confidence:

   a. 99%   **b.** 95%   **c.** 90%

10.4.11   A random sample of 91 observations from a population produced the following data:

$$\Sigma x = 2{,}734 \qquad \Sigma(x_i - \bar{x})^2 = 8{,}421$$

   a. Calculate the sample mean and sample standard deviation.

   b. Construct a 95% confidence interval for the population mean using the t distribution. Interpolate to find approximately the correct t value to use.

   c. Construct a 95% confidence interval for the population mean using the standard normal distribution. Compare with the answer from part **b.**

10.4.12   A random sample of 10 observations from a normal population produced a sample mean of 231.7 and a sample standard deviation of 9.3.

   a. Construct a 95% confidence interval for μ.

   b. Construct a 99% confidence interval for μ.

   c. Repeat parts **a** and **b** assuming the sample size is 121.

Statistical Applications

In each of the following problems, assume that the population from which the sample is selected is normal.

10.4.13   During a water shortage, a water company randomly sampled residential water meters to monitor daily water consumption. On one particular day, a sample of 30 meters showed a sample mean of $\bar{x} = 240$ gallons and a sample standard deviation of $s = 45$ gallons. Find a 90% confidence interval for the mean water consumption for the population.

10.4.14   To test a drug to be marketed for the treatment of an exotic virus, each patient was first given injections of the live virus until he or she was infected, and then treated with the drug. Because of the obvious dangers, only 15 volunteers could be obtained. One parameter of interest is the mean recovery time. The following recovery times (in days) were recorded:

$$
\begin{array}{ccccc}
10 & 11 & 12 & 17 & 22 \\
9 & 14 & 12 & 6 & 8 \\
12 & 9 & 14 & 11 & 18
\end{array}
$$

Find a 95% confidence interval for the mean recovery time.

10.4.15 An examination of the records for a random sample of 10 motor vehicles in a large fleet reveals the following operating costs (in cents per mile): 25.3, 27.3, 26.5, 27.0, 22.5, 23.5, 29.1, 26.8, 26.7, and 30.9. Construct a 90% confidence interval for μ. Assume that operating costs are normally distributed.

10.4.16 A chief of police is concerned about the speed of cars traveling over a stretch of highway on which there have been many accidents. A random sample of 36 automobiles showed an average speed of 59.4 miles per hour on this stretch and a sample standard deviation of 4.8 miles per hour. Find 90% and 99% confidence intervals for the mean speed of all cars traveling on this stretch of the highway.

10.4.17 A car rental company is interested in the amount of time its vehicles are out of operation for repair work. A random sample of 12 cars showed that, over the past year, the numbers of days each had been inoperative were as follows:

15, 11, 19, 24, 6, 18, 20, 15, 18, 12, 14, 19

Assuming that the population distribution is normal, find a 95% confidence interval for the mean number of days that vehicles in the company's fleet are out of operation.

10.4.18 An economist wants to determine the average annual income of full-time truck drivers. A random sample of 25 truck drivers is taken. We obtain $\bar{x} = \$20,000$ and $s = \$6,000$. Calculate the following confidence intervals for μ:
a. 95%   **b.** 90%

10.4.19 A random sample of 16 college teachers was taken to determine how much time they spent each week preparing lectures. The average number of hours was $\bar{x} = 10$ with $s^2 = 9$. Calculate the following confidence intervals for μ:
a. 95%   **b.** 95% assuming $n = 25$

10.4.20 During the past 20 years, the average number of days of rainfall annually in a northwestern city was $\bar{x} = 147$ with $s = 10$. Calculate the following confidence intervals for μ:
a. 95%   **b.** 99%

10.4.21 The manager of a grocery store wants to estimate the mean daily demand for bread. The following data are available (number of loaves sold per day): 148, 169, 118, 171, 168, 125, 134, 99, 142, 121. Assume that this is a random sample of daily demand and assume that the population is normal. Construct a 90% confidence interval for average daily demand for bread.

10.4.22 Construct a 90% left-sided confidence interval for the population mean. Suppose the population is normal. Suppose the sample size is $n = 10$, and assume $\bar{x} = 23.5$ and $s = 3.6$.

10.5 Confidence Intervals for Proportions (Large Samples)

In many problems in economics and business, we want to estimate the proportion of population members possessing some characteristic. For example, we may want to estimate the proportion of households with gross incomes below the poverty level or the proportion of individuals who are unemployed. Let p denote the proportion of items in a population that possess a certain characteristic. To estimate p, we take a random sample of n observations from the population and count the number X of items in the sample that possess the characteristic. The sample proportion $\hat{p} = X/n$ is used to estimate the population proportion p.

From Chapter 9, we know that $\hat{p}$ is a random variable having approximately a normal distribution (provided $np \geq 5$ and $nq \geq 5$) with mean p and variance pq/n where $q = (1 - p)$. In addition, the standardized random variable

$$Z = \frac{\hat{p} - p}{\sqrt{pq/n}}$$

follows the standard normal distribution. In Section 10.2, we showed that for the standard normal variable Z

$$P(-z_{\alpha/2} < Z < z_{\alpha/2}) = 1 - \alpha$$

After substituting for Z, we obtain

$$P\left(-z_{\alpha/2} < \frac{\hat{p} - p}{\sqrt{pq/n}} < z_{\alpha/2}\right) = 1 - \alpha$$

After multiplying through by the denominator, we obtain

$$P(-z_{\alpha/2}\sqrt{pq/n} < \hat{p} - p < z_{\alpha/2}\sqrt{pq/n}) = 1 - \alpha$$

Finally, a bit of algebra yields

$$P(\hat{p} - z_{\alpha/2}\sqrt{pq/n} < p < \hat{p} + z_{\alpha/2}\sqrt{pq/n}) = 1 - \alpha$$

This last probability statement tells us that the probability is $(1 - \alpha)$ that the random interval

$$(\hat{p} - z_{\alpha/2}\sqrt{pq/n}, \ \hat{p} + z_{\alpha/2}\sqrt{pq/n})$$

contains the true population proportion p. Because p and q are unknown, we replace their values by their sample estimates $\hat{p}$ and $\hat{q}$. Thus we can say that the probability is approximately $(1 - \alpha)$ that the random interval

$$(\hat{p} - z_{\alpha/2}\sqrt{\hat{p}\hat{q}/n}, \ \hat{p} + z_{\alpha/2}\sqrt{\hat{p}\hat{q}/n})$$

contains the true population proportion p.

DEFINITION   **Confidence Interval for the Population Proportion p** ───────────────

Let p denote the population proportion. Suppose we take a large random sample of n observations and obtain the sample proportion $\hat{p}$. A **confidence interval for the population proportion** having level of confidence $100(1 - \alpha)\%$ is given by

$$(\hat{p} - z_{\alpha/2}\sqrt{\hat{p}\hat{q}/n}, \ \hat{p} + z_{\alpha/2}\sqrt{\hat{p}\hat{q}/n})$$

where $z_{\alpha/2}$ is the value of the standard normal variable such that

$$P(Z > z_{\alpha/2}) = \alpha/2$$

EXAMPLE 10.17 **Confidence Interval for a Proportion** ————————————

A government study is designed to estimate the proportion of families having annual incomes below \$25,000 per year. A random sample of 500 families is contacted. In the sample, 200 families had annual incomes below \$25,000. Construct a 95% confidence interval for the population proportion p.

SOLUTION   The sample proportion is $\hat{p} = x/n = 200/500 = .4$ and $\hat{q} = (1 - \hat{p}) = .6$. For a 95% confidence interval, we have $\alpha = .05$ and $\alpha/2 = .025$. We obtain $z_{\alpha/2} = 1.96$. The 95% confidence interval for p is given by

$$(\hat{p} - z_{\alpha/2}\sqrt{\hat{p}\hat{q}/n}, \ \hat{p} + z_{\alpha/2}\sqrt{\hat{p}\hat{q}/n})$$
$$(.4 - 1.96\sqrt{(.4)(.6)/500}, \ .4 + 1.96\sqrt{(.4)(.6)/500})$$
$$(.357, .443)$$

This means that we can be 95% confident that the true population proportion is between .357 and .443.

One-Sided Confidence Intervals for the Population Proportion

Suppose we need a one-sided confidence interval that has either no upper or no lower limit. For example, suppose we take a random sample of n observations from some population having unknown proportion p. Suppose we wish to find the lower confidence limit LCL such that the probability is $(1 - \alpha)$ that p exceeds LCL. The one-sided interval (LCL, 1.00) is a *left-sided confidence interval.* The lower confidence limit is given by LCL $= \hat{p} - z_\alpha\sqrt{\hat{p}(1 - \hat{p})/n}$, where $\hat{p}$ is the sample proportion. Suppose we wish to find the upper confidence limit UCL such that the probability is $(1 - \alpha)$ that p is less than UCL. The one-sided interval (0, UCL) is a *right-sided confidence interval.* The upper confidence limit is given by UCL $= \hat{p} + z_\alpha\sqrt{\hat{p}(1 - \hat{p})/n}$.

EXAMPLE 10.18 **One-Sided Confidence Interval for a Population Proportion** ———————

Construct a right-sided 95% confidence interval for the proportion of defective items produced by a machine if 16 items are found to be defective in a random sample of 100 items.

SOLUTION   We have $n = 100$, $\hat{p} = {}^{16}\!/_{100} = .16$, and $(1 - \alpha) = .95$. Thus, $\alpha = .05$. From Table A.5, we obtain $z_{.05} = 1.645$. The desired upper confidence limit is

$$\text{UCL} = \hat{p} + z_\alpha\sqrt{\hat{p}(1 - \hat{p})/n} = .16 + 1.645\sqrt{.16(.84)/100} = .2203$$

The 95% right-sided confidence interval for p is (0, .2203). This means that we can be 95% confident that the population proportion is less than .2203.

Tips for Problem Solving

> To construct a confidence interval for the population proportion, follow these steps:
>
> **1.** Record the sample size n, the level of confidence $100(1 - \alpha)$, and the sample proportion $\hat{p}$. Record $\hat{q} = (1 - \hat{p})$.
>
> **2.** Find the Z scores $\pm z_{\alpha/2}$.
>
> **3.** Calculate the confidence interval
>
> $$(\hat{p} \pm z_{\alpha/2}\sqrt{\hat{p}\hat{q}/n})$$

Exercises

Statistical Concepts

10.5.1 State the formula for calculating a confidence interval for the population proportion.

10.5.2 To construct a confidence interval for the population proportion, what assumption is made concerning the sample size?

10.5.3 *True or false:* Let LCL and UCL denote the lower and upper boundaries of a confidence interval for the population proportion. The width of the confidence interval is $W = UCL - LCL$. Explain.

10.5.4 *True or false:* Let LCL and UCL denote the lower and upper confidence limits of a confidence interval for the population proportion. Then the sample proportion is $(LCL + UCL)/2$. Explain.

10.5.5 The width of a confidence interval for the population proportion depends on three factors. What are they, and how do they affect the width of a confidence interval?

10.5.6 Suppose a confidence interval for the population proportion having level of confidence $(1 - \alpha)$ has been constructed. Suppose the width of the confidence interval is $W = .08$ and the sample size is $n = 100$. How large should the sample size be to decrease the width of the confidence interval to $W = .04$?

10.5.7 *True or false:* Other things being equal, if we increase the sample size by a factor of 10, the width of a confidence interval for the population proportion will be multiplied by .2. Explain.

10.5.8 Explain how to construct a left-sided confidence interval for the population proportion.

10.5.9 *True or false:* Suppose we construct a two-sided and a left-sided confidence interval for the population proportion. In each case, suppose the level of confidence is $(1 - \alpha)$. The lower confidence limit for the two-sided interval exceeds the lower confidence limit for the left-sided interval. Explain.

Statistical Drills

10.5.10 *True or false:* Estimate the population proportion in the following cases:
   a. At a chemical plant, a simple random sample of 50 employees is taken. Of these, 32 have been to a doctor within the past year. Estimate the proportion of all employees who have been to a doctor.
   b. The governor wants to raise taxes to improve the state's system of public parks. In a random sample of 800 residents, 150 favor raising taxes. Estimate the proportion of all residents who favor raising taxes for this purpose.
   c. In a random sample of 382 coal miners, 107 favor going on strike for higher wages. Estimate the proportion of all coal miners who favor going on strike.

10.5.11 A random sample of size $n = 400$ yielded $\hat{p} = .34$.
  a. Is the sample size large enough to apply the methods discussed in this section for constructing a confidence interval for p? Explain.
  b. Construct a 95% confidence interval for p.

10.5.12 A random sample of size n yielded $\hat{p} = .34$. Construct a 95% confidence interval for p based on the following sample sizes:
  a. 20   **b.** 40   **c.** 80   **d.** 160   **e.** 320
  f. Plot these confidence intervals on a graph to show how the width of a confidence interval is affected by the sample size.

10.5.13 A random sample of size $n = 100$ yielded $\hat{p} = .64$. Construct a confidence interval for p based on the following level of confidence:
  a. 99%   **b.** 95%   **c.** 90%   **d.** 80%   **e.** 50%
  f. Plot these confidence intervals on a graph to show how the width of a confidence interval is affected by the level of confidence.

10.5.14 Suppose we take a random sample of $n = 100$ observations and obtain the sample proportion $\hat{p} = .73$. Construct a left-sided confidence interval for the population mean having the following levels of confidence:
  a. 99%   **b.** 95%   **c.** 90%

10.5.15 Suppose we take a random sample of $n = 64$ observations and obtain the sample proportion $\hat{p} = .29$. Construct a right-sided confidence interval for the population mean having the following levels of confidence:
  a. 99%   **b.** 95%   **c.** 90%

Statistical Applications

10.5.16 In a sample of 900 voters, 400 prefer candidate A. Find a 90% confidence interval for p.

10.5.17 Suppose that 55% of the people in a random sample of n people favor a law requiring stricter enforcement of traffic laws. Find a 95% confidence interval for p for the following sample sizes:
  a. $n = 100$
  b. $n = 400$
  c. $n = 1,600$
  d. Discuss how the width of the confidence interval changes when the sample size is multiplied by 4.

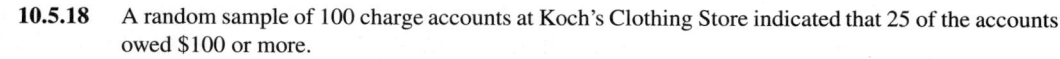

10.5.18 A random sample of 100 charge accounts at Koch's Clothing Store indicated that 25 of the accounts owed $100 or more.
  a. Find an 80% confidence interval for the proportion of charge accounts owing $100 or more.
  b. Find a 95% confidence interval for p.

10.5.19 A private consulting firm is given a governmental grant to study underreporting of income on federal income tax returns. The firm takes a random sample of 160 taxpayers and determines that 64 of them underreported their income on their tax return.
  a. Construct a 90% confidence interval for the proportion of all taxpayers who underreport their income.
  b. Construct a 99% confidence interval.

10.5.20 One hundred economists were randomly selected to get information about their reading habits. In the sample, 75 said that they read the *Wall Street Journal* regularly.
  a. Construct a 90% confidence interval for the proportion of all economists who read the *Wall Street Journal*.
  b. Construct a 98% confidence interval for p.

10.5.21   In a random sample of 100 students at a particular college, 60 indicated that they favored having the option of receiving pass–fail grades for elective courses. Obtain a 95% confidence interval for the proportion of the population of students who favor pass–fail grades for elective courses. Does this confidence interval contain the value $p = .5$? Explain why this particular value might be of interest.

10.5.22   An airline wants to determine the proportion of passengers that bring only carry-on luggage. In a random sample of 200 passengers, 44 passengers have only carry-on luggage. Find a 95% confidence interval for the proportion of passengers who have only carry-on luggage.

10.5.23   A random sample of 400 faculty members at a certain university contained 120 people who believed that the university should curtail its investments in South Africa. On the basis of this sample information, an analyst calculated the confidence interval (.25, .35) for the population proportion of faculty members favoring curtailment. What is the level of confidence of this interval?

10.5.33   Last year, a random sample of 1,025 student apartments indicated that the mean repair cost at the end of the year was $184.56; the sample standard deviation was $61.23. Construct a 98% confidence interval for the mean repair cost of all student apartments.

10.6   Determining the Sample Size

So far we have developed techniques for finding confidence intervals for a population mean or a population proportion based on a sample of information. At the conclusion of such a process, the researcher may think that the resulting confidence interval is too wide—that there is too much uncertainty about the parameter being estimated. For a fixed level of confidence, the only way to obtain a narrower confidence interval is to increase the sample size.

Sometimes the researcher will fix in advance the desired width of the confidence interval and then choose a sample size large enough to obtain it. In this section, we show how this is done. First, we discuss confidence intervals for the mean, and then for the population proportion.

Confidence Intervals for the Mean

Suppose an individual is interested in estimating the mean of a population having a known variance σ^2. How large a sample must be taken if the investigator wants the probability to be $(1 - \alpha)$ that the sampling error $|\bar{X} - \mu|$ is less than some amount D?

Recall that a $100(1 - \alpha)\%$ confidence interval for μ is given by

$$(\bar{x} - z_{\alpha/2}\sigma/\sqrt{n}, \ \bar{x} + z_{\alpha/2}\sigma/\sqrt{n})$$

This confidence interval is centered at the sample mean $\bar{x}$ and extends a distance

$$D = \frac{z_{\alpha/2}\sigma}{\sqrt{n}}$$

on each side of the sample mean, so D is half the width of the confidence interval. The investigator wants to fix D in advance and use a large enough sample size to guarantee that the confidence interval does not extend more than D units from the sample mean. If we solve this equation for n, we obtain

$$\sqrt{n} = \frac{z_{\alpha/2}\sigma}{D}$$

By squaring both sides of this equation, we obtain

$$n = \frac{z_{\alpha/2}^2 \sigma^2}{D^2}$$

Therefore choosing a sample of this size guarantees that the confidence interval extends a distance D on each side of the sample mean.

Sample size for a specific confidence interval for the mean

Suppose we take a random sample from a normal population with known variance σ^2. Then a $100(1 - \alpha)\%$ confidence interval for the population mean extends a distance D on each side of the sample mean if the number of observations is

$$n = \frac{z_{\alpha/2}^2 \sigma^2}{D^2}$$

The number of observations in a sample must be an integer, of course, so if the value of n in the preceding formula is not an integer, we round up to the next whole number to guarantee that our confidence interval does not exceed the required width. Note that specifying a smaller D requires a larger sample.

When σ^2 is unknown, as will usually be the case, it may be necessary to estimate its value by taking a small sample called a *pilot sample*. Otherwise, it may be necessary to make an educated guess.

EXAMPLE 10.19   **Sample Size Required to Estimate a Mean** ───────────────────

An economist wants to estimate the mean annual income of households in a particular congressional district. It is assumed that the population standard deviation is $\sigma = \$4,000$. The economist wants the probability to be .95 that the sample mean will be within $D = \$500$ of the true mean μ. How large a sample is required?

SOLUTION   We have $(1 - \alpha) = .95$, $z_{\alpha/2} = 1.96$, $D = 500$, and $\sigma = \$4,000$. We obtain

$$n = \frac{z_{\alpha/2}^2 \sigma^2}{D^2} = \frac{1.96^2 (4,000^2)}{500^2} = 245.86$$

To satisfy the requirement, a sample of at least 246 observations is needed.

Confidence Intervals for a Proportion

We proceed in exactly the same way if we want to determine the sample size needed to estimate a proportion p with a specified precision D at a certain level of confidence. At the level of confidence $(1 - \alpha)$, a confidence interval for p is given by

$$(\hat{p} - z_{\alpha/2}\sqrt{\hat{p}\hat{q}/n}, \ \hat{p} + z_{\alpha/2}\sqrt{\hat{p}\hat{q}/n})$$

This confidence interval is centered at the sample proportion $\hat{p}$ and extends a distance

$$D = z_{\alpha/2}\sqrt{\hat{p}\hat{q}/n}$$

on each side of the sample proportion. The distance D is half the width of the confidence interval. Now suppose that the investigator wants to fix D in advance. If we solve this equation for n, we obtain

$$\sqrt{n} = \frac{z_{\alpha/2}\sqrt{\hat{p}\hat{q}}}{D}$$

After squaring both sides, we obtain

$$n = \frac{z_{\alpha/2}^2 \hat{p}\hat{q}}{D^2}$$

This equation cannot be used directly because it involves the sample proportion $\hat{p}$, which will not be known at the outset of the investigation. If we have an initial estimate of p, substitute this value for $\hat{p}$ in the equation. If there is no such estimate available, substitute .5 for $\hat{p}$ in the formula. The product $\hat{p}\hat{q}$ cannot exceed .25 (the value when $\hat{p} = .5$), so the largest possible value for n is

$$n = \frac{.25\, z_{\alpha/2}^2}{D^2}$$

This equation shows the largest sample needed so that the probability is $(1 - \alpha)$ that the sampling error will be less than or equal to some amount D.

Sample size for a specific confidence interval for the proportion

> Suppose we take a random sample from some population. Then a $100(1 - \alpha)\%$ confidence interval for the population proportion extends at most a distance D on each side of the sample proportion if the number of observations is
>
> $$n = \frac{.25\, z_{\alpha/2}^2}{D^2}$$
>
> If an estimate of p exists, use the formula
>
> $$n = \frac{z_{\alpha/2}^2 \hat{p}\hat{q}}{D^2}$$

EXAMPLE 10.20   **Sample Size Required to Estimate a Proportion** ————————————

A polling agency is interested in predicting what proportion of votes a certain presidential candidate will receive. How large a sample must be taken if the agency wants to be 95% confident that its estimate is within .03 of the correct value?

SOLUTION   We have $(1 - \alpha) = .95$, $z_{\alpha/2} = 1.96$, and $D = .03$. We use $\hat{p} = \hat{q} = .5$ for lack of a better estimate. We obtain

$$n = \frac{.25\, z_{\alpha/2}^2}{D^2}$$

$$= \frac{.25(1.96^2)}{.03^2} = 1{,}067.1$$

The polling agency should sample 1,068 voters to obtain an estimate that is within .03 of the population proportion. If an estimate $\hat{p}$ of the population proportion had been available, we would have used it in the formula for n.

The sample size formula contains no information about the cost of obtaining additional sample information. Eventually a point is reached where the cost of obtaining additional information outweighs the benefits obtained from that information.

Exercises

Statistical Concepts

10.6.1   State the formula for calculating the sample size needed to construct a confidence interval for the population mean μ. You want the probability to be $(1 - \alpha)$ that your estimate is within a distance D of μ.

10.6.2   State the formula for calculating the sample size needed to construct a confidence interval for the population proportion p. You want the probability to be $(1 - \alpha)$ that your estimate is within a distance D of p.

10.6.3   You want to construct a confidence interval for μ that has width W and level of confidence $(1 - \alpha)$. Explain what happens to the required sample size if you make the desired width $W/2$.

Statistical Drills

10.6.4   Suppose you wish to be 95% confident that the sample mean is within $D = 2.1$ units of the population mean μ. Suppose you know that the population standard deviation is $\sigma = 8$. How many observations should be included in your sample?

10.6.5   Suppose you want to estimate a population mean with a 95% confidence interval of width $W = .2$ unit. Suppose you know that the population standard deviation is $\sigma = .5$ unit. How many observations should be included in your sample?

10.6.6   Suppose you want to construct a 95% confidence interval for the mean of a normal population. Suppose you know the population standard deviation is $\sigma = 8.4$ units.
   a. To see how increases in the sample size cause the width of the confidence interval to decrease, calculate the width W of the confidence interval for $n = 16, 25, 64, 100, 400$.
   b. Plot the width of the confidence interval against the sample size. Draw a smooth curve through the plotted points and observe how the width decreases as n increases.

10.6.7   Suppose you want the probability to be .95 that your sample mean is within a distance $D = 2.3$ units of the population mean μ. You do not know the population standard deviation σ, but you have information concerning the range. Sometimes σ is estimated based on information about the range (R) of the data. Many statisticians assume that the range of the data is approximately equal to 6σ. That is, $R \approx 6\sigma$. Suppose the range is known to be approximately 12.
   a. Obtain an estimate of σ.
   b. Find the approximate sample size that will produce the desired confidence interval.
   c. Repeat parts **a** and **b** but use the approximate relationship $R \approx 4\sigma$.

10.6.8   Find the approximate sample size necessary to estimate a binomial proportion p correct to within .02 with probability equal to .95.
   a. Assume that you have information that p is approximately .64.
   b. Assume that you have no information about p.

10.6.9 Find the approximate sample size necessary to estimate a binomial proportion p correct to within .04 with probability equal to .95.
   a. Assume that you have information that p is approximately .64.
   b. Assume that you have no information about p.

10.6.10 The following is a 95% confidence interval for p: (.34, .46).
   a. Find the sample proportion.
   b. How large was the sample used to construct this interval?

Statistical Applications

10.6.11 Starting annual salaries for college graduates with business administration degrees are believed to have a standard deviation of approximately $1,800. A 95% confidence interval estimate of the mean annual starting salary is desired. How large a sample should be taken if we want to be 95% confident that the maximum sampling error is to be as follows?
   a. $500   **b.** $200

10.6.12 The percentage of defective products produced by a certain method is thought to be about 5%. How large a sample is required if we want to be
   a. 99% confident that p is within .01 of $\hat{p}$?
   b. 99% confident that p is within .02 of $\hat{p}$?
   c. 90% confident that p is within .02 of $\hat{p}$?

10.6.13 A survey is to be taken to determine the average age of purchasers of *Newsweek* magazine. A preliminary study yielded an estimated variance of $\sigma^2 = 64$. How large a sample is required if we want to be
   a. 95% confident that μ is within 4 units of $\bar{x}$?
   b. 95% confident that μ is within 2 units of $\bar{x}$?
   c. 99% confident that μ is within 2 units of $\bar{x}$?

10.6.14 Radio station WRKC has an evening sports talk show that gets a lot of publicity because the host of the show is very popular. The host wants a raise because the show is so popular. Before granting a raise, WRKC executives want to know what proportion of the public listens to the show. How large a sample is required if they want to be 95% confident that their estimate is within .02 of the true proportion? Assume that p is approximately .2.

10.6.15 The State Unemployment Commission wants to estimate the proportion of the labor force that was unemployed during any part of last year. How large a sample is required if the commission wants to be 90% confident that the estimate is within .01 of the true proportion? Assume that p is approximately .1.

10.6.16 A school system wants to order 1,000 new desks, some of which will be designed for left-handed people. The purchasing agent must know what proportion of the population is left-handed. How large a sample is required to be 90% confident that the estimate is within .02 of the true value? Assume that p is approximately .2.

10.6.17 A well-known bank credit card firm is interested in estimating the proportion of credit card holders that carry a nonzero balance at the end of the month and thus incur an interest charge. The desired precision for the proportion estimate is $\pm 3\%$ at a 99% confidence level.
   a. How large a sample should be recommended if it is anticipated that roughly 70% of the firm's cardholders carry a nonzero balance at the end of the month?
   b. How large a sample would be recommended if no planning value for the population proportion could be specified?

10.6.18 A sample of 300 people was asked to identify their major source of news; 180 stated that their major source was television.

a. Construct a 95% confidence interval for the proportion of the people in the population for whom television is the major source of news.

b. How large a sample would be necessary to estimate the population proportion with a sampling error of .05 or less at a 95% confidence level?

10.6.19 A nationwide survey of practicing physicians is to be undertaken to estimate the mean number of prescriptions written per day. The desired margin of sampling error is $\pm.5$ with a 99% confidence coefficient. A pilot study revealed that a reasonable planning value for the population standard deviation is 7.

a. How many physicians should be contacted in the survey to estimate μ?

b. If the desired confidence coefficient were lowered to 95%, would the required sample size be substantially reduced?

10.7   Confidence Intervals for the Difference of Two Means

Confidence Intervals for $(\mu_1 - \mu_2)$ When the Variances Are Known or the Sample Sizes Are Large

At times we may be interested in constructing confidence intervals for the difference between two population means $(\mu_1 - \mu_2)$. The techniques discussed here are based on the assumption that we have two independent random samples from two normal populations with known variances. Alternatively, the results hold at least approximately when the sample sizes are large even though the populations are not normal and the variances have been estimated. How large n_1 and n_2 have to be to make the results approximately correct varies from problem to problem. The approximation is fairly good provided n_1 and n_2 are both at least 30.

Confidence interval for $(\mu_1 - \mu_2)$ when variances are known or sample sizes are large—independent samples

> Suppose we have independent random samples of size n_1 and n_2 from two normal populations having unknown means μ_1 and μ_2 and known variances σ_1^2 and σ_2^2. If the observed sample means are $\bar{x}_1$ and $\bar{x}_2$, a $100(1 - \alpha)\%$ confidence interval for $(\mu_1 - \mu_2)$ is given by
>
> $$\left((\bar{x}_1 - \bar{x}_2) - z_{\alpha/2} \sqrt{\frac{\sigma_1^2}{n_1} + \frac{\sigma_2^2}{n_2}}, \; (\bar{x}_1 - \bar{x}_2) + z_{\alpha/2} \sqrt{\frac{\sigma_1^2}{n_1} + \frac{\sigma_2^2}{n_2}} \right)$$
>
> If the sample sizes are large, say, 30 or more, and the population variances are unknown, then to a good approximation, a $100(1 - \alpha)\%$ confidence interval for $(\mu_1 - \mu_2)$ is obtained by replacing the population variances in the previous confidence interval by the corresponding sample variances s_1^2 and s_2^2. For large sample sizes, this approximation will usually be adequate even if the population distributions are not normal.

EXAMPLE 10.21   **Confidence Interval for $(\mu_1 - \mu_2)$ with Large Sample Sizes** ————————

In a sex discrimination case, an employee alleged that a large corporation paid men more than women for comparable work. Let population 1 represent all male employees performing certain jobs and population 2 all females performing comparable jobs at the corporation. Independent samples are taken of $n_1 = 100$ males and $n_2 = 100$ females; the sample means are $\bar{x}_1 = \$20,600$ and $\bar{x}_2 = \$19,700$, and the sample standard deviations are $s_1 = \$3,000$ and $s_2 = \$2,500$. Construct a 95% confidence interval for $(\mu_1 - \mu_2)$.

SOLUTION   Because the sample sizes are large, we can use the sample variances in place of the population variances and obtain the confidence interval by using the formula

$$\left((\bar{x}_1 - \bar{x}_2) - z_{\alpha/2} \sqrt{\frac{s_1^2}{n_1} + \frac{s_2^2}{n_2}}, \ (\bar{x}_1 - \bar{x}_2) + z_{\alpha/2} \sqrt{\frac{s_1^2}{n_1} + \frac{s_2^2}{n_2}} \right)$$

We have $z_{\alpha/2} = 1.96$. We obtain $(\bar{x}_1 - \bar{x}_2) = (20,600 - 19,700) = 900$. The confidence interval is

$$\left(900 - 1.96 \sqrt{\frac{3,000^2}{100} + \frac{2,500^2}{100}}, \ 900 + 1.96 \sqrt{\frac{3,000^2}{100} + \frac{2,500^2}{100}} \right)$$

$$(\$134.60, \ \$1,665.40)$$

Because this interval contains only positive values, we can be quite confident that $(\mu_1 - \mu_2) > 0$. Thus, it is reasonable to assume that the mean salary for males exceeds the mean salary for females.

Confidence Intervals for $(\mu_1 - \mu_2)$ When Variances Are Unknown and Sample Sizes Are Small

We now consider the case when the sample sizes are small, the population variances are unknown, and the populations are normal. The problem of finding a confidence interval has been solved for the special case when the unknown variances are equal. A general solution has not been found for the case where the unknown variances are unequal.

If we assume that the population variances are equal, then it is useful to pool the data from the two random samples to obtain one estimate of the common population variance σ^2. Let $\bar{x}_1$ and $\bar{x}_2$ denote the observed sample means, and s_1^2 and s_2^2 the observed sample variances. Let s_p^2 denote the pooled estimate of the common population variance. The pooled variance s_p^2 is given by

$$s_p^2 = \frac{(n_1 - 1)s_1^2 + (n_2 - 1)s_2^2}{n_1 + n_2 - 2}$$

Since the population variances are unknown, we estimate them by the pooled estimate s_p^2. The pooled sample variance s_p^2 is a weighted average of the two sample variances s_1^2 and s_2^2 with the weights being $(n_1 - 1)/(n_1 + n_2 - 2)$ and $(n_2 - 1)/(n_1 + n_2 - 2)$.

Confidence interval for $(\mu_1 - \mu_2)$ when variances are unknown but equal—independent samples

> Suppose we have independent random samples of n_1 and n_2 observations from normal populations with means μ_1 and μ_2 and a common variance σ^2. If the observed sample means are $\bar{x}_1$ and $\bar{x}_2$, then a $100(1 - \alpha)\%$ confidence interval for $(\mu_1 - \mu_2)$ is based on the t distribution and is given by
>
> $$\left((\bar{x}_1 - \bar{x}_2) - t_{\alpha/2,\nu} \sqrt{\frac{s_p^2}{n_1} + \frac{s_p^2}{n_2}}, \ (\bar{x}_1 - \bar{x}_2) + t_{\alpha/2,\nu} \sqrt{\frac{s_p^2}{n_1} + \frac{s_p^2}{n_2}} \right)$$
>
> where the number of degrees of freedom is $\nu = (n_1 + n_2 - 2)$ and s_p^2 is the pooled estimate of the common variance.

EXAMPLE 10.22   **Confidence Interval for $(\mu_1 - \mu_2)$ with Small Sample Sizes**

Two manufacturing companies produce carbide drill tips that are used to cut holes in steel sheets. A customer wishing to know which drill tips have the longer life purchases independent samples of $n_1 = 20$ drill tips from company 1 and $n_2 = 15$ drill tips from company 2. The mean lifelengths of the drill tips are $\bar{x}_1 = 78$ minutes and $\bar{x}_2 = 84$ minutes. The population variances are unknown but assumed to be equal. The sample variances are $s_1^2 = 41$ and $s_2^2 = 36$. Construct a 95% confidence interval for $(\mu_1 - \mu_2)$.

SOLUTION   We obtain the pooled variance estimate s_p^2 as follows:

$$s_p^2 = \frac{(20 - 1)(41) + (15 - 1)(36)}{20 + 15 - 2} = \frac{1{,}283}{33} = 38.88$$

For a 95% confidence interval, we have $\alpha = .05$ and $\alpha/2 = .025$. The degrees of freedom parameter is $\nu = (n_1 + n_2 - 2) = (20 + 15 - 2) = 33$. For 33 degrees of freedom, we obtain approximately $t_{\alpha/2,\nu} \approx 2.04$. The 95% confidence interval for $(\mu_1 - \mu_2)$ is

$$\left((\bar{x}_1 - \bar{x}_2) - t_{\alpha/2,\nu} \sqrt{\frac{s_p^2}{n_1} + \frac{s_p^2}{n_2}}, \ (\bar{x}_1 - \bar{x}_2) + t_{\alpha/2,\nu} \sqrt{\frac{s_p^2}{n_1} + \frac{s_p^2}{n_2}} \right)$$

The left endpoint of the interval is

$$(78 - 84) - 2.04 \sqrt{38.88/20 + 38.88/15} = -10.34$$

The right endpoint is

$$(78 - 84) + 2.04 \sqrt{38.88/20 + 38.88/15} = -1.66$$

The 95% confidence interval extends from -10.34 to -1.66. This means that we can be 95% confident that the average lifelength of tips for company 2 exceeds the average lifelength of tips for company 1 by somewhere between 1.66 and 10.34 minutes.

Exercises

Statistical Concepts

10.7.1 *True or false:* Suppose a 95% confidence interval for $(\mu_1 - \mu_2)$ is (3.54, 5.61). Then the evidence supports the hypothesis that μ_1 exceeds μ_2. Explain.

10.7.2 Explain how to construct a confidence interval for the difference of two population means when the population standard deviations are unknown but the sample sizes are large.

10.7.3 Explain how to construct a confidence interval for the difference of two population means when the population standard deviations are unknown and the sample sizes are small. What assumption is made about the population standard deviations? What assumption is made about the population distributions?

10.7.4 *True or false:* Suppose a 95% confidence interval for $(\mu_1 - \mu_2)$ does not contain the value 0. Then the evidence supports the hypothesis that $\mu_1 \neq \mu_2$. Explain.

10.7.5 Suppose we want to construct a confidence interval for $(\mu_1 - \mu_2)$ and the sample sizes are small. Explain how to estimate the standard deviation of the sampling distribution of $(\bar{X}_1 - \bar{X}_2)$ based on the assumption that the two population standard deviations are equal.

Statistical Drills

10.7.6 Suppose that we have obtained independent random samples from two populations with means μ_1 and μ_2. The sample sizes, sample means, and sample standard deviations are as follows:

$$n_1 = 80 \quad n_2 = 80 \quad \bar{x}_1 = 96.3 \quad \bar{x}_2 = 94.2 \quad s_1 = 6.1 \quad s_2 = 5.7$$

a. Construct a 95% confidence interval for $(\mu_1 - \mu_2)$.
b. Construct a 99% confidence interval for $(\mu_1 - \mu_2)$.

10.7.7 Suppose that we have obtained independent random samples from two normal populations with means μ_1 and μ_2. The sample sizes, sample means, and sample standard deviations are as follows:

$$n_1 = 20 \quad n_2 = 20 \quad \bar{x}_1 = 27.2 \quad \bar{x}_2 = 29.3 \quad s_1 = 5.9 \quad s_2 = 4.8$$

It is assumed that the population standard deviations are equal.
a. Construct a 95% confidence interval for $(\mu_1 - \mu_2)$.
b. Construct a 99% confidence interval for $(\mu_1 - \mu_2)$.

Statistical Applications

10.7.8 A school board wants to determine how the mean IQ of the students at inner city schools compares with the mean IQ of students at suburban schools. Independent random samples of 80 students are obtained from each school location. At inner city schools, the sample mean IQ is 105 and the sample standard deviation is 11. At suburban schools, the sample mean IQ is 101 and the standard deviation is 9.
a. Compute a 90% confidence interval for the difference between the mean IQs.
b. Compute a 99% confidence interval for this difference.

10.7.9 A department store sends out monthly statements to its credit customers. In the past, it has not enclosed a preaddressed envelope for returning payments with these statements. In a random sample of 100 of these accounts, the mean time to payment was 11.4 days and the sample standard deviation was 3.3 days. As an experiment, the company sent preaddressed envelopes with the accounts of 100 randomly and independently selected customers. For this sample, the mean time to payment was 8.2 days and the standard deviation was 3.3 days. Find a 95% confidence interval for the difference between the two population means.

10.7.10 A farmer wants to determine whether different types of feed can influence the mean number of eggs that hens lay per month. In a random sample of 100 hens that ate feed 1, the average number of eggs per month was $\bar{x}_1 = 15.2$ and the variance was 4. In a random sample of 100 hens that ate feed 2, the average number of eggs per month was $\bar{x}_2 = 14$ and the variance was 4. Construct a 95% confidence interval for $(\mu_1 - \mu_2)$.

10.7.11 A national restaurant franchise buys bakery goods from two suppliers who specialize in making cherry pies. The manager of a franchise wants to know whether the two bakers are filling the pies with the same quantity of cherries on the average. It is assumed that the unknown population variances are equal and that the populations are normal. The manager takes a random sample of 40 cherry pies, 25 from the first baker and 15 from the second. The quantity of cherries in each pie is weighed. The results are: $\bar{x}_1 = 11.6$ ounces, $s_1 = 1.8$, $\bar{x}_2 = 12.2$ ounces, and $s_2 = 2.0$. Construct a 95% confidence interval for $(\mu_1 - \mu_2)$.

10.8 Confidence Intervals for the Difference of Two Population Proportions

In Section 10.5, we showed how to construct a confidence interval for a single population proportion, but sometimes we are interested in comparing two population proportions. For example, in a large corporation we might be interested in comparing the proportion of male college students who eventually graduate to the proportion of female students who eventually graduate. In this section, we show how to obtain confidence intervals for the difference between two population proportions when independent large samples are taken from each population.

To construct a confidence interval for $(p_1 - p_2)$, the difference between two population proportions, follow the procedure in the box.

Confidence interval for $(p_1 - p_2)$

> Let $\hat{p}_1$ denote the observed proportion of successes in a random sample of n_1 observations from a population with proportion p_1 successes, and let $\hat{p}_2$ denote the observed proportion of successes in an independent random sample of n_2 observations from a population with proportion p_2 successes. A $100(1 - \alpha)\%$ confidence interval for $(p_1 - p_2)$ is given by the interval
>
> $$\left((\hat{p}_1 - \hat{p}_2) - z_{\alpha/2} \sqrt{\frac{\hat{p}_1\hat{q}_1}{n_1} + \frac{\hat{p}_2\hat{q}_2}{n_2}}, \; (\hat{p}_1 - \hat{p}_2) + z_{\alpha/2} \sqrt{\frac{\hat{p}_1\hat{q}_1}{n_1} + \frac{\hat{p}_2\hat{q}_2}{n_2}} \right)$$
>
> This result holds provided $n_1 p_1 \geqslant 5$, $n_1 q_1 \geqslant 5$, $n_2 p_2 \geqslant 5$, and $n_2 q_2 \geqslant 5$.

EXAMPLE 10.23 **Confidence Interval for $(p_1 - p_2)$** ————————————————

An employee alleged that a large corporation was discriminating against older employees by illegally terminating their employment. The company replied that, because of a decline in sales, it was necessary to cut the labor force by terminating employees. The company agreed that many older employees were terminated but argued that many younger employees were also terminated. In court, the opposing attorneys argued about whether the proportions of young and old employees who were terminated were equal. In a random sample of 100 young employees, 35 were terminated. In an independent sample of 100 old employees, 30 were terminated. Thus, we have $\hat{p}_1 = .35$ and $\hat{p}_2 = .30$. Find a 99% confidence interval for $(p_1 - p_2)$, the difference between the two population proportions.

SOLUTION   We have $(1 - \alpha) = .99$, so $\alpha = .01$ and $\alpha/2 = .005$. We obtain $z_{\alpha/2} = 2.58$. The confidence interval is

$$\left((.35 - .30) - 2.58 \sqrt{\frac{(.35)(.65)}{100} + \frac{(.30)(.70)}{100}}, (.35 - .30) + 2.58 \sqrt{\frac{(.35)(.65)}{100} + \frac{(.30)(.70)}{100}} \right)$$

$$(-.12, .22)$$

Thus, we are 99% confident that the true difference between p_1 and p_2 is between $-.12$ and $.22$. Since this interval contains the value 0, we cannot conclude that the two population proportions are different.

Exercises

Statistical Concepts

10.8.1   State the formula for calculating a confidence interval for the difference between two population proportions.

10.8.2   Suppose you wish to construct a confidence interval for the difference of two population proportions. What assumptions are necessary to apply the techniques discussed in this section?

10.8.3   *True or false:*   Suppose a 95% confidence interval for $(p_1 - p_2)$ does not contain the value 0. This can be viewed as evidence against the hypothesis that $p_1 = p_2$. Explain.

Statistical Drills

10.8.4   Suppose that we have obtained independent random samples from two populations with proportions p_1 and p_2. The sample sizes and sample proportions are as follows:

$$n_1 = 80 \quad n_2 = 80 \quad \hat{p}_1 = .40 \quad \hat{p}_2 = .43$$

a. Construct a 95% confidence interval for $(p_1 - p_2)$.
b. Construct a 99% confidence interval for $(p_1 - p_2)$.

10.8.5   Suppose that we have obtained independent random samples from two populations with proportions p_1 and p_2. The sample sizes and sample proportions are as follows:

$$n_1 = 64 \quad n_2 = 25 \quad \hat{p}_1 = .50 \quad \hat{p}_2 = .44$$

a. Construct a 95% confidence interval for $(p_1 - p_2)$.
b. Construct a 99% confidence interval for $(p_1 - p_2)$.

10.8.6   Suppose that we have obtained independent random samples from two populations with proportions p_1 and p_2. The sample sizes and sample proportions are as follows:

$$n_1 = 400 \quad n_2 = 400 \quad \hat{p}_1 = .52 \quad \hat{p}_2 = .48$$

a. Construct a 95% confidence interval for $(p_1 - p_2)$.
b. Construct a 99% confidence interval for $(p_1 - p_2)$.

Statistical Applications

10.8.7 A TV executive is interested in determining whether the proportion of people who watch a late-night talk show is higher with the regular host or a guest host. In a random sample of 400 people, 175 watch the show when the regular host is on. In a random sample of 500 people, 185 watch the show when a guest host is on. Calculate a 95% confidence interval for $(p_1 - p_2)$.

10.8.8 In a random sample of 200 city residents, 150 thought that the condition of local roads was very poor. In a random sample of 160 tourists, 90 thought the condition of local roads was very poor. Find a 90% confidence interval for the difference between the two population proportions.

10.8.9 A city planner claims that homeowners tend to have closer ties to their community than do renters. Thus, homeowners are more willing to pay for good schools and recreational facilities than are renters. In a random sample of 120 homeowners, 50 stated that the local tax rates were too high and 70 stated that tax rates were "about right." In an independent random sample of 200 renters, 70 thought that the local tax rates were too high and 130 thought they were "about right."
  a. Find a 99% confidence interval for the difference in the population proportions who think taxes are too high.
  b. Do the data support the city planner's claim?

10.9 Computer Applications

When we construct a confidence interval for a population mean, we typically have a random sample of n observations from a population with an unknown variance. To construct a confidence interval, we need to calculate both the sample mean $\bar{X}$ and the sample standard deviation S. In Chapter 4, we showed that to obtain the sample mean and sample variance using the SPSSX program, we issue the DESCRIPTIVES command followed by the names of the variables whose means and standard deviations are desired.

For example, refer to the data in Table 2.2 of Chapter 2, which shows selected characteristics of a sample of 50 students in a statistics class. Suppose we wish to construct a 95% confidence interval for the mean SAT score and the mean GPA for the population of all statistics students. To obtain the sample mean and sample standard deviation, issue the command

---
DESCRIPTIVES SAT, GPA

Figure 10.11 shows the appropriate computer output. Here we see that the sample mean for the SAT score is $\bar{x} = 1,122.40$, and the sample standard deviation is $s = 152.92$. For GPA, the sample mean is 2.92 and the sample standard deviation is .39.

```
Number of valid observations (listwise) =       50.00

                                              Valid
Variable      Mean      Std Dev   Minimum   Maximum    N   Label

SAT         1122.40     152.92     880.00   1490.00    50
GPA            2.92        .39       2.29      3.80    50   COLLEGE GRADE POINT AVERAGE
```

FIGURE 10.11 *SPSSX-generated output showing the sample mean and sample standard deviation for SAT scores and GPA in Table 2.2*

Because the sample size is $n = 50$, there are $\nu = 49$ degrees of freedom. The appropriate value of the t statistic is $t_{\alpha/2,\nu} \approx 2.02$. (We have used 40 degrees of freedom to find the critical value because t values for $\nu = 49$ are not in Table A.6.) The appropriate confidence interval is

$$\left(\bar{x} - t_{\alpha/2}\frac{s}{\sqrt{n}}, \bar{x} + t_{\alpha/2}\frac{s}{\sqrt{n}}\right)$$

$$[1{,}122.40 \pm 2.02(152.92/\sqrt{50})]$$

$$(1{,}122.40 \pm 43.686)$$

The desired confidence interval is thus $(1{,}078.714, 1{,}166.086)$.

Exercises

10.9.1   Construct a 95% confidence interval for the mean GPA for all students in the population using the data in Table 2.2 of Chapter 2.

10.9.2   Construct a 95% confidence interval for the mean age for all employees at Computech using the data in Table 2.1 of Chapter 2.

10.9.3   Construct a 95% confidence interval for the mean number of years of seniority for all employees at Computech using the data in Table 2.1 of Chapter 2.

Chapter 10 Supplementary Exercises

10.S.1   The Cincinnati Reds took a random sample of 400 spectators at several baseball games to estimate the mean distance that spectators traveled to attend the game. The results were $\bar{x} = 20.4$ miles and $s = 7$. Find a 99% confidence interval for the mean.

10.S.2   The highway department wants to estimate the mean number of cars that cross a certain bridge into town between 8 A.M. and 9 A.M. on weekdays. Traffic was counted on 16 different days. The sample variance was $s^2 = 25{,}000$ and $\bar{x} = 1{,}200$. Calculate a 90% confidence interval for the mean. Assume the population is normal.

10.S.3   A market researcher wants to select one sample to estimate both μ, the average age of people living within 5 miles of a proposed shopping mall site, and p, the proportion of people within that 5-mile radius who are between 21 and 45 years of age. The researcher wants to estimate μ with a 95% confidence interval that is no more than 6 years wide and p with a 90% confidence interval of width no greater than .1. It is known from previous studies that the standard deviation of age in the population is 15 years, and it is believed that p is near .3. How large a sample is required to construct confidence intervals for both μ and p that satisfy the above specifications?

10.S.4   In a random sample of 900 new cars, 10% had defective parts. Construct a 95% confidence interval for the proportion of all new cars having defective parts.

10.S.5   On election day, an exit poll is taken of a random sample of 2,400 voters. Candidate A gets 60% of the votes in this sample.
   a. Construct a 95% confidence interval for the percentage of votes the candidate will receive in the election.
   b. Is this sufficient evidence to convince you that candidate A will win the election?
   c. If you were the producer of an election night news program, would you be willing to announce that candidate A has won the election (before the votes were counted)? Explain.

10.S.6 For many years beer has been sold in aluminum cans. Recently a wine producer has proposed selling wine in cans, but the producer fears that many wine drinkers will refuse to buy wine unless it comes in a bottle. The producer hires a marketing agency to sample customers' opinions. The producer wants an estimate of the proportion of wine drinkers who would refuse to buy wine in a can within 3% of the population proportion with probability .95. Tentatively, the true percentage is estimated to be near 80%. How large a sample will be required?

10.S.7 A random sample of 1,000 adults was taken in Phoenix to estimate the unemployment rate in that city. In the sample, 8% of the labor force was unemployed.
  a. Find a 95% confidence interval for the unemployment rate in Phoenix.
  b. How large a sample must be taken if we want the estimated unemployment rate to be within .01 of the true unemployment rate with probability .95? Assume p is about .08.

10.S.8 The basic source of labor market information is the monthly survey data collected by the Bureau of the Census for the Bureau of Labor Statistics. Approximately 50,000 randomly selected households are surveyed monthly to represent the civilian noninstitutional population. The results are subject to some sampling error, which should be taken into account when analyzing the data.
  a. If a sample of 50,000 yields an estimated unemployment rate of 8%, find a 95% confidence interval for the true unemployment rate.
  b. How large a sample is required if we want to be 95% confident that p is within .02 of $\hat{p}$? Use $\hat{p} = .08$.

10.S.9 The American Cancer Society wants to estimate the proportion of high school seniors who smoke regularly. In a random sample of 1,000 seniors, 200 smoke regularly. Construct the following confidence intervals for the population proportion:
  a. 90%   **b.** 99%

10.S.10 The Wilson Company is doing tests to determine whether yellow, white, or orange tennis balls are easiest to see. During testing, it became necessary to estimate what proportion of the male population is color-blind, because color-blind males have difficulty seeing the orange ball. In a random sample of 1,600 males, 240 were color-blind.
  a. Construct a 95% confidence interval for the proportion of the male population that is color-blind.
  b. Construct a 99% confidence interval for p.

10.S.11 A physical education teacher claims that left-handed people have an advantage in sports such as tennis, baseball, and basketball because most people are right-handed and thus right-handed people do not get to practice enough against left-handed opponents. The instructor obtains independent random samples of 400 professional athletes and 400 nonathletes and finds that 95 of the athletes and 50 of the nonathletes are left-handed.
  a. Construct a 95% confidence interval for the difference between the population proportions.
  b. Is this sample evidence strong enough to make you agree with the instructor? Explain.

10.S.12 To prevent erosion along the sides of many highways, the highway department has spent a large sum of money planting grass. Because the cost of labor and equipment is a substantial part of the department's budget, the department wants to make sure that the grass seed being planted will help prevent erosion. Before purchasing a certain brand of seed, the department examined a random sample of 600 seeds. In the sample, 500 seeds germinated.
  a. Construct a 95% confidence interval for p.
  b. The seed company has claimed that the probability of germination is at least .9. Do you agree? Explain.

10.S.13 Congress is debating an antipollution proposal that will increase the cost of manufacturing steel and ultimately reduce the level of employment in certain steel-producing states. A politician claims that people's opinions about the proposed bill differ depending on whether they live in steel-producing states. In a random sample of 200 people from steel-producing states, 80 favor the bill. In an independent sample of 200 people from non-steel-producing states, 120 favor the bill.

a. Calculate a 95% confidence interval for the difference between the population proportions.

b. Does this confidence interval contain the value 0? Explain the significance of this result.

10.S.14 A quality control engineer would like to estimate the proportion of defects being produced on an assembly line to within .04 with 99% confidence. How large a sample is required if it is believed that p is approximately .1?

10.S.15 A company supervisor wants to estimate the mean length of time it takes employees to install mufflers on new cars. It is believed from previous studies that the population standard deviation is $\sigma = 4$ minutes. How large a sample is required to be 95% confident that $\bar{X}$ is within 1.5 minutes of the true value of μ?

10.S.16 The Newberry Toy Store has received a shipment of several thousand dolls from a foreign supplier. In a random sample of 100 dolls, 18 are found to be of such poor quality that they cannot be sold.

a. Find a 95% confidence interval for the proportion of defectives in the entire shipment.

b. Find a 99% confidence interval for the proportion of defectives in the entire shipment.

10.S.17 Two different manufacturing methods, casting and die forging, can be used to make parts for a newly developed supersonic airplane. One hundred parts were made by each method, and each part was subjected to a severe stress test. In the test, 40 of the castings failed and 30 of the forged parts failed. Find a 95% confidence interval for the difference between the population proportions of the cast and the forged parts that would fail the same test.

10.S.18 A national corporation is debating hiring a certain controversial sports star as its spokesman for a particular product. A polling agency is hired to estimate the percentage of the public who have a favorable opinion of the proposed spokesman. The agency wants its estimate to be accurate to within .03 with 95% confidence. How large a sample is required if it is believed that about 60% of the public have a favorable opinion of the individual?

10.S.19 A congressman would like to know how voters feel about a particular issue. Approximately how many voters should the congressman survey to estimate the true proportion favoring this bill to within .05 with probability .99? Since you do not have prior knowledge about p, substitute $p = .5$ in the formula to find the sample size.

10.S.20 Find the value $t_{.05}$ such that $P(t > t_{.05}) = .05$, where t follows the t distribution having the following degrees of freedom:

a. 10   **b.** 20   **c.** 30   **d.** 40

e. Compare these values with the standard normal variable Z; that is, find $z_{.05}$ such that $P(Z > z_{.05}) = .05$.

10.S.21 What is the probability that a t score with 14 degrees of freedom will exceed 1.761?

10.S.22 Approximately what is the 90th percentile of a t distribution having 500 degrees of freedom?

10.S.23 Will a 95% confidence interval based on the t distribution be wider or narrower than a 95% confidence interval based on the standard normal distribution? Explain.

10.S.24 A random sample of 200 households is taken to estimate the average annual expenditures on magazines. The results are: $\bar{x} = \$52$ and $s^2 = 400$.

a. Construct a 99% confidence interval for the population mean.

b. Construct a 95% confidence interval for the population mean.

c. What is the appropriate number of degrees of freedom?

10.S.25 A large university is considering giving each faculty member a telephone answering machine rather than having calls forwarded to a secretary when the professor is not in his or her office. A budget committee wants to estimate the average amount of time that secretaries spend handling phone calls for absent faculty members. Because getting the information is relatively expensive, the sample size must be kept small. A random sample of 16 secretaries is observed. The results are $\bar{x} = 36$ minutes per day and $s^2 = 320$.

a. Construct a 90% confidence interval for the population mean.
b. Construct a 95% confidence interval for the population mean.

10.S.26 In a certain city, baseball fans complain that they do not like to attend games because of huge traffic jams after the game. The fans claim that it takes about 35 minutes to get out of the stadium parking lots following a game when the attendance exceeds 30,000. To determine whether the complaints of the fans are justified, a random sample of 150 fans is taken during various games when the attendance exceeds 30,000. The average time it takes these 150 fans to exit from the parking lot is $\bar{x} = 27$ minutes and $s^2 = 49$. Construct a 95% confidence interval for the mean departure time.

10.S.27 An executive for Tri-State Coal Incorporated wants to determine the average amount of coal mined per day by miners in a certain mine. A random sample of 10 miners produces the following tonnages of coal on a certain day: 8, 11, 9, 6, 12, 10, 13, 9, 11, 9.

a. Find the sample mean and sample variance.
b. Find a 90% confidence interval for the mean tonnage.
c. Find a 95% confidence interval for the mean tonnage.

10.S.28 Professor Johnson has published an article in a psychology journal claiming that flight control operators (FCOs) and stockbrokers suffer the most stress during their jobs. Dr. Johnson selected a random sample of 15 FCOs and 10 stockbrokers and performed a series of tests to measure the stress of each individual. For the FCOs, the average score was $\bar{x}_1 = 216$ and the standard deviation was $s_1 = 22$; for the stockbrokers, the sample mean was $\bar{x}_2 = 198$ and the standard deviation was $s_2 = 36$. Assume the populations are approximately normal and the population variances are equal. Construct the following confidence intervals for $(\mu_1 - \mu_2)$:

a. 90%   **b.** 95%

10.S.29 The U.S. Department of Agriculture wants to determine the average number of eggs that children under 14 years of age consume each year. A random sample of 900 such children is obtained. In this sample, the average number of eggs consumed was $\bar{x} = 86$, and the sample standard deviation was $s = 16$. Find the following confidence intervals for the population mean:

a. 70%   **b.** 95%

10.S.30 Random samples of weights were taken from two package-filling processes at the Adams Feed Corporation. The first sample consisted of $n_1 = 150$ packages, and the second consisted of $n_2 = 175$ packages. The sample means were $\bar{x}_1 = 100.8$ pounds and $\bar{x}_2 = 101.4$ pounds, and the sample standard deviations were $s_1 = .9$ pound and $s_2 = 1.1$ pounds. Construct the following confidence intervals for $(\mu_1 - \mu_2)$:

a. 90%   **b.** 99%

10.S.31 The Nevada Highway Commission wants to know the mean weight of commercial vehicles traveling on a particular section of interstate highway. An inspector takes a sample of 100 randomly selected trucks and weighs them. The mean gross weight is $\bar{x} = 14.7$ tons and the sample standard deviation is $s = 4.8$ tons.

a. Construct a 90% confidence interval for the mean gross weight of commercial vehicles traveling on that section of the highway.
b. Construct a 99% confidence interval for μ.

10.S.32 The Maine Transportation Department wants to compare the durability of two different types of paint used to paint center lines on highways. On a heavily traveled road, 40 lines were painted, 20 with each type of paint. After 3 weeks, reflectometer readings were obtained. The higher the readings, the better is the durability of the paint. From the two independent samples of data, the following information was obtained: $\bar{x}_1 = 12.3$, $s_1^2 = 3$, $\bar{x}_2 = 11.6$, and $s_2^2 = 3$. Assume that the populations are normal and that the population variances are equal. Construct the following confidence intervals for ($\mu_1 - \mu_2$):

a. 95%   **b.** 99%

10.S.33 A computer manufacturer wants to estimate the speed of its new printer. Assume the population is approximately normal. A random sample yielded the following measured speeds (in characters per second):

116	130	112	105	115	136	111	127	118	124	111	126
117	119	107	110	134	123	122	114	122	103	115	130

a. Construct a 95% confidence interval for the population mean.
b. Construct a 99% confidence interval for the population mean.

10.S.34 The Nielsen television ratings are based on a sample of 2,000 homes out of a population of 80,000,000 homes. Suppose Nielsen estimates that 30% of these homes were watching a certain program.
a. Construct a 99% confidence interval for the population proportion.
b. Estimate the number of households that watched the program.
c. Find a 99% confidence interval for the number of households that watched the program.

10.S.35 To monitor a particular industrial process, a manufacturer wishes to estimate the mean diameter of items produced by the process during the past 24 hours. How many items must be sampled to estimate the population mean to within .01 inch with 95% confidence? Previous studies indicate that the standard deviation of the process is approximately .12 inch.

10.S.36 Suppose you have selected a random sample of $n = 10$ items from a normal population. Compare the standard normal z values with the corresponding t values if you were constructing the following confidence intervals.
a. 90% confidence interval
b. 95% confidence interval
c. 99% confidence interval

10.S.37 The following random sample of 10 observations was obtained from a normal population: 11, 25, 23, 14, 9, 16, 10, 18, 22, 24.
a. Calculate the sample mean and sample standard deviation.
b. Construct a 95% confidence interval for the population mean μ.
c. Construct a 99% confidence interval for the population mean μ.
d. Use the sample mean and sample standard deviation from part **a.** Assume that the sample size is $n = 25$ rather than $n = 10$. Repeat parts **b** and **c.** How does increasing the sample size affect the width of the confidence intervals?

10.S.38 A major concern of American companies is the increasing cost of fringe benefits such as health insurance, vision care, and dental care. A random sample of 20 employers indicated that the sample mean cost of health care per worker was $350 per month and the standard deviation was $68.
a. Construct a 95% confidence interval for the population mean monthly cost of health care per worker.
b. What assumption did you make about the population distribution?

10.S.39 Shoplifting is becoming an increasing cost of doing business. To determine the magnitude of the problem, a manager of a department store hired a security firm to observe a random sample of shoppers. The security agents randomly selected 400 shoppers and observed them while they were in the store. Of the 400 shoppers, 25 were seen stealing. Construct a 95% confidence interval for the proportion of all the store's customers who are shoplifters.

10.S.40 A bill to change federal gun control legislation is being debated in Congress. Before voting on the issue, a senator would like to know how constituents feel about the issue. Approximately how many constituents should be surveyed to estimate the true proportion favoring this bill to within .03 with 95% confidence?

10.S.41 Suppose the manager of a convenience store wants to estimate the proportion of customers who are shoplifters. To do this, a sample of customers is selected and observed by security agents. Suppose experience suggests that 8% of customers are shoplifters. How many customers should be observed if it is desired to estimate the proportion of shoplifters to within .02 with 95% confidence?

10.S.42 In each of the following instances, determine whether it would be appropriate to use a z value or a t value when constructing a confidence interval for μ. If neither is appropriate, explain why.
   a. Random sample of size $n = 20$ from a normal distribution with unknown mean μ and known standard deviation σ
   b. Random sample of size $n = 20$ from a normal distribution with unknown mean μ and unknown standard deviation σ
   c. Random sample of size $n = 100$ from a normal distribution with unknown mean μ and unknown standard deviation σ
   d. Random sample of size $n = 10$ from a nonnormal distribution with unknown mean μ and known standard deviation σ
   e. Random sample of size $n = 10$ from a nonnormal distribution with unknown mean μ and unknown standard deviation σ
   f. Random sample of size $n = 100$ from a nonnormal distribution with unknown mean μ and known standard deviation σ
   g. Random sample of size $n = 100$ from a nonnormal distribution with unknown mean μ and unknown standard deviation σ

10.S.43 For states to remain eligible for federal highway funds, the federal government requires them to certify that they are enforcing the 55-mile-per-hour speed limit. If too many drivers exceed the speed limit, a state is in jeopardy of losing its federal funding. Suppose the State Highway Patrol conducts a radar survey at a selected site on a state highway.
   a. How large a sample is required if it is desired to estimate p, the proportion of vehicles that exceed 60 miles per hour to within .03 with 95% confidence? Suppose it is believed that p is approximately .45.
   b. How large a sample is required if it is desired to estimate μ, the mean speed of the vehicles, to within 2 miles per hour with 95% confidence? Suppose it is believed that the population standard deviation is approximately $\sigma = 6$ miles per hour.

10.S.44 Quality control engineers use control charts to monitor the quality of industrial output. Typically, a small sample of output is observed at regular intervals to determine whether the process is in control during the time period when each sample was obtained. Suppose a quality control engineer at an automobile manufacturing plant wants to monitor the thickness of engine gaskets. The engineer takes a sample of 10 gaskets per hour and examines the thickness of each. In one sample of 10 gaskets, the mean and standard deviation are $\bar{x} = .25$ inch and $s = .02$ inch.
   a. Construct a 99% confidence interval for the mean gasket thickness produced during the period in which the sample was selected. What assumptions did you make when constructing the confidence interval?
   b. The process is to be considered out of control if the population mean thickness differs from .24 inch. Based on the confidence interval constructed in part **a,** would you conclude that the population mean differs from .24 inch? Explain.

References

Fisher, Ronald A. *Statistical Methods for Research Workers.* 14th ed. New York: Hafner Press, 1970.

Freund, John E., and R. E. Walpole. *Mathematical Statistics.* 4th ed. Englewood Cliffs, N.J.: Prentice-Hall, 1987.

Handbook of Tables for Probability and Statistics. 2d ed. Cleveland, Ohio: Chemical Rubber Co., 1968.

Hoel, Paul G. *Elementary Statistics.* 4th ed. New York: Wiley, 1976.

Nie, Norman E., C. Hadlai Hull, Jean G. Jenkins, Karen Steinbrenner, and Dale H. Bent. *SPSS Statistical Package for the Social Sciences.* 2d ed. New York: McGraw-Hill, 1975.

Norusis, Marija J. *SPSSX Introductory Statistics Guide.* New York: McGraw-Hill, 1990.

Norusis, Marija J. *SPSSX Advanced Statistics Guide.* Chicago: SPSS, 1990.

Norusis, Marija J. *The SPSS Guide to Data Analysis.* Chicago: SPSS, 1990.

Ryan, Thomas A., Brian L. Joiner, and Barbara F. Ryan. *Minitab Handbook.* 2d ed. Boston: PWS-Kent, 1985.

Ryan, Thomas A., Brian L. Joiner, and Barbara F. Ryan. *Minitab Reference Manual.* University Park, Penn.: Minitab Project, 1985.

SAS Introductory Guide. 3d ed. Cary, N.C.: SAS Institute, 1985.

SAS Procedures Guide for Personal Computers. Version 6 ed. Cary, N.C.: SAS Institute, 1986.

SAS Statistics Guide for Personal Computers. Version 6 ed. Cary, N.C.: SAS Institute, 1986.

SAS User's Guide: Basics. Version 5 ed. Cary, N.C.: SAS Institute, 1985.

SAS User's Guide: Statistics. Version 5 ed. Cary, N.C.: SAS Institute, 1985.

SPSSX User's Guide. Chicago: SPSS, 1990.

Student. "The Probable Error of a Mean." *Biometrika* 6 (1908): 1–25. The crucial article on the t distribution.

11 Hypothesis Testing

Inferences about population parameters can be made in two ways: We can estimate the parameters and construct confidence intervals around the estimates, or we can make decisions about the parameters by testing hypotheses.

For example, suppose a traffic engineer wants to know the mean number of cars that pass through a particular intersection on weekday mornings between 8 A.M. and 9 A.M. The engineer would probably obtain data for a sample of days and calculate the sample mean to estimate the population mean μ. Finally, he or she could construct a confidence interval to place bounds on the possible sampling error.

In contrast, suppose the engineer knows from years of experience that the average number of cars passing through the intersection between 8 A.M. and 9 A.M. is 2,500. New traffic lights have been installed and the engineer wants to determine whether the new system increases the traffic flow. Thus, the engineer wants to test the hypothesis that the mean is still 2,500 cars per day against the alternative hypothesis that the mean is greater than 2,500. This is a typical example of hypothesis testing.

To determine whether the mean is 2,500 or greater than 2,500, the engineer observes the traffic flow on a random sample of, say, 25 days and computes the sample mean and sample standard deviation. Suppose the sample mean is $\bar{x} = 2,800$ cars and the sample standard deviation is $s = 250$. What should the engineer conclude? It is not impossible to observe a sample mean of 2,800 if the true mean is 2,500, but it is very unlikely if the sample standard deviation is only 250. If the mean were 2,500 or lower, the probability of observing a sample mean as high as 2,800 would be very small. Thus, the engineer concludes that the population mean exceeds 2,500.

The hypothesis being tested, that the mean is 2,500, is called the *null hypothesis*. The decision to reject this hypothesis is based on the value of the sample mean, which serves as a *test statistic*. The entire set of values that the test statistic can assume is divided into two subsets, one corresponding to the *rejection region* and the other to the *acceptance region*. For the engineer, the rejection region would contain values of $\bar{X}$ that are far above the hypothesized value $\mu = 2,500$. If the test statistic computed from a sample assumes a value in the rejection region, the null hypothesis is rejected in favor of the alternative hypothesis. If the test statistic falls in the acceptance region, then the null hypothesis is not rejected.

One problem in testing any hypothesis is deciding which values of the test statistic to assign to the rejection region and which to the acceptance region. For example, should the value $\bar{x} = 2{,}600$ be assigned to the acceptance or rejection region? The answer to this question depends on the risks you are willing to take and the costs of making a mistake.

For example, since the new traffic light system in this example is very expensive, the engineer definitely wants to avoid the mistake of concluding that the mean is greater than 2,500 if, in fact, it is not. This mistake, called a *Type I error,* occurs when we incorrectly reject a null hypothesis that is true. To avoid this mistake, the engineer would decide to reject the null hypothesis only if $\bar{X}$ far exceeds 2,500. Suppose the engineer uses the *decision rule* "Reject the null hypothesis if $\bar{X} \geq 2{,}600$." The value 2,600 is then the *critical value* of the test statistic that separates the rejection region from the acceptance region.

Even if the true value of μ were 2,500, it would still be possible (though unlikely) to obtain a sample mean exceeding 2,600. If this happened, the engineer would incorrectly reject the null hypothesis and commit a Type I error. The probability of such an error is called the *level of significance* of the test. By increasing the critical value to, say, 2,650, the engineer would decrease the probability of committing a Type I error but make it easier to commit a *Type II error,* which occurs when we accept the null hypothesis when it is false.

For example, suppose the new traffic light system actually increases the hourly traffic flow to 2,650 cars. If the true mean is 2,650, the probability is .5 that the sample mean will be less than 2,650. Thus, if the engineer used the critical value 2,650, the probability would be .5 that the null hypothesis would not be rejected. Thus, the probability of a Type II error would be .5.

To decrease this probability, the engineer could decrease the critical value to, say, 2,600. However, by doing so he or she would then raise the probability of a Type I error. Thus, the choice of an appropriate critical value plays a crucial role in hypothesis testing. We now discuss the elements of a hypothesis test in more detail and provide examples of various types of tests.

11.1   Concepts of Hypothesis Testing

A population parameter is a number that describes some characteristic of a population, and a hypothesis is a statement about the value or set of values that a parameter or group of parameters can take. Many types of hypotheses can be tested by statistical methods.

Frequently we are concerned about the value of a single parameter, such as a population mean, a population variance, or a population proportion. At other times, we may wish to test a hypothesis that the means of two different populations are equal or that the proportions of two different populations possessing a certain characteristic are equal. In these cases we are testing hypotheses that involve parameters of two populations. We can generalize this concept to the problem of testing a hypothesis about k population means or proportions.

In this chapter, we discuss how to test hypotheses about a single population mean, a single population proportion, and a single population variance. In Chapter 12, we discuss how to test hypotheses about the difference between two population means and two population proportions. In Chapters 13 and 14, we extend this theory to cover tests of hypotheses concerning three or more means or proportions. Also in Chapter 14, we discuss how to test hypotheses about the equality of two population variances.

The Null Hypothesis H_0 and the Alternative Hypothesis H_1

The purpose of hypothesis testing is to choose between two conflicting hypotheses about the possible value of a population parameter. The two conflicting hypotheses are called the *null hypothesis,* denoted H_0, and the *alternative hypothesis,* denoted H_1. These two hypotheses are mutually exclusive; that is, when one is true, the other must be false.

DEFINITIONS **Null Hypothesis and Alternative Hypothesis**

The **null hypothesis** is an assumption concerning the value of the population parameter being studied. The **alternative hypothesis** specifies an alternative set of possible values of the population parameter that are not specified in the null hypothesis.

In some hypotheses, only one particular value of the population parameter is specified; in other hypotheses, a range of values is specified. Conclusions about the validity of these hypotheses are based on information obtained from random samples of observations collected from the populations of interest.

Suppose some hypothesis has been formed about the parameter being studied. We will not reject this hypothesis, which is the null hypothesis, unless the sample data provide strong evidence that the hypothesis is false.

EXAMPLE 11.1 **Null Hypothesis About a Population Proportion**

In 1987, the Environmental Protection Agency reported that the emissions equipment had been tampered with on 20% of the cars and trucks in the United States. Suppose you want to test this claim because you think this proportion is too high. In this example, the null hypothesis would be written

$$H_0: \quad p = .20$$

where p represents the proportion of emissions systems that have been tampered with.

In Example 11.1, the null hypothesis is a statement about a population proportion p that may be true or false. We will not reject the hypothesis unless we obtain strong evidence indicating that it is false. Before performing the test, we must formulate an alternative hypothesis, denoted H_1, against which the null hypothesis is tested. For example, we could test the null hypothesis against the alternative hypothesis that the true population proportion is less than .20. In this case, the alternative hypothesis would be denoted

$$H_1: \quad p < .20$$

DEFINITIONS **Simple Hypothesis and Composite Hypothesis**

If a hypothesis states that a certain population parameter θ equals a single specific value, such as $\theta = \theta_0$, then the hypothesis is said to be a **simple hypothesis.** This simple null hypothesis can be expressed as follows:

$$H_0: \quad \theta = \theta_0$$

When the hypothesis contains a range of possible values for the parameter of interest, the hypothesis is said to be a **composite hypothesis**. Examples of composite null hypotheses and their respective alternative hypotheses about a parameter θ are as follows:

$$H_0: \quad \theta \geqslant \theta_0$$
$$H_1: \quad \theta < \theta_0$$

and

$$H_0: \quad \theta \leqslant \theta_0$$
$$H_1: \quad \theta > \theta_0$$

In Example 11.1, the null hypothesis, $H_0: p = .20$, is a simple null hypothesis, and the alternative hypothesis, $H_1: p < .20$, is a composite alternative hypothesis. This alternative hypothesis is called a *one-sided alternative hypothesis* because all possible values of p in H_1 fall below $p = .20$, the value specified in H_0.

EXAMPLE 11.2   **One-Sided Alternative Hypothesis**

Suppose that a government inspector is concerned about whether a particular cereal producer is putting an average of at least 32 ounces of cereal into each box as claimed. In this case, it would be natural for the null hypothesis to be that the average amount of cereal in the boxes is at least 32 ounces and for the alternative hypothesis to be that the company is putting less than 32 ounces of cereal into the boxes. The null hypothesis is thus

$$H_0: \quad \mu \geqslant 32$$

and the alternative hypothesis is

$$H_1: \quad \mu < 32$$

Because this alternative hypothesis lies entirely on one side of the null hypothesis, it is a one-sided alternative hypothesis.

On some occasions, we might want to test a simple null hypothesis against a two-sided alternative hypothesis.

EXAMPLE 11.3   **Two-Sided Alternative Hypothesis**

The Kemco Company has purchased several new and less expensive pumps to remove water from flooded mines and construction sites. Kemco wants to compare them to the standard model pumps that the company has been using. Kemco may or may not switch exclusively to the new model depending on how well it performs compared to the standard model. Kemco knows that the standard models remove 5,000 gallons of water per hour on the average, and wants to test whether the new model is different from the standard model. The null hypothesis is

$$H_0: \quad \mu = 5,000$$

where μ represents the mean amount of water pumped in an hour by the new machines. The alternative hypothesis is the *two-sided alternative*

$$H_1: \quad \mu \neq 5{,}000$$

In Example 11.3, H_1 is a two-sided alternative hypothesis because it contains values of μ on both sides of the point specified in H_0.

The following formats show three common sets of the null hypothesis H_0 and the alternative hypothesis H_1:

1. Test $H_0: \theta = \theta_0$ against $H_1: \theta \neq \theta_0$ (two-sided test).
2. Test $H_0: \theta \geq \theta_0$ against $H_1: \theta < \theta_0$ (one-sided test).
3. Test $H_0: \theta \leq \theta_0$ against $H_1: \theta > \theta_0$ (one-sided test).

Sometimes we are not interested in the inequalities stated in the null hypotheses in formats 2 and 3; instead these hypotheses might be expressed as follows:

4. Test $H_0: \theta = \theta_0$ against $H_1: \theta < \theta_0$.
5. Test $H_0: \theta = \theta_0$ against $H_1: \theta > \theta_0$.

Students frequently confuse which hypothesis to call the null hypothesis and which to call the alternative hypothesis. In many statistical applications, the null hypothesis should correspond to the assumption that no change occurs when some new process or technique is tried. This was the case in Example 11.3, where the null hypothesis claims that the new pumps remove exactly the same amount of water as the old pumps. This explains the origin of the name "null" hypothesis, indicating no change between an old and a new situation.

Some statisticians argue that the null hypothesis should be the hypothesis that the decision maker wants to disprove. That is, the null hypothesis should specify the value(s) of the population parameter that the researcher thinks does not represent the true value(s) of the parameter; the alternative hypothesis then specifies the values of the parameter that the researcher believes do hold. The researcher then tests whether the sample data lead to rejection of H_0 and acceptance of H_1. Another common practice is to assign no special meaning to either the null or the alternative hypothesis, but to let these hypotheses merely represent two different assumptions about the population parameter.

How do we decide whether the null hypothesis should be simple or composite? In some cases, the choice arises quite naturally from the statement of the problem, as illustrated in the following examples.

EXAMPLE 11.4   **Choosing a Null and Alternative Hypothesis** —————————————

The Better Business Bureau is investigating a meat company that sells ground beef in 5-pound packages. There have been many complaints that the company is short-weighting its customers. The null hypothesis would be

$$H_0: \quad \mu \geq 5$$

The Better Business Bureau (and a jury) will accept the null hypothesis unless the data strongly suggest that the null hypothesis is false. We have no interest in whether the mean is actually greater than 5. Thus, the null hypothesis could also be

$$H_0: \quad \mu = 5$$

As far as we are concerned, either the mean is 5 or the mean is less than 5. The alternative hypothesis would thus be

$$H_1: \quad \mu < 5$$

EXAMPLE 11.5   **Choosing a Null and Alternative Hypothesis**

A company produces light bulbs for use in traffic signals in New York City. The average life of these light bulbs is 2,000 hours. Because it is very expensive to replace bulbs, the city wants to purchase the lights having the longest expected life for a given price. A second company claims that it produces better bulbs at the same cost. The city wants to examine a sample of the new bulbs to test whether, on the average, they last more than 2,000 hours. Because the new bulbs cost the same as the old bulbs, the city would not switch to the new bulbs unless the evidence was strong that the new bulbs outlast the old. When testing the new bulbs, the null hypothesis would be

$$H_0: \quad \mu \leq 2,000$$

The alternative hypothesis would then be

$$H_1: \quad \mu > 2,000$$

EXAMPLE 11.6   **Choosing a Null and Alternative Hypothesis**

A company produces a product that is used regularly by 30% of the people in a country; that is, $p = .30$, where p represents the proportion of people who use the product. The company has used a standard advertising approach for many years and is reluctant to change its ads unless strong evidence exists that a new advertising campaign would increase the proportion of people who use the product. New ads are used in a sample of cities to determine whether they are more effective than the old. The company would test the null hypothesis

$$H_0: \quad p \leq .30$$

against the alternative hypothesis

$$H_1: \quad p > .30$$

In this example, we are comparing the effects of a new advertising format with the effects of the old. The null hypothesis represents the status quo, or the state of affairs without the proposed change. The alternative hypothesis represents the state of affairs if the change has the effect claimed by its proponents.

EXAMPLE 11.7   **Choosing a Null and Alternative Hypothesis**

Under controlled laboratory conditions, a car door painted with standard paint will rust in 5 years, on the average, when subjected to a certain amount of dampness. A scientist claims that by using a new type of paint on the door, the rust can be retarded. The car company will not accept the scientist's claim unless the evidence strongly supports it.

Thus, the null hypothesis would be

$$H_0: \quad \mu \leqslant 5$$

and the alternative hypothesis would be

$$H_1: \quad \mu > 5$$

where μ is the average number of years before rust develops on treated car doors.

How the null and alternative hypotheses are formulated can have important consequences, because, in classical hypothesis testing, we do not reject the null hypothesis unless we have strong evidence against it. Thus, the null hypothesis is in a favored position relative to the alternative hypothesis. The null hypothesis has the status of a maintained hypothesis that will not be rejected because it is assumed to be true *unless the sample data provide strong contrary evidence.*

EXAMPLE 11.8   **Consequences of Choosing H_0 and H_1**

Consider the problem faced by officials of the Food and Drug Administration (FDA) when they are asked to approve a new medicine for sale to the public. There are two possibilities:

1. The medicine is beneficial.
2. The medicine is not beneficial.

Either of these possibilities could be chosen as the null, or maintained, hypothesis. The FDA uses the null hypothesis

H_0:   The medicine is not beneficial and should not be marketed.

The alternative hypothesis is thus

H_1:   The medicine is beneficial and should be marketed.

Thus, the FDA will not approve a new medicine unless the drug manufacturer can produce strong evidence that the null hypothesis is false. Because of this method of choosing H_0 and H_1, getting a new medicine approved in the United States is quite difficult.

There are various costs and benefits from choosing H_0 and H_1 in this way. The potential benefit is that it is very difficult to get harmful medicines onto the market, but the potential cost is that beneficial drugs can be delayed or even prohibited from being marketed. This cost can be enormous in terms of the pain (or death) suffered by individuals who could have been cured by a drug if it had been approved.

Now consider what would happen if the FDA used the following null and alternative hypotheses:

H_0:   The medicine is beneficial and should be marketed.

H_1:   The medicine is not beneficial and should not be marketed.

In this case, it would be assumed that a drug was beneficial and would not be prohibited unless there were strong evidence that it was ineffective or harmful. There would be a benefit and a cost involved in this choice of H_0, as well. The benefit of this procedure would be that if a drug was beneficial, it would not be delayed from reaching the market and patients would be cured more quickly; the cost would be that harmful drugs could reach the market relatively easily.

Example 11.8 shows that when performing a test, we should keep in mind the potential costs and benefits associated with a decision. In the United States, officials for the FDA feel that the potential costs of approving a harmful drug are so high that they make it very difficult to get a new drug approved. This viewpoint is an important factor in determining which drugs get approval and how long it takes to get approval.

When testing a hypothesis, we would like to be able to calculate a probability such as

$$P(\text{Theory is true}|\text{Observed data})$$

This notation represents the probability that a theory is true given the available data. Instead, we actually calculate the conditional probability,

$$P(\text{Observed data}|\text{Theory is true})$$

That is, what we actually determine is the probability that the data would look as they do if the theory were true.

These are two very different probabilities, and recognizing the difference is crucial to understanding the meaning and limitations of classical hypothesis tests. A hypothesis test is an attempted proof by statistical contradiction. For any particular theory, statisticians can deduce that if the theory is true, a sample of data is likely to look a certain way and unlikely to look some other way. If the observed sample data are of the likely kind, then they are consistent with the theory and tend to confirm it. If, on the other hand, the data are of the unlikely kind, then the data are not consistent with the theory, and the theory is rejected.

Of course, we cannot be certain that we are correct in rejecting a theory because the unlikely may well happen. Note also that not rejecting the null hypothesis is a relatively weak conclusion, because there can often be many other theories that would be consistent with the data.

Decision Rules

Based on the information provided by the sample, we must make a decision concerning the null hypothesis. There are two possibilities, namely, rejecting the null hypothesis H_0 in favor of the alternative H_1 or not rejecting H_0. Occasionally statisticians say that the null hypothesis is "accepted," but it is more common to say that the null hypothesis is "not rejected."

In a sense, a statistical test is like a trial by jury. Like the defendant in a criminal case, the null hypothesis is on trial. Also like the defendant, the null hypothesis is given the benefit of the doubt and assumed to be true unless the sample data provide strong evidence indicating that it is false. Note that in a jury trial, two types of mistakes can be made: Innocent people can be convicted, and criminals can be acquitted. Similarly, in hypothesis testing, two types of mistakes are possible: We can reject a null hypothesis that is correct (Type I error), or we can accept a null hypothesis that is incorrect (Type II error).

Before arriving at a conclusion, we need some *decision rule* for when to reject H_0. The decision to reject the null hypothesis is based on a *test statistic,* such as a sample mean $\bar{X}$, a sample proportion $\hat{p}$, a Z score, or a t score. We show how to use these statistics to test hypotheses as we proceed through this chapter.

DEFINITION **Test Statistic**

A **test statistic** is a random variable whose value is used to determine whether we reject the null hypothesis.

DEFINITION **Decision Rule**

The **decision rule** specifies the set of values of the test statistic for which the null hypothesis H_0 is rejected in favor of H_1 and the set of values for which H_0 is accepted (i.e., not rejected).

The decision rule separates the set of possible values of the test statistic into two exhaustive and mutually exclusive regions, the *acceptance region* and the *rejection region* (or *critical region*).

DEFINITIONS **Rejection Region and Acceptance Region**

The **rejection region** of a test, also called the **critical region,** consists of all values of the test statistic for which H_0 is rejected. The **acceptance region** consists of all values of the test statistic for which H_0 is accepted (not rejected).

When the alternative hypothesis is one-sided, the entire rejection region falls on one side of the acceptance region. When the alternative hypothesis is two-sided, part of the rejection region lies on either side of the acceptance region.

DEFINITION **Critical Value**

The **critical value** of the test statistic is the value that separates the critical region from the acceptance region.

A one-sided alternative hypothesis has one critical value, whereas a two-sided alternative hypothesis has two. These ideas will be developed in more detail in Section 11.2.

Type I and Type II Errors

There are four possible outcomes of any hypothesis test, two of which are correct and two of which are incorrect (see Table 11.1 on page 400):

1. H_0 is true and we do not reject H_0 (correct decision).
2. H_0 is false and we reject H_0 (correct decision).
3. H_0 is true and we reject H_0 (incorrect decision—Type I error).
4. H_0 is false and we do not reject H_0 (incorrect decision—Type II error).

TABLE 11.1   *Four possible outcomes of a hypothesis test*

| | ACTUAL SITUATION | |
Decision	H_0 true	H_0 false
H_0 not rejected	Correct decision	Type II error
H_0 rejected	Type I error	Correct decision

The two types of error are given special names, and it is very important to identify them correctly. If we reject the null hypothesis when in fact it is true (outcome 3), we commit a *Type I error*. If we accept the null hypothesis (or fail to reject it) when it is actually false and some other hypothesis is true (outcome 4), we commit a *Type II error*. Table 11.1 shows the four possible situations in hypothesis testing. A Type I error occurs when the test statistic falls in the rejection region even though the null hypothesis is true. The probability of making a Type I error is called the *level of significance* of the test and is denoted by the symbol α (Greek alpha).

DEFINITIONS   **Type I and Type II Errors**

A **Type I error** occurs if we reject H_0 when H_0 is true. A **Type II error** occurs if we accept (do not reject) H_0 when H_0 is false.

DEFINITION   **Level of Significance**

The **level of significance** of a test is the probability that the test statistic falls in the critical region given that H_0 is true. The level of significance is denoted by the symbol α where

$$\begin{aligned} \alpha &= P(\text{Type I error}) \\ &= P(H_0 \text{ reject}|H_0 \text{ true}) \\ &= P(\text{Test statistic in rejection region}|H_0 \text{ true}) \end{aligned}$$

The notation "$P(H_0 \text{ reject}|H_0 \text{ true})$" denotes the probability that we reject the null hypothesis H_0 *given that* the null hypothesis H_0 is true. That is, the vertical line is read "given that."

DEFINITION   **Probability of a Type II Error**

The **probability of making a Type II error** is denoted by the symbol β (Greek beta). The probability β of making a Type II error is the probability that the test statistic falls in the acceptance region when the null hypothesis is false. In symbols, we have

$$\beta = P(\text{Test statistic in acceptance region}|H_0 \text{ false})$$

EXAMPLE 11.9   **Example of a Type I Error**

In 1987, *USA Today* reported that a secret government survey found that illegal gambling amounts to at least $200 per year per adult in the United States. Suppose you believe that this figure is too high. You take a random sample of n individuals to estimate the mean amount of illegal gambling per person. You want to test the null hypothesis

$$H_0: \quad \mu \geq \$200$$

against the alternative hypothesis

$$H_1: \quad \mu < \$200$$

Suppose that H_0 is in fact true but you obtain a sample mean $\bar{x}$ that is much lower than \$200. Based on this sample information, you reject H_0 in favor of H_1. In this case, you would be making a Type I error because you rejected H_0 when H_0 was true.

This Type I error occurred because of bad luck. Although H_0 is true, you were unlucky because you obtained a sample that led you to believe that H_0 was false.

EXAMPLE 11.10 **Example of a Type II Error**

In Example 11.9, suppose that the amount of illegal gambling is actually much less than \$200 per person per year. Thus, H_0 is false and H_1 is true. However, suppose you obtain a sample mean $\bar{x}$ that is close to \$200 and thus decide that the evidence is not strong enough to make you reject H_0. You would then be making a Type II error, because you would not reject H_0 when it was false.

Note that like the Type I error in Example 11.9, this Type II error occurred because of bad luck. Although H_0 was false, you were unlucky because you obtained a sample that led you to believe that H_0 was true.

Implications of Rejecting or Accepting the Null Hypothesis

When the test statistic falls in the rejection region, this does not *prove* that the null hypothesis is false. Rather, it indicates that we should have strong doubts concerning the truth of the null hypothesis. Either the null hypothesis is false, or we have observed a very unlikely event that should occur with a small probability (less than α). Similarly, when the test statistic falls in the acceptance region, this does not *prove* that the null hypothesis is true, but merely indicates that the evidence is not strong enough to make us reject the null hypothesis.

Because the null hypothesis is assumed to be true before the test, it is not rejected unless the data provide strong contrary evidence. Usually we choose some small number, such as .05 or .01, as the level of significance of the test, thereby guaranteeing that the probability is low that we will reject a true null hypothesis.

If we have only a small sample of data, it is not likely that we will be able to reject an incorrect null hypothesis unless the hypothesis is wildly in error. Thus, accepting a null hypothesis on the basis of a small sample of information does not necessarily say a great deal in favor of the hypothesis. As the number of sample observations increases, however, it becomes more likely that we will detect a false null hypothesis.

Cost Analysis and Practical Significance

Statistical significance is not at all the same as practical importance. A statistically significant result is one that cannot be reasonably explained by sampling error. Such a result may well be of little or no practical importance. Conversely, a very important result may not appear to be statistically significant because the sample size is too small to allow us to reject the null hypothesis.

Whenever the test statistic falls in the rejection region, we say that the test statistic is *statistically significant* and reject the null hypothesis. Note that the word *significant* here is used in a technical sense and does not imply that the finding is of any practical importance. For example, the manufacturer of a water pump used to drain flooded construction sites may perform a test and conclude that a new model pump produces a statistically significant increase in the average amount of water pumped per hour. In such a case we can feel quite confident that the new pump actually does pump more water than the old. However, the increase might be so small compared with the increased cost of the new pump that switching to the new pump is not economically justified.

Exercises

Statistical Concepts

11.1.1   Define null hypothesis and alternative hypothesis.

11.1.2   Explain the difference between a one-sided and a two-sided alternative hypothesis.

11.1.3   Explain what is meant by a test statistic. Give an example.

11.1.4   Why is it important to know the sampling distribution of the test statistic under the assumption that the null hypothesis is true?

11.1.5   Explain what is meant by a decision rule.

11.1.6   Explain what is meant by a critical value of a test statistic.

11.1.7   Explain what is meant by the level of significance of a test of a hypothesis.

11.1.8   Suppose you test some null hypothesis and reject the null hypothesis in favor of the alternative hypothesis. Does this mean that the alternative hypothesis is correct? Explain.

11.1.9   Define the two types of errors that can be made when testing some hypothesis.

11.1.10   *True or false:*   A null hypothesis (H_0) is a theory about the values of one or more population parameters. The theory generally represents the status quo, which we accept unless we get strong evidence against it.

11.1.11   *True or false:*   An alternative hypothesis (H_1) is a theory that contradicts the null hypothesis. The alternative hypothesis will be accepted when there is strong evidence leading us to reject the null hypothesis.

11.1.12   *True or false:*   The critical region or rejection region for a test is the set of values of the test statistic for which the alternative hypothesis will be rejected.

11.1.13   *True or false:*   When the null hypothesis is true, the probability that the test statistic will fall in the critical region is called the level of significance of the test.

11.1.14   *True or false:*   The probability that the test statistic will fall in the critical region, given that H_0 is true, represents the probability of making a Type II error.

11.1.15   *True or false:*   The probability of making a Type II error is denoted $1 - \alpha$.

11.1.16   *True or false:*   Suppose the probability of making a Type I error is denoted α and the probability of making a Type II error is denoted β. Then $\alpha + \beta = 1$.

11.1.17   *True or false:*   Suppose the probability of making a Type I error is denoted α and the probability of making a Type II error is denoted β. Then $\alpha + \beta$ must be less than 1.

11.1.18 *True or false:* Suppose that, regardless of the evidence, you always accept the null hypothesis that an accused individual is innocent. Then $\alpha = 1$. Explain.

11.1.19 *True or false:* Suppose that, regardless of the evidence, you always accept the null hypothesis that an accused individual is innocent. Then $\beta = 0$. Explain.

11.1.20 *True or false:* Suppose that, regardless of the evidence, you always reject the null hypothesis that an accused individual is innocent. Then $\alpha = 0$.

11.1.21 *True or false:* Suppose it would be a very costly mistake if you reject the null hypothesis when, in fact, H_0 is true. You should choose a very small level of significance. Explain.

11.1.22 The probability of a Type I error depends on the size and location of the rejection region. Why don't we decrease the size of the rejection region and try to make α as small as possible?

11.1.23 *True or false:* Decreasing α increases the probability of not rejecting the hypothesis when it is false and some alternative hypothesis is true. Explain.

11.1.24 *True or false:* Other things being equal, decreasing α increases β.

11.1.25 *True or false:* A Type I error can occur only when the statistician decides to reject the null hypothesis. Explain.

11.1.26 *True or false:* A Type II error can occur only when the statistician decides to accept the null hypothesis. Explain.

11.1.27 Medical tests are used to detect various potential illnesses. For example, chest x-rays are used to detect cancer, and blood tests are performed to detect the presence of the HIV virus. There are two potential errors that can be made when reporting the result of a medical test. A false positive refers to a positive test result when, in fact, the individual does not have the disease. A false negative is a negative test result for an individual who does have the disease. Suppose the null hypothesis is that the individual does not have the disease.

a. Explain the consequences of making a Type I error. Relate your answer to a situation where a person is being tested for AIDS or cancer.

b. Explain the consequences of making a Type II error.

c. In your opinion, which type of error has the more serious consequences?

d. In your opinion, is it more important to try to minimize α or β?

Statistical Applications

11.1.28 If we were to compare a one-tailed test and a two-tailed test for a particular population parameter, what difference would we find in the form of the null and alternative hypotheses?

11.1.29 When evaluating a loan applicant, a financial officer is faced with the problem of granting loans to people who are good risks and denying loans to people who appear to be poor risks. In effect, the financial officer is testing the null hypothesis

$$H_0: \quad \text{The applicant is a good risk.}$$

against the alternative hypothesis

$$H_1: \quad \text{The applicant is a poor risk.}$$

The officer commits a Type I error when she rejects an applicant who is actually a good risk; she commits a Type II error when she grants a loan to an applicant who is a poor risk. Discuss the selection of a significance level α in the following instances:

a. Lending money is tight, interest rates are high, and loan applicants are numerous.

b. Lending money is plentiful, interest rates are moderate, and there is intense competition for loan applicants.

11.1.30   A quality control inspector tests the dimensions of parts used in a machining operation. Specifications require that the mean diameter of the parts be 2 inches. If a sample leads the quality control inspector to believe that the diameters are too large or too small, the machine is shut down and readjusted. State the appropriate null and alternative hypotheses for determining whether the machine should be shut down.

11.1.31   For each of the following situations, indicate whether a correct decision has been made or whether a Type I or Type II error has occurred.

   a. The null and alternative hypotheses are as follows:

$$H_0: \text{ New system is no better than the old one.}$$
$$H_1: \text{ New system is better.}$$

   (1) Adopt new system when new one is better.
   (2) Retain old system when new one is better.
   (3) Retain old system when new one is not better.
   (4) Adopt new system when new one is not better.

   b. The null and alternative hypotheses are as follows:

$$H_0: \text{ Batch of transistors is of good quality.}$$
$$H_1: \text{ Batch of transistors is of poor quality.}$$

   (1) Reject good-quality batch.
   (2) Accept good-quality batch.
   (3) Reject poor-quality batch.
   (4) Accept poor-quality batch.

11.1.32   An experiment is carried out to determine whether a new anesthetic results in lower death rates than other anesthetics.
   a. What should the null hypothesis be? The alternative hypothesis?
   b. What are the consequences of a Type I error? A Type II error?

11.1.33   For each of the following test situations, specify the null and alternative hypotheses and describe the possible Type I and Type II errors:
   a. The mean donation per contributor to a certain charity was $11.65 before a new public relations program was initiated. A random sample of donations received after the new program went into effect is used to determine whether the mean contribution is now larger.
   b. Last year, the mean duration of marriages that ended in divorce or annulment in a certain state was 4.7 years. A sociologist wishes to test whether new divorce legislation has changed the mean duration, based on a random sample of divorce records filed since the legislation was enacted.
   c. A chemist has developed a process for reducing the amount of sulfur emissions at nuclear power plants. Random samples of emissions treated by this process are obtained to test whether the mean sulfur content of the emissions has decreased.

11.1.34   A computer software company has devised a new type of spreadsheet unlike any now on the market. The company's president thinks it is equally likely that this idea will be a sensation or a flop. If it does flop, there will be small losses, but if it is a sensation, the gains will be substantial. The president orders a marketing survey to test consumer interest.
   a. Should the null hypothesis be that the idea will be a sensation or a flop?
   b. What value of α would you recommend?

11.1.35   The random variable X denotes the number of successes obtained in a random sample of $n = 50$ trials of an experiment. The researcher would like to show that p, the unknown probability of success, is less than .6. State the null and alternative hypotheses to be tested.

11.1.36   The lie detector test is used to indicate whether a person is telling the truth based on measurements of the person's blood pressure, respiration, and perspiration rates. Many people have criticized its use, claiming that it can give wrong indications. To analyze its effectiveness, psychologists gave the lie detector test to 1,000 people,. of whom 800 were instructed to tell the truth and 200 were instructed to lie. At the conclusion of the test, the polygraph machine indicated that approximately 80 of the truth tellers were lying and that approximately 24 of the liars were telling the truth.

a. For this test, the null hypothesis is that the person is telling the truth. What is the probability of committing a Type I error?

b. What is the probability of committing a Type II error?

c. The power of a test is the probability of rejecting H_0, given that H_0 is false. Find the power of the lie detector test.

 11.1.37 Pharmaceutical companies spend billions of dollars on research and development of new medicines. To obtain permission from the Food and Drug Administration (FDA) to market the medicine, the manufacturer must provide evidence that the medicine is safe and effective.

a. The FDA assumes that a new medicine is unsafe or ineffective unless there is strong evidence indicating otherwise. State the null and alternative hypotheses.

b. Given the nature of the null and alternative hypotheses as described in part **a,** describe the consequences of a Type I error.

c. Describe the consequences of a Type II error.

d. Suppose the FDA wants to be very confident that unsafe medicines do not reach the market. Should the FDA be more concerned that α is very small or that β is very small? Explain.

11.1.38 The random variable X denotes the number of successes obtained in $n = 10$ trials of an experiment. The researcher would like to show that p, the unknown probability of success, is greater than .4.

a. State the null and alternative hypotheses to be tested.

b. Suppose the researcher decides to reject H_0 if $X \geq 8$. Let $\alpha = P(X \geq 8)$. Find α.

c. Suppose that, in fact, $p = .5$. The researcher will commit a Type II error if $X < 8$. Find β if $p = .5$.

d. Suppose that, in fact, $p = .6$. The researcher will commit a Type II error if $X < 8$. Find β if $p = .6$.

e. Suppose that, in fact, $p = .7$. The researcher will commit a Type II error if $X < 8$. Find β if $p = .7$.

f. Explain why the results differ in parts **c–e.**

11.2 Testing Hypotheses About a Population Mean When Variance Is Known

In this section, we use the methodology of Section 11.1 to test hypotheses about the mean of a normal population with a known population variance σ^2. In Section 11.3, we see that these assumptions can be relaxed when the sample size is large.

We begin by testing the simple null hypothesis that the population mean is equal to some specified value μ_0. The null hypothesis is denoted

$$H_0: \quad \mu = \mu_0$$

Suppose the alternative hypothesis is the one-sided alternative that the population mean is less than μ_0. Thus, the alternative hypothesis is

$$H_1: \quad \mu < \mu_0$$

Once we have solved this problem, the solutions to other problems with different hypotheses will follow as natural extensions.

It is natural to base a test concerning the population mean on the sample mean $\bar{X}$. If the population is normal (or if the sample size is large), then *when the null hypothesis is true,* the random variable $\bar{X}$ has a normal distribution with mean μ_0 and variance σ^2/n, as shown in Figure 11.1. We would reject H_0 in favor of H_1 only if the observed sample mean $\bar{x}$ was substantially less than μ_0. In Figure 11.1, we would reject H_0 in favor of H_1 when the observed sample mean fell in the left tail of the distribution to the left of a critical value, denoted $\bar{x}^*$. For this test, the rejection region is the set of all values of the sample mean less than $\bar{x}^*$, and the acceptance region is the set of all values of the sample mean greater than or equal to $\bar{x}^*$. In Figure 11.1, we need to find the value $\bar{x}^*$ such that the area to its left is α, the level of significance of the test.

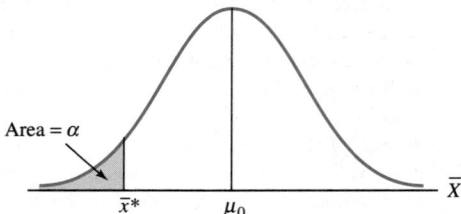

Area = α

$\bar{x}^*$ μ_0 $\bar{X}$

FIGURE 11.1   *Critical value $\bar{x}^*$ for a one-tailed test*

When we test H_0: $\mu = \mu_0$ against H_1: $\mu < \mu_0$, we use the decision rule:

Reject H_0 in favor of H_1 if and only if the observed sample mean $\bar{x}$ is less than the critical value $\bar{x}^*$.

When H_0 is true, the probability of observing a sample mean less than $\bar{x}^*$ is $P(\bar{X} < \bar{x}^*)$. We select $\bar{x}^*$ so that this probability is α, the level of significance of the test. That is, when H_0 is true, the probability of observing a sample mean less than $\bar{x}^*$ is α. If this event occurs, we will reject H_0 even though it is true. Thus, we have designed the test so that the probability of committing a Type I error equals the preassigned value α.

The Critical Value $\bar{x}^*$

If the population is normal with mean μ_0 and variance σ^2, then the random variable $\bar{X}$ follows a normal distribution with mean μ_0 and variance σ^2/n. Thus the random variable

$$Z = \frac{\bar{X} - \mu_0}{\sigma/\sqrt{n}}$$

follows the standard normal distribution $N(0, 1)$. If the area in the left tail of the distribution is α, we have

$$P(Z < -z_\alpha) = \alpha$$

When H_0 is true, we obtain

$$P\left(\frac{\bar{X} - \mu_0}{\sigma/\sqrt{n}} < -z_\alpha\right) = \alpha$$

After multiplying by $\sigma/\sqrt{n}$, we obtain

$$P(\bar{X} - \mu_0 < -z_\alpha\sigma/\sqrt{n}) = \alpha$$

or

$$P(\bar{X} < \mu_0 - z_\alpha\sigma/\sqrt{n}) = \alpha$$

Thus we find that the critical value $\bar{x}^*$ is the value

$$\bar{x}^* = \mu_0 - z_\alpha\sigma/\sqrt{n}$$

Critical value $\bar{x}^*$

When we test the null hypothesis

$$H_0: \quad \mu = \mu_0$$

against the one-sided alternative hypothesis

$$H_1: \quad \mu < \mu_0$$

the critical value $\bar{x}^*$ is

$$\bar{x}^* = \mu_0 - z_\alpha\sigma/\sqrt{n}$$

Whenever the observed sample mean $\bar{x}$ is less than $\bar{x}^*$, the observed Z score

$$z = \frac{\bar{x} - \mu_0}{\sigma/\sqrt{n}}$$

is less than $-z_\alpha$, and vice versa. Thus, the rule that we should reject H_0 if and only if the observed sample mean $\bar{x}$ is less than the critical value $\bar{x}^*$ is equivalent to the decision rule that we should reject H_0 if and only if the observed Z score is less than $-z_\alpha$. This situation is illustrated in Figure 11.2.

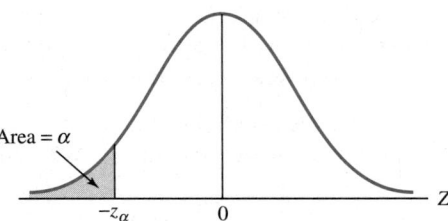

Area $= \alpha$

$-z_\alpha$ 0

FIGURE 11.2   *Critical value $-z_\alpha$ for a one-tailed test*

Procedure for Testing a Hypothesis About a Population Mean

The six-step procedure in the box on page 408 summarizes how to test the null hypothesis $H_0: \mu = \mu_0$ against the alternative hypothesis $H_1: \mu < \mu_0$ when the population is normal and the variance is known.

Testing the mean of a normal population with population variance known

Step 1: State H_0 and H_1.

Step 2: Select a level of significance α.

Step 3: Determine the critical value $-z_\alpha$ using Table A.5 in the Appendix.

Step 4: Obtain the sample of n observations and calculate the observed sample mean $\bar{x}$.

Step 5: Calculate the observed Z score

$$z = \frac{\bar{x} - \mu_0}{\sigma/\sqrt{n}}$$

Step 6: Reject H_0 in favor of H_1 if $z < -z_\alpha$. An equivalent decision rule is

Reject H_0 in favor of H_1 if $\bar{x} < \bar{x}^*$,

where $\bar{x}$ is the observed value of the sample mean and where

$$\bar{x}^* = \mu_0 - z_\alpha \sigma/\sqrt{n}$$

EXAMPLE 11.11

Testing a Hypothesis About a Mean of a Normal Population with Known Variance

Metaltech Industries manufactures carbide drill tips used in drilling oil wells. The life of a carbide drill tip is measured by how many feet can be drilled before the tip wears out. Metaltech claims that under typical drilling conditions, the life of a carbide tip follows a normal distribution with mean 32 feet and population variance 16. Suppose some customers disagree with Metaltech's claims and argue that Metaltech is overstating the mean. Metaltech agrees to examine a random sample of $n = 25$ carbide tips to test the null hypothesis

$$H_0: \quad \mu = 32$$

against the alternative hypothesis

$$H_1: \quad \mu < 32$$

If the null hypothesis is rejected, Metaltech has agreed to give customers a price rebate on past purchases. Suppose Metaltech decides to use a 5% level of significance and the observed sample mean is $\bar{x} = 29.5$. Test H_0 against H_1.

SOLUTION   To perform the test, we follow the six-step testing procedure.

Step 1: State the null and alternative hypotheses. We have

$$H_0: \quad \mu = 32 \quad \text{and} \quad H_1: \quad \mu < 32$$

Step 2: The level of significance is $\alpha = .05$.

Step 3: From Table A.5, we obtain $-z_\alpha = -z_{.05} = -1.645$.

Step 4: Based on a sample of 25 observations, the sample mean is $\bar{x} = 29.5$ feet.

Step 5: Calculate the observed Z score:

$$z = \frac{\bar{x} - \mu_0}{\sigma/\sqrt{n}} = \frac{29.5 - 32}{4/\sqrt{25}} = -3.125$$

Step 6:   Compare the observed Z score $z = -3.125$ with the critical value $-z_\alpha = -1.645$. We reject H_0 in favor of H_1 because the observed Z score is less than the critical value and falls in the rejection region. This situation is illustrated in Figure 11.3.

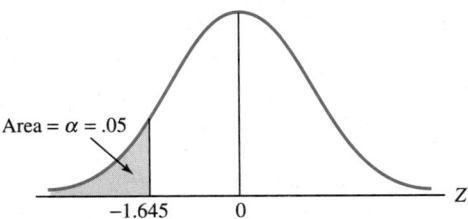

Area $= \alpha = .05$

-1.645 0 Z

FIGURE 11.3   *Critical value* $-z_{.05} = -1.645$
associated with the test in Example 11.11

EXAMPLE 11.12   **Calculation of the Critical Value** $\bar{x}*$

In Example 11.11, we showed that the critical value of Z was -1.645, so the rejection region consisted of all Z scores less than -1.645. Suppose we decide to use the sample mean as the test statistic rather than the Z score. Let us find the critical value $\bar{x}*$.

SOLUTION   We have

$$\begin{aligned}
\bar{x}* &= \mu_0 - z_\alpha \sigma/\sqrt{n} \\
&= 32 - 1.645(4/\sqrt{25}) \\
&= 30.684
\end{aligned}$$

This means that rejecting H_0 when $z < -1.645$ is equivalent to rejecting H_0 when the observed sample mean is less than 30.684. Because the observed sample mean is $\bar{x} = 29.5$, we reject H_0 in favor of H_1.

In Example 11.11, the fact that we rejected H_0 in favor of H_1 does not mean that we have *proved* that the true population mean is less than 32. Of course, the observed sample mean $\bar{x} = 29.5$ does cast doubt on the truth of the null hypothesis. In the sample of 25 observations, Metaltech obtained a Z score that was substantially less than the critical value -1.645. Metaltech now has two options: (1) It can continue to assume that the null hypothesis $H_0: \mu = 32$ is true, or (2) it can reject H_0 in favor of the alternative hypothesis that $\mu < 32$. Which course of action seems more reasonable? If Metaltech continues to support the null hypothesis, then it must admit that it has obtained a very unlikely sample of observations. That is, if H_0 were true, we would be very unlikely to obtain a sample of observations yielding a Z score less than -1.645. When H_0 is true, we should get a Z score less than -1.645 only 5% of the time. In other words, only 5% of all possible samples from the population would yield a Z score less than the critical value -1.645 if $\mu = 32$. However, if H_0 were *false* (so $\mu < 32$), it would not be so unusual to get a Z score less than -1.645. Thus, if we choose $\alpha = .05$, it is reasonable to reject the null hypothesis $H_0: \mu = 32$ in favor of $H_1: \mu < 32$ in Example 11.11.

Testing a Composite Null Hypothesis

Now consider the difference between testing the simple null hypothesis $H_0: \mu = \mu_0$ against $H_1: \mu < \mu_0$ and testing the composite null hypothesis $H_0: \mu \geq \mu_0$ against $H_1: \mu < \mu_0$ at a significance level α. If the population mean is exactly μ_0, the level of significance of each test will be exactly α. That is, if $\mu = \mu_0$, then the probability that the test statistic will fall in the rejection region is exactly α. On the other hand, if in fact $\mu > \mu_0$, then we are even less likely to incorrectly reject H_0. That is, if $\mu > \mu_0$, then the probability that the test statistic will fall in the rejection region is *less* than α and the probability of committing a Type I error is less than α.

When we test the simple hypothesis $H_0: \mu = \mu_0$ against $H_1: \mu < \mu_0$, the level of significance of the test is exactly α. When we test $H_0: \mu \geq \mu_0$ against $H_1: \mu < \mu_0$, the level of significance of the test is less than or equal to α. Specifically, the significance level is exactly α if $\mu = \mu_0$, and the significance level is less than α if $\mu > \mu_0$.

Testing a composite null hypothesis

> To test the composite null hypothesis
>
> $$H_0: \quad \mu \geq \mu_0$$
>
> against the one-sided alternative hypothesis
>
> $$H_1: \quad \mu < \mu_0$$
>
> follow the same six steps as when testing the simple null hypothesis $H_0: \mu = \mu_0$ against $H_1: \mu < \mu_0$.

Now consider the problem of testing the simple null hypothesis

$$H_0: \quad \mu = \mu_0$$

against the one-sided alternative hypothesis

$$H_1: \quad \mu > \mu_0$$

In this case, we would doubt the validity of the null hypothesis if the sample mean was substantially *greater than* the hypothesized value μ_0. Once again, if the population is normal and the variance is known, the random variable

$$Z = \frac{\bar{X} - \mu_0}{\sigma/\sqrt{n}}$$

follows the standard normal distribution. Because Z is a standard normal variable, we have

$$P(Z > z_\alpha) = \alpha$$

After substituting for Z, we obtain the decision rules given in the accompanying box. For this test, the rejection region consists of all observed Z scores greater than z_α. Equivalently, the rejection region consists of all values of the sample mean greater than the critical value $\bar{x}^*$.

Testing a simple null hypothesis: normal population with known variance

---

To test the null hypothesis

$$H_0: \quad \mu = \mu_0$$

against the alternative hypothesis

$$H_1: \quad \mu > \mu_0$$

use the decision rule

$$\text{Reject } H_0 \text{ in favor of } H_1 \text{ if } z > z_\alpha.$$

An equivalent decision rule is

$$\text{Reject } H_0 \text{ in favor of } H_1 \text{ if } \bar{x} > \bar{x}^*,$$

where the critical value is $\bar{x}^* = \mu_0 + z_\alpha \sigma / \sqrt{n}$.

When the alternative hypothesis is one-sided, the entire rejection region is placed under one tail of the distribution of the test statistic. When the alternative hypothesis is

$$H_1: \quad \mu > \mu_0$$

the entire rejection region is placed under the right tail of the distribution. When the alternative hypothesis is

$$H_1: \quad \mu < \mu_0$$

the entire critical region is placed under the left tail of the distribution.

The most frequently used levels of significance are 10%, 5%, and 1%. The following table shows the most common critical values z_α and $-z_\alpha$ used in one-tailed tests:

Level of significance α	CRITICAL Z SCORE	
	$H_1: \quad \mu < \mu_0$	$H_1: \quad \mu > \mu_0$
10%	$-z_{.10} = -1.28$	$z_{.10} = 1.28$
5%	$-z_{.05} = -1.645$	$z_{.05} = 1.645$
1%	$-z_{.01} = -2.33$	$z_{.01} = 2.33$

EXAMPLE 11.13 **Testing a Hypothesis About a Mean of a Normal Population with Known Variance**

A teachers' union is on strike for higher wages. The union claims that the mean salary for teachers is at most $25,000 per year, but a state legislator thinks that it is higher than $25,000 per year. The legislator does not want to reject the union's claim, however, unless the evidence is very strong against it. Because of this, a 1% level of significance is chosen. The legislator wants to test the null hypothesis

$$H_0: \quad \mu \leq \$25{,}000$$

against the one-sided alternative hypothesis

$$H_1: \quad \mu > \$25{,}000$$

Assume that salaries follow a normal distribution and that the population standard deviation is known to be $\sigma = \$4,000$. A random sample of $n = 100$ teachers is obtained, and the sample mean is $\bar{x} = \$27,000$. Perform the test.

SOLUTION   To perform the test, follow the six-step testing procedure. For level of significance $\alpha = .01$, the critical value is $z_\alpha = z_{.01} = 2.33$. Because the observed sample mean is $\bar{x} = \$27,000$, the Z score is

$$z = \frac{\bar{x} - \mu_0}{\sigma/\sqrt{n}} = \frac{27,000 - 25,000}{4,000/\sqrt{100}} = 5.0$$

Because $z = 5.0$ exceeds the critical value $z_\alpha = 2.33$ and falls in the rejection region, we reject H_0 in favor of H_1. This situation is illustrated in Figure 11.4.

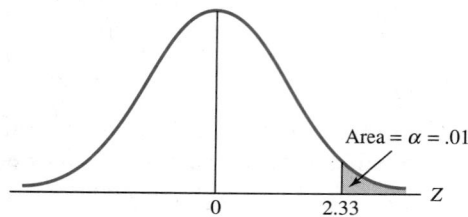

FIGURE 11.4   *One-tailed test for Example 11.13*

A Two-Tailed Test of the Population Mean

Now consider the test of the simple null hypothesis

$$H_0: \quad \mu = \mu_0$$

against the two-sided alternative hypothesis

$$H_1: \quad \mu \neq \mu_0$$

In performing this test, it is assumed that we have no reason to suspect that the population mean is on one side or the other of the hypothesized value μ_0. Thus we would reject H_0 if the observed sample mean was much greater than μ_0 or much less. Once again, if the population is normal with variance σ^2, then the Z score

$$Z = \frac{\bar{X} - \mu_0}{\sigma/\sqrt{n}}$$

follows the standard normal distribution. For a level of significance α, we have

$$P(Z < -z_{\alpha/2}) = \alpha/2 \quad \text{and} \quad P(Z > z_{\alpha/2}) = \alpha/2$$

Thus, the probability that Z exceeds $z_{\alpha/2}$ or is less than $-z_{\alpha/2}$ is α. We reject the null hypothesis if the observed Z score is less than $-z_{\alpha/2}$ or greater than $z_{\alpha/2}$.

Performing a two-tailed test of a mean of a normal population

Assume the population is normal and the variance is known. To test the null hypothesis

$$H_0: \quad \mu = \mu_0$$

against the two-sided alternative hypothesis

$$H_1: \quad \mu \neq \mu_0$$

select a level of significance α and find the critical values $-z_{\alpha/2}$ and $z_{\alpha/2}$ using Table A.5 in the Appendix. Obtain the sample of n observations and calculate the observed sample mean $\bar{x}$. Then calculate the observed Z score

$$z = \frac{\bar{x} - \mu_0}{\sigma/\sqrt{n}}$$

and use the decision rule

Reject H_0 in favor of H_1 if $z < -z_{\alpha/2}$ or if $z > z_{\alpha/2}$.

An equivalent decision rule is

Reject H_0 if $\bar{x} < \bar{x}_1^*$ or if $\bar{x} > \bar{x}_2^*$,

where the critical values that separate the acceptance region from the rejection region are

$$\bar{x}_1^* = \mu_0 - z_{\alpha/2}\sigma/\sqrt{n} \quad \text{and} \quad \bar{x}_2^* = \mu_0 + z_{\alpha/2}\sigma/\sqrt{n}$$

Thus, if the alternative hypothesis is two-sided, we perform a *two-tailed test* because part of the rejection region is placed in each tail of the distribution of the test statistic. This is illustrated in Figure 11.5.

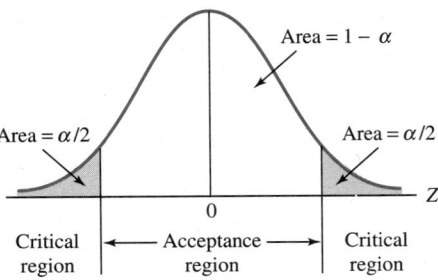

FIGURE 11.5   *Critical values $-z_{\alpha/2}$ and $z_{\alpha/2}$ for a two-tailed test*

As with one-tailed tests, the most commonly used values for the level of significance α are 10%, 5%, and 1%. The following table shows the most commonly used critical values $z_{\alpha/2}$ and $-z_{\alpha/2}$ when performing a two-tailed test:

Level of signifi- cance α	CRITICAL Z SCORE
	H_1: $\mu \neq \mu_0$
10%	$-z_{.05} = -1.645$ and $z_{.05} = 1.645$
5%	$-z_{.025} = -1.96$ and $z_{.025} = 1.96$
1% ·	$-z_{.005} = -2.58$ and $z_{.005} = 2.58$

EXAMPLE 11.14   **A Two-Tailed Test of a Mean of a Normal Population with Known Variance** ⎯⎯⎯

Members of Congress regularly tour foreign countries on fact-finding missions. A newspaper claims that daily expenditures of individuals on such junkets are normally distributed with mean $300 and standard deviation $\sigma = \$50$. To test this claim, a random sample of 25 members of Congress is obtained. Because we have no reason to suspect that the true mean is greater than $300 rather than less than $300, it is natural to test the simple null hypothesis

$$H_0: \quad \mu = 300$$

against the composite two-sided alternative hypothesis

$$H_1: \quad \mu \neq 300$$

Suppose the sample mean is $\bar{x} = \$260$. Test the null hypothesis using a 5% level of significance.

SOLUTION   For level of significance $\alpha = .05$, the two critical values are $z_1 = -z_{\alpha/2} = -1.96$ and $z_2 = z_{\alpha/2} = 1.96$. Thus the rejection region contains all Z scores less than -1.96 or greater than 1.96. The observed Z score is

$$z = \frac{\bar{x} - \mu_0}{\sigma/\sqrt{n}} = \frac{260 - 300}{50/\sqrt{25}} = -4$$

The test statistic falls in the rejection region because the observed Z score $z = -4$ is less than the critical value $z_1 = -1.96$, and so we reject H_0 in favor of H_1. See Figure 11.6. It appears that the population mean is less than the $300 claimed in the newspaper.

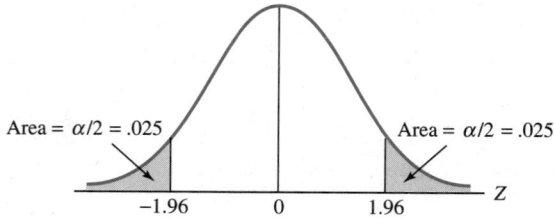

Area = $\alpha/2 = .025$ Area = $\alpha/2 = .025$

-1.96 0 1.96 Z

FIGURE 11.6   *Two-tailed test for Example 11.14*

EXAMPLE 11.15   **Finding the Critical Values of $\bar{X}$ for a Two-Tailed Test** ⎯⎯⎯

For Example 11.14, find the critical values of $\bar{X}$ that separate the critical regions from the acceptance region.

SOLUTION   Because we have used a two-tailed test, there are two critical values of $\bar{X}$. We obtain

$$\bar{x}_1^* = \mu_0 - z_{\alpha/2}\sigma/\sqrt{n} = 300 - 1.96(50/\sqrt{25}) = 280.40$$

and

$$\bar{x}_2^* = \mu_0 + z_{\alpha/2}\sigma/\sqrt{n} = 300 + 1.96(50/\sqrt{25}) = 319.60$$

See Figure 11.7.

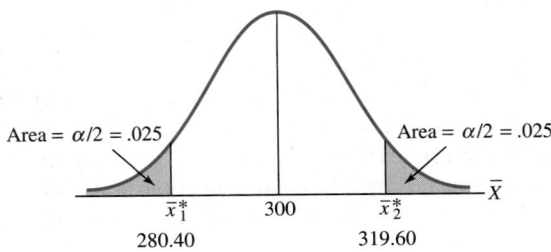

Area = $\alpha/2$ = .025   Area = $\alpha/2$ = .025

$\bar{x}_1^*$ 300 $\bar{x}_2^*$

280.40   319.60

FIGURE 11.7 *Critical values for the two-tailed test in Example 11.15*

In Example 11.14, the decision rule "Reject H_0 if the Z score is less than $z_1 = -1.96$ or greater than $z_2 = 1.96$" is equivalent to the decision rule "Reject H_0 if the observed sample mean $\bar{x}$ is less than $\bar{x}_1^* = 280.40$ or greater than $\bar{x}_2^* = 319.60$." Because we obtained $\bar{x} = 260$ in our sample, which falls in the rejection region, we reject H_0.

Two-Sided Hypothesis Tests and Confidence Intervals

There is a close connection between performing a two-tailed test with a level of significance α and calculating a confidence interval having level of confidence $(1 - \alpha)$. As already stated, for a two-tailed test using level of significance α, the acceptance region consists of all values of the sample mean between

$$\bar{x}_1^* = \mu_0 - z_{\alpha/2}\sigma/\sqrt{n} \quad \text{and} \quad \bar{x}_2^* = \mu_0 + z_{\alpha/2}\sigma/\sqrt{n}$$

Thus we accept H_0 if and only if the observed sample mean $\bar{x}$ is less than $z_{\alpha/2}\sigma/\sqrt{n}$ units from the hypothesized value μ_0. A $100(1 - \alpha)\%$ confidence interval for the population takes the form

$$(\bar{x} - z_{\alpha/2}\sigma/\sqrt{n}, \ \bar{x} + z_{\alpha/2}\sigma/\sqrt{n})$$

where $\bar{x}$ is the observed sample mean. This confidence interval contains all values of μ that are less than $(z_{\alpha/2}\sigma/\sqrt{n})$ units from $\bar{x}$. Thus, the $100(1 - \alpha)\%$ confidence interval for μ contains all possible values of μ_0 that would not be rejected if we performed a two-tailed test to test $H_0: \mu = \mu_0$ against $H_1: \mu \neq \mu_0$ using level of significance α.

A Further Comment on Statistical Significance Versus Practical Significance

As we mentioned at the end of the previous section, statistical significance and practical significance are not the same. If a result is statistically significant, then it is difficult to explain the result as occurring because of random chance. The fact that a result is statistically significant does not mean that it is important or of practical significance.

Consider the following hypothetical example based on a real-life occurrence. Suppose that vocational specialists believe that there are differences between the vocabulary abilities of children from urban areas and children from rural areas. A vocabulary test containing 200 words is designed. A student gets 2 points for knowing the meaning of the word. A student gets 1 point for being partially correct. In the past, scores achieved by children from urban areas have had a mean of $\mu = 148$ with standard deviation $\sigma = 10$. The same vocabulary test is given to a random sample of $n = 6{,}400$ rural children. The sample mean for the rural children is $\bar{x} = 147$. Is this 1-point difference statistically significant? To determine this, we test the null hypothesis $H_0\colon \mu = 148$ against the alternative hypothesis $H_1\colon \mu < 148$. We obtain the test statistic z as follows:

$$z = \frac{\bar{x} - \mu}{\sigma/\sqrt{n}} = \frac{148 - 147}{10/\sqrt{6{,}400}} = 8.00$$

Even if we choose $\alpha = .001$, this value falls in the critical region, so we would reject H_0 in favor of H_1. It follows that the difference between the vocabulary abilities of urban children and rural children is definitely statistically significant. The difference is too large to be attributed to chance.

Does this mean that the result is of any practical importance? Should we petition the state and the federal government to begin pouring more money into rural schools to improve the vocabulary abilities of rural children? Of course not. To assess the practical significance of the result, we need to use common sense. Suppose there is indeed a 1-point difference in the mean scores of urban and rural children. This is equivalent to a partial understanding of one word in a test involving 200 words. To most people, this difference is not large enough to require any remedial action. In fact, most people would say that the difference is so small that no action whatsoever is needed. The result indicates that urban children appear to be statistically different from rural children. The difference is so small, however, that it is not important.

The z statistic that is used to test a hypothesis about a population mean takes the form $(\bar{x} - \mu_0)/(\sigma/\sqrt{n})$, where μ_0 is the hypothesized value of the mean. For a given value of $(\bar{x} - \mu_0)$ and a given value of σ, the absolute value of z increases as n increases. It follows that, if the sample size n is very large, the test result will be statistically significant even if the difference $|\bar{x} - \mu_0|$ is very small. A large sample is useful because it enables us to detect very small deviations from the null hypothesis. Because the test statistic z represents $|\bar{x} - \mu_0|$ divided by the standard error $\sigma/\sqrt{n}$, if the sample size n is very large, even a very small difference $|\bar{x} - \mu_0|$ will yield a large and statistically significant value of z.

When the z statistic is statistically significant, the evidence supports the hypothesis that H_0 is not exactly correct. This says absolutely nothing about whether the discrepancy we have observed is of any practical importance.

Exercises

Statistical Concepts

11.2.1 Suppose you want to perform a one-sided test of hypothesis about the mean of a normal population having known standard deviation. Suppose you choose the level of significance to be α. Explain how to find the critical value z_α.

11.2.2 Suppose you want to perform a one-sided test of hypothesis about a population mean. Explain why it might be important to know that the population follows a normal distribution.

11.2.3 Explain the connection between a two-tailed test of a population mean using level of significance α and a confidence interval having level of confidence $(1 - \alpha)$.

11.2.4 Suppose you want to perform a two-sided test of hypothesis about the mean of a normal population using level of significance α. Suppose $\bar{x}$ lies less than $z_{\alpha/2}\sigma/\sqrt{n}$ units from μ. Show that a confidence interval having level of confidence $(1 - \alpha)$ contains the population mean μ, and show that you will not reject the null hypothesis.

11.2.5 *True or false:* Suppose you are testing the null hypothesis $H_0: \mu \leq 127$ against $H_1: \mu > 127$. H_0 will not be rejected if $\bar{X}$ is less than 127. Explain.

11.2.6 *True or false:* Suppose you are testing the null hypothesis $H_0: \mu \leq 127$ against $H_1: \mu > 127$. H_0 will be rejected only if $\bar{X}$ is significantly greater than 127. Explain.

11.2.7 Explain the difference between a one-tailed test and a two-tailed test of a population mean. How do you recognize when a test should be one-sided or two-sided?

11.2.8 *True or false:* A test result that is highly statistically significant must be very important. Explain.

11.2.9 Other things being equal, which value provides stronger evidence in favor of the null hypothesis: $z = 1.24$ or $z = 6.24$?

11.2.10 Can a result that is found to be statistically significant still be due to random chance? Explain.

11.2.11 *True or false:* With a large sample, even a small difference from what is assumed in H_0 can be statistically significant. Explain.

Statistical Drills

11.2.12 Suppose we use a standard normal Z score as our test statistic when performing some test. For each of the following situations, graph the sampling distribution of Z and sketch the location of the rejection region. The location of the critical region consists of all Z scores such that
  a. $z > 1.96$ **b.** $z > 1.645$ **c.** $z > 2.575$
  d. $z < -1.645$ or $z > 1.645$
  e. $z < -1.96$ or $z > 1.96$
  f. $z < -2.575$ or $z > 2.575$
  g. For each of the rejection regions listed in parts **a**–**f**, what is the probability of a Type I error?

11.2.13 A random sample of 100 observations from a normal population with standard deviation 50 yielded a sample mean of 109.
  a. Test the null hypothesis that $\mu = 98$ against the alternative hypothesis that $\mu > 98$. Let $\alpha = .05$. Interpret the results of the test.
  b. Test the null hypothesis that $\mu = 98$ against the alternative hypothesis that $\mu > 98$. Let $\alpha = .01$. Interpret the results of the test.
  c. Compare the results of the two tests you conducted in parts **a** and **b**. Explain why the results differ.

11.2.14 Suppose we take a random sample of n observations from a normal population with known standard deviation $\sigma = 40$. We obtain the sample mean $\bar{x} = 136$. Let $\alpha = .05$. Test $H_0: \mu \leq 130$ against $H_1: \mu > 130$. Suppose the sample size is as follows:

a. 25   **b.** 40   **c.** 100   **d.** 400

e. Explain why you make different decisions as the sample size increases.

f. In each case in parts **a–d,** find the observed Z score. Why does the Z score increase as the sample size increases?

g. In each case, find the critical value of $\bar{X}$ that separates the acceptance region from the critical region. Explain why these critical values change as the sample size changes.

11.2.15   Suppose we take a random sample of n observations from a normal population with known standard deviation $\sigma = 40$. We obtain the sample mean $\bar{x} = 116$. Let $\sigma = .05$. Test $H_0: \mu = 110$ against $H_1: \mu \neq 110$. Suppose the sample size is as follows:

a. 25   **b.** 40   **c.** 100   **d.** 400

e. Explain why you make different decisions as the sample size increases.

f. In each case in parts **a–d,** find the observed Z score. Why does the Z score change as the sample size increases?

g. In each case, find the critical value of $\bar{X}$ that separates the acceptance region from the critical region. Explain why these critical values change as the sample size changes.

h. In each case, construct a confidence interval for μ having level of confidence $(1 - \alpha)$. Explain how these confidence intervals can be used to test H_0 against H_1.

11.2.16   Let X be a normal random variable with known standard deviation $\sigma = 40$. Suppose we take a random sample of $n = 25$ observations and obtain the sample mean $\bar{x} = 136$.

a. Construct a confidence interval for the population mean having level of confidence 99%. Test $H_0: \mu = 130$ against $H_1: \mu \neq 130$ using $\alpha = .01$. Find the critical value of Z and the observed Z score. Show how to test the hypothesis by using the confidence interval.

b. Construct a confidence interval for the population mean having level of confidence 95%. Test $H_0: \mu = 130$ against $H_1: \mu \neq 130$ using $\alpha = .05$. Find the critical value of Z and the observed Z score. Show how to test the hypothesis by using the confidence interval.

Statistical Applications

11.2.17   Records indicate that the amounts paid per month by customers for long-distance telephone calls have a mean of $18.75 and a standard deviation of $7.80.

a. If a random sample of 50 long-distance phone bills is taken, what is the probability that the sample mean is greater than $20?

b. If a random sample of 100 bills is taken, what is the probability that the sample mean is greater than $20?

c. Suppose a random sample of 100 bills during a given month produced a sample mean of $21.25. Does this indicate that the mean level of the amounts billed per month for long-distance telephone calls has increased from $18.75? Test using $\alpha = .05$.

11.2.18   The manager of a fast-food restaurant claims that the mean expenditure per customer is $2.00 and the standard deviation is $.60. A new menu has been introduced at the restaurant, and the manager wants to test whether the mean expenditure has changed. The manager takes a random sample of 100 customers to test $H_0: \mu = \$2.00$ against $H_1: \mu \neq \$2.00$. Suppose the manager uses the decision rule "Accept H_0 if $\$1.90 \leq \bar{X} \leq \$2.10$; otherwise, reject H_0."

a. What is the test statistic?   **b.** What is the acceptance region?

c. What is the critical region?   **d.** What are the critical values of the test statistic?

e. Calculate α, the level of significance of the test.

11.2.19   Repeat Exercise 18 using the decision rule "Accept H_0 if $\$1.95 \leq \bar{X} \leq \$2.05$."

a. What is the test statistic?

b. What is the acceptance region?

c. What is the critical region?

d. What are the critical values of the test statistic?

e. Calculate α, the level of significance of the test.

f. How did the change in the decision rule affect the level of significance of the test?

11.2.20   Repeat Exercise 18 assuming that the sample size is 200 rather than 100.

a. What is the test statistic?

b. What is the acceptance region?

c. What is the critical region?

d. What are the critical values of the test statistic?

e. Calculate α, the level of significance of the test.

f. How did the change in the sample size affect the level of significance of the test?

11.2.21   The job placement center at a university claims that the mean starting salary of college graduates is $21,000 and the standard deviation is $2,500. A random sample of 100 college graduates is taken to test this claim. Use the decision rule "Accept H_0 if $20{,}500 \leq \bar{X} \leq \$21{,}500$; otherwise, reject H_0."

a. Is this a one-tailed or two-tailed test?

b. What is the test statistic?

c. What is the null hypothesis?

d. Given the decision rule, what do you think is the alternative hypothesis?

e. State the critical region, the acceptance region, and the critical values of the test statistic.

f. What is the distribution of the test statistic if H_0 is true?

g. Calculate α, the level of significance of this test.

11.2.22   A tire manufacturer claims that its tires have a mean life of at least 35,000 miles. A random sample of 16 of these tires is tested and the sample mean is 33,000 miles. Assume the population standard deviation is 3,000 miles and the lives of tires are approximately normally distributed. Test the manufacturer's claim using a 5% level of significance. Did you perform a one-tailed test? Why?

11.2.23   In Pennsylvania, automobiles must be inspected every 12 months. Many consumers want the state legislature to repeal this law, claiming that the law gives auto mechanics an opportunity to collect unnecessary inspection fees from the public. A lobbyist for the auto mechanics claims that the average inspection bill is at most $\mu = \$25$. A coalition of consumers claims the average is larger than $25. A random sample of 30 inspection bills is obtained, and the sample mean is $\bar{x} = \$31.86$. Assume the population is approximately normal with variance $\sigma^2 = 64$. Use a 5% level of significance and test the lobbyist's claim. State H_0 and H_1. Did you perform a one-tailed test? Why or why not?

11.2.24   A random sample of 100 observations from a normal population produced a sample mean of 231.7 and a sample standard deviation of 9.3. Because the sample size is large, it is reasonable to use the techniques discussed in this section even though the true standard deviation is unknown. Use the sample standard deviation as an estimate of σ and test the hypothesis H_0: $\mu = 225$ against H_1: $\mu \neq 225$ using $\alpha = .01$. Find the critical value of z and the observed Z score. Show how to test the hypothesis by constructing a confidence interval.

11.3   *p*-Value: Interpretation and Use

As described in this chapter, classical hypothesis testing involves choosing a test statistic and a level of significance α and then finding the associated rejection and acceptance regions. Difficulty arises, however, because in many problems the choice of α is completely arbitrary, particularly when it is difficult or impossible to determine the costs associated with Type I or Type II errors. In another approach to hypothesis testing, the smallest level at which the observed test statistic would be significant in a particular direction is reported.

Consider testing any one-tailed hypothesis. The observed value of the test statistic can be used to compute a tail probability called the *p-value,* or *observed level of significance.*

DEFINITION   ***p*-Value** ───

The ***p*-value** of a test is the probability of obtaining a value of the test statistic as extreme as or more extreme than the observed sample value when the null hypothesis H_0 is true.

The *p*-value answers the question, "If the null hypothesis is true, what is the probability that a random sample will yield the observed test statistic or one whose value is further from the expected value?" If this probability is very small, then we reject the null hypothesis by concluding that the discrepancy is too large to be explained by chance alone. The observed difference is then said to be statistically significant. If the *p*-value is not small, then we do not reject the null hypothesis, because the observed discrepancy may well be due to chance, that is, sampling error.

The *p*-value is the smallest level of significance α at which a null hypothesis can be rejected. The smaller the *p*-value, the more doubt is cast on the validity of the null hypothesis.

Calculating a *p*-value

To calculate a *p*-value, follow these steps:

Step 1:   Record the observed sample mean $\bar{x}$, the hypothesized value of the mean μ_0, the sample size n, and the population standard deviation σ.

Step 2:   Record the standard deviation of the sampling distribution of $\bar{X}$. This standard deviation is $\sigma/\sqrt{n}$, *not* σ.

Step 3:   Sketch a normal distribution showing the sampling distribution of the random variable $\bar{X}$. On the graph, denote the mean as μ_0. Locate the value $\bar{x}$ and shade the area under the curve beyond $\bar{x}$. This area is the one-tailed *p*-value.

Step 4:   To determine this area, calculate the Z score associated with the value $\bar{x}$.

Step 5:   Sketch a standard normal distribution and find the area under the curve beyond this Z score. This area is the one-tailed *p*-value.

Step 6:   If the test is a two-tailed test, double the *p*-value in step 5.

Step 7:   Examine your sketch and check that the *p*-value you have calculated is reasonable.

EXAMPLE 11.16   **Calculation of a *p*-Value** ─────────────────────────────

In Example 11.11, we performed a one-tailed test to test H_0: $\mu = 32$ against the alternative hypothesis H_1: $\mu < 32$. The observed Z score was $z = -3.125$. Calculate the *p*-value of the test.

SOLUTION   The *p*-value of the test is the probability of obtaining a Z score more extreme than the observed Z score of -3.125. From Table A.5, we obtain

$$p\text{-value} = P(Z < -3.125) = .5 - .4991 = .0009$$

This situation is illustrated in Figure 11.8. This *p*-value indicates that if H_0 is true, then in repeated samples we should obtain a Z score as small as or smaller than the one actually observed only 9 times out of 10,000, that is, with probability .0009. Because this value is extremely small, we have serious doubts about the validity of the null hypothesis.

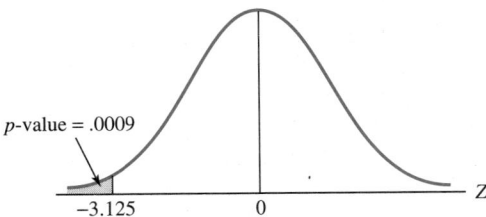

<p-value = .0009>

FIGURE 11.8   *p-value of the test described in Example 11.16*

In Example 11.11, we used a 5% level of significance and rejected H_0. Similarly, if we had used a 1% level of significance, the critical value of Z would have been -2.33, and the observed Z score would still have fallen in the rejection region. In fact, if we had used any level of significance greater than .0009, the observed Z score of -3.125 would have fallen in the rejection region. It follows that the *p*-value .0009 is the smallest level of significance at which the null hypothesis would have been rejected.

In statistical decision making, a *p*-value conveys more information than a report of whether the observed test statistic is statistically significant at some preselected level of significance α. For example, a *p*-value of, say, .002 is considerably more informative than the report that the test statistic is "significant at the .05 level." If we know that the *p*-value is .002, we know that we would have rejected H_0 not only at the .05 level of significance but also at the .005 level of significance. In fact, a *p*-value of .002 tells us that we would have rejected H_0 at all levels of significance exceeding .002.

When the *p*-value of a test is reported, the decision of whether to reject the null hypothesis is left up to the reader. Suppose a reported *p*-value is .04. A researcher who wants to use a 1% level of significance will not reject the null hypothesis, whereas a researcher who wants to use a 5% level of significance will.

It is especially useful to report a *p*-value when we do not have any specific reason for choosing a particular level of significance or when we have little or no information concerning the costs and consequences of committing a Type I or Type II error.

There is no general agreement on how to define *p*-values when the alternative hypothesis is two-sided. One approach is to report the one-tailed *p*-value and mention that the two-tailed *p*-value is presumably nearly twice as large as the one-tailed *p*-value. This is reasonable when the sampling distribution of the test statistic under H_0 is symmetric, because this corresponds to the idea that the standard two-sided test is a combination of two one-sided tests, each at level of significance $\alpha/2$. When the sampling distribution of the test statistic is not symmetric, doubling the p-value to obtain a two-sided p-value is less reasonable.

When publishing the results of a statistical test, many scientists use p-values rather than referring to levels of significance. Rather than selecting α and then conducting a test as explained in Section 11.2, the scientist reports the observed value of the test statistic that was used and the p-value associated with this observed value. This procedure allows the reader of the report to make his or her own decision to accept or reject the null hypothesis based on the reported p-value. The p-value is often referred to as the observed significance level of the test. One advantage of reporting test results as p-values is that readers are able to draw their own conclusions about the reported hypothesis test by choosing their own value for α and comparing it to the reported p-value. Also, the p-value provides an indication of how unusual the observed test statistic is. The p-value of the test represents the probability, assuming that H_0 is true, of obtaining a test statistic as extreme as or more extreme than that which was actually obtained from the sample data.

The following procedure describes how to decide whether to reject a null hypothesis by using a p-value:

1. Choose the maximum value of α that you are willing to tolerate.
2. If the p-value is less than the chosen value of α, then reject the null hypothesis.

Recall that the value of the Z statistic depends on the sample size n. Other things being equal, the larger the value of n, the larger the absolute value of Z. The p-value of a test is the area in the tail of the distribution beyond the value of Z, so the p-value of the test depends on the sample size. Even a small difference $(\bar{x} - \mu)$ can produce a large Z value and a very small (or statistically significant) p-value. A statistically significant p-value means that the observed result is difficult to explain by random chance. A statistically significant p-value does not necessarily mean that the observed result is important. Similarly, an important result may not be detected if the sample size is very small.

Exercises

Statistical Concepts

11.3.1 When a p-value is smaller than .01, some statisticians say the result is called very significant. When the p-value is between .01 and .05, the result is called significant. When the p-value is greater than .10, the result is considered not significant.

a. Suppose you want to test some hypothesis using $\alpha = .01$ as the level of significance. Suppose a researcher states that the p-value is very significant. Would this lead you to accept or reject the null hypothesis?

b Suppose you want to test some hypothesis using $\alpha = .01$ as the level of significance. Suppose a researcher states that the p-value is greater than .10. Would this lead you to accept or reject the null hypothesis?

c. Suppose you want to test some hypothesis using $\alpha = .05$ as the level of significance. Suppose a researcher states that the p-value is less than .01. Would this lead you to accept or reject the null hypothesis?

11.3.2 *True or false:* Suppose you wish to test $H_0: \mu \leq \mu_0$ against $H_1: \mu > \mu_0$. The p-value is the area to the right of the observed Z score.

11.3.3   *True or false:*   Suppose you wish to test $H_0: \mu = \mu_0$ against $H_1: \mu \neq \mu_0$. Suppose the observed Z score is positive. The *p*-value is twice the area to the left of the observed Z score.

11.3.4   *True or false:*   In a left-tailed test, the *p*-value is the area to the left of the test statistic if the test statistic is negative.

11.3.5   *True or false:*   In a two-tailed test, the *p*-value is twice the area to the right of a positive test statistic or to the left of a negative test statistic.

11.3.6   *True or false:*   To interpret a reported *p*-value, you can use the following rule: For a given level of significance α, reject the null hypothesis if and only if the *p*-value is less than α. Explain.

11.3.7   *True or false:*   A *p*-value represents the probability that the null hypothesis is true. Explain.

11.3.8   *True or false:*   A small *p*-value provides evidence supporting the alternative hypothesis. Explain.

11.3.9   *True or false:*   If we reject H_0 and the *p*-value is very small, this means that we feel strongly about rejecting H_0. It does *not* always mean that the population mean is very far away from the value stated in the null hypothesis. Explain.

11.3.10   *True or false:*   Suppose the *p*-value is very small. This means that there is strong evidence to reject the claim that $\mu = \mu_0$. If the sample size is very large, we could obtain a small *p*-value even if the true population mean is very close to μ_0.

11.3.11   Frequently, computers are used to perform hypothesis tests. Usually, the user specifies the kind of test to be carried out, that is, one- or two-tailed, and the distribution to be used. Data are entered and read by the program, and the analysis is done. Results are usually reported in a table that gives the value of the estimate, the standard deviation, the standard error of the estimator, the computed test statistic value, and the *p*-value. Suppose a computer program reports a *p*-value of .0000. Interpret this result. Does this mean that the *p*-value is exactly 0?

11.3.12   *True or false:*   The larger the *p*-value associated with a test of hypothesis, the stronger the support for the null hypothesis.

11.3.13   *True or false:*   The *p*-value of a test is the probability of getting a test statistic as extreme as or more extreme than the observed one. The probability is calculated based on the assumption that the null hypothesis is false. Explain.

11.3.14   *True or false:*   The larger the *p*-value, the stronger the evidence against the null hypothesis.

11.3.15   *True or false:*   The *p*-value of a test depends on the observed data, but the critical values of a test do not. Explain.

11.3.16   *True or false:*   If the *p*-value of a test is .01, then the probability is only .01 that the null hypothesis is true. Explain.

Statistical Drills

11.3.17   An experiment is conducted and the random variable X, the number of successes, denotes the number of successes obtained in $n = 10$ trials. The researcher would like to test $H_0: p \leq .4$ against $H_1: p > .4$.
 a. Suppose the researcher decides to reject H_0 if $X \geq 8$. Let $\alpha = P(X \geq 8)$. Find α.
 b. Suppose the researcher obtains 6 successes in 10 trials. Find the *p*-value associated with $X = 6$.
 c. Compare the *p*-value in part **b** with α in part **a**. What decision should the researcher make?

11.3.18   For each α and observed significance level combination, determine whether the null hypothesis would be accepted or rejected.
 a. $\alpha = .05$; *p*-value $= .07$
 b. $\alpha = .01$; *p*-value $= .07$
 c. $\alpha = .05$; *p*-value $= .03$
 d. $\alpha = .01$; *p*-value $= .02$
 e. $\alpha = .01$; *p*-value $= .007$

11.3.19   Explain the difference between a statistically significant result and a result that has practical significance.

11.3.20   A statistician wanted to test H_0: $\mu \leq 66$ against H_1: $\mu > 66$. The statistician obtained a p-value equal to .07. What is the smallest value of α for which the null hypothesis would be rejected?

11.3.21   A researcher wanted to test H_0: $\mu \geq 83$ against H_1: $\mu < 83$. The test results yielded the test statistic $z = -2.14$. Find the p-value for the test.

11.3.22   A scientist wanted to test H_0: $\mu = 77$ against H_1: $\mu \neq 77$. The test results yielded the test statistic $z = 1.75$. Find the p-value for the test.

11.3.23   A manager wanted to test H_0: $\mu \geq 114$ against H_1: $\mu < 114$. The statistician reported that the p-value of the test was .05. Find the value of the test statistic Z.

11.3.24   A personnel director wanted to test H_0: $\mu = 77$ against H_1: $\mu \neq 77$. The statistician reported that the p-value of the test was .05 and indicated that the Z score was positive. Find the value of the test statistic z.

11.3.25   A statistician wanted to test H_0: $\mu \geq 93$ against H_1: $\mu < 93$. The statistician reported that $n = 64$, $\sigma = 24$, and $\bar{X} = 88$. Find the p-value of the test.

11.3.26   A statistician wanted to test H_0: $\mu \leq 35$ against H_1: $\mu > 35$. The statistician reported that $n = 81$, $\sigma = 18$, and $\bar{X} = 39$. Find the p-value of the test.

Statistical Applications

11.3.27   At a state university, the SAT scores of entering students have had a mean of 1000 and a standard deviation of 180. Each year a sample of applications is taken to see whether the examination scores are at the same level as in previous years. Test the null hypothesis H_0: $\mu = 1000$ against a two-sided alternative hypothesis. A sample of 200 students in this year's class had a sample mean score of 980. Use a .05 level of significance.
a. Test this hypothesis by constructing the confidence interval.
b. Test this hypothesis by using a standardized test statistic.
c. What is the p-value for this test? Show how to perform the test by using the p-value.

11.3.28   A nationwide hamburger chain claims that its Super Burgers contain an average of $\mu = 6$ ounces of beef. The standard deviation is $\sigma = .7$ ounce. In a random sample of 50 Super Burgers, the average weight of beef was 5.7 ounces.
a. Use a 1% level of significance and test H_0: $\mu \geq 6$ against H_1: $\mu < 6$.
b. Find the one-tailed p-value of the test. Show how to perform the test by using the p-value.

11.3.29   Just before Mother's Day, the Florists' Association places commercials on television claiming that the average cost of a dozen roses is at most $18. A television station checks a random sample of 16 florists and finds that the average cost of a dozen roses is $23. Suppose the population standard deviation is $3 and assume that the population of prices is approximately normal.
a. Use a 1% level of significance and test H_0: $\mu \leq \$18$ against H_1: $\mu > \$18$.
b. Find the one-tailed p-value of the test. Can you show how to perform the test by using the p-value?

11.3.30   A firm claims that its flashlight batteries will operate continuously for 100 hours on the average. You select a random sample of 12 of those batteries to test H_0: $\mu = 100$ against H_1: $\mu \neq 100$ using a 5% level of significance. In your sample, you obtain $\bar{x} = 92$. Assume that the population is normal with variance $\sigma^2 = 600$.
a. Do you reject H_0?
b. Find the two-tailed p-value of the test. Show how to perform the test by using the p-value.

11.3.31   During the past decade, many companies have introduced flextime, which allows employees to choose their own work schedules, subject to various management restrictions. Supposedly flextime boosts employee morale and reduces absenteeism. The Bryce Manufacturing Company observed that, during the years prior to the introduction of flextime, the mean number of days of missed work was 6.8. Management takes a random sample of 100 employees who have flextime arrangements. For these 100 employees, the sample mean was 5.4 days of missed work and the sample standard deviation was 2.8 days. Should management conclude that flextime has reduced absenteeism, or does the reduction appear to be due to random chance? Choose an appropriate level of significance and perform the test. Also, find the *p*-value of the test.

11.4 Testing Hypotheses About a Population Mean with Large Sample Sizes

In Section 11.2, we showed how to test a null hypothesis about a mean of a normal population with a known variance σ^2. Often, however, we need such a test when the population is not normal or the variance is not known. Fortunately, if the sample size is large, say, 30 or more, then we can use all the methods described in Section 11.2, with only slight modifications, to test hypotheses under these less rigorous circumstances.

The tests in Section 11.2 rely on the fact that when the population distribution is normal with mean μ_0 and variance σ^2, the sample mean $\overline{X}$ follows a normal distribution with mean μ_0 and variance σ^2/n. According to the Central Limit Theorem, the sampling distribution of $\overline{X}$ for large samples is approximately normal even though the population is not normal. In addition, when the sample size is large, we can replace the unknown standard deviation σ in the formula for the Z score by the sample standard deviation S and still obtain an approximately standard normal variable.

A large-sample test of the population mean

> Suppose it is desired to test a simple or composite null hypothesis about the population mean against some appropriate alternative hypothesis. *If the sample size is large,* the test procedures developed in Section 11.2 can be used even when the population is not normal and the population variance is unknown, provided we replace σ in all the formulas by the observed sample standard deviation s.

EXAMPLE 11.17 **Two-Tailed Test of a Population Mean Using a Large Sample**

The Knight Metalworking Company produces metal ducts that are used in heating and air-conditioning systems in skyscrapers. Several holes have to be drilled in each duct so that they can be connected together with small screws. When a drill press machine is working properly, the holes have a mean diameter of 10 millimeters; the population standard deviation is unknown. When the drilling machine is not operating properly (mainly because of human error and incorrect gauge settings), the holes are either too big or too small. In either case, the metal sheets are defective and cannot be used. The owner of the company wishes to test whether a particular machine is working properly. In this case, it would be natural to test the simple null hypothesis

$$H_0: \quad \mu = 10$$

against the composite two-sided alternative hypothesis

$$H_1: \quad \mu \neq 10$$

A two-sided alternative hypothesis is appropriate because the machine would be operating improperly and need to be adjusted if the sample mean were either much greater than or much less than $\mu = 10$. A random sample of $n = 100$ holes is examined. The observed sample mean is $\bar{x} = 9.6$ millimeters, and the observed sample standard deviation is $s = 1$ millimeter. Test whether the machine is operating properly using a 5% level of significance.

SOLUTION   Because the sample size is large, the procedure described in Section 11.2 can be used if we substitute the sample standard deviation for the population standard deviation when calculating the Z score. There are two critical values of Z, denoted z_1 and z_2. For $\alpha = .05$, the two critical values are $z_1 = -z_{\alpha/2} = -1.96$ and $z_2 = z_{\alpha/2} = 1.96$. Thus the rejection region contains all Z scores less than -1.96 or greater than 1.96. The observed Z score is

$$z = \frac{\bar{x} - \mu_0}{s/\sqrt{n}} = \frac{9.6 - 10}{1/\sqrt{100}} = -4$$

Since the observed Z score falls in the rejection region (it is less than the critical value $z_1 = -1.96$), we reject H_0 in favor of H_1. See Figure 11.9.

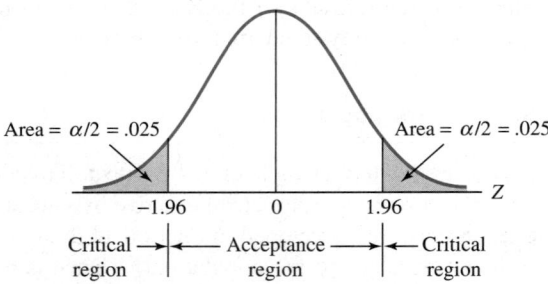

FIGURE 11.9   *Two-tailed test described in Example 11.17*

EXAMPLE 11.18   **Finding the Critical Values of $\bar{X}$ When Variance Is Unknown**

In Example 11.17, find the critical values of $\bar{X}$ that separate the critical region from the acceptance region.

SOLUTION   Because we have used a two-tailed test, there are two critical values of $\bar{X}$. We obtain

$$\bar{x}_1^* = \mu_0 - z_{\alpha/2}s/\sqrt{n} = 10 - 1.96(1/\sqrt{100}) = 9.804$$

and

$$\bar{x}_2^* = \mu_0 + z_{\alpha/2}s/\sqrt{n} = 10 + 1.96(1/\sqrt{100}) = 10.196$$

The decision rule "Reject H_0 if Z is less than $z_1 = -1.96$ or greater than $z_2 = 1.96$" is equivalent to the decision rule "Reject H_0 if $\bar{X}$ is less than $\bar{x}_1^* = 9.804$ or greater than

$\bar{x}_2^* = 10.196$." In our sample, we obtained $\bar{x} = 9.6$, which falls in the rejection region; thus we reject H_0. See Figure 11.10.

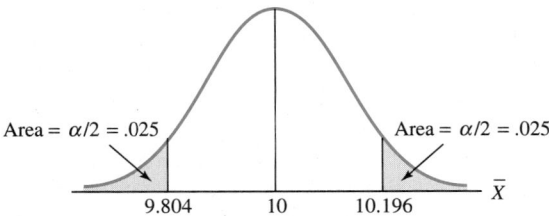

Area = $\alpha/2$ = .025   Area = $\alpha/2$ = .025

9.804   10   10.196 $\bar{X}$

FIGURE 11.10 *Critical values for the two-tailed test described in Example 11.18*

Exercises

Statistical Concepts

11.4.1 *True or false:* Suppose the population standard deviation σ is unknown, as is usually the case. Suppose we use the sample standard deviation s in place of σ when finding the observed Z score to test some hypothesis about a population mean. The resulting test statistic will be approximately correct if the sample size is large, say, $n \geq 30$. Explain.

11.4.2 *True or false:* Suppose the population standard deviation σ is unknown and the sample size is large, but assume the population is not necessarily normal. Suppose we use the sample standard deviation s in place of σ when finding the observed Z score to test some hypothesis about a population mean. The resulting test statistic will be approximately correct. Explain.

Statistical Drills

11.4.3 A researcher wanted to test H_0: $\mu = 95$ against H_1: $\mu \neq 95$, using $\alpha = .05$. In a sample of $n = 100$ observations, the sample mean was $\bar{X} = 98$ and the sample standard deviation was $s = 13.3$. Find the critical value of Z. Find the observed value of Z. Do you reject H_0?

11.4.4 A scientist wanted to test H_0: $\mu \leq 118$ against H_1: $\mu > 118$, using $\alpha = .05$. In a sample of $n = 64$ observations, the sample mean was $\bar{X} = 120$ and the sample standard deviation was $s = 7.6$. Find the critical value of Z. Find the observed value of Z. Do you reject H_0?

11.4.5 An accountant wanted to test H_0: $\mu = 66$ against H_1: $\mu \neq 66$, using $\alpha = .01$. In a sample of $n = 80$ observations, the sample mean was $\bar{X} = 60$ and the sample standard deviation was $s = 24.0$. Find the critical value of Z. Find the observed value of Z. Do you reject H_0?

11.4.6 An attorney wanted to test H_0: $\mu \leq 56$ against H_1: $\mu > 56$, using $\alpha = .01$. In a sample of $n = 60$ observations, the sample mean was $\bar{X} = 59$ and the sample standard deviation was $s = 8.6$. Find the critical value of Z. Find the observed value of Z. Do you reject H_0?

11.4.7 For each of the following situations, $n = 100$ and $\alpha = .05$. In each case, state whether H_0 should be accepted or rejected.
  a. H_0: $\mu \geq 24$; H_1: $\mu < 24$; $\bar{x} = 23.0$; $s = 4.7$
  b. H_0: $\mu \leq 14.9$; H_1: $\mu > 14.9$; $\bar{x} = 15.3$; $s = 1.5$
  c. H_0: $\mu = 167$; H_1: $\mu \neq 167$; $\bar{x} = 169$; $s = .92$

11.4.8 Suppose you wish to test the null hypothesis H_0: $\mu = 200$ against the one-sided alternative hypothesis H_1: $\mu > 200$. Use the decision rule "Reject H_0 if the sample mean of a random sample of 100 items is more than 208." Assume the population standard deviation is 80.

a. Express the decision rule in terms of Z.

b. For this decision rule, find α.

Statistical Applications

11.4.9   A small company that produces a sweetened cereal for children claims that its cereal boxes contain 16 ounces of cereal on the average. One of the company's filling machines has been readjusted recently, and the company wishes to test H_0: $\mu \geqslant 16$ against H_1: $\mu < 16$ using a 1% level of significance. The plant manager randomly selects 100 boxes of this cereal and obtains $\bar{x} = 15.8$ ounces with $s = .3$ ounce. Should we reject H_0?

11.4.10   Last year, the mean number of books borrowed per cardholder at a major university was 18.5 books per semester. A random sample of 100 cardholders showed the following results for this semester: $\bar{x} = 19.54$ books with $s = 6.31$ books. The library administration would like to know whether this semester's mean usage has changed from last semester's. Perform the appropriate two-tailed test using $\alpha = .05$.

11.4.11   A random sample of 100 observations from a population produced the following data:

$$\Sigma x = 1,600 \qquad \Sigma(x_i - \bar{x})^2 = 6,336$$

a. Calculate the sample mean and sample standard deviation.

b. Construct a 95% confidence interval for the population mean using the standard normal distribution.

c. Test H_0: $\mu = 19$ against H_1: $\mu \neq 19$ using $\alpha = .05$.

d. Explain the connection between your answers to parts **b** and **c**.

11.4.12   A random sample of 80 observations from a population produced the following data:

$$\Sigma x = 2,880 \qquad \Sigma(x_i - \bar{x})^2 = 25,596$$

a. Calculate the sample mean and sample standard deviation.

b. Test H_0: $\mu \geqslant 38$ against H_1: $\mu < 38$ using $\alpha = .05$.

c. Calculate the p-value of the test. Show how to perform the test in part **b** by using the p-value.

11.5 Testing Hypotheses About the Mean of a Normal Population with Unknown Variance

In this section, we consider the problem of how to test hypotheses about a population mean μ when the population is normal, the population variance is unknown, and the sample size is small ($n < 30$). Because the variance σ^2 is unknown and the sample size is small, the procedures discussed in Sections 11.2 and 11.3 are not appropriate. Instead, we use a test based on the Student t distribution rather than the standard normal distribution.

In Chapter 10, we showed that if the population is normal with mean μ_0, then the random variable

$$t = \frac{\bar{X} - \mu_0}{S/\sqrt{n}}$$

follows a Student t distribution with $\nu = (n - 1)$ degrees of freedom, where $\bar{X}$ is the sample mean and S is the sample standard deviation based on a random sample of n observations.

Suppose we want to test the null hypothesis

$$H_0: \quad \mu = \mu_0 \quad \text{or} \quad H_0: \quad \mu \leqslant \mu_0$$

against the one-sided alternative hypothesis

$$H_1: \quad \mu > \mu_0$$

As before, a sample mean that greatly exceeds μ_0 would cast doubt on the validity of the null hypothesis. Let $t_{\alpha,\nu}$ denote the critical value of the t distribution such that

$$P(t > t_{\alpha,\nu}) = \alpha$$

If the null hypothesis is true, the t score

$$t = \frac{\bar{X} - \mu_0}{S/\sqrt{n}}$$

follows the Student t distribution with $\nu = (n - 1)$ degrees of freedom, and the t score should exceed the critical value $t_{\alpha,\nu}$ with probability α. If the level of significance of the test is α, we reject H_0 in favor of H_1 whenever the observed t score exceeds $t_{\alpha,\nu}$, where the observed t score is the value

$$t = \frac{\bar{x} - \mu_0}{s/\sqrt{n}}$$

where $\bar{x}$ is the observed sample mean and s is the observed sample standard deviation.

The above discussion pertains to a one-tailed test where the rejection region falls in the right tail of the distribution. Alternatively, we can perform a one-tailed test where the rejection region is in the left tail, or a two-tailed test. Thus, there are three different cases, depending on the form of H_0 and H_1. The summary in the accompanying box describes how to test a hypothesis about the population mean when the population is normal and when the population variance is unknown.

**Testing the mean of a normal population
with unknown variance and a small sample**

Case 1:   To test the null hypothesis

$$H_0: \quad \mu = \mu_0 \qquad \text{or} \qquad H_0: \quad \mu \leq \mu_0$$

against the one-sided alternative hypothesis

$$H_1: \quad \mu > \mu_0$$

calculate the observed t score

$$t = \frac{\bar{x} - \mu_0}{s/\sqrt{n}}$$

and use the decision rule

Reject H_0 in favor of H_1 if $t > t_{\alpha,n-1}$.

Case 2:   To test the null hypothesis

$$H_0: \quad \mu = \mu_0 \qquad \text{or} \qquad H_0: \quad \mu \geq \mu_0$$

(*continued*)

against the one-sided alternative hypothesis

$$H_1: \quad \mu < \mu_0$$

use the decision rule

Reject H_0 in favor of H_1 if $t < -t_{\alpha,n-1}$.

Case 3:   To test the null hypothesis

$$H_0: \quad \mu = \mu_0$$

against the two-sided alternative hypothesis

$$H_1: \quad \mu \neq \mu_0$$

use the decision rule

Reject H_0 in favor of H_1 if $t < -t_{\alpha/2,n-1}$ or if $t > t_{\alpha/2,n-1}$.

EXAMPLE 11.19   **Small-Sample Test of the Mean of a Normal Population with Unknown Variance**

The manufacturer of a popular low-priced car claims that total repair costs resulting from low-speed car crashes follow a normal distribution with mean at most $200. A consumer agency thinks that this claim might be false and decides to test it. The null hypothesis is the manufacturer's claim

$$H_0: \quad \mu \leq \$200$$

and the alternative hypothesis is

$$H_1: \quad \mu > \$200$$

The consumer agency does not want to reject the manufacturer's claim unless the evidence against H_0 is very strong. Thus, the agency decides to use a 1% level of significance. Because it is very costly to test-crash a sample of cars, the agency test-crashes a random sample of only $n = 9$ cars. Use the following sample data to determine whether the null hypothesis should be rejected:

$$x_1 = \$245 \qquad x_2 = \$305 \qquad x_3 = \$175 \qquad x_4 = \$250 \qquad x_5 = \$280$$
$$x_6 = \$160 \qquad x_7 = \$250 \qquad x_8 = \$195 \qquad x_9 = \$210$$

SOLUTION   The observed sample mean is

$$\bar{x} = \frac{\Sigma x_i}{n} = \frac{2,070}{9} = 230$$

The observed sample variance is

$$s^2 = \frac{\Sigma(x_i - \bar{x})^2}{n - 1} = \frac{18,700}{8} = 2,337.5$$

The observed standard deviation is $s \approx \$48.35$. We use a one-tailed test and a 1% level of significance. The appropriate number of degrees of freedom is $(n - 1) = 8$, and the critical value is $t_{.01,8} = 2.896$. The rejection region therefore consists of all t scores exceeding 2.896. We obtain the observed t score

$$t = \frac{\bar{x} - \mu_0}{s/\sqrt{n}} = \frac{230 - 200}{48.35/\sqrt{9}} \approx 1.86$$

We do not reject H_0 because $t = 1.86$ falls in the acceptance region. See Figure 11.11.

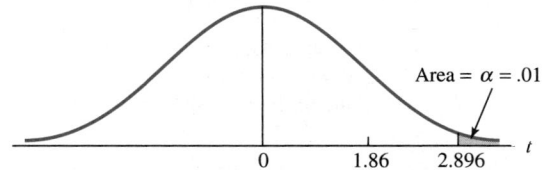

Area $= \alpha = .01$

0   1.86   2.896 t

FIGURE 11.11   *Critical value for the test described in Example 11.19*

We can use the table of critical values for the t distribution to obtain approximate p-values when performing a test that utilizes the t statistic as the test statistic. For example, in Example 11.19, the observed value of t was 1.86 and the appropriate number of degrees of freedom was 8. From Table A.6 in the Appendix, we see that $P(t > 1.860) = .05$. Thus, the p-value of the test is .05.

As another example, suppose the observed value of the t statistic had been $t = 1.49$. From Table A.6 in the Appendix, we see that $P(t > 1.860) = .05$ and $P(t > 1.397) = .10$. From this, we conclude that the p-value of the test is between .05 and .10.

Finally, suppose the observed value of the t statistic had been $t = 1.16$. From Table A.6 in the Appendix, we see that $P(t > 1.397) = .10$. It follows that $P(t > 1.16)$ must exceed .10. From this, we conclude that the p-value of the test is greater than .10.

The interpretation of the p-value for a t test is the same as for a z test based on the standard normal distribution.

Exercises

Statistical Concepts

11.5.1   Describe the assumptions that are made to use the t distribution to test a hypothesis about a population mean.

11.5.2   In each of the following cases, draw a graph of the t distribution and find the location of the rejection region. In each case, find the level of significance of the test.
a. $t > 1.476$ where df $= 5$
b. $t < -1.895$ where df $= 7$
c. $t < -2.120$ or $t > 2.120$ where df $= 16$

11.5.3   In each of the following cases, draw a graph of the t distribution and find the location of the rejection region. In each case, find the level of significance of the test.
a. $t > 2.201$ where df $= 11$
b. $t < -1.753$ where df $= 15$
c. $t < -2.093$ or $t > 2.093$ where df $= 19$

Statistical Drills

11.5.4   The following sample of eight values was randomly selected from a normal population: 11, 19, 28, 10, 34, 43, 22, 40.
   a. Calculate the sample mean and sample standard deviation.
   b. Test H_0: $\mu \geq 27.5$ against H_1: $\mu < 27.5$. Let $\alpha = .05$.
   c. Calculate the approximate p-value of the test. Show how to perform the test in part **b** by using the p-value.

11.5.5   The following sample of 10 values was randomly selected from a normal population: 21, 24, 18, 17, 54, 41, 26, 36, 53, 14.
   a. Calculate the sample mean and sample standard deviation.
   b. Test H_0: $\mu = 30$ against H_1: $\mu \neq 30$. Let $\alpha = .05$.
   c. Calculate the approximate p-value of the test. Show how to perform the test in part **b** by using the p-value.

11.5.6   The following sample of six values was randomly selected from a normal population: 122, 141, 108, 120, 142, 113.
   a. Calculate the sample mean and sample standard deviation.
   b. Test H_0: $\mu \geq 130$ against H_1: $\mu < 130$. Let $\alpha = .01$.
   c. Calculate the approximate p-value of the test. Show how to perform the test in part **b** by using the p-value.

11.5.7   A random sample of 19 observations produced the following information:

$$\Sigma\, x_i = 760 \qquad \Sigma\, (x_i - \bar{x})^2 = 1{,}800$$

   a. Calculate the sample mean and sample standard deviation.
   b. Test the null hypothesis that $\mu \geq 47$ against the alternative hypothesis that $\mu < 47$. Use $\alpha = .05$.
   c. Calculate the approximate p-value of the test. Show how to perform the test in part **b** by using the p-value.

Statistical Applications

11.5.8   A trucking firm believes that its mean weekly loss from damaged shipments is $1,800 or less. A sample of 15 weeks of operation shows a sample mean weekly loss of $2,000 and a sample standard deviation of $500. Use a .05 level of significance to test the trucking firm's claim. Assume that the distribution of weekly losses is approximately normal.

11.5.9   A contractor assumes that construction workers are idle 75 minutes per day or less. A sample of 25 construction workers had a mean idle time of 84 minutes per day. The sample standard deviation was 20 minutes. Assume the population is approximately normal.
   a. What is the two-tailed p-value associated with the sample result?
   b. Using a .05 level of significance, test the hypothesis H_0: $\mu \leq 75$.

11.5.10   For each of the following situations, state whether the test is right- or left-tailed and find the critical value z_α. For each sample result, calculate the Z score and indicate whether the null hypothesis should be accepted or rejected. Finally, for each outcome, determine the approximate probability of obtaining a value of the test statistic as rare as or rarer than the one you obtained by finding the tail area under the normal curve corresponding to the computed Z.
   a. H_0: $\mu \leq 100$; H_1: $\mu > 100$; $\alpha = .05$; $n = 100$; $\bar{x} = 103$; $s = 15$
   b. H_0: $\mu \geq 15$; H_1: $\mu < 15$; $\alpha = .01$; $n = 100$; $\bar{x} = 13.8$; $s = 2$
   c. H_0: $\mu \geq .8$; H_1: $\mu < .8$; $\alpha = .10$; $n = 36$; $\bar{x} = .6$; $s = .6$

11.5.11   In a study of risk-taking behavior, a researcher gave each subject a lottery ticket whereby the subject could win $500 with probability .1 and $0 with probability .9. By subsequent questioning, the researcher determined the minimum price at which the subject would sell the lottery ticket. Since the expected value of the lottery ticket was $50, the researcher wanted to test whether the mean minimum selling price was $50. Sixteen subjects participated in the study, and the sample results were $\bar{x} = \$41.25$ and $s = \$4.52$. The researcher believes that the minimum selling prices are approximately normally distributed. Conduct a two-sided test using $\alpha = .05$.

11.5.12   A company has produced a new type of steel-belted radial tire that it claims will last at least 40,000 miles on the average. You take a random sample of 16 tires and obtain $\bar{x} = 35,000$ miles and $s = 3,000$ miles. Assume that the population is approximately normal.
a. Test the company's claim using a 5% level of significance.
b. Did you perform a one-tailed test? Why or why not?

11.5.13   Suppose 12 tax returns are randomly sampled by the Internal Revenue Service. The charitable deduction claimed on each return is as follows:

$350 $310 $345 $290 $330 $300
 380   310   320   280   285   290

Assume that the population of values is approximately normal.
a. Calculate the sample mean and sample standard deviation.
b. Test $H_0: \mu = 320$ against $H_1: \mu \neq 320$. Let $\alpha = .05$.
c. Calculate the approximate p-value of the test. Show how to perform the test in part **b** by using the p-value.

11.6   Tests of the Population Proportion

Frequently we want to test hypotheses concerning the proportion of members of a population possessing some particular attribute. Such tests are based on the sample proportion $\hat{p}$. In Chapter 9, we showed that if the population proportion is p, the sample proportion $\hat{p}$ follows approximately a normal distribution with mean p and variance pq/n, where $q = (1 - p)$, provided $np \geq 5$ and $nq \geq 5$. Throughout the remainder of this section, it is always assumed that $np \geq 5$ and $nq \geq 5$, so that the distribution of $\hat{p}$ is approximately normal.

Suppose we want to test the null hypothesis

$$H_0: \quad p = p_0 \quad \text{or} \quad H_0: \quad p \geq p_0$$

against the alternative hypothesis

$$H_1: \quad p < p_0$$

where p_0 is some hypothesized value for the population proportion. If H_0 is true, then the sample proportion is approximately normal with mean p_0 and variance $p_0 q_0/n$, and the Z score

$$Z = \frac{\hat{p} - p_0}{\sqrt{p_0 q_0/n}}$$

approximately follows the standard normal distribution. Naturally, values of $\hat{p}$ far less than p_0 would cast doubt on the validity of the null hypothesis. If $\hat{p}$ is far less than p_0, the observed Z score will be far less than 0. This implies that the rejection region should be placed under the left tail of the standard normal distribution. This is precisely the same argument used in developing a test for the population mean.

Suppose the level of significance of the test is selected to be α. Then we reject H_0 if $Z < -z_\alpha$, where $-z_\alpha$ is the value such that

$$P(Z < -z_\alpha) = \alpha$$

The three possible cases—when the alternative hypothesis H_1 specifies values of p less than p_0, when H_1 specifies values greater than p_0, and when H_1 is two-sided—are discussed in the accompanying box.

Testing hypotheses about the population proportion

Suppose we want to test the null hypothesis that the population proportion is p_0 using a level of significance α. To perform the test, we take a sample of size n from the population and calculate the sample proportion $\hat{p}$. The tests are based on the assumption that $np_0 \geq 5$ and $nq_0 \geq 5$. If these conditions hold, then the random variable

$$Z = \frac{\hat{p} - p_0}{\sqrt{p_0 q_0 / n}}$$

approximately follows the standard normal distribution. Let z denote the observed value of the random variable Z after substituting the observed sample proportion $\hat{p}$ into the previous formula.

   Case 1:   To test the null hypothesis

$$H_0: \quad p = p_0 \qquad \text{or} \qquad H_0: \quad p \geq p_0$$

against the one-sided alternative hypothesis

$$H_1: \quad p < p_0$$

use the decision rule

Reject H_0 in favor of H_1 if $z < -z_\alpha$.

This is equivalent to the rule

Reject H_0 in favor of H_1 if $\hat{p} < \hat{p}^*$,

where the critical value is $\hat{p}^* = p_0 - z_\alpha \sqrt{p_0 q_0 / n}$.

   Case 2:   To test the null hypothesis

$$H_0: \quad p = p_0 \qquad \text{or} \qquad H_0: \quad p \leq p_0$$

against the alternative hypothesis

$$H_1: \quad p > p_0$$

use the decision rule

Reject H_0 in favor of H_1 if $z > z_\alpha$.

This is equivalent to the rule

$$\text{Reject } H_0 \text{ in favor of } H_1 \text{ if } \hat{p} > \hat{p}^*,$$

where the critical value is $\hat{p}^* = p_0 + z_\alpha \sqrt{p_0 q_0/n}$.

   Case 3:   To test the null hypothesis

$$H_0: \quad p = p_0$$

against the two-sided alternative hypothesis

$$H_1: \quad p \neq p_0$$

use the decision rule

$$\text{Reject } H_0 \text{ in favor of } H_1 \text{ if } z < -z_{\alpha/2} \text{ or if } z > z_{\alpha/2}.$$

This is equivalent to the rule

$$\text{Reject } H_0 \text{ if } \hat{p} < \hat{p}_1^* \text{ or if } \hat{p} > \hat{p}_2^*,$$

where the critical values are

$$\hat{p}_1^* = p_0 - z_{\alpha/2}\sqrt{p_0 q_0/n} \quad \text{and} \quad \hat{p}_2^* = p_0 + z_{\alpha/2}\sqrt{p_0 q_0/n}$$

When we test a hypothesis about a population proportion, *p*-values are calculated in the same way as they are when we test hypotheses about a population mean. Thus, the *p*-value of the test denotes the probability of obtaining a sample proportion as extreme as or more extreme than the value that was actually observed.

EXAMPLE 11.20   **Testing a Population Proportion** ─────────────────────────────

A politician claims that at least 60% of the population favors a strict gun control proposal that would severely restrict ownership of guns. A local hunting club claims that p is much less than 60%. You take a sample of size 100 to test the null hypothesis

$$H_0: \quad p = .6$$

against the one-sided alternative hypothesis

$$H_1: \quad p < .6$$

Suppose that 55 people in the sample favor the legislation. If we use a 5% level of significance, what decision should be made?

SOLUTION   If H_0 is true, then $\hat{p}$ has a normal distribution with mean $p = .6$ and variance $pq/n = (.6)(.4)/100 = .0024$. If we use a one-tailed test at the 5% level of significance, the critical region consists of all values of Z less than $-z_\alpha = -z_{.05} = -1.645$. In our sample, we obtained $\hat{p} = x/n = 55/100 = .55$. The observed Z score is

$$z = \frac{\hat{p} - p_0}{\sqrt{p_0 q_0/n}} = \frac{.55 - .60}{\sqrt{(.60)(.40)/100}} = -1.02$$

We do not reject H_0 because the Z score, -1.02, exceeds the critical value $-z_\alpha = -1.645$ and falls in the acceptance region. See Figure 11.12.

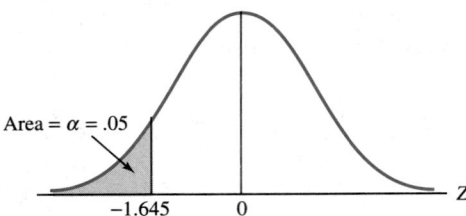

Area = $\alpha = .05$

$-1.645 \qquad 0$

Z

FIGURE 11.12 *Critical value for the test described in Example 11.20*

Alternatively, we can calculate the critical value

$$\hat{p}^* = p_0 - z_\alpha \sqrt{p_0 q_0/n} = .6 - 1.645\sqrt{(.6)(.4)/100} \approx .519$$

We do not reject H_0 because the observed sample proportion $\hat{p} = .55$ exceeds the critical value .519.

For this test, the p-value is

$$p\text{-value} = P(Z < -1.02) = .1539$$

Exercises

Statistical Concepts

11.6.1   State the assumptions that are needed to use the standard normal distribution to test a hypothesis about a population proportion.

11.6.2   Use the following information about the sample size and the hypothesized value of the population proportion p to determine whether the sample size is large enough to use the normal approximation methodology presented in this section to conduct a test of the null hypothesis H_0: $p = p_0$.
   a. $n = 100$; $p_0 = .10$
   b. $n = 50$; $p_0 = .10$
   c. $n = 500$; $p_0 = .05$
   d. $n = 25$; $p_0 = .10$

11.6.3   Use the following information about the sample size and the hypothesized value of the population proportion p to determine whether the sample size is large enough to use the normal approximation methodology presented in this section to conduct a test of the null hypothesis H_0: $p = p_0$.
   a. $n = 20$; $p_0 = .20$
   b. $n = 20$; $p_0 = .50$
   c. $n = 500$; $p_0 = .01$
   d. $n = 25$; $p_0 = .30$

Statistical Drills

11.6.4   Suppose a random sample of size $n = 20$ is obtained. A scientist wants to test H_0: $p = .6$ against H_1: $p \neq .6$. The scientist decides to reject H_0 if $X \leq 8$ or if $X \geq 16$.
   a. Use the table for the binomial distribution (Table A.2 in the Appendix) and find $P(X \leq 8)$ and $P(X \geq 16)$.

b. The level of significance of this test is the sum of the two probabilities calculated in part **a.** Find the level of significance of the test.

c. Suppose that, in fact, $p = .5$. We will commit a Type II error (and incorrectly accept H_0) if X takes a value from 9 to 15. Find the probability of committing a Type II error. That is, if $p = .5$, find $P(9 \leq X \leq 15)$.

11.6.5 Suppose a random sample of size $n = 20$ is obtained. A scientist wants to test $H_0: p = .5$ against $H_1: p \neq .5$. The scientist decides to reject H_0 if $X \leq 4$ or if $X \geq 16$.

a. Use the table for the binomial distribution (Table A.2 in the Appendix) and find $P(X \leq 4)$ and $P(X \geq 16)$.

b. The level of significance of this test is the sum of the two probabilities calculated in part **a.** Find the level of significance of the test.

c. Suppose that, in fact, $p = .7$. We will commit a Type II error (and incorrectly accept H_0) if X takes a value from 5 to 15. Find the probability of committing a Type II error. That is, if $p = .7$, find $P(5 \leq X \leq 15)$.

Statistical Applications

11.6.6 A magazine claims that at least 25% of its subscribers are college graduates. In a random sample of 200 subscribers, 41 are college graduates.

a. Use a 5% level of significance to test the validity of the magazine's claim.

b. Using the sample results, develop a 90% confidence interval for the proportion of the population that are college graduates.

c. Use the interval estimate of part **b** to test the magazine's claim about its subscribers. What is your conclusion?

11.6.7 A producer of a certain brand of coffee claims that at least 20% of all coffee drinkers prefer its product over the major competing brand. To test the validity of this claim, an individual samples 200 coffee drinkers and finds that 43 prefer the producer's brand. Test the producer's claim using $\alpha = .05$.

11.6.8 In checking the reliability of a bank's records, auditing firms sometimes ask a sample of the bank's customers to confirm the accuracy of their savings account balances as reported by the bank. An auditing firm is interested in estimating the proportion of a bank's savings accounts on whose balances the bank and the customer disagree. Of 400 savings account customers questioned by the auditors, 30 said their balance disagreed with that reported by the bank.

a. Estimate the actual proportion of the bank's savings accounts on whose balances the bank and customer disagree using a 95% confidence level.

b. The bank claims that the true fraction of accounts on which there is disagreement is no more than .05. You, as an auditor, doubt this claim. Does the sample provide evidence that the true fraction of accounts subject to disagreement exceeds .05? Use $\alpha = .05$ to perform the test.

11.6.9 A manufacturer of videodiscs believes that, under adequate quality control, no more than 6% of the discs should be returned as faulty. For a random sample of 250 sales of these discs, it was found that 22 were returned as faulty. Test at the 10% significance level the null hypothesis that the population percentage of discs returned as faulty is at most 6%.

11.6.10 A psychology text states that 10% of the population is left-handed. You do not know whether the proportion of left-handers is more or less than .10. You obtain a random sample of 200 people to test the book's claim using a 5% level of significance.

a. What decision would you make if, in your sample, 38 people were left-handed?

b. Why would it be appropriate to use a two-tailed test in this problem?

11.6.11 An economist states that 10% of a city's labor force is unemployed. You do not know whether the economist's estimate is too high or too low. Thus, you want to test $H_0: p = .10$ against $H_1: p \neq .10$

using a 5% level of significance. You obtain a random sample of 400 people in the labor force, of whom 27 are unemployed. Would you reject H_0?

11.6.12   A state legislature says that it is going to decrease its funding of the state university because, according to its sources, 36% of the graduates move out of the state within 3 years of graduation. As a faculty member at the university, you want to show that the proportion of graduates who move out of state is less than .36. You decide to test H_0: $p = .36$ against H_1: $p < .36$ using a 5% level of significance. You obtain a random sample of 160 graduates and find that 40 moved out of state within 3 years of graduation. Would you reject H_0? Use $\alpha = .05$.

11.7   Measuring the Power of a Test (Optional)

As discussed in Section 11.1, when testing any null hypothesis H_0 against an alternative hypothesis H_1, two types of errors can be committed. A Type I error occurs when we reject H_0 when it is true. The probability of a Type I error, denoted α and called the level of significance of the test, measures the probability that the test statistic falls in the critical region given that H_0 is true. A Type II error occurs when H_0 is false but we do not reject it. The probability of a Type II error, denoted by β, is the probability that the test statistic falls in the acceptance region when H_0 is false. This probability depends on the actual situation and varies with the value of the population parameter being tested.

For example, suppose the null hypothesis states that a population parameter θ equals some specific value θ_0. That is, the null hypothesis is

$$H_0: \quad \theta = \theta_0$$

If the true value of the population parameter θ was different from, but close to, θ_0, then it would not be unusual to obtain a test statistic that falls in the acceptance region; in such a case it would be relatively easy to accept the null hypothesis incorrectly. Thus, the probability of making a Type II error would be large. On the other hand, if the true value of θ was extremely different from θ_0, it would be very unlikely that the test statistic would fall in the acceptance region, and the probability of making a Type II error would be small.

It is important to know the probability of making a Type II error, so that if we accept the null hypothesis, we have some idea how likely such a decision would be if the null hypothesis is false. If the probability of making a Type II error is large, accepting the null hypothesis does not necessarily provide much evidence that the null hypothesis is true. Thus, when β is large, we cannot distinguish very well between the null hypothesis and various alternative hypotheses.

The probability β of making a Type II error is a conditional probability. We have

$$\beta = P(\text{Type II error})$$
$$= P(H_0 \text{ not rejected}|H_0 \text{ false})$$
$$= P(\text{Test statistic in acceptance region}|H_0 \text{ false})$$

The quantity $(1 - \beta)$ measures the probability that the test statistic falls in the rejection region when H_0 is false. Thus $(1 - \beta)$ measures the probability that the test correctly leads us to reject H_0 when H_0 is false. This probability $(1 - \beta)$ is called the *power of the test*. The reason for this name is that one can measure how good, or powerful, a test is by determining how frequently the test will reject H_0 when H_0 is actually false.

DEFINITION **Power of a Test** ────────────────────────────────

Let β denote the probability of making a Type II error. The probability $(1 - \beta)$ is called the **power of the test.** Thus,

$$1 - \beta = P(H_0 \text{ reject}|H_0 \text{ false})$$
$$= P(\text{Test statistic falls in rejection region}|H_0 \text{ false})$$

The power of the test measures the probability that the test correctly leads us to reject H_0 when H_0 is false.

──

Suppose we want to calculate the power of a test of the null hypothesis

$$H_0: \quad \mu = \mu_0$$

against the one-sided alternative hypothesis

$$H_1: \quad \mu > \mu_0$$

where the level of significance is α. This is a one-sided test where the rejection region lies entirely in the right tail of the distribution. We obtain a random sample of n observations from a normal population with a known variance σ^2. To find the power of the test, we first need to determine the critical value $\bar{x}^*$ that separates the acceptance region from the rejection region. In Section 11.2, we showed that the critical value is

$$\bar{x}^* = \mu_0 + z_\alpha \sigma/\sqrt{n}$$

Suppose the value of the population mean is actually μ_1 rather than the hypothesized value μ_0. We will correctly reject H_0 in favor of H_1 if the observed sample mean exceeds the critical value $\bar{x}^*$. Thus, the power of the test is the probability that the sample mean $\bar{X}$ exceeds $\bar{x}^*$, *given that $\mu = \mu_1$.* We obtain the rule described in the accompanying box.

Calculating the power of a test with rejection region in the right tail

Suppose we have a random sample of n observations from a normal population with a known variance σ^2. Suppose we want to test the null hypothesis

$$H_0: \quad \mu = \mu_0$$

against the one-sided alternative hypothesis

$$H_1: \quad \mu > \mu_0$$

where the level of significance is α. Suppose the mean is μ_1 rather than the hypothesized value μ_0. The *power of the test* is

$$\text{Power} = 1 - \beta = P(\bar{X} > \bar{x}^*)$$
$$= P(\bar{X} > \mu_0 + z_\alpha \sigma/\sqrt{n})$$

(continued)

where $\bar{X}$ follows the normal distribution with mean μ_1 and variance σ^2/n. We obtain

$$\text{Power} = P\left(Z > \frac{\bar{x}^* - \mu_1}{\sigma/\sqrt{n}}\right)$$

where $\bar{x}^* = \mu_0 + z_\alpha \sigma/\sqrt{n}$.

EXAMPLE 11.21   **Calculating the Power of a Right-Tailed Test**

In Example 11.13, we tested the null hypothesis that the mean salary of teachers was $25,000. We tested

$$H_0: \quad \mu \leq \$25,000$$

against the one-sided alternative hypothesis

$$H_1: \quad \mu > \$25,000$$

It was assumed that the population was normal with standard deviation $\sigma = \$4,000$. The sample size was $n = 100$, and the level of significance was .01. Find the power of the test if, in fact, the true population mean was $26,000.

SOLUTION   For $\alpha = .01$, the critical value of Z is $z_\alpha = 2.33$. Thus the critical value $\bar{x}^*$ is

$$\begin{aligned}
\bar{x}^* &= \mu_0 + z_\alpha \sigma/\sqrt{n} \\
&= 25,000 + 2.33(4,000/\sqrt{100}) = 25,932
\end{aligned}$$

The power of the test is calculated as follows:

$$\begin{aligned}
\text{Power} &= P(\bar{X} > \bar{x}^*) \\
&= P(\bar{X} > 25,932) \\
&= P\left(Z > \frac{\bar{x}^* - \mu_1}{\sigma/\sqrt{n}}\right) \\
&= P\left(Z > \frac{25,932 - 26,000}{4,000/\sqrt{100}}\right) \\
&= P(Z > -.17) = .5675
\end{aligned}$$

The rejection region consists of all values of $\bar{X}$ exceeding $25,932, so the power of the test is the probability that $\bar{X}$ exceeds $25,932 when $\mu = \$26,000$. The shaded area in Figure 11.13, which shows the power of the test, represents the probability that we will correctly reject H_0 when, in fact, $\mu = \$26,000$.

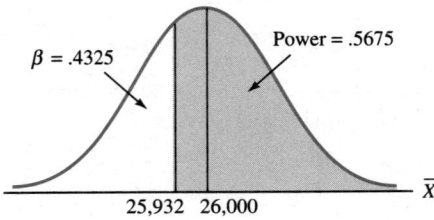

FIGURE 11.13   *Power of test described in Example 11.21*

Since the power of the test is $(1 - \beta) = .5675$, it follows that $\beta = .4325$. Thus, for $\mu = \$26,000$, the probability is .5675 that we will reject H_0 in favor of H_1; consequently, the probability is .4325 that we will incorrectly accept H_0.

In Example 11.21, the rejection region fell in the right tail of the distribution. Now consider the case when the rejection region falls in the left tail. We obtain the rule described in the accompanying box.

Calculating the power of a test with rejection region in the left tail

> Suppose we have a random sample of n observations from a normal population with a known variance σ^2. Suppose we want to test the null hypothesis
>
> $$H_0: \quad \mu = \mu_0$$
>
> against the one-sided alternative hypothesis
>
> $$H_1: \quad \mu < \mu_0$$
>
> where the level of significance is α. Suppose the mean is μ_1 rather than the hypothesized value μ_0. The power of the test is the probability that the sample mean $\bar{X}$ is less than the critical value $\bar{x}^*$ where
>
> $$\bar{x}^* = \mu_0 - z_\alpha \sigma / \sqrt{n}$$
>
> We have
>
> $$\text{Power} = P(\bar{X} < \bar{x}^*)$$
> $$= P(\bar{X} < \mu_0 - z_\alpha \sigma / \sqrt{n})$$
>
> where $\bar{X}$ follows a normal distribution with mean μ_1 and variance σ^2 / n. Equivalently, we obtain
>
> $$\text{Power} = P(\bar{X} < \bar{x}^*)$$
> $$= P\left(Z < \frac{\bar{x}^* - \mu_1}{\sigma / \sqrt{n}}\right)$$
>
> where $\bar{x}^* = \mu_0 - z_\alpha \sigma / \sqrt{n}$.

EXAMPLE 11.22 **Calculating the Power of a Left-Tailed Test** —————————————

In Example 11.11, we tested the null hypothesis

$$H_0: \quad \mu = 32$$

against the one-sided alternative hypothesis

$$H_1: \quad \mu < 32$$

It was assumed that the population was normal with variance $\sigma^2 = 16$. The sample size was $n = 25$, and the level of significance was .05. Find the power of the test if, in fact, $\mu = 29.5$.

SOLUTION   For $\alpha = .05$, the critical value of Z is $z_\alpha = 1.645$. For the critical value $\bar{x}^*$, we obtain

$$\begin{aligned}
\bar{x}^* &= \mu_0 - z_\alpha \sigma / \sqrt{n} \\
&= 32 - 1.645(4/\sqrt{25}) = 30.684
\end{aligned}$$

This is the same critical value calculated in Example 11.12.

The power of the test is

$$\begin{aligned}
\text{Power} &= P(\bar{X} < \bar{x}^*) \\
&= P(\bar{X} < 30.684) \\
&= P\left(Z < \frac{\bar{x}^* - \mu_1}{\sigma / \sqrt{n}}\right) \\
&= P\left(Z < \frac{30.684 - 29.5}{4/\sqrt{25}}\right) \\
&= P(Z < 1.48) = .9306
\end{aligned}$$

The shaded area in Figure 11.14 represents the power of the test when $\mu = 29.5$. Under these conditions, the probability is .9306 that we will correctly reject H_0 in favor of H_1, and so the probability of making a Type II error is $\beta = .0694$.

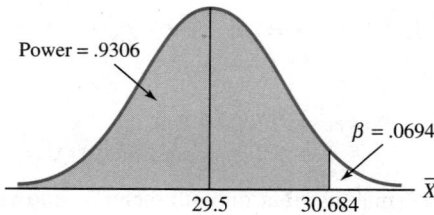

Power = .9306

$\beta = .0694$

29.5   30.684 $\bar{X}$

FIGURE 11.14   *Power of test described in Example 11.22*

In the accompanying box, we consider the power of the test of a null hypothesis

$$H_0: \quad \mu = \mu_0$$

against the two-sided alternative hypothesis

$$H_1: \quad \mu \neq \mu_0$$

Calculating the power of a test with rejection regions in both tails

Suppose we have a random sample of n observations from a normal population with a known variance σ^2. Suppose we want to test the null hypothesis

$$H_0: \quad \mu = \mu_0$$

against the two-sided alternative hypothesis

$$H_1: \quad \mu \neq \mu_0$$

where the level of significance is α. The power of the test is the probability that the sample mean $\bar{X}$ is less than $\bar{x}_1^*$ or greater than $\bar{x}_2^*$ when $\mu = \mu_1$, where $\bar{x}_1^*$ and $\bar{x}_2^*$ are the critical values given by

$$\bar{x}_1^* = \mu_0 - z_{\alpha/2}\sigma/\sqrt{n} \quad \text{and} \quad \bar{x}_2^* = \mu_0 + z_{\alpha/2}\sigma/\sqrt{n}$$

That is,

$$\begin{aligned} \text{Power} &= P(\bar{X} < \bar{x}_1^*) + P(\bar{X} > \bar{x}_2^*) \\ &= P(\bar{X} < \mu_0 - z_{\alpha/2}\sigma/\sqrt{n}) + P(\bar{X} > \mu_0 + z_{\alpha/2}\sigma/\sqrt{n}) \end{aligned}$$

where $\bar{X}$ follows a normal distribution with mean μ_1 and variance σ^2/n. Equivalently, we obtain

$$\begin{aligned} \text{Power} &= P(\bar{X} < \bar{x}_1^*) + P(\bar{X} > \bar{x}_2^*) \\ &= P\left(Z < \frac{\bar{x}_1^* - \mu_1}{\sigma/\sqrt{n}}\right) + P\left(Z > \frac{\bar{x}_2^* - \mu_1}{\sigma/\sqrt{n}}\right) \end{aligned}$$

EXAMPLE 11.23   **Calculating the Power of a Two-Tailed Test**

In Example 11.14, we tested the null hypothesis

$$H_0: \quad \mu = \$300$$

against the two-sided alternative hypothesis

$$H_1: \quad \mu \neq \$300$$

It was assumed that the population was normal with standard deviation $\sigma = 50$. The sample size was $n = 25$, and the level of significance was .05. Find the power of the test if, in fact, $\mu = \$270$.

SOLUTION   For $\alpha = .05$, we obtain $z_{\alpha/2} = 1.96$. Because we have performed a two-tailed test, there are two critical values of $\bar{X}$, calculated in Example 11.15 as $\bar{x}_1^* = \$280.40$ and $\bar{x}_2^* = \$319.60$.

The power of the test is

$$\begin{aligned} \text{Power} &= P(\bar{X} < \bar{x}_1^*) + P(\bar{X} > \bar{x}_2^*) \\ &= P(\bar{X} < 280.40) + P(\bar{X} > 319.60) \\ &= P\left(Z < \frac{\bar{x}_1^* - \mu_1}{\sigma/\sqrt{n}}\right) + P\left(Z > \frac{\bar{x}_2^* - \mu_1}{\sigma/\sqrt{n}}\right) \\ &= P\left(Z < \frac{280.40 - 270}{50/\sqrt{25}}\right) + P\left(Z > \frac{319.60 - 270}{50/\sqrt{25}}\right) \\ &= P(Z < 1.04) + P(Z > 4.96) \\ &= .8508 + .0000 = .8508 \end{aligned}$$

The shaded area in Figure 11.15 (page 444) represents the power of the test when, in fact, $\mu = \$270$. Under these circumstances, the probability is .8508 that we will correctly reject H_0 in favor of H_1, and so the probability of making a Type II error is $\beta = .1492$.

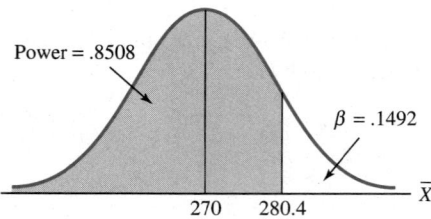

FIGURE 11.15   *Power of test described in Example 11.23*

The Power Curve

Suppose we are testing the null hypothesis H_0: $\mu = \mu_0$ and the true value of the mean is μ_1 rather than the hypothesized value μ_0. Other things being equal, we will be more likely to correctly reject H_0 when μ_1 is far from μ_0 than when μ_1 is relatively close to μ_0. The next example illustrates this point.

EXAMPLE 11.24   **Calculating the Power of a Test for Different Values of μ** ——————

Suppose a car manufacturer claims that when a certain model is driven at a speed of 50 miles per hour on a test track, the mileage follows a normal distribution with mean 30 miles per gallon and standard deviation 4 miles per gallon. A consumer advocate thinks that the manufacturer is overestimating average mileage. The advocate decides to test the null hypothesis H_0: $\mu = 30$ against the one-sided alternative hypothesis H_1: $\mu < 30$. Suppose the consumer advocate tests a sample of $n = 25$ cars and uses a 5% level of significance. Find the power of the test if the true mean is $\mu_1 = 28.5$.

SOLUTION   For $\alpha = .05$, we obtain $z_\alpha = 1.645$. We calculate the critical value

$$\bar{x}^* = \mu_0 - z_\alpha \sigma / \sqrt{n}$$
$$= 30 - 1.645(4/\sqrt{25}) = 28.684$$

The power of the test is

$$\text{Power} = P(\bar{X} < \bar{x}^*) = P(\bar{X} < 28.684)$$

when $\mu = 28.5$. To calculate this probability, we calculate the Z score

$$z = \frac{\bar{x}^* - \mu_1}{\sigma / \sqrt{n}} = \frac{28.684 - 28.5}{4/\sqrt{25}} = .23$$

We obtain

$$\text{Power} = P(Z < .23) = .5910$$

Thus, the probability is only .5910 that the null hypothesis will be rejected if the true mean is 28.5 miles per gallon, and the probability is .4090 that the null hypothesis will be incorrectly accepted. This shows that the test is not very powerful in testing the null hypothesis H_0: $\mu = 30$ against the alternative hypothesis H_1: $\mu < 30$ when $\mu = \mu_1 = 28.5$. The power of the test is illustrated in Figure 11.16.

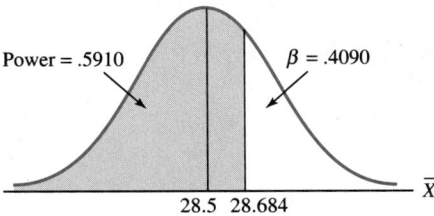

FIGURE 11.16 *Power of test for Example 11.24 when $\mu_1 = 28.5$*

Now let us repeat the entire procedure for the case when the true value of the mean is $\mu = 28.0$. The power of the test is now

$$\text{Power} = P(\bar{X} < \bar{x}^*) = P(\bar{X} < 28.684)$$

given that $\mu = 28.0$. We obtain the Z score

$$z = \frac{\bar{x}^* - \mu_1}{\sigma/\sqrt{n}} = \frac{28.684 - 28.0}{4/\sqrt{25}} \approx .86$$

and

$$\text{Power} = P(Z < .86) = .8051$$

This probability is the shaded area in Figure 11.17. A comparison of Figures 11.16 and 11.17 shows that the power of the test increases as the true mean μ_1 moves away from $\mu_0 = 30$.

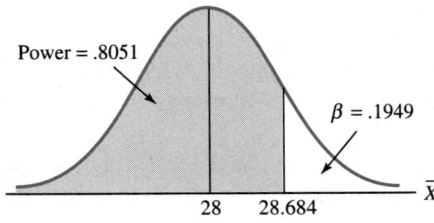

FIGURE 11.17 *Power of test for Example 11.24 when $\mu = 28.0$*

The power of the test in Example 11.24 can be calculated for any specific value of μ. The following table shows the power of the test associated with different values of μ when the sample size is $n = 25$:

μ	29.5	29.0	28.5	28.0	27.5	27.0	26.5
Power	.1539	.3446	.5910	.8051	.9306	.9826	.9968

DEFINITION   **Power Curve** ───────────────────────────────────

A **power curve** is a graph showing the power of the test associated with different values of the parameter of interest.

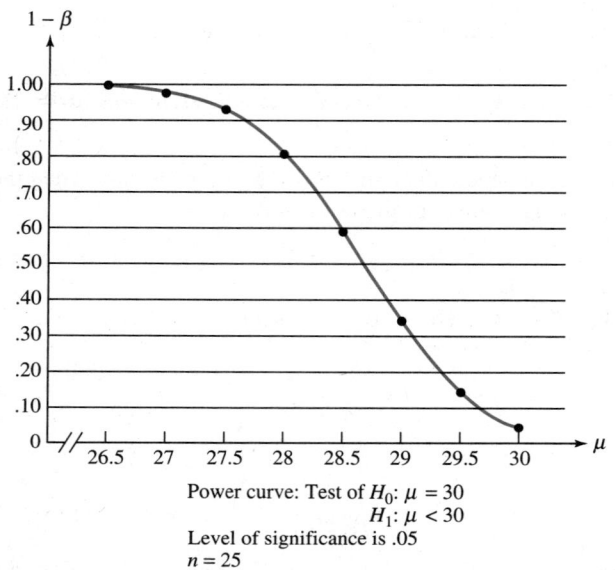

FIGURE 11.18   *Power curve for Example 11.24*

The data are graphed in Figure 11.18. The height of the curve above any value μ shows the probability that we will reject H_0 if the true mean is that particular value. The curve rises as μ moves away from the hypothesized value $\mu_0 = 30$. The higher the power curve, the more powerful the test and the more likely it is that we will correctly reject a false null hypothesis.

Other things equal, the more powerful the test, the better. We can increase the power of a test by increasing the sample size. The following example illustrates this point.

EXAMPLE 11.25   **Effect of Increased Sample Size on the Power of a Test** ──────────────

Let us repeat Example 11.24 using a larger sample size of $n = 64$.

SOLUTION   As in Example 11.24, for $\alpha = .05$, the critical value of Z is $z_\alpha = 1.645$, but the critical value for $\bar{x}*$ now becomes

$$\bar{x}* = \mu_0 - z_\alpha \sigma/\sqrt{n}$$
$$= 30 - 1.645(4/\sqrt{64}) = 29.1775$$

This change is due to the increase in the sample size, which decreases the value $\sigma/\sqrt{n}$. The power of the test is

$$\text{Power} = P(\overline{X} < \overline{x}^*) = P(\overline{X} < 29.1775)$$

when the true value of μ is 28.5. To calculate this probability, we calculate the Z score as follows:

$$Z = \frac{\overline{x}^* - \mu_1}{\sigma/\sqrt{n}} = \frac{29.1775 - 28.5}{4/\sqrt{64}} \approx 1.36$$

We then obtain

$$\text{Power} = P(Z < 1.36) = .9131$$

Thus, the probability is .9131 that the null hypothesis will be rejected if the true mean is 28.5 miles per gallon, as illustrated in Figure 11.19. This shows that the test is fairly powerful in detecting that the null hypothesis is false when, in fact, $\mu = 28.5$. Recall from Example 11.24 that when $\mu = 28.5$ and the sample size is $n = 25$, the power of the test is .5910. Increasing the sample size increases the probability of detecting a false null hypothesis.

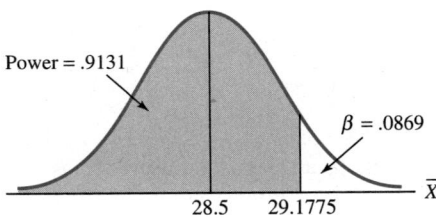

Power = .9131

$\beta = .0869$

$\overline{X}$

28.5   29.1775

FIGURE 11.19 *Power of test for Example 11.25*

As in Example 11.24, the power of a test can be calculated for different values of μ_1 based on a sample size of $n = 64$ rather than $n = 25$. Table 11.2 shows the power of the test associated with different values of μ and n. For any specific value $\mu = \mu_1$, the power of the test when $n = 64$ exceeds the power of the test when $n = 25$. Table 11.2 also shows that the power is greater when $n = 100$ than when $n = 64$. The power curves for these three values of n are shown in Figure 11.20 (page 448).

TABLE 11.2 *Power of test in Example 11.25 for different values of μ and n*

		n	
μ	**25**	**64**	**100**
29.5	.1539	.2578	.3446
29.0	.3446	.6406	.8051
28.5	.5910	.9131	.9826
28.0	.8051	.9909	.9996
27.5	.9306	.9996	1.0000
27.0	.9826	1.0000	1.0000
26.5	.9968	1.0000	1.0000

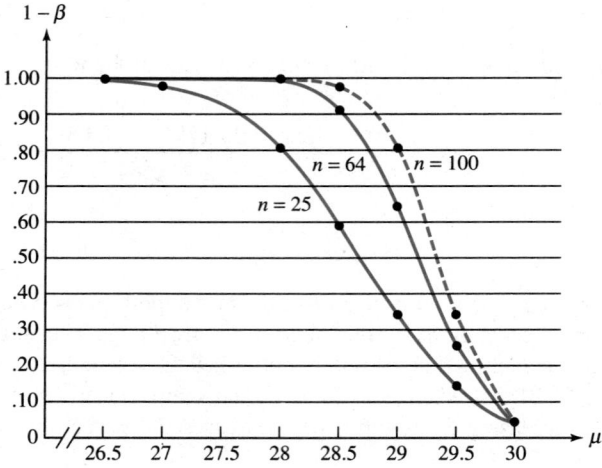

FIGURE 11.20   *Power curves for Example 11.25 for different sample sizes*

Properties of the Power Curve

The power of a test depends on four things, as follows:

1. The distance between the true mean μ_1 and the hypothesized mean μ_0, which is $|\mu_1 - \mu_0|$
2. The level of significance α
3. The population variance σ^2
4. The sample size n

Let's discuss these points one by one.

For given n, α, and σ^2, the farther the true mean μ_1 is from the hypothesized mean μ_0, the greater the power of the test. Thus, as $|\mu_1 - \mu_0|$ increases, we are more likely to detect that the true mean is not μ_0.

For given n, σ^2, and μ_1, the smaller the significance level α, the smaller the power of the test. This is true because reducing α reduces the size of the rejection region, thus making it more difficult to reject the null hypothesis.

For given n, α, and μ_1, the larger the population variance σ^2, the lower the power of the test. It is more difficult to detect that the null hypothesis is false when there is greater variability in the population because increasing σ moves the critical value $\bar{x}^*$ farther out into the tail of the distribution and reduces the size of the rejection region. Consequently, this reduces the power of the test.

For given α, σ^2, and μ_1, the greater the sample size n, the higher the power of the test. The more information we have, the easier it becomes to detect any deviation from the null hypothesis.

Because the level of significance is usually set at some low value such as .05 or .01, it is relatively unlikely that a *true* null hypothesis will be rejected. In fact, a low level of significance makes it unlikely that we will reject the null hypothesis *even when it is false.* This is especially the case when the sample size is small or when the true value of the population parameter is close to that specified in the null hypothesis. Thus, the fact that we do not reject the null hypothesis does not necessarily mean that the null hypothesis is true, nor does it necessarily provide strong support for the null hypothesis.

Choosing a small value for α means that the power of the test will be small in many situations. Occasionally this is a reasonable procedure to follow. For example, in some manufacturing situations, we may not want to make costly alterations to a production process, retool a plant, or change a chemical formula unless we have very strong evidence that the change will be beneficial. On the other hand, there are many situations where the null hypothesis should not be accorded special status. In some circumstances, treating the null and alternative hypotheses more equally may be more appropriate. For example, if the costs of Type I and Type II errors are approximately equal, then the optimal strategy is to balance the probabilities of making these two types of errors. When the costs of errors are not known, however, there is no optimal procedure for balancing the costs of Type I and Type II errors.

A potential problem arises when we have a very large sample of data. As the sample size increases, the power of the test increases. By constantly increasing the sample size, we can make the power of the test approach 1 and virtually guarantee that a false null hypothesis will be rejected. Of course, we want to reject H_0 when H_0 is false, but occasionally people *misinterpret* the results of the statistical decision. It is customary for researchers to report that the results of a test are "statistically significant" and that the null hypothesis has been rejected. If the statistical results are significant, we are led to conclude that the true situation differs from that stated in the null hypothesis.

When n is large and the power of the test approaches 1, it becomes relatively easy to detect even small deviations from the null hypothesis. In this situation, we must avoid the assumption that, just because we have rejected the null hypothesis, the difference between the null hypothesis and the true situation is of practical significance. This may or may not be the case. For instance, the null hypothesis may indeed be false, but the true situation may be so close to the null hypothesis that the difference is of no practical significance. In this situation, a statistical result that is *statistically* significant is not *practically* significant. This issue is illustrated in the next example.

EXAMPLE 11.26   **Statistical Versus Practical Significance**

The newspaper *USA Today* stated that, in the United States, the mean annual salary of male lawyers is $47,635 per year. A law firm in Texas wants to test whether the average salary of male lawyers in Texas is the same as the national average. Thus, the law firm will reject the *USA Today* claim if it obtains a sample mean that is either much higher or much lower than $47,635. Thus a two-tailed test is appropriate, and the law firm wants to test

$$H_0: \quad \mu = \$47,635$$

against the two-sided alternative hypothesis

$$H_1: \quad \mu \neq \$47,635$$

Assume that the population is normal with standard deviation $\sigma = \$10,000$, and use a 5% level of significance. Now consider two extreme cases: (1) The sample size is very small, $n = 9$; and (2) the sample size is very large, $n = 2,500$. Find the acceptance region of the test statistic $\bar{X}$ in both cases.

SOLUTION   Because the test is two-sided, the critical values of the test are

$$\bar{x}_1^* = \mu_0 - z_{\alpha/2}\sigma/\sqrt{n} \quad \text{and} \quad \bar{x}_2^* = \mu_0 + z_{\alpha/2}\sigma/\sqrt{n}$$

where $\mu_0 = \$47,635$, $z_{\alpha/2} = 1.96$, and $\sigma = \$10,000$. When $n = 9$, the acceptance region is any value of $\bar{X}$ between $\$41,101.67$ and $\$54,168.33$. When $n = 2,500$, the acceptance region is any value of $\bar{X}$ between $\$47,243$ and $\$48,027$. Thus when $n = 9$, we would not reject H_0 even if we obtained $\bar{x} = \$54,000$, an amount almost $\$7,000$ higher than the hypothesized mean. Here the true mean could be quite different from the hypothesized mean and it still might be very difficult to detect this difference.

When $n = 2,500$, we would reject the null hypothesis if we obtained the sample mean $\bar{x} = \$48,028$. This sample mean is only $\$393$ from the hypothesized mean of $\$47,635$, which to most people is of little practical significance. Although the difference would be *statistically* significant, it would not be *practically* significant.

Power Curve for Two-Sided Tests

When the alternative hypothesis is two-sided, part of the rejection region falls in each tail of the distribution. To calculate the power of the test, we have to calculate the probability that the test statistic falls in either rejection region given that H_0 is false, as illustrated in the next example.

EXAMPLE 11.27   **Calculating the Power of the Test When the Alternative Hypothesis Is Two-Sided**

A marketing report states that the annual per capita expenditures on cereal follow a normal distribution with mean $\mu = \$60$ and standard deviation $\sigma = \$15$. We want to test

$$H_0: \quad \mu = 60$$

against the two-sided alternative hypothesis

$$H_1: \quad \mu \neq 60$$

using a 5% level of significance and a random sample of $n = 100$ customers. Find the power curve associated with this test.

SOLUTION   For $\alpha = .05$, we obtain $z_{\alpha/2} = 1.96$. The critical values of the test are

$$\bar{x}_1^* = \mu_0 - z_{\alpha/2}\sigma/\sqrt{n} = 57.06$$

and

$$\bar{x}_2^* = \mu_0 + z_{\alpha/2}\sigma/\sqrt{n} = 62.94$$

We accept H_0 if ($57.06 \leq \bar{X} \leq 62.94$), as shown in Figure 11.21. The power of the test is the probability that $\bar{X}$ falls in the rejection region when H_0 is indeed false. Suppose that the true value of the mean is $\mu = 58$. We then obtain

$$\begin{aligned} \text{Power} &= P(\bar{X} < \bar{x}_1^*) + P(\bar{X} > \bar{x}_2^*) \\ &= P(\bar{X} < 57.06) + P(\bar{X} > 62.94) \end{aligned}$$

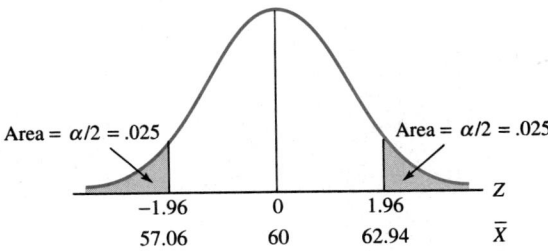

FIGURE 11.21 *Acceptance region for Example 11.27*

This probability is shown as the shaded area in Figure 11.22. To calculate the shaded area, we obtain the Z scores using $\mu = 58$ as follows:

$$z_1 = \frac{57.06 - 58}{15/\sqrt{100}} = -.63 \quad \text{and} \quad z_2 = \frac{62.94 - 58}{15/\sqrt{100}} = 3.29$$

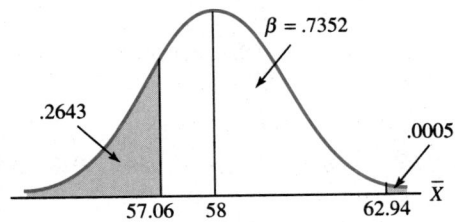

FIGURE 11.22 *Power of the test for Example 11.27*

We then obtain

$$\begin{aligned} \text{Power} &= P(\bar{X} < 57.06) + P(\bar{X} > 62.94) \\ &= P(Z < -.63) + P(Z > 3.29) \\ &= .2643 + .0005 = .2648 \end{aligned}$$

To get other points on the power curve, assign μ other values and repeat the process. The shaded areas in Figure 11.23 (page 452) show the power of the test for several values of μ.

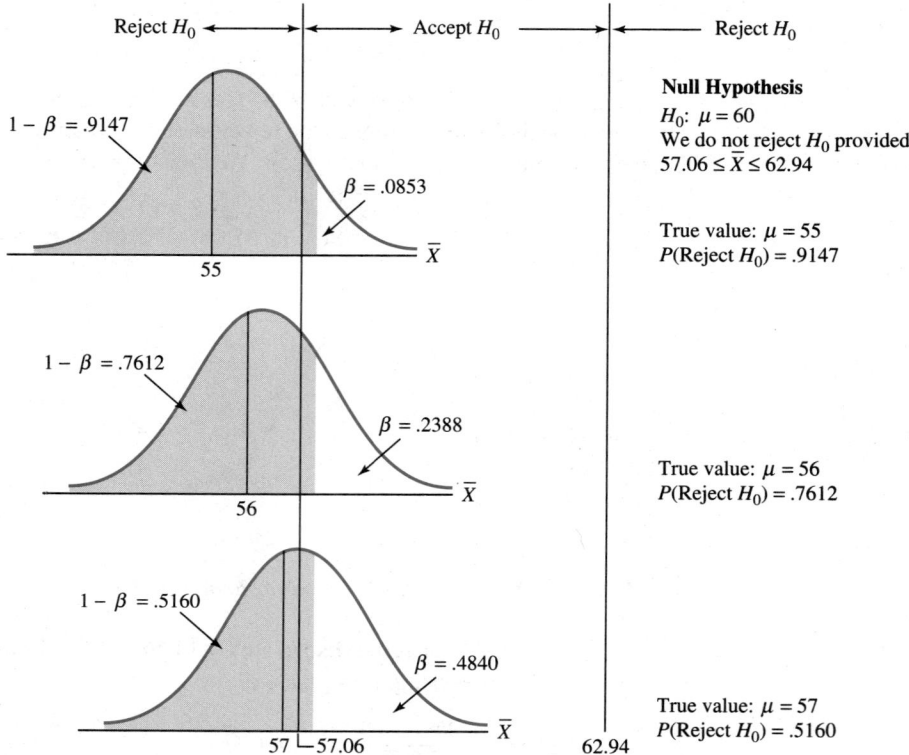

FIGURE 11.23 *Power of test for different values of μ in Example 11.27*

The data in the following table show the power of the test for various values of μ:

μ	55	56	57	58	59	60	61	62	63	64	65
Power	.9147	.7612	.5160	.2648	.1028	α = .0500	.1028	.2648	.5160	.7612	.9147

Because we used a two-tailed test, the acceptance region extends an equal distance (2.94 units) from $\mu = 60$ in each direction. Consequently, the power curve for this problem is symmetric about $\mu = 60$. This symmetry is present in any two-tailed test of the mean. Because the null hypothesis H_0 is true when $\mu = 60$, it is not possible to commit a Type II error when $\mu = 60$. When H_0 is true, the probability that the test statistic falls in the rejection region is not the power of the test but rather it is α, the level of significance of the test. When the true value of μ is extremely close to 60, the probability that the test statistic falls in the rejection region is approximately equal to α, the level of significance. Thus, the power of the test always exceeds α and approaches α as μ approaches 60. Figure 11.24, the power curve associated with the preceding data, illustrates that the power curve is symmetric about $\mu = 60$.

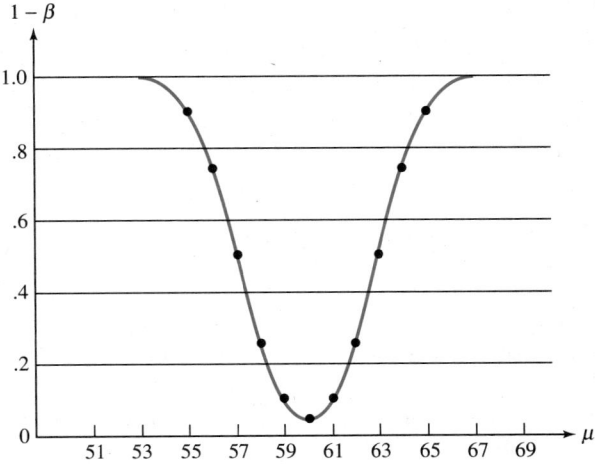

FIGURE 11.24   *Power curve for Example 11.27*

Tips for Problem Solving

Suppose you are testing the null hypothesis

$$H_0: \quad \mu = \mu_0$$

To calculate the power of a test, follow these steps:

1. Record the sample size n, the level of significance α, the hypothesized mean μ_0, and the population standard deviation σ.
2. Record the standard deviation of $\bar{X}$. This standard deviation is $\sigma/\sqrt{n}$, *not* σ.
3. Sketch the normal distribution showing the sampling distribution of $\bar{X}$. This distribution has mean μ_0 and standard deviation $\sigma/\sqrt{n}$. For a one-tailed test, shade an area equal to α in the appropriate tail. For a two-tailed test, shade an area of $\alpha/2$ in each tail.
4. Sketch a standard normal distribution and shade the tail area as in step 3. Find the critical values of the Z scores associated with the shaded tail area(s).
5. Find the critical value(s) of $\bar{X}$ associated with the critical value(s) of Z.
6. State the critical region and the acceptance region in terms of the critical values of $\bar{X}$.
7. Record the new hypothesized value of the population mean, denoted μ_1.
8. Sketch the new normal distribution showing the hypothesized sampling distribution of $\bar{X}$. This distribution has mean μ_1 and standard deviation $\sigma/\sqrt{n}$. Shade the areas corresponding to the critical region found in step 5.
9. Evaluate the shaded area(s) obtained in step 8. This is the power of the test.

Calculating the Power of the Test of a Population Proportion

The following procedures summarize how to calculate the power of the test of a hypothesis about a population proportion p using a one-sided or two-sided test:

Case 1:   Suppose we take a random sample of n observations and calculate the sample proportion $\hat{p}$ to test the null hypothesis

$$H_0:   p = p_0$$

against the one-sided alternative hypothesis

$$H_1:   p > p_0$$

where the significance level is α. If $np_0 \geq 5$ and $nq_0 \geq 5$, then $\hat{p}$ has an approximately normal distribution. Because the alternative hypothesis lies entirely to the right of $p = p_0$, the entire rejection region is placed in the right tail of the sampling distribution of $\hat{p}$ and consists of all values of the sample proportion exceeding the critical value $\hat{p}*$ where

$$\hat{p}* = p_0 + z_\alpha \sqrt{p_0 q_0/n}$$

The power of the test equals the probability that the test statistic falls in the rejection region when H_0 is false, so we obtain

$$\text{Power} = P(\hat{p} > \hat{p}*)$$
$$= P(\hat{p} > p_0 + z_\alpha \sqrt{p_0 q_0/n})$$

when $p = p_1$. Equivalently, we obtain

$$\text{Power} = P\left(Z > \frac{\hat{p}* - p_1}{\sqrt{p_1 q_1/n}}\right)$$

Case 2:   Suppose we are testing the null hypothesis

$$H_0:   p = p_0$$

against the alternative hypothesis

$$H_1:   p < p_0$$

where the significance level is α and n is the sample size. If we use $\hat{p}$ as a test statistic, then the rejection region consists of all values of the sample proportion less than the critical value $\hat{p}*$, where

$$\hat{p}* = p_0 - z_\alpha \sqrt{p_0 q_0/n}$$

In this case, the entire rejection region is placed in the left tail of the sampling distribution of $\hat{p}$ because the alternative hypothesis $H_1: p < p_0$ lies entirely to the left of $p = p_0$. In this case, we obtain

$$\text{Power} = P(\hat{p} < \hat{p}*)$$
$$= P(\hat{p} < p_0 - z_\alpha \sqrt{p_0 q_0/n})$$

when $p = p_1$. Equivalently, we obtain

$$\text{Power} = P\left(Z < \frac{\hat{p}* - p_1}{\sqrt{p_1 q_1/n}}\right)$$

Case 3:   Suppose we are testing the null hypothesis

$$H_0: \quad p = p_0$$

against the two-sided alternative hypothesis

$$H_1: \quad p \neq p_0$$

where the significance level is α and n is the sample size. If we use $\hat{p}$ as a test statistic, then the two critical values are

$$\hat{p}_1^* = p_0 - z_{\alpha/2}\sqrt{p_0 q_0/n}$$

and

$$\hat{p}_2^* = p_0 + z_{\alpha/2}\sqrt{p_0 q_0/n}$$

We obtain

$$\text{Power} = P(\hat{p} < \hat{p}_1^*) + P(\hat{p} > \hat{p}_2^*)$$

when $p = p_1$. Equivalently, we obtain

$$\text{Power} = P\left(Z < \frac{\hat{p}_1^* - p_1}{\sqrt{p_1 q_1/n}}\right) + P\left(Z > \frac{\hat{p}_2^* - p_1}{\sqrt{p_1 q_1/n}}\right)$$

EXAMPLE 11.28   **Power Curve for a One-Tailed Test of a Proportion** ——————————————

Last year, 40% of the TV audience watched the late-night news on Channel 2. Hoping to increase its audience share, the station expands its sports coverage. If the new format does not increase the station's share of the audience during a trial period, the station will return to its old format. Because the station will not retain the expanded sports format unless $p > .4$, the station wants to test the null hypothesis

$$H_0: \quad p \leq .4$$

against the one-sided alternative hypothesis

$$H_1: \quad p > .4$$

Suppose we take a random sample of $n = 100$ viewers and perform a one-tailed test using a 5% level of significance. Find the power curve associated with this test.

SOLUTION   We have $z_\alpha = 1.645$, $n = 100$, and $p_0 = .4$. The rejection region consists of all values of the sample proportion $\hat{p}$ exceeding the critical value $\hat{p}^*$, where

$$\hat{p}^* = p_0 + z_\alpha\sqrt{p_0 q_0/n}$$
$$= .4 + 1.645\sqrt{(.4)(.6)/100} \approx .48$$

The rejection region consists of all values of $\hat{p}$ such that $\hat{p} > .48$.

Now assume that the new news format actually increases the station's market share to .5. Then the sample proportion $\hat{p}$ is approximately normally distributed with mean .5 and variance $(.5)(.5)/100$. (Note that the variance is different when $p = .5$ than when $p = .4$.) The power of the test is given by

$$\text{Power} = P(\hat{p} > .48 | p = .5)$$

This situation is shown in Figure 11.25 (page 456).

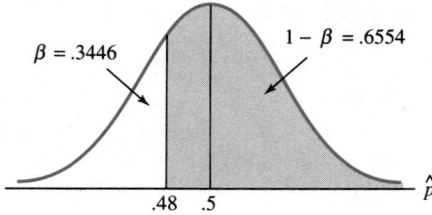

FIGURE 11.25   *Power of test in Example 11.28*

We obtain the Z score

$$z = \frac{.48 - .5}{\sqrt{(.5)(.5)/100}} = -.40$$

Finally, we get

$$P(\hat{p} > .48) = P(Z > -.40) = .6554$$

To find other points on the power curve, we assume other values for p, as calculated in the following table:

p		.40	.42	.45	.48	.50	.55	.60
Power	$(\alpha = .0500)$	.1093	.2709	.5000	.6554	.9207	.9929	

Using these data, we obtain the power curve shown in Figure 11.26. The power curve shows that this test is relatively powerful when p exceeds .5 but not when p is less than .5.

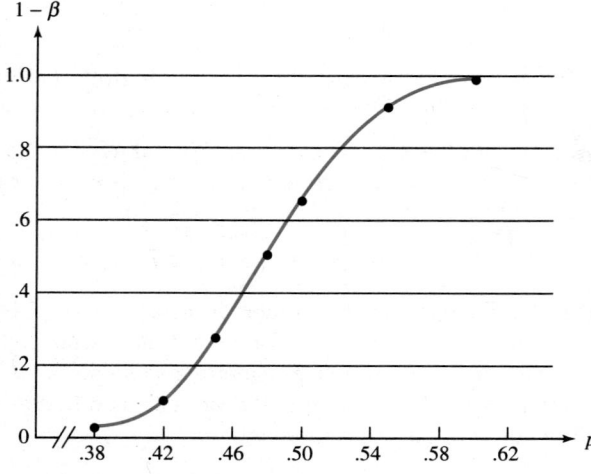

FIGURE 11.26   *Power curve for Example 11.28*

EXAMPLE 11.29 **Effect of Increasing Sample Size in Example 11.28** ————————————

In Example 11.28, we showed that when $p = .50$, the power of the test is .6554. This value depends on the fact that $\alpha = .05$ and $n = 100$. Suppose we increase the sample size to $n = 400$ and keep $\alpha = .05$. Show that the power will increase.

S O L U T I O N   To show this relationship, we must find a new rejection region based on a sample of 400. We have $z_\alpha = 1.645$, $n = 400$, and $p_0 = .4$. The rejection region consists of all values of the sample proportion $\hat{p}$ exceeding the critical value $\hat{p}*$ where

$$\hat{p}* = p_0 + z_\alpha \sqrt{p_0 q_0 / n}$$
$$= .4 + 1.645 \sqrt{(.4)(.6)/400} \approx .44$$

When $n = 100$, the rejection region is $\hat{p} > .48$. In the present case, when $n = 400$, the rejection region is $\hat{p} > .44$. The power of the test is

$$\text{Power} = P(\hat{p} > .44 | p = .5)$$

We obtain the Z score

$$z = \frac{.44 - .5}{\sqrt{(.5)(.5)/400}} = -2.40$$

which gives us

$$P(\hat{p} > .44) = P(Z > -2.40) = .9918$$

Increasing the sample size from 100 to 400 increases the power of the test from .6554 to .9918. For a fixed value of α, we can always increase the power of the test by increasing n.

Exercises

Statistical Concepts

11.7.1   Explain the difference between α and β.

11.7.2   Explain the relationship between β, the probability of committing a Type II error, and the power of a test.

11.7.3   How can we increase the power of a test without increasing the sample size?

11.7.4   Suppose you have constructed a power curve for either a one-tailed test or a two-tailed test. Explain the meaning of the "power" at the point corresponding to the value of the parameter assumed under the null hypothesis.

11.7.5   Explain how the sample size and the level of significance of a test influence the power of a test.

11.7.6   *True or false:*   For a fixed sample size, α and β are inversely related; as one increases, the other decreases.

11.7.7   *True or false:*   For a fixed value of α and β, increasing the sample size provides more information on which to base the decision and hence reduces β.

11.7.8   *True or false:*   The probability of a Type I error and the power of the test measure the risk of making an incorrect decision.

11.7.9   *True or false:*   Suppose a statistician has decided to reject the null hypothesis. Then the statistician cannot commit a Type II error. Explain.

11.7.10   *True or false:*   To calculate α, you find the probability that the test statistic will fall in the acceptance region, given that the null hypothesis is false. Explain.

11.7.11   *True or false:*   To calculate β, you find the probability that the test statistic will fall in the rejection region, given that the null hypothesis is false. Explain.

Statistical Drills

11.7.12   In the past, assembly line workers have been able to produce 900 or more units of output per hour, on the average. A quality control engineer wants to test H_0: $\mu \geqslant 900$ against H_1: $\mu < 900$ using $\alpha = .025$. The engineer determines hourly output during a random sample of $n = 25$ hours. Assume that hourly output follows a normal distribution with standard deviation $\sigma = 150$.
  a. Graph the sampling distribution of $\bar{X}$ assuming H_0 is true.
  b. Find the critical value of $\bar{X}$ that separates the critical region from the acceptance region.
  c. Now suppose that $\mu = 880$, rather than $\mu = 900$. Find the probability that the hypothesis test will incorrectly fail to reject H_0. That is, find β.
  d. If $\mu = 880$, find the probability that the test will correctly reject the null hypothesis. That is, find the power of the test.
  e. Repeat part **c** using $\mu = 860$. Compare β in the two situations. Explain the difference.
  f. Repeat part **d** using $\mu = 860$. Compare the power of the test in the two situations. Explain the difference.

11.7.13   A researcher wants to test H_0: $\mu \leqslant 760$ against H_1: $\mu > 760$ using $\alpha = .025$. Suppose the scientist takes a random sample of $n = 100$ observations from a normal population whose standard deviation is $\sigma = 50$.
  a. Graph the sampling distribution of $\bar{X}$ assuming H_0 is true.
  b. Find the critical value of $\bar{X}$ that separates the critical region from the acceptance region.
  c. Now suppose that $\mu = 770$, rather than $\mu = 760$. Find the probability that the hypothesis test will incorrectly fail to reject H_0. That is, find β.
  d. If $\mu = 770$, find the probability that the test will correctly reject the null hypothesis. That is, find the power of the test.
  e. Repeat part **c** using $\mu = 765$. Compare β in the two situations. Explain the difference.
  f. Repeat part **d** using $\mu = 765$. Compare the power of the test in the two situations. Explain the difference.

11.7.14   An investigator wants to test H_0: $\mu \geqslant 900$ against H_1: $\mu < 900$ using $\alpha = .025$. The investigator takes a random sample of $n = 25$ observations from a normal population whose standard deviation is $\sigma = 60$.
  a. Graph the sampling distribution of $\bar{X}$ assuming H_0 is true.
  b. Find the critical value of $\bar{X}$ that separates the critical region from the acceptance region.
  c. Now suppose that $\mu = 890$, rather than $\mu = 900$. Find the probability that the hypothesis test will incorrectly fail to reject H_0. That is, find β.
  d. If $\mu = 890$, find the probability that the test will correctly reject the null hypothesis. That is, find the power of the test.
  e. Repeat part **c** using $n = 100$. Compare β in the two situations. Explain the difference.
  f. Repeat part **d** using $n = 100$. Compare the power of the test in the two situations. Explain the difference.

11.7.15   A plant manager wants to test H_0: $\mu \leqslant 80$ against H_1: $\mu > 80$ using $\alpha = .05$. The manager takes a random sample of $n = 100$ observations from a uniform population whose standard deviation is $\sigma = 40$.
  a. Graph the sampling distribution of $\bar{X}$ assuming H_0 is true. Note that the population from which the sample is taken is not normal.
  b. Find the critical value of $\bar{X}$ that separates the critical region from the acceptance region.

c. Now suppose that $\mu = 85$, rather than $\mu = 80$. Find the probability that the hypothesis test will incorrectly fail to reject H_0. That is, find β.

d. If $\mu = 85$, find the probability that the test will correctly reject the null hypothesis. That is, find the power of the test.

e. Repeat part **c** using $n = 50$. Compare β in the two situations. Explain the difference.

f. Repeat part **d** using $n = 50$. Compare the power of the test in the two situations. Explain the difference.

11.7.16 A businessman wants to test $H_0: \mu = 280$ against $H_1: \mu \neq 280$ using $\alpha = .05$. The businessman takes a random sample of $n = 64$ observations from a normal population whose standard deviation is $\sigma = 24$.

a. Graph the sampling distribution of $\bar{X}$ assuming H_0 is true.

b. Find the critical values of $\bar{X}$ that separate the critical region from the acceptance region.

c. Graph the sampling distribution of $\bar{X}$ assuming $\mu = 283$.

d. Find the probability that the hypothesis test will incorrectly fail to reject H_0. That is, find β.

e. If $\mu = 283$, find the probability that the test will correctly reject the null hypothesis. That is, find the power of the test.

f. Find β for each of the following values of the population mean: 289, 286, 283, 277, 274, 270. Plot each value of β against the corresponding population mean. Plot β on the vertical axis and μ on the horizontal axis. Draw a curve through the six points.

g. Use the data from part **f** and plot the power of the test at each value of μ. Plot the power on the vertical axis against μ on the horizontal axis.

h. Explain the relationship between the power of a test and the distance between the true mean μ and μ_0, the assumed mean under H_0.

11.7.17 A scientist wants to test $H_0: \mu \geq 1{,}700$ against $H_1: \mu < 1{,}700$ using $\alpha = .025$. The scientist takes a random sample of $n = 100$ observations from a normal population whose standard deviation is $\sigma = 200$.

a. Graph the sampling distribution of $\bar{X}$ assuming H_0 is true.

b. Find the critical value of $\bar{X}$ that separates the critical region from the acceptance region.

c. Graph the sampling distribution of $\bar{X}$ assuming $\mu = 1{,}690$.

d. Find the probability that the hypothesis test will incorrectly fail to reject H_0. That is, find β.

e. If $\mu = 1{,}690$, find the probability that the test will correctly reject the null hypothesis. That is, find the power of the test.

f. Find β for each of the following values of the population mean: 1,695; 1,690; 1,680; 1,670; 1,660; 1,650; 1,640. Plot each value of β against the corresponding population mean. Plot β on the vertical axis and μ on the horizontal axis. Draw a curve through the seven points.

g. Use the data from part **f** and plot the power of the test at each value of μ. Plot the power on the vertical axis against μ on the horizontal axis.

h. Explain the relationship between the power of a test and the distance between the true mean μ and μ_0, the assumed mean under H_0.

i. Repeat parts **a–h** for a sample size of $n = 25$. Compare the power of the test for different values of μ based on using the different sample sizes. Explain how an increase in the sample size influences the power of the test.

11.7.18 An economist wants to test $H_0: \mu \geq 600$ against $H_1: \mu < 600$ using $\alpha = .025$. The individual takes a random sample of $n = 25$ observations from a normal population whose standard deviation is $\sigma = 100$.

a. Graph the sampling distribution of $\bar{X}$ assuming H_0 is true.

b. Find the critical value of $\bar{X}$ that separates the critical region from the acceptance region.

c. Graph the sampling distribution of $\bar{X}$ assuming $\mu = 590$.

d. Find the probability that the hypothesis test will incorrectly fail to reject H_0. That is, find β.

   e. If $\mu = 590$, find the probability that the test will correctly reject the null hypothesis. That is, find the power of the test.

   f. Find β for each of the following values of the population mean: 595, 590, 580, 570, 560, 550, 540. Plot each value of β against the corresponding population mean. Plot β on the vertical axis and μ on the horizontal axis. Draw a curve through the seven points.

   g. Use the data from part **f** and plot the power of the test at each value of μ. Plot the power on the vertical axis against μ on the horizontal axis.

   h. Repeat parts **a–g** but change the standard deviation to $\sigma = 50$. Compare the power of the test for different values of μ based on using the different standard deviations. Explain how a decrease in the standard deviation influences the power of the test.

Statistical Applications

11.7.19   At a supermarket, the average number of register mistakes per day per clerk was 18 and the standard deviation was 5. The owner of the supermarket purchased new cash registers in an effort to decrease the number of errors. The manager then took a random sample of 100 clerks on randomly selected days using the new registers to test $H_0: \mu = 18$ against $H_1: \mu \neq 18$.

   a. Find the critical values of the sample mean $\overline{X}$ such that the level of significance is 5%.

   b. Find the critical values of the sample mean $\overline{X}$ such that the level of significance is 1%.

   c. Suppose the sample size is 36 rather than 100. Find the critical values of the sample mean $\overline{X}$ such that the level of significance is 5%.

   d. In part **c,** suppose the population mean is $\mu = 16$. Find the power of the test.

   e. In part **a,** suppose that $\mu = 16$. Find the power of the test.

   f. Compare and explain the differences between answers to parts **d** and **e.**

11.7.20   With traditional advertising by mail, magazine sales in a community have had an average of 26 per day and a standard deviation of 5. A new sales technique is devised in which customers are given samples of the magazines and a certificate entitling them to a reduced price on a meal at a local restaurant. The technique is tried for 100 days. We want to test $H_0: \mu = 26$ against $H_1: \mu > 26$.

   a. Design a one-tailed test having a 5% level of significance. Find the critical value $\overline{x}^*$ that separates the acceptance region from the critical region.

   b. If $\overline{x} = 27$, what decision would you make in part **a**?

   c. Suppose the population mean is $\mu = 27$. Find the power of the test.

11.7.21   The quality control department of a food processing firm wants to test whether the average weight per package of a certain food is at least 25 ounces. Past experience indicates that the standard deviation of weight is .3 ounce. A random sample of 144 packages is obtained.

   a. Design a decision rule to test $H_0: \mu \geq 25$ against $H_1: \mu < 25$ such that $\alpha = 5\%$.

   b. If the true population mean is $\mu = 24.8$, find the power of the test.

   c. Repeat part **a** assuming the sample size is 36.

   d. In part **c,** assume the population mean is 24.8. Find the power of the test.

11.7.22   A business executive thinks that opening a Japanese restaurant would be profitable if at least 5% of the people in the community are interested in dining there. A random sample of 500 people in the community is obtained.

   a. Design a rule to test $H_0: p \geq .05$ against $H_1: p < .05$ using a 5% level of significance.

   b. If $\hat{p} = .04$, what decision would you make?

   c. Find the power of the test if the actual proportion is $p = .035$.

11.7.23   The McCormick Food Company sells 10-ounce cans of peas. The cans always contain several ounces of water. The quality control department wants to test whether the average weight of peas in the cans is at least 7 ounces. It is assumed that the standard deviation is $\sigma = .2$ ounce. A random sample of 100 cans of peas is tested.

a. Design a decision rule to test H_0: $\mu \geqslant 7$ against H_1: $\mu < 7$ using a 1% level of significance.
b. If in the sample $\bar{x} = 6.7$, what decision should be made?
c. Find the power of the test if $\mu = 6.9, 6.8, 6.7,$ and 6.6 ounces.

11.8 The Chi-Square Distribution

So far we have discussed how to test hypotheses about a population mean or a population proportion, but in many situations we are interested in testing a hypothesis about the population variance. To test such a hypothesis, we must compute a test statistic that depends on the sample variance S^2. If the null hypothesis is true and the population follows a normal distribution, then the test statistic has a sampling distribution that follows the **chi-square distribution.** Thus, to test a hypothesis about a population variance, we must know how to find the critical region of a chi-square distribution. In this section, we discuss the properties of the chi-square distribution and show how to calculate areas under the chi-square density function.

Like the t distribution, the chi-square distribution is a continuous probability distribution that depends on only one parameter, the degrees of freedom (ν), and consists of a family of distributions, one for each value of ν. Just as random variables that follow the t distribution are denoted by the letter t, random variables that follow the chi-square distribution are denoted by the symbol χ^2 (Greek chi squared).

Characteristics of the chi-square distribution

1. The random variable χ^2 varies from 0 to ∞ and is never negative.
2. As with all probability density functions, the density function of χ^2 is always nonnegative, and the area beneath the curve is 1.
3. For small values of ν, the chi-square distribution is skewed to the right, but as ν increases the curve becomes approximately symmetric. For $\nu = 1$ and $\nu = 2$, the curve is decreasing.
4. For large values of ν, say, 20 or more, the distribution is closely approximated by a normal curve.
5. The mean of the chi-square distribution is ν.
6. The variance of the chi-square distribution is 2ν. Thus, the mean and variance depend on the degrees of freedom.

Figure 11.27 (page 462) shows some chi-square distributions for different degrees of freedom. As the number of degrees of freedom increases, the curve becomes more symmetric. Because each value of ν yields a separate distribution, as with the t distribution, presenting detailed tables showing the areas under all the curves is not feasible. Instead, we have a table (Table A.7 in the Appendix) showing the critical values of χ^2 for different degrees of freedom, as we had for the t distribution. To test hypotheses with the chi-square distribution, we need to find the critical values of the chi-square distribution having a right tail area of α, where α is the level of significance of the test. The critical value depends on the degrees of freedom. The examples that follow indicate how to calculate the critical value for a chi-square distribution having ν degrees of freedom.

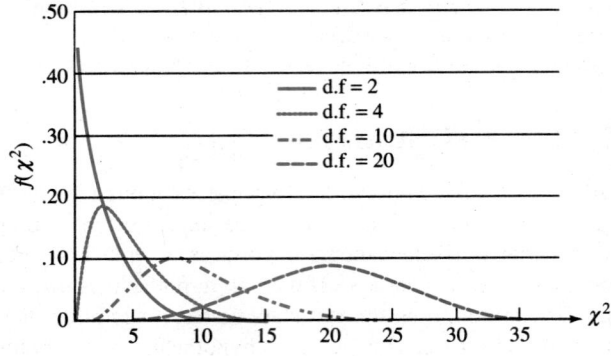

FIGURE 11.27 *Some chi-square distributions with different degrees of freedom*

Critical value $\chi^2_{\alpha,\nu}$

Let the random variable χ^2 follow the chi-square distribution having ν degrees of freedom. The expression $\chi^2_{\alpha,\nu}$ denotes the value such that

$$P(\chi^2 > \chi^2_{\alpha,\nu}) = \alpha$$

That is, if χ^2 follows a chi-square distribution with ν degrees of freedom, then the area under the density function to the right of $\chi^2_{\alpha,\nu}$ is α and the probability that χ^2 exceeds $\chi^2_{\alpha,\nu}$ is α.

EXAMPLE 11.30   **Finding a Critical Value for a Chi-Square Distribution**

Figure 11.28 shows the graph of the chi-square distribution with $\nu = 6$ degrees of freedom. Find the critical value of χ^2 such that the shaded area is .05.

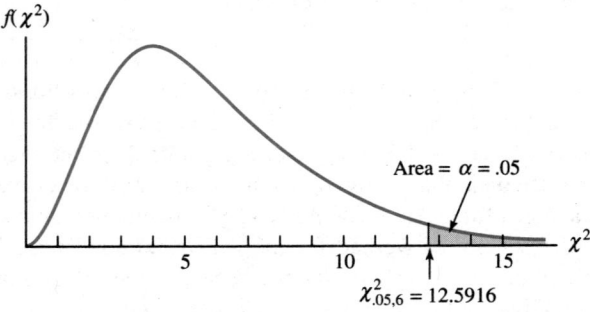

FIGURE 11.28   *Chi-square distribution having 6 degrees of freedom*

SOLUTION   Turn to Table A.7. Proceed down the column entitled Degrees of Freedom to $\nu = 6$. Now move to the right until you reach the column labeled .05. The result is $\chi^2_{.05,6} = 12.5916$. Thus, we know

$$P(\chi^2 > 12.5916) = .05$$

if χ^2 has 6 degrees of freedom.

EXAMPLE 11.31   **Finding a Critical Value for a Chi-Square Distribution**

Find the value of χ^2 such that the shaded area in Figure 11.28 is .01.

SOLUTION   Once again we have $\nu = 6$ degrees of freedom. We seek the value $\chi^2_{.01,6}$ in Table A.7. When $\nu = 6$, we find that $\chi^2_{.01,6} = 16.8119$.

To construct confidence intervals for a population variance and to perform two-tailed tests, we must find critical values of χ^2 such that the area in *each* tail of the distribution is $\alpha/2$. Because the chi-square distribution is not symmetric, the critical value for the left tail must be obtained independently of that for the right tail. The critical value such that the area in the left tail is $\alpha/2$ is denoted $\chi^2_{(1-\alpha/2),\nu}$ because the area to its right is $(1 - \alpha/2)$.

Exercises

Statistical Concepts

11.8.1   Describe the characteristics of the chi-square distribution.

11.8.2   *True or false:*   Suppose a random variable X has the chi-square distribution with 8 degrees of freedom. The mean of X is $(8 - 1) = 7$.

11.8.3   *True or false:*   Suppose a random variable X has the chi-square distribution with 6 degrees of freedom. The mean of X is 6, and the variance of X is 6.

11.8.4   *True or false:*   As the number of degrees of freedom gets large, say, df $\geq$ 20, the chi-square distribution begins to approximate a normal distribution with mean equal to 0.

11.8.5   *True or false:*   As the number of degrees of freedom gets large, say, df $\geq$ 20, the chi-square distribution begins to approximate a normal distribution with mean equal to the degrees of freedom.

11.8.6   *True or false:*   Suppose a random variable X has the chi-square distribution with 4 degrees of freedom. Then $P(X \leq 0) = 0$. Explain.

Statistical Drills

11.8.7   Suppose that χ^2 has $\nu = 8$ degrees of freedom. Find the values of χ^2 such that the right tail has the following areas:
    a. .05   **b.** .01   **c.** .10

11.8.8   Let the random variable χ^2 have a chi-square distribution with ν degrees of freedom. Find the value $\chi^2_{\alpha,\nu}$ such that $P(\chi^2 > \chi^2_{\alpha,\nu}) = \alpha$ where the values of α and ν are as follows:
    a. $\alpha = .01$; $\nu = 4$ **b.** $\alpha = .05$; $\nu = 7$

11.8.9   Suppose that χ^2 has $\nu = 6$ degrees of freedom. Find the values of χ^2 such that the left tail has the following areas:

    a. .01   **b.** .05   **c.** .10

11.8.10   Find the critical value of a chi-square distribution with 11 degrees of freedom having an area of $\alpha = .05$ to its right.

11.8.11   Find the critical value of a chi-square distribution with 50 degrees of freedom having an area of $\alpha = .05$ to its right.

11.9   Testing Hypotheses and Constructing Confidence Intervals About a Population Variance

To estimate a population variance σ^2, we take a random sample of size n and calculate the sample variance S^2 using the formula

$$S^2 = \sum_{i=1}^{n} \frac{(X_i - \bar{X})^2}{n - 1}$$

Because S^2 varies from sample to sample, S^2 is a random variable and has a sampling distribution. A variance can never be negative, so the probability distribution of S^2 begins at $S^2 = 0$. Since the mean of this probability distribution is the true value σ^2, the sample variance S^2 is an unbiased estimator of σ^2. The exact form of the sampling distribution of S^2 depends on the probability distribution of the population.

In this discussion, we assume that a random sample is selected from a normal population. The results will still hold approximately when the parent population is nonnormal, provided the deviations from normality are not too severe.

It can be shown that if the population is normal, the random variable

$$\chi^2 = \frac{(n - 1)S^2}{\sigma^2}$$

follows the chi-square distribution with $\nu = (n - 1)$ degrees of freedom. We use this result when testing hypotheses about σ^2 and constructing confidence intervals for σ^2.

Testing Hypotheses About σ^2

Let X be a random variable that follows a normal distribution with mean μ and variance σ^2. Suppose we want to test the hypothesis that the variance σ^2 equals some specific value σ_0^2 using level of significance α. The test is based on the sample variance S^2 and the test statistic $(n - 1)S^2/\sigma_0^2$, where σ_0^2 is the hypothesized variance. Let $\chi^2_{\alpha/2,\nu}$ be the value of the chi-square variable such that the area to its right is $\alpha/2$. It follows that $\chi^2_{1-\alpha/2,\nu}$ is the value of the chi-square variable such that the area to its right is $(1 - \alpha/2)$ and the area to its left is $\alpha/2$. This idea is illustrated in Figure 11.29. Given the value of α, we can find $\chi^2_{1-\alpha/2,\nu}$ and $\chi^2_{\alpha/2,\nu}$ for the chi-square distribution having ν degrees of freedom by using Table A.7.

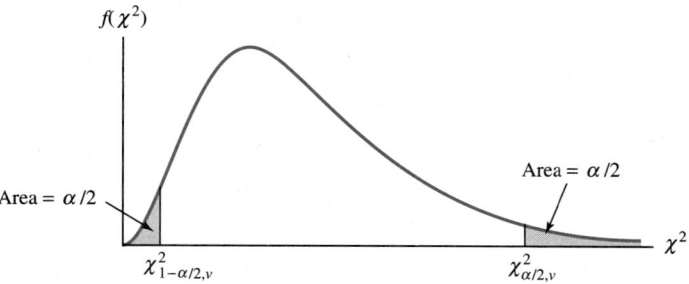

FIGURE 11.29 *Critical region for the chi-square statistic when performing a two-tailed test about a population variance*

Suppose we want to test the null hypothesis

$$H_0: \quad \sigma^2 = \sigma_0^2$$

against the two-sided alternative hypothesis

$$H_1: \quad \sigma^2 \neq \sigma_0^2$$

If H_0 is true, then the following probability statement holds:

$$P\left(\chi_{1-\alpha/2,\nu}^2 \leq \frac{(n-1)S^2}{\sigma_0^2} \leq \chi_{\alpha/2,\nu}^2\right) = \alpha$$

Using level of significance α, we would accept the null hypothesis if $(n-1)S^2/\sigma_0^2$ was between $\chi_{1-\alpha/2,\nu}^2$ and $\chi_{\alpha/2,\nu}^2$, as shown in Figure 11.29. If $(n-1)S^2/\sigma_0^2$ was less than $\chi_{1-\alpha/2,\nu}^2$ or greater than $\chi_{\alpha/2,\nu}^2$, then we would reject H_0 in favor of H_1.

For a one-tailed test, the entire rejection region is placed in one tail of the chi-square distribution, and we reject H_0 in favor of H_1 if the test statistic $\chi^2 = (n-1)S^2/\sigma_0^2$ falls in the rejection region. The procedure for testing a hypothesis about the population variance is summarized in the accompanying box.

Testing hypotheses about the variance of a normal population

Suppose we have a random sample of n observations from a normal population with variance σ_0^2. Let s^2 denote the observed sample variance. The following tests have level of significance α:

Case 1: To test the null hypothesis

$$H_0: \quad \sigma^2 = \sigma_0^2 \quad \text{or} \quad H_0: \quad \sigma^2 \leq \sigma_0^2$$

against the one-sided alternative hypothesis

$$H_1: \quad \sigma^2 > \sigma_0^2$$

calculate the test statistic

$$\chi^2 = (n-1)s^2/\sigma_0^2$$

(*continued*)

and use the decision rule

Reject H_0 in favor of H_1 if $\chi^2 > \chi^2_{\alpha, n-1}$.

Case 2:   To test the null hypothesis

$$H_0: \quad \sigma^2 = \sigma_0^2 \quad \text{or} \quad H_0: \quad \sigma^2 \geq \sigma_0^2$$

against the one-sided alternative hypothesis

$$H_1: \quad \sigma^2 < \sigma_0^2$$

use the decision rule

Reject H_0 in favor of H_1 if $\chi^2 < \chi^2_{1-\alpha, n-1}$.

Case 3:   To test the null hypothesis

$$H_0: \quad \sigma^2 = \sigma_0^2$$

against the two-sided alternative hypothesis

$$H_1: \quad \sigma^2 \neq \sigma_0^2$$

use the decision rule

Reject H_0 in favor of H_1 if $\chi^2 > \chi^2_{\alpha/2, n-1}$ or if $\chi^2 < \chi^2_{1-\alpha/2, n-1}$.

EXAMPLE 11.32   **Testing a Hypothesis About a Population Variance** ————————

Suppose we take a random sample of $n = 26$ observations from a normal population to test the null hypothesis

$$H_0: \quad \sigma^2 = 150$$

against the two-sided alternative hypothesis

$$H_1: \quad \sigma^2 \neq 150$$

using a 10% level of significance. Suppose the observed sample variance is $s^2 = 175$. Perform the test.

SOLUTION   The test statistic is

$$\frac{(n-1)s^2}{\sigma_0^2} = \frac{(25)(175)}{150} = 29.167$$

For a two-tailed test with $\alpha = .10$, we place an area of .05 in each tail of the distribution. From Table A.7 using $(n-1) = 25$ degrees of freedom, we obtain the critical values $\chi^2_{.95,25} = 14.61$ and $\chi^2_{.05,25} = 37.65$. Because 29.167 lies between 14.61 and 37.65, we do not reject the null hypothesis. See Figure 11.30.

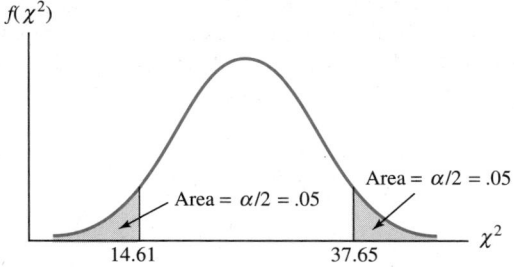

FIGURE 11.30   *Chi-square distribution for Example 11.32*

Confidence Intervals for the Population Variance

To construct confidence intervals for a population variance, we again make use of the test statistic $(n - 1)S^2/\sigma^2$ and the chi-square probability distribution. To construct a confidence interval having level of confidence $(1 - \alpha)$, we utilize the probability statement

$$P\left(\chi^2_{1-\alpha/2,\nu} \leq \frac{(n - 1)S^2}{\sigma^2} \leq \chi^2_{\alpha/2,\nu}\right) = 1 - \alpha$$

This is equivalent to the statement

$$P\left(\frac{\chi^2_{1-\alpha/2,\nu}}{(n - 1)S^2} \leq \frac{1}{\sigma^2} \leq \frac{\chi^2_{\alpha/2,\nu}}{(n - 1)S^2}\right) = 1 - \alpha$$

Finally, we obtain

$$P\left(\frac{(n - 1)S^2}{\chi^2_{\alpha/2,\nu}} \leq \sigma^2 \leq \frac{(n - 1)S^2}{\chi^2_{1-\alpha/2,\nu}}\right) = 1 - \alpha$$

Formula for a confidence interval for σ^2

> Suppose we have a random sample of n observations from a normal population with unknown variance σ^2. Let the observed sample variance be s^2. A confidence interval for σ^2 having level of confidence $(1 - \alpha)$ is given by
>
> $$\left(\frac{(n - 1)s^2}{\chi^2_{\alpha/2,\nu}}, \frac{(n - 1)s^2}{\chi^2_{1-\alpha/2,\nu}}\right)$$

EXAMPLE 11.33   **Establishing a Confidence Interval for the Population Variance** ————————

A new radar device is intended to enable police officers to measure the speed of passing cars instantaneously. To determine the reliability of the device, 61 cars were driven past an inspection point at precisely 50 miles per hour, and each car's speed was estimated by using the radar device. It is assumed that the population of readings is normal. Suppose the observed sample variance was $s^2 = 5$. Construct a 95% confidence interval for σ^2.

SOLUTION　The desired level of confidence is $(1 - \alpha) = .95$. Thus, each tail of the chi-square distribution should contain an area equal to .025. By using Table A.7 and 60 degrees of freedom, we obtain $\chi^2_{.975,60} = 40.48$ and $\chi^2_{.025,60} = 83.30$. This yields a 95% confidence interval of

$$\left(\frac{(60)(5)}{83.30}, \frac{(60)(5)}{40.48} \right)$$

or

$$(3.60, 7.41)$$

Thus, we are 95% confident that the true value of σ^2 lies between 3.60 and 7.41.

If the variance is 3.60, then the standard deviation is $\sqrt{3.60} = 1.90$. Similarly, if the variance is 7.41, then the standard deviation is 2.72. Thus, a 95% confidence interval for the standard deviation is $(1.90, 2.72)$. If the radar gun readings are approximately normally distributed, then about 95% of the observations should lie within 1.96σ units of the mean. If the population standard deviation is as large as 2.72, then approximately 95% of the radar readings should be within $(1.96)(2.72) = 5.33$ units of the true value. However, a radar measurement that is in error by as much as 5.33 miles per hour would be unsatisfactory for highway patrols, so a better measuring device is needed. On the other hand, knowing that the device measures the speed correctly within about 5.33 miles per hour indicates that the police should feel safe in giving tickets to anyone who is clocked going more than 5.33 miles per hour above the speed limit.

Exercises

Statistical Concepts

11.9.1　*True or false:*　When testing a hypothesis about a population variance, we use the chi-square distribution.

11.9.2　What assumptions are made about the population of data when testing a hypothesis about a population variance?

11.9.3　Let $\chi^2 = \dfrac{(n - 1)s^2}{\sigma^2}$. What is $E(\chi^2)$?

11.9.4　When using the chi-square test statistic to test a hypothesis about a population variance, what is the appropriate number of degrees of freedom to use?

Statistical Drills

11.9.5　The following sample of 10 observations was obtained from a normal population: 26.2, 31.3, 28.4, 32.5, 33.3, 23.4, 27.9, 28.0, 30.1, 33.6.
　a.　Calculate the sample mean and sample variance.
　b.　Construct a 95% confidence interval for the population variance.
　c.　Let $\alpha = .05$. Test $H_0: \sigma^2 \geq 15$ against $H_1: \sigma^2 < 15$.

11.9.6　The following sample of six observations was obtained from a normal population: .72, 1.04, 1.14, .95, .88, 1.23.
　a.　Calculate the sample mean and sample variance.
　b.　Construct a 95% confidence interval for the population variance.
　c.　Let $\alpha = .05$. Test $H_0: \sigma^2 = .20$ against $H_1: \sigma^2 \neq .20$.

11.9.7   Suppose a sample of eight observations was obtained from a normal population. The sample standard deviation was $s = 11.2$.
  a. Construct a 90% confidence interval for the population variance.
  b. Let $\alpha = .05$. Test $H_0: \sigma^2 \leq 100$ against $H_1: \sigma^2 > 100$.

11.9.8   Suppose a sample of 26 observations was obtained from a normal population. The sample standard deviation was $s = 20.5$.
  a. Construct a 90% confidence interval for the population variance.
  b. Let $\alpha = .05$. Test $H_0: \sigma^2 \geq 500$ against $H_1: \sigma^2 < 500$.

Statistical Applications

11.9.9   A machine at the Romano Drill Bit Company makes ¼-inch ball bearings. When the machine is operating properly, the variance of the diameter of the bearings is at most .0004. In a random sample of 41 bearings, the sample variance was $s^2 = .0007$. Use a 1% level of significance and test $H_0: \sigma^2 \leq .0004$ against $H_1: \sigma^2 > .0004$.

11.9.10   To be accepted in professional tournaments, tennis balls must pass rigorous tests proving that there is a minimum variance in liveliness from one ball to another. When the balls are dropped from a specified height, they must rebound to a height of 4 feet on the average. The brand of ball satisfying this requirement with the least variance will be selected for use on the pro tour. A new brand of ball, the Australian Gold, is produced. In a random sample of 91 balls, the height of each bounce had a sample mean of 4 feet and a sample variance of .36.
  a. Find a 90% confidence interval for the population variance.
  b. Find a 95% confidence interval for the population variance.
  c. The ball currently used on the tour has a population variance $\sigma^2 = .46$. The Australian Gold ball will be used if its variance is smaller. For the Australian Gold ball, test the null hypothesis $H_0: \sigma^2 \geq .46$ against $H_1: \sigma^2 < .46$ with a 5% level of significance.

11.9.11   Uniformity of product is an important quality characteristic that is closely monitored in quality control applications. In 1992, General Motors faced a recall of more than 100,000 automobiles because of flaws in the car paint. This recall had an eventual cost of several million dollars. Suppose that when a painting process is working correctly, the thickness of paint on car doors is a random variable having a normal distribution with mean $\mu = 6$ millimeters and variance $\sigma^2 = .8$. Suppose the paint thickness is measured on a random sample of 50 car doors. The sample variance is $s^2 = 2.3$. Select an appropriate level of significance. Can we conclude that the variance of the process has changed?

11.10   Computer Applications

When we test a null hypothesis about a population mean, we typically obtain a random sample of n observations from a population having an unknown variance. As described previously, to test the null hypothesis $H_0: \mu = \mu_0$ where μ_0 is some specified value of μ, we use the test statistic

$$t = \frac{\bar{X} - \mu_0}{S/\sqrt{n}}$$

To perform the test, we must calculate both the sample mean $\bar{X}$ and the sample standard deviation S. When the sample size is large, these tasks can be tedious. In Chapter 4, we showed that to obtain the sample mean and sample variance using the SPSSX program, we issue the DESCRIPTIVES command followed by the names of the variables whose means and standard deviations are desired.

For example, refer to the data in Table 2.2 of Chapter 2, showing the selected characteristics of a sample of 50 students in a statistics class. Suppose we wish to test the null hypothesis that in the population the mean SAT score is 1000 using a two-tailed test and a 5% level of significance. To obtain the sample mean and sample standard deviation for the variables SAT and GPA, issue the command

---

DESCRIPTIVES SAT, GPA

The computer output in Figure 11.31 shows that the sample mean for SAT scores in the Table 2.2 data is $\bar{x} = 1122.40$ and the sample standard deviation is $s = 152.92$.

Variable	Mean	Std Dev	Minimum	Maximum	Valid N	Label
SAT	1122.40	152.92	880.00	1490.00	50	
GPA	2.92	.39	2.29	3.80	50	COLLEGE GRADE POINT AVERAGE

FIGURE 11.31   *SPSSX-generated output showing the descriptive statistics for variable SAT in Table 2.2*

We then obtain the t statistic

$$t = \frac{1{,}122.400 \ - \ 1{,}000}{152.92/\sqrt{50}} = 5.66$$

Since there are $\nu = 49$ degrees of freedom, the critical value of the t statistic is $t_{\alpha/2,\nu} \approx 2.021$. We have used 40 degrees of freedom to find the critical value because values for $\nu = 49$ are not in the table. Alternatively, because the sample size is large, we can use the critical value $z_{.025} = 1.96$ from the standard normal distribution. The observed value of the test statistic, 5.66, exceeds the critical value and falls in the critical region. Thus, based on the sample of 50 observations, we reject the null hypothesis.

Exercises

11.10.1   For the data in Table 2.2 of Chapter 2, test the hypothesis H_0: $\mu = 3.0$ for the GPA against H_1: $\mu \neq 3.0$ for all students in the population. Let $\alpha = .05$.

11.10.2   For the data in Table 2.1 of Chapter 2 on the sample of 120 employees at Computech, test the hypothesis H_0: $\mu = 33.0$ for the mean age for all employees against H_1: $\mu \neq 33.0$. Let $\alpha = .05$.

11.10.3   For the data in Table 2.1 of Chapter 2, test the hypothesis that H_0: $\mu = 12$ for the number of years of seniority for all employees at Computech against H_1: $\mu \neq 12$. Let $\alpha = .05$.

Chapter 11 Supplementary Exercises

11.S.1   A company sells breakfast cereal in 16-ounce boxes. In a sample of 100 boxes of this cereal, the average weight was 15.89 ounces and the sample standard deviation was $s = .2$ ounce. Test H_0: $\mu = 16$ against H_1: $\mu < 16$ using a 5% level of significance.

11.S.2 To make a reasonable profit, an airline must average at least 58 passengers per flight on a certain route. In a sample of 100 flights, the average number of passengers was 57 and the sample variance was 36. Test H_0: $\mu \geq 58$ against H_1: $\mu < 58$ using a 1% level of significance.

11.S.3 A university publication claims that students spend an average of at most $110 per semester for textbooks. Suppose you take a random sample of 100 students and obtain $\bar{x} = \$130$. Assume the sample standard deviation is $40. Test H_0: $\mu \leq \$110$ against H_1: $\mu > \$110$ using a 5% level of significance.

11.S.4 In Exercise 3, find the power of the test if $\mu = \$120$.

11.S.5 In Exercise 3, find the power of the test if $\mu = \$115$.

11.S.6 A newspaper advertisement claims that at least 80% of the people who wear contact lenses experience no difficulty. You take a random sample of 400 people who have purchased contact lenses and find that 280 of them have experienced no difficulty. Test H_0: $p = .80$ against H_1: $p < .80$ using a 1% level of significance. Explain why we are using a one-tailed test.

11.S.7 In Exercise 6, find the one-tailed p-value of the test.

11.S.8 In Exercise 6, find the power of the test if $p = .75$. If $p = .70$.

11.S.9 One hundred students are polled to determine whether a certain controversial politician should be invited to lecture on campus. In the poll, 63 students favored inviting the speaker. Test H_0: $p = .5$ against H_1: $p > .5$ using a 5% level of significance.

11.S.10 In a random sample of 400 record albums produced by the Royal Record Company, 22 were defective. Is this sufficient evidence for concluding that the percentage of albums having flaws exceeds 4%? Use a 1% level of significance.

11.S.11 In a random sample of 200 walnut panels, 32 had major flaws. Is this sufficient evidence for concluding that the proportion of walnut panels that contain major flaws is greater than 10%? Use a 5% level of significance.

11.S.12 Last year, 60% of the patients at a hospital rented a television. For a trial period, the rental agency raised the fee for a random sample of 200 patients. In the sample, 90 patients decided to rent the TV. Is this sufficient evidence to convince the rental agency that the proportion of renters will be less than .6 if the price is raised for everyone? Use a 5% level of significance.

11.S.13 A sample of 1,600 voters is polled to estimate the percentage of votes a candidate will receive. We wish to test H_0: $p \leq .5$ against H_1: $p > .5$ using a 5% level of significance.
  a. Find the critical value of $\hat{p}$ associated with this test.
  b. Find the power of the test if $p = .55, .54, .53, .52$, and $.51$.
  c. Plot the power curve.
  d. Repeat parts **a, b,** and **c** using a sample size $n = 2,500$.

11.S.14 A campus coffee machine is supposed to put 8 ounces of coffee into a cup on the average. A random sample of 10 cups of coffee is taken. Test H_0: $\mu = 8$ against H_1: $\mu < 8$ using the t test and a 5% level of significance. The sample data are as follows:

  6.8   7.6   8.1   8.3   7.7   8.2   7.9   7.5   7.7   7.9

  a. Find the sample mean.
  b. Find the sample standard deviation.
  c. Perform the test. Assume the population is normal.

11.S.15 A union bricklayer is supposed to be able to lay 100 bricks per hour on the average. An apprentice is suspected of working significantly slower than average. The apprentice is checked during 10 randomly selected hours, and the number of bricks laid each hour is recorded as follows:

  92   88   104   97   78   89   106   89   92   101

a. Find the sample mean.

b. Find the sample standard deviation.

c. Use a one-tailed *t* test and a 5% level of significance to test whether the bricklayer is working significantly slower than average. Assume the population is normal.

11.S.16 The Douglas Drug Company claims that its medication takes effect within 12 minutes. The following data show the times that 10 patients required before feeling the effects of the medication:

$$8.1 \quad 11.6 \quad 9.4 \quad 13.3 \quad 16.2 \quad 10.3 \quad 14.6 \quad 15.3 \quad 13.6 \quad 15.2$$

It is assumed that the population is normally distributed.

a. Find the sample mean.

b. Find the sample standard deviation.

c. Use a 5% level of significance and test $H_0: \mu \le 12$ minutes against $H_1: \mu > 12$ minutes.

11.S.17 A sales representative for the Kaminski Fish Company claims that at least 50% of the American population eat fish regularly. In a random sample of 240 people, 106 said that they eat fish regularly.

a. Calculate the sample proportion who eat fish regularly.

b. Use a 5% level of significance to test the sales claim. Did you use a one-tailed test? Why or why not?

11.S.18 In a best-selling book, the author states that the average cost of a funeral is $800 or more. To test this claim, an investigator examines a random sample of 20 funerals. The sample mean funeral cost was $\bar{x} = \$600$ and the sample standard deviation was $s = \$108$. Assume that the population is approximately normal. Use a 1% level of significance and test $H_0: \mu \ge \$800$ against $H_1: \mu < \$800$.

11.S.19 In the past, 30% of the viewing audience watched the 11 o'clock news on Channel 7. To increase its share of the audience, Channel 7 hired some new newscasters. After 1 month, a random sample of 500 viewers were questioned. The executives will claim that their share has increased if at least 175 of the viewers in the sample were watching Channel 7.

a. What is the level of significance of this test?

b. If $p = .35$, what is the probability of incorrectly concluding that the proportion watching Channel 7 has not increased?

11.S.20 An executive for the Midwest Barge Line claims that barges that pass through lock 16 on the Mississippi River carry on the average at least 800 tons of coal. In a random sample of eight barges, the following tonnages were measured:

$$780 \quad 760 \quad 775 \quad 805 \quad 780 \quad 770 \quad 790 \quad 810$$

a. Find the sample mean.

b. Find the sample standard deviation.

c. Use a 5% level of significance and test $H_0: \mu \ge 800$ against $H_1: \mu < 800$. Assume the population is normal.

11.S.21 Last year, an airline found that the number of no-shows on the 10:00 A.M. flights to Chicago averaged 3.1. On a random sample of 20 of these flights during the last 3 months, the average number of no-shows was $\bar{x} = 2.4$ and the standard deviation was $s = .7$. Let $\alpha = .01$. Can we conclude the average number of no-shows per flight has declined? Assume the population is normal.

11.S.22 The Cottrell Soap Company has produced a new dishwashing liquid that it thinks is better than competitive products on the market. The company's market research department has distributed the new product to 400 randomly selected households. The board of directors has indicated that the product will be marketed if more than 25% of the public prefer the product over its competitors. Design the decision rule so that the probability of incorrectly concluding that the product should be marketed is no greater than 5%. The decision rule will be based on the proportion of individuals in the random sample who prefer the product.

11.S.23 Last year, Alston County received $180,000 in federal grants because the census showed that 6% of the county's families were below the poverty level. The county will receive $200,000 in federal aid this year provided the proportion below the poverty level this year is at least 6%. A research firm takes a random sample of 1,000 families in the county and determines that 52 were below the poverty level. Use a 5% level of significance. Should we conclude that the proportion of families below the poverty level is less than 6%?

11.S.24 Suppose that in Exercise 23 the true proportion of families in Alston County who are below the poverty level is .063. What is the probability that in a sample of 1,000 families we would get 52 or fewer impoverished families?

11.S.25 It is claimed that at most 25% of Americans are financially independent at age 65 and do not have to continue to work or become dependent on friends, relatives, or charity. Suppose you wish to test the validity of this claim by conducting a survey among 400 senior citizens. How many people in the sample would have to indicate financial independence to reject the claim? Use a 5% level of significance.

11.S.26 A doctor claims that a new serum is effective in preventing the common cold. A sample of 144 people were injected with the serum and observed over the winter. Suppose that 90 survived the winter without a cold. From prior information, it is known that the probability of surviving the winter without a cold is .5 when the serum is not used. Given the results of this experiment, what conclusions would you make regarding the effectiveness of the serum?

11.S.27 A market study for a new industrial product indicates that the firm should launch the product if the mean number of units sold per customer in the first solicitation of the firm's customers is more than 3.0. It is believed that the standard deviation of the number of units purchased in initial orders will be $\sigma = 1.8$ units. It is desired to test the null hypothesis that the mean number of orders in the population is less than or equal to 3.0 per customer against the alternative hypothesis that the mean number is greater than 3.0. In a random sample of 100 customers included in the first solicitation, the following decision rule is adopted: "Accept H_0 if $\bar{X} \leq 3.10$ and reject H_0 if $\bar{X} > 3.10$." Calculate the power of the test for μ values of 3.05, 3.10, 3.15, 3.20, 3.25, 3.30, 3.35, and 3.40. Sketch the power curve for this decision rule.

11.S.28 Last year, 4% of the customers at a restaurant indicated that they were dissatisfied with their food or service. Management initiates a training program for waiters and gives pay incentives to improve service. A random sample of 1,600 customers was then asked to rate the quality of the food and service. In the sample, 36 people were dissatisfied. Use this information to test whether the proportion of dissatisfied customers has declined. Use a 1% level of significance.

11.S.29 "Tests of hypotheses focus on the probability of incorrectly rejecting a particular hypothesis, the null hypothesis. By setting up the test to have a low probability of a Type I error, the researcher ensures that the test is unlikely to lead to the incorrect rejection of this particular hypothesis when it is true." Do you agree or disagree with this argument?

11.S.30 Suppose there are serious consequences of incorrectly rejecting a particular hypothesis.
  a. Should this hypothesis be the null hypothesis or the alternative hypothesis?
  b. Should α be large or small?
  c. Give an example of such a situation.

11.S.31 It is known that the random variable X has the continuous density function

$$f(x) = \begin{cases} 1/\theta & \text{for } 0 \leq X \leq \theta \\ 0 & \text{otherwise} \end{cases}$$

but it is not known whether the value of θ is 1 or 2. You will have just a single observation of X on which to base your choice between $H_0: \theta = 1$ and $H_1: \theta = 2$.

a. You decide to accept $\theta = 1$ if $X \leqslant .8$ and to accept $\theta = 2$ if $X > .8$. What are the probabilities of Type I and Type II errors?

b. What are the probabilities of Type I and Type II errors if you decide to accept $\theta = 1$ if $X \leqslant .6$ and to accept $\theta = 2$ if $X > .6$?

c. Which of these two rules is more appropriate if the consequences of a Type II error are more serious than those of a Type I error?

11.S.32   The Nutritious Foods Company claims that there are on average at least 4 ounces of raisins in a box of its cereal. A consumer group wants to test the null hypothesis H_0: $\mu \geqslant 4$. The group must choose a sample size and a level of significance. The following three possibilities are considered:

$$n = 25; \quad \alpha = .05$$
$$n = 25; \quad \alpha = .01$$
$$n = 100; \quad \alpha = .05$$

a. Which design has the smallest probability of falsely rejecting the company's claim?

b. Does the first or second design have the smaller probability of incorrectly accepting the company's claim? Why?

c. Does the first or third design have the smaller probability of incorrectly accepting the company's claim? Why?

 11.S.33   According to the U.S. Department of Commerce, approximately 20% of all automobiles on the road in the United States at a certain time are made in Japan. An autoworkers union that wants to limit imports believes that the proportion of Japanese cars on the road during the period in question is higher than 20%. Suppose a random sample of 4,000 cars is observed, 981 of which are made in Japan. Choose an appropriate value for α and conduct the hypothesis test. State whether you believe the reported figure.

 11.S.34   The Environmental Protection Agency monitors the materials that a chemical plant dumps into a river. No citations are issued as long as the mean level of the effluent of a certain toxic substance is no greater than 7.0 parts per million (ppm). In a random sample of $n = 75$ instances of dumping, the sample mean is 7.8 ppm and the sample standard deviation is $s = 2.1$ ppm. Should a citation be issued? Select a reasonable value for α and perform the test.

11.S.35   The publisher of a magazine is told that the average age of subscribers to the magazine is 36 years. To test this claim, a random sample of 150 subscribers is taken. The sample mean age is 32.4 years and the sample standard deviation is 8.3 years. Let $\alpha = .05$. Perform a two-sided test.

In the following problems, first determine whether the required test is one-tailed or two-tailed, and then solve the problem. You will have to make your own decisions: what kind of test to use, whether it is one- or two-tailed, which is the appropriate distribution, and in many cases you will have to choose your own level of significance α. As an aid in deciding whether a test is one-tailed or two-tailed, always ask yourself: Under what conditions do I want to take an action? The answer to this question will indicate the alternative hypothesis. (But remember that the equal sign must be in the null hypothesis.)

 11.S.36   In quality control applications, the null and alternative hypotheses are frequently specified as

    H_0:   The production process is performing in a satisfactory manner.
    H_1:   The production process is not performing in a satisfactory manner.

A manufacturer produces brake pads for new automobiles. The process is designed to produce brake pads with a mean thickness of .25 inch. To investigate whether the process is operating correctly, a random sample of 40 brake pads was obtained from the last hour's production. Their thicknesses (in inches) are listed here:

.246	.253	.249	.250	.248	.251	.248	.254
.253	.251	.252	.255	.256	.253	.254	.252
.253	.254	.253	.254	.256	.257	.256	.251
.249	.253	.257	.252	.256	.255	.252	.254
.250	.252	.251	.252	.253	.252	.255	.254

a. Do the data provide sufficient evidence to conclude that the process is not operating satisfactorily? Select an appropriate level of significance α. State the null hypothesis and the alternative hypothesis.

b. What is the approximate observed significance level associated with this test?

11.S.37 The first applications of quality control testing were primarily in the manufacturing sector, but many companies in the service sector have begun to use quality control techniques to monitor their services. For example, suppose the manager of a drive-through fast-food restaurant wants to assure customers that the mean waiting time will be 180 seconds or less. A random sample of 30 drive-through customers is selected daily, and the waiting times are recorded. The sample mean is $\bar{x} = 195$ seconds and the sample standard deviation is $s = 24$ seconds.

a. Do the data provide sufficient evidence to conclude that the process is not operating satisfactorily? Select an appropriate level of significance α. State the null hypothesis and the alternative hypothesis.

b. What is the approximate observed significance level associated with this test?

References

Fisher, Ronald A. *Statistical Methods for Research Workers.* 14th ed. New York: Hafner Press, 1970.

McClosky, Donald N. "The Loss Function Has Been Mislaid: The Rhetoric of Significance Tests." *American Economic Review.* May 1985, pp. 201–205.

Morrison, Donald, and Ramon E. Henkel. *The Significance Test Controversy.* Chicago: Aldine, 1970.

Neyman, Jerzy, and Egon S. Pearson. "On the Use and Interpretation of Certain Test Criteria for Purposes of Statistical Hypotheses." *Biometrika,* 1928, pp. 175–294.

Neyman, Jerzy, and Egon S. Pearson. "On the Problem of the Most Efficient Tests of Statistical Hypotheses." *Philosophical Transactions of the Royal Society,* 1933, pp. 289–337.

Norusis, Marija J. *SPSSX Introductory Statistics Guide.* New York: McGraw-Hill, 1990.

Norusis, Marija J. *SPSSX Advanced Statistics Guide.* Chicago: SPSS, 1990.

Norusis, Marija J. *The SPSS Guide to Data Analysis.* Chicago: SPSS, 1986.

Popper, Karl R. *The Logic of Scientific Discovery.* New York: Harper, 1965.

Ryan, Thomas A., Brian L. Joiner, and Barbara F. Ryan. *Minitab Handbook.* 2d ed. Boston: PWS-Kent, 1985.

Ryan, Thomas A., Brian L. Joiner, and Barbara F. Ryan. *Minitab Reference Manual.* University Park, Penn.: Minitab Project, 1985.

SAS Introductory Guide. 3d ed. Cary, N.C.: SAS Institute, 1985.

SAS Procedures Guide for Personal Computers. Version 6 ed. Cary, N.C.: SAS Institute, 1986.

SAS Statistics Guide for Personal Computers. Version 6 ed. Cary, N.C.: SAS Institute, 1986.

SAS User's Guide: Basics. Version 5 ed. Cary, N.C.: SAS Institute, 1985.

SAS User's Guide: Statistics. Version 5 ed. Cary, N.C.: SAS Institute, 1985.

SPSSX User's Guide. Chicago: SPSS, 1988.

12

Tests of Hypotheses Involving Two Populations

In this chapter, we discuss how to test hypotheses concerning the means μ_1 and μ_2, proportions p_1 and p_2, and the variances σ_1^2 and σ_2^2 of two populations.

Let θ_1 and θ_2 denote unknown population parameters from two populations. To test hypotheses about θ_1 and θ_2, we take random samples of size n_1 and n_2 from the two populations and calculate sample estimates $\hat{\theta}_1$ and $\hat{\theta}_2$. We then use these estimates to help us decide whether to reject some null hypothesis about the respective population parameters.

The hypotheses that will be considered in this chapter are as follows:

1. Test $H_0: \theta_1 - \theta_2 = D_0$ against $H_1: \theta_1 - \theta_2 \neq D_0$.
2. Test $H_0: \theta_1 - \theta_2 \leq D_0$ against $H_1: \theta_1 - \theta_2 > D_0$.
3. Test $H_0: \theta_1 - \theta_2 \geq D_0$ against $H_1: \theta_1 - \theta_2 < D_0$.

In most cases, the value of D_0 is hypothesized to be 0, but this does not have to be the case. From Chapter 11, you will recognize test 1 as a two-sided test and tests 2 and 3 as one-sided tests.

When $D_0 = 0$ is the hypothesized value of the difference $\theta_1 - \theta_2$, testing the null hypothesis $H_0: \theta_1 = \theta_2$ against the two-sided alternative hypothesis $H_1: \theta_1 \neq \theta_2$ is equivalent to testing $H_0: \theta_1 - \theta_2 = 0$ against $H_1: \theta_1 - \theta_2 \neq 0$.

12.1 Tests for the Differences of Means

To test $H_0: \mu_1 - \mu_2 = D_0$, take random samples of n_1 observations from population 1 and n_2 observations from population 2. (It is not necessary that n_1 and n_2 be equal, but that is frequently the case.) The samples should be independent of one another. This means, for example, that the samples should not contain the same experimental units. (Examples of observations in two samples that are dependent on one another would be the before-diet and after-diet weights of a sample of individuals and the car mileages achieved with gasoline A and with gasoline B using the same cars. Testing the difference between two population means when the samples are dependent is discussed in Section 12.3.) When the samples are independent, the test statistic used to test H_0 will be either a Z score or a t score, depending on whether the population variances are known or must be estimated.

To determine a critical region for the test statistic, we must know the sampling distribution of the test statistic. The test statistic will depend on the random variable $(\bar{X}_1 - \bar{X}_2)$. Values of $(\bar{X}_1 - \bar{X}_2)$ that are not close to D_0 would support the alternative hypothesis that D_0 is not the correct value for the difference $(\mu_1 - \mu_2)$.

If the two populations are both normally distributed or if the sample sizes are large (i.e., $n_1 \geq 30$ and $n_2 \geq 30$), then the random variable $(\bar{X}_1 - \bar{X}_2)$ has an approximately normal distribution with mean $(\mu_1 - \mu_2)$ and variance $[\sigma_1^2/n_1 + \sigma_2^2/n_2]$. Thus, if D_0 is the correct value for the difference between the two means, then the random variable

$$Z = \frac{(\bar{X}_1 - \bar{X}_2) - D_0}{\sqrt{\sigma_1^2/n_1 + \sigma_2^2/n_2}}$$

has a standard normal distribution.

The accompanying box describes how to perform a test concerning the difference between two population means when the population variances are known and the samples are independent.

Testing the difference between population means using large independent samples

Suppose we have independent random samples of n_1 and n_2 observations from two normal distributions with means μ_1 and μ_2 and variances σ_1^2 and σ_2^2, respectively. If the observed sample means are $\bar{x}_1$ and $\bar{x}_2$, then the following tests have significance level α:

Case 1:   To test the null hypothesis

$$H_0: \quad \mu_1 - \mu_2 = D_0 \quad \text{or} \quad H_0: \quad \mu_1 - \mu_2 \leq D_0$$

against the one-sided alternative hypothesis

$$H_1: \quad \mu_1 - \mu_2 > D_0$$

calculate the test statistic

$$z = \frac{(\bar{x}_1 - \bar{x}_2) - D_0}{\sqrt{\sigma_1^2/n_1 + \sigma_2^2/n_2}}$$

and use the decision rule

Reject H_0 in favor of H_1 if $z > z_\alpha$.

Case 2:   To test the null hypothesis

$$H_0: \quad \mu_1 - \mu_2 = D_0 \quad \text{or} \quad H_0: \quad \mu_1 - \mu_2 \geq D_0$$

against the one-sided alternative hypothesis

$$H_1: \quad \mu_1 - \mu_2 < D_0$$

use the decision rule

Reject H_0 in favor of H_1 if $z < -z_\alpha$.

(continued)

Case 3:   To test the null hypothesis

$$H_0: \quad \mu_1 - \mu_2 = D_0$$

against the two-sided alternative

$$H_1: \quad \mu_1 - \mu_2 \neq D_0$$

use the decision rule

Reject H_0 in favor of H_1 if $z > z_{\alpha/2}$ or if $z < -z_{\alpha/2}$.

If the population variances are unknown but the sample sizes n_1 and n_2 are large (say, 30 or more), then, to a good approximation, tests for the difference between population means can be performed by replacing the population variances by the observed sample variances s_1^2 and s_2^2. For large sample sizes, these approximations are good even when the population distributions are not normal.

In most tests, the value of D_0 is chosen to be 0. In this case, the null hypothesis states that the means of the two populations are equal.

EXAMPLE 12.1   **Testing the Difference Between Two Means (Large Samples)** ———————

Procter & Gamble (P&G) has made huge profits by selling Crest Toothpaste. P&G showed that brushing your teeth with toothpaste that contained fluoride helped prevent tooth decay, and Crest was the first toothpaste to contain fluoride. Now, of course, almost all toothpastes contain fluoride. Suppose you want to test the claim that the use of fluoride toothpaste helps prevent tooth decay by testing a group of 10-year-old children, all of whom have identical dental records at the start of the test. Let population 1 denote children who are told to use nonfluoridated toothpaste for a year, and let population 2 denote children who use fluoridated toothpaste for a year. Let μ_1 and μ_2 denote the mean number of cavities at the end of the year for children in each of the two populations. We want to test the null hypothesis

$$H_0: \quad \mu_1 - \mu_2 = 0$$

against

$$H_1: \quad \mu_1 - \mu_2 > 0$$

Because the population variances are unknown, we will estimate them by using the sample variances.

Suppose that in a random sample of 100 children who used the nonfluoridated toothpaste, the sample mean was $\bar{x}_1 = 4.8$ cavities and the sample variance was $s_1^2 = 1.1$, and in an independent random sample of 120 children who used fluoridated toothpaste, we obtained $\bar{x}_2 = 3.6$ cavities and $s_2^2 = .9$. Because the sample sizes are large, we can use the Z score when performing the test, replacing the population variances in the test statistic by the sample variances. We will use a 5% level of significance.

SOLUTION   For a one-tailed test using a level of significance $\alpha = .05$, the critical value of the test statistic is $z_\alpha = 1.645$. The observed value of the test statistic is

$$z = \frac{(\bar{x}_1 - \bar{x}_2) - D_0}{\sqrt{s_1^2/n_1 + s_2^2/n_2}}$$

where

$$s_1^2/n_1 + s_2^2/n_2 = 1.1/100 + .9/120 = .0185$$

We obtain the Z score

$$z = \frac{(\bar{x}_1 - \bar{x}_2) - 0}{\sqrt{.0185}} = \frac{4.8 - 3.6}{.136} \approx 8.82$$

We reject H_0 because $z = 8.82$ falls in the critical region.

In Example 12.1, we found a statistically significant difference between the two sample means. Children who used the fluoridated toothpaste got fewer cavities than children who did not. Is this result of any practical significance? That is, are the results strong enough to make you switch from nonfluoridated to fluoridated toothpaste? Most parents and children would probably say that the results are of practical significance. A 25% reduction in the expected number of cavities is probably sufficient enticement to get people to switch to fluoridated toothpaste (provided the price difference is not too great). The fact that nearly all toothpastes now contain fluoride shows that people do believe that this result is of both statistical and practical significance.

Exercises

Statistical Concepts

12.1.1 To test a hypothesis about the difference between two population means as described in this section, what assumptions are made about the population distributions and the population variances? How does your answer change if the sample sizes are large rather than small?

12.1.2 *True or false:* To test a hypothesis about the difference between two population means, the two population variances must be equal. Explain.

12.1.3 *True or false:* To test a hypothesis about the difference between two population means, we can substitute the sample variances for the population variances and obtain approximately correct results, provided the sample sizes are large, say, 30 or more. Explain.

12.1.4 *True or false:* If both populations are large, we can use the Z score as our test statistic to test a hypothesis about the difference between two population means, even though neither population is normal. Explain.

Statistical Drills

12.1.5 Assume we obtain independent samples from normal populations. Assume $\sigma_1^2 = 20$ and $\sigma_2^2 = 30$; $n_1 = n_2 = 10$; $\bar{x}_1 = 68$ and $\bar{x}_2 = 60$. Let $\alpha = .05$. Test $H_0: \mu_1 - \mu_2 = 0$ against $H_1: \mu_1 - \mu_2 \neq 0$.

12.1.6 Assume we obtain independent samples from normal populations. Assume $s_1^2 = 45$ and $s_2^2 = 30$; $n_1 = n_2 = 100$; $\bar{x}_1 = 68$ and $\bar{x}_2 = 60$. Let $\alpha = .05$. Test $H_0: \mu_1 - \mu_2 = 0$ against $H_1: \mu_1 - \mu_2 \neq 0$.

12.1.7 Assume we obtain independent samples from normal populations. Assume $s_1^2 = .87$ and $s_2^2 = .64$; $n_1 = 48$ and $n_2 = 63$; $\bar{x}_1 = 4.56$ and $\bar{x}_2 = 4.71$. Let $\alpha = .01$. Test $H_0: \mu_1 - \mu_2 \geq 0$ against $H_1: \mu_1 - \mu_2 < 0$.

12.1.8   Assume we obtain independent samples from normal populations. Assume $\sigma_1^2 = 5.4$ and $\sigma_2^2 = 6.2$; $n_1 = 12$ and $n_2 = 15$; $\bar{x}_1 = 7.8$ and $\bar{x}_2 = 7.2$. Let $\alpha = .01$. Test $H_0: \mu_1 - \mu_2 \leq 0$ against $H_1: \mu_1 - \mu_2 > 0$.

Statistical Applications

12.1.9   A superintendent of schools wants to determine whether children in public schools and in private schools read equally well. A reading test is given to 100 public school children at age 8 and 100 private school children at age 8. The sample statistics are $\bar{x}_1 = 206$, $s_1^2 = 450$; $\bar{x}_2 = 191$, $s_2^2 = 450$. Test $H_0: \mu_1 - \mu_2 = 0$ against $H_1: \mu_1 - \mu_2 \neq 0$ using a 5% level of significance.

12.1.10   A major producer of ice cream wants to increase its sales and its share of the market. It wants to spend most of its advertising expenditures on the segment of the market that consumes the most ice cream. To determine whether males or females eat more ice cream on the average, we take a random sample of 100 men (sample 1) and 100 women (sample 2), and determine how much ice cream each person eats during a year. We obtain $\bar{x}_1 = 12$ quarts, $s_1^2 = 8$; $\bar{x}_2 = 14$ quarts, $s_2^2 = 8$. Test $H_0: \mu_1 - \mu_2 = 0$ against $H_1: \mu_1 - \mu_2 \neq 0$ using a 1% level of significance.

12.1.11   The public transportation agency wants to determine whether the average income of people who ride the bus to work every day is the same as the average income of people who use the bus only occasionally or never. Random samples of size 100 are taken from each population. Assume that each population has a standard deviation of $1,800. We obtain $\bar{x}_1 = \$23,800$ and $\bar{x}_2 = \$25,000$. Test $H_0: \mu_1 - \mu_2 = 0$ against $H_1: \mu_1 - \mu_2 \neq 0$ using a 1% level of significance.

12.1.12   Suppose that the National Association of Truck Drivers is interested in improving the public image of truck drivers. They decide to test whether the average speed of trucks traveling on an interstate highway is significantly greater than the average speed of cars. In random samples of 100 cars and 200 trucks checked at one point on the highway, the average speed of the cars is $\bar{x}_1 = 52$ miles per hour and the average speed of the trucks is $\bar{x}_2 = 54$ miles per hour. Assume that the population variances are $\sigma_1^2 = 25$ and $\sigma_2^2 = 16$. Test $H_0: \mu_1 - \mu_2 = 0$ against $H_1: \mu_1 - \mu_2 < 0$ using a 5% level of significance.

12.1.13   A psychologist thinks that, other things being equal, people who wear glasses tend to have fewer automobile accidents than people who do not. She reasons that people who wear glasses are conscious of their vision problem and drive more carefully and conservatively. Random samples of 400 people who wear glasses and 400 people who do not wear glasses are obtained. The people are of approximately the same age and drive approximately the same number of miles per year. During a 12-month period, people wearing glasses had an average of $\bar{x}_1 = .72$ accident with sample standard deviation $s_1 = .3$. People not wearing glasses had an average of $\bar{x}_2 = .82$ accident with standard deviation $s_2 = .3$. Does this evidence support the psychologist's claim? Use a 5% level of significance.

12.2   Tests of Differences of Means Using Small Samples from Normal Populations When the Population Variances Are Equal but Unknown

Suppose we want to test a hypothesis about the difference between the means of two normal populations when the population variances σ_1^2 and σ_2^2 are equal but unknown. If the sample sizes are large, we can replace the unknown population variances by their sample estimates and follow the procedures developed in Section 12.1. However, when the sample sizes are small, we use a t statistic as the test statistic to obtain critical values from the t distribution. The populations are assumed to be normally distributed and the samples must be independent of one another. Because the t test is relatively robust, the test results are approximately correct when the population distributions are not exactly normal provided they do not deviate too far from normal.

The procedure in the accompanying box describes how to test a null hypothesis about the difference between two population means when the sample sizes are small. The test is based on the following assumptions:

1. The two populations are normal (or approximately normal).
2. The population variances are unknown but are assumed to be equal.
3. The samples are independent of one another.

We show how to test the hypothesis that $\sigma_1^2 = \sigma_2^2$ in Section 12.6. This test involves use of the F distribution, which will be discussed in Section 12.5.

Testing the difference between the means of two normal populations using small independent samples when the population variances are equal but unknown

Suppose we have independent random samples of n_1 and n_2 observations from normal distributions with means μ_1 and μ_2 and a common variance. Suppose the observed sample variances are s_1^2 and s_2^2. Let the observed sample means be $\bar{x}_1$ and $\bar{x}_2$, and let s_p^2 denote the observed pooled sample variance. The pooled estimate of the common population variance is calculated as follows:

$$s_p^2 = \frac{\Sigma(x_{i1} - \bar{x}_1)^2 + \Sigma(x_{i2} - \bar{x}_2)^2}{n_1 + n_2 - 2}$$

where x_{i1} is the ith observation from sample 1 and x_{i2} is the ith observation from sample 2. This is equivalent to the estimator

$$s_p^2 = \frac{(n_1 - 1)s_1^2 + (n_2 - 1)s_2^2}{n_1 + n_2 - 2}$$

The following tests have significance level α:

Case 1:   To test the null hypothesis

$$H_0: \quad \mu_1 - \mu_2 = D_0 \quad \text{or} \quad H_0: \quad \mu_1 - \mu_2 \leq D_0$$

against the one-sided alternative hypothesis

$$H_1: \quad \mu_1 - \mu_2 > D_0$$

calculate the test statistic

$$t = \frac{(\bar{x}_1 - \bar{x}_2) - D_0}{\sqrt{s_p^2/n_1 + s_p^2/n_2}}$$

and use the decision rule

Reject H_0 in favor of H_1 if $t > t_{\alpha,\nu}$.

(continued)

The random variable t follows the t distribution with $\nu = (n_1 + n_2 - 2)$ degrees of freedom.

Case 2:   To test the null hypothesis

$$H_0: \quad \mu_1 - \mu_2 = D_0 \quad \text{or} \quad H_0: \quad \mu_1 - \mu_2 \geqslant D_0$$

against the one-sided alternative hypothesis

$$H_1: \quad \mu_1 - \mu_2 < D_0$$

use the decision rule

$$\text{Reject } H_0 \text{ in favor of } H_1 \text{ if } t < -t_{\alpha,\nu},$$

where the number of degrees of freedom is $\nu = (n_1 + n_2 - 2)$.

Case 3:   To test the null hypothesis

$$H_0: \quad \mu_1 - \mu_2 = D_0$$

against the two-sided alternative hypothesis

$$H_1: \quad \mu_1 - \mu_2 \neq D_0$$

use the decision rule

$$\text{Reject } H_0 \text{ in favor of } H_1 \text{ if } t > t_{\alpha/2,\nu} \text{ or if } t < -t_{\alpha/2,\nu},$$

where the number of degrees of freedom is $\nu = (n_1 + n_2 - 2)$.

EXAMPLE 12.2   **Testing the Difference Between Two Means (Small Samples)** ───────

A market research firm wishes to know whether the mean number of hours of TV viewing per week is the same for teenage boys (population 1) as for teenage girls (population 2). They want to test $H_0: \mu_1 - \mu_2 = 0$ against $H_1: \mu_1 - \mu_2 \neq 0$ using a 5% level of significance. The unknown population variances are assumed to be equal. The following data were obtained: $n_1 = 20$, $\bar{x}_1 = 24.5$, $s_1^2 = 64$; $n_2 = 12$, $\bar{x}_2 = 28.7$, $s_2^2 = 71$.

SOLUTION   Because the sample sizes are small, we use the t score as our test statistic. Because the level of significance is $\alpha = .05$, there are $\nu = (n_1 + n_2 - 2) = 30$ degrees of freedom, and we are performing a two-tailed test, the critical values of the test statistic are $\pm t_{.025} = \pm 2.042$. Using the observed sample means 24.5 and 28.7 and the pooled estimate of the variance

$$s_p^2 = \frac{(19)(64) + (11)(71)}{20 + 12 - 2} = 66.57$$

we calculate the observed t statistic

$$t = \frac{(24.5 - 28.7) - 0}{\sqrt{(66.57/20) + (66.57/12)}} = \frac{-4.20}{\sqrt{8.88}} = -1.41$$

Thus, we do not reject H_0 because the test statistic falls in the acceptance region. See Figure 12.1.

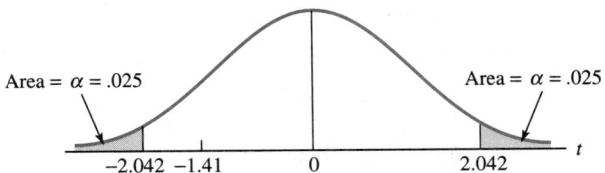

FIGURE 12.1   *The t distribution for Example 12.2*

Does this result seem reasonable? In the samples, the boys watched TV an average of 24.5 hours per week, whereas the girls averaged 28.7 hours. Because the sample sizes are relatively small and the pooled estimate of the variance is fairly large, our estimates of the population means are not very precise. Consequently, the observed difference of 4.2 hours in the sample means is not sufficient evidence to convince us that the two population means are different.

When the population variances are unknown and unequal, we cannot pool the data to obtain a single estimate of the variance. If the sample sizes are large, the appropriate test statistic is the Z score described in Section 12.1.

Formula for testing difference of means (unequal variances and large samples)

$$z = \frac{(\bar{x}_1 - \bar{x}_2) - (\mu_1 - \mu_2)}{\sqrt{(s_1^2/n_1) + (s_2^2/n_2)}}$$

When the sample sizes are small and the population variances are unknown and unequal, neither the t test nor the z test is theoretically appropriate. Nevertheless, many people still resort to one of these two tests for lack of a better alternative. However, an alternative test based on the ranks of the observations will be described in Chapter 21.

The results in this section are based on the assumption that the two samples used to calculate $\bar{x}_1$ and $\bar{x}_2$ are independent of one another. When the samples are not independent, we should use the methods discussed in Section 12.3 on paired difference tests.

As an example, observe how the t test leads to the wrong conclusion in the following problem.

EXAMPLE 12.3   **Inappropriate Use of the *t* Test**

Suppose 10 automobiles are randomly selected. One gallon of gasoline 1 is placed into each car, and each car is driven around a racetrack at a constant speed until all the gas is used. The process is then repeated using gasoline 2. The data in Table 12.1 (page 484) show the distances (in miles) traveled by each car using gasoline 1 and gasoline 2. A researcher would be interested in testing whether, in a large population of cars, the mean mileages are equal when using the two brands of gasoline. We obtain $\bar{x}_1 = 18.3$ miles, $s_1^2 = 89.79$; $\bar{x}_2 = 20.4$ miles, $s_2^2 = 86.27$.

TABLE 12.1 *Car distances for Example 12.3*

Car	Gasoline 1	Gasoline 2	Difference D_0
1	14	16	−2
2	21	24	−3
3	19	20	−1
4	11	15	−4
5	15	17	−2
6	16	19	−3
7	8	10	−2
8	32	33	−1
9	37	39	−2
10	10	11	−1
Total	183	204	−21
$\bar{x}$	18.3	20.4	−2.1
s^2	89.79	86.27	.9889

Suppose we incorrectly used the t test to test

$$H_0: \quad \mu_1 - \mu_2 = 0$$

against the alternative hypothesis

$$H_1: \quad \mu_1 - \mu_2 \neq 0$$

using $\alpha = .05$. The pooled estimate of the variance is

$$s_p^2 = \frac{(9)(89.79) + (9)(86.27)}{10 + 10 - 2} = 88.03$$

We obtain the test statistic

$$t = \frac{\bar{x}_1 - \bar{x}_2}{\sqrt{(s_p^2/n_1) + (s_p^2/n_2)}} = \frac{18.3 - 20.4}{\sqrt{(88.03/10) + (88.03/10)}} = -.50$$

For $\nu = 18$ degrees of freedom and $\alpha = .05$, the critical values of t are $\pm t_{.025,18} = \pm 2.101$. Because the observed value $t = -.50$ falls in the acceptance region for this test, we would not reject H_0.

However, if we compare the mileages car by car, we see that every car traveled farther using gasoline 2 than using gasoline 1. This provides very strong evidence that H_0 is false. Using the t test here leads to the wrong conclusion because the assumption of independent random samples has been violated. In Section 12.3, we will explain the appropriate procedure for performing this test.

Exercises

Statistical Concepts

12.2.1 Suppose we wish to test a hypothesis about the difference between the means of two populations using independent samples of data. It is assumed that the two population variances are equal and unknown. Explain how to perform the test if the sample sizes are small and the populations are approximately normal.

12.2.2 Suppose we wish to test a hypothesis about the difference between the means of two normal populations using independent samples of data. It is assumed that the two population variances are unknown. Explain how to perform the test if the sample sizes are large. Do we need to make any assumption about the normality of the populations? Do we need to assume that the unknown variances are equal?

12.2.3 Suppose we wish to test a hypothesis about the difference between the means of two normal populations using independent samples of data from the two populations. It is assumed that the two population variances are equal and unknown. The sample sizes are small. *True or false:*   The appropriate test statistic to use is the Z score. Explain.

12.2.4 Suppose we wish to test a hypothesis about the difference between the means of two nonnormal populations using independent samples of data from the two populations. It is assumed that the two population variances are unequal and unknown. The sample sizes are large. *True or false:*   The appropriate test statistic to use is the Z score. Explain.

12.2.5 Suppose we wish to test a hypothesis about the difference between the means of two normal populations using independent samples of data from the two populations. It is assumed that the two population variances are equal and unknown. The sample sizes are small. *True or false:*   The appropriate test statistic to use is the t score. Explain.

12.2.6 Suppose we wish to test a hypothesis about the difference between the means of two normal populations using independent samples of data from the two populations. It is assumed that the two population variances are equal and unknown. The sample sizes are small. *True or false:*   The appropriate number of degrees of freedom for the t statistic is $(n_1 + n_2)$. Explain.

Statistical Drills

12.2.7 Assume both populations are normal and we obtain independent samples of data from the two populations. Assume $s_1^2 = 18$ and $s_2^2 = 24$; $n_1 = n_2 = 10$; $\bar{x}_1 = 68$ and $\bar{x}_2 = 63$. Let $\alpha = .05$.
a. Find the pooled estimate of the population variance.
b. Determine the appropriate number of degrees of freedom for the test statistic t.
c. Test $H_0: \mu_1 - \mu_2 = 0$ against $H_1: \mu_1 - \mu_2 \neq 0$.

12.2.8 Assume both populations are normal and we obtain independent samples of data from the two populations. Assume $s_1^2 = 50$ and $s_2^2 = 60$; $n_1 = n_2 = 15$; $\bar{x}_1 = 112$ and $\bar{x}_2 = 106$. Let $\alpha = .05$.
a. Find the pooled estimate of the population variance.
b. Determine the appropriate number of degrees of freedom for the test statistic t.
c. Test $H_0: \mu_1 - \mu_2 \leq 0$ against $H_1: \mu_1 - \mu_2 > 0$.

12.2.9 Assume both populations are normal and we obtain independent samples of data from the two populations. Assume $s_1^2 = .87$ and $s_2^2 = .64$; $n_1 = 8$ and $n_2 = 16$; $\bar{x}_1 = 4.56$ and $\bar{x}_2 = 4.71$. Let $\alpha = .01$.
a. Find the pooled estimate of the population variance.
b. Determine the appropriate number of degrees of freedom for the test statistic t.
c. Test $H_0: \mu_1 - \mu_2 \geq 0$ against $H_1: \mu_1 - \mu_2 < 0$.

Statistical Applications

12.2.10 A time-and-motion study is conducted to test whether the mean length of time required to perform a certain task is the same for the employees on the day shift (population 1) and employees on the night shift (population 2). The data are as follows: $n_1 = 10$, $\bar{x}_1 = 26$, $s_1^2 = 64$; $n_2 = 8$, $\bar{x}_2 = 29$, $s_2^2 = 50$. Use a 5% level of significance and a two-tailed test. Assume the populations are approximately normal, the population variances are equal, and the samples are independent.

12.2.11 To test the effectiveness of a new fertilizer, a farm was divided into 50 plots of equal area. The soil quality of all plots was approximately the same. The new fertilizer was applied to 25 plots, and the old fertilizer was applied to the remaining 25. We obtain the following data: $\bar{x}_1 = 22$ bushels of wheat per plot; $s_1^2 = 4$; $\bar{x}_2 = 24$ bushels of wheat per plot, $s_2^2 = 6$. Test $H_0: \mu_1 - \mu_2 = 0$ against $H_1: \mu_1 - \mu_2 < 0$ using a 5% level of significance. Assume the population variances are equal, the populations are approximately normal, and the samples are independent.

12.3   Tests for Differences of Means of Paired Samples

When the observations from two populations occur in pairs or are related, the samples are not independent and the tests described in Sections 12.1 and 12.2 are inappropriate.

Examples of observations that occur in pairs are sales at stores before and after an advertising campaign and reading rates of individuals before and after taking a speed reading course. In these cases, rather than having two independent random samples, we have one random sample of n pairs of observations. Hypotheses concerning the differences between the means μ_1 and μ_2 are tested by treating the differences of the n pairs as a random sample from a population of such differences.

Let x_{i1} be the ith observation from population 1, and let x_{i2} denote the ith observation from population 2. Because the observations x_{i1} and x_{i2} occur in pairs, we have a random sample of n pairs of observations: $(x_{11}, x_{12}), (x_{21}, x_{22}), \ldots, (x_{n1}, x_{n2})$. For each pair, we calculate the difference d_i as follows:

$$d_i = x_{i1} - x_{i2} \qquad i = 1, 2, \ldots, n$$

Let $\mu_d = \mu_1 - \mu_2$ denote the mean of the population of differences, and let μ_0 denote some hypothesized value. The null hypothesis is

$$H_0: \quad \mu_d = \mu_0$$

and the alternative hypothesis takes one of the forms

$$H_1: \quad \mu_d \neq \mu_0 \qquad H_1: \quad \mu_d < \mu_0 \qquad \text{or} \qquad H_1: \quad \mu_d > \mu_0$$

To test H_0 against H_1, we first calculate the sample mean and sample variance of the observed differences $d_1, d_2, \ldots, d_n$. We obtain the sample mean

$$d = \frac{\sum\limits_{i=1}^{n} d_i}{n}$$

and the sample variance

$$s_d^2 = \frac{\sum\limits_{i=1}^{n} (d_i - \bar{d})^2}{n - 1}$$

$$= \frac{\sum\limits_{i=1}^{n} d_i^2 - n\bar{d}^2}{n - 1}$$

Suppose we have obtained a random sample of differences $d_1, d_2, \ldots, d_n$ from a population of differences that has a normal distribution with mean μ_0 and variance σ_d^2. Then the t statistic

$$t = \frac{\bar{d} - \mu_0}{s_d/\sqrt{n}}$$

follows the t distribution with $\nu = (n - 1)$ degrees of freedom. Values of t close to 0 support the null hypothesis, and values of t far from 0 support the alternative hypothesis.

The sample variance s_d^2 is an estimate of the population variance σ_d^2. If σ_d^2 is known, then we use σ_d^2 rather than s_d^2 and the Z score rather than the t score as the test statistic. In addition, we use the Z score when the sample size is large, even if the population variance is estimated. The procedure described in the accompanying box shows how to test the difference of two means when we have matched pairs of observations.

Testing the difference between population means of matched pairs

Suppose we have a random sample of n matched pairs of observations from populations with means μ_1 and μ_2. Let d_i be the ith difference

$$d_i = x_{i1} - x_{i2}$$

where x_{i1} is the ith observation from population 1 and x_{i2} is the ith observation from population 2. Let $\bar{d}$ and s_d denote the observed sample mean and standard deviation, respectively, for the n differences $d_1, d_2, \ldots, d_n$. If the population of differences is normal, then the following tests have significance level α:

Case 1: To test the null hypothesis

$$H_0: \quad \mu_1 - \mu_2 = \mu_0 \quad \text{or} \quad H_0: \quad \mu_1 - \mu_2 \leq \mu_0$$

against the one-sided alternative hypothesis

$$H_1: \quad \mu_1 - \mu_2 > \mu_0$$

calculate the test statistic

$$t = \frac{\bar{d} - \mu_0}{s_d/\sqrt{n}}$$

and use the decision rule

Reject H_0 in favor of H_1 if $t > t_{\alpha,\nu}$,

where the random variable t follows the t distribution with $\nu = (n - 1)$ degrees of freedom.

Case 2: To test the null hypothesis

$$H_0: \quad \mu_1 - \mu_2 = \mu_0 \quad \text{or} \quad H_0: \quad \mu_1 - \mu_2 \geq \mu_0$$

against the one-sided alternative hypothesis

$$H_1: \quad \mu_1 - \mu_2 < \mu_0$$

(continued)

use the decision rule

$$\text{Reject } H_0 \text{ in favor of } H_1 \text{ if } t < -t_{\alpha,\nu}.$$

Case 3:   To test the null hypothesis

$$H_0: \quad \mu_1 - \mu_2 = \mu_0$$

against the two-sided alternative hypothesis

$$H_1: \quad \mu_1 - \mu_2 \neq \mu_0$$

use the decision rule

$$\text{Reject } H_0 \text{ in favor of } H_1 \text{ if } t > t_{\alpha/2,\nu} \text{ or if } t < -t_{\alpha/2,\nu}.$$

When we want to test the null hypothesis that the two population means are equal, we set μ_0 equal to 0 in the formulas.

If the sample size is large, say, 30 or more, we can replace the t statistic by the Z score and use the standard normal distribution to obtain critical values for the test. Also, if the sample size is large, we can relax the assumption that the population distribution of paired differences is normal.

EXAMPLE 12.4   **Testing a Difference Between Means of Matched Pairs** ————————

Use the data in Table 12.1 to test the null hypothesis $H_0: \mu_1 - \mu_2 = 0$ against the two-sided alternative hypothesis $H_1: \mu_1 - \mu_2 \neq 0$ described in Example 12.3. That is, test $H_0: \mu_d = 0$ against $H_1: \mu_d \neq 0$. Use $\alpha = .05$.

SOLUTION   Because each car was driven twice, once with gasoline 1 and once with gasoline 2, the observations are paired and are not independent. Thus we should use a paired difference test. We will use a two-tailed test with the level of significance set at $\alpha = .05$. For $\nu = (n - 1) = 9$ degrees of freedom and $\alpha = .05$, the critical values of t are $\pm t_{.025,9} = \pm 2.262$. From Table 12.1, we obtain $\bar{d} = -2.1$, $s_d^2 = .9889$, and $s_d = .9944$.

The appropriate test statistic is

$$t = \frac{-2.1 - 0}{.9944/\sqrt{10}} = -6.68$$

The observed value $t = -6.68$ falls in the rejection region, so we reject H_0. As a result, we conclude that the mean of the population of differences is nonzero. This is equivalent to concluding that the average mileage using gasoline 1 is different from the average mileage using gasoline 2.

Exercises

Statistical Concepts

12.3.1   Suppose we wish to test a hypothesis about the difference between the population means of matched pairs. It is assumed that the population of paired differences is normal. *True or false:*   The t statistic has $(n_1 + n_2 - 2)$ degrees of freedom. Explain.

12.3.2 Suppose we wish to test a hypothesis about the difference between the population means of matched pairs. It is assumed that the population of paired differences is normal. Explain how to perform the test.

12.3.3 Suppose we wish to test a hypothesis about the difference between the population means of matched pairs. It is not assumed that the population of paired differences is normal, but the sample size is large. Explain how to perform the test.

Statistical Drills

12.3.4 In a random sample of $n = 20$ matched pairs, the sample mean difference was $\bar{d} = 3.4$ and the sample standard deviation was $s_d = 1.3$. Let $\alpha = .05$.
  a. Determine the appropriate number of degrees of freedom for the test statistic t.
  b. Test $H_0: \mu_d = 0$ against $H_1: \mu_d \neq 0$.

12.3.5 In a random sample of $n = 15$ matched pairs, the sample mean difference was $\bar{d} = 22.1$ and the sample standard deviation was $s_d = 14.3$. Let $\alpha = .01$.
  a. Determine the appropriate number of degrees of freedom for the test statistic t.
  b. Test $H_0: \mu_d \leq 0$ against $H_1: \mu_d > 0$.

12.3.6 The following data show measurements on a random sample of $n = 5$ matched pairs.

Period 1	126.3	135.4	157.8	113.8	116.4
Period 2	133.1	145.6	126.2	145.0	104.2

  a. Calculate $\bar{d}$ and s_d.
  b. Determine the appropriate number of degrees of freedom for the test statistic t.
  c. Let $\alpha = .05$. Test $H_0: \mu_d = 0$ against $H_1: \mu_d \neq 0$.

Statistical Applications

12.3.7 A publishing company wants to test whether secretaries type faster using word processor brand 1 or brand 2. Twelve secretaries are tested on each word processor, and their speeds in words per minute are recorded. Use a paired difference test and a 5% level of significance to test $H_0: \mu_1 - \mu_2 = 0$ against $H_1: \mu_1 - \mu_2 \neq 0$ given the following data on typing speed in words per minute:

Secretary	1	2	3	4	5	6	7	8	9	10	11	12
Brand 1	66	73	55	50	60	66	78	45	52	65	57	48
Brand 2	60	70	53	56	60	62	71	41	50	61	55	44

12.3.8 A tire manufacturer claims that its tires last at least as long as those of a competitor. Fifteen tires of each brand are randomly chosen. One tire of each type is placed on the rear of 15 different cars, which are driven until no tire tread remains. Use a 5% level of significance and a paired difference test to test $H_0: \mu_1 - \mu_2 \leq 0$ against $H_1: \mu_1 - \mu_2 > 0$ given the following results of tire mileage in thousands of miles:

Car	1	2	3	4	5	6	7	8	9	10	11	12	13	14	15
Brand 1	42	36	54	50	40	33	44	50	41	31	37	34	33	33	30
Brand 2	40	35	55	45	37	31	42	45	43	30	33	35	29	28	27

12.3.9 Suppose that a psychologist thinks that age influences IQ. Suppose that a random sample of 100 middle-age persons whose IQs had been tested at age 16 were tested again. Subtracting their earlier scores from their new scores resulted in a mean difference of $\bar{x} = 6$ points and a standard deviation of the sample of differences of $s = 7$ points. Using $\alpha = .01$ as the significance level, the psychologist wishes to test the null hypothesis $H_0: \mu_d \leq 0$ against $H_1: \mu_d > 0$. Perform the appropriate test. Does it appear that IQ improves with age?

12.4 Tests Concerning Differences of Proportions

Frequently we are interested in testing the hypothesis that the proportion of individuals who possess a certain characteristic in population 1 is the same as that in population 2. Let p_1 and p_2 denote such proportions of populations 1 and 2, respectively. Suppose we want to test the null hypothesis that the difference between p_1 and p_2 is some value D_0 (i.e., $H_0: p_1 - p_2 = D_0$).

To test this hypothesis, we take a sample of size n_1 from population 1 and calculate $\hat{p}_1$, the proportion in this sample possessing the characteristic. Then we do likewise for population 2 and calculate $\hat{p}_2$. If H_0 is true and if $n_1 p_1 \geq 5$, $n_1 q_1 \geq 5$, $n_2 p_2 \geq 5$, and $n_2 q_2 \geq 5$, then the random variable $(\hat{p}_1 - \hat{p}_2)$ is approximately normally distributed with mean D_0 and variance

$$\frac{p_1 q_1}{n_1} + \frac{p_2 q_2}{n_2}$$

The null hypothesis typically is $H_0: p_1 - p_2 = 0$. Because p_1 and p_2 are unknown (only their difference D_0 is specified), we cannot find the variance of $(p_1 - p_2)$ without first obtaining estimates of p_1 and p_2. When the null hypothesis assumes that $p_1 = p_2$, a reasonable way to estimate the variance is to pool the two samples together to obtain one estimate of $p = p_1 = p_2$. To estimate p, we pool the data from the two samples and use the following formula:

$$\hat{p} = \frac{x_1 + x_2}{n_1 + n_2}$$

where x_1 and x_2 are the numbers of observations in the first and second sample, respectively, that possess the characteristic of interest. To test the null hypothesis $H_0: p_1 - p_2 = 0$, follow the procedure described in the accompanying box.

Testing the equality of two population proportions

Let $\hat{p}_1$ denote the sample proportion of successes in a random sample of n_1 observations from population 1, and let $\hat{p}_2$ denote the sample proportion of successes observed in an independent random sample of n_2 observations from population 2. If it is hypothesized that the population proportions are equal, an estimate of the common proportion is given by

$$\hat{p} = \frac{n_1 \hat{p}_1 + n_2 \hat{p}_2}{n_1 + n_2} = \frac{x_1 + x_2}{n_1 + n_2}$$

If the sample sizes are large, the following tests have significance level α:

Case 1:   To test the null hypothesis

$$H_0: \quad p_1 - p_2 = 0 \quad \text{or} \quad H_0: \quad p_1 - p_2 \leq 0$$

against the one-sided alternative hypothesis

$$H_1: \quad p_1 - p_2 > 0$$

calculate the test statistic

$$z = \frac{\hat{p}_1 - \hat{p}_2}{\sqrt{\hat{p}\hat{q}/n_1 + \hat{p}\hat{q}/n_2}} = \frac{\hat{p}_1 - \hat{p}_2}{\sqrt{\hat{p}\hat{q}\left(\dfrac{n_1 + n_2}{n_1 n_2}\right)}}$$

where $\hat{p}$ is the pooled estimate of p. Use the decision rule

Reject H_0 in favor of H_1 if $z > z_\alpha$.

   Case 2:   To test the null hypothesis

$$H_0: \quad p_1 - p_2 = 0 \quad \text{or} \quad H_0: \quad p_1 - p_2 \geq 0$$

against the one-sided alternative hypothesis

$$H_1: \quad p_1 - p_2 < 0$$

use the decision rule

Reject H_0 in favor of H_1 if $z < -z_\alpha$.

   Case 3:   To test the null hypothesis

$$H_0: \quad p_1 - p_2 = 0$$

against the two-sided alternative hypothesis

$$H_1: \quad p_1 - p_2 \neq 0$$

use the decision rule

Reject H_0 in favor of H_1 if $z > z_{\alpha/2}$ or if $z < -z_{\alpha/2}$.

EXAMPLE 12.5   **Testing the Equality of Two Population Proportions** ————————————

In a sample of 400 products produced by machine 1, 23 were defective, and in a sample of 400 products produced by machine 2, 17 were defective. Test $H_0: p_1 - p_2 = 0$ against $H_1: p_1 - p_2 \neq 0$ using a 5% level of significance.

SOLUTION   We use a two-tailed test. For $\alpha = .05$, the critical values are $\pm z_{.025} = \pm 1.96$. To calculate the estimated variance, we obtain the pooled proportion

$$\hat{p} = \frac{x_1 + x_2}{n_1 + n_2} = \frac{23 + 17}{400 + 400} = \frac{40}{800} = .05$$

Thus, we estimate the variance of $(\hat{p}_1 - \hat{p}_2)$ as

$$\frac{\hat{p}\hat{q}}{n_1} + \frac{\hat{p}\hat{q}}{n_2} = \frac{(.05)(.95)}{400} + \frac{(.05)(.95)}{400} = .0002375$$

We have $\hat{p}_1 = 23/400 = .0575$ and $\hat{p}_2 = 17/400 = .0425$, and so we obtain the test statistic

$$z = \frac{.0575 - .0425}{\sqrt{.0002375}} \approx \frac{.015}{.015} = 1$$

Because the observed value $z = 1$ falls in the acceptance region, we do not reject H_0. See Figure 12.2. Our decision not to reject H_0 means that the evidence is not strong enough to lead us to conclude that one machine produces a higher proportion of defectives than the other.

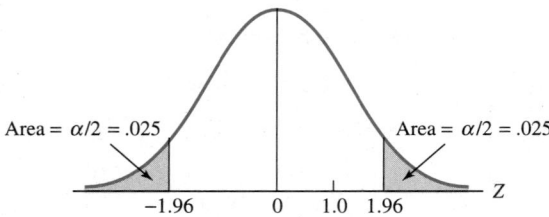

FIGURE 12.2   *The Z distribution for Example 12.5*

Exercises

Statistical Concepts

12.4.1   Suppose we want to test the null hypothesis $H_0: p_1 - p_2 = 0$. Explain how to estimate the variance $(p_1 q_1/n_1 + p_2 q_2/n_2)$ based on a pooled estimate of p.

12.4.2   Suppose we want to test the null hypothesis $H_0: p_1 - p_2 = 0$. To get a pooled estimate of the parameter p, we use the formula $\hat{p} = (x_1 + x_2)/(n_1 + n_2)$. Explain why we do not use the estimator $\hat{p} = (\hat{p}_1 + \hat{p}_2)/2$. [*Hint:*   Suppose $n_1 = 5,000$ and $n_2 = 50$.]

12.4.3   Suppose we want to test the null hypothesis $H_0: p_1 - p_2 = 0$. *True or false:*   The estimated variance of the random variable $(\hat{p}_1 - \hat{p}_2)$ is given by $(\hat{p}_1 \hat{q}_1/n_1 + \hat{p}_2 \hat{q}_2/n_2)$. Explain.

Statistical Drills

12.4.4   We take two independent random samples and obtain the following data: $n_1 = 100$, $n_2 = 100$, $\hat{p}_1 = .46$, $\hat{p}_2 = .41$. Let $\alpha = .05$.
a.   Calculate a pooled estimate of p.
b.   Test $H_0: p_1 - p_2 = 0$ against $H_1: p_1 - p_2 \neq 0$.

12.4.5   We take two independent random samples and obtain the following data: $n_1 = 120$, $n_2 = 240$, $x_1 = 24$, $x_2 = 36$. Let $\alpha = .05$.
a.   Calculate a pooled estimate of p.
b.   Test $H_0: p_1 - p_2 \leq 0$ against $H_1: p_1 - p_2 > 0$.

12.4.6   We take two independent random samples and obtain the following data: $n_1 = 200$, $n_2 = 250$, $x_1 = 80$, $x_2 = 90$. Let $\alpha = .01$.
a.   Calculate a pooled estimate of p.
b.   Test $H_0: p_1 - p_2 \leq 0$ against $H_1: p_1 - p_2 > 0$.

Statistical Applications

12.4.7 A pharmaceutical company wants to test whether aspirin 1 is as effective as aspirin 2. A random sample of 400 people with headaches are given aspirin 1, and 260 report that they feel better within 1 hour. Another independent random sample of 400 people with headaches are given aspirin 2, and 252 report that they feel better within 1 hour. Test $H_0: p_1 - p_2 = 0$ against $H_1: p_1 - p_2 \neq 0$ using a 5% level of significance.

12.4.8 In a sample poll of 100 voters from district 1, 60 favored a certain political candidate. In a sample poll of 100 voters from district 2, 40 favored the candidate. Test $H_0: p_1 - p_2 = 0$ against H_1: $p_1 - p_2 > 0$ using a 1% level of significance.

12.4.9 Let p_1 denote the percentage of people who were unemployed in March, and let p_2 denote the percentage of people who were unemployed in April. Suppose that during late March, the government instituted policies designed to lower the unemployment rate. We want to test whether the policies were effective. That is, we want to test $H_0: p_1 - p_2 = 0$ against $H_1: p_1 - p_2 > 0$ using, say, a 5% level of significance. In March, in a random sample of 1,000 people, 75 were unemployed. During April, in an independent random sample of 1,000 people, 65 were unemployed. Do you reject H_0?

12.4.10 A researcher desires to test whether the proportion of children born with defects is the same for mothers over 30 years of age and for mothers under 30 years of age. In a random sample of 500 babies whose mothers were over 30, 60 had birth defects. In a random sample of 400 babies whose mothers were under 30, 40 had birth defects. Test $H_0: p_1 - p_2 = 0$ against $H_1: p_1 - p_2 \neq 0$ using a 5% level of significance.

12.4.11 The Russell Construction Company purchases bricks from two different suppliers. In a random sample of 400 bricks from the first supplier, 26 were defective. In a random sample of 400 bricks from the second supplier, 44 were defective. Test the hypothesis that the defect rates are the same against the alternative hypothesis $H_1: p_1 - p_2 < 0$. Use a 5% level of significance.

12.5 The *F* Distribution

Occasionally it is necessary to compare two population variances. To test for the equality of two population variances, we calculate the ratio of the observed sample variances s_1^2/s_2^2. If this ratio is approximately equal to 1, we have no reason to doubt the hypothesis that the population variances σ_1^2 and σ_2^2 are equal. On the other hand, a very large or very small value for s_1^2/s_2^2 would provide evidence that the population variances are different.

How large or small must the ratio s_1^2/s_2^2 be before we reject the null hypothesis that the population variances are equal? To answer this question, we need to study the sampling distribution of the random variable S_1^2/S_2^2. When independent random samples are drawn from two normal populations with equal variances, then S_1^2/S_2^2 possesses a sampling distribution that is called the **F distribution.** We must be able to find critical values of the F distribution to test hypotheses about the equality of two population variances and to construct confidence intervals for the ratio σ_1^2/σ_2^2. In this section, we discuss the F distribution. In Section 12.6, we show how to perform tests and construct confidence intervals for a pair of population variances.

The F distribution is named in honor of the British statistician Sir Ronald Fisher (1890–1962), who began studying the distribution in the 1920s. Like the normal, t, and chi-square distributions, the F distribution is actually a family of distributions. For the F distribution, two parameters ν_1 and ν_2, called the *numerator* and *denominator degrees of freedom,* determine each different distribution. The random variable F is a continuous variable that can take any nonnegative value (it can never be negative because variances can never be negative). The F distribution is not symmetric; rather it is skewed to the right for small values of ν_1 and ν_2. Some typical F distributions for various combinations of ν_1 and ν_2 are shown in Figure 12.3.

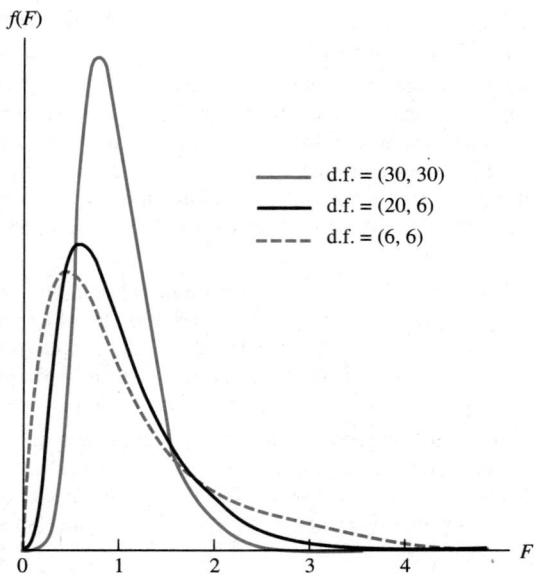

FIGURE 12.3   *Three examples of the F distribution*

When using the F distribution to test hypotheses, we must be able to find the critical values of F such that the area in the right and/or left tail of the distribution is some prespecified level of significance α. Suppose we want to find the critical value of F having ν_1 and ν_2 degrees of freedom such that the area in the right tail of the distribution is α. The critical values of the F statistic for different (ν_1, ν_2) combinations and various levels of significance α are provided in Table A.8 of the Appendix, which contains critical values for the levels of significance $\alpha = .05$ and $\alpha = .01$.

Critical value of F

Let the random variable F follow the F distribution with ν_1 and ν_2 degrees of freedom. Let α denote the area in the right tail of the F distribution. The critical value F_{α,ν_1,ν_2} is the value such that

$$P(F > F_{\alpha,\nu_1,\nu_2}) = \alpha$$

EXAMPLE 12.6   **Critical Value of the *F* Distribution** ────────────────────────────

Find the critical value of *F* such that the right tail of the distribution contains 5% of the area under the curve where *F* has numerator degrees of freedom $\nu_1 = 2$ and denominator degrees of freedom $\nu_2 = 9$.

SOLUTION   Refer to Table A.8 in the Appendix. Find the value located at the intersection of the column representing $\nu_1 = 2$ degrees of freedom and the row representing $\nu_2 = 9$ degrees of freedom. We obtain $F_{.05,2,9} = 4.26$, shown in Figure 12.4.

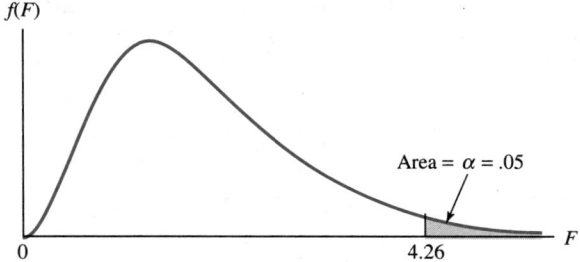

FIGURE   12.4   *The F distribution for Example 12.6*

──

EXAMPLE 12.7   **Critical Value of the *F* Distribution** ────────────────────────────

Find the critical value of *F* such that the right tail has area .01 where $\nu_1 = 2$ and $\nu_2 = 9$.

SOLUTION   Refer to the second portion of Table A.8, which shows critical values associated with $\alpha = .01$. Find the value located at the intersection of the column representing $\nu_1 = 2$ degrees of freedom and the row representing $\nu_2 = 9$ degrees of freedom. We obtain $F_{.01,2,9} = 8.02$.

──

Occasionally we may need critical values of *F* for values of ν_1 and ν_2 that are not given in Table A.8. For example, suppose we need the value $F_{.05,11,15}$. We can usually get a good estimate of the required *F* value by interpolation. From Table A.8, we obtain $F_{.05,10,15} = 2.54$ and $F_{.05,12,15} = 2.48$. By interpolation, we obtain $F_{.05,11,15} = 2.51$. A more detailed table would show that $F_{.05,11,15} = 2.51$, so in this case interpolation works very well.

In most statistics books, including this one, the tables for the *F* distribution show only the critical values for the right tail. Fortunately, we can determine the critical values for the left tail of the *F* distribution from the right-tail critical values. To find the value of *F* such that the area in the left tail is α, we use the fact that the lower-tail critical value, denoted by $F_{1-\alpha,\nu_1,\nu_2}$, can be determined from the upper-tail critical value F_{α,ν_2,ν_1} by using the relationship

$$F_{1-\alpha,\nu_1,\nu_2} = \frac{1}{F_{\alpha,\nu_2,\nu_1}}$$

Notice that in this formula, the degrees of freedom are reversed. Thus, $F_{1-\alpha,\nu_1,\nu_2}$ is the value of F having ν_1 and ν_2 degrees of freedom such that the area in the lower, or left, tail is α and the area in the right tail is $(1 - \alpha)$. On the other hand, F_{α,ν_2,ν_1} is the value of F having ν_2 and ν_1 degrees of freedom such that the area in the upper, or right, tail is α.

EXAMPLE 12.8   **Critical Values of the F Distribution**

Suppose the degrees of freedom of an F statistic are $\nu_1 = 8$ and $\nu_2 = 4$, respectively. Find the critical values of F such that each tail of the F distribution has area .05.

SOLUTION   We seek the two critical values $F_{.95,8,4}$ and $F_{.05,8,4}$. From Table A.8, we obtain $F_{.05,8,4} = 6.04$, but we still need to find $F_{.95,8,4}$. First we obtain the value $F_{.05,4,8} = 3.84$; then we obtain

$$F_{.95,8,4} = \frac{1}{F_{.05,4,8}} = \frac{1}{3.84} = .26$$

If the F statistic has $\nu_1 = 8$ and $\nu_2 = 4$ degrees of freedom, then

$$P(.26 \leq F \leq 6.04) = .90$$

This relationship is illustrated in Figure 12.5.

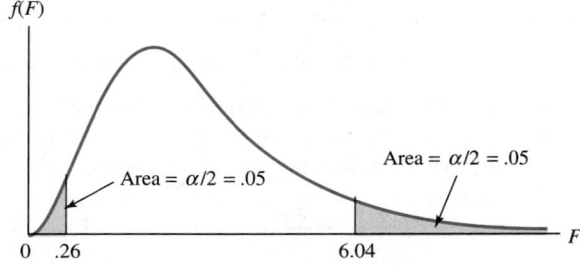

FIGURE 12.5 *The F distribution for Example 12.8*

Exercises

Statistical Concepts

12.5.1   *True or false:* The F distribution is symmetric about 0. Explain.

12.5.2   *True or false:* The F distribution varies from $-\infty$ to $+\infty$. Explain.

12.5.3   *True or false:* Like the t distribution and the chi-square distribution, the F distribution depends on a single parameter called the degrees of freedom. Explain.

12.5.4   *True or false:* Like the normal distribution, the F distribution depends on a pair of parameters called the mean and variance. Explain.

12.5.5   *True or false:* The F distribution is used to test the null hypothesis that two population variances are equal. Explain.

Statistical Drills

12.5.6 Using Table A.8 of the Appendix, find the following critical values:
a. $F_{.05,9,15}$ **b.** $F_{.01,9,15}$ **c.** $F_{.05,15,9}$

12.5.7 Find the critical value of F such that the left tail of the distribution has area .01 when the numerator and denominator degrees of freedom are the following:
a. 5 and 4   **b.** 4 and 5   **c.** 5 and 5

12.5.8 Find the critical value of F such that the left tail of the distribution has area .05 when the numerator and denominator degrees of freedom are the following:
a. 5 and 4   **b.** 4 and 5   **c.** 5 and 5

12.6   Testing Hypotheses and Constructing Confidence Intervals for Two Population Variances

Testing Hypotheses of Two Population Variances

Suppose we want to test the null hypothesis that the variances of two normal populations are equal. That is, we want to test the null hypothesis

$$H_0: \quad \sigma_1^2 = \sigma_2^2$$

against one of the alternative hypotheses

$$H_1: \quad \sigma_1^2 \neq \sigma_2^2 \qquad H_1: \quad \sigma_1^2 > \sigma_2^2 \qquad H_1: \quad \sigma_1^2 < \sigma_2^2$$

The null hypothesis $H_0: \sigma_1^2 = \sigma_2^2$ is equivalent to the null hypothesis

$$H_0: \quad \sigma_1^2 / \sigma_2^2 = 1$$

To test H_0 against H_1, we take random samples of size n_1 and n_2 from the two populations and calculate the sample variances S_1^2 and S_2^2. If the two populations are normally distributed and if $\sigma_1^2 = \sigma_2^2$ (so that H_0 is true), then the random variable

$$F = S_1^2 / S_2^2$$

follows the F distribution with numerator degrees of freedom $\nu_1 = (n_1 - 1)$ and denominator degrees of freedom $\nu_2 = (n_2 - 1)$.

When the alternative hypothesis requires a two-tailed test ($H_1: \sigma_1^2 \neq \sigma_2^2$), the rejection region will be equally divided between the lower and upper tails of the F distribution. Thus, to test H_0 against H_1 using a significance level α, we must find the two critical values of F such that the area in each tail of the F distribution equals $\alpha/2$.

The discussion in the accompanying box shows how to test hypotheses concerning the equality of two variances.

Testing the equality of variances of two normal populations

Let s_1^2 and s_2^2 be observed sample variances from independent random samples of n_1 and n_2 observations from normal populations with variances σ_1^2 and σ_2^2.

  Case 1: To test the null hypothesis

$$H_0: \quad \sigma_1^2 = \sigma_2^2 \qquad \text{or} \qquad H_0: \quad \sigma_1^2 \leq \sigma_2^2$$

(continued)

against the one-sided alternative hypothesis

$$H_1: \quad \sigma_1^2 > \sigma_2^2$$

calculate the test statistic

$$F = s_1^2/s_2^2$$

and use the decision rule

Reject H_0 in favor of H_1 if $F > F_{\alpha,\nu_1,\nu_2}$,

where $\nu_1 = (n_1 - 1)$ and $\nu_2 = (n_2 - 1)$. To perform the test when the alternative hypothesis is $H_1: \sigma_1^2 < \sigma_2^2$, just reverse the definitions of populations 1 and 2 and proceed as described previously.

Case 2:   To test the null hypothesis

$$H_0: \quad \sigma_1^2 = \sigma_2^2$$

against the two-sided alternative hypothesis

$$H_1: \quad \sigma_1^2 \neq \sigma_2^2$$

use the decision rule

Reject H_0 in favor of H_1 if $F < F_{1-\alpha/2,\nu_1,\nu_2}$ or if $F > F_{\alpha/2,\nu_1,\nu_2}$.

EXAMPLE 12.9   **Testing the Equality of Variances from Two Normal Populations**

An employee for the Metropolitan Bank thinks that the starting weekly salaries of newly hired male and female MBA graduates follow normal distributions. The employee wants to test the null hypothesis that the variance σ_1^2 among the men's starting salaries is the same as the variance σ_2^2 among the women's starting salaries. The employee obtains a random sample of 10 recently hired men and 7 recently hired women and obtains the sample variances $s_1^2 = 275$ and $s_2^2 = 225$, respectively. Test the null hypothesis $H_0: \sigma_1^2 = \sigma_2^2$ against the alternative hypothesis $H_1: \sigma_1^2 \neq \sigma_2^2$ using a 10% level of significance.

S O L U T I O N   The degrees of freedom are $\nu_1 = (n_1 - 1) = 9$ and $\nu_2 = (n_2 - 1) = 6$. For a two-tailed test using $\alpha = .10$, we obtain the critical values

$$F_{.05,9,6} = 4.10$$

and

$$F_{.95,9,6} = \frac{1}{F_{.05,6,9}} = \frac{1}{3.37} = .30$$

The value of the test statistic is given by

$$F = s_1^2/s_2^2 = 275/225 = 1.22$$

We do not reject H_0, because $F = 1.22$ falls in the acceptance region. See Figure 12.6. Our decision not to reject H_0 means that the evidence is not strong enough to lead us to conclude that the variance of salaries of male employees is different from the variance of salaries of female employees.

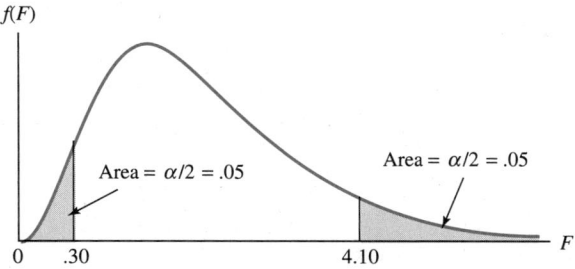

FIGURE 12.6 *The F distribution for Example 12.9*

Confidence Intervals for σ_1^2/σ_2^2

Suppose we take independent samples of size n_1 and n_2 from normal populations having variances σ_1^2 and σ_2^2. To construct a confidence interval for the ratio σ_1^2/σ_2^2, we utilize the fact that the random variable

$$F = \frac{S_1^2/\sigma_1^2}{S_2^2/\sigma_2^2}$$

has the F distribution with numerator degrees of freedom $\nu_1 = (n_1 - 1)$ and denominator degrees of freedom $\nu_2 = (n_2 - 1)$, and we utilize the probability statement

$$P\left(F_{1-\alpha/2,\nu_1,\nu_2} \leq \frac{S_1^2/\sigma_1^2}{S_2^2/\sigma_2^2} \leq F_{\alpha/2,\nu_1,\nu_2}\right) = 1 - \alpha$$

Algebraic manipulation yields the following equivalent probability statement:

$$P\left[\frac{S_1^2}{S_2^2}\left(\frac{1}{F_{\alpha/2,\nu_1,\nu_2}}\right) \leq \frac{\sigma_1^2}{\sigma_2^2} \leq \frac{S_1^2}{S_2^2}\left(\frac{1}{F_{1-\alpha/2,\nu_1,\nu_2}}\right)\right] = 1 - \alpha$$

Formula for a confidence interval for σ_1^2/σ_2^2

> A confidence interval for σ_1^2/σ_2^2 having level of confidence $(1 - \alpha)$ is given by
>
> $$\frac{s_1^2}{s_2^2}\left(\frac{1}{F_{\alpha/2,\nu_1,\nu_2}}\right), \quad \frac{s_1^2}{s_2^2}\left(\frac{1}{F_{1-\alpha/2,\nu_1,\nu_2}}\right)$$

EXAMPLE 12.10   **Constructing a Confidence Interval for the Ratio of Two Variances**

Two machines fill containers with fluid. The populations of fill amounts are assumed to be normal. Independent random samples are obtained of 21 containers filled by machine 1 and 16 containers filled by machine 2. The observed sample variances are $s_1^2 = .25$ and $s_2^2 = .08$. Construct a 90% confidence interval for σ_1^2/σ_2^2.

SOLUTION   The degrees of freedom are $\nu_1 = 20$ and $\nu_2 = 15$, and the critical values of F are

$$F_{.05,20,15} = 2.33$$

and

$$F_{.95,20,15} = \frac{1}{F_{.05,15,20}} = \frac{1}{2.20} = .45$$

The desired confidence interval is

$$\frac{.25}{.08}\left(\frac{1}{2.33}\right), \frac{.25}{.08}\left(\frac{1}{.45}\right)$$

Thus, we are 90% confident that the ratio σ_1^2/σ_2^2 lies in the interval (1.34, 6.94).

Exercises

Statistical Concepts

12.6.1   Suppose we want to test the null hypothesis H_0: $\sigma_1^2 = \sigma_2^2$. Suppose we have two independent samples from normal populations. *True or false:*   The appropriate test statistic is $F = s_1^2 - s_2^2$. Explain.

12.6.2   Suppose we want to test the null hypothesis H_0: $\sigma_1^2 \leq \sigma_2^2$ against H_1: $\sigma_1^2 > \sigma_2^2$. Suppose we obtain two independent samples from normal populations and calculate the test statistic $F = s_1^2/s_2^2$. *True or false:*   The appropriate decision rule is to reject H_0 if $F > F_{\alpha,\nu_1,\nu_2}$. Explain.

Statistical Drills

12.6.3   Suppose we take independent random samples from two normal populations. We obtain the following information: $n_1 = 16$, $n_2 = 21$, $s_1^2 = 134.6$, $s_2^2 = 164.3$. Test the null hypothesis H_0: $\sigma_1^2 = \sigma_2^2$ against H_1: $\sigma_1^2 \neq \sigma_2^2$. Let $\alpha = .10$.
   a. What are the appropriate numbers of degrees of freedom?
   b. Calculate the value of the F statistic.
   c. Determine the critical values of the test statistic.
   d. Do you reject the null hypothesis?
   e. Construct a 90% confidence interval for σ_1^2/σ_2^2.

12.6.4   Suppose we take independent random samples from two normal populations. We obtain the following information: $n_1 = 10$, $n_2 = 10$, $s_1^2 = 274.6$, $s_2^2 = 389.3$. Test the null hypothesis H_0: $\sigma_1^2 \geq \sigma_2^2$ against H_1: $\sigma_1^2 < \sigma_2^2$. Let $\alpha = .10$.
   a. What are the appropriate numbers of degrees of freedom?
   b. Calculate the value of the F statistic.
   c. Determine the critical value of the test statistic.
   d. Do you reject the null hypothesis?
   e. Construct a 90% confidence interval for σ_1^2/σ_2^2.

12.6.5   Suppose we take independent random samples from two normal populations. We obtain the following information: $n_1 = 25$, $n_2 = 30$, $s_1^2 = 421.3$, $s_2^2 = 554.3$. Test the null hypothesis H_0: $\sigma_1^2 = \sigma_2^2$ against H_1: $\sigma_1^2 \neq \sigma_2^2$. Let $\alpha = .10$.
   a. What are the appropriate numbers of degrees of freedom?
   b. Calculate the value of the F statistic.
   c. Determine the critical values of the test statistic.
   d. Do you reject the null hypothesis?
   e. Construct a 90% confidence interval for σ_1^2/σ_2^2.

Statistical Applications

In the following problems, assume that all populations are normal.

12.6.6 Two methods of filling boxes of cereal produce the same package fill weight on the average. The second method is faster, but some people suspect that this method produces a greater variance in fill amounts than the first method does. A random sample of 31 packages filled by method 1 is obtained, and the observed sample variance is $s_1^2 = .4$. A random sample of 26 packages filled by method 2 yields a sample variance of $s_2^2 = .6$.
a. Let $\alpha = .05$. Test $H_0: \sigma_1^2 = \sigma_2^2$. Use a one-tailed test.
b. Construct a 90% confidence interval for σ_1^2 / σ_2^2.

12.6.7 Suppose two machines produce 1-inch-diameter pipes with different degrees of precision. A random sample of 20 pipes is taken from each machine, yielding observed sample standard deviations of $s_1 = .002$ and $s_2 = .005$.
a. Let $\alpha = .02$. Should we conclude that σ_1^2 is different from σ_2^2? Use a two-tailed test.
b. Determine a 90% confidence interval for σ_1^2 / σ_2^2.

12.7   Computer Applications

Refer to the data in Table 2.2 of Chapter 2, of selected characteristics of a sample of statistics students. Suppose it is desired to test the null hypothesis that in the population of students, the mean SAT score for the males is the same as the mean SAT score for the females. In Section 9.8, we showed how to use the BREAKDOWN command in an SPSSX program to request the sample mean and sample standard deviation for a variable, such as SAT, broken down according to the values of another variable, such as SEX. The appropriate command would be

---
BREAKDOWN TABLES = SAT BY SEX

When the sample size is large, we obtain the appropriate sample means and sample standard deviations and substitute them into the Z score

$$Z = \frac{(\bar{X}_1 - \bar{X}_2) - (\mu_1 - \mu_2)}{\sqrt{S_1^2/n_1 + S_2^2/n_2}}$$

This statistic is approximately a standard normal variable when the sample sizes are large.

Exercises

12.7.1 Use the SPSSX program and the BREAKDOWN command to find the sample mean and sample standard deviation for the variable SALARY broken down according to the variable SEX using the data in Table 2.1 of Chapter 2. Let $\alpha = .05$ and test the null hypothesis that the mean salaries of males and females are equal. Use a two-tailed t test based on the assumption that the population variances are equal and the populations are normal.

12.7.2 Use the SPSSX program and the BREAKDOWN command to find the sample mean and sample standard deviation for the variable SALARY broken down according to the variable DEGREE using the data in Table 2.1 of Chapter 2.

a. Let $\alpha = .05$ and test the null hypothesis that the mean salaries of high school graduates and college graduates are equal. Use an appropriate one-tailed test.

b. Let $\alpha = .05$ and test the null hypothesis that the mean salaries of college graduates and individuals with postgraduate degrees are equal. Use an appropriate one-tailed test.

12.7.3   Use the SPSSX program and the BREAKDOWN command on the data in Table 2.1 of Chapter 2 to find the sample mean and sample standard deviation for the variable SENIOR broken down according to the variable DIV. Let $\alpha = .05$ and test the null hypothesis that the mean seniorities of individuals in the office and of individuals in the sales division are equal. Use an appropriate one-tailed test.

12.7.4   Use the SPSSX program and the BREAKDOWN command on the data in Table 2.2 of Chapter 2 to find the sample mean and sample standard deviation for the variable GPA broken down according to the variable SEX. Let $\alpha = .05$ and test the null hypothesis that the mean GPA scores of males and of females are equal. Use an appropriate two-tailed test.

12.7.5   Use the SPSSX program and the BREAKDOWN command on the data in Table 2.2 of Chapter 2 to find the sample mean and sample standard deviation for the variable SAT broken down according to the variable CLASS. Let $\alpha = .05$ and test the null hypothesis that the mean SAT scores of juniors and of seniors are equal. Use an appropriate two-tailed test.

12.7.6   Use the SPSSX program and the BREAKDOWN command on the data in Table 2.2 of Chapter 2 to find the sample mean and sample standard deviation for the variable GPA broken down according to the variable MAJOR. Let $\alpha = .05$ and test the null hypothesis that the mean GPA scores of business majors and of math majors are equal. Use an appropriate two-tailed test.

STATISTICS IN ACTION: CASE STUDY

Paint Problems on the Mellon Bank Building

The case study at the end of Chapter 4 described part of a lawsuit that the management of Mellon Bank filed against 20 local and national companies concerning flaws in the paint on the Mellon Bank Building.

A company that specializes in the chemical analysis of paint on buildings and bridges was hired to collect a sample of data. Data were obtained from a sample of 30 points where the paint had cracked and peeled (the "failure points") and from a sample of 247 points where the paint had not cracked or peeled (the "nonfailure points"). The following data refer to the total thickness of the paint at these 30 failure points and 247 nonfailure points. All values are in millimeters (mm).

	30 failure points	247 non-failure points
Mean total thickness	.1355	.0805
Standard deviation	.0643	.0272
Range	.3000	.1600

Total thickness	Relative frequency	Relative frequency
.0300–.0599	0.0%	26.7%
.0600–.0999	16.7%	43.4%
.1000–.1299	50.0%	25.9%
.1300–.1699	10.0%	4.0%
.1700–.3600	23.3%	0.0%

The data indicate that the mean thickness of the paint at the failure points was 68% greater than the mean thickness at the nonfailure points. This information supported the hypothesis that the thickness of the paint was an important determinant of paint failure. The relative frequency distributions also provided evidence supporting the hypothesis that the thickness of the paint was an important determinant of paint failure. Only 16.7% of the failure points had a thickness of .0999 mm or less, whereas 70.1% of the nonfailure points had a thickness of .0999 mm or less. Also, 23.3% of the failure points had a thickness of .1700 mm or more, whereas none of the nonfailure points had a thickness of .1700 mm or more.

This evidence suggests that excessive paint thickness was a major determinant of paint failure. The large standard deviation of paint thickness suggests that the paint thickness varied considerably from one spot to another. This suggests that there was a lack of quality control in the painting process.

Let μ_F denote the mean paint thickness for the population of failure points, and let μ_N denote the mean paint thickness for the population of nonfailure points. Let us perform a formal test to test the null hypothesis $H_0: \mu_F - \mu_N \leq 0$ against the alternative hypothesis $H_1: \mu_F - \mu_N > 0$. We do not want to claim that the paint thicknesses are different unless we have strong evidence supporting this conclusion, so we choose $\alpha = .01$ as the level of significance. We use the following information: $n_F = 30$, $n_N = 247$, $\bar{x}_F = .1355$, $\bar{x}_N = .0805$, $s_F^2 = .004134$, $s_N^2 = .000740$.

Because the sample sizes are relatively large, we use the Z score as the test statistic. If the null hypothesis is true, the Z score follows the standard normal distribution, at least approximately. We obtain

$$z = \frac{\bar{x}_F - \bar{x}_N}{\sqrt{\dfrac{s_F^2}{n_F} + \dfrac{s_N^2}{n_N}}}$$

$$= \frac{.1355 - .0805}{\sqrt{\dfrac{.004134}{30} + \dfrac{.000740}{247}}} = 4.635$$

Based on a one-tailed test, the critical value of z is 2.33. Our z statistic of 4.635 far exceeds the critical value. Thus, we reject the null hypothesis that the population means are equal.

According to the *Allegheny Bulletin*, the case was settled out of court for $13.1 million. This case study provides a classic example of how elementary descriptive statistics can help us make important decisions about data sets.

Chapter 12 Supplementary Exercises

12.S.1   In a TV commercial, company 1 claims that its golf balls are livelier than those made by company 2. A mechanical device hits 200 balls made by company 1 and 200 balls made by company 2. We measure how far each ball travels. The sample results are $\bar{x}_1 = 212$ yards, $s_1^2 = 81$; $\bar{x}_2 = 208$ yards, $s_2^2 = 81$. Test $H_0: (\mu_1 - \mu_2) = 0$ against $H_1: (\mu_1 - \mu_2) \neq 0$ using a 5% level of significance.

12.S.2   A sports publication claims that the average income of people who attend hockey games (population 1) is higher than the average income of people who attend pro football games. In a random sample of 200 hockey fans, we obtain $\bar{x}_1 = \$35,000$ with $s_1^2 = 9,000,000$. In a random sample of 200 football fans (population 2), we obtain $\bar{x}_2 = \$32,500$ with $s_2^2 = 9,000,000$. Test $H_0: (\mu_1 - \mu_2) = 0$ against $H_1: (\mu_1 - \mu_2) > 0$ using a 5% level of significance.

12.S.3   The Environmental Protection Agency wants to test the null hypothesis that Denver and Los Angeles have the same amount of air pollution. The air quality in each city is measured on 20 randomly selected days. In Denver, the sample mean was $\bar{x}_1 = 85$ and the sample variance was $s_1^2 = 1,000$. In Los Angeles, the sample mean was $\bar{x}_2 = 92$ with $s_2^2 = 800$. Test $H_0: (\mu_1 - \mu_2) = 0$ against $H_1:$

$(\mu_1 - \mu_2) \neq 0$ using a 5% level of significance. Assume the populations are normal with equal variances and use a t test.

12.S.4   The army is interested in determining whether method 1 or method 2 is a better way to teach soldiers. In a random sample of 1,000 soldiers taught using method 1, 700 pass the marksmanship test after 3 days of training; in an independent random sample of 1,000 soldiers taught by method 2, 740 pass the test after 3 days of training. Test $H_0: (p_1 - p_2) = 0$ against $H_1: (p_1 - p_2) \neq 0$ using a 5% level of significance.

12.S.5   A woman is running for governor of a southern state. Pollsters argue that men are biased against female candidates and that women are biased in favor of female candidates. In a sample of 100 women, 54 say that they favor a particular female candidate. In an independent sample of 100 men, 46 say that they favor this candidate. Test $H_0: (p_1 - p_2) = 0$ against $H_1: (p_1 - p_2) \neq 0$ using a 1% level of significance.

12.S.6   On a sample of 50 plots of land, a farmer plants wheat and uses fertilizer made by company 1; on another 50 plots, fertilizer by company 2 is used. When the wheat is harvested, the farmer measures how many bushels of wheat are obtained on each plot. The average yield on the plots using company 1 fertilizer was $\bar{x}_1 = 141$ with sample variance 400, and on the plots using company 2 fertilizer, the sample mean was $\bar{x}_2 = 127$ with sample variance 400. Test $H_0: (\mu_1 - \mu_2) = 0$ against $H_1: (\mu_1 - \mu_2) > 0$ using a 5% level of significance. Assume the population variances are equal, the populations are normal, and the samples are independent.

12.S.7   In an agricultural experiment to determine the effects of a particular insecticide, a field was planted with corn. Half of the plants were sprayed with the insecticide, and half were unsprayed. Several weeks later, independent random samples of 200 sprayed plants and 200 unsprayed plants were examined. The number of plants in each sample was as follows:

	Sprayed	Unsprayed
Healthy	131	111
Not healthy	69	89

If the significance level is set at .05, does the evidence indicate that a higher proportion of sprayed than of unsprayed plants were healthy? Use a one-tailed test.

12.S.8   The "fog index" is used to measure the reading difficulty of a written text. The fog index is the sum of the average number of words per sentence plus the percentage of words with three or more syllables. A random sample of six pages of material from the *Wall Street Journal* had the following fog indices:

    61 45 59 38 42 55

An independent random sample of six pages from *Sports Illustrated* had the following fog indices:

    40 36 29 28 35 42

Use a 5% level of significance and test the null hypothesis that the population mean fog indices are the same against the alternative that the population mean is higher for the *Wall Street Journal* than for *Sports Illustrated*.

12.S.9   We want to determine whether a new cover improves the distance that a golf ball will travel. Random samples of 200 new balls and 200 old balls are hit by a mechanical device. After measuring how far each ball travels, we obtain $\bar{x}_1 = 208$, $s_1^2 = 1,000$; $\bar{x}_2 = 198$, $s_2^2 = 800$. Use a 5% level of significance and test the null hypothesis $H_0: (\mu_1 - \mu_2) = 0$ against $H_1: (\mu_1 - \mu_2) > 0$. Assume that the population variances are equal and that the populations are normal.

12.S.10 In random samples of 200 males and 200 females, the proportions of individuals who attend church regularly were calculated. Suppose that 90 males and 110 females claimed that they attend church regularly. Test the null hypothesis that the population proportions are equal. Use $\alpha = .05$.

12.S.11 The braking ability of two new compact cars was being studied. The cars were driven 50 miles per hour when the brakes were applied, and the distances required to stop were measured. Ten cars of each type were studied. The sample means and sample standard deviations were $\bar{x}_1 = 138$ feet, $s_1 = 11$ feet; $\bar{x}_2 = 144$ feet, $s_2 = 13$ feet. Test $H_0: (\mu_1 - \mu_2) = 0$ against $H_1: (\mu_1 - \mu_2) \neq 0$ using a 1% level of significance. Assume the population variances are equal, the populations are normal, and the samples are independent.

12.S.12 Two independent samples of 80 equally intelligent schoolchildren are taught to read by different methods: The first group memorizes words, and the second tries to learn words by pronouncing the sounds of the letters. After several months of instruction, a reading test is given to all of the children. The test results were $\bar{x}_1 = 70$, $s_1 = 8$; $\bar{x}_2 = 63$, $s_2 = 7$. Test $H_0: \mu_1 - \mu_2 = 0$ against $H_2: \mu_1 - \mu_2 \neq 0$ using a 1% level of significance. Assume the population variances are equal, the populations are normal, and the samples are independent.

 12.S.13 In a random sample of 200 men, 120 favored the death penalty for certain crimes. In an independent random sample of 300 women, 250 favored the death penalty for the same crimes. Can we conclude that a greater proportion of women than men favor the death penalty? Use a 5% level of significance.

12.S.14 A new energy conservation bill is being considered in the U.S. Senate. In a random sample of 200 West Virginia residents, 160 were against the bill. In a random sample of 300 Maine voters, 200 favored the bill and 100 were against it. Let $\alpha = .05$. Can it be concluded that the proportion of voters who are against the bill is the same in Maine and West Virginia?

 12.S.15 A governmental agency hired a doctor to investigate the impact of a lead smelter on the level of lead in the blood of children living near the smelter. Ten of these children were chosen at random, and the following lead levels (micrograms per 100 milliliters of blood) were measured:

    18   16   21   14   17
    19   22   24   15   18

An independent sample of 7 children living in an area relatively free from possible lead pollution was also obtained; their blood samples had the following lead levels:

    9   13   8   15   17   12   11

a. State the null hypothesis and the alternative hypothesis.
b. With $\alpha = .05$, what do you conclude?

 12.S.16 A study was made of 10,000 men who had vasectomies, and each was matched to a man of approximately the same age, race, and marital status who had not been vasectomized. Over the 15-year period covered by the study, 220 of the vasectomized men and 310 of the nonvasectomized men died.

a. Let $\alpha = .05$ and test the null hypothesis that the two population proportions are equal. Does it appear to you that vasectomized men have a lower death rate?
b. Do you think the vasectomy causes the lower death rate? Could it be that men who choose to have a vasectomy are healthier than men who do not? Can you think of any possible explanation for these results?

 12.S.17 Food distributors use a system of quality control inspections to ensure that the contents of food containers meet various standards. Usually, differences in the mean weight of the contents of containers produced on different production lines or at different plants can be corrected by changing machine settings. On the other hand, differences in the variances of the weights of output produced on different assembly lines are more difficult to correct because much of the variation in weight may be due to random or unknown causes.

To monitor the weight of fruit in cans of output, random samples of output are collected from two production lines. The following data show the drained weights (in ounces) of the contents of cans of peaches filled by two different filling machines.

DRAINED WEIGHT OF PEACHES (IN OZ.)

Machine A

22.0	22.5	22.5	24.0	23.5	20.5	22.5	22.5	23.0	21.5
20.0	20.5	23.0	22.0	21.5	21.0	22.0	22.0	23.0	22.0
22.5	19.5	22.5	22.0	21.0	23.0	23.5	21.0	22.0	20.0
19.0	20.0	22.0	20.5	22.5	21.5	20.5	19.0	19.5	19.5
21.0	22.5	20.0	22.0	22.0	21.5	23.0	22.0	23.0	18.5
20.0	19.5	21.0	20.0	20.5	19.0	21.0	21.0	21.0	20.5
19.5	20.5	21.0	20.5	21.0	20.0	21.5	24.0	23.0	20.0

Machine B

22.5	19.5	21.0	21.5	21.0	21.5	20.5	22.0	21.5	23.5
19.0	21.5	23.0	21.0	23.5	21.0	20.5	19.5	22.0	21.0
20.0	23.5	24.0	20.5	21.5	22.0	20.5	21.0	22.5	20.0
19.0	20.5	21.0	20.5	22.5	21.5	25.0	21.0	19.0	21.0
22.5	22.0	23.0	22.0	23.5	22.5	22.0	22.0	19.5	20.5
18.5	22.0	22.5	21.0	21.5	21.5	20.5	20.5	16.5	21.5
24.0	22.0	17.5	21.0	22.5	19.5	22.5	15.5	20.0	22.5
22.0	17.5	21.0	22.0	23.5	22.0	22.0	20.5	24.0	21.5

a. For each machine, make a tally sheet of these measurements and arrange them in a frequency distribution.

b. Construct and plot the frequency distributions showing the class boundaries, class midpoints, and observed frequencies. Does it appear that the two distributions are approximately normal and have equal means?

c. Calculate the sample mean for each set of data.

d. Calculate the sample variance and sample standard deviation for each set of data.

e. Let $\alpha = .05$. Test $H_0: \mu_1 - \mu_2 = 0$ against $H_1: \mu_1 - \mu_2 \neq 0$.

f. Test the null hypothesis $H_0: \sigma_1^2 = \sigma_2^2$ against $H_1: \sigma_1^2 \neq \sigma_2^2$. Let $\alpha = .10$.

g. For each sample of data, determine the proportion of weights that are less than 20.0.

h. Let $\alpha = .05$. Test $H_0: p_1 - p_2 = 0$ against $H_1: p_1 - p_2 \neq 0$.

12.S.18 A small radio transmitting set is designed so that it can be used to generate a certain automatic signal. The time duration of this signal is one of the specified quality characteristics of the set. Random samples of output from two production lines were examined to determine whether there were any differences in the output of the two lines.

DURATION OF AUTOMATIC SIGNAL

Machine A

390	393	395	405	420	376	381	381	383	401
380	387	395	397	407	377	383	387	390	393
393	395	403	405	414	376	388	395	397	400
387	400	400	403	410	391	392	394	397	405
390	391	395	401	405	379	391	393	394	410

Machine B

390	397	400	406	428	380	382	389	391	399
362	363	372	375	377	357	360	368	370	372
390	395	395	397	406	382	399	401	406	406
390	395	395	400	410	381	390	394	397	399
367	389	398	401	435	362	378	396	400	429

a. For each machine, make a tally sheet of these measurements and arrange them in a frequency distribution.

b. Construct and plot the frequency distributions showing the class boundaries, class midpoints, and observed frequencies. Does it appear that the two distributions are approximately normal and have equal means?

c. Calculate the sample mean for each set of data.

d. Calculate the sample variance and sample standard deviation for each set of data.

e. Let $\alpha = .05$. Test $H_0: \mu_1 - \mu_2 = 0$ against $H_1: \mu_1 - \mu_2 \neq 0$.

f. Test the null hypothesis $H_0: \sigma_1^2 = \sigma_2^2$ against $H_1: \sigma_1^2 \neq \sigma_2^2$. Let $\alpha = .10$.

g. For each sample of data, determine the proportion of measurements that are less than 400.

h. Let $\alpha = .05$. Test $H_0: p_1 - p_2 = 0$ against $H_1: p_1 - p_2 \neq 0$.

12.S.19 Salk polio vaccine was tested by implementing a random double-blind experiment. A sample of approximately 400,000 children were the subjects of the experiment. Half the children were selected randomly to be given the vaccine, the other half were given a placebo (a pill having no medicinal effect). The test was double-blind because the patients did not know whether they were receiving the vaccine or the placebo. Similarly, until the test was concluded, the evaluators of the test did not know which children received the vaccine and which received the placebo. Assume that, of the 200,000 children who received the placebo (group 1), 142 contracted polio. Of the 200,000 who received the vaccine (group 2), 57 contracted polio. Test the null hypothesis $H_0: p_1 - p_2 = 0$ against $H_1: p_1 - p_2 \neq 0$. Let $\alpha = .01$.

a. Find $\hat{p}_1$ and $\hat{p}_2$. Be very careful to avoid round-off error.

b. Find the pooled estimate of p.

c. Perform the test.

d. Does it appear that the Salk vaccine was successful in lowering the polio rate?

e. Find the difference $(\hat{p}_1 - \hat{p}_2)$. In absolute terms, this difference is very small. Is this difference of any practical importance?

12.S.20 A "negative income tax" system is a system that gives supplemental income to people with low incomes instead of taking tax revenue away from them. A potential problem with a negative income tax system is that it may cause low-income individuals to stop working. A scientific study was carried out in New Jersey to test this proposition. A random sample of $n_1 = 400$ low-income families was selected to be the control group — they were not placed on the negative income tax system. A random sample of $n_2 = 400$ low-income families was selected to be the treatment group—they were put on the negative income tax system. All 800 families were observed for 3 years. Suppose that, during the 3-year period, the head of household in group 1 (the control group) worked an average of $\bar{x}_1 = 6,820$ hours with $s_1 = 3,600$. During the 3-year period, the head of household in group 2 (the treatment group) worked an average of $\bar{x}_2 = 6,240$ hours with $s_2 = 3,400$. Let $\alpha = .01$. Is the difference between the sample means statistically significant? What is your conclusion concerning the effect of the negative income tax?

12.S.21 A major problem haunting society is that, when convicts are released from prison, there is a high rate of recidivism; that is, the released prisoners return to a life of crime. Some sociologists argue that released prisoners tend to have little or no money, and thus they have a strong incentive to return to a life of crime. Perhaps the recidivism rate could be decreased if released prisoners were provided with income supplements during their first few months after being released from prison. The U.S. Department of Labor initiated a scientific study in Texas and Georgia to test this proposition. The following hypothetical example is based on the actual results.

Assume that 900 released prisoners were randomly selected. Of these, $n_1 = 600$ were randomly selected to be given income supplements. The remaining $n_2 = 300$ released prisoners were given no income supplements. Of the 600 ex-prisoners who were given supplemental income, 47.1% were rearrested within 1 year after being released. Of the 300 ex-prisoners who were not given income supplements, 45.4% were rearrested within 1 year. Let $\alpha = .01$. Does it appear that income supplements reduce recidivism?

During the first year after being released, the ex-prisoners who were given income supplements worked an average of 15.6 weeks with standard deviation $s_1 = 14.2$ weeks. The ex-prisoners who were not supported worked an average of 25.6 weeks with $s_2 = 17.1$ weeks. Does it appear that income support reduces the amount that the ex-prisoners worked?

References

Cochran, William G., and G. M. Cox. *Experimental Designs.* 2d ed. New York: Wiley, 1957.

Fisher, Ronald A. *Statistical Methods for Research Workers.* 14th ed. New York: Hafner, 1970.

Nie, Norman E., C. Hadlai Hull, Jean Jenkins, Karin Steinbrenner, and Dale H. Bent. *SPSS Statistical Package for the Social Sciences.* 2d ed. New York: McGraw-Hill, 1975.

Norusis, Marija J. *SPSSX Introductory Statistics Guide.* New York: McGraw-Hill, 1990.

Norusis, Marija J. *SPSSX Advanced Statistics Guide.* Chicago: SPSS, 1990.

Norusis, Marija J. *The SPSS Guide to Data Analysis.* Chicago: SPSS, 1986.

Ryan, Thomas A., Brian L. Joiner, and Barbara F. Ryan. *Minitab Handbook.* 2d ed. Boston: PWS-Kent, 1985.

Ryan, Thomas A., Brian L. Joiner, and Barbara F. Ryan. *Minitab Reference Manual.* University Park, Pa.: Minitab Project, 1985.

SAS Introductory Guide. 3d ed. Cary, N.C.: SAS Institute, 1985.

SAS Procedures Guide for Personal Computers. Version 6 ed. Cary, N.C.: SAS Institute, 1986.

SAS Statistics Guide for Personal Computers. Version 6 ed. Cary, N.C.: SAS Institute, 1986.

SAS User's Guide: Basics. Version 5 ed. Cary, N.C.: SAS Institute, 1985.

SAS User's Guide: Statistics. Version 5 ed. Cary, N.C.: SAS Institute, 1985.

SPSSX User's Guide. Chicago: SPSS, 1988.

13

Chi-Square Tests

In Chapters 11 and 12, we showed how to test hypotheses about a single population mean, a single population proportion, the difference between two means, and the difference between two proportions. In this chapter, we discuss methods for testing hypotheses about a whole set of proportions. In addition, we show how to test the null hypothesis that two qualitative random variables, such as the gender of an applicant and admittance to medical school, are independent of one another. Because all the tests discussed in this chapter use a test statistic that follows the chi-square distribution, they are called chi-square tests. To perform these tests, we need to find the critical values that separate the acceptance region and the critical region using the chi-square distribution.

13.1 The Chi-Square Goodness-of-Fit Test

In many situations in business and economics, we want to test the null hypothesis that a sample of data was selected from a population having certain characteristics. For example, we may wish to test the null hypothesis that the distribution of incomes or ages in a certain city is the same as it was in the last census. Usually the population is divided into several, say, K, categories, and the null hypothesis states that the proportions of observations in categories $1, 2, \ldots, K$ are $p_1, p_2, \ldots, p_K$. To test the null hypothesis, we determine how many observations in a sample of size n would be expected to fall in each category if the null hypothesis were true, and we compare these expected frequencies to the frequencies actually observed in the sample. The comparison is done by performing a *chi-square goodness-of-fit test*, which was developed in 1900 by the British statistician Karl Pearson (1857–1936). The following example introduces the chi-square goodness-of-fit test.

EXAMPLE 13.1   **Testing a Hypothesis About a Set of Population Proportions**

The Department of Commerce classifies households in the United States according to income. The publication *Current Population Reports* (Series P-60, No. 184 [Washington, D.C.: Bureau of the Census]) gives the following data on distribution of income in the United States, as reported in 1992.

Income	Proportion of households
$0–$12,499	.20
$12,500–$29,999	.30
$30,000–$74,999	.40
$75,000 or higher	.10

The mayor of Joliet, Illinois, wants to test whether the distribution of income in Joliet is the same as that in the entire United States because if incomes in Joliet are lower than in the United States, Joliet will be eligible for additional federal revenue-sharing funds. The null hypothesis is that the distribution of income in Joliet, Illinois, is the same as in the United States. If the null hypothesis is true, then the proportions of households in Joliet that fall into each of the four income categories will be .20, .30, .40, and .10. Thus, the null hypothesis is

$$H_0: \quad p_1 = .20, \ p_2 = .30, \ p_3 = .40, \ p_4 = .10$$

where p_i refers to the proportion of households falling into the ith income category. The alternative hypothesis is

$$H_1: \quad \text{At least one of the four proportions in } H_0 \text{ is incorrect.}$$

To test H_0 against H_1, we obtain a random sample of $n = 200$ households in Joliet and count the number of observations that fall into each category. These four frequencies, called the *observed frequencies,* are denoted o_1, o_2, o_3, and o_4. If the null hypothesis is true, we expect to obtain the following income distribution in Joliet:

Income	Proportion of households	Expected frequency (sample size = 200)
$0–$12,499	.20	40
$12,500–$29,999	.30	60
$30,000–$74,999	.40	80
$75,000 or higher	.10	20

The four *expected frequencies* are denoted e_1, e_2, e_3, and e_4. Large discrepancies between the observed frequencies and the expected frequencies cast doubt on the validity of the null hypothesis.

As a measure of the extent of discrepancy between the observed and expected frequencies, we calculate the chi-square statistic. In Example 13.1, there are four proportions in the null hypothesis, and the test statistic would be calculated as follows:

$$\chi^2 = \frac{(o_1 - e_1)^2}{e_1} + \frac{(o_2 - e_2)^2}{e_2} + \frac{(o_3 - e_3)^2}{e_3} + \frac{(o_4 - e_4)^2}{e_4}$$

If each observed frequency o_i is close to the corresponding expected frequency e_i, then the test statistic χ^2 is close to 0 and we do not reject H_0. In contrast, large differences between o_i and e_i lead to large values of χ^2 and cast doubt on the validity of the null hypothesis. As with any other test statistic, the problem at this point is to determine how large the test statistic χ^2 should be before we want to reject the null hypothesis.

Characteristics of the goodness-of-fit test

The chi-square goodness-of-fit test is used when testing hypotheses in experiments involving the following characteristics:

1. Every unit in the population falls into exactly one of K categories or cells, denoted $C_1, C_2, \ldots, C_K$.

2. The hypothesized proportion of items in the population that are members of category i is p_i. Thus, if an item is selected randomly, the probability that the item belongs to category i is p_i.
3. The null hypothesis H_0 states that the K proportions equal the K specific values $p_1, p_2, \ldots, p_K$.
4. The alternative hypothesis states that at least one of the proportions specified in H_0 is incorrect.
5. Because every observation belongs to one and only one category, the K proportions sum to 1; that is,

$$p_1 + p_2 + \cdots + p_K = 1$$

To test H_0, we take a random sample of n observations and determine how many observations belong to each of the K categories. These **observed frequencies**, denoted by $o_1, o_2, \ldots, o_K$, must sum to n:

$$o_1 + o_2 + \cdots + o_K = n$$

To test H_0, we compare the observed frequencies with what would be expected if the null hypothesis were true.

DEFINITION   **Expected Frequency**

For the ith category, the **expected frequency** is denoted e_i. If the sample contains n observations, the expected frequency for category i is

$$e_i = np_i \qquad i = 1, 2, \ldots, K$$

The sum of the expected frequencies is n because

$$
\begin{aligned}
e_1 + e_2 + \cdots + e_K &= np_1 + np_2 + \cdots + np_K \\
&= n(p_1 + p_2 + \cdots + p_K) = n
\end{aligned}
$$

The Chi-Square Test Statistic

To test the null hypothesis, we examine how close the observed frequencies $o_1, o_2, \ldots, o_K$ are to the expected frequencies $e_1, e_2, \ldots, e_K$. When they are close, the null hypothesis is supported; that is, the data provide a *good fit* to the proposed model, hence the name "goodness-of-fit" test. If the observed frequencies are quite different from the expected frequencies, the null hypothesis becomes doubtful. The larger the differences between o_i and e_i, the more doubtful we are that the null hypothesis is true.

DEFINITION   **Chi-Square Test Statistic**

Suppose each observation must fall in exactly one of K categories. The chi-square goodness-of-fit test is based on the following **chi-square test statistic**:

$$\chi^2 = \frac{(o_1 - e_1)^2}{e_1} + \frac{(o_2 - e_2)^2}{e_2} + \cdots + \frac{(o_K - e_K)^2}{e_K}$$

If the null hypothesis is true and the sample size is large so that each of the expected frequencies $e_1, e_2, \ldots, e_K$ is 5 or larger, the chi-square test statistic approximately follows a chi-square distribution with $(K - 1)$ degrees of freedom. If the expected frequency for any cell is less than 5, that particular cell should be combined with another cell or the sample size should be increased.

The degrees of freedom parameter $\nu = (K - 1)$ indicates the number of proportions p_i that can be chosen freely. Because the sum of the K proportions must be 1, only $(K - 1)$ of the proportions are free. Once any $(K - 1)$ proportions are specified, the last proportion is determined and thus no longer is free.

The chi-square goodness-of-fit test is summarized in the accompanying box.

The chi-square goodness-of-fit test

We observe a random sample of n observations, where each observation falls into exactly one of K categories. In the population, the hypothesized proportion of observations in category i is denoted by p_i. Assume we want to test the null hypothesis

$$H_0: \quad p_1, p_2, \ldots, p_K \text{ are equal to a set of prespecified values}$$

against the alternative hypothesis

$$H_1: \quad \text{At least one of the proportions is not correct}$$

Assume the level of significance of the test is α. The observed frequencies are $o_1, o_2, \ldots, o_K$. The expected frequencies are obtained using the equation $e_i = np_i$, and each expected frequency should be at least 5.

The test statistic is

$$\chi^2 = \sum_{i=1}^{K} \frac{(o_i - e_i)^2}{e_i}$$

and is distributed as chi-square with $\nu = (K - 1)$ degrees of freedom. The decision rule is

$$\text{Reject } H_0 \text{ in favor of } H_1 \text{ if } \chi^2 > \chi^2_{\alpha,\nu},$$

where $\chi^2_{\alpha,\nu}$ is the critical value such that

$$P(\chi^2 > \chi^2_{\alpha,\nu}) = \alpha$$

EXAMPLE 13.2   **Using the Chi-Square Goodness-of-Fit Test** ─────────────────

In Example 13.1, a random sample of 200 households in Joliet, Illinois, was obtained to test the null hypothesis

$$H_0: \quad p_1 = .20, \ p_2 = .30, \ p_3 = .40, \ p_4 = .10$$

where p_i refers to the proportion of households falling into the ith income category. The observed frequencies are shown in Table 13.1 along with the expected frequencies. Test the null hypothesis using a 1% level of significance.

SOLUTION There are $\nu = (K - 1) = 3$ degrees of freedom. For level of significance $\alpha = .01$, the critical value is $\chi^2_{.01,3} = 11.34$. The expected frequencies, shown in

TABLE 13.1 *Observed and expected frequencies for Example 13.2*

Income category i	o_i	e_i	$o_i - e_i$	$(o_i - e_i)^2$	$(o_i - e_i)^2/e_i$
1	55	40	15	225	5.625
2	65	60	5	25	.417
3	72	80	−8	64	.800
4	8	20	−12	144	7.200
Total	200	200	0		14.042

$$\chi^2 = \Sigma[(o_i - e_i)^2/e_i]$$
$$= 5.625 + .417 + .800 + 7.200 = 14.042$$

Table 13.1, are calculated as follows: $e_1 = np_1 = 200(.20) = 40$, $e_2 = np_2 = 200(.30) = 60$, and so forth. The chi-square test statistic is

$$\chi^2 = \frac{(55 - 40)^2}{40} + \frac{(65 - 60)^2}{60} + \frac{(72 - 80)^2}{80} + \frac{(8 - 20)^2}{20}$$
$$= 5.625 + .417 + .800 + 7.200 = 14.042$$

Note that this value is equal to the sum of the last column in Table 13.1. We reject H_0 because $\chi^2 = 14.042$ exceeds the critical value $\chi^2_{\alpha,\nu} = 11.34$, as illustrated in Figure 13.1. The data provide strong evidence that the income distribution in Joliet is different from the income distribution in the United States.

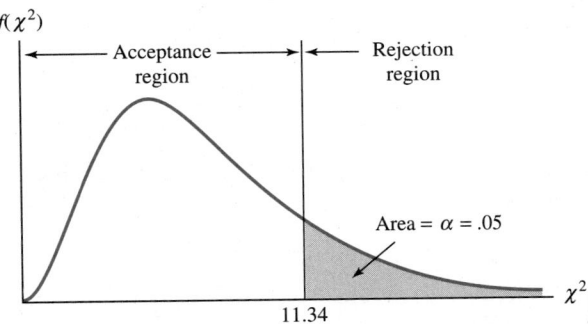

FIGURE 13.1 *Chi-square distribution for Example 13.2*

Testing Whether a Random Variable Has a Poisson Distribution

The chi-square goodness-of-fit test can be used to test whether there is a significant difference between an observed frequency distribution and any theoretical probability distribution. If the hypothesized distribution is a discrete distribution such as the Poisson distribution, the test can be carried out as described in Example 13.3.

EXAMPLE 13.3   **Testing Whether a Random Variable Has a Poisson Distribution**

It is hypothesized that the number of breakdowns per month of a computer system at a major university follows a Poisson distribution with mean $\mu = 2$. The data in Table 13.2 show the observed number of breakdowns per month during a sample of 100 months. Use a 5% level of significance and test the null hypothesis

H_0:   The population distribution of breakdowns is Poisson with mean $\mu = 2$

against the alternative hypothesis

H_1:   The population does not follow a Poisson distribution with mean $\mu = 2$

TABLE 13.2   *Observed and expected frequencies for Example 13.3*

Breakdowns	o_i	e_i	$o_i - e_i$	$(o_i - e_i)^2$	$(o_i - e_i)^2/e_i$
0	14	13.5	.5	.25	.019
1	20	27.1	-7.1	50.41	1.860
2	34	27.1	6.9	47.61	1.757
3	22	18.0	4.0	16.00	.889
4	5	9.0	-4.0	16.00	1.778
5 or more	5	5.3	$-.3$	.09	.017
Total	100	100.0			6.320

SOLUTION   Before we can determine the expected frequencies using the formula $e_i = np_i$, we must find the probabilities p_i. For the Poisson distribution, we have

$$P(X = x) = \frac{e^{-\mu}\mu^x}{x!}$$

From this formula (or from Table A.4), using $\mu = 2.0$, we obtain the following probabilities:

x	0	1	2	3	4	5 or more
$P(X = x)$	.1353	.2707	.2707	.1804	.0902	.0527

The expected frequencies e_i are obtained by using the rule $e_i = np_i$ where $n = 100$. For example, the expected frequency of $X = 3$ is

$$np_i = 100(.1804) = 18.04$$

The expected frequencies e_i are also shown in Table 13.2 (the expected frequencies have been rounded off to simplify the calculations). Note that each expected frequency exceeds 5.

There are $K = 6$ categories, so there are $\nu = (K - 1) = 5$ degrees of freedom. For $\alpha = .05$, the critical value is $\chi^2_{.05,5} = 11.07$. The chi-square test statistic is

$$\chi^2 = \sum_{i=1}^{6} \frac{(o_i - e_i)^2}{e_i}$$

$$= \frac{(14 - 13.5)^2}{13.5} + \frac{(20 - 27.1)^2}{27.1} + \cdots + \frac{(5 - 5.3)^2}{5.3} = 6.32$$

The test statistic falls in the acceptance region because the value 6.32 is less than the critical value 11.07. Hence, the null hypothesis is not rejected. Because the chi-square test statistic falls in the acceptance region, the evidence is not strong enough to lead us to conclude that the distribution of the number of computer breakdowns per month is different from a Poisson distribution with $\mu = 2$.

Testing Whether a Population Has a Normal Distribution

When using the chi-square goodness-of-fit test to determine whether a hypothesized distribution is continuous, such as the normal distribution, we must first divide the range of the theoretical density function into a number of intervals, or cells, and we use the areas above these intervals to represent the theoretical cell probabilities p_i. The expected cell frequencies are determined from the equation $e_i = np_i$, where n is the sample size. We then test whether the observed frequencies o_i differ significantly from the expected frequencies e_i by calculating the chi-square test statistic.

The following example illustrates how to test the null hypothesis that a sample of observations was selected from a normal population having a specific mean and variance.

EXAMPLE 13.4 **Testing Whether a Population Has a Normal Distribution with Given Mean and Variance**

The manufacturer of robots used in car assembly claims that the time required for the robots to perform a certain task follows a normal distribution with mean $\mu = 5$ seconds and variance $\sigma^2 = .04$. To test this hypothesis, we obtain a random sample of $n = 200$ observations. Use a 5% level of significance and test the null hypothesis that the times required to complete the task are distributed $N(5, .04)$ against the alternative hypothesis that the population is not $N(5, .04)$.

SOLUTION First we divide the range of the hypothesized distribution into K intervals. Suppose we choose $K = 10$ and select the 10 intervals so that each category has probability equal to .10, as shown in Figure 13.2. (The fact that each category has the same probability is unimportant.)

We want to determine the values $x_1, x_2, \ldots, x_9$ such that each shaded area and each unshaded area has probability .10. In Figure 13.2, the value x_5 is the mean of the distribution, so $x_5 = 5$ seconds.

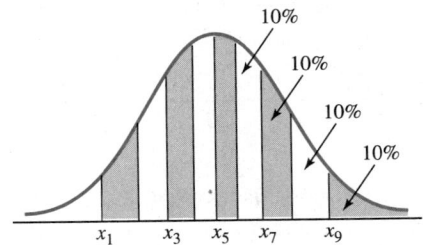

FIGURE 13.2 *Ten equal areas under the normal curve for $N(5, .04)$*

Now let us obtain the value of, say, x_7. In Figure 13.2, the area between x_5 and x_7 is .20. We obtain the value x_7 by first finding the standard normal variate such that the area between 0 and the variate is .20. From Table A.5, this standard normal variate is $z_7 \approx .52$ (see Figure 13.3). We then obtain x_7 by solving the equation

$$z_7 = .52 = \frac{x_7 - 5}{\sqrt{.04}}$$

The solution is $x_7 = 5.104$.

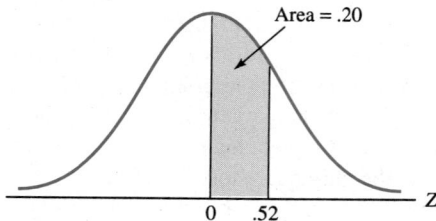

FIGURE 13.3 *Area under the standard normal curve for Example 13.4*

Similarly, to find the value of, say, x_1, we observe that the area from x_1 to x_5 is .40. The standard normal variate associated with x_1 is $z_1 = -1.28$. Thus, we obtain

$$z_1 = -1.28 = \frac{x_1 - 5}{\sqrt{.04}} \quad \text{and} \quad x_1 = 4.744$$

By proceeding in a similar manner, we can find the values of $x_1, x_2, \ldots, x_9$ as follows:

$$
\begin{array}{lll}
x_1 = 4.744 & x_2 = 4.832 & x_3 = 4.896 \\
x_4 = 4.950 & x_5 = 5.000 & x_6 = 5.050 \\
x_7 = 5.104 & x_8 = 5.168 & x_9 = 5.256
\end{array}
$$

These results indicate that we expect 10% of the observations to be less than 4.744, 10% to be between 4.744 and 4.832, and so forth. The expected frequency for each cell is

$$e_i = np_i = 200(.10) = 20$$

The expected frequencies are shown in Table 13.3 along with the observed frequencies based on the sample of 200 observations.

Because there are $K = 10$ categories, there are $\nu = (K - 1) = 9$ degrees of freedom. For $\alpha = .05$, the critical value is $\chi^2_{.05,9} = 16.92$. The chi-square test statistic is

$$\chi^2 = \frac{(18 - 20)^2}{20} + \frac{(23 - 20)^2}{20} + \cdots + \frac{(15 - 20)^2}{20}$$
$$= 4.10$$

We do not reject H_0 because $\chi^2 = 4.10$ is less than 16.92 and thus falls in the acceptance region. The evidence is not strong enough to lead us to conclude that the distribution of the time required to complete a task is different from a normal distribution with mean 5 and variance .04.

TABLE 13.3 *Observed and expected frequencies for Example 13.4*

Time (in seconds)	o_i	e_i	$o_i - e_i$	$(o_i - e_i)^2$	$(o_i - e_i)^2/e_i$
Less than 4.744	18	20	-2	4	.20
4.744–4.832	23	20	3	9	.45
4.832–4.896	19	20	-1	1	.05
4.896–4.950	22	20	2	4	.20
4.950–5.000	24	20	4	16	.80
5.000–5.050	17	20	-3	9	.45
5.050–5.104	18	20	-2	4	.20
5.104–5.168	23	20	3	9	.45
5.168–5.256	21	20	1	1	.05
5.256 or more	15	20	-5	25	1.25
Total	200	200	0		4.10

In the goodness-of-fit tests discussed thus far, the appropriate number of degrees of freedom was $\nu = (K - 1)$, where K is the number of categories. This is because the only constraint placed on the expected frequencies was that they had to sum to n, the total number of observations. When either of the population parameters (the mean or variance) has been estimated from the sample of data, the number of degrees of freedom must be reduced by the number of parameters that have been estimated.

Exercises

Statistical Concepts

13.1.1 *True or false:* The appropriate number of degrees of freedom to use when performing the chi-square goodness-of-fit test is $(n - 1)$, where n is the sample size. Explain.

13.1.2 *True or false:* When the chi-square goodness-of-fit test is performed, each cell frequency should be at least 5. Explain.

13.1.3 *True or false:* When the chi-square goodness-of-fit test is performed, the minimum value of chi-square is 0 and the occurrence of such a value would indicate a perfect fit. Explain.

13.1.4 Describe the characteristics of the goodness-of-fit test.

13.1.5 *True or false:* When the chi-square goodness-of-fit test is performed, the critical value of the chi-square test statistic increases as the number of cells increases. Explain.

13.1.6 Explain why the typical term in the chi-square statistic is $(o_i - e_i)^2/e_i$ rather than $(o_i - e_i)^2$.

Statistical Drills

13.1.7 Suppose you are performing a chi-square goodness-of-fit test. Find the critical value of the chi-square test statistic in each of the following situations:
  a. Number of cells = 6; $\alpha = .01$ **b.** Number of cells = 6; $\alpha = .05$
  c. Number of cells = 6; $\alpha = .10$ **d.** Number of cells = 9; $\alpha = .01$
  e. Number of cells = 9; $\alpha = .05$ **f.** Number of cells = 9; $\alpha = .10$

13.1.8 A random sample of $n = 500$ observations was obtained. Each observation fell into one of $K = 5$ cells. The numbers of observations in each cell were as follows:

Cell	1	2	3	4	5
n_i	125	104	89	92	90

Test the hypothesis H_0: $p_1 = p_2 = p_3 = p_4 = p_5 = .20$. Let $\alpha = .05$.
a. What is the appropriate number of degrees of freedom?
b. Determine the critical value of the chi-square test statistic.
c. Find each expected frequency.
d. Calculate the chi-square test statistic.
e. Perform the test.

13.1.9   A random sample of $n = 400$ observations was obtained. Each observation fell into one of $K = 4$ cells. The numbers of observations in each cell were as follows:

Cell	1	2	3	4
n_i	45	78	210	67

Test the hypothesis H_0: $p_1 = .10$, $p_2 = .20$, $p_3 = .50$, $p_4 = .20$. Let $\alpha = .05$.
a. What is the appropriate number of degrees of freedom?
b. Determine the critical value of the chi-square test statistic.
c. Find each expected frequency.
d. Calculate the chi-square test statistic.
e. Perform the test.

Statistical Applications

13.1.10   Suppose we toss a fair die 240 times and count how many times each value (1 through 6) occurs. We obtain the following results:

Value	1	2	3	4	5	6
Observed frequency	36	47	35	42	44	36

a. Find the expected frequencies if the die is fair.
b. Calculate the value of the chi-square statistic.
c. Find the critical value $\chi^2_{\alpha,\nu}$ to test the hypothesis that the die is fair using a 5% level of significance.
d. Test the hypothesis that the die is fair using $\alpha = .05$.

13.1.11   At the last census, 20% of the people in a town were less than 16 years of age, 20% were aged 16 to 30, 25% were aged 31 to 45, 15% were aged 46 to 60, and 20% were over 60. The planning commission wants to determine whether the age structure of the town has changed since the last census. A random sample of 200 residents of the town yields the following data:

Age (in years)	Less than 16	16–30	31–45	46–60	Over 60
Observed frequency	30	31	37	50	52

Test the hypothesis that the age structure of the town has not changed since the last census. Use a 1% level of significance.
a. Find the expected frequencies if the age structure has not changed.
b. Calculate the value of the chi-square statistic.
c. Find the critical value $\chi^2_{\alpha,\nu}$ to test the hypothesis that the age structure has not changed using a 1% level of significance.
d. Test the hypothesis using $\alpha = .01$.

13.1.12 A civil service exam is given to 200 job applicants. In recent years, grades have followed a normal distribution with mean 66.5 and variance 100. The grades of the job applicants are distributed as follows:

Grade	Under 49.5	49.5–59.5	59.5–69.5	69.5–79.5	79.5 and over
Observed frequency	22	38	56	50	34

    a. Graph the observed frequencies on a histogram. Does the histogram appear to have the shape of a normal distribution with mean 66.5 and variance 100?

    b. Test the hypothesis that the data came from a normal distribution with mean 66.5 and variance 100. Use a 5% level of significance.

13.1.13 A gambler claims that horses having starting positions near the rail have an advantage in a race because they have a shorter distance to run. A sample of 120 races is examined, each of which involved six horses. If starting position is not a factor, then each starting position should have produced 20 winners on the average. The following data show the number of winners from each starting position:

Starting position	1	2	3	4	5	6
Number of winners	25	22	18	19	21	15

Use the chi-square test to determine whether starting position is a significant factor. Assume a 5% level of significance.

13.1.14 Over a long period of time, the grades at a university have been distributed as follows: 15% A's, 20% B's, 40% C's, 20% D's, and 5% F's. Last semester a random sample of 1,000 grades showed 190 A's, 220 B's, 370 C's, 180 D's, and 40 F's. Determine at a 5% level of significance whether the grading pattern has changed.

13.1.15 Over a long period of time at a suburban restaurant, 60% of the customers ordered coffee, 20% ordered tea, 15% ordered milk, and 5% ordered cola. The owners opened a new restaurant in the business section of town, where they think that the drinking preferences of the customers may be different. In a random sample of 1,000 customers at the new restaurant, 550 ordered coffee, 240 ordered tea, 150 ordered milk, and 60 ordered cola. Determine at a 5% level of significance whether drinking patterns are different in town than at the suburban restaurant.

13.1.16 An investor is planning to buy some property in the heart of town to build a parking garage. The profitability of the project depends on the parking patterns of the potential customers at the garage. The investor assumes that 40% of the customers will stay less than one hour, 20% will stay from 1 to 2 hours, 15% will stay from 2 to 4 hours, and 25% will stay more than 4 hours. At a nearby lot, a random sample of 200 customers produced the following distribution:

Time of parking (in hours)	0–1	1–2	2–4	4 or more
Observed frequency	100	35	25	40

Should we reject the investor's assumptions? Use a 1% level of significance.

13.1.17 The safety engineer at a manufacturing firm claims that most plant accidents occur near quitting time, when employees are getting tired and are anxious to go home. Plant management will introduce new safety measures, such as an additional afternoon coffee break, if the engineer can prove his claim. The engineer examines a random sample of 80 plant accidents. The times of occurrence of the accidents were as follows:

Time	Number of accidents
8 a.m.–10 a.m.	16
10 a.m.–noon	16
Noon–1 p.m. (lunch)	—
1 p.m.–3 p.m.	21
3 p.m.–5 p.m.	27

Use a 5% level of significance and test whether the accidents are uniformly distributed throughout the day.

13.2   Tests of Independence and Contingency Tables

Suppose we have a random sample of n observations from some population where each observation is cross-classified according to two qualitative characteristics, or variables. An important application of the chi-square distribution involves testing the null hypothesis that, in such a population, the value taken by one variable is independent of the value taken by the other variable. Such a test is called a *chi-square test of independence.* For example, a random sample of employees is classified according to education and occupation. The data show whether an individual has a grade school, high school, or college education and whether the individual's occupation is professional, blue collar, or other. It seems reasonable to conjecture that the two variables Education and Occupation are not independent. As another example, suppose that a sample of individuals is classified according to whether they live in Canada, the United States, or Mexico and according to whether they prefer to watch hockey, baseball, or soccer on television. Once again, it seems reasonable to conjecture that the two variables Nationality and Favorite sport are not independent.

EXAMPLE 13.5   **Testing a Hypothesis About the Independence of Variables** ————————

A company planning a television advertising campaign wants to determine which TV shows its target audience watches. Suppose the company wants to test the null hypothesis

H_0:   Choice of TV program that an individual watches is independent
of the individual's income

against the alternative hypothesis

H_1:   Income and choice of TV program are not independent

In a random sample of $n = 500$ people, each person is classified into one of three income categories (low, medium, or high). In addition, each individual is classified according to which TV program the person watched (a hockey game, a movie, or the news).

The sample data are presented in Table 13.4. This table, which is a two-way classification table containing three rows and three columns, is referred to as a 3×3 *contingency table.* In general, if one of the variables contains H categories and the other variable contains K categories, then we obtain an $H \times K$ contingency table. Each combination of two attributes represents a *cell* in the contingency table.

TABLE 13.4 *Contingency table for Example 13.5*

	TYPE OF TV SHOW			
Income	Hockey	Movie	News	Total
Low	143	70	37	250
Medium	90	67	43	200
High	17	13	20	50
Total	250	150	100	500

In a contingency table, the value in row i and column j is denoted o_{ij} and called an *observed frequency*. For example, the value in row 3, column 2 of Table 13.4 is $o_{32} = 13$. This observed frequency represents the number of people who watched the movie on TV and had a high income. This contingency table contains nine cells.

Calculating Expected Frequencies

If the contingency table contains H rows and K columns, then there are $H \times K$ observed frequencies. Corresponding to the $H \times K$ contingency table is another $H \times K$ table containing the expected frequency for each cell based on the assumption that the null hypothesis is true. To compute the expected frequencies under the null hypothesis of independence, we must first determine the probability that an observation will fall into a particular cell if H_0 is true.

Let the K column categories be denoted $C_1, C_2, \ldots, C_K$, and let the column totals be denoted $c_1, c_2, \ldots, c_K$. Similarly, let the H row categories be denoted $R_1, R_2, \ldots, R_H$, and let the H row totals be denoted $r_1, r_2, \ldots, r_H$. We use the following rule: If events R_i and C_j are independent, then

$$P(R_i \cap C_j) = P(R_i)P(C_j)$$

Although we do not know the true population probabilities $P(R_i \cap C_j)$, $P(R_i)$, and $P(C_j)$, we can estimate them by using our sample of n observations to calculate relative frequencies. Because the number of observations that fall in row i is r_i, the relative frequency of row i is r_i/n. We use this value to estimate $P(R_i)$. Similarly, we use the relative frequency c_j/n to estimate the probability $P(C_j)$.

We want to test the null hypothesis

$$H_0: \quad P(R_i \cap C_j) = P(R_i)P(C_j) \quad \text{for all } i \text{ and } j$$

against the alternative hypothesis

$$H_1: \quad P(R_i \cap C_j) \neq P(R_i)P(C_j) \quad \text{for at least one } i \text{ and } j$$

Let e_{ij} denote the expected frequency in row i, column j. The expected frequency e_{ij} is

$$e_{ij} = nP(R_i \cap C_j)$$

If H_0 is true, this becomes

$$e_{ij} = nP(R_i)P(C_j)$$

If we use $(r_i/n)(c_j/n)$ to estimate the unknown probability $P(R_i \cap C_j) = P(R_i)P(C_j)$, the expected frequency becomes

$$e_{ij} = n(r_i/n)(c_j/n) = r_i c_j/n$$

Formula for expected frequencies

The estimated expected frequency for the cell in row i and column j is

$$e_{ij} = \frac{r_i c_j}{n}$$

where r_i is the number of observations in row i and c_j is the number of observations in column j.

The Chi-Square Test Statistic

A test of the null hypothesis that the variables are independent of one another is based on the magnitudes of the differences between the observed frequencies and the expected frequencies. Large differences between o_{ij} and e_{ij} provide evidence that the null hypothesis is false. The test is based on the following chi-square test statistic:

$$\chi^2 = \sum_{i=1}^{H} \sum_{j=1}^{K} \frac{(o_{ij} - e_{ij})^2}{e_{ij}}$$

We again impose the rule that each of the expected cell frequencies must be at least 5. If the null hypothesis is true, this chi-square test statistic approximately follows the chi-square distribution. The appropriate number of degrees of freedom represents the number of expected frequencies that can be chosen freely, provided the row and column totals of the expected frequency table are identical to the row and column totals of the observed frequency table. If we know the expected frequencies in the first $(H - 1)$ rows and $(K - 1)$ columns of the $H \times K$ table, then the remaining expected frequencies are uniquely determined. It follows that the appropriate number of degrees of freedom is

$$\nu = (H - 1)(K - 1)$$

Performing the chi-square test of independence

We take a random sample of n observations and classify each observation according to two variables. In the population, let $P(R_i)$ denote the probability that an observation possesses attribute R_i and let $P(C_j)$ denote the probability that an observation possesses attribute C_j. The null hypothesis states that the two classification schemes are independent of one another. We wish to test the null hypothesis

$$H_0: \quad P(R_i \cap C_j) = P(R_i)P(C_j) \quad \text{for } i = 1, 2, \ldots, H; \quad j = 1, 2, \ldots, K$$

against the alternative hypothesis

$$H_1: \quad P(R_i \cap C_j) \neq P(R_i)P(C_j) \quad \text{for some } i \text{ and } j$$

Let o_{ij} denote the number of observations in the cell in row i and column j of a contingency table. The expected frequency for this cell is e_{ij}, which is estimated using the formula

$$e_{ij} = r_i c_j / n$$

where r_i is the observed row total for row i and c_j is the observed column total for column j. Each expected frequency should be at least 5.

The chi-square test statistic is

$$\chi^2 = \sum_{i=1}^{H} \sum_{j=1}^{K} \frac{(o_{ij} - e_{ij})^2}{e_{ij}}$$

which has a chi-square distribution with $\nu = (H - 1)(K - 1)$ degrees of freedom. To test H_0 against H_1 using a level of significance α, use the decision rule

Reject H_0 in favor of H_1 if $\chi^2 > \chi^2_{\alpha,\nu}$.

EXAMPLE 13.6   **Using the Chi-Square Test of Independence**

Table 13.4 showed the observed frequencies relating to the incomes and choice of TV show for a random sample of $n = 500$ individuals described in Example 13.5. Let $\alpha = .05$ and test the null hypothesis that income and choice of TV program are independent of one another.

SOLUTION   The observed row and column frequencies are $r_1 = 250$, $r_2 = 200$, $r_3 = 50$, $c_1 = 250$, $c_2 = 150$, and $c_3 = 100$.

Table 13.5 contains the estimated expected cell frequencies along with the observed frequencies. For example, the expected frequency in row 2, column 3 is $e_{23} = 40$. This expected frequency is calculated as follows:

$$e_{23} = \frac{r_2 c_3}{n} = \frac{(200)(100)}{500} = 40$$

The other estimated expected frequencies are calculated in a similar manner.

TABLE 13.5 *Contingency table for Example 13.6*

Income	TYPE OF TV SHOW			Total
	Hockey	Movie	News	
Low				
o_{ij}	143	70	37	250
e_{ij}	125	75	50	250
Medium				
o_{ij}	90	67	43	200
e_{ij}	100	60	40	200
High				
o_{ij}	17	13	20	50
e_{ij}	25	15	10	50
Total				
o_{ij}	250	150	100	500
e_{ij}	250	150	100	500

All of the expected frequencies exceed 5, so the use of the chi-square test is appropriate. The chi-square test statistic is

$$\chi^2 = \frac{(143 - 125)^2}{125} + \frac{(70 - 75)^2}{75} + \frac{(37 - 50)^2}{50} + \frac{(90 - 100)^2}{100}$$
$$+ \frac{(67 - 60)^2}{60} + \frac{(43 - 40)^2}{40} + \frac{(17 - 25)^2}{25} + \frac{(13 - 15)^2}{15}$$
$$+ \frac{(20 - 10)^2}{10} = 21.174$$

The appropriate number of degrees of freedom is

$$\nu = (H - 1)(K - 1) = (3 - 1)(3 - 1) = 4$$

For $\alpha = .05$, the critical value is $\chi^2_{.05,4} = 9.49$. We reject the null hypothesis that program choice and income level are independent because $\chi^2 = 21.174$ falls in the rejection region. See Figure 13.4.

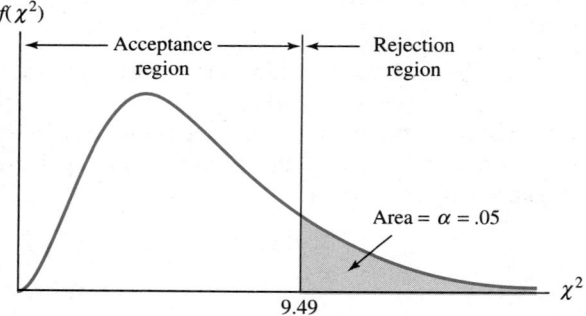

FIGURE 13.4   *Chi-square distribution for Example 13.6*

EXAMPLE 13.7   **Using the Chi-Square Test of Independence**

The manager of a bank is examining the mortgage payments made by customers of the bank. A payment is classified as good if it arrives on time and as delinquent if it arrives late or is not paid. In addition, the customer's income is classified as low, medium, or high. Table 13.6 shows the number of good and delinquent payments according to the person's income category. Use a 5% level of significance and test the null hypothesis

H_0:   The probability of a payment being good or delinquent
is independent of the person's income

against the alternative hypothesis

H_1:   The probability of a payment being good or delinquent
is not independent of the person's income

TABLE 13.6 *Observed frequencies for Example 13.7*

Payment	INCOME LEVEL			Total
	Low	Medium	High	
Good	45	50	65	160
Delinquent	5	20	15	40
Total	50	70	80	200

SOLUTION   Our first task is to estimate the expected frequencies. We use the rule $e_{ij} = r_i c_j / n$. For example, $e_{23} = (40)(80)/(200) = 16$. The expected frequencies are shown in Table 13.7.

TABLE 13.7 *Expected frequencies for Example 13.7*

Payment	INCOME LEVEL			Total
	Low	Medium	High	
Good	40	56	64	160
Delinquent	10	14	16	40
Total	50	70	80	200

All of the expected frequencies exceed 5, so the use of the chi-square test is appropriate. The chi-square test statistic is

$$\chi^2 = \frac{(45 - 40)^2}{40} + \frac{(5 - 10)^2}{10} + \frac{(50 - 56)^2}{56} + \frac{(20 - 14)^2}{14}$$

$$+ \frac{(65 - 64)^2}{64} + \frac{(15 - 16)^2}{16} = 6.418$$

The appropriate number of degrees of freedom is

$$\nu = (H - 1)(K - 1) = (2 - 1)(3 - 1) = 2$$

For $\nu = 2$ degrees of freedom and $\alpha = .05$, the critical value is $\chi^2_{.05,2} = 5.99$. We reject H_0 because $\chi^2 = 6.418$ exceeds the critical value of 5.99, as shown in Figure 13.5.

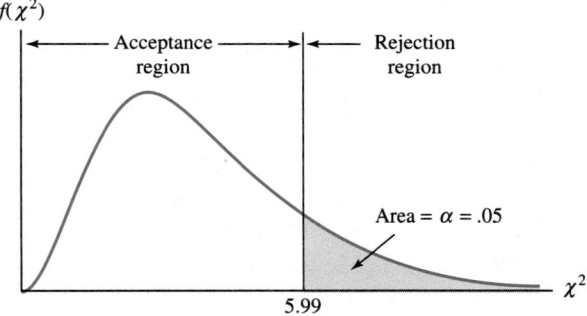

FIGURE 13.5 *Chi-square distribution for Example 13.7*

Does this result seem reasonable? That is, based on the observed frequencies in Table 13.6, does it appear that the payment of the loan depends on the level of income? In the entire sample of 200 mortgages, 40 (or 20%) were delinquent. In Table 13.6, only 5 of 50 (or 10%) of the low-income individuals were delinquent, whereas 20 of 70 (or 29%) medium-income individuals were delinquent. Thus it appears that in our sample too many medium-income people and too few low-income people were delinquent to claim that the probability of being delinquent is independent of income.

Some Comments

The chi-square test for independence is useful in helping to determine whether a relationship exists between two variables, say, between choice of TV program and income or between expenditures on a certain type of product and income, but it does not enable us to estimate or predict the values of one variable based on the value of the other. If it is determined that dependence does exist between two quantitative variables, then the techniques of regression analysis developed in Chapters 15, 16, and 17 are useful in helping us find a mathematical formula that expresses the nature of the mathematical relationship.

If we think two quantitative variables Y and X are related, we might want to estimate some functional relationship $Y = f(X)$ or test the hypothesis that the variables Y and X are related in a certain way.

Finally, when we perform the chi-square test, the critical region of the test statistic is placed in the right tail of the distribution; thus when the chi-square test statistic is large, the null hypothesis is rejected. Some statisticians have argued that part of the critical region should be placed in the left tail of the distribution, so that the null hypothesis is rejected if the computed value of χ^2 is too close to 0. They argue that values of χ^2 close to 0 indicate that the observed and expected frequencies agree "too well," possibly because the data have been incorrectly gathered, miscalculated, or erroneously reported.

Actually, small values of χ^2 should not be interpreted as evidence that the null hypothesis is false. Instead, they should prompt the investigator to examine the data carefully to see whether the observed and expected frequencies have been reported correctly. If no sources of error or bias can be found, the null hypothesis should not be rejected.

Exercises

Statistical Concepts

13.2.1   When constructing a contingency table, explain how to calculate a theoretical cell probability.

13.2.2   *True or false:*   In a contingency table, a cell probability is the product of the respective row and column probabilities. Explain.

13.2.3   When a contingency table is constructed, explain how to calculate an expected cell frequency.

13.2.4   *True or false:*   When a contingency table is constructed, the sum of the expected cell frequencies is equal to the sum of the observed cell frequencies. Explain.

13.2.5   *True or false:*   When a contingency table is constructed, the appropriate number of degrees of freedom is $(HK - 1)$. Explain.

13.2.6   *True or false:*   When a contingency table is constructed, the null hypothesis is that all cells are equally likely. Explain.

13.2.7   Explain why the critical region for the chi-square test statistic when we perform a test of independence is placed in the right tail of the distribution.

13.2.8   Suppose we perform a chi-square test of independence using $\alpha = .05$. How does the critical value of the chi-square test statistic vary as the number of rows and columns increases? Does the critical value get larger or smaller?

13.2.9   *True or false:*   Suppose we perform a chi-square test of independence using $\alpha = .05$. How does the critical value of the chi-square test statistic vary as the number of observations increases?

Statistical Drills

13.2.10   Suppose we perform a chi-square test of independence using $\alpha = .05$. Suppose we have $n = 200$ observations, $H = 6$ rows, and $K = 5$ columns. Suppose the observed value of the test statistic is $\chi^2 = 40.4$.
 a. Determine the appropriate number of degrees of freedom.
 b. Determine the critical value of the chi-square test statistic.
 c. Perform the test.
 d. For a chi-square test, the p-value of the test is the probability of obtaining a value of chi-square greater than or equal to the observed value. Use the chi-square table and find (approximately) the p-value of the test.

13.2.11   A sample of 250 observations was classified into the following 3×2 contingency table. Do the data support the null hypothesis that the two methods of classification are independent? Use $\alpha = .05$.

	CLASSIFICATION II	
Classification I	**A**	**B**
A	33	53
B	22	45
C	56	41

 a. Determine the appropriate number of degrees of freedom.
 b. Find the expected cell frequencies.
 c. Determine the critical value of the chi-square test statistic.
 d. Perform the test.
 e. What is the approximate p-value of the test?

13.2.12   A sample of 500 observations was classified into the following 4×3 contingency table. Do the data support the null hypothesis that the two methods of classification are independent? Use $\alpha = .05$.

	CLASSIFICATION II		
Classification I	**A**	**B**	**C**
A	27	38	62
B	34	54	39
C	46	31	44
D	43	35	47

 a. Determine the appropriate number of degrees of freedom.
 b. Find the expected cell frequencies.

   c. Determine the critical value of the chi-square test statistic.

   d. Perform the test.

   e. What is the approximate *p*-value of the test?

13.2.13   In the following situation, decide whether to accept or reject the null hypothesis that the variables are independent. The variables are Age of person and the person's Favorite brand of soft drink. Age has five categories and there are six brands of soft drink. The value of chi-square is 19.2. Let $\alpha = .05$.

13.2.14   In the following situation, decide whether to accept or reject the null hypothesis that the variables are independent. The variables are Newspaper read and the person's Political affiliation. Newspaper has three categories and there are three categories for political affiliation. The value of chi-square is 16.1. Let $\alpha = .01$.

13.2.15   In the following situation, decide whether to accept or reject the null hypothesis that the variables are independent. The variables are how the person would rate the Performance of the president (good, fair, poor) and the person's Political affiliation. There are three categories for political affiliation. The value of chi-square is 19.2. Let $\alpha = .01$.

Statistical Applications

13.2.16   We take a sample of 500 people, all of whom drive approximately 10,000 miles per year. The following data classify the individuals according to age and the number of auto accidents each has had during the last 3 years:

Number of accidents	AGE (IN YEARS)		
	Under 30	30–40	Over 40
0	61	109	180
1	27	25	48
More than 1	12	16	22

Test the hypothesis that the number of accidents is independent of the age of the driver using a 5% level of significance.

   a. Find the expected frequencies based on the assumption that the number of accidents is independent of age.

   b. Find the appropriate number of degrees of freedom to perform the test.

   c. Calculate the chi-square test statistic.

   d. What is the critical value $\chi^2_{\alpha,\nu}$ if $\alpha = .05$?

   e. Perform the test.

13.2.17   A recent Supreme Court decision dealt with the legality of abortion. A national women's organization has claimed that men and women tend to have different views concerning abortion. A random sample of men and women responded as follows:

Gender	RESPONSE		
	In favor	Opposed	Undecided
Men	86	74	40
Women	119	65	22

Test the hypothesis that there is no difference between men's and women's views on abortion. Use a 5% level of significance.

13.2.18   A standard argument of students in all universities is that grades in certain departments tend to be higher or lower than grades in other departments. To test this claim, samples of grades from three departments were obtained. Students' grades were distributed as follows:

Department	A	B	C	D	F
Math	50	140	200	180	20
Economics	40	50	90	60	12
History	60	100	180	110	25

Test the hypothesis that there is no difference in the grading policies of the three departments. Use a 5% level of significance.

13.2.19   A survey was taken to determine which TV network's evening news people prefer. Responses were classified according to the viewer's age for advertising purposes. Use a 5% level of significance and determine whether age and network preference are independent.

Age (in years)	ABC	NBC	CBS
Under 20	30	20	20
20–30	60	70	80
30–40	100	110	70
Over 40	110	100	80

13.2.20   A survey of 200 stockholders of a corporation was taken to determine whether they favored distributing the profits as dividends or reinvesting the profits in the firm. Test whether the opinions are independent of the number of shares owned. Use a 5% level of significance.

Number of shares owned	Pay dividends	Reinvest	Indifferent
0–99	35	17	8
100–249	31	23	6
250 or more	34	30	16

13.2.21   The Department of Transportation wants to determine whether the occurrence of fatal accidents is dependent upon the weight of the car. The following data show the outcomes of 300 accidents:

Type of accident	WEIGHT OF CAR (IN POUNDS)		
	0–1,599	1,600–2,299	2,300 or more
Fatal	72	20	18
Nonfatal	128	40	22

Let $\alpha = .05$ and test whether the variables are independent of one another.

13.2.22   A congressman sent questionnaires to 600 voters seeking their opinions on certain critical issues. One of the congressman's aides has warned him not to attach too much importance to the results because they do not represent a random sample of all voters' opinions since more poor than rich people tend to return such questionnaires. Let $\alpha = .01$ and test whether the return of the questionnaire is independent of income.

Questionnaire returned	INCOME (IN DOLLARS)		
	0–9,999	10,000–29,999	30,000 and over
Yes	90	50	60
No	110	150	140

13.3   Computer Applications

By using a CROSSTABS command with the appropriate options, we can get the SPSSX program to perform a chi-square test of independence for us. As we showed in Chapter 4, a CROSSTABS command can be used to generate an $H \times K$ contingency table, which shows the observed frequencies for each cell. By using the OPTIONS subcommand with the CROSSTABS command, we can get the program to print out the expected cell frequencies, the value of the chi-square test statistic, and the p-value associated with the chi-square test statistic. Thus, the problem of performing a chi-square test of independence can be reduced to entering the data into the SPSSX program and inserting the appropriate CROSSTABS and OPTIONS commands.

Refer to the data in Table 2.1 in Chapter 2, the characteristics of a random sample of 120 employees at the Computech Corporation. Suppose we want to test the null hypothesis that the division in which an individual is employed is independent of the individual's academic degree. To obtain the two-way contingency table containing the observed frequencies o_{ij}, issue the command

```
CROSSTABS TABLES = DEGREE BY DIV
```

To obtain the expected frequencies e_{ij} as well, issue the additional command

```
OPTIONS 14
```

Finally, to obtain the value of the chi-square test statistic and the associated p-value, issue the command

```
STATISTICS 1
```

Figure 13.6 shows the SPSSX computer output obtained by issuing the three commands

```
CROSSTABS TABLES = DEGREE BY DIV
OPTIONS 14
STATISTICS 1
```

on the data in Table 2.1. In the figure, each cell in the contingency table contains two numbers. The top entry is the observed cell frequency o_{ij}, and the second entry is the

```
                    DIV                         Page 1 of 1
             Count |
             Exp Val |OFFICE   MANUFACT SALES
                    |         URING
                    |  1.00|    2.00|    3.00| Row
DEGREE       -------+--------+--------+--------+  Total
      1.00   |   35  |   53  |    1  |     89
  HIGH SCHOOL |  28.2 |  43.0 |  17.8 |   74.2%
             +--------+--------+--------+
      2.00   |    3  |    5  |   15  |     23
  COLLEGE    |   7.3 |  11.1 |   4.6 |   19.2%
             +--------+--------+--------+
      3.00   |    0  |    0  |    8  |      8
  POST-GRAD  |   2.5 |   3.9 |   1.6 |    6.7%
             +--------+--------+--------+
            Column     38      58      24      120
            Total     31.7%   48.3%   20.0%   100.0%

       Chi-Square                Value          DF       Significance
  ---------------------      -----------       ----      ------------

  Pearson                      81.21947          4          .00000
  Likelihood Ratio             79.42636          4          .00000
  Mantel-Haenszel test for     49.34632          1          .00000
      linear association

  Minimum Expected Frequency -    1.600
  Cells with Expected Frequency < 5 -      4 OF      9 ( 44.4%)


  Number of Missing Observations:   0
```

FIGURE 13.6   *SPSSX-generated contingency table cross-tabulating DEGREE and DIV from data in Table 2.1*

expected cell frequency e_{ij} assuming that H_0 is true. For example, the observed frequency for the first row and second column is $o_{12} = 53$ and the expected frequency is $e_{12} = 43.0$. The observed frequency represents the number of individuals in the sample who have a high school degree and are employed in the manufacturing division.

The value of the chi-square test statistic, $\chi^2 = 81.21947$, is shown at the bottom of the computer output, along with the appropriate number of degrees of freedom, 4. The number shown under the word Significance, .0000, is the p-value associated with $\chi^2 = 81.21947$. That is, if χ^2 follows the chi-square distribution with 4 degrees of freedom, then $P(\chi^2 > 81.21947) = .0000$. The chi-square statistic provides extremely strong evidence that academic degree and division of employment are not independent.

Figure 13.7 (page 532) shows the corresponding computer output obtained from the SAS program. To obtain the frequency distribution, issue the SAS commands

```
PROC FREQ;
TABLES DEGREE * DIV;
```

Figure 13.8 (pages 532–533) shows the corresponding computer program and output obtained from the Minitab program. Note that the output is almost identical to the SAS output.

TABLE OF DEGREE BY DIV

DEGREE	DIV			
FREQUENCY PERCENT ROW PCT COL PCT	1	2	3	TOTAL
1	35 29.17 39.33 92.11	53 44.17 59.55 91.38	1 0.83 1.12 4.17	89 74.17
2	3 2.50 13.04 7.89	5 4.17 21.74 8.62	15 12.50 65.22 62.50	23 19.17
3	0 0.00 0.00 0.00	0 0.00 0.00 0.00	8 6.67 100.00 33.33	8 6.67
TOTAL	38 31.67	58 48.33	24 20.00	120 100.00

FIGURE 13.7 *SAS-generated contingency table cross-tabulating*
DEGREE and DIV from data in Table 2.1

```
MTB〉
READ 'COMPUTEC.DAT' INTO C1-C8
     120 ROWS READ
ROW   C1   C2   C3   C4   C5     C6   C7   C8
  1    0    1    1   35    1   1575   16    5
  2    1    1    1   31    2   1980   10    3
  3    1    1    1   51    2   2480   28    2
  4    0    1    1   57    1   1925   30    2
MTB〉
NAME C1 = 'SEX', C2 = 'RACE', C3 = 'DEGREE', C4 = 'AGE'
MTB〉
NAME C5 = 'DIV', C6 = 'SALARY', C7 = 'SENIOR', C8 = 'EXEMPT'
MTB〉
TABLE C3,C5;
SUBC〉
COUNTS;
SUBC〉
ROWPERCENTS;
```

FIGURE 13.8 *Minitab-generated contingency table cross-tabulating*
DEGREE and DIV from data in Table 2.1

```
SUBC〉
COLPERCENTS;
SUBC〉
TOTPERCENTS
ROWS:  DEGREE           COLUMNS:  DIV
```

	1	2	3	ALL
1	35	53	1	89
	39.33	59.55	1.12	100.00
	92.11	91.38	4.17	74.17
	29.17	44.17	0.83	74.17
2	3	5	15	23
	13.04	21.74	65.22	100.00
	7.89	8.62	62.50	19.17
	2.50	4.17	12.50	19.17
3	0	0	8	8
	—	—	100.00	100.00
	—	—	33.33	6.67
	—	—	6.67	6.67
ALL	38	58	24	120
	31.67	48.33	20.00	100.00
	100.00	100.00	100.00	100.00
	31.67	48.33	20.00	100.00

```
CELL CONTENTS –
COUNT
                       % OF ROW
                       % OF COL
                       % OF TBL
MTB 〉
STOP
```

FIGURE 13.8 *Continued*

Exercises

Use the SPSSX program and a CROSSTABS command on the data in Table 2.1 of Chapter 2 to complete Exercises 1–3.

13.3.1 Test the null hypothesis that the variables SEX and DEGREE are independent for the employees at the Computech Corporation.
a. What is the value of the chi-square test statistic?
b. What is the appropriate number of degrees of freedom?
c. What is the critical value of the test statistic if we use a 5% level of significance?
d. Do you reject the null hypothesis of independence?

13.3.2 Test the null hypothesis that the variables RACE and DEGREE are independent for the employees at the Computech Corporation.

a. What is the value of the chi-square test statistic?
b. What is the appropriate number of degrees of freedom?
c. What is the critical value of the test statistic if we use a 5% level of significance?
d. Do you reject the null hypothesis of independence?

13.3.3   Test the null hypothesis that the variables SEX and DIV are independent for the employees at the Computech Corporation.
a. What is the value of the chi-square test statistic?
b. What is the appropriate number of degrees of freedom?
c. What is the critical value of the test statistic if we use a 5% level of significance?
d. Do you reject the null hypothesis of independence?

Use the SPSSX program and a CROSSTABS command on the data in Table 2.2 of Chapter 2 to complete Exercises 4 and 5.

13.3.4   Test the null hypothesis that the variables SEX and MAJOR are independent for the students in the statistics class.
a. What is the value of the chi-square test statistic?
b. What is the appropriate number of degrees of freedom?
c. What is the critical value of the test statistic if we use a 5% level of significance?
d. Do you reject the null hypothesis of independence?

13.3.5   Test the null hypothesis that the variables CLASS and MAJOR are independent for the students in the statistics class.
a. What is the value of the chi-square test statistic?
b. What is the appropriate number of degrees of freedom?
c. What is the critical value of the test statistic if we use a 5% level of significance?
d. Do you reject the null hypothesis of independence?

STATISTICS IN ACTION: CASE STUDY

Testing for Age Discrimination

The use of statistical evidence has become commonplace in age discrimination litigation. In this case study, we discuss several statistical tests that can be used by economists and statisticians when attempting to support or refute the hypothesis that discrimination has occurred.

The basic antidiscrimination portion of Title VII (section 703) of the Civil Rights Act of 1964 as amended by the Equal Employment Act of 1972 provides:

It shall be unlawful employment practice for any employer (1) to fail or refuse to hire or to discharge any individual, or otherwise to discriminate against any individual with respect to his compensation, terms, conditions, or privileges of employment because of such individual's race, color, religion, sex or national origin; or (2) to limit, segregate or classify its employees or applicants for employment in any manner which would deprive or tend to deprive any individual of employment opportunities or otherwise adversely affect his status as an employee, because of such individual's race, color, religion, sex or national origin.

Age discrimination is included under the Age Discrimination in Employment Act (ADEA) of 1967. Discrimination against the physically and mentally disabled is prohibited under the Americans With Disabilities Act (ADA) of 1990.

The ADEA is designed to "promote employment of older persons based on their ability rather than age; to prohibit arbitrary age discrimination in employment; to help employers and workers find ways of meeting problems arising from the impact of age on employment." The ADEA applies to all private employers engaged in commerce who have 20 or more employees, states and their agencies and subdivisions, labor organizations that operate in industries affecting commerce, and employment agencies.

At a minimum, the ADEA applies to employment decisions explicitly based on age. It is rare, however, to find a situation where an employer has explicitly stated that a particular individual would not be hired or would be terminated because he or she was over 40 years of age.

Although it is rare that statistical evidence, by itself, is sufficient to establish that discrimination has occurred, the utilization of statistical analysis in discrimination cases has become commonplace. One way of showing statistical evidence of age discrimination is to compare the age distribution of persons treated favorably to the age distribution of persons treated unfavorably.

The statistical question is whether employee age was related to the employer's decision-making process with respect to staff reduction. We must compare what actually occurred with what we would expect to occur if employee age were totally unrelated to the decision-making process.

One of the first tasks for the statistician is to properly define the appropriate employee population at issue. In many situations, decisions concerning reductions in staff are made based on choosing various positions or job classifications for elimination, examining seniority within the job classification, or examining job performance within the job classification. It also may be relevant to consider the organizational unit or department in which the employee works and the training and experience that the employee has.

The population of employees that is analyzed should be alike with regard to these factors. For example, suppose an employer has established that it has a valid reason to have a staff reduction in, say, its manufacturing division. The appropriate population to study would be employees in the manufacturing division. Thus, for example, employees in the marketing division would be excluded from the analysis.

Under the null hypothesis, it is assumed that the employees to be terminated were picked without respect to age from the relevant population of employees. The null hypothesis, denoted H_0, states that age was not a factor in determining which employees would be terminated. The alternative hypothesis, denoted H_1, states that age was a factor in determining which employees would be terminated.

As an example, the data in Table 13.8 show the age classification of 1,013 employees of a large corporation according to whether they were retained or terminated. (The data are from an actual case that was tried in Pennsylvania.) Each employee has been classified into one of two different age categories and into one of two different employment status categories.

TABLE 13.8
Observed frequencies of the number of employees terminated and retained by age group: Under 40 versus 40 and over

Age	Retained employees	Terminated employees	
Under 40	481	28	509
40 and over	458	46	504
Totals	939	74	1,013
Proportion	.9269	.0731	

We observe that 92.69% of the labor force have been retained, and 7.31% of the labor force have been terminated. If the employees who were terminated were selected without regard to age, then we should observe that, except for chance deviations, approximately 92.69% of the employees under age 40 should have been retained and approximately 92.69%

of the employees age 40 and over should have been retained. Similarly, about 7.31% of the employees under age 40 should have been terminated and approximately 7.31% of the employees age 40 and over should have been terminated.

The question of interest is whether the differences in retention and termination rates between the younger and older employees could be reasonably attributed to chance, or whether the disparity is large enough to indicate that some factor other than chance (such as age) influenced the selection of the individuals to be terminated.

If terminations were independent of age, then we would expect to find that 92.69% of the employees in each age class were retained and 7.31% of the employees in each age class were terminated. Thus, under H_0, we would expect that, of the 509 employees under age 40, 471.82 would be retained and 37.18 would be terminated. Similarly, under H_0, we would expect that, of the 504 employees age 40 and over, 467.18 would be retained and 36.82 would be terminated. These expected frequencies are shown in Table 13.9.

The chi-square statistic based on a 2×2 table has $(2 - 1)(2 - 1) = 1$ degree of freedom. If we use $\alpha = .05$, the critical value of the chi-square statistic is 3.84. Similarly, if we use $\alpha = .01$, the critical value is 6.63.

The observed chi-square statistic calculated in Table 13.10 is $\chi^2 = 4.91$. This value exceeds the critical value $\chi^2_{.05,2} = 3.84$, if we use $\alpha = .05$. But the observed value of the test statistic does not exceed the critical value $\chi^2_{.01,2} = 6.63$ if we use $\alpha = .01$. That is, we obtain

$$.01 < P(\chi^2 > 4.91) < .05$$

Reasonable people will disagree about how rare an event has to be before we decide that it did not occur because of chance. In this situation, the chi-square statistic falls between the two proposed benchmarks based on a 5% or 1% level of significance. The observed value of χ^2 is significant at the 5% level but not at the 1% level.

In this case, the jury eventually sided with the employer and awarded no damages to the terminated employees. The jurors believed that the employer had justifiable non–age-related reasons for selecting the employees who were terminated.

This case study provides a classic example of how the chi-square test is used in age discrimination cases.

TABLE 13.9

Expected frequencies of number of employees terminated and retained by age group: Under 40 versus 40 and over

Age	Retained employees	Terminated employees	
Under 40	471.82	37.18	509
40 and over	467.18	36.82	504
Totals	939.00	74.00	1,013
Proportion	.9269	.0731	

TABLE 13.10   *Calculation of the chi-square statistic*

Row i	Column j	Observed frequency o_{ij}	Expected frequency e_{ij}	Difference $o_{ij} - e_{ij}$	Contribution to chi-square $(o_{ij} - e_{ij})^2/e_{ij}$
1	1	481	471.82	9.18	.1786
1	2	28	37.18	−9.18	2.2666
2	1	458	467.18	−9.18	.1804
2	2	46	36.82	9.18	2.2888
Total					4.9144

Chapter 13 Supplementary Exercises

13.S.1 In past years, 40% of students majored in humanities, 35% majored in social sciences, and 25% majored in natural sciences. In a recent survey of the student body, the following distribution of students was obtained: humanities, 341; social sciences, 382; and natural sciences, 277. Test the null hypothesis that there has been no change in the distribution of students' choice of major. Use a 5% level of significance.

13.S.2 An executive thinks that more employees are absent on Mondays and Fridays than on other days. A sample was obtained of 1,000 employees who were absent for just 1 day during a particular week. The results were as follows: Monday, 277; Tuesday, 162; Wednesday, 127; Thursday, 126; and Friday, 308. Test the hypothesis that an employee is equally likely to be absent on any weekday. Use a 5% level of significance.

13.S.3 A group of 500 children was asked to identify their favorite colors. A toy manufacturer wants to know whether the color preferences of boys and girls differ. We obtain the following data:

	FAVORITE COLOR			
Gender	Yellow	Red	Blue	Brown
Boy	107	85	33	25
Girl	93	65	67	25

Test the hypothesis that gender and favorite color are independent. Use a 1% level of significance.

13.S.4 In the sample of 500 children in Exercise 3, 200 children favored yellow, 150 favored red, 100 favored blue, and 50 favored brown. Test the hypothesis that, in the total population, these colors are favored according to the following proportions: 30% favor yellow, 30% favor red, 30% favor blue, and 10% favor brown. Use a 5% level of significance.

13.S.5 In a random sample of 485 voters, each individual was asked whether he or she thought inflation or unemployment was a more serious problem. The individuals were also classified by party affiliation. The results were as follows:

Party	Inflation	Unemployment
Republican	162	58
Democrat	98	67
Other	70	30

Use a 5% level of significance and test whether political party affiliation and perceived problem are independent.

13.S.6 A baseball player claims that he hits better during night games than during day games. Last season, the player's performance was as follows:

	Day games	Night games
Base hits	60	90
Outs	150	180

Do the data substantiate the player's claim? Use a 5% level of significance.

13.S.7 A pollster wants to test whether political party affiliation is independent of religious affiliation. A sample of voters yielded the following results:

Religion	Democrat	Republican	Other
Catholic	250	175	75
Protestant	400	300	100
Jewish	100	50	50
Other	150	75	75

Use a 5% level of significance. What decision should the pollster make?

13.S.8 In a sample of students, the distribution of grades was as follows:

Department	A	B	C	F
Social sciences	140	160	100	75
Natural sciences	160	190	90	75
Humanities	300	350	110	50

Use a 5% level of significance, and test whether the grading scale is independent of the department.

13.S.9 What is the appropriate number of degrees of freedom when we wish to test that a sample of data was obtained from a normal distribution if the number of intervals is 12 and the mean and standard deviation are unknown?

13.S.10 The manager of a restaurant claims that dinner bills are approximately normally distributed with mean $\mu = \$50$ and standard deviation $\sigma = \$10$. Let $\alpha = .05$ and use the following data to test the manager's claim:

Dinner bill (in dollars)	Frequency
20–29.99	28
30–49.99	52
50–59.99	58
60–69.99	36
70–79.99	20
80–89.99	16

13.S.11 A psychologist wants to test whether a relationship exists between mental ability and manual dexterity. An IQ test is given to a random sample of individuals. Then each person is timed while performing certain tasks. Let $\alpha = .05$. Test the null hypothesis that the two variables are independent of one another.

| Time (in seconds) | IQ | | |
	89 or less	90–110	111 or more
0–29	18	92	78
30–59	27	145	80
60–90	16	78	50

13.S.12   An economist wants to test whether the starting salaries of college graduates are independent of a graduate's major. A random sample of 400 students at the state university is selected. To control for ability, all the students in the sample have IQs between 115 and 120. Let $\alpha = .05$. Do the accompanying data indicate that starting salary and undergraduate major are independent?

Starting salary (in dollars)	Math	English	Economics	Business
		SUBJECT		
0–12,999	26	40	25	20
13,000–15,999	44	40	50	60
16,000 or more	30	20	25	20

13.S.13   The chef at Antonelli's Restaurant claims that men and women order different types of meals. Use the following data from a random sample of customers at Antonelli's to test whether choice of meal is independent of gender with $\alpha = .05$:

Gender	Meat	Fish	Fowl
Male	280	40	50
Female	86	42	42

13.S.14   In the banking industry, mortgage loans are classified as current if all payments are up to date, late if the mortgage payment is 1 to 45 days overdue, and delinquent if the payment is more than 45 days overdue. Nationwide, the proportions in each category are .80, .15, and .05, respectively. In a random sample of 1,000 mortgages from the Ashton National Bank, 750 are current, 190 are late, and 60 are delinquent. Use a 5% level of significance and test whether this bank is significantly different from the industry norm.

13.S.15   In an April 22, 1987, decision (*McClesky* v. *Kemp*, 55 LW 4537), the U.S. Supreme Court upheld, by a 5 to 4 margin, Georgia's death penalty. The challenge had concerned a black man's conviction in the killing of a white policeman during a 1978 robbery. The Court was given evidence purporting to show a disparity in the imposition of the death penalty in Georgia depending on the race of the murder victim and, to a lesser extent, the race of the defendant. The Supreme Court upheld the death penalty despite statistical evidence of racial disparity in its application. The crucial evidence in the trial was the following contingency table:

Death penalty	Black/ black	Black/ white	White/ black	White/ white
		RACE OF DEFENDANT/RACE OF VICTIM		
Yes	18	50	2	58
No	1,420	178	62	687

a. Find P(Death penalty$|$Victim black).
b. Find P(Death penalty$|$Victim white). Compare with part **a.**
c. Find P(Death penalty$|$Defendant black).
d. Find P(Death penalty$|$Defendant white). Compare with part **c.**
e. Find P(Death penalty$|$Defendant black and victim white).
f. Find P(Death penalty$|$Defendant white and victim white). Compare with part **e.**
g. Perform a chi-square test of independence to determine whether the variable Death penalty is independent of the variables Race of defendant and Race of victim. Use a 5% level of significance.

13.S.16   The computer center at a major university keeps records of the number of times per day the computer system is down. The following data show the number of breakdowns per day for a random sample of 100 days:

Breakdowns	0	1	2	3	4	5	6	7
Observed frequency	34	25	18	12	6	2	2	1

a. Test the null hypothesis that the distribution of breakdowns is Poisson. Let $\alpha = .05$.
b. In part **a**, what did you use as the mean?
c. What is the appropriate number of degrees of freedom to use?

13.S.17   Respondents to a questionnaire were cross-classified according to the amounts of time spent reading newspapers and watching television daily. The sample results are shown in the following table:

Hours watching TV	HOURS READING NEWSPAPER		
	Less than .5	.5 to 1.0	More than 1.0
Less than 2.5	10	16	9
2.5 or more	8	10	5

a. Let $\alpha = .05$. Test the null hypothesis that the number of hours spent watching TV and reading the newspaper are independent of one another.
b. What is the appropriate number of degrees of freedom?

13.S.18   The following data show the number of households that paid all or part of their credit card bills for the previous month according to the age of the head of the household:

Age (in years)	CHARGES PAID	
	All	Part
Under 31	200	200
31–40	210	190
41–50	240	160
51–60	300	100
Over 60	240	160

Let $\alpha = .05$. Test to see whether paying all or part of the charges is independent of the age of the head of the household.

13.S.19   The annual income tax payments of 1,000 randomly selected individuals are distributed as follows:

Payment	Frequency
Less than $1,000	50
$1,000 to under $1,500	70
$1,500 to under $2,000	160
$2,000 to under $2,500	240
$2,500 to under $3,000	250
$3,000 to under $4,000	150
$4,000 or more	80

Test the hypothesis that the distribution is normal with a mean of $2,500 and a standard deviation of $1,000. Use $\alpha = .05$.

13.S.20 In a study of social mobility, the occupations of a random sample of employees and of their fathers were recorded. These occupations were then ranked by status. The results are as follows:

	WORKER'S STATUS		
Father's status	**Upper**	**Middle**	**Lower**
Upper	572	390	161
Middle	351	720	433
Lower	118	327	440

a. If there was no social mobility whatsoever, what would the data look like?
b. If there was completely free social mobility, what would you expect the data to look like?
c. Perform a chi-square test of the null hypothesis that status of the worker and status of the father are independent. Use $\alpha = .05$.

13.S.21 The following 4×2 table of data shows a cross-classification of employees based on age and employment status. The data represent a reclassification of the data presented in Table 13.8. Let $\alpha = .01$ and test the null hypothesis that age and employment status are independent.

Age	Retained employees	Terminated employees	Totals
Under 40.0	481	28	509
40.0–49.9	304	15	319
50.0–59.9	127	21	148
60.0–69.9	27	10	37
Totals	939	74	1,013
Proportion	.9269	.0731	

a. Determine the appropriate number of degrees of freedom.
b. Find the expected cell frequencies.
c. Determine the critical value of the chi-square test statistic.
d. Perform the test.
e. What is the approximate p-value of the test?

13.S.22 In the production of an electrical device operated by a thermostatic control, random samples of control switches were tested to determine the "on" temperature at which the thermostat actually operated under a given setting. Results of the test were as follows:

"ON" TEMPERATURE AT WHICH THERMOSTATIC SWITCH OPERATES

54	56	56	56	55	51	52	54	56	54
54	52	50	57	55	56	55	56	53	50
53	54	57	56	52	53	52	58	55	54
52	55	54	55	56	56	53	53	54	55
55	52	53	52	53	50	54	53	55	55
57	54	53	52	53	52	52	54	53	55
54	53	55	52	52	59	55	54	51	55
58	53	57	56	51	58	57	56	51	59
55	55	55	56	53	54	57	54	55	54
54	53	56	53	55	53	53	57	54	53
58	57	57	56	59	59	57	56	59	59
56	58	58	58	58	57	55	57	58	59

a. Make a tally sheet of these measurements and arrange them in a frequency distribution. Use the classes 49.5–51.5, 51.5–53.5, 53.5–57.5, and 57.5–59.5.

b. Construct and plot the frequency distribution showing the class boundaries, class midpoints, and observed frequencies. Does it appear that the distribution is approximately normal?

c. Calculate the sample mean.

d. Calculate the sample variance and sample standard deviation.

Let $\alpha = .05$. Test the null hypothesis that the population distribution is normal.

e. Determine the appropriate number of degrees of freedom.

f. Find the expected cell frequencies.

g. Determine the critical value of the chi-square test statistic.

h. Perform the test.

i. What is the approximate p-value of the test?

13.S.23   A small radio transmitting set is designed to generate a certain automatic signal. The duration of this signal is one of the specified quality characteristics of the set. The following data show a random sample of output from a specific production line.

DURATION OF AUTOMATIC SIGNAL

390	393	395	405	420	376	381	381	383	401
380	387	395	397	407	377	383	387	390	393
393	395	403	405	414	376	388	395	397	400
387	400	400	403	410	391	392	394	397	405
390	391	395	401	405	379	391	393	394	410
390	397	400	406	428	380	382	389	391	399
362	363	372	375	377	357	360	368	370	372
390	395	395	397	406	382	399	401	406	406
390	395	395	400	410	381	390	394	397	399
367	389	398	401	435	362	378	396	400	429

a. Make a tally sheet of these measurements and arrange them in a frequency distribution.

b. Construct and plot the frequency distribution showing the class boundaries, class midpoints, and observed frequencies. Does it appear that the distribution is approximately normal?

c. Calculate the sample mean.

d. Calculate the sample variance and sample standard deviation.

Let $\alpha = .05$. Test the null hypothesis that the population distribution is normal.

e. Determine the appropriate number of degrees of freedom.

f. Find the expected cell frequencies.

g. Determine the critical value of the chi-square test statistic.

h. Perform the test.

i. What is the approximate p-value of the test?

13.S.24   A quality control engineer examined the output of four machines and classified each item of output as acceptable or defective. Test the null hypothesis that the proportion defective is independent of the machine.

	MACHINE			
Status	**A**	**B**	**C**	**D**
Good	284	361	504	802
Defective	28	37	52	61

    a. Determine the appropriate number of degrees of freedom.
    b. Find the expected cell frequencies.
    c. Determine the critical value of the chi-square test statistic.
    d. Perform the test.
    e. What is the approximate p-value of the test?

13.S.25   Many states have initiated lotteries and other gambling arrangements to raise revenue for various projects. None of these lotteries is a "fair game" because in each case the state takes in more revenue than it pays out. Thus, on the average, the typical participant in the lottery loses money. Many people have criticized these lotteries because they especially hurt poor people. There is a substantial amount of evidence indicating that individuals with lower incomes are more likely to play the lottery than individuals with higher incomes. The following data show the incomes earned by a sample of individuals classified according to whether they played the lottery at least once during the last year. Test the null hypothesis that playing the lottery is independent of income.

Played	ANNUAL INCOME (IN $1,000S)		
lottery	Under 20	20–40	Over 40
Yes	57	62	15
No	196	359	172

    a. Determine the appropriate number of degrees of freedom.
    b. Find the expected cell frequencies.
    c. Determine the critical value of the chi-square test statistic.
    d. Perform the test.
    e. What is the approximate p-value of the test?

13.S.26   A city has three major TV stations. In a poll taken last year, concerning whether people watched the 11 o'clock news, the distribution of responses was as follows: Channel 2, 29%; Channel 4, 18%; Channel 11, 14%; Other, 39%. A recent poll produced the following responses. Test the null hypothesis that the distribution of responses has remained the same.

Channel	2	4	11	Other
Number of viewers	282	141	152	32)

    a. Determine the appropriate number of degrees of freedom.
    b. Find the expected cell frequencies.
    c. Determine the critical value of the chi-square test statistic.
    d. Perform the test.
    e. What is the approximate p-value of the test?

13.S.27   Accounts receivable for large companies are classified as current, moderately late, very late, and uncollectible. Suppose that in the chemical industry, the percentages falling in these categories are 60%, 20%, 14%, and 6%, respectively. The Johnson Chemical Company reviews a sample of 1,000 accounts receivable, with 618, 230, 108, and 44 falling in the respective classes. Let $\alpha = .05$. Are these values in agreement with the industry ratios?

13.S.28   A sales manager wants to determine whether the proportion of people who drink diet cola is the same in different regions of the country. A sample of individuals shows that diet drinks are favored by 90 of 300 people in the North, 100 of 360 people in the South, 40 of 100 people in the Midwest, and 120 of 500 people in the West. Let $\alpha = .05$. Perform the test.

13.S.29 An engineer wants to determine, at the 1% level of significance, whether the proportion of defective pieces produced is the same for three machines. A sample of output indicates that machine A had 26 defectives in 900 pieces, machine B had 30 defectives in 800 pieces, and machine C had 22 defectives in 700 pieces. Perform the test.

13.S.30 A lawyer wants to check whether the race and gender of jurors in a county in Alabama chosen for murder trials reflects the race and gender makeup of the eligible jurors in the county. A sample of 200 jurors revealed the following frequencies:

White male: 80   White female: 70
Black male: 35   Black female: 15

In the county, the proportion of eligible jurors in each category was as follows:

White male: 35%   White female: 30%
Black male: 20%   Black female: 15%

Let $\alpha = .05$. Perform the test. If you reject the null hypothesis, explain which groups are overrepresented and which groups are underrepresented among the jurors.

13.S.31 An insurance adjuster believes that the population of yearly automobile accidents per driver follows a Poisson distribution. The following data show the performance of a sample of drivers. Find the mean number of accidents per driver. Let $\alpha = .05$. Perform the test.

Number of accidents per driver	Observed frequency
0	105,739
1	8,462
2	876
3	32
Total	115,109

13.S.32 A quality control engineer wants to determine (using $\alpha = .05$) whether the diameters of precision bolts can be described by a normal curve with mean $\mu = .450$ inch and standard deviation $\sigma = .01$. Perform the test based on the following sample of diameters.

Diameter	Frequency
Under .420	13
.420–under .430	58
.430–under .440	110
.440–under .450	345
.450–under .460	299
.460–under .470	126
.470–under .480	72
.480 and over	14

References

Cochran, William G. "The χ^2 Test of Goodness of Fit." *Annals of Mathematical Statistics* 23 (1952): 315–345.

Freund, John E., and R. E. Walpole. *Mathematical Statistics*. 4th ed. Englewood Cliffs, N.J.: Prentice-Hall, 1987.

Handbook of Tables for Probability and Statistics. 2d ed. Cleveland: Chemical Rubber Co., 1968.

Hoel, Paul G. *Elementary Statistics*. 4th ed. New York: Wiley, 1976.

Neter, John, William Wasserman, and G. A. Whitmore. *Fundamental Statistics for Business and Economics*. 4th ed. Boston: Allyn and Bacon, 1973.

Norusis, Marija J. *SPSSX Introductory Statistics Guide*. New York: McGraw-Hill, 1990.

Norusis, Marija J. *SPSSX Advanced Statistics Guide*. Chicago: SPSS, 1990.

Norusis, Marija J. *The SPSS Guide to Data Analysis*. Chicago: SPSS, 1986.

Pearson, Karl. "On the Criterion That a Given System of Deviations from the Probable in the Case of a Correlated System of Variables Is Such That It Can Be Reasonably Supposed to Have Arisen from Random Sampling." *Philosophical Magazine*. 5th series (1900), pp. 157–175.

Ryan, Thomas A., Brian L. Joiner, and Barbara F. Ryan. *Minitab Handbook*. 2d ed. Boston: PWS-Kent, 1985.

Ryan, Thomas A., Brian L. Joiner, and Barbara F. Ryan. *Minitab Reference Manual*. University Park, Penn.: Minitab Project, 1985.

SAS Introductory Guide. 3d ed. Cary, N.C.: SAS Institute, 1985.

SAS Procedures Guide for Personal Computers. Version 6 ed. Cary, N.C.: SAS Institute, 1986.

SAS Statistics Guide for Personal Computers. Version 6 ed. Cary, N.C.: SAS Institute, 1986.

SAS User's Guide: Basics. Version 5 ed. Cary, N.C.: SAS Institute, 1985.

SAS User's Guide: Statistics. Version 5th ed. Cary, N.C.: SAS Institute, 1985.

SPSSX User's Guide. Chicago: SPSS, 1988.

14 Analysis of Variance

In Chapter 12, we explained how to test the null hypothesis that the means of two different populations are equal to one another. Frequently we want to test the null hypothesis that three or more population means are equal. For example, we may want to examine the gasoline mileage of several different brands of automobiles, the incomes of workers in different occupations, or the costs of production using several different processes.

When testing for differences in the means of more than two populations, we usually do not proceed by considering all combinations of two populations at a time and testing for differences in each pair. First, such an approach would require several tests rather than just one. Second, if each individual test were conducted using a level of significance of, say, $\alpha = .05$, then the overall level of significance would be higher than .05. For example, if

$$H_0: \quad \mu_1 = \mu_2 = \mu_3$$

were true, and if the three null hypotheses $\mu_1 = \mu_2$, $\mu_1 = \mu_3$, and $\mu_2 = \mu_3$ were each tested using a .05 level of significance, then the probability of accepting each single hypothesis would be .95, and the probability of accepting all three hypotheses would be at least $.95^3 = .857$. (If the three tests were independent, this probability would be exactly $.95^3$.) Consequently, the probability of rejecting the null hypothesis when it was true could be as large as .143, which is substantially greater than .05.

Thus, we want to test *simultaneously* for differences among the means of all the populations, and we want the joint level of significance of the test to be α. To perform this test, we make use of the F distribution and use a method called **analysis of variance,** or **ANOVA**.

14.1 The One-Factor ANOVA Model

Suppose we obtain samples of data from K populations and want to test the null hypothesis

$$H_0: \quad \mu_1 = \mu_2 = \cdots = \mu_K$$

against the alternative hypothesis

$$H_1: \quad \text{At least two of the means differ.}$$

We might consider comparing sample means two at a time, repeatedly applying the t test described in Section 12.2. However, if $K = 5$, then there would be $_5C_2 = 10$ different pairs of means to test. Furthermore, if the null hypothesis were true and each test used α as the level of significance, the probability would be much greater than α that we would reject H_0 at least once. For a true null hypothesis, the more tests we perform, the greater the probability that we will reject H_0.

EXAMPLE 14.1   **Comparing Fuel Consumption of Three Makes of Automobile** ——————

There is a great deal of interest in comparing the mileage ratings of different makes of automobiles. Other things being equal, customers buy the car that gets the best mileage. Suppose we wish to compare the mean fuel consumption for $K = 3$ different makes of automobile, car 1, car 2, and car 3. Suppose five cars of each make are selected randomly and the gasoline mileage (in miles per gallon) is recorded for each car. Table 14.1 shows the results.

TABLE 14.1   *Gasoline mileage results for Example 14.1*

	Car 1	Car 2	Car 3
	18.2	19.8	21.2
	19.4	21.0	21.8
	19.6	20.0	22.4
	19.0	20.8	22.0
	18.8	20.4	21.6
Sample mean	19.0	20.4	21.8
Sample standard deviation	.548	.510	.447
Sample variance	.300	.260	.200

From Table 14.1, we see that the three sample means are $\bar{x}_1 = 19.0$, $\bar{x}_2 = 20.4$, and $\bar{x}_3 = 21.8$. These sample means all differ, but this would be the case even if all three population means were identical. The appropriate question is "Do the observed differences in the sample means provide evidence against the null hypothesis

$$H_0: \quad \mu_1 = \mu_2 = \mu_3$$

or can the observed differences be attributed to chance?" That is, we want to determine whether the differences between the sample means are large enough to convince us that the population means differ.

Before discussing how to perform an analysis of variance (or ANOVA) test, let us present an intuitive explanation of how analysis of variance works. From Table 14.1, the differences between the sample means are

$$\bar{x}_1 - \bar{x}_2 = 19.0 - 20.4 = -1.4$$
$$\bar{x}_1 - \bar{x}_3 = 19.0 - 21.8 = -2.8$$
$$\bar{x}_2 - \bar{x}_3 = 20.4 - 21.8 = -1.4$$

These differences (or their squares) provide a measure of the amount of variation *between* the three samples. We want to test whether these differences are due to chance or are indicative of true differences in the population means.

For example, the absolute difference between the sample means for cars 1 and 2 is 1.4 miles per gallon. Is this difference large enough to make us believe that μ_1 and μ_2 differ? The answer depends, at least in part, on whether $\bar{x}_1$ and $\bar{x}_2$ are good estimates of μ_1 and μ_2. If the sample variances s_1^2 and s_2^2 are small, then $\bar{x}_1$ and $\bar{x}_2$ should be fairly good estimates of μ_1 and μ_2, and a relatively small difference between the sample means could indicate that the population means differ. On the other hand, if s_1^2 and s_2^2 are relatively large, then the sample means $\bar{x}_1$ and $\bar{x}_2$ might not be very close to the population values μ_1 and μ_2, and even a sizable difference between the sample means would not necessarily indicate that μ_1 and μ_2 differ.

The sample variances (or standard deviations) measure the amount of variation present in each sample. If the amount of variation *within* each sample is small, then sizable differences between the sample means provide evidence of differences in the population means. If the sample variances are large, however, then observed differences in the sample means do not necessarily provide much evidence against the null hypothesis. Thus testing for the equality of means involves finding the amount of variation that exists *between* the samples and comparing this to the variation *within* samples. Hence the name "analysis of variance."

Figure 14.1 is a **dot diagram** showing the mileage data in Table 14.1. Each dot represents a sample observation. In Figure 14.1, the observations for car 1 are clustered together about the sample mean $\bar{x}_1 = 19.0$, the observations for car 2 about the sample mean $\bar{x}_2 = 20.4$, and the observations for car 3 about the sample mean $\bar{x}_3 = 21.8$. Figure 14.1 strongly suggests that the mileage data come from three populations having different means.

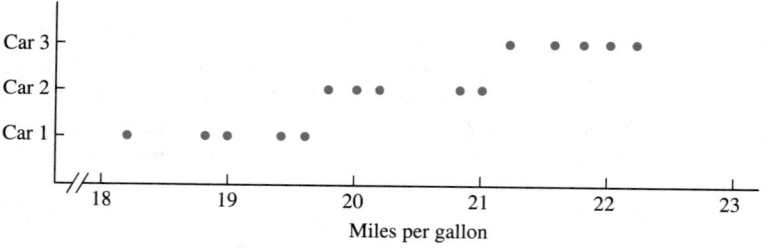

FIGURE 14.1   *Dot diagram for data in Table 14.1*

Now consider the data in Table 14.2, the mileage for three different brands of cars. The sample means in Table 14.2 are identical to those in Table 14.1, and so the variation *between* the samples in Table 14.2 is identical to the variation between the samples in Table 14.1. In Table 14.2, however, all the sample standard deviations are much larger than the corresponding values in Table 14.1. Thus, the variation *within* each sample is much greater in Table 14.2. For example, in Table 14.1 the standard deviation of gasoline mileage for car 1 is $s_1 = .548$, and in Table 14.2 the standard deviation for car 1 is $s_1 = 3.26$.

TABLE 14.2 *New gasoline mileage for Example 14.1*

	Car 1	Car 2	Car 3
	17.0	24.2	26.0
	20.4	22.0	19.8
	24.0	17.8	24.4
	15.8	16.2	16.0
	17.8	21.8	22.8
Sample mean	19.0	20.4	21.8
Sample standard deviation	3.26	3.29	3.97
Sample variance	10.66	10.84	15.76

Figure 14.2 shows the dot diagram for the mileage data in Table 14.2. In the figure, the observations for each car are less concentrated than in Figure 14.1. Thus it is not so obvious that the three populations are different; we would not be greatly surprised if all three samples came from the same population.

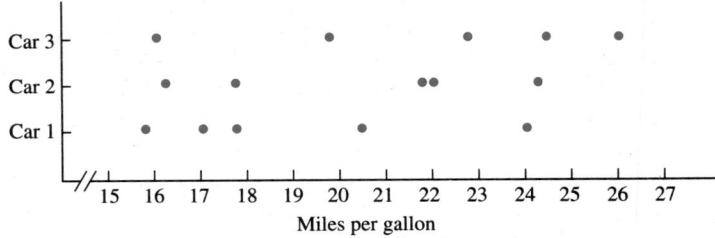

FIGURE 14.2 *Dot diagram for data in Table 14.2*

Terminology in Experimental Design

When data have been obtained according to certain sampling procedures, they may contain more information about the population means than could be obtained by using simple random sampling. The procedure used for obtaining sample data is called the design of the experiment. The development of experimental design originated with agricultural studies, where experiments have to be carefully designed because it takes an entire growing season to get a single observation. Thus it is very important to get as much information as possible out of small samples.

DEFINITIONS **Dependent Variable and Independent Variable**

The variable being studied is called the **dependent variable** or **response variable.** A variable that influences the dependent variable is called an **independent variable** or a **factor.**

In Example 14.1, the *dependent variable* is gasoline mileage. In Example 14.1, it is conjectured that gasoline mileage might depend on the make of car. Thus the make of car is an *independent variable,* or *factor,* that affects the values of the dependent variable. A factor may be either a quantitative or qualitative variable. In Example 14.1, Make of car is a qualitative variable that takes three different values. Another study could be performed to test the fuel consumption of cars of different ages. In this case, the factor Age of car would be a quantitative variable. In a study of crop yields using different brands of fertilizer, the factor would be qualitative, whereas in a study of crop yields using different quantities of fertilizer, the factor would be quantitative.

DEFINITIONS   **Factor Level and Treatment** ──────────────────────────────

A **factor level** is a particular value of a factor. If only a single factor is being considered, each factor level is called a **treatment;** if more than one factor is being considered, each combination of factor levels is called a **treatment.**

───

In Example 14.1, there is only one independent variable or factor, Make of car, that influences mileage, so the analysis of the differences in the three population means is called *one-way ANOVA.* There are three factor levels or treatments being considered— car 1, car 2, and car 3. Studies can also be performed using two or more factors. In such cases, a combination of different factor levels is called a *treatment.* For example, suppose it is conjectured that gasoline mileage depends on Make of car and Brand of gasoline. Each combination of a specific brand of car and a specific brand of gasoline would represent a treatment.

Example 14.1 represents what is called a *completely randomized design.* In a completely randomized design, the means are compared for K different treatments based on independent random samples of $n_1, n_2, \ldots, n_K$ observations drawn from populations associated with treatments $1, 2, \ldots, K$, respectively.

DEFINITION   **Completely Randomized Design** ──────────────────────────

A **completely randomized design** compares K population means by drawing independent random samples from each of the K populations.

───

The Logic Behind ANOVA

As stated earlier, the heart of the ANOVA technique is that variability in the sample data is divided into two components: the variation within each sample and the variation between the means of different samples. To perform the analysis of variance test, we compare the within-sample variation to the between-sample variation by computing an F statistic. If the appropriate assumptions hold, the F statistic will follow the F distribution, which was discussed in Section 12.5.

Format for one-way analysis of variance

We have K independent random samples of $n_1, n_2, \ldots, n_K$ observations from K populations. Each population is assumed to be normally distributed with a common variance σ^2. The K population means are denoted $\mu_1, \mu_2, \ldots, \mu_K$. One-way analysis of variance is used to test the null hypothesis

$$H_0: \quad \mu_1 = \mu_2 = \cdots = \mu_K$$

against the alternative hypothesis

$$H_1: \quad \text{At least two of the means differ.}$$

The Sample Observations

There are K populations, and the sample size for the jth sample is denoted n_j. We denote the sample observations by

$$x_{ij} \quad \text{where } i = 1, 2, \ldots, n_j \text{ and } j = 1, 2, \ldots, K$$

The symbol x_{ij} denotes the ith observation from the jth sample. The sample observations can be displayed as in Table 14.3.

TABLE 14.3 *Sample observations from independent random samples from K populations*

	POPULATION		
1	2	$\cdots$	K
x_{11}	x_{12}	$\cdots$	x_{1K}
x_{21}	x_{22}	$\cdots$	x_{2K}
$\vdots$	$\vdots$	$\vdots$	$\vdots$
$x_{n_1 1}$	$x_{n_2 2}$	$\cdots$	$x_{n_K K}$

Thus, we have a random sample of n_1 observations from population 1, a sample of n_2 observations from population 2, and so on up to a sample of n_K observations from population K. (The sample sizes do not have to be equal.) The total sample size is thus

$$n = n_1 + n_2 + \cdots + n_K$$

The Sample Means

Let $\bar{x}_1, \bar{x}_2, \ldots, \bar{x}_K$ denote the sample means of the K samples, and let $\bar{x}$ denote the average value of all n observations. The jth sample mean $\bar{x}_j$ is calculated as follows:

$$\bar{x}_j = \frac{\sum_{i=1}^{n} x_{ij}}{n_j} \quad \text{for } j = 1, 2, \ldots, K$$

In Table 14.1, the three sample means are $\bar{x}_1 = 19.0$, $\bar{x}_2 = 20.4$, and $\bar{x}_3 = 21.8$.

The null hypothesis states that the three population means are equal to some common value μ. To estimate this common value, we calculate the sample mean of all the sample observations. This overall sample mean, which is the sum of all the sample values divided by n, is denoted $\bar{x}$. Thus, the overall sample mean is calculated using the following formula:

$$\bar{x} = \sum_{j=1}^{K} \sum_{i=1}^{n_j} \frac{x_{ij}}{n} = \sum_{j=1}^{K} \frac{n_j \bar{x}_j}{n}$$

For the data in Table 14.1, the overall mean is

$$\bar{x} = \frac{(5)(19.0) + (5)(20.4) + (5)(21.8)}{15} = 20.4$$

Sums of Squares in ANOVA

As mentioned, the test of equality of means is based on comparing the variability within the samples to the variability between the samples. The *total sum of squares,* or the *sum of squares total,* measures the total variation in the entire sample of data and is denoted SST. SST is the sum of the squared deviations of all n observations x_{ij} about the overall mean $\bar{x}$.

DEFINITION   **Total Sum of Squares**

The **total sum of squares,** denoted SST, is calculated as

$$SST = \sum_{j=1}^{K} \sum_{i=1}^{n_j} (x_{ij} - \bar{x})^2$$

ANOVA partitions this total sum of squares into two components, the variation between the K samples and the variation within each sample. First, consider the variation within the K samples. The variability within the first sample is measured as the sum of the squared deviations of the observations about their sample mean $\bar{x}_1$. Denote this sum of squares within sample 1 as SS_1, that is,

$$SS_1 = \sum_{i=1}^{n_1} (x_{i1} - \bar{x}_1)^2$$

From Table 14.1, we obtain

$$SS_1 = (18.2 - 19.0)^2 + (19.4 - 19.0)^2 + (19.6 - 19.0)^2$$
$$+ (19.0 - 19.0)^2 + (18.8 - 19.0)^2$$
$$= 1.20$$

Similarly, for the sample from the second population, we obtain

$$SS_2 = \sum_{i=1}^{n_2} (x_{i2} - \bar{x}_2)^2$$

From Table 14.1, we obtain

$$SS_2 = (19.8 - 20.4)^2 + (21.0 - 20.4)^2 + (20.0 - 20.4)^2$$
$$+ (20.8 - 20.4)^2 + (20.4 - 20.4)^2$$
$$= 1.04$$

Finally, we obtain

$$SS_3 = (21.2 - 21.8)^2 + (21.8 - 21.8)^2 + (22.4 - 21.8)^2$$
$$+ (22.0 - 21.8)^2 + (21.6 - 21.8)^2$$
$$= .80$$

DEFINITION   **Sum of Squares Within**

The total variation within each of the K samples, denoted SSW, is called the **sum of squares within** and is calculated

$$SSW = \sum_{i=1}^{n_1} (x_{i1} - \bar{x}_1)^2 + \sum_{i=1}^{n_2} (x_{i2} - \bar{x}_2)^2 + \cdots + \sum_{i=1}^{n_K} (x_{iK} - \bar{x}_K)^2$$
$$= \sum_{j=1}^{K} \sum_{i=1}^{n_j} (x_{ij} - \bar{x}_j)^2$$
$$= SS_1 + SS_2 + \cdots + SS_K$$

Within the jth sample, the observations x_{ij} differ from the jth sample mean $\bar{x}_j$ due to unexplained random variation. Thus, the sum of squares within is also called the **sum of squares due to error** or the **error sum of squares,** and is frequently denoted SSE.

Thus, for the data in Table 14.1, the sum of squares within is

$$SSW = SS_1 + SS_2 + SS_3 = 1.20 + 1.04 + .80 = 3.04$$

Next we need to measure the variability between the samples. This is based on the squared differences

$$(\bar{x}_j - \bar{x})^2 \quad \text{for } j = 1, 2, \ldots, K$$

Each of these squared deviations is weighted by the number of observations in the corresponding sample so that samples with more observations are given greater weights. Then we calculate the *between-group sum of squares.*

DEFINITION   **Between-Group Sum of Squares**

The total variation between the K samples is called the **between-group sum of squares** or the **sum of squares between,** denoted SSB, and is calculated as follows:

$$SSB = n_1(\bar{x}_1 - \bar{x})^2 + n_2(\bar{x}_2 - \bar{x})^2 + \cdots + n_K(\bar{x}_K - \bar{x})^2$$
$$= \sum_{j=1}^{K} n_j(\bar{x}_j - \bar{x})^2$$

The between-group sum of squares is sometimes called the **treatment sum of squares,** denoted SSTR.

For the data in Table 14.1, the between-group sum of squares is

$$SSB = (5)(19.0 - 20.4)^2 + (5)(20.4 - 20.4)^2 + (5)(21.8 - 20.4)^2$$
$$= 19.6$$

It can be shown algebraically that the total sum of squares always equals the sum of the between-group sum of squares plus the within-group sum of squares. Thus, we obtain the fundamental ANOVA identity in the accompanying box.

Sum-of-squares identity

$$\text{SST} = \text{SSB} + \text{SSW}$$

The sum-of-squares identity shows how the total variation in the sample data can be decomposed into variation within samples and variation between samples. For the data in Table 14.1, the total sum of squares is

$$\text{SST} = \text{SSB} + \text{SSW} = 19.6 + 3.04 = 22.64$$

Calculating the Mean Square Within

The test of the equality of the population means is based on three assumptions:

■ *Assumption 1*:   The observations X_{ij} are independent.
■ *Assumption 2*:   For each population, the variance of X_{ij} is σ^2.
■ *Assumption 3*:   For each population, X_{ij} has a normal distribution.

The ANOVA test is based on different ways of estimating the unknown variance σ^2. An unbiased estimate of σ^2 can be obtained using the sample data from population 1 as follows:

$$s_1^2 = \frac{\sum_{i=1}^{n_1} (x_{i1} - \bar{x}_1)^2}{n_1 - 1}$$

Similarly, another unbiased estimate of σ^2 can be obtained using the sample data from the second population as follows:

$$s_2^2 = \frac{\sum_{i=1}^{n_2} (x_{i2} - \bar{x}_2)^2}{n_2 - 1}$$

This procedure can be repeated using any of the K samples. If H_0 is true, then $\bar{x}_1, \bar{x}_2, \ldots,$ $\bar{x}_K$ are all unbiased estimates of the common mean μ, and $s_1^2, s_2^2, \ldots, s_K^2$ are all unbiased estimates of the variance σ^2. Each of these K estimates of σ^2 is inefficient, however, because none utilizes the entire sample of data. A better estimate of the common variance σ^2 of the K populations can be obtained by using the entire sample of $(n_1 + n_2 + \cdots + n_K)$ observations and calculating the pooled estimate

$$s_p^2 = \frac{(n_1 - 1)s_1^2 + (n_2 - 1)s_2^2 + \cdots + (n_K - 1)s_K^2}{(n_1 + n_2 + \cdots + n_K - K)}$$

where

$$s_j^2 = \frac{\sum\limits_{i=1}^{n_j} (x_{ij} - \bar{x}_j)^2}{n_j - 1}$$

The pooled estimate s_p^2 is a weighted average of the sample variances s_j^2, where the weight given to the sample variance s_j^2 is

$$\frac{n_j - 1}{n_1 + n_2 + \cdots + n_K - K} \qquad i = 1, 2, \ldots, K$$

and the weights sum to 1. Observe that

$$(n_j - 1)s_j^2 = \sum\limits_{i=1}^{n_j} (x_{ij} - \bar{x}_j)^2 = SS_j$$

Thus, the numerator of s_p^2 is the same as the sum of squares within (SSW), and we can write

$$s_p^2 = \frac{SSW}{n - K}$$

This pooled estimate s_p^2 is an unbiased estimate of σ^2 even when the population means differ. In one-way ANOVA, the estimate s_p^2 is called the *mean square within* and is denoted MSW.

DEFINITION   **Mean Square Within** ————————————————————————————————

The **mean square within,** denoted MSW, is

$$MSW = \frac{SSW}{n - K}$$

The denominator $(n - K)$ is the *degrees of freedom* for MSW. The mean square within is also called the **within-group mean square** and the **mean square error,** and is sometimes denoted MSE.

——

For the data in Table 14.1,

$$MSW = SSW/(n - K) = 3.04/(15 - 3) = .253$$

Calculating the Mean Square Between

Another way of estimating σ^2 utilizes the overall mean $\bar{x}$, the K sample means $\bar{x}_1$, $\bar{x}_2, \ldots, \bar{x}_K$, and the squared deviations $(\bar{x}_j - \bar{x})^2$. If the populations are all normal, then the sampling distribution of $\bar{X}_j$ is normal with mean μ and variance $Var(\bar{X}_j) = \sigma^2/n_j$. That is, the sampling distribution of $\bar{X}_j$ is $N(\mu, \sigma^2/n_j)$. Assume for the moment that all the sample sizes are equal, so that $n_1 = n_2 = \cdots = n_K$. If H_0 is true, then the random variables $\overline{X}_1, \overline{X}_2, \ldots, \overline{X}_K$ all have the same probability distribution. Since these random variables are independent, they represent a random sample of K observations selected from a normal distribution with mean μ and variance σ^2/n_j. Thus, the sample variance of the $\overline{X}_j$'s can be used to estimate the population variance σ^2/n_j. To estimate $\text{Var}(\overline{X}_j) = \sigma^2/n_j$, we use the formula

$$S_{\overline{X}_j}^2 = \frac{\sum\limits_{j=1}^{K} (\overline{X}_j - \overline{X})^2}{K-1}$$

The observed sample variance $s_{\overline{X}_j}^2$ is an unbiased estimate of $\sigma_{\overline{X}_j}^2 = \sigma^2/n_j$, where n_j is the common sample size, so $n_j s_{\overline{X}_j}^2$ is an unbiased estimate of σ^2. However, $n_j s_{\overline{X}_j}^2$ is equal to $\text{SSB}/(K-1)$ in the special case when all sample sizes are equal, so we have

$$n_j s_{\overline{X}_j}^2 = \frac{\sum\limits_{j=1}^{K} n_j (\overline{x}_j - \overline{x})^2}{K-1} = \frac{\text{SSB}}{K-1}$$

The estimator $\text{SSB}/(K-1)$ is called the *mean square between* and is denoted MSB. For the data in Table 14.1,

$$\text{MSB} = 19.6/(3-1) = 9.80$$

Thus, when the null hypothesis that all means are equal is true, $\text{SSB}/(K-1)$ provides an unbiased estimate of σ^2. SSB depends on the sum

$$\sum_{j=1}^{K} n_j (\overline{x}_j - \overline{x})^2$$

If H_0 is false, this sum increases, and then $n_j s_{\overline{X}_j}^2$ tends to overestimate σ^2. When the means μ_j are not all equal, $n_j s_{\overline{X}_j}^2$ is not an unbiased estimator of σ^2. It can be shown that the expected value of $n_j s_{\overline{X}_j}^2$ is

$$E(n_j s_{\overline{X}_j}^2) = \sigma^2 + \frac{\sum\limits_{j=1}^{K} n_j (\mu_j - \mu)^2}{K-1}$$

where $\mu = \Sigma \mu_j / K$. Hence, if H_0 is false, then $n_j S_{\overline{X}_j}^2$ will be a biased estimator of σ^2, and the bias will be positive. Also, the bias increases as $(\mu_j - \mu)^2$ increases.

DEFINITION   **Mean Square Between** ————————————————————————————

The **mean square between** is

$$\text{MSB} = \frac{\text{SSB}}{K-1}$$

The denominator $(K-1)$ is called the *degrees of freedom* for MSB. The mean square between is also called the **between-group mean square** and the **mean square due to treatments** and is sometimes denoted MSTR.

When the population means are *not* equal, MSB is *not* an unbiased estimator of σ^2. If the means differ, the expected value of MSB *exceeds* σ^2. Thus, the ratio of MSB to MSW provides information concerning whether the population means are equal.

The ANOVA test statistic F

In one-way ANOVA, the appropriate test statistic is

$$F = \frac{\text{MSB}}{\text{MSW}}$$

The ANOVA Test Statistic

If the null hypothesis is true, MSB and MSW both provide unbiased estimates of σ^2, and F should not be significantly different from 1. Thus, when F is close to 1, the data do not provide evidence against H_0. In contrast, if the population means differ, MSB tends to overestimate σ^2 whereas MSW remains an unbiased estimator of σ^2. Consequently, when H_0 is false, the F statistic tends to exceed 1. Thus, large values of F provide evidence against the null hypothesis.

If the three assumptions hold and if the null hypothesis of equality of means is true, the F statistic follows the F distribution with numerator degrees of freedom $(K - 1)$ and denominator degrees of freedom $(n - K)$.

Explanation of the Degrees of Freedom

The number of degrees of freedom for SSB is $(K - 1)$, or 1 less than the number of populations. The appropriate number of degrees of freedom for SST is $(n - 1)$, and the number of degrees of freedom for SSW is $(n - K)$.

Formula for degrees of freedom

Sums of Squares: $\text{SST} = \text{SSB} + \text{SSW}$
Degrees of Freedom: $(n - 1) = (K - 1) + (n - K)$

These values for the degrees of freedom can be explained as follows: It is always the case that the sum of the deviations from the overall sample mean is 0. This puts one constraint on the deviations from the overall sample mean. For SST, only $(n - 1)$ of the n deviations from the overall mean are free. The last deviation can always be determined from the first $(n - 1)$ deviations. Thus, SST has $(n - 1)$ degrees of freedom.

SSB depends on the K squared deviations $(\bar{x}_j - \bar{x})^2$. But once $\bar{x}$ and the sample sizes are known, only $(K - 1)$ of the sample means $\bar{x}_j$ are free. Once the first $(K - 1)$ sample means and $\bar{x}$ are known, the last sample mean can be determined from the equation

$$\bar{x} = \frac{n_1 \bar{x}_1 + n_2 \bar{x}_2 + \cdots + n_K \bar{x}_K}{n}$$

Thus, SSB has $(K - 1)$ degrees of freedom.

SSW depends on the squared deviations of the x_{ij}'s from the appropriate sample means $\bar{x}_j$. In each sample, the deviations from the appropriate sample mean $\bar{x}_j$ sum to 0, so in each sample one of the deviations is not free. Thus, one observation in each sample can be determined from the other $(n_j - 1)$ observations and thus is not free. Since one deviation is not free in each of the K samples, only $(n - K)$ of the n deviations from the K sample means are free. Thus, the sum of squares within SSW has $(n - K)$ degrees of freedom.

Performing a One-Way ANOVA Test

The procedure for a one-way analysis of variance test for a completely randomized design is presented in the accompanying box.

One-way ANOVA test

Suppose we wish to test the null hypothesis

$$H_0: \quad \mu_1 = \mu_2 = \cdots = \mu_K$$

against the alternative hypothesis

$$H_1: \quad \text{At least two means are different.}$$

It is assumed that the K populations are normally distributed with a common variance σ^2.

Step 1:   Obtain K independent random samples of $n_1, n_2, \ldots, n_K$ observations from the K populations. Let n denote the total sample size; that is,

$$n = n_1 + n_2 + \cdots + n_K$$

Step 2:   Calculate MSW and MSB as follows:

$$\text{MSW} = \frac{\text{SSW}}{n - K} \quad \text{and} \quad \text{MSB} = \frac{\text{SSB}}{K - 1}$$

Step 3:   Calculate the F statistic

$$F = \frac{\text{MSB}}{\text{MSW}}$$

To test H_0 against H_1 using level of significance α, use the rule

$$\text{Reject } H_0 \text{ in favor of } H_1 \text{ if } F > F_{\alpha, K-1, n-K}.$$

The critical value $F_{\alpha, K-1, n-K}$ is determined using the F distribution with numerator degrees of freedom $(K - 1)$ and denominator degrees of freedom $(n - K)$ such that

$$P(F > F_{\alpha, K-1, n-K}) = \alpha$$

EXAMPLE 14.2   **A One-Way ANOVA Test**

Use the data in Table 14.1 and test

$$H_0:\quad \mu_1 = \mu_2 = \mu_3$$

against the alternative hypothesis

$$H_1:\quad \text{At least two means differ.}$$

Let $\alpha = .05$.

SOLUTION   For the data in Table 14.1, we have shown that the mean squares within and between are MSW = .253 and MSB = 9.80. The observed value of the test statistic is

$$F = \frac{\text{MSB}}{\text{MSW}} = \frac{9.80}{.253} = 38.735$$

The critical value is obtained using the F distribution with numerator degrees of freedom $(K - 1) = 2$ and denominator degrees of freedom $(n - K) = 12$. The critical value is

$$F_{.05,2,12} = 3.89$$

The observed test statistic ($F = 38.735$) far exceeds the critical value (3.89) and thus falls in the critical region, so we reject H_0. (See Figure 14.3.) Now, refer back to the dot diagram in Figure 14.1, which shows the sample observations from the three populations. The ANOVA test confirms our suspicion that the data came from three populations having different means. The one-way ANOVA test says that there is too much variation between samples relative to the variation within each sample for us to believe that the population means are equal. Figure 14.1 supports this conclusion.

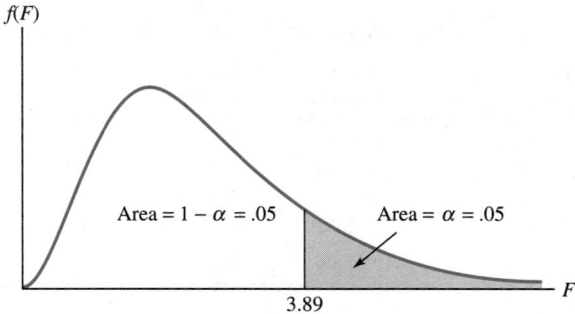

FIGURE 14.3   *Critical region of the one-way ANOVA test in Example 14.2*

Finding a *p*-Value

The *p*-value of this test is the probability of obtaining a value of F greater than or equal to 38.735, where there are 2 numerator degrees of freedom and 12 denominator degrees of freedom. In Table A.8 in the Appendix, the critical value of F associated with 2 and 12 degrees of freedom and right-tail probability of .01 is 6.93. So $P(F > 6.93) = .01$. It follows that $P(F > 38.735)$ is much less than .01. Thus, we know that the p-value of the test is much lower than .01.

Suppose the observed value of F had been 4.23. Refer to Table A.8 in the Appendix, and find the critical value of F associated with 2 and 12 degrees of freedom and right-tail probability of .05. The corresponding value of F is 3.89. This tells us that $P(F > 3.89) = .05$. Now refer to Table A.8 and find the critical value of F associated with 2 and 12 degrees of freedom and right-tail probability of .01. The corresponding value of F is 6.93. This tells us that $P(F > 6.93) = .01$. From this information, we conclude that $P(F > 4.23)$ is between .01 and .05. Thus, we know that the p-value of the test is between .01 and .05.

The One-Way ANOVA Table

The various computations needed for one-way ANOVA are usually presented in a standard format called an *analysis of variance table*. Such a format is shown in Table 14.4. The one-way ANOVA table shows the three sums of squares SST, SSB, and SSW; their respective degrees of freedom; the respective mean squares; and the appropriate value of the F statistic. Table 14.5 is the one-way ANOVA table for the data in Table 14.1.

TABLE 14.4 *Format of a one-way ANOVA table*

Source of variation	Sum of squares	Degrees of freedom	Mean square	F
Between-group	SSB	$K - 1$	$MSB = SSB/(K - 1)$	MSB/MSW
Within-group	SSW	$n - K$	$MSW = SSW/(n - K)$	
Total	SST	$n - 1$		

TABLE 14.5 *ANOVA table for mileage data in Table 14.1 and Example 14.2*

Source of variation	Sum of squares	Degrees of freedom	Mean square	F
Between-group (make of car)	19.60	2	9.800	38.735
Within-group (random error)	3.04	12	.253	
Total	22.64	14		

EXAMPLE 14.3   **Another One-Way ANOVA Test**

Use the data in Table 14.2 and test

$$H_0: \quad \mu_1 = \mu_2 = \mu_3$$

against the alternative hypothesis

$$H_1: \quad \text{At least two means differ.}$$

Let $\alpha = .05$.

SOLUTION   For the data in Table 14.2, the sum of squares within is

$$SSW = (n_1 - 1)s_1^2 + (n_2 - 1)s_2^2 + (n_3 - 1)s_3^2$$
$$= (4)(10.66) + (4)(10.84) + (4)(15.76)$$
$$= 149.04$$

The mean square within is

$$MSW = SSW/(n - K) = 149.04/12 = 12.42$$

and the overall mean is $\bar{x} = 20.4$.

The sum of squares between is

$$SSB = n_1(\bar{x}_1 - \bar{x})^2 + n_2(\bar{x}_2 - \bar{x})^2 + n_3(\bar{x}_3 - \bar{x})^2$$
$$= (5)(19.0 - 20.4)^2 + (5)(20.4 - 20.4)^2 + (5)(21.8 - 20.4)^2$$
$$= 19.6$$

The between-group mean square is

$$MSB = SSB/(K - 1) = 19.6/2 = 9.8$$

The observed F statistic is

$$F = \frac{MSB}{MSW} = \frac{9.80}{12.42} = .79$$

The critical value of the test statistic is obtained using the F distribution with numerator degrees of freedom $(K - 1) = 2$ and denominator degrees of freedom $(n - K) = 12$. The critical value is

$$F_{.05,2,12} = 3.89$$

Because the observed test statistic (.79) is less than the critical value (3.89) and thus falls in the acceptance region, we do not reject H_0.

Now refer back to the dot diagram in Figure 14.2. Although the three sample means are the same as in Example 14.2, the conclusion is reversed because the within-sample variation is much greater for these data than for the data in Table 14.1. Table 14.6 shows the corresponding one-way ANOVA table.

TABLE 14.6   *ANOVA table for mileage data in Table 14.2 and Example 14.3*

Source of variation	Sum of squares	Degrees of freedom	Mean square	F
Between-group (make of car)	19.60	2	9.80	.79
Within-group (random error)	149.04	12	12.42	
Total	168.64	14		

Assumptions Underlying the ANOVA Test

The one-way ANOVA test requires that the populations have normal distributions; otherwise the distribution of the F statistic would not exactly follow the F distribution. The use of the F distribution will be approximately correct, however, if the distributions of the populations do not differ too much from normality; the accuracy of the approximation improves as the distributions approach normality.

The test also depends on the assumption that the variances of the different populations are equal. If the variances are not equal, the test is inappropriate. The idea underlying the F test is that if the means of the populations are equal, we can pool all the data together and treat the observations as if they had all been obtained from the same population. If the population variances differ, this argument does not hold. Thus, the assumption that each population has the same variance is important.

Exercises

Statistical Concepts

14.1.1   Describe how a dot diagram can be useful in detecting differences in population means.

14.1.2   *True or false:*   In ANOVA, the dependent variable being studied is called a factor and the independent variable is called a treatment. Explain.

14.1.3   *True or false:*   In ANOVA, a completely randomized design involves comparing the means of K populations by drawing independent random samples from each of the K populations. Explain.

14.1.4   State and explain the null and alternative hypotheses that are being tested in analysis of variance.

14.1.5   *True or false:*   In ANOVA, the test of equality of means is based on comparing the variability *within* the samples to the variability *between* the samples. This is why the procedure is called analysis of *variance*. Explain.

14.1.6   The test of equality of a set of population means is based on three assumptions. State each assumption and explain what it means.

14.1.7   *True or false:*   The mean square within (MSW) is SSW/$(K - 1)$, where $(K - 1)$ is called the degrees of freedom for MSW. Explain.

14.1.8   *True or false:*   The mean square between (MSB) is SSB/$(n - 1)$, where $(n - 1)$ is called the degrees of freedom for MSB. Explain.

14.1.9   State the appropriate numbers of degrees of freedom for SST, SSB, and SSW.

14.1.10   *True or false:*   In one-way ANOVA, the appropriate test statistic is $F = $ MSW/MSB. Explain.

14.1.11   *True or false:*   In one-way ANOVA, the appropriate test statistic is $F = $ MSB/$(K - 1)$. Explain.

14.1.12   *True or false:*   In one-way ANOVA, the appropriate test statistic can be written as

$$F = \frac{\text{SSW}/(n - K)}{\text{SSB}/(K - 1)}$$

Explain.

14.1.13   *True or false:*   In one-way ANOVA, the critical region for the F statistic is placed in both tails of the distribution. Explain.

Statistical Drills

14.1.14   The following data show the results of a completely randomized design experiment in which each of two treatments was replicated five times.

Sample 1	Sample 2
5.1	7.0
6.1	7.4
7.3	8.5
6.0	6.4
4.2	6.7

a. Draw a dot diagram. Does it appear that the data came from populations having different means?
b. Calculate the sample mean for each sample.
c. Calculate SST, SSW, and SSB.
d. Calculate MSW and MSB.
e. Calculate the test statistic $F = $ MSB/MSW. What are the numerator and denominator degrees of freedom?
f. Perform the ANOVA test for significant differences among the population means. Use $\alpha = .05$.
g. What is the approximate p-value for the test? Does this p-value agree with your test results?

14.1.15   The accompanying data show the results of a completely randomized design experiment in which each of four treatments was replicated five times.

Sample 1	Sample 2	Sample 3	Sample 4
5.4	7.3	9.5	8.1
6.3	7.8	7.1	7.2
7.8	8.8	7.6	6.3
6.6	5.2	6.9	5.8
5.2	6.5	8.3	7.5

a. Draw a dot diagram. Does it appear that the data came from populations having different means?
b. Calculate the sample mean for each sample.
c. Calculate SST, SSW, and SSB.
d. Calculate MSW and MSB.
e. Calculate the test statistic $F = $ MSB/MSW. What are the numerator and denominator degrees of freedom?
f. Perform the ANOVA test for significant differences among the population means. Use $\alpha = .05$.
g. What is the approximate p-value for the test? Does this p-value agree with your test results?

14.1.16   An experiment was conducted in a completely randomized design format. A sample of 10 observations was obtained from each of 4 populations. Some of the relevant data are shown in the following table.

Source	d.f.	SS	MS	F
Between	_____	_____	63.2	_____
Within	_____	107.5	_____	
Total	_____	_____		

a. Find the appropriate numbers of degrees of freedom for each sum of squares.
b. Find SSB and SST.
c. Find MSW.
d. Calculate the F statistic for testing for significant differences among the population means.
e. Let $\alpha = .05$. Find the critical value of the test statistic.
f. Perform the test.

14.1.17   An experiment was conducted in a completely randomized design format. A sample of 16 observations was obtained from each of 5 populations. Some of the relevant data are shown in the following table.

Source	d.f.	SS	MS	F
Between	____	____	97.3	____
Within	____	149.1	____	
Total	____	____		

a. Find the appropriate numbers of degrees of freedom for each sum of squares.
b. Find SSB and SST.
c. Find MSW.
d. Calculate the F statistic for testing for significant differences among the population means.
e. Let $\alpha = .05$. Find the critical value of the test statistic.
f. Perform the test.

Statistical Applications

14.1.18   An agronomist at the Department of Agriculture wants to test the effects of three different brands of fertilizer on the yields of tomato plants: 5 plants are fertilized with brand 1, 4 plants with brand 2, and 6 plants with brand 3. The yields of the tomato plants in pounds are as follows:

Brand 1:   5.4, 4.8, 4.6, 5.2, 5.9
Brand 2:   4.7, 4.4, 3.7, 4.2
Brand 3:   6.1, 6.0, 6.2, 4.7, 5.2, 6.0

Calculate the following:
a. Sample means for the three brands of fertilizer
b. SST, SSB, SSW
c. MSB, MSW
d. F
e. Degrees of freedom for SSB and SSW
f. Present the relevant information in an ANOVA table.
g. Test the null hypothesis that the three population means are equal.

14.1.19   A study was undertaken to determine whether the mean cost of a market basket of food is the same in four cities. Five stores were sampled in each city, and the cost of the market basket of goods was recorded as follows:

Store	New York 1	Boston 2	Chicago 3	Detroit 4
1	75	80	81	68
2	78	79	79	74
3	82	83	84	75
4	79	77	80	73
5	74	77	77	75

Calculate the following:
a. $\bar{x}_1, \bar{x}_2, \bar{x}_3, \bar{x}_4$
b. SST, SSB, SSW

   c. MSB, MSW
   d. F
   e. Degrees of freedom for SSB and SSW
   f. Test the null hypothesis that the four population means are equal. Let $\alpha = .05$.

14.1.20   During rush hour, policemen are assigned to various downtown intersections to speed up the flow of traffic and reduce congestion. To determine the intersections most in need of police supervision, the number of automobiles passing through four intersections was recorded for a 30-minute period on 6 days. The data are as follows:

Day	Exit 1	Exit 2	Exit 3	Exit 4
1	400	470	360	320
2	470	490	400	410
3	430	420	425	440
4	490	470	455	400
5	510	500	450	460
6	390	360	370	380

   a. Calculate the mean number of cars passing through each intersection.
   b. Calculate SST, SSB, and SSW.
   c. Let $\alpha = .05$. Test whether the population mean is the same at each intersection.

14.1.21   The pollution index for five cities was taken on 8 randomly selected days. The data were as follows:

City	$\bar{x}_i$	s_i^2
1	108	25
2	111	28
3	116	26
4	104	32
5	109	36

   Use $\alpha = .05$ and test $H_0: \mu_1 = \mu_2 = \mu_3$.

14.2   The Statistical Model for One-Way ANOVA

In this section, we describe the one-way ANOVA model and provide additional insight into how the one-way ANOVA test works.

One-way ANOVA model for a completely randomized design

> Let X_{ij} denote the ith observation selected from the jth population where $i = 1$, $2, \ldots, n_j$ and $j = 1, 2, \ldots, K$. Let μ_j denote the jth population mean. Let e_{ij} denote the amount by which X_{ij} differs from μ_j, so that
>
> $$e_{ij} = X_{ij} - \mu_j$$
>
> (*continued*)

The mathematical model for a one-way ANOVA model is

$$X_{ij} = \mu_j + e_{ij}$$

where e_{ij} represents a random effect associated with the ith observation from the jth population.

The random effects e_{ij} are assumed to be normally and independently distributed with mean 0 and homogeneous variance σ^2. That is,

1. $E(e_{ij}) = 0$ for all i and j.
2. $\text{Var}(e_{ij}) = \sigma^2$ for all i and j.
3. The e_{ij}'s are independently distributed.
4. The e_{ij}'s are normally distributed.

The equation

$$X_{ij} = \mu_j + e_{ij}$$

states that the random variable X_{ij} consists of two components: the jth population mean μ_j plus a random amount by which X_{ij} differs from μ_j.

The mean value of population j can be denoted

$$\mu_j = \mu + t_j \quad j = 1, 2, 3, \ldots, K$$

where μ denotes the overall mean in all K populations, and

$$t_j = \mu_j - \mu$$

denotes the treatment effect associated with being in population j. The null hypothesis states that all the population means are equal to a common value μ (with no subscript):

$$H_0: \quad \mu_1 = \mu_2 = \cdots = \mu_K = \mu$$

This is equivalent to the null hypothesis that the treatment effects are all 0:

$$H_0: \quad t_1 = t_2 = \cdots = t_K = 0$$

If the null hypothesis is true, there is no treatment effect, and each value X_{ij} consists of the common effect μ and a random error e_{ij}.

Derivation of the Sum-of-Squares Identity

We have the following relationships:

$$X_{ij} = \mu_j + e_{ij} = \mu + t_j + e_{ij} \quad i = 1, 2, \ldots, n_j; j = 1, 2, \ldots, K$$

$$t_j = \mu_j - \mu$$

$$e_{ij} = X_{ij} - \mu_j$$

Rearranging the first equation yields

$$X_{ij} - \mu = t_j + e_{ij}$$

Substituting for t_j and e_{ij} yields

$$X_{ij} - \mu = (\mu_j - \mu) + (X_{ij} - \mu_j)$$

If we replace μ by its estimate $\bar{x}$ and μ_j by its estimate $\bar{x}_j$, we obtain the sample identity

$$x_{ij} - \bar{x} = (\bar{x}_j - \bar{x}) + (x_{ij} - \bar{x}_j)$$

Squaring both sides of this equation yields

$$(x_{ij} - \bar{x})^2 = (\bar{x}_j - \bar{x})^2 + (x_{ij} - x_j)^2 + 2(\bar{x}_j - \bar{x})(x_{ij} - \bar{x}_j)$$

If this equation is summed over all i and j, the last term on the right sums to 0, and we obtain

$$\sum_{j=1}^{K} \sum_{i=1}^{n_j} (x_{ij} - \bar{x})^2 = \sum_{j=1}^{K} \sum_{i=1}^{n_j} (x_j - \bar{x})^2 + \sum_{j=1}^{K} \sum_{i=1}^{n_j} (x_{ij} - \bar{x}_j)^2$$

The first term to the right of the equals sign can be simplified to give the following formula, which shows the relationship among the three sums of squares SST, SSB, and SSW:

$$\sum_{j=1}^{K} \sum_{i=1}^{n_j} (x_{ij} - \bar{x})^2 = \sum_{j=1}^{K} n_j(\bar{x}_j - \bar{x})^2 + \sum_{j=1}^{K} \sum_{i=1}^{n_j} (x_{ij} - \bar{x}_j)^2$$

or

$$\text{SST} = \text{SSB} + \text{SSW}$$

A Graphical Explanation of ANOVA

In one-way ANOVA, the F test statistic is the ratio of the two mean squares MSB and MSW. When the null hypothesis that the population means are all equal is true, MSB and MSW are both unbiased estimators of the variance σ^2. When the null hypothesis is false, the mean square MSB tends to overestimate σ^2. This relationship can be shown graphically.

Figure 14.4 (page 568) depicts $K = 3$ populations of data. Each population is assumed to have a normal distribution with the same variance σ^2. If H_0 is true, the three populations are identical, and we can treat all the observations as one sample of size $(n_1 + n_2 + n_3)$ from a single population having mean μ and variance σ^2.

To estimate the variance σ^2, we could pool all the sample data and obtain one pooled sample variance. Alternatively, we could obtain the sample variance for each of the three samples and estimate the pooled variance by taking a weighted average of the three sample variances. If H_0 is true, both of these methods should yield unbiased estimates of the population variance.

Now consider the case when H_0 is false, so that at least two of the population means are different. This situation is illustrated in Figure 14.5 (page 568), which shows three normal distributions having equal variances but different means. A weighted average of the three sample variances will still yield an unbiased estimate of the variance σ^2.

Now consider what happens if we pool the data from the three populations. Figure 14.6 (page 569) shows how the total variation in the data remains constant if we pool populations having equal means and variances, whereas the total variation in the data increases if we pool populations having equal variances but different means. The term *analysis of variance* is derived from the fact that when testing H_0 against H_1, we are in a sense testing whether the variance of each individual population is the same as the variance of the pooled data after the populations have been pooled.

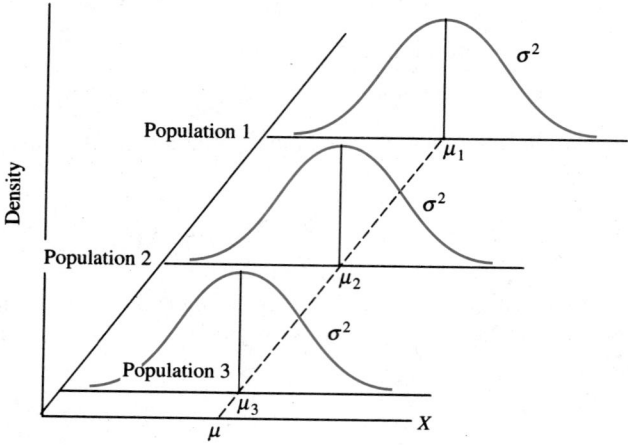

The null hypothesis is true:

(a) Populations are normal.
(b) Populations have the same variance σ^2.
(c) Populations have the *same* mean ($\mu_1 = \mu_2 = \mu_3 = \mu$).

FIGURE 14.4   *Three normal distributions having equal means*

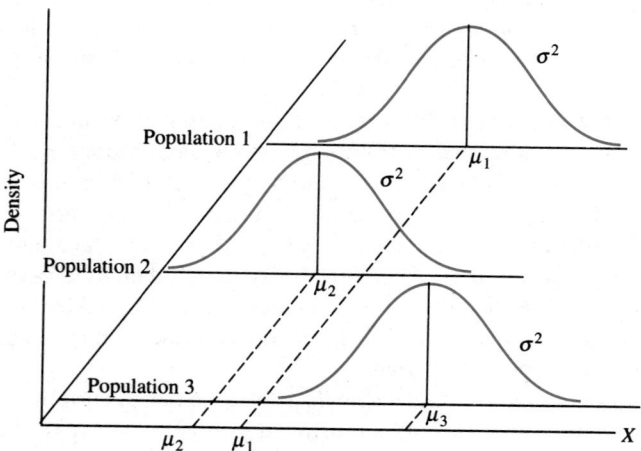

The null hypothesis is false:

(a) Populations are normal.
(b) Populations have the same variance σ^2.
(c) Populations have *different* means.

FIGURE 14.5   *Three normal distributions having different means*

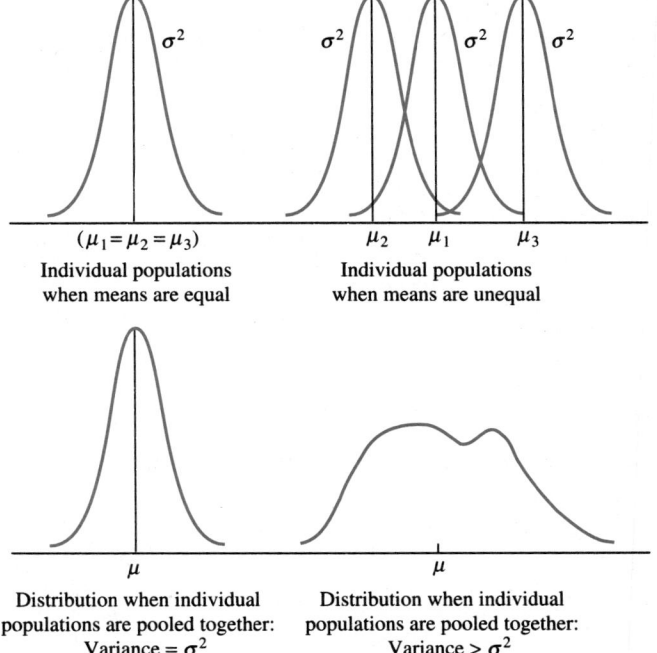

FIGURE 14.6 *Pooling data from normal distributions with equal means and pooling data from normal distributions with different means*

Equivalence of the *t* Test and One-Way ANOVA When Testing H_0: $\mu_1 = \mu_2$ Against H_1: $\mu_1 \neq \mu_2$

Suppose we have independent random samples of n_1 and n_2 observations from two populations and we wish to test the null hypothesis H_0: $\mu_1 = \mu_2$ against the alternative hypothesis H_1: $\mu_1 \neq \mu_2$. It is assumed that both populations are normal and have the same unknown variance σ^2. In Section 12.2, we showed how to test H_0 against H_1 by using the *t* statistic

$$t = \frac{\bar{x}_1 - \bar{x}_2}{\sqrt{s_p^2(1/n_1 + 1/n_2)}}$$

where s_p^2 is the pooled estimate of the variance

$$s_p^2 = \frac{\Sigma(x_{i1} - \bar{x}_1)^2 + \Sigma(x_{i2} - \bar{x}_2)^2}{n_1 + n_2 - 2}$$

It is also possible to perform the test using the one-way ANOVA test based on the test statistic

$$F = \frac{\text{MSB}}{\text{MSW}}$$

where the number of populations is $K = 2$. We will show that when $K = 2$, the one-way ANOVA test is equivalent to the t test discussed in Section 12.2 because the F statistic used in one-way ANOVA equals the square of the t statistic; that is, $F = t^2$. This can be shown as follows.

When there are only two populations, the F statistic is

$$F = \frac{MSB}{MSW} = \frac{SSB/(K - 1)}{SSW/(n - K)}$$

$$= \frac{[n_1(\bar{x}_1 - \bar{x})^2 + n_2(\bar{x}_2 - \bar{x})^2]/(2 - 1)}{[\Sigma(x_{i1} - \bar{x}_1)^2 + \Sigma(x_{i2} - \bar{x}_2)^2]/(n_1 + n_2 - 2)}$$

If we square the t statistic, we obtain

$$t^2 = \frac{(\bar{x}_1 - \bar{x}_2)^2}{s_p^2(1/n_1 + 1/n_2)} = \frac{n_1 n_2}{n_1 + n_2}\left[\frac{(\bar{x}_1 - \bar{x}_2)^2}{s_p^2}\right]$$

To show that $F = t^2$, we use the following identity:

$$\bar{x} = \frac{n_1\bar{x}_1 + n_2\bar{x}_2}{n_1 + n_2}$$

From this we obtain

$$\bar{x}_1 - \bar{x} = \frac{n_2(\bar{x}_1 - \bar{x}_2)}{n_1 + n_2}$$

and

$$\bar{x}_2 - \bar{x} = \frac{n_1(\bar{x}_2 - \bar{x}_1)}{n_1 + n_2}$$

Substituting these two equations into the formula for SSB yields

$$SSB = n_1(\bar{x}_1 - \bar{x})^2 + n_2(\bar{x}_2 - \bar{x})^2$$

$$= \frac{n_1 n_2^2(\bar{x}_1 - \bar{x}_2)^2 + n_2 n_1^2(\bar{x}_2 - \bar{x}_1)^2}{(n_1 + n_2)^2}$$

$$= \frac{n_1 n_2}{n_1 + n_2}(\bar{x}_1 - \bar{x}_2)^2$$

Next we obtain

$$\frac{SSW}{n_1 + n_2 - 2} = \frac{\Sigma(x_{i1} - \bar{x}_1)^2 + \Sigma(x_{i2} - \bar{x}_2)^2}{n_1 + n_2 - 2}$$

$$= s_p^2$$

Substitution for SSB and SSW/$(n_1 + n_2 - 2)$ in the formula for F gives the desired result of $F = t^2$.

Furthermore, when the numerator degrees of freedom is $\nu_1 = 1$ and the denominator degrees of freedom is ν_2, the critical value of the F distribution will always be the square of the critical value of t having ν_2 degrees of freedom. This last statement is easy to verify by checking some values from the t and F tables in the Appendix. For example, if the level of significance of a two-tailed t test is $\alpha = .05$ and the number of degrees of freedom is $\nu_2 = 10$, then the critical values of t are ± 2.228 (where 2.5% is the area in each tail of the distribution). For the F test, we would use the F value for $\alpha = .05$ using degrees of freedom $\nu_1 = 1$ and $\nu_2 = 10$. This critical value is $F = 4.96$, which is equal to $t^2 = (2.228)^2$. This relationship is illustrated in the next example.

EXAMPLE 14.4 **Equivalence of One-Way ANOVA and the Two-Sample t Test** ——————

Suppose we want to test $H_0: \mu_1 = \mu_2$ against $H_1: \mu_1 \neq \mu_2$ using a 5% level of significance. Two independent random samples of size $n_1 = 21$ and $n_2 = 21$ are obtained, which have the following sample means and sample variances: $\bar{x}_1 = 30$, $s_1^2 = 8$; $\bar{x}_2 = 33$, $s_2^2 = 10$. It is assumed that both populations are normal and have equal variances.

SOLUTION For the t test, we first calculate the pooled estimate of the variance. We obtain

$$
\begin{aligned}
s_p^2 &= \frac{\Sigma(x_{i1} - \bar{x}_1)^2 + \Sigma(x_{i2} - \bar{x}_2)^2}{n_1 + n_2 - 2} \\
&= \frac{(n_1 - 1)s_1^2 + (n_2 - 1)s_2^2}{n_1 + n_2 - 2} \\
&= \frac{(20)(8) + (20)(10)}{40} \\
&= 9
\end{aligned}
$$

Next we calculate the t statistic as follows:

$$
\begin{aligned}
t &= \frac{\bar{x}_1 - \bar{x}_2}{\sqrt{s_p^2(1/n_1 + 1/n_2)}} \\
&= \frac{30 - 33}{\sqrt{9/21 + 9/21}} = -3.240
\end{aligned}
$$

The critical values of the t statistic are $t = \pm 2.021$. We reject H_0 because the t statistic is less than -2.021 and thus falls in the critical region.

For the ANOVA test, we use the F statistic

$$
\begin{aligned}
F &= \frac{\text{MSB}}{\text{MSW}} = \frac{\text{SSB}/(K - 1)}{\text{SSW}/(n - K)} \\
&= \frac{[n_1(\bar{x}_1 - \bar{x})^2 + n_2(\bar{x}_2 - \bar{x})^2]/(2 - 1)}{[\Sigma(x_{i1} - \bar{x}_1)^2 + \Sigma(x_{i2} - \bar{x}_2)^2]/(n_1 + n_2 - 2)}
\end{aligned}
$$

To calculate F, we need to find the overall mean $\bar{x}$. We obtain

$$
\begin{aligned}
\bar{x} &= \frac{n_1\bar{x}_1 + n_2\bar{x}_2}{n_1 + n_2} \\
&= \frac{(21)(30) + (21)(33)}{42} = 31.5
\end{aligned}
$$

We have already shown that $SSW/(n - K) = s_p^2$. Thus, the denominator of the F statistic is 9, and the F statistic becomes

$$F = \frac{[n_1(\bar{x}_1 - \bar{x})^2 + n_2(\bar{x}_2 - \bar{x})^2]/(2 - 1)}{[\Sigma(x_{i1} - \bar{x}_1)^2 + \Sigma(x_{i2} - \bar{x}_2)^2]/(n_1 + n_2 - 2)}$$

$$= \frac{[(21)(30 - 31.5)^2 + (21)(33 - 31.5)^2]/1}{9}$$

$$= 10.5$$

Note that $F = 10.5 = (-3.240)^2 = t^2$.

The critical value of F having degrees of freedom $\nu_1 = 1$ and $\nu_2 = 40$ is $F = 4.08$, which is the square of the critical value of t. We reject H_0 because $F = 10.5$ falls in the critical region. This example illustrates that when only two means are being tested, the one-way ANOVA test and the t test yield equivalent results.

Exercises

Statistical Concepts

14.2.1 The mathematical model for a one-way ANOVA model is $X_{ij} = \mu_j + e_{ij}$, where e_{ij} represents a random effect. What four assumptions are made about the random effects?

14.2.2 *True or false:* Suppose three populations are normal and have the same variance σ^2 but have different means. If we pool the data from the three populations, the pooled variance will be larger than σ^2. Explain.

14.2.3 *True or false:* Suppose we have independent random samples from two normal populations and wish to test $H_0: \mu_1 - \mu_2 = 0$. The F statistic used in ANOVA will be the square of the t statistic. Explain.

Statistical Drills

14.2.4 The accompanying data show the results of a completely randomized design experiment in which each of two treatments was replicated five times.

Sample 1	Sample 2
4.2	7.0
6.9	6.4
5.3	8.5
4.7	4.4
4.3	5.7

a. Draw a dot diagram. Does it appear that the data came from populations having different means?
b. Calculate the sample mean for each sample.
c. Calculate SST, SSW, and SSB.
d. Calculate MSW and MSB.
e. Calculate the test statistic $F = MSB/MSW$. What are the numerator and denominator degrees of freedom?
f. Perform the ANOVA test for significant differences among the population means. Use $\alpha = .05$.

g. Calculate the pooled estimate of the variance.
h. Calculate the test statistic t and show that $F = t^2$.
i. Let $\alpha = .05$. Use the t test and perform a two-tailed test for equivalence of means.

14.2.5 The following data show the results of a completely randomized design experiment in which each of two treatments was replicated 10 times.

Sample 1	Sample 2
16.3	17.5
27.3	12.2
19.2	13.4
25.7	24.1
18.2	31.2
8.7	17.8
14.5	10.2
25.8	15.7
25.7	17.2
17.3	26.4

a. Draw a dot diagram. Does it appear that the data came from populations having different means?
b. Calculate the sample mean for each sample.
c. Calculate SST, SSW, and SSB.
d. Calculate MSW and MSB.
e. Calculate the test statistic $F = \text{MSB/MSW}$. What are the numerator and denominator degrees of freedom?
f. Perform the ANOVA test for significant differences among the population means. Use $\alpha = .05$.
g. Calculate the pooled estimate of the variance.
h. Calculate the test statistic t and show that $F = t^2$.
i. Let $\alpha = .05$. Use the t test and perform a two-tailed test for equivalence of means.

Statistical Applications

MKTG

14.2.6 A marketing representative wants to determine whether the mean income of subscribers to *Sports Illustrated* is the same as that of *Newsweek* subscribers. A random sample of $n_1 = 15$ subscribers to *Sports Illustrated* and an independent random sample of $n_2 = 15$ subscribers to *Newsweek* are obtained. The incomes (in thousands of dollars) are as follows:

Sports Illustrated: 24, 34, 32, 45, 13, 17, 24, 27, 31, 19, 20, 42, 13, 41, 25
Newsweek: 41, 37, 65, 15, 26, 35, 52, 27, 28, 47, 36, 44, 68, 21, 33

a. Calculate the sample means and sample variances.
b. Calculate s_p^2, the pooled estimate of the variance.
c. Calculate the t statistic based on the sample data.
d. Find the critical value of t using $\alpha = .05$.
e. Do you accept or reject H_0?

14.2.7 Use the data in Exercise 6 and perform a one-way ANOVA test using $\alpha = .05$.
a. Calculate the observed value of the F statistic.
b. Verify that this value is the square of the t statistic obtained in part **c** of Exercise 6.
c. Calculate the critical value of F using $\alpha = .05$.
d. Verify that the critical value of F is the square of the critical value of t.
e. Do you accept or reject H_0?

Computer Applications

Statistical computer packages perform all the arithmetic calculations necessary to perform a one-way analysis of variance test. The following example uses computer output generated by the SPSSX statistical program, but other computer packages, including Minitab and SAS, generate similar outputs.

EXAMPLE 14.5   ## Using SPSSX for One-Way ANOVA

The cost of maintaining and repairing a car can be an important factor in determining what make of car a customer will buy. The annual maintenance and repair costs are recorded for each car in samples of 20 cars of each of three makes. Use the data in Table 14.7 and the SPSSX computer package to test the null hypothesis

$$H_0: \quad \mu_1 = \mu_2 = \mu_3$$

against the alternative hypothesis

$$H_1: \quad \text{The means are not all equal.}$$

Use a 5% level of significance. The populations are assumed to be normal and the population variances to be equal.

TABLE 14.7   *Annual automobile repair costs for Example 14.5*

Car	Make 1	Make 2	Make 3
1	$280	$ 90	$110
2	164	80	92
3	301	67	84
4	114	86	86
5	88	43	96
6	76	119	128
7	123	172	61
8	94	74	184
9	204	27	58
10	89	62	67
11	53	80	82
12	40	128	110
13	27	104	90
14	104	32	64
15	33	60	80
16	27	40	66
17	43	27	75
18	36	18	48
19	40	24	57
20	61	50	63

SOLUTION   The following instructions perform a one-way ANOVA test using the SPSSX program:

```
TITLE 'ANNUAL MAINTENANCE COSTS'
DATA LIST FREE/CAR COST
BEGIN DATA
1 280
2  90
3 110
1 164
2  80
3  92
    .
    .
    .
END DATA
ONEWAY COST BY CAR(1,3)
STATISTICS ALL
FINISH
```

When we input the data, each observation is coded as a 1, a 2, or a 3, depending on the make of car. Figure 14.7 shows the SPSSX computer output. The output shows the three sample means and standard deviations to be $\bar{x}_1 = 99.85$, $s_1 = 80.3782$; $\bar{x}_2 = 69.15$, $s_2 = 39.9648$; $\bar{x}_3 = 85.05$, $s_3 = 31.0339$. For the entire sample of 60 observations, the overall sample mean and standard deviation are $\bar{x} = 84.6833$, $s = 55.3612$. The output also shows a 95% confidence interval for each of the means.

- O N E W A Y -

```
        Variable  COST      ESTIMATED ANNUAL COST OF REPAIRS
      By Variable  CAR       TYPE OF CAR

                             ANALYSIS OF VARIANCE
```

SOURCE	D.F.	SUM OF SQUARES	MEAN SQUARES	F RATIO	F PROB.
BETWEEN GROUPS	2	9428.9333	4714.4667	1.5678	.2174
WITHIN GROUPS	57	171398.0500	3006.9833		
TOTAL	59	180826.9833			

GROUP	COUNT	MEAN	STANDARD DEVIATION	STANDARD ERROR	MINIMUM	MAXIMUM	95 PCT CONF INT FOR MEAN	
Grp 1	20	99.8500	80.3782	17.9731	27.0000	301.0000	62.2318 TO	137.4682
Grp 2	20	69.1500	39.9648	8.9364	18.0000	172.0000	50.4459 TO	87.8541
Grp 3	20	85.0500	31.0339	6.9394	48.0000	184.0000	70.5257 TO	99.5743
TOTAL	60	84.6833	55.3612	7.1471	18.0000	301.0000	70.3820 TO	98.9847
FIXED EFFECTS MODEL			54.8360	7.0793			70.5073 TO	98.8594
RANDOM EFFECTS MODEL				8.8642			46.5432 TO	122.8234

```
RANDOM EFFECTS MODEL - ESTIMATE OF BETWEEN COMPONENT VARIANCE      85.3742


Tests for Homogeneity of Variances

    Cochrans C = Max. Variance/Sum(Variances) = .7162, P =  .000 (Approx.)
    Bartlett-Box F =                            9.364 , P =  .000
    Maximum Variance / Minimum Variance         6.708
```

F I G U R E 14.7 *SPSSX-generated output for Example 14.5*

The sums of squares, mean squares, and F statistic are SSB = 9,428.9333, MSB = 4,714.4667, SSW = 171,398.0500, MSW = 3,006.9833, and F = 1.5678.

The computer printout shows .2174 as the value F PROB. F PROB. is the p-value, or observed significance level, associated with F = 1.5678. That is, if the random variable F follows the F distribution with 2 and 57 degrees of freedom, then $P(F > 1.5678)$ = .2174. Because this p-value exceeds the 5% level of significance of the test, the F statistic falls in the acceptance region; we would not reject the hypothesis that the population group means are equal. Because the observed F statistic falls in the acceptance region, the evidence is not strong enough to make us conclude that the mean repair costs of the three types of cars are different.

Exercises

Use the data in Table 2.2 of Chapter 2 to complete Exercises 1–3.

14.3.1   Test the null hypothesis that the mean SAT scores are equal for students having different majors.
   a. Find the sample mean and sample standard deviation for each major.
   b. Find SST, SSB, and SSW.
   c. Find MSB and MSW.
   d. Find the F statistic.
   e. Let α = .05. Test the null hypothesis that the population means are equal.

14.3.2   Test the null hypothesis that the mean grade point averages are the same for students having different majors.
   a. Find the sample mean and sample standard deviation for each major.
   b. Find SST, SSB, and SSW.
   c. Find MSB and MSW.
   d. Find the F statistic.
   e. Let α = .05. Test the null hypothesis that the population means are equal.

14.3.3   Test the null hypothesis that the mean SAT scores are the same for freshmen, sophomores, juniors, and seniors.
   a. Find the sample mean and sample standard deviation for each class.
   b. Find SST, SSB, and SSW.
   c. Find MSB and MSW.
   d. Find the F statistic.
   e. Let α = .05. Test the null hypothesis that the population means are equal.

Use the data in Table 2.1 of Chapter 2 to complete Exercises 4–6.

14.3.4   Test the null hypothesis that the mean seniority is the same for employees working in different divisions.
   a. Find the sample mean and sample standard deviation for each division.
   b. Find SST, SSB, and SSW.
   c. Find MSB and MSW.
   d. Find the F statistic.
   e. Let α = .05. Test the null hypothesis that the population means are equal.

14.3.5   Test the null hypothesis that the mean salary is the same for employees working in different divisions.
   a. Find the sample mean and sample standard deviation for each division.
   b. Find SST, SSB, and SSW.

    c. Find MSB and MSW.

    d. Find the F statistic.

    e. Let $\alpha = .05$. Test the null hypothesis that the population means are equal.

14.3.6   Test the null hypothesis that the mean salary is the same for employees having different academic degrees.

    a. Find the sample mean and sample standard deviation for each academic degree.

    b. Find SST, SSB, and SSW.

    c. Find MSB and MSW.

    d. Find the F statistic.

    e. Let $\alpha = .05$. Test the null hypothesis that the population means are equal.

Chapter 14 Supplementary Exercises

14.S.1   Suppose there are four classes in a statistics course. All use the same text and have the same exams, but each has a different teacher. Use a 1% level of significance to test whether the test scores of the students in the following table indicate that the teacher has an effect on a student's score:

	TEACHER ($K = 4$)			
Student	**1**	**2**	**3**	**4**
1	8	9	8	9
2	6	4	10	8
3	4	7	8	9
4	5	8	6	7
5	7		6	10
6			5	7

14.S.2   Two different brands of golf balls are hit by a mechanical device, and the distance each ball travels is recorded. The experiment is performed 100 times and yields the following information: $\bar{x}_1 = 200$ yards, $\bar{x}_2 = 204$ yards, and SSE $= 20,000$.

    a. Calculate SSB and SST.

    b. Calculate MSB, MSE, and F.

    c. Use a 5% level of significance and test whether the average distances of the balls are equal. Use the F test.

    d. Perform the test in part **c** using the t test.

14.S.3   Over the years, a controversy has developed over whether the resilience of baseballs used in the major leagues has changed. Five balls that were made for use in each of the years 1960, 1965, 1970, and 1975 are randomly selected. Each ball is dropped onto a cement floor from a height of 30 feet, and the height of the ball's bounce in feet is recorded in the following table:

	YEAR			
Observation	**1960**	**1965**	**1970**	**1975**
1	9.2	9.3	8.8	9.0
2	9.0	9.1	8.9	8.8
3	9.3	9.3	8.7	9.3
4	8.9	9.0	9.2	8.8
5	8.8	9.1	9.0	8.9

Use a 5% level of significance to test whether the resiliencies of the balls are equal.

14.S.4   Twenty babies are randomly selected for an experiment to determine whether different brands of baby food affect a child's weight. The babies are separated into five groups, and each group is fed a different diet. The weight gained in pounds by each baby after being on the diet for 4 months is recorded in the following table:

			TYPE OF DIET		
Baby	1	2	3	4	5
1	10	9	12	6	11
2	11	6	11	8	9
3 .	7	8	8	9	10
4	8	7	10	8	7

Use a 1% level of significance to test whether the type of diet affects weight gain.

14.S.5   A plastics manufacturer tests the tensile strength of three different types of polyethylene material. A random sample of five products is taken for each material type. The data in pounds per square inch are as follows:

Product	Type 1	Type 2	Type 3
1	380	385	365
2	400	435	415
3	410	375	380
4	390	410	400
5	420	430	410

a. Present the relevant information in an ANOVA table.
b. Let $\alpha = .05$. Determine whether the mean tensile strengths of the types of materials differ significantly.

14.S.6   Four packaging materials were tested for moisture retention by storing the same food product in each for 3 days and then determining the moisture loss. Each type of package was used to wrap six food samples. The moisture losses are given in the accompanying table.
a. Construct the ANOVA table.
b. Let $\alpha = .01$. Determine whether the mean moisture losses differ significantly.

Material 1	Material 2	Material 3	Material 4
27	18	18	22
29	22	17	25
21	19	23	29
24	23	20	27
25	25	25	26
20	20	17	25

14.S.7   The accompanying data show the ratio of gross profits to total sales for a random sample of firms classified by asset size. At the 5% level of significance, can we conclude that mean profit ratios are the same for different-size firms?

PROFIT-TO-SALES RATIO		
Small firms	**Medium firms**	**Large firms**
.10	.17	.22
.06	.16	.02
.22	.22	.08
.15	.06	.12
.19	.04	.14
.13	.08	.10

14.S.8   A police department wanted to compare three makes of cars before ordering an entire fleet. Each car was driven for 20,000 miles, and the cost of operation per mile was noted. Because purchase prices were equal, they were ignored. The accompanying data show the costs of operation per mile (in cents).

OPERATING COSTS		
Make A	**Make B**	**Make C**
9.2	9.0	8.6
9.0	10.2	8.4
9.6	9.6	9.6
8.6	9.8	9.0
8.2	8.6	9.2
	11.2	
	10.4	

a. Construct an ANOVA table.
b. Let $\alpha = .01$. Can we conclude that mean operating costs differ?

14.S.9   The manager of a Burger World store suspects that the average number of noontime customers may depend on the day of the week. To determine the number of employees to use on any given day, the manager wants to know whether the daily averages differ. The accompanying data show the number of customers at noon on randomly selected days. Let $\alpha = .05$ and test $H_0: \mu_1 = \mu_2 = \mu_3 = \mu_4 = \mu_5$.

Monday	Tuesday	Wednesday	Thursday	Friday
116	97	89	98	114
126	122	111	107	118
146	84	116	109	124
108	128	104	114	142
96	106	112	102	106
130	114	98	104	132

14.S.10   Three different automatic printing machines are used to print the designs on cereal boxes. Observations were taken randomly to determine how many boxes were being printed per minute by each machine. Each machine was checked during 5 randomly selected minutes. Test the hypothesis that the mean outputs per minute for the machines are equal. Use $\alpha = .05$. The following data show the number of boxes printed by each machine during the randomly selected minutes:

	MACHINE		
Sample	1	2	3
1	64	68	70
2	62	66	72
3	65	67	66
4	67	69	72
5	66	67	68

14.S.11 The Midwest Rent-a-Car Company owns four maintenance and repair facilities. All of Midwest's cars are sent to these shops when they need repairs. The director of maintenance thinks that the four facilities do not do an equally good job of repairing the cars, so he takes 20 identical cars and has his mechanics create identical mechanical problems in each car. Five different cars are sent to each of the four repair shops, and the repair times (in hours) are as follows:

	REPAIR SHOP			
Sample	1	2	3	4
1	4.2	5.2	6.2	5.4
2	5.4	5.8	5.4	5.8
3	4.8	6.2	6.0	6.6
4	3.8	6.0	6.0	4.8
5	4.4	5.6	5.8	4.8

a. Calculate the sample means.
b. Calculate the overall mean.
c. Calculate the F statistic.
d. Test $H_0: \mu_1 = \mu_2 = \mu_3 = \mu_4$. Use $\alpha = .05$.

14.S.12 A college professor wants to determine the best way to present an important topic to his class. He has the following three choices: He can lecture (condition 1), he can lecture plus assign supplementary reading (condition 2), or he can show a film and assign the same supplementary reading as in condition 2 (condition 3). He decides to do an experiment to evaluate the three options. He solicits 27 volunteers from his class and randomly assigns 9 to each of the 3 conditions. The following scores (percentage correct) on an exam were obtained from these 27 students:

Condition 1	92	86	87	76	80	87	92	83	84
Condition 2	86	93	97	81	94	89	98	90	91
Condition 3	81	80	72	82	83	89	76	88	83

a. What is the overall null hypothesis?
b. What is the conclusion? Use $\alpha = .05$.

14.S.13 To test whether memory changes with age, a researcher conducts an experiment on four groups differing in age, with six subjects in each group. Assume all individuals are in good health and the groups do not differ by IQ, education, or any other possible contaminating variable. Each subject is shown a series of nonsense syllables (like DUT or FAM) at a rate of one syllable every 4 seconds. Each series is shown twice, and the subjects are asked to write down as many syllables as they can remember. The number of syllables remembered by each subject grouped by age is as follows:

	AGE (IN YEARS)		
30	**40**	**50**	**60**
14	12	17	13
13	15	14	10
15	16	14	7
17	11	9	8
12	12	13	6
10	18	15	9

Does age have an effect on memory? Use $\alpha = .05$.

14.S.14 An important quality characteristic of a certain new product is a measure of strength that can be tested only by a rather costly destructive test. Random samples of 10 items are tested from each of three machines, machines A, B, and C, with the following results:

	TEST VALUES		
Sample	**A**	**B**	**C**
1	1,200	1,380	1,270
2	1,300	1,140	1,370
3	1,260	1,290	1,320
4	1,500	1,450	1,520
5	1,660	1,350	1,550
6	1,440	1,290	1,210
7	1,420	1,380	1,510
8	1,110	1,280	1,200
9	1,110	1,080	1,170
10	1,250	1,470	1,290

Let $\alpha = .05$. Can we reject the null hypothesis that the machines produce equally strong products?

14.S.15 A quality control engineer wants to test whether five machines that produce aircraft fittings are producing output with the same mean pitch of threads. A random sample of 20 items was taken from the output of each of the five machines. The data, presented in the accompanying table, are expressed in units of .0001 in. in excess of .40 in. Let $\alpha = .05$. Test the null hypothesis that the five population means are equal.

Sample number	MACHINE					Sample number	MACHINE				
	A	**B**	**C**	**D**	**E**		**A**	**B**	**C**	**D**	**E**
1	36	35	34	33	32	11	34	38	35	34	38
2	31	31	34	32	30	12	36	38	39	39	40
3	30	30	32	30	32	13	36	40	35	26	33
4	32	33	33	32	35	14	36	35	37	34	44
5	32	34	37	37	35	15	30	37	33	34	35
6	32	32	31	33	33	16	28	31	33	33	33
7	33	33	36	32	31	17	33	30	34	33	35
8	23	33	36	35	36	18	27	28	29	27	30
9	43	36	35	24	31	19	35	36	29	27	32
10	36	35	36	41	41	20	33	35	35	39	36

14.S.16 A quality control engineer at a food processing plant inspects the drained weight (in ounces) of standard-grade tomatoes present in size no. 3 cans. There are four production lines. The engineer wants to test whether the mean weights of tomatoes from the four lines are equal. Let $\alpha = .05$. Perform the test using the following data.

Sample number	PRODUCTION LINE				Sample number	PRODUCTION LINE			
	A	B	C	D		A	B	C	D
1	22.0	22.5	22.5	24.0	16	21.5	20.5	22.0	21.5
2	20.5	22.5	22.5	23.0	17	19.0	21.5	23.0	21.0
3	20.0	20.5	23.0	22.0	18	21.0	20.5	19.5	22.0
4	21.0	22.0	22.0	23.0	19	20.0	23.5	24.0	20.5
5	22.5	19.5	22.5	22.0	20	22.0	20.5	21.0	22.5
6	23.0	23.5	21.0	22.0	21	19.0	20.5	21.0	20.5
7	19.0	20.0	22.0	20.5	22	21.5	25.0	21.0	19.0
8	21.5	20.5	19.0	19.5	23	22.5	22.0	23.0	22.0
9	21.0	22.5	20.0	22.0	24	22.5	22.0	22.0	19.5
10	21.5	23.0	22.0	23.0	25	18.5	22.0	22.5	21.0
11	20.0	19.5	21.0	20.0	26	21.5	20.5	20.5	16.5
12	19.0	21.0	21.0	21.0	27	24.0	22.0	17.5	21.0
13	19.5	20.5	21.0	20.5	28	19.5	22.5	15.5	20.0
14	20.0	21.5	24.0	23.0	29	22.0	17.5	21.0	22.0
15	22.5	19.5	21.0	21.5	30	22.0	20.0	20.5	24.0

14.S.17 A certain automatic filling operation fills metal containers with a sticky plastic compound. Each container is supposed to contain 20 lb of compound. Once a container is filled, it is not possible to empty it to find the exact weight of its contents because some of the compound adheres to the sides and bottom of the container. For this reason, indirect methods of analyzing filling weights were used in the following study. The purpose of the study was to find out whether more overfill was being used to satisfy the weight specification or whether there was not enough overfill.

One part of the study dealt with the weights of the containers themselves. The containers were purchased from four different producers. A random sample of 20 containers was taken from the most recent shipment of containers from each vendor. Each container was weighed, and the following results (in ounces) were obtained. Let $\alpha = .05$. Test the null hypothesis that the mean weights of the containers made by different vendors are equal.

Sample number	VENDOR				Sample number	VENDOR			
	A	B	C	D		A	B	C	D
1	13.0	14.3	15.1	14.3	11	15.1	12.2	14.1	13.2
2	13.5	14.3	14.3	13.5	12	12.3	14.5	14.3	14.1
3	14.3	13.5	13.3	14.4	13	14.8	13.3	14.2	13.8
4	15.1	14.3	14.4	13.1	14	15.4	14.4	14.9	13.9
5	14.5	12.2	14.1	14.8	15	15.3	12.2	14.7	14.5
6	13.4	13.4	14.9	14.4	16	14.1	13.7	14.1	14.3
7	12.3	13.1	14.2	13.7	17	14.4	14.1	13.4	14.4
8	15.5	13.8	12.7	12.1	18	14.7	13.1	12.5	14.3
9	14.3	14.3	13.3	14.1	19	15.7	13.7	14.8	13.2
10	15.8	13.1	14.2	13.1	20	13.7	13.8	14.2	14.8

References

Cochran, W. G., and G. M. Cox. *Experimental Designs.* 2d ed. New York: Wiley, 1957.

Fisher, Ronald A. *The Design of Experiments.* 9th ed. New York: Hafner Press, 1971.

Fisher, Ronald A. *Statistical Methods for Research Workers.* 14th ed. New York: Hafner Press, 1970.

Fisher, Ronald A., and W. A. Mackenzie. "Studies in Crop Variation." *Journal of Agricultural Science,* 1923, pp. 311–320.

Neter, John, and William Wasserman. *Applied Linear Statistical Models.* Homewood, Ill.: Irwin, 1974.

Nie, Norman E., C. Hadlai Hull, Jean G. Jenkins, Karin Steinbrenner, and Dale H. Bent. *SPSS Statistical Package for the Social Sciences.* 2d ed. New York: McGraw-Hill, 1975.

Norusis, Marija J. *SPSSX Introductory Statistics Guide.* New York: McGraw-Hill, 1990.

Norusis, Marija J. *SPSSX Advanced Statistics Guide.* Chicago: SPSS, 1990.

Norusis, Marija J. *The SPSS Guide to Data Analysis.* Chicago: SPSS, 1986.

Ryan, Thomas A., Brian L. Joiner, and Barbara F. Ryan. *Minitab Reference Manual.* University Park, Penn.: Minitab Project, 1985.

Ryan, Thomas A., Brian L. Joiner, and Barbara F. Ryan. *Minitab Handbook.* 2d ed. Boston: PWS-Kent, 1985.

SAS Introductory Guide. 3d ed. Cary, N.C.: SAS Institute, 1985.

SAS Procedures Guide for Personal Computers. Version 6 ed. Cary, N.C.: SAS Institute, 1986.

SAS Statistics Guide for Personal Computers. Version 6 ed. Cary, N.C.: SAS Institute, 1986.

SAS User's Guide: Basics. Version 5 ed. Cary, N.C.: SAS Institute, 1985.

SAS User's Guide: Statistics. Version 5 ed. Cary, N.C.: SAS Institute, 1985.

Scheffé, Henry. *The Analysis of Variance.* New York: Wiley, 1959.

SPSSX User's Guide. Chicago: SPSS, 1988.

15 Regression and Correlation

Stochastic Relationships and Scatter Diagrams

In many instances, the values taken by one variable Y are influenced by or related to the values taken by some other variable X, and we are interested in exploring the relationship between these two variables. If a relationship exists, we may be able to use values of X to estimate or predict values of Y.

For example, we may be interested in determining what relationship (if any) exists between a person's consumption expenditures (Y) and the person's annual income (X), between the annual sales of a corporation (Y) and annual advertising expenditures (X), or between the rate of inflation (Y) and the change in the money supply (X).

In each of the preceding examples, the variable Y represents the *dependent variable* and the variable X, the *independent variable*. We construct a mathematical model to describe the hypothesized relationship between Y and X. For example, we may think that there is an approximately linear relationship between a person's consumption (Y) and the person's income (X). To see whether this hypothesis is true, it would be natural to obtain a sample of data and examine whether the observed data points lie close to a straight line.

Scatter Diagrams

When the hypothesized model contains only one independent variable (X), the appropriateness of some hypothesized mathematical model can be examined by plotting the observed data on a graph called a *scatter diagram*.

DEFINITION **Scatter Diagram** ─────────────────────────────

Suppose we have a sample of n pairs of values (x_i, y_i). A **scatter diagram** is a graph showing the n sample observations (x_1, y_1), (x_2, y_2), . . . , (x_n, y_n).

─────────────────────────────

On a scatter diagram, the dependent variable is always plotted on the vertical axis and the independent variable on the horizontal axis. A scatter diagram is useful for *revealing the form of the relationship* (if any) that exists between two variables. Various scatter diagrams are shown in Figure 15.1. The data in plots (b) and (d) can be described quite well by linear equations because the observed points cluster more or less around straight lines. The data in plot (a) are better described by an exponential, or quadratic, equation. No type of equation appears to describe the data in plot (c) well; there does not appear to be any discernible relationship between the two variables. A scatter diagram can also be useful for *revealing the strength of the relationship* between two variables. A linear equation fits the data much better in plots (b) and (d) than in plot (c), for example.

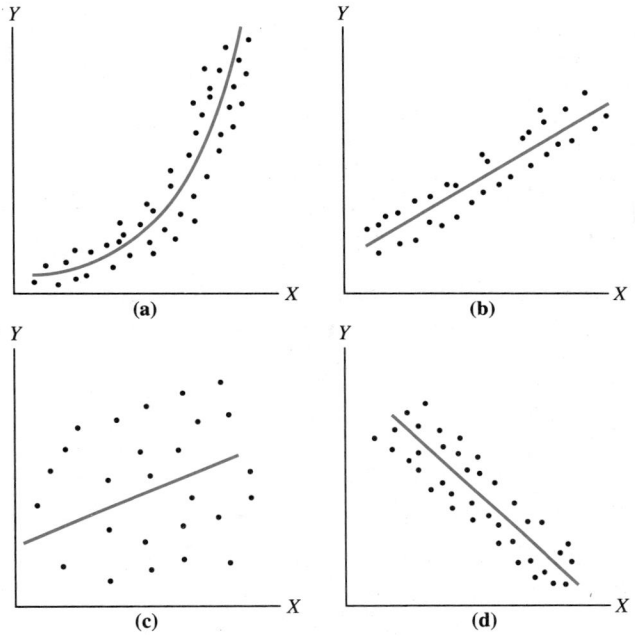

FIGURE 15.1 *Scatter diagrams*

By inspecting a scatter diagram, we can get information to help answer four important questions about the data:

1. Does there appear to be a simple mathematical relationship between Y and X? That is, do the points tend to cluster about some straight line or other simple curve?
2. Is the relationship between Y and X positive or negative? The relationship between Y and X is *positive* (or *direct*) if increases in X tend to be associated with increases in Y, and *negative* (or *inverse*) if increases in X tend to be associated with decreases in Y.
3. Is the relationship between Y and X approximately linear or nonlinear? A relationship between Y and X is approximately linear if the plotted points tend to lie close to some straight line. The relationship is approximately nonlinear if the plotted points tend to lie close to some nonlinear function.
4. How strong is the relationship between Y and X? The relationship between Y and X is relatively strong if the plotted points tend to lie close to some line or simple curve. In general, the closer the points lie to the line or curve, the stronger the relationship between Y and X.

Besides linear relationships, various other functions are useful in fitting models to data, including polynomial functions and logarithmic functions. If the scatter diagram indicates that a straight line describes the relationship between Y and X well, we assume that Y and X are approximately related to one another according to the equation

$$Y = \beta_0 + \beta_1 X$$

where the coefficients β_0 and β_1 are unknown numbers. This is the equation of a straight line where β_0 is the *Y-intercept* of the line and β_1 is the *slope* of the line. The coefficient β_1 measures the change in Y associated with a 1-unit increase in X. The coefficient β_0 measures the value of Y when $X = 0$. Our task is to estimate the unknown coefficients β_0 and β_1 and determine whether the resultant straight line provides a good explanation of the relationship between Y and X.

In a *deterministic relationship,* for any given value of X, the value of the variable Y is a constant whose value depends on X and can be determined with certainty.

DEFINITIONS   **Deterministic Relationship and Stochastic Relationship** ─────────────

A relation $Y = f(X)$ is **deterministic** if each value of X is paired with one and only one value of Y, and **stochastic** if each value of X is associated with a whole probability distribution of values of Y.

EXAMPLE 15.1   **Examples of Deterministic Relationships** ─────────────────

The following relationships are deterministic:

a. A company sells computers for \$960 each. Let X denote the number of computers sold during a given month, and Y the total monthly revenue. Then Y and X are related according to the deterministic relationship

$$Y = 960X$$

This model is deterministic because, when a value of X is substituted into the equation, the value of Y is determined exactly and no allowance need be made for error.

b. An individual rents a car for use on a business trip. The rental fee is \$100 plus \$.20 per mile. Let X denote the number of miles driven, and let Y denote the total cost of renting the car. Then Y and X are related according to the deterministic relationship

$$Y = 100 + .20X$$

In a *stochastic relationship,* for any given value of X, the variable Y is a random variable whose value depends on X but cannot be predicted with certainty. Nevertheless, knowledge of X helps predict the value of Y. A stochastic model contains a probabilistic, or random, component that is added to the deterministic portion of the model to account for the random or unexplained error of prediction.

EXAMPLE 15.2   **Examples of Stochastic Relationships** ─────────────────

The following relationships are stochastic:

a. Let X denote annual income for a household, and let Y denote the household's annual expenditures on recreation and entertainment. For a given value $X = x$, the value of Y

cannot be predicted exactly, since other factors influence a family's expenditures on recreation. In general, however, as X increases, Y also increases; thus, knowledge of X is helpful in predicting Y. A probabilistic model relating expenditures for the ith household Y_i to income of the ith household $X = x_i$ is given by the equation

$$Y_i = \beta_0 + \beta_1 x_i + e_i$$

where e_i is assumed to be a random error variable with expected value equal to 0 and variance equal to σ_e^2. As an example, we might have the hypothesis that expenditures on recreation and entertainment are related to household income according to the equation

$$Y_i = 250 + .1x_i + e_i$$

where e_i is the random factor that makes expenditures greater or less than $(250 + .1x_i)$. Thus, for a given value $X = x_i$, there is an entire probability distribution of possible values of Y.

b. Let the random variable Y_i denote the income that will be received by the main performer at the ith rock concert, and let $X = x_i$ denote the observed number of spectators at the ith concert. Suppose the performer will receive \$50,000 plus 20% of all concession sales. Assume the typical spectator spends about \$3.00 on concessions, plus or minus some random amount. Thus, the performer would receive approximately \$.60 per spectator in income from concessions. Thus, total income from the ith concert is a random variable (Y_i) that is related to the number of spectators (x_i) according to the stochastic relation

$$Y_i = 50,000 + .60x_i + e_i$$

For a given value of $X = x_i$, total revenue cannot be predicted exactly, and there is an entire distribution of possible values of Y associated with each value of X.

There are numerous situations in economics and business where a dependent variable Y is stochastically related to some independent variable X. Our goal is to express Y as some function of X and possibly a random error variable. Some possibilities are

$$Y_i = \beta_0 + \beta_1 x_i + e_i \quad \text{and} \quad Y_i = \beta_0 + \beta_1 x_i + \beta_2 x_i^2 + e_i$$

There are numerous other possibilities, including models that contain other explanatory variables besides the variable X. These models will be discussed in Chapters 16, 17, and 18.

In this chapter, we will restrict ourselves to the model

$$Y_i = \beta_0 + \beta_1 x_i + e_i$$

where β_0 and β_1 are unknown coefficients that have to be estimated.

Goals of Regression Analysis

The technique of estimating stochastic relationships and analyzing the estimators is called **regression analysis.** In regression analysis, we estimate a *mathematical model* that explains the relationship between a dependent variable Y and one or more independent variables. The mathematical model consists of a hypothesized equation and some assumptions that together provide a simplified or idealized representation of behavior observed in the real world. In construction of the model, two problems must be addressed: (1) In many applications in business and economics, the precise mathematical form of any underlying relationship will not be known, and (2) in many cases, no deterministic relationship holds *exactly* in the real world. Thus, in general, our mathematical models will be only approximations of real-world behavior.

In this chapter, we discuss the *simple linear regression model,* where we explain the value of the dependent variable Y based on the known value of an explanatory variable X. When there are two or more explanatory variables in the equation to explain Y, the mathematical model is called a *multiple regression model.*

Reasons for using regression analysis

> There are at least four reasons for using regression analysis:
>
> **1.** Regression analysis enables us to *quantify* a theory about how the variables Y and X are related.
> **2.** Regression analysis enables us to *test* a theory about the relationship of a variable Y to a variable X.
> **3.** Regression analysis enables us to *measure the strength of the relationship* between Y and X.
> **4.** Regression analysis enables us to *predict,* or forecast, the value of Y, given the value of some independent variable X.

The following discussion provides some examples of how regression analysis can be used to quantify theories, test theories, measure the strength of relationships, and forecast values of Y.

1. *Quantifying a theory:*   According to economic theory, expenditures on entertainment and recreation tend to increase as income increases. An economist would want to know *by how much* expenditures would increase if income increased by, say, $10,000. If we estimate a stochastic relationship between expenditures and income, we have quantified the theory.

2. *Testing theories:*   It is hypothesized that annual sales of a corporation are positively related to advertising expenditures. Is this theory true? By gathering data and constructing a model showing how sales and advertising expenditures are related, we can eventually test the hypothesis that advertising expenditures influence sales.

3. *Measuring the strength of relationships:*   Annual incomes tend to increase with years of education, but many other variables also influence an individual's income. How important is education as a determinant of income? Is income closely related to years of education? Regression analysis enables us to estimate the relationship between income and education and also to measure the strength of the relationship.

4. *Forecasting:*   An executive thinks that annual sales are influenced by the dollar amount of advertising expenditures. Regression analysis can be used to estimate an equation explaining the relationship between annual sales and advertising expenditures. Once the relationship between advertising expenditures and sales has been estimated, the equation can be used to forecast annual sales for any given level of advertising expenditures.

EXAMPLE 15.3   **A Simple Linear Regression Model** ————————————————————————

A large corporation is planning to open a nationwide chain of sporting goods stores. The corporation hires a marketing agency to perform a market analysis to help determine optimal locations for the stores. One of the factors that the marketing agency is asked to study is the relationship between the explanatory variable family income (X) and the dependent variable household expenditure on recreation (Y).

Suppose the marketing agency contacted a random sample of 20 families and obtained the data in Table 15.1 on weekly expenditures on recreation and on weekly income. (In actual practice, hundreds of families might be contacted.) The observations (x_1, y_1), $(x_2, y_2), \ldots, (x_{20}, y_{20})$ are plotted in the scatter diagram in Figure 15.2 (page 590). (On the scatter diagram, a straight line has been drawn that is the *estimated* line that best describes the sample data. This idea will be described in more detail in Section 15.2.)

We would expect expenditures on recreation to depend on income and perhaps on other factors as well. Specifically, we would expect families with relatively high incomes to have relatively high expenditures on recreation, and families with relatively low incomes to have relatively low expenditures on recreation. The scatter diagram shows that this pattern seems to hold for the sample of 20 data points.

TABLE 15.1 *Data for Example 15.3*

Family	Weekly expenditure on recreation Y	Weekly income X
1	$90	$900
2	60	800
3	45	600
4	70	650
5	30	300
6	40	350
7	50	400
8	55	700
9	85	950
10	75	700
11	90	800
12	70	750
13	60	500
14	40	300
15	25	250
16	65	700
17	60	600
18	50	450
19	35	250
20	35	200

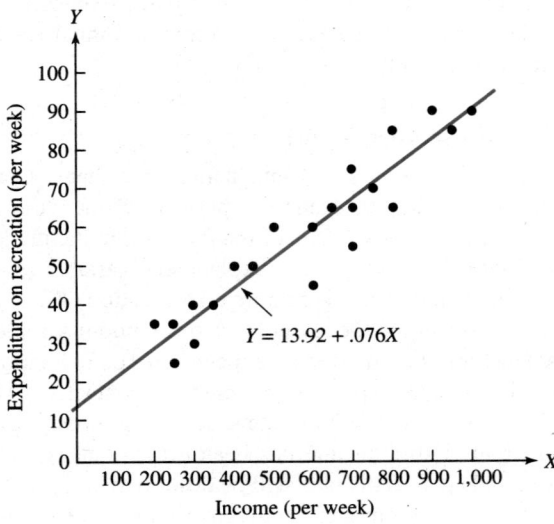

FIGURE 15.2   *Scatter diagram for Example 15.3*

The relationship between expenditures on recreation and family income in Example 15.3 is stochastic rather than deterministic. Thus, it is not reasonable to associate just a single level of expenditure on recreation with a particular level of family income. Rather, it is more realistic to think of a whole distribution of possible levels of expenditures on recreation for each possible level of family income.

Exercises

Statistical Concepts

15.1.1 Suppose Y and X are related according to the equation

$$Y = \beta_0 + \beta_1 X$$

Explain why the coefficient β_0 is called the intercept and β_1 is called the slope.

15.1.2 Explain the difference between a deterministic and a stochastic relationship.

15.1.3 Give three examples of deterministic relationships and three examples of stochastic relationships.

15.1.4 Draw a scatter diagram showing a nonlinear positive stochastic relationship between Y and X.

15.1.5 Draw a scatter diagram showing a nonlinear positive deterministic relationship between Y and X.

15.1.6 Draw a scatter diagram showing a linear positive stochastic relationship between Y and X where the relationship is weak.

15.1.7 *True or false:* Suppose Y and X are related according to the deterministic equation $Y = 5 - 3X$. A 1-unit increase in X is associated with a 3-unit increase in Y. Explain.

15.1.8 *True or false:* Suppose the logarithm of Y and the logarithm of X are associated according to the equation $\ln(Y) = 3 + 5\ln(X)$. A 1-unit increase in X is associated with a 5-unit increase in Y. Explain.

Statistical Drills

15.1.9 The equation of a straight line is $y = 7 + 3x$.
   a. Graph this line.
   b. What is the slope?
   c. What is the intercept?

15.1.10 The equation of a straight line is $y = 5 - 2x$.
   a. Graph this line.
   b. What is the slope?
   c. What is the intercept?

15.1.11 The equation of a straight line is $y = 3x$.
   a. Graph this line.
   b. What is the slope?
   c. What is the intercept?

15.1.12 The accompanying data are pairs of values of Y and X.

x_i	y_i
6.1	17.4
7.3	18.5
6.0	16.4
4.2	15.7
5.1	16.2
8.2	19.7

   a. Plot the data on a scatter diagram. Does there appear to be a relationship between Y and X?
   b. Is the relationship between Y and X stochastic or deterministic?
   c. Does the relationship between Y and X appear to be approximately linear or nonlinear?
   d. Does the relationship between Y and X appear to be strong or weak?
   e. Does the relationship between Y and X appear to be positive or negative?

15.1.13 The accompanying data are pairs of values of Y and X.

x_i	y_i
6.1	7.4
7.3	5.6
6.0	4.3
4.2	5.5
5.1	6.4
8.2	7.5

   a. Plot the data on a scatter diagram. Does there appear to be a relationship between Y and X?
   b. Is the relationship between Y and X stochastic or deterministic?
   c. Does the relationship between Y and X appear to be approximately linear or nonlinear?
   d. Does the relationship between Y and X appear to be strong or weak?
   e. Does the relationship between Y and X appear to be positive or negative?

15.1.14   The accompanying data are pairs of values of Y and X.

x_i	y_i
2.0	9.0
3.0	14.0
4.0	21.0
5.0	30.0
6.0	41.0
7.0	54.0
8.0	69.0

a. Plot the data on a scatter diagram. Does there appear to be a relationship between Y and X?
b. Is the relationship between Y and X stochastic or deterministic?
c. Does the relationship between Y and X appear to be approximately linear or nonlinear?
d. Does the relationship between Y and X appear to be strong or weak?
e. Does the relationship between Y and X appear to be positive or negative?

15.1.15   The accompanying data are pairs of values of Y and X.

x_i	y_i
6.1	74
7.3	86
6.0	73
4.2	55
5.1	64
8.2	95

a. Plot the data on a scatter diagram. Does there appear to be a relationship between Y and X?
b. Is the relationship between Y and X stochastic or deterministic?
c. Does the relationship between Y and X appear to be approximately linear or nonlinear?
d. Does the relationship between Y and X appear to be strong or weak?
e. Does the relationship between Y and X appear to be positive or negative?

15.1.16   Let $Y = 12 - 3X$.
a. If $x_1 = 1$, find y_1.
b. If $x_2 = 3$, find y_2.
c. Plot the points (x_1, y_1) and (x_2, y_2).
d. Graph the line $Y = 12 - 3X$.

15.1.17   Graph each of the following equations:
a. $Y = 4 + 2X$ b. $Y = 4 - 2X$ c. $Y = 2 + 2X$ d. $Y = 2 + 3X$

Statistical Applications

15.1.18   Suppose that a company installs and repairs copy machines. The company studied the relationship between repair costs for a sample of six machines and the number of pages copied by each machine. The goal is to identify machines whose costs are too high relative to their copying volumes. The repair costs in dollars and the number of pages copied (in thousands) for the six machines are as follows:

Machine	1	2	3	4	5	6
Repair cost	85	120	70	165	125	90
Pages copied	900	1,350	550	850	1,500	800

a. Which variable is the dependent variable and which is the independent variable? Why?
b. Make a scatter diagram of these observations.
c. Does the maintenance cost of any machine seem out of line?

15.2 The Simple Linear Regression Model

In the simple linear regression model, it is assumed that for every value of X, there is an entire distribution of values of Y. We use the symbol $E(Y_i|X = x_i)$, or more simply $E(Y_i|x_i)$, to denote the conditional expected value of the random variable Y_i when the variable X takes the specific value x_i.

The simple linear regression model states that the relationship between the expected value of Y_i and the value x_i can be expressed as

$$E(Y_i|x_i) = \beta_0 + \beta_1 x_i$$

where β_0 and β_1 are unknown parameters that have to be estimated. This means that the mean value of Y_i for a given value x_i lies on a straight line whose intercept is β_0 and whose slope is β_1. See Figure 15.3.

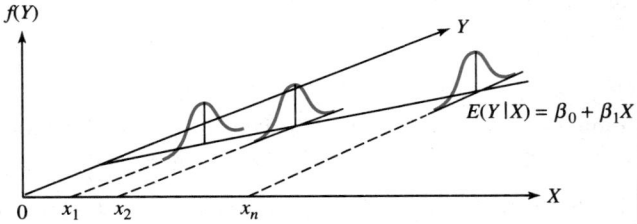

FIGURE 15.3 *The population regression model*

DEFINITION **Population Regression Line**

The equation

$$E(Y_i|X = x_i) = \beta_0 + \beta_1 x_i$$

is called the **population regression line**. The equation states that for a given value $X = x_i$, the expected value of Y_i is $(\beta_0 + \beta_1 x_i)$.

The model states that the *mean values of the subpopulations of Y* associated with different values of X lie on a straight line. In Figure 15.3, there is a whole population of values of Y associated with each value of X, and the mean of each distribution falls on the population regression line. Although the distributions have different means, it is assumed that the distributions have the same variance σ^2.

Because the population parameters β_0 and β_1 are unknown, the equation of the population regression line is unknown and has to be estimated using a sample of observations (x_i, y_i). Once we have estimated the parameters β_0 and β_1, we obtain an estimated regression line, which is called the *sample regression line*.

Because the observed sample points (x_i, y_i) do not lie exactly on the population regression line, the observed value y_i will differ from the expected value $E(Y_i|x_i)$. The quantity

$$e_i = Y_i - E(Y_i|X = x_i) = Y_i - (\beta_0 + \beta_1 x_i)$$

measures the amount by which the random value Y_i differs from the expected value of Y_i, given that $X = x_i$. The quantity e_i is called a *random error term.*

DEFINITION   **Random Error Term**

In the population regression model, the **random error term**, denoted e_i, is the difference between the random variable Y_i and the mean value of Y_i given that $X = x_i$. That is,

$$\begin{aligned} e_i &= Y_i - E(Y_i|X = x_i) \\ &= Y_i - (\beta_0 + \beta_1 x_i) \end{aligned}$$

The random error e_i is also called a **random disturbance term**.

Rearranging the equation defining the random error term (see the accompanying box) yields

$$Y_i = \beta_0 + \beta_1 x_i + e_i$$

In Example 15.3, we can think of a family's expenditures on recreation (Y_i) as being composed of the sum of the following two components:

1. $(\beta_0 + \beta_1 x_i)$, the systematic component that measures the mean value of Y_i given $X = x_i$
2. e_i, which represents a deviation from the systematic component

The component e_i reflects the multitude of factors other than income that influence expenditures on recreation, such as the size of the family, the ages of the family members, and so forth.

Given the sample of observations (x_i, y_i) for $i = 1, 2, \ldots, n$, we want to estimate the parameters β_0 and β_1.

DEFINITION   **Sample Regression Line**

The estimated equation

$$\hat{y}_i = b_0 + b_1 x_i$$

is called the **sample regression line**. The value b_0 is the sample estimate of the population parameter β_0, and the value b_1 is the sample estimate of the population parameter β_1. The value $\hat{y}_i$ is called the *fitted value* of Y_i or the *predicted value* of Y_i when $X = x_i$.

In Section 15.3, we show the formulas that are used to determine the sample estimates b_0 and b_1 for a given sample of data. Using the data in Table 15.1, we obtained the sample regression line

$$\hat{y}_i = 13.92 + .076x_i$$

This line is plotted on the scatter diagram in Figure 15.2 on page 590. Note that although the observations (x_i, y_i) do not lie exactly on the line, they do lie relatively close to it. This leads us to believe that the regression model describes fairly accurately the relationship between expenditures on recreation and family income.

DEFINITION   **Residuals**

For the given value $X = x_i$, the difference between the actual value y_i and the fitted value $\hat{y}_i$ is denoted $\hat{e}_i$. The values $\hat{e}_1, \hat{e}_2, \ldots, \hat{e}_n$ are called the **residuals** and are obtained using the equation

$$\hat{e}_i = y_i - \hat{y}_i$$
$$= y_i - (b_0 + b_1 x_i)$$

In Figure 15.2, the distances between the sample points and the estimated line are the residuals. Figure 15.4 shows the relationship among y_i, $\hat{y}_i$, and $\hat{e}_i$. As can be seen, the residual will be positive if the sample observation (x_i, y_i) lies above the sample regression line and negative if it lies below. If (x_i, y_i) lies exactly on the line, the residual will be 0 and the predicted value $\hat{y}_i$ and the actual value y_i will be exactly the same. That is, the residual $\hat{e}_i$ will be 0 if $y_i = \hat{y}_i$. If each of the residuals is close to 0, then each predicted value $\hat{y}_i$ will be close to the actual value y_i, and each of the observed points (x_i, y_i) will lie close to the estimated line $\hat{y}_i = b_0 + b_1 x_i$.

An analysis of the residuals can tell us a lot about whether the estimated model is satisfactory. A good model will have small residuals with no regular patterns evident in a graph of the residuals versus the dependent or independent variable. Residual analysis is discussed in detail in Chapter 18.

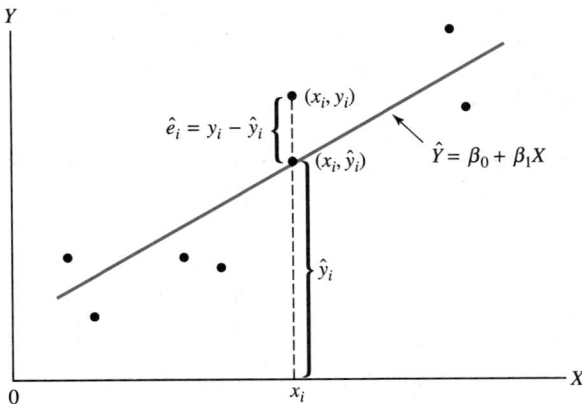

FIGURE 15.4   *Scatter diagram showing the relationship among y_i, $\hat{y}_i$, and the residual $\hat{e}_i$*

Basic Assumptions of the Simple Linear Regression Model

The basic assumptions of the simple linear regression model are summarized in the accompanying box.

Basic assumptions of the simple linear regression model

> In the simple linear regression model, it is assumed that the dependent variable Y_i is related to the observed value of the independent variable $X = x_i$ according to the equation
>
> $$Y_i = \beta_0 + \beta_1 x_i + e_i \qquad i = 1, 2, \ldots, n$$
>
> where the random error terms e_i are assumed to have the following properties, which are called the *basic assumptions of the simple linear regression model:*
>
> 1. *Normality:*   For any value x_i, the error term has a normal distribution.
> 2. *Zero mean:*   For any value x_i, $E(e_i) = 0$.
> 3. *Homoscedasticity:*   The variance of e_i, denoted σ_e^2, is the same for all values of X.
> 4. *No serial correlation:*   The error terms are independent of one another.
> 5. *Independence of e_i and x_i:*   The error terms e_i are independent of the values of the independent variable X.

The first three assumptions state that for each value x_i, the error term e_i is a normally distributed random variable with 0 mean and variance σ_e^2.

The assumption concerning homoscedasticity means that every disturbance has the same variance. This assumption rules out, for example, the possibility that the dispersion of the disturbances is greater for high values of X than for low values. This means that the variance for each subpopulation of values of Y is the same regardless of the value of X. When the assumption of equal variances is violated, some of the optimal properties of our estimators will no longer hold. For example, our estimates of variances become biased and our tests of hypotheses lose their validity. This is discussed in more detail in Chapter 18.

The assumption of independence of the error terms means that a disturbance in one time period is unrelated to disturbances at other time periods. A violation of this assumption indicates that something systematic has been omitted from the model, in which case estimates of variances will be biased and tests of hypotheses invalid.

The assumptions underlying the regression model are used in making inferences about the values of the three unknown regression parameters β_0, β_1, and the disturbance variance σ_e^2.

Figure 15.5 shows a population regression line and an estimated sample regression line. The lines differ because, in general, b_0 and b_1 will differ from the true values of β_0 and β_1.

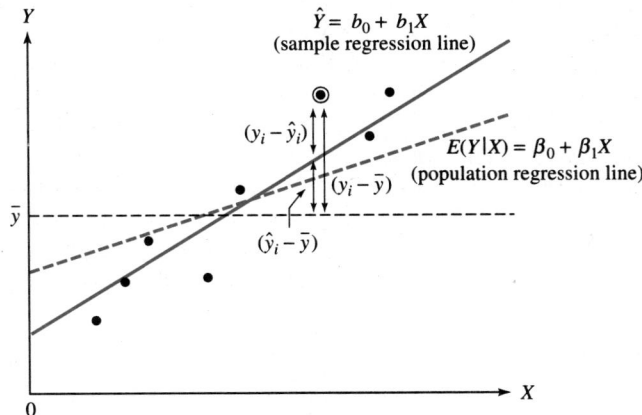

FIGURE 15.5 *A population and sample regression line*

Exercises

Statistical Concepts

15.2.1 Explain what is meant by the population regression line. Draw a graph illustrating the population regression line.

15.2.2 *True or false:* Suppose Y and X follow a deterministic linear relationship. Then the population and sample regression lines will be identical. Explain.

15.2.3 *True or false:* Suppose Y and X follow a deterministic linear relationship. Then the residuals will all be 0. Explain.

15.2.4 *True or false:* The residual represents the difference between the fitted value of Y_i and the actual value of Y_i. Explain.

15.2.5 State the five basic assumptions of the simple linear regression equation.

15.2.6 *True or false:* The observed values y_i fall on the population regression line. Explain.

15.2.7 *True or false:* The residuals measure the deviations between the predicted values $\hat{y}_i$ and the mean value $\bar{y}$. Explain.

15.2.8 Suppose a sample regression equation is

$$\hat{y}_i = 10 + 3.5x_i$$

a. The first observation is ($x_1 = 10$, $y_1 = 50$). Find the first residual.
b. The second observation is ($x_2 = 14$, $y_2 = 58$). Find the second residual.

15.3 Method of Least Squares

In this section, we show how to use the sample observations (x_i, y_i) to estimate the population parameters β_0 and β_1. Because the sample observations do not fall exactly on a straight line, many different lines can be drawn through the plotted data points. The problem is to select the line that, in some sense, best describes the data.

To avoid individual judgment in constructing the sample regression line, we need a definition of what is the best-fitting line. It might seem reasonable to define the best line as the line that produces the smallest residuals $\hat{e}_i$ on the average. The problem with this method is that some of the residuals are positive and some are negative. As a result, the sum of the residuals can be 0 even though none of the individual residuals is close to 0. To avoid this problem, we define the best line as the line for which the sum of the squared residuals is the smallest. This procedure is called the **method of least squares.**

The method of least squares produces a line that minimizes the sum of the squared vertical distances from the observed data points to the line. Any other line has a larger sum. In Figure 15.2 (page 590), the least-squares line has been superimposed on the scatter diagram.

The Least-Squares Estimates b_0 and b_1

Because the derivation of the formulas for the least-squares coefficients b_0 and b_1 requires the use of calculus, it is not included here. By definition, the least-squares line is the line that minimizes the residual sum of squares ($\hat{e}_1^2 + \hat{e}_2^2 + \cdots + \hat{e}_n^2$).

DEFINITION   **Residual Sum of Squares** ───────────────────────────────

The **residual sum of squares**, denoted SSE (an acronym for "sum of squared errors"), is given by

$$\text{SSE} = \sum_{i=1}^{n} \hat{e}_i^2 = \sum_{i=1}^{n} (y_i - \hat{y}_i)^2 = \sum_{i=1}^{n} (y_i - b_0 - b_1 x_i)^2$$

This is algebraically equal to

$$\text{SSE} = \sum_{i=1}^{n} y_i^2 - b_0 \sum_{i=1}^{n} y_i - b_1 \sum_{i=1}^{n} x_i y_i$$

Given the sample points (x_i, y_i), the least-squares estimates of the population coefficients β_0 and β_1 are the values b_0 and b_1 that minimize SSE, which are given in the accompanying box.

Formula for the slope of the sample regression line

Either of the following two formulas can be used to obtain the value of b_1, the **slope** of the sample regression line:

$$b_1 = \frac{\sum_{i=1}^{n} x_i y_i - n\bar{x}\bar{y}}{\sum_{i=1}^{n} x_i^2 - n\bar{x}^2}$$

or

$$b_1 = \frac{n \sum_{i=1}^{n} x_i y_i - \left(\sum_{i=1}^{n} x_i \right) \left(\sum_{i=1}^{n} y_i \right)}{n \sum_{i=1}^{n} x_i^2 - \left(\sum_{i=1}^{n} x_i \right)^2}$$

The advantage of the second formula is that it avoids the potential round-off error that might result from calculating $\bar{x}$ and $\bar{y}$.

When calculating and reporting the estimated coefficients, it is important to maintain as many significant digits as possible. Rounding off the coefficient early in an analysis can lead to substantial errors later on.

Formula for the intercept of the sample regression line

The value of b_0, the **intercept** of the sample regression line, is

$$b_0 = \bar{y} - b_1 \bar{x}$$

where b_1 is the slope of the sample regression line.

Residuals Sum to Zero

For the sample regression line, the residuals will sum to 0. That is,

$$\Sigma \, \hat{e}_i = 0$$

The proof is as follows. The sum of the residuals is

$$\Sigma \, \hat{e}_i = \Sigma \, y_i - \Sigma \, \hat{y}_i$$
$$= \Sigma \, y_i - \Sigma(b_0 + b_1 x_i)$$
$$= \Sigma \, y_i - nb_0 - b_1 \Sigma \, x_i$$

If we divide both sides of this equation by n, we obtain

$$(\Sigma \, \hat{e}_i)/n = \bar{y} - b_0 - b_1 \bar{x}$$

Replacing b_0 by its value ($\bar{y} - b_1 \bar{x}$), we obtain

$$(\Sigma \, \hat{e}_i)/n = 0$$

which proves that $\Sigma \, \hat{e}_i = 0$.

Because the residuals sum to 0, we can check our estimation results. If the residuals do not sum to 0, then we know that we have made an error in calculating our estimates. Many computer packages print out the sum of the residuals. Occasionally the sum will be slightly different from 0 in the fourth or fifth decimal place. This small discrepancy can usually be attributed to round-off error.

The arithmetic calculations involved in obtaining all the appropriate estimates and test statistics associated with the simple linear regression model become very tedious as the number of observations increases. In addition, when there is more than one independent variable in the model, the computational burden increases tremendously. Fortunately, many computer programs exist that make it very easy to obtain the appropriate estimates and test statistics without doing any arithmetic computations.

In this book, whenever any arithmetic computations are required to illustrate some point, the number of observations will be kept small to minimize the number of computations. The reader should note, however, that it is not unusual to estimate actual regression models in which there are hundreds or even thousands of sample observations.

EXAMPLE 15.4   A Linear Regression Model

A stockbroker thinks that the selling prices of individual stocks are approximately linearly related to the annual dividends paid by the stocks. The data for a random sample of 10 stocks are shown in Table 15.2 and are plotted in Figure 15.6. Calculate the sample regression line and then find the predicted value of Y given that $X = 13$.

TABLE 15.2   *Data for Example 15.4*

Stock i	Annual dividend (in dollars) x_i	Price (in dollars) y_i	x_iy_i	x_i^2	y_i^2
1	13	115	1,495	169	13,225
2	4	45	180	16	2,025
3	12	100	1,200	144	10,000
4	5	50	250	25	2,500
5	6	55	330	36	3,025
6	8	85	680	64	7,225
7	3	40	120	9	1,600
8	4	50	200	16	2,500
9	5	45	225	25	2,025
10	7	70	490	49	4,900
Sums	67	655	5,170	553	49,025

SOLUTION   To find the estimates b_0 and b_1, we need the sums Σx_i, Σy_i, Σx_iy_i, and Σx_i^2, which are calculated in Table 15.2. The sum Σy_i^2 is calculated for later use. The sample means are $\bar{x} = 6.7$ and $\bar{y} = 65.5$. The slope estimate is

$$b_1 = \frac{\Sigma x_iy_i - n\bar{x}\bar{y}}{\Sigma x_i^2 - n\bar{x}^2}$$

$$= \frac{5,170 - 10(6.7)(65.5)}{553 - 10(6.7)(6.7)}$$

$$= 7.5072$$

and the intercept is

$$b_0 = \bar{y} - b_1\bar{x}$$
$$= 65.5 - (7.5072)(6.7)$$
$$= 15.2017$$

The sample regression equation is thus

$$\hat{y}_i = 15.2017 + 7.5072x_i$$

This equation can be used to predict the price of a stock based on its annual dividend. For example, suppose a stock's dividend is $13. The predicted selling price of this stock would be

$$\hat{y} = 15.2017 + 7.5072(13) = 112.80$$

To show how to calculate a residual, let us calculate the first residual $\hat{e}_1$ in Example 15.4. The first stock paid a dividend of $x_1 = \$13$, and the actual selling price was $y_1 = \$115$. Since the predicted price of a stock that pays a $13 dividend is $112.80, the first residual is

$$\hat{e}_1 = y_1 - \hat{y}_1 = 115.00 - 112.80 = 2.20$$

The sample data and the sample regression line are shown in Figure 15.6.

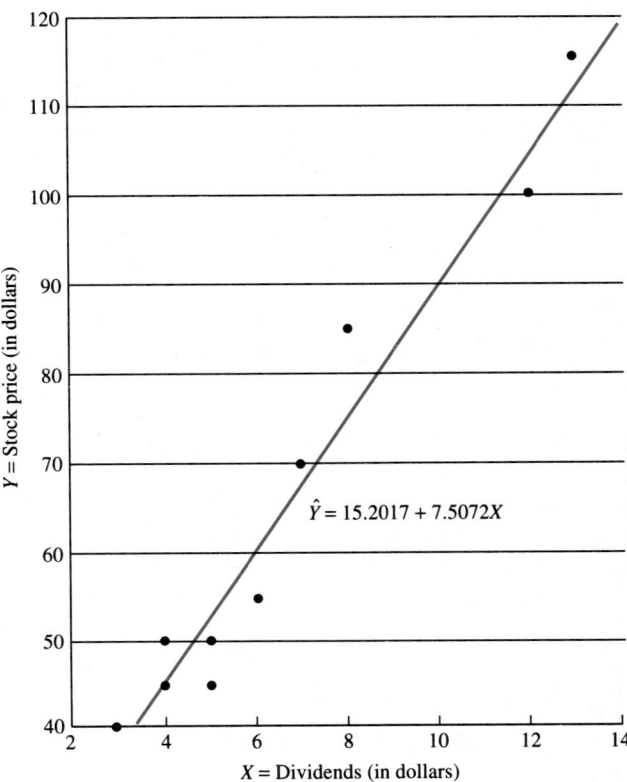

FIGURE 15.6 *Scatter diagram of data in Table 15.2*

Table 15.3 shows the observed values y_i, the predicted values $\hat{y}_i$, and the residuals $\hat{e}_i$ for the stock prices in Example 15.4. Note that except for round-off error, the residuals sum to 0. Later we shall want to calculate $\Sigma \, \hat{e}_i^2$, so the values $\hat{e}_i^2$ are also shown.

TABLE 15.3   *Observed values, predicted values, and residuals for Example 15.4*

i	y_i	$\hat{y}_i$	$\hat{e}_i$	$\hat{e}_i^2$
1	115	112.7953	2.2047	4.8607
2	45	45.2305	$-.2305$	.0531
3	100	105.2881	-5.2881	27.9640
4	50	52.7377	-2.7377	7.4950
5	55	60.2449	-5.2449	27.5090
6	85	75.2593	9.7407	94.8812
7	40	37.7233	2.2767	5.1834
8	50	45.2305	4.7695	22.7481
9	45	52.7377	-7.7377	59.8720
10	70	67.7521	2.2479	5.0531
	$\overline{655}$	654.9994	.0006	255.6196

Estimation of σ_e^2

The simple linear regression model contains three unknown parameters, β_0, β_1, and σ_e^2. We still need to obtain an estimate for σ_e^2. To do so, we utilize the sum of squared errors SSE. The following formula shows how to calculate an unbiased estimate of the population variance σ_e^2, which is denoted s_e^2:

Formula for unbiased estimate of σ_e^2 in the simple linear regression model

$$s_e^2 = \text{SSE}/(n - 2)$$

In this formula, the denominator $(n - 2)$ is the degrees of freedom of SSE, where n is the sample size and 2 represents the number of coefficients to be estimated in the regression line, namely, β_0 and β_1.

Degrees of freedom for multiple regression model

> In the multiple regression model, which includes several explanatory variables, the degrees of freedom of SSE is n minus the number of coefficients to be estimated in the regression line.

For example, in the regression model

$$E(Y_i|x_i) = \beta_0 + \beta_1 x_i$$

SSE has $(n - 2)$ degrees of freedom because there are two coefficients to be estimated. In the regression model

$$Y_i = \beta_0 + \beta_1 x_{i1} + \beta_2 x_{i2} + e_i$$

SSE has $(n - 3)$ degrees of freedom because there are 3 coefficients to be estimated, β_0, β_1, and β_2. In the *multiple regression model*

$$Y_i = \beta_0 + \beta_1 x_{i1} + \beta_2 x_{i2} + \cdots + \beta_K x_{iK} + e_i$$

SSE has $[n - (K + 1)]$ degrees of freedom because there are $(K + 1)$ coefficients to be estimated, $\beta_0, \beta_1, \ldots, \beta_K$.

Standard Error of the Regression

The square root of σ_e^2 is the standard deviation σ_e and represents the standard deviation of the subpopulation of Y's associated with a given value $X = x$. This standard deviation measures the amount of variation, or scatter, about the population regression line. The smaller the value of σ_e, the more concentrated are the values of Y about the population regression line and the closer the points (x_i, y_i) lie to it. The sample statistic used to estimate σ_e is denoted s_e and is called the *estimated standard error of the regression*.

DEFINITION **Estimated Standard Error of the Regression** ————————————

For the simple linear regression model, the **estimated standard error of the regression** is defined as

$$s_e = \sqrt{\frac{\Sigma(y_i - \hat{y}_i)^2}{n - 2}} = \sqrt{\frac{\Sigma \hat{e}_i^2}{n - 2}} = \sqrt{\frac{SSE}{n - 2}}$$

When we perform calculations by hand, the following expression can be used to calculate s_e:

Formula for s_e

$$s_e = \sqrt{\frac{\Sigma y_i^2 - b_0 \Sigma y_i - b_1 \Sigma x_i y_i}{n - 2}}$$

EXAMPLE 15.5 **Calculation of s_e** ————————————

Calculate the estimated standard error of the regression s_e for Example 15.4.

SOLUTION From Table 15.3, we obtain

$$SSE = \Sigma \hat{e}_i^2 = 255.6196$$

Alternatively, we can obtain SSE from the formula

$$\begin{aligned}SSE &= \Sigma y_i^2 - b_0 \Sigma y_i - b_1 \Sigma x_i y_i \\ &= 49{,}025 - (15.2017)(655) - (7.5072)(5{,}170) \\ &= 255.663\end{aligned}$$

(The slight difference in the two values of SSE is due to round-off error.)

An unbiased estimate of σ_e^2 is given by

$$s_e^2 = \frac{\text{SSE}}{n-2} = \frac{255.6196}{8} = 31.952$$

and the estimated standard error of the regression is

$$s_e = 5.653$$

This means that the subpopulation of stock prices associated with a given dividend $X = x_i$ has an estimated standard deviation of \$5.65.

Rounding Inaccuracies

Whenever you use a hand calculator to estimate a regression equation, you should carry as many significant digits as possible. For example, the estimate b_0 depends on $\bar{y}$, $\bar{x}$, and b_1. If any of these three quantities have been rounded off, then the estimate b_0 will be affected. Similarly, because all the residuals $\hat{e}_i$ depend on b_0 and b_1, rounding b_0 or b_1 will affect all the residuals. This, in turn, will affect the calculation of SSE and s_e.

There are many equivalent algebraic formulas that can be used to calculate various least squares estimates. Many of these formulas involve the following sums of squares and cross-products.

Alternative formulas

Various sums of squares and cross-products are useful in regression analysis:

$$\text{SS}_{yy} = \Sigma(y_i - \bar{y})^2 = \Sigma y_i^2 - n\bar{y}^2$$

$$\text{SS}_{xx} = \Sigma(x_i - \bar{x})^2 = \Sigma x_i^2 - n\bar{x}^2$$

$$\text{SS}_{xy} = \Sigma(x_i - \bar{x})(y_i - \bar{y}) = \Sigma x_i y_i - n\bar{x}\bar{y}$$

$$b_1 = \frac{\text{SS}_{xy}}{\text{SS}_{xx}}$$

Also,

$$\text{SS}_{yy} = \Sigma y_i^2 - \frac{(\Sigma y_i)^2}{n}$$

$$\text{SS}_{xx} = \Sigma x_i^2 - \frac{(\Sigma x_i)^2}{n}$$

$$\text{SS}_{xy} = \Sigma x_i y_i - \frac{(\Sigma x_i)(\Sigma y_i)}{n}$$

The residual sum of squares can be calculated as follows:

$$\text{SSE} = \Sigma(y_i - \hat{y}_i)^2 = \text{SS}_{yy} - \frac{(\text{SS}_{xy})^2}{\text{SS}_{xx}}$$

$$= \text{SS}_{yy} - b_1\text{SS}_{xy}$$

Tips for Problem Solving

1. If you estimate a simple linear regression model using a hand calculator, always calculate the following five sums:

$$\Sigma x_i \quad \Sigma y_i \quad \Sigma x_i y_i \quad \Sigma x_i^2 \quad \Sigma y_i^2$$

It will become evident later in this chapter why all five sums are useful.

2. Be careful to avoid round-off errors. Carry out your calculations to at least six decimal places if possible.

3. Always graph the data on a scatter diagram. A glance at such a graph should reveal whether a linear regression model will provide a good approximation to the relationship between Y and X.

4. Always plot the estimated sample regression line on the scatter diagram. If the plotted line seems unreasonable, check your arithmetic for calculation errors.

Exercises

Statistical Concepts

15.3.1 Explain what is meant by the residual sum of squares.

15.3.2 *True or false:* For the simple linear regression model, the random disturbance terms sum to 0. Explain.

15.3.3 *True or false:* For the simple linear regression model, every residual will be 0 if the data fall exactly on a straight line. Explain.

15.3.4 Let σ_e^2 denote the variance of the random error terms. *True or false:* The value of this variance is an indicator of whether the data lie close to or far from the population regression line. Explain.

15.3.5 *True or false:* In the simple linear regression model, the number of degrees of freedom of SSE is $(n - 2)$. Explain.

15.3.6 Let $SS_{xx} = \Sigma(x_i - \bar{x})^2$. Let $SS_{xy} = \Sigma(x_i - \bar{x})(y_i - \bar{y})$.
a. Prove that $SS_{xx} = \Sigma x_i^2 - n\bar{x}^2$.
b. Prove that $SS_{xy} = \Sigma x_i y_i - n\bar{x}\bar{y}$.
c. Prove that $b_1 = SS_{xy}/SS_{xx}$.

15.3.7 *True or false:* The sample regression line always passes through the point $(\bar{x}, \bar{y})$. Explain.

Statistical Drills

15.3.8 The accompanying data are pairs of values of Y and X.

x_i	y_i
6.1	17.4
7.3	18.5
6.0	16.4
4.2	15.7
5.1	16.2

a. Plot the data on a scatter diagram. Does there appear to be a relationship between Y and X?
b. Find Σx_i, Σy_i, $\Sigma x_i y_i$, Σx_i^2, and Σy_i^2.

c. Calculate $\bar{x}$ and $\bar{y}$.
d. Calculate the estimate b_1.
e. Calculate the estimate b_0.
f. Calculate the fitted values $\hat{y}_i$.
g. Graph the sample regression line $\hat{y}_i = b_0 + b_1 x_i$ on the scatter diagram.
h. Calculate the residuals $\hat{e}_i = y_i - \hat{y}_i$.
i. Calculate SSE.
j. Calculate s_e, the estimated standard error of the regression.

15.3.9 The accompanying data are pairs of values of Y and X.

x_i	y_i
2.0	11.0
3.0	16.3
4.0	16.7
5.0	21.1
6.0	26.8

a. Plot the data on a scatter diagram. Does there appear to be a relationship between Y and X?
b. Find Σx_i, Σy_i, $\Sigma x_i y_i$, Σx_i^2, and Σy_i^2.
c. Calculate $\bar{x}$ and $\bar{y}$.
d. Calculate the estimate b_1.
e. Calculate the estimate b_0.
f. Calculate the fitted values $\hat{y}_i$.
g. Graph the sample regression line $\hat{y}_i = b_0 + b_1 x_i$ on the scatter diagram.
h. Calculate the residuals $\hat{e}_i = y_i - \hat{y}_i$.
i. Calculate SSE.
j. Calculate s_e, the estimated standard error of the regression.

15.3.10 The accompanying data are pairs of values of Y and X.

x_i	y_i
6.1	2.9
7.3	3.5
6.0	3.0
4.2	2.0
5.1	2.6
8.2	4.3

a. Plot the data on a scatter diagram. Does there appear to be a relationship between Y and X?
b. Find Σx_i, Σy_i, $\Sigma x_i y_i$, Σx_i^2, and Σy_i^2.
c. Calculate $\bar{x}$ and $\bar{y}$.
d. Calculate the estimate b_1.
e. Calculate the estimate b_0.
f. Calculate the fitted values $\hat{y}_i$.
g. Graph the sample regression line $\hat{y}_i = b_0 + b_1 x_i$ on the scatter diagram.
h. Calculate the residuals $\hat{e}_i = y_i - \hat{y}_i$.
i. Calculate SSE.
j. Calculate s_e, the estimated standard error of the regression.

15.3.11 Suppose you estimate a sample regression line using $n = 25$ observations. You obtain SSE = 16.8. Find s_e.

15.3.12 Suppose you are given the following information:

$$n = 10 \quad \Sigma x_i = 34 \quad \Sigma y_i = 30 \quad \Sigma x_i y_i = 28 \quad \Sigma x_i^2 = 251 \quad \Sigma y_i^2 = 348$$

a. Find the equation of the sample regression line.
b. Find SSE.
c. Find s_e.

15.3.13 Suppose you are given the following information:

$$n = 20 \quad \Sigma x_i = 14 \quad \Sigma y_i = 43 \quad \Sigma x_i y_i = 328 \quad \Sigma x_i^2 = 521 \quad \Sigma y_i^2 = 434$$

a. Find the equation of the sample regression line.
b. Find SSE.
c. Find s_e.

Statistical Applications

15.3.14 Let x_i denote the percentage change in advertising expenditures for a firm during the ith sales period, and let y_i denote the eventual percentage change in sales revenue during the next period. The data are as follows:

x_i	8	10	7	6	9	5	6	7	5	6
y_i	6	9	5	5	10	8	7	9	7	8

a. Plot the data on a scatter diagram.
b. Find Σx_i, Σy_i, $\Sigma x_i y_i$, and Σx_i^2.
c. Find $\bar{x}$ and $\bar{y}$.
d. Find b_1.
e. Find b_0.
f. Graph the sample regression line $\hat{y}_i = b_0 + b_1 x_i$.
g. If $X = 11$, predict Y.
h. Find the residuals $\hat{e}_i$.

15.3.15 For a certain type of automobile, yearly repair costs in dollars (Y) are approximately linearly related to the age in years (X) of the car. The following table shows these data for 10 cars:

Repair costs Y	Age X
80	2
99	3
79	1
138	7
170	10
140	8
114	4
83	1
94	2
110	5

a. Plot the data on a scatter diagram.
b. Determine the linear regression of Y on X and graph it.
c. Estimate the repair costs of a 6-year-old car and a 3-year-old car.

15.3.16   The selling price of a used car is inversely (or negatively) related to the age of the car. That is, as the age increases, the selling price tends to decrease. The following table shows data for 10 cars of a certain make and model:

Selling price (in dollars) Y	Age (in years) X
980	5
1,760	3
1,100	5
600	8
2,100	2
1,600	3
1,400	4
710	7
800	6
1,800	3

a. Plot the data.
b. Find the equation of the sample regression line and graph it.
c. Estimate the selling price of a 4-year-old car.

15.3.17   A utility company believes that (except for July and August, when air conditioners are used) the per-customer usage of electricity in kilowatt-hours per month is inversely (or negatively) related to the average monthly temperature. The following data show the average monthly temperature in degrees Fahrenheit (computed as the average temperature at noon during the month) and the per-customer usage of electricity for a northeastern city for a sample of 10 months:

Electricity usage Y	Average monthly temperature X
1,000	18
420	50
400	55
705	30
550	45
850	25
1,020	17
670	35
610	38
560	42

a. Plot the data.
b. Find the equation of the regression line and graph it.
c. Estimate per-customer usage of electricity during a month when the average temperature is 40 degrees.

15.4   Explanatory Power of a Linear Regression Equation

After we have estimated the sample regression line, we want to determine whether the line provides a good fit and we want to measure the goodness of fit. The underlying population regression model states that the random variable Y_i is related to x_i by the equation

$$Y_i = \beta_0 + \beta_1 x_i + e_i$$

According to this model, Y varies for two reasons. One source of variability in Y values is due to variation in the X values. This variability can be explained by the regression model. The other source of variability in Y values is due to the random error e. This variability is unexplained by the regression model.

The ith residual $\hat{e}_i$ is defined by the equation

$$\hat{e}_i = y_i - \hat{y}_i$$

We can rearrange this equation to obtain

$$y_i = \hat{y}_i + \hat{e}_i$$

After subtracting the sample mean $\bar{y}$ from both sides of this equation, we obtain

$$(y_i - \bar{y}) = (\hat{y}_i - \bar{y}) + \hat{e}_i$$

Squaring all these deviations and summing them over the index i, we obtain

$$\Sigma(y_i - \bar{y})^2 = \Sigma(\hat{y}_i - \bar{y})^2 + \Sigma \hat{e}_i^2 + 2 \Sigma(\hat{y}_i - \bar{y})\hat{e}_i$$

It can be shown that the last sum on the right is 0, provided the regression line contains the constant b_0. Thus, we obtain

$$\Sigma(y_i - \bar{y})^2 = \Sigma(\hat{y}_i - \bar{y})^2 + \Sigma \hat{e}_i^2$$

These three sums of squares are denoted SST, SSR, and SSE, respectively. Thus, we obtain

$$\text{SST} = \text{SSR} + \text{SSE}$$

where SST is a measure of the total variation in the dependent variable Y, SSR is a measure of the total variation in the fitted values $\hat{y}_i$, and SSE measures the variation in Y that is unexplained by the model.

Sum of squares decomposition

The equation

$$\sum_{i=1}^{n} (y_i - \bar{y})^2 = \sum_{i=1}^{n} (\hat{y}_i - \bar{y})^2 + \sum_{i=1}^{n} \hat{e}_i^2$$

can be expressed as

Total variation = Explained variation + Unexplained variation

(continued)

or

$$SST = SSR + SSE$$

where

Total sum of squares: $$SST = \sum_{i=1}^{n} (y_i - \bar{y})^2$$

Regression sum of squares: $$SSR = \sum_{i=1}^{n} (\hat{y}_i - \bar{y})^2$$

Sum of squared errors: $$SSE = \sum_{i=1}^{n} \hat{e}_i^2$$

Because SST, SSR, and SSE are all sums of squares, each sum must be nonnegative. It follows that $SST \geq SSR$ and $SST \geq SSE$.

The equation

$$SST = SSR + SSE$$

has an important interpretation. For a given value of SST, the smaller the value of SSE, the smaller the deviations of the observed values of Y from the sample regression line. Thus, the smaller the value of SSE, the better the sample regression line fits the data.

To measure how well a sample regression line fits the data, we divide the total variation in the dependent variable (SST) into two components: (1) the variation in Y that can be explained by the sample regression line (SSR), and (2) the variation in Y that cannot be explained by the sample regression line (SSE).

Coefficient of Determination R^2

The equation

$$SST = SSR + SSE$$

can be expressed as

$$SSR = SST - SSE$$

If we divide both sides of this equation by SST, we obtain

$$\frac{SSR}{SST} = 1 - \frac{SSE}{SST}$$

This gives us the coefficient of determination R^2, described in the accompanying box.

DEFINITION **Coefficient of Determination** ─────────────────────────

The ratio SSR/SST, denoted by the symbol R^2, is called the **coefficient of determination** and is equal to the following:

$$R^2 = \frac{SSR}{SST} = 1 - \frac{SSE}{SST}$$

SSR measures the variation in Y that is explained by the regression equation, whereas SST measures the total variation of Y in the sample. Thus, the ratio SSR/SST measures the *proportion* of the variation in Y that is explained by the regression equation. Because SST $\geq$ SSR, it follows that

$$0 \leq R^2 \leq 1$$

The coefficient of determination R^2 provides a measure of the relative amount of variation in Y explained by the regression line. The higher the value of R^2, the greater the explanatory power of the regression equation. If SSE is small relative to SST, then R^2 will be close to 1, and the regression equation explains most of the variation in Y. If every observation falls exactly on the sample regression line, then each residual will be 0. In this case, SSE = 0 and $R^2 = 1$, and we have a perfect fit.

On the other hand, if SSE/SST is close to 1, then the residuals account for a large part of the variation in Y and R^2 will be close to 0. When R^2 is close to 0, the regression line does not explain the dependent variable Y well. In the extreme case, SSE = SST and SSR = 0. When SSR = 0, every predicted value $\hat{y}_i$ is equal to $\bar{y}$ and there is no variation in the predicted values. That is, when SSR = 0, the predicted values do not change as X changes. Thus, $R^2 = 0$ (or $R^2 \approx 0$) indicates that the independent variable X does not influence the dependent variable.

The use of the symbol R^2 is based on the fact that in the simple linear regression model, R^2 equals the square of the sample correlation coefficient between Y and X, which is denoted by the symbol R. The correlation coefficient R is discussed in Section 15.9.

An alternative formula for R^2 is shown in the accompanying box.

Alternative formula for R^2

The sample variances of X and Y are

$$s_X^2 = \frac{\Sigma(x_i - \bar{x})^2}{n - 1} = \frac{\Sigma x_i^2 - n\bar{x}^2}{n - 1}$$

and

$$s_Y^2 = \frac{\Sigma(y_i - \bar{y})^2}{n - 1} = \frac{\Sigma y_i^2 - n\bar{y}^2}{n - 1}$$

Then

$$R^2 = b_1^2 \frac{s_X^2}{s_Y^2}$$

The sample variances are positive values, so $R^2 = 0$ if and only if $b_1 = 0$. If $b_1 = 0$, then the regression line is a horizontal line and the predicted values of Y do not change for different values of X.

EXAMPLE 15.6   **Calculating the Coefficient of Determination** ———————————

In Example 15.4, we obtained the following sample regression equation relating fitted stock prices ($\hat{y}_i$) to cash dividends (x_i):

$$\hat{y}_i = 15.2017 + 7.5072x_i$$

In Example 15.5, we obtained all the residuals and SSE. Calculate R^2 for this model.

SOLUTION   Using the data from Tables 15.2 and 15.3, we obtain

$$\text{SST} = \Sigma(y_i - \bar{y})^2 = \Sigma \, y_i^2 - n\bar{y}^2$$
$$= 49{,}025 - 10(65.5)^2$$
$$= 6{,}122.5$$

From Table 15.3, we obtain SSE = 255.6196. Finally, we obtain

$$R^2 = 1 - \frac{\text{SSE}}{\text{SST}}$$

$$= 1 - \frac{255.6196}{6{,}122.5}$$

$$= .9582$$

Alternatively, we could calculate

$$s_X^2 = \frac{\Sigma \, x_i^2 - n\bar{x}^2}{n - 1}$$

$$= \frac{553 - 10(6.7)^2}{9} = 11.567$$

and

$$s_Y^2 = \frac{\Sigma \, y_i^2 - n\bar{y}^2}{n - 1} = \frac{49{,}025 - 10(65.5)^2}{9} = 680.278$$

We can then calculate R^2 using the formula

$$R^2 = b_1^2 \left(\frac{s_X^2}{s_Y^2} \right)$$

$$= (7.5072)^2 \left(\frac{11.567}{680.278} \right)$$

$$= .9582$$

The R^2 value indicates that, in the sample, dividends explain more than 95% of the total variation in stock prices. Figure 15.6 on page 601 shows that most of the sample observations lie very close to the sample regression line.

——

Tips for Problem Solving

1. The value of R^2 is one of the most important statistics associated with a regression model. R^2 is a descriptive measure of the strength of the regression relationship.

2. The total variation in Y is SST $= \Sigma(y_i - \bar{y})^2$.
3. The variation in Y explained by the model is SSR $= \Sigma(\hat{y}_i - \bar{y})^2$.
4. The variation in Y unexplained by the model is SSE $= \Sigma(y_i - \hat{y}_i)^2$.

5.
$$R^2 = \frac{\text{SSR}}{\text{SST}} = 1 - \frac{\text{SSE}}{\text{SST}}; \quad 0 \leqslant R^2 \leqslant 1$$

R^2 measures the proportion of the total variation in Y that is explained by the regression relationship between Y and X. The higher the value of R^2, the better the fit and the higher our confidence in the regression equation.

Exercises

Statistical Concepts

15.4.1 *True or false:* If Y and X are related according to a deterministic linear relationship, then SST will be 0. Explain.

15.4.2 *True or false:* If Y and X are related according to a deterministic linear relationship, then SSR will be 1. Explain.

15.4.3 *True or false:* If Y and X are related according to a deterministic linear relationship, then SSE will be 0 and SST = SSR. Explain.

15.4.4 *True or false:* If Y and X are related according to a deterministic linear relationship, then the coefficient of determination will be 1. Explain.

15.4.5 *True or false:* The residual e_i is defined as $e_i = y_i - b_0 + b_1 x_i$. Explain.

15.4.6 *True or false:* SSE must be nonnegative. Explain.

15.4.7 Can SSE ever be 0? Explain why or why not.

15.4.8 Can SSE ever be equal to SST? Explain why or why not.

15.4.9 *True or false:* s_Y^2 and s_e^2 are the same thing. Explain.

15.4.10 *True or false:* Suppose a sample regression line has slope $b_1 = 0$. Then SSR = 0. Explain.

15.4.11 *True or false:* Suppose a sample regression line has slope $b_1 = 0$. Then SSE = SST and $R^2 = 0$. Explain.

15.4.12 *True or false:* Suppose a sample regression line has slope $b_1 = 0$. Then SSE = SST and $R^2 = 1$. Explain.

15.4.13 *True or false:* There is no clear answer to the following question: How high must R^2 be before we can feel confident in using a regression model? Explain.

Statistical Drills

15.4.14 The following data are pairs of values of Y and X.

x_i	y_i
6.1	17.4
7.3	18.5
6.0	16.4
4.2	15.7
5.1	16.2

    a. Plot the data on a scatter diagram.
    b. Find the estimated sample regression line.
    c. Graph the sample regression line $\hat{y}_i = b_0 + b_1 x_i$ on the scatter diagram.
    d. Calculate SST, SSR, and SSE.
    e. Calculate R^2 and interpret its value. Does this value of R^2 indicate that the data lie close to the sample regression line? Examine your graph.

15.4.15   The following data are pairs of values of Y and X.

x_i	y_i
2.0	11.0
3.0	16.3
4.0	16.7
5.0	21.1
6.0	26.8

    a. Plot the data on a scatter diagram.
    b. Find the estimated sample regression line.
    c. Graph the sample regression line $\hat{y}_i = b_0 + b_1 x_i$ on the scatter diagram.
    d. Calculate SST, SSR, and SSE.
    e. Calculate R^2 and interpret its value. Does this value of R^2 indicate that the data lie close to the sample regression line? Examine your graph.

15.4.16   The following data are pairs of values of Y and X.

x_i	y_i
6.1	2.9
7.3	3.5
6.0	3.0
4.2	2.0
5.1	2.6
8.2	4.3

    a. Plot the data on a scatter diagram.
    b. Find the estimated sample regression line.
    c. Graph the sample regression line $\hat{y}_i = b_0 + b_1 x_i$ on the scatter diagram.
    d. Calculate SST, SSR, and SSE.
    e. Calculate R^2 and interpret its value. Does this value of R^2 indicate that the data lie close to the sample regression line? Examine your graph.

15.4.17   Suppose you obtain the following sample of four data points:

x_i	0	1	0	1
y_i	0	0	1	1

    a. Plot these four points on a scatter diagram.
    b. Without doing any calculations, what is the equation of the sample regression line?
    c. Perform the arithmetic calculations, and find the sample regression line.
    d. Find the residuals and calculate their sum.
    e. Calculate SST, SSR, and SSE.
    f. Calculate R^2 and interpret its value.

Statistical Applications

15.4.18 In a certain industry, costs are believed to be approximately linearly related to output. The following data show the cost of production (in thousands of dollars) and the number of units produced for several firms in the industry:

Cost Y	Units produced X
140	40
158	50
150	45
195	80
215	100
212	90
180	75
165	60

a. Plot the data.
b. Estimate the regression line of Y on X.
c. Find R^2 and s_e, and interpret both.
d. Predict Y if $X = 85$.

15.4.19 Does a high value of R^2 imply that two variables are causally related? Explain.

15.5 Estimating Variances of Estimated Coefficients in the Regression Model

The estimators b_0 and b_1 are random variables whose values change from sample to sample. Thus, b_0 and b_1 have sampling distributions. Their variances are denoted $\sigma_{b_0}^2$ and $\sigma_{b_1}^2$. In general, these variances are unknown and must be estimated. Our estimates of the true variances are denoted $s_{b_0}^2$ and $s_{b_1}^2$. Similarly, the symbols σ_{b_0} and σ_{b_1} denote the true standard deviations of b_0 and b_1, and s_{b_0} and s_{b_1} represent the estimated standard deviations of b_0 and b_1.

If the basic assumptions of the simple linear regression model hold, then the variances of b_0 and b_1 are given by the following formulas:

Formulas for the true variances of b_0 and b_1

$$\sigma_{b_0}^2 = \frac{\sigma_e^2 \sum_{i=1}^{n} x_i^2}{n \left(\sum_{i=1}^{n} x_i^2 - n\bar{x}^2 \right)}$$

and

$$\sigma_{b_1}^2 = \frac{\sigma_e^2}{\sum_{i=1}^{n} x_i^2 - n\bar{x}^2} = \frac{\sigma_e^2}{\sum_{i=1}^{n} (x_i - \bar{x})^2}$$

The variances of b_0 and b_1 depend on σ_e^2, the variance of the random error terms. The value σ_e^2 will generally be unknown and will have to be estimated. This implies that the variances of b_0 and b_1 also have to be estimated. To estimate them, we replace σ_e^2 in the preceding equations by its unbiased estimate s_e^2. To obtain an unbiased estimate of the variance of b_1, we use the following formula:

Unbiased estimate of the variance of b_1

$$s_{b_1}^2 = \frac{s_e^2}{\sum_{i=1}^{n} x_i^2 - n\bar{x}^2} = \frac{s_e^2}{\sum_{i=1}^{n}(x_i - \bar{x})^2}$$

DEFINITION   **Estimated Standard Error of b_1** ——————————————————

The estimated standard deviation of b_1, called the **estimated standard error of b_1**, is denoted s_{b_1} and is calculated as follows:

$$s_{b_1} = \frac{s_e}{\sqrt{\sum x_i^2 - n\bar{x}^2}} = \frac{s_e}{\sqrt{\sum(x_i - \bar{x})^2}}$$

The estimated standard error of b_1 is used in testing hypotheses about the value of the population parameter β_1 and in constructing confidence intervals for the value of β_1. Observe that s_{b_1} decreases as the sum $\Sigma(x_i - \bar{x})^2$ increases. That is, as the amount of variation in the X's increases, our estimate of the slope of the regression line becomes more precise.

In a similar fashion, s_{b_0} is called the *estimated standard error of b_0* and is defined in the accompanying box.

DEFINITION   **Estimated Standard Error of b_0** ——————————————————

The estimated standard deviation of b_0, called the **estimated standard error of b_0**, is denoted s_{b_0} and is calculated as follows:

$$s_{b_0} = \frac{s_e\sqrt{\sum x_i^2}}{\sqrt{n(\sum x_i^2 - n\bar{x}^2)}} = \frac{s_e\sqrt{\sum x_i^2}}{\sqrt{n\,\Sigma(x_i - \bar{x})^2}}$$

EXAMPLE 15.7   **Estimating the Standard Errors of the Estimated Coefficients** ————————————

Using the data in Table 15.4 on average weekly expenditures (in dollars) and average weekly after-tax income (in dollars), estimate the coefficients β_0 and β_1 in the model

$$E(Y_i|x_i) = \beta_0 + \beta_1 x_i$$

Also calculate R^2 and the estimated standard errors of the estimated coefficients. The data are graphed in Figure 15.7.

TABLE 15.4   *Data for Example 15.7*

Income (in dollars) x_i	Expenditures (in dollars) y_i	$x_i y_i$	x_i^2	y_i^2
400	350	140,000	160,000	122,500
300	250	75,000	90,000	62,500
350	325	113,750	122,500	105,625
400	370	148,000	160,000	136,900
200	180	36,000	40,000	32,400
300	270	81,000	90,000	72,900
375	330	123,750	140,625	108,900
380	350	133,000	144,400	122,500
325	300	97,500	105,625	90,000
400	360	144,000	160,000	129,600
3,430	3,085	1,092,000	1,213,150	983,825

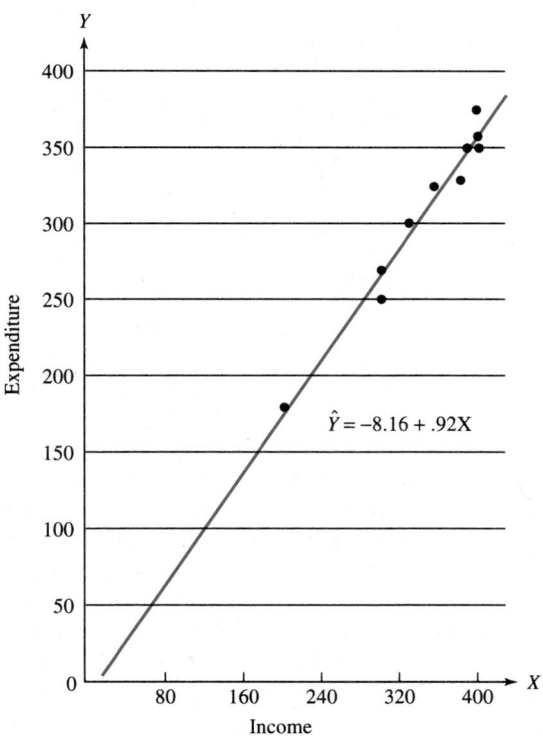

FIGURE 15.7   *Scatter diagram of data in Table 15.4*

SOLUTION   From the data in Table 15.4, we obtain $\bar{x} = 3{,}430/10 = 343$ and $\bar{y} = 3{,}085/10 = 308.5$. Therefore,

$$b_1 = \frac{\displaystyle\sum_{i=1}^{10} x_i y_i - 10\bar{x}\bar{y}}{\displaystyle\sum_{i=1}^{10} x_i^2 - 10\bar{x}^2}$$

$$= \frac{33{,}845}{36{,}660} = .923213$$

$$b_0 = \bar{y} - b_1\bar{x} = -8.16217$$

Thus, we obtain the sample regression equation

$$\hat{y}_i = -8.16217 + .923213x_i$$

Next, we obtain

$$\text{SST} = \sum_{i=1}^{10}(y_i - \bar{y})^2 = \sum_{i=1}^{10} y_i^2 - 10\bar{y}^2 = 32{,}102.50$$

$$\text{SSE} = \Sigma\, y_i^2 - b_0\, \Sigma\, y_i - b_1\, \Sigma\, x_i y_i = 856.367$$

$$s_e^2 = \frac{\text{SSE}}{n-2} = 107.046 \quad s_e = 10.346$$

$$R^2 = 1 - \frac{\text{SSE}}{\text{SST}} = .9733$$

$$s_{b_0}^2 = 354.236 \quad s_{b_0} = 18.8211$$

$$s_{b_1}^2 = .0029 \quad s_{b_1} = .054$$

To evaluate the results, it is useful to note the following facts:

1. R^2 is close to 1. This indicates that, in the sample, the variable income explains almost all the variation in expenditures. Thus, the linear equation provides a very good fit to the data.

2. The estimated standard error of the regression is $s_e = 10.346$. This means that the estimated standard deviation of the subpopulation of Y's associated with a given value of X is 10.346. For a variable with a normal distribution, we should expect approximately 68% of the observations to lie within 1 standard deviation of the mean and about 95% of the observations to lie within 2 standard deviations of the mean. Thus we would expect Y to lie within 10.346 units of the regression line about 68% of the time and within 20.693 units of the line 95% of the time.

3. The coefficient $b_1 = .923213$ indicates that every \$1.00 increase in income tends to be associated with about a \$.92 increase in expenditure. The estimated standard error of b_1 is $s_{b_1} = \$.054$.

Exercises

Statistical Concepts

15.5.1   *True or false:*   Other things being equal, the smaller the value of $s_{b_1}^2$, the more confidence we have that our estimate b_1 is close to the population parameter β_1. Explain.

15.5.2   *True or false:*   The value of s_{b_1} provides an estimate of the standard deviation of the sampling distribution of b_1. Explain.

Statistical Drills

15.5.3   Suppose you are given the following information:

$$n = 10 \quad \Sigma x_i = 34.3 \quad \Sigma y_i = 30.9 \quad \Sigma x_i y_i = 10{,}328 \quad \Sigma x_i^2 = 12{,}521 \quad \Sigma y_i^2 = 10{,}434$$

a. Find the equation of the sample regression line.
b. Find SST, SSR, and SSE.
c. Find s_e.
d. Find the estimated standard error for b_1.
e. Find the estimated standard error for b_0.

15.5.4   Suppose you are given the following information:

$$n = 10 \quad \Sigma x_i = 12 \quad \Sigma y_i = 97 \quad \Sigma x_i y_i = 93 \quad \Sigma x_i^2 = 19 \quad \Sigma y_i^2 = 7{,}461$$

a. Find the equation of the sample regression line.
b. Find SST, SSR, and SSE.
c. Find s_e.
d. Find the estimated standard error for b_1.
e. Find the estimated standard error for b_0.

Statistical Applications

15.5.5   The managers of the Ajax Corporation want to determine whether profits are linearly related to sales revenue. Let y_i denote profits (in millions of dollars) at the Ajax Corporation in year i, and let x_i denote sales (in millions) in year i. The data are as follows:

Y	1	4	7	2	5	4	9	8
X	8	12	33	10	22	16	45	38

Use the model

$$E(Y_i|x_i) = \beta_0 + \beta_1 x_i$$

a. Estimate β_0, β_1, and σ_e^2, and interpret the values.
b. Calculate SST, SSE, and SSR.
c. Calculate R^2 and s_e, and interpret the values.
d. Estimate the variances of b_0 and b_1.
e. Discuss possible violations of the basic assumptions.

15.5.6   It is hypothesized that states with high unemployment rates tend to decrease in population and states with low unemployment rates tend to gain population. Let x_i denote the percent unemployment rate of state i, and let y_i denote the net influx in thousands of migrants into state i. (If y_i is negative, the state suffered a net loss.) The data are as follows:

Influx Y	Unemployment rate X
270	4
50	6
−400	12
−30	9
−5	7
45	5
73	6
−10	9
−80	10

a. Estimate the population regression equation and interpret the result.
b. Find R^2 and s_e, and interpret the values.
c. Estimate the variances of b_0 and b_1.

15.5.7   Sam's Pizza has several stores in the suburbs around a city. Sam wants to know how the demand for his pizzas depends on the prices he charges. Occasionally he raises or lowers his prices to see how the demand increases or decreases. The following data show the prices in dollars Sam charged during certain months and the number of large pepperoni pizzas sold during those months:

Pizzas sold, Y	490	480	480	400	410	440
Price, X	4.75	5.00	5.25	5.50	5.20	5.30

a. Estimate the population regression equation and interpret the result.
b. Calculate R^2 and interpret the value.
c. Estimate the variances of b_0 and b_1.

15.5.8   A market analyst for the Weber Refrigerator Company has visited various appliance stores in a city to get data on the selling prices of different brands of refrigerators. Use the following data to help determine whether the selling price in dollars is linearly related to the volume in cubic feet of the refrigerator:

Price Y	Volume X
320	15.8
580	21.3
669	24.4
480	18.2
525	19.5
420	17.5
400	16.3
550	21.0

a. Estimate the population regression equation and interpret the result.
b. Estimate the variances of b_0 and b_1.
c. Calculate R^2 and interpret the value.

Hypothesis Testing in the Linear Regression Model

We frequently are interested in testing whether knowledge of an independent variable X is useful in explaining the values of Y. For example, we may want to test whether a linear relationship exists between X and Y. If X and Y are not linearly related, then in the population regression line

$$E(Y_i|x_i) = \beta_0 + \beta_1 x_i$$

we should have $\beta_1 = 0$. If $\beta_1 = 0$, the values of X are of no use in predicting Y. In Chapters 16, 17, and 18, we will examine more complex potential relationships between dependent and independent variables. If Y is linearly related to X, then $\beta_1 \neq 0$. Thus, we are usually interested in testing the null hypothesis

$$H_0: \quad \beta_1 = 0$$

against the alternative hypothesis

$$H_1: \quad \beta_1 \neq 0$$

At times, we may use a one-sided null hypothesis such as

$$H_0: \quad \beta_1 \geq 0 \qquad \text{or} \qquad H_0: \quad \beta_1 \leq 0$$

or a one-sided alternative hypothesis such as

$$H_1: \quad \beta_1 > 0 \qquad \text{or} \qquad H_1: \quad \beta_1 < 0$$

For example, if we reject $H_0: \beta_1 = 0$ in favor of the hypothesis $H_1: \beta_1 > 0$, then we are saying that the sample data support the hypothesis that the slope of the population regression line is positive. The following examples show some of the different hypotheses that we might want to test.

EXAMPLE 15.8 **A One-Sided Test** ⸻

Let y_i be the annual income of individual i, and x_i be the years of formal education of individual i. The relationship between Y and X is assumed to have the form

$$E(Y_i|x_i) = \beta_0 + \beta_1 x_i$$

It is reasonable to assume that either education has no effect on income or its effect is positive. Thus, we would test the null hypothesis

$$H_0: \quad \beta_1 = 0$$

against the one-sided alternative hypothesis

$$H_1: \quad \beta_1 > 0$$

EXAMPLE 15.9 **Another One-Sided Test** ⸻

Let x_i be the annual tuition charged at a state university during year i, and let y_i be the number of students who enroll during year i. Again we assume that

$$E(Y_i|x_i) = \beta_0 + \beta_1 x_i$$

It is reasonable to assume either that tuition cost has no effect on the number of students who enroll or that tuition cost and the number of students who enroll are negatively related. Thus, we would test the null hypothesis

$$H_0: \quad \beta_1 = 0$$

against the one-sided alternative hypothesis

$$H_1: \quad \beta_1 < 0$$

EXAMPLE 15.10   **A Two-Sided Test**

Suppose we want to test whether the grade point average (GPA) of a college student is linearly related to the income of the student's parents. Let y_i be the GPA of student i, and let x_i be the income of the parents of student i. We assume

$$E(Y_i|x_i) = \beta_0 + \beta_1 x_i$$

Because we have no reason a priori to think that β_1 is positive rather than negative, we would test the null hypothesis

$$H_0: \quad \beta_1 = 0$$

against the two-sided alternative hypothesis

$$H_1: \quad \beta_1 \neq 0$$

The parameter β_1 represents the slope of the population regression line and measures the change in the expected value of Y associated with a 1-unit change in X. If $\beta_1 = 0$, then the expected value of Y does not change as X changes, and the population regression line will be a horizontal line (see Figure 15.8). We can also test hypotheses about β_0, but usually there is much more interest in testing hypotheses about β_1. If we reject the hypothesis that $\beta_1 = 0$, then we are saying that the values of X are helpful in predicting Y.

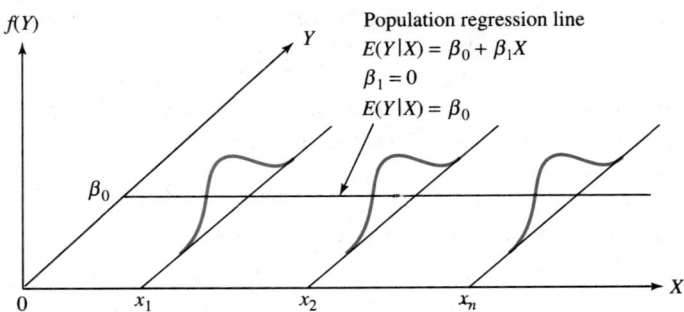

FIGURE 15.8   *A horizontal regression line*

EXAMPLE 15.11   **Another Two-Sided Test**

The null hypothesis does not always have the form

$$H_0: \quad \beta_1 = 0$$

although this is by far the most frequent case. Suppose an economist claims that, in the United States, annual income is linearly related to years of education and the slope of the population regression line is approximately $\beta_1 = \$2,000$. That is, the economist claims that, in the United States, an increase of 1 year in education tends to be associated with an increase of approximately \$2,000 in annual income. Suppose we want to test the null hypothesis that the slope of the population regression line is $\beta_1 = \$2,000$. Then we would test the null hypothesis

$$H_0: \quad \beta_1 = \$2,000$$

against the two-sided alternative hypothesis

$$H_1: \quad \beta_1 \neq \$2,000$$

Tests of hypotheses concerning the value of β_1 are based on the fact that if the basic assumptions of the simple linear regression model hold, then the random variable

$$t = \frac{b_1 - \beta_1}{s_{b_1}}$$

follows the Student t distribution with $(n - 2)$ degrees of freedom. The procedure in the accompanying box shows how to test a hypothesis about the slope of the population regression line.

Tests concerning the slope of the population regression line

Case 1: Suppose we want to test the null hypothesis

$$H_0: \quad \beta_1 = \beta_1^* \quad \text{or} \quad H_0: \quad \beta_1 \leq \beta_1^*$$

against the one-sided alternative hypothesis

$$H_1: \quad \beta_1 > \beta_1^*$$

where β_1^* is some hypothesized value of β_1. Suppose the level of significance of the test is α. The test is based on the t statistic

$$t = \frac{b_1 - \beta_1^*}{s_{b_1}}$$

If H_0 is false, then the t statistic tends to be greater than 0. Thus, large positive values of the t statistic provide evidence against the null hypothesis. We use the decision rule

Reject H_0 in favor of H_1 if $t > t_{\alpha,n-2}$,

where $t_{\alpha,n-2}$ is the critical value of the t distribution having $(n - 2)$ degrees of freedom such that

$$P(t > t_{\alpha,n-2}) = \alpha$$

(continued)

Case 2:   Suppose we want to test the null hypothesis

$$H_0: \quad \beta_1 = \beta_1^* \quad \text{or} \quad H_0: \quad \beta_1 \geq \beta_1^*$$

against the one-sided alternative hypothesis

$$H_1: \quad \beta_1 < \beta_1^*$$

Use the decision rule

Reject H_0 in favor of H_1 if $t < -t_{\alpha, n-2}$.

Case 3:   To test the null hypothesis

$$H_0: \quad \beta_1 = \beta_1^*$$

against the two-sided alternative hypothesis

$$H_1: \quad \beta_1 \neq \beta_1^*$$

use the decision rule

Reject H_0 in favor of H_1 if $t < -t_{\alpha/2, n-2}$ or if $t > t_{\alpha/2, n-2}$.

The most frequently performed test involves testing the null hypothesis $H_0: \beta_1 = 0$ against one of the alternative hypotheses

$$H_1: \quad \beta_1 \neq 0, \qquad H_1: \quad \beta_1 > 0, \qquad \text{or} \qquad H_1: \quad \beta_1 < 0$$

To perform the test, the test statistic is simply the t statistic $t = b_1/s_{b_1}$.

EXAMPLE 15.12   **Testing the Null Hypothesis H_0: $\beta_1 = 0$**

Refer to Example 15.7 where we used data concerning Y_i = Average weekly expenditures of the ith individual and X_i = Average weekly after-tax income of the ith individual to obtain the sample regression line

$$\hat{y}_i = -8.16217 + .923213x_i \qquad R^2 = .9733$$
$$(18.82) \qquad\quad (.054) \qquad\qquad n = 10$$

The values in parentheses are estimated standard deviations. It is quite obvious that expenditures depend on income. Thus, if we formally tested the null hypothesis that $\beta_1 = 0$, the results should overwhelmingly reject H_0. To show this, let us test the null hypothesis

$$H_0: \quad \beta_1 = 0$$

against the one-sided alternative hypothesis

$$H_1: \quad \beta_1 > 0$$

using a 1% level of significance.

SOLUTION   We have $n = 10$ observations, so the number of degrees of freedom is $(n - 2) = 8$. For $\alpha = .01$, we obtain the critical value $t_{.01,8} = 2.896$. We have $b_1 = .923213$ and $s_{b_1} = .054$. We calculate the t statistic

$$t = \frac{b_1 - 0}{s_{b_1}} = \frac{.923213}{.054} = 17.10$$

We reject the null hypothesis that $\beta_1 = 0$ because the t statistic falls in the critical region. In regression analysis, it is quite common to get results like this where the absolute value of the t statistic is very large. This provides very strong evidence that the population regression coefficient is nonzero.

EXAMPLE 15.13

Testing the Null Hypothesis H_0: $\beta_1 = .90$ ――――――――――――――

Suppose an economist analyzing the data in Example 15.7 argues that the marginal propensity to consume is .90 when using net income. That is, the economist claims that if after-tax income increases by $1.00, the average increase in consumption is $.90. Let us use a 5% level of significance, and test the null hypothesis

$$H_0: \quad \beta_1 = .90$$

against the two-sided alternative hypothesis

$$H_1: \quad \beta_1 \neq .90$$

SOLUTION We have $\alpha = .05$ and $(n - 2) = 8$ degrees of freedom. Thus, we have $\alpha/2 = .025$. The critical values of t are $t_{.025,8} = 2.306$ and $-t_{.025,8} = -2.306$. Thus, we will reject H_0 in favor of H_1 if $t < -2.306$ or if $t > 2.306$. We calculate the t statistic

$$t = \frac{b_1 - \beta_1^*}{s_{b_1}} = \frac{.923213 - .90}{.054} = .4298$$

We do not reject H_0 because $t = .4298$ falls in the acceptance region. This means that the evidence is not strong enough for us to conclude that the true value of the marginal propensity to consume out of additional income differs from .90.

Testing hypotheses about the intercept

Let β_0^* be some hypothesized value of β_0. To test hypotheses about the intercept of the population regression line, calculate the t statistic

$$t = \frac{b_0 - \beta_0^*}{s_{b_0}}$$

and proceed exactly as when testing hypotheses about the slope of the population regression line. When H_0 is true, this t statistic follows the t distribution with $(n - 2)$ degrees of freedom. To perform the test, we compare this t statistic with the critical values of the t distribution using α as the level of significance.

In most cases, there is little or no interest in testing hypotheses about the intercept of the regression equation. Occasionally we may be interested in testing that the intercept is 0, which would indicate that the regression line passes through the origin. The following example shows how to perform the test.

EXAMPLE 15.14

Testing a Null Hypothesis About the Intercept ――――――――――――

For the data in Example 15.7, use a 5% level of significance and test the null hypothesis

$$H_0: \quad \beta_0 = 0$$

against the two-sided alternative hypothesis

$$H_1: \quad \beta_0 \neq 0$$

SOLUTION   We have $\alpha = .05$ and $(n - 2) = 8$ degrees of freedom. We obtain $\alpha/2 = .025$, which we use to calculate the critical values $t_{.025,8} = 2.306$ and $-t_{.025,8} = -2.306$. The acceptance region for H_0 is $-2.306 \leqslant t \leqslant 2.306$. We have $b_0 = -8.16217$ and $s_{b_0} = 18.82$, and so we obtain the t statistic

$$t = \frac{b_0 - 0}{s_{b_0}} = \frac{-8.16217}{18.82} = -.4337$$

We do not reject H_0 because $t = -.4337$ falls in the acceptance region. This means that the evidence is not strong enough for us to conclude that the true value of the intercept of the consumption function differs from 0.

Tips for Problem Solving

> **1.** The most frequently performed test in regression analysis is the test of the null hypothesis $H_0: \beta_1 = 0$ against the alternative hypothesis $H_1: \beta_1 \neq 0$. Some statisticians call this a model utility test because the test helps determine whether the model is useful in explaining values of Y.
> **2.** Rejection of H_0 indicates that the evidence supports the hypothesis that changes in the variable Y are associated with changes in the variable X. Rejection of H_0 indicates that the model is useful in helping to explain values of Y.
> **3.** Nonrejection of H_0 indicates that there is no strong evidence that changes in Y are associated with changes in X. Nonrejection of H_0 indicates that the model is not useful in helping to explain values of Y.
> **4.** To perform the test, you need to know the estimated slope b_1 and its estimated standard deviation s_{b_1}. The test statistic is $t = b_1/s_{b_1}$, which follows the t distribution with $(n - 2)$ degrees of freedom provided the basic assumptions hold.
> **5.** As a general rule, values of t much larger than 2 lead to rejection of H_0.

Exercises

Statistical Concepts

15.6.1   Explain why we are interested in testing the null hypothesis $H_0: \beta_1 = 0$.

15.6.2   *True or false:*   Suppose we cannot reject the null hypothesis $H_0: \beta_1 = 0$. This proves that Y is not linearly related to X. Explain.

15.6.3   *True or false:*   Suppose we want to test the null hypothesis $H_0: \beta_1 = 0$. The appropriate test statistic is $t = b_1/s_e$. Explain.

15.6.4   Suppose you want to test the null hypothesis $H_0: \beta_1 = 0$ in the simple linear regression model. Statisticians and economists frequently say that a t statistic greater than 2 is statistically significant. Can you explain the reasoning behind this statement?

Statistical Drills

15.6.5   Suppose you are given the following information:

$$n = 21 \quad b_1 = 4.53 \quad s_{b_1} = 3.21$$

Let $\alpha = .05$. Suppose we want to test $H_0: \beta_1 = 0$ against $H_1: \beta_1 \neq 0$.

a. What is the appropriate number of degrees of freedom for the t statistic?
b. Find the critical values for the t statistic.
c. Perform the test.

15.6.6   Suppose you are given the following information:

$$n = 15 \quad b_1 = -6.71 \quad s_{b_1} = 8.38$$

Let $\alpha = .01$. Suppose we want to test $H_0: \beta_1 \geq 0$ against $H_1: \beta_1 < 0$.

a. What is the appropriate number of degrees of freedom for the t statistic?
b. Find the critical values for the t statistic.
c. Perform the test.

15.6.7   Suppose you are given the following information concerning a simple linear regression model:

$$n = 10 \quad b_1 = 5.6 \quad b_0 = 12.4 \quad s_{b_1} = 3.4 \quad s_{b_0} = 1.2$$

Let $\alpha = .05$. Suppose we want to test $H_0: \beta_1 = 0$ against $H_1: \beta_1 \neq 0$.

a. What is the appropriate number of degrees of freedom for the t statistic?
b. Find the critical values for the t statistic.
c. Perform the test.
Suppose we want to test $H_0: \beta_0 = 0$ against $H_1: \beta_0 \neq 0$.
d. What is the appropriate number of degrees of freedom for the t statistic?
e. Find the critical values for the t statistic.
f. Perform the test.

15.6.8   Suppose you are given the following information concerning a simple linear regression model:

$$n = 21 \quad b_1 = 25.6 \quad b_0 = 62.4 \quad s_{b_1} = 4.4 \quad s_{b_0} = 41.2$$

Let $\alpha = .05$. Suppose we want to test $H_0: \beta_1 = 0$ against $H_1: \beta_1 \neq 0$.

a. What is the appropriate number of degrees of freedom for the t statistic?
b. Find the critical values for the t statistic.
c. Perform the test.
Suppose we want to test $H_0: \beta_1 = 60$ against $H_1: \beta_1 \neq 60$.
d. What is the appropriate number of degrees of freedom for the t statistic?
e. Find the critical values for the t statistic.
f. Perform the test.

Statistical Applications

15.6.9   Refer to Exercise 5 at the end of Section 15.5.
a. Test $H_0: \beta_1 = 0$ against $H_1: \beta_1 > 0$ using a 1% level of significance.
b. Test $H_0: \beta_0 = 0$ against $H_1: \beta_0 \neq 0$ using a 1% level of significance.
c. Test $H_0: \beta_1 = .2$ against $H_1: \beta_1 \neq .2$ using a 1% level of significance.
d. Test $H_0: \beta_1 = .25$ against $H_1: \beta_1 \neq .25$ using a 10% level of significance.

15.6.10   It is claimed that a judge tends to give shorter sentences to wealthy people than to poor people for the same crime. The following data show the length of jail sentences in days given to eight individuals who committed similar crimes and their respective incomes in thousands of dollars:

Jail sentence Y	Income X
10	45
30	6
25	12
20	20
15	30
20	22
25	25
15	25

a. Estimate the linear regression equation.
b. Test H_0: $\beta_1 = 0$ against H_1: $\beta_1 < 0$ using a 5% level of significance.

15.6.11   The manager of Discount Distributing Company claims that sales of soda pop at her store are not very responsive to changes in price. The price of the best-selling brand has changed six times during the last 6 months. The following data show the prices charged in dollars and the number of cases of soda pop sold during the 20 days following each price change:

Cases sold, Y	600	550	560	500	520	540
Price per case, X	4.25	5.25	4.75	5.50	5.00	4.50

a. Estimate the linear regression equation.
b. Let $\alpha = .05$. Test H_0: $\beta_1 = 0$ against H_1: $\beta_1 < 0$.
Do you agree with the manager's claim?

15.6.12   A teaching tennis pro claims that a player can increase the speed of his or her serve by stringing the racquet tighter. To prove this point, the pro takes 10 identical racquets and has them strung at different tensions from 56 to 65 pounds. A playing pro then serves five times with each racquet, and the average speed of the serve is recorded in miles per hour. The data are as follows:

Average speed	Tension
98	56
102	57
100	58
101	59
104	60
107	61
104	62
108	63
109	64
110	65

a. Estimate the population regression equation.
b. Let $\alpha = .05$. Test the pro's conjecture.

Confidence Intervals for the Regression Coefficients

A confidence interval for the population parameter β_1 is based on the fact that, if the basic assumptions of the simple linear regression model hold, the t statistic

$$t = \frac{b_1 - \beta_1}{S_{b_1}}$$

follows the Student t distribution with $(n - 2)$ degrees of freedom. A confidence interval having level of confidence $(1 - \alpha)$ is based on the probability statement

$$P(-t_{\alpha/2,\nu} < t < t_{\alpha/2,\nu}) = 1 - \alpha$$

or equivalently

$$P\left(-t_{\alpha/2,\nu} \leq \frac{b_1 - \beta_1}{S_{b_1}} \leq t_{\alpha/2,\nu}\right) = 1 - \alpha$$

where the number of degrees of freedom is $\nu = (n - 2)$. By rearranging terms, we can obtain the equation

$$P(b_1 - t_{\alpha/2,\nu} S_{b_1} < \beta_1 < b_1 + t_{\alpha/2,\nu} S_{b_1}) = 1 - \alpha$$

This probability statement means that, in repeated sampling, the random interval

$$(b_1 - t_{\alpha/2,\nu} S_{b_1}, \, b_1 + t_{\alpha/2,\nu} S_{b_1})$$

will contain the true value β_1 approximately $100(1 - \alpha)\%$ of the time, where the value $t_{\alpha/2,\nu}$ is obtained from the t distribution having $\nu = (n - 2)$ degrees of freedom. For our specific values of b_1 and s_{b_1}, we cannot be sure that β_1 lies in that interval. If we were to repeat the estimation procedure many times with many samples of data, the confidence interval would contain the true value of β_1 approximately $100(1 - \alpha)\%$ of the time.

Confidence interval for the slope

The interval

$$(b_1 - t_{\alpha/2,\nu} S_{b_1}, \, b_1 + t_{\alpha/2,\nu} S_{b_1})$$

is called a $100(1 - \alpha)\%$ confidence interval for β_1.

By proceeding in a similar fashion, it is possible to obtain the formula for a confidence interval for the intercept of the population regression line, which is shown in the accompanying box.

Confidence interval for the intercept

A confidence interval for the regression intercept β_0 having level of confidence $(1 - \alpha)$ is given by

$$(b_0 - t_{\alpha/2,\nu} S_{b_0}, \, b_0 + t_{\alpha/2,\nu} S_{b_0})$$

EXAMPLE 15.15   **A Confidence Interval for the Slope**

Refer to the data in Example 15.7. We obtained the results

$$\hat{y}_i = -8.16217 + .923213x_i \qquad R^2 = .9733$$
$$\quad\;\;(18.82) \qquad\;\;\; (.054) \qquad\qquad n = 10$$

where the values in parentheses are estimated standard errors of the corresponding esti-mated coefficients. Construct a 90% confidence interval for the slope β_1.

S O L U T I O N   We have $\alpha/2 = .05$ and the number of degrees of freedom is $(n - 2) =$ 8. We obtain $t_{\alpha/2,\nu} = t_{.05,8} = 1.860$. The desired confidence interval is

$$(b_1 - t_{\alpha/2,\nu}\, s_{b_1},\; b_1 + t_{\alpha/2,\nu}\, s_{b_1})$$

or

$$[.923213 - 1.860(.054),\; .923213 + 1.860(.054)]$$

Thus, the confidence interval is $(.82277, 1.02365)$. In this example, the coefficient β_1 measures the marginal propensity to consume (MPC) out of disposable income. The con-fidence interval indicates that it is not unreasonable to suspect that the MPC could be as low as .82. On the other hand, a value as large as 1.02 is unreasonable, because this would indicate that a \$1.00 increase in disposable income leads to a \$1.02 increase in expendi-tures. It is unreasonable to suspect that the MPC exceeds 1.00.

EXAMPLE 15.16   **Effect of Cigarette Smoking by Expectant Mothers on the IQ of the Child**

On March 15, 1987, the newspaper *USA Today* reported the results of a study in which researchers in Baltimore determined that 3-year-old children whose mothers smoked dur-ing pregnancy scored an average of 5 points lower on an IQ test than children of women who did not smoke while pregnant. Suppose that after reading this report, you decide to examine the relationship between the number of cigarettes smoked per day by an expec-tant mother and the IQ of her child at age 3. Let x_i denote the average number of cigarettes smoked per day by the ith expectant mother, and let y_i denote the IQ of the ith child at age 3. Suppose all of the women in your study have approximately the same IQ, and suppose that all of the fathers also have approximately the same IQ so that differences in the IQs of the children cannot be attributed to differences in the IQs of the parents. It is hypothesized that differences in the IQs of the children can be attributed to the number of cigarettes smoked per day by the expectant mother and other random factors.

The hypothesized model is

$$E(Y_i|x_i) = \beta_0 + \beta_1 x_i$$

We obtain a random sample of $n = 20$ observations showing the IQs of children at age 3 and the number of cigarettes smoked per day by the child's mother during pregnancy. The estimated results are as follows:

$$\hat{y}_i = 104 - .60x_i \qquad R^2 = .17$$
$$\quad\;\;(1.2) \quad\; (.15) \qquad\;\; s_e = 7.8$$

The values in parentheses are the estimated standard errors of the estimated coefficients. Analyze the regression results.

SOLUTION   The slope of the sample regression line is $b_1 = -.60$. This indicates that an increase of one cigarette per day is associated with a decrease of .6 point in IQ. The coefficient $b_0 = 104$ indicates that the predicted IQ for children whose mothers do not smoke ($x_i = 0$) is 104.

It is natural to test the null hypothesis that the number of cigarettes smoked by the expectant mother does not influence the IQ of the child. The alternative hypothesis is that increased smoking causes the child's IQ to decrease. Test the null hypothesis

$$H_0: \quad \beta_1 = 0$$

against the one-sided alternative hypothesis

$$H_1: \quad \beta_1 < 0$$

Suppose the level of significance is chosen to be $\alpha = .05$. If the null hypothesis is true, then the t statistic follows the t distribution with $(n - 2) = 18$ degrees of freedom. The critical value of the test statistic is $t_{.05,18} = -1.734$. The observed value of the test statistic is

$$t = \frac{b_1}{s_{b_1}} = \frac{-.60}{.15} = -4.00$$

This value exceeds the critical value and thus falls in the rejection region, so we reject the null hypothesis that $\beta_1 = 0$. The sample regression line provides very strong evidence that the number of cigarettes smoked by the expectant mother is related to the IQ of the child.

Once we have rejected H_0, we would want to construct a confidence interval for the population coefficient β_1. Suppose we decide to construct a confidence interval for β_1 having level of confidence $(1 - \alpha) = .95$. The critical value of t is $t_{.025,18} = 2.101$. The confidence interval utilizes the estimated coefficient $b_1 = -.60$ and the estimated standard error $s_{b_1} = .15$. The 95% confidence interval for β_1 is

$$(b_1 - t_{\alpha/2,\nu} \, s_{b_1}, \; b_1 + t_{\alpha/2,\nu} \, s_{b_1})$$

or

$$[-.60 - 2.101(.15), \; -.60 + 2.101(.15)]$$

The desired confidence interval extends from $-.9152$ to $-.2849$. This means that we can be 95% confident that the true value of β_1 falls within this interval. If the population regression coefficient is as low as $-.9152$, then an increase of $X = 10$ cigarettes per day would lead to a decrease in IQ of 9.152 points. If the population regression slope is $-.2849$, then an increase of $X = 10$ cigarettes per day would lead to a decrease in IQ of 2.849 points.

A researcher would also be interested in observing whether the number of cigarettes smoked by the expectant mother explained most of the variation in the IQs of the children at age 3. (Recall that the IQs of all the mothers were approximately equal and the IQs of all the fathers were approximately equal, so that parents' IQs do not explain variations in the children's IQs.) The coefficient of determination is $R^2 = .17$, indicating that the number of cigarettes smoked by the expectant mother explains 17% of the observed variation in the IQs of the children. Alternatively, we could say that the model does not explain 83% of the variation in the IQs of the children.

A researcher would also examine the estimated standard error of the regression, $s_e = 7.8$. This indicates that the standard deviation of the error term is estimated to be 7.8. From the Empirical Rule, which was discussed in Chapter 4, we know that about 95% of the observations for a bell-shaped distribution should lie within 2 standard deviations of the mean. Using $s_e = 7.8$, about 95% of the IQs should lie within 15.6 units of the population regression line for a given value of X. Because an error of 15.6 units in an estimate of a child's IQ is relatively large, the unexplained random factors that influence a child's IQ are relatively sizable. This reinforces the information obtained by examining the value of R^2.

After we examine these data, the general conclusion should be that cigarette smoking by the expectant mother apparently is related to the child's IQ. An increase of one cigarette per day smoked by the expectant mother is associated with roughly a .6-unit decrease in the child's IQ, but the decrease could be as small as a .2849-unit decrease in IQ or as large as a .9152-unit decrease in IQ. The data are not highly concentrated about a straight line because the value of R^2 is only .17. This means that 83% of the variability in the IQs of the children is unexplained. It should be kept in mind, however, that if the number of cigarettes smoked by the expectant mother explains 17% of the variation in IQs of the children, then this can be a sizable and important influence on IQ.

Exercises

Statistical Concepts

15.7.1   *True or false:*   Suppose we cannot reject the null hypothesis $H_0: \beta_1 = 0$ using $\alpha = .05$. The 95% confidence interval for β_1 will contain the value 0. Explain.

15.7.2   *True or false:*   Suppose we are constructing a 95% confidence interval for β_1 in the simple linear regression model. Other things being equal, a wide confidence interval is preferable to a narrow one. Explain.

15.7.3   Suppose we have constructed a 99% confidence interval for β_1 in the simple linear regression model. Assume this confidence interval contains the value 0. What can you say about testing $H_0: \beta_1 = 0$?

15.7.4   *True or false:*   Suppose we have constructed 99% confidence intervals for β_1 and β_0 in the simple linear regression model. Assume these confidence intervals are very narrow. We can infer that R^2 is close to 1. Explain.

15.7.5   Suppose we have estimated a simple linear regression model. Suppose we obtain $b_1 = 4.78$ and $R^2 = 1.00$. What can we determine about the 95% confidence interval for β_1?

15.7.6   *True or false:*   Suppose a 95% confidence interval for β_1 contains the value 0. This implies that Y and X are not causally related. Explain.

15.7.7   Examine the formula for s_{b_1}, and explain how the values of x_i influence the value of s_{b_1} and consequently affect the width of confidence intervals for β_1.

15.7.8   Other things being equal, would you prefer $\Sigma(x_i - \bar{x})^2$ to be large or small when constructing a confidence interval for β_1?

15.7.9   Sometimes it is possible to design a regression study by setting the values of the independent variable X at different levels and then observing the corresponding values of the dependent variable Y. If you were designing such a study, how would you recommend selecting the values of X?

15.7.10   Suppose a statistician constructs a 95% confidence interval for β_1, and this confidence interval does not contain the value 0. *True or false:*   If we test $H_0: \beta_1 = 0$ against $H_1: \beta_1 \neq 0$ using $\alpha = .05$, we will reject H_0. Explain.

Statistical Drills

15.7.11   Suppose you are given the following information:

$$n = 11 \qquad b_1 = 4.82 \qquad s_{b_1} = 3.58$$

   a. Construct a 95% confidence interval for β_1 and interpret the result.
   b. Construct a 99% confidence interval for β_1.
   c. Let $\alpha = .05$. Suppose we want to test $H_0: \beta_1 = 0$ against $H_1: \beta_1 \neq 0$. Perform the test and interpret the result.

15.7.12   Suppose you are given the following information:

$$n = 15 \qquad b_1 = -6.71 \qquad s_{b_1} = 8.38$$

   a. Construct a 95% confidence interval for β_1 and interpret the result.
   b. Construct a 99% confidence interval for β_1.
   c. Let $\alpha = .05$. Suppose we want to test $H_0: \beta_1 = 0$ against $H_1: \beta_1 \neq 0$. Perform the test and interpret the result.

15.7.13   Suppose you are given the following information concerning a simple linear regression model:

$$n = 10 \quad b_1 = 5.6 \quad b_0 = 12.4 \quad s_{b_1} = 3.4 \quad s_{b_0} = 1.2$$

   a. Construct a 95% confidence interval for β_1 and interpret the result.
   b. Construct a 99% confidence interval for β_1.
   c. Let $\alpha = .05$. Suppose we want to test $H_0: \beta_1 = 0$ against $H_1: \beta_1 \neq 0$. Perform the test and interpret the result.

Statistical Applications

15.7.14   Refer to Exercise 5 in Section 15.5. Construct 90%, 95%, and 99% confidence intervals for β_0 and β_1.

15.7.15   The unemployment rate in a midwestern city tends to depend on the number of employees working at a large steel mill located in the city, as shown in the following table:

Unemployment rate (in percent) Y	Employees (in thousands) X
4	14
6	11
12	8
8	10
5	12
9	9
11	7
10	8
14	7
9	8

a. Plot the data.

b. Estimate the population regression equation, and plot the sample regression line on the scatter diagram. Interpret the result.

c. Calculate a 95% confidence interval for β_1 and interpret the result.

15.7.16   The Acme Corporation manufactures color televisions. It is believed that the number of color televisions sold per month is linearly related to the amount of money spent on advertising. The following data show the number of color TVs sold (in thousands) and total advertising expenditures (in thousands of dollars) during 8 months:

Number sold Y	Advertising expenditure X
62	22
50	15
58	17
64	20
68	22
64	23
70	25
72	26

a. Plot the data.

b. Estimate the population regression equation and plot the sample regression line. Interpret.

c. Construct a 95% confidence interval for β_1.

d. Test $H_0: \beta_1 = 0$ against $H_1: \beta_1 > 0$. Let $\alpha = .05$. Interpret the result.

15.7.17   The Quick-In-Quick-Out Food Store specializes in giving fast service to customers who are in a hurry and only want to buy a few items. The manager thinks that the daily sales revenue during the spring is related to temperature. He thinks that the higher the temperature, the more customers the store has and the higher the revenue. Use the following data on daily revenue (in thousands of dollars) and noon temperature (in degrees Fahrenheit) to test this hypothesis:

Daily revenue, Y	2.6	3.4	2.5	3.2	3.3	2.7
Temperature, X	58	66	63	40	52	60

a. Estimate the population regression equation. Interpret the result.

b. Construct a 99% confidence interval for β_1.

c. Let $\alpha = .01$. Test $H_0: \beta_1 = 0$ against $H_1: \beta_1 > 0$. Interpret the result.

15.8   Prediction Using the Regression Model

One of the goals of regression analysis is to obtain an equation that will enable us to predict Y for a given value of X. There are two different prediction problems of interest:

1. We may want to estimate or predict the *actual value* of the random variable Y_i from the equation $Y_i = \beta_0 + \beta_1 x_i + e_i$ when the independent variable takes the value x_i.

2. We may want to estimate the *conditional mean* $E(Y_i | x_i)$ that results when the independent variable takes the value x_i. This conditional mean represents the *average value* of the random variable Y_i when the independent variable is fixed at x_i.

To predict Y for a given value of X, we use the equation

$$\hat{y}_i = b_0 + b_1 x_i$$

We use this equation whether we are predicting an individual Y value or predicting the mean value of Y associated with a given value of X.

Formula to predict Y_i and $E(Y_i|x_i)$

The best point estimate of Y_i and $E(Y_i|x_i)$ is given by

$$\hat{y}_i = b_0 + b_1 x_i$$

Although the prediction of an individual value Y_i and the estimate of the mean value $E(Y_i|x_i)$ are the same for a given value x_i, the sampling errors associated with the two predictions are different. This is reasonable because the estimate of the mean value of Y_i does not require an estimate of the random error e_i.

Effects of Sampling Error

In general, the estimated value $\hat{y}_i$ will differ from the mean value $E(Y_i|x_i) = (\beta_0 + \beta_1 x_i)$ because b_0 and b_1 are only estimates of the unknown coefficients β_0 and β_1. Thus, the sampling error involved in predicting the mean value of Y for a given X is due to the sampling error involved in estimating β_0 and β_1.

The prediction of an individual value Y_i is subject to sampling error because we use $(b_0 + b_1 x_i)$ as our prediction of the actual value y_i, which is actually $(\beta_0 + \beta_1 x_i + e_i)$. We have sampling error because b_0 and b_1 are estimates of β_0 and β_1 and because we have used 0 as our estimate of the random error e.

Confidence Intervals for Predictions

Just as we construct confidence intervals for the mean of a population based on a sample of values, it is natural to construct confidence intervals for the mean value of Y_i and for an individual value of Y_i based on our estimated value $\hat{y}_i = b_0 + b_1 x_i$.

To construct confidence intervals for the mean value of Y associated with a given value of X, say, $X = x_p$, we use the sampling distribution of $\hat{Y}_p$. If the basic assumptions of the simple linear regression model hold, then $\hat{Y}_p$ is normally distributed with mean

$$E(\hat{Y}_p|x_p) = \beta_0 + \beta_1 x_p$$

and variance

$$\text{Var}(\hat{Y}_p) = \sigma_e^2 \left(\frac{1}{n} + \frac{(x_p - \bar{x})^2}{\Sigma\, x_i^2 - n\bar{x}^2} \right)$$

Figure 15.9 (page 636) shows the characteristics of the sampling distribution of $\hat{Y}_p$. The formulas in the following boxes show how to construct confidence intervals for $E(Y_p|x_p)$ and for an individual value of Y_p.

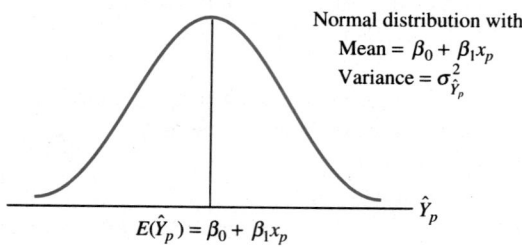

FIGURE 15.9   *Sampling distribution of $\hat{Y}_p$*

To estimate $\text{Var}(\hat{Y}_p)$, we replace σ_e^2 by its estimate $s_e^2 = \text{SSE}/(n - 2)$. We obtain the estimated variance of $\hat{Y}_p$.

Formula to estimate the variance of $\hat{Y}_p$

$$s_{\hat{Y}_p}^2 = s_e^2 \left(\frac{1}{n} + \frac{(x_p - \bar{x})^2}{\sum x_i^2 - n\bar{x}^2} \right)$$

Formula for a confidence interval for $E(Y_p|x_p)$

Given a specific value x_p, a confidence interval for the mean value of Y_p having level of confidence $(1 - \alpha)$ is given by

$$\hat{Y}_p \pm t_{\alpha/2, \nu}\, s_e \sqrt{\frac{1}{n} + \frac{(x_p - \bar{x})^2}{\sum x^2 - n\bar{x}^2}}$$

where the number of degrees of freedom is $\nu = (n - 2)$.

Formula for a confidence interval for Y_p

Given a specific value x_p, a confidence interval for the actual value of Y_p having level of confidence $(1 - \alpha)$ is given by

$$\hat{Y}_p \pm t_{\alpha/2, \nu}\, s_e \sqrt{1 + \frac{1}{n} + \frac{(x_p - \bar{x})^2}{\sum x^2 - n\bar{x}^2}}$$

The confidence interval for an individual Y is always wider than the confidence interval for the mean value of Y. This is reasonable because the mean value of Y does not involve the random error e. For a given level of confidence, the wider the confidence interval, the greater the uncertainty surrounding our prediction. We make the following observations concerning the width of a confidence interval for a prediction of Y_i or $E(Y_i|x_i)$:

1. Other things being equal, the larger the sample size n, the smaller the width of the confidence interval. This reflects the fact that we become more certain about a prediction as we obtain more sample information.

2. Other things being equal, the larger the value s_e, the greater the width of the confidence interval. Recall that s_e^2 is an estimate of σ_e^2, the variance of the error terms. The error terms represent the discrepancy between the observed values of the dependent variables and their expected values. The standard deviation of the error s_e is an estimate of the standard deviation of the population of Y_i values associated with the value x_i. A large value s_e indicates that the distribution of Y values is not highly concentrated about the population regression line, thus making a precise estimate of the population regression line difficult to obtain.

3. The quantity $(\Sigma\ x_i^2 - n\bar{x}^2) = \Sigma(x_i - \bar{x})^2 = (n - 1)s_x^2$, where s_x^2 is the sample variance of the X values. Thus, $(\Sigma x_i^2 - n\bar{x}^2)$ is a multiple of the sample variance of X and can be used as a measure of the amount of variation in the sample of observations on the independent variable. A large variance implies that we have information for a wide variety of values of X. This fact makes it easier to determine how the variable Y depends on the variable X. Consequently, our estimates of the slope and intercept of the population regression line improve and our predictions become more precise. As a result, the larger the value of $(\Sigma\ x_i^2 - n\bar{x}^2)$, the narrower the confidence intervals for our predictions.

4. The larger the value of $(x_p - \bar{x})^2$, the wider the confidence intervals for the predictions. This means that our predictions become less precise the farther x_p is from the sample mean of the independent variable. This result reflects the fact that, if our sample data are centered at $\bar{x}$, we would have more confidence in a prediction of Y_p when x_p is close to $\bar{x}$ than when it is far from $\bar{x}$.

EXAMPLE 15.17 Confidence Interval for a Prediction

Suppose the annual repair cost of an automobile Y_i is approximately linearly related to the age of the car x_i. A sample of 15 cars was used to estimate the equation

$$\hat{y}_i = 50 + 25x_i$$

Assume that $s_e = 30$, $\bar{x} = 5$, and $(\Sigma x^2 - n\bar{x}^2) = 50$. Let us find 95% confidence intervals for the mean value of Y_p for cars aged $x_p = 1, 2, 3, 4, 5, 6, 7, 8,$ and 9 years.

SOLUTION   We have $\alpha = .05$ and $(n - 2) = 13$ degrees of freedom. The critical value of the t distribution is $t_{.025,13} = 2.160$. We obtain the predicted values and confidence intervals shown in Table 15.5 (page 638). The sample regression line with the confidence intervals is shown in Figure 15.10. Note that the width of the confidence interval increases as x_p gets farther away from $\bar{x} = 5$.

TABLE 15.5   *Confidence intervals for $E(Y_p|x_p)$ in Example 15.17*

x_p	$\hat{y}_p$	$\sqrt{\dfrac{1}{n} + \dfrac{(x_p - \bar{x})^2}{\Sigma\, x^2 - n\bar{x}^2}}$	$\hat{y}_p \pm t_{.025}s_e \sqrt{\dfrac{1}{n} + \dfrac{(x_p - \bar{x})^2}{\Sigma\, x^2 - n\bar{x}^2}}$
1	75	.62	(34.82, 115.18)
2	100	.50	(67.59, 132.41)
3	125	.38	(100.37, 149.63)
4	150	.29	(131.20, 168.80)
5	175	.26	(158.15, 191.85)
6	200	.29	(181.20, 218.80)
7	225	.38	(200.37, 249.63)
8	250	.50	(217.59, 282.41)
9	275	.62	(234.82, 315.18)

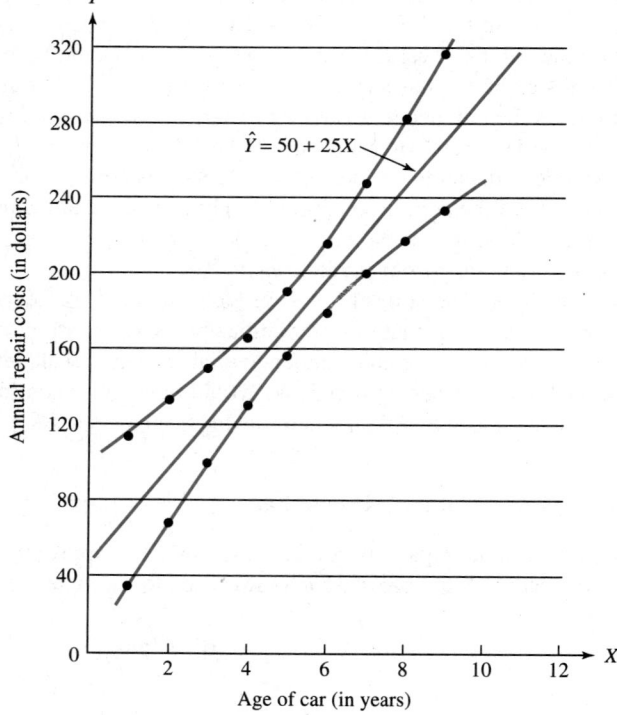

FIGURE 15.10   *Regression line with confidence intervals for $E(Y_i|x_i)$ for Example 15.17*

Some Cautionary Comments

Be careful when forecasting the value of a dependent variable if the value of the independent variable lies beyond the sample range. For example, suppose that the sample regression line indicates that annual repair costs for cars aged 1 to 5 years are related to the age of the car according to the equation $\hat{y}_i = 50 + 25x_i$, where y_i denotes the predicted annual repair costs of car i and x_i denotes the age of car i. We should hesitate to use this equation to forecast annual repair costs for a car that is, say, 10 years old. The relationship between Y and X might shift or become nonlinear outside the sample range 0 to 5 years. The process of forecasting the dependent variable when the independent variable falls outside the sample range is called *extrapolation*. Extrapolation is risky because the sample data provide no evidence that the relationship between Y and X is linear beyond the range of the data.

Another potential pitfall is a prediction based on a sample regression line that has been estimated using outdated data. The less recent the data used in estimating a sample regression line, the more likely it is that there will have been a change in the population regression line.

Finally, after estimating a regression line, examine the residuals to see whether any of the basic assumptions have been violated. It also is useful to examine the scatter diagram. Such an examination is probably the easiest way to find evidence of a nonlinear relationship between Y and X.

Exercises

Statistical Concepts

15.8.1   *True or false:*   If $s_e \neq 0$, then a 95% confidence interval for the actual value of Y_i will be wider than a 95% confidence interval for $E(Y_i|x_i)$. Explain.

15.8.2   *True or false:*   Suppose we construct a 95% confidence interval for the actual value of Y_p associated with the specific value x_p. The larger the value of $(x_p - \bar{x})^2$, the wider the confidence interval. Explain.

15.8.3   *True or false:*   Suppose we construct a 95% confidence interval for the actual value of Y_p associated with the specific value x_p. The larger the value of $\Sigma(x_i - \bar{x})^2$, the wider the confidence interval. Explain.

15.8.4   Explain what is meant by extrapolation, and discuss why it is a questionable procedure.

Statistical Drills

15.8.5   Suppose you are given the following information concerning a sample regression equation:

$$\hat{y}_i = 20 + 45x_i$$
$$s_e = 6 \quad \bar{x} = 20 \quad n = 30 \quad \Sigma x_i^2 - n\bar{x}^2 = 64$$

a. Construct a 95% confidence interval for the actual value Y_p associated with $x_p = 10$.
b. Construct a 95% confidence interval for the mean value $E(Y_p|x_p)$ associated with $x_p = 10$.
c. Construct a 95% confidence interval for the actual value Y_p associated with $x_p = 20$. Compare this with your answer to part a.
d. Construct a 95% confidence interval for the mean value $E(Y_p|x_p)$ associated with $x_p = 20$. Compare this with your answer to part b.

15.8.6   Suppose you are given the following information concerning a sample regression equation:

$$\hat{y}_i = 2 + 5x_i$$
$$\text{SSE} = 132 \quad \bar{x} = 42 \quad n = 30 \quad \Sigma x_i^2 - n\bar{x}^2 = 124$$

a. Construct a 95% confidence interval for the actual value Y_p associated with $x_p = 38$.
b. Construct a 95% confidence interval for the mean value $E(Y_p|x_p)$ associated with $x_p = 38$.

c. Construct a 95% confidence interval for the actual value Y_p associated with $x_p = 34$. Compare this with your answer to part **a**.

d. Construct a 95% confidence interval for the mean value $E(Y_p|x_p)$ associated with $x_p = 34$. Compare this with your answer to part **b**.

15.8.7 Suppose you are given the following information concerning a sample regression equation:

$$\hat{y}_i = -12 + 26x_i$$

$$SSE = 100 \quad \bar{x} = 30 \quad n = 10 \quad \Sigma x_i^2 - n\bar{x}^2 = 36$$

a. Construct a 95% confidence interval for the actual value Y_p associated with $x_p = 25$.

b. Construct a 95% confidence interval for the mean value $E(Y_p|x_p)$ associated with $x_p = 25$.

c. Construct a 95% confidence interval for the actual value Y_p associated with $x_p = 35$. Compare this with your answer to part **a**.

d. Construct a 95% confidence interval for the mean value $E(Y_p|x_p)$ associated with $x_p = 35$. Compare this with your answer to part **b**.

Statistical Applications

15.8.8 The manager of a gas station has changed the price of a gallon of gas numerous times and noted how sales changed in response. Let y_i denote the gallons sold on the ith day, and let x_i denote price charged in dollars on the ith day. The data are as follows:

Gallons sold Y	Price X
7,500	1.12
7,200	1.14
6,800	1.16
7,350	1.14
6,750	1.18
5,700	1.24
6,250	1.22
6,200	1.20
5,950	1.22
6,850	1.16

a. Plot the data and estimate the population regression equation.

b. Predict Y if $x = \$1.20$.

c. Find 95% confidence intervals for Y and for the mean value of Y if $x = \$1.20$.

d. Predict Y if $x = \$1.26$.

e. Find 95% confidence intervals for Y and for the mean value of Y if $x = \$1.26$.

15.8.9 The following data show the number of pairs of shoes produced and the costs of production in thousands of dollars during several weeks at a shoe factory:

Weekly cost Y	Number of pairs X
9.6	1,860
8.7	1,760
8.6	1,720
(*continued*)	

Weekly cost Y	Number of pairs X
7.9	1,640
8.2	1,700
10.0	1,900
9.7	1,850

a. Plot the data and estimate the population regression equation.

b. Find a 95% confidence interval for Y and the mean value of Y if $x = 1,700$.

15.8.10 Suppose you obtained the sample regression line $\hat{y} = 3 + 2x$. Suppose the number of observations is 10 and you obtained $\bar{x} = 6$, $s_e = 1$, and $(\Sigma\, x^2 - n\bar{x}^2) = 20$.

a. Plot the regression line.

b. Find 95% confidence intervals for the mean values of Y_p if $x_p = 2, 3, 4, 5, 6, 7, 8, 9$, and 10.

c. Plot the endpoints of these confidence intervals on the graph with the regression line, and draw a smooth curve through these points.

15.9   The Correlation Coefficient

Often we are not as interested in a mathematical equation relating two variables as we are in the strength of the linear relationship between the variables Y and X. In such cases, the statistical technique known as **correlation analysis** can be used. In correlation analysis, it is assumed that both Y and X are random variables.

The Population Correlation Coefficient, ρ

Suppose we want to find a suitable way of measuring the strength of the relationship between Y and X. One way of doing this is by measuring how well a regression line fits a sample of data and using the coefficient of determination R^2. Using the correlation coefficient provides a second way.

Suppose we are studying two random variables Y and X that vary together in a joint distribution. In the population of (X, Y) values, there is a mean value of X, denoted μ_X, and a mean value of Y, denoted μ_Y. The population variances of X and Y are denoted σ_X^2 and σ_Y^2. Recall that σ_X^2 measures the average value of $(X - \mu_X)^2$ in the population, and similarly for σ_Y^2. That is,

$$\sigma_X^2 = E[(X - \mu_X)^2] \quad \text{and} \quad \sigma_Y^2 = E[(Y - \mu_Y)^2]$$

Given a sample of data, we estimate σ_X^2 by calculating the sample variance s_X^2, where

$$s_X^2 = \frac{\Sigma(x_i - \bar{x})^2}{n - 1}$$

We use a corresponding approach to calculate the sample variance for Y.

One way of measuring whether two quantitative variables are related is to calculate a covariance between the variables. If we have a population of observations, we calculate a *population covariance*; otherwise, we calculate a *sample covariance*. The population and sample covariances are defined in the accompanying boxes.

DEFINITION   **Population Covariance**

The **population covariance**, denoted by the symbol σ_{XY}, measures the average value of $(X - \mu_X)(Y - \mu_Y)$ in the population of (X, Y) values. The population covariance is given by

$$\sigma_{XY} = \text{Cov}(X, Y) = E[(X - \mu_X)(Y - \mu_Y)]$$

DEFINITION   **Sample Covariance**

Given a sample of (X, Y) values, we use the **sample covariance** to estimate σ_{XY}. The sample covariance, denoted s_{XY}, is calculated using the formula

$$s_{XY} = \frac{\Sigma(x_i - \bar{x})(y_i - \bar{y})}{n - 1}$$

A problem with using the covariance to measure the strength of the relationship between X and Y is that the magnitudes (but not the signs) of σ_{XY} and s_{XY} depend on the units used to measure X and Y. For example, if we are studying the relationship between the circumference of a tree X and the height of the tree Y, then $\text{Cov}(X, Y)$ will be much larger if X and Y are measured in inches than if they are measured in feet. This scaling problem is especially common in economics and business, where dollar values are frequently expressed in thousands, millions, or billions.

Note that the standardized score $(X - \mu_X)/\sigma_X$ does not depend on the dimensions used to measure X, and similarly for $(Y - \mu_Y)/\sigma_Y$. It follows that the *standardized covariance*

$$E\left[\frac{(X - \mu_X)(Y - \mu_Y)}{\sigma_X \sigma_Y}\right]$$

does not depend on the units used to measure X and Y.

DEFINITION   **Population Correlation Coefficient**

The standardized covariance, called the **population correlation coefficient** and denoted by the Greek letter ρ (rho), is defined by the formula

$$\rho = \frac{\sigma_{XY}}{\sigma_X \sigma_Y}$$

It can be shown that for any population of (X, Y) values, it is always the case that

$$-1 \leq \rho \leq 1$$

That is, ρ cannot exceed $+1$ or be less than -1.

The sign of ρ is the same as the sign of σ_{XY} and depends on the average of the cross-products $(X - \mu_X)(Y - \mu_Y)$. The amounts $(X - \mu_X)$ and $(Y - \mu_Y)$ measure the amounts by which X and Y are above or below average. If X and Y are both above average or both below average, then the cross-product $(X - \mu_X)(Y - \mu_Y)$ will be positive. If the average cross-product is positive, then the covariance σ_{XY} is positive and ρ is positive; in such a case, we say that X and Y are positively correlated. If X is above average when Y is below average or vice versa, then the cross-product $(X - \mu_X)(Y - \mu_Y)$ will be negative and ρ will be negative. Then we say that X and Y are negatively correlated.

Three values of ρ are of special interest. When $\rho = 0$, then X and Y are not linearly related, and we say that X and Y are uncorrelated. When all the values of X and Y lie exactly on a straight line having a positive slope, then $\rho = 1$. If all values of X and Y lie exactly on a straight line having a negative slope, then $\rho = -1$. If the values of X and Y lie close to a straight line having a positive (or negative) slope, then ρ will be close to $+1$ (or -1). This shows that ρ measures the strength, or closeness, of the linear relation between two variables.

We thus have the following results:

1. A correlation of -1 implies perfect negative linear association.

2. A correlation of 1 implies perfect positive linear association.

3. The larger the absolute value of the population correlation coefficient, the stronger the linear association between the random variables.

4. A correlation of 0 implies no *linear* association. In particular, if X and Y are independent random variables, then $\rho = 0$. On the other hand, $\rho = 0$ does not necessarily imply that X and Y are independent of one another, because they may be related in a nonlinear way.

Figures 15.11, 15.12, and 15.13 (page 644) show that (X, Y) populations can be quite different but still have the same correlation coefficient. The correlation coefficient

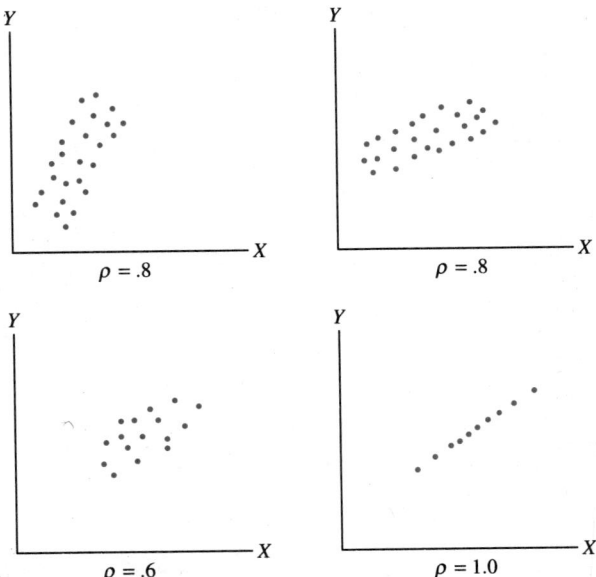

FIGURE 15.11 *Examples of positive correlation*

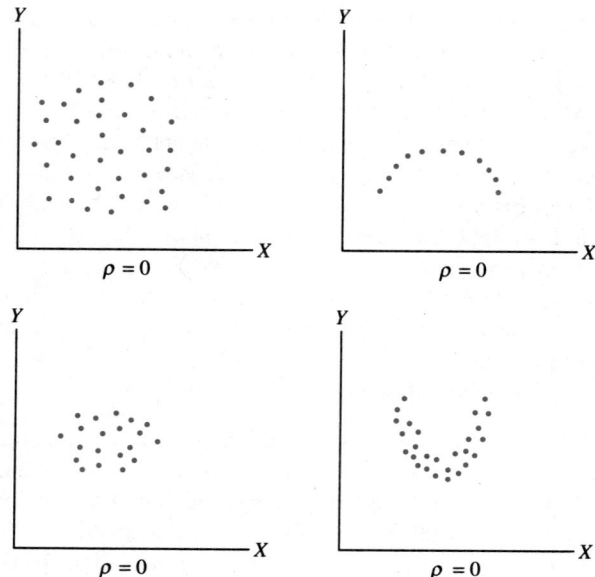

FIGURE 15.12   *Examples of zero correlation*

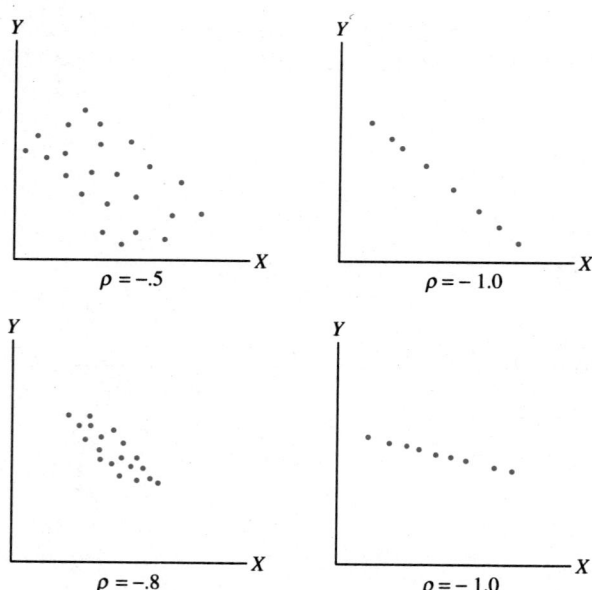

FIGURE 15.13   *Examples of negative correlation*

indicates whether the data lie close to a straight line. Two different sets of data might be scattered about two population regression lines, one line having slope $\beta_1 = .1$ and the other line having slope $\beta_1 = 100$, and in each case the correlation coefficient could be, say, .7.

The Sample Correlation Coefficient

To estimate the population correlation coefficient ρ, we calculate the sample correlation coefficient, denoted by the symbol R. The sample correlation coefficient R measures the strength of the linear relationship of the sample data (x_i, y_i). Recall that the population correlation coefficient is

$$\rho = \frac{\sigma_{XY}}{\sigma_X \sigma_Y}$$

To estimate ρ, we substitute the sample estimates s_{XY}, s_X, and s_Y for their population counterparts.

DEFINITION

Sample Correlation Coefficient

The **sample correlation coefficient**, denoted R, is obtained by the formula

$$R = \frac{s_{XY}}{s_X s_Y}$$

To calculate R, the following formula can be used:

Computational formula for R

$$R = \frac{\sum x_i y_i - n\bar{x}\bar{y}}{\sqrt{\sum x_i^2 - n\bar{x}^2}\,\sqrt{\sum y_i^2 - n\bar{y}^2}}$$

Independence of Unit of Measurement

For the population correlation coefficient and the sample correlation coefficient, interchanging the definitions of the two variables X and Y in the formulas does not change the results. Thus, when calculating the correlation coefficient, it does not matter which variable is denoted as X and which as Y.

The correlation coefficient is not expressed in any unit of measure and does not change when a different unit of measure is used for either of the variables. For example, we will get the same correlation coefficient if some variable X that measures distance is recorded in yards or centimeters or if some variable Y that measures time is recorded in seconds or hours.

Relationship Between Sample Correlation Coefficient and Slope of Sample Regression Line

The sample correlation coefficient is related to the slope of the sample regression line. The slope of the sample regression line can be calculated as

$$b_1 = \frac{\sum(x_i - \bar{x})(y_i - \bar{y})}{\sum(x_i - \bar{x})^2}$$

If we divide both numerator and denominator by $(n - 1)$, we obtain

$$b_1 = \frac{s_{XY}}{s_X^2}$$

Recall that the sample correlation coefficient can be expressed as

$$R = \frac{s_{XY}}{s_X s_Y}$$

Thus we obtain the following equations:

Relationship between R and b_1

$$b_1 = R \frac{s_Y}{s_X} \quad \text{and} \quad R = b_1 \frac{s_X}{s_Y}$$

where s_Y and s_X are the sample standard deviations of X and Y.

Because s_Y and s_X are both positive, it follows that $R = 0$ if and only if $b_1 = 0$. That is, the sample correlation coefficient is 0 if and only if the sample regression line is a horizontal line having zero slope. When the sample regression line is horizontal, this indicates that knowledge of X does not help in predicting Y. When $R = 0$, this implies that, for the sample data, X is uncorrelated with Y. Thus, saying X and Y are uncorrelated is basically the same as saying that X and Y are not linearly related.

Also, because both s_Y and s_X are positive, the sign of R is the same as the sign of b_1. R will be positive when the sample regression line has a positive slope and negative when the sample regression line has a negative slope.

As with the population correlation coefficient ρ, it is possible that $R = 0$ and that X and Y are related in a nonlinear fashion. Thus, $R = 0$ indicates only the absence of a linear relationship between X and Y and does not imply anything about nonlinear relationships.

Recall that the coefficient of determination R^2 measures the goodness of fit of the sample regression line. In the simple linear regression model, the coefficient of determination R^2 is the square of the sample correlation coefficient R. Since $0 \le R^2 \le 1$ and since R is the square root of R^2, it follows that

$$-1 \le R \le 1$$

EXAMPLE 15.18

Calculation of a Sample Correlation Coefficient ———————————————

The data in Table 15.6 show the depth X in thousands of feet and the cost of drilling Y in thousands of dollars for a sample of oil wells. Calculate the correlation coefficient between the depth of the well and the cost of drilling.

SOLUTION   We obtain $\bar{x} = 5$ and $\bar{y} = 10$. We then calculate R as follows:

$$R = \frac{\sum x_i y_i - n\bar{x}\bar{y}}{\sqrt{\sum x_i^2 - n\bar{x}^2} \sqrt{\sum y_i^2 - n\bar{y}^2}}$$

$$= \frac{355 - 6(5)(10)}{\sqrt{184 - 6(5)(5)} \sqrt{698 - 6(10)(10)}}$$

$$= .953$$

TABLE 15.6   *Depth of oil wells*

Well i	Depth x_i	Cost y_i	x_iy_i	x_i^2	y_i^2
1	3	6	18	9	36
2	5	12	60	25	144
3	1	3	3	1	9
4	6	13	78	36	169
5	8	14	112	64	196
6	7	12	84	49	144
	30	60	355	184	698

This sample correlation coefficient is quite close to 1 and indicates that the data lie close to a straight line having a positive slope. The graph of the data in Figure 15.14 indicates that the drilling cost and the depth of the well are positively correlated.

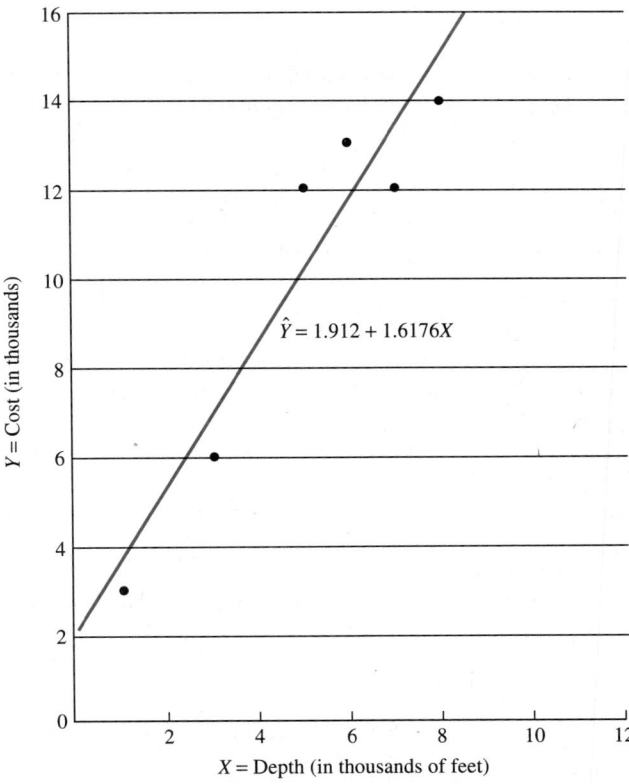

FIGURE 15.14   *Scatter diagram of data in Example 15.18*

By using the data in Table 15.6, we can obtain the sample regression line

$$\hat{y}_i = 1.912 + 1.6176x_i$$

The value of the coefficient of determination is $R^2 = .91$, which is the square of the sample correlation coefficient.

Testing Hypotheses About the Population Correlation Coefficient

The sample correlation coefficient R can be interpreted as an estimate of the population correlation coefficient ρ, where X and Y are both random variables having a joint probability distribution. If we use this interpretation, then it is natural to develop hypotheses about the value of ρ in the population and to test these hypotheses by making use of the estimate R. By far the most frequently performed test is of the null hypothesis that the population correlation coefficient is 0. Note that since $b_1 = R(s_Y/s_X)$, testing $\rho = 0$ is equivalent to testing that $\beta_1 = 0$, where β_1 is the slope of the population regression line.

Tests of the null hypothesis $H_0: \rho = 0$ are based on the assumption that X and Y follow a bivariate normal distribution. Suppose we have a random sample (x_1, y_1), $(x_2, y_2), \ldots, (x_n, y_n)$ of observations from the bivariate normal distribution. When $\rho = 0$, the t statistic

$$t = R/s_R$$

follows the Student t distribution with $(n - 2)$ degrees of freedom, where s_R is the estimated standard deviation of R. We calculate s_R by the formula

$$s_R = \sqrt{\frac{1 - R^2}{n - 2}}$$

The appropriate tests are then derived as indicated in the accompanying box.

Testing the null hypothesis $H_0: \rho = 0$

Let R be the sample correlation coefficient, calculated from a random sample of n pairs of observations from a joint normal distribution. The tests of the null hypothesis

$$H_0: \quad \rho = 0$$

described subsequently have significance level α and are based on the test statistic

$$t = R/s_R$$

Case 1:   To test H_0 against the one-sided alternative hypothesis

$$H_1: \quad \rho > 0$$

use the decision rule

$$\text{Reject } H_0 \text{ in favor of } H_1 \text{ if } t > t_{\alpha,\nu}$$

where $t_{\alpha,\nu}$ is the critical value such that

$$P(t > t_{\alpha,\nu}) = \alpha$$

where the random variable t follows a Student t distribution with $\nu = (n - 2)$ degrees of freedom.

      Case 2:   To test H_0 against the one-sided alternative hypothesis

$$H_1: \quad \rho < 0$$

use the decision rule

      Reject H_0 in favor of H_1 if $t < -t_{\alpha,\nu}$.

      Case 3:   To test H_0 against the two-sided alternative

$$H_1: \quad \rho \neq 0$$

use the decision rule

      Reject H_0 in favor of H_1 if $t < -t_{\alpha/2,\nu}$ or if $t > t_{\alpha/2,\nu}$.

EXAMPLE 15.19   **Testing the Null Hypothesis $H_0: \rho = 0$**

For the data in Example 15.18, the sample correlation between the depth of an oil well and the cost of drilling was $R = .953$. It is reasonable to suspect that there is a positive relationship between these two variables. Thus we would want to test the null hypothesis

$$H_0: \quad \rho = 0$$

against the one-sided alternative hypothesis

$$H_1: \quad \rho > 0$$

SOLUTION   Because we have $n = 6$ observations and $\alpha = .05$, we obtain the t statistic

$$t = \frac{.953}{\sqrt{(1 - .953^2)/(6 - 2)}} = 6.29$$

We use a one-sided test. For $\alpha = .05$, the critical value of t having $(n - 2) = 4$ degrees of freedom is $t_{.05,4} = 2.132$. We reject H_0 in favor of H_1 because the t statistic exceeds the critical value and falls in the rejection region of the test.

Inappropriate Use of Correlation Analysis

If correlation analysis is used indiscriminately, conclusions can be misleading or even false. A common mistake is to assume that correlation implies causation. However, the fact that X and Y are correlated does not imply that X causes Y or vice versa. It may be that X causes Y, that Y causes X, that both X and Y are influenced by some other variable, or that the correlation between X and Y is *spurious*.

DEFINITION   **Spurious Correlation**

Two variables are said to be **spuriously correlated** when their correlation is nonzero and there is no reason to believe that the variables are related to one another.

An example of spurious correlation can be illustrated by calculating the sample correlation coefficient between the annual number of crimes committed in New York City and the average salary of full professors at Harvard University during the years 1950 to 1970. These variables are highly positively correlated, but there is no apparent reason to believe that one variable is connected to the other. In this case, the correlation coefficient does not measure causality.

Another problem occurs when X and Y are, say, negatively related, but correlation analysis indicates that the effect of X on Y is positive. The next example illustrates this.

EXAMPLE 15.20   **An Inappropriate Use of Correlation Analysis** ————————————————

An important theory in economics states that, other things being equal, the higher the price of a commodity, the less the demand for it. Thus, if other things were equal, we would expect to find that the prices of tickets to professional football games and the number of tickets sold were negatively correlated. However, the data show that since 1960 both prices of tickets and the number of tickets sold per year have steadily increased. Thus, the data reveal a positive correlation between the price of tickets and the quantity sold. Does this evidence support the hypothesis that the law of demand is false? Does the sample evidence support the hypothesis that, to sell more tickets, professional football teams should increase prices? Did the increase in prices cause the increase in quantity sold?

Using correlation analysis on sales of tickets and the price of tickets is inappropriate because we are neglecting other important variables that affect the number of tickets sold. Since 1960, sales have increased for several reasons: The total population of the United States has increased, so there are more potential ticket buyers; incomes have increased, so people have more money to spend on football tickets; the number of professional teams has increased; the popularity of professional football has increased; and so forth. Once these things are taken into account, the law of demand still holds true. That is, once population, income, number of teams, and other pertinent factors are considered, we still find that the higher the price of tickets, the lower the quantity sold.

The following argument shows that the law of demand holds (as prices increase, the quantity demanded decreases) and that the sample correlation coefficient is giving misleading results. Assume that, at the present time, 50,000 people are willing to pay, say, $10 each to see a certain game. Consider what would happen if tomorrow the home team announced that the price of each ticket had been increased to, say, $15. Immediately after the announcement, the number of individuals who would want to buy a ticket would decrease and fewer than 50,000 tickets would be sold. This shows that, if other factors remain constant, the quantity of football tickets demanded and the price of football tickets are negatively related.

It is important not to misinterpret the correlation coefficient. A nonzero population correlation coefficient means that there is a statistical relationship between the variables Y and X. It does not mean that changes in X *cause* changes in Y. It means that the variables Y and X vary together in a predictable way, for whatever reason. The previous examples do not illustrate the failure of correlation analysis, just a failure to interpret correlation correctly.

The problem with correlation analysis is that it ignores all the variables that may influence Y other than the single variable X. In economics and business, there are many cases where it is unreasonable to assume that a variable Y is a function of just a single variable X. A more appropriate way to determine the true relationship between Y and X may be to construct a multiple regression model that includes all the important variables influencing Y. Such models are discussed in Chapter 16.

Nevertheless, the correlation coefficient can be useful in finding potential variables to insert into a multiple regression model. There might be a relatively high correlation between Y and some variable X_1 and between Y and some other variable X_2. If this is the case, it might be that a regression model should contain both variables X_1 and X_2 as independent variables.

Multiple regression is a statistical technique that enables the investigator to discover how Y is influenced by a whole set of other variables, and thus it overcomes many of the problems inherent in simple correlation analysis.

Correlation and Statistical Significance

Suppose we use a 5% level of significance and reject the null hypothesis H_0: $\rho = 0$ in favor of the alternative hypothesis H_1: $\rho > 0$. Frequently the statistician will say that the sample correlation coefficient R is "significantly different from 0 at the 5% level of significance." This does not mean that variable X explains a significant amount of the variation in Y. The word *significant* is being used with two different meanings here.

For example, if we use a 5% level of significance and the sample is large enough ($n > 383$), then a sample correlation coefficient of $R = .1$ will be significantly different from 0 if we perform a formal statistical test using the t statistic. Although a sample correlation coefficient of $R = .1$ may be statistically significant (i.e., it is significantly different from 0), it may be of no practical significance. When $R = .1$, this means that $R^2 = .01$ and thus X explains only 1% of the variation in Y, leaving 99% of the variation in Y still unexplained.

Finally, it should be noted that if a variable Y is correlated with each of two other variables, say, X_1 and X_2, this does not imply that those two variables together in a regression equation will explain significantly more of the variation in Y than just one of them alone. It might be that X_1 and X_2 are highly correlated with one another and contribute virtually the same information for explaining or predicting Y.

Tips for Problem Solving

1. The first step in calculating the correlation coefficient is to calculate the following five sums:

$$\Sigma x_i \quad \Sigma y_i \quad \Sigma x_i^2 \quad \Sigma y_i^2 \quad \Sigma x_i y_i$$

2. Values of R close to $+1$ indicate that the observations lie near a straight line with a positive slope.
3. Values of R close to -1 indicate that the observations lie near a straight line with a negative slope.
4. The relative strength of the linear relationship between Y and X is indicated by the absolute value of R. Thus, $R = -.93$ indicates a relatively strong nega-

(continued)

tive linear relationship; $R = .24$ indicates a relatively weak positive linear relationship.

5. There is no perfect guideline for deciding when a linear relationship is "strong" or "weak." Recall, however, that R^2 measures the proportion of the variation in Y that is explained by the independent variable X.

Exercises

Statistical Concepts

15.9.1   *True or false:*   Let s_{xy} denote the sample covariance. It must be that $0 \leq s_{xy}$. Explain.

15.9.2   *True or false:*   Let R denote the sample correlation coefficient. It must be that $0 \leq R \leq 1$. Explain.

15.9.3   *True or false:*   A correlation of 1 implies that the slope of the sample regression line is $b_1 = 1$. Explain.

15.9.4   *True or false:*   The larger the absolute value of the population correlation coefficient, the steeper the population regression line. Explain.

15.9.5   *True or false:*   A correlation of 0 implies that X and Y are independent of one another. Explain.

15.9.6   *True or false:*   If $b_1 > 0$, then $R > 0$. Explain.

15.9.7   *True or false:*   If $\rho > 0$, then $R > 0$. Explain.

15.9.8   Explain how to test the null hypothesis H_0: $\rho = 0$. What assumption is required concerning the joint distribution of X and Y?

15.9.9   Explain what is meant by spurious correlation.

15.9.10   *True or false:*   The determination of causality cannot be directly answered in the context of correlation or regression analysis. Explain.

15.9.11   *True or false:*   Suppose that variables Y and X both tend to increase as the population increases. Then Y and X will tend to be positively correlated even though there may be no causal relationship between Y and X. Explain.

15.9.12   *True or false:*   Suppose we are testing the null hypothesis H_0: $\rho = 0$. Depending on the sample size, it is possible that we will not reject H_0 in a situation when $R = .54$, and we will reject H_0 in a situation when $R = .05$. Explain.

15.9.13   A major logical error that people frequently commit is called the "post hoc, ergo propter hoc" fallacy. A literal translation of this Latin term is "After this, therefore because of this." The fallacy is this: The fact that event B occurred after event A does not imply that event A caused event B. For example, sociologists have noted that in recent years there has been a rapid rise in the number of incidents of violence shown on television. At the same time, there has been a rapid rise in the crime rate per person in major urban areas. Now people are petitioning the government to set some standards that will limit the amount of violence shown on TV. Regardless of how you feel about violence on TV, analyze the data in a logical way. Do the data indicate that increased violence on TV causes more crime? Explain. Is it possible to prove that increased violence on TV causes crime? Suppose you estimate a regression model that shows that the crime rate appears to be linearly related to the incidence of violence on TV. Does this prove that violence on TV causes crime?

Statistical Drills

15.9.14   Use the following data:

x_i	4	6	3	9	8
y_i	11	15	9	21	19

a. Plot the data on a scatter diagram.
b. Estimate the sample linear regression line. Plot this line on the scatter diagram.
c. Calculate the sample correlation coefficient.
Replace the y_i data with the following values:

y_i	7	9	6	12	11

d. Plot the data on a scatter diagram.
e. Estimate the sample linear regression line. Plot this line on the scatter diagram.
f. Calculate the sample correlation coefficient.
g. Compare the slopes of the two regression lines and the correlation coefficients. *True or false:* The correlation coefficient measures the strength of the linear relationship, not the steepness of the regression line. Explain.

15.9.15 Use the following data:

x_i	4	4	6	6	8	8
y_i	12	2	14	0	9	5

a. Plot the data on a scatter diagram.
b. Estimate the sample linear regression line. Plot this line on the scatter diagram.
c. Calculate the sample correlation coefficient.

15.9.16 Use the following data:

x_i	4	6	3	9	8
y_i	11	17	12	22	15

a. Plot the data on a scatter diagram.
b. Estimate the sample linear regression line. Plot this line on the scatter diagram.
c. Calculate the sample correlation coefficient. Does the magnitude of R indicate that the regression line provides a good fit to the data?

15.9.17 Use the following data:

x_i	4	6	3	9	8
y_i	110	170	120	220	150

These are the same data as in Exercise 16 except that the values of y_i have been multiplied by 10.
a. Plot the data on a scatter diagram.
b. Estimate the sample linear regression line. Plot this line on the scatter diagram. Compare the slope and intercept with the slope and intercept in Exercise 16. Do you see the relationship?
c. Calculate the sample correlation coefficient and compare it with the value obtained in Exercise 16. Does the magnitude of R tell us about the steepness of the slope of the regression line?

15.9.18 In a sample of $n = 25$ data points, the sample correlation coefficient was $R = .67$. Let $\alpha = .05$. Test $H_0: \rho \leqslant 0$ against $H_1: \rho > 0$.
a. What is the appropriate number of degrees of freedom?
b. What is the critical value of the test statistic t?
c. Perform the test.

15.9.19 In a sample of $n = 8$ data points, the sample correlation coefficient was $R = .5$. Let $\alpha = .05$. Test $H_0: \rho \leqslant 0$ against $H_1: \rho > 0$.
a. What is the appropriate number of degrees of freedom?

b. What is the critical value of the test statistic t?

c. Perform the test.

d. *True or false:* If the sample size is very small, say, 8 or less, it is relatively difficult to reject the hypothesis that $\rho = 0$. Explain.

e. Assume that $n = 25$. Now perform the test.

15.9.20   We obtain a sample of $n = 8$ data points, and calculate the sample correlation coefficient. Let $\alpha = .05$. We wish to test $H_0: \rho \leqslant 0$ against $H_1: \rho > 0$.

a. What is the appropriate number of degrees of freedom?

b. What is the critical value of the test statistic t?

c. How large must R be for us to reject the null hypothesis? That is, find the value of R that makes the test statistic t equal to the critical value.

d. Repeat parts **a–c** using $n = 62$.

15.9.21   In a sample of $n = 12$ data points, the sample correlation coefficient was $R = -.6$. Let $\alpha = .05$. Test $H_0: \rho \geqslant 0$ against $H_1: \rho < 0$.

a. What is the appropriate number of degrees of freedom?

b. What is the critical value of the test statistic t?

c. Perform the test.

Statistical Applications

15.9.22   Calculate the sample correlation coefficient for the following data concerning grades of eight students on math and economics tests. Interpret the results.

| Math, X | 89 | 86 | 74 | 90 | 46 | 82 | 68 | 94 |
| Physics, Y | 82 | 72 | 80 | 96 | 68 | 89 | 74 | 90 |

Plot the data on a scatter diagram.

15.9.23   The following data show the total number of priests and ministers in a certain town and the total expenditures on alcohol in thousands of dollars in that town between 1970 and 1995:

Year	Expenditure on alcohol Y	Number of clergy X
1970	480	64
1975	500	69
1980	535	74
1985	565	85
1990	610	106
1995	640	114

a. Calculate the sample correlation coefficient between X and Y. Interpret the results.

b. Do you think that increases in X cause increases in Y, increases in Y cause increases in X, or neither? How do you explain the high value of R?

15.9.24   The accompanying data show the size of various houses (in square feet) and the annual heating expenditures in dollars in those houses during the last year:

| Heating expenditures, Y | 260 | 345 | 420 | 235 | 400 | 280 |
| Size of house, X | 800 | 1,000 | 1,600 | 750 | 1,700 | 850 |

Calculate the sample correlation coefficient and interpret the results.

15.9.25   A sample of 18 values of X and Y yielded a correlation coefficient of $R = .5$. Test $H_0: \rho = 0$ against $H_1: \rho \neq 0$ using a 5% level of significance.

15.9.26   A correlation coefficient based on a sample of size 27 was computed to be .4. Can we conclude that ρ differs from 0 at a significance level of 5%? Of 1%?

15.9.27   Suppose that the Goodyear Tire Company collects data on the relationship between a car's speed and the distance it travels before stopping (after the brakes are applied). Would you expect the correlation coefficient to be positive or negative? Is there a causal relationship in this case?

15.9.28   If a regression line can explain 36% of the variation in the dependent variable, what is the correlation coefficient?

15.10   Computer Applications

In actual practice, statisticians almost always use computers for regression and correlation analysis. Many different computer programs are available that will generate all of the information we have discussed in this chapter. The following example illustrates how to obtain and interpret computer output using the SPSSX computer program.

EXAMPLE 15.21   **Using SPSSX for Regression Analysis** ————————————————————

We will use the SPSSX computer program to obtain the sample regression line and various other useful statistical information for the data presented in Table 15.1 on the weekly expenditures on recreation and weekly incomes for a sample of 20 families, described in Example 15.3. The data and the sample regression line were plotted in Figure 15.2 (page 590).

Following is the complete SPSSX set of instructions for reading the data, generating a scatter diagram showing the pairs of values of the variables EXPEND and INCOME, and printing out the sample regression equation that explains EXPEND as a linear function of INCOME:

```
TITLE 'EXPENDITURE STUDY'
DATA LIST FREE/EXPEND INCOME
BEGIN DATA

     90      900
     60      800
      :        :
      :        :
     35      200

END DATA
SCATTERGRAM EXPEND WITH INCOME
REGRESSION VARIABLES = EXPEND INCOME/
    DEPENDENT = EXPEND/
    ENTER INCOME
FINISH
```

```
* * * *   M U L T I P L E   R E G R E S S I O N   * * * *
```

Listwise Deletion of Missing Data

Equation Number 1   Dependent Variable..   EXPEND   EXPENDITURES ON RECREATION

Block Number 1. Method: Enter   INCOME

Variable(s) Entered on Step Number 1..   INCOME   WEEKLY INCOME

		Analysis of Variance			
Multiple R	.91067		DF	Sum of Squares	Mean Square
R Square	.82932				
Adjusted R Square	.81984	Regression	1	6016.72585	6016.72585
Standard Error	8.29416	Residual	18	1238.27415	68.79301
		F =	87.46130	Signif F =	.0000

------------------ Variables in the Equation ------------------

Variable	B	SE B	Beta	T	Sig T
INCOME	.076379	.008167	.910671	9.352	.0000
(Constant)	13.918919	4.916352		2.831	.0111

End Block Number   1   All requested variables entered.

FIGURE 15.15 *SPSSX-generated output for Example 15.21*

The computer output from this set of instructions, shown in Figure 15.15, is explained as follows:

Estimated coefficients: The estimated slope b_1 and the estimated intercept b_0 are found in the column labeled B. We have $b_1 = .076379$ and $b_0 = 13.918919$.

Estimated standard deviations: The estimated standard deviations of b_1 and b_0 are found in the column labeled SE B. We obtain $s_{b_1} = .008167$ and $s_{b_0} = 4.916352$.

t statistics: The t statistics used to test the null hypotheses $H_0: \beta_0 = 0$ and $H_0: \beta_1 = 0$ are printed in the column labeled T. The t statistic used to test $H_0: \beta_1 = 0$ is $t = 9.352$. This is $t = b_1/s_{b_1} = .076379/.008167$. The t statistic used to test $H_0: \beta_0 = 0$ is $t = 2.831$. This is $t = b_0/s_{b_0} = 13.918919/4.916352$.

p-values: In the table, the values found in the column labeled Sig T refer to observed significance levels, or *p*-values, for the t statistics. For the slope coefficient, the observed significance level is Sig T $= .0000$. This means that for the t distribution having $(n - 2) = 18$ degrees of freedom, the combined area to the right of $t = 9.352$ and to the left of $t = -9.352$ is .0000. For the intercept (constant) coefficient, the observed significance level is Sig T $= .0111$. This means that, for the t distribution having $(n - 2) = 18$ degrees of freedom, the combined area to the right of $t = 2.831$ and to the left of $t = -2.831$ is .0111. Having these *p*-values printed out is especially convenient, because we do not need to refer to tables of the t distribution to test the null hypothesis that a population regression coefficient is 0.

The SPSSX program also prints out what is called an analysis of variance table. This table contains information concerning the sample correlation coefficient *R,* the value of R^2, the regression sum of squares (SSR), the residual sum of squares (SSE), and the estimated standard error of the regression (s_e). The following values and their interpretations are printed in the table.

Sample correlation coefficient: The sample correlation coefficient, which equals the square root of R^2, is shown in the row labeled Multiple R as $R = .91067$.

Coefficient of determination: The coefficient of determination is shown in the row labeled R Square as $R^2 = .82932$.

Standard error: The estimated standard deviation of the error terms s_e is shown in the row labeled Standard Error as 8.29416.

Regression sum of squares: The regression sum of squares SSR is shown to be 6,016.72585.

Residual sum of squares: The residual sum of squares SSE is shown to be 1,238.27415 and has $(n - 2) = 18$ degrees of freedom. Recall that an unbiased estimate of the variance of the error terms is given by $s_e^2 = \text{SSE}/(n - 2) = 1{,}238.27415/18 = 68.79301$. After taking the square root, we obtain $s_e = 8.29416$, which agrees with the value given above.

F statistic: The statistic F is used to test the null hypothesis that all of the independent variables considered together do not help explain the value of the dependent variable. In the simple linear regression model, the value of the F statistic is the square of the t statistic used to test the null hypothesis H_0: $\beta_1 = 0$. Note that $F = 87.46130 = (9.352)^2 = t^2$.

Figure 15.16 shows the SAS computer output for Example 15.21. Compare the regression coefficients, t statistics, s_e, R^2, SSE, and other statistics with the corresponding output from the SPSSX program.

The SAS System

General Linear Models Procedure

Dependent Variable: EXPEND

Source	DF	Sum of Squares	Mean Square	F Value	Pr > F
Model	1	6016.72585141	6016.72585141	87.46	0.0001
Error	18	1238.27414859	68.79300825		
Corrected Total	19	7255.00000000			

	R-Square	C.V.	Root MSE		EXPEND Mean
	0.829321	14.67992	8.29415507		56.50000000

Source	DF	Type I SS	Mean Square	F Value	Pr > F
INCOME	1	6016.72585141	6016.72585141	87.46	0.0001

Source	DF	Type III SS	Mean Square	F Value	Pr > F
INCOME	1	6016.72585141	6016.72585141	87.46	0.0001

Parameter	Estimate	T for H0: Parameter=0	Pr > |T|	Std Error of Estimate
INTERCEPT	13.91891892	2.83	0.0111	4.91635194
INCOME	0.07637862	9.35	0.0001	0.00816702

FIGURE 15.16 *SAS-generated output for Example 15.21*

Exercises

Use the data in Table 2.3 of Chapter 2 to complete Exercises 1–4.

15.10.1 Test the hypothesis that the average teachers' salary in a state (X_2) is linearly related to per capita income in the state (X_5).
   a. Estimate the simple linear regression model that explains SALARY as a function of INCOME.
   b. Does this equation seem to do a good job of explaining the variable SALARY? Generate a scatter diagram of the data.
   c. Find the value of R^2 and s_e.
   d. Test the null hypothesis that the slope of the population regression line is 0 against the alternative hypothesis that it is positive. Let $\alpha = .05$.

15.10.2 Test the hypothesis that the average teachers' salary in a state (X_2) is linearly related to tax revenue per capita in the state (X_3).
   a. Estimate the simple linear regression model that explains SALARY as a function of REVENUE.
   b. Does this equation seem to do a good job of explaining the variable SALARY? Generate a scatter diagram of the data.
   c. Find the value of R^2 and s_e.
   d. Test the null hypothesis that the slope of the population regression line is 0 against the alternative hypothesis that it is positive. Let $\alpha = .05$.

15.10.3 Test the hypothesis that tax revenue per capita in a state (X_3) is linearly related to per capita income in the state (X_5).
   a. Estimate the simple linear regression model that explains REVENUE as a function of INCOME.
   b. Does this equation seem to do a good job of explaining the variable REVENUE? Generate a scatter diagram of the data.
   c. Find the value of R^2 and s_e.
   d. Test the null hypothesis that the slope of the population regression line is 0 against the alternative hypothesis that it is positive. Let $\alpha = .05$.

15.10.4 Test the hypothesis that the average hourly earnings in manufacturing in a state (X_6) is linearly related to the percentage of employees in the state who are unionized (X_1).
   a. Estimate the simple linear regression model that explains PCTUNION as a function of WAGE.
   b. Does this equation seem to do a good job of explaining the variable PCTUNION? Generate a scatter diagram of the data.
   c. Find the value of R^2 and s_e.
   d. Test the null hypothesis that the slope of the population regression line is 0 against the alternative hypothesis that it is positive. Let $\alpha = .05$.

STATISTICS IN ACTION: CASE STUDY

Paint Problems on the Mellon Bank Building

The case study at the end of Chapter 4 described a lawsuit filed by the management of Mellon Bank against 20 local and national companies concerning flaws that existed in one of its buildings. The case study in Chapter 4 considered problems concerned with paint peeling on the exterior of the building. The outer walls of the Mellon Bank Building consist of more than 2,000 metal panels with dimensions 6 ft by 10 ft. Each panel was painted with a thin coat of primer and a thicker layer of topcoat paint. After a few years, the paint began to chip and crack at numerous places on the building.

A lawsuit arose concerning who, if anyone, was responsible for the defective paint job. Various theories were proposed concerning why the paint was peeling. In this case study, we examine the theory that the total thickness of the primer coat and the topcoat was too great.

A company specializing in the chemical analysis of paint on buildings and bridges was hired to collect a sample of data. Data were obtained from a sample of 30 points where the paint had cracked and peeled (the "failure points") and from a sample of 247 points where the paint had not cracked or peeled (the "nonfailure points"). At each of the 247 nonfailure points, a "cut test" was applied. The cut test consists of using a metal scraper to make a series of scratches in the paint in a tic-tac-toe design. Then a special piece of tape is placed over the scratch. As the tape is removed, some of the paint is also removed. The amount of paint that is found on the tape is an indicator of the adhesion quality of the paint. The less paint on the tape, the better the adhesion quality of the paint. Let X_i denote the total paint thickness at the ith sample point, and let Y_i denote the amount of paint on the tape after applying the cut test. The variable Y_i assumed values between 0.0 and 5.0. A value close to 0 indicates that almost no paint was removed during the cut test. A value near 5.0 indicates that nearly all the paint on the building was removed during the cut test. If points with a large total paint thickness have worse adhesion qualities, the correlation coefficient between the variables Y and X should be positive. On the other hand, if adhesion quality of the paint is unrelated to paint thickness, the sample correlation coefficient between Y and X should not be significantly different from 0. Based on the random sample of 247 points, the sample correlation coefficient between Y and X was $R = .5993$. Suppose we test the null hypothesis $H_0: \rho = 0$ against $H_1: \rho > 0$ using a 1% level of significance ($\alpha = .01$). We obtain the t statistic

$$t = \frac{R}{s_R} = \frac{.5993}{\sqrt{\dfrac{1 - .3592}{245}}} = 11.72$$

Based on a 1% level of significance and with $(n - 2) = 245$ degrees of freedom, the critical value of t is 2.326, so we reject H_0 in favor of H_1. To four decimal places, the p-value of the test is .0000. These results strongly support the conclusion that the thicker the paint, the worse the adhesion qualities of the paint.

As more evidence, consider the following information, which shows that paint thickness on the tape was positively related to paint thickness on the building. In the sample, Y had a value between 0.0 and 1.50 in 42 cases; for these 42 cases, the mean value of X was .0590 millimeter. Y had a value between 1.50 and 3.50 in 153 cases; for these 153 cases, the mean value of X was .0770 millimeter. Finally, Y had a value between 3.50 and 5.00 in 52 cases; for these 52 cases, the mean value of X was .1082 millimeter. These results indicate that as the paint thickness increased, more paint came off the building during the cut test.

Many additional tests were performed on the paint to rule out other potential causes of paint failure. The evidence strongly supported the hypothesis that excessive paint thickness was a major determinant of paint failure and poor paint adhesion.

As reported in the *Allegheny Bulletin,* the case was settled out of court for $13.1 million. This case study provides a classic example of how correlation analysis can help us make important decisions about relationships between variables.

Chapter 15 Supplementary Exercises

15.S.1 Graph the following straight lines:
  a. $Y = 2 + 2X$ **b.** $Y = 3 + 2X$ **c.** $Y = 3 - 2X$ **d.** $Y = 3 - .5X$

15.S.2 We are interested in determining whether the selling price of a used car of a specific model is related to the age of the car. We take a sample of eight cars of the same model and obtain the following data:

Age (in years), X	2	4	5	3	1	3	5	1
Selling price (in dollars), Y	1,600	800	800	1,300	2,000	1,100	600	1,800

a. Plot the data (x_i, y_i) on a scatter diagram.
b. Obtain the least-squares estimates b_0 and b_1.
c. Graph the line $\hat{y}_i = b_0 + b_1 x_i$.
d. Calculate R, the correlation coefficient. Using a 5% level of significance, test $H_0: \rho = 0$ against $H_1: \rho \neq 0$.
e. Calculate R^2.

15.S.3 The accompanying data show the annual incomes (X) of 10 families and their annual expenditures on entertainment and recreation (Y), all expressed in thousands of dollars. It is hypothesized that expenditures on entertainment and recreation are linearly related to incomes.

Recreation expenditures, Y	2.0	4.2	3.4	4.9	6.6	5.9	4.8	2.8	5.1	8.6
Income, X		11	16	14	26	45	32	24	18	25	50

a. Plot the data.
b. Estimate and graph the linear regression line.
c. Calculate R^2.
d. Calculate the correlation coefficient between X and Y.
e. Using a 5% level of significance, test whether the population correlation coefficient is 0.

15.S.4 The U.S. Government publishes numerous documents and sells them to the public. The following data show the selling prices in dollars of various documents and their lengths in pages:

Price, Y	8	19	14	16	10	12	16
Number of pages, X	240	890	430	620	330	425	535

a. Plot the data.
b. Estimate the regression equation $\hat{y}_i = b_0 + b_1 x_i$.
c. Calculate the coefficient of determination R^2.
d. Predict the selling price for a 700-page book.

15.S.5 A professional football analyst had a theory that teams that threw many passes during a game tended to win fewer games than teams that seldom passed. The following data show the number of games won by various teams and the average number of passes they threw per game throughout the season:

Games won, Y	13	10	6	4	7	9	5	8
Passes per game, X	18	26	34	32	27	29	32	28

a. Plot the data.
b. Calculate the correlation coefficient.
c. Test $H_0: \rho = 0$ against $H_1: \rho < 0$. Let $\alpha = .05$.
d. Can you think of any reason that X and Y would be negatively correlated?

15.S.6 What is the correlation coefficient between Y and X in the following situations:
a. The variable X is a constant.
b. X always exceeds Y by 10 units.
c. X is always twice as large as Y.
d. X and Y are related by the equation $Y = 1 + 3X$.

15.S.7 Whether intelligence is inherited or determined mainly by environmental factors has long been debated. A random sample of 8-year-old children is selected, and the children and their parents are given IQ tests. The results of the test follow:

Child's IQ, Y	101	111	122	103	95	99	89	103
Average of parents' IQs, X	105	122	116	98	88	108	102	105

a. Plot the data.
b. Calculate the correlation coefficient and test $H_0: \rho = 0$ against $H_1: \rho > 0$. Let $\alpha = .05$.
c. Estimate and graph the linear regression equation.
d. Predict a child's IQ if the average IQ of the parents is 112.

15.S.8 A baseball player and the team's general manager are discussing the player's contract for the next season. The player wants a raise because he claims that people come to watch him play. The manager claims that people do not pay to watch stars play but to see the team win. To support his theory, the manager offers the following data concerning percentage of games won and average per game attendance in thousands during recent years:

Year	Average attendance Y	Percent games won X
1986	22	64
1987	20	60
1988	14	48
1989	18	56
1990	10	42
1991	11	46
1992	17	55
1993	23	65
1994	25	68

a. Plot the data.
b. Calculate the correlation coefficient. Test $H_0: \rho = 0$ against $H_1: \rho > 0$. Let $\alpha = .05$.
c. Estimate the sample regression line and graph it.

15.S.9 Suppose that we calculate the sample regression equation for a given set of sample data and obtain $b_0 = 2.0$ and $b_1 = 0$. Explain which of the following will be true and which false:
a. SSE $= 0$ **b.** $R^2 = 0$ **c.** $R = 0$ **d.** SSR $=$ SST
e. SSR $=$ SSE **f.** SSR $= 0$ **g.** SSE $=$ SST

 15.S.10 Let Y represent the resale price of the ith used motorcycle and X represent the mileage the motorcycle has traveled. Suppose we have the following data showing the selling prices in dollars and mileages in thousands of a sample of six motorcycles:

Y	100	400	600	500	300	300
X	30	8	5	7	9	i2

a. Estimate β_0, β_1, and σ_e^2 for the linear regression model.
b. Calculate R^2 and s_e.
c. Estimate the variances of b_0 and b_1.

d. Test $H_0: \beta_1 = 0$ against $H_1: \beta_1 < 0$ using a 5% level of significance.
e. Test $H_0: \beta_0 = 0$ against $H_1: \beta_0 \neq 0$ using a 5% level of significance.
f. Calculate a 95% confidence interval for β_1.

15.S.11 A company believes that its sales are linearly related to its advertising expenditures and obtained the following data expressed in millions of dollars:

Sales, Y	500	200	400	900	860	940	770
Advertising expenditures, X	60	35	45	80	75	85	70

a. Estimate the population regression equation.
b. Calculate R^2.
c. Calculate a 95% confidence interval for β_1.
d. Predict sales if advertising expenditures equal 100.
e. Calculate a 95% confidence interval for the mean value of Y if $X = 100$.

15.S.12 The owner of a large firm that manufactures furniture believes that annual national expenditures on furniture are linearly related to national personal disposable income. Use the accompanying data, expressed in billions of dollars, from the *Economic Report of the President* and perform the following tasks:
a. Plot the data on a scatter diagram.
b. Estimate and graph the population regression line.
c. Calculate R^2 and s_e.
d. Estimate the variances of b_0 and b_1.
e. Test $H_0: \beta_1 = 0$ against $H_1: \beta_1 > 0$ using a 5% level of significance.
f. Construct a 95% confidence interval for β_1.
g. Estimate expenditures on furniture if disposable income is $900 billion.
h. Construct a 95% confidence interval for the mean value of Y if $X = 900$.

Expenditures on furniture, Y	20	18	22	24	26	30	30	30	38	40
Personal disposable income, X	350	364	385	404	438	473	511	546	591	634

15.S.13 You are given the following data on sales (in tens of thousands of dollars) and advertising expenditures (in thousands of dollars):

Sales, Y	31	40	25	30	20	26
Advertising expenditures, X	5	11	3	4	3	5

a. Calculate the regression equation of sales on advertising expenditures.
b. Find 95% confidence intervals for β_0 and β_1.
c. Use a 5% level of significance and test $H_0: \beta_1 = 0$ against $H_1: \beta_1 > 0$.
d. Predict sales if $6,000 is spent on advertising.
e. Calculate a 95% confidence interval for the mean value of Y if $X = 6$.

15.S.14 The following data show the number of minutes required to type reports of varying lengths:

Typing time, Y	60	130	100	290	175
Number of pages, X	10	20	15	50	30

a. Estimate the linear regression line of Y on X.
b. Find a 95% confidence interval for β_1.
c. Predict how long it will take to type a 40-page report.
d. Find a 95% confidence interval for the actual value of Y given that $X = 40$.

15.S.15 The Metropolitan Fuel Company has collected data to study the relationship between the temperature at noon in degrees Fahrenheit and the gallons of fuel oil consumed in single-family dwellings on that day. To predict the supplies of fuel oil that will be needed on cold days, the company must estimate the relationship between temperature and consumption of fuel oil. The following table shows a sample of the data collected by Metro executives:

Fuel oil consumption, Y	7.2	6.7	6.3	5.1	4.7	3.5	2.8	1.2
Temperature at noon, X	12	18	21	33	38	46	52	70

a. Plot the data and estimate the linear regression line. Draw this line on the scatter diagram.
b. Construct a 95% confidence interval for the coefficient β_1.
c. Predict average consumption if the daily temperature at noon is 60° F.
d. Calculate R^2. Does this value indicate that the linear model is appropriate for explaining the values of Y?
e. Construct a 95% confidence interval for the mean value of Y given that $X = 70°$ F.

15.S.16 The Jones Rustproofing Company operates a chain of outlets in Chicago. The company rustproofs automobiles. Management believes that the number of customers in a quarter of the year can be predicted relatively accurately by using a linear regression model in which the explanatory variable is the number of new automobile registrations in Chicago in the previous quarter. The following data show the number of customers in hundreds during the last eight quarters and the number of new car registrations in thousands for each previous quarter.

Customers per quarter, Y	7.1	8.2	6.3	9.1	8.7	6.4	5.2	8.1
New cars registered, X	14.4	17.1	11.9	20.2	17.0	14.0	11.1	15.2

a. Plot the data on a scatter diagram.
b. Calculate the correlation coefficient between Y and X.
c. Test $H_0: \rho = 0$ against $H_1: \rho > 0$ using a 5% level of significance.
d. Based on your answer to part **c**, does it appear that using X will be helpful in predicting Y?
e. Estimate the linear regression equation.
f. If $X = 12$ in some quarter, what would you predict for Y in the following quarter?
g. Construct 95% confidence intervals for b_0 and b_1.
h. Let $\alpha = .05$. Test $H_0: \beta_1 = 0$ against $H_1: \beta_1 > 0$.
i. Assume that $X = 15$ in some quarter. Predict Y for the following quarter.
j. Construct a 95% confidence interval for $E(Y|X = 15)$.
k. Construct a 95% confidence interval for Y given that $X = 15$.
l. Note that in the sample the values of X vary between 11.1 and 20.2. Would you be more confident in a forecast of Y for a period when $X = 15$ than for a period when $X = 25$? Why or why not?

15.S.17 Economists have long debated whether changes in government spending (ΔG) or changes in the money supply (ΔM) cause changes in gross national product (ΔGNP). Suppose we want to examine this issue. The accompanying data from the *Economic Report of the President* show the level of GNP, the level of government spending, and the money supply (measured as the sum of currency, demand deposits, travelers checks, other checkable deposits, money market mutual funds, savings, and small time deposits).

a. Calculate the year-to-year change in each variable. Denote these annual changes as ΔGNP_t, ΔG_t, and ΔM_t.
b. Plot ΔGNP_t versus ΔG_t on a scatter diagram.
c. Estimate the linear regression equation $\Delta GNP_t = \beta_0 + \beta_1 \Delta G_t + e_t$.

d. Plot the sample regression equation on the scatter diagram.
e. Find R^2. Is this value large enough to make you believe that changes in GNP are caused (or explained by) changes in G?
f. Plot ΔGNP_t versus ΔM_t on a scatter diagram.
g. Estimate the linear regression equation $\Delta GNP_t = \beta_0 + \beta_1 \Delta M_t + e_t$.
h. Plot the sample regression equation on the scatter diagram.
i. Find R^2. Is this value large enough to make you believe that changes in GNP are caused (or explained by) changes in M?

Year	GNP	G	M
1	2,416	573	628
2	2,484	567	713
3	2,608	571	805
4	2,744	565	861
5	2,729	573	908
6	2,695	581	1,023
7	2,827	580	1,163
8	2,959	589	1,286
9	3,115	604	1,389
10	3,192	609	1,498
11	3,187	621	1,630
12	3,249	630	1,792
13	3,166	642	1,952
14	3,279	649	2,186
15	3,490	675	2,374
16	3,585	721	2,567
17	3,677	748	2,805

Note: All data measured in billions of 1982 dollars.

15.S.18   Plutonium has been produced in the Pacific Northwest since the 1940s, and some radioactive wastes have leaked into the Columbia River. A study of cancer incidence in nine communities bordering the river compared an exposure index (X) and the cancer mortality rate per 100,000 residents (Y). The results are as follows:

Exposure index, X	8.34	6.41	3.41	3.83	2.57	11.64	1.25	2.49	1.62
Cancer mortality, Y	210.3	177.9	129.9	162.3	130.1	207.5	113.5	147.1	137.5

a. Calculate the sample correlation coefficient and test whether the population correlation coefficient is 0. Use $\alpha = .05$.
b. Estimate the simple linear regression model.
c. Calculate R^2 and plot the data.

15.S.19   In a study involving a large sample of U.S. males aged 21 to 54, there was a positive correlation between income and blood pressure, a positive correlation between income and age, and a positive correlation between blood pressure and age. Do you think any of these variables is causing the other to change? Can you explain the results?

15.S.20   An agent for the Internal Revenue Service claims that there is approximately a linear relationship between people's wage income (X) and their income earned from interest and dividends (Y).

Interest and dividend income (thousands of dollars) Y	Wage income (thousands of dollars) X
2.3	38.3
3.7	45.4
5.6	67.7
.4	16.3
1.1	18.1
.3	23.4
.1	21.2
5.2	59.0
7.1	83.2
4.8	48.6
3.2	54.2
4.1	65.7
.7	31.0

a. Use the accompanying data and find the sample regression line that expresses Y as a linear function of X.

b. Plot the data on a scatter diagram.

c. Find R^2. Based on this value of R^2, would you feel confident in predicting the interest and dividend income of a household with a wage income of \$35,000?

d. Let $\alpha = .05$. Test the null hypothesis that wage income does not help explain interest and dividend income.

 15.S.21 The sales manager for a large corporation has to set sales goals for all the salespeople in the division. People who exceed their goals are eligible for year-end bonuses, so it is important that the goals be set in a reasonable manner. To set the goals for second-year employees, the sales manager argues that sales during the person's second year (Y) are linearly related to that person's sales during the first year (X). The sales manager gathers historical data from past employees to see whether their second-year sales were related to their first-year sales.

Sales in second year (thousands of dollars) Y	Sales in first year (thousands of dollars) X
134.6	110.6
104.7	110.2
213.7	156.8
98.3	82.3
112.4	101.2
78.4	56.7
145.0	109.8
152.3	125.6
69.5	78.1
103.7	92.2

a. Use the accompanying data and find the sample regression line that expresses Y as a linear function of X.

b. Plot the data on a scatter diagram.

c. Find R^2. Based on this value of R^2, would you feel confident in setting a sales goal for a person who had sales of $120,000 last year?

d. Let $\alpha = .05$. Test the null hypothesis that X does not help explain Y.

References

Ames, Edward, and Stanley Reiter. "Distributions of Correlation Coefficients in Economic Time Series." *Journal of the American Statistical Association* 56 (1961): 637–656.

Draper, Norman R., and Harry Smith. *Applied Regression Analysis.* 2d ed. New York: Wiley, 1981.

Goldberger, Arthur. *Econometric Theory.* New York: Wiley, 1964.

Griffiths, William E., R. Carter Hill, and George G. Judge. *Learning and Practicing Econometrics.* New York: Wiley, 1993.

Gujarati, D. *Basic Econometrics.* 2d ed. New York: McGraw-Hill, 1988.

Johnston, John. *Econometric Methods.* New York: McGraw-Hill, 1972.

Judge, G. G., R. C. Hill, W. E. Griffiths, and T. C. Lee. *The Theory and Practice of Econometrics.* 2d ed. New York: Wiley, 1985.

Judge, G. G., R. C. Hill, W. E. Griffiths, H. Lutkepohl, and T. C. Lee. *Introduction to the Theory and Practice of Econometrics.* 2d ed. New York: Wiley, 1988.

Klein, L. *An Introduction to Econometrics.* Englewood Cliffs, N.J.: Prentice-Hall, 1962.

Kmenta, Jan. *Elements of Econometrics.* 2d ed. New York: Macmillan, 1986.

Neter, John, and William Wasserman. *Applied Linear Statistical Models.* 2d ed. Homewood, Ill.: Irwin, 1985.

Neter, John, William Wasserman, and G. A. Whitmore. *Applied Statistics.* 3d ed. Boston: Allyn and Bacon, 1988.

Nie, Norman E., C. Hadlai Hull, Jean G. Jenkins, Karin Steinbrenner, and Dale H. Bent. *SPSS Statistical Package for the Social Sciences.* 2d ed. New York: McGraw-Hill, 1975.

Norusis, Marija J. *SPSSX Introductory Statistics Guide.* New York: McGraw-Hill, 1990.

Norusis, Marija J. *SPSSX Advanced Statistics Guide.* Chicago: SPSS, 1990.

Norusis, Marija J. *The SPSS Guide to Data Analysis.* Chicago: SPSS, 1986.

Pindyck, R., and D. Rubinfeld. *Econometric Models and Forecasts.* New York: McGraw-Hill, 1981.

Ryan, Thomas A., Brian L. Joiner, and Barbara F. Ryan. *Minitab Handbook.* 2d ed. Boston: PWS-Kent, 1985.

Ryan, Thomas A., Brian L. Joiner, and Barbara F. Ryan. *Minitab Reference Manual.* University Park, Penn.: Minitab Project, 1985.

SAS Introductory Guide. 3d ed. Cary, N.C.: SAS Institute, 1985.

SAS Procedures Guide for Personal Computers. Version 6 ed. Cary, N.C.: SAS Institute, 1986.

SAS Statistics Guide for Personal Computers. Version 6 ed. Cary, N.C.: SAS Institute, 1986.

SAS User's Guide: Basics. Version 5 ed. Cary, N.C.: SAS Institute, 1985.

SAS User's Guide: Statistics. Version 5 ed. Cary, N.C.: SAS Institute, 1985.

Schmidt, P. *Econometrics.* New York: Marcel Dekker, 1976.

SPSSX User's Guide. Chicago: SPSS, 1988.

Theil, H. *Principles of Econometrics.* New York: Wiley, 1971.

Wonnacott, T. J., and R. J. Wonnacott. *Regression: A Second Course in Statistics.* New York: Wiley, 1981.

16 Multiple Regression Models

In Chapter 15, we discussed the simple linear regression model, where the values taken by a dependent variable are related to the values taken by one independent variable. In this chapter, we discuss multiple regression models, where the values taken by a dependent variable are related to the values taken by several independent variables. In the first section, we discuss how to estimate and analyze multiple regression models. In later sections, we study several specialized topics involving multiple regression analysis.

Some of the exercises in this chapter will involve small samples of data so that the computations will not be too burdensome and can be solved by hand. Some problems have large data sets, however, and require the use of a computer. Although it is not necessary to use a computer to understand the material presented in this chapter, you should learn how to use a computer regression package to duplicate the results of some examples in this chapter.

16.1 Models with Two Explanatory Variables

Multiple regression is a generalization of the simple linear regression analysis, discussed in Chapter 15. Simple regression analysis enables us to analyze a relationship between a dependent variable and a single explanatory variable. The same ideas can be extended to analyze relationships between a dependent variable and two or more explanatory variables. If knowledge of one variable X helps us predict the value of Y, then it is natural to consider whether knowledge of several variables $X_1, X_2, \ldots, X_K$ enables us to provide an even better prediction of the value of Y. The symbol x_{i1} denotes the ith observation on the variable X_1; similarly, x_{i2} denotes the ith observation on the variable X_2, and in general x_{ij} denotes the ith observation on the jth independent variable X_j.

In multiple regression analysis, the dependent variable Y is related to two or more explanatory variables according to some hypothesized model. Usually it is assumed that the dependent variable is linearly related to the explanatory variables, but this does not have to be the case.

Regression model with two explanatory variables

> The regression model with two explanatory variables is given by
> $$Y_i = \beta_0 + \beta_1 x_{i1} + \beta_2 x_{i2} + e_i \qquad i = 1, 2, \ldots, n$$
> *(continued)*

where x_{i1} represents the ith observation on the explanatory variable X_1 and x_{i2} denotes the ith observation on the second explanatory variable X_2. As for the simple linear regression model, the following assumptions are made:

1. e_i $(i = 1, 2, \ldots, n)$ is normally distributed.
2. $E(e_i) = 0, i = 1, 2, \ldots, n$
3. $\text{Var}(e_i) = \sigma_e^2, i = 1, 2, \ldots, n$
4. e_i and e_j are independent of one another for all i and j.
5. The errors e_i and the x_{ij} values are independent.

In the multiple regression model, we need the following additional assumption:

6. The variables X_1 and X_2 are not perfectly linearly related. That is, it is not possible to find a set of numbers c_0, c_1, and c_2 such that

$$c_0 + c_1 x_{i1} + c_2 x_{i2} = 0$$

for every $i = 1, 2, \ldots, n$.

For given values of x_{i1} and x_{i2}, the mean value of the random variable Y_i lies on the *population regression plane* given by

$$E(Y_i | x_{i1}, x_{i2}) = \beta_0 + \beta_1 x_{i1} + \beta_2 x_{i2}$$

Assumption 6 rules out a condition called *perfect multicollinearity,* in which one of the explanatory variables can be expressed exactly as a linear function of the other explanatory variables and a constant term. If this occurs in the two-variable model, then it is not possible to determine whether changes in Y are related to changes in X_1 or changes in X_2 because the latter two variables are perfectly linearly related. Whenever one changes, the other changes in a perfectly predictable way.

There are four unknown parameters in the population regression model, $\beta_0, \beta_1, \beta_2,$ and σ_e^2. The coefficient β_0 is the constant term. The coefficient of the first explanatory variable x_{i1} is β_1 and measures the increase in the mean value of Y_i associated with a 1-unit increase in x_{i1} for a fixed value of the second variable x_{i2}. The coefficient of the second explanatory variable x_{i2} is β_2 and measures the increase in the mean value of Y_i associated with a 1-unit increase in x_{i2} for a fixed value of the first variable x_{i1}.

Sample regression equation

For the model containing two explanatory variables, the sample regression equation is

$$\hat{y}_i = b_0 + b_1 x_{i1} + b_2 x_{i2}$$

where $\hat{y}_i$ is the fitted, or predicted, value of the random variable Y_i, and $b_0, b_1,$ and b_2 are the estimated values of the parameters $\beta_0, \beta_1,$ and β_2.

The ith residual $\hat{e}_i$ is given by

$$\hat{e}_i = y_i - \hat{y}_i$$
$$= y_i - b_0 - b_1 x_{i1} - b_2 x_{i2} \qquad i = 1, 2, \ldots, n$$

The least-squares method says we should choose the estimates b_0, b_1, and b_2 so as to minimize the residual sum of squares SSE where

$$\text{SSE} = \sum_{i=1}^{n} \hat{e}_i^2 = \sum_{i=1}^{n} (y_i - \hat{y}_i)^2$$
$$= \sum_{i=1}^{n} (y_i - b_0 - b_1 x_{i1} - b_2 x_{i2})^2$$

For convenience, we define the following totals:

$$T_{11} = \Sigma x_{i1}^2 - n\bar{x}_1^2 \qquad\qquad T_{1Y} = \Sigma x_{i1} y_i - n\bar{x}_1 \bar{y}$$
$$T_{12} = \Sigma x_{i1} x_{i2} - n\bar{x}_1 \bar{x}_2 \qquad T_{2Y} = \Sigma x_{i2} y_i - n\bar{x}_2 \bar{y}$$
$$T_{22} = \Sigma x_{i2}^2 - n\bar{x}_2^2 \qquad\qquad T_{YY} = \Sigma y_i^2 - n\bar{y}^2$$

(The value T_{YY} will be needed later but is not needed to find the estimated coefficients.)
We obtain

$$b_1 = \frac{T_{22} T_{1Y} - T_{12} T_{2Y}}{\Delta} \qquad\qquad (1)$$

$$b_2 = \frac{T_{11} T_{2Y} - T_{12} T_{1Y}}{\Delta} \qquad\qquad (2)$$

where $\Delta = T_{11} T_{22} - T_{12}^2$. Once we obtain b_1 and b_2, we can calculate b_0 as follows:

$$b_0 = \bar{y} - b_1 \bar{x}_1 - b_2 \bar{x}_2 \qquad\qquad (3)$$

Thus, the computational procedure is as follows:

1. Obtain all the means, $\bar{y}$, $\bar{x}_1$, and $\bar{x}_2$.
2. Obtain all the sums of squares and sums of cross-products, Σy_i^2, Σx_{i1}^2, Σx_{i2}^2, $\Sigma x_{i1} x_{i2}$, $\Sigma x_{i1} y_i$, and $\Sigma x_{i2} y_i$.
3. Obtain T_{11}, T_{12}, T_{22}, T_{1Y}, T_{2Y}, and T_{YY}.
4. Find b_1 and b_2 from equations (1) and (2).
5. Substitute these into equation (3) to obtain b_0.

It is much easier to express the formulas for computing the regression coefficients using matrices, but we will not pursue the matrix approach here, since a knowledge of matrix algebra is not assumed for this book. (The interested reader should consult an econometrics textbook, such as William E. Griffiths, R. Carter Hill, and George G. Judge, *Learning and Practicing Econometrics,* New York: Wiley, 1993.)

Figure 16.1 (page 670) shows a hypothetical regression plane fitted to a hypothetical sample of observations. The sample regression plane is the plane that minimizes the squared deviations from the plane in the vertical direction.

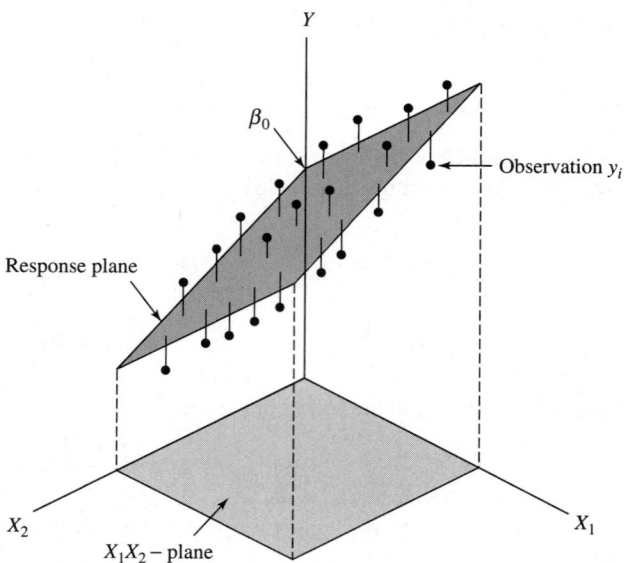

FIGURE 16.1   *A hypothetical regression plane*

EXAMPLE 16.1   **Estimating a Multiple Regression Model**

It is reasonable to suspect that gasoline mileage for a car is determined mainly by the car's weight and engine size. An automotive engineer gathered information for a sample of $n = 10$ different cars. The data in Table 16.1 show y_i, the gasoline mileage (in miles per gallon) of the ith car; x_{i1}, the engine size of the ith car (in hundreds of cubic inches); and x_{i2}, the weight of the ith car (in tons). Estimate the regression plane and predict the mileage for a car that has a 240-cubic-inch engine (i.e., a 2.4-hundred-cubic-inch engine) and weighs .9 ton.

TABLE 16.1   *Data for Example 16.1*

Miles per gallon Y	Engine size X_1	Weight X_2	X_1Y	X_2Y	X_1^2	X_2^2	X_1X_2	Y^2
32	1.0	1.0	32.0	32.0	1.00	1.00	1.00	1,024
27	1.4	1.3	37.8	35.1	1.96	1.69	1.82	29
22	2.1	1.5	46.2	33.0	4.41	2.25	3.15	484
14	3.2	1.7	44.8	23.8	10.24	2.89	5.44	196
21	2.2	1.4	46.2	29.4	4.84	1.96	3.08	441
19	2.4	1.7	45.6	32.3	5.76	2.89	4.08	361
32	1.4	1.2	44.8	38.4	1.96	1.44	1.68	1,024
18	2.3	1.7	41.4	30.6	5.29	2.89	3.91	324
38	1.2	.8	45.6	30.4	1.44	.64	.96	1,444
35	1.3	.9	45.5	31.5	1.69	.81	1.17	1,225
258	18.5	13.2	429.9	316.5	38.59	18.46	26.29	7,252

SOLUTION   Typically, we would use a computer to solve this problem. For purposes of illustration, we used a hand calculator to calculate the sums shown in Table 16.1. We have

$$\Sigma \, y_i = 258 \qquad \Sigma \, x_{i1} = 18.5 \qquad \Sigma \, x_{i2} = 13.2 \qquad \Sigma \, x_{i1}x_{i2} = 26.29$$

$$\Sigma \, x_{i1}^2 = 38.59 \qquad \Sigma \, x_{i2}^2 = 18.46 \qquad \Sigma \, x_{i1}y_i = 429.9 \qquad \Sigma \, x_{i2}y_i = 316.5$$

and the means $\bar{y} = 25.8$, $\bar{x}_1 = 1.85$, and $\bar{x}_2 = 1.32$. The sums of squares of deviations from the respective means are as follows:

$$T_{11} = 4.365 \qquad T_{12} = 1.87 \qquad T_{22} = 1.036$$

$$T_{1Y} = -47.4 \qquad T_{2Y} = -24.06 \qquad T_{YY} = 595.6 \qquad \Delta = 1.02524$$

Finally, we compute the estimated coefficients:

$$b_1 = \frac{(1.036)(-47.4) - (1.87)(-24.06)}{1.02524} = -4.01291$$

$$b_2 = \frac{(4.365)(-24.06) - (1.87)(-47.4)}{1.02524} = -15.98055$$

$$b_0 = 25.8 - (-4.0129)(1.85) - (-15.98055)(1.32) = 54.3182$$

We obtain the estimated equation (after rounding) of

$$\hat{y}_i = 54.3182 - 4.0129x_{i1} - 15.9806x_{i2}$$

The predicted mileage for a car that has a 2.4-hundred-cubic-inch engine and weighs .9 ton is obtained by substituting the values $x_{i1} = 2.4$ and $x_{i2} = .9$ into the estimated equation. The predicted value is then

$$\hat{y} = 54.3182 - 4.0129(2.4) - 15.9806(.9) = 30.3047$$

or about 30.3 miles per gallon.

The computer output for this problem as it appears from the SPSSX program is shown in Figure 16.2 (page 672). The regression coefficients computed by the program are shown in the bottom half of the figure in the column labeled B. The printout indicates that the estimated coefficients are as follows:

$$b_0 = \text{Constant} = 54.318218$$
$$b_1 = \text{Coefficient of Size of engine} = -4.012914$$
$$b_2 = \text{Coefficient of Weight of car} = -15.980551$$

All these values agree with the values just calculated. The rest of the information shown in the computer output will be explained later in this chapter.

* * * * M U L T I P L E   R E G R E S S I O N * * * *

Listwise Deletion of Missing Data

Equation Number 1   Dependent Variable..   MILEAGE   MILES PER GALLON

Block Number 1. Method: Enter   SIZE   WEIGHT

Variable(s) Entered on Step Number 1.. WEIGHT   WEIGHT IN TONS
                                    2.. SIZE   SIZE OF ENGINE IN HUNDREDS OF CUBIC INCH

```
Multiple R           .98230      Analysis of Variance
R Square             .96492                         DF     Sum of Squares    Mean Square
Adjusted R Square    .95489      Regression          2         574.70418      287.35209
Standard Error      1.72775      Residual            7          20.89582        2.98512

                                 F =   96.26158      Signif F =  .0000
```

------------------ Variables in the Equation ------------------

```
Variable            B           SE B        Beta          T    Sig T

WEIGHT        -15.980551      3.565004    -.666491     -4.483   .0029
SIZE           -4.012914      1.736792    -.543538     -2.311   .0541
(Constant)     54.318218      2.484910               21.859   .0000
```

End Block Number   1   All requested variables entered.

FIGURE 16.2 *SPSSX-generated output for Example 16.1*

There are many different algebraic formulas that can be used to calculate the least squares estimates. In the multiple regression model, different authors adopt different notation systems. Many of the least squares formulas involve the following sums of squares and cross-products.

Some useful formulas

The following formulas are useful in the multiple regression model containing two explanatory variables X_{i1} and X_{i2}. Some authors use the notation SS to denote sum of squares.

$$T_{YY} = \text{SS}_{yy} = \Sigma(y_i - \bar{y})^2 = \Sigma y_i^2 - n\bar{y}^2$$

$$T_{11} = \text{SS}_{11} = \Sigma(x_{i1} - \bar{x}_1)^2 = \Sigma x_{i1}^2 - n\bar{x}_1^2$$

$$T_{22} = \text{SS}_{22} = \Sigma(x_{i2} - \bar{x}_2)^2 = \Sigma x_{i2}^2 - n\bar{x}_2^2$$

$$T_{1Y} = \text{SS}_{1y} = \Sigma(x_{i1} - \bar{x}_1)(y_i - \bar{y}) = \Sigma x_{i1}y_i - n\bar{x}_1\bar{y}$$

$$T_{2Y} = \text{SS}_{2y} = \Sigma(x_{i2} - \bar{x}_2)(y_i - \bar{y}) = \Sigma x_{i2}y_i - n\bar{x}_2\bar{y}$$

$$T_{12} = \text{SS}_{12} = \Sigma(x_{i1} - \bar{x}_1)(x_{i2} - \bar{x}_2) = \Sigma x_{i1}x_{i2} - n\bar{x}_1\bar{x}_2$$

$$b_1 = \frac{\text{SS}_{22}\text{SS}_{1y} - \text{SS}_{12}\text{SS}_{2y}}{\text{SS}_{11}\text{SS}_{22} - \text{SS}_{12}^2}$$

$$b_2 = \frac{\text{SS}_{11}\text{SS}_{2y} - \text{SS}_{12}\text{SS}_{1y}}{\text{SS}_{11}\text{SS}_{22} - \text{SS}_{12}^2}$$

$$b_0 = \bar{y} - b_1\bar{x}_1 - b_2\bar{x}_2$$

(continued)

> Unless the number of observations is very small, we recommend using a computer to generate multiple regression results.

Exercises

Statistical Concepts

16.1.1 Suppose Y and X are related according to the equation

$$Y = \beta_0 + \beta_1 X_1 + \beta_2 X_2$$

Interpret the coefficients β_1 and β_2. For a given value of X_1, suppose the value of X_2 increases by 1 unit. By how much will Y change?

16.1.2 State the assumptions underlying the multiple regression model with two explanatory variables.

16.1.3 *True or false:* In the multiple regression model, the estimated coefficient b_0 equals the mean value of the dependent variable. Explain.

16.1.4 In the multiple regression model, suppose the variable $X_2 = X_1^2$. Suppose we obtain the estimated model

$$y_i = 10 + 8x_i + 3x_i^2$$

Such an equation, where Y is related to X and X^2, is called a *parabola*. This is a special case of a multiple regression model where one explanatory variable is the square of another explanatory variable.
  a. Plot this equation for the values $x_i = 0, 1, 2, 3, 4, 5$. Do the values trace out a straight line?
  b. Change the coefficient of x_i^2 from 3 to -3. Now plot the equation.
  c. How does the sign of the coefficient of x_i^2 affect the shape of the parabola?

Statistical Drills

16.1.5 Suppose you obtain the following information:

$$n = 10 \quad \Sigma y_i = 260 \quad \Sigma x_{i1} = 20 \quad \Sigma x_{i2} = 15 \quad \Sigma x_{i1}x_{i2} = 28 \quad \Sigma x_{i1}^2 = 45$$
$$\Sigma x_{i2}^2 = 25 \quad \Sigma x_{i1}y_i = 428 \quad \Sigma x_{i2}y_i = 310$$

  a. Compute the regression estimates b_1, b_2, and b_0.
  b. Write the equation of the sample regression plane.
  c. Assume we have $x_{i1} = 2.4$ and $x_{i2} = 4.1$. Use the estimated regression model and predict the actual value of Y_i.
  d. What is the estimated intercept in this model?
  e. By how much does the predicted value of Y_i change when x_{i1} increases by 1 unit?
  f. By how much does the predicted value of Y_i change when x_{i2} increases by 1 unit?
  g. By how much does the predicted value of Y_i change when both x_{i1} and x_{i2} increase by 1 unit?

16.1.6 In a multiple regression model, the following coefficient estimates were obtained:

$$b_0 = 14.5 \quad b_1 = 5.6 \quad b_2 = 11.7$$

  a. Write the equation of the estimated multiple regression model.
  b. Suppose we have $x_{i1} = 14$ and $x_{i2} = 5$. Predict the value of Y_i.
  c. Suppose we have $x_{i1} = 15$ and $x_{i2} = 5$. Predict the value of Y_i.
  d. Suppose we have $x_{i1} = 14$ and $x_{i2} = 6$. Predict the value of Y_i.

16.1.7   In a multiple regression model, the following coefficient estimates were obtained:

$$b_0 = -10 \quad b_1 = 4.5 \quad b_2 = -6.0$$

a. Write the equation of the estimated multiple regression model.
b. Suppose we have $x_{i1} = 8$ and $x_{i2} = 4$. Predict the value of Y_i.
c. Suppose we have $x_{i1} = 9$ and $x_{i2} = 4$. Predict the value of Y_i.
d. Suppose we have $x_{i1} = 8$ and $x_{i2} = 5$. Predict the value of Y_i.

16.1.8   The accompanying data are values of X_{i1}, X_{i2}, and Y_i.

x_{i1}	x_{i2}	y_i
3.0	6.0	14.0
5.0	4.0	24.0
7.0	5.0	29.0
9.0	8.0	40.0
8.0	4.0	30.0

a. Estimate the multiple linear regression model.
b. Predict Y_i given $x_{i1} = 6.0$ and $x_{i2} = 4.0$.

16.1.9   The accompanying data are values of X_i and Y_i.

x_i	y_i
1.0	9.0
3.0	28.0
2.0	15.0
6.0	86.0
5.0	63.0

a. Plot the data on a scatter diagram.
b. Does it appear that a parabola would fit the data better than a straight line?
c. Let x_{i1} denote the values x_i. Let x_{i2} denote the values x_i^2. Estimate the parabolic regression model

$$\hat{y}_i = b_0 + b_1 x_i + b_2 x_i^2$$

d. Predict Y_i given $x_i = 6.0$.

Statistical Applications

16.1.10   It is hypothesized that a student's grade point average in college is related to the student's college entrance scores on math and English tests. Estimate the model

$$Y_i = \beta_0 + \beta_1 x_i + \beta_2 z_i + e_i$$

where Y = grade point average, X = math score, and Z = English score. The data are shown in the accompanying table.

Y	X	Z
2.5	320	470
3.1	545	530
3.8	690	570

(continued)

Y	X	Z
3.5	570	570
2.7	460	490
3.0	540	550
3.7	710	680
3.2	600	590
2.4	490	520

16.1.11 The number of tennis racquets sold per year by a nationwide distributor depends on the price of the racquet and on the number of tennis players. Estimate the multiple regression equation using the data in the accompanying table.

Quantity sold (in thousands) Y	Price (in dollars) X	Number of players (in millions) Z
40	20	2.0
60	20	3.0
50	40	3.2
60	30	4.0
80	25	4.0
64	44	4.5
72	30	5.1
53	34	4.4

16.1.12 Suppose the price of a house in thousands of dollars (Y) depends on the size of the house in thousands of square feet (X) and the area of the yard in thousands of square feet (Z). The data are shown in the accompanying table.

Price Y	Size of house X	Area of yard Z
48	2.2	4.3
54	2.1	5.2
35	1.3	2.4
89	3.3	7.8
59	3.2	3.8
67	2.8	4.0
76	2.9	5.0
51	2.5	4.8

a. Calculate the correlation coefficient between price and the size of the house.
b. Estimate the multiple regression equation relating price to the size of the house and the area of the yard.
c. Calculate the correlation between the size of the house and the area of the yard.

16.1.13 A researcher is trying to evaluate the potential benefits of a proposed irrigation project on crop yields in a developing country. The only variables affecting output in bushels per acre (Y) are total rainfall in inches during the growing season (R) and average temperature in degrees Fahrenheit during the growing season (T). Use the accompanying data and estimate the model

$$Y_i = \beta_0 + \beta_1 r_i + \beta_2 t_i + e_i$$

Year	Yield Y	Total rainfall R	Average temperature T
1987	55	19	69
1988	65	17	78
1989	80	21	67
1990	75	17	80
1991	70	19	75
1992	50	18	70
1993	60	20	73
1994	65	21	71

16.2   The General Multiple Regression Model

The theory of least squares can be extended to cases in which the regression equation contains more than two explanatory variables.

Multiple regression model

> The multiple regression model is
>
> $$Y_i = \beta_0 + \beta_1 x_{i1} + \beta_2 x_{i2} + \cdots + \beta_K x_{iK} + e_i$$
>
> where
>
> Y_i is a random variable representing the ith value of the dependent variable Y
> $\beta_0, \beta_1, \ldots, \beta_K$ are unknown population coefficients
> x_{i1} is the ith observed value of the first independent variable X_1
> x_{i2} is the ith observed value of the second independent variable X_2
>
> .
> .
> .
>
> x_{iK} is the ith observed value of the Kth independent variable X_K
> e_i is the ith value of the unobservable random disturbance term

The same basic assumptions are made for the multiple regression model as for the two-explanatory-variable regression model discussed in Section 16.1. These basic assumptions are repeated in the accompanying box.

Basic assumptions of the multiple regression model

> The following basic assumptions are made about the error terms e_i and the values of the independent variables $X_1, X_2, \ldots, X_K$:

1. *Normality:* For any set of values of the independent variables, the error term e_i is a normally distributed random variable.
2. *Zero mean:* For any set of values of the independent variables, $E(e_i) = 0$.
3. *Homoscedasticity:* The variance of e_i, denoted σ_e^2, is the same for all values of the independent variables.
4. *No serial correlation:* The error terms are independent of one another for $i \neq j$.
5. *Independence of e_i and x_{ij}:* The error terms e_i are independent of the values of the independent variables. The independent variables are either fixed numbers or random variables that are independent of the error terms. If the x_{ij}'s are random variables, then all inferences are carried out conditionally on the observed values of the x_{ij}'s.
6. *No perfect multicollinearity:* It is not possible to find a set of numbers c_0, $c_1, \ldots, c_K$ such that

$$c_0 + c_1 x_{i1} + c_2 x_{i2} + \cdots + c_K x_{iK} = 0$$

for every $i = 1, 2, \ldots, n$.

The purpose of the assumption of no perfect multicollinearity is to exclude independent variables that can be determined exactly as a linear function of other independent variables. For example, if our model contains the variables X_1, X_2, and X_3, then the last assumption rules out a case such as

$$x_{i3} = d_0 + d_1 x_{i1} + d_2 x_{i2}$$

for $i = 1, 2, \ldots, n$. Note that if X_3 could be perfectly explained in terms of X_1 and X_2, then the variable X_3 would provide no information that was not already included in the variables X_1 and X_2. In such a case, we would not be able to determine the separate effect that X_3 has on the dependent variable. As a practical matter, it is safe to assume that the last assumption is not violated.

Population regression equation

The population regression equation is

$$E(Y_i | x_{i1}, x_{i2}, \ldots, x_{iK}) = \beta_0 + \beta_1 x_{i1} + \beta_2 x_{i2} + \cdots + \beta_K x_{iK}$$

and shows the mean value of Y_i associated with the given values $x_{i1}, x_{i2}, \ldots, x_{iK}$ of the explanatory variables.

The parameter β_0 is the constant term, or Y-intercept, and measures the mean value of Y_i when all the independent variables are set to 0. The parameter β_1 measures the change in the mean value of Y_i corresponding to a 1-unit increase in the value of x_{i1}, when all the other independent variables are held constant; the parameter β_2 measures the change in the mean value of Y_i corresponding to a 1-unit increase in the value of x_{i2}, when all other independent variables are held constant, and so forth. For example, if we increase x_{i2} by 1 unit, we obtain

$$E(Y_i|x_{i1}, x_{i2} + 1, \ldots, x_{iK}) = \beta_0 + \beta_1 x_{i1} + \beta_2(x_{i2} + 1) + \cdots + \beta_K x_{iK}$$

After subtracting, we obtain

$$E(Y_i|x_{i1}, x_{i2} + 1, \ldots, x_{iK}) - E(Y_i|x_{i1}, x_{i2}, \ldots, x_{iK}) = \beta_2$$

DEFINITION   **Random Error Term** ──────────────────────────────

In the population regression model, the **random error term,** denoted e_i, is the difference between the value of the random variable Y_i and the expected value $E(Y_i|x_{i1}, x_{i2}, \ldots, x_{iK})$. We obtain

$$e_i = Y_i - E(Y_i|x_{i1}, x_{i2}, \ldots, x_{iK})$$
$$= Y_i - (\beta_0 + \beta_1 x_{i1} + \beta_2 x_{i2} + \cdots + \beta_K x_{iK})$$

After rearranging terms in the equation for the random error term e_i, we obtain the equivalent expression

$$Y_i = \beta_0 + \beta_1 x_{i1} + \beta_2 x_{i2} + \cdots + \beta_K x_{iK} + e_i$$

where, for any values of the independent variables, the mean value of e_i is 0.

The population parameters $\beta_0, \beta_1, \ldots, \beta_K$ are unknown and are estimated using a sample of n observations on the dependent variable Y and the K independent variables $X_{i1}, X_{i2}, \ldots, X_{iK}$. Once we have estimated the parameters $\beta_0, \beta_1, \ldots, \beta_K$, we obtain an estimated regression equation, which is called the *sample regression equation.*

DEFINITION   **Least-Squares Estimates** ──────────────────────────────

The **least-squares estimates** of the population parameters $\beta_0, \beta_1, \ldots, \beta_K$ are the values $b_0, b_1, \ldots, b_K$ that minimize the residual sum of squares SSE, where

$$SSE = \Sigma(y_i - \hat{y}_i)^2 = \Sigma(y_i - b_0 - b_1 x_{i1} - b_2 x_{i2} - \cdots - b_K x_{iK})^2$$

An alternative formula for SSE is

$$SSE = \Sigma y_i^2 - b_0 \Sigma y_i - b_1 \Sigma x_{i1} y_i - b_2 \Sigma x_{i2} y_i - \cdots - b_K \Sigma x_{iK} y_i$$

DEFINITION   **Sample Regression Equation** ──────────────────────────────

The estimated equation

$$\hat{y}_i = b_0 + b_1 x_{i1} + b_2 x_{i2} + \cdots + b_K x_{iK}$$

is called the **sample regression equation.** The value b_0 is the sample estimate of the population parameter β_0, the value b_1 is the sample estimate of the population parameter β_1, and so forth. The value $\hat{y}_i$ is called the *fitted value of Y_i* or the *predicted value of Y_i.*

Many hand calculators can produce the estimated coefficients in a simple linear regression model that contains only one explanatory variable. For the multiple regression model, however, the estimated coefficients are almost always obtained by using a computer. Even when the model contains only two explanatory variables, the arithmetic calculations necessary to obtain the estimated coefficients can be quite tedious.

If we estimate the multiple regression model by the method of least squares and if the model contains a constant term, then we obtain

$$\Sigma \, \hat{e}_i = 0$$

that is, the sum of the residuals is always 0 provided the sample regression model contains a constant term b_0.

The Gauss–Markov Theorem

The estimated coefficients $b_0, b_1, \ldots, b_K$ are random variables whose values change from sample to sample. If the basic assumptions hold, the least-squares estimates have several important properties described by the Gauss–Markov theorem, which is summarized in the following box. By virtue of the Gauss–Markov theorem, the least-squares estimators are said to be *best linear unbiased estimators.*

The Gauss–Markov theorem

> If basic assumptions 2 through 6 (page 677) hold, the method of least squares provides estimates $b_0, b_1, \ldots, b_K$, which have the following properties:
>
> **1.** The estimated coefficients are *unbiased* estimates of the true population parameters $\beta_0, \beta_1, \ldots, \beta_K$.
> **2.** The estimators $b_0, b_1, \ldots, b_K$ have the *minimum variances* among the class of linear unbiased estimators.

If, in addition to assumptions 2 through 6, we add basic assumption 1 (normality of the errors) to our model, then the sampling distributions of the least-squares estimators are normal. The assumption of normality is required if we want to determine the sampling distributions of our estimators for testing hypotheses and constructing confidence intervals.

To test hypotheses and to construct confidence intervals, we need an estimate of σ_e^2. From Chapter 15, recall that in the simple linear regression model an unbiased estimator of σ_e^2 is $s_e^2 = \text{SSE}/(n - 2)$, where $(n - 2)$ is the degrees of freedom of the model. An unbiased estimator of σ_e^2 is defined in an analogous manner in the multiple regression model. In the formula for s_e^2, the denominator is n minus the number of coefficients that have to be estimated. In the multiple regression model, there are $(K + 1)$ coefficients $\beta_0, \beta_1, \ldots, \beta_K$ to be estimated, and so the denominator in the formula for s_e^2 is

$$n - (K + 1) = n - K - 1$$

Thus the formula for s_e^2 is

$$s_e^2 = \text{SSE}/(n - K - 1)$$

which is an unbiased estimator of σ_e^2. The value

$$\nu = n - K - 1$$

is the degrees of freedom of the regression equation. In a sense, we use up $(K + 1)$ degrees of freedom before calculating SSE because we first estimate the $(K + 1)$ coefficients β_0, $\beta_1, \ldots, \beta_K$.

Estimated Standard Error of the Regression

The estimated standard error of the regression, denoted s_e, is an estimate of σ_e, the standard deviation of the error terms, and is simply the square root of s_e^2:

DEFINITION   **Estimated Standard Error of the Regression**

The **estimated standard error of the regression** is

$$s_e = \sqrt{\frac{\text{SSE}}{n - K - 1}}$$

EXAMPLE 16.2   **Calculation of s_e**

Use the data in Table 16.1 along with the estimated coefficients b_0, b_1, and b_2 from Example 16.1 to determine SSE and s_e for the model estimated in Example 16.1 for predicting gasoline mileage.

SOLUTION   We obtain

$$
\begin{aligned}
\text{SSE} &= \Sigma\, y_i^2 - b_0\, \Sigma\, y_i - b_1\, \Sigma\, x_{i1} y_i - b_2\, \Sigma\, x_{i2} y_i \\
&= 7{,}252 - 54.3182(258) - (-4.0129)(429.9) - (-15.9806)(316.5) \\
&= 20.8958
\end{aligned}
$$

An unbiased estimate of the variance of the errors is

$$
\begin{aligned}
s_e^2 &= \text{SSE}/(n - K - 1) \\
&= 20.8958/(10 - 3) = 2.9851
\end{aligned}
$$

and the estimated standard error of the regression is

$$s_e = \sqrt{2.9851} = 1.7277$$

Based on the Empirical Rule, for a normal population, approximately 95% of the error terms should be less than $2s_e$ (or 3.4554) units from the estimated plane and approximately 95% of the sample observations should lie within $2s_e = 3.46$ units of the estimated plane.

Reading SPSSX Computer Output

Refer again to the SPSSX computer output shown in Figure 16.2. At the top of the output is the analysis of variance, or ANOVA table. In the ANOVA table, the residual sum of squares (SSE) is one of the entries in the column labeled Sum of Squares. From the

SPSSX computer output, we see that the residual sum of squares is SSE = 20.8958. The estimated standard error of the regression (s_e) is reported next to the label Standard Error. From the SPSSX output, we obtain the standard error s_e = 1.7277, which agrees with the value calculated previously. The value s_e^2 is reported under the column Mean Square in the row labeled Residual. From the ANOVA table, we obtain s_e^2 = 2.9851. The computer-generated values for SSE and s_e agree with the values we calculated by hand.

Tips for Problem Solving

1. If you estimate a multiple linear regression model by hand, always calculate the residual sum of squares and the standard error of the regression. If you use a computer, be sure to record the value of s_e.

2. Be careful to avoid round-off errors. Carry out your calculations to at least six decimal places if possible. Rounding off the estimated coefficients will affect the value of SSE and s_e.

3. Always examine the value of s_e. The value of s_e provides evidence concerning the typical distance between the actual value of y_i and the corresponding predicted value in your set of observations.

4. Large values of s_e indicate that the "typical" residual is large in absolute value. This could indicate that predictions made using the estimated model may not be very accurate.

5. How large a value of s_e you might be willing to tolerate depends on what it is you are trying to predict. For example, an error of $300 in predicting an individual's weekly income would be totally unacceptable, but an error of $300 in predicting annual revenues of a department store would be an extremely accurate prediction.

Exercises

Statistical Concepts

16.2.1   *True or false:*   Suppose you use the two explanatory variables X_1 and X_2 in your multiple regression model. Suppose the values of these variables always obey the equation

$$x_{i1} = 10 + 2x_{i2}$$

Then the variables X_1 and X_2 are perfectly collinear. Explain.

16.2.2   The Gauss–Markov theorem states that, under certain conditions, our estimated coefficients have minimum variances. Explain why we want our estimated coefficients to have small variances.

16.2.3   What is meant by the standard error of the regression? What does it estimate?

16.2.4   *True or false:*   The standard error of the regression is another name for SST and measures the total variation in the dependent variable. Explain.

16.2.5   *True or false:*   The residual represents the difference between the fitted value of Y_i and the predicted value of Y_i. Explain.

16.2.6   What does it mean to say that the least squares estimators are best?

16.2.7   *True or false:*   Other things being equal, we would prefer a model with a small value of s_e to a model with a large value of s_e. Explain.

16.2.8   **a.** Suppose you estimated a regression model to predict weekly expenditures on food per person. Would you be pleased if your model yielded an estimated standard error of $s_e = \$300$?

  b. Suppose you estimated a regression model to predict annual federal tax revenues. Would you be pleased if your model yielded an estimated standard error of $s_e = \$3$ million?

  c. Do your answers to parts **a** and **b** indicate that the value of s_e must be examined and interpreted in relation to the values assumed by the dependent variable? Explain.

Statistical Drills

16.2.9   In a multiple regression model, the following coefficient estimates were obtained:

$$b_0 = -10 \qquad b_1 = 4.5 \qquad b_2 = -6.0$$

  a. Write the equation of the estimated multiple regression model.

  b. Suppose we have $n = 25$ observations and SSE = 480. What is the appropriate number of degrees of freedom for this regression model?

  c. Find the estimated standard error of the regression.

16.2.10   In a multiple regression model, the following coefficient estimates were obtained:

$$b_0 = 6.0 \qquad b_1 = 4.0 \qquad b_2 = 5.0$$

  a. Write the equation of the estimated multiple regression model.

  b. Suppose we have $n = 16$ observations. What is the appropriate number of degrees of freedom for this regression model?

  c. Suppose we have $\Sigma y_i^2 = 800$, $\Sigma y_i = 50$, $\Sigma x_{i1} y_i = 60$, $\Sigma x_{i2} y_i = 20$. Find the estimated standard error of the regression.

Statistical Applications

16.2.11   State the assumptions of the classic normal linear regression model. Give an example of how each of these assumptions can be violated.

16.2.12   Explain what is meant by the statement that b_0 and b_1 are unbiased estimators of β_0 and β_1. What assumptions are needed if b_0 and b_1 are to be unbiased estimators of β_0 and β_1?

16.2.13   The following model was fitted to a sample of supermarkets to explain profit levels:

$$Y_i = \beta_0 + \beta_1 x_{i1} + \beta_2 x_{i2} + \beta_3 x_{i3} + e_i$$

where

$$Y_i = \text{Profits in thousands of dollars}$$
$$X_{i1} = \text{Food sales in tens of thousands of dollars}$$
$$X_{i2} = \text{Nonfood sales in tens of thousands of dollars}$$
$$X_{i3} = \text{Store size in thousands of square feet}$$

The estimated regression coefficients were $b_1 = .031$, $b_2 = .089$, and $b_3 = .487$. Interpret these estimates.

16.2.14   The accompanying data show product sales (Y), expenditure on advertising (X_1), and expenditure on salespeople (X_2), all measured in thousands of dollars for a sample of 12 sales districts.

Y	X_1	X_2
132	18	10
148	25	11
112	19	6
160	24	16
100	15	7
178	26	17

(continued)

Y	X_1	X_2
161	25	14
128	16	12
139	17	12
144	23	12
159	22	14
138	15	15

a. Compute the multiple regression of Y on X_1 and X_2.
b. Calculate SSE and s_e. Interpret your results.

16.3 Measuring Goodness of Fit

Just as in the simple linear regression model, we can decompose the total variation in the dependent variable (SST) into two components: (1) the variation in Y that can be explained by the sample regression equation, denoted SSR, and (2) the variation in Y that cannot be explained by the sample regression equation, denoted SSE.

Decomposition of sum of squares

> The total variation SST is decomposed into the explained variation SSR and the unexplained variation SSE,
>
> $$\Sigma(y_i - \bar{y})^2 = \Sigma(\hat{y}_i - \bar{y})^2 + \Sigma \hat{e}_i^2$$
>
> or
>
> $$SST = SSR + SSE$$
>
> where
>
> Total sum of squares: $SST = \Sigma(y_i - \bar{y})^2 = \Sigma y_i^2 - n\bar{y}^2$
> Regression sum of squares: $SSR = \Sigma(\hat{y}_i - \bar{y})^2$
> Residual sum of squares: $SSE = \Sigma \hat{e}_i^2$

As in the simple linear regression model, the *coefficient of multiple determination* is denoted R^2 and is calculated using the following formula:

$$R^2 = \frac{SSR}{SST}$$

$$= 1 - \frac{SSE}{SST}$$

As in simple linear regression, the coefficient of multiple determination R^2 measures the proportion of the variation in the dependent variable that is explained by the independent variables, and it must be that

$$0 \leqslant R^2 \leqslant 1$$

Values of R^2 close to 1 indicate that the independent variables explain most of the variation in Y and that the sample data tend to lie near the estimated regression equation. Values of R^2 close to 0 indicate that a large proportion of the variation in Y is unexplained. If $R^2 = 1$, then SSE = 0 and every observation exactly satisfies the estimated regression equation. In this case, we have a perfect fit.

EXAMPLE 16.3

Calculation of R^2

Calculate R^2 for the data in Table 16.1, referring also to Examples 16.1 and 16.2.

SOLUTION   From the sums in Table 16.1, we obtain

$$\text{SST} = T_{YY} = \Sigma\, y_i^2 - n\bar{y}^2 = 7{,}252 - 10(25.8)^2 = 595.6$$

In Example 16.2, we found that the sum of squared residuals SSE is 20.8958. We obtain

$$R^2 = 1 - \frac{\text{SSE}}{\text{SST}} = 1 - \frac{20.8958}{595.6}$$
$$= .9649$$

This value indicates that our multiple regression equation explains approximately 96% of the variation in gasoline mileage.

Refer again to the SPSSX computer output shown in Figure 16.2. In the ANOVA table, the regression sum of squares (SSR) is given as one of the entries in the column labeled Sum of Squares next to the label Regression and has a value of 574.70418. Also, from the ANOVA table, the residual sum of squares (SSE) is given as 20.89582, and the value of R^2, reported next to the label R Square, is listed as .96492. This agrees with the value we computed by hand.

Adjusted Coefficient of Determination $\bar{R}^2$

Recall that

$$R^2 = 1 - \frac{\text{SSE}}{\text{SST}}$$

where

$$\text{SST} = \sum_{i=1}^{n} (y_i - \bar{y})^2 \quad \text{and} \quad \text{SSE} = \sum_{i=1}^{n} \hat{e}_i^2$$

The SST remains constant when another explanatory variable is added to a model because such an addition has no effect on the sum $\Sigma(y_i - \bar{y})^2$. On the other hand, adding an additional explanatory variable to the model causes SSE to decline provided the estimated coefficient of the new variable is not exactly 0 and thus causes the value of R^2 to increase. To this extent, then, the value of R^2 depends on the number of explanatory variables included in the model. This causes a problem when we try to compare the goodness of fit of two models that have the same dependent variable but different numbers of explanatory variables.

Another measure of goodness of fit takes into account the number of explanatory variables included in an equation. The measure, called the **adjusted R square** and de-noted $\bar{R}^2$, is calculated as shown in the accompanying box.

Formula for the adjusted R square

> The formula for the **adjusted R square** is
>
> $$\bar{R}^2 = 1 - \frac{\text{SSE}/(n - K - 1)}{\text{SST}/(n - 1)}$$
>
> $$= 1 - \frac{n - 1}{n - K - 1}(1 - R^2)$$

The term

$$s_Y^2 = \text{SST}/(n - 1)$$

measures the sample variance of the Y's, which is the total variance to be explained. The term

$$s_e^2 = \text{SSE}/(n - K - 1)$$

is an unbiased estimate of σ_e^2, the variance of the error terms, which is the unexplained variance. By substituting, we obtain

$$\bar{R}^2 = 1 - (s_e^2/s_Y^2)$$

Thus, $\bar{R}^2$ measures the proportion of the variation in the Y's that is explained by the X's. If $\bar{R}^2$ is close to 1, the model explains a large proportion of the total variation in the Y's.

Unlike R^2, $\bar{R}^2$ can decrease when an additional explanatory variable is added to a model. Although SSE will decrease when an extra explanatory variable is included, so will the term $(n - K - 1)$. Thus the term $s_e^2 = \text{SSE}/(n - K - 1)$ can increase or decrease.

Some statisticians object to using the statistic $\bar{R}^2$ as a measure of goodness of fit because it can take negative values whenever the term

$$\frac{n - 1}{n - K - 1}(1 - R^2)$$

exceeds 1. This can occur if R^2 is close to 0 and K is large.

EXAMPLE 16.4　**Calculation of the Adjusted R Square** ———————————————————

Refer to Example 16.3 and Figure 16.2, which shows the SPSSX computer output for Example 16.1. Calculate $\bar{R}^2$.

SOLUTION　We have SST = 595.6, SSE = 20.8958, $n = 10$, and $K = 2$. We obtain

$$\bar{R}^2 = 1 - \frac{20.8958/7}{595.6/9} = .95489$$

In the ANOVA table of Figure 16.2, the value of $\bar{R}^2$ is reported next to the label Adjusted R Square as .95489.

Tips Concerning $\bar{R}^2$

> 1. $R^2 = 1 - \dfrac{\text{SSE}}{\text{SST}}$
>
> 2. R^2 depends on both SST and SSE. R^2 will be close to 1 when SSE is small *relative* to SST.
>
> 3. A large value of SST indicates that there is a large amount of dispersion or variation in the observed values of Y. If SST is very large, it is possible for SSE (and s_e) to be relatively large and still yield a value of R^2 close to 1.
>
> 4. Adding an explanatory variable to a model causes no change in SST, but will cause the value of SSE to decrease provided the estimated coefficient is not exactly equal to zero (a very unlikely occurrence). Thus, including an additional explanatory variable in a model causes R^2 to increase regardless of whether the additional variable truly belongs in the model.
>
> 5. $\bar{R}^2$, the adjusted R^2, takes into account the number of explanatory variables in the model. Inclusion of an additional explanatory variable in a model will cause the adjusted R^2 to decrease if the decrease in the number of degrees of freedom is not offset by a sufficient decrease in SSE.

Exercises

Statistical Concepts

16.3.1   *True or false:* Suppose we have estimated two different regression models explaining different dependent variables. Each model contains a constant term and two explanatory variables. In each model, $n = 25$ and SSE $= 100$. In model 1, SST $= 400$, whereas in model 2, SST $= 500$. These models yield different values for R^2 but identical values for s_e. Explain.

16.3.2   *True or false:* The adjusted R^2 must decrease when an additional explanatory variable is introduced into a model. Explain.

16.3.3   *True or false:* Suppose our regression model provides a perfect fit. Then $R^2 = 1$, but the adjusted R^2 will be less than 1. Explain.

Statistical Drills

16.3.4   Calculate $\bar{R}^2$ under the following conditions:
   a. $R^2 = .60$, $n = 25$, $K = 3$ **b.** $R^2 = .10$, $n = 25$, $K = 3$
   c. $R^2 = .10$, $n = 15$, $K = 5$ **d.** $R^2 = .05$, $n = 15$, $K = 5$

16.3.5   You estimate a multiple regression model and calculate the value of R^2. Then you decide to reestimate the model using an additional explanatory variable that has an estimated coefficient that is nonzero.
   a. Explain why it must be the case that R^2 will be larger for the model with the additional variable.
   b. Discuss what happens to the value of $\bar{R}^2$. Does it necessarily increase? Decrease?

16.3.6   In Exercise 10 of Section 16.1, a model was estimated that explained a student's grade point average as a function of entrance scores on a math test (X) and on an English test (Z).
   a. Calculate R^2 and $\bar{R}^2$ for the estimated multiple regression equation.
   b. Omit the variable Z from the equation and reestimate the model.
   c. Calculate R^2 and $\bar{R}^2$ for the reestimated model, and compare the results with the results in part **a**. Do the changes in R^2 agree with the conclusions from Exercise 5?

    d. Reestimate the model but omit the variable X and keep Z as an explanatory variable.

    e. Calculate R^2 and $\bar{R}^2$ for the reestimated model, and compare the results with the results in part **a.** Do the changes in R^2 agree with the conclusions from Exercise 5?

    f. Compare these results with the results in part **c.** Can you make any prediction about what will happen to R^2 if we simultaneously add a new variable to a model and delete a variable from a model?

16.3.7 In Exercise 13 of Section 16.1, a model was estimated that explained crop yield as a function of rainfall (R) and temperature (T).

    a. Calculate R^2 and $\bar{R}^2$ for the estimated multiple regression equation.

    b. Omit the variable T from the equation and reestimate the model.

    c. Calculate R^2 and $\bar{R}^2$ for the reestimated model, and compare the results with the results in part **a.** Do the changes in R^2 agree with the conclusions from Exercise 5?

    d. Reestimate the model but omit the variable R and keep T as an explanatory variable.

    e. Calculate R^2 and $\bar{R}^2$ for the reestimated model, and compare the results with the results in part **a.** Do the changes in R^2 agree with the conclusions from Exercise 5?

    f. Compare these results with the results in part **c.**

16.3.8 Consider the following information from a multiple regression equation:

$$\Sigma\, \hat{e}_i^2 = 90$$

$$\text{Values of } Y = 11,\ 13,\ 15,\ 18,\ 23,\ 26,\ 28,\ 38,\ 45$$

    a. Find SST, SSE, and SSR.   **b.** Find R^2.   **c.** Find s_e.

16.4   Confidence Intervals and Tests of Hypotheses Concerning the Regression Coefficients

To construct confidence intervals for the true population parameters $\beta_0, \beta_1, \ldots, \beta_K$, it is necessary to estimate the standard deviations of the estimated coefficients $b_0, b_1, \ldots, b_K$ and to make use of the t distribution. These estimated standard deviations are called the estimated standard errors of the estimated coefficients.

DEFINITION   **Estimated Standard Error of b_i** ——————————————————————

The estimated standard deviation of b_i is called the **estimated standard error of b_i** and is denoted by the symbol s_{b_i}.

——

    The estimated standard error of each regression coefficient has the same interpretation as the estimated standard error for a regression coefficient in the simple linear regression model, which was discussed in Section 15.5. The symbol s_{b_i} denotes a specific value of the random variable S_{b_i}.

    The construction of confidence intervals and the testing of hypotheses are based on the fact that if the basic assumptions hold, then the random variable

$$t = \frac{b_i - \beta_i}{S_{b_i}}$$

is distributed as the Student t with $(n - K - 1)$ degrees of freedom.

The formulas for calculating s_{b_i} are quite complicated and will not be given here. Fortunately, computer software packages report the values for s_{b_i} along with the estimated coefficients b_i.

Formula for confidence interval for β_i

A $100(1 - \alpha)\%$ confidence interval for a coefficient β_i takes the form

$$(b_i - t_{\alpha/2,\nu}s_{b_i}, \; b_i + t_{\alpha/2,\nu}s_{b_i})$$

where s_{b_i} is the estimated standard deviation of b_i and $t_{\alpha/2,\nu}$ is the critical value of the t distribution that has $\nu = (n - K - 1)$ degrees of freedom such that

$$P(t > t_{\alpha/2,\nu}) = \alpha/2$$

EXAMPLE 16.5   **Construction of Confidence Intervals for Regression Coefficients** ──────────

In Example 16.1, we obtained the sample regression equation

$$\hat{y}_i = 54.3182 - 4.0129x_{i1} - 15.9806x_{i2}$$

where Y is the gasoline mileage, X_1 is the engine size in hundreds of cubic inches, and X_2 is the weight of the car in tons. Let us construct 95% confidence intervals for each of the regression coefficients.

SOLUTION   We have $n = 10$ observations and $K = 2$ explanatory variables in the model, so the appropriate number of degrees of freedom is $\nu = n - K - 1 = 7$. To construct a 95% confidence interval, we need the critical values from the t distribution having 7 degrees of freedom. We obtain $t_{.025,7} = 2.365$.

From the computer printout for Example 16.1 (reproduced here in Figure 16.3), we obtain the estimated standard deviations

$$s_{b_0} = 2.485 \qquad s_{b_1} = 1.737 \qquad s_{b_2} = 3.565$$

The desired confidence intervals are

β_0: $54.318 \pm 2.365(2.485)$ or $(48.441, 60.195)$

β_1: $-4.013 \pm 2.365(1.737)$ or $(-8.121, .095)$

β_2: $-15.981 \pm 2.365(3.565)$ or $(-24.412, -7.550)$

Recall that the variable X_1 is engine size, so the coefficient b_1 measures the estimated influence of engine size on gasoline mileage when weight remains unchanged. We would expect this influence to be negative—that is, the larger the engine, the lower the gasoline mileage. However, the 95% confidence interval for β_1 contains the value 0. This implies that if we performed a two-sided test using $\alpha = .05$, we would not reject H_0: $\beta_1 = 0$.

The confidence interval for β_2 is fairly wide, which indicates that our sample of data has not provided a very precise estimate of the effect of car weight on gas mileage. Nevertheless, the confidence interval for β_2 does not contain the value 0, so we can feel quite confident that increasing a car's weight does have a negative effect on gas mileage, when engine size is held constant.

```
                                 * * * *   M U L T I P L E   R E G R E S S I O N   * * * *

Listwise Deletion of Missing Data

Equation Number 1    Dependent Variable..    MILEAGE    MILES PER GALLON

Block Number  1.  Method:  Enter      SIZE     WEIGHT

Variable(s) Entered on Step Number  1..    WEIGHT    WEIGHT IN TONS
                                    2..    SIZE      SIZE OF ENGINE IN HUNDREDS OF CUBIC INCH

Multiple R            .98230       Analysis of Variance
R Square              .96492                          DF    Sum of Squares    Mean Square
Adjusted R Square     .95489       Regression          2         574.70418      287.35209
Standard Error       1.72775       Residual            7          20.89582        2.98512

                                   F =      96.26158      Signif F =   .0000

------------------- Variables in the Equation -------------------

Variable          B          SE B        Beta        T    Sig T

WEIGHT     -15.980551     3.565004    -.666491    -4.483    .0029
SIZE        -4.012914     1.736792    -.543538    -2.311    .0541
(Constant)  54.318218     2.484910                21.859    .0000

End Block Number   1    All requested variables entered.
```

FIGURE 16.3 *SPSSX-generated output for Example 16.1*

Hypothesis Testing

The procedure in the accompanying box describes how to test the null hypothesis that the parameter β_i is equal to some specific value β_i^*.

Tests of hypotheses concerning the regression coefficients

Let β_i^* denote some hypothesized value of the regression coefficient β_i. If the basic assumptions hold, then the appropriate test statistic is

$$t = \frac{b_i - \beta_i^*}{s_{b_i}}$$

Case 1:   Suppose we wish to test the null hypothesis

$$H_0:\ \beta_i = \beta_i^* \qquad \text{or} \qquad H_0:\ \beta_i \leq \beta_i^*$$

against the one-sided alternative hypothesis

$$H_1:\ \beta_i > \beta_i^*$$

using a level of significance α. Use the decision rule

Reject H_0 in favor of H_1 if $t > t_{\alpha,\nu}$,

where the number of degrees of freedom of the Student t distribution is $\nu = (n - K - 1)$.

Case 2:   To test the null hypothesis

$$H_0:\ \beta_i = \beta_i^* \qquad \text{or} \qquad H_0:\ \beta_i \geq \beta_i^*$$

(continued)

against the one-sided alternative hypothesis

$$H_1: \quad \beta_i < \beta_i^*$$

use the decision rule

Reject H_0 in favor of H_1 if $t < -t_{\alpha,\nu}$.

Case 3:   To test the null hypothesis

$$H_0: \quad \beta_i = \beta_i^*$$

against the two-sided alternative hypothesis

$$H_1: \quad \beta_i \neq \beta_i^*$$

use the decision rule

Reject H_0 in favor of H_1 if $t < -t_{\alpha/2,\nu}$ or if $t > t_{\alpha/2,\nu}$.

An important special case arises when the hypothesized value of an individual parameter is 0. If $\beta_j = 0$, then a change in the value of the independent variable (X_j) will not affect $E(Y_i)$, the expected value of the dependent variable. For example, in the regression model

$$E(Y_i|x_{i1}, x_{i2}, \ldots, x_{iK}) = \beta_0 + \beta_1 x_{i1} + \beta_2 x_{i2} + \cdots + \beta_K x_{iK}$$

suppose the true value of the parameter β_1 is 0. Then, given that the variables $X_2, X_3, \ldots,$ X_K are also to be used, information on the variable X_1 contributes nothing further toward explaining the behavior of the dependent variable and the mean value of Y_i is the same for all values of X_1.

EXAMPLE 16.6   **Testing Hypotheses About a Regression Coefficient** ————————————————————

In Example 16.1, it is reasonable to conjecture either that, for fixed weight, engine size has no effect on gasoline mileage or that the larger the engine, the lower the mileage. Thus, it is natural to test the null hypothesis

$$H_0: \quad \beta_1 = 0$$

against the one-sided alternative hypothesis

$$H_1: \quad \beta_1 < 0$$

Test this hypothesis using a 5% level of significance.

SOLUTION   We have $\alpha = .05$, $n = 10$, and $K = 2$. The appropriate number of degrees of freedom is $\nu = (10 - 2 - 1) = 7$. The critical value of t having 7 degrees of freedom is $-t_{.05,7} = -1.895$. The hypothesized value is $\beta_1^* = 0$. To test H_0, we calculate the t statistic

$$t = \frac{b_1 - 0}{s_{b_1}} = \frac{-4.0129}{1.737} = -2.31$$

In Figure 16.3, the observed t statistic for the variable SIZE (X_{i1} in Example 16.1), shown in the column labeled T, is $t = -2.311$, which corresponds to the value just calculated.

Because this observed t statistic is less than the critical value, it falls in the rejection region. Thus, if we use $\alpha = .05$, we reject H_0 in favor of H_1. However, if we had used the slightly different level of significance $\alpha = .025$, then the critical value of the test would have been $-t_{.025} = -2.365$. In this case, the observed t statistic would have fallen in the acceptance region and we would not have rejected H_0.

Under the column Sig T and in the row for SIZE is the value .0541. This is the two-tailed p-value associated with $t = -2.311$, indicating that the combined area in the tails of the t distribution to the left of -2.311 and to the right of 2.311 is .0541.

In the model, the variable X_2 measures the weight of the car. We would expect heavy cars to get lower gasoline mileage than light cars, so it is natural to test $H_0: \beta_2 = 0$ against $H_1: \beta_2 < 0$. The appropriate t statistic is

$$ t = \frac{b_2}{s_{b_2}} = \frac{-15.9806}{3.565} = -4.48 $$

This t statistic falls in the critical region if we use a 5% level of significance, so we reject the null hypothesis. The computer printout in Figure 16.3 shows that the two-tailed p-value associated with $t = -4.48$ are .0029, so this t statistic provides extremely strong evidence that the population correlation coefficient β_2 is negative.

In Example 16.6, we showed that if we performed a one-tailed test using a 2.5% level of significance, we would accept the null hypothesis that $\beta_1 = 0$, but if we used a 5% level of significance we would reject the null hypothesis that $\beta_1 = 0$. The question becomes "Should we keep the variable X_1 in the model or remove it and express gasoline mileage solely as a function of car weight?" This is a difficult question to answer, and the decision depends partly on the opinions of the investigator. Remember that statistical evidence is a guide to decision making, but it is not a replacement for common sense and sound judgment. Most people would probably keep X_1 in the model because its p-value is fairly low and because there are strong theoretical reasons for believing that engine size affects gasoline mileage. On the other hand, we should not lose sight of the fact that a major reason for carrying out a statistical analysis is to determine whether the facts support these intuitive feelings. In Section 16.5, when we discuss *multicollinearity,* we shall get more insight into why it is difficult to determine whether $\beta_1 = 0$.

Using the F Test to Test $H_0: \beta_1 = \beta_2 = \cdots = \beta_K = 0$

The t test can be used to test a hypothesis about any single coefficient in the regression model, but testing a hypothesis about β_1 and then testing a hypothesis about β_2 is not the same as performing a *joint test* about β_1 and β_2 simultaneously. To test whether our model explains a significant portion of the variation in Y, we need to perform a joint test of the regression coefficients. In the general regression model, we may want to test the null hypothesis

$$ H_0: \beta_1 = \beta_2 = \cdots = \beta_K = 0 $$

against the alternative hypothesis

H_1: At least one of the coefficients $\beta_1, \beta_2, \ldots, \beta_K$ is nonzero.

Sequentially performing a series of t tests, each at the 5% level of significance, on each of the coefficients $\beta_1, \beta_2, \ldots, \beta_K$ is not the same as jointly testing the null hypothesis that all K coefficients are 0. For example, when we perform two separate t tests on the coefficients β_1 and β_2, each at the 5% level of significance, the joint level of significance will not be .05. If the probability of rejecting a true null hypothesis that $\beta_1 = 0$ is .05 and the probability of rejecting a true null hypothesis that $\beta_2 = 0$ is also .05, the joint probability of rejecting *both* $\beta_1 = 0$ and $\beta_2 = 0$ is less than .05 and depends on the correlation between the sample estimators b_1 and b_2. (If the estimators were independent, the joint probability of rejecting both hypotheses if they were in fact true would be .05 $\times$ .05 = .0025.) The correct method for testing all the coefficients jointly is based on an F statistic, which follows the F distribution. The procedure is described in the accompanying box.

Testing the null hypothesis $H_0: \beta_1 = \beta_2 = \cdots = \beta_K = 0$

Consider the multiple regression model

$$E(Y_i|x_{i1}, x_{i2}, \ldots, x_{iK}) = \beta_0 + \beta_1 x_{i1} + \beta_2 x_{i2} + \cdots + \beta_K x_{iK}$$

Suppose the basic assumptions hold, and we wish to test the joint null hypothesis

$$H_0: \quad \beta_1 = \beta_2 = \cdots = \beta_K = 0$$

against the alternative hypothesis

$$H_1: \quad \beta_1, \beta_2, \ldots, \beta_K \text{ are not all } 0$$

using a level of significance α. To perform the test, calculate the F statistic

$$F = \frac{\text{SSR}/K}{\text{SSE}/(n - K - 1)}$$

The validity of the null hypothesis would be in doubt if the regression sum of squares was large compared with the error sum of squares. Hence, the null hypothesis would be rejected for large values of the test statistic F.

To test H_0 against H_1, use the decision rule

$$\text{Reject } H_0 \text{ in favor of } H_1 \text{ if } F > F_{\alpha, K, (n-K-1)}$$

where $F_{\alpha, K, (n-K-1)}$ is the value such that

$$P(F > F_{\alpha, K, (n-K-1)}) = \alpha$$

and where the test statistic F follows the F distribution with degrees of freedom K and $(n - K - 1)$.

EXAMPLE 16.7   **Testing the Joint Hypothesis $H_0: \beta_1 = \beta_2 = 0$** —————————

From the computer output in Figure 16.3 we see that the estimated equation for Example 16.1 is

$$\hat{y}_i = 54.3182 - 4.0129 x_{i1} - 15.9806 x_{i2}$$

where Y is the gas mileage, X_1 is the engine size, and X_2 is the weight of the car. Use a 5% level of significance and test the joint null hypothesis

$$H_0: \quad \beta_1 = \beta_2 = 0$$

against the alternative hypothesis

$$H_1: \quad \beta_1 \text{ and } \beta_2 \text{ are not both } 0.$$

SOLUTION   From the ANOVA table in Figure 16.3, we see that SSE (the residual sum of squares) is 20.8958. Thus, we obtain

$$SSE/(n - K - 1) = 20.8958/(10 - 3) = 2.985$$

In the ANOVA table, the quantity $SSE/(n - K - 1)$ is called the *residual mean square*. The regression sum of squares is SSR = 574.7042 and we obtain

$$SSR/K = 574.7042/2 = 287.3521$$

In the ANOVA table, the quantity SSR/K is called the *regression mean square*. We calculate the F statistic by finding the ratio of the regression mean square (MSR) and the residual mean square (MSE). We obtain

$$F = MSR/MSE = 287.3521/2.985 = 96.26$$

This F statistic can be obtained directly from the ANOVA table. In the table, we see that the F statistic is

$$F = 96.26158$$

Large values of F indicate that the null hypothesis is false. For $\alpha = .05$, the critical value of F having $K = 2$ and $(n - K - 1) = (10 - 2 - 1) = 7$ degrees of freedom is $F_{.05,2,7} = 4.74$. We reject H_0 in favor of H_1 because the F statistic of 96.26 far exceeds the critical value.

In Figure 16.3, the value Signif F represents the *p*-value associated with $F = 96.26$ and is shown to be .0000. This provides extremely strong evidence that H_0 is false.

Comments on the F Test

Example 16.7 represents a typical regression model using economic or business data. When we build a model to explain some economic or business variable, it is almost always the case that we will reject

$$H_0: \quad \beta_1 = \beta_2 = \cdots = \beta_K = 0$$

in favor of H_1. Rejection of H_0 means that, *as a group*, the explanatory variables do help explain the values of the dependent variable Y, but this result alone does not imply that we have a good model or that the estimated equation will yield good predictions of Y. Nor does it mean that the estimated equation provides a good fit to the data. In general, a good model explains most of the variation in Y and satisfies the statistical assumptions. In practice, most statisticians rely more on R^2, s_e, and the individual t tests than on the overall F test when assessing the quality of some regression model.

The F statistic is related to the value of R^2 according to the following equation:

$$F = \frac{\text{SSR}/K}{\text{SSE}/(n - K - 1)} = \frac{(n - K - 1)R^2}{K(1 - R^2)}$$

A small value of F strongly indicates that our model is inadequate, but, in general, a small value of F will be accompanied by a small value of R^2; consequently, the F statistic does not provide much additional information about the quality of a model that cannot be learned from inspection of R^2 alone.

Tips for Problem Solving

1. When reporting the results from estimating a multiple regression model, be sure to indicate the estimated standard error of each coefficient and/or the appropriate t ratio.

2. Some statisticians report standard errors and some report t ratios, so be sure to label your results to avoid confusion.

3. If you report the estimated coefficients and their standard errors, get in the habit of mentally estimating the ratio b_i/s_{b_i}. If this ratio is small, say, less than 2, there is some question whether the variable belongs in the model as an important explanatory variable, given the other variables in the model.

4. If you report the estimated coefficients and their t ratios, get in the habit of examining whether the t ratio exceeds 2. Large absolute values of the t ratio provide evidence against the null hypothesis $H_0: \beta_i = 0$.

5. Large values of the F statistic

$$F = \frac{\text{SSR}/K}{\text{SSE}/(n - K - 1)}$$

lead us to reject the joint hypothesis $H_0: \beta_1 = \cdots = \beta_K = 0$.

Exercises

Statistical Concepts

16.4.1   *True or false:*   Suppose we estimate a multiple regression model. Suppose the model contains three explanatory variables and a constant term. The appropriate number of degrees of freedom to use when performing the t test is $(n - 3)$. Explain.

16.4.2   *True or false:*   Suppose we wish to test $H_0: \beta_i = 0$ against $H_1: \beta_i \neq 0$ using $\alpha = .05$. If the 95% confidence interval for β_i does not contain 0, then we will reject H_0. Explain.

16.4.3   *True or false:*   Suppose we want to test the null hypothesis $H_0: \beta_1 = \beta_2 = \cdots = \beta_K = 0$. The appropriate test statistic is a t statistic with $(n - K - 1)$ degrees of freedom. Explain.

16.4.4   Explain why it is relevant to test the null hypothesis $H_0: \beta_1 = \beta_2 = \cdots = \beta_K = 0$. If we do not reject H_0, what does this tell us about our regression model?

16.4.5   Explain why we typically do not test the null hypothesis $H_0: \beta_0 = \beta_1 = \beta_2 = \cdots = \beta_K = 0$. Explain why we typically would not want to include the constant term β_0 in the null hypothesis.

16.4.6   *True or false:*   Suppose we perform an F test and reject the null hypothesis $H_0: \beta_1 = \beta_2 = \cdots = \beta_K = 0$. This implies that our regression model provides a good fit to the data. Explain.

16.4.7   *True or false:*   Suppose that $\beta_3 > \beta_5$. When we estimate the regression model, we will obtain $b_3 > b_5$. Explain.

16.4.8   *True or false:*   Suppose that $\beta_3 > \beta_5 > 0$. When we estimate the regression model, it is possible that we could reject H_0: $\beta_5 = 0$ and not reject H_0: $\beta_3 = 0$. Explain.

16.4.9   The F statistic used to test the null hypothesis H_0: $\beta_1 = \beta_2 = \cdots = \beta_K = 0$ is given by

$$F = \frac{\text{SSR}/K}{\text{SSE}/(n - K - 1)}$$

a. Prove that this F statistic is equivalent to

$$F = \left(\frac{n - K - 1}{k}\right)\left(\frac{R^2}{1 - R^2}\right)$$

b. *True or false:*   For a given value of K, F tends to increase as R^2 increases. Explain.

16.4.10   *True or false:*   When performing the F test to test the null hypothesis H_0: $\beta_1 = \beta_2 = \cdots = \beta_K = 0$, suppose we reject H_0. We can conclude that each of the coefficients $\beta_1, \beta_2, \ldots, \beta_K$ is different from 0. Explain.

Statistical Drills

16.4.11   We obtain the following estimated regression model:

$$\hat{y}_i = 4.61 + 5.76x_{i1} - 11.71x_{i2} + 19.23x_{i3}$$
$$(1.21) \quad (2.43) \quad\quad (7.82) \quad\quad (6.14)$$

The values in parentheses are estimated standard deviations. The sample size was $n = 20$.
a. What is the appropriate number of degrees of freedom associated with a t test to test a single coefficient?
b. Let $\alpha = .05$. What is the critical value of the t statistic to test H_0: $\beta_1 = 0$ against H_1: $\beta_1 \neq 0$?
c. Perform the test.
d. Let $\alpha = .05$. Test H_0: $\beta_2 \geq 0$ against H_1: $\beta_2 < 0$.
e. Suppose we have SSR $= 531.02$ and SSE $= 45.76$. Let $\alpha = .05$. Test the null hypothesis H_0: $\beta_1 = \beta_2 = \beta_3 = 0$ against H_1: $\beta_1, \beta_2, \beta_3$ are not all 0. What are the appropriate degrees of freedom of the F statistic? Perform the test.

16.4.12   We obtain the following regression results from a computer printout:

$$\text{SSE} = 52.60 \qquad \text{SSR} = 1,742.12 \qquad n = 25$$

The regression model contains four explanatory variables plus a constant term. Let $\alpha = .05$.
a. Test the null hypothesis H_0: $\beta_1 = \beta_2 = \beta_3 = \beta_4 = 0$ against H_1: $\beta_1, \beta_2, \beta_3, \beta_4$ are not all 0. What are the appropriate degrees of freedom of the F statistic? Perform the test.
b. What is the value of R^2 for this regression model?

Statistical Applications

16.4.13   You have obtained the following sample regression equation:

$$\hat{y}_i = 2.304 + 3.412x_{i1} - 4.657x_{i2}$$
$$(.87) \quad\quad (1.23) \quad\quad (3.78)$$

The values in parentheses are estimated standard deviations. The sample size was $n = 25$.
a. Let $\alpha = .05$ and test H_0: $\beta_1 = 0$ against H_1: $\beta_1 > 0$.
b. In part **a,** what is the appropriate number of degrees of freedom?
c. Let $\alpha = .05$ and test H_0: $\beta_2 = 0$ against H_1: $\beta_2 < 0$.

16.4.14   In Exercise 13, suppose $R^2 = .86$. Let $\alpha = .05$.
 a. Calculate the F statistic used to test the joint null hypothesis $H_0: \beta_1 = \beta_2 = 0$ against the alternative hypothesis that at least one of these two population regression coefficients is nonzero.
 b. What are the appropriate numbers of degrees of freedom for the F statistic?
 c. What is the critical value for the F statistic?
 d. Do you reject the null hypothesis?

16.4.15   Consider the following sample regression results:

$$\hat{y}_i = \begin{array}{cccc} 15.4 & + \ 2.20 x_{i1} & + \ 48.14 x_{i2} & \qquad R^2 = .355 \\ (6.14) & (.42) & (5.21) & \qquad n = 27 \end{array}$$

The numbers in parentheses are the estimated standard errors of the sample regression coefficients.
 a. Construct a 95% confidence interval for β_1.
 b. Would you reject $H_0: \beta_2 = 0$ at the 5% level of significance?
 c. Construct a 99% confidence interval for β_2.
 d. Would you reject $H_0: \beta_2 = 0$ at the 1% level of significance? Explain how this answer could be obtained from the results in part **c**.

16.5   The Multicollinearity Problem

Determining the final form of a multiple regression equation is not an easy task. In most situations, numerous variables are likely candidates for inclusion in any regression model. The **multicollinearity problem** occurs when two or more explanatory variables are highly correlated. If two explanatory variables X_j and X_k are highly correlated, then Y will be explained about equally well by an equation containing only X_j or only X_k as by an equation containing both X_j and X_k. The reason is that most of the information about Y learned by examining X_k can be learned about Y by examining X_j, because X_j and X_k tend to move together. Furthermore, if X_j and X_k tend to move together, it is difficult to separate their effects on Y, and it will not be possible to obtain very reliable estimates of the coefficients β_j and β_k (i.e., estimates having small variances). A high degree of multicollinearity between X_j and X_k is undesirable because the estimates of the regression coefficients β_j and β_k will tend to have large standard errors and the confidence intervals for the true values of β_j and β_k will tend to be very wide; thus we cannot be very sure about the correct values of these coefficients.

The main purpose of multiple regression analysis is to separate and estimate these effects, but the researcher should realize that if both X_j and X_k influence Y and if X_j and X_k tend to move together, then determining how variations in Y are related to variations in X_j or in X_k alone will always be difficult.

If there is an *exact* linear relationship between two or more explanatory variables, then the condition of *exact,* or *perfect,* multicollinearity exists. In this case, the least squares estimation procedure breaks down, and we do not obtain unique coefficient estimates. As an example of perfect multicollinearity, suppose every employee in a certain city pays a 5% wage tax. Let x_{i1} denote wages earned by the ith individual. Let x_{i2} denote taxes paid by the ith individual. The variables X_{i1} and X_{i2} are perfectly correlated. One variable can be perfectly predicted from knowledge of the other. Thus, wages and tax revenue are perfectly collinear. Now suppose we create a regression model in which expenditure on entertainment by individuals is believed to be a function of wage income and tax expenditure. Because the explanatory variables are perfectly collinear, it is impossible to separate the effects of the two variables.

Usually, perfect collinearity does not exist in regression models. In practice, the variables in most economic regression models are not exactly collinear, but frequently the variables tend to be approximately collinear. When two or more of the explanatory variables are approximately collinear, the variances of the estimated regression coefficients tend to be large.

A commonly used rule of thumb is that a correlation coefficient greater than .8 between a pair of explanatory variables indicates that a potentially serious multicollinearity problem may exist, and we may not be able to get very precise coefficient estimates.

When two or more of the explanatory variables are highly correlated with one another, it becomes difficult to determine how each individual variable affects the dependent variable. As a consequence, the individual regression estimates become less precise. Although the estimated coefficients remain unbiased, their standard deviations will be large. Large standard errors indicate that sampling variability is high and that confidence intervals will be wide. The information about the regression coefficients that is provided by the sample data is relatively imprecise.

The t statistic used to test whether a particular regression coefficient is 0 is the ratio of the estimated coefficient divided by that coefficient's standard error. When the standard errors of the estimated coefficients are large, it is common for the t statistics to be small. This leads us to conclude that the estimated regression coefficients are not significantly different from 0. This result occurs even though the regression model yields a high value of R^2 and the overall F statistic is statistically significant.

The heart of the multicollinearity problem is that, although the explanatory variables may help explain the dependent variable, they do not provide sufficient information to allow us to get precise estimates of their individual effects on the dependent variable.

Individual coefficient estimates can be very sensitive to the inclusion or exclusion of other explanatory variables. If X_1 and X_2 are highly collinear, deleting one of the variables from the model will tend to cause a large change in the coefficient of the other variable, although the value of R^2 may hardly change at all. The influence of one of the explanatory variables on the dependent variable has been shifted to the other explanatory variable.

If we have a problem of near perfect, but not perfect, multicollinearity, the least squares estimator is still the best linear unbiased estimate. Sometimes even this *best* estimate may not be very good.

When we try to predict values of Y, the multicollinearity problem may not be so important. Example 16.1 presented a multiple regression model in which gasoline mileage was related to the two explanatory variables Size of engine and Weight of car. Usually, Size of engine and Weight of car tend to be highly positively correlated, making it difficult to determine how each of these variables affects mileage and whether changes in mileage are due to changes in engine size or in car weight. When predicting Mileage (Y), however, it should make little difference whether our model contains just Size of engine, just Weight of car, or both, provided these variables move together. If only Size of engine is included in the model, then the variations in Y caused by the variable Weight of car will be attributed to the variable Size of engine.

This example shows the heart of the multicollinearity problem. If Size of engine and Weight of car are highly correlated, determining how each affects Mileage will be extremely difficult. On the other hand, our predictions of Mileage will be approximately the same if we use as explanatory variables only Size of engine, only Weight of car, or both. A serious problem arises, however, if we want to predict Mileage for cars where Size of engine and Weight of car are not correlated. In this case, we can get very different estimates depending on what we use as explanatory variables.

Multicollinearity in Macroeconomic Models

The multicollinearity problem is especially troublesome when estimating macroeconomic models, because most aggregate economic variables tend to move relatively closely together. For example, some economists argue that changes in Gross National Product (GNP) are caused primarily by changes in the money supply (M). Estimating a regression equation using GNP as the dependent variable and M as the explanatory variable, they obtain high values of R^2, regression coefficients with the correct signs, and large t statistics. Therefore, they argue that the data tend to support their hypothesis. On the other hand, many other economists argue that changes in GNP are caused primarily by changes in government spending (G). With a regression equation using GNP as the dependent variable and G as an explanatory variable, these economists also obtain a high value for R^2, coefficients having the correct signs, and large t statistics. They, too, argue that the data support their hypothesis.

Suppose an arbitrator is asked to decide who is correct. The arbitrator suggests estimating a third equation that includes both G and M as explanatory variables. This equation will also have a high value of R^2, coefficients with the expected signs, and large t statistics. However, the estimated coefficients in the third equation will differ from those in the first two equations, and the standard errors of the estimated coefficients will be much larger than in either of the models containing a single explanatory variable. This indicates that we have a multicollinearity problem. Because the explanatory variables M and G tend to move together over time, separating their effects on GNP is difficult.

The available data are not adequate enough to enable us to determine which model (if any) is correct. If we had additional data from periods when G and M moved in opposite directions, it would be easier to choose the correct model and to determine precisely the effects of G and M on GNP. In most economic and business problems, however, we are not so fortunate to find data that enable us to choose between competing models easily.

Multicollinearity is a problem because explanatory variables in economics and business tend to be correlated. A regression coefficient is usually interpreted as a measure of the change in the mean value of Y when the value of the particular explanatory variable is increased by 1 unit while the values of all other explanatory variables remain unchanged. This idea is nice theoretically, but in practice it may not be possible to change certain economic variables while holding the other variables fixed. Usually we do not obtain data from a controlled experiment where the investigator is free to select the desired values for the explanatory variables.

Multicollinearity is not an error in a regression model. Instead, multicollinearity is a data problem. The purpose of regression analysis is to determine the separate effects of the independent variables $X_1, X_2, \ldots, X_K$ on Y, but when the independent variables tend to move together, it is extremely difficult to get good estimates of these effects.

We do not test for the presence or absence of multicollinearity, since in economic models some degree of multicollinearity is always present. What we want to do is examine the *degree* of multicollinearity so that we can better interpret the various regression coefficients.

The Correlation Matrix

A good way to examine the degree of multicollinearity is to calculate the sample correlation coefficients between each possible pair of explanatory variables X_j and X_k. The sample correlation coefficient between, say, X_3 and X_4 is denoted R_{34}. When some correlation coefficient such as R_{34} is close to $+1$ or -1, we know that X_3 and X_4 tend to move together, and it will be difficult to estimate their separate effects on Y.

EXAMPLE 16.8 **The Multicollinearity Problem**

In Example 16.1, gasoline mileage (Y) was related to the engine size of a car and the weight of a car. This hypothesized relationship seems reasonable, but the estimated coefficient for Size of engine was of questionable statistical significance. One possible explanation for this lack of statistical significance is that the sample correlation between Size of engine (X_1) and Weight of car (X_2) is $R_{12} = .879$. Thus, it is possible that some of the effect of X_1 on Mileage has been shifted to X_2. The high correlation between X_1 and X_2 explains why the estimated coefficient b_1 is barely statistically significant.

Summary of the multicollinearity problem

> **1.** The multicollinearity problem occurs when two or more explanatory variables are highly intercorrelated so that it becomes difficult to obtain precise coefficient estimates.
> **2.** A potential collinearity problem is indicated by a large sample correlation coefficient between any pair of explanatory variables.
> **3.** When a multicollinearity problem exists, estimated coefficients may have very large standard errors even though the value of R^2 may be large.
> **4.** Because of the large standard errors, it may be difficult to reject the null hypothesis that a particular regression coefficient is 0.
> **5.** When a severe multicollinearity problem exists, estimated coefficients may have the wrong signs or implausible magnitudes.

Exercises

Statistical Concepts

16.5.1 Explain what is meant by the multicollinearity problem.

16.5.2 *True or false:* Let X_1 = square feet of space in a house and X_2 = number of rooms in the house. The variables X_1 and X_2 tend to be highly positively correlated, and a regression model that contains both X_1 and X_2 as explanatory variables will have a multicollinearity problem. Explain.

16.5.3 *True or false:* The presence of the multicollinearity problem causes the least squares estimators to be biased. Explain.

16.5.4   *True or false:*   The presence of the multicollinearity problem causes the least squares estimators to be less precise. That is, the variances of the estimates tend to increase, and we are less confident about the true values of the regression coefficients. Explain.

16.5.5   *True or false:*   When two or more explanatory variables are correlated with one another, the value of R^2 will be low. Explain.

16.5.6   *True or false:*   When two or more explanatory variables are correlated with one another, the predictions from our regression model will be very imprecise. Explain.

16.5.7   *True or false:*   One way of identifying a multicollinearity problem is to examine the correlation coefficients between pairs of explanatory variables. Explain.

Statistical Applications

16.5.8   Refer to Exercise 10 in Section 16.1 where a model was estimated that explained a student's grade point average (Y) as a function of the student's entrance exam scores on a math test (X) and an English test (Z).

a. Calculate the sample correlation coefficient between Y and X. What will be the sign of the estimated regression coefficient of X if we estimate a simple linear regression model having Y as a function of X? Verify this result by actually estimating the equation.

b. Calculate the sample correlation coefficient between Y and Z. What will be the sign of the estimated regression coefficient of Z if we estimate a simple linear regression model having Y as a function of Z? Verify this result by actually estimating the equation.

c. Calculate the sample correlation coefficient between Z and X. Can you predict what will happen to the estimated regression coefficients in parts **a** and **b** if we estimate the multiple regression model having Y as a function of both X and Z? Verify this result by actually estimating the equation.

16.5.9   You have estimated a model containing three explanatory variables. You then add a fourth explanatory variable to the model. The value of R^2 increases by only a small amount, and the estimated coefficient of the new explanatory variable is not statistically significant. Could multicollinearity be a problem?

16.5.10   You have estimated a model containing three explanatory variables. You then add a fourth explanatory variable to the model. The value of R^2 increases by only a small amount and the estimated coefficient of the new explanatory variable is statistically significant. The coefficients of the other variables change substantially, and all of their standard deviations increase substantially. Could multicollinearity be a problem?

16.5.11   You have estimated a model containing three explanatory variables. You then add a fourth explanatory variable to the model. The correlation coefficient between the fourth explanatory variable and the third explanatory variable is .98, but the sample correlation is .02 between X_4 and X_2 and .04 between X_4 and X_1. Could multicollinearity be a problem?

16.5.12   You have estimated a model containing three explanatory variables. You then add a fourth explanatory variable to the model. The correlation coefficient between the fourth explanatory variable and the third explanatory variable is .98, and the sample correlation is .92 between X_4 and X_2 and .94 between X_4 and X_1. Could multicollinearity be a problem?

16.5.13   Refer to Exercise 13 in Section 16.1 and Exercise 7 in Section 16.3.

a. Find the sample correlation coefficient between rainfall (R) and temperature (T).

b. Predict what you think would happen to the estimated coefficient of T if the variable R was omitted from the model.

c. Compare your prediction in part **b** with the actual result obtained in Exercise 7 in Section 16.3.

16.5.14 A researcher has obtained the following sample regression equation:

$$\hat{y}_i = 23.45 + 32.74x_{i1} + 62.09x_{i2}$$

The sample correlation between X_1 and X_2 is .95. What will happen to the estimated coefficient of X_1 if the variable X_2 is removed from the model and the equation is reestimated?

16.5.15 An economist wants to estimate the following production function relating to output for a given firm during period t:

$$Q_t = \beta_0 + \beta_1 L_t + \beta_2 K_t + e_t$$

where

$$L_t = \text{Dollars of labor employed in period } t$$
$$K_t = \text{Dollars of capital expenditure in period } t$$

The firm's budget is such that the firm always spends $80,000 per year for capital and labor.
a. Is there a multicollinearity problem?
b. Can the equation be estimated using annual data?

16.6   Computer Applications

It is almost a necessity to use a statistical computer package when performing multiple regression analysis. As the numbers of explanatory variables and observations increase, it becomes extremely time-consuming to calculate the sample regression coefficients, the standard deviations, and other associated test statistics. The next example shows how to use the SPSSX program to estimate a multiple regression model.

EXAMPLE 16.9 **Estimating a Multiple Regression Model Using SPSSX** ───────────

There are substantial differences among the states in the average salary paid to teachers in the public elementary and secondary schools. It is natural to hypothesize that the average salary of elementary and secondary school teachers in state i, Y_i, is influenced by the state and local tax revenue per capita in state i, x_{i1}. Suppose you think that it is possible that the percentage of the labor force that is unionized in state i, x_{i2}, and the income per capita in state i, x_{i3}, also influence Y_i.

Since salaries are paid out of tax revenue, it is reasonable to conjecture that the coefficient of x_{i1} should be positive. In addition, it might be expected that highly unionized states tend to pay higher salaries than other states and that states with high average incomes might tend to pay higher salaries. Thus it might be conjectured that both β_2 and β_3 are positive.

The conjectured model is thus

$$E(Y_i|x_{i1}, x_{i2}, x_{i3}) = \beta_0 + \beta_1 x_{i1} + \beta_2 x_{i2} + \beta_3 x_{i3}$$

The following SPSSX instructions were used to generate the regression output for this regression equation, shown in Figure 16.4 (page 702).

```
TITLE  'SALARY STUDY'
DATA LIST FREE/SALARY REVENUE PCTUNION INCOME
BEGIN DATA

    22934    2729    18.2    10673
    41480    8349    30.4    18187
       .
       .
       .
    27244    5440    15.6    13223

END DATA
REGRESSION VARIABLES = SALARY REVENUE PCTUNION INCOME/
      DEPENDENT = SALARY/
       ENTER REVENUE, PCTUNION, INCOME
FINISH
```

```
                              * * * *   M U L T I P L E   R E G R E S S I O N   * * * *

Listwise Deletion of Missing Data

Equation Number 1    Dependent Variable..   SALARY   AVERAGE TEACHER SALARY IN 1

Block Number  1.  Method:  Enter       REVENUE  PCTUNION INCOME

Variable(s) Entered on Step Number  1..    INCOME    PERSONAL INCOME PER CAPITA IN 1985
                                    2..    PCTUNION  % UNION MEMBERSHIP IN 1982
                                    3..    REVENUE   STATE AND LOCAL REVENUE PER CAPITA IN 19

Multiple R            .89492        Analysis of Variance
R Square              .80088                         DF     Sum of Squares      Mean Square
Adjusted R Square     .78817        Regression        3     699673630.26439   233224543.42146
Standard Error   1923.86399         Residual         47     173958874.36306     3701252.64602

                                    F =      63.01233     Signif F =  .0000

------------------ Variables in the Equation ------------------

Variable            B          SE B         Beta        T  Sig T

INCOME          .445013      .197333      .229615     2.255  .0288
PCTUNION     149.303607    41.585449      .266464     3.590  .0008
REVENUE        2.176366      .398163      .549522     5.466  .0000
(Constant)  7501.498262  1757.305539                  4.269  .0001

End Block Number   1   All requested variables entered.
```

FIGURE 16.4 *SPSSX-generated output for unconstrained model in Example 16.9*

The command

REGRESSION VARIABLES = SALARY REVENUE PCTUNION INCOME/

tells the computer that you want to perform regression analysis and names the variables that you will be working with. The command

DEPENDENT = SALARY/

tells the computer to estimate a regression equation whose dependent variable is SALARY, and the command

---

ENTER REVENUE, PCTUNION, INCOME

---

tells the computer to use the three variables REVENUE, PCTUNION, and INCOME as explanatory variables.

Exercises

16.6.1 Refer to the data in Table 2.3 of Chapter 2.
  a. Estimate the multiple regression equation that explains per capita income as a function of the percentage of employees who are unionized and average hourly earnings in manufacturing.
  b. Find the value of R^2. Does this model seem to do a reasonably good job of explaining per capita income?
  c. Find s_e.
  d. Test the null hypothesis that the coefficient of the variable Hourly earnings is 0. Use a 5% level of significance. What is the p-value of the test?

16.6.2 Refer to the data in Table 2.3 of Chapter 2.
  a. Estimate the multiple regression equation that explains tax revenue per capita as a function of per capita income, percentage of employees who are unionized, and average hourly earnings in manufacturing.
  b. Find the value of R^2. Does this model seem to do a reasonably good job of explaining tax revenues per capita?
  c. Find s_e.
  d. Test the null hypothesis that the coefficient of the variable Hourly earnings is 0. Use a 5% level of significance. What is the p-value of the test?
  e. Construct a 95% confidence interval for the coefficient of the variable Hourly earnings.

16.6.3 Refer to the data in Table 2.3 of Chapter 2.
  a. Estimate the multiple regression equation that explains tax revenue per capita as a function of per capita income and average hourly earnings in manufacturing.
  b. Find the value of R^2. Does this model seem to do a reasonably good job of explaining tax revenues per capita?
  c. Find s_e.
  d. Test the null hypothesis that the coefficient of the variable Hourly earnings is 0. Use a 5% level of significance. What is the p-value of the test?
  e. Compare the values of R^2 here and in Exercise 2. Observe that R^2 is larger for the model that contains an additional explanatory variable.
  f. Compare the estimated regression coefficients in Exercise 2 with the corresponding estimates here. Observe how all the values change when a variable is added to or deleted from the model.
  g. Compare the estimated standard deviations for the regression coefficients in Exercise 2 with the corresponding standard deviations here. Once again observe how all the standard deviations change when a variable is added to or deleted from the model.

STATISTICS IN ACTION: CASE STUDY

Estimating the Value of Coal Used by a Nationwide Steel Company

In 1989, the Internal Revenue Service filed suit against a nationwide steel company (we'll call it The Steel Company to preserve anonymity), claiming that the company was taking too large a deduction for the cost of coal used in producing steel. The dispute arose as follows.

The Steel Company mined its own coal reserves, which it used in the production of steel. According to IRS regulations, the value of the coal could be deducted as a coal depletion allowance before determining company profits and taxes. Because the coal was not purchased on the open market, a dispute arose about the appropriate price to attribute to a ton of coal.

Coal mined from different seams has different values because each seam of coal has unique physical and chemical attributes that make the coal better or worse as a fuel source. For example, the moisture content, ash content, and sulphur content of the coal all affect its price. The IRS gathered data concerning the physical and chemical characteristics of the coal purchased on the open market and estimated a multiple regression model designed to explain the market price of coal as a linear function of the explanatory variables Volatility, Ash content, and Sulphur content.

The IRS substituted the characteristics of the coal used by The Steel Company into its model and arrived at a predicted market price for the company's coal. According to the IRS, The Steel Company had overvalued its coal and had made too large a cost deduction. The IRS claimed that The Steel Company overestimated its coal depletion allowance and, as a result, it underestimated its profits. Thus, the IRS argued that The Steel Company paid too little in corporate income taxes. According to the IRS, The Steel Company owed the federal government more than $25 million in additional income taxes.

The Steel Company wanted to examine the IRS's model to see whether the IRS's argument was valid. It appeared that the IRS's model for predicting the market price of the coal contained major flaws that made their results invalid.

The most important factor in obtaining an adequate model is to start with an adequate theory. Thus, before building a model to predict the price of coal, we must have some knowledge about what factors increase or decrease the value of coal.

Based on this knowledge, it appeared that the equation estimated by the IRS was inadequate because price was expressed as a linear function of only the variables Volatility, Ash content, and Sulphur content. The leading authorities in the coal and steel industries support the hypothesis that the price of coal depends on many more factors in addition to these three. Thus, the IRS model suffers from what is called the error of omitted variables.

In particular, the IRS model contained no information about the following important variables: the moisture content of the coal, the free swelling index, the Giesler plasticity index, the petrographic composition of the coal, the vitrinoid reflectance index of the coal, and the degree of oxidation. (These variables indicate how hot the coal will burn, how it expands or contracts during burning, how it will affect the brick linings of the furnace, and so forth.)

By applying some standard statistical tests, the IRS should have realized that its model suffered from the omitted variables problem and was totally inadequate. The model estimated by the IRS was as follows:

$$\text{Price} = 59.12 - .47 \, (\text{Volatility})$$
$$- .19 \, (\text{Ash content})$$
$$+ 4.45 \, (\text{Sulphur content})$$

In this model, the coefficient of the variable Sulphur is 4.45. This estimated coefficient is *positive*. This means that the IRS model predicts that, other things being equal, the coal becomes more valuable as its sulphur content increases. This result is totally contrary to economic theory and observed facts. Increased sulphur content decreases the value of coal. If the IRS model were theoretically correct, the estimated coefficient of the variable Sulphur should be negative, not positive. This incorrect sign is an obvious indicator of an inadequate model.

The IRS failed to report the value of R^2 for their estimated model. After some investigation, it was learned that the value of R^2 for the IRS model was .059. Thus, the IRS model explained only 5.9% of the observed variation in coal prices. Even if we ignored the omitted variables problem and the incorrect sign of the coefficient of the variable Sulphur content, no expert in the coal business would feel comfortable in predicting potential coal prices based on a model that provided such a poor fit to the data. The value of R^2 is so low that it is clear that additional variables must be included in the model to adequately explain variations in coal prices.

The IRS also failed to report the value for s_e for their estimated model. After some investigation, it was learned that the standard error of estimate was $s_e = \$6.43$. The value of s_e influences the width of a confidence interval for the predicted value of the dependent variable associated with any set of explanatory variables. The 95% confidence interval for actual coal price would be wider than $\hat{y} \pm 2s_e$. We have $2s_e = \$12.86$, so typical prediction errors will be extremely large. A prediction error of $12.86 in the price of a ton of coal would be totally unacceptable.

In summary, the IRS model suffered from the error of omitted variables; it contained a coefficient having a theoretically incorrect sign; it provided a very poor fit based on the low value for R^2; and it provided very poor forecasts based on the large value of s_e.

These arguments were presented to the attorneys at the IRS. In addition, reports of leading chemical engineers and mining experts were included supporting these conclusions. After reviewing these arguments, the IRS decreased The Steel Company's tax bill by over $25,000,000 and the case was dropped.

This case study provides a classic example of some of the typical mistakes that people make when constructing and estimating a multiple regression model. Keep in mind, however, that it usually is much easier to criticize someone else's model (as was done here) than it is to create a model of one's own that passes the scrutiny of a host of potential critics.

Chapter 16 Supplementary Exercises

 16.S.1   As the age of an automobile increases, its resale value decreases. The resale value also decreases as the mileage of the car increases. An auto insurance appraiser has obtained data from the sales of numerous used cars. The appraiser has estimated the following regression equation from data on a certain car model:

$$\hat{y}_i = 5,462.1 - 412.9x_{i1} - 52.3x_{i2}$$

where

$$Y = \text{Resale price of car}$$
$$X_1 = \text{Age of car (in years)}$$
$$X_2 = \text{Mileage of car (in thousands)}$$

In the sample, cars varied from 1 to 6 years in age and from 5,000 to 50,000 in mileage.
a. Predict the selling price of a 3-year-old car driven 30,000 miles.

b. Predict the selling price of a 10-year-old car driven 90,000 miles.

c. Does this problem show the danger of extrapolating far outside the range of the observed data?

16.S.2 The accompanying data show the output of sugar beets in tons, the mean July temperature in degrees Fahrenheit, and the rainfall in July in inches at a certain farm during a 10-year period.

Production of sugar beets Y	Mean July temperature X_1	July rainfall X_2
470	62	3.3
520	64	4.2
560	63	3.2
510	60	3.8
500	64	3.1
550	61	4.0
630	62	4.3
640	64	3.7
650	61	3.0
620	58	4.3

a. Estimate the regression equation $E(Y_i|x_{i1}, x_{i2}) = \beta_0 + \beta_1 x_{i1} + \beta_2 x_{i2}$.

b. Calculate R^2. Is this value large enough to make you think that X_1 and X_2 can be used to make good predictions of Y?

c. Let $\alpha = .05$. Test $H_0: \beta_1 = 0$ and $H_0: \beta_2 = 0$.

16.S.3 In a study of production costs at 62 coal mines, data were obtained on costs per ton (Y). Costs were related to degrees of mechanization (X_1), measure of geological difficulty (X_2), and percentage of absenteeism (X_3). The following estimates were obtained from a computer printout:

Variable	Coefficient	Standard deviation
Constant	8.1	2.0
X_1	3.2	.8
X_2	4.1	.4
X_3	2.2	1.3

a. Test $H_0: \beta_1 = 0$ against $H_1: \beta_1 > 0$. Use $\alpha = .05$.

b. Test $H_0: \beta_3 = 0$ against $H_1: \beta_3 > 0$. Use $\alpha = .05$.

c. Construct a 95% confidence interval for β_3.

d. Construct a 95% confidence interval for β_2.

16.S.4 Consider the following estimated sample regression equation:

$$\hat{y}_i = 12 + 6x_{i1} - 3x_{i2}$$

Determine which of the following statements are true, which are false, and which are indeterminate. Explain your answer.

a. When x_{i2} increases by 1 unit, $\hat{y}_i$ increases by 3 units.

b. For a given x_{i2}, when x_{i1} increases by 1 unit, $\hat{y}_i$ increases by 6 units.

c. Y is more strongly correlated with X_1 than with X_2 because the coefficient of X_1 is larger.

d. It is possible that the coefficient 6 might not be significantly different from 0 and that the coefficient -3 could be significantly different from 0 if we tested each coefficient using $\alpha = .05$.

16.S.5 The executive vice-president of a large corporation has been studying the annual salaries of the corporation's salespeople. A regression equation is estimated having the following variables:

$$Y = \text{Annual salary of employee (in thousands of dollars)}$$
$$X_1 = \text{Employee's number of years of education}$$
$$X_2 = \text{Employee's number of years seniority}$$
$$X_3 = \text{Value of equipment sold by employee (in thousands of dollars)}$$

The estimated equation is

$$\hat{y}_i = 16 + .5x_{i1} + 1.2x_{i2} + .06x_{i3}$$

a. Interpret each of the four regression coefficients.
b. Suppose a salesperson has 16 years of education and 5 years of seniority, and sold $100,000 worth of equipment. Predict this person's salary.
c. Should X_3 be dropped from the model because its coefficient is so small?
d. Is X_2 the most important explanatory variable because its coefficient is the largest?

16.S.6 The sales manager of Norbert's Cracker Company thinks that the company's yearly sales revenues are related to annual advertising expenditures and to the number of sales representatives employed by the company. From data for the last 10 years, the following regression results are obtained for the dependent variable of sales revenues (in thousands of dollars):

Variable	Coefficient	Standard deviation	t
Constant	147.6	10.2	14.5
Advertising (in thousands of dollars)	65.4	4.1	16.0
Number of salespeople	32.3	9.6	3.4

a. Suppose the company has 20 sales representatives and spends $10,000 on advertising. Predict the sales revenue.
b. Construct a 95% confidence interval for $\beta_0, \beta_1,$ and β_2.
c. Let $\alpha = .05$. Test $H_0: \beta_2 = 0$ against $H_1: \beta_2 > 0$.

References

Draper, Norman R., and Harry Smith. *Applied Regression Analysis.* 2d ed. New York: Wiley, 1981.
Goldberger, Arthur. *Econometric Theory.* New York: Wiley, 1964.
Griffiths, William E., R. Carter Hill, and George G. Judge. *Learning and Practicing Econometrics.* New York: Wiley, 1993.
Gujarati, D. *Basic Econometrics.* 2d ed. New York: McGraw-Hill, 1988.
Intriligator, Michael. *Econometric Models, Techniques and Applications.* Englewood Cliffs, N.J.: Prentice-Hall, 1978.
Johnston, John. *Econometric Methods.* New York: McGraw-Hill, 1972.
Judge, G. G., R. C. Hill, W. E. Griffiths, and T. C. Lee. *The Theory and Practice of Econometrics.* 2d ed. New York: Wiley, 1985.
Judge, G. G., R. C. Hill, W. E. Griffiths, H. Lutkepohl, and T. C. Lee. *Introduction to the Theory and Practice of Econometrics.* 2d ed. New York: Wiley, 1988.
Klein, L. *An Introduction to Econometrics.* Englewood Cliffs, N.J.: Prentice-Hall, 1962.
Kmenta, Jan. *Elements of Econometrics.* 2d ed. New York: Macmillan, 1986.
Nie, Norman E., C. Hadlai Hull, Jean G. Jenkins, Karin Steinbrenner, and Dale H. Bent. *SPSS Statistical Package for the Social Sciences.* 2d ed. New York: McGraw-Hill, 1975.

Norusis, Marija J. *SPSSX Introductory Statistics Guide.* New York: McGraw-Hill, 1983.

Norusis, Marija J. *SPSSX Advanced Statistics Guide.* Chicago: SPSS, 1990.

Norusis, Marija J. *The SPSS Guide to Data Analysis.* Chicago: SPSS, 1990.

Pindyck, Robert S., and Daniel L. Rubinfeld. *Econometric Models and Economic Forecasts.* 2d ed. New York: McGraw-Hill, 1981.

Ryan, Thomas A., Brian L. Joiner, and Barbara F. Ryan. *Minitab Reference Manual.* University Park, Penn.: Minitab Project, 1985.

Ryan, Thomas A., Brian L. Joiner, and Barbara F. Ryan. *Minitab Handbook.* 2d ed. Boston: PWS-Kent, 1985.

SAS Introductory Guide. 3d ed. Cary, N.C.: SAS Institute, 1985.

SAS Procedures Guide for Personal Computers. Version 6 ed. Cary, N.C.: SAS Institute, 1986.

SAS Statistics Guide for Personal Computers. Version 6 ed. Cary, N.C.: SAS Institute, 1986.

SAS User's Guide: Basics. Version 5 ed. Cary, N.C.: SAS Institute, 1985.

SAS User's Guide: Statistics. Version 5 ed. Cary, N.C.: SAS Institute, 1985.

Schmidt, P. *Econometrics.* New York: Marcel Dekker, 1976.

SPSSX User's Guide. Chicago: SPSS, 1988.

Theil, Henri. *Principles of Econometrics.* New York: Wiley, 1971.

Wonnacott, Thomas H., and Ronald J. Wonnacott. *Regression: A Second Course in Statistics.* New York: Wiley, 1981.

17 Special Topics in Multiple Regression Analysis

17.1 Models Involving Polynomials

Many different mathematical functions can be used to describe relationships between economic variables. Only a few are used in practice, however, and generally the formulas used are not too complicated.

The linear relationship discussed in Chapter 15 is a special case of a larger class of functions called the *class of polynomial functions*. The general equation for the class of polynomial functions is

$$Y = \beta_0 + \beta_1 X + \beta_2 X^2 + \beta_3 X^3 + \cdots$$

If the highest exponent is 2, we have a *second-degree polynomial*, called a *parabola*. If the highest exponent is 3, we have a *third-degree polynomial*, or *cubic polynomial*. Higher-degree polynomials are seldom used. Figure 17.1 shows three representative polynomial curves.

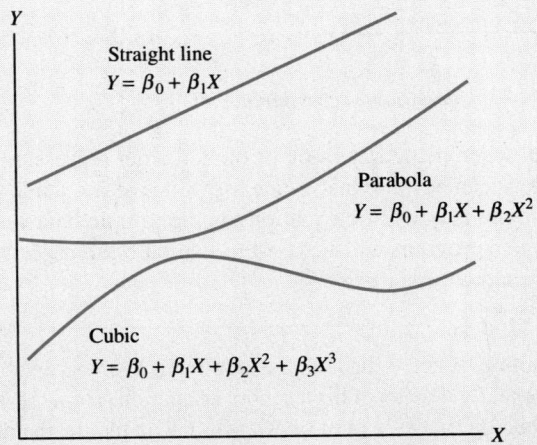

FIGURE 17.1 *Examples of various polynomials*

DEFINITION   **Second-Degree Polynomial Regression Model**

The **second-degree polynomial regression model** (also called the *quadratic regression model*) is based on the equation

$$Y_i = \beta_0 + \beta_1 x_i + \beta_2 x_i^2 + e_i$$

where e_i is a random variable that satisfies all the basic assumptions. For a given value x_i, the mean value of Y_i is given by

$$E(Y_i|x_i) = \beta_0 + \beta_1 x_i + \beta_2 x_i^2$$

Figure 17.2 shows a situation in which the values of Y are scattered about a second-degree polynomial, or quadratic, regression curve. The sample quadratic regression equation is

$$\hat{y}_i = b_0 + b_1 x_i + b_2 x_i^2$$

where the coefficients b_0, b_1, and b_2 are the estimated values of the population parameters β_0, β_1, and β_2, respectively. To obtain the estimates b_0, b_1, and b_2, we proceed exactly as when estimating the regression model with two explanatory variables, where we use X as the first explanatory variable and X^2 as the second.

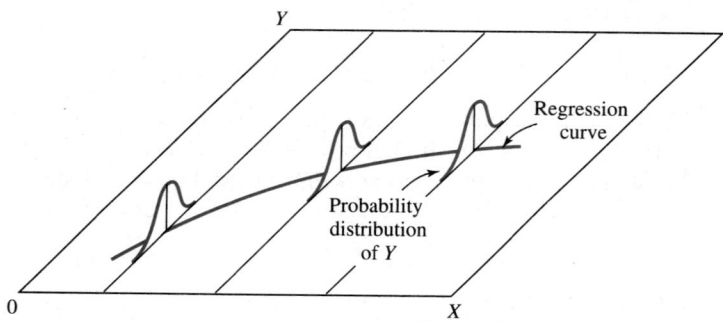

FIGURE 17.2 *A quadratic regression model*

Be careful when using quadratic or higher-order regression models to make predictions. Predictions can be quite inaccurate for values of X outside the range of the observed data, even though the quadratic equation fits the sample data well. As in Chapter 16, the SPSSX computer program will be used to estimate the regression models that are discussed in this chapter.

EXAMPLE 17.1   **A Quadratic Model**

The data in Table 17.1 show the average income in 1985 earned by white males of various ages who had exactly 4 years of high school education (from the 1986 *Current Population Survey* published by the Bureau of the Census). The plot of the data in Figure 17.3 shows that the relationship between income and age is nonlinear. Let us estimate the quadratic regression model

$$E(Y_i|x_i, x_i^2) = \beta_0 + \beta_1 x_i + \beta_2 x_i^2$$

TABLE 17.1 *Data for Example 17.1 on average 1985 income of white male high school graduates*

Group	Age group	Average age X	Income Y
1	18–24	21	$14,108
2	25–29	27	20,523
3	30–34	32	22,471
4	35–39	37	26,017
5	40–44	42	27,314
6	45–49	47	28,093
7	50–54	52	27,288
8	55–59	57	26,706
9	60–64	62	25,951

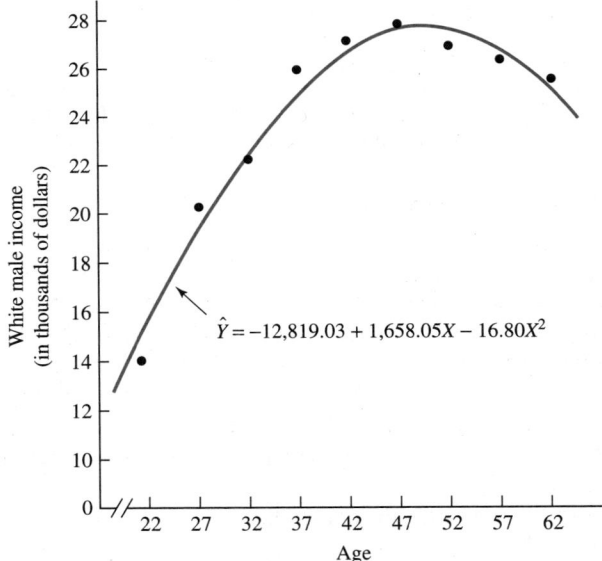

$$\hat{Y} = -12,819.03 + 1,658.05X - 16.80X^2$$

FIGURE 17.3 *Income versus age in 1985 for white male high school graduates (Example 17.1)*

SOLUTION The SPSSX computer output is shown in Figure 17.4 (page 712). The sample regression equation, which is graphed on the scatter diagram in Figure 17.3, is

$$\hat{y}_i = -12,819.03 + 1,658.05x_i - 16.80x_i^2 \qquad R^2 = .98535$$

t statistics: $\quad(-5.65)\qquad\quad(14.30)\qquad(-12.19)$

The values in parentheses are the t statistics used to test the null hypothesis that the corresponding coefficient is 0. Let us test the null hypothesis

$$H_0: \quad \beta_2 = 0$$

```
                                    * * * *   M U L T I P L E   R E G R E S S I O N   * * * *

Listwise Deletion of Missing Data

Equation Number 1      Dependent Variable..   INCOME   INCOME

Block Number  1.  Method:  Enter      AGE       AGESQ

Variable(s) Entered on Step Number   1..    AGESQ
                                     2..    AGE        AGE

Multiple R              .99265        Analysis of Variance
R Square                .98535                             DF      Sum of Squares        Mean Square
Adjusted R Square       .98047        Regression           2       162921784.80527     81460892.40263
Standard Error      635.36570         Residual             6         2422137.41695       403689.56949

                                      F =      201.79093        Signif F =    .0000

------------------ Variables in the Equation ------------------

Variable            B          SE B        Beta        T    Sig T

AGESQ        -16.802749      1.378766   -4.314698   -12.187   .0000
AGE         1658.048471    115.976280    5.061611    14.296   .0000
(Constant)  -12819.02566  2269.258525               -5.649    .0013

End Block Number   1    All requested variables entered.
```

FIGURE 17.4   *SPSSX-generated output for Example 17.1, explaining income as a function of age*

against the one-sided alternative hypothesis

$$H_1: \quad \beta_2 < 0$$

Suppose we use a 5% level of significance. There are $n = 9$ observations and $K = 2$ explanatory variables. The t statistic has $\nu = (n - K - 1) = 6$ degrees of freedom. For $\alpha = .05$, the critical value of t is $-t_{.05,6} = -1.943$. The computer output shows that the t statistic for b_2 is $t = -12.19$. Because this value falls in the critical region, we reject H_0 in favor of H_1. Alternatively, the computer printout shows that the two-tailed p-value for this t statistic is .0000, so the evidence is quite strong that the population coefficient β_2 is negative.

Figure 17.3 shows that the curve fits the data quite well. The computer output reveals that $R^2 = .98535$, indicating that the sample regression equation explains more than 98% of the variation in income. We can be confident that predictions made with this model should be very good as long as the X value is in the range of the observed data. That is, the model should predict income accurately if the age variable is, say, 30, but not if it is 75.

Suppose we want to predict the average income of white males who are age 30. We substitute $x = 30$ into the sample regression equation and obtain the predicted income

$$\hat{y} = -12,819.03 + 1,658.05(30) - 16.80(30)^2$$
$$= \$21,802.47$$

Figure 17.4 shows that the estimated standard error of the regression is $s_e = \$635.36570$, which is relatively small when one considers that we are predicting the average income of white males based on age; that is, an error of \$635 would be considered a relatively small error.

Finally, note that the F statistic used to test the joint hypothesis $H_0: \beta_1 = \beta_2 = 0$ is $F = 201.79093$. Because the p-value for this value of F is .0000, we can certainly reject the null hypothesis.

Exercises

Statistical Concepts

17.1.1 Suppose Y and X are related according to the quadratic equation

$$Y = \beta_0 + \beta_1 X + \beta_2 X^2$$

Explain how the shape of the quadratic curve differs for $\beta_2 > 0$ and for $\beta_2 < 0$.

17.1.2 Suppose Y and X are related according to the quadratic equation

$$Y = \beta_0 + \beta_1 X + \beta_2 X^2$$

Explain how the location of the quadratic curve differs for $\beta_0 > 0$ and for $\beta_0 < 0$.

Statistical Drills

17.1.3 Suppose we obtain the estimated model

$$\hat{y}_i = 10 + 6x_i + 2x_i^2$$

a. Plot this equation for the values $x_i = 0, 1, 2, 3, 4, 5$.
b. Change the coefficient of x_i^2 from 2 to -2. Now plot the equation.
c. How does the sign of the coefficient of x_i^2 affect the shape of the parabola?

17.1.4 Suppose we obtain the estimated regression model

$$\hat{y}_i = b_0 + b_1 x_i + b_2 x_i^2$$

Suppose that the value of x_i increases by 1 unit. By how much will $\hat{y}_i$ change?

Statistical Applications

17.1.5 When a daily newspaper is published, the number of errors made by a typesetter tends to be related to the speed at which the typesetter is working. The following data show the number of errors made in an hour (Y) and the speed of the typesetter (X) in lines set per minute:

Y	9	34	15	50	28	5	66	44	23	17
X	25	40	30	50	40	20	60	45	35	30

a. Plot the data on a scatter diagram. Does a linear or quadratic relationship seem to describe the relationship between Y and X better?
b. Estimate the linear regression model

$$E(Y_i|x_i) = \beta_0 + \beta_1 x_i$$

c. Estimate the quadratic regression model

$$E(Y_i|x_i, x_i^2) = \beta_0 + \beta_1 x_i + \beta_2 x_i^2$$

d. Let $\alpha = .05$. Test $H_0: \beta_2 = 0$ against $H_1: \beta_2 \neq 0$.
e. Plot the estimated equations on the scatter diagram.

17.1.6   An architect claims that the cost of constructing an office building is related to the area of the floor space in the building according to the quadratic regression model

$$E(Y_i|x_i, x_i^2) = \beta_0 + \beta_1 x_i + \beta_2 x_i^2 + e_i$$

where Y is the cost per square foot in dollars and X is the floor space in 100,000s of square feet. The data are as follows:

Y	16	19	22	26	28	29	30	33	36	40
X	2.6	3.4	4.3	4.5	5.0	6.2	6.8	7.2	8.4	9.7

a. Plot the data on a scatter diagram. Does a linear or quadratic relationship seem to describe the relationship between Y and X better?

b. Estimate the linear regression model, and find the adjusted R^2.

c. Estimate the quadratic regression model, and find the adjusted R^2.

d. Plot the estimated equations on a scatter diagram.

e. Let $\alpha = .05$. Test $H_0: \beta_2 = 0$ against $H_1: \beta_2 > 0$.

f. Predict the cost of constructing a building that is 200 feet by 200 feet and has six floors. Use the equation estimated in part **b** and then the estimated equation in part **c**. Which prediction do you think is better?

17.2   Dummy Variables in Regression Models

Some observed phenomena are qualitative rather than quantitative and thus cannot be measured on a continuous scale. The presence or absence of some particular quality may be important in explaining the value of some dependent variable Y, however, and we may want to take this into account when constructing regression models. For example, an individual's income might depend on whether the person possesses a college degree, an individual's expected expenditure on recreation might depend on whether the person is male or female, and the value of a car might depend on whether it has air conditioning.

Until now, in the multiple regression model

$$E(Y_i|x_{i1}, x_{i2}, \ldots, x_{iK}) = \beta_0 + \beta_1 x_{i1} + \beta_2 x_{i2} + \cdots + \beta_K x_{iK}$$

we have implicitly assumed that the X variables were all measured on some continuous scale. We can greatly increase the scope of the regression model by incorporating qualitative variables into it in the form of *dummy variables*.

DEFINITION   **Dummy Variables** ─────────────────────────────────────

Dummy variables are specially constructed variables that indicate the presence or absence of some characteristic. They assume a value of 1 or 0, depending on whether a certain characteristic is present.

───

In this section, we incorporate dummy variables in the multiple regression model and explain how to interpret the estimated coefficients of the dummy variables.

Some Uses of Dummy Variables

Dummy variables may be used to represent and compare factors such as the following:

1. *Temporal effects:* Examples include wartime versus peacetime, holiday season versus non-holiday season, summer versus nonsummer, strike period versus non-strike period, and different quarters of the year.

2. *Spatial effects:* Examples include north versus south, urban versus rural, city A versus city B, developed versus underdeveloped countries, and farm versus nonfarm communities.

3. *Qualitative variables:* Examples include male versus female, college graduate versus noncollege graduate, skilled versus unskilled employee, married versus single, renter versus homeowner, employed versus unemployed, and white versus nonwhite.

4. *Broad groupings of quantitative variables:* Income over $50,000 versus income under $50,000, age over 25 versus age under 25, 3 or more children versus fewer than 3 children, and sales less than $1 million per year versus sales greater than $1 million per year.

Base case and the coefficient of a dummy variable

> The **base case** refers to any observation for which the dummy variable is equal to 0. The coefficient of a dummy variable measures the difference between being in the base case and not being in the base case.

The following example shows a model that contains a dummy variable and explains how to interpret the estimated coefficients.

EXAMPLE 17.2   **Dummy Variables—The Ph.D. Effect** ————————————————

Suppose the salaries of employees at a particular research institute depend on seniority (or number of years employed at the institute), whether the employee has a Ph.D. degree, and other random factors.

Suppose we express the relationship between the variables in terms of the following multiple regression model:

$$E(Y_i|x_i, z_i) = \beta_0 + \beta_1 x_i + \beta_2 z_i$$

where

$$Y = \text{Salary in dollars}$$
$$X = \text{Years of seniority}$$
$$Z = \begin{cases} 1 & \text{if individual has a Ph.D.} \\ 0 & \text{if individual does not have a Ph.D.} \end{cases}$$

We assume that all the basic assumptions of the multiple regression model hold.

Suppose the ith individual has seniority of x_i years and does not have a Ph.D. Thus, the variable Z assumes the value 0. The expected salary would be

$$E(Y_i|x_i, z_i = 0) = \beta_0 + \beta_1 x_i + \beta_2(0)$$
$$= \beta_0 + \beta_1 x_i$$

A person who has a Ph.D. and the same seniority of x_i years would have an expected salary of

$$E(Y_i|x_i, z_i = 1) = \beta_0 + \beta_1 x_i + \beta_2(1)$$
$$= \beta_0 + \beta_1 x_i + \beta_2$$

Thus, for a non-Ph.D. with x_i years of seniority, we obtain

$$E(Y_i|x_i, \text{Non-Ph.D.}) = E(Y_i|x_i, z_i = 0) = \beta_0 + \beta_1 x_i$$

For a Ph.D. with x_i years of seniority, we obtain

$$E(Y_i|x_i, \text{Ph.D.}) = E(Y_i|x_i, z_i = 1) = \beta_0 + \beta_1 x_i + \beta_2$$

Suppose β_0 is \$15,000, β_1 is \$1,000, and β_2 is \$2,500. The expected salary of a non-Ph.D. with x_i years of seniority would be

$$E(Y_i|x_i, z_i = 0) = 15,000 + 1,000 x_i$$

The constant $\beta_0 = 15,000$ represents the starting salary, and the coefficient $\beta_1 = 1,000$ represents the annual salary increment.

The expected salary of a Ph.D. with x_i years of seniority would be

$$E(Y_i|x_i, z_i = 1) = 15,000 + 1,000 x_i + 2,500$$

The coefficient $\beta_2 = 2,500$ represents the effect of having the Ph.D. ($z_i = 1$) as opposed to not having the Ph.D. ($z_i = 0$) and indicates that a person who has a Ph.D. earns, on the average, \$2,500 more per year than a person with the same seniority who does not have a Ph.D. Thus testing the hypothesis that $\beta_2 = 0$ is equivalent to testing the hypothesis that there is no difference between the salaries of Ph.D.'s and the salaries of non-Ph.D.'s.

Figure 17.5 illustrates the relationship between salary, years of seniority, and educational level. The regression model is

$$E(Y_i|x_i, z_i) = 15,000 + 1,000 x_i + 2,500 z_i$$

where

$$Z = \begin{cases} 1 & \text{if individual has a Ph.D.} \\ 0 & \text{otherwise} \end{cases}$$

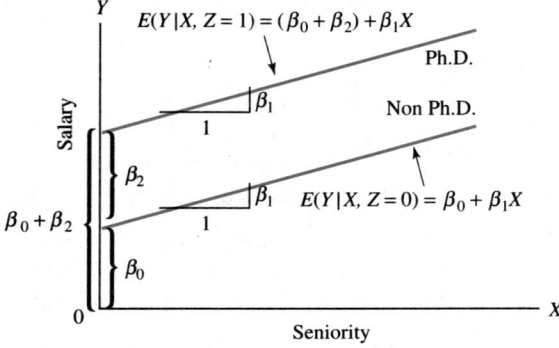

FIGURE 17.5 *Illustrating the Ph.D. effect using a dummy variable*

The expected salary of a non-Ph.D. is determined by

$$E(Y_i|x_i, z_i = 0) = 15,000 + 1,000x_i$$

This is the base case, represented by the lower line on Figure 17.5. The expected salary of a Ph.D. is determined by the equation

$$E(Y_i|x_i, z_i = 1) = 15,000 + 1,000x_i + 2,500$$

This is the upper line on Figure 17.5. Having a Ph.D. shifts the intercept of the line from \$15,000 to \$17,500, although the slope of the line remains 1,000.

We see that adding the dummy variable Z to the model allows the value of the intercept parameter to be different for people with a Ph.D. degree than for people without a Ph.D. degree. In Example 17.2, adding the dummy variable Z to the economic model creates a *parallel shift* in the relationship between income and years of seniority.

A dummy variable like variable Z is incorporated into a regression model to capture a shift in the intercept as the result of some qualitative factor. In Example 17.2, it is reasonable to expect that the coefficient β_2 would be positive, indicating that, at any level of seniority, the person with the Ph.D. degree earns more than the person without the Ph.D. degree.

The properties of the least squares coefficient estimators are not affected by the fact that one of the explanatory variables consists of only 0's and 1's. The coefficient of the dummy variable (β_2 in Example 17.2) is treated like the other coefficients in the model. We can construct a confidence interval based on the estimated coefficient b_2. We also can perform a test of statistical significance just as we did when examining the coefficients of continuous explanatory variables. A test of the null hypothesis $H_0: \beta_2 = 0$ will be a test of whether possessing the Ph.D. degree has a significant effect on the level of income. If $\beta_2 = 0$, then having a Ph.D. degree has no effect on the level of income.

EXAMPLE 17.3 Effect of Air Conditioning on a Car's Resale Price

Suppose we want to estimate the coefficients in a multiple regression model in which the resale value of a certain model car depends on the age of the car and on whether the car has air conditioning. Let Y be the resale price in dollars, and X_1 be the age of the car in years; let X_2 be 1 if air conditioning is present and 0 if it is not.

The data in Table 17.2 (on page 718) refer to a sample of 20 used cars of a certain make and model. Estimate the multiple regression model.

S O L U T I O N   There are 7 cars that have air conditioning and 13 that do not. The relevant computer output is shown in Figure 17.6.

We obtain the estimated regression equation

$$\hat{y}_i = 5,005.22 - 1,005.79x_{i1} + 448.39x_{i2} \qquad R^2 = .996222$$
$$t \text{ statistics:} \quad (120.46) \quad (-62.93) \quad (12.27)$$

The coefficient $b_1 = -1,005.79$ indicates that an increase of 1 year in the car's age tends to be associated with a decrease of \$1,005.79 in the car's value. The coefficient $b_2 = 448.39$ measures the effect of an air conditioner. A car with air conditioning

TABLE 17.2   *Data for Example 17.3 on used cars*

Car	Price Y	Age X_1	Dummy X_2
1	4,000	1	0
2	3,050	2	0
3	4,350	1	1
4	3,900	1	0
5	1,950	3	0
6	3,000	2	0
7	1,400	4	1
8	4,500	1	1
9	2,950	2	0
10	900	4	0
11	3,600	2	1
12	4,100	1	0
13	2,100	3	0
14	1,000	4	0
15	2,400	3	1
16	4,000	1	0
17	4,400	1	1
18	2,900	2	0
19	4,450	1	1
20	2,050	3	0

DEPENDENT VARIABLE: PRICE
SUM OF SQUARED RESIDUALS = 100,598
STANDARD ERROR OF THE REGRESSION = 76.9254
R-SQUARED = .996222
ADJUSTED R-SQUARED = .995777
F STATISTIC (2, 17) = 2241.18

VARIABLE	ESTIMATED COEFFICIENT	STANDARD DEVIATION	t STATISTIC
CONSTANT	$b_0 = 5005.22$	41.5501	120.462
AGE	$b_1 = -1005.79$	15.9829	-62.929
DUMMY	$b_2 = 448.387$	36.5542	12.266

FIGURE 17.6   *Computer output for predicting used car prices in Example 17.3*

$(x_{i2} = 1)$ is worth \$448.39 more than a car of the same age that does not have air conditioning $(x_{i2} = 0)$.

The price of a car without air conditioning would be predicted using the equation

$$\hat{y}_i = 5,005.22 - 1,005.79x_{i1} + 448.39(0)$$
$$= 5,005.22 - 1,005.79x_{i1}$$

and the price of a car with air conditioning would be predicted using the equation

$$\hat{y}_i = 5,005.22 - 1,005.79x_{i1} + 448.39(1)$$
$$= 5,453.61 - 1,005.79x_{i1}$$

The t statistics are all significant at the 5% level of significance, which supports the hypotheses that β_0, β_1, and β_2 are all nonzero. The value $R^2 = .996222$ indicates that the regression model fits the data very well. Since the estimated standard error of the regression is $s_e = 76.9254$, the Empirical Rule indicates that about 95% of the observations should lie within $2s_e = \$153.85$ of the regression equation. The large value of the F statistic indicates that we would reject the null hypothesis $H_0: \beta_1 = \beta_2 = 0$.

The data and the two regression equations are plotted in Figure 17.7. The lower line refers to cars without air conditioning, and the upper line to cars with air conditioning. The vertical distance between the two lines is 448.397.

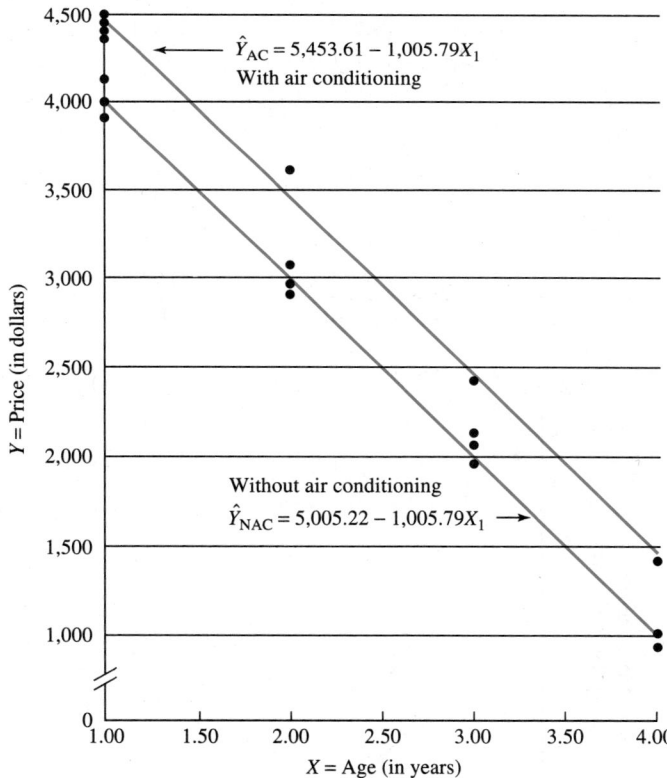

FIGURE 17.7   *Resale prices of used cars versus age of the car for Example 17.3*

EXAMPLE 17.4   **Effect of Gender on Income of an Employee**

The data in Table 17.3 on page 720 (from the *Current Population Survey, 1986,* by the Bureau of the Census) show the average annual incomes earned in 1985 by white males and white females of various ages who had exactly 4 years of high school education. The data are plotted in Figure 17.8. In Example 17.1, we used the data for the males to estimate a quadratic equation that explained income as a function of age and (age)^2. Now let us estimate a model that explains the income of both males and females.

TABLE 17.3   *Data on income, age, and gender for Example 17.4*

Group	Age group	Average age	S^a	Income
1	18–24	21	1	$14,108
2	25–29	27	1	20,523
3	30–34	32	1	22,471
4	35–39	37	1	26,017
5	40–44	42	1	27,314
6	45–49	47	1	28,093
7	50–54	52	1	27,288
8	55–59	57	1	26,706
9	60–64	62	1	25,951
10	18–24	21	0	11,154
11	25–29	27	0	14,517
12	30–34	32	0	16,215
13	35–39	37	0	16,199
14	40–44	42	0	15,777
15	45–49	47	0	15,906
16	50–54	52	0	16,217
17	55–59	57	0	15,755
18	60–64	62	0	15,240

[a] S is 1 if individuals are male and 0 if individuals are female.

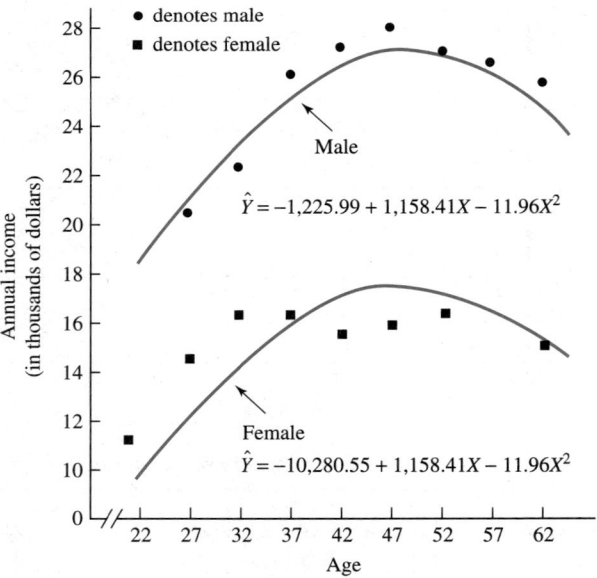

FIGURE 17.8   *Income versus age for males and females for Example 17.4*

From the scatter diagram, income appears to depend on age and on gender. In addition, it appears that the relationship between income and age is nonlinear. Let us estimate the multiple regression model

$$E(Y_i|x_i, x_i^2, s_i) = \beta_0 + \beta_1 x_i + \beta_2 x_i^2 + \beta_3 s_i + e_i$$

where

$$Y = \text{Income}$$
$$X = \text{Age}$$
$$S = \begin{cases} 1 & \text{if individual is male} \\ 0 & \text{if individual is female} \end{cases}$$

SOLUTION   The SPSSX computer output is shown in Figure 17.9. The estimated regression equation is

$$\hat{y}_i = -10,280.55 + 1,158.41x_i - 11.96x_i^2 + 9,054.56s_i \qquad R^2 = .92048$$
$$t \text{ statistics: } (-2.28) \qquad (5.05) \qquad (-4.39) \qquad (10.82)$$

The value $R^2 = .92048$ indicates that the model explains more than 92% of the variation in average income. The base case is $S = 0$, where the individual is a white female. Thus the population regression coefficient β_3 indicates the difference in income between a white male of a certain age and a white female of the same age. The estimated coefficient of variable S is $b_3 = 9,054.56$, which indicates that, on the average, a white male earns approximately $9,054.56 more than a white female of the same age.

```
* * * *   M U L T I P L E   R E G R E S S I O N   * * * *

Listwise Deletion of Missing Data

Equation Number 1     Dependent Variable..   INCOME    INCOME

Block Number  1.  Method:  Enter      AGE       AGESQ     SEX


Variable(s) Entered on Step Number  1..     SEX       GENDER
                                    2..     AGESQ
                                    3..     AGE       AGE


Multiple R           .95942     Analysis of Variance
R Square             .92048                      DF     Sum of Squares      Mean Square
Adjusted R Square    .90344     Regression        3     511156373.90450   170385457.96817
Standard Error   1775.95953     Residual         14      44156451.70661     3154032.26476

                                F =     54.02147     Signif F =   .0000


------------------ Variables in the Equation ------------------

Variable            B          SE B         Beta         T    Sig T

SEX          9054.555556   837.195353     .815088    10.815   .0000
AGESQ         -11.955059     2.725116   -2.368984    -4.387   .0006
AGE          1158.413808   229.225790    2.728950     5.054   .0002
(Constant) -10280.55246   4504.654889                -2.282   .0386


End Block Number   1   All requested variables entered.
```

FIGURE 17.9 *SPSSX-generated output for Example 17.4, explaining income as a function of age and gender*

Figure 17.8 shows quite clearly that white males earn more than white females of the same age. Suppose we wanted to put this statement to a formal test. To test the null hypothesis that white males and white females of the same age earn the same annual income, let us test the null hypothesis

$$H_0: \quad \beta_3 = 0$$

against the one-sided alternative hypothesis

$$H_1: \quad \beta_3 > 0$$

Let us use a 1% level of significance. To perform the test, the appropriate number of degrees of freedom is $(n - K - 1) = (18 - 3 - 1) = 14$. The critical value of the test statistic is $t_{.01,14} = 2.624$. Because the SPSSX output shows that the t statistic for b_3 is $t = 10.815$, we reject the null hypothesis. Alternatively, the computer output shows that the p-value associated with $t = 10.815$ is .0000.

To predict the average income for white males of a given age, we substitute $s_i = 1$ into the sample regression equation. We obtain

$$\hat{y}_i = -10{,}280.55 + 1{,}158.41x_i - 11.96x_i^2 + 9{,}054.56(1)$$

or

$$\hat{y}_i = -1{,}225.99 + 1{,}158.41x_i - 11.96x_i^2$$

To predict the average income for white females of a given age, we substitute $s_i = 0$ into the sample regression equation and obtain

$$\hat{y}_i = -10{,}280.55 + 1{,}158.41x_i - 11.96x_i^2$$

These two curves are plotted in Figure 17.8.

For females, the intercept is $-10{,}280.55$. For males, the intercept is $-1{,}225.99 = (-10{,}280.55 + 9{,}054.56)$. The difference between these two intercepts of 9,054.56 estimates that a white male of a given age earns, on the average, \$9,054.56 more than a white female of the same age.

Could it be argued that the data in Example 17.4 provide evidence of sex discrimination in the United States labor market, or are there other reasons that might explain why men get paid more than women? Recall that in Chapter 15 we showed that if we omit important explanatory variables from a model, then it is possible to obtain misleading results. The data in Example 17.4 might indicate that there is sex discrimination in the job market, but other factors should also be considered.

As an extreme example, suppose that there are only two types of jobs for male and female high school graduates, office clerk and truck driver. Suppose that truck drivers earn about \$10,000 per year more than clerks. Furthermore, suppose that 90% of the males are truck drivers and 10% are clerks, whereas 10% of the females are truck drivers and 90% are clerks. Finally, assume that male and female truck drivers are paid identical salaries and that male and female clerks are paid identical salaries. A regression model that ignores the individual's occupation would indicate that the average male earns much more than the average female, but this result would be misleading because males and females who do identical work are paid identical salaries.

Regression Models with More Than One Dummy Variable

Frequently it is necessary to include more than one dummy variable in a regression model. In the following example, we attempt to explain the level of income based on the age, gender, and race of the individual. Because the variables Gender and Race are both qualitative variables, we must construct two dummy variables in our model.

EXAMPLE 17.5   **A Regression Model with Two Dummy Variables** ──────────────

In Example 17.4, we estimated a regression model to explain the average incomes of white males and females who have 4 years of high school education. The *Current Population Survey* also reports data concerning the incomes earned by black high school graduates. Let us introduce a second dummy variable R into the model to take into account the race of the individual. The population regression model becomes

$$E(Y_i|x_i,\ x_i^2,\ s_i,\ r_i) = \beta_0 + \beta_1 x_i + \beta_2 x_i^2 + \beta_3 s_i + \beta_4 r_i$$

where

$$Y = \text{Income}$$
$$X = \text{Age}$$
$$S = \begin{cases} 1 & \text{if individual is male} \\ 0 & \text{if individual is female} \end{cases}$$
$$R = \begin{cases} 1 & \text{if individual is white} \\ 0 & \text{if individual is black} \end{cases}$$

According to our definitions of the dummy variables S and R, the base case is $S = 0$ and $R = 0$, which refers to a black female. The data are in Table 17.4 (page 724).

SOLUTION   The computer output is shown in Figure 17.10 (page 725). The estimated regression equation is

$$\hat{y}_i = -9,465.50 + 1,010.39x_i - 10.21x_i^2 + 7,159.50s_i + 2,965.94r_i$$
$$t \text{ statistics:}\quad (-2.81)\qquad (5.92)\qquad (-5.03)\qquad (11.48)\qquad (4.76)$$

The value $R^2 = .87886$ indicates that this expanded model fits the data fairly well.

The coefficient of interest in this problem is the coefficient of the dummy variable R. The computer output shows that $b_4 = 2,965.94$, which indicates that, in our sample data, whites of a given gender and age earn $2,965.94 more than blacks of the same age and gender.

To test the null hypothesis that whites of a certain age and gender earn the same annual income as blacks of the same age and gender, let us test the null hypothesis

$$H_0:\ \beta_4 = 0$$

against the one-sided alternative hypothesis

$$H_1:\ \beta_4 > 0$$

using a 1% level of significance. The appropriate number of degrees of freedom is $(n - K - 1) = (36 - 4 - 1) = 31$, and so the critical value of the test statistic is $t_{.01,31}$

= 2.457. The computer output shows that the t statistic for b_4 is $t = 4.756$. We reject the null hypothesis because the t statistic falls in the critical region. In addition, the computer output shows that the p-value for $t = 4.756$ is .0000. We thus reject the null hypothesis and conclude that the variable R does help explain income.

TABLE 17.4 *Data on income, age, gender, and race for Example 15.4*

Group	Age group	Average age	S^a	R^b	Income
1	18–24	21	1	1	$14,108
2	25–29	27	1	1	20,523
3	30–34	32	1	1	22,471
4	35–39	37	1	1	26,017
5	40–44	42	1	1	27,314
6	45–49	47	1	1	28,093
7	50–54	52	1	1	27,288
8	55–59	57	1	1	26,706
9	60–64	62	1	1	25,951
10	18–24	21	0	1	11,154
11	25–29	27	0	1	14,517
12	30–34	32	0	1	16,215
13	35–39	37	0	1	16,199
14	40–44	42	0	1	15,777
15	45–49	47	0	1	15,906
16	50–54	52	0	1	16,217
17	55–59	57	0	1	15,755
18	60–64	62	0	1	15,240
19	18–24	21	1	0	11,698
20	25–29	27	1	0	16,809
21	30–34	32	1	0	17,420
22	35–39	37	1	0	19,657
23	40–44	42	1	0	21,393
24	45–49	47	1	0	21,772
25	50–54	52	1	0	20,423
26	55–59	57	1	0	22,775
27	60–64	62	1	0	22,775
28	18–24	21	0	0	9,573
29	25–29	27	0	0	12,696
30	30–34	32	0	0	14,180
31	35–39	37	0	0	16,548
32	40–44	42	0	0	15,763
33	45–49	47	0	0	14,471
34	50–54	52	0	0	15,379
35	55–59	57	0	0	14,366
36	60–64	62	0	0	14,366

[a] S is 1 if individuals are male and 0 if individuals are female.
[b] R is 1 if individuals are white and 0 if individuals are black.

```
                              * * * *   M U L T I P L E   R E G R E S S I O N   * * * *

Listwise Deletion of Missing Data

Equation Number 1     Dependent Variable..   INCOME     INCOME

Block Number  1.  Method:  Enter      AGE       AGESQ      SEX       RACE


Variable(s) Entered on Step Number  1..     RACE      RACE
                                    2..     AGESQ
                                    3..     SEX       GENDER
                                    4..     AGE       AGE


    Multiple R          .93748    Analysis of Variance
    R Square            .87886                       DF    Sum of Squares     Mean Square
    Adjusted R Square   .86323    Regression          4    787257142.50738  196814285.62684
    Standard Error  1870.91324    Residual           31    108509807.13151    3500316.35908

                                  F =      56.22757    Signif F =   .0000


------------------ Variables in the Equation ------------------

Variable          B          SE B        Beta        T     Sig T

RACE       2965.944444    623.637747    .297294     4.756   .0000
AGESQ       -10.206705      2.029974  -2.252072    -5.028   .0000
SEX        7159.500000    623.637747    .717640    11.480   .0000
AGE        1010.388895    170.753284   2.650371     5.917   .0000
(Constant) -9465.504363  3370.033065               -2.809   .0085


End Block Number   1   All requested variables entered.
```

F I G U R E   1 7 . 1 0   *SPSSX-generated output for Example 17.5, explaining income*
                            as a function of age, gender, and race

Qualitative Variables with More Than Two Classes

In Examples 17.2–17.5, the dummy variables classified an observation into one of two categories. Many times a qualitative characteristic has more than two possible classifications. For example, we may construct a model to explain the incomes earned by individuals who have different qualitative characteristics. We may want to take into account each person's current marital status (single never married, married, single divorced, single widowed), level of educational attainment (no high school diploma, high school diploma, college degree, graduate degree), region of residence (North, South, Midwest, Southwest, Far West), occupation (plumber, electrician, carpenter, accountant, teacher), and type of residence (apartment, townhouse, trailer, house). The qualitative variables Marital status, Level of educational attainment, Region of residence, Occupation, and Type of residence are inherently different from quantitative variables such as Income, Age, and Years of seniority. Quantitative variables take obvious numerical values. There is no obvious numerical value that the qualitative variables take and the effect of the qualitative variables on the dependent variable is measured by introducing a series of dummy variables.

For example, suppose we have a theory that an individual's expenditure on recreation E_i depends on the individual's wage income W_i, the individual's gender, and the individual's type of residence. Furthermore, we assume that the factors gender and type of residence affect only the intercept of the equation. To take into account the gender of the individual, we define the dummy variable S as follows:

$$S_i = \begin{cases} 1 & \text{if } i\text{th individual is female} \\ 0 & \text{if } i\text{th individual is male} \end{cases}$$

The variable Gender is a qualitative variable that takes only two possible values. Thus, only one dummy variable is needed to measure the presence or absence of a particular characteristic.

Suppose that type of residence is a qualitative variable that takes one of four possible values: apartment, townhouse, trailer, or house. To take into account the type of residence of the individual, we introduce a series of three dummy variables R_1, R_2, and R_3 as follows:

$$R_{i1} = \begin{cases} 1 & \text{if } i\text{th individual lives in an apartment} \\ 0 & \text{otherwise} \end{cases}$$

$$R_{i2} = \begin{cases} 1 & \text{if } i\text{th individual lives in a townhouse} \\ 0 & \text{otherwise} \end{cases}$$

$$R_{i3} = \begin{cases} 1 & \text{if } i\text{th individual lives in a trailer} \\ 0 & \text{otherwise} \end{cases}$$

Type of residence takes one of four possible values, so three dummy variables are needed to measure the effect of the presence or absence of a characteristic. It is not necessary to define a dummy variable for every possible state of a qualitative variable. To measure the effect of how expenditures of males differ from expenditures of females, we need one, not two dummy variables. To measure the effect of how type of residence affects expenditures, we need three, not four dummy variables. The state of the qualitative variable that is not represented by a dummy variable is called the *base case*. The coefficient of any dummy variable measures the difference between being in the base case and being in any other specific state or category. Thus, in a regression model that includes the three types of residence dummies, the base case (the state that does not have an explicit dummy variable) is the case when an individual lives in a house. The coefficient of the dummy variable R_{i3} measures the difference between the intercept for a person who lives in a house and a person who lives in a trailer.

This point can be better illustrated by examining the basic regression model. The assumption that the intercept varies according to the individual characteristics can be explained in terms of the following model:

$$E_i = \beta_{i0} + \beta_1 W_i + e_i \qquad i = 1, 2, \ldots, n$$

The subscript i on the constant term (or intercept term) is used to indicate that the intercept may be different for different individuals. If we assume that the intercept changes according to the gender and type of residence of the individual, the intercept term can be expressed as

$$\beta_{i0} = \beta_0 + \gamma S_i + \delta_1 R_{i1} + \delta_2 R_{i2} + \delta_3 R_{i3}$$

where the coefficients γ, δ_1, δ_2, and δ_3 are parameters that measure the effects of the qualitative characteristics on the intercept in the basic regression model. If we substitute this equation into the basic model, we obtain

$$E_i = \beta_0 + \gamma S_i + \delta_1 R_{i1} + \delta_2 R_{i2} + \delta_3 R_{i3} + \beta_1 W_i + e_i \qquad i = 1, 2, \ldots, n$$

The effect of each qualitative state on the dependent variable can be observed by enumerating each particular state. For simplicity, we omit the random error term. We obtain

$$E_i = \beta_0 + \beta_1 W_i \qquad\qquad \text{for a male who lives in a house}$$
$$E_i = \beta_0 + \gamma + \beta_1 W_i \qquad\quad \text{for a female who lives in a house}$$
$$E_i = \beta_0 + \delta_1 + \beta_1 W_i \qquad\quad \text{for a male who lives in an apartment}$$
$$E_i = \beta_0 + \gamma + \delta_1 + \beta_1 W_i \qquad \text{for a female who lives in an apartment}$$
$$E_i = \beta_0 + \delta_2 + \beta_1 W_i \qquad\quad \text{for a male who lives in a townhouse}$$
$$E_i = \beta_0 + \gamma + \delta_2 + \beta_1 W_i \qquad \text{for a female who lives in a townhouse}$$
$$E_i = \beta_0 + \delta_3 + \beta_1 W_i \qquad\quad \text{for a male who lives in a trailer}$$
$$E_i = \beta_0 + \gamma + \delta_3 + \beta_1 W_i \qquad \text{for a female who lives in a trailer}$$

Observe that the coefficient γ is absent in each relationship involving a male. This is because, when the individual is a male, $S_i = 0$. Observe that the coefficient δ_1 appears only when a person lives in an apartment. This is because, except when the individual lives in an apartment, $R_{i1} = 0$. Similarly, the coefficient δ_2 appears only when a person lives in a townhouse, and the coefficient δ_3 appears only when the individual lives in a trailer.

In this model, the base case is represented by males who live in a house. The regression model completely enumerates the eight possible categories of gender and type of residence, with each case having a different intercept. Only one dummy variable for gender and three for type of residence are required for this complete enumeration of the eight possibilities.

By introducing the one dummy variable for gender and three dummy variables for type of residence, we can examine such questions as whether there is a significant difference between expenditures of men and women, and whether the type of residence affects a person's expenditures. If the statistical model satisfies the basic assumptions, the coefficients in the model can be estimated by least squares, and the various coefficients can be tested for statistical significance. For example, we can test whether there is a significant difference in the expenditures of men and women by testing the null hypothesis $H_0: \gamma = 0$ against the alternative hypothesis $H_1: \gamma \neq 0$ based on a t test. To test whether a person living in an apartment has different recreation expenditures than a person living in a house, we test the null hypothesis $H_0: \delta_1 = 0$ against the alternative hypothesis $H_1: \delta_1 \neq 0$ based on a t test. To test whether a person living in a townhouse has different recreation expenditures than a person living in a house, we test the null hypothesis $H_0: \delta_2 = 0$ against the alternative hypothesis $H_1: \delta_2 \neq 0$ based on a t test, and so forth. Finally, we can perform a joint test that any set of coefficients are all simultaneously 0 by performing the F test. For example, we might want to test the null hypothesis $H_0: \delta_1 = \delta_2 = \delta_3 = 0$ against the alternative hypothesis $H_1: \delta_1, \delta_2, \delta_3$ are not all 0. Rejection of H_0 indicates that at least one of the types of residence matters when explaining expenditures.

When using quarterly data, it is customary to include dummy variables in the multiple regression model to represent the different quarters or seasons of the year. Because there are $m = 4$ quarters, we need to create $(m - 1) = 3$ dummy variables. Suppose we select the fourth quarter to be the base case. We could define the three dummy variables as follows:

$$Q_1 = \begin{cases} 1 & \text{if observation refers to the first quarter} \\ 0 & \text{otherwise} \end{cases}$$

$$Q_2 = \begin{cases} 1 & \text{if observation refers to the second quarter} \\ 0 & \text{otherwise} \end{cases}$$

$$Q_3 = \begin{cases} 1 & \text{if observation refers to the third quarter} \\ 0 & \text{otherwise} \end{cases}$$

For an observation from the fourth quarter, we have $q_{i1} = 0$, $q_{i2} = 0$, and $q_{i3} = 0$. For an observation from the first quarter, we have $q_{i1} = 1$, $q_{i2} = 0$, $q_{i3} = 0$. For an observation from the second quarter, the values of the dummy variables are $q_{i1} = 0$, $q_{i2} = 1$, $q_{i3} = 0$. If the observation is from the third quarter, the values are $q_{i1} = 0$, $q_{i2} = 0$, and $q_{i3} = 1$. The coefficient of variable Q_1 measures the difference in the dependent variable as a result of its being in quarter 1 rather than in quarter 4, the base case, and similarly for Q_2 and Q_3.

A dummy variable takes on only the values 0 and 1. We *do not* create a dummy variable that takes on the values, say, 0, 1, and 2, or 1, 2, and 3. The two values 0 and 1 work like a light switch that is either "on" or "off." A value of 0 means that a certain condition is false, and a value of 1 means that the condition is true.

Number of dummy variables

> If a qualitative variable falls into one of m classes, then $(m - 1)$ dummy variables are required to represent the variable in the multiple regression model.

Suppose that every observation falls into exactly one of $m = 3$ categories. In the regression model, we must create $(m - 1) = 2$ dummy variables, which we will call D_1 and D_2. Let's label the three categories 0, 1, and 2, and call category 0 the base category. If an observation falls in category 0, then set $d_{i1} = 0$ and $d_{i2} = 0$. If an observation falls in category 1, we set $d_{i1} = 1$ and $d_{i2} = 0$. If an observation falls in category 2, we set $d_{i1} = 0$ and $d_{i2} = 1$. If we use this framework, then each observation is classified as follows:

$$d_{i1} = \begin{cases} 1 & \text{if and only if the } i\text{th observation falls in category 1} \\ 0 & \text{otherwise} \end{cases}$$

$$d_{i2} = \begin{cases} 1 & \text{if and only if the } i\text{th observation falls in category 2} \\ 0 & \text{otherwise} \end{cases}$$

This idea can easily be extended to cover observations that fall into exactly one of m categories, where m is any value greater than or equal to 2.

EXAMPLE 17.6   **Using Dummy Variables to Represent Three Categories** ————————

A research economist for a large bank has been examining the selling prices of corporate and municipal bonds and thinks that the selling prices of bonds depend on four criteria:

the interest rate paid by the bond, the interest rate paid on Treasury bills on the day that the bond is sold, the bond's Moody's rating, and whether the income from the bond is tax exempt. The economist obtains data on 15 different bonds at various times. We define the following variables:

P_i = Price of the ith bond in dollars

I_i = Interest rate of the ith bond in percentage points

T_i = Interest rate on Treasury bills in percentage points on day bond was sold

$$E_i = \begin{cases} 1 & \text{if income from the } i\text{th bond is tax exempt} \\ 0 & \text{if income from the } i\text{th bond is not tax exempt} \end{cases}$$

Suppose that each bond is rated AAA, AA, or A. There are $m = 3$ categories, so it is necessary to construct $(m - 1) = 2$ dummy variables. Suppose we call these two dummy variables R_1 and R_2. We define the values taken by these dummy variables as follows:

$$R_{i1} = \begin{cases} 1 & \text{if the } i\text{th bond is rated AAA} \\ 0 & \text{otherwise} \end{cases}$$

$$R_{i2} = \begin{cases} 1 & \text{if the } i\text{th bond is rated AA} \\ 0 & \text{otherwise} \end{cases}$$

It follows that for a bond with an A rating, we have $R_{i1} = 0$ and $R_{i2} = 0$, which represents the base case.

The multiple regression model is

$$P_i = \beta_0 + \beta_1 I_i + \beta_2 T_i + \beta_3 E_i + \beta_4 R_{i1} + \beta_5 R_{i2} + e_i$$

The economist hypothesizes the following: β_1 should be positive because bonds paying high interest rates should be worth more than bonds paying low interest rates, other things being equal; β_2 should be negative because the demand for corporate bonds will decline when the interest rate on government bonds increases; β_3 should be positive because, when the income from the bond is tax exempt, the bond is more valuable than when the income is taxable, other things being equal; and β_4 and β_5 should both be positive because AAA and AA bonds should be more valuable than A bonds, other things being equal.

Each bond has a stated value of $100, but its actual selling price varies as the other variables change. The actual selling prices and the other data are shown in Table 17.5 (page 730). Figure 17.11 shows the relevant computer output.

The estimated regression equation is

$$\hat{P}_i = 93.615 + .764I_i - 1.357T_i + 6.166E_i + 7.213R_{i1} + 6.931R_{i2}$$
$$t \text{ statistics:} \quad (14.706) \quad (.904) \quad (-1.885) \quad (2.564) \quad (2.252) \quad (1.954)$$

All of the estimated coefficients have the appropriate sign, and all of the t statistics are statistically significant at a 5% level of significance. The value $R^2 = .80825$ indicates that the regression equation explains approximately 81% of the variation in bond prices.

The coefficient of E_i, 6.166, indicates that the price of a tax-exempt bond ($E_i = 1$) is approximately $6.17 more than the price of a non-tax-exempt bond, other things being equal. The coefficient of R_{i1}, 7.213, indicates that the price of an AAA bond is approxi-

TABLE 17.5   *Data for bonds in Example 17.6*

Bond	Price P_i	Interest rate I_i	Treasury bill rate T_i	Tax status E_i[a]	AAA status R_{i1}[b]	AA status R_{i1}[c]
1	89	6	8	0	0	0
2	98	8	8	0	0	1
3	100	8	6	0	1	0
4	100	10	7	1	0	1
5	97	5	6	0	1	0
6	105	9	7	1	1	0
7	108	10	8	1	1	0
8	97	6	4	1	0	0
9	89	5	5	0	0	0
10	87	9	9	0	1	0
11	100	9	6	0	0	1
12	86	6	10	0	0	0
13	90	7	7	0	0	0
14	106	9	7	1	0	1
15	103	11	6	0	1	0

[a] E_i is 1 if bond is tax exempt and 0 if it is not.
[b] R_{i1} is 1 if bond has AAA rating and 0 otherwise.
[c] R_{i2} is 1 if bond has AA rating and 0 otherwise.

```
* * * *   M U L T I P L E    R E G R E S S I O N   * * * *

Listwise Deletion of Missing Data

Equation Number 1    Dependent Variable..   PRICE

Block Number  1.  Method:  Enter      I        T        E        R1        R2

Variable(s) Entered on Step Number  1..    R2         aa dummy
                                    2..    T          treasury bill rate
                                    3..    E          tax exempt dummy
                                    4..    I          interest rate
                                    5..    R1         aaa dummy


Multiple R           .89903      Analysis of Variance
R Square             .80825                          DF     Sum of Squares     Mean Square
Adjusted R Square    .70172      Regression           5          588.40532       117.68106
Standard Error      3.93834      Residual             9          139.59468        15.51052

                                 F =       7.58718      Signif F =   .0047


------------------ Variables in the Equation ------------------

Variable          B          SE B        Beta        T     Sig T

R2          6.930957      3.546328     .439955     1.954   .0824
T          -1.357273       .720047    -.288682    -1.885   .0921
E           6.166209      2.404593     .417246     2.564   .0305
I            .763545       .845038     .203542      .904   .3898
R1          7.213173      3.203508     .507238     2.252   .0509
(Constant) 93.614943      6.365676                14.706   .0000


End Block Number   1   All requested variables entered.
```

FIGURE 17.11   *SPSSX-generated output for Example 17.6, explaining bond prices as a function of other interest rates and bond rating*

mately \$7.21 more than the price of an A-rated bond. Similarly, the coefficient of R_{i2} indicates that the price of an AA-rated bond is approximately \$6.93 more than the price of an A-rated bond. The difference between these coefficients, ($7.21 - $6.93) = $.28, measures the difference in price between an AAA-rated bond and an AA-rated bond.

Let us predict the selling price of a tax-exempt, AA-rated bond where the interest rate is $I_i = 6\%$ and the Treasury bill rate is $T_i = 8\%$. Because the bond is tax exempt, we use $E_i = 1$. Because the bond is AA-rated, we set R_{i1} equal to 0 and R_{i2} equal to 1. We obtain

$$\hat{P}_i = 93.615 + .764(6) - 1.357(8) + 6.166(1) + 7.213(0) + 6.931(1)$$
$$= \$100.44$$

The Dummy Variable Trap: Creating Too Many Variables

Suppose we want to estimate a regression model where each observation falls into exactly one of m categories. For example, if we think that income depends on age and on gender, each observation can be classified according to whether the individual is male or female. Thus there are $m = 2$ categories. A common error is to construct a multiple regression model that contains a constant term, some explanatory variables, and m rather than $(m - 1)$ dummy variables. This mistake of creating one too many dummy variables is called the **dummy variable trap.**

To classify observations into m categories we need $(m - 1)$, not m, dummy variables. When the observation falls into the base case, each dummy variable takes the value 0. The coefficient of the jth dummy variable measures the effect of being in category j as opposed to being in the base category. If we do not specify a base case, the coefficients will lose this interpretation.

If we create m rather than $(m - 1)$ dummy variables for a qualitative variable that contains m categories, the least-squares estimation procedure breaks down and will not yield unique estimates of the regression coefficients if the equation contains a constant term.

Exercises

Statistical Concepts

17.2.1 *True or false:* A dummy variable is an explanatory variable used to measure the effect of a specific value of a qualitative variable. Explain.

17.2.2 Suppose a statistician obtained the following estimated regression model

$$E_i = 400 + .25W_i + 25S_i$$

where S_i is a dummy variable indicating the gender of the ith individual and $S_i = 1$ if the individual is a male. Suppose another statistician estimated the model using the same data but defined $S_i = 1$ if the individual is a female.
  a. What would happen to the constant term?
  b. What would happen to the coefficient of W_i?
  c. What would happen to the coefficient of S_i?

17.2.3 *True or false:* The effect of a dummy variable is to cause a parallel shift in the regression line by adjusting the intercept term. Explain.

17.2.4   *True or false:*   Suppose a qualitative variable can take one of four possible values. To introduce this variable into the regression model, we can define a dummy variable as follows:

$$D_i = \begin{cases} 0 & \text{if observation falls in category 0} \\ 1 & \text{if observation falls in category 1} \\ 2 & \text{if observation falls in category 2} \\ 3 & \text{if observation falls in category 3} \end{cases}$$

Explain.

17.2.5   *True or false:*   To avoid the dummy variable trap, for each qualitative variable, we define one less dummy variable than the number of categories that exist. Explain.

Statistical Drills

17.2.6   Express the formula for a regression model containing a constant term, one quantitative explanatory variable, one qualitative variable having two levels, and one qualitative variable having three levels.

17.2.7   In a multiple regression model, the following model was obtained:

$$E_i = 625 + .05W_i + 125S_i$$

where

$$E_i = \text{Annual expenditure on recreation}$$
$$W_i = \text{Annual wage income}$$
$$S_i = \begin{cases} 1 & \text{if the person is male} \\ 0 & \text{if the person is female} \end{cases}$$

a. Write the estimated model for a male. Plot this line.
b. Write the estimated model for a female. Plot this line.
c. Predict expenditure for a male whose wage income is $25,000.
d. Predict expenditure for a female whose wage income is $25,000.

Statistical Applications

17.2.8   Suppose data from a large corporation are used to estimate a regression model that explains the corporation's salary structure. The estimated equation is

$$\hat{y}_i = 10,000 + 1,000x_{i1} + 1,000x_{i2} + 3,000x_{i3} - 2,000x_{i4}$$

where

$$y_i = \text{Salary of individual } i$$
$$x_{i1} = \text{Years of service of individual } i$$
$$x_{i2} = \begin{cases} 1 & \text{if highest degree is B.A. for individual } i \\ 0 & \text{otherwise} \end{cases}$$
$$x_{i3} = \begin{cases} 1 & \text{if individual } i \text{ has master's or Ph.D.} \\ 0 & \text{otherwise} \end{cases}$$
$$x_{i4} = \begin{cases} 1 & \text{if individual } i \text{ is a female} \\ 0 & \text{male} \end{cases}$$

a. Suppose a female Ph.D. has been employed for 4 years. Predict her salary.
b. Suppose a male with no college degree has been employed 5 years. Predict his salary.
c. Suppose a new male recruit who has a B.A. is hired. Predict his salary.
d. What is the difference in starting salaries between a B.A. and a Ph.D. recruit of the same gender?
e. Interpret the coefficient of x_{i4}. How would the coefficient change if we let x_{i4} be 1 for a male and 0 otherwise?

17.2.9 Suppose a regression equation has been estimated to explain housing values in a city. The estimated equation is

$$\hat{y}_i = 20{,}000 - 200(\text{age of house}) + 2{,}000(\text{number of rooms})$$
$$- 500(\text{miles from city hall}) + 4{,}000x_{i1} - 1{,}000x_{i2} + 2{,}000x_{i3}$$

where

$$x_{i1} = \begin{cases} 1 & \text{if house is in eastern part of city} \\ 0 & \text{otherwise} \end{cases}$$

$$x_{i2} = \begin{cases} 1 & \text{if house is in western part of city} \\ 0 & \text{otherwise} \end{cases}$$

$$x_{i3} = \begin{cases} 1 & \text{if house is in northern part of city} \\ 0 & \text{otherwise} \end{cases}$$

a. A 7-room, 5-year-old house is 4 miles from the city hall and in the eastern part of town. Estimate its value in dollars.
b. Which part of town is most desirable?
c. Which part of town is least desirable?
d. Which part of town is the base case?
e. Rank the four sections of town in terms of desirability.
f. A 5-room house is 6 miles from the city hall, 8 years old, and in the southern part of town. Estimate its value.

17.2.10 Explain how three dummy variables whose values are denoted q_{i1}, q_{i2}, and q_{i3} can be used in a regression model designed to forecast quarterly sales of a firm. The dummy variables should represent the quarterly effect or quarterly deviation from the base quarter.

17.2.11 The following regression model was obtained to explain weekly sales in thousands of dollars at a fast-food outlet:

$$\hat{y}_i = 10.1 - 4.4x_{i1} + 6.9x_{i2} + 14.3x_{i3}$$

where

$$y_i = \text{Weekly sales at the } i\text{th outlet}$$
$$x_{i1} = \text{Number of competitors within 1 mile of the } i\text{th outlet}$$
$$x_{i2} = \text{Population in thousands within 1 mile of the } i\text{th outlet}$$
$$x_{i3} = \begin{cases} 1 & \text{if drive-up window is present at the } i\text{th outlet} \\ 0 & \text{otherwise} \end{cases}$$

a. What is the expected amount of sales attributable to the drive-up window?
b. Predict sales for a store with two competitors, a population of 8,000 within 1 mile, and no drive-up window.
c. Predict sales for a store with one competitor, a population of 3,000 within 1 mile, and a drive-up window.

17.2.12 Each of the following variables is proposed as an explanatory variable in a regression model. For which of the variables would it be necessary to create one or more dummy variables? If one or more dummy variables should be created, explain how to do it.
a. Age of consumer   **b.** Income of household   **c.** Gender of employee
d. Price of a product   **e.** Season of the year   **f.** Country of birth
g. Highest academic degree of an employee   **h.** Years of seniority of an employee
i. Marital status of an employee

17.2.13 Consider the following sample regression results:

$$\hat{y}_i = \begin{array}{cccc} 17,000 + 1,000s_i + 3,000x_{i1} + 5,000x_{i2} \\ (2,121.7) \quad (167.1) \quad\;\; (365.7) \quad\;\; (562.3) \end{array} \quad \begin{array}{l} R^2 = .78 \\ n = 45 \end{array}$$

where

$$y_i = \text{Annual salary of the } i\text{th employee}$$
$$s_i = \text{Years of seniority of the } i\text{th employee}$$
$$x_{i1} = \begin{cases} 1 & \text{if B.A. is highest degree of the } i\text{th employee} \\ 0 & \text{otherwise} \end{cases}$$
$$x_{i2} = \begin{cases} 1 & \text{if M.S. or higher is highest degree of the } i\text{th employee} \\ 0 & \text{otherwise} \end{cases}$$
$$x_{i3} = \begin{cases} 1 & \text{if high school or lower is highest degree} \\ 0 & \text{otherwise} \end{cases}$$

The values in parentheses are estimated standard errors.

a. Explain why the variable X_3 is not included in the equation.

b. Are the signs of the coefficients plausible? Explain.

c. Predict the salary of an employee with a B.A. degree and 10 years of seniority.

d. Predict the salary of an employee with a Ph.D. degree and 10 years of seniority.

e. Predict the salary of an employee with a high school degree and 10 years of seniority.

f. At the 5% level of significance, would you conclude that salaries for employees with a B.A. degree are higher than salaries for employees with a high school degree, assuming the same amount of seniority?

g. Suppose you reestimate the model and replace the variable X_1 by the variable X_3. What would the new estimated equation be?

h. Suppose you reestimate the model and replace the variable X_2 with the variable X_3. What would the new estimated equation be?

17.3   Estimating Equations in Logarithmic Form

Frequently economic variables are related by some nonlinear functional relationship. One such relationship, the polynomial relationship, was discussed in Section 17.1. Another important functional relationship is the *logarithmic* relationship.

Suppose we have n sample observations on the variables X and Y, and it is hypothesized that values of Y are approximately related to values of X according to the equation

$$Y = AB^X$$

where A and B are unknown population parameters. If we take the natural logarithm of both sides of the equation, we obtain

$$\ln Y = \ln A + (\ln B)X$$

If we let

$$Y' = \ln Y, \quad \beta_0 = \ln A, \quad \text{and} \quad \beta_1 = \ln B$$

we obtain the equivalent expression

$$Y' = \beta_0 + \beta_1 X$$

If we add an error term to the equation to indicate that the values of Y' are not perfectly explained by the values of X, we obtain the population regression model

$$Y_i' = \beta_0 + \beta_1 x_i + e_i$$

The coefficients β_0 and β_1 can be estimated by the method of least squares, provided we use $Y' = \ln Y$ in all the formulas in place of Y. If we apply the method of least squares to the transformed model, we obtain the estimated coefficients b_0 and b_1. We obtain estimates for A and B in the equation $Y = AB^X$ by taking antilogarithms of b_0 and b_1.

The transformation of the equation

$$Y = AB^X$$

into the equation

$$Y' = \beta_0 + \beta_1 X$$

is called the *semilogarithmic transformation*. In this model, the variable Y is transformed, but the variable X retains its original values.

Another exponential function that is sometimes used to describe the relationship between X and Y is the equation

$$Y = AX^B$$

If we take the natural logarithm of both sides of this equation, we obtain

$$\ln Y = \ln A + B(\ln X)$$

By redefining the variables, we obtain the linear equation

$$Y' = \beta_0 + \beta_1 X'$$

where

$$Y' = \ln Y, \quad \beta_0 = \ln A, \quad \beta_1 = B, \quad \text{and} \quad X' = \ln X$$

Once again we add an error term to the model to indicate that the values of Y' are not perfectly explained by the values of X', and we obtain the population regression model

$$Y_i' = \beta_0 + \beta_1 X_i' + e_i$$

Once again the parameters β_0 and β_1 can be estimated by using the least-squares method, provided we use the logs of the original values of Y and X as data. The transformation of the equation

$$Y = AX^B$$

into the equivalent equation

$$Y' = \beta_0 + \beta_1 X'$$

is called the *double logarithmic transformation*.

EXAMPLE 17.7   **Using the Double Logarithmic Transformation** ——————————————————

Suppose an analyst for the Environmental Protection Agency is studying the relationship between the speed at which a car travels (S) and the amount of gasoline consumed per mile (G). It is believed that the variables are approximately related in a nonlinear way according to the equation

$$G = AS^B$$

where G represents gasoline consumption measured in miles per gallon and S represents the average speed of the car in miles per hour. A car is driven around a track at different speeds under carefully controlled conditions; the results are presented in Table 17.6 and plotted in Figure 17.12. We want to obtain the least-squares estimates of the population coefficients.

TABLE 17.6   *Data on gasoline mileage and car speed for Example 17.7*

Miles per gallon G	Average speed S	ln G Y	ln S X	XY	X^2
36	30	3.584	3.401	12.188	11.568
30	35	3.401	3.555	12.092	12.640
25	40	3.219	3.689	11.874	13.608
23	45	3.135	3.807	11.936	14.491
20	50	2.996	3.912	11.719	15.304
19	55	2.944	4.007	11.799	16.059
17	60	2.833	4.094	11.600	16.764
16	65	2.773	4.174	11.574	17.426
14	70	2.639	4.248	11.212	18.050
13	75	2.565	4.317	11.074	18.641
		30.089	39.206	117.069	154.549

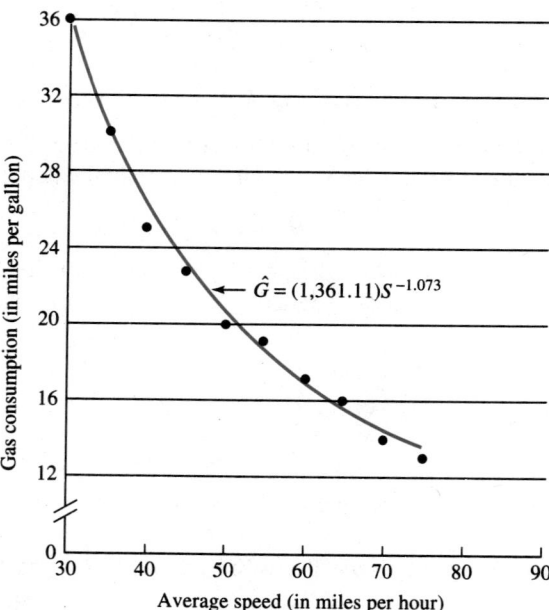

FIGURE 17.12   *Gas consumption versus average speed for Example 17.7*

We transform the equation $G = AS^B$ by taking natural logarithms and obtain

$$\ln G = \ln A + B \ln S$$

We define $Y = \ln G$, $X = \ln S$, $\beta_0 = \ln A$, and $\beta_1 = B$. After adding an error term, we obtain the population regression equation

$$Y_i = \beta_0 + \beta_1 x_i + e_i$$

SOLUTION To show the types of calculations that are involved, we have solved the problem using a hand calculator. The necessary calculations are shown in Table 17.6. First, we obtain the values $y_i = \ln G_i$ and $x_i = \ln S_i$. Next, we obtain $\bar{x} = 39.206/10 = 3.9206$ and $\bar{y} = 30.089/10 = 3.0089$. We obtain b_1 and b_0 from the equations

$$b_1 = \frac{\Sigma x_i y_i - n\bar{x}\bar{y}}{\Sigma x_i^2 - n\bar{x}^2} = -1.073$$

$$b_0 = \bar{y} - b_1\bar{x} = 7.216$$

The sample regression equation is

$$\hat{y}_i = 7.216 - 1.073x_i$$

or

$$\ln \hat{G}_i = 7.216 - 1.073 \ln S_i$$

After taking antilogarithms, we obtain the estimated coefficient

$$\hat{A} = e^{7.216} = 1{,}361.11$$

Also we have

$$\hat{B} = b_1 = -1.073$$

In estimated form, the original model can be written as

$$\hat{G} = 1{,}361.11 S^{-1.073}$$

Thus, if a car was driven at a speed of 68 miles per hour, the predicted gas mileage would be

$$\hat{G} = 1{,}361.11(68)^{-1.073} = 14.71 \text{ miles per gallon}$$

Exercises

17.3.1 Assume that sales of a firm are related to advertising expenditures according to the equation

$$S = AB^X$$

where S is sales in thousands of dollars and X is advertising expenditures in thousands of dollars. The data are as follows:

S	70	90	140	200	152
X	10	12	16	20	17

a. Plot the data on a scatter diagram.
b. Plot $\ln S$ versus X.

c. Estimate the equation

$$\ln S_i = \beta_0 + \beta_1 x_i + e_i$$

d. Plot the estimated equation on the scatter diagram.

17.3.2   A theory states that the yield Y in pounds of a certain agricultural product is related to the amount X of fertilizer in pounds used according to the model

$$\ln Y_i = \beta_0 + \beta_1 \ln x_i + e_i$$

The data are as follows:

Y	6	8	16	19	26	47	63
X	2	3	4	5	6	8	10

a. Plot $\ln Y$ versus $\ln X$ on a scatter diagram.
b. Estimate the population regression model.
c. Plot the estimated equation on the scatter diagram.
d. Predict the yield if 7 pounds of fertilizer are used.

17.3.3   A chemist studied the concentration of a solution (Y) at various time intervals (X). Fifteen identical solutions were prepared and then examined after various time delays. The results follow:

Solution	1	2	3	4	5	6	7	8	9	10	11	12	13	14	15
X	9	9	9	7	7	7	5	5	5	3	3	3	1	1	1
Y	.07	.09	.08	.16	.17	.21	.49	.58	.53	1.22	1.15	1.07	2.84	2.57	3.10

a. Plot the data.
b. Estimate the simple linear regression model and plot the sample regression line.
c. Calculate R^2 and s_e.
d. Test the null hypothesis that β_1 is 0. Let $\alpha = .05$.
e. What transformations might you try to achieve linearity?

17.3.4   Refer to Exercise 3.
a. Let the dependent variable be $Y' = \ln Y$ and obtain the estimated regression function for the transformed data.
b. Plot the estimated regression line and the transformed data. Does the regression line appear to provide a good fit to the transformed data?
c. Obtain the residuals and plot them against the fitted values. What do your plots show?

17.4   Stepwise Regression

When constructing a multiple regression model, it is frequently difficult to determine which explanatory variables to include. Researchers often estimate various models using different sets of explanatory variables, examine the t statistics of the estimated coefficients, and then delete those variables with insignificant t statistics. This is a questionable procedure because a variable can have a coefficient whose t statistic is statistically significant in one formulation of the model but not in another formulation.

Because most economic variables are highly correlated with one another, it is difficult to determine which variables belong in a model. To use an extreme example, suppose that a researcher has data on a dependent variable Y and on six potential explanatory variables X_1, X_2, X_3, X_4, X_5, and X_6. Because the researcher does not know which variables to include in the model, he or she decides to estimate all possible regression models. That is, Y is regressed on each combination of the X_j variables taken one at a time, two at a time, three at a time, and so on. There are a total of 63 different such combinations of the explanatory variables. Suppose that in each equation the researcher examines the signs of the estimated coefficients and their t statistics and rejects equations containing estimated coefficients whose signs do not agree with some underlying theory or whose t statistics are insignificant. At this stage, the researcher probably still has several nonrejected equations. Estimating all possible regressions is an example of a practice called "data mining," whereby a researcher performs numerous statistical procedures on a given set of data and reports only the most favorable results.

To obtain a suitable model, some researchers use a technique known as **stepwise regression.** Suppose that we have data on a dependent variable Y and six proposed explanatory variables X_1, X_2, X_3, X_4, X_5, and X_6. A stepwise regression computer program works in the following way. First, the computer estimates each of the simple linear regression models

$$Y_i = \beta_0 + \beta_j x_{ij} + e_i \qquad j = 1, 2, \ldots, 6$$

The computer selects and prints the results for the equation having the highest R^2. Suppose this equation is

$$\hat{y}_i = b_0 + b_3 x_{i3}$$

Next, the program estimates each of the equations

$$Y_i = \beta_0 + \beta_3 x_3 + \beta_j x_{ij} + e_i \qquad j = 1, 2, 4, 5, 6$$

and prints the results for the equation having the highest R^2. Note that in this second step, variable X_3 remains in the model; the computer includes as the second explanatory variable the one that increases R^2 the most. Suppose that, after the second step, the estimated model is

$$\hat{y}_i = b_0 + b_3 x_{i3} + b_5 x_{i5}$$

In the next step, each of the equations

$$Y_i = \beta_0 + \beta_3 x_{i3} + \beta_5 x_{i5} + \beta_j x_{ij} + e_i \qquad j = 1, 2, 4, 6$$

is estimated. Again the equation having the highest R^2 is selected. Suppose this equation is

$$\hat{y}_i = b_0 + b_3 x_{i3} + b_5 x_{i5} + b_6 x_{i6}$$

Now the stepwise regression routine examines whether any of the X_j variables already in the model should be dropped. For example, it is possible that X_5 and X_6 together explain Y perfectly but by themselves neither explains Y as well as does X_3. After X_5 and X_6 have been included in the equation, X_3 would no longer be needed, and it would drop out in the next round.

At each step, the stepwise regression routine examines each variable not in the equation and selects the variable that increases R^2 the most. Then the routine reexamines each variable that has been included in the equation and determines whether any of them should be dropped. At each stage, the criterion used for adding or dropping a variable is the resultant effect on R^2. The procedure stops when no variable can be added that causes R^2 to increase by more than a prespecified amount R_1^2 and when no variable can be deleted that causes R^2 to decrease by less than some other prespecified amount R_2^2, where $R_1^2 > R_2^2$. The stepwise regression routine is available in many standard regression packages, including the SPSSX program and the SAS program.

A problem with stepwise regression is that at each step, all the estimated coefficients, standard deviations, and t statistics change. As a result, in one step an estimated coefficient b_j could be significantly different from 0 according to the t test, whereas at the next step the new estimated coefficient b_j may not be significantly different from 0.

Some researchers use stepwise regression because it removes the biases and opinions of the model builder from the task of selecting what variables to include in the model. Other researchers condemn stepwise regression for exactly the same reasons, claiming that the form of the model should be based on theoretical considerations.

The following example shows in detail how a stepwise regression program works.

EXAMPLE 17.8   **Using Stepwise Regression to Build a Model Explaining the Dow Jones Industrial Average** ―――――――――――――――――――――――――

Suppose we wish to estimate a model that explains the value of the dependent variable Y = DOWJONES, which is the Dow Jones Industrial Average at the end of a particular year. This variable could depend on many potential explanatory variables. For example, it might be thought that DOWJONES depends on the level of the Gross National Product (GNP) during the year: When GNP increases, it might be expected that DOWJONES also increases. It might also be conjectured that DOWJONES depends on the level of the Consumer Price Index because when the general price level increases, the prices of stocks should also increase. Other potential explanatory variables that will be considered here are the level of federal government spending, the level of national defense spending, the national unemployment rate, and the level of interest rates in the national economy.

It might be that one, some, or all of these variables might help explain the value of the Dow Jones Industrial Average. Thus it might be wise to let the data help us decide which variables to include in the final model. This is exactly the type of problem that stepwise regression is helpful in solving.

The data in Table 2.4 in Chapter 2 show annual values from 1960 through 1986 for the following variables:

$$Y = \text{Dow Jones Industrial Average (DOWJONES)}$$
$$X_1 = \text{Gross National Product (GNP)}$$
$$X_2 = \text{Federal government spending (TOTALGOV)}$$
$$X_3 = \text{Federal government defense spending (DEFENSE)}$$
$$X_4 = \text{National unemployment rate (URATE)}$$
$$X_5 = \text{Consumer Price Index (CPI)}$$
$$X_6 = \text{6-month Treasury bill rate (TBILL)}$$

All data were taken from the 1987 *Economic Report of the President*. To request stepwise regression in the SPSSX program, the following three commands were issued:

REGRESSION VARIABLES = DOWJONES GNP TOTALGOV DEFENSE
    URATE CPI TBILL/
    DEPENDENT = DOWJONES/
    STEPWISE

SOLUTION   Figure 17.13 shows the first step of the SPSSX stepwise regression output. A brief explanation of the steps follows.

```
* * * *   M U L T I P L E   R E G R E S S I O N   * * * *
Equation Number 1    Dependent Variable..  DOWJONES   DOW JONES INDUSTRIAL AVER

  Descriptive Statistics are printed on Page   24

Block Number  1.  Method: Stepwise    Criteria   PIN  .2000   POUT  .2500

Variable(s) Entered on Step Number  1..   DEFENSE    TOTAL GOV. EXPENDITURE ON DEFENSE

Multiple R           .83418     Analysis of Variance
R Square             .69586                    DF    Sum of Squares     Mean Square
Adjusted R Square    .68369     Regression      1    999643.51633    999643.51633
Standard Error    132.19909     Residual       25    436914.97450     17476.59898

                               F =     57.19897     Signif F =  .0000

------------------ Variables in the Equation ------------------       ------------ Variables not in the Equation ------------

Variable          B         SE B       Beta       T   Sig T       Variable    Beta In  Partial  Min Toler      T   Sig T

DEFENSE     2.860302    .378197    .834182    7.563  .0000        GNP        -.903639 -.379705   .053700   -2.011  .0557
(Constant) 600.408660  48.954657             12.265  .0000        TOTALGOV   -2.570624 -.411574  .007796   -2.212  .0367
                                                                  URATE       -.343628 -.482634   .599975   -2.700  .0125
                                                                  CPI        -1.299957 -.589886   .062626   -3.579  .0015
                                                                  TBILL       -.420853 -.595200   .608330   -3.629  .0013
```

FIGURE 17.13 *SPSSX-generated output for step 1 of stepwise regression for Example 17.8, explaining the Dow Jones Industrial Average as a function of defense spending*

Step 1:   The variable DEFENSE is the first to be included in the model. That is, it is the single best predictor of DOWJONES. The sample regression equation is

$$\text{DOWJONES} = 600.41 + 2.86 \text{ DEFENSE} \qquad R^2 = .69586$$
$$t \text{ statistics:} \quad (12.27) \quad (7.56)$$

When just DEFENSE is used to explain DOWJONES, we have $R^2 = .69586$. That is, by itself, DEFENSE explains almost 70% of the total variation in DOWJONES; no other variable explains DOWJONES as well. The coefficient of DEFENSE is positive, which is reasonable because we would expect the Dow Jones Industrial Average to increase as defense spending increases. The computer output shows a t statistic of 7.563 for the coefficient of DEFENSE, which has a p-value of .0000.

Step 2:   The SPSSX output for the second step is given in Figure 17.14 (page 742). The variable TBILL is included in the second step, because it creates the greatest marginal improvement in explaining DOWJONES. The model relating DOWJONES to DEFENSE and TBILL is

$$\text{DOWJONES} = 720.00 + 3.76 \text{ DEFENSE} - 33.75 \text{ TBILL} \qquad R^2 = .80361$$
$$t \text{ statistics:} \quad (13.9) \quad (9.46) \qquad\qquad (-3.63)$$

For this model, R^2 is .80361 and the adjusted R^2 is .78724, which means that the two variables DEFENSE and TBILL explain approximately 80% of the variation in DOW-JONES. Thus, the inclusion of the variable TBILL causes R^2 to increase from .70 to .80.

* *

```
Variable(s) Entered on Step Number  2..    TBILL     INTERST ON 6 MONTH T-BILLS

Multiple R              .89644      Analysis of Variance
R Square                .80361                         DF    Sum of Squares      Mean Square
Adjusted R Square       .78724      Regression          2     1154426.34782     577213.17391
Standard Error      108.42281       Residual           24      282132.14301      11755.50596

                                    F =      49.10152      Signif F =   .0000


Equation Number 1    Dependent Variable..   DOWJONES   DOW JONES INDUSTRIAL AVER

------------------ Variables in the Equation ------------------      ------------ Variables not in the Equation ------------

Variable         B          SE B        Beta        T    Sig T      Variable    Beta In   Partial   Min Toler       T    Sig T

DEFENSE      3.763413     .397686    1.097566     9.463   .0000      GNP        -.008880  -.003590    .032102     -.017   .9864
TBILL      -33.749187    9.300849    -.420853    -3.629   .0013      TOTALGOV   -.287633  -.043724    .004538     -.210   .8356
(Constant) 719.998142   51.944337                13.861   .0000      URATE      -.283143  -.488529    .469473    -2.685   .0132
                                                                     CPI        -.740329  -.291832    .030517    -1.463   .1569
```

FIGURE 17.14   *SPSSX-generated output for step 2 of stepwise regression for Example 17.8, explaining the Dow Jones Industrial Average as a function of defense spending and the Treasury bill rate*

In this model, the t statistics for the coefficients of DEFENSE and TBILL are 9.463 and -3.629. Because the p-values for these t statistics are each less than .002, we can feel very confident that both of the population regression coefficients are nonzero. Also note that both coefficients have the correct sign. That is, we would expect increases in defense spending to have a positive effect on the Dow Jones Industrial Average and increases in interest rates to have a negative effect. Also observe that the coefficient of the variable DEFENSE changes from 2.86 in the first step to 3.76 in the second step.

Step 3:   The SPSSX computer output for the third step is shown in Figure 17.15. The variable URATE (the national unemployment rate) is included in the model on the third step because it creates the next greatest marginal improvement in explaining DOW-JONES. The model relating DOWJONES to DEFENSE, TBILL, and URATE is

$$\text{DOWJONES} = 879.53 + 4.28 \text{ DEFENSE} - 30.15 \text{ TBILL} - 39.94 \text{ URATE} \qquad R^2 = .85048$$
$$t \text{ statistics:} \quad (11.7) \quad (10.6) \qquad\qquad (-3.59) \qquad\qquad (-2.69)$$

Including URATE causes R^2 to increase from .80361 to .85048. In this model, the t statistics for the coefficients of DEFENSE, TBILL, and URATE are 10.610, -3.590, and -2.685, respectively. Since the p-values for these t statistics are all less than .02, we can feel confident that all three of the population regression coefficients are nonzero. Also note that all three coefficients have the correct sign. We would expect increases in the unemployment rate (URATE) to have a negative effect on the Dow Jones Industrial Average.

```
Variable(s) Entered on Step Number  3..    URATE     THE RATE OF UNEMPLOYMENT

Multiple R          .92221      Analysis of Variance
R Square            .85048                          DF    Sum of Squares    Mean Square
Adjusted R Square   .83097      Regression           3    1221760.24652    407253.41551
Standard Error    96.63878      Residual            23     214798.24431      9339.05410

                                F =      43.60757      Signif F =   .0000

----------------- Variables in the Equation ------------------       ------------ Variables not in the Equation -------------

Variable            B          SE B       Beta       T   Sig T       Variable     Beta In   Partial   Min Toler     T   Sig T

DEFENSE       4.281039      .403493   1.248528   10.610  .0000       GNP          .718034   .293629    .025004    1.441  .1637
TBILL       -30.146025     8.397881   -.375921   -3.590  .0015       TOTALGOV    1.468848   .228159    .003608    1.099  .2836
URATE       -39.935204    14.872723   -.283143   -2.685  .0132       CPI          .546984   .156410    .012226     .743  .4655
(Constant)  879.526760    75.321584             11.677  .0000
```

FIGURE 17.15   *SPSSX-generated output for step 3 of stepwise regression for Example 17.8, explaining the Dow Jones Industrial Average as a function of defense spending, the Treasury bill rate, and the national unemployment rate*

When the SPSSX computer program was used on the data in Table 2.4, the stepwise regression procedure stopped at this stage when the default value was $\alpha = .05$. The program stopped because no remaining explanatory variable when inserted into the equation had a coefficient whose p-value was less than .05. To increase the number of steps in this example and to show a few other interesting features of stepwise regression, the default p-value was increased to .65. Thus the program continued to insert values into the equation as long as the p-value of the test was less than .65. This is a very lenient criterion; an estimated regression coefficient could be less than 1 standard deviation from 0 and still be included in the model.

Step 4:   The SPSSX computer output for the fourth step is shown in Figure 17.16. The variable GNP (Gross National Product) is included at this step. The model relating DOWJONES to DEFENSE, TBILL, URATE, and GNP is

```
* * * *   M U L T I P L E   R E G R E S S I O N   * * * *

Equation Number 1    Dependent Variable..   DOWJONES   DOW JONES INDUSTRIAL AVER

Variable(s) Entered on Step Number  4..    GNP     GROSS NATIONAL PRODUCT

Multiple R          .92918      Analysis of Variance
R Square            .86337                          DF    Sum of Squares    Mean Square
Adjusted R Square   .83853      Regression           4    1240279.73501    310069.93375
Standard Error    94.45508      Residual            22     196278.75582      8921.76163

                                F =      34.75434      Signif F =   .0000

----------------- Variables in the Equation ------------------       ------------ Variables not in the Equation -------------

Variable            B          SE B       Beta       T   Sig T       Variable     Beta In    Partial   Min Toler      T   Sig T

DEFENSE       2.320975     1.416454    .676892    1.639  .1155       TOTALGOV    -.230646   -.021585    .001197     -.099  .9221
TBILL       -39.989073    10.679317   -.498664   -3.745  .0011       CPI        -4.208678   -.400755    .001239    -2.004  .0581
URATE       -51.093365    16.471012   -.362254   -3.102  .0052
GNP            .143127      .099342    .718034    1.441  .1637
(Constant)  974.915445    99.011538              9.846  .0000
```

FIGURE 17.16   *SPSSX-generated output for step 4 of stepwise regression for Example 17.8, explaining the Dow Jones Industrial Average as a function of defense spending, the Treasury bill rate, the national unemployment rate, and Gross National Product*

$$\text{DOWJONES} = 974.92 + 2.32 \text{ DEFENSE} - 39.99 \text{ TBILL}$$
$$t \text{ statistics:} \quad (9.85) \quad (1.64) \quad\quad\quad (-3.74)$$

$$- 51.09 \text{ URATE} + .14 \text{ GNP} \quad R^2 = .86337$$
$$(-3.10) \quad\quad\quad (1.44)$$

For this model R^2 is .86337 and the adjusted R^2 is .83853. Thus the inclusion of GNP increased R^2 from .85048 to .86337. The t statistic for the coefficient of GNP, 1.441, has a p-value of .1637, so there is some doubt as to whether the population coefficient for GNP is nonzero. However, the signs for the estimated coefficients of the four explanatory variables all agree with theoretical considerations.

Step 5:   The SPSSX computer output for the fifth step is shown in Figure 17.17. The variable CPI (Consumer Price Index) is included here, resulting in the model

$$\text{DOWJONES} = 1,115.21 + 5.24 \text{ DEFENSE} - 10.17 \text{ TBILL}$$
$$t \text{ statistics:} \quad (9.59) \quad (2.66) \quad\quad\quad (-.57)$$

$$+ 10.55 \text{ URATE} + .70 \text{ GNP} - 11.80 \text{ CPI} \quad R^2 = .88531$$
$$(.31) \quad\quad\quad (2.39) \quad (-2.00)$$

The t statistic for the coefficient of CPI, -2.004, has a p-value of .0581, so there is some doubt about whether the population coefficient for CPI is nonzero. In addition, the coefficients of all the other explanatory variables have changed from the previous step, and the coefficient of the unemployment rate (URATE) has changed in sign from negative to positive. Also note that the p-values for the coefficients of TBILL and URATE have increased to .5767 and .7622, respectively. Now it appears that neither of these estimated coefficients is significantly different from 0. Thus we have an example of estimated coefficients that were statistically significant at one step of the analysis but not at a later step.

```
Variable(s) Entered on Step Number   5..    CPI       CONSUMER PRICE INDEX

Multiple R            .94091        Analysis of Variance
R Square              .88531                          DF     Sum of Squares      Mean Square
Adjusted R Square     .85801        Regression         5     1271803.02769     254360.60554
Standard Error      88.57482        Residual          21      164755.46314       7845.49824

                                    F =    32.42122       Signif F =   .0000

------------------ Variables in the Equation ------------------        ------------ Variables not in the Equation ------------

Variable          B           SE B        Beta        T     Sig T      Variable    Beta In  Partial  Min Toler       T    Sig T

DEFENSE       5.241340    1.971519    1.528591     2.659   .0147       TOTALGOV    .096860  .009865    .001190      .044   .9652
TBILL       -10.168219   17.933604    -.126798     -.567   .5767
URATE        10.547305   34.412241     .074781      .306   .7622
GNP            .699234     .292653     3.507898    2.389   .0263
CPI         -11.797944    5.885736   -4.208678    -2.004   .0581
(Constant) 1115.205912  116.271144                 9.591   .0000
```

FIGURE 17.17   *SPSSX-generated output for step 5 of stepwise regression for Example 17.8, explaining the Dow Jones Industrial Average as a function of defense spending, the Treasury bill rate, the national unemployment rate, Gross National Product, and the Consumer Price Index*

Step 6:   The SPSSX computer output for the sixth step is shown in Figure 17.18. Here the variable URATE (unemployment rate) is *removed* from the model, leaving the model

$$\text{DOWJONES} = 1{,}106.66 + 4.78 \text{ DEFENSE} - 14.51 \text{ TBILL}$$

t statistics: (10.01)   (3.82) (-1.35)

$$+ .63 \text{ GNP} - 10.19 \text{ CPI} \qquad R^2 = .88480$$

(3.51) (-3.94)

The stepwise regression program stops at this stage because no additional explanatory variable has a regression coefficient whose t statistic satisfies the .65 p-value criterion.

```
                        * * * *   M U L T I P L E   R E G R E S S I O N   * * * *
Equation Number 1    Dependent Variable..   DOWJONES   DOW JONES INDUSTRIAL AVER

Variable(s) Removed on Step Number  6..    URATE      THE RATE OF UNEMPLOYMENT

Multiple R           .94064         Analysis of Variance
R Square             .88480                         DF      Sum of Squares      Mean Square
Adjusted R Square    .86385         Regression       4        1271066.01054     317766.50263
Standard Error     86.73169         Residual        22         165492.48029       7522.38547

                                    F =      42.24278        Signif F =   .0000


------------------ Variables in the Equation ------------------        ------------ Variables not in the Equation ------------

Variable           B         SE B        Beta         T    Sig T        Variable    Beta In  Partial  Min Toler       T   Sig T

DEFENSE       4.781237    1.251455    1.394406     3.821  .0009         TOTALGOV     .084672  .008606   .001190      .039  .9689
TBILL       -14.507965   10.777061    -.180914    -1.346  .1919         URATE        .074781  .066734   .001239      .306  .7622
GNP            .629274     .179343    3.156926     3.509  .0020
CPI         -10.185895    2.586787   -3.633612    -3.938  .0007
(Constant) 1106.664365  110.533066                10.012  .0000


End Block Number   1   PIN =     .200 Limits reached.
```

FIGURE 17.18   *SPSSX-generated output for step 6 of stepwise regression for Example 17.8, explaining the Dow Jones Industrial Average as a function of defense spending, the Treasury bill rate, Gross National Product, and the Consumer Price Index*

This example of stepwise regression illustrates the problems that can occur when data are highly collinear. Variables that are important at one step of the analysis can be insignificant at a later step. Because GNP, total government spending, and federal defense spending are all highly correlated with one another, it is difficult to separate their individual effects on DOWJONES.

By examining the correlations between potential explanatory variables in a stepwise regression model, it is sometimes possible to anticipate some of the estimation problems. This idea is discussed in the next section.

17.5   The Correlation Matrix

In addition to stepwise regression, there is another useful computer tool called the **correlation matrix** that helps us select explanatory variables for a regression model.

Suppose we wish to build a regression model to explain some dependent variable Y using a set of K potential explanatory variables $X_1, X_2, \ldots, X_K$. Before estimating any regression equations, it is useful to know the sample correlation coefficients between Y and each explanatory variable X_i. In addition, it is useful to know the sample correlation coefficient between each potential explanatory variable X_i and every other potential explanatory variable X_j. Many computer programs, including SPSSX and SAS, routinely calculate all these correlation coefficients. The resulting set of sample correlation coefficients is presented in a rectangular array called a sample correlation matrix.

DEFINITION   **Sample Correlation Matrix** ───────────────────────────────

Suppose we have a sample of observations on K variables $X_1, X_2, \ldots, X_K$. The **sample correlation matrix** shows the sample correlation between each pair of variables X_i and X_j. We denote the sample correlation between X_i and X_j as R_{ij}.

EXAMPLE 17.9   **A Sample Correlation Matrix** ───────────────────────────

Let us generate a sample correlation matrix for the data in Table 2.3 of Chapter 2, which show the values in 1985 for the 50 states and the District of Columbia for the following variables:

PCTUNION_i = Percentage of employees who are unionized in state i

SALARY_i = Average annual salary of secondary school teachers in state i

REVENUE_i = State and local tax revenue per capita in state i

EXPEND_i = Expenditures per elementary and secondary school pupil in state i

INCOME_i = Income per capita in state i

WAGE_i = Average hourly wage in manufacturing employment in state i

The following commands generated the sample correlation matrix shown in Figure 17.19 using the SPSSX program

---

REGRESSION DESCRIPTIVES = CORR/
      VARIABLES = PCTUNION, SALARY, REVENUE, EXPEND, INCOME, WAGE/

Figure 17.19 also shows the mean and standard deviation for each variable.

```
                                  * * * *   M U L T I P L E   R E G R E S S I O N   * * * *

Listwise Deletion of Missing Data

               Mean     Std Devi  Label

PCTUNION     19.731        7.460  % UNION MEMBERSHIP IN 1982
SALARY    24357.784     4180.030  AVERAGE TEACHER SALARY IN 1986
REVENUE    3684.529     1055.438  STATE AND LOCAL REVENUE PER CAPITA IN 19
EXPEND     2509.275     1479.569  EXPENDITURES PER PUPIL IN 1986
INCOME    13238.804     2156.789  PERSONAL INCOME PER CAPITA IN 1985
WAGE          9.407        1.245  AVERAGE HOURLY EARNINGS IN MANUFACTURING

N of Cases =     51

Correlation:

              PCTUNION    SALARY    REVENUE     EXPEND     INCOME       WAGE

PCTUNION        1.000      .613       .438       .326       .460       .706
SALARY           .613     1.000       .840       .747       .767       .562
REVENUE          .438      .840      1.000       .797       .755       .425
EXPEND           .326      .747       .797      1.000       .474       .413
INCOME           .460      .767       .755       .474      1.000       .373
WAGE             .706      .562       .425       .413       .373      1.000
```

FIGURE 17.19   *SPSSX-generated sample correlation matrix for Example 17.9*

Several interesting observations can be made about the values in the sample correlation matrix.

The largest correlation coefficient in the table is .840, which represents the sample correlation between SALARY and REVENUE. This means that the single variable that best explains variations in teachers' salaries is the variable REVENUE. Also observe that SALARY is highly correlated with EXPEND and INCOME. The sample correlation between SALARY and PCTUNION is .613, which indicates that teachers tend to earn more in heavily unionized states.

The sample correlation between PCTUNION and INCOME is .460. This means that average annual income tends to be higher in heavily unionized states. The square of this correlation coefficient is .2116, which means that, in a linear regression model, the variable PCTUNION explains 21.16% of the state-by-state variation in annual income. Although unionization is positively correlated with income per capita, the correlation is relatively low. This implies that the single variable PCTUNION does not account for income in states very well.

The positive sample correlation between INCOME and WAGE of .373 indicates that states that pay high wages to manufacturing employees tend to have high per capita incomes. Although the sample correlation between INCOME and WAGE is positive, it is not very high, and the single variable WAGE does not explain well why some states have high average incomes and others have low average incomes.

The sample correlation between PCTUNION and WAGE is .706. Most people probably think that states that are heavily unionized pay higher manufacturing wages than states that are not; the data tend to support this belief.

Also, note that the sample correlation between EXPEND and REVENUE is .797. This tells us that states that spend the most money per pupil on elementary and secondary school students also tend to be the states that have the highest tax revenue per capita.

Finally, every variable is perfectly correlated with itself. Thus, in the correlation matrix, every value on the main diagonal of the correlation matrix is 1.00. In addition, the correlation between X_i and X_j is the same as the correlation between X_j and X_i, so the correlation matrix is symmetric. That is, the value in row i and column j is the same as the value in row j and column i.

The correlation matrix can be useful in helping to select variables to include in a regression model. Variables that are highly correlated with a dependent variable are obvious candidates for inclusion in a regression equation. In addition, the correlation matrix can be helpful in detecting multicollinearity. If the sample correlation between X_i and X_j is relatively large, it may be difficult to determine the separate effects of X_i and X_j on a dependent variable Y.

EXAMPLE 17.10 **Examining the Sample Correlation Matrix** —————————————

Let us examine the sample correlation matrix in Figure 17.19 and see how this information can help us build a regression model. The variables are as defined in Example 17.9. It is clear that EXPEND will depend on the variable SALARY, and vice versa. It is not clear whether increases in EXPEND cause increases in SALARY, or vice versa. Suppose we ignore the variable SALARY and build a regression model to explain the variable

EXPEND. We are interested in finding variables that help explain why different states have different expenditures on education per pupil; in other words, EXPEND is our dependent variable. Although we could use stepwise regression to help build a suitable model, first let us examine the sample correlation matrix and see whether we can predict in advance what will happen when we apply the stepwise regression procedure.

SOLUTION   In Figure 17.19, examine the sample correlation coefficient between the dependent variable EXPEND and each of the potential explanatory variables. The largest of these sample correlation coefficients is .797, so REVENUE will be the first variable to enter the regression equation in the stepwise regression procedure. The next variable to enter the regression in the stepwise procedure is the one that adds most to explaining EXPEND after taking into account the presence of REVENUE. We cannot determine this with complete accuracy from the sample correlation matrix, but we can get some clues. After REVENUE, EXPEND is most highly correlated with the variable INCOME. Thus it is reasonable to conjecture that INCOME would enter the regression model on the second step. INCOME is also highly correlated with REVENUE ($R = .755$), however, so INCOME might not explain much of the variation in EXPEND that has not already been explained by REVENUE.

Now examine some of the other potential candidates for entry into the regression model. The sample correlation between PCTUNION and WAGE is .706. Because this value is relatively high, these two variables together might not explain much more of the variation in EXPEND than either of the variables individually. In addition, neither WAGE nor PCTUNION is highly correlated with EXPEND, so it is likely that the stepwise regression procedure will not insert either variable.

Now let us see whether the correlation matrix helped us predict which variables would enter the regression model by using the stepwise regression procedure.

EXAMPLE 17.11   **Using Stepwise Regression to Explain a State's Expenditures per Pupil** ————

We will use the SPSSX stepwise regression procedure to build a model to explain EXPEND as a function of the variables PCTUNION, REVENUE, INCOME, and WAGE.

Step 1:   The computer output for the first step is shown in Figure 17.20. The variable REVENUE is the first to be included in the model. That is, it is the single best predictor of EXPEND. (By examining the correlation matrix, we already knew this.) The model representing this relationship is

$$\text{EXPEND} = -1{,}606.52 + 1.117 \text{ REVENUE} \qquad R^2 = .63495$$
$$t \text{ statistics:} \qquad (-3.47) \qquad (9.23)$$

When just REVENUE is used to explain EXPEND, the value of R^2 is .63495. That is, by itself, REVENUE explains about 63% of the total variation in EXPEND. The sign of the coefficient of REVENUE is positive, which agrees with economic theory. The computer output shows a t statistic of 9.232 for the estimated coefficient of REVENUE, which has a p-value of .0000.

```
* * * *   M U L T I P L E   R E G R E S S I O N   * * * *
```

Equation Number 1   Dependent Variable..   EXPEND   EXPENDITURES PER PUPIL IN 1

  Descriptive Statistics are printed on Page   42

Block Number 1. Method: Stepwise   Criteria   PIN .0500   POUT .1000

Variable(s) Entered on Step Number 1..   REVENUE   STATE AND LOCAL REVENUE PER CAPITA IN 19

```
Multiple R          .79684      Analysis of Variance
R Square            .63495                        DF    Sum of Squares      Mean Square
Adjusted R Square   .62750      Regression         1    69499117.52665    69499117.52665
Standard Error   903.02386      Residual          49    39957152.63022      815452.09449

                                F =      85.22771     Signif F =   .0000
```

```
----------------- Variables in the Equation ------------------        ------------ Variables not in the Equation ------------

Variable           B        SE B       Beta       T    Sig T          Variable    Beta In  Partial  Min Toler      T    Sig T

REVENUE     1.117049     .120999    .796837   9.232    .0000          PCTUNION    -.027704 -.041224  .808294    -.286   .7762
(Constant) -1606.523721 463.409636           -3.467    .0011          SALARY       .264852  .238098  .295025    1.698   .0959
                                                                      INCOME      -.297863 -.323076  .429468   -2.365   .0221
                                                                      WAGE         .090117  .134977  .818957     .944   .3500
```

FIGURE 17.20 *SPSSX-generated output for step 1 of stepwise regression for Example 17.11, explaining expenditures per pupil as a function of tax revenue per capita*

Step 2:   The SPSSX computer output for the second step is shown in Figure 17.21. Here the variable INCOME enters the model, because it offers the greatest marginal improvement in explaining EXPEND. The model relating EXPEND to REVENUE and INCOME is

$$\text{EXPEND} = -63.457 + 1.432 \ \text{REVENUE} - .204 \ \text{INCOME} \qquad R^2 = .67305$$
$$t \text{ statistics:} \quad (-.08) \qquad (8.11) \qquad (-2.37)$$

Variable(s) Entered on Step Number 2..   INCOME   PERSONAL INCOME PER CAPITA IN 1985

```
Multiple R          .82040      Analysis of Variance
R Square            .67305                        DF    Sum of Squares      Mean Square
Adjusted R Square   .65943      Regression         2    73669782.12424    36834891.06212
Standard Error   863.45343      Residual          48    35786488.03262      745551.83401

                                F =      49.40621     Signif F =   .0000
```

```
----------------- Variables in the Equation ------------------        ------------ Variables not in the Equation ------------

Variable           B        SE B       Beta       T    Sig T          Variable    Beta In  Partial  Min Toler      T    Sig T

REVENUE     1.432447     .176545   1.021823   8.114    .0000          PCTUNION     .021110  .032377  .408661     .222   .8252
INCOME       -.204336    .086393   -.297863  -2.365    .0221          SALARY       .464285  .409010  .253732    3.073   .0035
(Constant) -63.456882  788.657058            -.080     .9362          WAGE         .109785  .173090  .405502    1.205   .2343
```

FIGURE 17.21 *SPSSX-generated output for step 2 of stepwise regression for Example 17.11, explaining expenditures per pupil as a function of tax revenue per capita and income per capita*

Including INCOME increases R^2 from .63495 to .67305, but the coefficient of INCOME has the wrong sign. Based on economic theory, we would expect increases in income per capita to have a positive effect on expenditures per pupil. The fact that the estimated coefficient of INCOME is negative shows that we have a multicollinearity problem. From the correlation matrix in Figure 17.19, we see that INCOME is positively correlated with EXPEND ($R = .474$), but INCOME is much more highly correlated with REVENUE ($R = .755$) than with EXPEND. After taking into account the effect of REVENUE on EX-

PEND, the effect of the variable INCOME on EXPEND is not clear. On theoretical grounds, we would reject the model resulting from the second step of the stepwise regression procedure, since it is highly implausible that increases in income per capita would lead to decreases in expenditures per pupil.

At this stage, the stepwise regression procedure stops because none of the remaining explanatory variables has a *p*-value less than .05 when entered into the model.

This example should be studied carefully. Quite frequently in applied research, estimated coefficients will have the wrong sign. In this example, the wrong sign can be explained by the fact that the effect of INCOME on EXPEND has already been accounted for by the inclusion of the variable REVENUE.

At this stage, the stepwise regression procedure stops because none of the remaining explanatory variables has a *p*-value less than .05 when entered into the model.

Exercises

Statistical Concepts

17.5.1   Explain what is meant by data mining.

17.5.2   Explain what is meant by stepwise regression. Explain how it is performed.

17.5.3   *True or false:*   A useful way of trying to predict which variables will enter into a stepwise regression model is to examine the sample correlation matrix. Explain.

Statistical Applications

17.5.4   Refer to the data in Table 2.4 in Chapter 2, which were used in Example 17.8 to predict the Dow Jones Industrial Average (DOWJONES).
   a. Generate a correlation matrix using the SPSSX commands

   REGRESSION DESCRIPTIVES = CORR/
      VARIABLES = GNP, TOTALGOV, DEFENSE, URATE, CPI, TBILL, DOWJONES/

---

   b. Examine these sample correlation coefficients. Which variable has the highest correlation with the Dow Jones Industrial Average?
   c. Which variable has the highest correlation with the Gross National Product?
   d. Which variable has the highest correlation with the Consumer Price Index?
   e. Find the correlation between the Consumer Price Index and the Treasury bill rate.
   f. Find the correlation between the Dow Jones Industrial Average and the Consumer Price Index.

17.5.5   Refer to the data in Table 2.3 in Chapter 2, which were used in Examples 17.9 and 17.10.
   a. Generate a correlation matrix using the SPSSX commands

   REGRESSION DESCRIPTIVES = CORR/
      VARIABLES = PCTUNION, SALARY, REVENUE, EXPEND, INCOME, WAGE/

---

   b. Examine these sample correlation coefficients. Which variable has the highest correlation with the variable SALARY?
   c. Which variable has the highest correlation with REVENUE?
   d. Which variable has the highest correlation with INCOME?
   e. Find the correlation between SALARY and REVENUE and between SALARY and INCOME.
   f. Find the correlation between WAGE and PCTUNION.

Computer Applications

To perform the types of statistical analyses described in this chapter, using a computer is almost mandatory. In this chapter, we discussed five different statistical procedures:

1. Polynomial regression
2. Regression with dummy variables
3. Logarithmic regression
4. Stepwise regression
5. The correlation matrix

All of these procedures are relatively easy to carry out using SPSSX or SAS. For example, polynomial regression is nothing more than multiple regression where one of the explanatory variables is the square (or cube or higher power) of another explanatory variable. To estimate a polynomial regression equation, we first instruct the computer to create a new variable that equals the original variable raised to the appropriate power. In SPSSX, this is done via a COMPUTE command. For example, suppose we wish to create a new variable (named SPEEDSQ) that is the square of the variable SPEED. Issue the command

```
COMPUTE SPEEDSQ = SPEED * SPEED
```

In SPSSX, an asterisk represents a multiplication sign. To create a variable (named SPEEDCUB) that is SPEED raised to the third power, issue the command

```
COMPUTE SPEEDCUB = SPEED * SPEED * SPEED
```

To estimate the polynomial regression model, proceed exactly as in any other multiple regression model and treat the newly created variables as any other explanatory variable.

To estimate a model that includes dummy variables, we can use an IF command to create the appropriate dummy variables. If the variable is dichotomous, such as gender, then the values of the variable should be coded as 0 and 1 and the dummy variable is ready for use. If the variable is multinomial, such as academic class, then the appropriate dummy variables have to be created. For example, examine the data in Table 2.2 in Chapter 2, where the variable CLASS is coded as 1, 2, 3, or 4 to indicate whether a student is a freshman, sophomore, junior, or senior. Suppose we use freshman as the base case. Then we need to create three dummy variables to represent the sophomores, the juniors, and the seniors. To create a variable (named JUNIOR) that takes the value 1 if a student is a junior and 0 otherwise, issue the command

```
IF (CLASS EQ 3) JUNIOR = 1
```

The two additional commands

```
IF (CLASS EQ 2) SOPH = 1
IF (CLASS EQ 4) SENIOR = 1
```

will create the other two required dummy variables. At this stage, we have created three new variables whose values consist of 0's and 1's in the appropriate places.

To estimate a model that includes the logarithm of some variable, we again need to use a COMPUTE command to create the appropriate variable. For example, to create a variable (named LOGSPEED) that is the natural logarithm of the variable SPEED, issue the command

---

COMPUTE LOGSPEED = LN(SPEED)

In Section 17.4, we discussed how to obtain computer output for stepwise regression by adding the statement

---

STEPWISE

---

at the end of the regression commands.

In Section 17.5, we showed how to obtain a correlation matrix by issuing the command

---

DESCRIPTIVES = CORR/

---

along with the other regression commands.

Chapter 17 Supplementary Exercises

$$ **17.S.1** College students were studied to determine the relationship between annual expenditure on clothing and total annual family income. The following equation was obtained:

$$\text{Expenditure}_i = 85 + .02(\text{family income})_i + 175s_i + 45x_{i1} + 22x_{i2}$$

where

$$s_i = \begin{cases} 1 & \text{if female} \\ 0 & \text{if male} \end{cases}$$

$$x_{i1} = \begin{cases} 1 & \text{if student lives in Northeast} \\ 0 & \text{otherwise} \end{cases}$$

$$x_{i2} = \begin{cases} 1 & \text{if student lives in Midwest} \\ 0 & \text{otherwise} \end{cases}$$

a. Explain the meaning of the term $175s_i$ in the equation. Is it reasonable that the coefficient of s_i is positive?

b. A male student from Maine has a family income of \$40,000. Estimate his annual expenditure on clothing.

c. For male residents of the Midwest, graph the relationship between expenditures on clothing and family income. Do the same for female residents of the Midwest.

17.S.2 The cost accountant for TLG Industries has estimated a quadratic regression equation to explain the number of refrigerators produced annually by the firm in terms of the total number of workers employed. The estimated equation is

$$\hat{y}_t = 372.6 + 984.3x_t - 4.7x_t^2$$

where

$$\hat{y}_t = \text{Predicted number of washing machines produced in year } t$$
$$x_t = \text{Number of full-time employees in year } t \text{ at TLG Industries}$$

a. Plot the sample regression equation.
b. Predict output if the firm has 40 full-time employees.
c. Can you show that at some point $\hat{y}_t$ will level off and then decrease when x_t increases?
d. What is $\hat{y}_t$ if $x_t = 0$? Is this reasonable?
e. Do the answers to parts c and d imply that the equation is useless in predicting Y? Is it possible for the equation to be useful over a limited range of values of X?

17.S.3 A government study of eating habits contained the following regression results:

Variable	Coefficient	Standard deviation
Constant	78.4	3.9
X_1: Log food price index	-17.3	4.2
X_2: Log income per capita	.07	.003

The dependent variable is the logarithm of food expenditures per capita. There were 200 observations in the study. Construct a 95% confidence interval for the following:
a. β_1 b. β_2

17.S.4 This problem requires the use of a computer. The Bardol Health Club has franchises in 15 different cities. The marketing manager believes that the profit Y_i that the ith franchise earned last year depended on four factors: x_{i1}, the average income (in thousands of dollars) in the community where the ith franchise is located; x_{i2}, the total population (in 100,000s) within 15 miles of the ith franchise; x_{i3}, annual advertising expenditures (in thousands of dollars) of the ith franchise; and x_{i4}, which equals 1 if there is a swimming pool and 0 if there is not. The data are shown in the accompanying table.

Franchise	Profits Y	Income X_1	Population X_2	Advertising X_3	Status X_4
1	94	34	4.3	21	1
2	22	17	1.0	4	0
3	44	23	2.5	8	0
4	81	33	4.0	15	1
5	74	26	3.1	14	1
6	54	26	2.0	9	0
7	112	40	4.2	20	1
8	28	19	1.1	7	0
9	56	25	2.1	8	0
10	77	29	3.8	16	1
11	62	25	2.9	14	0
12	80	31	3.7	19	1
13	48	24	2.2	10	0
14	50	24	2.4	9	0
15	40	22	3.1	7	0

a. Estimate the population regression model

$$Y_i = \beta_0 + \beta_1 x_{i1} + \beta_2 x_{i2} + \beta_3 x_{i3} + \beta_4 x_{i4} + e_i$$

b. Let $\alpha = .05$. Test $H_0: \beta_1 = 0$ against $H_1: \beta_1 > 0$.

c. Let $\alpha = .05$. Test $H_0: \beta_2 = 0$ against $H_1: \beta_2 > 0$.

d. Let $\alpha = .05$. Test $H_0: \beta_3 = 0$ against $H_1: \beta_3 > 0$.

e. Calculate s_e.

f. Calculate R^2. Given the value of R^2, would you have much faith in forecasts generated by this model?

g. A certain franchise plans to spend $x_{i3} = \$12,000$ on advertising next year. Assume that for this franchise $x_{i1} = \$22,000$, $x_{i2} = 216,000$, and $x_{i4} = 1$. Forecast profits for this franchise.

h. Calculate the correlation coefficient between X_1 and X_3. If this correlation coefficient is large, then multicollinearity could make it difficult to get good estimates of the coefficients of X_1 and X_3.

17.S.5   A tax consultant studied the relationship between selling price and assessed valuation of one-family residential dwellings in a large tax district. The accompanying data represent a random sample of nine recent sales transactions of one-family dwellings located on corner lots and another random sample of 14 recent sales of one-family dwellings not located on corner lots. In the data, both selling price (Y) and assessed valuation (X) are expressed in thousands of dollars. Assume that the error term variances in the two populations are equal. Estimate the regression model

$$Y_i = \beta_0 + \beta_1 x_{i1} + \beta_2 x_{i2} + e_i$$

where x_{i2} is 1 if the house is on a corner lot and 0 otherwise.

	CORNER LOTS				NONCORNER LOTS		
Lot	Y	X_1	X_2	Lot	Y	X_1	X_2
1	56.2	17.5	1.0	10	31.2	10.0	0.0
2	42.5	12.5	1.0	11	36.9	13.8	0.0
3	68.6	20.0	1.0	12	41.0	15.0	0.0
4	54.8	16.0	1.0	13	51.8	19.5	0.0
5	50.0	15.0	1.0	14	48.0	17.0	0.0
6	47.5	14.7	1.0	15	33.3	12.5	0.0
7	56.9	17.5	1.0	16	38.0	14.5	0.0
8	34.0	12.3	1.0	17	35.9	12.8	0.0
9	39.0	11.5	1.0	18	32.0	12.0	0.0
				19	44.3	16.0	0.0
				20	29.0	10.0	0.0
				21	46.1	17.0	0.0
				22	30.0	10.8	0.0
				23	42.0	15.0	0.0

a. Plot the sample data for the two groups on one graph, using different symbols for the two samples. Does the regression relation appear to be the same for the two populations?

b. Let $\alpha = .05$. Use a two-tailed test and test whether β_2 is 0.

c. Plot the estimated regression functions for the two groups and describe the differences between them.

d. Prepare a residual plot for each sample. Does the assumption of equal error term variances appear to be reasonable here?

17.S.6 The accompanying data show Gross National Product per capita in U.S. dollars (Y), the population in millions (X_1), land in thousands of square kilometers (X_2), percentage of literate adults in the population (X_3), and the percentage of the economy attributed to agriculture (X_4) for a sample of 34 countries.
  a. Compute the multiple regression of Y on X_1, X_2, X_3, and X_4.
  b. Find the sample correlation matrix for the five variables.
  c. Examine the sample correlation matrix and predict which variables will enter first, second, and third using stepwise regression.
  d. Estimate Y as a function of the other four variables using stepwise regression.

Y	X_1	X_2	X_3	X_4
90	1.2	144	22	55
130	4.4	26	23	25
160	4.9	1,267	8	47
190	16.4	945	66	45
240	1.3	30	40	30
270	1.5	1,031	17	26
300	133.5	2,027	62	31
140	5.6	118	25	47
240	8.1	587	58	40
320	7.9	1,001	44	32
420	43.8	541	82	27
450	5.1	753	39	14
550	18.3	447	28	21
730	2.8	407	80	20
840	5.0	49	60	20
1,110	17.0	2,382	35	8
1,160	10.6	757	88	10
1,360	116.1	8,512	76	12
1,960	21.7	256	85	16
2,850	3.6	21	88	7
1,240	2.1	51	88	21
860	5.9	164	38	17
690	7.5	322	20	25
430	5.2	196	10	28
2,880	3.2	70	98	22
6,130	7.5	84	99	5
7,590	9.8	31	99	2
8,550	4.0	324	99	6
8,520	220.0	9,363	99	3
7,340	14.1	7,687	100	5
6,680	2.8	1,760	45	3
910	9.6	115	96	21
1,580	21.6	238	98	31
3,150	34.7	313	98	16

References

Draper, Norman R., and Harry Smith. *Applied Regression Analysis.* 2d ed. New York: Wiley, 1981.

Goldberger, Arthur. *Econometric Theory.* New York: Wiley, 1964.

Griffiths, William E., R. Carter Hill, and George G. Judge. *Learning and Practicing Econometrics.* New York: Wiley, 1993.

Gujarati, D. *Basic Econometrics.* 2d ed. New York: McGraw-Hill, 1988.

Johnston, John. *Econometric Methods.* New York: McGraw-Hill, 1972.

Judge, G. G., R. C. Hill, W. E. Griffiths, and T. C. Lee. *The Theory and Practice of Econometrics.* 2d ed. New York: Wiley, 1985.

Judge, G. G., R. C. Hill, W. E. Griffiths, H. Lutkepohl, and T. C. Lee. *Introduction to the Theory and Practice of Econometrics.* 2d ed. New York: Wiley, 1988.

Klein, L. *An Introduction to Econometrics.* Englewood Cliffs, N.J.: Prentice Hall, 1962.

Kmenta, Jan. *Elements of Econometrics.* 2d ed. New York: Macmillan, 1986.

Lovell, Michael C. "Data Mining." *Review of Economics and Statistics* 65 (1983): 1–12.

Nie, Norman E., C. Hadlai Hull, Jean G. Jenkins, Karin Steinbrenner, and Dale H. Bent. *SPSS Statistical Package for the Social Sciences.* 2d ed. New York: McGraw-Hill, 1975.

Norusis, Marija J. *SPSSX Introductory Statistics Guide.* New York: McGraw-Hill, 1990.

Norusis, Marija J. *SPSSX Advanced Statistics Guide.* Chicago: SPSS, 1990.

Norusis, Marija J. *The SPSS Guide to Data Analysis.* Chicago: SPSS, 1986.

Pindyck, Robert S., and Daniel L. Rubinfeld. *Econometric Models and Economic Forecasts.* 2d ed. New York: McGraw-Hill, 1981.

Ryan, Thomas A., Brian L. Joiner, and Barbara F. Ryan. *Minitab Reference Manual.* University Park, Penn.: Minitab Project, 1985.

Ryan, Thomas A., Brian L. Joiner, and Barbara F. Ryan. *Minitab Handbook.* 2d ed. Boston: PWS-Kent, 1985.

SAS Introductory Guide. 3d ed. Cary, N.C.: SAS Institute, 1985.

SAS Procedures Guide for Personal Computers. Version 6 ed. Cary, N.C.: SAS Institute, 1986.

SAS Statistics Guide for Personal Computers. Version 6 ed. Cary, N.C.: SAS Institute, 1986.

SAS User's Guide: Basics. Version 5 ed. Cary, N.C.: SAS Institute, 1985.

SAS User's Guide: Statistics. Version 5 ed. Cary, N.C.: SAS Institute, 1985.

Schmidt, P. *Econometrics.* New York: Marcel Dekker, 1976.

SPSSX User's Guide. Chicago: SPSS, 1988.

Suits, Daniel. "Dummy Variables: Mechanics vs. Interpretation." *Review of Economics and Statistics* 66(1984): 177–180.

Theil, Henri. *Principles of Econometrics.* New York: Wiley, 1971.

Wonnacott, Thomas H., and Ronald J. Wonnacott. *Regression: A Second Course in Statistics.* New York: Wiley, 1981.

18 Residual Analysis and Violations of the Basic Assumptions

In the previous chapters, we showed how to estimate and interpret the parameters β_0, $\beta_1, \ldots, \beta_K$ and σ_e^2 in the population regression equation

$$E(Y_i | x_{i1}, x_{i2}, \ldots, x_{iK}) = \beta_0 + \beta_1 x_{i1} + \beta_2 x_{i2} + \cdots + \beta_K x_{iK}$$

In addition, we showed that the coefficient of determination R^2 and the estimated standard error of the regression s_e can be used to measure the goodness of fit of the estimated model. Finally, we showed how to use the correlation matrix and stepwise regression to help select an appropriate model.

In deriving our results, we made use of the basic assumptions concerning the error terms and the explanatory variables. When the basic assumptions hold, the least-squares estimators of the regression coefficients are normally distributed and unbiased and have minimum variance among the class of linear unbiased estimators. In addition, the estimated variances are unbiased estimators of the corresponding population parameters.

In this chapter, we discuss methods of testing to determine whether the basic assumptions hold, and we describe how the properties of the least-squares estimators are affected when any of the basic assumptions are violated. In addition, we develop alternative methods of estimation when the consequences of such a violation are serious.

There are many reasons that a specific model can be inadequate and many ways to check or test for the inadequacies. In this chapter, we cover many of these reasons, show how to test for violations of the basic assumptions, and give procedures for improving inadequate models.

18.1 Searching for Model Inadequacies

The process of checking the adequacy of a proposed model involves many steps. Some of the things that should be examined are the following:

1. *Signs of coefficients:* Examine each estimated coefficient and determine whether the sign of the estimated coefficient agrees with theoretical expectations. When an estimated coefficient has the wrong sign, there is strong evidence that the model is inadequate and should be changed.
2. *Standard deviations of coefficients:* Examine the estimated standard deviation of each estimated coefficient. Coefficients with large standard deviations have not been estimated with much precision.

3. *t statistics and p-values:*   The t statistic $t = b_i/s_{b_i}$ is used to test the null hypothesis that the population regression coefficient is 0. When the t statistic is close to 0 or when the p-value is large, we do not have strong evidence that the true population regression coefficient is nonzero. This raises the question of whether the variable should be included in the model.

4. *R^2 and adjusted R^2:*   Examine the values of R^2 and the adjusted R^2 to get an idea of how well the explanatory variables explain variation in the dependent variable. Values of R^2 close to 0 indicate that the explanatory variables explain only a small proportion of the variation in the dependent variable.

5. *Estimated standard error of the regression:*   Examine s_e to get an idea of the absolute size of the typical residual. If s_e is small, the sample data tend to lie close to the estimated equation. Approximately 95% of the observations should lie within 2 standard errors of the regression line. Large standard errors indicate that predictions typically will not be very accurate.

6. *Residual analysis:*   Examine the residuals to discover whether any of the basic assumptions have been violated. Plot the residuals against the sample values of the independent variables, against the fitted values of the dependent variable ($\hat{y}_i$), or against time if applicable. In addition, a histogram of the residuals may reveal a violation of the normality assumption. If there are no violations of the basic assumptions, the residual plots should show no distinct patterns of variation.

Using scatter plots to detect violations of the basic assumptions can pose problems because interpreting the plots is based on subjective judgment rather than well-defined statistical tests. In addition, unless the violation is quite obvious, interpreting the plots usually requires considerable skill and experience. In the remainder of this chapter, we discuss potential violations of the basic assumptions, and we describe graphical techniques and formal statistical tests for detecting such violations.

Once we determine that some basic assumption fails to hold, we have to decide whether to adjust the least-squares model in some way or to use a different estimation technique. When an assumption fails to hold, it is sometimes possible to transform the model so that the basic assumptions will hold in the transformed model.

When a test indicates a violation, a common practice is to transform the model and reestimate the regression coefficients using an estimation method other than ordinary least squares that takes the violation into account. However, there is a statistical problem involved with estimating a model, testing it, revising it, and reestimating. If we construct confidence intervals and test for the significance of individual regression coefficients after we have performed tests and revised the model, the results are conditional on the outcome of the original tests. In statistical literature, this is known as *pretest bias*. When this is done, the usual significance levels or confidence intervals are not the same as when no preliminary testing has been done.

If the regression model adequately describes real-world behavior, the residuals should possess properties that tend to confirm the assumptions we have made; at the very least, they should not lead us to reject the assumptions. When we examine the residuals, we should ask ourselves, "Do the residuals make it appear that the regression model is inadequate or that the assumptions are wrong?"

Searching for Outliers

Before performing any formal statistical tests, it is useful to examine scatter diagrams of the original data and the residuals to see whether there are any obvious errors. Occasionally data are reported incorrectly or typing errors are made when inputting the data into the computer. Frequently such errors can be spotted by examining a scatter diagram or by searching for residuals that are much larger or much smaller than most of the others.

DEFINITIONS **Outlier and Outlier Residual** ─────────────────────────────

An **outlier** is any observation that is quite different in magnitude from other observations. An **outlier residual** is a residual that is much larger in absolute value than most of the other residuals.

For practical purposes, an outlier can be considered to be any observation that lies more than $3s_e$ units from the estimated regression equation.

───

EXAMPLE 18.1 **Using a Scatter Diagram to Detect an Outlier** ─────────────────────────

Consider the data plotted in Figure 18.1. This is a scatter diagram showing the average salary of elementary and secondary school teachers (Y) against the average tax revenue

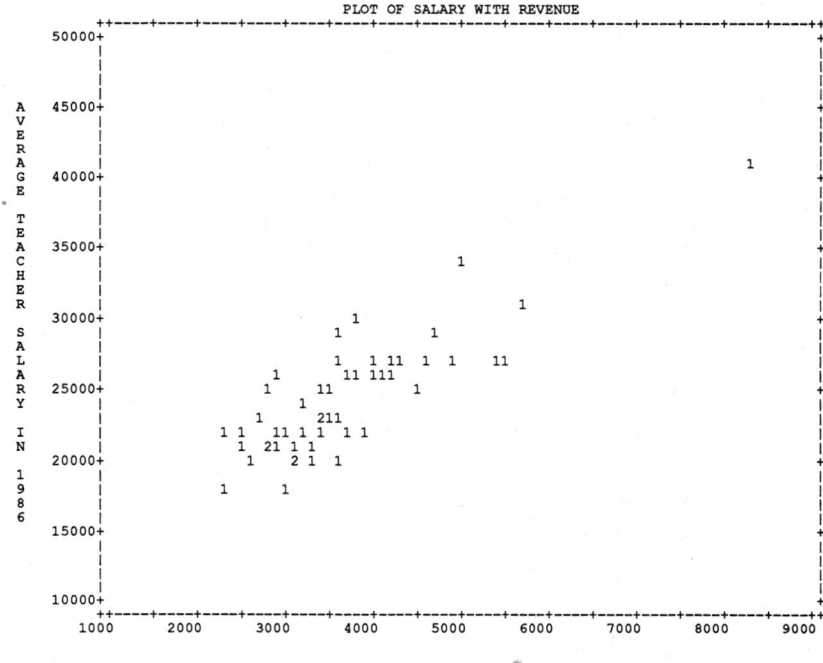

FIGURE 18.1 *Teachers' salaries versus tax revenue per capita in the 50 states and District of Columbia; Alaska is an outlier*

per capita (X) for each of the 50 states and the District of Columbia. The data were presented in Chapter 2 in Table 2.3. The scatter diagram reveals one very obvious outlier—the point in the upper right corner of the graph, which represents the data for the state of Alaska.

In Alaska, the average salary of teachers is approximately 25% higher than in any other state. In addition, Alaska has much higher tax revenue per capita than any other state. Including the data point for Alaska will have a strong effect on the estimated regression coefficients, because the estimated regression line is found by minimizing the squared residuals. To minimize SSE, the sample regression line will have to pass close to the Alaska data point.

Standardized Residuals

Another way of detecting outliers is by calculating *standardized residuals*. Because the standardized residuals have a mean of 0 and a standard deviation of 1, the relative magnitudes of the residuals are easier to judge when they are divided by s_e. For example, the fact that a particular residual is, say, 6,253.21 provides little information. However, if you know that its standardized value is 4.2, you know that this residual is much larger than most residuals in absolute value.

DEFINITION   **Standardized Residual** ———————————————————————————————

The ith **standardized residual** is denoted by the symbol SR_i and is defined as

$$SR_i = \hat{e}_i / s_e$$

where $\hat{e}_i$ is the ith residual and s_e is the estimated standard error of the regression.

———

Many computer programs, including the SPSSX program, plot both the residuals and the standardized residuals along with their calculated values. If the standardized residuals are approximately normally distributed, the Empirical Rule tells us that approximately 95% of the standardized residuals should be between -2.00 and $+2.00$ and more than 99% of the standardized residuals should be between -3.00 and $+3.00$. Thus standardized residuals less than -3 or greater than $+3$ should not occur with great frequency.

One of the easiest ways to detect a violation of the basic assumptions is to plot the observations (x_i, y_i) on a scatter diagram or to plot each residual $\hat{e}_i$ versus the observation y_i, versus $\hat{y}_i$, versus x_{ij}, and so forth. There are many different residual plots that can be generated. Throughout this chapter, these plots are used to detect violations of the basic assumptions.

Exercises

Statistical Concepts

18.1.1   Explain what is meant by residual analysis. What is an outlier residual? What is a standardized residual?

18.1.2   Explain how to examine the signs of estimated regression coefficients to detect possible model inadequacy.

18.1.3   Explain what is indicated by small t statistics associated with estimated regression coefficients.

18.1.4   What do small values of R^2 indicate? Explain.

18.1.5   Explain what is signaled by a large estimated standard error of the regression.

Statistical Applications

18.1.6   *True or false:*   In a regression model, the variable with the largest estimated coefficient is the most important. Explain.

18.1.7   *True or false:*   In a regression model, the variable with the largest observed t statistic is the most important variable. Explain.

 18.1.8   A realtor studied the relation between household electricity consumption (Y) and the number of rooms in the house (X). The realtor estimated a linear regression model and obtained the following residuals:

House	1	2	3	4	5	6	7	8	9	10
X	2	3	4	5	5	5	6	7	7	8
$\hat{e}$	3.2	2.9	−1.7	−2.0	−2.3	−1.2	−.9	.5	.7	.8

a. Plot the residuals $\hat{e}_i$ against x_i.

b. What problem appears to be present here? Might a transformation alleviate this problem?

\$\$ 18.1.9   A sociologist estimated a linear regression model to relate per capita earnings (Y) to the average number of years of education (X) in 12 different cities. The fitted values $\hat{Y}$ and the standardized residuals (SR) follow:

City	1	2	3	4	5	6	7	8	9	10	11	12
$\hat{Y}$	9.9	9.3	10.2	9.6	10.2	12.4	14.3	9.6	9.2	15.6	11.2	13.1
SR	−1.12	.81	−.76	.43	.65	−.17	1.62	1.79	−.53	−3.78	.74	.32

a. Plot the standardized residuals against the fitted values. What does the plot suggest?

b. Do you detect an outlier?

18.2   Violations of Basic Assumption 1: Nonnormal Errors

The first basic assumption states that the error terms follow a normal distribution. In this section, we discuss potential problems when this basic assumption fails to hold.

Basic assumption 1

> Each of the random error terms e_i is selected from a normal distribution with mean 0 and variance σ_e^2.

Consequences of Nonnormality of the Error Terms

If the normality assumption does not hold, the least-squares estimators of the regression coefficients are still the best linear unbiased estimators. In other words, even with non-normal errors, the least-squares estimators are unbiased and have the smallest variance among all linear unbiased estimators of the respective parameters. However, all confidence intervals and tests of hypotheses concerning the regression coefficients do depend on the assumption of normality.

Without the assumption of normality, the sampling distributions of the least-squares estimators $b_0, b_1, \ldots, b_K$ are not normal and the t statistics used for testing hypotheses about the population regression coefficients do not follow the t distribution. Strictly speaking, therefore, the confidence intervals and hypothesis tests that we have developed would not apply. Fortunately, however, if the distribution of the error terms does not differ too radically from normality, the use of the t distribution to construct confidence intervals and to test hypotheses is still approximately correct.

In many cases, as the sample size increases, the sampling distributions of the least-squares estimators approach normality even when the error terms are not normal. This implies that when the sample sizes are relatively large, our confidence intervals will be approximately correct and tests of hypotheses will have approximately the correct level of significance.

Detecting Nonnormality of Error Terms

The SPSSX computer program can produce a histogram of the residuals and the standardized residuals. If the errors are normally distributed and the sample size is fairly large, the histogram of the standardized residuals should appear like a standard normal distribution.

```
HISTOGRAM
STUDENTIZED RESIDUAL
  N EXP N      (  * = 2 CASES.    . : = NORMAL CURVE)
  3  0.37   OUT **
  1  0.73  3.00 *
  3  1.85  2.66 :*
  4  4.23  2.33 *:
 10  8.65  2.00 ***:*
 14 15.85  1.66 ********.
 21 26.01  1.33 ***********.
 31 38.23  1.00 ****************
 48 50.34  0.66 ************************.
 55 59.38  0.33 ****************************.
 63 62.74  0.00 ******************************.*
 64 59.38 -0.33 ****************************:**
 62 50.34 -0.66 ************************:******
 44 38.23 -1.00 *****************.***
 28 26.01 -1.33 **************:*
 14 15.85 -1.66 ********.
  7  8.65 -2.00 ***:
  1  4.23 -2.33 *.
  1  1.85 -2.66 :
  0  0.73 -3.00
  0  0.37   OUT
```

FIGURE 18.2   *SPSSX-generated histogram of residuals*

Figure 18.2 is an SPSSX printout of a histogram of standardized residuals.

18.3   Violations of Basic Assumption 2: Nonzero Mean

The second basic assumption states that each error term is a random variable selected from a distribution having a mean of 0.

Basic assumption 2

Each of the random error terms e_i is a random variable selected from a distribution having a mean of 0.

The assumption that the mean of each of the error terms is 0 implies that the specification of the population regression equation is

$$E(Y_i|x_{i1}, x_{i2}, \ldots, x_{iK}) = \beta_0 + \beta_1 x_{i1} + \beta_2 x_{i2} + \cdots + \beta_K x_{iK}$$

If the mean of the error term is not 0, but, say, μ_i, we have

$$E(Y_i|x_{i1}, x_{i2}, \ldots, x_{iK}) = \beta_0 + \beta_1 x_{i1} + \beta_2 x_{i2} + \cdots + \beta_K x_{iK} + \mu_i$$

The consequences of violating this basic assumption depend on the nature of the mean μ_i. If μ_i differs for each observation, the intercept of the equation becomes $(\beta_0 + \mu_i)$ and differs for each observation. Thus, the mean value of Y_i changes not only because of changes in the values of the independent variables $x_{i1}, x_{i2}, \ldots, x_{iK}$ but also for other reasons. In other words, the relationship between Y_i and $x_{i1}, x_{i2}, \ldots, x_{iK}$ has not been correctly specified. For example, if we have omitted the important explanatory variable Z, it might be that the mean μ_i is

$$\mu_i = \beta_{K+1} z_i$$

Consequences of a Nonzero Mean of the Error Terms

If the errors do not have zero means, we have either omitted some important independent variables or used an incorrect functional expression. In either case, we have estimated the wrong model. As a result, the least-squares estimates of the regression coefficients will be biased. Omitting an important variable and estimating an incorrect functional form are examples of what are called *specification errors*.

Detecting a Nonzero Mean of the Error Terms

Omitted Variables

The omission of an important variable can frequently be detected by estimating different regression models using different sets of independent variables. It is customary to use the t statistic to examine the statistical significance of each estimated coefficient to determine which variables should be included in a model.

The discussion of stepwise regression in Section 17.4 showed that adding or deleting an explanatory variable causes the t statistics to change for all of the other estimated regression coefficients. An estimated coefficient that appears to be statistically significant in one formulation of a model may not be statistically significant in another formulation. Unless there are strong theoretical reasons for keeping a variable in a model, it is customary to remove any variable whose t statistic is not significant at some prespecified level of significance.

Detecting whether an important variable has been omitted from a model is frequently a trial-and-error process. Various models are estimated and the results are examined to determine whether the coefficients have the appropriate signs and magnitudes. Tests are performed to determine whether the estimated coefficients are statistically significant, and the values of R^2 and s_e are examined to see whether the model provides a good fit. If the model seems to be lacking in some way, it is customary to include additional independent variables or to estimate some alternative functional form. There is no easy way to determine whether an important explanatory variable has been omitted from the model except to include the variable in a model and examine the estimated coefficient of the variable.

Sometimes the estimated coefficients of a regression model do not seem to make sense. For example, sometimes economic theory tells us that a particular coefficient should be positive but the estimated coefficient is negative. Because such a result violates the economic theory, we would reject this model and search for a better one. Generally an estimated coefficient with the wrong sign indicates that the model has not been specified properly. Perhaps some other important explanatory variables have been omitted or perhaps the wrong functional form has been estimated.

Occasionally incorrect signs or unusual values for estimated coefficients can be attributed to *multicollinearity*. If two explanatory variables X_1 and X_2 are highly correlated, then it is difficult to determine their separate influences on the dependent variable Y. Because this is a data problem rather than a specification error, there is little that can be done about it. When two explanatory variables X_1 and X_2 are highly correlated, it is difficult to get precise estimates of their coefficients β_1 and β_2.

For example, consider what happens when variables X_1 and X_2 are highly positively correlated with each other and are both positively correlated to the dependent variable Y. In advance, we do not know whether Y is a function of just X_1, just X_2, or both X_1 and X_2. There are three different models that we could estimate. Suppose we estimate all three different models and obtain the following three sample regression equations:

$$\hat{y}_i = b_0 + b_1 x_{i1} + b_2 x_{i2} \tag{1}$$
$$\hat{y}_i = c_0 + c_1 x_{i1} \tag{2}$$
$$\hat{y}_i = d_0 + d_1 x_{i2} \tag{3}$$

What would we expect to find, given that X_1 and X_2 are highly positively correlated? The estimated coefficient c_1 in equation (2) would be larger than the estimated coefficient b_1 in equation (1), because in the second equation some of the influence that X_2 has on Y has been attributed to the lone explanatory variable X_1. If the correct regression model contains both X_1 and X_2, that is, if the correct model is

$$Y_i = \beta_0 + \beta_1 x_{i1} + \beta_2 x_{i2} + e_i$$

then c_1 will tend to overestimate the true effect of X_1 on Y. For the same reason, the estimated coefficient d_1 will exceed the estimated coefficient b_2 and d_1 will tend to overestimate the true effect of X_2 on Y. Consequently, the estimates c_1 and d_1 will be biased estimates that tend to overestimate the true coefficients β_1 and β_2, respectively.

Thus, because of the high correlation between X_1 and X_2, it is difficult to get precise estimates of β_1 and β_2. The standard errors of the regression coefficients b_1 and b_2 will be large, causing the confidence intervals for the true values of β_1 and β_2 to be fairly wide. Although we have obtained the best estimates that we can, these estimates may not be very good.

When the explanatory variables are highly correlated, the precision of the estimates is affected. The standard errors of the estimated coefficients will tend to be large, indicating that the estimates are not very reliable.

One way to check for multicollinearity is to find the sample correlation coefficients between all pairs of potential explanatory variables in the model. When the sample correlation coefficient between X_i and X_j is large in absolute value, it may be difficult to separate the effects of these two variables on the dependent variable Y.

Incorrect Functional Form

The classical regression model states that the variables are related according to the equation

$$E(Y_i|x_{i1}, x_{i2}, \ldots, x_{iK}) = \beta_0 + \beta_1 x_{i1} + \beta_2 x_{i2} + \cdots + \beta_K x_{iK}$$

If the expected value of Y_i is related to the explanatory variables according to some other form of equation, we have made a specification error. For example, perhaps the logarithm of Y is related to the explanatory variables according to some equation, or perhaps $E(Y_i)$ is related to X^2 as well as X, or perhaps $E(\ln Y_i)$ is related to $\ln X$. If we estimate the model

$$Y_i = \beta_0 + \beta_1 x_{i1} + \beta_2 x_{i2} + e_i$$

when the variables are related according to some other equation, we have estimated an incorrect functional form. (Strictly speaking, when we omit an important explanatory variable, we estimate an incorrect functional form; however, we have already discussed this case.) One way to detect an incorrect functional form is to plot the observations (x_i, y_i) on a scatter diagram. This procedure works fine when there is only one independent variable, but other procedures have to be used when there are many explanatory variables.

Another way of discovering an incorrect functional form is to plot the residuals $\hat{e}_i$ versus the observations of the explanatory variables x_{ij}; that is, construct a scatter diagram showing the points $(x_{ij}, \hat{e}_i)$. Suppose that we plot the points $(x_{ij}, \hat{e}_i)$ and obtain a scatter diagram where the points lie in a shaded area, as in Figure 18.3. This pattern occurs quite frequently when analyzing economic and business data. In the scatter diagram, small and large values of X tend to be associated with positive residuals, whereas middle values of X tend to be associated with negative residuals. Because the residuals should be randomly scattered, such patterns in a residual plot usually indicate that one of the assumptions has been violated. The pattern in Figure 18.3 indicates that we tend to overestimate Y for

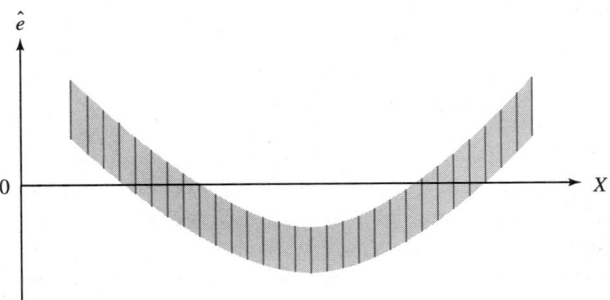

FIGURE 18.3 *Plot of residuals versus an explanatory variable X*

middle values of X and to underestimate Y when X is very small or very large. This is an indication that we have fitted a straight line to data better described by a curvilinear relationship, such as that illustrated in Figure 18.4.

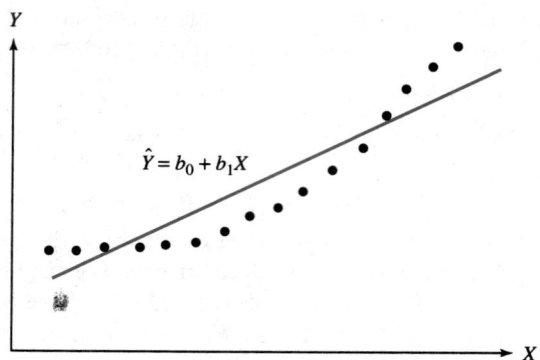

FIGURE 18.4 *Fitting a straight line to curvilinear data*

When a model has several explanatory variables, each one should be plotted against Y and against the residuals. If the residual plot indicates that the relationship between Y and an explanatory variable X_j is curvilinear, we should try a different functional form such as a polynomial. These types of models were discussed in Section 17.1.

EXAMPLE 18.2   **Detecting a Nonlinear Relationship** ———————————————————

The data in Table 18.1 show before-tax income (X) and federal income taxes paid (Y) for families covered in a recent census. Let us construct a model to explain taxes as a function of a family's income.

S O L U T I O N   Suppose we estimate the linear model

$$Y_i = \beta_0 + \beta_1 x_i + e_i$$

The SPSSX computer output for this model is shown in Figure 18.5. The sample regression equation is

$$\hat{y}_i = -2{,}189.33 + .234768x_i \qquad R^2 = .9609$$
$$t \text{ statistics:} \quad (-5.85) \qquad (20.43)$$

Income explains approximately 96% of the variation in Y ($R^2 = .9609$) so the data fall close to a straight line. The t statistics for both coefficients are very large in absolute value and both have a p-value of .0000, so the evidence is very strong that both population regression coefficients are nonzero. At this stage, the computer output might lead us to believe that we have a fairly good model explaining tax expenditures as a linear function of income.

TABLE 18.1 *Data for Example 18.2*

Taxes paid Y	Before-tax income X
$ 128	$ 3,750
329	6,250
519	8,750
801	11,250
1,085	13,750
1,464	16,250
1,903	18,750
2,389	21,250
2,847	23,750
3,369	26,250
3,830	28,750
4,454	31,250
5,048	33,750
5,650	36,250
6,294	38,750
7,259	42,500
8,829	47,500
11,158	55,000
15,767	67,500

★ ★ ★ ★ M U L T I P L E   R E G R E S S I O N ★ ★ ★ ★

Listwise Deletion of Missing Data

Equation Number 1   Dependent Variable..   TAX   FEDERAL INCOME TAX

Block Number 1. Method: Enter   INCOME

Variable(s) Entered on Step Number 1..   INCOME   BEFORE TAX INCOME

		Analysis of Variance			
Multiple R	.98025		DF	Sum of Squares	Mean Square
R Square	.96090	Regression	1	292839012.69294	292839012.69294
Adjusted R Square	.95860	Residual	17	11916851.09654	700991.24097
Standard Error	837.25220				

F =   417.74989   Signif F =   .0000

------------------ Variables in the Equation ------------------

Variable	B	SE B	Beta	T	Sig T
INCOME	.234768	.011486	.980254	20.439	.0000
(Constant)	-2189.330098	374.218973		-5.850	.0000

End Block Number   1   All requested variables entered.

FIGURE 18.5 *SPSSX-generated output for Example 18.2, explaining tax expenditures as a linear function of before-tax income*

Now let us examine some scatter diagrams and computer-generated plots of the residuals to search for clues on how to improve the model. The data are plotted in Figure 18.6. From the scatter diagram, it is quite obvious that the relationship between Y and X is nonlinear.

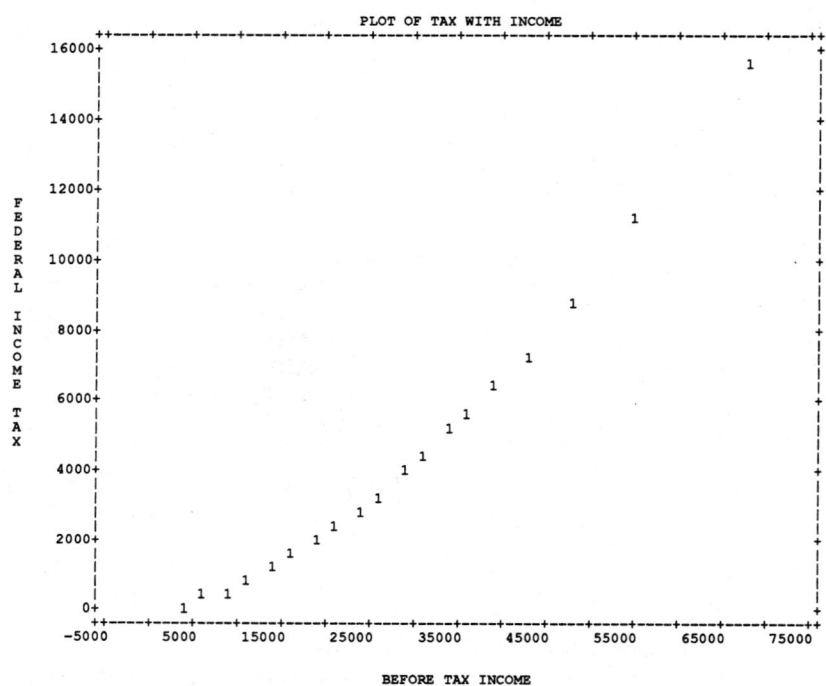

FIGURE 18.6   *SPSSX-generated scatter diagram showing tax expenditures versus before-tax income*

Figure 18.7 shows the standardized residuals SR_i plotted against the observed standardized x_i values. This scatter diagram reveals that the residuals associated with low values of X are positive, the residuals associated with middle values of X are negative, and the residuals associated with large values of X are positive.

Figure 18.8 shows the standardized residuals SR_i plotted against the standardized values of $\hat{y}_i$. This scatter diagram reveals basically the same information as the scatter diagram of $\hat{e}_i$ plotted against x_i. The residuals are positive for small values $\hat{y}_i$, negative for middle values, and positive for very large predicted values $\hat{y}_i$.

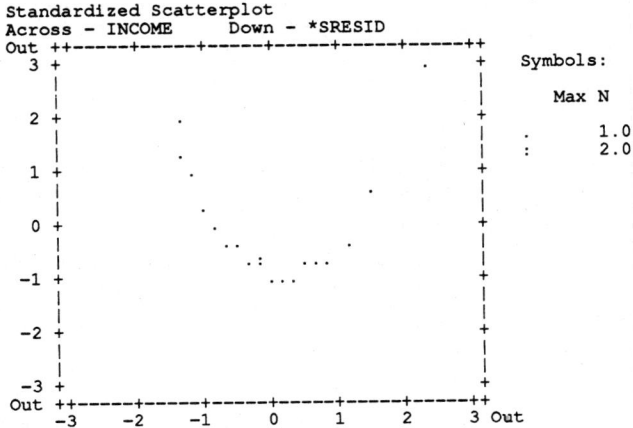

FIGURE 18.7 *SPSSX-generated scatter diagram showing standardized values of $(x_i, \hat{e}_i)$ for Example 18.2*

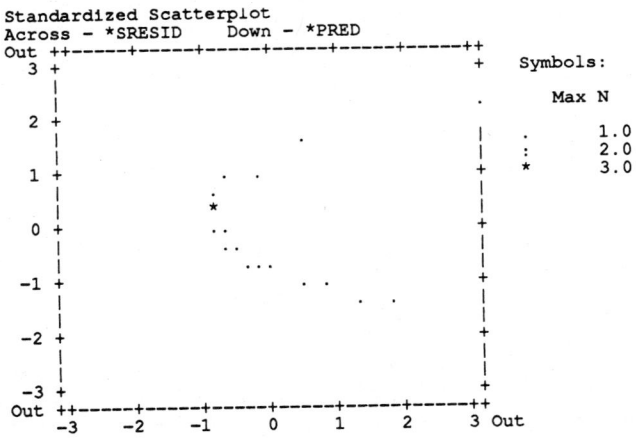

FIGURE 18.8 *SPSSX-generated scatter diagram showing standardized values of $(\hat{y}_i, \hat{e}_i)$ for Example 18.2*

From the scatter diagrams and plots of the residuals, the relationship between tax expenditures and income is better explained by a nonlinear equation. Let us estimate the nonlinear model

$$Y_i = \beta_0 + \beta_1 x_i + \beta_2 x_i^2 + e_i$$

The computer output for this model is shown in Figure 18.9. The estimated model is

$$\hat{y}_i = -312.97 + .076x_i + .0000024x_i^2 \qquad R^2 = .99979$$

$$t \text{ statistics:} \quad (-7.02) \quad (25.02) \quad (54.46)$$

In the computer output, the coefficient of the variable x_i^2 is written as 2.41280E–06. The symbol E–06 tells us to move the decimal point six places to the left. Thus, the coefficient of x_i^2 is $b_2 = .0000024128$. The quadratic equation explains more than 99.9% of the variation in taxes paid ($R^2 = .99979$). Each estimated regression coefficient has a t statistic whose p-value is .0000, which supports the hypothesis that each population regression coefficient is nonzero.

```
* * * *   M U L T I P L E   R E G R E S S I O N   * * * *

Listwise Deletion of Missing Data

Equation Number 1    Dependent Variable..   TAX    FEDERAL INCOME TAX

Block Number  1.  Method:  Enter      INCOME    INCSQ

Variable(s) Entered on Step Number  1..    INCSQ
                                    2..    INCOME     BEFORE TAX INCOME

Multiple R            .99990      Analysis of Variance
R Square              .99979                        DF    Sum of Squares      Mean Square
Adjusted R Square     .99976      Regression          2   304691929.36914   152345964.68457
Standard Error      63.21314      Residual           16       63934.42033        3995.90127

                                  F =   38125.55775      Signif F =   .0000

------------------ Variables in the Equation ------------------

Variable           B           SE B        Beta        T     Sig T

INCSQ       2.41280E-06    4.4301E-08     .691370    54.463   .0000
INCOME          .076066        .003040    .317608    25.020   .0000
(Constant)  -312.971650     44.555543              -7.024   .0000

End Block Number   1   All requested variables entered.
```

FIGURE 18.9   *SPSSX-generated output for Example 18.2, explaining tax expenditures as a quadratic function of before-tax income*

Exercises

Statistical Concepts

18.3.1   One of the basic assumptions of the multiple regression model is that the error terms follow a normal distribution. *True or false:*   If the error terms do not follow a normal distribution, then the estimated regression coefficients will be unbiased, but they will not be best. Explain.

18.3.2   One of the basic assumptions of the multiple regression model is that the error terms follow a normal distribution. *True or false:*   If the error terms do not follow a normal distribution, then the estimated regression coefficients will be unbiased and they will not be best, but the t tests and confidence intervals will not be exactly correct. Explain.

18.3.3 One of the basic assumptions of the multiple regression model is that the error terms have a mean of 0. *True or false:* If the error terms do not have a mean of 0, then the estimated regression coefficients will be unbiased. Explain.

18.3.4 One of the basic assumptions of the multiple regression model is that the error terms have a mean of 0. *True or false:* If the error terms do not have a mean of 0, then the population regression model has been misspecified. Explain.

18.3.5 Explain how a plot of the residuals versus the explanatory variables can be used to detect a possible nonlinear relationship between the dependent and explanatory variables.

Statistical Applications

18.3.6 Suppose the variables Y, X, and Z are exactly related according to the equation

$$Y = 4.0 + 2.2X + 4.2Z$$

In a sample of data, the sample correlation coefficient between X and Z is .94. Suppose we make a specification error and estimate the model

$$Y_i = \beta_0 + \beta_1 x_i + e_i$$

a. Will we obtain an unbiased estimate of β_1? Explain.

b. Can we determine whether the bias in part **a** will be positive or negative? Will we tend to overestimate or underestimate the true coefficient of X?

c. Suppose the sample mean of the variable Z is 10. What is the mean of the set of random error terms?

18.3.7 *True or false:* In a multiple regression model, if the t statistic for an estimated coefficient is not significantly different from 0, the variable should be deleted from the model because we know that the population regression coefficient is 0. Explain.

18.3.8 *True or false:* We can never be certain that our regression model correctly explains why the dependent variable changes. Explain.

18.3.9 Suppose the variables Y, X, and X^2 are exactly related according to the equation

$$Y = 2.0 + 1.2X + 2.5X^2$$

a. Plot this relationship.

b. Suppose we obtain a sample of data, and all the values of X are positive. We make a specification error and estimate the model

$$Y_i = \beta_0 + \beta_1 x_i + e_i$$

Discuss what the sample regression line will look like compared with the correct quadratic model.

c. What will happen if we use the estimated model to forecast values of Y?

18.4   Violations of Basic Assumption 3: Heteroscedasticity

The third basic assumption states that each of the error terms is a random variable selected from a distribution having the same variance σ_e^2. If the error terms all have the same variance, the errors are said to be *homoscedastic.* The word *homoscedastic* is derived from two Greek words that mean "having the same spread." If the error terms do not have equal variances, they are said to be *heteroscedastic,* from two Greek words meaning "having different spreads." In this section, we discuss the potential problems when this assumption fails to hold.

Basic assumption 3

> Each of the random variables $e_1, e_2, \ldots, e_n$ is a random variable selected from a distribution with finite variance σ_e^2. The variance σ_e^2 is the same for all values of the explanatory variables.

This assumption is violated if the dispersion of points about the true regression line varies with the magnitude of the explanatory variables. Figure 18.10 shows a situation where the errors are homoscedastic; that is, each of the probability distributions shown has the same variance, σ_e^2, regardless of the value of X.

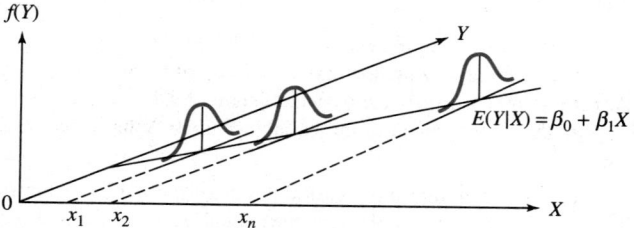

FIGURE 18.10   *Population regression line with homoscedastic errors*

Figure 18.11 shows a population regression model in which the errors are heteroscedastic. In the figure, the mean of each distribution falls on the regression line, but the variance of the distributions increases as X increases. If the errors are heteroscedastic, subpopulations of Y values associated with different values of X will have different variances. In economic and business data, variances frequently increase as the explanatory variables increase in magnitude, as illustrated in Figure 18.11.

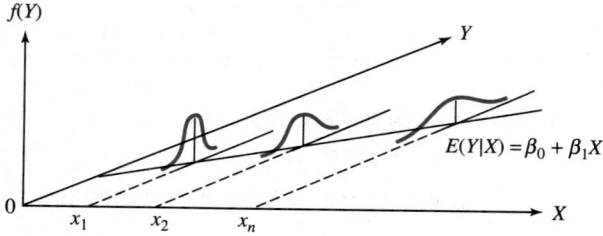

FIGURE 18.11   *Population regression line with heteroscedastic errors*

For example, suppose we have data on the annual income and annual consumption expenditures of individual families, and we formulate a model in which consumption expenditures are explained as a function of income. In this case, the assumption of homoscedasticity may not be very plausible because we would expect less variation in consumption for low-income families than for high-income families. At low incomes, the typical level of consumption is low and the variation about this level is relatively small. Consumption cannot fall too far below the average level, because this would mean near starvation for the family. In addition, consumption cannot rise too far above the average, because the family's assets and credit position would not allow it. These constraints are generally less binding for families with higher incomes. This argument implies that the consumption expenditures of families whose annual incomes are, say, $15,000 per year are likely to be more highly concentrated than the consumption expenditures of families whose annual incomes are, say, $80,000 per year. The appropriate model in this case would contain heteroscedastic error terms.

Consequences of Heteroscedasticity of Error Variances

If the errors do not have equal variances, the least-squares estimates $b_0, b_1, \ldots, b_K$ will still be unbiased estimates of the population parameters $\beta_0, \beta_1, \ldots, \beta_K$, but the estimates will not be efficient. In addition, the least-squares estimate of σ_e^2 and the estimates of the variances of the estimated coefficients b_i will be biased. If we know the nature of the heteroscedasticity, there exists an alternative linear unbiased estimator that yields estimates whose variances are smaller than the least-squares variances.

The fact that all the estimated variances are biased invalidates all the t tests and F tests used to test hypotheses about the values of the population parameters and invalidates all the confidence intervals for the population parameters.

Because heteroscedasticity of the error variances makes our tests and confidence intervals invalid, it is important to detect the presence of heteroscedasticity. Most tests for homoscedasticity are designed to test the assumption of equal error variances against some relatively simple alternative, such as the variance σ_i^2 of the ith error term e_i is related to the value of one of the independent variables, say, x_{ij}. For example, when trying to explain family consumption as a function of family income, it is reasonable to believe that the variance of the error terms increases as the independent variable of family income increases.

Detecting Heteroscedasticity of the Error Terms

Graphical Techniques

In some cases, plotting the residuals $\hat{e}_i$ against the values of the individual independent variables x_{ij} or against the predicted values $\hat{y}_i$ can be useful in detecting the presence of heteroscedasticity. For example, examine Figure 18.12, which shows a set of residuals $\hat{e}_i$ plotted against the values of an independent variable X. In the figure, there does not appear to be any systematic relationship between the magnitudes of the residuals and the values of X. Thus the scatter diagram provides no evidence that the variances of the error terms are not constant. If the variances are all equal, the residuals should fall approximately within a rectangle, as illustrated in Figure 18.12.

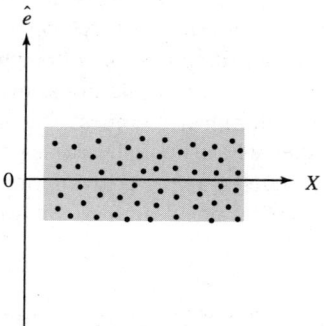

FIGURE 18.12   *Scatter diagram of $(x_i, \hat{e}_i)$ showing homoscedastic residuals*

In Figure 18.13, the amount of variation in the residuals tends to increase with the values of X. This suggests that the variance of the error terms is related to the magnitude of the variable X. If the error variance is not constant for all values of X and if the variance increases as X increases, the residuals will tend to lie in a funnel, as illustrated in Figure 18.13.

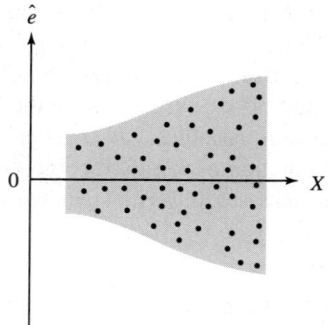

FIGURE 18.13   *Scatter diagrams of $(x_i, \hat{e}_i)$ showing heteroscedastic residuals*

Except in relatively obvious situations, interpreting a scatter plot of residuals depends on subjective judgment. Next we consider a formal procedure for testing the null hypothesis that the errors are homoscedastic against the alternative hypothesis that the variance of the error term is related to the magnitude of one of the independent variables.

The Goldfeld–Quandt Test

The Goldfeld–Quandt test is designed to test the null hypothesis that errors are homoscedastic against the alternative hypothesis that the errors are not homoscedastic. It is used when the observations can be arranged in order of supposed increasing variance of the error terms. This can be done, for example, when we think that the variance is related to the values of some explanatory variable, say, x_{ij}. The null hypothesis is

$$H_0: \quad \sigma_1^2 = \sigma_2^2 = \cdots = \sigma_n^2$$

where σ_i^2 is the variance of the ith error term. The test is described in the accompanying box.

Goldfeld–Quandt test for homoscedastic errors

Suppose we wish to test the null hypothesis

$$H_0: \quad \sigma_1^2 = \sigma_2^2 = \cdots = \sigma_n^2$$

against the alternative hypothesis

$$H_1: \quad \text{Var}(e_i) = \sigma^2 x_{ij}^2$$

where σ^2 is some unknown constant. Let the level of significance of the test be α.

To perform the test, arrange the sample observations according to the magnitudes of the variable X_j and split the sample of observations into three subsamples containing n_1, p, and n_2 observations. Let n_1 denote the number of observations in the first subsample, p the number of observations in the second subsample, and n_2 the number of observations in the third subsample. Typically $n_1 = n_2$ and p is much smaller. The middle set of p observations is omitted from the analysis.

Estimate the proposed regression model twice. In one equation, use only the first n_1 observations; in the other equation, use only the last n_2 observations. Denote the estimated variance from the first regression equation as s_1^2 and that from the second regression equation as s_2^2. Calculate the F statistic

$$F = s_2^2/s_1^2$$

If H_0 is true, then s_2^2 should be approximately equal to s_1^2, and the ratio s_2^2/s_1^2 should be approximately equal to 1. If H_0 is false, the F should be large because s_2^2 should exceed s_1^2.

To perform the test, use the decision rule

Reject H_0 in favor of H_1 if $F > F_{\alpha,\nu_1,\nu_2}$

where F_{α,ν_1,ν_2} is the critical value of the F distribution such that

$$P(F > F_{\alpha,\nu_1,\nu_2}) = \alpha \qquad \text{(continued)}$$

When H_0 is true, the F statistic follows the F distribution with numerator degrees of freedom $\nu_1 = (n_2 - K - 1)$ and denominator degrees of freedom $\nu_2 = (n_1 - K - 1)$, where K denotes the number of explanatory variables in the model not counting the constant term.

The reason for omitting a set of observations from the middle of the sample is to make the variances in the two remaining subsamples more disparate. The best number of observations to omit is not clear. The power of the test is increased by increasing the difference between the "small" variances and the "large" variances, but at the same time, the power is reduced because fewer observations are used in performing the test. Based on experiments, Goldfeld and Quandt recommend dropping about one-sixth of the observations. The arbitrariness in choosing p, the number of observations to be omitted, represents a rather unsatisfactory aspect to the test, because it might be possible for a statistician to tailor the results of the test to agree with his or her wishes by a clever choice of p.

EXAMPLE 18.3   Testing for Heteroscedasticity of the Error Variances

The data in Table 18.2 show the annual incomes and the years of education for a sample of 28 adult males who work for a certain company and have the same number of years of seniority. Figure 18.14 shows a scatter diagram of the data. The scatter diagram shows that the dispersion of annual income tends to increase as the years of education increase. Thus, it is reasonable to suspect that the errors are heteroscedastic.

TABLE 18.2   *Data on annual income and years of education for Example 18.3*

Years of education	Number of individuals	Annual income (in thousands of dollars)
8	6	14, 16, 18, 20, 22, 24
12	6	20, 24, 28, 32, 36, 40
14	4	22, 28, 38, 50
16	6	24, 30, 38, 42, 50, 56
18	6	26, 34, 44, 52, 56, 64

The hypothesized relationship between income (Y) and years of education (X) is

$$Y_i = \beta_0 + \beta_1 x_i + e_i$$

Let us estimate the population regression equation and perform the Goldfeld–Quandt test for heteroscedasticity.

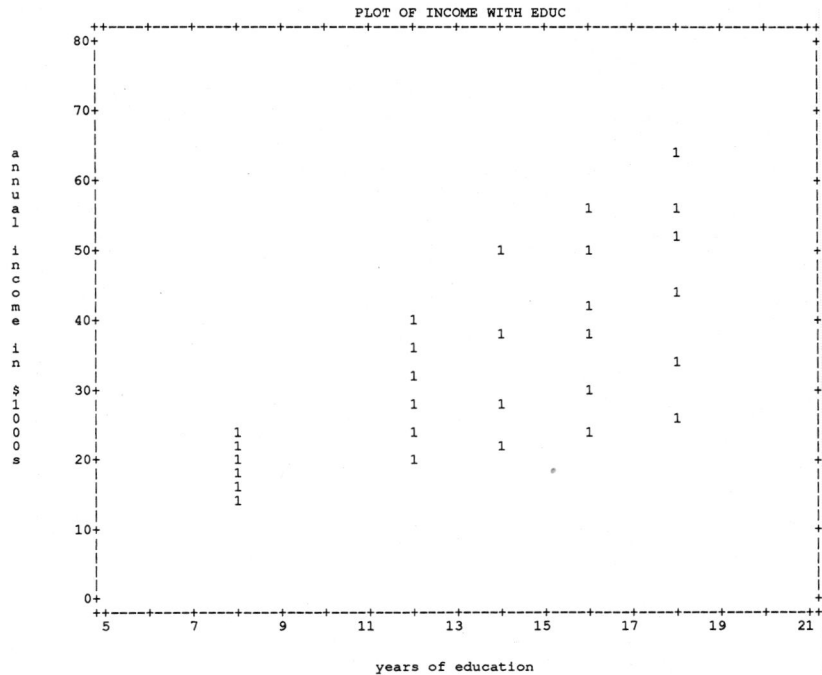

FIGURE 18.14 *SPSSX-generated scatter diagram showing income versus years of education*

SOLUTION   The SPSSX computer output from estimating this model is given in Figure 18.15 (page 778). The sample regression equation using all 28 observations is

$$\hat{y}_i = -2.333 + 2.667 x_i \qquad R^2 = .499$$
$$t \text{ statistics:} \quad (-.32) \qquad (5.09)$$

If the errors are heteroscedastic, the estimated standard errors, the t statistics, and the p-values shown in the computer output are all invalid.

   Figure 18.16 (page 778) shows a computer-generated scatter diagram of the residuals $\hat{e}_i$ plotted against the values of the explanatory variable X. (In the graph, the residuals and the values of X have all been standardized by subtracting the mean and dividing by the standard deviation.) Note how the residuals tend to have more dispersion for large values of X than for small values of X. This provides evidence supporting the hypothesis that the errors are heteroscedastic.

   Let us use the Goldfeld–Quandt test to test the null hypothesis

$$H_0: \quad \sigma_1^2 = \sigma_2^2 = \cdots = \sigma_n^2$$

```
* * * *   M U L T I P L E   R E G R E S S I O N   * * * *

Listwise Deletion of Missing Data

Equation Number 1    Dependent Variable..   INCOME   annual income in $1000s

Block Number  1.  Method:  Enter      EDUC

Variable(s) Entered on Step Number  1..    EDUC       years of education

Multiple R           .70651      Analysis of Variance
R Square             .49915                       DF      Sum of Squares      Mean Square
Adjusted R Square    .47989      Regression        1         2523.42857       2523.42857
Standard Error      9.86836      Residual         26         2532.00000         97.38462

                                 F =     25.91198     Signif F =   .0000

------------------ Variables in the Equation ------------------

Variable            B         SE B       Beta       T      Sig T

EDUC          2.666667     .523864     .706507    5.090    .0000
(Constant)   -2.333333    7.350113               -.317    .7534

End Block Number   1   All requested variables entered.
```

FIGURE 18.15 *SPSSX-generated output for Example 18.3, explaining income as a function of years of education*

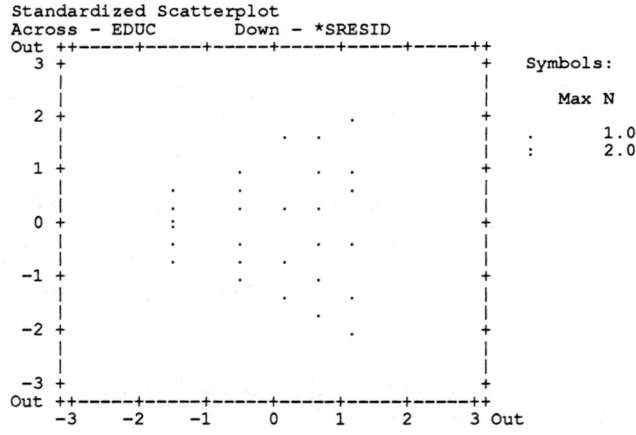

FIGURE 18.16 *SPSSX-generated scatter diagram showing* $(x_i, \hat{e}_i)$ *for Example 18.3*

against the alternative hypothesis that the variances increase as X increases. As suggested by the Goldfeld–Quandt test, the data have been placed in ascending order according to the magnitude of X. Suppose we let $p = 4$ and discard the middle 4 observations, corresponding to individuals with 14 years of education. We use the first $n_1 = 12$ observations to estimate the regression equation and then repeat the process using the last $n_2 = 12$ observations.

Figure 18.17 shows the SPSSX computer output for the regression equation estimated by using the first $n_1 = 12$ observations. The estimated regression equation is

$$\hat{y}_i = -3.000 + 2.750x_i \qquad R^2 = .509$$
$$t \text{ statistics:} \quad (-.35) \qquad (3.22)$$

To perform the Goldfeld–Quandt test, we need the value s_1^2 (s_e^2). This value, listed in the computer output as the residual mean square, is 35.00. Alternatively, recall that s_1^2 is the square of the estimated standard error of the regression. The computer output lists the estimated standard error of the regression as $s_1 = 5.91608$. Thus, we obtain $s_1^2 = 35.00$.

```
* * * *   M U L T I P L E    R E G R E S S I O N   * * * *

Listwise Deletion of Missing Data

Equation Number 1     Dependent Variable..    INCOME    annual income in $1000s

Block Number  1.  Method:  Enter         EDUC

Variable(s) Entered on Step Number  1..      EDUC      years of education

Multiple R              .71352      Analysis of Variance
R Square                .50912                       DF    Sum of Squares    Mean Square
Adjusted R Square       .46003      Regression        1       363.00000      363.00000
Standard Error        5.91608       Residual         10       350.00000       35.00000

                                    F =    10.37143     Signif F =   .0092

------------------ Variables in the Equation ------------------

Variable          B        SE B        Beta        T    Sig T

EDUC         2.750000     .853913     .713524    3.220   .0092
(Constant)  -3.000000    8.708234               -.345   .7376

End Block Number   1    All requested variables entered.
```

FIGURE 18.17   *SPSSX-generated regression output using first*
$n_1 = 12$ observations from Example 18.3

Figure 18.18 (page 780) shows the computer output for the regression equation estimated using only the last $n_2 = 12$ observations. The estimated regression equation is

$$\hat{y}_i = -8.000 + 3.000x_i \qquad R^2 = .059$$
$$t \text{ statistics:} \quad (-.12) \qquad (.79)$$

and s_2^2 has the value 172.80.

To apply the Goldfeld–Quandt test, calculate the ratio

$$F = s_2^2/s_1^2 = 172.80/35.00 = 4.937$$

If the null hypothesis of homoscedasticity is true, this F value follows the F distribution with numerator degrees of freedom $\nu_1 = (n_2 - K - 1) = 10$ and denominator degrees of freedom $\nu_2 = (n_1 - K - 1) = 10$. Suppose we wish to test the null hypothesis that the errors are homoscedastic using level of significance $\alpha = .01$. The critical value of the F statistic is $F_{.01,10,10} = 4.85$. Because the actual F statistic, 4.937, exceeds the critical value, we would reject the null hypothesis that the errors are homoscedastic.

```
                     * * * *   M U L T I P L E   R E G R E S S I O N   * * * *

Listwise Deletion of Missing Data

Equation Number 1    Dependent Variable..    INCOME    annual income in $1000s

Block Number  1.  Method:  Enter    EDUC


Variable(s) Entered on Step Number  1..    EDUC    years of education


Multiple R            .24254    Analysis of Variance
R Square              .05882                      DF    Sum of Squares    Mean Square
Adjusted R Square    -.03529    Regression         1         108.00000      108.00000
Standard Error      13.14534    Residual          10        1728.00000      172.80000

                                F =       .62500    Signif F =   .4475


------------------ Variables in the Equation ------------------

Variable            B         SE B       Beta        T    Sig T

EDUC        3.000000     3.794733    .242536     .791   .4475
(Constant) -8.000000    64.621978               -.124   .9039


End Block Number  1    All requested variables entered.
```

FIGURE 18.18 *SPSSX-generated regression output using last $n_2 = 12$ observations from Example 18.3*

Because we have multiple observations of Y for each given value of X, there is yet another way to show that the errors are homoscedastic. There are six individuals who have $X = 8$ years of education. For these six individuals, the sample variance of Y associated with $X = 8$ is $s^2 = 14$. Similarly, for the individuals with 12 years of education, the sample variance is $s^2 = 56$. The sample variance for the six values of Y associated with $X = 16$ is $s^2 = 144$, and the sample variance for the six values of Y associated with $X = 18$ is $s^2 = 201.6$. Observe that the sample variance of incomes for individuals with 18 years of education is more than 14 times greater than the sample variance of incomes for individuals who have only 8 years. Thus we see that income depends on years of education, and we also see that the amount of variation in income increases as the level of education increases.

Correcting for Heteroscedasticity of Error Terms Using Weighted Least Squares

When tests indicate that the error terms have unequal variances, estimating the regression coefficients by the method of least squares may not be the most appropriate procedure. To get (1) efficient estimates of the coefficients, (2) unbiased estimates of the variances of the estimated coefficients, and (3) valid t tests and confidence intervals, it is necessary to take into account the fact that the error terms have unequal variances. An alternative estimation procedure called *weighted least squares,* denoted WLS, can be used for this purpose.

The difficulty with the method of weighted least squares is that the values of σ_i^2 up to a factor of proportionality must be known. In most cases, these values will not be known and the weighted-least-squares method will not be feasible. Instead we must estimate the variances σ_i^2 from the sample and estimate the regression coefficients by using what is called *feasible weighted least squares.*

Weighted-least-squares method of regression

Suppose that the error terms are not homoscedastic and that, in fact, the variances of the error terms are directly proportional to the squared value of some variable Z; that is,

$$\sigma_i^2 \propto z_i^2 \quad \text{for } i = 1, 2, \ldots, n$$

where the values z_i are known. The variable Z may or may not be one of the independent variables in the regression equation. When the variances of the error terms follow this particular relationship, the weighted-least-squares estimation procedure will yield the best linear unbiased estimates of the population regression coefficients $\beta_0, \beta_1, \ldots, \beta_K$.

To obtain the weighted-least-squares estimates of the population regression coefficients, rather than estimating the equation

$$Y_i = \beta_0 + \beta_1 x_{i1} + \beta_2 x_{i2} + \cdots + \beta_K x_{iK} + e_i$$

we estimate the regression equation

$$Y_i/z_i = \beta_0(1/z_i) + \beta_1(x_{i1}/z_i) + \beta_2(x_{i2}/z_i) + \cdots + \beta_K(x_{iK}/z_i) + u_i$$

and the error term $u_i = e_i/z_i$ will have a constant variance.

In the weighted-least-squares regression model, the dependent variable is Y_i/z_i rather than Y_i. The values of the independent variables are $1/z_i, x_{i1}/z_i, x_{i2}/z_i, \ldots, x_{iK}/z_i$ rather than $1, x_{i1}, x_{i2}, \ldots, x_{iK}$. (Unless Z is one of the explanatory variables X_j, there is no intercept term in the weighted-least-squares model, whereas if $Z = X_K$, say, then the coefficient β_K will appear to be the constant term.) The coefficients in the weighted-least-squares model are estimated by applying the method of least squares to the transformed data. The resulting estimates $b_0, b_1, \ldots, b_K$ are the best linear unbiased estimates of the parameters $\beta_0, \beta_1, \ldots, \beta_K$, and the corresponding estimated variances are unbiased estimates of the true variances. In addition, the t tests, p-values, and confidence intervals will be theoretically valid.

EXAMPLE 18.4   **Estimation by the Method of Weighted Least Squares** ———————

Example 18.3 contained data showing the years of education and annual incomes of 28 individuals. Based on the Goldfeld–Quandt test, we rejected the null hypothesis that the errors are homoscedastic.

Suppose the variances of the error terms follow the equation

$$\sigma_i^2 = \sigma^2 x_i^2 \quad \text{for } i = 1, 2, \ldots, n$$

To obtain efficient estimates of the regression coefficients, apply the method of weighted least squares. Transform the original model

$$Y_i = \beta_0 + \beta_1 x_i + e_i$$

to the weighted-least-squares model

$$Y_i/x_i = \beta_0(1/x_i) + \beta_1(x_i/x_i) + u_i$$

or

$$Y_i/x_i = \beta_0(1/x_i) + \beta_1 + u_i$$

where $u_i = e_i/x_i$. The SPSSX computer output is shown in Figure 18.19, where the dependent variable INVINC is Y_i/x_i and the explanatory variable INVEDUC represents the inverse of the education variable (i.e., INVEDUC = $1/x_i$). The estimated equation is

$$\hat{y}_i/x_i = -2.264(1/x_i) + 2.661 \qquad R^2 = .008$$
$$t \text{ statistics:} \qquad (-.46) \qquad\qquad (6.44)$$

```
* * * *   M U L T I P L E   R E G R E S S I O N   * * * *

Listwise Deletion of Missing Data

Equation Number 1    Dependent Variable..    INVINC

Block Number  1.  Method:  Enter      INVEDUC

Variable(s) Entered on Step Number  1..    INVEDUC

Multiple R           .08990      Analysis of Variance
R Square             .00808                    DF    Sum of Squares    Mean Square
Adjusted R Square   -.03007      Regression     1           .09189         .09189
Standard Error       .65859      Residual      26         11.27713         .43374

                                 F =    .21185      Signif F =   .6491


------------------ Variables in the Equation ------------------

Variable            B          SE B       Beta        T    Sig T

INVEDUC      -2.263625     4.918022   -.089901     -.460   .6491
(Constant)    2.661434      .413336                6.439   .0000

End Block Number   1   All requested variables entered.
```

FIGURE 18.19 *SPSSX-generated regression output for feasible weighted least squares*

Multiply both sides of this equation by x_i to obtain the WLS model

$$\hat{y}_i = -2.264 + 2.661x_i$$

Now let us compare the estimated coefficients and the estimated standard deviations for the two models:

Model	b_0	s_{b_0}	b_1	s_{b_1}
Least squares	−2.333	7.350	2.667	.524
Weighted least squares	−2.264	4.918	2.661	.413

The WLS estimated coefficients are slightly different from the least-squares estimated coefficients, and the WLS standard deviations are smaller than the least-squares standard deviations. The WLS standard deviations are theoretically valid (assuming that the variance structure follows the assumptions made earlier) and lead to valid t tests and confidence intervals, whereas the least-squares standard deviations are not theoretically valid.

For the WLS model, the value $R^2 = .00808$ is quite small, but this does not necessarily mean that the model is of no use. Recall that we are trying to explain values of Y_i, not Y_i/x_i. To measure goodness of fit, it is not appropriate to examine the residuals from the model containing Y_i/x_i as the dependent variable. Instead, substitute the estimated coefficients from the WLS model back into the original (untransformed) model and calculate the residuals. Then calculate R^2 using the relationship $R^2 = (1 - \text{SSE/SST})$.

Heteroscedasticity of the Dependent Variable

Another form of heteroscedasticity called *dependent-variable heteroscedasticity* has been proposed. This involves the assumption that the variance of the error term is proportional to the squared mean of Y_i. That is, we assume

$$\sigma_i^2 = \sigma^2[E(Y_i)]^2 = \sigma^2(\beta_0 + \beta_1 x_i)^2$$

To obtain the feasible weighted-least-squares estimates, we follow a two-step procedure. First, we estimate the equation

$$Y_i = \beta_0 + \beta_1 x_i + e_i$$

by the method of least squares to obtain the fitted values $\hat{y}_i$. Then we transform the data to obtain the weighted-least-squares model

$$Y_i/\hat{y}_i = \beta_0(1/\hat{y}_i) + \beta_1(x_i/\hat{y}_i) + u_i$$

where $u_i = e_i/\hat{y}_i$. The estimated coefficients from this model are not the optimal weighted-least-squares estimates, because we have used the estimated weights $\hat{y}_i$ rather than the correct weights $(\beta_0 + \beta_1 x_i)$, which are unknown.

Exercises

Statistical Concepts

18.4.1   One of the basic assumptions of the multiple regression model is that the error terms are homoscedastic. Explain what this means.

18.4.2   One of the basic assumptions of the multiple regression model is that the error terms are homoscedastic. *True or false:*   If the error terms are not homoscedastic, then the estimated regression coefficients will be unbiased, but they will not be best. Explain.

18.4.3   *True or false:*   If the error terms are not homoscedastic, then the estimated variances of the estimated regression coefficients will be biased. Explain.

18.4.4   Explain how to use a graphical technique to detect the possible absence of homoscedasticity.

18.4.5   Explain how to perform the Goldfeld–Quandt test to detect the possible absence of homoscedasticity.

18.4.6   Explain what is meant by dependent-variable heteroscedasticity.

Statistical Applications

18.4.7   A doctor estimated a linear regression model to explain the relation between the concentration of a drug in plasma (Y) and the dosage of the drug in grams (X) given to the patient. The residuals and the dosages are as follows:

Dosage	1	2	3	4	5	6	7	8	9
X	.15	.64	.81	.23	.77	.98	.48	.83	.99
$\hat{e}$	.50	2.10	−3.40	.30	−1.70	4.20	−.60	2.60	−4.00

Plot the residuals $\hat{e}$ against X. What conclusions do you draw from the plot?

18.4.8   The following data show the number of students in 12 different sections of a computer class (X) and the weekly cost of computer time for the class in dollars (Y):

Section	1	2	3	4	5	6	7	8	9	10	11	12
X	16	14	22	10	14	17	10	13	19	12	18	11
Y	77	70	85	50	62	70	52	63	88	57	81	54

a. Fit a linear regression function by ordinary least squares.

b. Obtain the residuals and prepare a plot of the residuals against X. What does the residual plot suggest?

c. Assume that $\sigma_i^2 = kx_i^2$ and use weighted least squares to fit the linear regression function. Are the regression coefficients similar to those obtained in part **a** with ordinary least squares? How do the standard deviations of the regression coefficients compare?

d. What transformation of the variables was used?

18.5   Violations of Basic Assumption 4: Serial Correlation

The fourth basic assumption of the multiple regression model states that the error terms are independent of one another (or are uncorrelated with one another). If the error terms are not independent of one another, they are said to be *serially correlated* or *autocorrelated*. In this section, we discuss the problems that arise when this basic assumption fails to hold.

Basic assumption 4

> Let e_i and e_j be the random error terms associated with the random variables Y_i and Y_j. The error terms e_i and e_j are assumed to be independent of one another. We have $E(e_i e_j) = 0$ for $i \neq j$.

The problem of serially correlated error terms arises when we are studying *time series* data. For example, data concerning annual savings and annual income for one family for 10 consecutive years are time series data. Data from 15 different families concerning their savings and their income in 1 specific year are cross-sectional data. A third possibility exists. Suppose we obtain data concerning savings and income for each of the 15 families for a period of 10 years. Then we have a study involving *pooled* time series and cross-sectional data. Such a data set is called *panel data.*

DEFINITIONS **Time Series and Cross-Sectional Data**

Time series data refer to observations that can be ordered chronologically. **Cross-sectional** data refer to observations concerning many different individuals at one point in time.

With time series data, the order in which the observations are made is important. To emphasize that the data are time series data, it is customary to use the subscript t rather than i to denote different observations. Thus, y_t denotes the observed value of the random variable Y during time period t, y_{t-1} denotes the observed value of Y during period $(t - 1)$, and so forth. Serially correlated errors can occur in time series studies because random events that influence Y in period $(t - 1)$ can have lingering effects that influence Y in the following period t. If this is the case, we say that the random disturbances e_{t-1} and e_t are serially correlated.

With cross-sectional data, the ordering of the observations is irrelevant. There is little reason to expect that the random errors associated with different observations are correlated. Thus, serially correlated errors are not a concern when the model involves cross-sectional data.

With time series data, however, successive errors often tend to be positively correlated. That is, positive errors tend to be followed by positive errors, and negative errors tend to be followed by negative errors, because random events that cause positive errors in one period have lasting effects that also cause positive errors in the next period. In time series models, serially correlated errors can arise because an important explanatory variable has been omitted from the model.

The error term represents a summary of a large number of random factors that enter the relationship under study but are not measurable or have been omitted from the model for some reason. In a time series model, it would be reasonable to suspect that the effect of these factors in one period carries over somewhat to the following periods.

In general, the shorter the period between observations, the greater the likelihood of serially correlated error terms. Thus we would be more likely to find serial correlation in a model dealing with monthly or quarterly observations than in a model dealing with annual observations.

Consequences of Serially Correlated Data

The following are some of the problems caused by the presence of serially correlated errors:

1. The sample regression coefficients $b_0, b_1, \ldots, b_K$ are unbiased estimates of the population parameters $\beta_0, \beta_1, \ldots, \beta_K$, but the estimates are not efficient. That is, if the nature of the serial correlation is known, there is another estimation technique that results in unbiased estimators having smaller variances.

2. The least-squares estimate of σ_e^2 is biased. Typically, if the error terms are positively correlated, the true variance will be underestimated.

3. The estimated variances of the sample regression coefficients are biased. Typically, if the errors are positively correlated, the true variances will be underestimated.
4. Because the estimated variances are biased, the tests of hypotheses involving the t and F distributions and the confidence intervals for the population parameters are invalid. The t statistics do not follow the t distribution, and the associated p-values will be incorrect.

When estimating any model using time series data, it is important to test whether the errors are serially correlated. The presence of serially correlated errors can indicate that something systematic has been omitted from the model, that an incorrect functional form has been estimated, or that an important explanatory variable has been omitted.

Generation of Serially Correlated Errors

When all the basic assumptions hold, each error term represents an independent random drawing from a normal distribution having 0 mean and variance σ_e^2. When the error terms are serially correlated, the drawings are no longer independent. In this case, it is necessary to postulate an alternative hypothesis concerning the generation of the error terms.

From basic assumption 2, we have

$$E(e_i) = 0 \quad i = 1, 2, \ldots, n$$

so we can write the covariance between e_i and e_j as

$$\begin{aligned} \text{Cov}(e_i, e_j) &= E\{[e_i - E(e_i)][e_j - E(e_j)]\} \\ &= E(e_i e_j) \end{aligned}$$

The null hypothesis that the errors are not serially correlated implies that

$$\text{Cov}(e_i, e_j) = E(e_i e_j) = 0 \quad \text{for } i \neq j$$

By far, the most frequently postulated alternative hypothesis states that the error terms follow what is known as a *first-order autoregressive scheme,* which is described in the accompanying box.

DEFINITION | **First-Order Autoregressive Scheme** ————————————————

Suppose the error term e_t is generated according to the equation

$$e_t = \rho e_{t-1} + u_t$$

where u_t is a random variable having a normal distribution with mean 0 and variance σ_u^2. It is assumed that all the random variables u_t and u_s are independent of one another for $t \neq s$. In addition, it is assumed that $-1 < \rho < 1$. Then the random variable e_t is said to follow a **first-order autoregressive scheme.**

If e_t follows a first-order autoregressive scheme, each error term e_t is equal to a proportion ρ of the preceding error term plus a new random effect represented by u_t. The amount ρe_{t-1} measures the amount of the previous error term that is carried over to the current period, whereas u_t represents new error effects created during the current period. The coefficient ρ is called the *population correlation coefficient* between e_t and e_{t-1}.

By successive substitutions for $e_{t-1}, e_{t-2}, \ldots, e_1$, we obtain

$$
\begin{aligned}
e_t &= \rho e_{t-1} + u_t \\
&= \rho(\rho e_{t-2} + u_{t-1}) + u_t \\
&= \rho^2 e_{t-2} + \rho u_{t-1} + u_t \\
&= \rho^2(\rho e_{t-3} + u_{t-2}) + \rho u_{t-1} + u_t \\
&= \rho^3 e_{t-3} + \rho^2 u_{t-2} + \rho u_{t-1} + u_t \\
&\quad \vdots \\
&= \rho^{t-1} e_1 + \rho^{t-2} u_2 + \cdots + \rho^2 u_{t-2} + \rho u_{t-1} + u_t
\end{aligned}
$$

Thus, each error term e_t is generated as a function of the random effects $u_2, u_3, \ldots, u_t$ and the initial error term e_1. To make the description of the autoregressive process complete, it is assumed that

$$
e_1 = \frac{u_1}{\sqrt{1 - \rho^2}}
$$

Thus, e_1 is normally distributed with mean 0 and variance $\sigma_u^2/(1 - \rho^2)$. This form for e_1 greatly simplifies certain derivations in the regression model. The variance of e_t is described in the accompanying box.

Variance of e_t in a first-order autoregressive scheme

> If e_t follows the first-order autoregressive scheme, the variance of e_t is given by
>
> $$\sigma_e^2 = \text{Var}(e_t) = \sigma_u^2/(1 - \rho^2)$$
>
> for $t = 1, 2, \ldots$. (For a more detailed discussion, see John Johnston, *Econometric Methods* [New York: McGraw-Hill, 1972].)

In addition, it can be shown that the covariance between e_t and e_{t-1} is

$$
E(e_t e_{t-1}) = \rho \sigma_e^2
$$

In general, the covariance between e_t and e_{t-s} is

$$
E(e_t e_{t-s}) = \rho^s \sigma_e^2
$$

The correlation between e_t and e_{t-1} is ρ, whereas the correlation between e_t and e_{t-s} is ρ^s. The parameter ρ measures the degree of correlation between the two random variables e_t and e_{t-1}. If ρ is close to 1 in absolute value, there is a high degree of relationship between e_t and e_{t-1}, and the effect of the error term e_{t-1} will linger for many periods and take a long time to die out. When ρ is positive, there is a tendency for positive error terms to follow positive error terms and negative error terms to follow negative error terms. When

ρ is negative, there is a tendency for the sign to change for successive error terms, so that positive error terms tend to follow negative error terms and vice versa.

When $\rho = 0$, we have

$$e_t = u_t$$
$$\text{Var}(e_t) = \sigma_e^2 = \sigma_u^2$$

If $\rho = 0$, all the basic assumptions about the error terms hold because the random variables u_t are normally and independently distributed with 0 mean and constant variance.

Detecting Serial Correlation of Error Terms

Graphical Techniques

The discussion of the first-order autoregressive scheme indicated that successive error terms tend to have the same sign when ρ is positive and opposite signs when ρ is negative. When $\rho = 0$, there is no pattern in the behavior of successive values of e_t, because successive values of e_t are independent of one another. To get some information about the process that generated the error terms, we have to rely on the residuals $\hat{e}_t$ because the true error terms are unknown.

To detect serially correlated errors, it is useful to plot the residuals against time. If the errors are uncorrelated, the residuals should be randomly scattered about 0. If they are positively correlated (as is frequently the case), there should be a tendency for positive residuals to follow positive residuals, and negative residuals to follow negative residuals. Thus, a plot would tend to exhibit a wavelike pattern. If the errors are negatively correlated (a much less frequent occurrence), negative residuals should tend to follow positive residuals and vice versa, giving the plot a very choppy appearance.

Figure 18.20 shows two different plots of residuals versus time. In these two plots, the residuals are positively correlated, indicating the presence of positive serial correlation among the error terms. If the residuals look like those in plot 1, an appropriate equation to estimate would be $Y_t = \beta_0 + \beta_1 x_t + \beta_2 x_t^2 + e_t$, where $\beta_2 > 0$. If the residuals look like those in plot 2, the same quadratic equation should be estimated, and we would expect to find $\beta_2 < 0$.

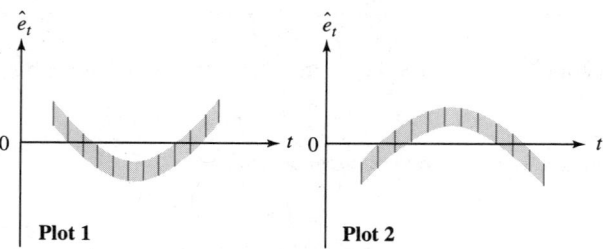

Plot 1   **Plot 2**

FIGURE   18.20   *Residual plots showing positive serial correlation*

Figure 18.21 shows a typical series of noncorrelated residuals $\hat{e}_t$ plotted against time. There is no systematic pattern in this time series. In contrast, Figure 18.22 shows residuals with strong positive serial correlation. Positive residuals tend to be followed by positive residuals and negative residuals by negative residuals.

If there is strong negative serial correlation, positive residuals tend to be followed by negative residuals and negative residuals by positive residuals. If $\rho < 0$, there will be many more changes of sign between successive residuals than when the error terms are not autocorrelated.

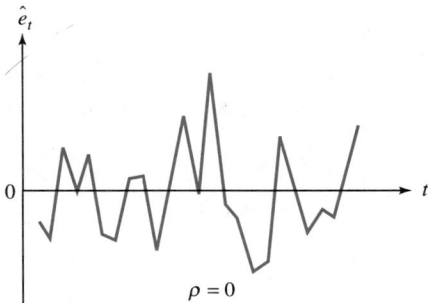

$\rho = 0$

FIGURE 18.21   *Residual plot showing no serial correlation*

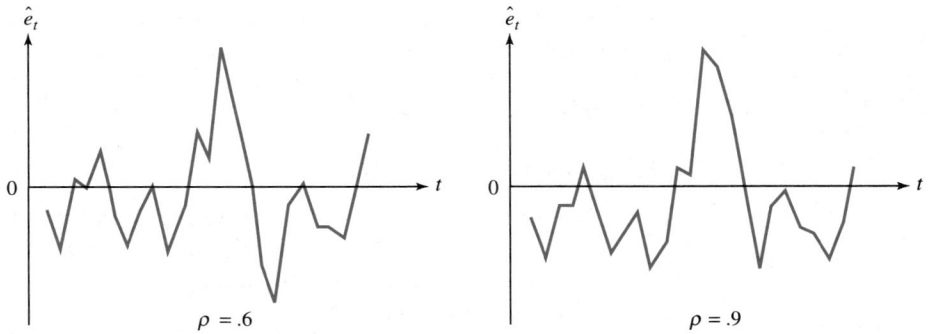

$\rho = .6$ $\rho = .9$

FIGURE 18.22   *Residual plots showing strong positive serial correlation*

Graphing the residuals is a useful way to detect glaring errors in a model. For example, if we estimate a straight line when the data really follow a quadratic equation, a time plot of the residuals will generally indicate this. In many cases, however, the presence of serially correlated errors is not so obvious, and we have to rely on various statistical tests rather than on graphs.

We now discuss how to estimate the parameter ρ and how to test the null hypothesis $H_0: \rho = 0$.

Estimation of ρ

Suppose we think that the error terms follow the first-order autoregressive scheme

$$e_t = \rho e_{t-1} + u_t$$

We want to test the null hypothesis

$$H_0: \quad \rho = 0$$

against one of the possible alternative hypotheses

$$H_1: \quad \rho \neq 0, \qquad H_1: \quad \rho > 0, \qquad \text{or} \qquad H_1: \quad \rho < 0$$

There are several ways to estimate the parameter ρ and several tests to test H_0. One estimator of ρ is defined in the accompanying box.

Estimator for ρ

The estimated value of ρ, denoted $\hat{\rho}$, is defined by the equation

$$\hat{\rho} = \frac{\displaystyle\sum_{t=2}^{n} \hat{e}_t \hat{e}_{t-1}}{\displaystyle\sum_{t=2}^{n} \hat{e}_{t-1}^2}$$

There are two popular tests for testing H_0 against H_1. For lack of a better name, we will call one test the *asymptotic test,* because the test is valid asymptotically—that is, when the sample size gets very large. The other test is called the *Durbin–Watson test* in honor of the two men, James Durbin and Geoffrey Watson, who developed it.

Testing for Serial Correlation

The Asymptotic Test

To test $H_0: \rho = 0$ against, say, $H_1: \rho > 0$, we use the fact that if H_0 is true and *the sample size is large,* $\hat{\rho}$ is approximately normally distributed with mean 0 and variance $1/n$, where n denotes the sample size. Thus, as the sample size increases, the sampling distribution of the standardized test statistic

$$Z = \hat{\rho}/\sqrt{1/n} = \hat{\rho} \sqrt{n}$$

approaches the standard normal distribution.

The asymptotic test

Suppose we wish to test the null hypothesis

$$H_0: \quad \rho = 0$$

against the one-sided alternative hypothesis

$$H_1: \quad \rho > 0$$

using a level of significance α. Calculate the test statistic

$$z = \hat{\rho}/\sqrt{1/n} = \hat{\rho} \sqrt{n}$$

and use the decision rule

Reject H_0 in favor of H_1 if $z > z_\alpha$.

> To test H_0 against H_1: $\rho < 0$, use the decision rule
>
> Reject H_0 in favor of H_1 if $z < -z_\alpha$.
>
> To test H_0 against H_1: $\rho \neq 0$, use the decision rule
>
> Reject H_0 in favor of H_1 if $z < -z_{\alpha/2}$ or if $z > z_{\alpha/2}$.

The Durbin–Watson Test

The test that is used most often to test for the presence of serially correlated error terms is called the **Durbin–Watson test**. Most computer programs will compute the Durbin–Watson test statistic d, which is defined as follows:

$$d = \frac{\sum\limits_{t=2}^{n} (\hat{e}_t - \hat{e}_{t-1})^2}{\sum\limits_{t=1}^{n} \hat{e}_t^2}$$

If we expand the numerator, we obtain

$$d = \frac{\sum\limits_{t=2}^{n} \hat{e}_t^2 - 2 \sum\limits_{t=2}^{n} \hat{e}_t \hat{e}_{t-1} + \sum\limits_{t=2}^{n} \hat{e}_{t-1}^2}{\sum\limits_{t=1}^{n} \hat{e}_t^2}$$

We then obtain the approximate relationship

$$d \approx 2 - 2\hat{\rho}$$

Thus, the following approximate relations hold:

- If $\hat{\rho} \approx 1$, then $d \approx 0$.
- If $\hat{\rho} \approx 0$, then $d \approx 2$.
- If $\hat{\rho} \approx -1$, then $d \approx 4$.

The value of d must lie between 0 and 4, with values of d near 2 supporting the null hypothesis of no serial correlation, values near 0 indicating positive serial correlation, and values near 4 indicating negative serial correlation.

The test statistic d is a random variable whose value varies from sample to sample. The sampling distribution of d is quite complicated and depends on the values assumed by the explanatory variables. As a result, the sampling distribution of d changes from one problem to another, and no single distribution can be tabulated to provide critical values. However, Durbin and Watson found bounds that hold for all data sets, and these bounds have been tabulated.

To test for positive serial correlation, the critical region falls in the left tail of the d distribution. Let d_α denote the critical value d_α such that the probability of obtaining a d statistic less than d_α is equal to the level of significance α. That is,

$$P(d < d_\alpha) = \alpha$$

Our decision rule is

$$\text{Reject } H_0: \rho = 0 \text{ in favor of } H_1: \rho > 0 \text{ if } d < d_\alpha.$$

For a given value α, the critical value d_α is unknown, but Durbin and Watson found that this critical value must lie between two values called the *lower bound* $d_{\alpha,L}$ and the *upper bound* $d_{\alpha,U}$. Thus we have, for a given level of significance α,

$$d_{\alpha,L} \leq d_\alpha \leq d_{\alpha,U}$$

To perform the test, we compare the test statistic d to the two known values $d_{\alpha,L}$ and $d_{\alpha,U}$ rather than to the unknown value d_α. If $d < d_{\alpha,L}$, then $d < d_\alpha$ and we should reject H_0. Similarly, if $d > d_{\alpha,U}$, then $d > d_\alpha$ and d falls in the acceptance region. If d falls between $d_{\alpha,L}$ and $d_{\alpha,U}$, we cannot tell whether d is less than or greater than the correct critical value d_α, and the test is inconclusive.

The procedure for performing the Durbin–Watson test is summarized in the accompanying box.

The Durbin–Watson test

Consider the population regression model

$$Y_t = \beta_0 + \beta_1 x_{t1} + \beta_2 x_{t2} + \cdots + \beta_K x_{tK} + e_t$$

where

$$e_t = \rho e_{t-1} + u_t$$

where u_t follows the basic assumptions. Suppose we wish to test the null hypothesis

$$H_0: \quad \rho = 0$$

against the one-sided alternative hypothesis

$$H_1: \quad \rho > 0$$

using a level of significance α. Calculate the Durbin–Watson statistic:

$$d = \frac{\sum_{t=2}^{n} (\hat{e}_t - \hat{e}_{t-1})^2}{\sum_{t=1}^{n} \hat{e}_t^2}$$

Use the decision rule

$$\text{Reject } H_0 \text{ if } d < d_{\alpha,L}.$$

$$\text{Accept } H_0 \text{ if } d > d_{\alpha,U}.$$

$$\text{Make no conclusion if } d_{\alpha,L} < d < d_{\alpha,U}.$$

The lower and upper bounds $d_{\alpha,L}$ and $d_{\alpha,U}$, respectively, depend on the sample size n, the number of explanatory variables included in the regression equation K, and the level of significance of the test α. (Tables for the Durbin–Watson test showing appropriate values of $d_{\alpha,L}$ and $d_{\alpha,U}$ can be found in Table A.13 in the Appendix for $\alpha = .05$.)

Occasionally, we want to test against the alternative of negative autocorrelation; that is,

$$H_1: \quad \rho < 0$$

The appropriate test is precisely the same as for positive autocorrelation except that it is based on the statistic $(4 - d)$ rather than d. This quantity is then compared with tabulated values $d_{\alpha,L}$ and $d_{\alpha,U}$.

EXAMPLE 18.5 **Performing the Durbin–Watson Test** ————————————

Suppose we wish to test $H_0: \rho = 0$ against $H_1: \rho > 0$ at the 5% level of significance using the Durbin–Watson test. Assume that there are 20 observations and $K = 1$ explanatory variable. Find the lower and upper bounds $d_{\alpha,L}$ and $d_{\alpha,U}$.

SOLUTION   We have $\alpha = .05$, $n = 20$, and $K = 1$. From Table A.13 in the Appendix, the appropriate values are $d_{.05,L} = 1.20$ and $d_{.05,U} = 1.41$. Thus, we should reject H_0 if $d < 1.20$, not reject H_0 if $d > 1.41$, and withhold judgment if $1.20 \le d \le 1.41$. Figure 18.23 illustrates the Durbin–Watson test.

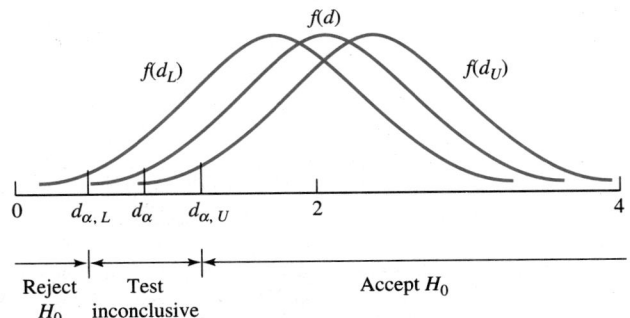

FIGURE 18.23 *Distribution of the Durbin–Watson d statistic*

In most studies concerned with estimating regression equations from time series data, the value of the d statistic is presented along with the other estimates. If the tests indicate serially correlated errors, one response is to reestimate the equation using a more advanced estimation technique. Another approach is to reexamine the specification of the regression model, since serial correlation may indicate the presence of some unexplained systematic influence on the dependent variable.

EXAMPLE 18.6 **Testing for Serial Correlation Using the Durbin–Watson Test** ————

The data in Table 2.4 of Chapter 2 show the values of the Dow Jones Industrial Average (Y) and the Gross National Product (X) for 1960 through 1986. Figure 18.24 (page 794) provides a scatter diagram of the data.

Suppose it is hypothesized that the values of Y are related to the values of X according to the linear model

$$Y_t = \beta_0 + \beta_1 x_t + e_t$$

Estimate the linear regression model and test for the presence of positive serial correlation in the error terms.

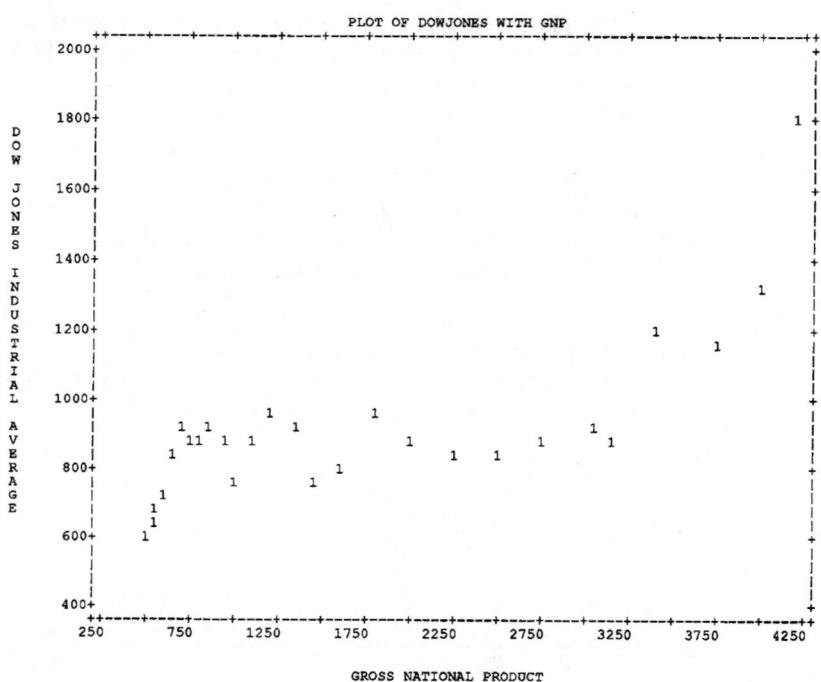

FIGURE 18.24　*SPSSX-generated scatter diagram showing the Dow Jones Industrial Average versus Gross National Product*

SOLUTION　The SPSSX computer output for this model is shown in Figure 18.25. The sample regression line is

$$\hat{y}_t = 648.438 + .152x_t, \quad R^2 = .582$$
$$t \text{ statistics: } (11.93) \quad (5.90) \quad d = .72$$

Figure 18.26 shows a portion of the SPSSX computer output concerning the residuals. The graph shows a plot of the standardized residuals versus time. The printed values show the actual values of the Dow Jones Industrial Average, the predicted values $\hat{y}_t$, and the residuals $\hat{e}_t$. In the figure, negative residuals tend to follow negative residuals, and positive residuals tend to follow positive residuals. For example, the first four residuals are negative, the next six residuals are positive, and so forth. The serpentine pattern of residuals leads us to suspect that the error terms are positively serially correlated.

Now use the Durbin–Watson test to test the null hypothesis

$$H_0: \quad \rho = 0$$

Listwise Deletion of Missing Data

Equation Number 1   Dependent Variable..   DOWJONES   DOW JONES INDUSTRIAL AVER

Block Number 1. Method: Enter   GNP

Variable(s) Entered on Step Number 1..   GNP   GROSS NATIONAL PRODUCT

```
Multiple R              .76295     Analysis of Variance
R Square                .58209                      DF    Sum of Squares    Mean Square
Adjusted R Square       .56538     Regression        1       836209.26588   836209.26588
Standard Error     154.96441       Residual         25       600349.22495    24013.96900

                                   F =    34.82179    Signif F =  .0000
```

------------------ Variables in the Equation ------------------

```
Variable            B         SE B      Beta       T    Sig T

GNP           .152080    .025772    .762950    5.901   .0000
(Constant)  648.437897  54.373382            11.926   .0000
```

End Block Number   1   All requested variables entered.

Residuals Statistics:

```
              Min       Max     Mean   Std Dev   N

*PRED    726.8046 1288.5264 916.7263  179.3374  27
*RESID  -245.5624  504.2336   .0000   151.9551  27
*ZPRED    -1.0590    2.0732   .0000    1.0000   27
*ZRESID   -1.5846    3.2539   .0000     .9806   27
```

FIGURE 18.25 *SPSSX-generated regression output for Example 18.6,*
explaining the Dow Jones Industrial Average
as a function of Gross National Product

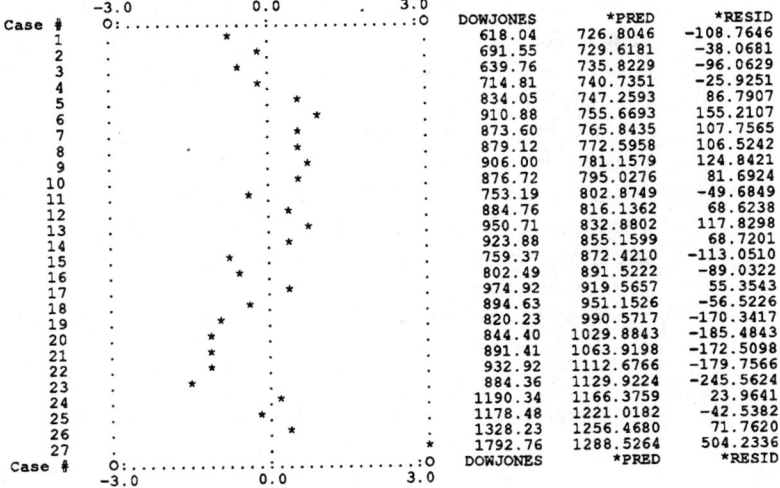

Equation Number 1   Dependent Variable..   DOWJONES   DOW JONES INDUSTRIAL AVER

Casewise Plot of Standardized Residual

*: Selected   M: Missing

```
            -3.0        0.0    .    3.0
Case #   O:................:O   DOWJONES     *PRED      *RESID
   1     .        *     .      618.04     726.8046   -108.7646
   2     .         *.   .      691.55     729.6181    -38.0681
   3     .       * .    .      639.76     735.8229    -96.0629
   4     .        *.    .      714.81     740.7351    -25.9251
   5     .          *   .      834.05     747.2593     86.7907
   6     .          . * .      910.88     755.6693    155.2107
   7     .          .*  .      873.60     765.8435    107.7565
   8     .          .*  .      879.12     772.5958    106.5242
   9     .          . * .      906.00     781.1579    124.8421
  10     .          .*  .      876.72     795.0276     81.6924
  11     .       * . .  .      753.19     802.8749    -49.6849
  12     .         .*   .      884.76     816.1362     68.6238
  13     .         . *  .      950.71     832.8802    117.8298
  14     .         .*   .      923.88     855.1599     68.7201
  15     .      * . .   .      759.37     872.4210   -113.0510
  16     .      * .     .      802.49     891.5222    -89.0322
  17     .         .*   .      974.92     919.5657     55.3543
  18     .       * . .  .      894.63     951.1526    -56.5226
  19     .      * .     .      820.23     990.5717   -170.3417
  20     .     * .      .      844.40    1029.8843   -185.4843
  21     .     * .      .      891.41    1063.9198   -172.5098
  22     .     * .      .      932.92    1112.6766   -179.7566
  23     .    * .       .      884.36    1129.9224   -245.5624
  24     .        . *   .     1190.34    1166.3759     23.9641
  25     .        *.    .     1178.48    1221.0182    -42.5382
  26     .         . *  .     1328.23    1256.4680     71.7620
  27     .         .    *     1792.76    1288.5264    504.2336
Case #   O:................:O   DOWJONES     *PRED      *RESID
            -3.0        0.0        3.0
```

FIGURE 18.26 *SPSSX-generated output showing the residuals from Example 18.6*

against the one-sided alternative hypothesis

$$H_1: \quad \rho > 0$$

using a 5% level of significance.

The value of the Durbin–Watson test statistic is $d = .72139$. There are $n = 27$ observations and $K = 1$ explanatory variable, namely, GNP. From Table A.13, for $\alpha = .05$, the two critical values for the Durbin–Watson test are $d_{.05,L} = 1.32$ and $d_{.05,U} = 1.47$. Since the calculated value $d = .72139$ is far below the lower critical value, we reject the null hypothesis of independence and conclude that the error terms are serially correlated.

Correcting for Serial Correlation of the Error Terms

When the error terms are serially correlated, we have several possible actions:

1. Search for additional explanatory variables in hopes that the error terms will then be serially uncorrelated.
2. Try other functional forms, such as a polynomial regression model or a logarithmic regression model.
3. Use the least-squares estimates even though we have evidence that the errors are serially correlated.
4. Use a different estimation technique that takes serial correlation into account.

The first two approaches are always recommended. Serially correlated errors indicate that there is something systematic in the error terms that the model is not explaining. The theory should be reexamined to try to find the source of these patterns. Sometimes use of a trend variable t can be helpful in reducing serial correlation. (The trend variable will be discussed in detail in Chapter 19.)

When the first two approaches fail, we still have the latter two options. We can continue to use the unbiased but inefficient least-squares estimates, or we can use another estimation technique that takes the serial correlation into account. The problem with using the least-squares estimates is that all of the variance estimates are biased, thus invalidating all confidence intervals and tests of hypotheses. In the remainder of this section, we discuss the Prais–Winsten estimation technique, which is designed to correct for serially correlated error terms.

The Prais–Winsten (PW) estimator is named after the two men, S. J. Prais and C. B. Winsten, who proposed it in 1954 (S. J. Prais and C. B. Winsten, "Trend Estimators and Serial Correlation," Cowles Commission Discussion Paper No. 383 [Chicago: 1954]). The idea behind the PW estimator is to transform the original model containing the serially correlated errors into a new model containing independent error terms that satisfy the basic assumptions.

The PW estimator works as follows. The original equation

$$Y_t = \beta_0 + \beta_1 x_t + e_t$$

can be multiplied by ρ and lagged by one period to obtain

$$\rho Y_{t-1} = \rho \beta_0 + \rho \beta_1 x_{t-1} + \rho e_{t-1}$$

If we subtract the second equation from the first, we obtain

$$Y_t - \rho Y_{t-1} = (1 - \rho)\beta_0 + \beta_1(x_t - \rho x_{t-1}) + e_t - \rho e_{t-1}$$

But since $e_t - \rho e_{t-1} = u_t$, we can write

$$Y_t - \rho Y_{t-1} = (1 - \rho)\beta_0 + \beta_1(x_t - \rho x_{t-1}) + u_t$$

for $t = 2, 3, \ldots, n$. In this way, we have replaced the serially correlated error terms e_t by the independent error terms u_t. However, we appear to have lost one observation in this process. To bring back u_1, we consider the first observation

$$Y_1 = \beta_0 + \beta_1 x_1 + e_1$$

Recall that

$$e_1 = \frac{u_1}{\sqrt{1 - \rho^2}}$$

After substituting for e_1, we obtain

$$Y_1 = \beta_0 + \beta_1 x_1 + \frac{u_1}{\sqrt{1 - \rho^2}}$$

We multiply both sides of this equation by $\sqrt{1 - \rho^2}$ to obtain

$$Y_1\sqrt{1 - \rho^2} = \beta_0\sqrt{1 - \rho^2} + \beta_1 x_1\sqrt{1 - \rho^2} + u_1$$

The Prais–Winsten estimator is obtained by estimating the model

$$Y_t^* = \beta_0 w_t^* + \beta_1 x_t^* + u_t$$

where, for $t = 1$,

$$Y_1^* = Y_1\sqrt{1 - \rho^2}, \quad w_1^* = \sqrt{1 - \rho^2}, \quad \text{and} \quad x_1^* = x_1\sqrt{1 - \rho^2}$$

and for $t = 2, 3, \ldots, n$,

$$Y_t^* = Y_t - \rho Y_{t-1}, \quad w_t^* = 1 - \rho, \quad \text{and} \quad x_t^* = x_t - \rho x_{t-1}$$

In the transformed equation, there are two explanatory variables, w_t^* and x_t^*, and no constant term. Note that w_t^* has the same value for all observations except the first one. This transformation is known as the *Prais–Winsten transformation*.

By applying least squares to the transformed model, we obtain best linear unbiased estimates. That is, our estimates have all the nice properties possessed by the least-squares estimators when the basic assumptions hold. (These optimal properties are based on the assumption that the correct value of ρ is used when making the PW transformation.)

In general, the value of ρ will be unknown and must be estimated. There are various ways of estimating ρ, the most popular estimator being

$$\hat\rho = \frac{\displaystyle\sum_{t=2}^{n} \hat{e}_t \hat{e}_{t-1}}{\displaystyle\sum_{t=2}^{n} \hat{e}_{t-1}^2}$$

Another estimator of ρ can be obtained by making use of the Durbin–Watson statistic d and the approximate relationship

$$d \approx 2 - 2\hat{\rho}$$

which by rearranging yields

$$\hat{\rho} = 1 - d/2$$

When we replace the true value ρ by an estimate $\hat{\rho}$ in the PW transformation, the resulting estimates are called the *feasible PW estimates*.

EXAMPLE 18.7   **Using the Prais–Winsten Estimator** ────────────────────────

Use the results from Example 18.6 and obtain the feasible Prais–Winsten estimator for the model relating the value of the Dow Jones Industrial Average to Gross National Product.

S O L U T I O N   Using the result $d = .72139$ from Example 18.6, we obtain the estimated value of ρ,

$$\hat{\rho} = 1 - d/2 = .639305$$

After making the PW transformation, we obtain the PW estimated model

$$\hat{y}_t^* = 570.807w_t^* + .207x_t^*$$

The SPSSX computer results are shown in Figure 18.27.

```
                      * * * *  MULTIPLE REGRESSION THROUGH THE ORIGIN  * * * *

Listwise Deletion of Missing Data

Equation Number 1    Dependent Variable..   YSTAR

Block Number  1.  Method:  Enter      WSTAR    XSTAR

Variable(s) Entered on Step Number  1..    XSTAR
                                    2..    WSTAR

Multiple R           .95295        Analysis of Variance
R Square             .90812                        DF    Sum of Squares      Mean Square
Adjusted R Square    .90076        Regression       2     3888665.43485    1944332.71742
Standard Error    125.45325        Residual        25      393462.93751      15738.51750

                                   F =    123.53976     Signif F =  .0000

------------------- Variables in the Equation -------------------

Variable              B        SE B        Beta       T    Sig T

XSTAR           .206925     .047705     .450444    4.338   .0002
WSTAR        570.807268  107.728273     .550235    5.299   .0000

End Block Number   1   All requested variables entered.
```

FIGURE 18.27   *SPSSX-generated output for the feasible Prais–Winsten estimator in Example 18.7*

Table 18.3 shows the estimated coefficients and estimated standard deviations when the model is estimated by least squares and by the feasible Prais–Winsten technique. The feasible PW slope estimate is approximately 33% larger than the least-squares slope estimate, and the feasible PW estimated standard deviations are much larger than the (biased) least-squares estimates.

TABLE 18.3 *Least-squares estimates and feasible Prais–Winsten estimates of regression model for Dow Jones Industrial Average problem*

Estimation technique	b_0	s_{b_0}	b_1	s_{b_1}
Least squares	648.438	54.37	.152	.026
Feasible PW	570.807	107.73	.207	.048

Recall that the purpose of using the PW estimation technique is to correct for serially correlated errors to get unbiased variance estimates, correct t statistics, correct p-values, correct confidence intervals, and efficient estimates. However, because we have used an estimated value of ρ rather than the true value, the feasible PW estimates do not necessarily possess the optimal properties of the PW estimator obtained from using the correct ρ. Nevertheless, economic research has indicated that, in general, the feasible PW estimates are preferable to the least-squares estimates, even though they do not necessarily possess the optimal properties of the true PW estimates.

Exercises

Statistical Concepts

18.5.1 Explain the difference between time series data and cross-sectional data.

18.5.2 For each of the following data sets, classify the data as time series or cross-sectional.
  a. Incomes of all employees of IBM during 1993
  b. Annual profits earned by IBM from 1980 to 1993
  c. Grade point averages of all the students in an economics course
  d. Annual Gross National Product of the United States from 1980 to 1993
  e. Prime interest rate on December 31 for each year from 1980 to 1993
  f. Interest rate on credit card balances charged by 12 national lenders on December 31, 1993

18.5.3 *True or false:* Serial correlation tends to be a problem when we are analyzing time series data. Explain.

18.5.4 *True or false:* If the error terms are not independent of one another, then the estimated regression coefficients will be unbiased, but they will not be best. Explain.

18.5.5 *True or false:* If the error terms are not independent of one another, then the standard t tests and confidence intervals will be invalid. Explain.

18.5.6 *True or false:* If the error terms are not independent of one another, then the estimated variance of the error terms will be biased. Explain.

18.5.7 Explain how to use a graphical technique to detect the possible presence of serial correlation.

18.5.8 Suppose the error terms follow a first-order autoregressive scheme. Explain how to obtain an estimate of ρ.

18.5.9 Suppose it is suspected that the error terms follow a first-order autoregressive scheme. Explain how to perform the Durbin–Watson test to test the null hypothesis $H_0: \rho = 0$.

18.5.10   Suppose it is believed that the error terms in a regression model are serially correlated. List four possible courses of action.

18.5.11   Explain how to perform the Prais–Winsten estimation technique to correct for serially correlated error terms.

18.5.12   *True or false:*   A value of the Durbin–Watson statistic close to 0 provides evidence that the error terms may be negatively serially correlated. Explain.

18.5.13   *True or false:*   A value of $\hat{\rho}$ close to 0 provides evidence that the error terms may be negatively serially correlated. Explain.

18.5.14   *True or false:*   A value of $\hat{\rho}$ close to 4 provides evidence that the error terms may be negatively serially correlated. Explain.

Statistical Drills

18.5.15   Suppose we estimate a regression model containing a constant term and two explanatory variables. The sample size is $n = 25$. Suppose we obtain the Durbin–Watson statistic $d = 1.85$. Let $\alpha = .05$.
a. Find the two critical values d_L and d_U.
b. Test $H_0: \rho \leq 0$ against $H_1: \rho > 0$.

18.5.16   Suppose we estimate a regression model containing a constant term and two explanatory variables. The sample size is $n = 30$. Suppose we obtain the Durbin–Watson statistic $d = 1.45$. Let $\alpha = .05$.
a. Find the two critical values d_L and d_U.
b. Test $H_0: \rho \leq 0$ against $H_1: \rho > 0$.

18.5.17   Suppose we estimate a regression model containing a constant term and two explanatory variables. The sample size is $n = 25$. Suppose we obtain the estimated value $\hat{\rho} = .30$. Let $\alpha = .05$. Test $H_0: \rho \leq 0$ against $H_1: \rho > 0$ using the asymptotic test.

Statistical Applications

18.5.18   The General Nutrient Corporation (GNC) sells various types of vitamin supplements in a regional market. GNC wants to predict its sales by using national sales figures for vitamin supplements. (To make future predictions, GNC relies on national predictions published by the industry's trade association.) The accompanying data show quarterly GNC sales in millions of dollars (Y) and national sales in millions of dollars (X) for a 5-year period.

Quarter	t	GNC sales Y	Industry sales X
Year 1			
1	1	20.96	127.3
2	2	21.40	130.0
3	3	21.96	132.7
4	4	21.52	129.4
Year 2			
1	5	22.39	135.0
2	6	22.76	137.1
3	7	23.48	141.2
4	8	23.66	142.8
Year 3			
1	9	24.10	145.5
2	10	24.01	145.3
3	11	24.54	148.3
4	12	24.30	146.4

Quarter	t	GNC sales Y	Industry sales X
Year 4			
1	13	25.00	150.2
2	14	25.64	153.1
3	15	26.36	157.3
4	16	26.98	160.7
Year 5			
1	17	27.52	164.2
2	18	27.78	165.6
3	19	28.24	168.7
4	20	28.78	171.7

a. Plot the data.
b. Find the sample regression line and plot it.
c. Calculate s_e.
d. Calculate the estimated variances for b_0 and b_1.
e. Find the residuals and plot them versus time.
f. Calculate the Durbin–Watson test statistic.
g. Let $\alpha = .05$. Test for the presence of positive serial correlation. What are the critical values $d_{\alpha,L}$ and $d_{\alpha,U}$?

18.5.19 Suppose it is believed that the errors in Exercise 18 follow the first-order autoregressive model

$$e_t = \rho e_{t-1} + u_t$$

a. Estimate ρ.
b. Calculate the transformed values $(y_t - \hat{\rho} y_{t-1})$ and $(x_t - \hat{\rho} x_{t-1})$.
c. Estimate β_0 and β_1 using the feasible Prais–Winsten estimator.
d. Calculate the estimated variances of b_0 and b_1. Compare these estimates with the estimates obtained in Exercise 18.

18.5.20 A researcher estimates a regression model that contains three explanatory variables and a constant, based on a random sample of 30 observations. The researcher wants to perform the Durbin–Watson test to test for positive serial correlation using a 5% level of significance. Find $d_{.05,L}$ and $d_{.05,U}$.

18.5.21 Repeat Exercise 20 if the model contains two explanatory variables and the sample size is $n = 25$.

18.6 Violations of Basic Assumption 5: Correlation Between Errors and Explanatory Variables

The fifth basic assumption of the multiple regression model states that the error terms are independent of the explanatory variables. In this section, we discuss potential problems when the fifth basic assumption fails to hold.

Basic assumption 5

> Each of the random variables $e_1, e_2, \ldots, e_n$ is assumed to be independent of the explanatory variables $X_1, X_2, \ldots, X_K$.

The fifth basic assumption means that when the explanatory variable takes the value x_{ij}, nothing can be inferred about the value of e_i. Violations of this assumption usually occur in one of two ways. The first case involves a simultaneous equation model, and the second involves a model containing the lagged value of the dependent variable.

For the first case, consider the following highly simplified model of income determination:

$$\text{Consumption function:} \quad C_t = \beta_0 + \beta_1 Y_t + e_t \tag{1}$$
$$\text{GNP or income identity:} \quad Y_t = C_t + I_t + G_t \tag{2}$$

where

$$C_t = \text{Aggregate consumption in time period } t$$
$$Y_t = \text{Aggregate income in time period } t$$
$$I_t = \text{Total investment in time period } t$$
$$G_t = \text{Total government spending in time period } t$$
$$e_t = \text{Random disturbance in time period } t$$

I_t and G_t are assumed to be nonstochastic and are determined outside the model.

We want to estimate the parameters β_0 and β_1 based on a sample of data for C and Y. In this example, the values of Y_t and e_t are not independent, so basic assumption 5 is violated. We can show that Y_t is not independent of e_t by substituting equation (1) into equation (2). We obtain

$$Y_t = \beta_0 + \beta_1 Y_t + e_t + I_t + G_t$$

or

$$Y_t = \frac{\beta_0}{1 - \beta_1} + \frac{1}{1 - \beta_1} I_t + \frac{1}{1 - \beta_1} G_t + \frac{1}{1 - \beta_1} e_t$$

which shows that Y_t depends on e_t.

The second way that basic assumption 5 can be violated is when the regression model contains a lagged dependent variable. For example, suppose that in a model the dependent variable Y_t depends on Y_{t-1}, the value of Y in the previous period (that is, at time $t - 1$) and on random effects. Suppose the value of Y at time t is determined by

$$Y_t = \beta_0 + \beta_1 Y_{t-1} + e_t \tag{3}$$

In this model, the explanatory variable Y_{t-1} is the lagged value of the dependent variable. At time t, the value of Y_{t-1} is already determined. If e_t is independent of $e_1, e_2, \ldots, e_{t-1}$, then e_t and Y_{t-1} will be independent. The problem is that Y_{t-1} is not independent of e_{t-1}, which means that the explanatory variable Y_{t-1} is not independent of *all* the error terms. It is assumed that equation (3) holds for past time periods, so if we lag the model one time period, we see that Y_{t-1} was determined by

$$Y_{t-1} = \beta_0 + \beta_1 Y_{t-2} + e_{t-1}$$

which shows that Y_{t-1} is dependent on e_{t-1}. In equation (3) we say that Y_{t-1} and e_t are *contemporaneously uncorrelated,* but the explanatory variable Y_{t-1} is not independent of e_{t-1}, e_{t-2}, and so forth. For basic assumption 5 to hold, the explanatory variable must be independent of *all* the error terms. Thus assumption 5 is violated whenever one or more lagged values of the dependent variable are used as explanatory variables.

Basic assumption 5 is needed to ensure that the least-squares estimators b_0 and b_1 are unbiased estimators of β_0 and β_1. When it is violated, the least-squares estimates are biased. Much of econometric theory concerns the problem of developing estimation techniques that produce estimators having good properties when basic assumption 5 is violated. Most econometrics texts treat these techniques in detail.

Consequences of Explanatory Variables Being Dependent on Error Terms

If basic assumption 5 fails to hold, the least-squares estimates $b_0, b_1, \ldots, b_K$ are not unbiased estimates of the population regression coefficients $\beta_0, \beta_1, \ldots, \beta_K$. In addition, all the estimates of the variances of the estimated coefficients are biased. Thus, the confidence intervals and tests of hypotheses are invalid.

Models containing lagged dependent variables are called *autoregressive models* and are treated in more detail in Chapter 19.

18.7   Interpreting the Results of a Regression Model

The following example shows how to interpret the results of an estimated regression model. The example is based on a study of turnover in the labor market published by James F. Ragan, Jr. ("Turnover in the Labor Market: A Study of Quit and Layoff Rates," *Economic Review,* Federal Reserve Bank of Kansas City, May 1981, pp. 13–22).

Ragan obtained the following results for the U.S. economy from 1950 through 1979.

$$\ln Y_t = 4.47 \; - \; .34(\ln x_{t1}) \; + \; 1.22(\ln x_{t2}) \; + \; 1.22(\ln x_{t3})$$

t statistics: $\quad$ (4.28) $\quad$ (−5.31) $\qquad$ (3.46) $\qquad\qquad$ (3.10)

$$+ \; .80(\ln x_{t4}) \; - \; .0054 x_{t5} \qquad \bar{R}^2 = .5370$$

$\qquad\qquad$ (1.10) $\qquad$ (−3.09)

where

Y = Quit rate in manufacturing, defined as number of people leaving jobs voluntarily per 100 employees

x_1 = Adult male unemployment rate

x_2 = Percentage of employees younger than 25

$x_3 = N_{t-1}/N_{t-4}$ = Ratio of manufacturing employment in quarter $(t - 1)$ to that in quarter $(t - 4)$

x_4 = Percentage of female employees

x_5 = Time trend (1 = first quarter of 1950)

Let us analyze these regression results.

1. *Type of data.* We have time series data extending from the first quarter of 1950 to the fourth quarter of 1979.

2. *Number of observations.* We have four observations per year for 30 years from 1950 through 1979, yielding a total of 120 observations. This should be a sufficient

number of observations to prevent a few outliers from having a distorting effect on the regression results.

3. *Form of equation and signs of the coefficients.* The model is expressed in logarithmic form. The dependent variable represents the logarithm of the voluntary quit rate. Does it appear that the signs of the estimated coefficients are correct?

■ *Coefficient of* ln x_{t1}: The coefficient $-.34$ is negative. This is correct because as the adult male unemployment rate increases, we would expect employees to become more reluctant to voluntarily quit their jobs in hopes of finding a better position elsewhere. When the unemployment rate increases, we would expect the voluntary quit rate to decrease.

■ *Coefficient of* ln x_{t2}: The coefficient 1.22 is positive. This is correct because young employees are more likely than older employees to voluntarily quit jobs. When the percentage of employees younger than 25 increases, we would expect the voluntary quit rate to increase.

■ *Coefficient of* ln x_{t3}: The coefficient 1.22 is positive. Variable x_{t3} measures the ratio N_{t-1}/N_{t-4} where N_t represents manufacturing employment during period t. When the ratio is increasing, employment in the manufacturing sector of the economy has increased relative to the four previous quarters. Thus, an increase in x_{t3} indicates that manufacturing employment is increasing. This would make employees more likely to voluntarily quit a low-paying job in hopes of finding a higher-paying job in the manufacturing sector of the economy. Thus, the positive coefficient for ln x_{t3} is theoretically correct. When employment is increasing in the manufacturing sector of the economy, we would expect the voluntary quit rate to increase.

■ *Coefficient of* ln x_{t4}: The coefficient .80 is positive. Variable x_{t4} measures the percentage of women employees during period t. Data indicate that female employees are more likely than male employees to voluntarily quit their jobs (partly because of pregnancy and child rearing). Thus, the positive coefficient for ln x_{t4} is theoretically correct. When the percentage of female employment increases, we would expect the voluntary quit rate to increase.

■ *Coefficient of* x_{t5}: The coefficient $-.0054$ is negative. Variable x_{t5} measures a time trend that takes the values $t = 1, 2, \ldots, 120$. Data indicate that the overall voluntary quit rate has slowly been declining over time. Thus, the negative coefficient for x_{t5} is correct. The model predicts that, as time passes, we would expect the voluntary quit rate to decrease. Note that variable x_{t5} is not expressed in logarithmic form.

4. *Magnitudes of the estimated coefficients.* It is difficult to theoretically predict the exact size of any of the coefficients, but we would expect that none of the coefficients would be large in absolute value. For example, consider the coefficient of ln x_{t2}, which is 1.22. This coefficient indicates that, when the percentage of employees younger than age 25 increases by 1 percentage point, the voluntary quit rate increases by approximately 1.22 percentage points. Although this exact value cannot be justified through a theoretical argument, we can rule out coefficients that are negative or that are much larger than 1.22. For example, it would not be reasonable to obtain a coefficient of 10.5. Such a coefficient would indicate that a 1-percentage-point increase in the percentage of employees under age 25 would lead to an increase in the voluntary quit rate of 10.5 percentage points. Such an effect would appear to be much too large. Similar arguments can be made concerning each of the estimated regression coefficients. None of the estimated coefficients has a value that appears to be unreasonable.

5. *Magnitudes of the t statistics.* The model contains a constant term and five explanatory variables. For each regression coefficient, it is customary to test the null hypothesis $H_0: \beta_j = 0$ against the alternative hypothesis that the coefficient is nonzero. Typically, a one-sided test is performed utilizing the t statistic. The model contains six estimated coefficients, and we have used $n = 120$ observations, so the appropriate number of degrees of freedom is $120 - 6 = 114$. To test the null hypothesis $H_0: \beta_j = 0$ against a one-tailed alternative hypothesis, we compare the observed value of the t statistic to the critical value of t obtained from the statistical tables based on 114 degrees of freedom. Suppose we use a 5% level of significance. The critical value of the t statistic would be approximately $t = 1.661$. Except for the coefficient of the variable $\ln x_{t4}$, all of the t statistics exceed 1.661 in absolute value. Thus, except for the coefficient of $\ln x_{t4}$, we reject the null hypothesis that the population regression coefficient is 0. An argument could be made that, since its coefficient is not statistically significant, the variable $\ln x_{t4}$ should be omitted from the model. Ragan believed that he had strong theoretical reasons for retaining the variable, so he did not delete it from his model.

6. *Magnitude of R^2.* Ragan reports the value $\bar{R}^2$ that represents R^2 adjusted for the number of coefficients in the model. The value $\bar{R}^2 = .5370$ indicates that the model explains about 53.7% of the variation in the logarithm of quit rates between the first quarter of 1950 and the fourth quarter of 1979. From another point of view, we could say that the model leaves unexplained about 46.3% of the total variation in the logarithm of quarterly quit rates.

7. *Standard error of estimate and the Durbin–Watson statistic.* These values are not reported. Since the model utilizes time series data, it would be useful to test for the presence of serially correlated error terms by using the Durbin–Watson test statistic. Detection of positive serial correlation could be an indication that the author has omitted an important explanatory variable. If a researcher wanted to use this model to predict future quit rates, it would be useful to know the value of the standard error of estimate (s_e). This value is necessary to construct confidence intervals for forecasts.

18.8   Computer Applications

To test the assumptions of the multiple regression model, we usually resort to analyzing the residuals. In the SPSSX program, we can analyze the residuals by issuing one or both of the commands RESIDUALS and CASEWISE.

To illustrate how these commands are used, suppose we want to estimate the model

$$Y_i = \beta_0 + \beta_1 x_i + e_i$$

and analyze the residuals. Issue the REGRESSION command to generate the sample regression line. To obtain the Durbin–Watson statistic used to test for serial correlation, issue the command

```
RESIDUALS = DEFAULT, DURBIN/
```

To obtain a listing of all the actual values of the dependent variable along with the predicted values and the residuals, issue the command

```
CASEWISE = DEFAULT, ALL/
```

The following commands produce the estimated regression line, the Durbin–Watson statistic, a printout of the actual and predicted values of Y, and the residuals:

```
REGRESSION VARIABLES = Y, X/
         DEPENDENT = Y/
         ENTER X/
         RESIDUALS = DEFAULT, DURBIN/
         CASEWISE = DEFAULT, ALL/
```

Exercises

18.8.1   Refer to the time series data in Table 2.4 in Chapter 2, which show data values for the variables GNP, TOTALGOV, DEFENSE, URATE, CPI, TBILL, and DOWJONES from 1960 through 1986.
   a. Generate a scatter diagram of DOWJONES versus GNP. Do these two variables appear to be linearly related?
   b. Estimate the linear model where DOWJONES is a function of GNP. Examine the value of R^2. Does GNP appear to do a good job of explaining the value of DOWJONES?
   c. Print out the residuals and obtain the Durbin–Watson statistic.
   d. Examine the residual plot and determine whether GNP appears to explain DOWJONES adequately.
   e. Let $\alpha = .05$ and test the null hypothesis that the error terms are not serially correlated against the alternative hypothesis that the errors are positively correlated.

18.8.2   Refer to Exercise 1.
   a. Generate a scatter diagram of TBILL versus CPI. Do these two variables appear to be linearly related?
   b. Estimate the linear model where TBILL is a function of CPI. Examine the value of R^2. Does CPI appear to do a good job of explaining the value of TBILL?
   c. Print out the residuals and obtain the Durbin–Watson statistic.
   d. Examine the residual plot and determine whether CPI appears to explain TBILL adequately.
   e. Let $\alpha = .05$ and test the null hypothesis that the error terms are not serially correlated against the alternative hypothesis that the errors are positively correlated.

Chapter 18 Supplementary Exercises

18.S.1   At many universities, incoming freshmen are required to take a placement test in mathematics. The director of admissions selected 20 students randomly and recorded their placement scores (X_1) and their grade point averages at the end of the freshman year (Y). The data are as follows:

Student	Y	X_1	Student	Y	X_1
1	3.1	5.5	11	2.0	4.9
2	2.3	4.8	12	2.9	5.4
3	3.0	4.7	13	2.3	5.0
4	1.9	3.9	14	3.2	6.3
5	2.5	4.5	15	1.8	4.6
6	3.7	6.2	16	1.4	4.3
7	3.4	6.0	17	2.0	5.0
8	2.6	5.2	18	3.8	5.9
9	2.8	4.7	19	2.2	4.1
10	1.6	4.4	20	1.5	4.7

  a. Estimate the simple linear regression model.
  b. Plot the data and the estimated sample regression line.
  c. Predict the grade point average for a student whose entrance test score is 4.0.
  d. Find s_e and R^2.
  e. Find the estimated standard deviations for each coefficient.
  f. Obtain a 99% confidence interval for β_1. Does this confidence interval contain 0?
  g. Let $\alpha = .01$. Test the null hypothesis H_0: $\beta_1 = 0$. Is it appropriate to perform a one-tailed test here?
  h. Obtain a 95% confidence interval for the mean value of Y_i given that $x_{i1} = 4.0$.

18.S.2   Refer to Exercise 1. Find all the residuals.
  a. Do the residuals sum to 0?
  b. Plot the residuals against the fitted values $\hat{y}_i$. Does this plot indicate any shortcomings of the linear regression model?

18.S.3   Refer to Exercise 1. The following data show an IQ score (X_2) and a high school grade average (X_3) for each of the 20 students:

Student	X_2	X_3	Student	X_2	X_3
1	105	2.9	11	123	3.2
2	113	2.8	12	114	3.3
3	118	3.1	13	120	3.4
4	107	2.4	14	132	2.6
5	110	3.0	15	122	3.0
6	125	2.4	16	110	2.8
7	115	3.5	17	119	3.3
8	121	3.1	18	109	3.4
9	117	3.1	19	116	2.6
10	111	2.9	20	108	2.7

a. Plot the residuals versus each of these variables. Does it appear that the model could be improved by including either or both of these potential explanatory variables?

b. Estimate the model

$$Y_i = \beta_0 + \beta_1 x_{i1} + \beta_2 x_{i2} + e_i$$

c. Estimate the model

$$Y_i = \beta_0 + \beta_1 x_{i1} + \beta_2 x_{i3} + e_i$$

d. Estimate the model

$$Y_i = \beta_0 + \beta_1 x_{i1} + \beta_2 x_{i2} + \beta_3 x_{i3} + e_i$$

e. Compare the results from parts **b–d**. Note the changes in R^2, in the estimated coefficients, in their estimated standard deviations, and in the significance of the t statistics.

18.S.4 The accompanying monthly data are the amount of assets in millions of constant dollars (Y) and the number of clients in thousands (X) at a small bank for 20 months. A simple linear regression model is believed to be appropriate, but positively autocorrelated error terms may be present.

Month	Y	X	Month	Y	X
1	2.20	2.52	11	2.84	3.73
2	2.04	2.17	12	2.87	3.80
3	2.07	2.23	13	2.75	3.57
4	2.22	2.52	14	2.75	3.58
5	2.11	2.30	15	2.69	3.44
6	2.24	2.52	16	2.33	2.72
7	2.48	3.02	17	2.48	3.01
8	2.48	3.01	18	2.52	3.11
9	2.73	3.53	19	2.79	3.62
10	2.69	3.46	20	2.79	3.62

a. Fit a simple linear regression model by ordinary least squares and obtain the residuals. Also obtain the estimated variances of the regression coefficients.

b. Plot the residuals against time, and explain whether you find any evidence of positive autocorrelation.

c. Conduct a formal test for positive autocorrelation using the Durbin–Watson test and a significance level of $\alpha = .01$. Is the residual analysis in part **b** in accord with the test result?

18.S.5 Refer to Exercise 4.

a. Obtain a point estimate of the autocorrelation parameter ρ using the formula

$$\hat{\rho} = \frac{\sum\limits_{t=2}^{n} \hat{e}_t \hat{e}_{t-1}}{\sum\limits_{t=2}^{n} \hat{e}_{t-1}^2}$$

b. Obtain an alternative estimate as $\hat{\rho} = 1 - d/2$, where d is the Durbin–Watson statistic.

c. Use the estimate in part **b** and obtain the feasible Prais–Winsten estimates for the population regression coefficients. Compare these estimates with the estimates obtained in Exercise 4.

d. Estimate the variances of the sample regression coefficients and compare with the estimates obtained in Exercise 4.

18.S.6 Suppose you want to estimate the following consumption function:

$$C_t = \beta_0 + \beta_1 I_t + \beta_2 W_t + e_t$$

where

$$C_t = \text{Consumption in year } t$$
$$I_t = \text{Income in year } t$$
$$W_t = \text{Wealth in year } t$$

Assume that $E(e_t) = 0$ and $E(e_t^2) = (C_t^2)\sigma^2$. Transform the model into one in which the disturbance term is homoscedastic and describe the steps required to estimate it.

18.S.7 Apply the Goldfeld–Quandt test to the accompanying data and comment on the results. (Drop the middle six observations.)

Observation	Y	X	Observation	Y	X
1	29.5	49.6	11	115.1	140.2
2	31.8	55.7	12	121.5	150.7
3	38.4	63.3	13	137.4	167.3
4	49.8	75.4	14	154.3	181.3
5	54.7	78.3	15	172.0	207.9
6	73.2	96.7	16	196.5	235.0
7	80.4	109.3	17	216.6	258.9
8	87.2	115.8	18	224.7	227.7
9	89.1	118.8	19	252.0	297.6
10	106.3	133.0	20	268.6	309.4

18.S.8 Consider the following regression equation:

$$\hat{y}_t = 34 + 28 x_t \qquad n = 40$$
$$\quad\;\; (3) \quad\;\; (6) \qquad d = 1.40$$

The numbers in parentheses are estimated standard errors of the coefficients.
  a. Does positive autocorrelation exist at the 5% level of significance?
  b. Is the coefficient of X significantly different from 0 at the 5% level of significance? Use a one-tailed test.

18.S.9 Suppose the following regression model has a serially correlated disturbance term

$$Y_t = \beta_0 + \beta_1 x_t + e_t$$

where $e_t = \rho e_{t-1} + u_t$. The first three observations in the data set are as follows:

Y	140	204	310
X	12	15	20

  a. If you estimate this model by ordinary least squares (OLS), would $E(b_1) = \beta_1$?
  b. Using OLS to estimate, would $\text{Var}(b_1)$ be the minimum for linear and unbiased estimators of β_1?
  c. Would the estimated variance of b_1 be unbiased?
  d. Write the appropriate transformation for estimating this model using the Prais–Winsten technique.

e. Suppose $\rho = .5$. Write the Prais–Winsten transformation and calculate the values for the three observations above.

18.S.10   You are given the model

$$Y_t = \beta_0 + \beta_1 x_t + e_t$$

with the following data:

Observation	Y	X
1	2.4	1
2	2.5	2
3	2.5	3
4	1.6	4
5	3.7	5
6	5.8	6
7	6.9	7
8	7.0	8
9	11.1	9
10	11.2	10
11	11.3	11
12	13.4	12
13	16.5	13
14	11.5	14
15	12.6	15

a. Estimate the simple linear regression model.
b. Calculate the Durbin–Watson statistic and test for positive first-order autocorrelation at the .05 significance level.

18.S.11   Using a series of 25 annual observations, a student estimated the following regression model:

$$\hat{y}_t = -509 + .89x_{t1} + 17.27x_{t2} - 21.20x_{t3} - .522x_{t4} + 7.36x_{t5}$$
$$(23.4) \quad (3.24) \quad\quad (53.46) \quad\quad (.175) \quad\quad (5.69)$$
$$R^2 = .885 \quad d = 1.046$$

where

Y = Yearly Dow Jones Industrial Average
X_1 = Ratio of annual corporate profit to annual corporate sales
X_2 = Index of industrial production
X_3 = Corporate bond yield
X_4 = Disposable income per capita
X_5 = Consumer Price Index

The figures in parentheses below the estimated coefficients are the estimated standard errors.
a. Discuss the sign and statistical significance of each coefficient.
b. Do any of the coefficients appear to have the wrong sign?
c. Which coefficients appear to be statistically insignificant?
d. Examine the Durbin–Watson statistic. What does it indicate?

18.S.12   The accompanying data show the number of ski tickets in thousands sold last season (Y) on a sample of 100 days at 10 different ski resorts. The data also show for each resort the number of miles of nonexpert trails (X_1) and the chair lift capacity in skiers per hour (X_2).

Resort	Tickets sold Y	Miles of trails X_1	Lift capacity X_2
1	21,929	12.5	2,300
2	5,729	4.5	1,100
3	25,897	14.5	3,200
4	9,786	5.2	1,500
5	32,411	15.8	3,900
6	7,331	4.7	1,300
7	10,524	8.3	1,900
8	46,572	23.6	5,600
9	35,486	18.2	4,300
10	13,165	7.7	1,900

a. Find the correlation between Y and X_1.   **b.** Find the correlation between Y and X_2.

c. Find the correlation between X_1 and X_2.

d. Estimate the simple linear regression model where Y is a function of X_1.

e. Estimate the simple linear regression model where Y is a function of X_2.

f. Estimate the multiple regression model where Y is a function of X_1 and X_2.

g. Comment on the differences in the estimated coefficients in parts **d–f**. Can you explain these differences?

h. Comment on the differences in the estimated standard deviations in parts **d–f**. Can you explain these differences?

i. Comment on the differences in the values of R^2 in parts **d–f**. Can you explain these differences?

 18.S.13 The accompanying data show total expenditures on education as a percentage of Gross National Product (Y), per capita income (X_1), median educational attainment in years of the population over 25 years of age (X_2), and the ratio of the population age 0 to 14 to the total population (X_3).

Y	X_1	X_2	X_3
2.1	1,160	5.71	.30
4.5	2,310	8.04	.29
4.5	1,570	3.18	.24
8.5	2,750	9.12	.30
4.3	1,360	5.71	.24
6.2	2,080	4.38	.26
3.4	2,360	4.39	.25
4.4	1,200	7.23	.21
4.6	1,100	5.36	.31
5.6	1,590	7.57	.33
4.4	1,410	3.62	.24
4.2	1,420	5.85	.24
7.2	1,760	4.42	.27
5.9	2,160	5.20	.26
4.1	2,700	5.38	.23

a. Find the correlation matrix.

b. Estimate the multiple regression model

$$Y_i = \beta_0 + \beta_1 x_{i1} + \beta_2 x_{i2} + \beta_3 x_{i3} + e_i$$

c. Find the residuals. What can you learn by examining the residuals from the fitted model?

d. Do you find any evidence of heteroscedasticity? Do you see any outliers?

e. Do you observe any multicollinearity problems?

f. Is it reasonable to test for serially correlated error terms in this problem? Why or why not?

References

Durbin, James R. "Testing for Serial Correlation in Least Squares Regression When Some of the Regressors Are Lagged Dependent Variables." *Econometrica* 38 (1970):410–421.

Durbin, James R., and Geoffrey S. Watson. "Testing for Serial Correlation in Least Squares Regression." *Biometrika* (1950):409–428; (1951):159–178; (1971):1–20.

Goldberger, Arthur. *Econometric Theory.* New York: Wiley, 1964.

Goldfeld, Stephen M., and Richard E. Quandt. "Some Tests for Homoscedasticity." *Journal of the American Statistical Association* (1965):539–547.

Griffiths, William E., R. Carter Hill, and George G. Judge. *Learning and Practicing Econometrics.* New York: Wiley, 1993.

Gujarati, D. *Basic Econometrics.* 2d ed. New York: McGraw-Hill, 1988.

Intriligator, Michael. *Econometric Models, Techniques and Applications.* Englewood Cliffs, N.J.: Prentice-Hall, 1978.

Johnston, John. *Econometric Methods.* New York: McGraw-Hill, 1972.

Judge, George G., W. E. Griffiths, R. Carter Hill, Helmut Lutkepohl, and Tsoung-Chao Lee. *The Theory and Practice of Econometrics.* 2d ed. New York: Wiley, 1985.

Judge, G. G., R. C. Hill, W. E. Griffiths, H. Lutkepohl, and T. C. Lee. *Introduction to the Theory and Practice of Econometrics.* 2d ed. New York: Wiley, 1988.

Kadiyala, Koteswara R. "A Transformation Used to Circumvent the Problem of Autocorrelation." *Econometrica* 36 (1968):93–96.

Klein, L. *An Introduction to Econometrics.* Englewood Cliffs, N.J.: Prentice Hall, 1962.

Kmenta, Jan. *Elements of Econometrics.* 2d ed. New York: Macmillan, 1986.

Maddala, G. *Econometrics.* New York: McGraw-Hill, 1977.

Maddala, G. *Introduction to Econometrics.* 2d ed. New York: Macmillan, 1992.

Nie, Norman E., C. Hadlai Hull, Jean G. Jenkins, Karin Steinbrenner, and Dale H. Bent. *SPSS Statistical Package for the Social Sciences.* 2d ed. New York: McGraw-Hill, 1975.

Norusis, Marija J. *SPSSX Introductory Statistics Guide.* New York: McGraw-Hill, 1990.

Norusis, Marija J. *SPSSX Advanced Statistics Guide.* Chicago: SPSS, 1990.

Norusis, Marija J. *The SPSS Guide to Data Analysis.* Chicago: SPSS, 1986.

Pindyck, Robert S., and Daniel L. Rubinfeld. *Econometric Models and Economic Forecasts.* 2d ed. New York: McGraw-Hill, 1981.

Ryan, Thomas A., Brian L. Joiner, and Barbara F. Ryan. *Minitab Handbook.* 2d ed. Boston: PWS-Kent, 1985.

Ryan, Thomas A., Brian L. Joiner, and Barbara F. Ryan. *Minitab Reference Manual.* University Park, Penn.: Minitab Project, 1985.

SAS Introductory Guide. 3d ed. Cary, N.C.: SAS Institute, 1985.

SAS Procedures Guide for Personal Computers. Version 6 ed. Cary, N.C.: SAS Institute, 1986.

SAS Statistics Guide for Personal Computers. Version 6 ed. Cary, N.C.: SAS Institute, 1986.

SAS User's Guide: Basics. Version 5 ed. Cary, N.C.: SAS Institute, 1985.

SAS User's Guide: Statistics. Version 5 ed. Cary, N.C.: SAS Institute, 1985.

Schmidt, P. *Econometrics.* New York: Marcel Dekker, 1976.

SPSSX User's Guide. Chicago: SPSS, 1988.

Theil, Henri. *Principles of Econometrics.* New York: Wiley, 1971.

19 Time Series Analysis I: Estimation of the Trend Component

The study of business and economic activity often requires the analysis of data that have been collected over a period of time. Any series of observations that can be arranged chronologically is called a **time series,** and the study of such series is called **time series analysis.** Business executives, economists, and government officials use time series analysis in forecasting sales, tax revenues, government expenditures, and so on.

Following are four major reasons for performing time series analysis:

1. *To forecast the value of a dependent variable:* For example, the seller of any product has to forecast demand for the product to know how many items to stock in inventory, investors in the bond market need to forecast interest rates, and investors in the stock market need to forecast stock prices and dividends.

2. *To describe or explain seasonal patterns:* For example, we might be interested in describing how department store sales fluctuate from season to season.

3. *To quantify theories:* Most economic theories can help us predict the signs of coefficients but not their magnitudes. For example, economic theory tells us that increases in interest rates have a negative effect on car sales, but it says nothing about the magnitude of this effect. We can use regression analysis and time series analysis to estimate the magnitude.

4. *To test theories:* Proposed economic theories often relate the value of a certain variable to the value of some other variable. We can use regression analysis and time series analysis to estimate the population regression coefficients in a model and to test hypotheses about the values of these coefficients.

19.1 Components of a Time Series

In general, the fluctuations in an economic time series result from four different components:

1. Trend
2. Seasonal variation
3. Cyclical variation
4. Irregular, or random, variation (sometimes called the random error term)

In this chapter, we describe popular models used to determine the trend component of a time series and describe techniques used to determine the seasonal component.

Not all of these components need be present in every time series, and some are more important in some series than in others. In fact, our goal is to verify the presence of these components and to determine and quantify their importance. In this section, we describe each of these components and provide examples of variables possessing these components individually or in combination.

The Trend

Trend ─────────────────────────────────

The **trend** is the long-term movement in a time series.

Many economic variables, such as the Gross Domestic Product, total government spending, and the sales of new cars, have generally increased over the years. This does not mean that each series has always moved upward from month to month or from year to year but that the long-term trend has been upward over a period of many years. Of course, not all trends are upward. For example, the total number of people employed on farms in the United States has steadily declined since 1947. Whether upward or downward, the trend of a time series is represented by a smooth curve.

Table 19.1 shows annual data for four variables containing a strong trend component, which are graphed in Figures 19.1–19.4 (page 816–817). Examine the graphs and note the different trends. Figure 19.1 shows that U.S. Gross Domestic Product follows a relatively smooth, nonlinear upward trend. In Figure 19.2, gross domestic investment also follows a nonlinear trend, but, unlike for the GDP, the growth has been irregular. Figure 19.3 shows that the U.S. population has followed a very smooth, approximately linear trend since 1961. Finally, Figure 19.4 shows that the farm population has followed an irregular negative trend since 1961. Each of the four variables has a different trend, illustrating the fact that trends may be positive or negative, approximately linear or nonlinear, regular or irregular.

Seasonal Variation

Seasonal Variation ─────────────────────────────

Seasonal variation represents fluctuations in a time series that tend to repeat in a regular way year after year.

Many variables measured monthly or quarterly show seasonal variation, which is often caused by the effects of the weather (hence the term *seasonal*). Energy consumption, travel and vacation expenditures, and farm output are all affected by changes in the weather. Holidays such as Christmas or Easter cause other seasonal effects. Still others are caused by schools being in session and income tax season.

TABLE 19.1 *Trended variables, 1961–1994*

Year	GDP (in billions)	Gross investment (in billions)	Population (in millions)	Farm population (in millions)
1961	531.8	77.9	183.7	14.8
1962	571.6	87.9	186.5	14.3
1963	603.1	93.4	189.2	13.4
1964	648.0	101.7	191.9	13.0
1965	702.7	118.0	194.3	12.4
1966	769.8	130.4	196.6	11.6
1967	814.3	128.0	198.7	10.9
1968	889.3	139.9	200.7	10.5
1969	959.9	155.2	202.7	10.3
1970	1,010.7	150.3	205.1	9.7
1971	1,097.2	175.5	207.7	9.4
1972	1,207.0	205.6	209.9	9.6
1973	1,349.6	243.1	212.0	9.5
1974	1,458.6	245.8	213.9	9.3
1975	1,585.9	226.0	216.0	8.9
1976	1,768.4	286.4	218.0	8.3
1977	1,974.1	358.3	220.2	6.2
1978	2,232.7	434.0	222.6	6.5
1979	2,488.6	480.2	225.1	6.2
1980	2,708.0	467.6	227.8	6.1
1981	3,030.6	558.0	230.0	5.8
1982	3,149.6	503.4	232.2	5.6
1983	3,405.0	546.7	234.3	5.8
1984	3,777.2	718.9	236.3	5.8
1985	4,038.7	714.5	238.5	5.4
1986	4,268.6	717.6	240.7	5.2
1987	4,539.9	749.3	242.8	5.0
1988	4,900.4	793.6	245.0	5.0
1989	5,250.8	832.3	247.3	4.8
1990	5,546.1	808.9	249.9	4.6
1991	5,724.8	744.8	252.6	4.6
1992	6,020.2	788.3	255.4	4.4
1993	6,343.3	882.0	258.1	4.4
1994	6,736.9	1,037.5	260.7	4.2

Source: Economic Report of the President, 1995 (Washington, D.C.: U.S. Government Printing Office, 1995).

For example, the demand for air conditioners is much higher in the summer than in the winter, and so one would expect data on factory shipments of air conditioners to show a pronounced, predictable seasonal pattern with most shipments just before the start of summer and only a few shipments during the winter. A seasonal index can be calculated that indicates how each month departs from what would be expected on the basis of trend alone.

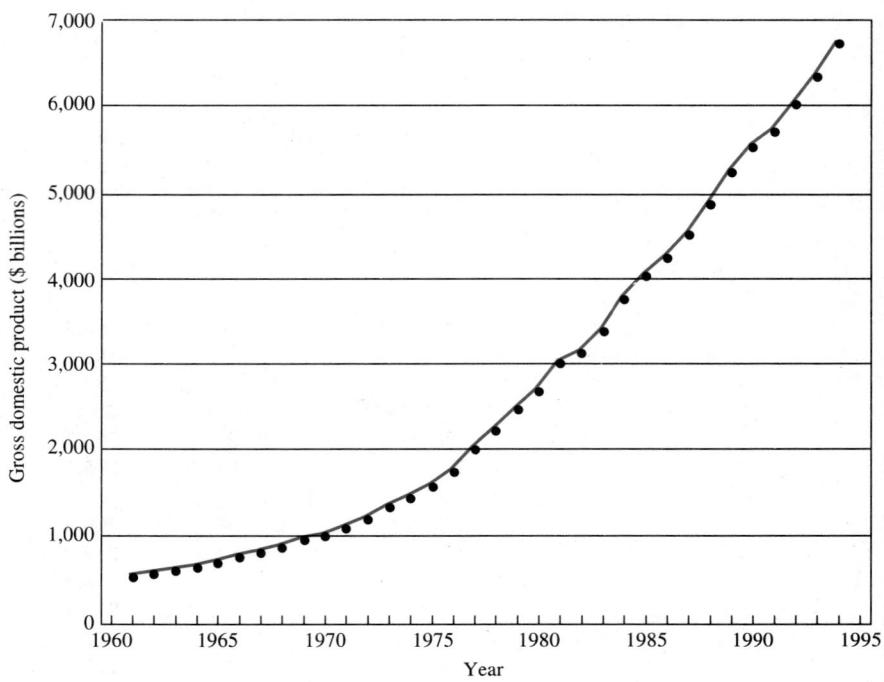

FIGURE 19.1   *Gross Domestic Product, 1961–1994*

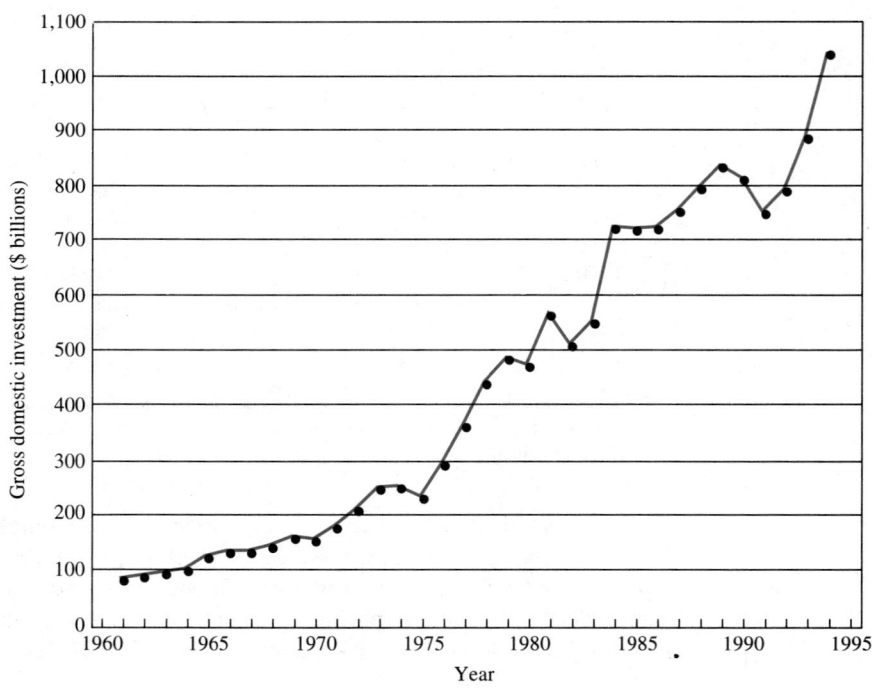

FIGURE 19.2   *Gross domestic investment, 1961–1994*

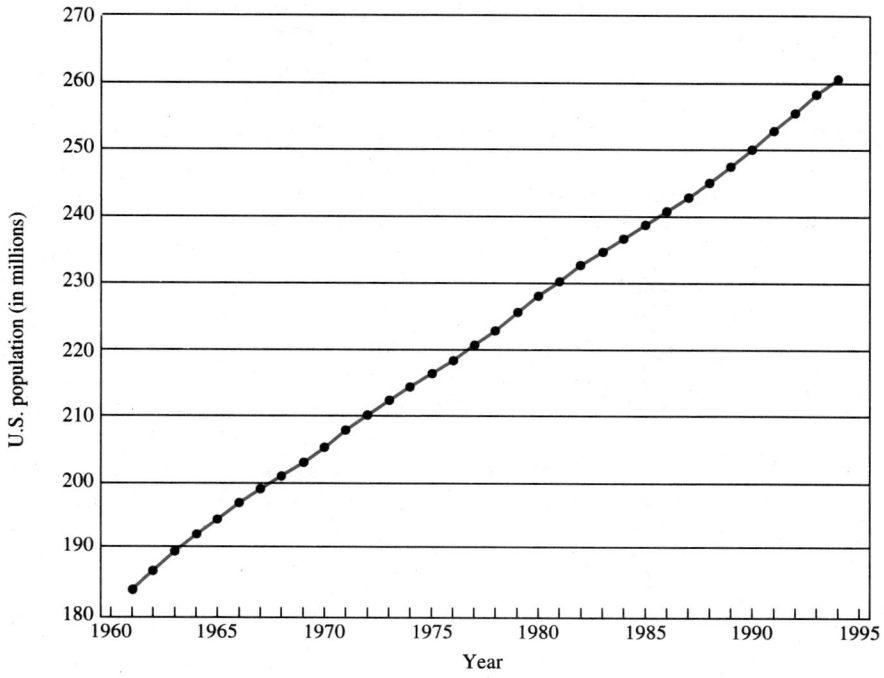

FIGURE 19.3 *Population of the United States, 1961–1994*

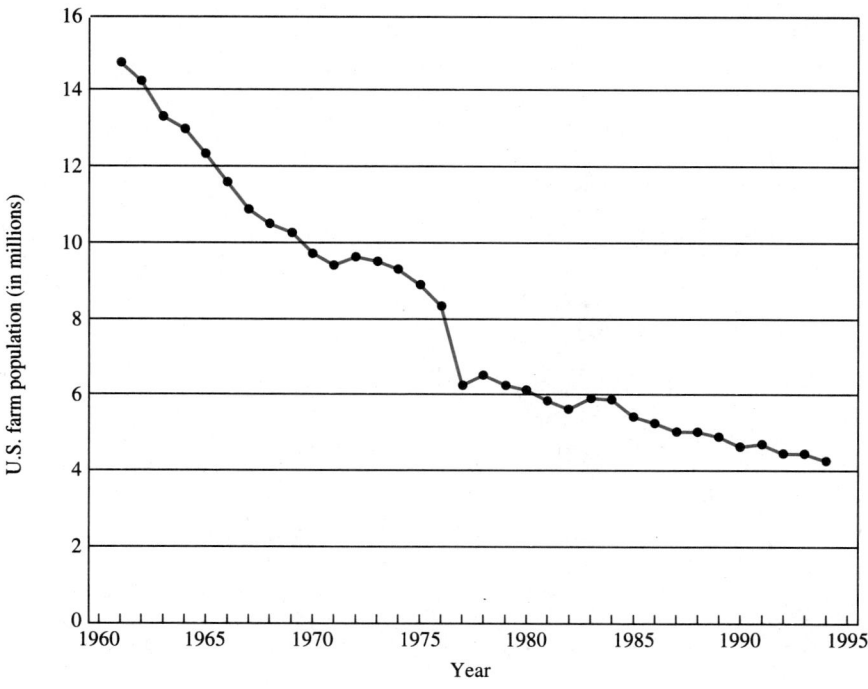

FIGURE 19.4 *U.S. farm population, 1961–1994*

A knowledge of seasonal variation is often useful in interpreting data. For example, suppose that unemployment increased by 2% between May and July of 1987. Before taking any action, government officials would be likely to ask, "To what extent is this increase due to seasonal factors rather than to a downturn in economic conditions?" In other words, even if there were no change in the trend and cyclical factors had no effect, would one expect an increase of this magnitude because of seasonal factors alone?

Usually it is much easier to make comparisons over time when data have been seasonally adjusted. For example, total unemployment always increases during the summer when schools let out and students begin to look for jobs. Similarly, unemployment always decreases during the Christmas season, when stores increase their sales staffs. To make valid comparisons over time, we should remove these seasonal effects. Otherwise, we might be inclined to take actions to effect a change in some variable when no such action is called for.

Table 19.2 shows data on three monthly variables containing seasonal components: monthly retail sales, monthly production of electric power, and monthly shipments of air

TABLE 19.2 *Seasonal variables, 1991–1994*

	Retail sales (in billions)	Electric power (in millions of kilowatt-hours)	Shipments of air conditioners (in thousands)
1991			
January	133.94	248	159
February	131.20	211	185
March	152.50	221	496
April	151.14	209	532
May	162.81	234	613
June	156.91	248	447
July	157.58	272	171
August	162.70	268	63
September	149.21	234	12
October	154.90	223	22
November	158.57	221	31
December	184.77	234	76
1992			
January	141.27	244	103
February	142.28	218	227
March	153.84	225	523
April	158.17	211	545
May	164.92	220	557
June	163.46	237	380
July	164.78	266	243
August	165.26	255	106
September	159.50	235	5
October	168.13	221	24
November	166.41	221	17
December	203.56	244	104

TABLE 19.2 (*continued*)

	Retail sales (in billions)	Electric power (in millions of kilowatt-hours)	Shipments of air conditioners (in thousands)
1993			
January	147.81	246	134
February	144.40	225	236
March	164.02	235	478
April	169.72	211	453
May	175.48	222	440
June	174.95	250	536
July	177.16	282	512
August	176.39	279	68
September	170.39	237	36
October	175.71	224	43
November	180.56	226	55
December	217.92	246	85
1994			
January	154.60	262	
February	155.81	225	
March	184.21	232	
April	181.77	215	
May	187.15	228	
June	190.12	264	
July	185.81	278	
August	193.78	275	
September	185.93	238	
October	189.71	228	
November	194.74	225	
December	233.33	243	

Source: Survey of Current Business (Washington, D.C.: U.S. Department of Commerce, 1995).

conditioners. These variables are graphed in Figures 19.5–19.7 (pages 820–821), and each exhibits a different seasonal behavior and a different long-term behavior.

Figure 19.5 shows monthly retail sales from January 1991 through December 1994. Because the curve rises to the right, sales are positively trended over time. Since 1991, the trend has been approximately linear. However, note the seasonal peak each December and the seasonal trough each February.

Figure 19.6 shows the monthly production of electric power from fuel for the same time period. Once again the curve has a positive slope and thus a positive trend. Unlike retail sales, however, the production of electric power tends to have two peaks each year, one during the middle of winter, when electric power is needed to produce heat, and the other during the middle of summer, when electric power is needed to produce air conditioning.

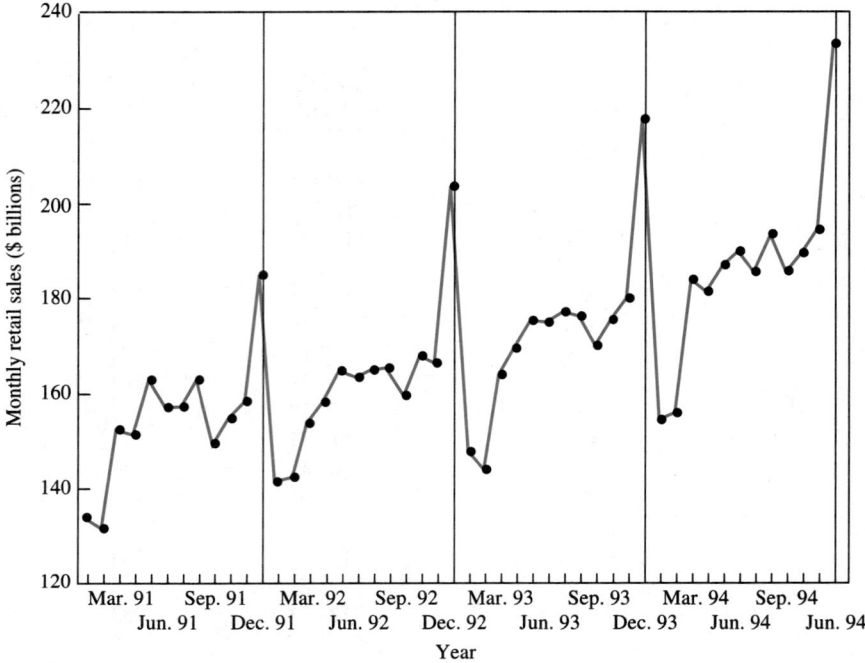

FIGURE 19.5   *Monthly U.S. retail sales, 1991–1994*

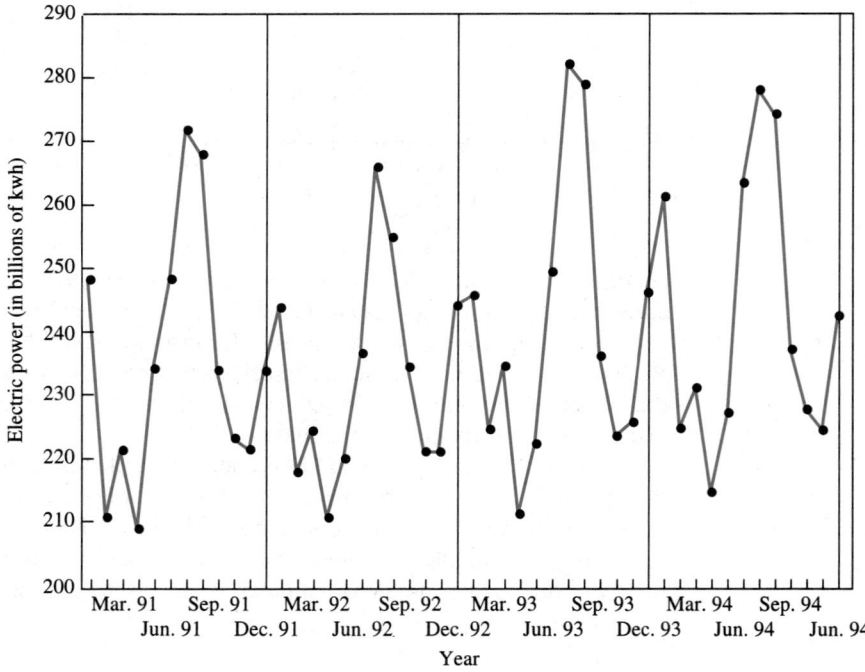

FIGURE 19.6   *Monthly production of electric power, 1991–1994*

Figure 19.7 shows the monthly factory shipments of air conditioners for January 1991 through December 1993. This variable does not appear to have any positive or negative trend, but there is a seasonal pattern. Shipments peak prior to the warm-weather months and die out as the temperature drops. This seasonal pattern indicates that stores increase their inventories a few months before the peak demand period and do not replenish their inventories until the following spring.

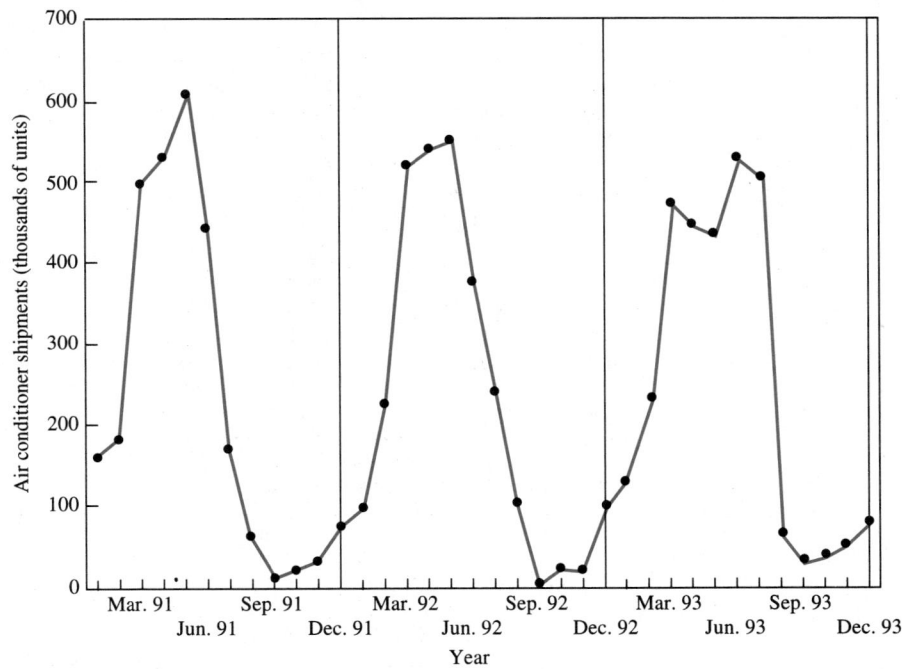

FIGURE 19.7   *Monthly shipments of air conditioners, 1991–1993*

Cyclical Variation

Cyclical Variation

Cyclical variation refers to fluctuations about a long-term trend that are attributable to changing business or economic conditions.

Because cyclical variations vary greatly in length and magnitude, removing their effect can be quite difficult. Economic activity in the United States has always exhibited an irregular cyclical pattern with periods of economic expansion followed by periods of recession and vice versa. These fluctuations, often called *business cycles,* differ from seasonal fluctuations because they cover longer periods of time, are brought about by different causes, and are less predictable. An index of aggregate economic activity showing stages of prosperity, recession, recovery, and prosperity is an important example of a time series variable containing a cyclical component.

Table 19.3 shows the number of new private housing starts from 1961 through 1994; the data are plotted in Figure 19.8. The graph shows that a cyclical component is present.

TABLE 19.3   *New private housing starts, 1961–1994*

Year	Units (in thousands)	Year	Units (in thousands)
1961	1,313.0	1978	2,020.3
1962	1,462.9	1979	1,745.1
1963	1,603.2	1980	1,292.2
1964	1,528.8	1981	1,084.2
1965	1,472.8	1982	1,062.2
1966	1,164.9	1983	1,703.0
1967	1,291.6	1984	1,749.5
1968	1,507.6	1985	1,741.8
1969	1,466.8	1986	1,805.4
1970	1,433.6	1987	1,620.5
1971	2,052.2	1988	1,488.1
1972	2,356.6	1989	1,376.1
1973	2,045.3	1990	1,192.7
1974	1,337.7	1991	1,013.9
1975	1,160.4	1992	1,199.7
1976	1,537.5	1993	1,287.6
1977	1,987.1	1994	1,453.1

Source: Economic Report of the President, 1995 (Washington, D.C.: U.S. Government Printing Office, 1995).

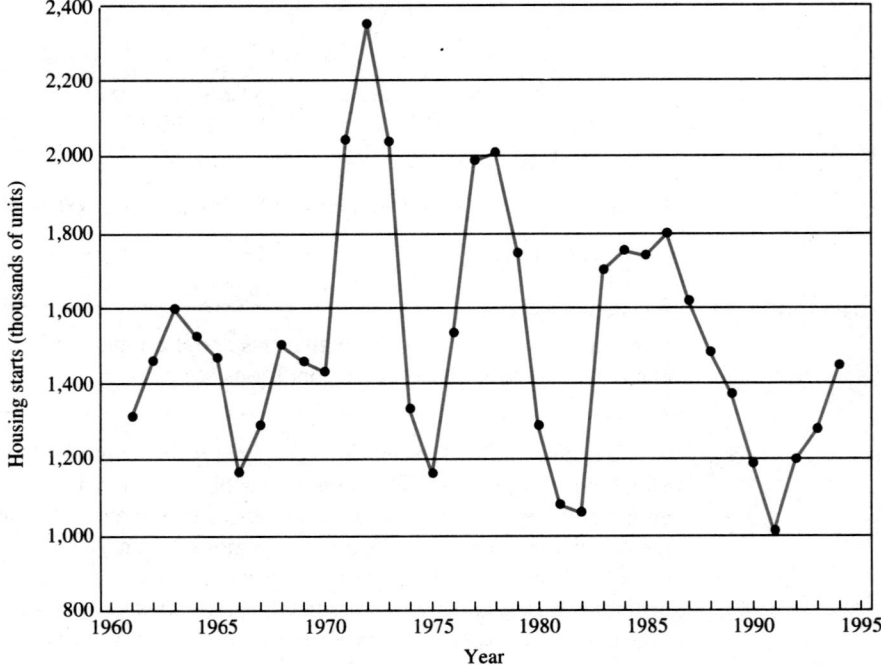

FIGURE 19.8   *Private housing starts, 1961–1994*

Since 1961, there has been no long-run trend, and annual variations have been quite irregular. Because there is no regular pattern in the graph of housing starts, it is difficult to produce a good forecast of housing starts by using a time series model.

Irregular (Random) Variation

DEFINITION   **Irregular Variation**

Irregular variation, or **random variation,** is variation in a time series that is unpredictable, taking place randomly at various points of time. Such variation is not accounted for by trend, seasonal, or cyclical factors.

The irregular component of a time series is not systematic and is caused by unpredictable influences. Irregular variation is conceptually similar to the random error term in regression analysis. The random error term and irregular variation both represent movements that are unexplained by the model.

Relationships Between Time Series Components

It is of interest to identify the influence of the four components on a particular time series and to determine their relationships with one another. We do this by constructing a time series model, which usually takes one of two forms: additive or multiplicative. These two types of models are described in the accompanying boxes.

Additive time series model

In the **additive time series model,** the value of the dependent variable Y can be represented as the sum of the four time series components. Thus, the additive model takes the form

$$Y = T + S + C + I$$

where T is the trend factor, S is the seasonal variation, C is the cyclical variation, and I is the irregular variation.

In the additive model, each of the four components is measured in the same units as the dependent variable Y, and the components S, C, and I are measured as deviations from the trend value T.

Multiplicative time series model

In the **multiplicative time series model,** the value of the dependent variable Y can be represented as the product of the four time series components. Thus, the multiplicative model takes the form

$$Y = T \times S \times C \times I$$

where T is the trend factor, S is the seasonal variation, C is the cyclical variation, and I is the irregular variation.

(continued)

> In the multiplicative model, only one of the components, usually the trend, is expressed in the same unit of measure as the dependent variable Y. The other components are expressed as percentage deviations from the trend.

In the additive model, the deviations from the trend are measured in absolute terms; in the multiplicative model, the deviations from the trend are measured in percentages. For example, suppose we have data showing monthly sales at a department store. If sales tend to increase year after year, there is a long-term upward trend. In general, sales will be highest during December (because of Christmas) and lowest during February. We would be estimating this seasonal effect in absolute terms if our model estimated that sales during December tended to be, say, $1 million above the annual monthly average, and in percentage terms if our model estimated that sales during December tended to be, say, 40% above the annual monthly average.

The data in Tables 19.1, 19.2, and 19.3 show different patterns of behavior. The type of model used to explain the behavior of a time series variable depends on how frequently the data are gathered and on what pattern of behavior we are trying to describe. For instance, because annual data have no seasonal effects, we are concerned only with describing trend and cyclical effects. In contrast, monthly and quarterly data have not only trend and cyclical effects, but also seasonal effects. Daily data might also vary from day to day on a regular basis. For example, total sales at supermarkets tend to be high on Friday and Saturday and low on Monday and Tuesday. The same pattern holds for daily receipts at movie theaters and department stores and deposits into savings and checking accounts at banks.

In the remaining sections of this chapter, we discuss techniques for estimating the trend of a time series variable. In the next chapter, we discuss how to estimate the seasonal component.

Exercises

Statistical Concepts

19.1.1   Typical fluctuations in an economic time series result from four different components. Name and define each component.

19.1.2   *True or false:*   Trend is defined as the long-term upward movement in a time series. Explain.

19.1.3   *True or false:*   Seasonal and cyclical movements differ in that seasonal movements are more predictable and occur at regular intervals of time. Explain.

Statistical Drills

19.1.4   Plot the following monthly data, and identify time series components that appear to be present.

Time	1	2	3	4	5	6	7	8	9	10	11	12
Value	28	32	44	51	58	62	67	76	80	75	68	54

Time	13	14	15	16	17	18	19	20	21	22	23	24
Value	38	44	60	71	79	85	92	106	114	105	98	80

Time	25	26	27	28	29	30	31	32	33	34	35	36
Value	58	64	70	81	89	95	99	113	121	115	103	97

19.1.5 Plot the following annual data, and identify time series components that appear to be present.

Time	1	2	3	4	5	6	7	8
Value	15	19	33	39	58	66	77	86

19.1.6 Plot the following quarterly data, and identify time series components that appear to be present.

Time	1	2	3	4	5	6	7	8	9	10	11	12
Value	28	32	44	51	26	35	42	56	24	30	46	55

Time	13	14	15	16	17	18	19	20	21	22	23	24
Value	27	32	41	53	29	33	40	50	20	35	44	52

Statistical Applications

19.1.7 Consider the following time-dependent variables:

Monthly department store sales
Monthly production of new automobiles
Monthly average temperature in New York City
Monthly sales of ski equipment
Annual profits of ski manufacturers
Annual housing starts
Monthly sales of cigarettes
Monthly sales of bread
Daily receipts from newspaper sales
Daily recordings of temperature at noon in New York City
Hourly recordings of temperature in New York City
Monthly, weekly, or daily number of births of children in the United States

a. State which time series contain a trend component and explain why.
b. State which time series contain a seasonal component and explain why.
c. State which time series contain a cyclical component.

19.1.8 Mention some time series that contain the following:
a. A trend component   **b.** A seasonal component   **c.** A cyclical component

19.2 The Linear Trend Model

Usually the first step in analyzing a time series is to estimate the trend component T. If the amount of increase or decrease in a series from one time period to another is fairly constant, a straight line may describe the trend appropriately. At times, however, other types of curves may be needed to represent the trend component.

One of the most popular ways of forecasting values of a variable is to project a constant increase or decrease each period. One model for such a forecast is the *linear trend model;* the model is estimated by the *sample trend line.* Both concepts are defined in the accompanying boxes.

DEFINITION **Linear Trend Model**

The **linear trend model** is represented by the equation

$$Y_t = \beta_0 + \beta_1 t + e_t$$

where

Y_t = Value of the dependent variable Y during time period t

t = tth unit of time

e_t = Random movement unexplained by the trend variable during time period t

---

DEFINITION **Sample Trend Line**

The estimated regression equation

$$\hat{y}_t = b_0 + b_1 t$$

is called the **sample trend line,** where b_0 and b_1 are the least-squares estimates of the population parameters β_0 and β_1.

Given the observations $y_1, y_2, \ldots, y_n$ and the values $t = 1, 2, \ldots, n$, we estimate β_0 and β_1 using the formulas

$$b_1 = \frac{\sum\limits_{t=1}^{n} ty_t - n\bar{t}\,\bar{y}}{\sum\limits_{t=1}^{n} t^2 - n\bar{t}^2} \quad \text{and} \quad b_0 = \bar{Y} - b_1\bar{t}$$

where $\bar{t} = \sum t/n$ and $\bar{y} = \sum y_t/n$. This estimation of the population parameters results in the *sample trend line.*

A linear trend model is exactly the same as the simple linear regression model discussed in Chapter 15, where the explanatory variable X is replaced by the variable t. Thus, every formula in Chapter 15 holds for the linear trend model if we replace X everywhere by the variable t.

EXAMPLE 19.1 **Estimating the Linear Trend Model**

The data in Table 19.4 show annual sales of retail stores from 1981 through 1993 and are graphed in Figure 19.9. Let us estimate the coefficients in the linear trend model and forecast retail sales in 1997 and 1998. (Because the observations are annual data, there are no seasonal effects.)

SOLUTION   We use the explanatory variable Time (represented by t), where t equals 1 in 1981, 2 in 1982, and so on. To estimate the linear trend model, we need the sums $\sum y_t$, $\sum t$, $\sum ty_t$, and $\sum t^2$. These are calculated in Table 19.4. The least-squares estimates are then

$$b_1 = \frac{\sum ty_t - n\bar{t}\,\bar{y}}{\sum t^2 - n\bar{t}^2}$$

$$= \frac{156.84 - 13(91/13)(20.11/13)}{819 - 13(91/13)^2} = .088297$$

and

$$b_0 = (\sum y_t/n) - b_1(\sum t/n)$$
$$= (20.11/13) - .088297(91/13) = .928846$$

TABLE 19.4   *Computations to estimate the trend line*

Year	Retail sales (in trillions)	Period t	tY	t^2
1981	$1.04	1	1.04	1
1982	1.08	2	2.16	4
1983	1.17	3	3.51	9
1984	1.29	4	5.16	16
1985	1.37	5	6.85	25
1986	1.45	6	8.70	36
1987	1.54	7	10.78	49
1988	1.66	8	13.28	64
1989	1.76	9	15.84	81
1990	1.85	10	18.50	100
1991	1.86	11	20.46	121
1992	1.96	12	23.52	144
1993	2.08	13	27.04	169
Total	20.11	91	156.84	819

Source: *Survey of Current Business* (Washington, D.C.: U.S. Department of Commerce, 1995).

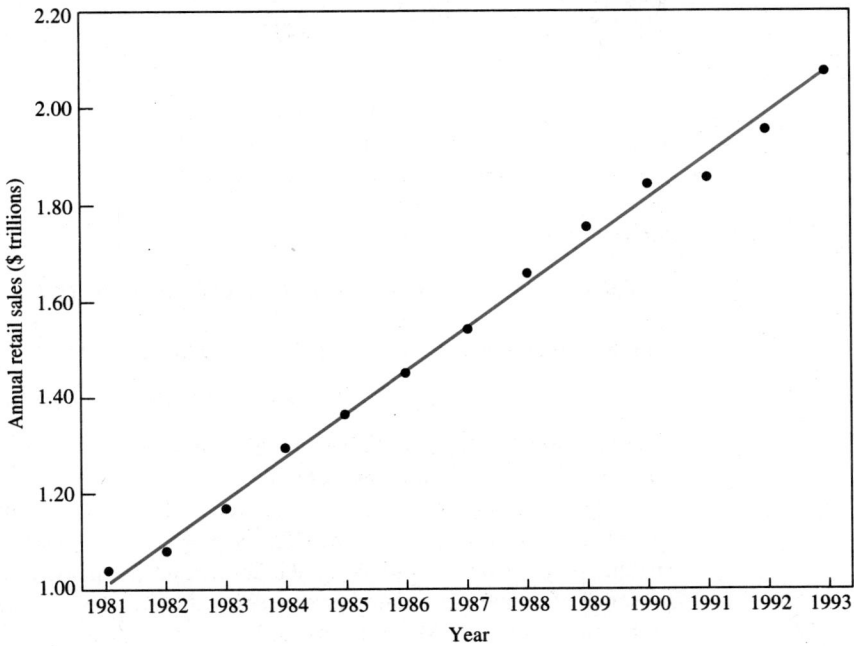

FIGURE 19.9   *Annual retail sales, 1981–1993*

The estimated trend equation is

$$\hat{y}_t = .928846 + .088297t \qquad R^2 = .99457$$

$$t \text{ statistics:} \quad (59.482) \quad (44.880)$$

where $t = 1$ in 1981.

The sample trend line is plotted in Figure 19.9, and the SPSSX computer output for this model is shown in Figure 19.10. The trend variable t explains over 99% of the annual variation in Y ($R^2 = .99457$). The estimated standard error of the regression is $s = .02654$, which indicates that the typical residual is quite small.

The slope of the sample trend line is $b_1 = .088297$, so we predict that retail sales will increase by \$.088297 trillion, or \$88.3 billion, per year.

The value of the trend variable is $t = 1$ in 1981, so in 1997 the value of the explanatory variable is $t = 17$. The forecasted value for retail sales in 1997 would thus be

$$\hat{y}_{17} = .928846 + .088297(17) = 2.429895$$

```
* * * *   M U L T I P L E   R E G R E S S I O N   * * * *

Listwise Deletion of Missing Data

Equation Number 1    Dependent Variable..   SALES   ANNUAL RETAIL SALES

Block Number   1.  Method:  Enter      TIME

Variable(s) Entered on Step Number  1..    TIME

Multiple R          .99728      Analysis of Variance
R Square            .99457                         DF    Sum of Squares   Mean Square
Adjusted R Square   .99407      Regression          1          1.41893       1.41893
Standard Error      .02654      Residual           11           .00775        .00070

                                F =    2014.24796      Signif F =   .0000

------------------ Variables in the Equation ------------------

Variable            B           SE B        Beta        T     Sig T

TIME            .088297      .001967     .997281    44.880   .0000
(Constant)      .928846      .015616                59.482   .0000

End Block Number   1   All requested variables entered.
```

FIGURE 19.10 *SPSSX-generated output for Example 19.1*

and the forecasted value for retail sales in 1998 would be

$$\hat{y}_{18} = .928846 + .088297(18) = 2.518192$$

The data are in trillions of dollars, so the forecasted values for 1997 and 1998 are \$2.430 trillion and \$2.518 trillion, respectively.

Figure 19.11 shows the actual and predicted values for annual retail sales along with the residuals, as computed by the SPSSX program. The residual plot does not show any striking patterns that would indicate that the model is inadequate. To test for the possible presence of serially correlated error terms, examine the Durbin–Watson statistic. In this problem, the Durbin–Watson statistic is $d = 1.53551$, which falls in the acceptance region if we are testing $H_0: \rho = 0$ against $H_1: \rho > 0$ using a 5% level of significance.

```
                                      * * * *   M U L T I P L E   R E G R E S S I O N   * * * *

Equation Number 1    Dependent Variable..   SALES   ANNUAL RETAIL SALES


Casewise Plot of Standardized Residual

*: Selected    M: Missing

          -3.0            0.0            3.0
  Case #  O:...............:...............:O      SALES      *PRED       *RESID
       1  .                .         *     .       1.04       1.0171       .0229
       2  .           *    .               .       1.08       1.1054      -.0254
       3  .             *  .               .       1.17       1.1937      -.0237
       4  .                .   *           .       1.29       1.2820     7.9670E-03
       5  .                *               .       1.37       1.3703    -3.2967E-04
       6  .             *  .               .       1.45       1.4586    -8.6264E-03
       7  .              *.                .       1.54       1.5469    -6.9231E-03
       8  .                .    *          .       1.66       1.6352       .0248
       9  .                .      *        .       1.76       1.7235       .0365
      10  .                .      *        .       1.85       1.8118       .0382
      11  .        *       .               .       1.86       1.9001      -.0401
      12  .           *    .               .       1.96       1.9884      -.0284
      13  .                .  *            .       2.08       2.0767     3.2967E-03
  Case #  O:...............:...............:O      SALES      *PRED       *RESID
          -3.0            0.0            3.0


Residuals Statistics:

              Min       Max      Mean    Std Dev   N

  *PRED     1.0171    2.0767    1.5469     .3439   13
  *RESID    -.0401     .0382     .0000     .0254   13
  *ZPRED   -1.5407    1.5407     .0000    1.0000   13
  *ZRESID  -1.5112    1.4388     .0000     .9574   13

Total Cases =      13

Durbin-Watson Test =    1.53551
```

FIGURE 19.11 *SPSSX-generated actual and predicted values and residuals for Example 19.1*

EXAMPLE 19.2 **Forecasting Population Using a Linear Trend Model** ⎯⎯⎯⎯⎯⎯⎯⎯⎯⎯

Let Y_t denote the population of the United States in millions in year t, where t equals 1 in 1961, 2 in 1962, and so on through 34 in 1994. The data are shown in Table 19.1 (page 815) and plotted in Figure 19.3 (page 817). From the graph, a linear trend model appears to be appropriate to describe the behavior of Y_t. Estimate the linear trend model.

SOLUTION   The sample trend line is

$$\hat{y}_t = 182.3487 + 2.2631t \qquad R^2 = .999$$
$$t \text{ statistics: } (949.6) \qquad (236.4)$$

where $t = 1$ in 1961.

The SPSSX computer output is shown in Figure 19.12 (page 830). This trend model explains more than 99.9% of the variation in population. To forecast the population in 2000, we substitute $t = 40$ into the estimated equation and obtain

$$\hat{y}_{40} = 182.3487 + 2.2631(40) = 272.8727$$

Over a period of 5 to 10 years, the linear trend model should provide fairly accurate forecasts of the U.S. population, which has followed an essentially linear path during any

```
                                    * * * *   M U L T I P L E   R E G R E S S I O N   * * * *

Listwise Deletion of Missing Data

Equation Number 1    Dependent Variable..   POP    POPULATION

Block Number  1.  Method:  Enter      TIME

Variable(s) Entered on Step Number  1..    TIME

Multiple R              .99971      Analysis of Variance
R Square                .99943                           DF    Sum of Squares    Mean Square
Adjusted R Square       .99941      Regression            1       16760.53048    16760.53048
Standard Error          .54756      Residual             32           9.59422         .29982

                                    F =   55902.06586      Signif F =   .0000


------------------ Variables in the Equation ------------------

Variable              B         SE B       Beta         T   Sig T

TIME            2.263102     .009572     .999714    236.436   .0000
(Constant)    182.348663     .192032                949.576   .0000


End Block Number    1    All requested variables entered.
```

FIGURE 19.12   *SPSSX-generated output for Example 19.2*

one 20- to 30-year period. Over longer periods of time, however, the population of the United States has tended to follow a nonlinear path.

Table 19.5 shows the United States population from 1800 to 1990; the data are plotted in Figure 19.13. In Figure 19.3, we saw that the time path of population appeared to be linear from 1961 through 1994, but Figure 19.13 shows that the time path is definitely nonlinear over the longer period from 1800 through 1990. Figure 19.13 shows that a linear model might provide a reasonable representation of population growth during any particular 20- to 30-year period but not over longer periods. Thus, the use of the estimated linear trend equation to forecast population more than a few years in the future is not

TABLE 19.5   *U.S. population, 1800–1990*

Year	Population (in millions)	Year	Population (in millions)
1800	5.3	1900	76.0
1810	7.2	1910	92.0
1820	9.6	1920	105.7
1830	12.9	1930	122.8
1840	17.1	1940	131.7
1850	23.2	1950	150.7
1860	31.4	1960	178.5
1870	39.8	1970	203.3
1880	50.2	1980	226.5
1890	62.9	1990	248.7

Source: Statistical Abstract of the United States, 1994
(Washington, D.C.: U.S. Department of Commerce, 1994).

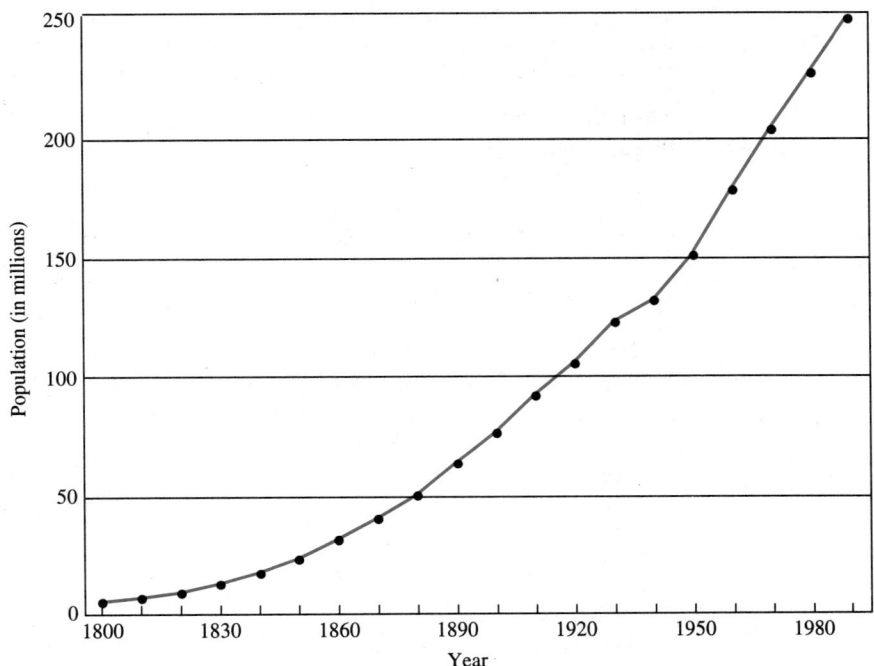

F I G U R E   1 9 . 1 3   *U.S. population, 1800–1990*

recommended. To describe the nonlinear trend in population, we could use a polynomial trend model, which is discussed in the next section.

When a trend variable is included as one of the explanatory variables in a multiple regression model, the model is a mixture of an explanatory model and a descriptive model. The pure trend model describes but does not explain broad movements in a variable over time. It is useful to consider why adding the trend variable to a multiple regression model seems to produce good results in many situations. For example, adding a trend variable to a regression model commonly increases the adjusted R^2 by a substantial amount. Why?

Refer to the omitted-variable problem discussed in Section 18.3. If one or more relevant explanatory variables are omitted from a model, the estimated coefficients will be biased. The nature and amount of the bias depend on how the included variables are correlated with the excluded variables and how the effects of the excluded variables can be shifted to the coefficients of the included variables. Because many of the variables omitted from a model may exhibit trends over time, a trend variable may pick up the effect of these omitted variables and thereby reduce the potential bias in the coefficients of the other variables included in the equation.

Exercises

Statistical Concepts

19.2.1   Projecting future values of a time series based on a trend model is called extrapolating. Explain why it is dangerous to extrapolate more than a few years into the future using a trend model.

19.2.2   *True or false:*   A potential weakness of any trend model for an economic variable is that it assumes that the future will mimic the past. If underlying economic conditions change, there is reason to believe that the trend model will not produce accurate forecasts. Explain.

Statistical Drills

19.2.3   Plot the following data.

Year	1	2	3	4	5	6	7	8
Value	13	16	18	25	25	29	36	40

a. Estimate a linear trend model.
b. Calculate R^2.
c. Forecast the values for years 9 and 10.
d. Do you feel confident that your forecast will be fairly accurate? Explain why or why not.

19.2.4   Plot the following data.

Year	1	2	3	4	5	6	7	8
Value	108	162	184	251	295	329	368	420

a. Estimate a linear trend model.
b. Calculate R^2.
c. Forecast the values for years 9 and 10.
d. Do you feel confident that your forecast will be fairly accurate? Explain why or why not.

19.2.5   Plot the following data.

Year	1	2	3	4	5	6	7	8
Value	78	162	144	251	233	329	568	418

a. Estimate a linear trend model.
b. Calculate R^2.
c. Forecast the values for years 9 and 10.
d. Do you feel confident that your forecast will be fairly accurate? Explain why or why not.

19.2.6   Plot the following monthly data.

Time	1	2	3	4	5	6	7	8	9	10	11	12
Value	28	32	44	51	58	62	67	76	80	75	68	54

Time	13	14	15	16	17	18	19	20	21	22	23	24
Value	38	44	60	71	79	85	92	106	114	105	98	80

Time	25	26	27	28	29	30	31	32	33	34	35	36
Value	58	64	70	81	89	95	99	113	121	115	103	97

a. Estimate a linear trend model.
b. Calculate R^2.
c. Do you feel confident that your forecast will be fairly accurate? Explain why or why not.

19.2.7   Plot the following annual data.

Time	1	2	3	4	5	6	7	8
Value	15	19	33	39	58	66	77	86

a. Estimate a linear trend model.
b. Calculate R^2.
c. Forecast the values for years 9 and 10.
d. Do you feel confident that your forecast will be fairly accurate? Explain why or why not.

19.2.8   Plot the following quarterly data.

Time	1	2	3	4	5	6	7	8	9	10	11	12
Value	28	32	44	51	26	35	42	56	24	30	46	55

Time	13	14	15	16	17	18	19	20	21	22	23	24
Value	27	32	41	53	29	33	40	50	20	35	44	52

a. Estimate a linear trend model.
b. Calculate R^2.
c. Do you feel confident that your forecast will be fairly accurate? Explain why or why not.

Statistical Applications

$$ **19.2.9**   We are given the accompanying data on the money supply (currency plus demand deposits) for the years 1981 to 1994.

Year	Period t	Currency plus demand deposits (in billions) Y	Year	Period t	Currency plus demand deposits (in billions) Y
1981	1	$436.3	1988	8	$ 787.4
1982	2	474.3	1989	9	794.7
1983	3	521.0	1990	10	826.4
1984	4	552.1	1991	11	897.7
1985	5	619.9	1992	12	1,024.8
1986	6	724.5	1993	13	1,128.4
1987	7	750.1	1994	14	1,147.6

Source: *Economic Report of the President, 1995* (Washington, D.C.: U.S. Government Printing Office, 1995).

a. Graph the data on a scatter diagram.   **b.** Estimate the trend line and graph it.
c. Calculate R^2.   **d.** Forecast Y for the year 2000.

19.2.10   The accompanying data show total population in agriculture for the years 1981 to 1994.

Year	Period t	Farm population (in millions) Y	Year	Period t	Farm population (in millions) Y
1981	1	6.05	1988	8	4.99
1982	2	5.85	1989	9	4.95
1983	3	5.63	1990	10	4.80
1984	4	5.79	1991	11	4.59
1985	5	5.75	1992	12	4.63
1986	6	5.36	1993	13	4.48
1987	7	5.23	1994	14	4.44

Source: *Economic Report of the President, 1995* (Washington, D.C.: U.S. Government Printing Office, 1995).

a. Graph the data on a scatter diagram.   **b.** Estimate the trend line and graph it.
c. Calculate R^2.   **d.** Forecast Y for 2000.

$$ **19.2.11**   The accompanying data show personal disposable income in the United States for the years 1981 to 1994.

Year	Period	Disposable income (in billions of 1987 dollars)	Year	Period	Disposable income (in billions of 1987 dollars)
1981	1	$2,795.8	1988	8	$3,404.3
1982	2	2,820.4	1989	9	3,464.9
1983	3	2,893.6	1990	10	3,524.5
1984	4	3,080.1	1991	11	3,538.5
1985	5	3,162.1	1992	12	3,648.1
1986	6	3,261.9	1993	13	3,704.1
1987	7	3,289.5	1994	14	3,835.4

Source: Economic Report of the President, 1995 (Washington, D.C.: U.S. Government Printing Office, 1995).

a. Estimate the trend line and graph it along with the data.
b. Forecast disposable income for 2000 and 2001.

   19.2.12   The accompanying data show personal expenditures on food and beverages.

Year	Period	Expenditures on food and beverages (in billions of 1987 dollars)	Year	Period	Expenditures on food and beverages (in billions of 1987 dollars)
1981	1	$446.6	1988	8	$513.4
1982	2	451.4	1989	9	515.0
1983	3	463.4	1990	10	523.9
1984	4	472.3	1991	11	518.8
1985	5	483.0	1992	12	514.7
1986	6	494.1	1993	13	524.0
1987	7	500.7	1994	14	535.2

Source: Economic Report of the President, 1995 (Washington, D.C.: U.S. Government Printing Office, 1995).

a. Plot the data; estimate the trend line and graph it.
b. Forecast personal expenditures on food and beverages for 1995.

19.2.13   The following data show the populations in millions of two counties:

Year	County 1	County 2	Year	County 1	County 2
1981	14.2	11.8	1985	13.4	13.6
1982	13.8	12.2	1986	13.0	13.9
1983	13.9	12.3	1987	13.2	14.4
1984	13.7	12.9	1988	13.5	14.6

a. Estimate the trend line for county 1 using $t = 1$ in 1981.
b. Use the equation from part **a** to predict the population of county 1 in 1998.
c. Plot the data for county 1 and graph the trend line.
d. Note that since 1986, the population in county 1 has been increasing. Do you think that the population in county 1 is adequately explained by a linear trend line? Do you have a lot of confidence in your answer to part **b**?
e. Estimate the trend line for county 2 using $t = 1$ in 1981.
f. Use the equation in part **e** to predict the population of county 2 in 1998.
g. Plot the data for county 2 and graph the trend line.
h. Do you have more confidence in your prediction in part **f** than in part **b**? Calculate R^2 for each equation. Does this help explain your preference?

19.2.14 The following data show the price of a share of stock in the Molleran Manufacturing Corporation on the last day of the month for 8 consecutive months:

Month	Price	Month	Price
January	$6.42	May	$10.14
February	7.27	June	11.61
March	8.42	July	12.50
April	9.10	August	13.42

a. Plot the data and estimate a linear trend line using $t = 1$ for January.
b. Predict the price of the stock on the last days of September, October, November, and December.
c. Calculate R^2. Is this value large enough to suggest that your predictions might be reliable?
d. Graph the trend line on the scatter diagram.

19.2.15 The accompanying data show the number of new business incorporations for the years 1981 to 1994.

Year	Period t	Incorporations (in thousands) Y	Year	Period t	Incorporations (in thousands) Y
1981	1	581	1988	8	685
1982	2	567	1989	9	677
1983	3	600	1990	10	647
1984	4	635	1991	11	628
1985	5	664	1992	12	667
1986	6	703	1993	13	706
1987	7	686	1994	14	730

Source: Economic Report of the President, 1995 (Washington, D.C.: U.S. Government Printing Office, 1995).

a. Plot the data on a scatter diagram.
b. Estimate the linear trend equation.
c. Plot the estimated line on the scatter diagram.
d. Calculate R^2.
e. Predict Y for 2000.

19.2.16   The accompanying data show the number of business failures for the years 1981 to 1994.

Year	Period *t*	Business failures (in thousands)	Year	Period *t*	Business failures (in thousands)
1981	1	17	1988	8	57
1982	2	25	1989	9	50
1983	3	31	1990	10	61
1984	4	52	1991	11	88
1985	5	57	1992	12	97
1986	6	62	1993	13	86
1987	7	61	1994	14	71

Source: Economic Report of the President, 1995 (Washington, D.C.: U.S. Government Printing Office, 1995).

a. Plot the data on a scatter diagram.   **b.** Estimate the linear trend equation.
c. Plot the estimated line on the scatter diagram.   **d.** Calculate R^2.
e. Predict Y for 2000.

19.3   The Polynomial Trend Model

When graphed over time, many economic variables approximately follow a quadratic curve rather than a straight line; in such cases, it is appropriate to fit a quadratic trend curve. Occasionally a cubic or higher-order curve is appropriate to describe the path of an economic variable over time. The general polynomial trend model is defined as follows.

DEFINITION   **Polynomial Trend Model** ————————————————————————————

The **polynomial trend model** is

$$Y_t = \beta_0 + \beta_1 t + \beta_2 t^2 + \cdots + \beta_s t^s + e_t$$

where

$$
\begin{aligned}
Y_t &= \text{Value of the dependent variable at time } t \\
t &= t\text{th unit of time} \\
s &= \text{Degree of the polynomial} \\
e_t &= \text{Random error term at time } t
\end{aligned}
$$

The linear trend model is a special case of the polynomial trend model in which $s = 1$. The *quadratic*, or *second-degree, trend model* is a special case of the polynomial trend model in which $s = 2$. For most variables studied in economics and business, s rarely exceeds 2 or 3.

Economic variables such as annual sales of a firm and annual profits often tend to follow a quadratic trend curve for a period of years. In addition, the time paths of many aggregate variables for the U.S. economy can be approximated by a quadratic trend curve. Naturally we have to be quite cautious in using these trend curves to predict future values because if the underlying forces that influence the dependent variable change in the future, the future path of the dependent variable will no longer follow the polynomial trend.

EXAMPLE 19.3 **Fitting a Second-Degree Polynomial**

The data for the U.S. Gross Domestic Product in Table 19.1 are plotted in Figure 19.1 on page 816. From Figure 19.1, it is obvious that a linear trend model would be inadequate, and a second-degree polynomial might be appropriate to explain the values of the GDP. Let us estimate the coefficients in the second-degree polynomial model

$$Y_t = \beta_0 + \beta_1 t + \beta_2 t^2 + e_t$$

where t equals 1 in 1961, 2 in 1962, and so on.

SOLUTION Estimating this model is equivalent to estimating a multiple regression model in which the first explanatory variable is $t = 1, 2, 3, \ldots, 34$ and the second explanatory variable is $t^2 = 1, 4, 9, \ldots, 1,156$. Quadratic polynomial models were discussed in Section 17.1.

Figure 19.14 shows the SPSSX computer output for the second-degree polynomial model. The estimated GDP polynomial trend curve is

$$GDP_t = 538.763 - 5.273t + 5.621t^2 \qquad R^2 = .998$$
$$t \text{ statistics: } \quad (11.5) \qquad (-.85) \qquad (32.8)$$

where t equals 1 in 1961, 2 in 1962, and so on.

```
* * * *   M U L T I P L E   R E G R E S S I O N   * * * *

Listwise Deletion of Missing Data

Equation Number 1    Dependent Variable..    GDP    GROSS DOMESTIC PRODUCT

Block Number  1.  Method:  Enter      T        TSQ

Variable(s) Entered on Step Number   1..     TSQ
                                     2..     T

Multiple R            .99911      Analysis of Variance
R Square              .99821                        DF      Sum of Squares       Mean Square
Adjusted R Square     .99810      Regression         2     127900266.15112    63950133.07556
Standard Error      85.91501      Residual          31        228823.04418        7381.38852

                                  F =     8663.69964      Signif F =    .0000


------------------ Variables in the Equation ------------------

Variable             B          SE B        Beta          T   Sig T

TSQ            5.620942      .171375     1.024937     32.799   .0000
T             -5.273181     6.183306     -.026649      -.853   .4003
(Constant)   538.763135    46.936853                  11.478   .0000

End Block Number    1    All requested variables entered.
```

FIGURE 19.14 *SPSSX-generated output for Example 19.3*

The computer output shows that the t statistic for b_2 is 32.8 and the p-value is .0000. Thus, the evidence is extremely strong that the population regression coefficient β_2 is nonzero. In addition, the R^2 value indicates that the model explains more than 99% of the annual variation in GDP. The data and the estimated second-degree polynomial are plotted in Figure 19.15 (page 838); note that the data fall extremely close to the estimated trend curve.

Now let us use the estimated polynomial equation to predict GDP in 2000. Based on $t = 1$ in 1961, the value of t in 2000 is $t = 40$. After we substitute $t = 40$ into the estimated equation, the predicted value of GDP (in billions of dollars) in 2000 is

$$GDP_{40} = 538.763 - 5.273(40) + (5.621)(40)^2 = 9{,}321.443$$

This forecast is based on the assumption that past trends will continue unchanged in the future. However, if economic conditions change, then another forecasting technique might be necessary. Also, the further into the future we forecast, the less confidence we have in our forecast.

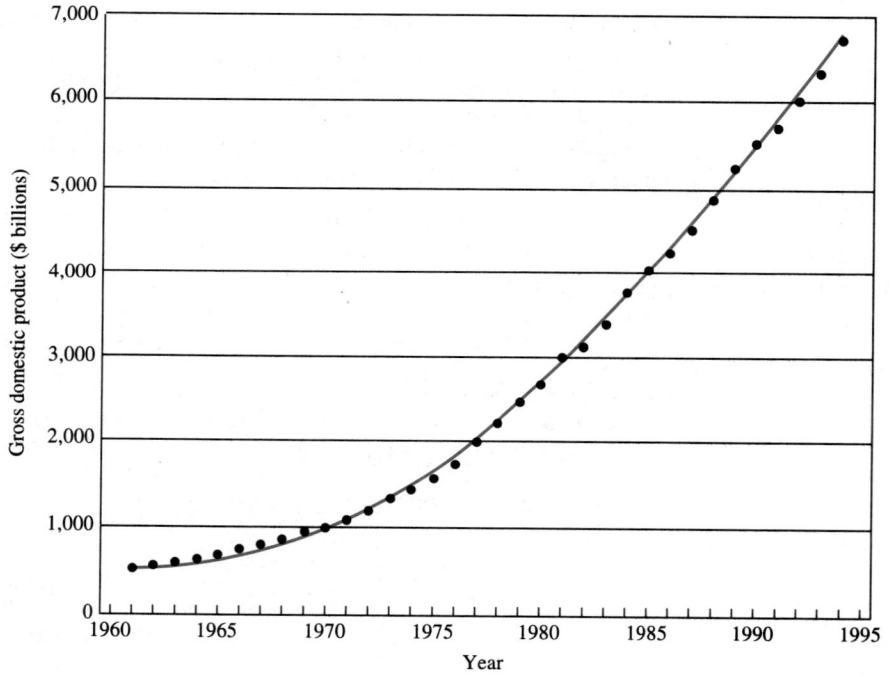

FIGURE 19.15   *Data and estimated polynomial model for Example 19.3*

Exercises

Statistical Concepts

19.3.1   *True or false:*   When choosing between a linear and a quadratic trend model, it is not reasonable merely to choose the model that yields the higher value of R^2 because the linear model can never have a higher value of R^2 than the quadratic model. Explain.

19.3.2   *True or false:*   If some variable grows at approximately a constant percentage rate each year, then a quadratic trend model would be preferred to a linear model. Explain.

19.3.3   *True or false:*   If some variable grows by approximately the same absolute amount each year, then a quadratic trend model would be preferred to a linear model. Explain.

Statistical Drills

19.3.4   Plot the following data.

Year	1	2	3	4	5	6	7	8
Value	5	7	12	18	27	40	55	79

a. Estimate a quadratic trend model.
b. Calculate R^2.
c. Forecast the values for years 9 and 10.
d. Do you feel confident that your forecast will be fairly accurate? Explain why or why not.

19.3.5   Plot the following data.

Year	1	2	3	4	5	6	7	8
Value	10	22	38	59	88	139	188	275

a. Estimate a quadratic trend model.
b. Calculate R^2.
c. Forecast the values for years 9 and 10.
d. Do you feel confident that your forecast will be fairly accurate? Explain why or why not.

19.3.6   Plot the following annual data.

Time	1	2	3	4	5	6	7	8
Value	13	16	21	28	37	48	61	76

a. Estimate a quadratic trend model.
b. Calculate R^2.
c. Forecast the values for years 9 and 10.
d. Do you feel confident that your forecast will be fairly accurate? Explain why or why not.

19.3.7   You have estimated a quadratic trend model and obtained the following sample regression equation:

$$Y_t \doteq 27.2 - 16.3t + 3.2t^2$$

a. Plot this quadratic curve.   **b.** Forecast Y_t for periods 5 and 6.   **c.** Forecast Y_t for period 20.

Statistical Applications

19.3.8   The accompanying table shows the index of production in the iron and steel industry for the years 1981 to 1994, where the index value for 1987 is 100. Plot the data.

Year	Period	Index (1987 = 100)	Year	Period	Index (1987 = 100)
1981	1	117.5	1988	8	108.7
1982	2	83.2	1989	9	107.2
1983	3	91.0	1990	10	106.5
1984	4	102.4	1991	11	98.7
1985	5	101.8	1992	12	101.9
1986	6	93.7	1993	13	106.9
1987	7	100.0	1994	14	114.2

Source: *Economic Report of the President, 1995* (Washington, D.C.: U.S. Government Printing Office, 1995).

a. Estimate the linear trend line.
b. Estimate the second-degree polynomial trend curve.
c. Forecast the values of the index in 1998, 1999, and 2000 by using the linear trend line and the polynomial trend curve.

19.3.9 Suppose a firm's sales are growing over time according to the equation

$$S_t = \beta_0 + \beta_1 t + \beta_2 t^2 + e_t$$

where S = sales and t = time period. Estimate this equation using the data in the accompanying table.

Period t	Sales (in millions)	Period t	Sales (in millions)
1	$10	5	$ 58
2	15	6	83
3	24	7	112
4	37	8	150

19.3.10 The following data show the annual sales in millions of dollars (Y) for the Bowen Candy Company during a 10-year period:

t	1	2	3	4	5	6	7	8	9	10
Y	1.2	1.6	1.9	2.5	3.1	3.8	4.7	5.7	7.0	8.0

a. Plot the data on a scatter diagram. Does a linear or quadratic relationship better describe the relationship between Y and t?
b. Estimate the linear trend model

$$Y_t = \beta_0 + \beta_1 t + e_t$$

c. Estimate the quadratic regression equation

$$Y_t = \beta_0 + \beta_1 t + \beta_2 t^2 + e_t$$

d. Calculate R^2 for each equation.
e. Test $H_0: \beta_2 = 0$ against $H_1: \beta_2 \neq 0$. Let $\alpha = .05$.
f. Plot the estimated equations on the scatter diagram.

19.3.11 An analyst for Ajax Records has estimated a quadratic regression equation to help forecast future revenues of the company. The estimated equation is

$$\hat{y}_t = 216.8 + 9.1t + .3t^2$$

where

$$\hat{y}_t = \text{Predicted revenue in year } t \text{ in thousands of dollars}$$
$$t = 1 \text{ in 1981, 2 in 1982, and so forth}$$

a. Plot the quadratic regression equation.
b. Predict revenue in 1996.
c. Predict revenue in 1997.

19.3.12 Another analyst for Ajax Records estimated the following cubic regression equation for forecasting revenues:

$$\hat{y}_t = 214.5 + 8.6t + .2t^2 + .1t^3$$

where

$$\hat{y}_t = \text{Predicted revenue in year } t \text{ in thousands of dollars}$$
$$t = 1 \text{ in 1981, 2 in 1982, and so forth}$$

a. Plot the cubic regression equation.
b. Predict revenue in 1996.
c. Predict revenue in 1997.

19.4 The Exponential Trend Model

The exponential model is useful for finding the trend in a variable that tends to grow at a fairly constant percentage rate each time period.

DEFINITION

The Exponential Trend Model ————————————————————

In the **exponential trend model,** it is assumed that Y_t can be explained by the equation

$$Y_t = c \, \exp(\beta_1 t + \beta_2 t^2 + \cdots + \beta_s t^s)$$

After taking the natural logarithm of both sides of this equation and adding a random error term, we obtain

$$\ln Y_t = \ln c + \beta_1 t + \beta_2 t^2 + \cdots + \beta_s t^s + e_t$$

or

$$\ln Y_t = \beta_0 + \beta_1 t + \beta_2 t^2 + \cdots + \beta_s t^s + e_t$$

where $\beta_0 = \ln c$.

———

Like the polynomial trend model, the exponential trend model can be used to describe movements in a time series variable that follows a relatively smooth but nonlinear path over time. As an example, let us estimate the exponential trend model for the stock of currency in the United States since 1961.

EXAMPLE 19.4

Estimating an Exponential Trend Model ————————————————————

Table 19.6 (page 842) shows the stock of currency in the United States from 1961 through 1994, and the data are plotted in Figure 19.16. The data follow a smooth, nonlinear path, and an exponential model seems to be a good candidate for describing the behavior of Y_t, the stock of currency at time t. (A quadratic trend model would also do a good job of describing Y_t.) Let us estimate the exponential trend model

$$Y_t = ce^{\beta_1 t}$$

SOLUTION After transforming the model by taking logarithms and adding a random error term, we obtain

$$\ln Y_t = \beta_0 + \beta_1 t + e_t$$

where $\beta_0 = \ln c$ and $\ln Y_t$ denotes the natural logarithm of the currency stock at time t. To estimate the coefficients, we use $\ln Y_t$ as the dependent variable and t as the explanatory

TABLE 19.6   *Stock of U.S. currency, 1961–1994*

Year	Currency (in billions) Y	ln Y	Year	Currency (in billions) Y	ln Y
1961	$29.3	3.38	1978	$ 96.0	4.56
1962	30.3	3.41	1979	104.8	4.65
1963	32.2	3.47	1980	115.4	4.75
1964	33.9	3.52	1981	122.6	4.81
1965	36.0	3.58	1982	132.5	4.89
1966	38.0	3.64	1983	146.2	4.98
1967	40.0	3.69	1984	156.1	5.05
1968	43.0	3.76	1985	167.9	5.12
1969	45.7	3.82	1986	180.7	5.20
1970	48.6	3.88	1987	196.9	5.28
1971	52.0	3.95	1988	212.2	5.36
1972	56.2	4.03	1989	222.6	5.41
1973	60.8	4.11	1990	246.7	5.51
1974	67.0	4.20	1991	267.1	5.59
1975	72.8	4.29	1992	292.2	5.68
1976	79.5	4.38	1993	321.4	5.77
1977	87.4	4.47	1994	353.6	5.87

Source: Economic Report of the President, 1995 (Washington, D.C.: U.S. Government Printing Office, 1995).

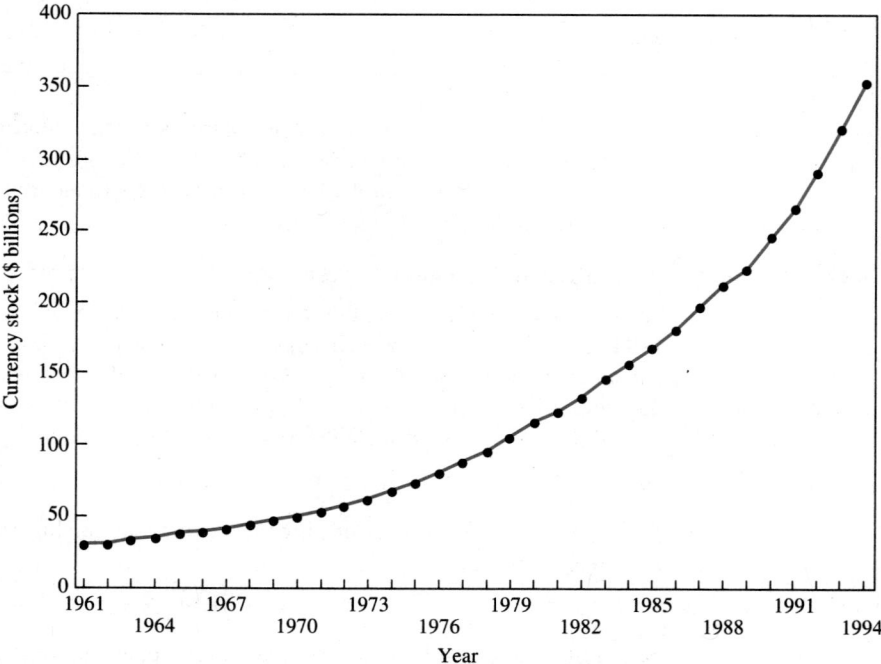

FIGURE 19.16   *Scatter diagram of data in Table 19.6*

variable where t equals 1 in 1961, 2 in 1962, and so forth. Table 19.6 also shows the values of ln Y_t. From Figure 19.17, which shows the values of ln Y_t plotted against t, it appears that ln Y_t can be described as a linear function of time.

The estimated model is

$$\ln \hat{y}_t = 3.170158 + .077778t \qquad R^2 = .996$$
$$t \text{ statistics:} \quad (193.4) \qquad (95.2)$$

where t equals 1 in 1961, 2 in 1962, and so on. The SPSSX computer output for this model is shown in Figure 19.18 (page 844). The computer output shows that the t statistic for b_1 is 95.171 and the p-value is .0000, so we can feel very confident that the population regression coefficient β_1 is nonzero. From R^2, we see that the model explains more than 99% of the variation in ln Y_t (not Y_t). From the plot of the estimated logarithmic equation in Figure 19.17, we can see that the estimated logarithmic equation explains the data almost perfectly.

Recalling that $b_0 = \ln \hat{c} = 3.170158$, we obtain the estimate for c of

$$\hat{c} = e^{3.170158} = 23.8112$$

This yields the exponential trend curve

$$\hat{y}_t = 23.8112e^{.077778t}$$

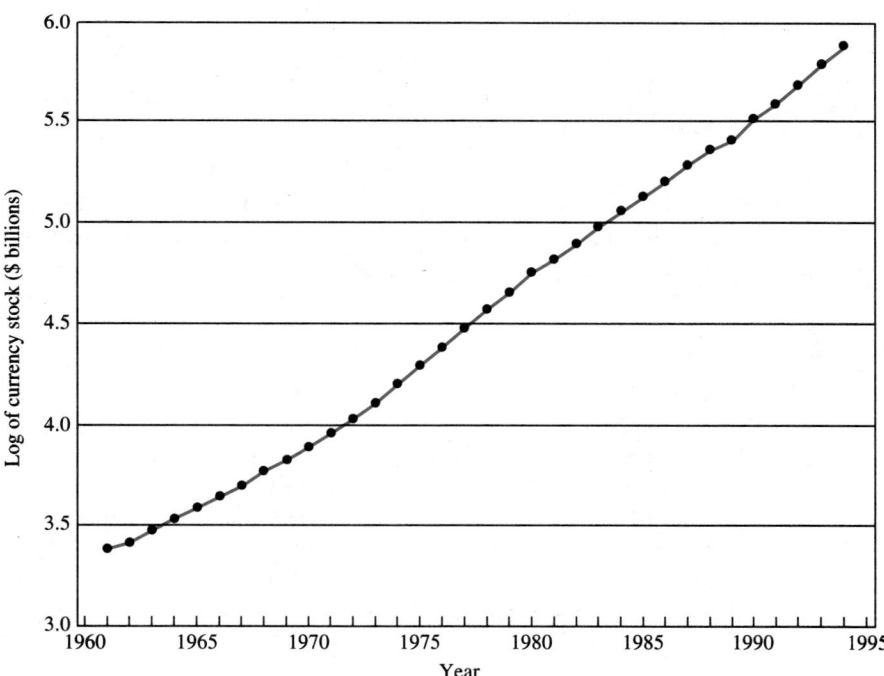

FIGURE 19.17 *Scatter diagram showing the logarithm of stock of currency versus time from Table 19.6*

```
* * * *   M U L T I P L E   R E G R E S S I O N   * * * *

Listwise Deletion of Missing Data

Equation Number 1    Dependent Variable..   LNY

Block Number  1.  Method:  Enter      T

Variable(s) Entered on Step Number  1..    T

Multiple R          .99824      Analysis of Variance
R Square            .99648                      DF   Sum of Squares   Mean Square
Adjusted R Square   .99637      Regression       1        19.79657      19.79657
Standard Error      .04675      Residual        32          .06994        .00219

                                F =    9057.57473      Signif F =  .0000


------------------ Variables in the Equation ------------------

Variable              B        SE B       Beta        T    Sig T

T               .077778  8.1724E-04    .998238   95.171   .0000
(Constant)     3.170158     .016396              193.352   .0000


End Block Number   1   All requested variables entered.
```

.FIGURE 19.18   *SPSSX-generated output for Example 19.4*

Now suppose we wanted to use this model to predict the value for the stock of currency in 1998. Based on $t = 1$ in 1961, the value of t in 1998 would be $t = 38$. The predicted value is then

$$\hat{y}_{38} = 23.8112e^{.077778(38)} = 457.4740$$

or approximately $457.5 billion.

In this example, we obtained an almost perfect fit ($R^2 = .996$) by using an exponential trend model. If a quadratic trend equation had been estimated instead, the estimated equation would have been

$$\hat{y}_t = 43.4188 - 3.5112t + .3508t^2$$

and the value of R^2 would have been .99408. This shows that if the exponential model provides a good description, then a quadratic trend model will probably provide a good description of the data also.

Exercises

Statistical Concepts

19.4.1 *True or false:*   When choosing between a quadratic trend model and an exponential trend model, it is not reasonable merely to choose the model that yields the higher value of R^2 because the two models do not use the same dependent variable. Explain.

19.4.2 *True or false:*   If some variable grows at approximately a constant percentage rate each year, then the exponential trend model would be preferred to a linear model. Explain.

Statistical Drills

19.4.3 Plot Y_t and $\ln Y_t$ against time t.

Year	1	2	3	4	5	6	7	8
Value	3	7	11	22	31	45	61	84

a. Estimate the exponential trend model

$$\ln Y_t = \beta_0 + \beta_1 t + e_t$$

b. Plot the estimated trend line on the scatter plot.
c. Calculate R^2.
d. Forecast the values for years 9 and 10.
e. Do you feel confident that your forecast will be fairly accurate? Explain why or why not.

19.4.4 Plot Y_t and ln Y_t against time t.

Year	1	2	3	4	5	6	7	8
Value	3	7	11	10	31	28	61	84

a. Estimate the exponential trend model

$$\ln Y_t = \beta_0 + \beta_1 t + e_t$$

b. Plot the estimated trend line on the scatter plot.
c. Calculate R^2.
d. Forecast the values for years 9 and 10.
e. Do you feel confident that your forecast will be fairly accurate? Explain why or why not.

Statistical Applications

$$ 19.4.5 The accompanying data show the total interest paid by federal, state, and local governments in the United States from 1981 to 1994.

Year	Period	Interest paid (in billions)	Year	Period	Interest paid (in billions)
1981	1	110.2	1988	8	229.9
1982	2	130.6	1989	9	251.0
1983	3	146.6	1990	10	269.6
1984	4	174.6	1991	11	283.9
1985	5	195.9	1992	12	282.3
1986	6	207.9	1993	13	279.3
1987	7	215.9	1994	14	286.0

Source: Economic Report of the President, 1995 (Washington, D.C.: U.S. Government Printing Office, 1995).

a. Estimate the linear trend line using $t = 1$ for 1981. Calculate R^2.
b. Forecast interest payments in 1998 and 1999 using the linear trend line.
c. Estimate the model ln $Y_t = \beta_0 + \beta_1 t + e_t$, where Y_t = interest payments in year t. Calculate R^2.
d. Forecast interest payments in 1995 and 1996 using the exponential trend line.

$$ 19.4.6 The accompanying data show the money supply (currency plus demand deposits) for the years 1981 to 1994.

Year	Period	Money supply (in billions)	Year	Period	Money supply (in billions)
1981	1	$436.3	1988	8	$ 787.4
1982	2	474.3	1989	9	794.7
1983	3	521.0	1990	10	826.4
1984	4	552.1	1991	11	897.7
1985	5	619.9	1992	12	1,024.8
1986	6	724.5	1993	13	1,128.4
1987	7	750.1	1994	14	1,147.6

Source: Economic Report of the President, 1995 (Washington, D.C.: U.S. Government Printing Office, 1995).

a. Estimate the linear trend line using $t = 1$ in 1981 and calculate R^2.
b. Estimate the money supply in 1995 and 1996 using the linear trend line.
c. Estimate the equation $\ln Y_t = \beta_0 + \beta_1 t + e_t$ and calculate R^2.
d. Estimate the money supply in 1998 and 1999 using the exponential trend line.

19.4.7 Suppose it is hypothesized that the Gross Domestic Product (GDP) can be explained by the model

$$\ln \mathrm{GDP}_t = \beta_0 + \beta_1 t + e_t$$

where t represents time as measured in the accompanying data. The *Economic Report of the President* gives the following data, expressed in billions of 1987 dollars:

Year	t	GDP (in billions of 1987 dollars)	Year	t	GDP (in billions of 1987 dollars)
1960	1	$1,970.8	1980	5	$3,776.3
1965	2	2,470.5	1985	6	4,279.8
1970	3	2,873.9	1990	7	4,897.3
1975	4	3,221.7			

a. Plot GDP versus time.
b. Plot ln GDP versus time.
c. Estimate and plot the equation $\ln \mathrm{GDP}_t = \beta_0 + \beta_1 t + e_t$.
d. Forecast GDP for 1995.

19.4.8 A marketing researcher studied annual sales of a product that had been introduced 10 years ago. The data were as follows, where t is the year (coded) and Y is sales in thousands of units:

t	1	2	3	4	5	6	7	8	9	10
y	98	135	162	178	221	232	283	300	374	395

a. Prepare a scatter plot of the data. Does a linear relation appear adequate here?
b. Use the transformation $Y'_t = \ln Y_t$ and obtain the estimated linear regression function for the transformed data.
c. Plot the estimated regression line and the transformed data. Does the regression line appear to be a good fit to the transformed data?
d. Obtain the residuals and plot them against the fitted values.
e. Express the estimated regression equation in the original units.

19.5   Autoregressive Forecasting Models

A frequently used time series model is the *autoregressive model*. This is a special case of the multiple regression model in which some or all of the explanatory variables are lagged values of Y_t.

DEFINITION   **Autoregressive Model**

The **autoregressive model** takes the form

$$Y_t = \beta_0 + \beta_1 y_{t-1} + \beta_2 y_{t-2} + \cdots + \beta_s y_{t-s} + e_t$$

The value y_{t-s} is called the *lagged value* of Y at time $(t - s)$. The *order* of the autoregressive model is s.

In an autoregressive equation, the value of Y at time t is explained by the values of Y during the previous s periods. As usual, a random error term e_t is included. This model might be criticized because it might be some set of variables $X_1, X_2, \ldots, X_K$ rather than $y_{t-1}, y_{t-2}, \ldots, y_{t-s}$ that causes Y to assume a specific value. On the other hand, if the true causal factors are $X_1, X_2, \ldots, X_K$ and they vary in a relatively systematic manner from one period to the next, then Y should also vary in a systematic manner from one period to the next. Thus, knowledge of the values $y_{t-1}, y_{t-2}, \ldots, y_{t-n}$ should provide us with information when forecasting y_t, y_{t+1}, and so on. In a sense, the lagged dependent variable Y_{t-1} can be thought of as being a proxy for the true causal variables $X_1, X_2, \ldots, X_K$.

If the time series we are trying to forecast follows a relatively smooth path over time, a low-order autoregressive model (i.e., an autoregressive model with a small value of s) can usually produce fairly accurate results. For example, since GDP follows a fairly smooth path over time (except for a few aberrations), a simple first-order or second-order autoregressive model produces a fairly accurate set of forecasts for most purposes.

A useful procedure for checking the forecasting ability of a model is to hold out the last few values of Y (e.g., y_n, y_{n-1}, and y_{n-2}) before estimating the model. After estimating the model, obtain the predicted values $\hat{y}_{n-2}, \hat{y}_{n-1}$, and $\hat{y}_n$ and compare them with the actual values. If the two sets of values are close to each other, then reestimate the model using all the data and use the reestimated model to forecast the future values y_{n+1}, y_{n+2}, and so forth.

One benefit of using an autoregressive equation for forecasting is that it is not necessary to predict the future values of the explanatory variables first. If Y_t is expressed as a function of an explanatory variable X, then it is necessary to know or forecast x_{t+1} to forecast Y_{t+1}. However, if Y_t is expressed as a function of y_{t-1}, we need only know or forecast Y_t to forecast Y_{t+1}. If a linear or quadratic trend model fits the data well, a low-order autoregressive equation will also provide a good fit.

EXAMPLE 19.5 **Estimating an Autoregressive Model**

The data in Table 19.7 show gross business savings from 1980 to 1994 in billions of dollars. Figure 19.19 (page 848) is a scatter diagram of the data. From the graph, it appears that knowing the value of gross business savings in period $(t - 1)$ is useful in predicting the value of gross business savings in period t. That is, it appears that Y_t can be explained as a function of y_{t-1}. Let us estimate the regression coefficients in the first-order autoregressive model

$$y_t = \beta_0 + \beta_1 Y_{t-1} + e_t$$

where Y_t = gross business savings in year t.

TABLE 19.7 *Autoregressive model for gross business savings Y_t*

Year	Y_t	Y_{t-1}	$Y_t Y_{t-1}$	Y_{t-1}^2
1980	346	—	—	—
1981	394	346	136,324	119,716
1982	417	394	164,298	155,236
1983	473	417	197,241	173,889
1984	521	473	246,433	223,729
1985	546	521	284,466	271,441

(*continued*)

TABLE 19.7   (*continued*)

Year	Y_t	Y_{t-1}	$Y_t Y_{t-1}$	Y_{t-1}^2
1986	534	546	291,564	298,116
1987	589	534	314,526	285,156
1988	647	589	381,083	346,921
1989	667	647	431,549	418,609
1990	691	667	460,897	444,889
1991	726	691	501,666	477,481
1992	753	726	546,678	527,076
1993	790	753	594,870	567,009
1994	850	790	671,500	624,100

Source: Survey of Current Business (Washington, D.C.:
U.S. Department of Commerce, 1995).

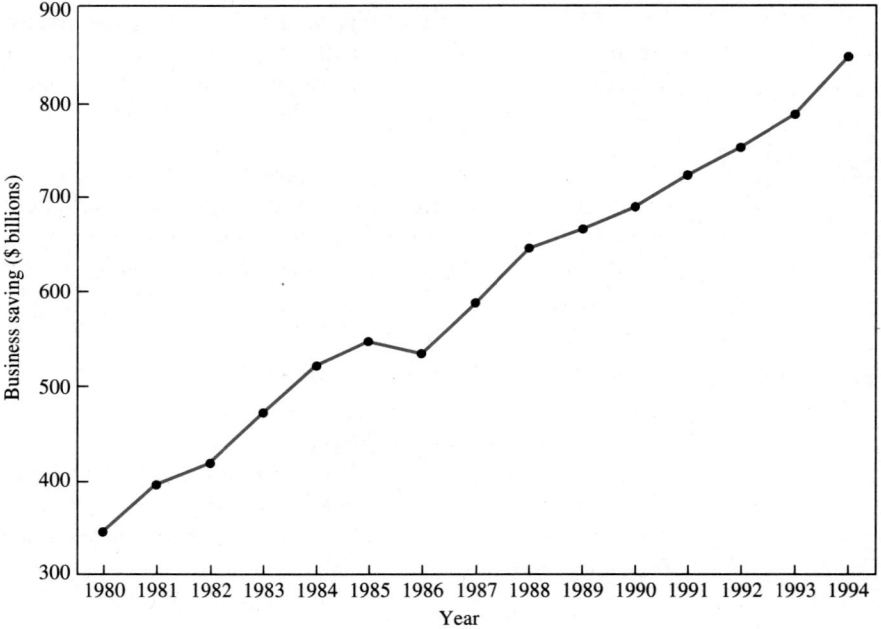

FIGURE 19.19   *Scatter diagram of data for gross business savings in Table 19.7*

SOLUTION   This is the same problem as estimating the simple linear regression
model

$$Y_t = \beta_0 + \beta_1 x_t + e_t$$

using $x_t = y_{t-1}$. In general, a computer would be used to estimate this model. The data
and the appropriate calculations are shown in Table 19.7.

In a sense, we lose the first observation of Y_t because of the lagged variable y_{t-1}. That
is, our first equation is

$$Y_{1981} = \beta_0 + \beta_1 y_{1980} + e_{1981}$$

We have 15 observations on Y from 1980 through 1994, but the number of observations in the regression model is $n = 14$. The sum of the y_i's is calculated using the data for 1981 through 1994.

To obtain the least-squares estimates with a hand calculator, we substitute y_{t-1} for x_t in the equations for the least-squares coefficients.

The SPSSX computer output is shown in Figure 19.20.

```
* * * *   M U L T I P L E   R E G R E S S I O N   * * * *

Listwise Deletion of Missing Data

Equation Number 1    Dependent Variable..   Y   GROSS BUSINESS SAVINGS

Block Number  1.  Method:  Enter      YLAG


Variable(s) Entered on Step Number  1..      YLAG


Multiple R            .98983      Analysis of Variance
R Square              .97976                          DF     Sum of Squares     Mean Square
Adjusted R Square     .97808      Regression           1       250576.52905    250576.52905
Standard Error       20.76693     Residual            12         5175.18524       431.26544

                                  F =      581.02623      Signif F =   .0000


------------------ Variables in the Equation ------------------

Variable              B          SE B       Beta         T   Sig T

YLAG             .993473      .041215    .989831    24.104   .0000
(Constant)     39.773372    24.466174               1.626   .1300


End Block Number    1    All requested variables entered.
```

FIGURE 19.20   *SPSSX-generated output for Example 19.5*

The estimated model is

$$\hat{y}_t = 39.773 + .993\, y_{t-1} \qquad R^2 = .97976$$
$$t \text{ statistics:} \quad (1.626) \qquad (24.10)$$

The t statistic for b_1 is $t = 24.10$, and the p-value for b_1 is .0000. The t statistic in an autoregressive model does not exactly follow the t distribution because one of the basic assumptions of the classical linear regression model has been violated: The error terms are not independent of the explanatory variables because of the presence of the lagged dependent variable. Nevertheless, this t statistic is so extreme that we can still feel quite confident that the population regression slope β_1 is nonzero.

The model explains more than 97% of the variation in Y, which is a common occurrence in autoregressive models. In an autoregressive model, the value of R^2 will be quite close to 1 whenever the trend of the dependent variable is relatively smooth, because it should be possible to get a good forecast of a value of Y when the previous value of Y is known.

Let us use the autoregressive equation to forecast gross business savings for 1995. From the autoregressive model, we obtain

$$\hat{y}_{1995} = 39.773 + .993\, y_{1994}$$

We substitute $y_{1994} = 850$ and obtain

$$\hat{y}_{1995} = 39.773 + .993(850) = 883.823$$

Recall that the original data are in billions of dollars. Thus, the forecast for 1995 would be $883.823 billion.

Exercises

Statistical Concepts

19.5.1 *True or false:* Suppose a variable, such as population of the United States, is well explained by a linear trend model. Then the variable will be well explained by an autoregressive model. Explain.

19.5.2 Suppose a variable is well explained by an autoregressive model. Does it follow that the variable will be well explained by a linear trend model? Consider a variable such as mean monthly temperature in a city before explaining your answer.

19.5.3 Suppose we have monthly sales data for some retail product such as women's sweaters. Explain why it might be useful to include a term such as y_{t-12} in the model. What does the variable y_{t-12} represent?

19.5.4 Suppose we have quarterly sales data for some retail product such as ski apparel. Explain why it might be useful to include a term such as y_{t-4} in the model. What does the variable y_{t-4} represent?

Statistical Drills

19.5.5 Consider the following eight annual observations.

Year	1	2	3	4	5	6	7	8
y_t	12.0	15.3	17.2	21.3	22.8	25.1	26.2	29.0

a. Plot the data.
b. Estimate the first-order autoregressive model, and interpret the results.
c. Calculate R^2 and interpret the value.
d. Forecast the value of Y_t for period 9.
e. Forecast the value of Y_t for period 10.
f. Does it appear that a linear trend model would provide a good explanation of the data? Estimate the linear trend model.
g. Use the linear trend model and forecast the value of Y_t for period 9. Compare with the forecast in part **d.**

19.5.6 Consider the following 10 annual observations.

Year	1	2	3	4	5	6	7	8	9	10
y_t	6.0	8.3	11.5	16.3	19.8	22.1	26.7	29.7	34.2	36.1

a. Plot the data.
b. Estimate the first-order autoregressive model, and interpret the results.
c. Calculate R^2 and interpret the value.
d. Forecast the value of Y_t for period 11.
e. Forecast the value of Y_t for period 12.
f. Does it appear that a linear trend model would provide a good explanation of the data? Estimate the linear trend model.
g. Use the linear trend model and forecast the value of Y_t for period 11. Compare with the forecast in part **d.**

Statistical Applications

19.5.7 Refer to the data in Exercise 5 in Section 19.4.
    a. Estimate the model $Y_t = \beta_0 + \beta_1 y_{t-1} + e_t$. Calculate R^2. Interpret your results.
    b. Forecast Y_t in 1995 and 1996.
    c. Compare your results with the results obtained in Section 19.4.

19.5.8 Refer to the data in Exercise 6 in Section 19.4.
    a. Estimate the model $Y_t = \beta_0 + \beta_1 y_{t-1} + e_t$. Calculate R^2. Interpret your results.
    b. Forecast Y_t in 1995 and 1996.
    c. Compare your results with the results obtained in Section 19.4.

19.5.9 Refer to the data in Exercise 7 in Section 19.4.
    a. Estimate the model $Y_t = \beta_0 + \beta_1 y_{t-1} + e_t$. Calculate R^2. Interpret your results.
    b. Forecast Y_t in 1995.
    c. Compare your results with the results obtained in Section 19.4.

19.5.10 The accompanying data show quarterly observations on profits (Y) and advertising expenditures (X) for a local corporation. Estimate the model

$$Y_t = \beta_0 + \beta_1 x_t + \beta_2 y_{t-1} + e_t$$

All data are in thousands of dollars.

t	Y	X	t	Y	X
1	116	42	11	168	63
2	120	43	12	173	66
3	124	44	13	182	68
4	132	47	14	190	72
5	139	50	15	191	74
6	145	52	16	195	77
7	150	54	17	201	80
8	153	56	18	204	83
9	157	59	19	210	85
10	161	60	20	214	87

19.6   Measuring Forecast Accuracy

The linear trend model is one of the most frequently used models for predicting or forecasting future values of some dependent variable Y and has certain advantages and disadvantages compared with other techniques for forecasting. Advantages include the following:

    1. The trend line is very easy to estimate. Because we need only a set of observations on Y, we have none of the problems involved in collecting data for explanatory variables.

    2. The trend line is easy to understand and interpret; nearly everyone intuitively understands how to forecast the value of some dependent variable Y by extrapolating into the future along a linear graph.

Disadvantages of the linear trend model include the following:

    1. By using a trend line to forecast, we are not really identifying the underlying factors that cause Y to change over time.

2. If the behavior of the underlying factors changes in the future, then the trend line might produce very poor forecasts.

The basic problem with using an estimated trend line to forecast the future is that the trend line may oversimplify a relatively complicated situation. For example, it is easy to forecast future sales if we assume that sales will continue to grow in the future just as they have in the past, but sometimes this assumption may be naive and unfounded. In fact, sales depend on income, people's tastes, total population, the prices and availability of competing products, advertising expenditures, and so forth. If the behavior of any of these explanatory variables changes in the future, then sales can be expected to change accordingly. However, the linear trend line does not explicitly take into account the behavior of these independent variables. Using the trend line for forecasting becomes more and more suspect as the forecast period moves farther and farther into the future, because, as time passes, it becomes more likely that the behavior of the true explanatory variables will change.

On the other hand, there is also a major disadvantage in forecasting future values of Y by using a multiple regression model containing many different explanatory variables. Suppose we think a variable Y depends on the K explanatory variables $X_1, X_2, \ldots, X_K$. To forecast future values of Y, we need to know the future values of $X_1, X_2, \ldots, X_K$. How do we get this information? In many cases we would have to forecast the future values of the X's (probably by using linear trend equations!).

EXAMPLE 19.6   **Difficulty of Forecasting the Future**

The data in Table 19.8 show small-denomination time deposits from 1961 to 1981 and are plotted in Figure 19.21. It appears that a quadratic trend model describes the behavior of Y appropriately. Let us estimate the quadratic trend model

$$Y_t = \beta_0 + \beta_1 t + \beta_2 t^2 + e_t$$

and use this model to forecast the values of Y for 1982 to 1986.

TABLE 19.8   *Small-denomination time deposits, 1961–1981*

Year	Deposits (in billions) Y_t	Year	Deposits (in billions) Y_t
1961	$ 14.8	1971	$189.7
1962	20.1	1972	231.6
1963	25.5	1973	265.8
1964	29.2	1974	287.9
1965	34.5	1975	337.9
1966	55.0	1976	390.8
1967	77.8	1977	445.7
1968	100.5	1978	521.5
1969	120.4	1979	635.3
1970	151.1	1980	730.2
		1981	825.1

Source: Economic Report of the President, 1987 (Washington, D.C.: U.S. Government Printing Office, 1987).

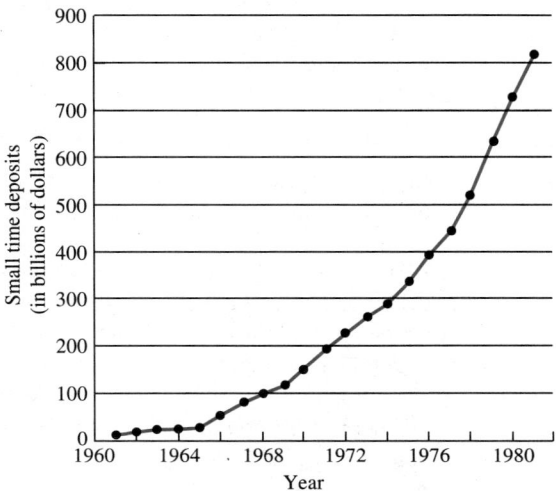

FIGURE 19.21 *Scatter diagram for data in Example 19.6*

SOLUTION We obtain the estimated quadratic equation

$$\hat{y}_t = 48.159 - 14.622\,t + 2.373\,t^2 \qquad R^2 = .993$$

t statistics: (3.027) (−4.39) (16.1)

The SPSSX computer output for this model is shown in Figure 19.22.

```
                             * * * *   M U L T I P L E   R E G R E S S I O N   * * * *

Listwise Deletion of Missing Data

Equation Number 1    Dependent Variable..   Y    SMALL TIME DEPOSITS

Block Number  1.  Method:  Enter      T        TSQ


Variable(s) Entered on Step Number  1..    TSQ
                                    2..    T


Multiple R          .99642        Analysis of Variance
R Square            .99286                            DF    Sum of Squares     Mean Square
Adjusted R Square   .99206        Regression           2      1213896.32873   606948.16436
Standard Error    22.02769        Residual            18         8733.94365      485.21909

                                  F =    1250.87445      Signif F =   .0000


------------------ Variables in the Equation ------------------

Variable            B          SE B       Beta        T    Sig T

TSQ           2.372952      .147071    1.348968    16.135   .0000
T           -14.622469     3.331529    -.366960    -4.389   .0004
(Constant)   48.159398    15.911930                 3.027   .0073


End Block Number   1   All requested variables entered.
```

FIGURE 19.22 *SPSSX-generated output for Example 19.6*

Because R^2 is .99286, the estimated model explains more than 99% of the variation in Y. Suppose we want to use this model to predict the value for small-denomination time deposits from 1982 to 1986. Since these values are already known, we can check our forecasts against the actual values.

Based on $t = 1$ in 1961, the value of t in 1982 would be $t = 22$. We obtain the predicted value

$$\hat{y}_{22} = 48.159 - 14.622(22) + 2.373(22)^2 = 875.007$$

Similarly, we can obtain the forecasted values for 1983 through 1986. The actual and forecasted values for small-denomination time deposits for 1982 through 1986 are shown in Table 19.9.

The actual values of Y for 1961 to 1986 are plotted in Figure 19.23. It is quite apparent that the long-term behavior of time deposits changed in 1982. At the end of 1981, an individual making forecasts for 1982 through 1986 would have thought that the forecasts would be very accurate. In fact, however, the forecasts are very inaccurate.

TABLE 19.9 *Actual and forecasted values for small-denomination time deposits*

Year	Actual value (in billions) Y	Forecasted value (in billions) $\hat{Y}$
1982	$852.8	$ 875.007
1983	785.2	967.170
1984	887.5	1,064.079
1985	880.3	1,165.734
1986	852.4	1,272.135

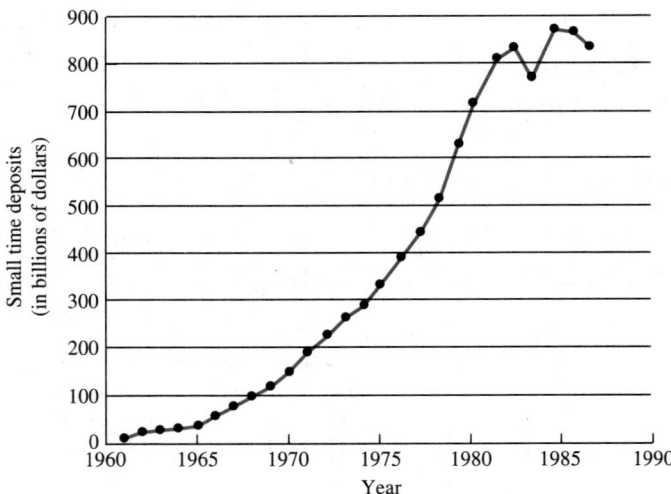

FIGURE 19.23 *Actual small-denomination time deposits versus time, 1961–1986*

Prior to 1982, time deposits traced out an extremely smooth path on the scatter diagram, but afterward the behavior of the variable became much more volatile and less predictable. Because of the recent change in the behavior of Y, any forecasts of future values of small-denomination time deposits based on data from 1961 to 1981 will be inaccurate.

The previous example points out the problems in forecasting the value of any variable. Trend models rely on the assumption that past trends will continue unchanged in the future. In Example 19.6, such was not the case. An observer might wonder what caused the sudden change in the behavior of small-denomination time deposits in 1982. A likely explanation for the slowdown in their growth after 1981 is the rapid growth of a new investment vehicle, Money Market Deposit Accounts (MMDAs), which did not exist before 1982. MMDAs grew from $0 in 1981 to $43.2 billion in 1982, $379.2 billion in 1983, and $570.7 billion in 1986. This factor is not accounted for in the quadratic trend model used in Example 19.6.

Three Measures of Forecast Accuracy

When we make our forecast, there is no way of knowing with certainty whether it will be accurate. For example, given the data on small-denomination time deposits for 1961 through 1981, most people would think that they could produce very accurate forecasts for 1982 through 1986. We have shown, however, that the forecasts generated by the quadratic trend model in Example 19.6 are very inaccurate. Only after the actual values have been observed can we determine whether we have made a good forecast.

We want a model that minimizes the differences between the forecast values and the actual values. We discuss the following methods of measuring the quality of a forecast:

1. Mean absolute forecast error (MAFE)
2. Mean absolute percentage error (MAPE)
3. Root mean squared error (RMSE)

All of these measures depend on the actual forecast errors $(y_t - \hat{y}_t)$.

The first method, the mean absolute forecast error, is defined as follows.

DEFINITION **Mean Absolute Forecast Error (MAFE)**

Suppose we have forecast the value of Y for m periods. The mean of the absolute forecast errors is called the **mean absolute forecast error** (MAFE). We obtain

$$\text{MAFE} = \frac{\Sigma |y_t - \hat{y}_t|}{m}$$

where y_t is the actual value of Y observed at time t and $\hat{y}_t$ is the forecast value of Y for time t.

An alternative way of measuring forecast error is in percentage terms. When the variable being forecast takes very large values, a forecast error that is quite large in absolute terms might be quite small in percentage terms. For example, Gross Domestic

Product currently exceeds $6 trillion per year. A forecast error of $1 billion is large in absolute terms, but quite small in percentage terms—less than 1%. The absolute percentage forecast error is defined as

$$\text{Absolute percentage forecast error} = \frac{|y_t - \hat{y}_t|}{y_t} \times 100$$

DEFINITION   **Mean Absolute Percentage Error (MAPE)**

Suppose we have forecast the value of Y for m periods. The mean of the absolute percentage forecast errors is called the **mean absolute percentage error** (MAPE). We obtain

$$\text{MAPE} = \frac{\Sigma[|y_t - \hat{y}_t|/y_t]}{m} \times 100$$

A third measure of forecast accuracy is called the root mean squared error, defined as follows.

DEFINITION   **Root Mean Squared Error (RMSE)**

The **root mean squared error** of a forecast is defined by the equation

$$\text{RMSE} = \sqrt{\frac{\Sigma(y_t - \hat{y}_t)^2}{m}}$$

where m is the number of time periods for which forecasts have been made.

The basic difference between MAFE and RMSE is that the latter penalizes extreme errors more heavily than does MAFE. For this reason RMSE is an appropriate measure of forecast error if the costs of making forecast errors increase more than proportionally to the size of the error. Regardless of which measure of forecast error is chosen, the model providing the smallest MAFE, MAPE, or RMSE is the most accurate.

Supposedly the statistics MAFE, MAPE, and RMSE are based on observed and forecast values at future points in time. When we are examining a model, however, it is usually impractical to fit the model to historical data and then wait for future actual values to become available in order to measure forecast accuracy. Thus the statistician will sometimes separate the historical data base into two sets, one for estimating the parameters in the model, which is then used to forecast the values of the remaining observations.

EXAMPLE 19.7   **Measuring Forecast Accuracy**

In Example 19.6, we actually had 26 observations on small-denomination time deposits from 1961 through 1986. We used the first 21 observations to estimate the model and the last 5 observations to measure the model's forecast accuracy. Calculate MAFE, MAPE, and RMSE.

SOLUTION   Refer to the actual and forecasted values for the variable Small-denomination time deposits, which are given in Table 19.9 (page 854). For example, in 1982 the absolute forecast error is

$$|y_t - \hat{y}_t| = |852.8 - 875.007| = \$22.207 \text{ billion}$$

and the absolute percentage forecast error is

$$\frac{|y_t - \hat{y}_t|}{y_t} \times 100 = \frac{|852.8 - 875.007|}{852.8} \times 100 = 2.6\%$$

For 1983, the absolute forecast error is $181.97 billion and the percentage forecast error is 23.2%. Thus, the forecast error is much larger in 1983 than in 1982.

   The absolute forecast errors, the absolute percentage forecast errors, and the squared forecast errors for 1982 through 1986 are shown in Table 19.10. By 1986, the absolute forecast error has grown to $419.735 billion and the absolute percentage forecast error to 49.2%. From 1982 through 1986, the MAFE is $217.185 billion, the MAPE is 25.5%, and the RMSE is $253.94 billion.

TABLE 19.10 *Forecast errors for Example 19.7*

SMALL TIME DEPOSITS (IN BILLIONS)

Year	Actual value Y	Forecast value $\hat{Y}$	Absolute forecast error	Absolute percentage error	Squared forecast error
1982	$852.8	$ 875.007	$ 22.207	2.6	$ 493.15
1983	785.2	967.170	181.970	23.2	33,113.08
1984	887.5	1,064.079	176.579	19.9	31,180.14
1985	880.3	1,165.734	285.434	32.4	81,472.57
1986	852.4	1,272.135	419.735	49.2	176,177.47
			1,085.925	127.3	322,436.41

Note: MAFE = 1,085.925/5 = 217.185; MAPE = 127.3/5 = 25.46%;
RMSE = $\sqrt{322,436.41/5}$ = 253.94.

Exercises

Statistical Concepts

19.6.1   *True or false:*   The mean absolute forecast error provides a measure of the typical percentage error we have made in forecasting values of Y. Explain.

19.6.2   *True or false:*   The mean absolute percentage error shows us how large our future forecast errors will be as a percentage of the actual value. Explain.

19.6.3   *True or false:*   The root mean squared error provides a measure of the typical in-sample forecast error that was made, but it does not tell us how big the error is relative to the magnitude of the variable we are trying to forecast. Explain.

19.6.4   Suppose you want to forecast the population of California for the next year. Would you expect the mean absolute forecast error to be larger or smaller than 5,000? Would you expect the mean absolute percentage error to be greater than or less than .25? Explain.

Statistical Drills

19.6.5   Consider the following eight annual observations.

Year	1	2	3	4	5	6	7	8
y_t	11.0	13.4	15.2	17.3	19.8	22.1	26.3	29.0

a. Plot the data.
b. Estimate the first-order autoregressive model, and interpret the results.
c. Forecast the value of Y_t for each period.
d. Calculate MAPE, MAFE, and RMSE. Interpret each value.

19.6.6   Consider the following 10 annual observations.

Year	1	2	3	4	5	6	7	8	9	10
y_t	6.0	8.5	10.1	13.3	14.8	16.1	19.7	24.7	25.2	27.1

a. Plot the data.
b. Estimate the first-order autoregressive model, and interpret the results.
c. Forecast the value of Y_t for each period.
d. Calculate MAPE, MAFE, and RMSE. Interpret each value.

Statistical Applications

19.6.7   Refer to the data in Exercise 5 in Section 19.4.
a. Estimate the model $Y_t = \beta_0 + \beta_1 y_{t-1} + e_t$.
b. Forecast the value of Y_t for each period.
c. Calculate MAPE, MAFE, and RMSE. Interpret each value.

19.6.8   Refer to the data in Exercise 6 in Section 19.4.
a. Estimate the model $Y_t = \beta_0 + \beta_1 y_{t-1} + e_t$.
b. Forecast the value of Y_t for each period.
c. Calculate MAPE, MAFE, and RMSE. Interpret each value.

19.6.9   Refer to the data in Exercise 7 in Section 19.4.
a. Estimate the model $Y_t = \beta_0 + \beta_1 y_{t-1} + e_t$.
b. Forecast the value of Y_t for each period.
c. Calculate MAPE, MAFE, and RMSE. Interpret each value.

19.7   Computer Applications

In this chapter, we discussed the following four models for estimating the long-term trend in a time series:

1. Linear trend model
2. Polynomial trend model
3. Exponential trend model
4. Autoregressive model

The SPSSX program can be used to estimate each of these models. To estimate the linear trend model, read in the n data values of the dependent variable along with the values 1 through n for the variable Time. Suppose we name the dependent variable Y and the explanatory variable TIME. To estimate the linear trend model

$$Y_t = \beta_0 + \beta_1 t + e_t$$

issue the following commands:

REGRESSION VARIABLES = Y, TIME/
    DEPENDENT = Y/
    ENTER TIME/

To estimate a quadratic trend model, we first have to create the squared term for Time. Thus to estimate the quadratic trend model

$$Y_t = \beta_0 + \beta_1 t + \beta_2 t^2 + e_t$$

issue the following commands:

COMPUTE TIMESQ = TIME * TIME
REGRESSION VARIABLES = Y, TIME, TIMESQ/
    DEPENDENT = Y/
    ENTER TIME, TIMESQ/

To estimate an exponential trend model, we first have to create the variable representing the natural logarithm of the dependent variable Y; let's call it LNY. If the model contains a squared term for the variable Time, we also have to create the appropriate variable. Then we estimate the model. To estimate the quadratic exponential trend model

$$\ln Y_t = \beta_0 + \beta_1 t + \beta_2 t^2 + e_t$$

issue the following commands:

COMPUTE LNY = LN(Y)
COMPUTE TIMESQ = TIME * TIME
REGRESSION VARIABLES = LNY, TIME, TIMESQ/
    DEPENDENT = LNY/
    ENTER TIME, TIMESQ/

To estimate an autoregressive trend model, we first have to create a new variable representing the lagged dependent variable. Suppose we name the new variable LAGY. To do this, we issue a LAG command and then we estimate the model. To estimate the single-period autoregressive model, issue the following commands:

COMPUTE LAGY = LAG(Y, 1)
REGRESSION VARIABLES = Y, LAGY/
    DEPENDENT = Y/
    ENTER LAGY/

Exercises

19.7.1   Refer to the data in Table 2.4 in Chapter 2, which show data values for the time series variables TOTALGOV and DEFENSE indicated from 1960 through 1986.
a. For each variable, generate a scatter diagram versus time.

b. For each variable, estimate the linear trend model and examine the value of R^2 for each variable.

c. For each variable, print out the residuals and obtain the Durbin–Watson statistic.

d. Examine the residual plot and determine whether a quadratic model seems to be preferable to a linear model.

e. For each model, let $\alpha = .05$ and test the null hypothesis that the error terms are not serially correlated against the alternative hypothesis that the errors are positively correlated.

19.7.2   Refer to Exercise 1.

a. For each variable, estimate the quadratic trend model.

b. For each variable, print out the residuals and obtain the Durbin–Watson statistic.

c. Examine the residual plot and determine whether a quadratic model seems to be suitable.

d. For each model, let $\alpha = .05$ and test the null hypothesis that the error terms are not serially correlated against the alternative hypothesis that the errors are positively correlated.

19.7.3   Refer to Exercise 1.

a. For each variable, estimate the first-order autoregressive model.

b. For each variable, print out the residuals and obtain the Durbin–Watson statistic.

c. Examine the residual plot and determine whether an autoregressive model appears to be suitable.

d. For each model, let $\alpha = .05$ and test the null hypothesis that the error terms are not serially correlated against the alternative hypothesis that the errors are positively correlated. Is this test strictly valid? Discuss.

STATISTICS IN ACTION: CASE STUDY

Estimating Lost Profits for a Regional Electrical Contractor

In 1989, an electrical contractor located in the northeastern United States was providing services to a large German construction firm that was building a major industrial complex in the United States. The German firm adopted numerous changes in the original contract, which led to huge cost overruns and long construction delays and bottlenecks. A dispute arose between the electrical contractor and the German firm concerning who was responsible for the cost overruns and the construction delays. Also, the German firm experienced a series of financial problems in Germany, which caused the firm to have a serious cash-flow problem.

The German firm failed to pay several million dollars in fees owed to the electrical contractor. The electrical contractor was forced to borrow money at high interest rates to pay its workers and to pay numerous bills. Eventually, the contractor had to lay off many skilled employees. Because of its precarious financial position (its debts exceeded its assets), the contractor could not get the insurance backing necessary to enable it to bid on other construction projects in the region. As a result, future business and future potential profits were lost. Employees continued to leave the firm, the firm lost its good reputation, and what had been a thriving business turned into a bankrupt firm.

The electrical contractor wanted to estimate the potential economic losses resulting from the actions allegedly undertaken by the German construction firm. One task was to estimate the potential contract revenue (and the potential profits) that the electrical contractor would have earned from 1989 through 1992 in the absence of the alleged actions undertaken by the German construction firm.

The past history of the electrical firm was examined. The company, established in the early 1920s, had experienced slow, but steady, growth until approximately 1970. Before 1970, most of the company's contracts were with small schools and small municipalities. During the 1970s and 1980s, revenues expanded greatly, the number of employees increased rapidly, and the firm began to expand its market to include large institutional, industrial, and health care facilities in the region. Also, during this period, the company began performing additional services in the fields of design, construction management, and engineering.

Before the dispute in 1989, the electrical contractor had a long history of steady and profitable growth. After the dispute, contract revenues and gross profits did not follow their former trends: They were much lower than they would have been without the alleged actions of the German firm.

The contract revenues earned by the electrical contractor declined each year from 1989 through 1992. One possible explanation for this decline is a lack of construction activity in the region. However, a review of construction activity indicated that this explanation was not valid. Rather, the data indicated that the electrical firm could have increased its contract revenues over previous amounts.

A linear trend model was estimated using actual contract revenue data from 1980 to 1988. The estimated linear trend equation was

Potential contract revenue
$$= \$2,647,778 + \$794,378 \times \text{Time}$$

where time = 0 in 1980, 1 in 1981, and so forth. The value of the coefficient of determination was $R^2 = .96$, so the equation provided a very good fit to past data. Based on past performance, it appeared reasonable to project potential contract revenues for 1989 through 1992 using the linear trend model. To project potential contract revenue for 1989–1992 without the alleged actions of the German firm, the values time = 9, 10, 11, and 12 were substituted into the estimated equation. For example, predicted contract revenues for 1990 (time = 10) amounted to \$10,591,558:

Potential contract revenue
$$= \$2,647,778 + \$794,378 \times 10$$
$$= \$10,591,558$$

In 1990, actual contract revenues were only \$2,569,539. It was argued that the shortfall of \$8,022,019 in contract revenues during 1990 could be attributed to the dispute with the German firm. Similar results were obtained for 1989, 1991, and 1992.

Next, it was argued that, from 1980 through 1988, gross profits at the electrical contracting firm were linearly related to the level of contract revenue. Thus, once the value of lost contract revenues had been projected, it was possible to project the value of lost potential gross profits.

The attorneys representing the electrical contractor argued that the German firm should reimburse the electrical contractor for its lost potential gross profits from 1989 through 1992 and, in addition, should make a payment to cover lost potential gross profits after 1992. It was argued that, from 1989 through 1992, the electrical company had lost contract revenues exceeding \$35,000,000 and lost profits exceeding \$5,000,000. After a long period of litigation, the case was settled and the German firm paid damages exceeding \$6,000,000.

This case provides a classic example of a situation where a linear trend line was helpful in projecting future values of an economic vari3able. Before using a trend curve to project future values, you should be certain that the curve provides a good fit to past data. Next you should try to show that it is reasonable to assume that the trend model would continue to apply in the future. Of course, there is no way to guarantee that the future will continue to mimic past behavior. In fact, the attorneys for the German firm argued that there was no reason to believe that behavior in 1989 through 1992 would be similar to behavior from 1980 to 1988.

A review of regional data, however, showed that there were no unexpected changes in basic economic conditions in the regional construction industry. Other competing firms that were similar in size to the electrical contracting firm continued to experience steady growth. Thus, there was no evidence to suggest that the electrical contracting firm would not continue to grow after 1988 in the same way that it grew prior to 1988.

At the present time, the electrical contracting firm is trying to make a comeback and has rehired some of its former employees.

However, the company still has not attained the level of its 1988 contract revenues.

Chapter 19 Supplementary Exercises

19.S.1   A physical fitness spa has been in existence for 8 years. The numbers of customers using the spa each year are as follows:

Year	Customers (in thousands)	Year	Customers (in thousands)
1988	18	1992	48
1989	23	1993	59
1990	30	1994	74
1991	40	1995	88

a. Plot the data and estimate the linear trend line.
b. Graph the trend line.
c. Estimate the numbers of customers for 1996 and 1997.
d. Estimate the autoregressive equation $Y_t = \beta_0 + \beta_1 y_{t-1} + e_t$.
e. Use the autoregressive equation to predict the numbers of customers for 1996 and 1997.
f. Estimate the semilogarithmic equation $\ln Y_t = \beta_0 + \beta_1 t + e_t$.
g. Use the semilogarithmic equation to predict the numbers of customers for 1996 and 1997.

19.S.2   The following data show the number of college students in a state for selected years:

Year	Students (in thousands)	Year	Students (in thousands)
1986	103	1991	111
1987	105	1992	112
1988	106	1993	114
1989	108	1994	116
1990	109	1995	118

a. Plot the data.
b. Fit a linear trend line and calculate R^2.
c. Predict the numbers of students for 1996 and 1997.
d. Estimate the equation $Y_t = \beta_0 + \beta_1 Y_{t-1} + e_t$ and find R^2.
e. Use part **d** to predict the number of students for 1996.

19.S.3   A real estate agent has been studying the prices of new homes in Baldwin County. Let Y_t denote the average selling price (in thousands of dollars) of new homes sold in Baldwin County in year t. The agent has obtained the following regression equation:

$$\hat{y}_t = 1.02 + 1.06 y_{t-1}$$

Suppose the average price of homes sold in 1995 was $140,000. Predict the average selling prices in 1996, 1997, and 1998.

19.S.4 The following data show the annual sales of a corporation:

Year	Sales (in millions)	Year	Sales (in millions)
1986	$320	1991	$560
1987	366	1992	620
1988	390	1993	640
1989	422	1994	690
1990	456	1995	740

a. Estimate a linear trend line for the sales.   **b.** Predict sales in 1996, 1997, and 1998.
c. Plot the data and graph the estimated line.   **d.** Calculate R^2, the coefficient of determination.

19.S.5 The following data show the gross sales for a petroleum company from 1986 to 1995:

Year	Gross sales (in billions)	Year	Gross sales (in billions)
1986	$3.9	1991	$5.3
1987	3.9	1992	5.5
1988	3.6	1993	6.4
1989	4.8	1994	7.0
1990	4.5	1995	7.1

a. Estimate the linear trend line.
b. Plot the data and graph the line.
c. Predict sales for the year 1998.
d. Estimate a semilogarithmic trend line and predict sales for 1998.

19.S.6 The following data show the population of animals on a sheltered game preserve:

Year	Population	Year	Population
1988	40	1992	800
1989	90	1993	1,900
1990	230	1994	4,200
1991	440	1995	8,800

a. Fit a quadratic curve to the data.
b. Plot the data and graph the estimated curve.
c. Predict the size of the population for 1996.

19.S.7 The accompanying data show the average gross weekly earnings of employees at a manufacturing firm for selected years.

Year	Average weekly earnings	Year	Average weekly earnings
1987	$331.71	1992	$390.96
1988	340.73	1993	410.11
1989	354.42	1994	428.49
1990	368.04	1995	452.11
1991	380.06		

a. Plot the data.
b. Fit a linear trend line.
c. Predict weekly earnings for 1996.
d. Fit the equation $\ln Y_t = \beta_0 + \beta_1 t + e_t$.
e. Predict weekly earnings for 1996 using part **d**.

19.S.8   The following data show the number of mobile homes shipped by a national producer for selected years:

Year	Units shipped	Year	Units shipped
1987	90,200	1992	217,300
1988	118,000	1993	240,360
1989	150,840	1994	317,950
1990	191,320	1995	412,690
1991	216,470		

a. Plot the data.
b. Fit a linear trend line.
c. Predict shipments for 1997 using the linear trend line.
d. Fit a semilog trend line.
e. Predict shipments for 1997 using the semilog trend line.

19.S.9   The following data show U.S. expenditures in billions of dollars on gasoline and oil for selected years:

Year	Period	Amount (in billions)	Year	Period	Amount (in billions)
1986	1	79.7	1991	6	102.9
1987	2	84.7	1992	7	105.5
1988	3	86.9	1993	8	105.6
1989	4	96.2	1994	9	107.3
1990	5	108.4			

Source: Economic Report of the President, 1995 (Washington, D.C.: U.S. Government Printing Office, 1995).

a. Plot the data.
b. Fit the linear trend line and predict expenditures for 1995 and 1996.
c. Estimate the quadratic trend curve.

19.S.10   The accompanying data refer to local government revenues from property taxes for a particular county.

Year	Revenues (in millions)	Year	Revenues (in millions)
1984	$16	1990	$25
1985	18	1991	26
1986	19	1992	28
1987	20	1993	31
1988	21	1994	34
1989	23	1995	38

a. Plot the data and fit the linear trend line.
b. Predict tax revenues for 1997 using the linear trend line.
c. Plot and fit the semilogarithmic trend line.
d. Predict tax revenues for 1997 using the semilogarithmic trend line.

19.S.11 The Martin Jewelry Company has hired an advertising agency to promote its products. The advertising agency used data on the company's monthly sales revenues and monthly advertising expenditures in recent months to generate the sample regression equation

$$\hat{y}_t = 13.20 + 2.10x_{t1} + 1.24x_{t2} + .62x_{t3}$$

where

$$Y = \text{Monthly sales revenues (in thousands of dollars)}$$
$$X_1 = \text{Monthly advertising expenditures (in hundreds of dollars)}$$
$$X_2 = X_1^2$$
$$X_3 = \text{Time period (1 in month 1, 2 in month 2, etc.)}$$

a. Suppose that in month 10 the Martin Jewelry Company intends to spend $500 on advertising. Predict sales revenues for that month.
b. Suppose that in month 10 the Martin Jewelry Company intends to spend $600 on advertising. Predict revenues for that month.
c. What is the change in predicted revenue when advertising expenditures are increased from $500 to $600?
d. Suppose that in month 11 the Martin Jewelry Company intends to spend $800 on advertising. Predict revenues for that month.
e. Suppose that in month 11 the Martin Jewelry Company intends to spend $900 on advertising. Predict revenues for that month.
f. What is the change in predicted revenue when advertising expenditures are increased from $800 to $900?
g. Why is the answer to part **f** different from the answer to part **c**?

MKTG **19.S.12** A marketing analyst studied annual sales of a product that had been introduced 10 years ago. The data were as follows, where t is the year (coded) and Y is sales in thousands of units:

t	1	2	3	4	5	6	7	8	9	10
Y	98	135	162	178	221	232	283	300	374	395

a. Prepare a scatter plot of the data. Does a linear relation appear adequate here?
b. Use the transformation $Y' = \sqrt{Y}$ and obtain the estimated linear regression function for the transformed data.
c. Plot the estimated regression line and the transformed data. Does the regression line appear to be a good fit to the transformed data?
d. Obtain the residuals and plot them against the fitted values. What does your plot show?
e. Express the estimated regression equation in the original units.

References

Box, George E. P., and Gwilym M. Jenkins. *Time Series Analysis, Forecasting, and Control.* San Francisco: Holden-Day, 1970.

Granger, Clive, and Paul Newbold. *Forecasting Economic Time Series.* London: Academic Press, 1977.

Griffiths, William E., R. Carter Hill, and George G. Judge. *Learning and Practicing Econometrics.* New York: Wiley, 1993.

Gujarati, D. *Basic Econometrics.* 2d ed. New York: McGraw-Hill, 1988.

Hoff, John C. *A Practical Guide to Box–Jenkins Forecasting.* Belmont, Calif.: Lifetime Learning, 1983.

Judge, George G., R. Carter Hill, William E. Griffiths, and Tsoung-Chao Lee. *The Theory and Practice of Econometrics.* 2d ed. New York: Wiley, 1985.

Judge, George G., R. Carter Hill, William E. Griffiths, Helmut Lutkepohl, and Tsoung-Chao Lee. *Introduction to the Theory and Practice of Econometrics.* 2d ed. New York: Wiley, 1988.

Makridakis, Spyros G. "A Survey of Time Series Analysis." *International Statistical Review* 44 (1976): 29–70.

Makridakis, Spyros, and Steven C. Wheelwright. *Forecasting: Methods and Applications.* New York: Wiley, 1978.

Morgenstern, Oskar. *On the Accuracy of Economic Observations.* 2d ed. Princeton, N.J.: Princeton University Press, 1963.

Nelson, Charles R. *Applied Time Series for Managerial Forecasting.* San Francisco: Holden-Day, 1973.

Nie, Norman E., C. Hadlai Hull, Jean G. Jenkins, Karin Steinbrenner, and Dale H. Bent. *SPSS Statistical Package for the Social Sciences.* 2d ed. New York: McGraw-Hill, 1975.

Norusis, Marija J. *SPSSX Introductory Statistics Guide.* New York: McGraw-Hill, 1990.

Norusis, Marija J. *SPSSX Advanced Statistics Guide.* Chicago: SPSS, 1990.

Norusis, Marija J. *The SPSS Guide to Data Analysis.* Chicago: SPSS, 1986.

Pankratz, Alan. *Forecasting with Univariate Box–Jenkins Models: Concepts and Cases.* New York: Wiley, 1983.

Pindyck, Robert S., and Daniel L. Rubinfeld. *Econometric Models and Economic Forecasts.* 3d ed. New York: McGraw-Hill, 1985.

Ryan, Thomas A., Brian L. Joiner, and Barbara F. Ryan. *Minitab Handbook.* 2d ed. Boston: PWS-Kent, 1985.

Ryan, Thomas A., Brian L. Joiner, and Barbara F. Ryan. *Minitab Reference Manual.* University Park, Penn.: Minitab Project, 1985.

SAS Introductory Guide. 3d ed. Cary, N.C.: SAS Institute, 1985.

SAS Procedures Guide for Personal Computers. Version 6 ed. Cary, N.C.: SAS Institute, 1986.

SAS Statistics Guide for Personal Computers. Version 6 ed. Cary, N.C.: SAS Institute, 1986.

SAS User's Guide: Basics. Version 5 ed. Cary, N.C.: SAS Institute, 1985.

SAS User's Guide: Statistics. Version 5 ed. Cary, N.C.: SAS Institute, 1985.

SPSSX User's Guide. Chicago: SPSS, 1988.

20 Time Series Analysis II: Estimation of the Seasonal Component

Many time series in economics contain a *seasonal component,* a regular fluctuation about the trend that repeats year after year. In some applications, we want to estimate the size of the seasonal component either in absolute terms or as a percentage of the trend value. In other cases, the forecaster may want to determine the long-term trend of the time series and remove any seasonal influence. When we remove the seasonal component from the time series, we obtain a new time series that is said to be *seasonally adjusted.* In this chapter, we discuss different ways of estimating the seasonal component and of obtaining a seasonally adjusted series.

20.1 Seasonal Adjustment by Using Moving Averages

In this section, we show how to use the *method of moving averages* to obtain a seasonally adjusted series. This method smooths out the effect of any seasonal or irregular component by replacing each observation y_t by a smoothed value s_t, where s_t is an average of y_t and some of the adjacent values y_{t-2}, y_{t-1}, y_{t+1}, y_{t+2}, and so forth.

The simplest moving average is called a *simple centered $(2m + 1)$-point moving average,* where we replace y_t by the average of y_t, the previous m values y_{t-m}, y_{t-m+1}, ..., y_{t-1}, and the succeeding m values y_{t+1}, y_{t+2}, ..., y_{t+m}. If the observations used in calculating the average are given different weights, we obtain a *weighted $(2m + 1)$-point moving average* as opposed to a simple moving average. The moving average is centered because y_t is the central value in the set of values used to calculate s_t.

DEFINITION **A Simple Centered $(2m + 1)$-Point Moving Average** ─────────

Let y_1, y_2, ..., y_n be n observations on a time series, and let m be some nonnegative integer. To calculate a **simple centered $(2m + 1)$-point moving average,** replace the value y_t by the smoothed value s_t, where

$$s_t = \frac{y_{t-m} + y_{t-m+1} + \cdots + y_{t-1} + y_t + y_{t+1} + y_{t+2} + \cdots + y_{t+m}}{2m + 1}$$

for $t = m + 1, m + 2, \ldots, n - m$.

───

In calculating s_t, there are $(2m + 1)$ terms in the numerator, and y_t is the central value. For example, suppose we want to compute a simple centered 5-point moving average. Let s_t denote the smoothed value of y_t. If the original data are denoted by y_1, $y_2, \ldots, y_n$, then a simple 5-point moving average would take the form

$$s_3 = \frac{y_1 + y_2 + y_3 + y_4 + y_5}{5}$$

$$s_4 = \frac{y_2 + y_3 + y_4 + y_5 + y_6}{5}$$

$$\vdots$$

$$s_{n-2} = \frac{y_{n-4} + y_{n-3} + y_{n-2} + y_{n-1} + y_n}{5}$$

The smoothed value s_3 is the average of five values, namely, y_3, the two previous values y_1 and y_2, and the two succeeding values y_4 and y_5. Note that the smoothed time series contains fewer observations than the original time series. In the 5-point moving average, the values s_1, s_2, s_{n-1}, and s_n cannot be calculated, and so we lose four observations, two at each end of the series. In general, when we compute a $(2m + 1)$-point moving average, we lose m observations at each end of the series. Thus, we lose a total of $2m$ observations.

Simple Versus Weighted Moving Averages

The simple centered $(2m + 1)$-point moving average gives an equal weight of $1/(2m + 1)$ to each original observation used to calculate the smoothed value. Thus, in a simple 5-point moving average, the smoothed value is obtained by giving a weight of $\frac{1}{5}$ to each of the five adjacent values.

The relative importance of some of the original observations can be increased by computing a **weighted moving average.** For example, in a 5-point weighted moving average, the weights might be $\frac{1}{8}$, $\frac{1}{4}$, $\frac{1}{4}$, $\frac{1}{4}$, and $\frac{1}{8}$, based on the assumption that the middle values in a series of observations should have greater weight than the end observations. If we use some other weighting system, an entirely different moving average will be determined. The only requirement is that the weights sum to 1.

You may be wondering why we use $(2m + 1)$ observations to form the moving average rather than, say, $2m$ observations. Because the value $(2m + 1)$ is an odd number, the moving-average value for time t will be centered at that time. If we computed the moving average over an even number of observations, the moving-average values would lie *between* the time points rather than *at* the time points.

Constructing a Centered Four-Quarter Moving Average

Suppose we wish to smooth quarterly data by calculating a four-quarter moving average. Because the moving average consists of the average of an even number of observations, the moving·average value will be centered in the middle of two observations. For example, the first value in the four-quarter moving average would be calculated as

$$s_{2.5} = \frac{y_1 + y_2 + y_3 + y_4}{4}$$

where $s_{2.5}$ is centered between times $t = 2$ and $t = 3$. Similarly, we could calculate

$$s_{3.5} = \frac{y_2 + y_3 + y_4 + y_5}{4}$$

To center the moving-average values, we would calculate the moving-average value s_3 as the average of $s_{2.5}$ and $s_{3.5}$; thus we would obtain

$$s_3 = \frac{s_{2.5} + s_{3.5}}{2}$$

$$= \frac{\dfrac{y_1 + y_2 + y_3 + y_4}{4} + \dfrac{y_2 + y_3 + y_4 + y_5}{4}}{2}$$

$$= \frac{y_1}{8} + \frac{y_2 + y_3 + y_4}{4} + \frac{y_5}{8}$$

DEFINITION   **A Centered Four-Quarter Moving Average**

Suppose we have the quarterly data $y_1, y_2, \ldots, y_n$. The **centered four-quarter moving average** at time t is denoted s_t, where

$$s_t = \frac{y_{t-2}}{8} + \frac{y_{t-1} + y_t + y_{t+1}}{4} + \frac{y_{t+2}}{8}$$

The values s_t can be calculated for $t = 3, 4, \ldots, (n - 2)$.

The formula shows that a centered four-quarter moving average is actually a weighted five-point moving average with weights ⅛, ¼, ¼, ¼, and ⅛.

Constructing a Centered 12-Month Moving Average

When we construct a centered 12-month moving average, the first two noncentered values would be

$$s_{6.5} = \frac{y_1 + y_2 + \cdots + y_{12}}{12}$$

and

$$s_{7.5} = \frac{y_2 + y_3 + \cdots + y_{13}}{12}$$

We obtain the centered value

$$s_7 = \frac{s_{6.5} + s_{7.5}}{2}$$

$$= \frac{y_1 + 2y_2 + \cdots + 2y_{12} + y_{13}}{24}$$

Thus, a centered 12-month moving average is actually a weighted 13-point moving average with weights ¹⁄₂₄, ¹⁄₁₂, ..., ¹⁄₁₂, and ¹⁄₂₄.

DEFINITION   **A Centered 12-Month Moving Average** —————————————————————

Suppose we have the monthly data $y_1, y_2, \ldots, y_n$. The **centered 12-month moving average** at time t is denoted s_t, where

$$s_t = \frac{y_{t-6}}{24} + \frac{y_{t-5} + y_{t-4} + \cdots + y_t + \cdots + y_{t+4} + y_{t+5}}{12} + \frac{y_{t+6}}{24}$$

The values s_t can be calculated for $t = 7, 8, \ldots, (n - 6)$.

———

One objective in constructing a moving average is to eliminate seasonal, cyclical, and irregular fluctuations. Frequently a time plot of the smoothed values will give the analyst a good idea of the long-term behavior of the variable being studied because a moving average dampens fluctuations in a time series and thus reveals systematic movement. Graphs showing original data and smoothed values make it easy to recognize seasonal effects.

Estimating the Seasonal Component and Constructing a Seasonal Index

The centered moving-average values can be useful for gaining information about the structure of a time series. These values have been smoothed so that they are mainly free from seasonal and irregular components. The series of moving averages forms the basis for many seasonal adjustment procedures. One important application of moving averages is estimating the seasonal component of a time series and constructing a seasonal index and a seasonally adjusted series.

The technique of seasonal adjustment called the *ratio-to-moving-average approach* is based on the assumption that the time series has a stable seasonal pattern year after year. For any given month or quarter in any year, the effect of seasonality is assumed to raise or lower the observation by some constant proportion from the moving-average value. The seasonal variations are estimated by a quantity called a *seasonal index*, which is defined as follows.

DEFINITION   **Seasonal Index** —————————————————————————————————

A **seasonal index** expresses the value of a time series variable in each month (or quarter) as a percentage of the trend or moving-average value for that month (or quarter).

———

For example, a seasonal index of 105 for some month (or quarter) means that, because of seasonal factors, the observation during the month (or quarter) in question is expected to be 5% above the trend, or moving-average, value. A seasonal index of 95 indicates that the monthly (or quarterly) value is expected to be 5% below the trend, or moving-average, value. The base value of a seasonal index is always 100, or 100%. The construction of a seasonal index is explained in the next box.

Constructing a seasonal index from a moving average

Suppose we have a time series of monthly or quarterly observations $y_1, y_2, \ldots , y_n$. To construct a seasonal index, perform the following steps:

1. Calculate a centered 12-month moving average s_t for each time period when using monthly data or a centered four-quarter moving average s_t when using quarterly data.
2. For each time period, express y_t as a percentage of s_t. That is, calculate the percentage

$$100\left(\frac{y_t}{s_t}\right)$$

3. To find the average seasonal effect for each month or quarter, calculate the sample mean of all percentages for that month or quarter. These sample means are the unadjusted seasonal index values.
4. Scale the monthly or quarterly index values so that the mean of all the monthly or quarterly index values is 100%.

The following example shows how to calculate a seasonal index for monthly data.

EXAMPLE 20.1   **Constructing a Seasonal Index** ───────────────────────────

The data in column 1 of Table 20.1 show the monthly shipments of air conditioners from 1983 through 1986; the data are plotted in Figure 20.1 (page 873). The figure shows that shipments follow a seasonal pattern with a peak each year in May or June and a trough in October. Let us construct a seasonal index by using the ratio-to-moving-average method.

TABLE 20.1 *Shipments of air conditioners, in thousands of units*

	(1)	(2)	(3)	(4) Seasonal	(5) Adjusted
Month	y_t	s_t	$100(y_t/s_t)$	index	series
1983					
January	88	—	—	67.45	130.5
February	130	—	—	94.10	138.2
March	309	—	—	192.81	160.3
April	259	—	—	190.51	136.0
May	300	—	—	219.80	136.5
June	265	—	—	196.43	134.9
July	306	170.42	180	105.29	290.6
August	108	179.25	60	34.87	309.7
September	58	192.88	30	21.72	267.0
October	32	209.33	15	12.83	249.4
November	52	229.67	23	19.74	263.4
December	98	257.63	38	44.42	220.6

(*continued*)

TABLE 20.1 *(continued)*

Month	(1) y_t	(2) s_t	(3) $100(y_i/s_t)$	(4) Seasonal index	(5) Adjusted series
1984					
January	168	268.63	63	67.45	249.1
February	262	261.13	100	94.10	278.4
March	504	257.83	195	192.81	261.4
April	459	257.25	178	190.51	240.9
May	588	257.38	228	219.80	267.5
June	648	257.96	251	196.43	329.9
July	187	260.29	72	105.29	177.6
August	47	262.63	18	34.87	134.8
September	40	264.33	15	21.72	184.2
October	36	268.13	13	12.83	280.6
November	51	272.67	19	19.74	258.4
December	113	264.83	43	44.42	254.4
1985					
January	209	254.50	82	67.45	309.9
February	277	254.71	109	94.10	294.4
March	530	255.96	207	192.81	274.9
April	524	255.83	205	190.51	275.1
May	632	254.83	248	219.80	287.5
June	416	254.33	164	196.43	211.8
July	171	251.08	68	105.29	162.4
August	68	243.58	28	34.87	195.0
September	49	234.83	21	21.72	225.6
October	24	227.50	11	12.83	187.1
November	39	217.13	18	19.74	197.6
December	113	210.04	54	44.42	254.4
1986					
January	131	218.42	60	67.45	194.2
February	175	227.21	77	94.10	186.0
March	422	229.75	184	192.81	218.9
April	456	232.63	196	190.51	239.4
May	451	234.54	192	219.80	205.2
June	427	234.38	182	196.43	217.4
July	361	—	—	105.29	342.9
August	89	—	—	34.87	255.2
September	89	—	—	21.72	409.8
October	53	—	—	12.83	413.1
November	56	—	—	19.74	283.7
December	92	—	—	44.42	207.1

Source: Survey of Current Business (Washington, D.C.: U.S. Department of Commerce, 1987).

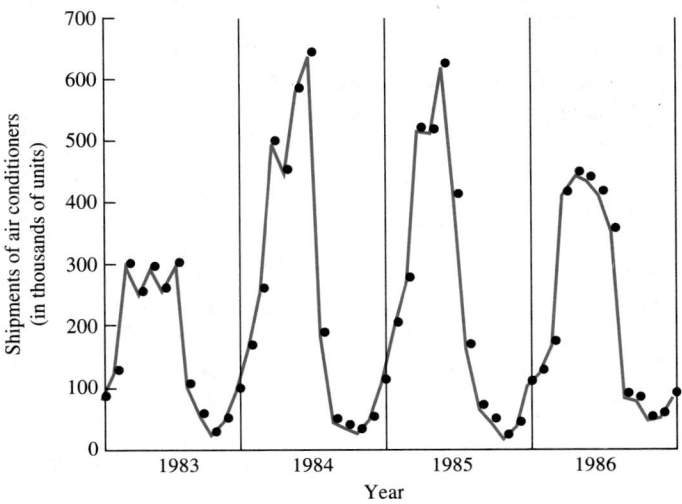

FIGURE 20.1   *Monthly shipments of air conditioners, 1983–1986*

SOLUTION   *Step* 1:   Calculate centered 12-month moving averages using the following equations:

$$s_7 = \frac{y_1 + 2y_2 + \cdots + 2y_{12} + y_{13}}{24} = \frac{4{,}090}{24} = 170.42$$

$$s_8 = \frac{y_2 + 2y_3 + \cdots + 2y_{13} + y_{14}}{24} = \frac{4{,}302}{24} = 179.25$$

$$\vdots$$

$$s_{42} = \frac{y_{36} + 2y_{37} + \cdots + 2y_{47} + y_{48}}{24} = \frac{5{,}625}{24} = 234.38$$

The centered moving averages s_t rounded off to two decimal places are shown in column 2 of Table 20.1. We lose 12 observations, 6 at each end of the series. This accounts for the blanks in Table 20.1.

   Step 2:   For each observation, express y_t as a percentage of the moving-average value s_t. That is, calculate $100(y_t/s_t)$ for $t = 7, 8, \ldots, 42$. For example, for July 1983, we have $t = 7$ and

$$100(y_7/s_7) = 100(306/170.42) \approx 180$$

Similarly, for August 1983 we have $t = 8$ and

$$100(y_8/s_8) = 100(108/179.25) \approx 60$$

After calculating $100(y_t/s_t)$ for the appropriate values of t, we have three percentages for each month, shown in column 3 in Table 20.1. For example, for January we have the three percentages 63, 82, and 60. Because these percentages differ, we still have some irregular variation in the data.

Step 3:   Calculate the mean percentage for each individual month. By averaging the percentages for each month, we eliminate some of the irregular variation. The objective in averaging is to obtain a value that represents the typical seasonal effect for each month. For January, we obtain the mean percentage

$$(63 + 82 + 60)/3 = 68.33$$

Similarly, for February we obtain

$$(100 + 109 + 77)/3 = 95.33$$

The 12 monthly averages are shown in Table 20.2.

Step 4:   We want the average monthly percentage index to be 100 so that, on the average, we have not increased or decreased the series. Thus, it may be necessary to scale the monthly mean percentages. This is done by multiplying each monthly average by the

TABLE 20.2   *Ratio-to-moving-average seasonal indexes for data in Table 20.1*

Month	Monthly average $100(y_t/s_t)$	Scaled seasonal index
January	68.33	67.45
February	95.33	94.10
March	195.33	192.81
April	193.00	190.51
May	222.67	219.80
June	199.00	196.43
July	106.67	105.29
August	35.33	34.87
September	22.00	21.72
October	13.00	12.83
November	20.00	19.74
December	45.00	44.42
Total	1,215.66	1,199.97

appropriate scale factor. For example, because the sum of our monthly means is 1,215.66, their average is 101.305. Because we want the average to be 100, we should adjust the monthly means by multiplying each one by the appropriate scaling factor. In this case, the appropriate scaling factor is

$$100/101.305 = 1,200/1,215.66 = .9871$$

Multiplying each monthly average by this scale factor will yield a new set of monthly averages whose mean is 100. For example, the scaled average for January is 68.33 × .9871 = 67.45.

The scaled monthly averages in Table 20.2 are the monthly seasonal indexes. Except for round-off error, these scaled averages have a mean of 100. If the sum of the original averages were close to 1,200 (instead of 1,215.66), step 4 could be omitted without substantially altering the results.

The seasonal index for January of 67.45 indicates that shipments of air conditioners in January tend to be 67.45% of the centered moving-average value for January. Similarly, the seasonal index for May of 219.80 means that shipments in May tend to be 219.80% of the moving-average value for May.

Seasonal Adjustment

After we have obtained the seasonal index for each month (or quarter), we can *seasonally adjust* a time series by dividing each original observation y_t by the seasonal index and multiplying the quotient by 100. The seasonally adjusted data show the monthly or quarterly trend components after the seasonal influence has been removed.

For example, in Example 20.1, the total shipments of air conditioners in January 1984 were 168. Because the seasonal index for January is 67.45, the seasonally adjusted value for January 1984 is

$$(168/67.45) \times 100 = 249.07$$

The series of seasonally adjusted values is shown in column 5 of Table 20.1 and plotted in Figure 20.2. Compare the seasonally adjusted values in Figure 20.2 with the raw data

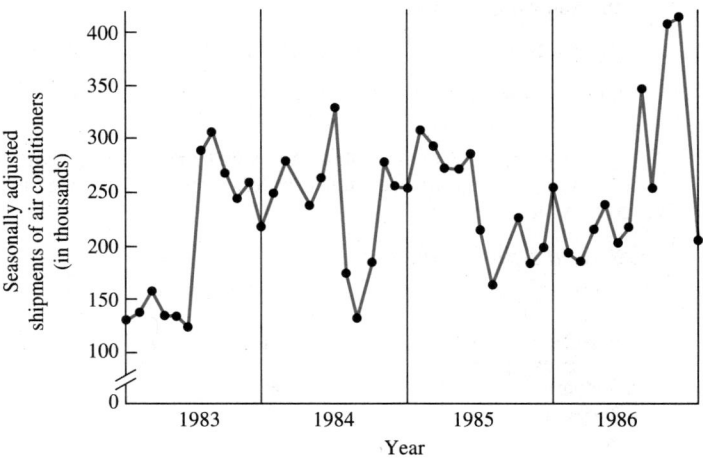

FIGURE 20.2 *Seasonally adjusted values for data in Figure 20.1*

plotted in Figure 20.1. Note the marked seasonal pattern in the raw data, which has been removed in the seasonally adjusted data.

DEFINITION   **Seasonally Adjusted Value** ─────────────────────────────────────

Given a time series $y_1, y_2, \ldots, y_n$, a series of **seasonally adjusted values** y_t is obtained by using the formula

$$\text{Seasonally adjusted value} = \frac{y_t}{\text{Seasonal index}} \times 100$$

Exercises

Statistical Concepts

20.1.1   Suppose we have monthly data on new car sales. Suppose we want to construct a simple centered 13-point moving average and find the smoothed value for January 1993. *True or false:* The smoothed value for January 1993 is simply the average of the monthly values for January 1992 through January 1993. Explain.

20.1.2   *True or false:* A problem with constructing a centered moving average is that values are lost at both ends of the series. Explain.

20.1.3   Explain the difference between a simple and a weighted moving average.

20.1.4   *True or false:* If a time series has a regular upward trend, then a forecast based on a moving-average model will tend to underestimate the correct value. Explain.

20.1.5   Which will produce a smoother curve, a 5-point or a 9-point centered moving average? Why?

20.1.6   Show that an unweighted centered moving average of order 4 is equivalent to a weighted moving average of order 5 with weights $\frac{1}{8}$, $\frac{2}{8}$, $\frac{2}{8}$, $\frac{2}{8}$, and $\frac{1}{8}$.

20.1.7   Explain whether a 12-month moving average removes the following components:
a. Trend   **b.** Seasonality

Statistical Drills

20.1.8   Use the following numbers:

    8   6   5   1   7   8   6   5   1   7   8   6   5   1   7   8   6   5   1   7   8   6   5   1   7

Compute an unweighted moving average of the following orders:
a. 2   **b.** 3   **c.** 4   **d.** 5   **e.** 6

20.1.9   Suppose a moving average of order 5 is calculated. How many numbers will be in the moving average if the original series contains the following numbers of observations?
a. 20   **b.** 25   **c.** 30

Statistical Applications

20.1.10   The following data show monthly sales in millions of dollars for a large corporation:

Month	Year 1	Year 2	Year 3
January	39	40	43
February	40	43	46
March	44	47	49
April	47	59	54
May	55	64	64
June	58	62	70
July	54	54	65
August	42	51	59
September	37	49	56
October	36	49	52
November	37	49	53
December	38	49	55

a. Smooth the data by using an unweighted 5-month moving average.
b. Plot the original and smoothed data.
c. Construct a seasonal index.

20.1.11 The following data show monthly retail sales of a department store in millions of dollars for 3 years.

Month	Year 1	Year 2	Year 3
January	2.7	2.5	3.3
February	2.9	3.1	3.4
March	3.2	3.6	4.2
April	3.9	4.4	4.1
May	4.4	4.7	5.2
June	5.1	4.8	5.8
July	4.4	3.9	5.4
August	3.9	3.8	4.6
September	3.6	3.4	4.5
October	4.2	4.5	5.2
November	5.1	5.2	6.0
December	6.3	7.1	8.3

a. Compute a seasonal index based on a 12-month moving average.
b. Plot the actual and the deseasonalized data.

20.2 Seasonal Adjustment by Using Dummy Variables

The ratio-to-moving-average method of seasonal adjustment measures the seasonal component of a time series as a percentage deviation from the moving-average value. Dummy variables can be used to measure seasonal components as absolute (or additive) deviations from the trend value rather than as a proportional (or percentage) deviation.

As shown in Section 17.2, if an observation can be classified into one of m categories, we can measure the effect of falling in any specific category by constructing $(m - 1)$ dummy variables. The coefficients of the dummy variables measure the effect of being in a specific category rather than in the base category. In time series models, any particular month or quarter can be picked as the base category, although dummy variables are more commonly used for quarterly data. With quarterly data, we need to construct only three dummy variables to measure differences between the base quarter and the other three quarters, whereas monthly data require 11 dummy variables. The next example shows how to use dummy variables to determine seasonal effects in a model containing quarterly data.

EXAMPLE 20.2 **Using Dummy Variables for Seasonal Adjustment** ———————————

The data in Table 20.3, the quarterly production of beer from 1983 through 1986, are plotted in Figure 20.3 (page 878). There appears to be a slightly positive linear trend and definite quarterly seasonal components. For example, beer production is always higher in quarters 2 and 3 (the warm-weather quarters) than in quarters 1 and 4 (the cold-weather quarters).

Let us construct a multiple regression model to forecast quarterly production. As explanatory variables, we use a linear trend variable t to explain the trend component and quarterly dummy variables Q_1, Q_2, and Q_3 to measure the seasonal components. Quarter 4 is the base case.

TABLE 20.3   *Quarterly beer production, 1983–1986 (in millions of barrels)*

Quarter	t	Y	COEFFICIENT		
			Q_1	Q_2	Q_3
1983: 1	1	46.11	1	0	0
2	2	52.18	0	1	0
3	3	52.48	0	0	1
4	4	41.38	0	0	0
1984: 1	5	46.62	1	0	0
2	6	53.71	0	1	0
3	7	50.81	0	0	1
4	8	41.09	0	0	0
1985: 1	9	46.72	1	0	0
2	10	55.06	0	1	0
3	11	50.84	0	0	1
4	12	40.61	0	0	0
1986: 1	13	47.42	1	0	0
2	14	55.31	0	1	0
3	15	50.65	0	0	1
4	16	43.12	0	0	0

Source: Survey of Current Business (Washington, D.C.: U.S. Department of Commerce, 1987).

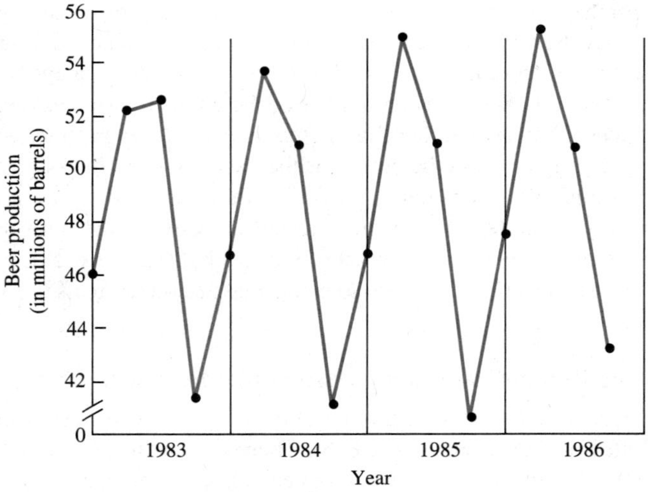

FIGURE 20.3   *Quarterly beer production, 1983–1986*

Table 20.3 shows the values of the explanatory variable t, where t equals 1 in quarter 1 of 1983, 2 in quarter 2 of 1983, and so on up to 16 in quarter 4 of 1986. Table 20.3 also shows the values for the three dummy variables Q_1, Q_2, and Q_3, where

$$Q_1 = \begin{cases} 1 & \text{if observation of quarter 1} \\ 0 & \text{otherwise} \end{cases}$$

$$Q_2 = \begin{cases} 1 & \text{if observation of quarter 2} \\ 0 & \text{otherwise} \end{cases}$$

$$Q_3 = \begin{cases} 1 & \text{if observation of quarter 3} \\ 0 & \text{otherwise} \end{cases}$$

In this example, there are $m = 4$ categories (or quarters), so we have constructed $(m - 1) = 3$ dummy variables. We have arbitrarily chosen the fourth quarter to be the base category, so the estimated coefficient of the dummy variable Q_1 indicates the difference between projected production in quarter 1 and projected production in quarter 4, and similarly for dummy variables Q_2 and Q_3.

SOLUTION   The SPSSX computer output is shown in Figure 20.4. The estimated regression model is

$$\hat{y}_t = 40.672 + .088t + 5.431Q_1 + 12.691Q_2 + 9.733Q_3 \qquad R^2 = .97191$$
t statistics: (55.68)   (1.61)   (7.87)   (18.2)   (14.1)

```
* * * *   M U L T I P L E   R E G R E S S I O N   * * * *

Listwise Deletion of Missing Data

Equation Number 1    Dependent Variable..   Y   QUARTERLY BEER PRODUCTION

Block Number  1.  Method:  Enter     T     Q1      Q2      Q3


Variable(s) Entered on Step Number  1..    Q3        Q3
                                    2..    T         TIME ELAPSED IN QUARTERS
                                    3..    Q2        Q2
                                    4..    Q1        Q1


Multiple R          .98586     Analysis of Variance
R Square            .97191                        DF    Sum of Squares    Mean Square
Adjusted R Square   .96170     Regression          4         361.09250       90.27313
Standard Error      .97400     Residual           11          10.43534         .94867

                               F =     95.15780     Signif F =   .0000


------------------ Variables in the Equation ------------------

Variable              B          SE B        Beta        T     Sig T

Q3             9.732813      .690868     .874588     14.088    .0000
T               .087812      .054448     .084004      1.613    .1351
Q2            12.690625      .697275    1.140376     18.200    .0000
Q1             5.430938      .707824     .488022      7.673    .0000
(Constant)    40.671875      .730497                 55.677    .0000


End Block Number   1   All requested variables entered.
```

FIGURE 20.4 *SPSSX-generated output for Example 20.2*

The coefficient of determination is $R^2 = .97191$, and the adjusted R^2 is .96170. Thus, the model explains approximately 96% of the quarterly variation in beer production. The estimated coefficient of each dummy variable has a very large t statistic. Each of these three t statistics has a p-value of .0000, indicating that there are definite quarterly seasonal effects. The t statistic for the coefficient of the trend variable t is 1.613, which is of questionable significance.

The graph shows that beer production tends to be lower in quarter 4 than in the other three quarters. Thus, we would expect the estimated coefficients of the three dummy variables to be positive, indicating that production in any of those quarters is higher than in the fourth quarter.

Does the time trend belong in the model? To answer this question, we could test the null hypothesis

$$H_0: \quad \beta_1 = 0$$

against the one-sided alternative hypothesis

$$H_1: \quad \beta_1 > 0$$

using, say, a 5% level of significance. We have $\nu = (n - K - 1) = (16 - 4 - 1) = 11$ degrees of freedom, and the critical value of t is $t_{.05,11} = 1.796$. This means that we would not reject the null hypothesis that the population regression coefficient of the trend variable t is 0. Although there might be a positive trend factor in beer production, the evidence is not extremely strong. Nevertheless, for purposes of illustration, we will retain the trend variable in the equation.

During the fourth quarter of any year we have $Q_1 = 0$, $Q_2 = 0$, and $Q_3 = 0$. For quarter 4, the estimated model becomes

$$\hat{y}_t = 40.672 + .088t$$

This represents a linear trend model having intercept 40.672 and slope .088.

During the first quarter of any year, we have $Q_1 = 1$, $Q_2 = 0$, and $Q_3 = 0$, and the estimated model becomes

$$\hat{y}_t = 40.672 + .088t + 5.431(1) = 46.103 + .088t$$

This represents a new trend line having intercept 46.103 and slope .088. For quarter 1, then, the trend line has the same slope as the trend line for quarter 4, but the intercept is 5.431 units higher than for the trend line of quarter 4. The coefficient of the dummy variable Q_1 measures the amount by which the trend line for the base case (quarter 4) should be shifted up or down to obtain the trend line for quarter 1. The coefficient 5.431 indicates that quarterly sales during quarter 1 tend to be 5.431 million barrels higher than during quarter 4 after we have taken the trend into account.

During the second quarter of any year, we have $Q_1 = 0$, $Q_2 = 1$, and $Q_3 = 0$, and the estimated model becomes

$$\hat{y}_t = 40.672 + .088t + 12.691 = 53.363 + .088t$$

The trend line for quarter 2 is 12.691 units above the trend line for quarter 4. As with the coefficient of Q_1, the coefficient of the dummy variable Q_2 measures the amount by which the trend line for the base case (quarter 4) should be shifted up or down to obtain the trend line for quarter 2. Similarly, the trend line for quarter 3 will be 9.733 units above the trend line for quarter 4.

In this example, all of the coefficients of the dummy variables are positive because the base case (quarter 4) was the quarter having the lowest production. Thus, the trend line for quarter 4 has to be raised to obtain the trend line for any other quarter.

EXAMPLE 20.3 **Forecasting with Dummy Variables**

Let us use the model developed in Example 20.2 to forecast beer production during each quarter of 1989.

SOLUTION For quarter 1 of 1989, the value of the trend variable is $t = 25$ and the values of the dummy variables are $Q_1 = 1$, $Q_2 = 0$, and $Q_3 = 0$. To forecast, we substitute these values into the estimated model and obtain

$$\hat{y}_{25} = 40.672 + .088(25) + 5.431(1) + 12.691(0) + 9.733(0)$$
$$= 48.303$$

For quarter 2 of 1989, t equals 26, and the values of the dummy variables are $Q_1 = 0$, $Q_2 = 1$, and $Q_3 = 0$. Substitute these values into the estimated model and obtain the forecast

$$\hat{y}_{26} = 40.672 + .088(26) + 5.431(0) + 12.691(1) + 9.733(0)$$
$$= 55.651$$

For quarter 3 of 1989, t equals 27, and the values of the dummy variables are $Q_1 = 0$, $Q_2 = 0$, and $Q_3 = 1$. We obtain the forecast

$$\hat{y}_{27} = 40.672 + .088(27) + 5.431(0) + 12.691(0) + 9.733(1) = 52.781$$

Finally, for quarter 4 of 1989, t equals 28, and the values of the dummy variables are $Q_1 = 0$, $Q_2 = 0$, and $Q_3 = 0$. We obtain the forecast

$$\hat{y}_{28} = 40.672 + .088(28) + 5.431(0) + 12.691(0) + 9.733(0) = 43.136$$

Exercises

Statistical Concepts

20.2.1 *True or false:* If we add dummy variables to a model to explain quarterly seasonality, we must create four dummy variables because there are four quarters. Explain.

20.2.2 *True or false:* If we add dummy variables to a model to explain quarterly seasonality, we must create one dummy variable that assumes the value 1, 2, 3, or 4, depending on the quarter. Explain.

20.2.3 Suppose we estimate a model to explain quarterly sales of a product that has peak sales during the Christmas season (quarter 4). Suppose we create three dummy variables to account for quarterly seasonality. Suppose the base case is quarter 4. Can you predict the signs of the coefficients of the three dummy variables? Will they all be positive? Negative? Explain.

Statistical Applications

20.2.4 Consider the following regression results:

$$\hat{y}_t = 23.4 + .14t + 6.2Q_2 + 5.3Q_3 + 18.3Q_4$$

where $Q_i = 1$ in the ith quarter of each year and 0 otherwise, and y_t represents total sales in thousands of dollars in period t for a department store.
a. Explain the meaning of each of the regression coefficients.
b. Is it plausible that each of the coefficients is positive?
c. Explain why the coefficient of Q_4 is larger than the coefficients of Q_2 and Q_3.

d. Suppose the variable Q_4 had been omitted from the model and a dummy variable for Q_1 had been included. Would the coefficient of t change?

e. Suppose the variable Q_4 had been omitted from the model and a dummy variable for Q_1 had been included. Would the constant term change? If so, how?

f. Forecast sales in period 15.

20.2.5   The accompanying data refer to quarterly sales (in millions of dollars) of children's toys by a nationwide chain:

Quarter	t	Y
Year 1: 1	1	2.8
2	2	3.2
3	3	3.1
4	4	7.4
Year 2: 1	5	2.4
2	6	3.4
3	7	3.5
4	8	8.0
Year 3: 1	9	3.1
2	10	3.5
3	11	3.7
4	12	9.6
Year 4: 1	13	4.0
2	14	4.2
3	15	4.3
4	16	11.7
Year 5: 1	17	4.2
2	18	5.1
3	19	5.3
4	20	14.5

a. Plot the data.

b. Estimate the model

$$Y_t = \beta_0 + \beta_1 t + \beta_2 Q_2 + \beta_3 Q_3 + \beta_4 Q_4 + e_t$$

where quarter 1 is the base case.

c. Estimate the model

$$Y_t = \beta_0 + \beta_1 t + \beta_2 Q_2 + \beta_3 Q_3 + \beta_4 Q_1 + e_t$$

where quarter 4 is the base case.

d. Compare the estimated constant terms in each equation. Can you explain why this change occurred?

e. Compare the coefficients of the trend variable in the two equations.

f. Compare the coefficients of Q_2 in the two equations. Can you explain why this change occurred?

g. Plot the relationship between sales and time separately for quarter 1, quarter 2, quarter 3, and quarter 4.

h. Forecast sales for each quarter of year 6.

Seasonal Adjustment by Using the Ratio-to-Trend Method

We estimated the seasonal component as a percentage deviation from the moving-average value in Section 20.1 and as an absolute deviation from a trend value in Section 20.2. A third method is called the *ratio-to-trend method,* which estimates the seasonal component as the percentage deviation from the trend.

Suppose that a given time series can be represented by the multiplicative model $Y_t = T_t \times S_t \times C_t \times I_t$. The ratio-to-trend method estimates the seasonal index (S) by removing the trend (T) from the series but not the cyclical (C) and irregular (I) variations. First, we obtain the monthly (or quarterly) trend values by fitting a least-squares trend line or some other curve. Let $\hat{y}_t$ denote the fitted trend value for time t. Trend can be eliminated from the original series by dividing each value y_t by the corresponding fitted value $\hat{y}_t\ (= T)$. Thus, we obtain

$$\frac{y_t}{\hat{y}_t} = \frac{y_t}{T_t} = S_t \times C_t \times I_t$$

The fluctuations in the product $(S \times C \times I)$ are assumed to represent mainly seasonal fluctuations. We remove much of the random effect $(C \times I)$ by averaging $(S \times C \times I)$ for each month or quarter.

Calculating the ratio-to-trend seasonal index

The ratio-to-trend seasonal index for monthly data is calculated as follows:

1. Estimate the sample trend line for the original data $y_1, y_2, \ldots, y_n$.
2. Obtain the fitted, or trend, values $\hat{y}_1, \hat{y}_2, \ldots, \hat{y}_n$.
3. Obtain the detrended values $y_1/\hat{y}_1, y_2/\hat{y}_2, \ldots, y_n/\hat{y}_n$.
4. Obtain the average of all values $y_i/\hat{y}_i$ corresponding to January. Do the same for every other month.
5. Multiply each average by 100 to convert it to a percentage. These 12 averages are the seasonal indexes for each of the 12 months.

EXAMPLE 20.4   **Constructing a Seasonal Index with the Ratio-to-Trend Method** ————

The data in Table 20.4 show the monthly sales of retail stores from 1983 through 1986 and are plotted in Figure 20.5 (page 884). The data are highly seasonal with a peak each year in December and a trough each year in February. There appears to be a positive long-term trend. Let us obtain a seasonal index for these data using the ratio-to-trend method.

SOLUTION   *Step* 1:   Estimate the linear trend model. The SPSSX computer program was used to obtain the estimated trend equation

$$\hat{y}_t = 92.055762 + .743047t \qquad R^2 = .55869$$
$$t \text{ statistics:} \qquad (33.6) \qquad (7.63)$$

where t equals 1 in January 1983, 2 in February 1983, and so on to 48 in December 1986.

TABLE 20.4 *Monthly retail sales, 1983–1986 (in billions of dollars)*

Month	1983	1984	1985	1986
January	$ 81.3	$ 93.1	$ 98.8	$105.6
February	78.9	93.7	95.6	99.7
March	93.8	104.3	110.2	114.2
April	94.0	104.3	113.1	115.7
May	97.8	111.3	120.3	125.4
June	100.6	112.0	115.0	120.4
July	99.6	106.6	115.5	120.7
August	100.2	110.7	121.1	124.1
September	98.0	103.9	114.2	124.6
October	100.7	109.2	116.1	123.1
November	103.9	113.3	118.6	120.8
December	125.7	131.8	139.5	151.5

Source: Survey of Current Business (Washington, D.C.: U.S. Department of Commerce, 1987).

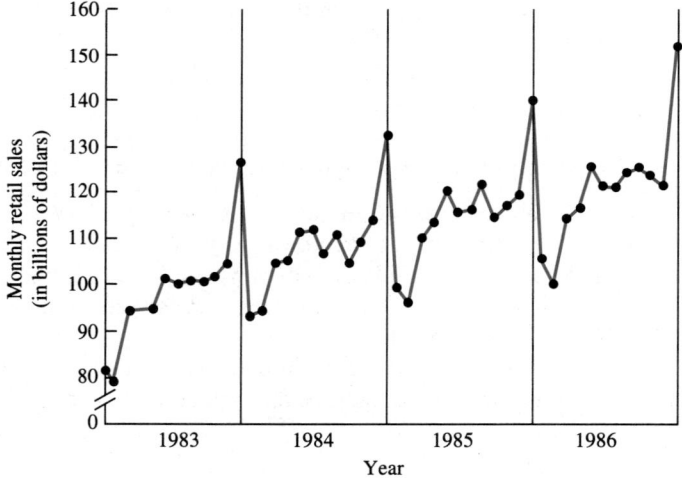

FIGURE 20.5 *Monthly retail sales, 1983–1986*

The computer output is shown in Figure 20.6. The relatively low value of $R^2 = .55869$ shows that there is a great deal of unexplained seasonal variation about the trend line.

Step 2: Calculate the fitted, or estimated, trend value $\hat{y}_t$ for each month. For example, in December 1984 we have $t = 24$ and the fitted value is

$$\hat{y}_t = 92.055762 + .743047(24) = 109.8889$$

Figure 20.7 (page 886) shows the actual values of y_t, the predicted values $\hat{y}_t$, and the residuals. For example, the predicted value for the 24th observation is $\hat{y}_{24} = 109.8889$ and the actual value is $y_{24} = 131.8$, giving a residual of 21.9111.

```
                          * * * *   M U L T I P L E   R E G R E S S I O N   * * * *

Listwise Deletion of Missing Data

Equation Number 1     Dependent Variable..    Y    RETAIL SALES

Block Number  1.  Method:   Enter      T


Variable(s) Entered on Step Number  1..     T         TIME ELAPSED IN MONTHS


Multiple R              .74746         Analysis of Variance
R Square                .55869                            DF     Sum of Squares    Mean Square
Adjusted R Square       .54910         Regression          1       5086.12033      5086.12033
Standard Error         9.34539         Residual           46       4017.47446        87.33640

                                       F =     58.23597       Signif F =   .0000


------------------- Variables in the Equation -------------------

Variable             B          SE B        Beta          T    Sig T

T                 .743047      .097369      .747458      7.631   .0000
(Constant)      92.055762     2.740495                  33.591   .0000


End Block Number   1   All requested variables entered.
```

FIGURE 20.6   *SPSSX-generated output for Example 20.4*

In the computer output, there are four very large positive residuals for observations 12, 24, 36, and 48, which are the residuals for the month of December.

Step 3:   Divide each actual observation y_t by the corresponding fitted value $\hat{y}_t$ to obtain the ratio r_t, where r_t is defined by the equation

$$r_t = \frac{y_t}{\hat{y}_t}$$

for $t = 1, 2, \ldots, 48$. The ratios r_t yield a series of proportions showing each monthly observation as a proportion of the corresponding trend value. For example, in December 1984 the actual value is $y_{24} = 131.8$ and the fitted value is $\hat{y}_{24} = 109.8889$, yielding a ratio of

$$r_{24} = \frac{131.8}{109.9} = 1.20$$

All of the monthly ratios are shown in Table 20.5 (page 887). All the ratios for January and February are less than 1, and all the ratios for December are greater than 1. Thus, observations for January and February are always below the trend line, and observations for December are always above the trend line.

Step 4:   Calculate the average (mean) ratio for each month. For example, for December the ratios are 1.24, 1.20, 1.17, and 1.19 for the 4 years studied. The average ratio for December is

$$\frac{1.24 + 1.20 + 1.17 + 1.19}{4} = 1.20$$

The mean ratio for each month is shown in Table 20.5 in column 5, labeled Average. Months with mean ratios exceeding 1 tend to fall above the trend line, whereas months with mean ratios less than 1 tend to fall below it.

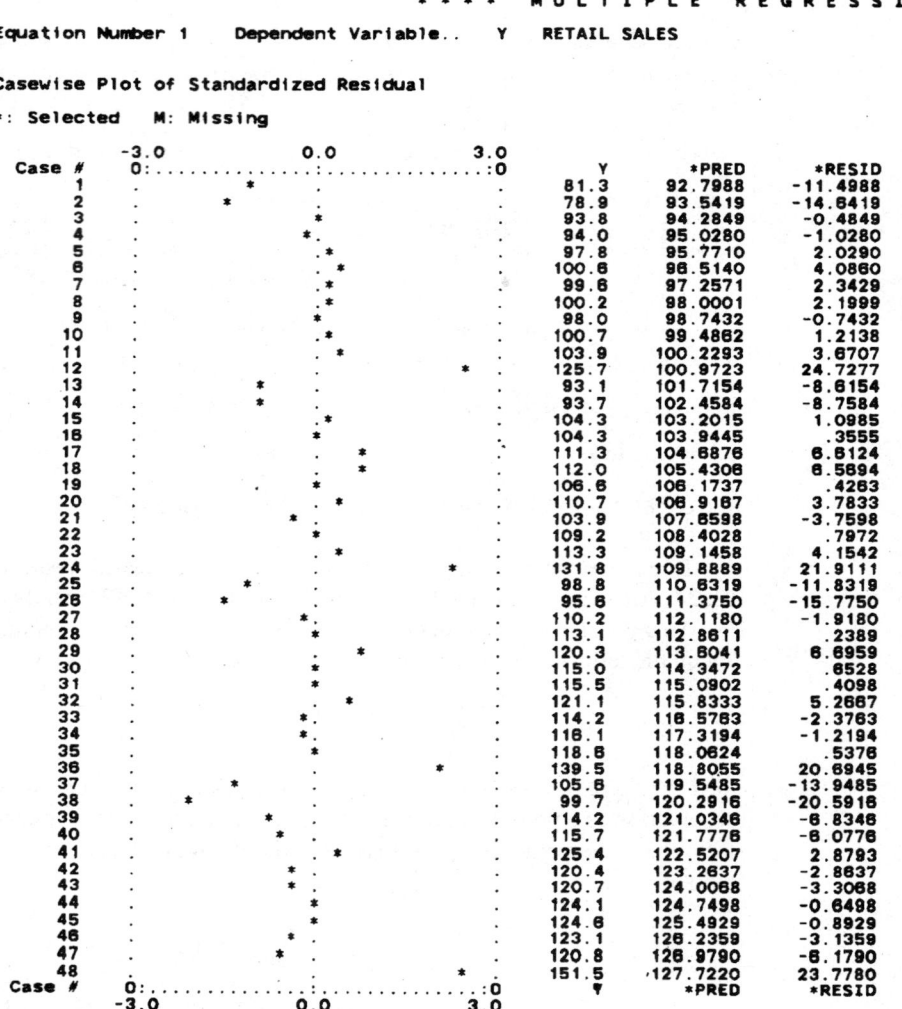

Equation Number 1   Dependent Variable..   Y   RETAIL SALES

Casewise Plot of Standardized Residual

*: Selected   M: Missing

Case #			Y	*PRED	*RESID
1			81.3	92.7988	-11.4988
2			78.9	93.5419	-14.6419
3			93.8	94.2849	-0.4849
4			94.0	95.0280	-1.0280
5			97.8	95.7710	2.0290
6			100.6	96.5140	4.0860
7			99.6	97.2571	2.3429
8			100.2	98.0001	2.1999
9			98.0	98.7432	-0.7432
10			100.7	99.4862	1.2138
11			103.9	100.2293	3.6707
12			125.7	100.9723	24.7277
13			93.1	101.7154	-8.6154
14			93.7	102.4584	-8.7584
15			104.3	103.2015	1.0985
16			104.3	103.9445	.3555
17			111.3	104.6876	6.6124
18			112.0	105.4306	6.5694
19			106.6	106.1737	.4263
20			110.7	106.9167	3.7833
21			103.9	107.6598	-3.7598
22			109.2	108.4028	.7972
23			113.3	109.1458	4.1542
24			131.8	109.8889	21.9111
25			98.8	110.6319	-11.8319
26			95.6	111.3750	-15.7750
27			110.2	112.1180	-1.9180
28			113.1	112.8611	.2389
29			120.3	113.6041	6.6959
30			115.0	114.3472	.6528
31			115.5	115.0902	.4098
32			121.1	115.8333	5.2667
33			114.2	116.5763	-2.3763
34			116.1	117.3194	-1.2194
35			118.6	118.0624	.5376
36			139.5	118.8055	20.6945
37			105.6	119.5485	-13.9485
38			99.7	120.2916	-20.5916
39			114.2	121.0346	-6.8346
40			115.7	121.7776	-6.0776
41			125.4	122.5207	2.8793
42			120.4	123.2637	-2.8637
43			120.7	124.0068	-3.3068
44			124.1	124.7498	-0.6498
45			124.6	125.4929	-0.8929
46			123.1	126.2359	-3.1359
47			120.8	126.9790	-6.1790
48			151.5	127.7220	23.7780

FIGURE 20.7   *SPSSX-generated actual values, predicted values, and residuals for Example 20.4*

Step 5:   Multiply each monthly mean ratio by 100 to obtain the monthly seasonal index. For example, the seasonal index for December is $1.20 \times 100 = 120.00$. The monthly seasonal indexes are shown in column 6 of Table 20.5.

The December seasonal index of 120.00 indicates that, on the average, sales during December are 20% above the trend line. Similarly, the seasonal index for February is 86.00, indicating that sales during February tend to be 14% below the trend line.

TABLE 20.5 *Seasonal indexes for data in Table 20.4 obtained by ratio-to-trend method*

Month	(1) 1983 $y_t/\hat{y}_t$	(2) 1984 $y_t/\hat{y}_t$	(3) 1985 $y_t/\hat{y}_t$	(4) 1986 $y_t/\hat{y}_t$	(5) Average	(6) Seasonal index
January	.88	.92	.89	.88	.8925	89.25
February	.84	.91	.86	.83	.8600	86.00
March	.99	1.01	.98	.94	.9800	98.00
April	.99	1.00	1.00	.95	.9850	98.50
May	1.02	1.06	1.06	1.02	1.0400	104.00
June	1.04	1.06	1.01	.98	1.0225	102.25
July	1.02	1.00	1.00	.97	.9975	99.75
August	1.02	1.04	1.05	.99	1.0250	102.50
September	.99	.97	.98	.99	.9825	98.25
October	1.01	1.01	.99	.98	.9975	99.75
November	1.04	1.04	1.00	.95	1.0075	100.75
December	1.24	1.20	1.17	1.19	1.2000	120.00

Forecasting with the Seasonal Index

Obtaining a forecast with the seasonal index and trend value

Let $\hat{y}_t$ be the estimated trend value of a variable in period t, and let SI_t be the seasonal index for period t. The forecast value for period t is $(\hat{y}_t \times SI_t)/100$.

EXAMPLE 20.5   **Forecasting with the Ratio-to-Trend Model** ————————————

Let us use the ratio-to-trend model to forecast retail sales in December 1988 based on the data in Table 20.4.

SOLUTION   First, calculate the forecast trend value. For December 1988, t is 72. From the estimated trend equation, we obtain

$$\hat{y}_{72} = 92.055762 + .743047(72) = 145.555$$

This is the estimated trend value. Now we must adjust this value for seasonality. To do this, we multiply the estimated trend value by the seasonal index for that month and divide by 100 to obtain the monthly forecast. For December, the seasonal index is 120.00 and the trend value is 145.555. The forecast value for December 1988 is thus

$$(\hat{y}_{72} \times SI_{72})/100 = (145.555 \times 120.00)/100$$
$$= 174.666$$

Exercises

Statistical Concepts

20.3.1   *True or false:*   The ratio-to-trend method estimates the seasonal component as the deviation of the actual value from the trend forecast. Explain.

20.3.2   *True or false:*   The ratio-to-trend method estimates the detrended values as $y_i - \hat{y}_i$, that is, as the deviation between the actual value and the forecast value. Explain.

20.3.3   *True or false:*   Using the ratio-to-trend method of seasonal adjustment, suppose a particular month has a seasonal index of 110.0. The values for this month tend to fall 110 units above the trend line. Explain.

Statistical Applications

20.3.4   Refer to the data in Exercise 10 of Section 20.1. Compute a seasonal index using the ratio-to-trend method and plot the actual and deseasonalized data. Compute forecasts for the next 12 values.

20.3.5   Refer to the data in Exercise 11 of Section 20.1. Compute a seasonal index using the ratio-to-trend method and plot the actual and deseasonalized data. Compute forecasts for the next 12 values.

20.4   Exponential Smoothing

Another method of smoothing data is called **exponential smoothing,** a forecasting procedure used when the variable being predicted has no regular seasonal component and no regular upward or downward trend. When a time series has no trend or seasonal component, the usual objective is to forecast the *level* of the time series. Thus, given the values $y_1, y_2, \ldots, y_n$, we want to predict the value y_{n+1}.

One way of predicting a value y_{n+1} is to use the last observation y_n. However, this is impractical because each observation y_t contains a random component, and if the random component is large, then y_t is not a good predictor of the level of the time series.

If y_t contains a substantial irregular component, it is reasonable to use several previous values of the time series to predict a value to average out the random component. In the most extreme case, we would take the average of all n observations to predict y_{n+1}. The problem with this procedure is that the predicted value gives each previous value equal weight. In many cases, it is more reasonable to give recent observations greater weight than older observations. In general, we expect recent data to be more reliable indicators of the future than distant data.

Exponential smoothing generates a forecast as a weighted average of current and past values. In calculating the weighted average, more recent values of the time series get greater weights than distant observations, as described in the next definition. The smoothed value s_t that is obtained is an estimate of the level of the series at time t, whereas the actual value y_t measures the level of the series *plus* some random factor.

DEFINITION   **Exponential Smoothing** ————————————————————————

Suppose we have observations $y_1, y_2, \ldots, y_n$. Let s_t denote the smoothed value at time t. The method of **exponential smoothing** calculates the smoothed values in the following way: In period 1, set s_1 equal to y_1. For all succeeding periods, the smoothed value s_t is determined using the equation

$$s_t = cy_t + (1 - c)s_{t-1}$$

where c is some constant that satisfies the inequality $0 \leqslant c \leqslant 1$.

If we lag the model one period, we obtain

$$s_{t-1} = cy_{t-1} + (1 - c)s_{t-2}$$

Substituting s_{t-1} into the original equation yields

$$s_t = cy_t + (1 - c)[cy_{t-1} + (1 - c)s_{t-2}]$$
$$= cy_t + c(1 - c)y_{t-1} + (1 - c)^2 s_{t-2}$$

Similarly, we can substitute for s_{t-2} and obtain

$$s_t = cy_t + c(1 - c)y_{t-1} + (1 - c)^2[cy_{t-2} + (1 - c)s_{t-3}]$$

If we continue substituting, we eventually obtain

$$s_t = c \sum_{j=0}^{t-2} (1 - c)^j y_{t-j} + (1 - c)^{t-1} y_1$$

This formula shows that the current smoothed value s_t depends on every past observation y_{t-j}. If the smoothing constant c is close to 1, the coefficients $(1 - c)^j$ die out quickly, and the smoothed value depends almost exclusively on recent values of y_t. When c is close to 0, the coefficients $(1 - c)^j$ die out much more slowly, and the smoothed value s_t is more heavily influenced by distant values of y_t than when c is near 1.

The choice of c depends on the characteristics of the series being smoothed. If the series is quite volatile, use a small value of c to give a large weight to the previous smoothed value s_{t-1} and only a small weight to the current value y_t. If the data are not too volatile, then the random component of the series is probably fairly small. In this case, y_t should get a large weight when we calculate s_t.

EXAMPLE 20.6 **Exponential Smoothing**

The data in Table 20.6 show the price of a share of stock on the first trading day of the month for a period of 8 months. Construct a smoothed series using the exponential smoothing technique.

TABLE 20.6 *Price of a share of stock*

Time	Observation	Smoothed value
1	16.4	16.400
2	17.6	16.880
3	16.1	16.568
4	15.5	16.141
5	17.2	16.565
6	16.8	16.659
7	18.1	17.235
8	17.0	17.141

SOLUTION   Let us smooth the series using $c = .4$. We obtain

$$s_1 = y_1 = 16.4$$
$$s_t = cy_t + (1 - c)s_{t-1} \quad \text{for } t = 2, 3, \ldots, 8$$
$$s_2 = (.4)(17.6) + (.6)(16.4) = 16.88$$
$$s_3 = (.4)(16.1) + (.6)(16.88) = 16.568$$

and so forth. The smoothed data are also shown in Table 20.6.

Forecasting with the Exponential Smoothing Model

The exponential smoothing model is used to average out the effects of random variation so we can estimate the level of a time series not subject to trend or seasonal components. The model can be used to forecast the future level of a time series as described in the accompanying box.

Forecasting with exponential smoothing

Let $y_1, y_2, \ldots, y_n$ be a set of observations of a time series with no regular seasonal or trend component. Let $\hat{s}_{n+j}$ denote the forecast level of the series in period $(n + j)$. We use the forecast

$$\hat{s}_{n+j} = s_n = cy_n + (1 - c)s_{n-1}$$

where s_n is the smoothed value for period n and c is the smoothing constant. Thus, the forecast value of the level of the series for all future periods is the last estimate of the level of the series.

EXAMPLE 20.7   **Forecasting with the Exponential Smoothing Model**

Use the data in Table 20.6 to forecast the stock price on the first trading day of months 9 and 10.

SOLUTION   In Example 20.6, the smoothed value for period 8 is

$$s_8 = 17.141$$

At the end of period 8, the forecasts for periods 9 and 10 are

$$\hat{s}_9 = s_8 = 17.141$$
$$\hat{s}_{10} = s_8 = 17.141$$

At the end of period 9, we would get a new smoothed value s_9, from which we would obtain the estimate for period 10. Thus, at the end of period 9, the updated forecast for period 10 would be

$$\hat{s}_{10} = s_9$$

Forecasting by exponential smoothing is very easy to do, but whether this model will outperform a carefully designed regression model is questionable. Its major advantage is its simplicity, but this simplicity is also its major drawback. Unless the time series to be forecast moves very smoothly, the forecasts generated via exponential smoothing will be inaccurate, because the method does not take into account underlying factors that cause the behavior of y_t to change over time. A well-designed regression model can frequently identify these factors and take them into account.

Two-Parameter Exponential Smoothing

Exponential smoothing based on the equation

$$s_t = cy_t + (1 - c)\, s_{t-1}$$

is called one-parameter exponential smoothing because it depends on the single parameter c. When there is an upward trend in the time series, one-parameter exponential smoothing will generally produce forecasts that are too low. This follows from the fact that, if $y_t > y_{t-1}$ and $c < 1$, then $s_t < y_t$. Two-parameter exponential smoothing, also called **Holt–Winters exponential smoothing,** is a smoothing technique that incorporates both an exponentially smoothed component S_t and a trend component T_t. The technique uses a set of equations to generate forecasts F_{t+1} for a time series that contains a trend. The exponentially smoothed component is estimated using the equation

$$S_t = \gamma Y_t + (1 - \gamma)\,(S_{t-1} + T_{t-1})$$

The trend component is estimated using the equation

$$T_t = \delta(S_t - S_{t-1}) + (1 - \delta)T_{t-1}$$

The forecast for period $(t + k)$ for $k = 1, 2, \ldots$ is

$$F_{t+k} = S_t + kT_t$$

The system of equations requires two smoothing constants γ and δ, each of which takes a value between 0 and 1. The value of γ determines the smoothness of the time series for S_t. Large values of γ make the series move much like Y_t; small values of γ give more weight to past values of the series and make the series move more like the trend component.

The trend component T_t is a weighted average of the previous trend component T_{t-1} and the change in the smoothed component $(S_t - S_{t-1})$. Small values of δ give more emphasis to past estimates of the trend component T_{t-1}, whereas large values of δ give more emphasis to the recent change in the smoothed component.

The calculation of forecasts using the two-parameter exponential smoothing method is summarized in the following box.

Calculation of forecasts using two-parameter exponential smoothing

> **1.** Select a smoothing constant γ between 0 and 1. Large values of γ give greater weight to current values of the time series.
>
> *(continued)*

2. Select a smoothing constant δ between 0 and 1. Large values of δ give greater weight to the more recent trend of the series and less to the past trend of the series.

3. Given the data $Y_1, Y_2, \ldots, Y_t$, calculate the start-up values S_2 and T_2 as follows:

$$S_2 = Y_2$$
$$T_2 = Y_2 - Y_1$$

4. Calculate all subsequent exponentially smoothed values S_t and trend components T_t as follows:

$$S_t = \gamma Y_t + (1 - \gamma)(S_{t-1} + T_{t-1})$$
$$T_t = \delta(S_t - S_{t-1}) + (1 - \delta)T_{t-1}$$

5. Calculate the forecast for period $(t + 1)$ as follows:

$$F_{t+1} = S_t + T_t$$

Calculate the forecast for period $(t + k)$ as follows:

$$F_{t+k} = S_t + kT_t$$

EXAMPLE 20.8   **Forecasting with Two-Parameter Exponential Smoothing**

The data in Table 20.7 show the monthly sales of a certain brand of computer. Use the two-parameter exponential smoothing model, and calculate the smoothed components S_t and the trend components T_t using the smoothing constants $\gamma = .8$ and $\delta = .5$. Calculate the 1-month-ahead forecasts.

TABLE 20.7   *Monthly sales (in thousands of dollars) for Example 20.8*

Month	Sales	Month	Sales	Month	Sales
1	29	9	60	17	88
2	32	10	72	18	89
3	37	11	75	19	94
4	48	12	83	20	95
5	44	13	86	21	100
6	49	14	87	22	102
7	57	15	86	23	99
8	54	16	90	24	104

SOLUTION   The smoothed components S_t, the trend components T_t, and the 1-month-ahead forecasts F_{t+1} are obtained as follows:

$$S_2 = Y_2 = 32$$
$$T_2 = Y_2 - Y_1 = 32 - 29 = 3.0$$
$$F_3 = S_2 + T_2 = 32 + 3.0 = 35.0$$
$$S_3 = .8Y_3 + (1 - .8)(S_2 + T_2) = .8(37) + .2(32 + 3.0) = 36.6$$

$$T_3 = .5(S_3 - S_2) + (1 - .5) \, T_2 = .5(36.6 - 32.0) + .5(3.0) = 3.8$$
$$F_4 = S_3 + T_3 = 36.6 + 3.8 = 40.4$$

$$S_4 = .8Y_4 + (1 - .8) \, (S_3 + T_3) = .8(48) + .2(36.6 + 3.8) = 46.5$$
$$T_4 = .5(S_4 - S_3) + (1 - .5) \, T_3 = .5(46.5 - 36.6) + .5(3.8) = 6.8$$
$$F_5 = S_4 + T_4 = 46.5 + 6.8 = 53.3$$

and so forth.

All the values of S_t, T_t, and F_t are given in Table 20.8 along with the forecast errors $Y_t - F_t$; all values have been rounded off. Figure 20.8 (page 894) is a plot of the values of Y_t and the forecast values F_t.

TABLE 20.8 *Forecasts of monthly sales (in thousands of dollars) for Example 20.8*

Month t	Actual sales Y_t	Smoothed sales S_t	Smoothed trend T_t	Forecast sales F_t	Forecast error $Y_t - F_t$
1	29	—	—	—	—
2	32	32.0	3.0	—	—
3	37	36.6	3.8	35.0	2.0
4	48	46.5	6.8	40.4	7.6
5	44	45.9	3.1	53.3	−9.3
6	49	49.0	3.1	49.0	.0
7	57	56.0	5.1	52.1	4.9
8	54	55.4	2.2	61.1	−7.1
9	60	59.5	3.2	57.7	2.3
10	72	70.1	6.9	62.7	9.3
11	75	75.4	6.1	77.0	−2.0
12	83	82.7	6.7	81.5	1.5
13	86	86.7	5.3	89.4	−3.4
14	87	88.0	3.3	92.0	−5.0
15	86	87.1	1.2	91.3	−5.3
16	90	89.7	1.9	88.3	1.7
17	88	88.7	.5	91.5	−3.5
18	89	89.0	.4	89.2	−.2
19	94	93.1	2.2	89.4	4.6
20	95	95.1	2.1	95.3	−.3
21	100	99.4	3.2	97.2	2.8
22	102	102.1	3.0	102.7	−.7
23	99	100.2	.5	105.1	−6.1
24	104	103.3	1.8	100.7	3.3

In Table 20.8, the trend component T_t provides a measure of the general upward movement per month in the time series. In the formula for T_t, the smoothing constant $\delta = .5$ gives an equal weight to the past trend value T_{t-1} and to the most recent trend $(S_t - S_{t-1})$. In the formula for S_t, the smoothing constant $\gamma = .8$ gives heavy weight to the most recent value of Y_t. If the original time series is fairly volatile, a smaller value of γ would yield a smoother series.

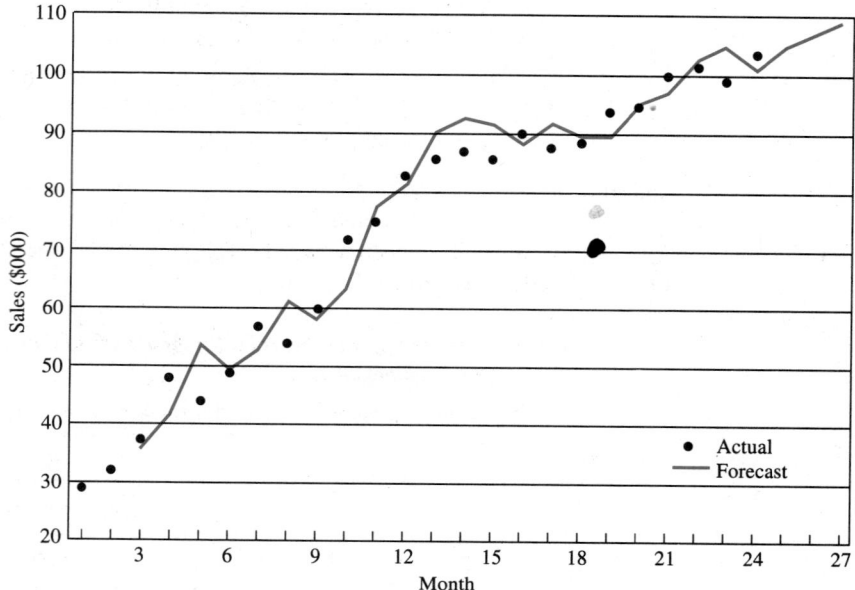

FIGURE 20.8 *Sales data and two-parameter exponentially smoothed forecasts*

The one-step-ahead forecast F_{t+1} is the sum of the last smoothed value S_t plus the estimated increase due to the trend T_t. The forecast for period $(t + 2)$ is again based on the final smoothed estimate S_t, but now we add 2 times the estimated increase due to the trend. Thus, the two-step-ahead forecast would be

$$F_{t+2} = S_t + 2T_t$$

For example, in Table 20.8, we listed the firm's monthly sales along with the smoothed and trend components for each month. The forecast for month 25 is

$$F_{25} = S_{24} + T_{24} = 103.3 + 1.8 = 105.1$$

The forecasts for months 26 and 27, plotted in Figure 20.8, are as follows:

$$F_{26} = S_{24} + 2T_{24} = 103.3 + 2(1.8) = 106.9$$
$$F_{27} = S_{24} + 3T_{24} = 103.3 + 3(1.8) = 108.7$$

In actual practice, different smoothing constants γ and δ can be tried in hopes of improving the forecasts and decreasing the forecast errors. This result is usually determined by trial and error.

The forecast errors provide a measure of the accuracy of the forecast and can be used to help select appropriate smoothing constants. As discussed in the previous chapter, two commonly used measures of forecast accuracy are the **mean absolute forecast error (MAFE)** and the **root mean squared error (RMSE)** of the forecasts. MAFE is the arithmetic mean of the absolute values of the forecast errors, whereas RMSE is the square root of the average of the squared forecast errors. The formulas for calculating MAFE and RMSE are given in the following box.

Measures of forecast accuracy

> 1. The **mean absolute forecast error (MAFE)** is the arithmetic mean of the absolute values of the forecast errors,
>
> $$\text{MAFE} = \frac{\Sigma |Y_t - F_t|}{n}$$
>
> where n is the number of forecasts used for the evaluation.
> 2. The **root mean squared error (RMSE)** is the square root of the average of the squared forecast errors,
>
> $$\text{RMSE} = \sqrt{\frac{\Sigma (Y_t - F_t)^2}{n}}$$

Table 20.8 lists 22 forecast errors, one for each of the periods 3 through 24. Because the first values of S_t and T_t are for period $t = 2$, the first forecast is for period $t = 3$. For period 3, the actual value is $Y_3 = 37.0$, the forecast is $F_3 = 35.0$, and the forecast error, denoted E_t, is

$$E_3 = 37.0 - 35.0 = 2.0$$

For period 4, the actual value is $Y_4 = 48.0$, the forecast is $F_4 = 40.4$, and the forecast error is

$$E_4 = 48.0 - 40.4 = 7.6$$

The mean absolute forecast error is obtained by calculating the arithmetic mean of the absolute forecast errors. The sum of the absolute forecast errors is 82.9. We obtain

$$\text{MAFE} = \frac{2.0 + 7.6 + \cdots + 3.3}{22} = \frac{82.9}{22} = 3.77$$

The RMSE is obtained as the square root of the mean of the squared forecast errors. For example, for period 3, the forecast error is $E_3 = 2.0$, and the squared forecast error is 4.0; for period 4, the forecast error is $E_4 = 7.6$, and the squared forecast error is 57.76. The sum of the squared forecast errors is 478.94. We obtain

$$\text{RMSE} = \sqrt{\frac{4.0 + 57.8 + \cdots + 10.9}{22}} = \sqrt{\frac{478.94}{22}} = 4.67$$

Different choices for γ and δ will lead to different values for MAFE and RMSE. One way of selecting the smoothing constants is to select the model with the smallest MAFE or the smallest RMSE. Of course, the model with the smallest MAFE or RMSE will not necessarily provide the best future forecasts. Also, the formulas for MAFE and RMSE give equal weights to all the forecast errors regardless of which period is forecast. Some forecasters might argue that it would be preferable to give higher weights to more recent forecast errors than to forecast errors for periods in the distant past.

Exercises

Statistical Concepts

20.4.1   Suppose we use the exponential smoothing model

$$s_t = cy_t + (1 - c)s_{t-1}$$

Suppose the values y_t are quite volatile. The time series of values of s_t will be more volatile if we use a large value of c rather than a small value of c. Explain.

20.4.2   *True or false:*   The purpose of the two-parameter exponential smoothing model is to allow the investigator to take into account an upward or downward trend. Explain.

20.4.3   Which will produce a smoother curve, a one-parameter exponentially smoothed curve using $c = .3$ or $c = .7$? Why?

20.4.4   *True or false:*   When we use one-parameter exponential smoothing, a very stable or smooth time series calls for a large value of c and an extremely volatile series calls for a small value of c. Explain.

Statistical Drills

20.4.5   Use the following data and construct a smoothed series using the one-parameter exponential smoothing technique:

Time	1	2	3	4	5	6	7	8
Value	9	7	12	8	10	14	16	13

  a. Plot the time series.
  b. Find the smoothed values using $c = .5$ and plot them.
  c. Find forecasts for periods 9, 10, and 11.
  d. Find the smoothed values using $c = .9$ and plot them.
  e. Find forecasts for periods 9, 10, and 11.

20.4.6   Use the following data and construct a smoothed series using the one-parameter exponential smoothing technique:

Time	1	2	3	4	5	6	7	8
Value	3.2	4.1	3.7	3.9	3.4	3.7	4.1	3.6

  a. Plot the time series.
  b. Find the smoothed values using $c = .2$ and plot them.
  c. Find forecasts for periods 9, 10, and 11.
  d. Find the smoothed values using $c = .8$ and plot them.
  e. Find forecasts for periods 9, 10, and 11.

20.4.7   Use the following data and construct a smoothed series using the two-parameter exponential smoothing technique:

Time	1	2	3	4	5	6	7	8
Value	3	6	8	8	10	14	13	19

  a. Plot the time series.
  b. Find the smoothed values and trend values $\gamma = .8$ and $\delta = .5$.

c. Find the forecasts for periods 3–8. Plot them.
d. Find the MAFE and RMSE.
e. Find forecasts for periods 9, 10, and 11.
f. Repeat steps **b–e** using $\gamma = .6$ and $\delta = .5$.

20.4.8 Use the following data and construct a smoothed series using the two-parameter exponential smoothing technique:

Time	1	2	3	4	5	6	7	8
Value	12	10	13	15	21	27	26	34

a. Plot the time series.
b. Find the smoothed values and trend values $\gamma = .8$ and $\delta = .5$.
c. Find the forecasts for periods 3–8. Plot them.
d. Find the MAFE and RMSE.
e. Find forecasts for periods 9, 10, and 11.
f. Repeat steps **b–e** using $\gamma = .6$ and $\delta = .5$.

Statistical Applications

20.4.9 The following data show monthly receipts in thousands of dollars of a new-car franchise during 1994 and 1995:

						MONTH						
Year	Jan.	Feb.	Mar.	Apr.	May	June	July	Aug.	Sep.	Oct.	Nov.	Dec.
1994	84	82	86	92	96	102	108	114	112	106	98	92
1995	90	90	94	100	110	110	114	120	122	116	112	106

a. Plot the data.
b. Let $c = .8$ and forecast receipts 1 month ahead from July 1994 to December 1995 using exponential smoothing.
c. Plot these forecasts and calculate the monthly errors in the forecast.

20.4.10 The following data show the number of gallons of gasoline sold per month (in hundreds of thousands) at a service station:

						MONTH						
Year	Jan.	Feb.	Mar.	Apr.	May	June	July	Aug.	Sep.	Oct.	Nov.	Dec.
1	1.8	1.8	1.9	2.1	2.3	2.5	2.8	3.0	2.7	2.5	2.3	2.2
2	2.2	2.3	2.4	2.5	2.7	2.7	3.1	3.2	2.8	2.8	2.4	2.4
3	2.5	2.6	2.6	2.7	2.9	3.0	3.2	3.3	3.1	2.9	2.7	2.5

a. Plot the data.
b. Estimate the trend line $Y_t = \beta_0 + \beta_1 t + e_t$.
c. Use the estimated trend line and predict Y for each month.
d. Calculate the residuals and plot them.
e. Let $c = .8$ and calculate forecasts 1 month ahead by exponential smoothing.
f. Plot these forecasts and calculate the forecast errors.
g. Plot these errors.

20.4.11   The following data show monthly sales in thousands of cases for a nationwide distributor of bottled spring water. Construct a smoothed series using the two-parameter exponential smoothing technique.

Month	1	2	3	4	5	6	7	8	9
Sales	5	6	14	22	23	32	46	46	42

Month	10	11	12	13	14	15	16	17	18
Sales	47	53	48	63	67	72	83	93	95

Month	19	20	21	22	23	24	25	26	27
Sales	94	88	87	90	91	95	101	109	108

Month	28	29	30	31	32	33	34	35	36
Sales	114	119	125	134	147	145	152	150	156

a. Plot the time series.
b. Find the smoothed values and trend values $\gamma = .8$ and $\delta = .6$.
c. Find the forecasts for periods 3–36. Plot them.
d. Find the MAFE and RMSE.
e. Find forecasts for periods 37, 38, and 39.
f. Repeat steps **b–e** using $\gamma = .6$ and $\delta = .5$.

Chapter 20 Supplementary Exercises

20.S.1   The accompanying data refer to monthly sales (in thousands of dollars) of a small clothing store.

Month	Year 1	Year 2	Year 3
January	13.4	14.4	14.5
February	14.6	15.9	14.6
March	17.4	16.8	16.4
April	16.3	15.3	15.8
May	17.6	17.9	18.2
June	16.4	16.4	17.0
July	14.9	15.1	16.4
August	16.8	17.0	17.4
September	16.6	16.0	16.6
October	22.3	20.8	21.5
November	23.8	22.1	21.3
December	25.2	24.4	25.8

a. Find the linear trend component.
b. Find the seasonal index using the ratio-to-trend method.
c. Seasonally adjust the data using the index from part **b.**
d. Plot the original data, the seasonally adjusted data, and the linear trend line.
e. Remove the seasonal component by computing a centered 12-month moving average of the original data.
f. Plot the original data and the smoothed data.
g. Estimate the linear trend line using the smoothed data.
h. Use parts **a** and **b** to forecast sales for each month of year 4.

20.S.2 The accompanying data are monthly car sales at a small distributor in Chicago.

Month	Year 1	Year 2	Year 3
January	31	33	34
February	33	34	34
March	41	40	39
April	49	53	49
May	56	59	59
June	66	64	67
July	71	73	72
August	74	75	76
September	63	61	62
October	56	54	55
November	50	46	47
December	42	40	39

   a. Calculate a seasonal index using the simple average method.
   b. Deseasonalize the data using the index in part **a.**
   c. Deseasonalize the data by calculating a centered 12-month moving average.

 20.S.3 The accompanying table shows quarterly sales data for a business firm in millions of dollars.

Quarter	Year 1	Year 2	Year 3	Year 4
1	2.6	3.6	4.9	7.2
2	2.9	4.1	5.8	7.5
3	3.5	4.8	6.4	8.2
4	5.7	6.9	8.8	10.7

   a. Compute a centered four-quarter moving average.
   b. Fit a logarithmic trend curve using the results from part **a.**
   c. Compute a quarterly index using the results from part **b** and the ratio-to-trend method.
   d. Forecast the trend values for year 5.
   e. Use parts **c** and **d** to forecast the actual values for year 5.

20.S.4 Usage of an indoor tennis facility has been increasing every year, but the number of customers fluctuates from month to month because of seasonal factors, as shown in the accompanying table.

Month	Year 1	Year 2	Year 3	Year 4	Year 5
January	400	500	620	760	860
February	450	510	650	800	910
March	400	470	600	720	810
April	380	450	510	660	710
May	380	420	450	610	620
June	190	200	220	310	320
July	100	200	220	280	300
August	100	180	210	260	260
September	120	230	280	320	330
October	180	290	410	450	470
November	270	450	520	590	610
December	310	480	620	680	710

a. Plot the data.

b. Fit a linear trend line and graph it.

c. Construct a centered 12-month moving average, and plot the smoothed values.

d. Construct a seasonal index using the method of simple averages.

e. Use the trend equation to predict the number of customers for each month of year 6. Adjust these values by using the seasonal index from part **d.** Plot these predictions.

f. Construct a seasonal index using the ratio-to-trend method.

g. Predict the number of customers for each month of year 6 using the results of parts **b** and **f.** Plot these predictions.

h. Construct a seasonal index using the ratio-to-moving-average method.

i. Predict the number of customers for each month of year 6 using the results of parts **b** and **h.** Plot these predictions.

References

Box, George E. P., and Gwilym M. Jenkins. *Time Series Analysis, Forecasting, and Control.* San Francisco: Holden-Day, 1970.

Granger, Clive, and Paul Newbold. *Forecasting Economic Time Series.* London: Academic Press, 1977.

Hoff, J. C. *A Practical Guide to Box–Jenkins Forecasting.* Belmont, Calif.: Lifetime Learning, 1983.

Makridakis, Spyros G. "A Survey of Time Series Analysis." *International Statistical Review* 44 (1976): 29–70.

Makridakis, Spyros, and Steven C. Wheelwright. *Forecasting: Methods and Applications.* New York: Wiley, 1978.

Moore, Geoffrey H., and Julius Shiskin. "Early Warning Signals for the Economy." In *Statistics: A Guide to the Unknown,* ed. by Judith M. Tanur *et al.* San Francisco: Holden-Day, 1972. A discussion of the history, nature, and reliability of leading, coincident, and lagging indicators of business cycles.

Morgenstern, Oskar. *On the Accuracy of Economic Observations.* 2d ed. Princeton, N.J.: Princeton University Press, 1963. Mandatory reading for anyone working with business and economic statistics.

Nelson, Charles R. *Applied Time Series for Managerial Forecasting.* San Francisco: Holden-Day, 1973. An excellent introduction.

Nie, Norman E., C. Hadlai Hull, Jean G. Jenkins, Karin Steinbrenner, and Dale H. Bent. *SPSS Statistical Package for the Social Sciences.* 2d ed. New York: McGraw-Hill, 1975.

Norusis, Marija J. *SPSSX Introductory Statistics Guide.* New York: McGraw-Hill, 1990.

Norusis, Marija J. *SPSSX Advanced Statistics Guide.* Chicago: SPSS, 1990.

Norusis, Marija J. *The SPSS Guide to Data Analysis.* Chicago: SPSS, 1986.

Pankratz, Alan. *Forecasting with Univariate Box–Jenkins Models: Concepts and Cases.* New York: Wiley, 1983.

Ryan, Thomas A., Brian L. Joiner, and Barbara F. Ryan. *Minitab Handbook.* 2d ed. Boston: PWS-Kent, 1985.

Ryan, Thomas A., Brian L. Joiner, and Barbara F. Ryan. *Minitab Reference Manual.* University Park, Penn.: Minitab Project, 1985.

SAS Introductory Guide. 3d ed. Cary, N.C.: SAS Institute, 1985.

SAS Procedures Guide for Personal Computers. Version 6 ed. Cary, N.C.: SAS Institute, 1986.

SAS Statistics Guide for Personal Computers. Version 6 ed. Cary, N.C.: SAS Institute, 1986.

SAS User's Guide: Basics. Version 5 ed. Cary, N.C.: SAS Institute, 1985.

SAS User's Guide: Statistics. Version 5 ed. Cary, N.C.: SAS Institute, 1985.

SPSSX User's Guide. Chicago: SPSS, 1988.

21 Some Nonparametric Tests

The tests developed in the previous chapters depend on a rather strict set of assumptions concerning the distribution of the random variable being studied. For example, to use the t statistic for testing the hypothesis that the mean of a population is equal to a specified value, the underlying population must at least approximate a normal distribution. Sometimes, however, the distribution of the underlying population does not meet such requirements. In these situations it is desirable to base our inferences on tests that are valid over a wide range of distributions of the parent population. Such tests are called **nonparametric,** or **distribution-free,** tests. Besides requiring few assumptions, nonparametric statistics may be used to analyze data consisting of rankings or ratings as well as quantitative measurements.

In this chapter, we discuss the following six nonparametric tests:

1. *Sign test:* This test is used to analyze samples of data (A_i, B_i) that occur in matched pairs. It tests the null hypothesis that, for a given pair, the probability that A_i exceeds B_i is $p = .5$ or that, in a certain population, half the individuals have a certain characteristic and half do not. The data may consist of numerical values, rankings, or preferences.

2. *The Mann–Whitney test:* This test is used to test the null hypothesis that two populations are identical, that their means are equal, or that their medians are equal. The data consist of numerical values or rankings from independent samples from two populations.

3. *The Wilcoxon signed-rank test:* This test is used to test the null hypothesis that two populations are identical, that their means are equal, or that their medians are equal. The data consist of numerical values or rankings from matched pairs of data from two populations.

4. *The Kruskal–Wallis test:* This test is used as an alternative to analysis of variance for testing the null hypothesis that several populations are identical, that their means are equal, or that their medians are equal. The data may consist of numerical values or rankings from independent samples from several populations.

5. *Runs test:* This test is used to test the null hypothesis that the sequence of occurrence of observations in a sample is random. The data may consist of either numerical values or qualitative observations.

6. *Rank correlation coefficient test:*   The rank correlation coefficient test is used to test the null hypothesis that there is no linear relationship between two variables when the variables X and Y represent rankings of two variables. It is used as an alternative to the correlation coefficient discussed in Section 15.9.

21.1   The Sign Test

The easiest nonparametric test to perform is called the **sign test**, which is used to analyze data (A_i, B_i) that occur in matched pairs. The null hypothesis states that the probability is .5 that observation A_i exceeds observation B_i. The data may consist of numerical values, rankings, or preferences. The sign test can be used as an alternative to the paired t test discussed in Section 12.3, where we tested the null hypothesis that the mean difference is 0. The sign test can be used in any situation where the null hypothesis states that half the observations in a population have a certain characteristic and the other half do not.

For example, suppose we ask 15 different customers to rate each of two products on a scale of 1 to 10. For each individual, we have two ratings or scores. Let A_i denote the ith individual's rating for product A and let B_i denote the ith individual's rating for product B. The sign test can be used to test the null hypothesis that, for the entire population of ratings by all individuals, there is no overall tendency to prefer one product over the other.

For each individual in the sample, record a plus sign $(+)$ if $(A_i - B_i)$ is greater than 0 and a minus sign $(-)$ if $(A_i - B_i)$ is less than 0. If the difference equals 0, eliminate the observation from the sample and decrease the sample size accordingly. The sign test is used to test the hypothesis that positive differences are just as likely as negative differences.

For the entire population, let p denote the probability of obtaining a plus sign on any selection. The null hypothesis is

$$H_0: \quad p = .5$$

and the alternative hypothesis takes one of the forms

$$H_1: \quad p \neq .5, \qquad H_1: \quad p > .5, \qquad \text{or} \qquad H_1: \quad p < .5$$

Let X denote the number of plus signs obtained in a random sample of n observations (after eliminating all ties). If H_0 is true, the random variable X follows the binomial distribution with $p = .5$. When the sample size is small (say, 10 or less), the binomial distribution is used to find the critical region of the test, as described in the accompanying box. When the sample size is 10 or more, the normal distribution can be used to find the critical region, because the random variable X approximately follows the normal distribution.

Performing the sign test with small samples

Suppose we want to test the null hypothesis

$$H_0: \quad p = .5$$

against the one-sided alternative hypothesis

$$H_1: \quad p > .5$$

using the level of significance α where the sample size is 10 or less. Let x_0 denote the observed number of positive values (or successes) in a sample of size n, where n is the sample size after eliminating all ties. Use the binomial distribution to calculate $P(X \geq x_0)$. Use the decision rule

Reject H_0 in favor of H_1 if $P(X \geq x_0) < \alpha$.

When the alternative hypothesis is H_1: $p < .5$, use the decision rule

Reject H_0 in favor of H_1 if $P(X \leq x_0) < \alpha$.

When the alternative hypothesis is H_1: $p \neq .5$, use the decision rule

Reject H_0 in favor of H_1 if $P(X \leq x_0) < \alpha/2$ or if $P(X \geq x_0) < \alpha/2$.

EXAMPLE 21.1   **Using the Sign Test with a Small Sample** —————————————————

Every summer the Coca-Cola and Pepsi-Cola companies conduct taste tests, where they ask random samples of people to taste the two drinks and select the one that tastes better. Suppose you blindfold a random sample of 12 people and have them taste and rate each drink from 1 to 10. Let A_i denote the rating given to Coca-Cola by the ith individual and B_i denote the rating given to Pepsi-Cola by the ith individual. Suppose you obtain the results given in Table 21.1. At a 5% level of significance, do the results indicate that one cola is preferred over the other?

TABLE 21.1   *Ratings of colas for Example 21.1*

Individual	Coca-Cola A_i	Pepsi-Cola B_i	Sign of $(A_i - B_i)$	
1	10	8	+	
2	7	5	+	
3	5	7	−	
4	6	6	0	delete
5	8	7	+	
6	8	5	+	
7	7	7	0	delete
8	6	3	+	
9	6	7	−	
10	5	3	+	
11	7	5	+	
12	8	6	+	

SOLUTION   After deleting the 2 ties from the sample, we are left with a sample of 10 observations, consisting of 8 plus values and 2 minus values. We want to test the null hypothesis

$$H_0: \quad p = .5$$

against the two-sided alternative hypothesis

$$H_1: \quad p \neq .5$$

where p is the proportion of the population who prefer Coca-Cola. We use a two-tailed test with each tail of the critical region having probability $\alpha/2 = .025$. For $x_0 = 8$, the binomial distribution yields

$$P(X = 8) = .0439, \quad P(X = 9) = .0098, \quad P(X = 10) = .0010$$

and

$$P(X \geq x_0) = P(X \geq 8) = .0439 + .0098 + .0010 = .0547$$

Thus, the probability of obtaining a result at least as extreme as the one we observed, if the null hypothesis is true, is .0547. Since .0547 exceeds .025, we do not reject the null hypothesis.

In Example 21.1, the effective sample size was $n = 10$. When the sample size is larger than 10, the normal distribution provides a good approximation to the sampling distribution of X; thus we can use the normal approximation to the binomial distribution when performing the sign test. When n exceeds 10, the sampling distribution of X is approximately normal with mean

$$\mu = np = .5n$$

and variance

$$\sigma^2 = npq = .25n$$

To perform the sign test when n exceeds 10, use the procedure described in the accompanying box.

Performing the sign test with large samples

Suppose we want to test the null hypothesis

$$H_0: \quad p = .5$$

against the one-sided alternative hypothesis

$$H_1: \quad p > .5$$

using level of significance α where the sample size exceeds 10. To perform the sign test, calculate the test statistic

$$z = \frac{x_0 - .5n}{\sqrt{.25n}}$$

where x_0 denotes the observed number of plus signs and n is the sample size after eliminating all ties. Use the decision rule

Reject H_0 in favor of H_1 if $z > z_\alpha$,

where the critical value z_α is obtained from the standard normal distribution table.

> When the alternative hypothesis is $H_1: p < .5$, use the decision rule
>
> Reject H_0 in favor of H_1 if $z < -z_\alpha$.
>
> When the alternative hypothesis is $H_1: p \neq .5$, use the decision rule
>
> Reject H_0 in favor of H_1 if $z < -z_{\alpha/2}$ or if $z > z_{\alpha/2}$.

EXAMPLE 21.2   **Using the Sign Test with a Large Sample**

A random sample of 110 customers were asked to test-drive two cars. Each customer was asked to rate the handling of each car. Suppose that 63 preferred car A, 37 preferred car B, and 10 were indifferent. Let p denote the probability that car A is preferred in the population. Test the null hypothesis

$$H_0: \quad p = .5$$

against the two-sided alternative hypothesis

$$H_1: \quad p \neq .5$$

using a 5% level of significance.

SOLUTION   Let the random variable X denote the number of individuals who prefer car A. After omitting ties, we have a sample of $n = 100$ observations. The number who prefer car A (a plus sign) is $x_0 = 63$. If H_0 is true, the random variable X is approximately normally distributed with mean $np = 50$ and variance $npq = 25$. For $\alpha = 5\%$, the critical values of the test statistic are $z_{.025} = 1.96$ and $-z_{.025} = -1.96$. The observed test statistic is

$$z = \frac{63 - 50}{\sqrt{25}} = 2.60$$

Because the observed Z score exceeds the critical value, we reject the null hypothesis in favor of $p \neq .5$.

Exercises

Statistical Concepts

21.1.1   Explain what is meant by a nonparametric test. When would a nonparametric test be preferred over the parametric tests discussed in previous chapters?

21.1.2   *True or false:*   The sign test can be applied to numerical values, rankings, or qualitative preferences. Explain.

21.1.3   What distribution is used when performing the sign test with small samples? With large samples? Explain.

21.1.4   *True or false:*   When performing the sign test with a small sample, the appropriate test statistic is the number of times that $(A - B)$ is positive, or the number of times that A is preferred to B. Explain.

21.1.5   *True or false:*   When performing the sign test with a small sample, the appropriate sample size is the total number of pairs of observations. Explain.

Statistical Drills

21.1.6   In a random sample of $n = 15$ individuals, 8 preferred product A, 4 preferred product B, and 3 were indifferent. Let $\alpha = .05$. Use the sign test and test $H_0: p = .5$ against $H_1: p \neq .5$.

21.1.7   In a random sample of $n = 23$ individuals, 8 preferred product A, 12 preferred product B, and 3 were indifferent. Let $\alpha = .05$. Use the sign test and test $H_0: p = .5$ against $H_1: p \neq .5$.

21.1.8   In a random sample of $n = 408$ individuals, 208 preferred product A, 192 preferred product B, and 8 were indifferent. Let $\alpha = .05$. Use the sign test and test $H_0: p = .5$ against $H_1: p \neq .5$.

Statistical Applications

21.1.9   A random sample of 112 people examine two television sets. Sixty-two people prefer the picture on television A, 38 prefer the picture on television B, and 12 are indifferent. Use the two-tailed sign test and a 5% level of significance to test whether the proportion of people in the population who prefer television A is .5.

21.1.10   Two TV commercials for a certain brand of automobile are shown to a random sample of 100 people: 48 prefer commercial 1, 42 prefer commercial 2, and 10 are indifferent. Use the two-tailed sign test and a 5% level of significance to test whether one commercial is preferred to the other.

21.1.11   An automobile manufacturer surveys a random sample of 120 dealers concerning what should be done to promote sales. Sixty favor increasing the amount of advertising, 40 favor reducing prices, and 20 are indifferent. Use the two-tailed sign test and a 5% level of significance to test whether one proposal is favored over the other.

    21.1.12   A congressman wants to know how the voters feel about a proposed tariff. In a random sample of 490 voters, 220 favor the tariff, 180 oppose it, and 90 are indifferent. Use the two-tailed sign test and a 5% level of significance to test whether one proposal is favored over the other.

21.2   The Mann–Whitney Test

In Section 21.1, we discussed the sign test, which is designed to determine whether significant differences exist between two populations based on a random sample of matched pairs of observations. In this section, we introduce the Mann–Whitney test, which is designed to determine whether significant differences exist when *independent random samples* are taken from two populations.

The test is named after Henry B. Mann and D. R. Whitney, who developed the test in 1947. It is used to test the null hypothesis that two population means or medians are identical when the data consist of ranks or when some of the assumptions underlying the *t* test do not apply. The Mann–Whitney test is thus an alternative to the *t* test for independent samples, which was discussed in Section 12.2.

The null hypothesis tested can take one of several forms:

1. The two samples have been selected from the same population.
2. The two samples have been selected from populations having identical means.
3. The two samples have been selected from populations having identical medians.

The test relies on the assumption that the two population distributions have the same shape and spread. Assume we have two independent samples containing n_1 and n_2 observations from two populations. When applying the Mann–Whitney test, the first step is to pool the data from the two samples into one set of observations. Next, rank these observations from the lowest to the highest score.

If the null hypothesis is true, the observations from the two samples should be randomly scattered throughout the ranking of the pooled data. If the data do not come from populations having identical medians, the sample observations from the population having the smaller mean or median will tend to have lower rankings, whereas the sample observations from the population having the larger mean or median will tend to have higher rankings. The statistic used in the Mann–Whitney test, called *U*, is described in the accompanying box.

Formula for calculating the Mann–Whitney *U* statistic

Let R_1 denote the sum of the ranks of the observations from population 1, and let R_2 denote the sum of the ranks of the observations from population 2. The Mann–Whitney *U* statistic is

$$U = n_1 n_2 + \frac{n_1(n_1 + 1)}{2} - R_1$$

If the null hypothesis is true and the sample sizes are large, the random variable *U* is approximately normally distributed with mean

$$\mu_U = \frac{n_1 n_2}{2}$$

and variance

$$\sigma_U^2 = \frac{n_1 n_2(n_1 + n_2 + 1)}{12}$$

The approximation to the normal distribution is adequate when n_1 and n_2 are both 10 or more.

To test the null hypothesis, we use the test statistic

$$z = \frac{u - \mu_U}{\sigma_U}$$

where *u* is the observed value of the random variable *U*. When the null hypothesis is true, the random variable *Z* approximately follows the standard normal distribution.

The procedure for performing the Mann–Whitney test with large samples is described in the box at the top of page 908.

Performing the Mann–Whitney test with large samples

> Suppose we have independent random samples of n_1 and n_2 observations from two populations. Pool the observations together and rank them. Let R_1 denote the sum of the ranks of the observations from the first population. To test the null hypothesis that the populations have equal means or medians at a level of significance α, calculate the test statistic
>
> $$z = \frac{u - \mu_U}{\sigma_U}$$
>
> where u is the observed value of the Mann–Whitney U statistic.
>
> *Case 1*:   If the alternative hypothesis states that the mean or median of population 1 exceeds the mean or median of population 2, use the decision rule
>
> $$\text{Reject } H_0 \text{ in favor of } H_1 \text{ if } z < -z_\alpha.$$
>
> *Case 2*:   If the alternative hypothesis states that the mean or median of population 1 is less than the mean or median of population 2, use the decision rule
>
> $$\text{Reject } H_0 \text{ in favor of } H_1 \text{ if } z > z_\alpha.$$
>
> *Case 3*:   If the alternative hypothesis states that the mean or median of population 1 differs from the mean or median of population 2, use the decision rule
>
> $$\text{Reject } H_0 \text{ in favor of } H_1 \text{ if } z < -z_{\alpha/2} \text{ or if } z > z_{\alpha/2}.$$

EXAMPLE 21.3   **Using the Mann–Whitney Test with a Large Sample**

In a sex discrimination suit against a large corporation, a female employee alleged that male employees at the corporation received larger pay raises than female employees. All the employees worked in the same department and had approximately identical job performance ratings. Suppose you wish to test the null hypothesis that the average pay raises are the same for males and females. You obtain the following data on the pay increases (in thousands of dollars) from a random sample of 20 males (population 1) and 15 females (population 2) in the department.

POPULATION 1 (MALES)									
3.06	2.11	3.08	2.94	2.78	2.62	2.73	2.69	2.51	2.91
3.16	3.19	3.42	2.66	2.98	2.80	2.16	2.22	3.10	2.55

POPULATION 2 (FEMALES)									
3.12	3.18	2.95	2.86	2.75	2.81	3.07	3.22	2.90	2.48
2.77	2.72	2.96	3.26	3.23					

The observations are ranked in Table 21.2. Use a one-tailed test and a 5% level of significance.

TABLE 21.2 *Rankings of pay raises for Example 21.3*

Ordered data	Sample	Rank for sample 1	Rank for sample 2
2.11	1	1	
2.16	1	2	
2.22	1	3	
2.48	2		4
2.51	1	5	
2.55	1	6	
2.62	1	7	
2.66	1	8	
2.69	1	9	
2.72	2		10
2.73	1	11	
2.75	2		12
2.77	2		13
2.78	1	14	
2.80	1	15	
2.81	2		16
2.86	2		17
2.90	2		18
2.91	1	19	
2.94	1	20	
2.95	2		21
2.96	2		22
2.98	1	23	
3.06	1	24	
3.07	2		25
3.08	1	26	
3.10	1	27	
3.12	2		28
3.16	1	29	
3.18	2		30
3.19	1	31	
3.22	2		32
3.23	2		33
3.26	2		34
3.42	1	35	
		$R_1 = 315$	$R_2 = 315$

SOLUTION   The sample sizes are $n_1 = 20$ and $n_2 = 15$. The sum of the ranks for males is $R_1 = 315$, and so the observed value of the U statistic is

$$U = 20(15) + \frac{20(20 + 1)}{2} - 315 = 195$$

If the null hypothesis is true, the random variable U is approximately normally distributed with mean

$$\mu_U = \frac{20(15)}{2} = 150$$

and variance

$$\sigma_U^2 = \frac{20(15)(20 + 15 + 1)}{12} = 900$$

The observed test statistic is

$$z = \frac{195 - 150}{\sqrt{900}} = 1.5$$

The rejection region is $z < -1.645$. Because the observed value of the test statistic falls in the acceptance region, we do not reject the null hypothesis that the means are equal.

If we thought that salary increases for males and females were normally distributed in Example 21.3, we could have used the t test to test the hypothesis that there was no difference in the mean increases. If increases are not approximately normally distributed, however, it is more appropriate to use a nonparametric test such as the U test. The t test is more powerful than the Mann–Whitney test and will detect true differences between two populations more often than will the Mann–Whitney test. This is because the t test uses more information from the data. The Mann–Whitney test substitutes ranks for the actual values of observations, thus losing useful information. In general, if the assumptions of the t test appear reasonable, it should be used. When the data are ranks or come from a markedly nonnormal distribution, the Mann–Whitney test is preferable.

If there are ties in observations, average the ranks of the tied observations and assign this averaged rank to each of the tied observations. For example, if the fourth, fifth, and sixth observations are all tied, assign the rank of 5 (the mean of 4, 5, and 6) to each of these observations. The next observation would receive a rank of 7. If the ninth and tenth observations are tied, assign the rank 9.5 to each; the next observation would receive a rank of 11.

The normal distribution does not necessarily give a good approximation to the probability distribution of U when n_1 or n_2 is less than 10. For such small values of n_1 or n_2, tables of probabilities for U are available, such as Table A.9 in the Appendix. The procedure in the accompanying box shows how to perform the Mann–Whitney test when the sample size is small.

Performing the Mann–Whitney test with small samples

Suppose we have independent random samples of n_1 and n_2 observations from two populations. Let population 1 denote the population with the smaller sample size. Suppose we wish to test the null hypothesis that the populations have equal means or medians using a level of significance α. Pool the observations from the two samples and calculate the Mann–Whitney U statistic. Let u_0 denote the observed value of U. Refer to Table A.9. For given values of n_1 and n_2, Table A.9 shows $P(U \le u_0)$. Use the decision rule

Reject H_0 in favor of H_1 if $P(U \le u_0) < \alpha$.

EXAMPLE 21.4 **Using the Mann–Whitney Test with a Small Sample**

Suppose we have $n_1 = 4$ observations from population 1 and $n_2 = 5$ observations from population 2. Use the following data and a 5% level of significance to test the null hypothesis that the data came from populations having identical means or medians:

Population	2	2	1	2	1	1	2	1	2
Observation	18	21	22	25	29	30	31	34	36
Rank	1	2	3	4	5	6	7	8	9

SOLUTION Population 1 is designated as the population with the smaller sample size. The sum of ranks for the observations from population 1 is

$$R_1 = 3 + 5 + 6 + 8 = 22$$

The observed value of the Mann–Whitney U statistic is

$$u_0 = 4(5) + \frac{4(4+1)}{2} - 22 = 8$$

From Table A.9, using $n_1 = 4$, $n_2 = 5$, and $u_0 = 8$, we obtain

$$P(U \le 8) = .3651$$

Because this probability exceeds α, we do not reject the null hypothesis.

Exercises

Statistical Concepts

21.2.1 *True or false:* The Mann–Whitney test is designed to test whether significant differences exist when independent random samples are obtained. Explain.

21.2.2 What is the test statistic when performing the Mann–Whitney test?

21.2.3 When performing the Mann–Whitney test, what is the sampling distribution of the test statistic? What is assumed about the sample sizes?

21.2.4 Explain how to perform the Mann–Whitney test when the sample sizes are small.

Statistical Drills

21.2.5 Two independent random samples were obtained from populations 1 and 2. It is assumed that the populations are not normal but have the same shape and spread. Use the Mann–Whitney test to decide whether the mean of population 1 exceeds the mean of population 2. Suppose the following data were obtained: $n_1 = 25$, $n_2 = 15$, $R_1 = 320$.
a. Compute the U test statistic. **b.** Let $\alpha = .05$. Perform the test.

21.2.6   Two independent random samples were obtained from populations 1 and 2. It is assumed that the populations are not normal but have the same shape and spread. Use the Mann–Whitney test to decide whether the median of population 1 equals the median of population 2. Suppose the following data were obtained: $n_1 = 7$, $n_2 = 8$, $R_1 = 46$.
a. Compute the U test statistic.   **b.** Let $\alpha = .05$. Perform the test.

21.2.7   The following data represent random samples obtained from two independent populations. Use the Mann–Whitney test to determine whether population 1 tends to lie to the right of population 2.

Sample 1	7	5	14	18	9	11
Sample 2	8	6	4	13	11	10

a. State H_0 and H_1.
b. Find the rejection region if α is approximately .05.
c. Rank the pooled samples of data, and find the value of the U test statistic.
d. Perform the test.

Statistical Applications

21.2.8   ?✓   A national pizza distributor conducted taste tests to compare its pizza with the local neighborhood favorite. In a particular neighborhood, seven individuals ate the national brand and six ate the local favorite, and each individual gave a rating from 1 to 20 for the pizza. Let $\alpha = .05$ and test the null hypothesis that there is no preference for one pizza over the other. The data follow:

Local pizza ratings	18	10	12	15	9	16	
National pizza ratings	16	5	7	11	19	11	13

21.2.9   An educator administers a reading test to a random sample of 13 boys and 12 girls. Their scores are as follows:

Boys' scores	62	87	55	63	71	80	91	78	96	84	59	77	88
Girls' scores	66	86	81	74	72	85	97	99	46	68	73	83	

Use the U test to determine whether boys' scores are significantly different from girls' scores. Use a 5% level of significance.

21.3   The Wilcoxon Signed-Rank Test

The sign test is used to test whether two populations have equal means or medians based on a matched pair of observations. However, it utilizes only the signs of the differences between pairs of observations and ignores the magnitudes of the differences. The Wilcoxon signed-rank test, developed by Frank Wilcoxon (1892–1965) in 1945, is an alternative to the sign test. The test requires as data not only the signs of the differences between pairs of observations, but also the ranks of the differences. Because it uses more information, the Wilcoxon signed-rank test is more powerful than the sign test. On the other hand, it is not as generally applicable as the sign test because at times the ranks of the differences of paired observations are not known. The Wilcoxon signed-rank test, like the sign test, utilizes data from two *related* samples. The samples may refer to before and after observations, husband–wife responses, an individual's scores on two tests, and so forth.

Suppose we obtain a sample of n observations from population 1 and a sample of n observations from population 2, where each observation in sample 1 is related to an observation in sample 2. The null hypothesis is that the two populations are identical. If the null hypothesis is true, we would expect half of the differences of paired observations to be negative and half to be positive. In addition, we would expect positive and negative differences of any given magnitude to occur with equal probability.

For each matched pair, calculate the absolute value of the difference. As in the sign test, differences of 0 are ignored. Rank the absolute values from lowest to highest and obtain the sum of the ranks for the differences that are positive. Denote this sum as T_p. Then sum the ranks of the differences that are negative and denote this number as T_n. If the two distributions are identical, we would expect these two rank sums to be approximately equal. The sum of all n rankings is

$$T_p + T_n = \frac{n(n+1)}{2}$$

and since we expect T_p to equal T_n, the mean value (or expected value) of T_p and of T_n is

$$\mu_{T_p} = \mu_{T_n} = \frac{n(n+1)}{4}$$

If T_p or T_n differs significantly from $n(n+1)/4$, the hypothesis that populations 1 and 2 have different distributions is supported. Note that

$$\left| T_p - \frac{n(n+1)}{4} \right| = \left| T_n - \frac{n(n+1)}{4} \right|$$

That is, if T_p is, say, 5 units above the mean, then T_n is 5 units below the mean. Thus, T_p and T_n differ from the mean by exactly the same amount but in opposite directions. It makes no difference, therefore, whether we use T_p or T_n to test the hypothesis that the distributions are identical.

Define the random variable T to be the smaller of T_p and T_n. That is,

$$T = \min(T_p, T_n)$$

Small values of T support the alternative hypothesis H_1 that the population distributions are not identical.

When n is large (say, $n \geq 15$), then T is approximately normally distributed with mean

$$\mu_T = \frac{n(n+1)}{4}$$

and variance

$$\sigma_T^2 = \frac{n(n+1)(2n+1)}{24}$$

and the random variable

$$Z = \frac{T - \mu_T}{\sigma_T}$$

approximately follows the standard normal distribution. The procedure in the accompanying box shows how to perform the Wilcoxon signed-rank test when the sample size is large ($n > 15$).

Performing the Wilcoxon signed-rank test with large samples

The Wilcoxon signed-rank test uses a random sample of matched pairs of observations. We wish to test the null hypothesis that the two population distributions are identical or the null hypothesis that the distribution of differences is centered at 0. Discard pairs for which the difference is 0 and rank the absolute differences in ascending order. Then calculate the sum of the ranks for positive differences and for negative differences. The observed Wilcoxon signed-rank test statistic t_0 is the smaller of these two sums. When the sample size exceeds 15, the observed test statistic

$$z = \frac{t_0 - \mu_T}{\sigma_T}$$

is distributed approximately as a standard normal variate.

Case 1: If the alternative hypothesis is one-sided, reject the null hypothesis if $z < -z_\alpha$, where α is the level of significance of the test.

Case 2: If the alternative hypothesis is two-sided, reject the null hypothesis if $z < -z_{\alpha/2}$, where α is the level of significance of the test.

Any tied values are given the same rank, which is the average rank of the tied values.

EXAMPLE 21.5   **Using the Wilcoxon Signed-Rank Test with a Large Sample** ────────

A soft drink bottler has produced a new drink using two different recipes, one of which is much sweeter. The bottler asks 20 individuals to taste both drinks and rate the drinks on a scale of 1 to 10, where 10 means the individual likes the drink very much. Use the Wilcoxon signed-rank test to test the null hypothesis that neither drink is preferred over the other. Use a two-tailed test and a 5% level of significance. The data are in Table 21.3.

SOLUTION   The sum of the ranks for the positive differences is $T_p = 154$ and for negative differences, $T_n = 17$. The Wilcoxon signed-rank test statistic is the smaller of these two values. Thus, $t_0 = 17$. The mean is

$$\mu_T = \frac{n(n + 1)}{4} = \frac{18(19)}{4} = 85.5$$

(Remember, $n = 18$ because we deleted 2 pairs of observations.) The variance is

$$\sigma_T^2 = \frac{n(n + 1)(2n + 1)}{24} = \frac{18(19)(37)}{24} = 527.25$$

TABLE 21.3 *Ratings of soft drinks for Example 21.5*

Individual	Drink A	Drink B	Difference	Sign of (A − B)	Rank of positives	Rank of negatives
1	10	6	4	+	13	
2	8	5	3	+	8.5	
3	6	2	4	+	13	
4	8	2	6	+	16	
5	7	4	3	+	8.5	
6	5	6	− 1	−		2
7	1	4	− 3	−		8.5
8	3	5	− 2	−		4.5
9	9	9	0	Omit		
10	7	8	− 1	−		2
11	4	2	2	+	4.5	
12	5	2	3	+	8.5	
13	8	1	7	+	18	
14	6	3	3	+	8.5	
15	8	2	6	+	16	
16	7	6	1	+	2	
17	4	1	3	+	8.5	
18	8	2	6	+	16	
19	9	5	4	+	13	
20	3	3	0	Omit		
					$T_p = 154$	$T_n = 17$

Finally, the observed test statistic is

$$z = \frac{t_0 - \mu_T}{\sigma_T} = \frac{17 - 85.5}{\sqrt{527.25}} = -2.98$$

Using a 5% level of significance, we have $\alpha/2 = .025$, and the critical value of the test statistic is $- z_{.025} = - 1.96$. We reject the null hypothesis because the observed value $z = - 2.98$ falls in the critical region.

When n is 15 or less, the normal distribution does not necessarily provide a good approximation to the distribution of the random variable T. For small values of n, tables of probabilities for T are available, such as Table A.10 in the Appendix. The procedure in the accompanying box shows how to perform the Wilcoxon signed-rank test when the sample size is small.

Performing the Wilcoxon signed-rank test with small samples

Suppose we have matched samples of n observations from two populations and wish to test the null hypothesis that the populations have identical distributions using a level of significance α. Let t_0 denote the observed value of T. For given values of n, Table A.10 in the Appendix shows the critical value T_α such that $P(T \leqslant T_\alpha) = \alpha$. Use the decision rule

$$\text{Reject } H_0 \text{ in favor of } H_1 \text{ if } t_0 < T_\alpha.$$

EXAMPLE 21.6   Using the Wilcoxon Signed-Rank Test with a Small Sample

Two makes of tires are tested on the rear wheels of six different cars. The number of miles traveled until a tire fails is recorded. Because one tire of each make is used on each car, the observations occur in matched pairs. Use the Wilcoxon signed-rank test and a 5% level of significance to test the null hypothesis

$$H_0: \quad \text{The population means are equal}$$

against the two-sided alternative hypothesis

$$H_1: \quad \text{The population means are not equal}$$

The data are in Table 21.4.

TABLE 21.4   *Tire mileage for Example 21.6*

Car	Tire A (in miles)	Tire B (in miles)	Difference (A − B)	Rank of positives	Rank of negatives
1	20,000	19,000	1,000	1	
2	24,600	23,000	1,600	2	
3	32,500	37,000	− 4,500		3
4	36,600	30,100	6,500	4	
5	37,200	25,500	11,700	5	
6	23,000	39,500	− 16,500		6
				$T_p = 12$	$T_n = 9$

SOLUTION   The sum of ranks for positive differences is $T_p = 12$; the sum of ranks for negative differences is $T_n = 9$. Thus, we have $t_0 = \min(T_p, T_n) = 9$. In Table A.10 using $\alpha = .05$ and a two-tailed test, we see that for $n = 6$ observations, the critical value of t_0 is $T = 1$. Because the value $T = 9$ exceeds the critical value, we do not reject the hypothesis that the tires are equally good.

Exercises

Statistical Concepts

21.3.1   *True or false:*   The Wilcoxon signed-rank test is applicable when the data occur in matched pairs. Explain.

21.3.2   *True or false:*   The Wilcoxon signed-rank test is designed to test whether significant differences exist when independent random samples are obtained. Explain.

21.3.3   *True or false:*   The Wilcoxon signed-rank test is more powerful than the sign test. Explain.

21.3.4   *True or false:*   The Wilcoxon signed-rank test is more generally applicable than the sign test. Explain.

21.3.5   When performing the Wilcoxon signed-rank test, what is the sampling distribution of the test statistic? What is assumed about the sample sizes?

21.3.6   *True or false:*   To perform the Wilcoxon signed-rank test, we must be able to rank the differences of paired samples. Explain.

21.3.7   *True or false:*   When performing the Wilcoxon signed-rank test, we discard pairs having a difference of 0.

21.3.8   *True or false:*   When performing the Wilcoxon signed-rank test, we rank the absolute values of the paired differences.

Statistical Drills

21.3.9   In a paired difference experiment, a random sample of $n = 8$ pairs of observations was obtained. Use the Wilcoxon signed-rank test to determine whether the two distributions are different. The accompanying data show the paired differences:

1.3	2.5	-3.5	-6.8	2.7	2.9	-3.1	4.2

a. State H_0 and H_1.
b. Rank the observed differences according to their absolute values.
c. Compute T_p and T_n.
d. Compute the Wilcoxon test statistic t_0.
e. Let $\alpha = .10$. Do the data support the conclusion that the two populations are different?

21.3.10   In a paired difference experiment, a random sample of $n = 10$ pairs of observations was obtained. Use the Wilcoxon signed-rank test to determine whether distribution A lies to the right of distribution B. The accompanying data show the paired differences (A $-$ B):

2.3	3.5	3.6	-1.8	2.7	2.4	-1.1	4.2	2.1	-1.2

a. State H_0 and H_1.
b. Rank the observed differences according to their absolute values.
c. Compute T_p and T_n.
d. Compute the Wilcoxon test statistic t_0.
e. Let $\alpha = .05$. Do the data support the conclusion that population A lies to the right of population B?

21.3.11   In a paired difference experiment, $n = 40$ randomly selected pairs were obtained. The sums of the ranks of the absolute differences were $T_p = 440$ and $T_n = 380$. Let $\alpha = .01$. Use the Wilcoxon signed-rank test to determine whether the two populations are different.

21.3.12   The following data are from a paired difference study. Let $\alpha = .05$. Use the Wilcoxon signed-rank test to determine whether the two populations are different.

Observation	1	2	3	4	5	6	7	8
Population 1	6.3	5.2	4.8	9.3	7.5	8.2	9.1	6.0
Population 2	5.8	5.1	4.9	7.2	6.0	8.4	8.0	5.4

Statistical Applications

21.3.13   A certain magazine is shown to 20 randomly selected married couples. Each person is asked to rate the magazine on a scale of 1 to 10. Perform the Wilcoxon signed-rank test using the accompanying data to test whether the husbands' tastes are the same as the wives'. Use a 5% level of significance.

Wife	7	2	4	3	5	9	2	1	3	2
Husband	9	6	2	8	6	8	8	6	2	6
Wife	4	5	4	3	5	6	3	6	2	1
Husband	5	6	8	9	9	5	4	9	3	7

21.3.14 Ten randomly selected companies are examined in 1990 and again in 1995 to determine whether a smaller percentage of income is being allocated to research and development (R&D). The following data show the percentage of income spent on R&D by the companies:

Company	1	2	3	4	5	6	7	8	9	10
1990	18	26	17	22	19	31	24	20	14	20
1995	11	24	19	20	22	25	17	23	13	16

Using $\alpha = .05$, test to see whether the percentages declined in 1985.

21.3.15 An advertising agency wishes to compare the opinions of seven randomly selected viewers before and after watching a TV special sponsored by a foreign car manufacturer. A series of questions is asked to measure the viewers' opinions of the firm (0 = very poor, 100 = very good). The following results were obtained:

Viewer	1	2	3	4	5	6	7
Before TV show	45	80	68	47	26	75	30
After TV show	72	81	60	58	20	87	45

Let $\alpha = .05$. Test for a significant difference between the opinions before and after the TV special.

21.4   The Kruskal–Wallis Test

In Chapter 14, one-way analysis of variance (ANOVA) was used to test the null hypothesis that J population means were equal to one another. The ANOVA test requires independent random samples from normal populations. An alternative test, the **Kruskal–Wallis test**, can be used when the normality assumption is questionable. Just as ANOVA is a generalization of a two-sample t test, the Kruskal–Wallis test is a generalization of the two-sample Mann–Whitney test.

The Kruskal–Wallis test, developed in 1952 by William H. Kruskal and W. Allen Wallis, is used to compare more than two population means or medians. The test is based on the ranks of the sample observations.

The null and alternative hypotheses are as follows:

H_0: The J population means (or medians) are all equal.
H_1: The J population means (or medians) are not all equal.

Alternatively, the Kruskal–Wallis test can be used to test

H_0: The J populations are identical.
H_1: The J populations are not identical.

The test relies on the assumption that all the population distributions have the same shape and spread.

To perform the test, obtain independent random samples from each of the J populations. Let n_j denote the size of the sample selected from the jth population. The total sample size is $n = \Sigma\, n_j$. Pool *all* n observations together and rank them from lowest to highest; assign the mean rank for tied observations. Let T_j denote the sum of the ranks of the n_j observations from population j. Let $\overline{T}_j = T_j/n_j$ denote the sample mean rank of the sample of observations from population j. If the values $\overline{T}_1, \overline{T}_2, \ldots, \overline{T}_J$ are not close to one another, the alternative hypothesis that the populations are not identical is supported.

Performing the Kruskal–Wallis test

> The Kruskal–Wallis test is used to test the null hypothesis
>
> $$H_0: \quad \text{The } J \text{ population means are equal}$$
>
> against the alternative hypothesis
>
> $$H_1: \quad \text{The } J \text{ population means are not equal}$$
>
> To perform the test, calculate the test statistic
>
> $$K = \frac{12}{n(n+1)} \left(\sum_{j=1}^{J} \frac{T_j^2}{n_j} \right) - 3(n+1)$$
>
> If H_0 is true, the sampling distribution of K is approximately chi-square with $\nu = (J - 1)$ degrees of freedom provided each sample contains five or more observations. The greater the differences in the central location of the J population distributions, the larger will be the value of K. Thus, large values of K lead us to reject the null hypothesis. The critical region is placed in the right tail of the chi-square distribution. To perform the test using a level of significance α, use the decision rule
>
> $$\text{Reject } H_0 \text{ in favor of } H_1 \text{ if } K > \chi_{\alpha,\nu}.$$

EXAMPLE 21.7 **Performing the Kruskal–Wallis Test** ————————————————

A corporation deciding which of $J = 4$ makes of automobile to order for its fleet examines independent random samples of five cars of each type. The operating cost per mile for each car was determined after each car was driven 15,000 miles. The data in Table 21.5 show operating costs per mile in cents as well as the ranks of operating costs. Use the Kruskal–Wallis test to determine whether the four distributions have equal means. Let $\alpha = .05$.

TABLE 21.5 *Operating costs and ranks of cars for Example 21.7*

CAR 1		CAR 2		CAR 3		CAR 4	
Cost	Rank	Cost	Rank	Cost	Rank	Cost	Rank
18.8¢	10	17.2¢	5	20.2¢	15	17.7¢	6
16.3	3	16.0	1	20.6	16	19.9	14
16.5	4	19.4	12	16.2	2	19.3	11
17.8	7	18.5	9	19.7	13	20.8	17
18.2	8	20.9	18	21.3	19	21.6	20
	$T_1 = 32$		$T_2 = 45$		$T_3 = 65$		$T_4 = 68$

SOLUTION   For each of the four samples, we sum the ranks and obtain $T_1 = 32$, $T_2 = 45$, $T_3 = 65$, and $T_4 = 68$. The test statistic is

$$K = \frac{12}{n(n+1)} \left(\sum_{j=1}^{J} \frac{T_j^2}{n_j} \right) - 3(n+1)$$

$$= \frac{12}{20(21)} \left(\frac{1{,}024}{5} + \frac{2{,}025}{5} + \frac{4{,}225}{5} + \frac{4{,}624}{5} \right) - 3(21)$$

$$= 4.99$$

When $\alpha = .05$, the critical value of the chi-square statistic having 3 degrees of freedom is $\chi^2_{.05,3} = 7.81$. Because K is smaller than 7.81, we do not reject the null hypothesis that the population distributions are identical.

Exercises

Statistical Concepts

21.4.1   *True or false:*   The Kruskal–Wallis test is applicable when the data occur in matched pairs. Explain.

21.4.2   *True or false:*   The Kruskal–Wallis test is designed to test whether significant differences exist when two independent random samples are obtained. Explain.

21.4.3   *True or false:*   The Kruskal–Wallis test is designed to test whether significant differences exist when more than two independent random samples are obtained. Explain.

21.4.4   *True or false:*   The Kruskal–Wallis test is used as an alternative to ANOVA when the assumption of normality is questionable. Explain.

21.4.5   When performing the Kruskal–Wallis test, what is the sampling distribution of the test statistic? What is assumed about the sample sizes?

21.4.6   *True or false:*   When performing the Kruskal–Wallis test, we rank the absolute values of the observations. Explain.

21.4.7   *True or false:*   When performing the Kruskal–Wallis test to test $J = 4$ means, the test statistic K approximately follows the chi-square distribution with 4 degrees of freedom. Explain.

Statistical Drills

21.4.8   Suppose we perform a Kruskal–Wallis test that J population means are equal. Find the critical region of the test statistic under the following conditions:
  a. $J = 5$, $\alpha = .05$ **b.** $J = 4$, $\alpha = .05$
  c. $J = 5$, $\alpha = .01$ **d.** $J = 6$, $\alpha = .01$

21.4.9   Suppose we obtain independent random samples of sizes $n_1 = 16$, $n_2 = 20$, $n_3 = 25$, $n_4 = 20$ from four independent nonnormal populations. The rank sums for each sample of data are $T_1 = 813$, $T_2 = 760$, $T_3 = 940$, $T_4 = 808$. Perform the Kruskal–Wallis test to determine whether the four populations are identical.
  a. State H_0 and H_1.
  b. Compute the Kruskal–Wallis test statistic K.
  c. What is the appropriate number of degrees of freedom of the test statistic?
  d. Let $\alpha = .05$. Do the data support the conclusion that the four populations are identical?

21.4.10   The following data represent random samples from three independent nonnormal populations. Let $\alpha = .05$. Use the Kruskal–Wallis test to determine whether the three populations are different.

Observation	1	2	3	4	5	6	7	8
Population 1	6.3	5.2	4.8	9.3	7.5	8.2	9.1	6.0
Population 2	5.8	5.1	4.9	7.2	6.0	8.4	8.0	5.4
Population 3	8.2	6.1	2.4	5.7	7.1	8.0	7.8	6.2

21.4.11 The following data represent random samples from four independent nonnormal populations. Let $\alpha = .05$. Use the Kruskal–Wallis test to determine whether the four populations are different.

Observation	1	2	3	4	5
Population 1	5.1	8.2	7.0	9.2	7.1
Population 2	9.2	6.3	8.2	8.1	6.5
Population 3	8.7	5.7	3.4	4.2	3.9
Population 4	7.0	5.1	2.8	5.7	6.3

Statistical Applications

21.4.12 Random samples of chickens were fed four different types of feed to observe the effects of the food on weight gain. A random sample of 32 chickens was obtained, and each type of feed was given to 8 chickens for an identical period of time. The weight gains of the chickens (in ounces) are shown in the accompanying table. Let $\alpha = .05$. Use the Kruskal–Wallis test to test the null hypothesis that the population mean weight gains are all equal.

Feed 1	Feed 2	Feed 3	Feed 4
13.0	14.2	16.3	19.2
15.1	13.1	18.4	17.1
17.2	11.0	20.2	16.8
11.3	10.8	19.3	18.3
14.7	8.9	17.2	19.4
18.1	14.2	14.8	21.2
16.2	12.3	15.1	15.3
15.3	11.7	14.2	17.4

21.4.13 An engineer at a manufacturing plant uses wire cord in a certain production process. He has access to four different kinds of wire cord, and he wants to determine whether the mean breaking strengths of the cords are the same. The data in the accompanying table show the force required in pounds to break each wire cord. Let $\alpha = .05$ and test whether the population means are equal using the Kruskal–Wallis test.

Wire A	Wire B	Wire C	Wire D
228	210	261	220
237	217	282	225
251	204	251	215
214	230	248	222
246	209	263	221
256		290	

21.4.14 An automotive engineer wants to test whether the mean times required to install spark plugs in new cars on an assembly line are the same for four different types of cars. The data in the accompanying table show the times (in seconds) for installation. Let $\alpha = .05$ and test whether the population means are equal using the Kruskal–Wallis test.

Car A	Car B	Car C	Car D
21.1	19.6	19.8	21.4
27.8	21.7	21.3	22.4
28.3	21.9	23.4	20.8
33.4	22.2	25.2	23.2
35.2	25.1	20.7	24.0

21.5 The Runs Test

When working with time series data, we frequently wish to test whether the order or sequence of occurrence of the observations in a sample is random. For example, suppose that on a typical day 10% of the products manufactured at a certain plant are found to be defective. If we selected products randomly and tested them, we would not expect to find any definite pattern, but if the products were selected sequentially as they came off the assembly line, we might expect a pattern to appear where good products tended to be followed by good products and defectives by defectives.

Alternatively, suppose we recorded the weights of packages after they were filled by a machine. If the process was in control, weights above the median and below the median should occur randomly. One way to detect whether the process is out of control is to observe the sequence of weights and determine whether the series of observations appears to be nonrandom. For example, if the process is out of control, we might expect above-average weights to follow above-average weights and below-average weights to follow below-average weights.

The **runs test** is designed to test whether a sequence of observations has unusual or unlikely patterns. Suppose we label each observation as a "success" or "failure." A success might denote that a product is satisfactory, the weight of a package is above the median, the diameter of a pipe is greater than the median, the income of a sample respondent is above the median, and so forth. We are interested in detecting any unusual patterns in the sequence of sample observations.

DEFINITION   **Run**
  _____

A **run** is a sequence of like observations.

  _____

For example, we might ask a sample of 20 voters whether they prefer candidate A or candidate B. If the election is expected to be very close, it would not be unusual to find that exactly 10 favor candidate A and 10 candidate B. It would be unusual, however, if the preferences were obtained in the sequence

A A A A A A A A A A B B B B B B B B B B

where the first 10 voters all favored A and the last 10 favored B. This sequence contains only two *runs,* a run of 10 A's and a run of 10 B's. Such an occurrence would lead us to suspect that the first observations were selected in a manner favorable to candidate A and the last 10 in a manner favorable to candidate B. In other words, we suspect nonrandomness because the number of runs is too small.

As another example, suppose a mechanical device is used to fill cans with oil. A sample of 20 filled cans is selected and the weight of each can is carefully measured. Denote a weight above the median by an A and a weight below the median by a B. It would seem unusual if the 20 observations occurred in the sequence

A B A B A B A B A B A B A B A B A B A B

where overfilled cans are followed by underfilled cans and vice versa. This sequence contains 20 runs. Here we suspect nonrandomness because the number of runs is too large.

To test for the existence of patterns, we use the runs test, which is based on the probability of obtaining the observed number of runs. The first step is to determine the number of runs in the data. The following example shows the number of runs in two sequences of observations.

EXAMPLE 21.8 **Number of Runs in a Sequence**

The following sequence contains two runs:

$$\underbrace{A \ \ A \ \ A \ \ A \ \ A}_{1} \quad \underbrace{B \ \ B \ \ B \ \ B \ \ B}_{2}$$

The following sequence contains five runs:

$$\underbrace{A \ \ A \ \ A}_{1} \quad \underbrace{B \ \ B}_{2} \quad \underbrace{A}_{3} \quad \underbrace{B \ \ B \ \ B}_{4} \quad \underbrace{A}_{5}$$

Let the random variable r denote the number of runs in the data. We reject the null hypothesis that the observations occur randomly if r is either very small or very large. To perform the test, we need n_A (the number of observations of type A) and n_B (the number of observations of type B). Usually the runs test for randomness is a two-tailed test. If n_A and n_B both exceed 10, the random variable r approximately follows the normal distribution with mean and variance

$$\mu_r = \frac{2n_A n_B}{n_A + n_B} + 1$$

$$\sigma_r^2 = \frac{2n_A n_B (2n_A n_B - n_A - n_B)}{(n_A + n_B)^2 (n_A + n_B - 1)}$$

The procedure in the accompanying box describes how to perform the runs test when the sample size is large.

Performing the runs test with large samples

Suppose we obtain a random sample of n observations that are ordered sequentially. We wish to test the null hypothesis

H_0: The sequence of A's and B's is random

against the alternative hypothesis

H_1:　The sequence of A's and B's is nonrandom

Let α be the level of significance. Let n_A and n_B denote the number of A's and B's, respectively. Let r_0 denote the observed number of runs. Calculate the test statistic

$$z = \frac{r_0 - \mu_r}{\sigma_r}$$

For large values of n_A and n_B, the sampling distribution of the random variable Z is approximately standard normal.

　　Case 1:　To test H_0 against the one-sided alternative that there are too many runs, use the decision rule

Reject H_0 in favor of H_1 if $z > z_\alpha$.

　　Case 2:　To test H_0 against the one-sided alternative that there are too few runs, use the decision rule

Reject H_0 in favor of H_1 if $z < -z_\alpha$.

　　Case 3:　To test H_0 against the two-sided alternative that there are either too many or too few runs, use the decision rule

Reject H_0 in favor of H_1 if $z < -z_{\alpha/2}$ or if $z > z_{\alpha/2}$.

We assume that the approximation to normality is good if $n_A > 10$ and $n_B > 10$.

EXAMPLE 21.9　**Using the Runs Test with a Large Sample**

A filling machine is set to put 16 ounces of oil into cans. Because of the viscosity of the oil, the weights differ from can to can. The weights of 50 cans of oil were obtained as they came off a filling machine. When the machine is in control, weights above (A) and below (B) the 16-ounce average occur randomly. The following sequence was observed:

A　A　B　B　B　A　B　A　A　A　B　A　B　B　B　A　A　A　A
B　B　B　B　B　B　A　A　A　B　A　B　A　B　B　B　B　A　A
A　A　B　B　B　A　B　A　B　B　B　B

Test the null hypothesis that the sequence of A's and B's is random. Use a 5% level of significance.

S O L U T I O N　There are 22 A's and 28 B's, so $n_A = 22$ and $n_B = 28$. There are $r_0 = 22$ runs. If the process is random, the random variable r is approximately normally distributed with mean

$$\mu_r = \frac{2n_A n_B}{n_A + n_B} + 1$$

$$= \frac{2(22)(28)}{22 + 28} + 1 = 25.64$$

and variance

$$\sigma_r^2 = \frac{2n_A n_B (2n_A n_B - n_A - n_B)}{(n_A + n_B)^2 (n_A + n_B - 1)}$$

$$= \frac{2(22)(28)[2(22)(28) - 22 - 28]}{(22 + 28)^2(22 + 28 - 1)} = \frac{1,456,224}{122,500} = 11.89$$

Let us use a two-tailed test to test the null hypothesis that the sequence is random against the two-sided alternative hypothesis that there are too few or too many runs. The observed test statistic is

$$z = \frac{22 - 25.64}{\sqrt{11.89}} = -1.056$$

For $\alpha = .05$, the two critical values are $\pm z_{.025} = \pm 1.96$. Since z falls in the acceptance region, we do not reject the null hypothesis that the process is random.

When n_A and n_B are 10 or less, Table A.11 in the Appendix shows critical values for r. To perform the runs test when n_A and n_B are small, use the procedure described in the accompanying box.

Performing the runs test with small samples

> We wish to test the null hypothesis
>
> $$H_0: \quad \text{The sequence of A's and B's is random}$$
>
> against the alternative hypothesis
>
> $$H_1: \quad \text{The sequence of A's and B's is nonrandom}$$
>
> Refer to Table A.11, which shows one-sided p-values for various values of r_0. To test H_0 against the one-sided alternative that there are too few runs, use Table A.11 and obtain $P(r \leq r_0)$, where r_0 is the observed number of runs. Use the decision rule
>
> $$\text{Reject } H_0 \text{ in favor of } H_1 \text{ if } P(r \leq r_0) > \alpha.$$
>
> To test H_0 against the one-sided alternative that there are too many runs, use the decision rule
>
> $$\text{Reject } H_0 \text{ in favor of } H_1 \text{ if } P(r \geq r_0) < \alpha.$$
>
> To test H_0 against the two-sided alternative that there are too many or too few runs, use the decision rule
>
> $$\text{Reject } H_0 \text{ in favor of } H_1 \text{ if } P(r \geq r_0) < \alpha/2 \text{ or if } P(r \leq r_0) > \alpha/2.$$

EXAMPLE 21.10   **Using the Runs Test with a Small Sample** ————————————

A ski resort has annual observations on total winter snowfall for 12 years. The median level of snowfall is obtained, and for each year we record whether the annual snowfall was above or below the median. The sequence of above- and below-average years was as follows:

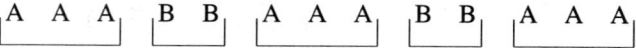

Test the null hypothesis that the sequence is random. Perform a two-tailed test with $\alpha = .05$.

SOLUTION   The numbers of A's and B's are $n_A = 8$ and $n_B = 4$. There are $r_0 = 5$ runs. To perform the test, refer to Table A.11. In the table, the value n_1 is always the smaller of n_A and n_B. For $n_1 = 4$ and $n_2 = 8$, we have $P(r \leq 5) = .279$. Because this value exceeds $\alpha/2 = .025$, we do not reject the null hypothesis that the sequence is random. The probability .279 indicates that if H_0 is true, we should observe 5 or fewer runs in a sample of 12 observations with probability .279. This p-value exceeds the level of significance of the test, so we do not reject H_0.

Exercises

Statistical Concepts

21.5.1   *True or false:*   The runs test is used to test the null hypothesis that a sequence of observations, say, successes and failures, has been produced in a random order. Explain.

21.5.2   Explain how to calculate the number of runs in a sequence of observations.

21.5.3   Explain how to perform the runs test using a large sample. What assumption is made about the sample size?

21.5.4   Explain how to perform the runs test using a small sample.

21.5.5   When performing the runs test using a large sample, what is the sampling distribution of the test statistic?

21.5.6   *True or false:*   When performing the runs test, the order in which the observations occur is of crucial importance. Explain.

Statistical Drills

21.5.7   We obtain a sample of $n = 50$ observations containing 20 successes and 30 failures. There are 17 runs in the sample. Use the runs test to test for randomness.
a. State H_0 and H_1.
b. Compute the expected number of runs.
c. Compute the standard deviation for the number of runs.
d. Compute the observed value of the test statistic z.
e. Let $\alpha = .05$. Perform a two-sided test.

21.5.8   We obtain a sample of $n = 80$ observations containing 50 successes and 30 failures. There are 37 runs in the sample. Use the runs test to test for randomness.
a. State H_0 and H_1.
b. Compute the expected number of runs.
c. Compute the standard deviation for the number of runs.
d. Compute the observed value of the test statistic z.
e. Let $\alpha = .05$. Perform a two-sided test.

21.5.9   We obtain a sample of $n = 16$ observations containing 8 successes and 8 failures. There are 11 runs in the sample. Use the runs test to test for randomness.
  a. State H_0 and H_1.   **b.** Let $\alpha = .05$. Perform a two-sided test.

Statistical Applications

21.5.10   A computer is used to generate digits from 0 to 9. The data that follow show the sequence of 40 observations:

$$8 \quad 1 \quad 5 \quad 7 \quad 1 \quad 3 \quad 7 \quad 8 \quad 6 \quad 7 \quad 2 \quad 9 \quad 3 \quad 2 \quad 9 \quad 6 \quad 5 \quad 6 \quad 6 \quad 6$$
$$5 \quad 6 \quad 1 \quad 1 \quad 3 \quad 2 \quad 3 \quad 4 \quad 7 \quad 4 \quad 8 \quad 4 \quad 3 \quad 5 \quad 6 \quad 1 \quad 5 \quad 9 \quad 3 \quad 7$$

Use the runs test and a 5% level of significance to test the following hypotheses:
  a. Values above 4.5 and values below 4.5 occur randomly.
  b. Odd values and even values occur randomly.

21.5.11   A basketball coach wants to test the hypothesis that, for a skilled basketball player, making or missing free throws occurs randomly. The alternative hypothesis is that some sort of learning process is involved, so that the runs test should show evidence of an unusual number of runs. Use a 5% level of significance. An S denotes a successful shot; an F denotes a failure, or missed shot. The results after 60 shots are as follows:

$$S \quad S \quad S \quad F \quad S \quad F \quad F \quad S \quad S \quad S \quad S \quad S \quad F \quad F \quad S \quad S \quad S \quad S \quad S \quad S$$
$$F \quad S \quad S \quad F \quad F \quad S \quad S \quad S \quad S \quad S \quad F \quad S \quad S \quad S \quad S \quad F \quad F \quad S \quad S \quad F$$
$$S \quad S \quad F \quad F \quad F \quad S \quad S \quad S \quad F \quad S \quad S \quad S \quad S \quad F \quad S \quad S \quad S \quad F \quad S \quad S$$

21.5.12   The runs test can be used instead of the Durbin–Watson test to determine whether the error terms in a regression model are serially correlated. In a regression model, positive (P) and negative (N) residuals occurred in the following sequence:

$$P \quad N \quad P \quad P \quad P \quad N \quad N \quad N \quad P \quad P \quad P \quad P \quad N \quad N \quad P \quad N \quad N \quad P \quad N \quad N \quad P \quad N \quad P \quad N \quad N \quad P \quad P \quad P$$

Let $\alpha = .05$ and test for randomness. Use a two-tailed test.
  a. How many runs are there?
  b. State the mean and variance of the random variable r.
  c. State the observed value of the test statistic z.
  d. Do you reject the null hypothesis that the sequence is random?

21.5.13   Suppose we compare the rate of return on a security to the average rate of return in the market. The random-walk hypothesis states that rates of return above and below the market average are completely random, following no pattern over time. The following data show the market-adjusted rate of return for a security for 30 consecutive days (negative values indicate below-market returns):

$$.05 \quad .07 \quad -.03 \quad -.02 \quad .06 \quad .02 \quad .04 \quad -.03 \quad .02 \quad .01 \quad -.02 \quad .03 \quad .04 \quad .02 \quad -.02$$
$$.03 \quad .01 \quad -.02 \quad -.03 \quad -.02 \quad .06 \quad -.10 \quad -.06 \quad .02 \quad .01 \quad -.03 \quad .01 \quad -.02 \quad -.02 \quad .04$$

Let $\alpha = .05$ and test for randomness. Use a two-tailed test.
  a. How many runs are there?
  b. State the mean and variance of the random variable r.
  c. State the observed value of the test statistic z.
  d. Do you reject the null hypothesis that the sequence is random?

21.6   Rank Correlation

The population correlation coefficient ρ, discussed in Chapter 15, measures whether above (or below) average values of a variable Y tend to occur with above (or below) average values of another variable X. To calculate the sample correlation coefficient R, it is necessary to know the magnitudes of the Y and X variables. At times the values of these variables are ranks rather than numerical magnitudes. The **population rank correlation coefficient**, denoted by the symbol ρ_s, measures the correlation between the ranks of two variables X and Y. To estimate the population rank correlation coefficient, we calculate the **sample rank correlation coefficient**, denoted R_s. The rank correlation coefficient was introduced by Charles E. Spearman (1863–1945), a British psychology professor, and is frequently called *Spearman's rank correlation coefficient* (hence the subscript "s").

Let x_i and y_i denote the ranks of the ith observation on variables X and Y in a sample of n observations. We would like to determine whether high (or low) rankings of Y tend to be associated with high (or low) rankings of X. The sample rank correlation coefficient R_s can be calculated just like the sample correlation coefficient R, as described in the accompanying box.

Formula for calculating the sample rank correlation coefficient

Suppose we have a sample of n pairs of ranks (x_i, y_i). Spearman's sample rank correlation coefficient can be calculated as follows:

$$R_s = \frac{\sum_{i=1}^{n}(x_i - \bar{x})(y_i - \bar{y})}{\sqrt{\sum_{i=1}^{n}(x_i - \bar{x})^2 \sum_{i=1}^{n}(y_i - \bar{y})^2}}$$

However, the sample rank correlation coefficient is usually calculated in another way that gives an equivalent value provided no two of the x_i's or y_i's are given the same rank. This alternative method is shown in the accompanying box.

Alternative formula for calculating the sample rank correlation coefficient

If there are no ties in ranks of the x_i's or y_i's, the sample rank correlation coefficient can be calculated as follows:

$$R_s = 1 - \frac{6\sum_{i=1}^{n} d_i^2}{n(n^2 - 1)}$$

where $d_i = (x_i - y_i)$ denotes the difference in ranks between observations x_i and y_i.

If $R_s = 1$, the ranks x_i and y_i are identical ($d_i = 0$ for all i). If $R_s = -1$, the ranks x_i are in exactly the opposite order of the ranks y_i. Large positive (or negative) values of R_s support the hypothesis that ρ_s is positive (or negative) in the population.

The sample rank correlation coefficient is a random variable whose value changes from sample to sample. If the population rank correlation coefficient ρ_s is 0 and the sample size is 30 or more, then the t statistic

$$t = \frac{R_s}{s_{R_s}}$$

approximately follows the Student t distribution with $\nu = (n - 2)$ degrees of freedom, where the observed sample standard deviation is

$$s_{R_s} = \sqrt{\frac{1 - R_s^2}{n - 2}}$$

The procedure described in the accompanying box shows how to test the null hypothesis that the population rank correlation coefficient is 0 when the sample size is large ($n \geqslant 30$).

Spearman's rank correlation test with large samples

Case 1: Suppose we wish to test the null hypothesis

$$H_0: \quad \rho_s = 0$$

against the one-sided alternative hypothesis

$$H_1: \quad \rho_s > 0$$

using level of significance α. Calculate the test statistic

$$t = \frac{R_s}{s_{R_s}}$$

If H_0 is true and n is 30 or larger, this t statistic is approximately distributed as Student's t with $\nu = (n - 2)$ degrees of freedom. To test H_0 against H_1, calculate the test statistic t and use the decision rule

Reject H_0 in favor of H_1 if $t > t_{\alpha,\nu}$.

Case 2: To test the null hypothesis

$$H_0: \quad \rho_s = 0$$

against the one-sided alternative hypothesis

$$H_1: \quad \rho_s < 0$$

use the decision rule

Reject H_0 in favor of H_1 if $t < -t_{\alpha,\nu}$.

Case 3: To test the null hypothesis

$$H_0: \quad \rho_s = 0$$

against the two-sided alternative hypothesis

$$H_1: \quad \rho_s \neq 0$$

use the decision rule

Reject H_0 in favor of H_1 if $t < -t_{\alpha/2,\nu}$ or if $t > t_{\alpha/2,\nu}$.

EXAMPLE 21.11

Testing for Rank Correlation with a Large Sample

A corporation uses private interviews and written tests to determine which job candidates to hire. Because interviews are much more expensive than written tests, the corporation would like to discontinue interviewing if the rankings of job candidates based on the written test are highly correlated with rankings based on personal interviews.

A random sample of 62 job candidates were ranked on an interview and again on a written test. The sample rank correlation between the two rankings is $R_s = .20$. Using $\alpha = .05$, test the null hypothesis

$$H_0: \quad \rho_s = 0$$

against the one-sided alternative hypothesis

$$H_1: \quad \rho_s > 0$$

SOLUTION The t statistic has $\nu = n - 2 = 60$ degrees of freedom. The critical value of t is $t_{.05,60} = 1.671$, and the observed test statistic is

$$t = \frac{.20}{\sqrt{\dfrac{1 - .20^2}{62 - 2}}} = 1.58$$

Because the observed test statistic $t = 1.58$ does not exceed the critical value $t_{.05,60} = 1.671$, we do not reject H_0.

If the sample size is smaller than 30, the t distribution does not generally serve as a useful approximation to the distribution of the statistic t. Instead, to test the null hypothesis that the population rank correlation coefficient is 0, refer to Table A.12, which shows the critical values of R_s for a one-tailed test where α has the values .05, .025, .01, and .005. The table shows the critical values $R_{s,\alpha}$ such that $P(R_s \geq R_{s,\alpha}) = \alpha$.

Spearman's rank correlation test with small samples

Case 1: Suppose we wish to test the null hypothesis

$$H_0: \quad \rho_s = 0$$

against the one-sided alternative hypothesis

$$H_1: \quad \rho_s > 0$$

using level of significance α. If $n \leq 30$, use the decision rule

Reject H_0 in favor of H_1 if $R_s > R_{s,\alpha}$.

Case 2: To test the null hypothesis

$$H_0: \quad \rho_s = 0$$

against the one-sided alternative hypothesis

$$H_1: \quad \rho_s < 0$$

use the decision rule

Reject H_0 in favor of H_1 if $R_s < -R_{s,\alpha}$.

Case 3: To test the null hypothesis

$$H_0: \quad \rho_s = 0$$

against the two-sided alternative hypothesis

$$H_1: \quad \rho_s \neq 0$$

use the decision rule

Reject H_0 in favor of H_1 if $R_s < -R_{s,\alpha/2}$ or if $R_s > R_{s,\alpha/2}$.

EXAMPLE 21.12   **Testing for Rank Correlation with a Small Sample** ────────────

Suppose we want to test the null hypothesis that the rankings given to 15 colleges by the general public are uncorrelated with the rankings given by college professors. Test $H_0: \rho_s = 0$ against $H_1: \rho_s > 0$ using $\alpha = .05$. The data are in Table 21.6.

TABLE 21.6   *Ratings of colleges for Example 21.12*

College	Ranking by public	Ranking by professors	d_i	d_i^2
1	4	5	−1	1
2	12	10	2	4
3	14	3	11	121
4	2	13	−11	121
5	1	2	−1	1
6	11	9	2	4
7	3	1	2	4
8	10	8	2	4
9	5	4	1	1
10	13	11	2	4
11	9	6	3	9
12	6	7	−1	1
13	15	15	0	0
14	7	12	−5	25
15	8	14	−6	36
				336

SOLUTION   Refer to Table A.12 to obtain the critical value of the test statistic. For $n = 15$ and $\alpha = .05$, we obtain the critical value $R_{s,.05} = .441$. The value of the sample rank correlation coefficient is

$$R_s = 1 - \frac{6 \sum\limits_{i=1}^{n} d_i^2}{n(n^2 - 1)}$$

$$= 1 - \frac{6(336)}{15(15^2 - 1)}$$

$$= 1 - .60$$

$$= .40$$

Because the calculated value $R_s = .40$ does not exceed the critical value, we do not reject the null hypothesis that $\rho_s = 0$.

Exercises

Statistical Concepts

21.6.1   Explain how the rank correlation coefficient R_s differs from the sample correlation coefficient R.

21.6.2   Let us compare the standard correlation coefficient ρ with the rank correlation coefficient ρ_s.
  a. *True or false:*   If $\rho = 1$, then $\rho_s = 1$. Explain your answer.
  b. *True or false:*   If $\rho_s = 1$, then $\rho = 1$. Explain your answer.

Statistical Drills

21.6.3   Suppose we have obtained a random sample of n pairs of observations of variables (x_i, y_i). Suppose the values of x_i and y_i have been ranked from lowest to highest, and we want to test whether the ranks of x_i tend to be positively related to the ranks of y_i.
  a. Suppose $n = 20$ and $\alpha = .05$. What is the critical value for Spearman's rank correlation coefficient?
  b. Suppose $n = 10$ and $\alpha = .01$. What is the critical value for Spearman's rank correlation coefficient?

21.6.4   In a random sample of $n = 15$ pairs of observations, the rank correlation coefficient was $R_s = .46$. Let $\alpha = .05$. Test the null hypothesis that the population correlation coefficient is 0 against the alternative hypothesis that the correlation coefficient is positive.

Statistical Applications

21.6.5   Gina Aldisert is an analyst for a chain of major appliance stores. She thinks that there is a positive association between people's preferences for washing machines and the manufacturer's suggested retail price. A potential customer is asked to rank six brands of washing machines. The accompanying data show the preference rankings along with the price. Do the data suggest an agreement between the person's preference rankings and the manufacturer's suggested retail prices?

Machine	1	2	3	4	5	6
Preference	2	3	1	4	6	5
Price ($)	386.99	415.99	420.00	345.25	338.80	368.95

     a. Obtain a listing of the ranks of the prices.

     b. Calculate R_s.

     c. Let $\alpha = .05$. Perform a one-tailed test. Does there appear to be a correlation between preferences and prices?

21.6.6 Ed Chidiac and Bryce Arendt are corporate credit officers for a major national bank. They provide advice concerning whether the bank should make loans to various medium-size businesses. Ed and Bryce work independently of one another and make separate reports concerning whether funds should be made available to various companies. Bill Weil is the manager of the credit policy division and is in charge of training all the employees of the division. Bill hopes that, after sufficient training, all of his employees will make the same decision when reviewing a loan. In addition, Ed would prefer that each employee rate companies the same way he does. Of course, complete unanimity is not always possible. For example, sometimes Ed or Bryce might recommend making a loan that Bill thinks is too risky. Occasionally, the opposite occurs. Bill decides to perform a small experiment to determine whether the rankings made by Ed and Bryce are positively correlated with his own rankings. A set of eight companies, which had applied for millions of dollars worth of loans, was analyzed. The accompanying data show the preference rankings for Ed, Bryce, and Bill.

Company	1	2	3	4	5	6	7	8
Ed's ranking	3	2	4	1	5	6	8	7
Bryce's ranking	4	2	1	5	6	3	7	8
Bill's ranking	2	3	1	4	6	5	8	7

     a. Let $\alpha = .05$. Do the data suggest an agreement between Bill's preference rankings and Ed's preference rankings?

     a. Let $\alpha = .05$. Do the data suggest an agreement between Bill's preference rankings and Bryce's preference rankings?

21.6.7 The following data show the annual salaries in thousands of dollars Y for a sample of company presidents and the gross sales in millions of dollars X of their companies.

Company	1	2	3	4	5	6	7	8	9	10	11	12
Y (in thousands)	$120	132	98	86	145	210	450	104	80	80	180	150
X (in millions)	$103	420	312	798	203	154	210	306	161	410	120	556

     a. Obtain the ranks of the two variables.

     b. Calculate the sample rank correlation coefficient R_s.

     c. Let $\alpha = .05$. Test $H_0: \rho_s = 0$ against $H_1: \rho_s > 0$.

     d. Calculate the sample correlation coefficient R. Is there much difference between R and R_s?

21.6.8 A political scientist wanted to examine the relationship between the rate of economic growth of a country and the stability of its government. Each of 10 countries was ranked on these two variables. Calculate R_s and test $H_0: \rho_s = 0$ using a 5% level of significance.

Country	A	B	C	D	E	F	G	H	I	J
Rank by Growth	2	6	1	8	3	5	10	4	9	7
Rank by Stability	1	4	3	6	2	10	8	5	9	7

21.6.9 An executive determines which college graduates to hire based on the students' grades on an achievement test and an interview. Because interviewing the students is very expensive, the executive wonders whether she could eliminate the interview and still choose the best candidates. The data for 12 candidates follow:

Candidate	1	2	3	4	5	6	7	8	9	10	11	12
Test Score	88	64	95	86	80	70	66	70	75	60	65	84
Interview Rank	3	10	1	2	6	7	8	9	4	12	11	5

a. Use the accompanying data and calculate the rank correlation coefficient R_s.
b. Test $H_0: \rho_s = 0$ using a 5% level of significance.

21.7   Computer Applications

Many statistical computer programs, including SPSSX, SAS, and Minitab, can perform the statistical tests described in this chapter. In this section, we show how to use the SPSSX program to perform these tests.

Mann–Whitney Test

The SPSSX computer program can be used to perform the Mann–Whitney test. The use of a computer is especially helpful when the sample sizes are large. The following example shows how to perform the Mann–Whitney test using the SPSSX computer program.

EXAMPLE 21.13   **Using SPSSX to Perform the Mann–Whitney Test** ——————————

Nutrition is thought to be an important component in cancer development in humans. For example, there is evidence that the per capita consumption of dietary fats is positively correlated with the incidence of colon cancer. Medical studies have examined the relationship between diet and tumor development in rats. One hypothesis of interest is whether the length of time until a tumor develops in rats fed diets high in saturated fats differs from that in rats fed diets high in unsaturated fats. If it is reasonable to assume that the time for tumor development is normally distributed, the two-sample t test can be used to test the null hypothesis that the population means are equal. Otherwise, the Mann–Whitney test should be used.

Table 21.7 shows the number of days until tumor development for a random sample of 20 rats fed a diet high in saturated fats and an independent random sample of 20 rats fed a diet high in unsaturated fats. Let us test the null hypothesis that the population means are equal by using the Mann–Whitney test and the SPSSX computer program.

SOLUTION   To perform the test using the SPSSX program, issue the command

```
NPAR TESTS M–W = TIME BY DIET(0,1)
```

This command tells the computer to perform a nonparametric Mann–Whitney test. The variable TIME measures the number of days until tumor development, and the variable DIET was recorded as 0 for rats on diets high in saturated fats and 1 for rats on diets high in unsaturated fats.

TABLE 21.7 *Number of days to tumor development in rats for Example 21.13*

Saturated diet		Unsaturated diet	
84	102	121	107
116	112	80	93
90	77	60	65
74	83	81	74
83	90	78	85
124	94	54	107
161	104	62	67
114	109	73	91
85	83	91	84
100	117	111	87

Figure 21.1 shows the output generated by the SPSSX program. The output shows that the mean rank is 24.98 for the rats fed saturated fats and 16.02 for the rats fed unsaturated fats. The U statistic has the value 110.5, and the value of the test statistic Z is $z = -2.4221$. This value of Z has a two-tailed p-value of .0154. If we were using a 5% level of significance, we would reject the null hypothesis that the two population means are equal.

```
- - - - - Mann-Whitney U - Wilcoxon Rank Sum W Test

      TIME         TIME ELAPSED IN DAYS
    by DIET        TYPE OF FOOD


    Mean Rank     Cases

       24.98        20   DIET = 0   SATURATED FATS
       16.02        20   DIET = 1   UNSATURATED FATS
                    --
                    40   Total

                                    Exact        Corrected for ties
        U             W           2-Tailed P       Z      2-Tailed P
      110.5         499.5          .0143        -2.4221      .0154
```

FIGURE 21.1 *SPSSX-generated output for the Mann–Whitney test for Example 21.13*

Wilcoxon Signed-Rank Test

The SPSSX computer program can be used to perform the Wilcoxon test, as shown in the following example.

EXAMPLE 21.14

Using SPSSX to Perform the Wilcoxon Signed-Rank Test

Every Monday morning for 20 weeks, two stockbrokers independently invested $1,000 in mutual funds. Every Friday afternoon, each broker recorded the profit earned during the week measured as the difference between the price on Friday afternoon and the price on Monday morning, as shown in Table 21.8. Let us use the Wilcoxon signed-rank test to test the null hypothesis that the average returns earned by the two stockbrokers are the same.

TABLE 21.8   *Weekly profits for Example 21.14*

Week	Broker 1	Broker 2	Week	Broker 1	Broker 2
1	5	7	11	8	4
2	4	7	12	6	5
3	6	5	13	10	11
4	11	8	14	14	12
5	2	4	15	12	10
6	8	4	16	7	8
7	3	7	17	9	12
8	6	7	18	6	8
9	9	12	19	8	6
10	5	4	20	5	7

SOLUTION   To perform the test, issue the command

NPAR TESTS WILCOXON = S1 WITH S2

In the computer statement, the variable S1 measures the weekly profit earned by the first stockbroker; the variable S2, that by the second. The computer output in Figure 21.2 shows that the mean rank of weekly profits was 10.44 for the first stockbroker and 10.55 for the second. For the Wilcoxon test, the Z score is $z = -.4107$. The two-tailed p-value associated with this value is .6813. If we tested the null hypothesis at a 5% level of significance, we would not reject the null hypothesis that the weekly profits earned by the two brokers have the same mean.

```
- - - - - Wilcoxon Matched-Pairs Signed-Ranks Test

      S1          STOCK 1
with S2           STOCK 2


   Mean Rank      Cases

      10.44          9   - Ranks (S2 LT S1)
      10.55         11   + Ranks (S2 GT S1)
                     0     Ties (S2 EQ S1)
                    --
                    20     Total

      Z =    -.4107              2-Tailed P =   .6813
```

FIGURE 21.2   *SPSSX-generated output for Wilcoxon signed-rank test for Example 21.14*

Kruskal–Wallis Test

The SPSSX computer program can be used to perform the Kruskal–Wallis test, as shown in the next example.

EXAMPLE 21.15   **Using SPSSX to Perform the Kruskal–Wallis Test**

Tourism is the largest revenue producer in the state of Colorado. The state intends to begin an advertising campaign to recruit more tourists, and it wants to spend its advertising dollars in states whose residents are most likely to visit Colorado. Before deciding where to advertise, government officials commissioned a statistical study to determine where most of Colorado's tourists come from and how much they spend in Colorado. As part of the study, the officials want to determine whether the average expenditures of tourists vary with their home state. The data in Table 21.9 show the total expenditures in Colorado by samples of tourists from four states. Let us use the Kruskal–Wallis test to test the null hypothesis that the population means are equal for the four states.

TABLE 21.9 *Expenditures in Colorado by home state of tourists for Example 21.15*

California	Texas	Illinois	Nevada
$564	$ 812	$550	$620
581	800	480	840
643	690	690	790
674	724	650	890
697	543	670	910
417	1,060	820	940
806	840	740	860
392	680	760	880
	700		780
	1,110		

SOLUTION The SPSSX output shown in Figure 21.3 was generated by executing the SPSSX computer statement

NPAR TESTS K–W = EXPENSE BY STATE(1,4)

In the printout, the variable EXPENSE measures the amount of tourist expenditures and the variable STATE was recorded as 1 for California, 2 for Texas, 3 for Illinois, and 4 for Nevada. The output shows the mean rank for each group, the number of observations in each group, the chi-square statistic uncorrected for ties, and the chi-square statistic and

```
- - - - - Kruskal-Wallis 1-Way Anova

    EXPENSE     AVERAGE AMOUNT SPENT
  by STATE      TOURISTS STATE OF RESIDENCE

  Mean Rank    Cases

       9.63        8    STATE = 1    CALIFORNIA
      21.10       10    STATE = 2    TEXAS
      13.56        8    STATE = 3    ILLINOIS
      25.94        9    STATE = 4    NEVADA

                 --

                 35    Total

                                        Corrected for ties
      Cases    Chi-Square  Significance   Chi-Square  Significance
        35       13.1694       .0043        13.1731       .0043
```

FIGURE 21.3 *SPSSX-generated output for Kruskal–Wallis test for Example 21.15*

its significance level (or *p*-value) corrected for ties. The corrected chi-square statistic is 13.1731, which has a *p*-value of .0043. If we tested the null hypothesis that the population means are equal using a 1% level of significance, we would reject the null hypothesis that the four population means are equal.

Runs Test

The SPSSX computer program can be used to perform the runs test, as shown in the next example.

EXAMPLE 21.16   **Using SPSSX to Perform the Runs Test**

The Huwalt Machine Tool Company produces pipes that are used in the engines of nuclear-powered submarines. Following are the diameters of a random sample of 50 pipes:

.617	.610	.619	.614	.609	.609	.617	.613
.614	.609	.615	.616	.608	.607	.608	.614
.621	.618	.622	.615	.611	.622	.616	.615
.621	.614	.613	.619	.606	.616	.618	.616
.617	.614	.614	.618	.613	.610	.613	.614
.617	.618	.607	.621	.616	.612	.613	.614
.614	.611						

When the production process is operating correctly, the diameters of successive pipes are independent of one another; when it is not, pipes with diameters above (below) average tend to follow other pipes with diameters that are above (below) average. That is, large pipes tend to follow large pipes, and small pipes tend to follow small pipes. When the production process is not operating correctly, costly adjustments have to be made. Use the runs test to test the null hypothesis that successive observations are independent of one another.

SOLUTION   The SPSSX computer output shown in Figure 21.4 was generated by executing the SPSSX computer statement

---
NPAR TESTS RUNS (MEAN) = D

In the printout, the variable D represents the diameter of the pipe. The code (MEAN) indicates that observations should be recorded as above or below average according to whether they do or do not exceed the sample mean. The figure shows that the mean

```
- - - - - Runs Test

    D           DIAMETER OF PIPE

        Runs:   22              Test value =  .61 (Mean)

        Cases:  27    LT Mean
                23    GE Mean              Z = -1.1046
                --
                50    Total      2-Tailed P =   .2693
```

FIGURE 21.4   *SPSSX-generated output for the runs test for Example 21.16*

diameter is $\bar{X} = .61$. The Z score is $z = -1.1046$, and the two-tailed p-value associated with this score is .2693. If we were testing the null hypothesis at a 5% level of significance, we would not reject the null hypothesis that successive diameters are independent of one another.

Chapter 21 Supplementary Exercises

21.S.1 A political scientist wishes to determine whether the political preference of homeowners is independent of that of their next-door neighbors. A sequence of 30 adjacent homeowners were interviewed, and R was recorded for Republicans and D for Democrats. The results are as follows:

R D R R R D D D D D R R R R R D D D D D R D R D D D R R D D

   a. Explain how the runs test can be used for this problem.
   b. State the null and alternative hypotheses.
   c. Complete the test using $\alpha = .05$.

21.S.2 A random sample of 28 pipes is selected from the output of a pipe-making machine, and the diameter of each pipe is recorded. Those with diameters smaller than the median are marked S, and those with diameters larger than the median are marked L. The data are as follows:

S L L L S S L L L S S L L L L L L L S S S S L S L S S S

   a. Explain how the runs test can be used to test whether the sequence of large and small diameters is random.
   b. State the null and alternative hypotheses.
   c. Complete the test using $\alpha = .05$.

21.S.3 The general manager of a chain of 30 food stores is interested in determining whether containers used for storing food in home freezers will sell better when displayed next to frozen foods or paper goods. The containers are displayed with frozen foods for a week and then with paper products for a week. The numbers of items sold are shown in the accompanying table.

Store	Frozen foods	Paper goods	Store	Frozen foods	Paper goods
1	40	60	16	20	29
2	75	40	17	49	60
3	25	28	18	32	22
4	20	30	19	15	32
5	9	19	20	80	121
6	11	27	21	17	11
7	16	37	22	26	19
8	28	28	23	33	27
9	20	33	24	19	23
10	16	19	25	27	29
11	14	19	26	44	40
12	41	60	27	52	46
13	14	19	28	17	27
14	7	12	29	29	33
15	28	20	30	32	19

    a. Let $\alpha = .05$ and use the sign test to determine whether location affects sales.

    b. Use the Wilcoxon test to determine whether location affects sales.

21.S.4 Two mixes, A and B, are used to make concrete beams. A sample of 12 beams made from mix A has an average strength of 5,094 psi (pounds per square inch), and a sample of 10 beams made from mix B has an average strength of 5,745 psi. The following data show the strength of the concrete beams (in psi):

Mix A		Mix B	
5,050	5,120	4,280	6,000
6,120	4,900	5,920	4,320
5,000	5,210	5,500	7,100
4,650	5,020	4,988	6,040
5,100	5,041	6,700	6,602
4,800	5,117		

Let $\alpha = .05$ and test whether mix B is better than mix A.

    a. Use the standard t test based on the assumption that the samples are independent and are obtained from normal populations.

    b. Use the Mann–Whitney test.

21.S.5 A university owns or leases several hundred cars for use on official business. A random sample of 10 repair bills for car 1 and 15 repair bills for car 2 are shown in the accompanying table. Let $\alpha = .05$. Use the Mann–Whitney test to determine whether one make is more expensive to repair than the other.

Car 1		Car 2		
$27.00	$106.81	$145.00	$25.00	$18.20
52.61	97.32	107.00	52.05	35.70
77.08	35.18	14.95	81.75	46.00
66.90	212.17	28.00	19.60	62.00
45.12	96.00	32.00	27.33	20.00

 21.S.6 A tax consultant is studying the ratio of assessed value to sales value for properties in two sections of a city. A sample of 11 sales from section 1 and 13 sales from section 2 is shown in the accompanying data. Let $\alpha = .05$ and test whether the assessment ratios are the same in the two sections of town.

Section 1		Section 2	
.54	.62	.45	.57
.66	.60	.49	.62
.48	.61	.60	.64
.71	.68	.58	.56
.49	.70	.59	.61
.56		.66	.67
		.47	

21.S.7 A food processor undertook a study to compare consumer opinions of two brands of catsup. In a random sample of 30 consumers, each tasted both brands and gave a rating from 0 to 10 (a rating of 10 signifying the best catsup). Let $\alpha = .05$. Do the following data indicate that one brand is preferred to the other?

Consumer	Catsup A	Catsup B	Consumer	Catsup A	Catsup B
1	6	8	16	5	5
2	5	7	17	7	6
3	8	6	18	8	9
4	4	5	19	6	5
5	7	9	20	1	3
6	9	2	21	5	8
7	1	9	22	7	6
8	2	4	23	2	5
9	4	8	24	1	9
10	9	7	25	5	2
11	8	9	26	9	8
12	6	6	27	6	5
13	1	4	28	4	7
14	5	3	29	8	6
15	3	7	30	1	4

21.S.8 Two brands of automobile tires were placed on the rear wheels of 10 cars. The number of miles in thousands before tire failure was recorded for each tire. The data are as follows:

Car	1	2	3	4	5	6	7	8	9	10
Tire 1	22	33	30	41	36	29	35	40	37	32
Tire 2	33	20	19	27	31	26	36	22	27	33

Let $\alpha = .05$.
a. Use the t test to determine whether the tires are equally good.
b. Use the sign test to test whether the tires are equally good.
c. Use the Wilcoxon signed-rank test to test whether the tires are equally good.

21.S.9 A corporation selects its executives from its sales force (S) or from its research department (R). During the past 4 years, 10 executives have been selected. The following sequence shows the order in which the 10 executives were selected:

R R S R S S S S R R

Does this sequence imply that the order of selection was nonrandom?

21.S.10 Two different brands of shingles were used to cover houses in a certain neighborhood. The following data show the life in years of service before the shingles had to be replaced:

Brand A   20   23   27   24   29   26   30   20   18
Brand B   16   20   26   18   27   19   22   18   20

Let $\alpha = .05$. Determine whether the shingles are equally good using the following tests:
a. t test   **b.** Mann–Whitney test

21.S.11   The advertising agent for a mail order company conducted an experiment to determine whether multicolored advertisements were more effective than black-and-white advertisements containing bonus coupons. In each of 10 cities, 500 households received a color ad and 500 received a black-and-white ad. The number of orders placed by the recipients of these ads is shown in the accompanying data. Let $\alpha = .05$. Use the following tests to determine whether the ads were equally effective:

a. Sign test

b. Wilcoxon signed-rank test

City	Color	Black and white
Cincinnati	106	97
Detroit	89	92
Boston	122	104
Chicago	144	112
Miami	85	94
Dallas	180	130
Houston	195	175
Atlanta	148	150
Denver	135	108
Pittsburgh	110	114

21.S.12   The cost of land per acre in thousands of dollars for land in two neighboring towns follows:

Town 1   42   47   46   45   38   40   51   48   49
Town 2   37   42   39   44   38   43

Let $\alpha = .05$. Use the following tests to test the null hypothesis that land prices are equal in the two towns:

a. Wilcoxon signed-rank test   **b.** *t* test

21.S.13   A securities analyst wants to investigate the association between the profitability of companies and their liquidity. He ranked 12 firms in an industry according to their profitability and liquidity as shown in the following data:

Firm	1	2	3	4	5	6	7	8	9	10	11	12
Profitability	8	1	9	3	10	12	2	11	7	5	4	6
Liquidity	4	5	6	1	9	2	12	10	3	7	11	8

a. Calculate the rank correlation coefficient using the ranks in the given data.

b. Test whether the correlation is nonzero using $\alpha = .05$.

21.S.14   Two loan officers rank their bank's loan applications according to their desirability, as shown in the following data:

Loan	1	2	3	4	5	6	7	8	9	10
Officer 1	2	3	10	4	5	9	1	6	8	7
Officer 2	1	2	6	3	5	10	4	7	9	8

a. Calculate the rank correlation coefficient.

b. Let $\alpha = .05$. Determine whether the correlation is significantly different from 0.

21.S.15 A leading toothpaste manufacturer advertises that in a recent medical study, 70% of the people tested had brighter teeth after using its Very Bright toothpaste than after using Brand X. The data upon which these statements were based were collected from a random sample of 10 individuals in an experiment where each individual used both toothpastes. Half of the individuals used Brand X for 3 weeks and then Very Bright for the same time period; the other half used Very Bright first and then Brand X. A brightness test was given at the end of each 3-week period. Thus, there were two scores for each individual, one following the use of Brand X and one following the use of Very Bright. The following are the scores (the higher, the brighter):

Subject	1	2	3	4	5	6	7	8	9	10
Very Bright	5	4	4	2	3	4	1	3	6	6
Brand X	4	3	2	3	1	1	3	4	5	4

a. Test the manufacturer's claim. What is the alternative hypothesis? What is the null hypothesis?
b. Using $\alpha = .05$, what do you conclude?
c. What error may you be making in your answer to part **b**?
d. Does the advertising seem misleading?

21.S.16 Lois White is a quality control engineer for a major Midwest chemical firm. The firm packages various products in plastic containers. For a certain product, the advertised volume is stated as 350 fluid ounces. To avoid shortchanging the customer, the firm inserts a mean of 352 ounces of material in the containers. Lois knows that, when a filling machine is not properly cleaned and oiled, it tends to get sticky and can either overfill or underfill the container. Testing for runs is one way of determining whether a filling machine is operating correctly. A sequence of containers was examined as they came off the production line. The volumes were determined with great accuracy; a recording was made showing whether the container contained more (M) or less (L) than 32 fluid ounces. The following data show the sequential results:

M L L M M M L L M L L L M M M L L M L M L L L M M M L L M M
L L M M M L L M L M L M M M L L L M L M M L L M M M M L L M L

Let $\alpha = .05$. Test the null hypothesis that the sequence is random.

References

Conover, W. J. *Practical Nonparametric Statistics.* 2d ed. New York: Wiley, 1980.

Gujarati, D. *Basic Econometrics.* 2d ed. New York: McGraw-Hill, 1988.

Kendall, M. G., and A. Stuart. *The Advanced Theory of Statistics.* Vol. 2, 4th ed. New York: Hafner Press, 1979.

Mann, H. B., and R. Whitney. "On a Test of Whether One of Two Random Variables Is Stochastically Larger Than the Other." *Annals of Mathematical Statistics* 18 (1947): 50–60.

Nie, Norman E., C. Hadlai Hull, Jean G. Jenkins, Karin Steinbrenner, and Dale H. Bent. *SPSS Statistical Package for the Social Sciences.* 2d ed. New York: McGraw-Hill, 1975.

Norusis, Marija J. *SPSSX Introductory Statistics Guide.* New York: McGraw-Hill, 1990.

Norusis, Marija J. *SPSSX Advanced Statistics Guide.* Chicago: SPSS, 1990.

Norusis, Marija J. *The SPSS Guide to Data Analysis.* Chicago: SPSS, 1986.

Ryan, Thomas A., Brian L. Joiner, and Barbara F. Ryan. *Minitab Handbook.* 2d ed. Boston: PWS-Kent, 1985.

Ryan, Thomas A., Brian L. Joiner, and Barbara F. Ryan. *Minitab Reference Manual.* University Park, Penn.: Minitab Project, 1985.

SAS Introductory Guide. 3d ed. Cary, N.C.: SAS Institute, 1985.

SAS Procedures Guide for Personal Computers. Version 6 ed. Cary, N.C.: SAS Institute, 1986.

SAS Statistics Guide for Personal Computers. Version 6 ed. Cary, N.C.: SAS Institute, 1986.

SAS User's Guide: Basics. Version 5 ed. Cary, N.C.: SAS Institute, 1985.

SAS User's Guide: Statistics. Version 5 ed. Cary, N.C.: SAS Institute, 1985.

Siegel, S. *Nonparametric Statistics for the Behavioral Sciences.* New York: McGraw-Hill, 1956.

Spearman, Charles E. "The Proof and Measurement of Association Between Two Things." *American Journal of Psychology* 15 (1904): 72–101.

SPSSX User's Guide. Chicago: SPSS, 1988.

Wald, A., and J. Wolfowitz. "On a Test of Whether Two Samples Are from the Same Population." *Annals of Mathematical Statistics* 2 (1940): 146–162.

Wilcoxon, Frank. "Individual Comparisons by Ranking Methods." *Biometrics Bulletin* 1, no. 6 (1945): 80–83.

22 Introduction to Statistical Decision Theory

Chapter 11 covered the classical theory of hypothesis testing. We saw that, at the conclusion of a test, we must make a statistical decision either to reject or not to reject the null hypothesis. Classical hypothesis testing tells us how to make statistical decisions in problems in which the goodness of a test is determined by α and β, the probabilities of making Type I and Type II errors. The classical procedure discussed in Chapter 11, however, does not take into account the costs and benefits associated with different decisions and different actions.

Before taking any action, a decision maker should consider what the potential costs or benefits from any action might be. If incorrectly rejecting a true null hypothesis would be a very costly mistake, we might require a great deal of evidence supporting the alternative hypothesis before we would reject the null hypothesis. On the other hand, if a Type I error is relatively inexpensive, we might reject the null hypothesis on the basis of relatively weak contradictory evidence. In this chapter we show how to account for these economic consequences.

22.1 Payoff Tables and Opportunity Loss Tables

The process of decision making involves selecting a single action from a set of several possible actions. First, however, we have to specify what actions are available. Examples of types of actions (or choices) include the following:

1. Which stocks, bonds, and mutual funds to invest in
2. Whether to build a manufacturing plant of small, medium, or large size
3. Whether to market a new product
4. Which method to use for producing a product

We will denote different actions as a_1, a_2, and so on. Associated with each action is a *payoff*, or consequence, resulting from that particular act.

Notation for actions

> Possible **actions** are labeled a_1, a_2, . . . , a_n and are assumed to form a mutually exclusive, collectively exhaustive set.

States of Nature

Generally, we do not know which action will be best unless we know with certainty the other factors that affect the payoff associated with each action. In decision making under uncertainty, the payoff from any action depends on factors beyond our control, which are called **states of nature**. These states of nature are denoted as s_1, s_2, and so on.

Notations for states of nature

> The **states of nature** in a decision problem are denoted s_1, s_2, . . . , s_k. No two states of nature can be in effect at the same time, and all possible states of nature must be included in the analysis.

In decision making under certainty, we know which state will occur, and we evaluate the possible payoffs for each possible action. Then we select the best action given the possible payoffs. In decision making under uncertainty, we do not know which particular outcome will follow from an action because the state of nature is unknown at the time the action is taken. These ideas are illustrated in the following example.

EXAMPLE 22.1   **Decision Making Under Uncertainty** ────────────────────

A farmer must decide which of two crops to plant. The possible actions are

$$a_1 = \{\text{Plant crop A}\}$$
$$a_2 = \{\text{Plant crop B}\}$$

The farmer's profits or losses depend not only on which crop is planted, but also on the weather, the crop's selling price at harvest time, labor costs, and so forth. Thus, the states of nature include all the factors affecting the farmer's final profit. The farmer cannot predict profit with certainty, because the eventual states of nature are unknown at the time the decision is made about which crop to plant.

───

In many cases, the number of possible states of nature affecting the payoff can be so large that the analysis of the problem is almost impossible. Under such circumstances, we have to reduce the number of possible states of nature to a workable number.

Payoffs and Payoff Tables

We assume that the payoff associated with taking a particular action when a particular state of nature occurs is known. In the real world, this assumption is often unrealistic. Usually many different states of nature are possible, each yielding a different set of payoffs for the various possible actions. In economics, it is customary to use the symbol Π to denote the payoff, or profit, from an action.

Notation for payoffs

> The symbol Π_{ij} denotes the **payoff** received when we take action a_i and state of nature s_j occurs. The payoffs may be positive, negative, or 0.

DEFINITION   **Payoff Table** ───

A **payoff table** shows the payoff Π_{ij} associated with each possible action a_i and each possible state of nature s_j.

───

A general payoff table is shown in Table 22.1. The payoff Π_{ij} is in row i and column j of the table, where rows indicate the possible actions and columns indicate the different states of nature. For example, Π_{32} represents the payoff when action a_3 is taken and state of nature s_2 occurs; it is found at the intersection of row 3 and column 2.

TABLE 22.1   *A payoff table*

	STATE OF NATURE			
Action	s_1	s_2	$\ldots$	s_k
a_1	Π_{11}	Π_{12}	$\ldots$	Π_{1k}
a_2	Π_{21}	Π_{22}	$\ldots$	Π_{2k}
$\vdots$	$\vdots$	$\vdots$	$\ddots$	$\vdots$
a_n	Π_{n1}	Π_{n2}	$\ldots$	Π_{nk}

EXAMPLE 22.2   **Decision Making Under Uncertainty** ─────────────────────────────

You may buy 100 shares of one of four different stocks, each of which is currently selling for $10 per share. None of the stocks will pay dividends next year. You want to buy the stock that will increase the most in price next year. Find the optimal action.

SOLUTION   This problem involves decision making under uncertainty because you do not know the value of the stocks next year. Prices depend on the general state of the economy, interest rates, inflation rates, and so forth. There are many possible states of nature, each yielding a different set of payoffs.

You think stock prices in 1 year depend on whether the inflation rate increases, is constant, or decreases (states of nature s_1, s_2, and s_3, respectively). For s_1, you think that the respective prices of the four stocks a year from now will be $5, $10, $15, and $20; for s_2, $9, $13, $7, and $16; for s_3, $12, $14, $16, and $9.

There is a different potential payoff, or profit, associated with each particular action, depending on which state of nature actually occurs. For example, if the inflation rate increases (s_1), the price of stock 4 will increase from $10 to 20, and action a_4 will yield a profit of $1,000. (At $10 per share, 100 shares of stock 4 would cost $1,000 today. If s_1 occurs, then the price of stock 4 will increase to $20 per share and the 100 shares will be worth $2,000, yielding a $1,000 profit.)

Table 22.2 shows the 12 potential payoffs associated with the four actions and the three states of nature described in this problem.

In the payoff table, no single action is obviously the best one. If the inflation rate increases (s_1), the optimal action is a_4 because it yields the largest profit ($1,000). Under s_2, action a_4 is again optimal; but if s_3 occurs, action a_3 is optimal.

TABLE 22.2 *Payoff table for Example 22.2*

	STATE OF NATURE		
Action	s_1 (Inflation rate increases)	s_2 (Inflation rate constant)	s_3 (Inflation rate decreases)
a_1 (Buy stock 1)	− $ 500	− $100	$200
a_2 (Buy stock 2)	0	300	400
a_3 (Buy stock 3)	500	− 300	600
a_4 (Buy stock 4)	1,000	600	− 100

Regardless of which state of nature occurs, a_2 always yields a larger payoff than a_1. This means that a_1 can never be the optimal action. Action a_1 is said to be *dominated* by action a_2.

DEFINITION   **Dominated Action** ——————————————————————————————————

An action a_i is said to be a **dominated action** if another action a_j exists such that, for every possible state of nature, the payoff for a_j is at least as large as the payoff for a_i. A dominated action will never be the optimal action and can be eliminated from the analysis without affecting the solution. Action a_j dominates action a_i if $\Pi_{jk} \geq \Pi_{ik}$ for all values of k.

EXAMPLE 22.3   **Decision Making Under Uncertainty** ——————————————————————

A grocer must decide each morning how many baskets of perishable fruit to purchase for sale during the day. He buys baskets of fruit for $3 each and sells them for $5 apiece, yielding a profit of $2 per unit. At the end of the day, all remaining units are sold to a senior citizens' group for $1 per unit (i.e., at a loss of $2 each). The grocer can order 5, 10, or 15 units, and the daily demand is 6, 11, or 14 units. Construct the payoff table for this problem.

SOLUTION   The possible actions are

$$a_1: \quad \{\text{Order 5 units}\}$$
$$a_2: \quad \{\text{Order 10 units}\}$$
$$a_3: \quad \{\text{Order 15 units}\}$$

The states of nature are

$$s_1: \quad \{\text{Demand of 6 units}\}$$
$$s_2: \quad \{\text{Demand of 11 units}\}$$
$$s_3: \quad \{\text{Demand of 14 units}\}$$

To show how to calculate some potential payoffs, suppose action a_2 is taken (the grocer orders 10 units at a total cost of $30). If state s_1 prevails, the grocer sells 6 units at $5 each and the remaining 4 units at $1 each for a total revenue of $34. Thus, the payoff associated with action a_2 and state of nature s_1 is $4 ($34 − $30). As another example, suppose action a_3 is taken and state of nature s_1 occurs. Then the grocer orders 15 units at a total cost of $45, sells 6 units at $5 each, and sells the remaining 9 units at $1 each, for a total revenue of $39. Thus, the payoff is $-\$6$ (i.e., a loss of $6). The other payoffs are shown in Table 22.3.

TABLE 22.3 *Payoff table for Example 22.3*

	STATE OF NATURE		
Action	s_1 (Demand of 6)	s_2 (Demand of 11)	s_3 (Demand of 14)
a_1 (Order 5)	$10	$10	$10
a_2 (Order 10)	4	20	20
a_3 (Order 15)	-6	14	26

In Example 22.3, if the grocer knew the demand on a given day, it would be easy to choose the optimal action. Typically, however, we do not know what state of nature will occur, so we need some criteria for choosing which action to take. In the next section, we discuss various criteria to use when choosing an optimal action.

Opportunity Loss

Another way of representing potential payoffs is by an opportunity loss table. Suppose we take action a_i and state of nature s_j occurs. Our payoff is Π_{ij}. First, we find the largest payoff associated with s_j. Suppose this payoff is Π_{mj}. If we had chosen action a_m, our payoff would have been Π_{mj} rather than Π_{ij}. The difference in payoffs is

$$L_{ij} = \Pi_{mj} - \Pi_{ij}$$

In a sense, by choosing the wrong action, we lost the opportunity to earn an additional L_{ij}.

DEFINITION

Opportunity Loss

Let Π_{mj} be the highest payoff associated with state of nature s_j. Let Π_{ij} denote the payoff associated with taking action a_i when state s_j occurs. For state s_j, the **opportunity loss** from taking action a_i is

$$L_{ij} = \Pi_{mj} - \Pi_{ij}$$

For a given state of nature s_j, the opportunity loss L_{ij} is the difference between the highest possible payoff associated with state s_j and the payoff actually realized for the act selected. For a given state of nature, the opportunity loss associated with the optimal action is 0 and the opportunity loss for other actions is nonnegative. The opportunity loss L_{ij} can be obtained from the payoff table by subtracting the payoff Π_{ij} from the largest entry in the same column.

Formula for opportunity loss

$$L_{ij} = (\max_k \Pi_{kj}) - \Pi_{ij}$$

EXAMPLE 22.4 **Constructing an Opportunity Loss Table**

Refer to Example 22.2 and Table 22.2. Construct the corresponding opportunity loss table.

SOLUTION We calculate the opportunity losses in the following way. Assume the state of nature is s_1. The maximum payoff is $\Pi_{41} = \$1,000$ by choosing action a_4. That is, $\max_k \Pi_{k1} = \Pi_{41} = \$1,000$ and the opportunity losses are

$$L_{11} = 1,000 - \Pi_{11} = 1,000 - (-500) = \$1,500$$
$$L_{21} = 1,000 - \Pi_{21} = 1,000 - 0 = \$1,000$$
$$L_{31} = 1,000 - \Pi_{31} = 1,000 - 500 = \$\,500$$
$$L_{41} = 1,000 - \Pi_{41} = 1,000 - 1,000 = \$\quad 0$$

By following a similar procedure for states s_2 and s_3, we obtain the opportunity losses shown in Table 22.4.

TABLE 22.4 *Opportunity loss table for Example 22.4*

	STATE OF NATURE		
Action	s_1	s_2	s_3
a_1	$1,500	$700	$400
a_2	1,000	300	200
a_3	500	900	0
a_4	0	0	700

Consider $L_{31} = \$500$. When s_1 occurs, we earn $500 by taking action a_3; but by taking a_4 we could have earned $1,000. Consequently, by taking action a_3, we lose the opportunity of earning an additional $500.

The opportunity loss table is sometimes called the *regret table*.

Exercises

Statistical Concepts

22.1.1 Explain what is meant by states of nature in a decision problem.

22.1.2 *True or false:* One action is said to dominate another action when the second action is never preferred to the first, regardless of the state of nature. Explain.

22.1.3 Define the opportunity loss associated with taking a certain action.

22.1.4 At baseball games, it is common for newspaper reporters to second-guess the manager. For example, suppose the manager brings in a relief pitcher late in a ball game and the next batter hits a home run to win the game. The reporters frequently criticize the manager's decision to bring in the relief pitcher. Is it possible that the manager made a smart decision, but because possible outcomes were uncertain, his choice was an unlucky one? Explain.

22.1.5 Is it necessarily the case that a good outcome was the result of a good decision, or is it possible that an unwise decision was made, but the decision maker got lucky?

22.1.6 Give an example of a situation where a good decision led to a bad outcome.

22.1.7 Give an example of a situation where a bad decision led to a good outcome.

Statistical Drills

22.1.8 A decision problem consists of three possible actions and three possible states of nature. Consider the following payoff table for this decision problem:

	STATE OF NATURE		
Action	s_1	s_2	s_3
a_1	10	-6	-8
a_2	5	-1	7
a_3	-4	-2	3

a. Is any action dominated by any other action? Explain why or why not.
b. Construct the opportunity loss table.

22.1.9 A decision problem consists of four possible actions and three possible states of nature. Consider the following payoff table for this decision problem:

	STATE OF NATURE		
Action	s_1	s_2	s_3
a_1	9	-8	-5
a_2	5	0	6
a_3	-4	-2	13
a_4	-2	9	3

a. Is any action dominated by any other action? Explain why or why not.
b. Construct the opportunity loss table.

Statistical Applications

22.1.10 An investor has the opportunity to buy four different stocks. Each stock costs $50 per share, and the investor will purchase 20 shares of one of the stocks and sell it 1 year later. If there is a recession (state s_1), the selling prices of the four stocks will be $40, $52, $58, and $45. If there is no recession (state s_2), the selling prices will be $53, $56, $54, and $60, respectively.
a. Determine the payoff table.
b. Are any of the actions dominated? If so, which one?
c. Determine the opportunity loss table.

22.1.11 The Bess Management Company (BMC) submits a proposal to the government to study the costs of repairing bridges in major eastern cities. BMC does not presently own a computer, but if it wins the contract, a large amount of computer work will be required. The company has the option of renting a computer for $45,000 today. If they wait a few months until the contract is awarded, the rent will be $72,000. If BMC does not win the contract, it will not rent a computer. If it does win the contract, the company's profits except for computer costs will be $160,000. The possible states of nature are s_1 (win the contract) and s_2 (not win the contract). The possible actions are a_1 (rent now) and a_2 (wait and possibly rent later). Determine the following:
a. The payoff table
b. The opportunity loss table
c. Whether any action is a dominated action

22.1.12 A common problem in business is deciding the quantity to order of perishable inventories—that is, items that lose the major portion of their economic value after a given date due to either obsolescence or spoilage. For example, if a newsstand orders too many papers and cannot sell them all, the excess papers have little value. Suppose a buyer for a large department store is trying to decide how many of a new style of bathing suit to order. Because of rapid changes in fashion, she does not want to order too many, but if she orders too few, she will lose profits for her department. The store pays $30 each for bathing suits and sells them for $50 each. Any bathing suits not sold at the end of the season will be sold for $15 each. The buyer believes that the department will sell 4, 5, 6, 7, or 8 dozen bathing suits. Bathing suits must be purchased by the department store in lots of 1 dozen.

a. Formulate the payoff table for this decision problem.
b. Convert the payoff table to an opportunity loss table.

22.2 Criteria for Making Decisions

Decision-making problems can be classified according to the degree of knowledge we have about the state of nature that will occur. In *decision making under certainty,* the state of nature that will occur is known in advance. In this case, the decision maker can determine which action to take to maximize profits. In decision making under uncertainty, the state of nature that will occur is unknown. However, there are two extreme cases for decision making under uncertainty. In one case, the probability that any state of nature will occur is known; in the other case, nothing is known about such probabilities. Probabilities that are known may be based on the relative frequency of occurrence of these states in the past or on theoretical grounds, or they may be nothing more than subjective estimates.

For most decision-making problems of interest, we have imperfect knowledge of the probability distribution of the various states of nature. In these problems, we use all available information to get a subjective estimate of the pertinent probabilities. Problems in decision making where the probabilities for the states of nature are unknown can be further divided into categories based on how much information we have concerning the probability distribution of the states of nature.

In most situations of interest in business and everyday life, we have to make some assumptions concerning the probability of occurrence of each state of nature. After assigning a probability to each possible state, we can calculate the expected, or average, payoff that we would obtain by taking any action. We make this calculation by weighting each possible payoff by its probability of occurrence. In decision making under uncertainty, a commonly used rule is to select the action that yields the highest expected payoff. This idea is discussed in more detail later.

The Decision-Making Process

The following steps should be used to solve a statistical decision problem:

1. List all the alternative choices, or actions, a_i that are available.
2. List each possible event, or state of nature, s_j that has an influence on the potential payoff.
3. For each state of nature, list the payoff obtained from following each action.
4. If possible, assign a probability to the occurrence of each state of nature.
5. State a criterion for selecting an optimal action.
6. Select the optimal action based on the criterion in step 5.

Various criteria can be used to determine the optimal action. Some different criteria are as follows:

1. The maximax, or Hurwicz, criterion
2. The maximin, or Wald, criterion
3. The minimax, or Savage, criterion
4. The Laplace, or strategy-of-insufficient-reason, criterion
5. The expected monetary value (EMV) criterion
6. The expected opportunity loss (EOL) criterion
7. The expected utility criterion

These criteria are discussed next, except for expected utility, which we discuss in Section 22.3.

Maximax Criterion

DEFINITION **Maximax Criterion**

Under the **maximax criterion**, we select the action with the largest possible payoff.

The maximax criterion is used by supreme optimists. Optimists assume that, of all the states of nature, the state that yields the maximum overall profit will occur. Then the optimist selects the action associated with this maximum profit.

EXAMPLE 22.5 **Maximax Criterion**

Find the optimal action in Example 22.2 based on the maximax criterion.

SOLUTION In Example 22.2 the maximax criterion would lead us to select action a_4 because $1,000 is the largest payoff in Table 22.2.

The maximax criterion is rarely used because it ignores the possibilities of large losses or small payoffs. Businesses on the brink of bankruptcy might adopt the maximax criterion if only a large payoff could save the company.

Maximin Criterion

DEFINITION **Maximin Criterion**

Under the **maximin criterion**, we find the worst possible payoff for each action and take the action yielding the largest (or maximum) of all these minimum payoffs (hence the name *maximin*).

Users of the maximin criterion are pessimists, assuming that no matter which action is taken, the worst possible state of nature will occur.

EXAMPLE 22.6 **Maximin Criterion**

Find the optimal action in Example 22.2 according to the maximin criterion.

SOLUTION   From the payoff table (Table 22.2), the worst payoff for a_1 is $-\$500$; for a_2, \$0; for a_3, $-\$300$; and for a_4, $-\$100$. The maximum of these four minimum payoffs is \$0, so the maximin criterion tells us to choose action a_2. By choosing a_2, the worst that could happen is a payoff of \$0; any other action could result in a loss.

The maximin criterion was developed by Abraham Wald and is sometimes called the Wald criterion. It is a conservative strategy that can be appropriate when a company is in financial difficulty and cannot afford a large loss. It selects actions that avoid large losses, but neglects large payoffs that are possible by choosing other actions.

This strategy is not as unrealistic as it may appear at first glance. In certain situations, it may be necessary for an individual or a business to avoid a large possible loss or to guarantee a minimum profit.

Minimax Criterion

DEFINITION   **Minimax Criterion** ────────────────────────

Under the **minimax criterion**, we find the maximum opportunity loss possible for each action and take the action with the smallest such loss (hence the name *minimax*).

The minimax criterion was developed by L. J. Savage. The minimax criterion is concerned with opportunity loss rather than profits, and the criterion tells us to minimize the largest potential opportunity loss from each course of action.

EXAMPLE 22.7   **Minimax Criterion** ────────────────────────

Find the optimal action in Example 22.4 based on the minimax criterion.

SOLUTION   The maximum opportunity loss for a_1 is \$1,500; for a_2, \$1,000; for a_3, \$900; and for a_4, \$700. The minimum of these maximum opportunity losses is \$700. Thus, under the minimax criterion, the optimal action is a_4.

The minimax criterion is a pessimistic strategy because it uses only the worst possible opportunity loss for each act.

Strategy of Insufficient Reason

If the decision maker has no information on the states of nature, a common assumption is that each state of nature is equally likely to occur. This idea was first developed by Pierre Laplace. Since each possible payoff is assumed to be equally likely, we find an expected payoff for each action by taking the arithmetic average of the possible payoffs associated with that action.

DEFINITION   **Laplace Criterion** ────────────────────────

Under the **Laplace criterion**, we take the action that has the largest expected payoff based on the assumption that every possible state of nature is equally likely.

EXAMPLE 22.8 **Laplace Criterion** ———————————————————————————————

Use the Laplace criterion to determine the optimal action in Example 22.3.

SOLUTION According to Table 22.3, the possible payoffs from action a_1 are $10, $10, and $10; the average is $10. The possible payoffs from action a_2 are $4, $20, and $20; the average is $44/3, or $14.67. The possible payoffs from action a_3 are $-$6, $14, and $26; the average payoff is $34/3 = $11.33. The Laplace criterion tells us to select action a_2 because its expected payoff is the largest.

——

Unlike the other criteria discussed so far, the Laplace criterion uses all the data in the payoff matrix, not just the largest or smallest value in a row or column. Naturally, this criterion is inappropriate if the states of nature are not approximately equally likely. If we have no information whatsoever about the probabilities of the states of nature, it is debatable whether the assumption that all the states are equally likely is wise.

Each of the criteria discussed thus far ignores any information concerning the probabilities of occurrence of each state of nature. Because we usually want to weight the possible payoffs or losses by their probability of occurrence, the use of the maximax, maximin, minimax, and Laplace criteria is limited.

Expected Monetary Value Criterion

The expected monetary value (EMV) criterion is sometimes called the Bayes criterion after the Reverend Thomas Bayes, an 18th-century minister and mathematician. To use the EMV criterion, we must be able to assign a probability to each state of nature, and the sum of these probabilities must be 1. These probabilities represent our feelings about the likelihood of the various states of nature. Frequently, however, no data are available to help us determine such probabilities. Nevertheless, if we have had experience in similar situations, the probabilities in those circumstances may be useful. Since most people are influenced in their choice of actions by their subjective feelings about the likelihood of different events, it is reasonable to incorporate these probabilities into the decision-making process. However, because of this subjective procedure, some statisticians reject the EMV criterion.

DEFINITION **Prior Probabilities** ———————————————————————————————————

The probabilities assigned to the various states of nature are called **prior probabilities**. We let $p_j = P(s_j)$ denote the probability that state of nature s_j occurs.

——

DEFINITION **Expected Monetary Value** ———————————————————————————————

The **expected monetary value** of action a_i, denoted EMV(a_i), is defined as

$$\text{EMV}(a_i) = \sum_{j=1}^{k} \Pi_{ij} p_j$$

where

$$\Pi_{ij} = \text{Payoff of action } a_i \text{ when state of nature } s_j \text{ occurs}$$
$$p_j = P(s_j) = \text{Probability that state of nature } s_j \text{ occurs } (j = 1, 2, \ldots, k)$$
$$k = \text{Number of possible states of nature}$$

DEFINITION   **Expected Monetary Value Criterion** ———————————————————

Under the **expected monetary value criterion**, we take the action with the highest expected monetary value.

EXAMPLE 22.9   **Expected Monetary Value Criterion** ———————————————

Suppose that, based on previous experience, the investor in Example 22.2 believes that the prior probabilities for the possible states of nature are as shown in Table 22.5. Find the optimal action based on the EMV criterion.

SOLUTION   From Table 22.2, the payoffs from action a_1 are $-\$500$ under s_1, $-\$100$ under s_2, and $\$200$ under s_3. To calculate the EMV from taking action a_1, these payoffs must be weighted by their probabilities. The expected monetary value of a_1 is

$$\text{EMV}(a_1) = (-500)(.5) + (-100)(.3) + (200)(.2) = -\$240$$

TABLE 22.5   *Prior probabilities for Example 22.9*

State of nature s_j	Prior probability $P(s_j)$
s_1 (Inflation rate increases)	.5
s_2 (Inflation rate constant)	.3
s_3 (Inflation rate decreases)	.2

The payoffs under a_2 are $\$0$, $\$300$, and $\$400$, so the EMV is

$$\text{EMV}(a_2) = (0)(.5) + (300)(.3) + (400)(.2) = \$170$$

Similarly, for actions a_3 and a_4, we obtain

$$\text{EMV}(a_3) = (500)(.5) + (-300)(.3) + (600)(.2) = \$280$$
$$\text{EMV}(a_4) = (1,000)(.5) + (600)(.3) + (-100)(.2) = \$660$$

By the EMV criterion, a_4 is the optimal action, because its expected payoff of $\$660$ is larger than the expected payoff from any other action.

Expected Opportunity Loss Criterion

DEFINITION **Expected Opportunity Loss**

The **expected opportunity loss** of action a_i, denoted $EOL(a_i)$, is defined as

$$EOL(a_i) = \sum_{j=1}^{k} L_{ij} p_j$$

where

L_{ij} = Opportunity loss if action a_i is taken and state of nature s_j occurs
$p_j = P(s_j)$ = Prior probability that s_j occurs ($j = 1, 2, \ldots, k$)
k = Number of possible states of nature

DEFINITION **Expected Opportunity Loss Criterion**

Under the **expected opportunity loss criterion**, we take the action having the smallest expected opportunity loss.

EXAMPLE 22.10 **Expected Opportunity Loss Criterion**

Using the prior probabilities in Table 22.5, find the optimal action for Example 22.4 based on the EOL criterion.

SOLUTION From Tables 22.4 and 22.5, the EOL for action a_1 is

$$EOL(a_1) = (1,500)(.5) + (700)(.3) + (400)(.2) = \$1,040$$

In a similar manner, we obtain

$$EOL(a_2) = (1,000)(.5) + (300)(.3) + (200)(.2) = \$630$$
$$EOL(a_3) = (500)(.5) + (900)(.3) + (0)(.2) \quad = \$520$$
$$EOL(a_4) = (0)(.5) + (0)(.3) + (700)(.2) \quad = \$140$$

Action a_4 has the minimum EOL. Recall that a_4 also had the largest EMV. In fact, the EMV and EOL criteria always choose the same action.

The last decision criterion, the expected utility criterion, will be discussed in Section 22.3.

Expected Payoff Using Perfect Information

If we knew that state s_j was going to occur, it would be easy to select the action a_i having the largest payoff Π_{ij}. That is, for a given value of j, we would maximize Π_{ij} over $i = 1$, $2, \ldots, n$. Suppose we knew that state s_j was going to occur. Let's denote the maximum payoff that we could obtain as $\max_i \Pi_{ij}$. For each state s_j, weight the maximum payoff by the probability that state s_j occurs. The sum of these weighted payoffs is called the *ex-*

pected payoff using perfect information (EPPI). EPPI represents the profit that would be realized under the following circumstances:

1. We could repeat the problem many times under identical conditions.
2. We knew with certainty on each trial which state s_j was going to occur.
3. The states of nature occurred with the assumed prior probabilities.

DEFINITION   **Expected Payoff from Perfect Information** ————————————

To calculate the **expected payoff from perfect information** (EPPI), multiply the highest payoff associated with each state by the state's probability of occurrence $P(s_j)$. EPPI is the sum of these products over all states of nature. Notationally, we have

$$\text{EPPI} = \sum_{j=1}^{k} P(s_j)(\max_i \Pi_{ij})$$

EXAMPLE 22.11   **Expected Payoff from Perfect Information** ————————————

Calculate the EPPI using the payoffs Π_{ij} given in Table 22.2 and the probabilities $P(s_j)$ given in Table 22.5.

S O L U T I O N   If s_1 occurs, the maximum payoff is \$1,000; this amount is $\max_i \Pi_{i1}$. If s_2 occurs, the maximum payoff is \$600; this is $\max_i \Pi_{i2}$. If s_3 occurs, the maximum payoff is \$600; this is $\max_i \Pi_{i3}$. Because the states s_1, s_2, and s_3 occur with probabilities .5, .3, and .2, respectively, we obtain

$$\text{EPPI} = .5(1,000) + .3(600) + .2(600) = \$800$$

This result indicates that if we knew in advance which state was going to occur and each state occurred with the assumed probabilities, and if on each trial we took the optimal action, then in the long run our payoff would be \$800 per trial.

Expected Value of Perfect Information

Before actually selecting a course of action, there is another option available. We can postpone action until we obtain additional information about the state of nature. Before taking an action, we should determine the cost of obtaining additional information and the potential value of this information. The *maximum* value of additional information about the state of nature is the same as the value of *perfect information* on which state of nature will occur.

The *expected value of perfect information* (EVPI) is the maximum amount of money that should be spent to obtain perfect information on which state of nature will occur. The EVPI measures the difference between the maximum payoff with perfect information (i.e., under conditions of certainty) and the expected payoff under uncertainty.

DEFINITION   **Expected Value of Perfect Information** ————————————

The expected payoff using perfect information is EPPI. Without perfect information, the best we can do is to maximize EMV. Denote the maximum value of EMV as EMV*. The difference between EPPI and EMV* represents the expected additional profit we would obtain if we had perfect information about which state was going to occur. This additional profit represents an upper bound for the amount of money we should be willing to spend to obtain perfect information. This amount is called the **expected value of perfect information** (EVPI). Notationally, we have

$$EVPI = EPPI - EMV*$$

$$= \sum_{j=1}^{k} P(s_j)(\max_i \Pi_{ij}) - \max_i \left[\sum_{j=1}^{k} P(s_j)\Pi_{ij} \right]$$

where EMV* is the maximum value of EMV.

EXAMPLE 22.12 **Expected Value of Perfect Information**

Using the payoffs in Table 22.2 and the probabilities in Table 22.5, calculate the EVPI.

SOLUTION In Example 22.11, we found that the expected profit using perfect information was EPPI = $800. In Example 22.9, we found that the maximum expected monetary value was EMV* = $660 by using action a_4. We obtain

$$EVPI = EPPI - EMV* = 800 - 660 = \$140$$

Another way of calculating EVPI utilizes the data in an opportunity loss table. In such a table, the entries L_{ij} represent the potential cost of not knowing which state of nature will occur. EVPI measures the maximum amount we should spend to obtain information that would enable us to avoid these opportunity losses. EVPI equals the EOL associated with the optimal decision.

EXAMPLE 22.13 **Calculating EVPI Using an Opportunity Loss Table**

Using Table 22.4, Example 22.4, and Example 22.10, calculate the EVPI.

SOLUTION In Example 22.4, the optimal action in stock purchasing was a_4, and from Example 22.10, we have EOL(a_4) = $140. Thus, the investor should be willing to spend up to $140 to obtain perfect information about next year's rate of inflation. This can be explained as follows. With no advance information concerning which state of nature is going to occur, a_4 is the optimal action. If we knew in advance that s_1 or s_2 was going to occur, the optimal action would still be a_4, so no opportunity is lost. On the other hand, if we knew in advance that s_3 was going to occur, the optimal action would be a_3, and the opportunity loss from choosing a_4 would be $700. Recall that the states of nature were assumed to occur with probabilities $P(s_1) = .5$, $P(s_2) = .3$, and $P(s_3) = .2$. Thus, by choosing a_4, there is a 20% chance of suffering an opportunity loss of $700. The EOL from choosing a_4 is 700(.2) = $140. Thus, the investor should be willing to pay up to $140 to determine whether s_3 is going to occur.

Some Comments About the Expected Monetary Value Criterion

The EMV criterion is the most frequently used criterion for making decisions. If the same problem occurs repeatedly under approximately identical conditions, then in the long run we maximize profits by choosing the action with the highest EMV. In some situations, however, a decision maker may not want to take the action with the largest EMV. For example, if the action having the highest EMV has large negative and large positive payoffs, a conservative (or risk-averting) decision maker might take some other action to avoid the possibility of large losses.

The EMV criterion uses only the average, or expected, payoff. It does not take into account the variability, or variance, of the possible payoffs. A prudent decision maker will frequently accept a smaller expected payoff if the variability of the possible payoffs is also smaller.

Exercises

Statistical Concepts

22.2.1   *True or false:*   The maximax criterion is used by extreme optimists and the maximin criterion is used by extreme pessimists. Explain.

22.2.2   Explain how the minimax criterion differs from the maximin criterion.

22.2.3   Explain the major drawback of the Laplace criterion.

22.2.4   *True or false:*   Under the expected monetary value criterion, we choose the action having the highest possible payoff. Explain.

22.2.5   *True or false:*   Under the expected monetary value criterion, we weight each possible payoff from an action by its probability of occurrence, and we choose the action having the highest expected payoff. Explain.

22.2.6   *True or false:*   A disadvantage of the EMV criterion is that it requires knowledge of the probabilities of occurrence of each state of nature. Explain.

22.2.7   *Explain:*   To be a rational decision maker, you first must identify your goals and then you must adopt some rule (or payoff method) that allows you to rank different actions according to how well they satisfy your goals.

22.2.8   *True or false:*   The maximax, minimax, and maximin criteria do not require knowledge of any prior probabilities. Explain.

Statistical Drills

22.2.9   A decision problem consists of three possible actions and three possible states of nature. Consider the following payoff table for this decision problem:

	STATE OF NATURE		
	s_1	s_2	s_3
Prior probability:	.2	.3	.5
Action			
a_1	10	-6	-8
a_2	5	-1	7
a_3	-4	-2	3

a. Find the optimal action based on the maximax criterion.
b. Find the optimal action based on the maximin criterion.
c. Find the optimal action based on the minimax criterion.
d. Find the optimal action based on the EMV criterion.

e. Find the optimal action based on the expected opportunity loss criterion.

f. Find the expected payoff using perfect information.

g. Find the expected value of perfect information.

22.2.10 A decision problem consists of four possible actions and three possible states of nature. Consider the following payoff table for this decision problem:

	STATE OF NATURE		
	s_1	s_2	s_3
Prior probability:	.4	.2	.4
Action			
a_1	9	-8	-5
a_2	5	0	6
a_3	-4	-2	13
a_4	-2	9	3

a. Find the optimal action based on the maximax criterion.

b. Find the optimal action based on the maximin criterion.

c. Find the optimal action based on the minimax criterion.

d. Find the optimal action based on the EMV criterion.

e. Find the optimal action based on the expected opportunity loss criterion.

f. Find the expected payoff using perfect information.

g. Find the expected value of perfect information.

Statistical Applications

22.2.11 Use the information in Exercise 10 of Section 22.1. As prior probabilities, use $P(\text{Recession}) = .2$ and $P(\text{No recession}) = .8$. Find the optimal action using the following criteria:

a. Maximax   **b.** Minimax   **c.** Maximin

d. EMV   **e.** EOL   **f.** Calculate the EVPI.

22.2.12 Use the information presented in Exercise 11 of Section 22.1. Assume the company's prior probabilities are $P(s_1) = .4$ and $P(s_2) = .6$. Find the optimal action using the following criteria:

a. Maximax   **b.** Minimax   **c.** Maximin

d. EMV   **e.** EOL   **f.** Calculate the EVPI.

22.2.13 The Bowen Hardware Store must decide how many cans of driveway sealer to order for the summer season. It takes a long time for an order to be delivered, so Mr. Bowen can place only one order early in the spring. The cost per can is \$12, and his selling price is \$18 per can. In October, any unsold cans are sold to the local school district for \$9 per can. Mr. Bowen has to decide whether to buy 100, 200, 300, or 400 cans. He thinks the demand will be 100, 200, 300, or 400 cans with probabilities .2, .5, .2, and .1, respectively. These probabilities are based on Mr. Bowen's past business experience.

a. Determine the payoff table for this decision problem.

b. Find the optimal action if we use the EMV criterion.

c. Find the EPPI.

d. Find the EVPI.

22.2.14 Mr. Anderson is in charge of inventory control at a large department store. He has to decide how many lawn mowers to order for the summer season. Because the lawn mowers are widely available, Mr. Anderson loses the customer to another store if he is out of stock when a customer orders one. It takes 1 week for Mr. Anderson to receive new supplies, so each Monday he decides how many lawn mowers to order for the following week. Suppose Mr. Anderson makes \$70 profit on every sale, and every unsold lawn mower costs him $15 in interest and storage costs. He has determined that the demand for lawn mowers approximately follows the Poisson distribution with mean $\mu = .9$. How many lawn mowers should Mr. Anderson order to maximize expected profits?

22.2.15   A woman is considering purchasing stock in one of the following three business lines: mining, oil refining, or computer software development. The expected profit in each line as a function of future inflation rates is given in the following payoff matrix:

Action	s_1 (Inflation up)	s_2 (Inflation down)	s_3 (Inflation unchanged)
a_1 (Mining)	140	510	290
a_2 (Oil refining)	420	210	250
a_3 (Software)	280	300	300

a. What is the maximin strategy?
b. What is the maximax strategy?
c. If the entrepreneur decides to invest in the mining industry, is she a pessimist or an optimist? Explain.

22.2.16   Refer to the payoff matrix in Exercise 15. Assume that the entrepreneur makes decisions according to the expected profit criterion.
a. Which activity will she choose if she believes that the following probabilities characterize the three states of nature: $P(s_1) = .3$, $P(s_2) = .3$, and $P(s_3) = .4$?
b. Which activity will she choose if it is given that $P(s_1) = .1$, $P(s_2) = .8$, and $P(s_3) = .1$?

22.2.17   Refer to Exercise 12 in Section 22.1. Use the payoff table drawn up in part **a** of that problem to do the following:
a. Find the action that would be prescribed by the maximax criterion.
b. Find the action that would be prescribed by the maximin criterion.

22.2.18   Refer to the payoff table for the decision problem in Exercise 12 of Section 22.1. The buyer has determined the following probability distribution for the number of bathing suits she can sell:

Dozens Sold	4	5	6	7	8
Probability	.25	.30	.20	.20	.05

How many dozen bathing suits should the buyer order according to the expected payoff criterion?

22.3   Utility Theory and the Expected Utility Criterion

Although people often use the expected monetary value criterion when deciding which action to take, there are situations where individuals clearly do not follow the EMV criterion. For example, almost all homeowners have some sort of insurance to protect themselves against fires, and drivers have insurance to protect themselves against traffic accidents. Historically, insurance companies have earned profits, which means that, on the average, they receive more money in premiums than they pay out in claims. Thus, on the average, people who buy insurance pay more in premiums than they receive in claims, so an individual who buys insurance can expect to lose money. Consequently, the expected monetary value from buying insurance is negative. If people followed the EMV criterion, no insurance would be purchased. The question, then, is "Why do people buy insurance?" or "Why do people pay more in premiums than they expect to receive in claims?"

One possible explanation is that in many cases people do not maximize EMV. Rather, they maximize their expected utility.

The Concept of Utility

DEFINITION **Utility**

A person's **utility** for a quantity of money or for an object is the amount of happiness or satisfaction that the person obtains by possessing and using the money or object.

The *utility* of a given item varies from person to person. Thus, a loaf of bread could have an extremely high utility to a poor person and a very low utility to a rich person. In addition, the utility of a second loaf of bread is not necessarily the same as the utility of the first loaf of bread. According to the theory of diminishing marginal utility, each extra unit of a good tends to increase utility by less than the previous unit did. If this theory is correct, then winning N dollars (where $N > 1$) causes our utility to increase by less than N times the utility of winning \$1. Similarly, losing N dollars causes our utility to decrease by more than N times the utility lost by losing \$1. Rather than choose the action that maximizes the expected monetary value, people actually choose the action that maximizes the expected utility. This is the **expected utility criterion**.

Because utility is strictly a subjective concept, the decision maker's personal values must be taken into consideration when solving a problem by the expected utility criterion. To use the expected utility criterion, the payoffs in a payoff table should not be measured in monetary values but in values indicating the personal utility derived from that monetary payoff.

Assumptions of Utility Theory

A utility index for a rational person is assumed to satisfy the following requirements:

1. There is a complete ranking of preferences. For any pair of outcomes A and B, the decision maker can determine whether A is preferred to B, B is preferred to A, or there is no preference for either.
2. Preferences are transitive. If outcome A is preferred to outcome B and if outcome B is preferred to outcome C, then A is preferred to C. If we are indifferent between A and B and we are indifferent between B and C, then we are also indifferent between A and C.
3. Suppose that actions A and B have the same potential payoffs, but action A has a higher probability of success than action B. Then action A is preferred to action B.
4. It is assumed that if A is preferred to B and B is preferred to C, then there exists a gamble that offers A with probability p and C with probability $(1 - p)$ such that the decision maker will be indifferent between taking the gamble and receiving B with certainty.

To analyze a decision problem in terms of utilities rather than monetary payoffs, we must transform the potential payoffs in a payoff table into utilities. To do this we must know the decision maker's utility function. If the decision maker's utility function is unknown, we must construct a utility index that will reveal the decision maker's preferences.

Constructing a Utility Index

The unit in which utility is measured is arbitrary, so we can scale the utility index in any way that is convenient. To construct a utility index, proceed as described in the accompanying box.

Constructing a utility index

> **1.** Let L denote the lowest monetary payoff in the payoff table and let H denote the highest monetary payoff. Assign utility 0 to payoff L and 100 to payoff H. That is, $U(L) = 0$ and $U(H) = 100$, where $U(X)$ denotes the utility associated with payoff X.
> **2.** Let Π be any payoff between L and H. Determine the probability p such that the decision maker is indifferent between receiving payoff Π with certainty and taking a gamble that pays H with probability p or L with probability $(1 - p)$.
> **3.** Define the utility of payoff Π to be $100p$. That is,
>
> $$U(\Pi) = 100p$$

The reason for defining utility as described in the box is that the expected utility from taking the gamble is

$$100p + 0(1 - p) = 100p$$

and the decision maker is indifferent between taking the gamble and receiving payoff Π for certain. Because the decision maker is indifferent, taking the gamble must have the same utility as receiving payoff Π for certain.

EXAMPLE 22.14   **Constructing a Utility Index** ————————————————————

Suppose a decision maker is confronted with a problem that has two possible actions a_1 and a_2 and two possible states of nature s_1 and s_2. The payoff associated with action a_i and state s_j is Π_{ij}. The payoff table is shown in Table 22.6. Let us construct a utility index based on this payoff table.

TABLE 22.6   *Payoff table for Example 22.14*

| Action | STATE OF NATURE | |
	s_1	s_2
a_1	− $400	$800
a_2	$400	$200

SOLUTION   The lowest payoff is $L = -\$400$ and the highest is $H = \$800$. The utilities assigned to these two payoffs are $U(L) = U(-400) = 0$ and $U(H) = U(800) = 100$. To find the utility associated with a payoff of $200, we ask the decision maker which of the following situations he prefers:

1. Receiving Π_{22} = \$200 with certainty, or
2. Receiving \$800 if a red card is drawn from a deck containing 1 red card and 99 black cards or losing \$400 if a black card is drawn.

Most people would choose option 1. Now ask the decision maker which of the following situations he would prefer:

1. Receiving Π_{22} = \$200 with certainty, or
2. Receiving \$800 if a red card is drawn from a deck containing 2 red cards and 98 black cards or losing \$400 if a black card is drawn.

Suppose the decision maker again chooses option 1. We continue this process, increasing the number of red cards and decreasing the number of black cards until eventually we get the decision maker to switch from choosing option 1 to choosing option 2. Suppose the decision maker switches from option 1 to option 2 provided there are 65 red cards and 35 black cards. This means the point of indifference occurs when the probability of winning the maximum payoff is p = .65; thus the utility associated with payoff Π_{22} = \$200 is

$$U(\$200) = .65 \times U(\$800) + .35 \times U(-\$400)$$
$$= .65 \times 100 + .35 \times 0 = 65$$

Next we repeat this process for payoff Π_{21} = \$400. Suppose the decision maker is indifferent between receiving \$400 with certainty and taking a gamble where he can gain \$800 with probability p = .85 or lose \$400 with probability .15. Then the utility associated with payoff Π_{21} = \$400 is

$$U(\$400) = .85 \times 100 + .15 \times 0 = 85$$

Now the utility index for payoffs of \$800, \$400, \$200, and −\$400 is known; it is plotted in Figure 22.1. By proceeding in a similar fashion, we can determine the value of the utility index for any payoff between H = \$800 and L = −\$400.

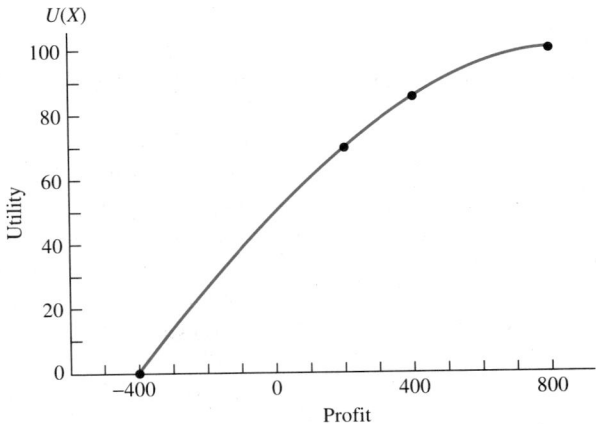

FIGURE 22.1 *Utility curve for Example 22.14*

Attitudes Toward Risk and the Shape of the Utility Curve

The shape of the utility curve in Figure 22.1 is important because it characterizes the decision maker's attitude toward risk. Observe that, as required of all utility curves, total utility increases as the potential payoff increases, but in this figure utility increases at a decreasing rate as the payoff increases. This illustrates *declining marginal utility*.

The shape of Figure 22.1 is typical of decision makers who are said to be *risk averse*. In Example 22.14, consider the expected monetary value from taking the gamble in which the decision maker gains $800 with probability $p = .65$ or loses $400 with probability .35. The expected monetary value of this gamble is

$$\text{EMV} = .65 \times 800 + .35 \times -400 = \$380$$

Recall that in Example 22.14, the decision maker indicated that he was indifferent between taking the gamble or receiving $200 with certainty. The fact that the expected monetary value of the gamble is $380 and that the decision maker would be willing to accept $200 rather than take the gamble indicates an aversion to risk taking.

The decision maker has revealed that he is willing to take a smaller monetary payment ($200) with certainty rather than take a gamble with a higher expected monetary value ($380).

The shape of the utility function can be used to describe the person's attitude toward risk. Individuals are classified as being risk averse, risk neutral, or risk seeking. The utility functions for each basic attitude are shown in Figure 22.2.

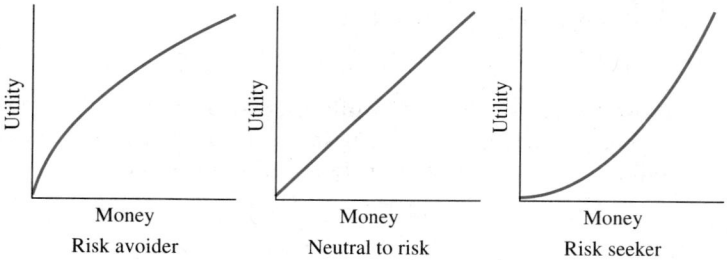

FIGURE 22.2 *Three types of utility curves showing attitudes toward risk*

Risk Aversion

Let X be a payoff that is less than Y. Whenever a decision maker prefers receiving the smaller payoff X with certainty to taking a gamble whose expected monetary value is Y, the person is said to be **risk averse**. This attitude is present when a person's marginal utility of money declines with larger amounts of money. The risk-averse individual tends to be conservative and will choose a certain fixed return over a gamble with a higher expected monetary value. Risk aversion is quite common, as shown by the fact that people buy insurance.

Risk Neutrality

When a person considers the utility of receiving a given amount of money with certainty to be equal to the utility of taking a gamble whose expected monetary value is the same amount, the person is said to be **risk neutral**. The utility function for a risk-neutral individual is a straight line. For these people, the marginal utility of money is constant; thus, every extra dollar of income increases utility by exactly the same amount as the previous dollar. Risk-neutral people behave as though they were maximizing expected monetary value. They would not buy insurance because the expected monetary payoff is negative; that is, the premium exceeds the expected claim.

When dealing with small sums of money, many people exhibit risk-neutral behavior. Thus, many people would be willing to take a gamble where they have an equal chance of winning or losing $1. For larger sums of money, however, their behavior becomes risk averse. Thus, few people would be willing to bet $10,000 on the toss of a coin. Risk neutrality is the implicit assumption made by those who use the expected monetary value criterion. For a person who is risk neutral, maximizing expected utility will yield the same optimal action as following the expected monetary value criterion.

Risk Seeking

When a person considers the utility of receiving X dollars with certainty to be less than the utility of taking a gamble whose expected monetary value is X, the person is said to be **risk seeking**. The utility function for a risk seeker increases at an increasing rate. For these people, the marginal utility of money is increasing, so every extra dollar of income increases utility by more than the previous dollar. Risk seekers will accept huge riches-or-ruin gambles because they are motivated by the possibility of obtaining a large payoff.

When the stakes are large, most people tend to be risk averse. When the stakes are small, these same people often tend to be risk neutral. Therefore a given person's utility curve can have portions that are risk averse, portions that are risk neutral, and portions that are risk seeking.

The St. Petersburg Paradox

An indication that people frequently maximize expected utility rather than expected monetary value is apparent in a famous problem called the St. Petersburg paradox. This problem illustrates a situation where people refuse to take a fair gamble based on the expected monetary value criterion. A solution to the St. Petersburg paradox was proposed by the Swiss mathematician Daniel Bernoulli (1700–1782), who studied gamblers at the casinos of St. Petersburg.

The St. Petersburg paradox involves the following game between persons A and B: A fair coin is tossed repeatedly until a head appears. If a head appears on the first toss, A pays B $1 and the game ends; if a head appears for the first time on the second toss, A pays B $2 and the game ends; if a head appears first on the third toss, A pays B $4 and the game ends. In general, the game ends when the first head appears and A pays B 2^{n-1} if the first head appears on the nth toss.

Bernoulli asked how much player B should be willing to pay player A for the privilege of playing this game to make the game an even gamble. To make the game fair, player B should pay player A the expected monetary value of the gamble. This value can be calculated as follows: The possible payments to B are $1, $2, $4, $8, and so forth depending on whether the first head appears on the first toss, the second toss, the third toss, the fourth toss, and so forth. The respective probabilities of these payments are $1/2$, $1/4$, $1/8$, and so forth. Thus the expected payment to B is the sum

$$1(1/2) + 2(1/4) + 4(1/8) + \cdots = .50 + .50 + .50 + \cdots = \infty$$

Even though the expected monetary value of the game is infinity, no person would be willing to pay an infinite sum of money for the privilege of playing this game (even though such a payment would make the gamble fair). Bernoulli argued that people making decisions under uncertainty do not attempt to maximize expected monetary value but try to maximize expected utility instead. He argued that the total utility of money (the overall welfare people derive from the possession of money) rises as people get more money, but the *marginal utility* of money (the increase in total utility associated with a $1 increase in the quantity of money) decreases as people get more money. According to the theory of declining marginal utility, although each extra dollar of income increases our total utility, the amount of increase is less than that caused by the previous dollar of income. It follows that the decrease in total utility from losing $1 exceeds the gain in total utility from winning $1.

Thus, any game in which there is an equal probability of gaining or losing a given amount is a fair game in expected monetary terms. In expected utility terms, the same game would be considered unfair if the person's utility function exhibited decreasing marginal utility. As a result, if people maximize expected utility, they will refuse to take fair gambles because, although the expected monetary value of the gamble is $0, the expected utility of the gamble is negative.

Expected Utility Criterion

DEFINITION   **Expected Utility Criterion** ─────────────────────────────────

According to the **expected utility criterion**, a decision maker should choose the action for which the expected utility is the highest.

To employ the expected utility criterion, we first have to determine the utility associated with each possible payoff. This can be done by assuming that the decision maker has some specific utility function or by constructing a utility index. Next, it is necessary to determine the action that has the highest expected utility. For each action a_i, the k possible payoffs $\Pi_{i1}, \Pi_{i2}, \ldots, \Pi_{ik}$ occur with probabilities $P(s_1), P(s_2), \ldots, P(s_k)$. The utilities assigned to these payoffs are $U(\Pi_{i1}), U(\Pi_{i2}), \ldots, U(\Pi_{ik})$. For any given action a_i, we obtain the expected utility payoff by weighting each potential utility by its probability of occurrence. This yields the *expected utility* of action a_i. A decision maker using the expected utility criterion determines the expected utility of each possible action and then selects the action that maximizes expected utility.

DEFINITION   **Expected Utility** ─────────────────────────────────

Suppose a decision maker has n possible actions $a_1, a_2, \ldots, a_n$ and there are k possible states of nature. Let Π_{ij} denote the payoff from action i if state j occurs. Let U_{ij} denote the utility corresponding to the ith action and jth state, and let p_j denote the probability of the *j*th state of nature occurring. The **expected utility** of action a_i, denoted $E[U(a_i)]$, is

$$E[U(a_i)] = \sum_{j=1}^{k} p_j U_{ij}$$

EXAMPLE 22.15   **Expected Utility Criterion**

In Example 22.14, assume that state s_1 occurs with probability .3 and state s_2 with probability .7. Find the optimal action based on the expected utility criterion.

SOLUTION   In terms of utility, the individual's payoff matrix is shown in Table 22.7. The expected utility of action a_1 is

$$E[U(a_1)] = (.3)(0) + (.7)(100) = 70$$

The expected utility of action a_2 is

$$E[U(a_2)] = (.3)(85) + (.7)(65) = 71$$

Under the expected utility criterion, the optimal action is a_2.

TABLE 22.7   *Utility payoff table for Example 22.14*

	STATE OF NATURE	
Action	s_1 $P(s_1) = .3$	s_2 $P(s_2) = .7$
a_1	0	100
a_2	85	65

Under the EMV criterion, we would use the payoffs in Table 22.6. The expected payoffs from actions a_1 and a_2 are

$$\text{EMV}(a_1) = (.3)(-400) + (.7)(800) = \$440$$
$$\text{EMV}(a_2) = (.3)(400) + (.7)(200)\quad = \$260$$

This example illustrates that, based on the expected utility criterion, the optimal action is a_2, but based on the expected monetary value criterion, the optimal action is a_1.

Exercises

Statistical Concepts

22.3.1   *True or false:*   A person's utility for an item can be measured by the price of the item. Explain.

22.3.2   Explain the differences among a person who is risk averse, a person who is risk neutral, and a person who is a risk seeker.

22.3.3   Explain the St. Petersburg paradox.

22.3.4   Explain what is meant by the expected utility criterion.

Statistical Drills

22.3.5   On a TV game show, a contestant is offered $8,000 in cash. Alternatively, the contestant is permitted to open one of two treasure chests and keep the contents. The contestant is told that one chest is empty and the other contains $12,000. The contestant elects to open a chest. Is this person risk averse, risk neutral, or a risk seeker?

22.3.6   A decision problem consists of three possible actions and three possible states of nature. Consider the following payoff table for this decision problem, where x_{ij} denotes the payoff associated with action i and state j:

	STATE OF NATURE		
	S_1	S_2	S_3
Prior probability:	.2	.3	.5
Action			
a_1	10	6	4
a_2	5	0	7
a_3	4	8	3

Consider the following utility function:

$$U(x_{ij}) = \frac{\sqrt{40x_{ij}}}{.2}$$

a. Find the expected utility associated with each action.
b. Find the optimal action based on the expected utility criterion.
c. Find the expected monetary value associated with each action.
d. Find the optimal action based on the EMV criterion.

22.3.7   A decision problem consists of four possible actions and three possible states of nature. Consider the following payoff table for this decision problem, where x_{ij} denotes the payoff associated with action i and state j:

	STATE OF NATURE		
	S_1	S_2	S_3
Prior probability:	.4	.2	.4
Action			
a_1	9	-7	-5
a_2	5	0	6
a_3	-4	-2	13
a_4	-2	9	3

Consider the following utility function:

$$U(x_{ij}) = \frac{\sqrt{35 + 5x_{ij}}}{.1}$$

a. Find the expected utility associated with each action.
b. Find the optimal action based on the expected utility criterion.
c. Find the expected monetary value associated with each action.
d. Find the optimal action based on the EMV criterion.

Statistical Applications

22.3.8 Suppose you are indifferent between receiving $200 with certainty and participating in a gamble in which you have a probability of .2 of receiving $2,000 and a probability of .8 of receiving nothing. You are also indifferent between receiving $1,000 with certainty and participating in a gamble in which you have a probability of .7 of receiving $2,000 and a probability of .3 of receiving nothing. Finally, you are indifferent between receiving $1,600 with certainty and participating in a gamble in which you have a probability of .95 of receiving $2,000 and a probability of .05 of receiving nothing.

a. Let $U(\$2,000) = 100$ and $U(\$0) = 0$. Find $U(\$200)$, $U(\$1,000)$, and $U(\$1,600)$.

b. Plot your utility function for money over the range $0 to $2,000.

22.3.9 Consider the utility function for money

$$U(X) = \frac{100\sqrt{X}}{\sqrt{120}}$$

and the following payoff table:

	STATE OF NATURE		
Action	S_1 $P(s_1) = .25$	S_2 $P(s_2) = .40$	S_3 $P(s_3) = .35$
a_1	$120	$25	$ 0
a_2	30	60	40

a. Use the utility function to convert the outcomes in the payoff table from monetary values to utility values.

b. Which action is optimal based on the expected monetary value criterion?

c. Which action is optimal based on the expected utility criterion?

22.3.10 A doctor is involved in a malpractice suit. She can either settle out of court for $300,000 or go to court. If she goes to court and loses, she must pay the plaintiff $825,000 plus $75,000 in court costs. If she wins in court, the plaintiff pays the court costs and the doctor pays $0.

a. Construct a payoff table for this decision problem.

b. The doctor's lawyer estimates the probability of winning to be .2. Use the expected monetary value criterion to decide whether the doctor should settle or go to court.

c. Suppose the doctor's utility function for money is given by

$$U(X) = \frac{\sqrt{X + 900,000}}{\sqrt{900,000}}$$

Convert the payoffs to utility values.

d. Use the expected utility criterion to decide whether the doctor should settle or go to court.

22.4 Decision Tree Analysis

Decision problems can also be solved using decision tree analysis. When the number of acts and states is small (say, fewer than four actions and fewer than four states), a decision tree is useful because it makes every possible action and every possible state of nature easy to see. When the number of actions or possible states of nature gets fairly large, a decision tree can become too large to be manageable. For example, a payoff table containing 10 actions and 10 possible states of nature would require a decision tree with 100 branches.

DEFINITION   **Decision Tree**

A **decision tree** is a graph showing the possible actions that can be taken, the possible states of nature and their probabilities, and the possible payoffs associated with each action and each state of nature.

Decision trees can be especially useful when the problem involves a sequence of decisions to be made or a sequence of states of nature that can occur.

Decision trees use the following conventions:

1. Points in the tree where a decision must be made about an action are called *decision nodes* (represented by rectangles).
2. Points where a state of nature occurs are called *state nodes* or chance nodes (represented by circles).
3. Each path leading from a decision node represents a different action.
4. Each path leading from a state node represents a different state of nature.
5. Near each branch leading from a state node, we record the prior probability of the occurrence of that specific state of nature. These probabilities must sum to 1.
6. At the end of each branch leading from a state node, we write the payoff that would occur at that point.
7. The branch probabilities are multiplied by the payoffs at the end of those branches. The sum of the products is then recorded in the circle from which these branches emanate. The value in the circle represents the EMV of an action.

We should examine all actions from a decision node and select the action resulting in the highest EMV. Other actions should be deleted (often denoted by drawing two small lines through the appropriate branches).

EXAMPLE 22.16   **Using a Tree Diagram**

The Burns Chemical Corporation must decide whether to invest in the development of a new quick-acting glue. The corporation has three actions open to it: $a_1 = \{$Do not invest$\}$, $a_2 = \{$Hire one chemist at a cost of \$40,000$\}$, and $a_3 = \{$Hire two chemists at a cost of \$70,000$\}$. If the product is developed successfully, Burns will be able to produce 80,000 units at a cost of \$1 each and sell them all for \$3 each. If development is unsuccessful, all the research costs will be lost. The probability that one chemist working alone can develop the product successfully is .3. If two chemists work together, the probability of successful development is .6. Let us construct a decision tree for this problem and determine the optimal action.

SOLUTION   Figure 22.3 shows the decision tree. Although the sequence of actions and states of nature occur from left to right on the tree, the problem is solved by going from right to left on the tree.

Emanating from the first rectangle, or decision node, are the three possible actions: $a_1 = \{$Hire no one$\}$, $a_2 = \{$Hire one chemist$\}$, and $a_3 = \{$Hire two chemists$\}$. The payoff from a_1 is \$0. If we hire one chemist, we proceed across the tree to the first circle, or state node, where the project either succeeds (with probability .3) or fails (with probability .7). If it fails, the initial research costs are lost, so the payoff is $-$\$40,000 (placed at the end

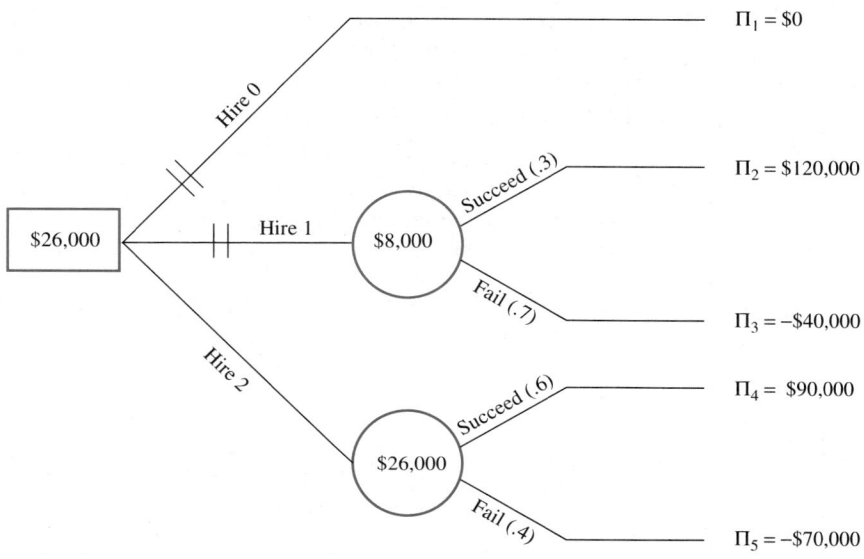

$\Pi_1 = \$0$

$\Pi_2 = \$120,000$

$\Pi_3 = -\$40,000$

$\Pi_4 = \$90,000$

$\Pi_5 = -\$70,000$

FIGURE 22.3 *Decision tree for chemical research decision, Example 22.16*

of the branch). If it succeeds, there is a net payoff of $120,000 ($160,000 in gross profit from the sale of 80,000 units minus $40,000 in research costs). If two chemists are hired and the project fails, the loss is $70,000; if two chemists are hired and the project succeeds, the firm earns $160,000 in gross profits. Subtracting the $70,000 in research costs yields a payoff of $90,000. The relevant data are placed on the decision tree.

To solve the problem, we proceed from right to left. The number in the circle at the end of the Hire 1 branch is the EMV of hiring one chemist. We obtain

$$\text{EMV(Hire one chemist)} = (120,000)(.3) + (-40,000)(.7) = \$8,000$$

The expected payoff from hiring two chemists is

$$\text{EMV(Hire two chemists)} = (90,000)(.6) + (-70,000)(.4) = \$26,000$$

The largest expected payoff, $26,000, occurs when two chemists are hired. This value is placed in the rectangle, and the other two actions are eliminated. The optimal action (EMV = $26,000) is for Burns to hire two chemists.

In the following example, a standard payoff table is inadequate because there are sequential actions, one of which is made after a state of nature is observed.

EXAMPLE 22.17   **Constructing a Decision Tree for Sequential Actions** ────────────

The T. J. Television Network has to decide which of two television series to produce for next season. The first series is a detective show with a 30% chance of a $300,000 profit and a 70% chance of an $80,000 loss. The second possibility is a comedy show with a 40% chance of a $200,000 profit and a 60% chance of a $100,000 loss. If the comedy series is successful, the network has the option of developing a second spin-off series using one of the stars from the first series. This spin-off would yield a profit of $80,000 with probability .2 or a loss of $40,000 with probability .8. Construct a decision tree for this problem and determine the optimal action.

SOLUTION Figure 22.4 shows the decision tree. The numbers by the triangles show losses or gains at the point where they occur. The payoffs at the right end of the tree branches represent the sum of all payoffs shown in the triangles along the path leading to the end points. We obtain

$$\text{EMV(Detective series)} = (300,000)(.3) + (-80,000)(.7) = \$34,000$$

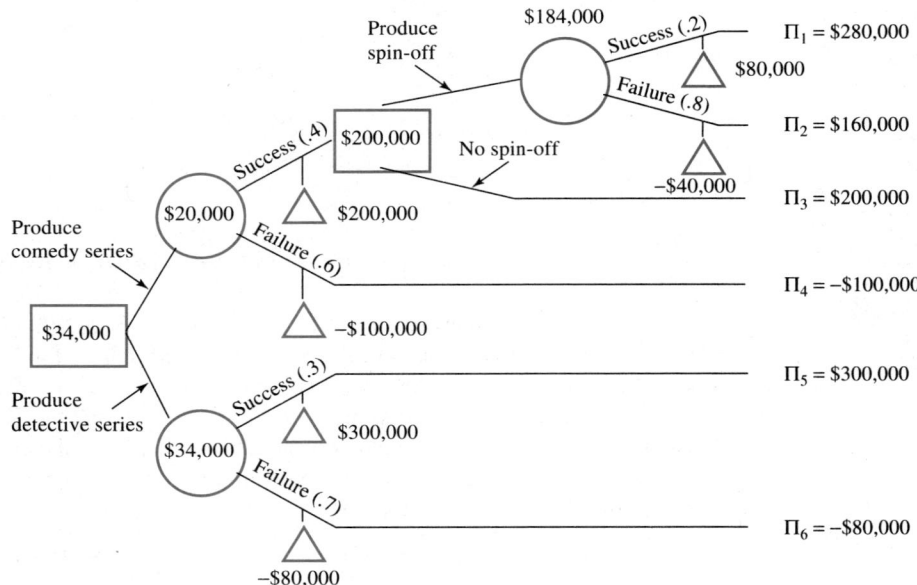

FIGURE 22.4 *Decision tree for television decision, Example 22.17*

The upper path of branches shows that if the comedy series and its spin-off are both produced and are both successful, the total payoff is $\Pi_1 = \$280,000$, the sum of the payoffs by the triangles along the branches leading to Π_1 ($200,000 + $80,000). Similarly, if the comedy series succeeds and the spin-off fails, the total payoff is $200,000 - \$40,000 = \$160,000$.

The state node at the end of the branch labeled Produce comedy series shows the EMV for the comedy series

$$\text{EMV(Comedy series)} = (200,000)(.4) + (-100,000)(.6) = \$20,000$$

where the number 200,000 is in the rectangle at the end of the branch labeled Success.

The action No spin-off yields a higher EMV than Produce spin-off. Similarly, the action Produce detective series yields a higher EMV than Produce comedy series. The optimal action is to produce the detective series, which has an expected payoff of $34,000.

Sensitivity Analysis

We should use payoff tables and decision trees with caution, since the prior probabilities and possible payoffs are only approximately correct. Sometimes relatively small changes in the state-of-nature probabilities or in the potential payoffs can lead to a different optimal action.

For example, if, in Example 22.17, we change the probability of a successful comedy series from .4 to .5, the EMV(comedy series) changes from $20,000 to $50,000. The optimal action would then be to produce the comedy rather than the detective series.

In Example 22.17, the optimal decision is sensitive to the prior probabilities. Before taking an action based on the EMV criterion, it is advisable to perform a sensitivity analysis by making modest changes in payoffs and state-of-nature probabilities to see whether the choice of optimal action is sensitive to any of these changes. If the choice is sensitive, it might be a good idea to invest some money in obtaining more reliable estimates.

Exercises

22.4.1 Use the data in Exercise 10 in Section 22.1.
  a. Construct the decision tree.
  b. Determine the optimal action based on the expected payoff.

22.4.2 Use the data in Exercise 11 in Section 22.1.
  a. Construct the decision tree.
  b. Determine the optimal action based on the expected payoff.

$$ **22.4.3** Mr. Watkins owns a plot of land on the outskirts of Orlando, Florida, and has to decide whether to sell the land now or wait a year and sell it then. The land value will increase if a shopping center is built nearby and decrease if a factory is built instead. The probability is .6 that the shopping center will be built and .4 that the factory will be built. If Watkins sells today, he will get $400,000; if he sells next year and a shopping center is built, he thinks he will get $600,000. If the factory is built instead, he will get $250,000.
  a. Construct the decision tree for this problem.
  b. Determine the optimal action based on the expected payoff.

Chapter 22 Supplementary Exercises

$$ **22.S.1** Mike Newman, just graduated from City College, has been offered a job in which the present value to Mike is $200,000. If Mike declines the job offer and goes to graduate school, the probability is .2 that he will get a Ph.D. degree and a job in which the present value is $350,000; the probability is .6 that he will get an M.A. degree and a job in which the present value is $280,000; and the probability is .2 that he will quit graduate school without a degree and get a job in which the present value is $140,000. Construct a decision tree and determine Mike's optimal action. Should he go to graduate school if he wants to maximize the expected present value of his income?

22.S.2 A grocer stocks a perishable product. Each item costs $3 and is sold for $5. Unsold items are a total loss. The grocer has to decide whether to stock 10, 20, or 30 units. The demand situation, in the grocer's opinion, is as follows: $P(\text{Demand} = 8) = .4$, $P(\text{Demand} = 17) = .5$, and $P(\text{Demand} = 31) = .1$.
  a. Construct payoff and opportunity loss tables.
  b. Determine the optimal action using the EMV criterion.

    c. Calculate the EVPI using the opportunity loss table.

    d. Construct a decision tree.

22.S.3 The National Oil Company has to decide whether to drill a well on a site in Alaska. Preliminary analysis indicates a 60% probability of striking oil on the site. A successful well will yield a net profit of $400,000, but a dry well will cause a net loss of $200,000.

    a. Construct a payoff table and a decision tree.

    b. Construct an opportunity loss table.

    c. Determine the optimal action using the EMV criterion.

    d. Calculate the EVPI.

22.S.4 A farmer must decide which one of three crops (c_1, c_2, or c_3) to plant. The potential profits depend on the weather. If the weather is good, the farmer thinks the net profit from crop c_1 will be $30,000; from c_2, $40,000; and from c_3, $50,000. If the weather is bad, each crop will yield a net loss of $10,000 for c_1, $15,000 for c_2, and $20,000 for c_3. The farmer thinks the probability of good weather is .6.

    a. Construct a payoff table and determine the optimal action.

    b. Determine the optimal action by using the minimax criterion.

22.S.5 Mary Anderson has $10,000 to invest for 1 year. She has three potential options: (a_1) she can buy a savings certificate that will yield $900; ($a_2$) she can buy stock that she thinks will sell in 1 year for a profit of $1,100 if there is no recession or a profit of $500 if there is a recession; (a_3) she can buy some real estate that she thinks will yield a profit of $3,000 if a new shopping center is built on the neighboring land, a profit of $800 if a park is built there, or a loss of $700 if the neighboring land is left undeveloped. Mary thinks the probability of a recession is .3; of a shopping center, .4; of a park, .3; and of no development, .3.

    a. Construct a decision tree.

    b. Determine the optimal action and find the expected payoff.

22.S.6 The Petrol Company has drilling rights on a large tract of land in Alaska. A preliminary analysis indicates that $P(s_1) = .3$, where s_1 means that oil exists on the land. The prior probability that the land does not contain oil is $P(s_2) = .7$, where s_2 means that oil is not present. The company has two possible actions: $a_1 = \{\text{Drill}\}$ or $a_2 = \{\text{Do not drill}\}$. The potential payoffs are as follows:

Action	s_1 = Oil	s_2 = No oil
a_1 = Drill	$1,000,000	− $500,000
a_2 = Do not drill	0	0

    a. Find the EMV of each action and determine the optimal action.

    b. Determine the EPPI.

    c. Determine the EVPI.

22.S.7 The following data show the profits (in millions of dollars) associated with five actions and six states of nature. Determine the best action under the following criteria:

Action	s_1	s_2	s_3	s_4	s_5	s_6
			STATE OF NATURE			
a_1	0	− 11	− 32	− 48	26	67
a_2	− 15	− 32	− 47	33	67	102
a_3	− 35	− 50	21	65	102	139
a_4	− 48	22	67	102	139	150
a_5	24	67	112	129	144	− 110

    a. Maximin

    b. Maximax

    c. Expected monetary value assuming $P(s_1) = .10$, $P(s_2) = .20$, $P(s_3) = .30$, $P(s_4) = .10$, $P(s_5) = .05$, and $P(s_6) = .25$.

22.S.8 The accompanying data show profits (in millions of dollars) associated with four actions and five states of nature. Determine the best action under the following criteria:

Action	STATE OF NATURE				
	s_1	s_2	s_3	s_4	s_5
a_1	88	67	58	-14	0
a_2	0	14	62	105	2
a_3	5	6	9	70	-5
a_4	-53	89	0	-12	102

    a. Maximin

    b. Maximax

    c. Expected monetary value assuming $P(s_1) = .2$, $P(s_2) = .3$, $P(s_3) = .2$, $P(s_4) = .1$, and $P(s_5) = .2$.

22.S.9 An investor is thinking of buying a bankrupt resort, which would cost $500,000 in interest fees annually plus an additional expense of $200,000 per year for overhead. At a contemplated net profit of $200 per guest, the probabilities are .1, .5, and .4, respectively, that 3,000 guests, 3,500 guests, or 4,000 guests per year will use the resort. Use a decision tree to determine the optimal action, given the desire to maximize expected monetary value and given the option of not making the investment at all.

References

Luce, Robert D., and Howard Raiffa. *Games and Decisions.* New York: Wiley, 1958.

Neumann, John von, and Oskar Morgenstern. *The Theory of Games and Economic Behavior.* Rev. ed. Princeton: Princeton University Press, 1953.

Raiffa, Howard. *Decision Analysis: Introductory Lectures on Choices Under Uncertainty.* Reading, Mass.: Addison-Wesley, 1968.

Savage, Leonard J. "The Theory of Statistical Decision." *Journal of the American Statistical Association,* March 1951, pp. 55–67.

Schlaifer, Robert. *Probability and Statistics for Business Decisions: An Introduction to Managerial Economics Under Uncertainty.* New York: McGraw-Hill, 1959.

Schlaifer, Robert. *Analysis of Decisions Under Uncertainty.* New York: McGraw-Hill, 1969.

23 Quality Control

23.1 Introduction

Quality control is a set of statistical techniques used in business and management to help producers attain and maintain consistently high-quality output. Quality control techniques help producers spot production problems as they occur—problems that may be due to worker fatigue, machine wear, faulty machine adjustment, or production design problems. Businesses save money when problems are quickly identified before numerous defective products are produced.

Originally, checking the quality of output involved 100% inspection of all the goods produced. In a modern production facility, the sheer amount of output produced makes this an inefficient and expensive way to sort good products from defective ones. Also, in certain cases, testing a product may destroy it, for example, when measuring the chemical content of medicines or the strength of wire cables. It would be unreasonable to inspect every unit of output in these instances.

In 1924, Dr. Walter A. Shewhart, an employee of Bell Laboratories, introduced statistical process control to detect changes in the output of manufacturing processes. Shewhart emphasized that variation is inevitable in any production process. He advocated the use of a graphical technique, the *control chart,* to track variation in industrial processes. The control chart helps the producer determine when a production process should be adjusted or corrected. The control chart is based on sample data obtained from the process at different points in time.

Although quality control had its origins in the 1920s, it was not utilized on a broad scale until World War II. During the war, the federal government recruited many statisticians and economists to design methods to improve the quality of industrial output used in the war effort. After World War II, emphasis on quality control declined in most countries. The Japanese, however, continued to emphasize quality control techniques, mainly because of the influence of the American statistician W. Edwards Deming. Partly as a result of Japan's success in manufacturing very-high-quality consumer products, interest in statistical quality control in the United States and other countries has increased markedly.

Statistical Quality Control

DEFINITION **Statistical Quality Control**

The application of statistical techniques to monitor the quality of a stream of output is called **statistical quality control (SQC).** Other names that are used frequently are **total quality management (TQM)** and **statistical process control (SPC).**

Basic to the idea of statistical quality control is the concept of a *process*.

DEFINITION **Process**

A **process** is any action or set of actions that results in the output of a particular good or service over time. A process is any ongoing procedure that produces output over time.

Some variation is present in the output of every process. No two items produced by a person or machine will be exactly identical. For example, no two engine bearings have exactly identical dimensions. Variation in output can affect the quality of the output, which may lead to dissatisfied customers, expensive repairs, and lost profits.

There are two ways to identify a production process that is producing defective or low-quality output. One procedure is to examine 100% of the final output to detect any defects. This can be very expensive because repairing or correcting a defective product can be nearly as costly as producing a new item. Also, thousands of defective products may be produced before the problem is noticed and the process is corrected.

Shewhart argued that it is more cost effective to monitor the crucial production processes while the goods or services are being produced so that potential problems can be identified quickly and the process can be adjusted immediately.

Quality control emphasizes the idea of prevention rather than repair, since it is less costly to prevent defects than it is to repair the defective products and afterward adjust the process.

Random Causes of Variation

The output of every repetitive process involves some uncontrollable random variation caused by the innumerable small events that occur when producing any product. This variation is referred to as *random variation,* or variation due to common causes. Even with careful attention to keeping conditions constant, some random variation in output will always be present and is inherent in the process.

DEFINITION **Random Causes of Variation**

Random variation is attributed to unknown causes that are considered to be inherent to the process. This variation cannot be reduced without redesigning the entire process.

Assignable Causes of Variation

Suppose a manufacturing process produces engine bearings whose diameters are slowly increasing over time because a machine setting is going out of adjustment. The change in the average bearing diameter over time caused by the defective machine setting is said to be due to an *assignable cause of variation.*

DEFINITION **Assignable Causes of Variation**

Assignable causes of variation are events that affect the output of a process. These events are not inherent to the design of the process.

Some factors found to be assignable causes of variation in manufacturing processes are as follows:

- Changes in the machines used in the process
- Changes in the quality or brand of raw materials used as inputs
- Changes in the quality of the employees
- Changes in temperature when the process is operating
- Changes in humidity when the process is operating
- Changes in the level of experience of the workers
- Changes in the educational background of the workers
- Changes in the amount and type of training of the workers
- Changes in output during the day shift and night shift
- Changes in quality of output on different days of the week
- Changes in quality before and after holidays, weekends, or vacations
- Changes in the lighting in the work environment
- Changes in the amount of noise and pollution present in the work environment

A process that is affected by assignable causes of variation is said to be **out of control.** A process that is affected only by random variation is said to be **in control.**

The goal of quality control analysis is to identify and remove assignable causes of variation and thereby reduce the total variation in output. Although quality control methods can help us identify *when* trouble has occurred, they do not necessarily indicate *why* the trouble has occurred. Sometimes, identifying the source of the problem may require engineering expertise or hands-on experience with the operation of the process.

Upper and Lower Specification Limits

In industry, many measurable characteristics (such as the dimensions of automobile parts or the quantities of ingredients in chemical compounds) must lie within some specified range. For example, bearing diameters might be required to fall between, say, .960 and .970 inch. The minimum and maximum acceptable values for a variable are called the specification limits.

DEFINITIONS   **Upper and Lower Specification Limits** ———————————————

The smallest acceptable value of a variable is called the **lower specification limit (LSL).** The largest acceptable value of a variable is called the **upper specification limit (USL).**

Specification limits may be two-sided with upper and lower limits or one-sided with either an upper or a lower limit.

Defective Items

DEFINITION   **Defective Item** ———————————————

A **defective item** is one that does not meet all of the specifications placed on it. Thus, an item is defective when one or more of its characteristics falls outside the specification limits.

Even though a process is in control, it can still be producing some defective output. For example, Figure 23.1 shows the probability distribution of a quantitative variable, such as the diameter of a bottle cap. In Figure 23.1, the upper and lower specification limits are plotted at the values LCL = .495 inch and USL = .505 inch. Because part of the probability distribution falls outside the specification limits, it is inevitable that some of the output will be defective.

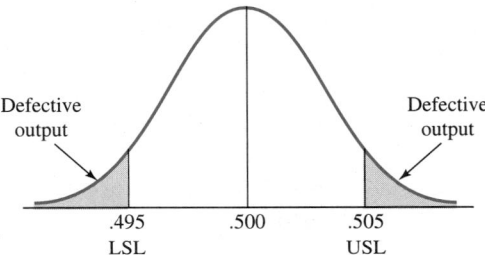

FIGURE 23.1   *Probability distribution of an output variable and its upper and lower specification limits*

Even though the production process may be in control, the process will not be capable of producing 100% acceptable product because there is too much random variation present. To ensure 100% acceptable output, the production process has to be redesigned to reduce this random variation. If the standard deviation of the probability distribution is sufficiently reduced, almost all of the output may fall within the specification limits.

Uniformity Versus Acceptability

Just as a process that is in control can produce defective output, a process that is out of control can produce acceptable output. Even though the output of a process is acceptable, it is possible that the process is changing in such a way that it is likely that future output will be defective. The changes in uniformity over time are an indicator that the production process is out of control. An assignable cause of variation influences the probability distribution of future output and could cause future output to be defective even though 100% of current output is acceptable.

The following example illustrates a situation where a process is out of control despite the fact that it has produced 100% acceptable output.

EXAMPLE 23.1   **Diameters of Bottle Caps** ─────────────────────────────────────

The data in Table 23.1 (page 982) show the diameters of a sample of 100 bottle caps that were obtained sequentially from a production process at a food distributing plant. The specification limits require that the caps must be between .560 inch and .595 inch in diameter to ensure a proper fit. Caps outside this range are considered defective and must be discarded.

Table 23.2 (page 982) shows the frequency distribution and relative frequency distribution of the diameters of the sample of 100 bottle caps listed in Table 23.1. Figure 23.2 (page 982) shows the histogram of the diameters of the bottle caps. The frequency distribution indicates that all of the diameters fall within the required specifications. Thus, all the output in the sample would be considered acceptable.

TABLE 23.1   *Diameters of Bottle Caps for Example 23.1*

.5821	.5843	.5765	.5811	.5784	.5833	.5826	.5827	.5834	.5791
.5782	.5802	.5771	.5835	.5845	.5832	.5811	.5837	.5824	.5823
.5838	.5848	.5844	.5852	.5834	.5842	.5868	.5859	.5832	.5849
.5824	.5835	.5827	.5858	.5868	.5858	.5844	.5858	.5862	.5867
.5875	.5869	.5872	.5869	.5866	.5858	.5865	.5862	.5875	.5858
.5866	.5856	.5868	.5879	.5887	.5845	.5847	.5855	.5844	.5872
.5867	.5853	.5865	.5876	.5889	.5846	.5847	.5856	.5847	.5876
.5863	.5867	.5858	.5873	.5887	.5846	.5851	.5853	.5866	.5874
.5889	.5864	.5862	.5870	.5883	.5859	.5890	.5912	.5879	.5878
.5892	.5912	.5910	.5877	.5887	.5911	.5919	.5917	.5883	.5875

TABLE 23.2   *Frequency Distribution of Diameters of Bottle Caps for Data in Table 23.1*

Diameter	Frequency	Relative frequency
.5760–.5779	2	.02
.5780–.5799	3	.03
.5800–.5819	3	.03
.5820–.5839	16	.16
.5840–.5859	29	.29
.5860–.5879	32	.32
.5880–.5899	9	.09
.5900–.5919	6	.06
Total	100	1.00

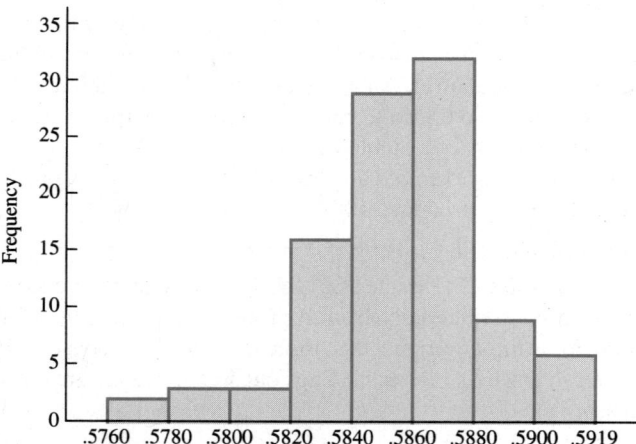

FIGURE 23.2   *Frequency distribution of diameters of bottle caps listed in Table 23.1*

Time Series Data and Time Plots

The output of any production process can be arranged chronologically to form a time series. It is always useful to plot the data versus time, because any unusual or systematic patterns in the sequence of observations become apparent. Frequently, a time plot will reveal more than a sophisticated statistical analysis.

Figure 23.3 shows the time plot of the bottle cap diameters listed in Table 23.1. To detect the potential presence of assignable causes of variation, it is useful to search for unusual patterns in the observed series of output.

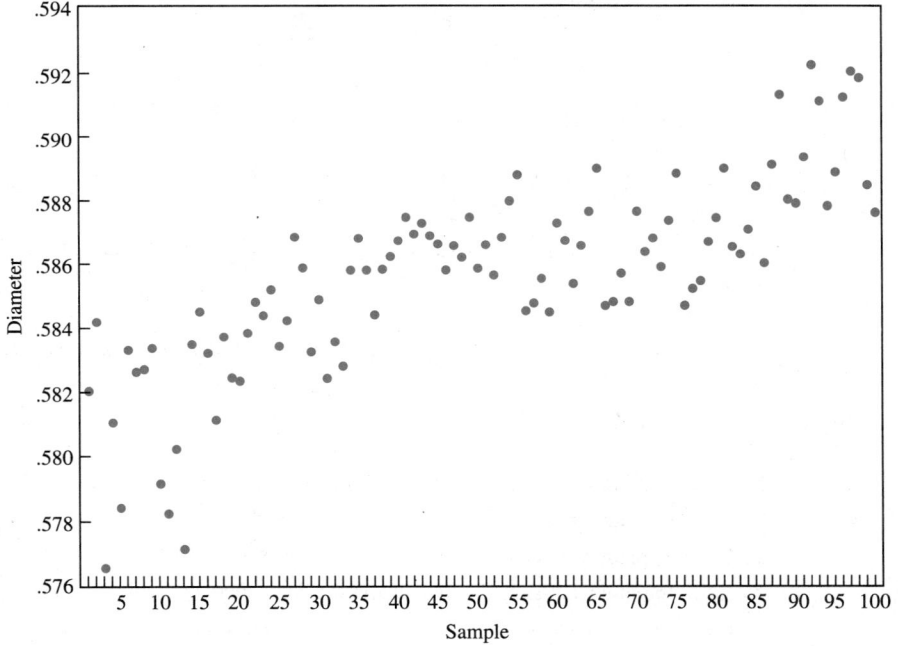

FIGURE 23.3 *Time plot of diameters of bottle caps listed in Table 23.1*

Figure 23.4 (page 984) shows the time plot of diameters along with a horizontal line, called the **centerline,** located at the mean of the observations. Also, successive points have been connected to help visualize patterns in the data.

Figure 23.4 shows a gradual upward trend in the time series of diameters. This should be a clue that something is affecting the process and causing the diameters to slowly increase. Perhaps a machine setting has gotten out of adjustment.

Even though all the values in the sample meet the producer's specification limits, this upward trend indicates that the process may soon start producing defective output. Observing the histogram alone would not have revealed this potential problem.

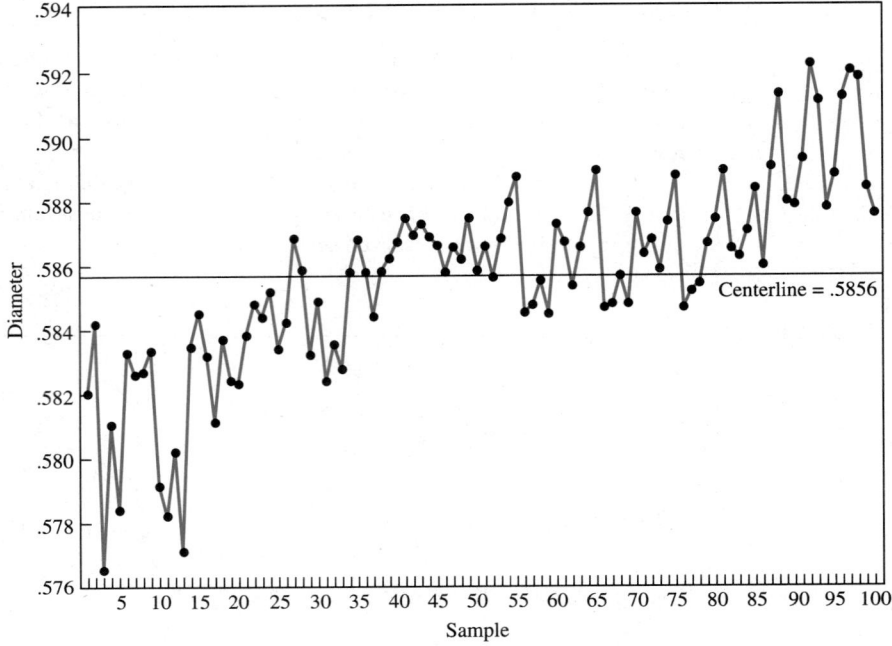

FIGURE 23.4 *Time plot of diameters of bottle caps with a centerline*

Exercises

23.1.1 Explain what is meant by an upper and a lower specification limit.

23.1.2 Explain what is meant by common causes of variation.

23.1.3 Explain what is meant by assignable causes of variation.

23.1.4 Explain what is meant by a process. Why is there variability in the output of every process?

23.1.5 List some sources of assignable causes of variation in manufacturing situations.

23.2 Some Patterns That Reveal Assignable Causes of Variation

To see how quality control techniques are used to search for assignable causes of variation, it is useful to think of the series of values in Table 23.1 as a set of observations selected from a series of probability distributions. The chronological series of observations can be thought of as being generated by random selections from a sequence of probability distributions, as illustrated in Figure 23.5. If the means of these probability distributions gradually increase over time, then we are likely to obtain an observed series of output like the one illustrated in Figure 23.4.

Statistical quality control techniques are designed to detect potential changes over time in the means and variances of this sequence of distributions. If all the distributions are identical (with identical means and variances), the process is in control. If the distributions change over time, the process is out of control.

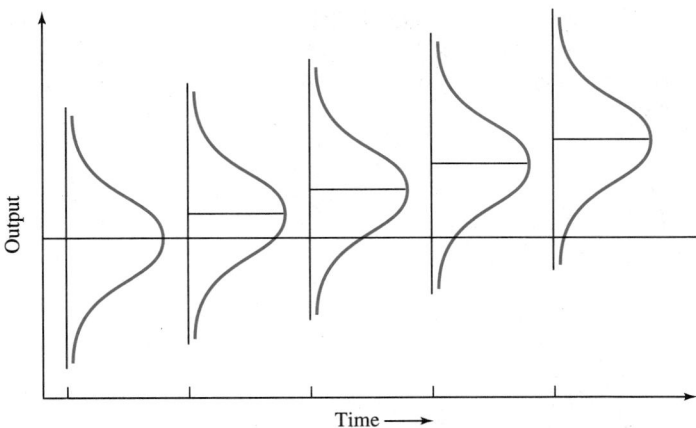

FIGURE 23.5 *A sequence of probability distributions with an upward trend in mean value*

Time Plots Showing Various Patterns of Output

The values plotted in Figure 23.4 provide an example of a time series of data generated by a process that is out of control because the mean level of output is gradually increasing over time. There are other patterns of output that occur frequently when processes have gone out of control because of assignable causes of variation.

Figure 23.6 shows the time plot of a series with a decreasing trend. Such a time plot could be generated by output obtained from a process where the mean level of output is shifting downward over time.

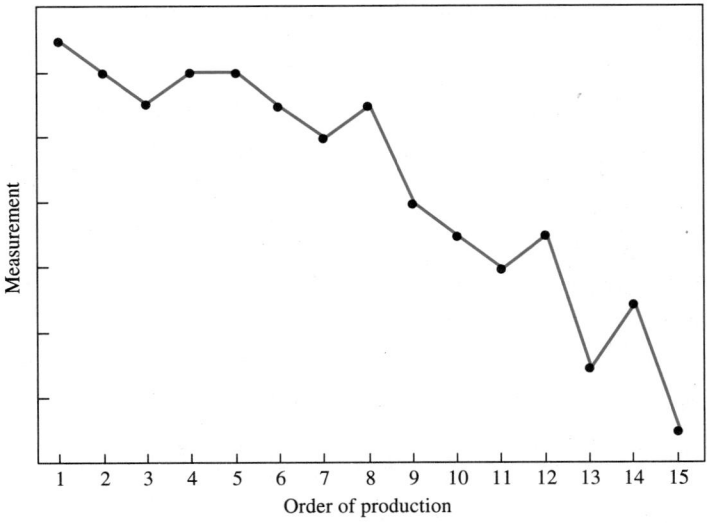

FIGURE 23.6 *Example of a time series with a decreasing trend*

Figure 23.7 shows the time plot of an oscillating sequence in which the observations alternate above and below the centerline. This type of sequence could be caused by a defective piece of machinery, such as a worn bearing or a valve that sticks at regular intervals.

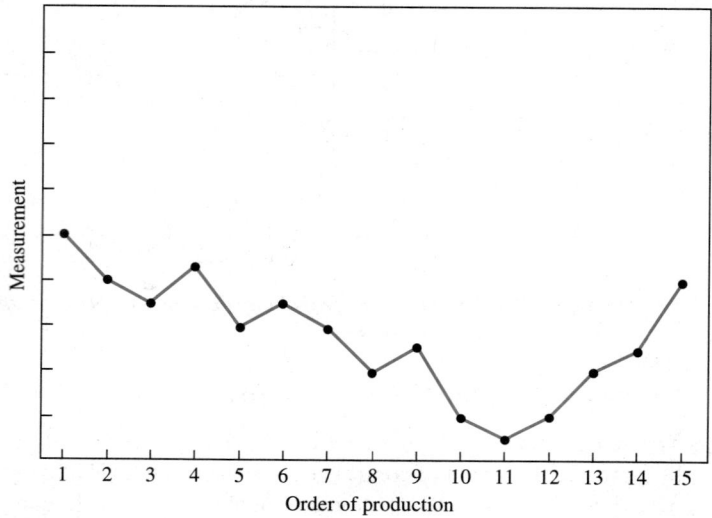

FIGURE 23.7 *Example of an oscillating time series*

Figure 23.8 shows the time plot of a series with an unchanging mean but an increasing variance. Such a series might occur when a production worker has become fatigued

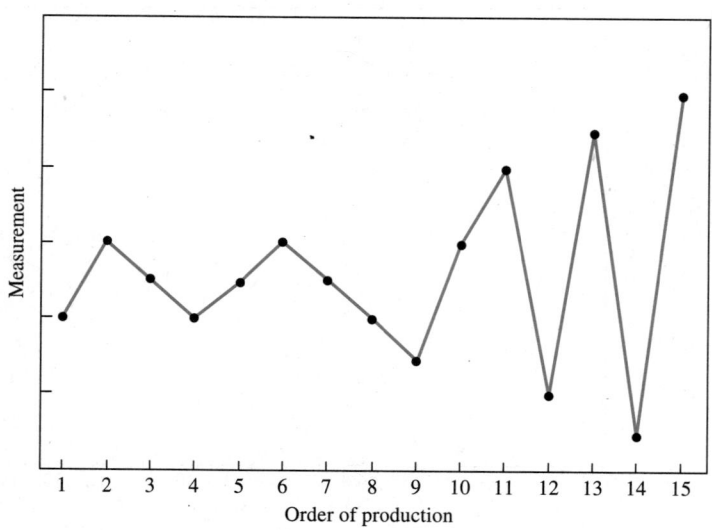

FIGURE 23.8 *Example of a time series with increasing variance*

and loses concentration. In Figure 23.8, the mean level of output may be acceptable, but the individual output varies greatly from one item to another.

Figure 23.9 illustrates a time series of probability distributions from which the time series in Figure 23.8 could have been obtained. Suppose that, over time, the output variable has a probability distribution with a constant mean but an increasing variance. The observed output over time can be thought of as the result of random selections from a sequence of probability distributions, as illustrated in Figure 23.9.

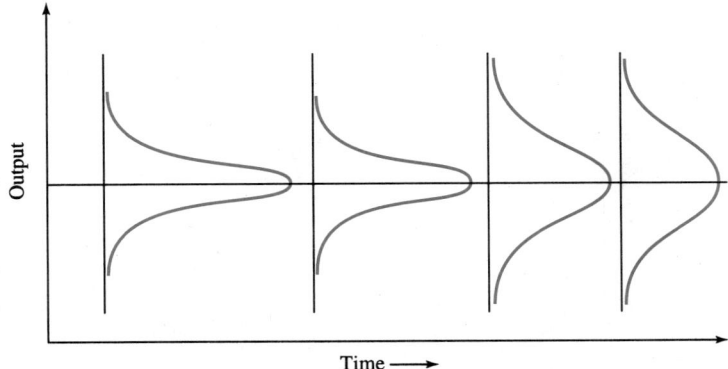

FIGURE 23.9 *A sequence of probability distributions with gradually increasing variances*

Figure 23.10 shows the time plot of a process whose output fluctuates in a fairly regular cycle. Such output could be caused by hourly changes in temperature or humidity,

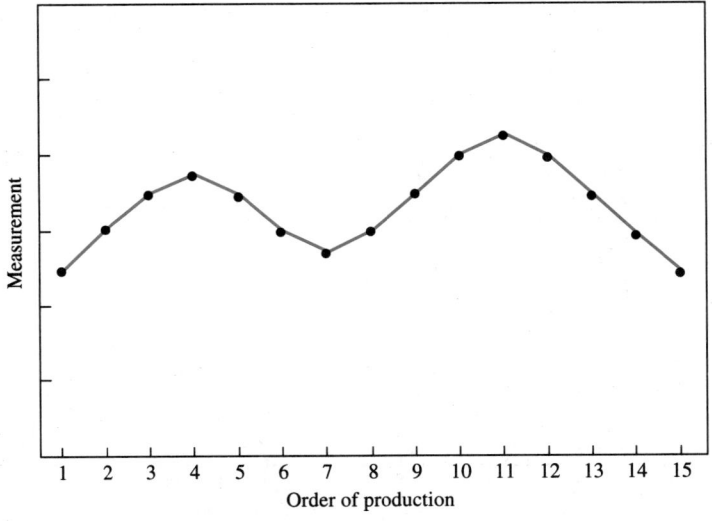

FIGURE 23.10 *Example of a cyclical time series*

changes in employees during different shifts, or changes in worker concentration before and after work breaks.

Figure 23.11 shows the time plot of a series affected by a one-time random shock or a freak occurrence. This might occur when a new, untrained employee operates a machine for one work period or when a single bad batch of input raw materials, such as an incorrect chemical compound, is used.

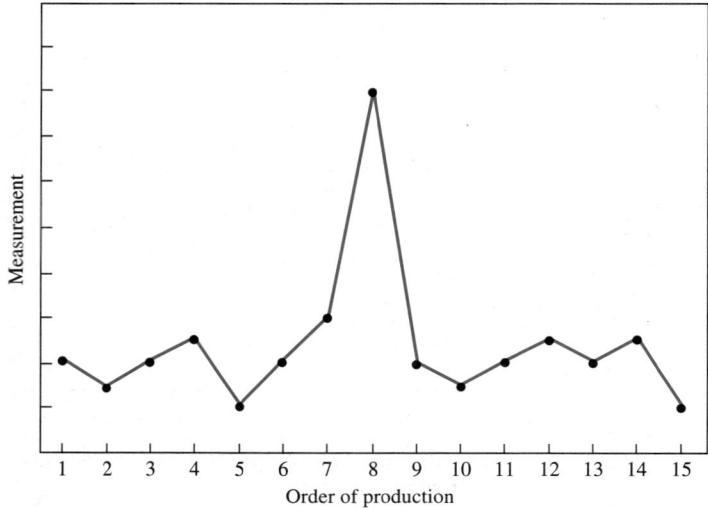

FIGURE   23.11   *Example of a time series affected by a one-period random shock*

Figure 23.12 shows the time plot of a series affected by a sudden change in the mean level of output. This could occur when a piece of machinery breaks and a new piece of

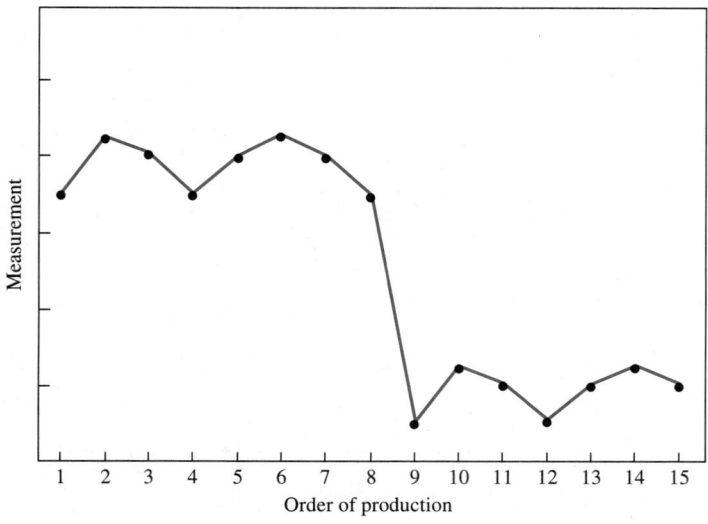

FIGURE   23.12   *Example of a time series with a shift in mean level of output*

machinery is introduced, when a worker is changed, when a shift is changed, or when the source of the input raw materials is changed.

Figure 23.13 illustrates how the time series in Figure 23.12 could have been obtained. At each point in time, the observed output over time can be thought of as the result of random selections from a sequence of probability distributions such as that illustrated in Figure 23.12. Figure 23.13 illustrates a situation where there was a permanent downward shift in the mean of these probability distributions. In such a case, we are likely to obtain an observed series of output such as that illustrated in Figure 23.12.

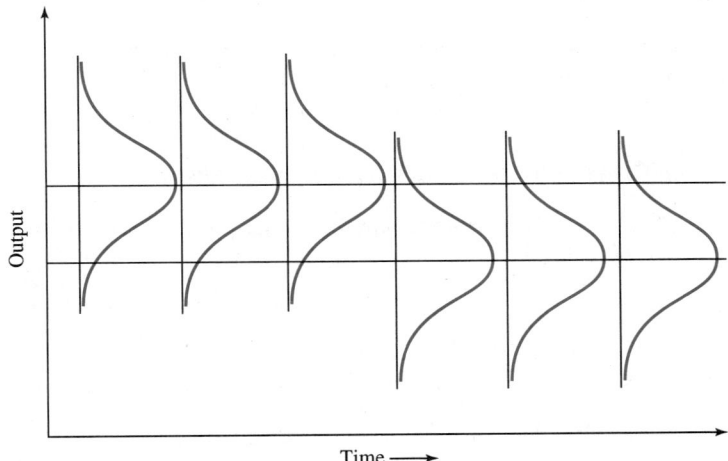

FIGURE 23.13 *A sequence of probability distributions with a permanent downward shift in process mean*

Exercises

Statistical Concepts

23.2.1 When analyzing the output from some manufacturing process, explain why it is important to construct both a frequency distribution and a time plot to show the sequence of output. What can be learned from the time plot that is not disclosed by the frequency distribution?

23.2.2 Draw time plots of output showing an upward trend, a changing variance, a cyclical mean, a sudden permanent shift in mean, and a one-time shift in mean.

Statistical Drills

23.2.3 Consider the following time series data showing the lengths of time in seconds an employee took to produce each of 20 products on an assembly line:

Product	1	2	3	4	5	6	7	8	9	10
Time Required	7.3	7.8	7.6	6.9	8.1	7.6	8.3	8.5	7.9	8.6
Product	11	12	13	14	15	16	17	18	19	20
Time Required	7.9	8.8	8.6	8.9	8.1	8.6	9.3	9.5	9.9	9.6

a. Construct a time series plot. Connect the plotted points and add a centerline.
b. Do you detect any particular pattern in the data?
c. Can you think of any explanation for this pattern?

23.2.4 The following data show number of units of output produced per hour during a period of 20 hours. Four employees rotate performing the job, with each employee working for 1 hour.

Hour	1	2	3	4	5	6	7	8	9	10
Output	17	25	31	11	18	28	34	13	18	26

Hour	11	12	13	14	15	16	17	18	19	20
Output	38	15	20	29	48	24	31	37	53	32

a. Construct a time series plot. Connect the plotted points and add a centerline.
b. Do you detect any particular pattern in the data?
c. Can you think of any explanation for this pattern?

23.3 $\bar{x}$-Chart to Monitor the Process Mean

Quality control techniques can be thought of as a set of procedures designed to test the null hypothesis

$$H_0: \quad \text{The process is in control}$$

against the alternative hypothesis

$$H_1: \quad \text{The process is out of control}$$

Control charts are the most commonly used statistical tools in quality control analysis. The most important control charts are the $\bar{x}$-chart (which is used to monitor the mean level of a process), the R-chart (which monitors process variability), and the p-chart (which monitors the proportion of defectives).

In this section, we discuss the $\bar{x}$-chart, which is used to monitor potential changes in the mean level of some variable. The $\bar{x}$-chart is typically used along with an R-chart, which monitors changes in the variability of some variable. These charts help us determine whether a process has gone out of control because the process mean or the process standard deviation has changed. The R-chart is explained in the next section.

The first step in constructing a control chart is to collect samples of data from the process being studied. Typically, a new sample of data is collected at each of k regular intervals, such as at the start of each hour, each shift, or some other systematic interval, where each sample contains n items. For each of the k samples, the sample mean $\bar{x}_i$ is calculated. The essential feature of an $\bar{x}$-chart is a plot of the sequence of k sample means ($\bar{x}_i$) against the sample number.

The theory behind the $\bar{x}$-chart is that, if the process is in control, the sequence of sample means should vary randomly about the process population mean μ and most of the sample means should fall within 3 standard deviations of the population mean; that is, most of the sample means should fall in the interval ($\mu - 3\sigma_{\bar{x}}, \mu + 3\sigma_{\bar{x}}$).

Any unusual value of a sample mean $\bar{x}_i$ or any unusual pattern in the plot of sample means serves as an indication that the process may have gone out of control. As long as the control chart signals that the process is operating normally, the process is allowed to continue. When the chart signals trouble, corrective action is taken.

Selection of Samples: Rational Subgrouping

Before collecting the data, both the frequency with which samples will be taken and the sample size n must be determined. To improve the chances of detecting any change in the process, the sample collection should be timed such that any suspected change in the process occurs *between* samples, not *within* samples. In an optimal situation, sample means obtained before the occurrence of some assignable cause for variation will be substantially different from those obtained after the change, an indication that something has happened in the interim to change the process.

Rational Subgrouping

DEFINITIONS   **Subgroups and Rational Subgroups**

Each of the k samples used in a quality control investigation is called a **subgroup.** The subgroups are called **rational subgroups** if they have been selected in such a way that the potential variation within each particular subgroup is minimized and the potential variation between different subgroups is maximized.

Choice of Sample Size (n) and Choice of Number of Subgroups (k)

Typically the sample size n is kept small because it is desirable to minimize the opportunity for variation *within* a subgroup. In industrial situations, the most frequently used value for subgroup size is $n = 4$ or $n = 5$. Most quality control engineers recommend that the number of subgroups should be at least $k = 20$.

When the process is in control, the process population mean μ is constant from sample to sample. The value of μ is unknown, so we estimate it by calculating the *grand mean,* the average of the k sample means.

DEFINITION   **Grand Mean $\bar{x}$**

The **grand mean** is denoted $\bar{x}$ (without a subscript) and is calculated as follows:

$$\bar{x} = \frac{\bar{x}_1 + \bar{x}_2 + \cdots + \bar{x}_k}{k}$$

where k denotes the number of samples, and $\bar{x}_i$ denotes the sample mean for the ith sample.

On the $\bar{x}$-chart, a horizontal centerline is plotted at the grand mean to represent the estimated mean of the process.

Theoretical Control Limits

On the $\bar{x}$-chart, additional horizontal lines are plotted 3 standard deviations above and below the centerline. These lines represent the **upper control limit (UCL)** and the **lower control limit (LCL)** of the control chart.

When determining the upper and lower control limits, it is important to use the standard deviation for the sample mean ($\sigma/\sqrt{n}$), not the standard deviation σ for an individual observation. The $\bar{x}$-chart is based on the k sample means $\bar{x}_i$ rather than on individual measurements x_i, so the relevant standard deviation is the standard deviation of $\bar{x}_i$, which is $\sigma/\sqrt{n}$, not σ.

If values of μ and σ were known, the theoretical upper and lower control limits would be determined as follows:

$$\text{Theoretical upper control limit:} \quad \mu + 3\frac{\sigma}{\sqrt{n}}$$

$$\text{Theoretical lower control limit:} \quad \mu - 3\frac{\sigma}{\sqrt{n}}$$

Estimation of the Upper and Lower Control Limits

To estimate the upper and lower control limits, we must estimate both the process mean μ and the quantity $3\sigma/\sqrt{n}$.

The process mean is estimated by the grand mean $\bar{x}$. If the process is in control, the grand mean provides an unbiased estimate of the process mean.

To estimate the quantity $3\sigma/\sqrt{n}$, we could rely on $3s/\sqrt{n}$, where s is the sample standard deviation of all the process observations. This was done in the early years of quality control. Quality control engineers found that shop workers had difficulty understanding and calculating the sample standard deviation. Subsequent statistical research relied on the series of k sample ranges to obtain reliable estimates of $3\sigma/\sqrt{n}$.

To estimate $3\sigma/\sqrt{n}$, we first compute the sample range R_i for each of the k subgroups. Next, obtain $\bar{R}$, the mean of the k sample ranges. Next, the mean range, $\bar{R}$, is multiplied by a constant A_2 and the value $A_2\bar{R}$ is used as an estimate of $3\sigma/\sqrt{n}$. The appropriate value of the constant A_2 depends on the sample size n. Values of A_2 are given in Table 23.3 and in Table A.14 in the Appendix for all sample sizes.

TABLE 23.3 *Values of A_2 Used when Constructing an $\bar{x}$-Chart*

Sample size n_i	Coefficient A_2	Sample size n_i	Coefficient A_2
2	1.880	14	.235
3	1.023	15	.223
4	.729	16	.212
5	.577	17	.203
6	.483	18	.194
7	.419	19	.187
8	.373	20	.180
9	.337	21	.173
10	.308	22	.167
11	.285	23	.162
12	.266	24	.157
13	.249	25	.153
		Over 25	$3\sigma/\sqrt{n}$

$A_2\bar{R}$ is used to estimate the quantity $3\sigma/\sqrt{n}$ when constructing upper and lower control limits on an $\bar{x}$-chart.

The estimated upper and lower control limits are obtained by substituting $\bar{x}$ for μ and substituting $A_2\bar{R}$ for $3\sigma/\sqrt{n}$ in the formulas for the theoretical control limits.

DEFINITION **Estimated Upper and Lower Control Limits** ————————————

$$\text{Estimated upper control limit:} \quad \bar{x} + A_2\bar{R}$$
$$\text{Estimated lower control limit:} \quad \bar{x} - A_2\bar{R}$$

where

$$\bar{R} = \frac{R_1 + R_2 + \cdots + R_k}{k}$$

Elements of an x̄-chart

The control chart for the mean is obtained by plotting the sequence of k sample means $(\bar{x}_i)$ against the sample number. In addition, horizontal lines are plotted to represent the centerline and the upper and lower control limits.

$$\text{Centerline:} \qquad \bar{x} = \frac{1}{k}\sum_{i=1}^{k}\bar{x}_i$$

$$\text{Upper control limit:} \qquad \text{UCL} = \bar{x} + A_2\bar{R}$$

$$\text{Lower control limit:} \qquad \text{LCL} = \bar{x} - A_2\bar{R}$$

Note: The value of A_2 depends on the sample size n and is given in Table 23.3 and in Table A.14 of the Appendix. It is recommended that k be at least 20.

The theory underlying the use of the control chart relies on the assumption that each sample mean $\bar{x}_i$ is a value selected from a distribution that is approximately normal with constant mean μ and standard deviation $\sigma/\sqrt{n}$. (Recall that the Central Limit Theorem tells us that the sampling distribution of the sample mean tends toward a normal distribution as the sample size increases. If the distribution of the individual observations is roughly mound shaped, the sampling distribution of the sample mean $\bar{x}_i$ will be approximately normal even when the sample size is as small as 4 or 5.)

The table showing areas under the standard normal distribution indicates that the probability that any normal random variable will fall more than 3 standard deviations above its mean is approximately .0013. Similarly, the probability is .0013 that it will fall more than 3 standard deviations below its mean. Thus, the probability is approximately .0026 (or less than 3 in 1,000) that an individual sample mean $\bar{x}_i$ will fall more than 3 standard deviations from the process mean μ.

If the process is in control and the sample means follow approximately a normal distribution, fewer than 3 in 1,000 sample means should fall outside the interval ($\mu - 3\sigma/\sqrt{n}$, $\mu + 3\sigma/\sqrt{n}$). When we observe even one sample mean $\bar{x}_i$ outside the upper or lower control limits, we suspect that the process is out of control. In such a case, the source of the problem should be sought and corrected.

The $\bar{x}$-chart should not be used without first constructing the corresponding R-chart, since the calculation of the upper and lower control limits on the $\bar{x}$-chart rely on the average sample range, $\bar{R}$. If the R-chart reveals that the variability of the process is out of control, the control limits on the $\bar{x}$-chart will not be correct. (In this section, we assume that the process variability is in control and remains constant.)

Rules for Pattern Analysis

Even if all the sample means fall between the control limits, the occurrence of an unusual sequence of observations can indicate that the process is out of control. Any pattern of output that appears to be nonrandom and systematic could be evidence that the process is out of control. For example, a sequence of gradually increasing sample means might indicate that the process is out of control.

Quality control engineers have developed many different rules or guidelines concerning what represents an "unusual" sequence of observations. Each of the guidelines illustrates an event that has a very low probability of occurring when the process is in control. Searching for unusual patterns in the time plot of data on the control chart is referred to as **pattern analysis.**

The rules in the accompanying box are frequently used by quality control engineers to help identify situations when a process may have gone out of control.

Rules for pattern analysis

Consider a process to be **out of control** if any of the following patterns is observed:

Pattern 1:   Any value falls more than 3 standard deviations from the centerline
Pattern 2:   Nine consecutive values fall on one side of the centerline
Pattern 3:   Six consecutive values are steadily increasing or decreasing
Pattern 4:   Fourteen consecutive values are alternating up and down
Pattern 5:   At least two out of three consecutive values fall more than 2 standard deviations above (below) the centerline
Pattern 6:   At least four out of five consecutive values fall more than 1 standard deviation above (below) the centerline
Pattern 7:   Fifteen consecutive values (on both sides of the centerline) fall within 1 standard deviation of the centerline
Pattern 8:   Eight consecutive values (on both sides of the centerline) fall more than 1 standard deviation from the centerline

Some quality control engineers rely solely on pattern 1 when monitoring a process. Based on pattern 1, a process is considered to be out of control if the control chart contains at least one value beyond the upper or lower control limits. Other analysts recommend applying all eight of the rules for pattern analysis. If the process is in control, each pattern listed in the box has a very small probability of occurring.

Before applying the rules for pattern analysis, it is useful to visualize a normal distribution that has been separated into eight "zones" or regions, four located above the mean (denoted zones 1A, 2A, 3A, and 4A) and four below the mean (denoted zones 1B, 2B,

3B, and 4B). Zone 1A denotes all values between μ and $(\mu + 1\sigma)$; that is, Zone 1A contains all values between the mean and 1 standard deviation above the mean. Zone 2A denotes all values between $(\mu + 1\sigma)$ and $(\mu + 2\sigma)$; that is, zone 2A contains all values falling more than 1 and less than 2 standard deviations above the mean. Zone 3A denotes all values between $(\mu + 2\sigma)$ and $(\mu + 3\sigma)$. Finally, zone 4A denotes all values more than 3 standard deviations above the mean. Zones 1B, 2B, 3B, and 4B represent similar regions below the mean.

If the data follow a normal distribution, the probability of obtaining an observation in any particular zone can be found by referring to the table showing areas under the standard normal distribution (Table A.5 in the Appendix). For example, the probability that a point falls in zone 1A is the same as the probability of getting a Z score between 0 and 1.0. This probability is .3413. The probability that a point falls in zone 2A is the same as the probability of getting a Z score between 1.0 and 2.0. This probability is $.4772 - .3413 = .1359$. The probability that a point falls in zone 3A is the same as the probability of getting a Z score between 2.0 and 3.0. This probability is $.4987 - .4772 = .0215$. Finally, the probability that a point falls in zone 4A is the same as the probability of obtaining a value greater than $z = 3$ from the standard normal distribution. This probability is $.5000 - .4987 = .0013$. Identical probabilities apply to zones 1B, 2B, 3B, and 4B. These probabilities are useful for determining the probability of occurrence of various observations and various sequences of observations.

Figure 23.14**a** shows the location of these zones on a horizontal line. Figure 23.14**b** shows the location of these zones on a standard normal curve.

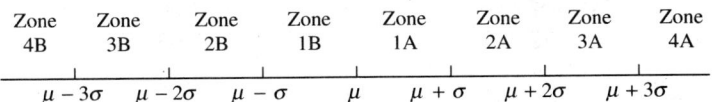

FIGURE 23.14a *Location of zones 1, 2, 3, and 4 on a horizontal line*

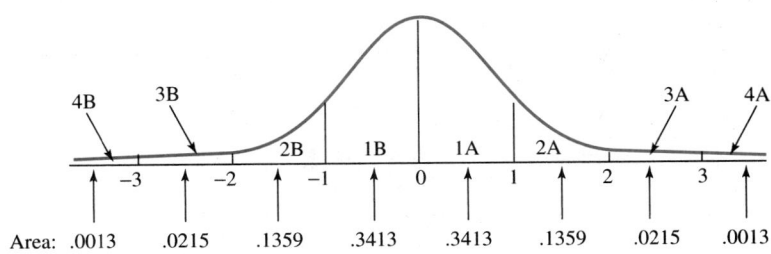

FIGURE 23.14b *Location of zones 1, 2, 3, and 4 on a standard normal curve*

To apply all the rules for pattern analysis, it is useful to show all these zones on the control chart. Figure 23.15 shows the form of a typical $x̄$-chart. The chart contains a horizontal centerline located at the sample mean of the variable being plotted. The additional horizontal lines are located 1, 2, and 3 standard deviations above and below the centerline. These lines separate the control chart into the eight zones previously described. Finally, the values of the k sample means are plotted on the control chart.

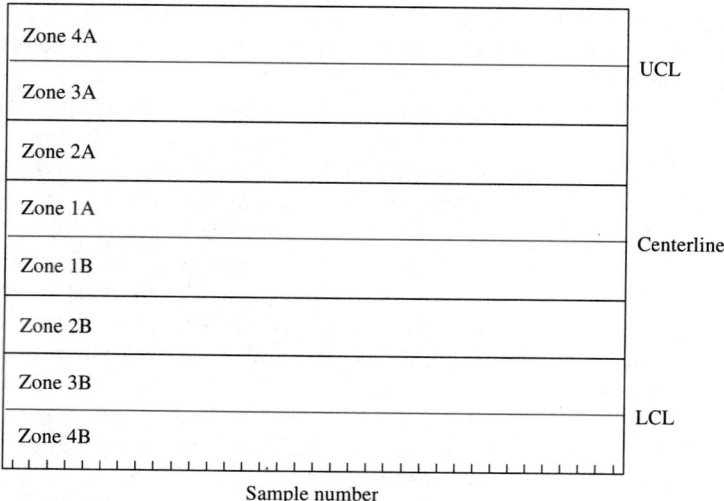

FIGURE 23.15   *Example of the format of a typical control chart*

Figure 23.16 shows an $\bar{x}$-chart where a process is considered to be out of control because a sample mean falls above the upper control limit or below the lower control limit.

Figures 23.17–23.23 show other patterns that might indicate that a process is out of control.

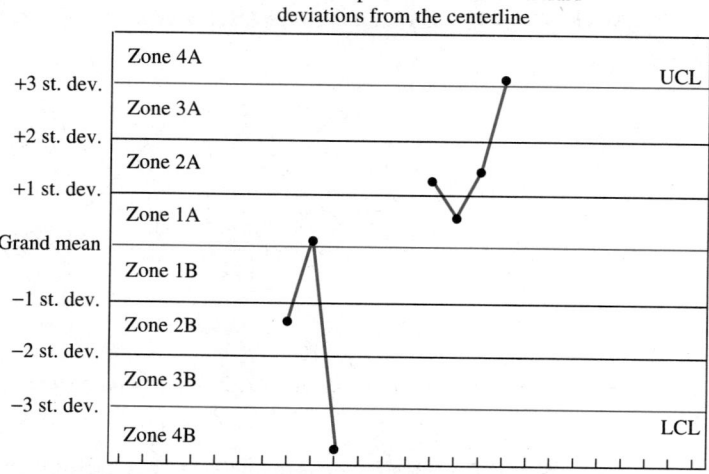

**FIGURE 23.16   *Example of pattern 1: At least one value
more than 3 standard deviations from the centerline***

Figure 23.17 shows a control chart where pattern 2 has occurred because nine consecutive sample means fall above the centerline. If successive values are independent, the probability of obtaining nine consecutive values on one side of the centerline is $(.5)^9 = .00195$, or less than 2 times in 1,000.

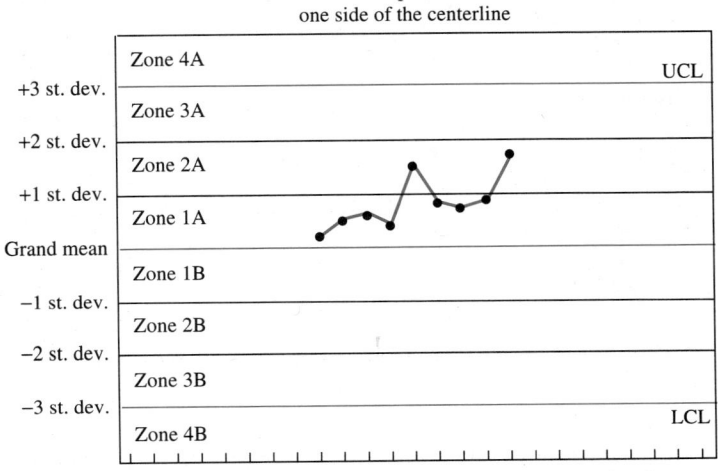

FIGURE 23.17 *Example of pattern 2: Nine consecutive values on one side of the centerline*

Figure 23.18 illustrates a situation where pattern 3 has occurred because there are six consecutive sample means steadily increasing or decreasing. Figure 23.19 (page 998)

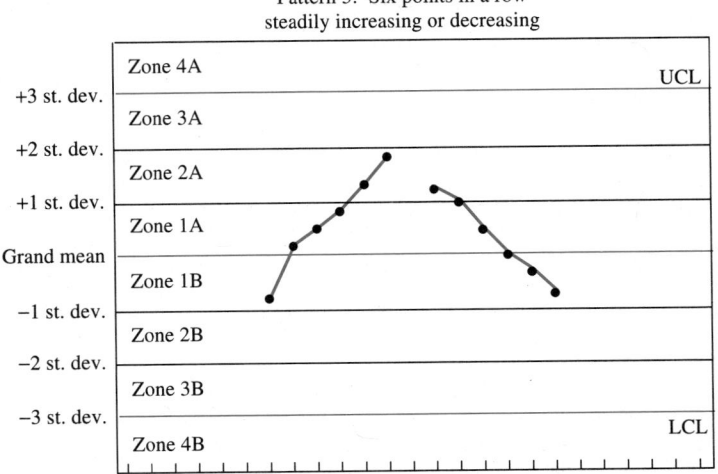

FIGURE 23.18 *Example of pattern 3: Six consecutive values steadily increasing or decreasing*

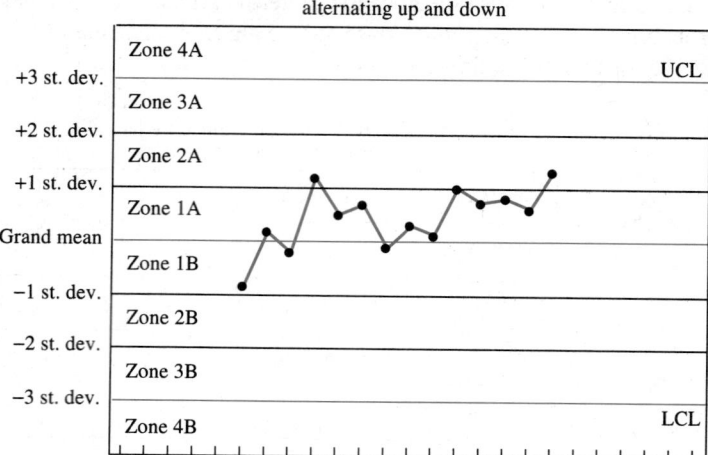

FIGURE 23.19 *Example of pattern 4: Fourteen consecutive values alternating up and down*

shows a control chart where pattern 4 has occurred because there are 14 consecutive sample means alternating up and down.

Figure 23.20 illustrates three situations where pattern 5 has occurred because there are at least two out of three consecutive sample means more than 2 standard deviations above (below) the centerline. When the process is in control, for a given sequence of three consecutive values, pattern 5 occurs with probability .003072, or less than 4 times in 1,000.

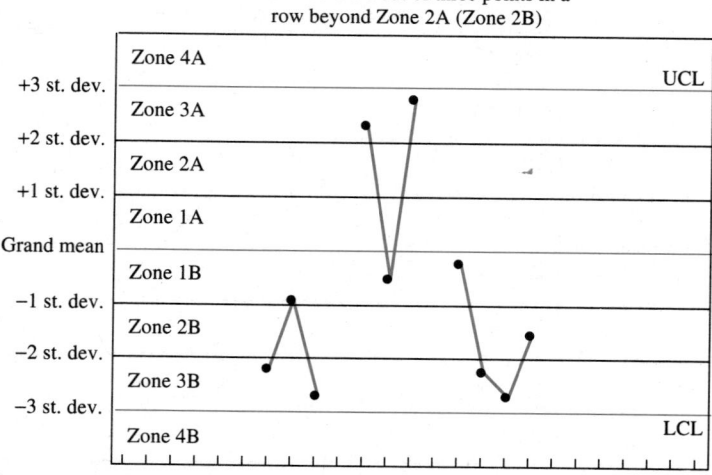

FIGURE 23.20 *Example of pattern 5: Two out of three consecutive values more than 2 standard deviations above (below) the centerline*

Figure 23.21 illustrates two situations where pattern 6 has occurred because there are at least four out of five consecutive sample means more than 1 standard deviation above (below) the centerline. When the process is in control, for a given sequence of five consecutive values, pattern 6 occurs with probability .005538, or less than 6 times in 1,000.

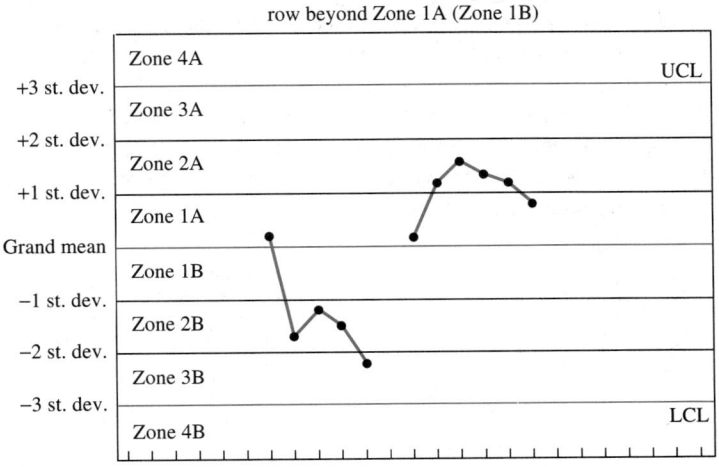

Figure 23.22 illustrates a situation where pattern 7 has occurred because there are 15 consecutive sample means less than 1 standard deviation from the centerline. When the

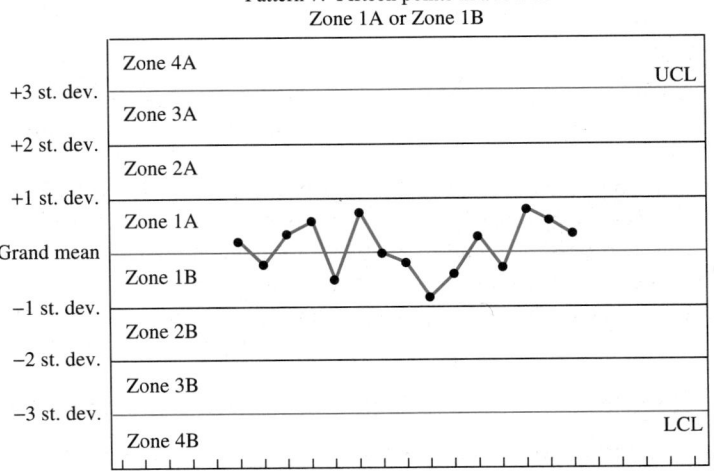

FIGURE 23.22 *Example of pattern 7: Fifteen consecutive values less than 1 standard deviation from the centerline*

process is in control, for a given sequence of 15 consecutive values, pattern 7 occurs with probability .00325, or less than 4 times in 1,000.

Pattern 7 deserves a special comment. When pattern 7 occurs, the output appears to be too uniform, rather than too variable. Thus, the occurrence of pattern 7 could be a signal that the process has changed in a favorable way toward producing more uniform output. Management should inspect the process to determine the cause of this fortunate occurrence and to see whether it can be permanently incorporated into the process. (Some quality control engineers regard the occurrence of pattern 7 as a potential indication that the data may have been "fudged" because the values are "too uniform.")

Figure 23.23 illustrates a situation where pattern 8 has occurred because there are eight consecutive sample means more than 1 standard deviation from the centerline. When a process is in control, the probability of occurrence of pattern 8 is .000103, or approximately 1 in 10,000.

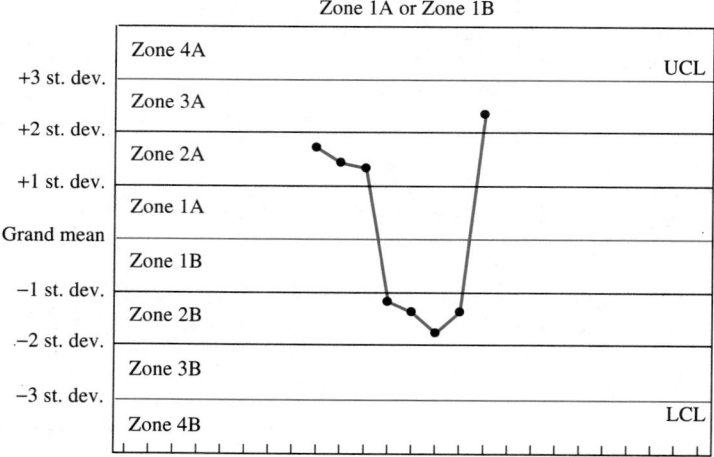

FIGURE 23.23   *Example of pattern 8: Eight consecutive values more than 1 standard deviation from the centerline*

In pattern analysis, any of the patterns described in the box on page 994 indicates that either a very rare event has occurred or the process is out of control.

If a process is in control, its future output can be predicted in the sense that we know its probability distribution. If a process is out of control, it is difficult to predict the future output of the process because its probability distribution is changing at different values in time.

Difference Between Specification Limits and Control Limits

It is important to keep in mind the distinction between control limits and specification limits. Specification limits are requirements that are specified by the producer or consumer of a product. The locations of the specification limits are determined by the demands of the producer or consumer; they are not determined by the amount of random variation present in the production process.

Upper and lower control limits are located 3 standard deviations from the estimated process mean, so the location of the control limits does not depend on any restrictions imposed by the manufacturer or consumer. Control limits are functions of the inherent variability present in the manufacturing process. The location of the upper and lower control limits depends on the magnitude of the standard deviation of the variable being analyzed.

The location of specification limits is determined by how the producer *wants* the process to perform, whereas the location of the control limits is determined by how the process *actually* performs. Specification limits are determined by the demands of consumers or producers, whereas control limits are determined by the natural variation present in the production process.

A process can be in control even though some of its output falls outside the specification limits. In such a situation, management has to find a new or better way of producing the product; that is, the process design must be changed to reduce the amount of random variation and to make the output more uniform.

Also, a process can be out of control even though all of its output falls within the specification limits. For example, consider a process where the mean level of output is gradually increasing, but where all the output observed to date falls within the specification limits. When a process is out of control, it does not mean that the output of the process is defective. It means that variability can be reduced by the elimination of an assignable cause of variation.

The optimal situation is to attain a state where a process both is in control and satisfies the tolerance specifications.

Constructing an x̄-Chart

The information in the following box summarizes how to construct and interpret an x̄-chart.

Constructing an x̄-chart: Summary

1. Collect at least $k = 20$ samples (subgroups) of data, each containing n observations. Usually $n = 4$ or $n = 5$.
2. When selecting the samples, use a method of rational subgrouping.
3. Calculate the sample mean $\bar{x}_i$ and sample range R_i for each sample.
4. Calculate the grand mean $\bar{\bar{x}}$, which is the average of the sample means, and calculate the average of the sample ranges, $\bar{R}$, as follows:

$$\bar{\bar{x}} = \frac{\bar{x}_1 + \bar{x}_2 + \cdots + \bar{x}_k}{k}$$

$$\bar{R} = \frac{R_1 + R_2 + \cdots + R_k}{k}$$

(*continued*)

where

$$k = \text{Number of samples (i.e., subgroups)}$$
$$\bar{x}_i = \text{Sample mean for the } i\text{th sample}$$
$$R_i = \text{Range of the } i\text{th sample}$$

5. Plot the centerline, the control limits, and the zone boundaries:

Centerline: $\bar{x}$

Upper control limit: $\bar{x} + A_2 \bar{R}$

Lower control limit: $\bar{x} - A_2 \bar{R}$

where A_2 is a constant that depends on the sample size n. Values for A_2 are given in Table 23.3 and in the Appendix in Table A.14 for samples of size $n = 2$ to $n = 25$.

6. Plot the k sample means $\bar{x}_i$ in chronological order versus the sample number.

7. Apply the rules for pattern analysis to determine whether the process is out of control.

8. If the process is deemed to be out of control, search for assignable causes of variation that may be affecting the process. These causes should be eliminated to bring the process under control.

9. Before using the control chart, monitor the variability of the process by constructing an R-chart.

EXAMPLE 23.2   **Constructing an $\bar{x}$-Chart** ———————————————————————

Suppose a production process at a food processing plant fills containers with fruit and syrup. An inspector wants to monitor the mean level of the weight of the fruit in the containers. At the beginning of every hour for $k = 25$ hours, the inspector selects five containers, drains the syrup, and weighs the remaining fruit. The sample data are shown in Table 23.4.

    a. Examine whether this sampling strategy follows the rules for rational subgrouping.

    b. Construct the $\bar{x}$-chart for the process.

    c. Analyze the data and decide whether the process is in control.

SOLUTION   **a.**   The sampling procedure is rational if it enables us to detect any process changes that may have occurred during the 25-hour period. Within each sample, the individual measurements were made consecutively so there is little chance that any change in the process occurred within the sample period. Since the samples were taken 1 hour apart, there should be sufficient time between samples to enable us to detect any changes that may have occurred in the filling process.

    b.   Table 23.4 shows the five observations in each of the 25 samples. The table also shows the 25 sample means $\bar{x}_i$ and the 25 sample ranges R_i. For example, for the first sample, the sample mean and sample range are as follows:

$$\bar{x}_1 = \frac{14 + 12 + 11 + 11 + 14}{5} = 12.4$$

$$R_1 = 14 - 11 = 3$$

TABLE 23.4 *Fill Weights of Fruit for Example 23.2*

Sample number (hour)	Sample observations (weight in ounces)					Sample mean $\bar{x}_i$	Sample range R_i
1	14	12	11	11	14	12.4	3
2	12	13	11	11	12	11.8	2
3	16	12	11	11	11	12.2	5
4	11	12	12	12	13	12.0	2
5	14	11	11	11	12	11.8	3
6	13	10	12	11	12	11.6	3
7	12	14	12	10	11	11.8	4
8	13	13	13	12	12	12.6	1
9	11	11	15	11	13	12.2	4
10	10	14	12	14	11	12.2	4
11	13	13	11	11	12	12.0	2
12	16	13	12	11	11	12.6	5
13	12	12	11	12	14	12.2	3
14	12	12	13	12	14	12.6	2
15	11	13	12	11	14	12.2	3
16	10	13	11	13	13	12.0	3
17	13	12	11	12	12	12.0	2
18	12	14	10	10	10	11.2	4
19	12	12	13	15	10	12.4	5
20	15	11	12	14	11	12.6	4
21	13	12	14	12	11	12.4	3
22	13	13	15	13	12	13.2	3
23	17	14	11	12	12	13.2	6
24	11	10	10	11	11	10.6	1
25	14	13	12	11	11	12.2	3
						$\bar{\bar{x}} = 12.16$	$\bar{R} = 3.20$

In addition, the table shows the grand mean $\bar{\bar{x}}$ (the average of the 25 sample means) and $\bar{R}$ (the average of the 25 sample ranges):

$$\bar{\bar{x}} = \frac{12.4 + 11.8 + \cdots + 12.2}{25} = 12.16$$

$$\bar{R} = \frac{3 + 2 + \cdots + 3}{25} = 3.20$$

On the control chart, the centerline is positioned at the grand mean $\bar{\bar{x}} = 12.16$.

To locate the upper and lower control limits, we need the constant A_2, which can be found in Table 23.3 and in Table A.14 in the Appendix. For $n = 5$, we obtain $A_2 = .577$. The upper and lower control limits are

UCL: $\bar{\bar{x}} + A_2\bar{R} = 12.16 + .577(3.20) = 14.0064$

LCL: $\bar{\bar{x}} - A_2\bar{R} = 12.16 - .577(3.20) = 10.3136$

Next, the 25 sample means are plotted on the control chart in chronological order. Figure 23.24 shows the $\bar{x}$-chart with the sample means, the centerline, and the control limits. Note that none of the sample means fall outside the control limits. Quality control engineers who apply only the first rule for pattern analysis would stop here and would say that the process appears to be in control.

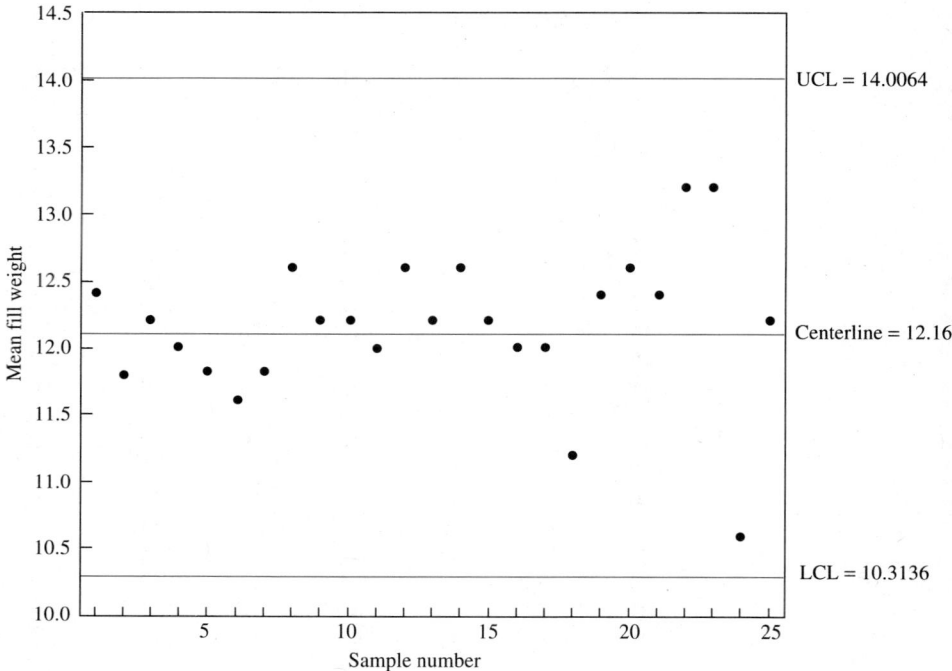

FIGURE 23.24 *$\bar{x}$-chart for Example 23.2 showing fill weights of fruit*

c. To apply all the rules for pattern analysis, we draw additional horizontal lines on the control chart. These lines are located 1 and 2 standard deviations above and below the grand mean.

The lines located 1 standard deviation above and below the grand mean are determined as follows:

■ Boundary between zones 1A and 2A:

$$\bar{x} + \frac{1}{3}(A_2\bar{R}) = 12.16 + \frac{1}{3}(.577)(3.20) = 12.7755$$

■ Boundary between zones 1B and 2B:

$$\bar{x} - \frac{1}{3}(A_2\bar{R}) = 12.16 - \frac{1}{3}(.577)(3.20) = 11.5445$$

The lines that are located 2 standard deviations above and below the grand mean are determined as follows:

■ Boundary between zones 2A and 3A:

$$\bar{x} + \frac{2}{3}(A_2\bar{R}) = 12.16 + \frac{2}{3}(.577)(3.20) = 13.3909$$

■ Boundary between zones 2B and 3B:

$$\bar{x} - \frac{2}{3}(A_2\bar{R}) = 12.16 - \frac{2}{3}(.577)(3.20) = 10.9291$$

Figure 23.25 shows a more detailed $\bar{x}$-chart. The lines separate the control chart into eight zones, four above the centerline and four below the centerline.

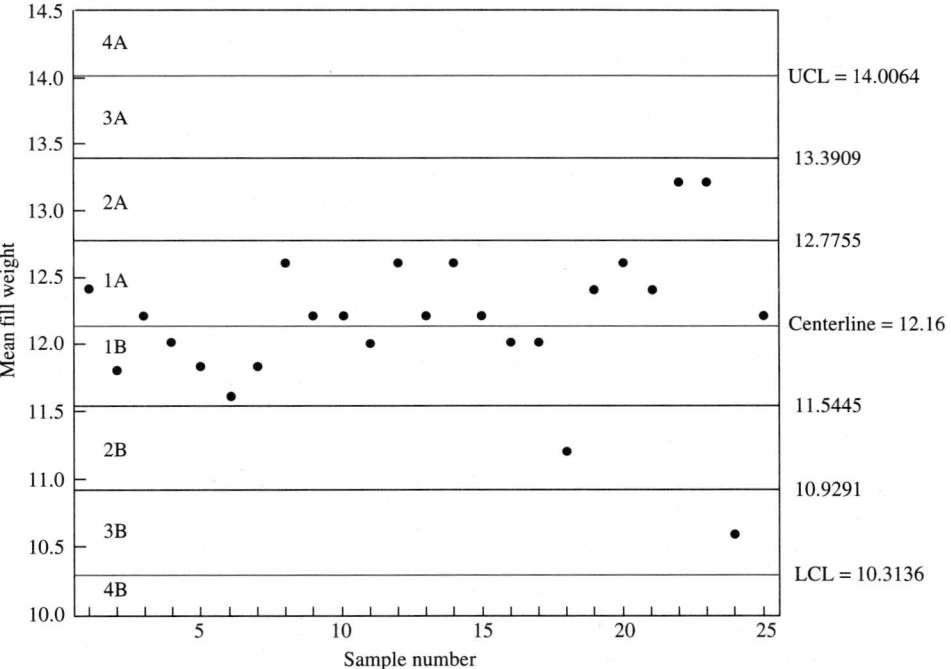

FIGURE 23.25   *Control chart for Example 23.2 showing additional zones*

At this stage, the chart would be considered a *trial $\bar{x}$-chart* because we have not yet established whether the process was in control during the period when the 25 samples were taken. Until we have established that the process was in control when the data were obtained, the centerline and control limits should be viewed as preliminary estimates.

Now, we examine the rules for pattern analysis.

Pattern 1:   None of the sample means falls outside the upper and lower control limits, so we have no reason to believe that the process was out of control when the samples were taken.

Pattern 2:   We never observe nine consecutive sample means on one side of the centerline, so we have no reason to believe that the process was out of control when the samples were obtained.

Pattern 3:   We never observe six consecutive sample means steadily increasing or decreasing.

Pattern 4:   We never observe 14 consecutive sample means alternating up and down.

Pattern 5:   We never observe two out of three consecutive sample means more than 2 standard deviations above (below) the centerline. (To fall more than 2 standard deviations from the centerline, the sample mean must be larger than 13.3909 or less than 10.9291.)

Pattern 6:   We never observe four out of five consecutive sample means more than 1 standard deviation above (below) the centerline. (To fall more than 1 standard deviation from the centerline, the sample mean must be greater than 12.7755 or less than 11.5445.)

Pattern 7:   We never observe 15 consecutive sample means within 1 standard deviation of the centerline. (To fall within 1 standard deviation of the centerline, the sample mean must be between 11.5445 and 12.7755.)

Pattern 8:   We never observe eight consecutive sample means more than 1 standard deviation from the centerline. (To fall more than 1 standard deviation from the centerline, the sample mean must be larger than 12.7755 or less than 11.5445.)

The sequence of sample means reveals no patterns that indicate that the process is out of control. Accordingly, we act as if the process was in control when the data were collected, and the trial centerline and trial control limits are now considered suitable for future monitoring of process output.

Remember that this conclusion assumes that the process variation has already been monitored on an *R*-chart. We will examine the process variation in the next section.

Monitoring the Process Based on Additional Data

Now, suppose we obtain additional sample data at a later time period from the same process. Let us examine the new data and determine whether the process remained in control during the later sample periods.

EXAMPLE 23.3   **Monitoring the Fill Weight of Fruit: Additional Data**

Suppose that each hour for 10 hours a sample of five containers of fruit was obtained from the process described in Example 23.2. After the syrup was drained, the weight of fruit in each container was recorded. All the new sample data, along with the 10 new sample means and new sample ranges, are shown in Table 23.5. Determine whether the process remained in control during the period when the new samples were obtained.

S O L U T I O N   Beginning with sample number 26, we plot the 10 new sample means on the original control chart. The new control chart is shown in Figure 23.26.

Next, we apply the rules for pattern analysis.

Pattern 1:   No values fall outside the control limits, so we still assume the process is in control.

Pattern 2:   We never observe nine consecutive sample means on one side of the centerline.

Pattern 3:   We do observe six consecutive sample means steadily decreasing (sample 27 through sample 33). This indicates that the process may have gone out of control between period 27 and period 33.

TABLE 23.5   *Additional Data on Fill Weights of Fruit*

Sample number (hour)	Sample observations (weight in ounces)					Sample mean $\bar{x}_i$	Sample range R_i
26	15	12	11	12	14	12.8	4
27	14	15	12	15	12	13.6	3
28	14	13	14	13	12	13.2	2
29	13	14	12	13	13	13.0	2
30	12	15	14	11	12	12.8	4
31	13	12	12	12	12	12.2	1
32	12	13	11	10	11	11.4	3
33	11	13	10	11	10	11.0	3
34	12	12	13	14	13	12.8	2
35	10	14	11	14	11	12.0	4

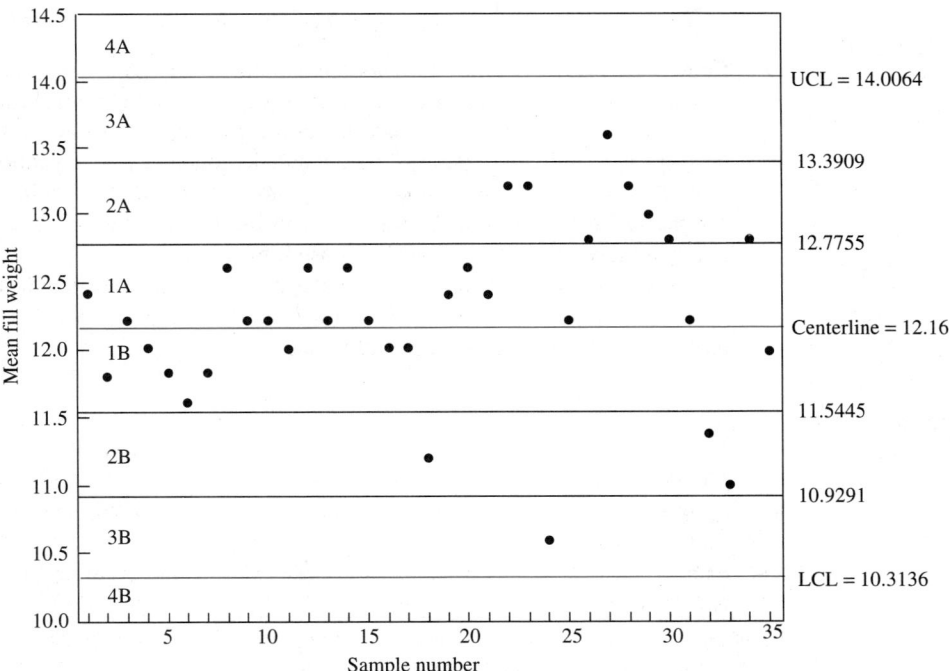

FIGURE 23.26   *x̄-chart for Example 23.3*
monitoring 10 additional sample values

Pattern 4:   We never observe 14 consecutive sample means alternating up and down.

Pattern 5:   We do not observe two out of three consecutive sample means more than 2 standard deviations above the centerline. (To fall more than 2 standard deviations from the centerline, the sample mean must be larger than 13.3909 or less than 10.9291.)

Pattern 6:   We do observe four out of five consecutive sample means more than 1 standard deviation above the centerline. (To fall more than 1 standard deviation above the centerline, the sample mean must be greater than 12.7755.) Samples 26 through 30 all fall more than 1 standard deviation above the centerline. This indicates that the process may have gone out of control between period 26 and period 30.

Pattern 7:   We never observe 15 consecutive sample means within 1 standard deviation of the centerline.

Pattern 8:   We never observe eight consecutive sample means more than 1 standard deviation from the centerline. (To fall more than 1 standard deviation from the centerline, the sample mean must be larger than 12.7755 or less than 11.5445.)

After applying all the rules for pattern analysis, we decide that the process may have gone out of control at some time between period 26 and period 30. Thus, we should begin searching for the source of the problem and try to determine what assignable causes of variation may have influenced the process during this time period.

Caution: $\bar{x}$-Charts Show Control Limits, Not Specification Limits

Remember that the $\bar{x}$-chart does not contain the upper and lower specification limits, USL and LSL. Instead, the $\bar{x}$-chart is concerned with statistical control limits. A control chart does not reveal whether the process can meet its specification limits. The specification limits refer to individual items of output and do not appear on the control chart. The values plotted on the $\bar{x}$-chart are sample means, not individual observations x_i. The fact that a sample mean falls within the specification limits does not imply that all the individual items in the sample also fall within those limits.

Specification limits are used to determine whether a particular item is acceptable or defective. Control limits are used to determine whether a process is operating in control or is subject to assignable causes of variation.

Exercises

Statistical Concepts

23.3.1   Describe the elements of a control chart, and explain how a control chart is used to detect whether a process is out of control.

23.3.2   Explain what is meant by pattern analysis, and explain how it is used to test whether a process is out of control.

23.3.3   *True or false:*   If any of the patterns listed in the rules for pattern analysis occurs, we know the process is out of control. Explain.

23.3.4   *True or false:*   If all the points on a control chart lie inside the upper and lower control limits, the process is in control. Explain.

23.3.5   Explain the difference between a specification limit and a control limit.

Statistical Drills

23.3.6   Suppose we are constructing an $\bar{x}$-chart. Find the value of A_2 for each of the following sample sizes:
a. $n = 4$ **b.** $n = 5$ **c.** $n = 6$

23.3.7   Suppose a manufacturing process is observed for $k = 25$ hours by taking a sample of size $n = 5$ items each hour. The grand mean is $\bar{x} = 27.5$ and the sample mean range is $\bar{R} = 4.7$.

a. Use this information to construct upper and lower control limits for an $\bar{x}$-chart.

b. Find the boundaries separating the zones on a control chart.

23.3.8   Suppose a manufacturing process is observed for $k = 20$ hours by taking a sample of size $n = 6$ items each hour. The grand mean is $\bar{\bar{x}} = 18.0$ and the sample mean range is $\bar{R} = 3.3$.

a. Use this information to construct upper and lower control limits for an $\bar{x}$-chart.

b. Find the boundaries for zones 1, 2, 3, and 4.

Statistical Applications

23.3.9   Twenty samples of size $n = 5$ were collected to construct an $\bar{x}$-chart. The following sample means and ranges were calculated for these data:

Sample number i	Sample mean $\bar{x}_i$	Sample range R_i	Sample number i	Sample mean $\bar{x}_i$	Sample range R_i
1	58.3	6.2	11	58.5	9.3
2	59.2	8.3	12	61.7	8.9
3	62.1	5.8	13	58.3	8.4
4	60.2	5.9	14	62.0	8.9
5	59.6	8.7	15	58.8	6.7
6	61.7	8.5	16	61.2	6.1
7	61.3	6.1	17	64.7	7.8
8	57.8	6.3	18	56.8	8.7
9	60.0	7.5	19	57.2	7.2
10	64.8	6.9	20	62.7	7.7

a. Find the grand mean.

b. Find the average of the sample ranges.

c. Find the upper and lower control limits.

d. Find the boundaries for zones 1, 2, 3, and 4.

e. Plot the 20 sample means on a control chart.

f. Use the rules for pattern analysis to determine whether the process was in control when the data were obtained. Does the chart indicate that the process is or is not under control?

23.3.10   A control chart is constructed to monitor the tensile strength in pounds of a certain yarn. Each subgroup size is 5. After 25 subgroups, we obtain the grand mean $\bar{\bar{x}} = 21.7$ and $\bar{R} = 3.8$.

a. Compute the values of the upper and lower control limits for the $\bar{x}$-chart.

b. Find the boundaries for zones 1, 2, 3, and 4.

23.3.11   A control chart is constructed to monitor the weight in ounces of the contents of a certain container. Each subgroup size is 10. The sample mean and sample range are calculated for each subgroup. After 20 subgroups, the sum of the sample means is 590.5 and the sum of the sample ranges is 6.5.

a. Compute the values of the upper and lower control limits for the $\bar{x}$-chart.

b. Find the boundaries for zones 1, 2, 3, and 4.

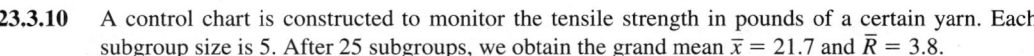

23.3.12   A control chart is constructed to monitor the shear strength in pounds of test spot welds. Each subgroup size is 3. The values of the sample mean and sample range are calculated for each subgroup. After 30 subgroups, the sample means sum to 12,900 and the sample ranges sum to 1,140.

a. Compute the values of the upper and lower control limits for the $\bar{x}$-chart.

b. Find the boundaries for zones 1, 2, 3, and 4.

23.3.13   The process for the manufacture of a certain yarn appears to be in control with respect to the quality characteristic Tensile strength. There is a single lower specification limit of $L = 18$ lb. Suppose the process mean is 20.3 lb and the process standard deviation is $\sigma = 1.2$ lb. What proportion of output do you expect to fail to meet its lower specification limit? Assume that the normal distribution is applicable.

23.3.14 In a statistically controlled container-filling process, the lower specification limit is $L = 32$ oz. It is desired to hold the overfill to as low a value as possible consistent with meeting this specification. In a sample of data, it was estimated that the process mean was 32.6 oz with standard deviation $\sigma = .25$ oz. Determine what percentage of the containers you would expect to contain less than 32 oz. Assume that the normal distribution is applicable. If it were permissible for 5% to be below 32 oz, what should the mean be corrected to from its present value?

23.3.15 The specified minimum strength for a weld is 370 lb. In a sample of data, it was estimated that the process mean was 422 lb with standard deviation $\sigma = 24.8$ lb. If the process is in statistical control and normally distributed, what can you conclude regarding its ability to meet this specification?

23.3.16 The resistance of an electrical component is specified as 78 ± 6 ohms. Suppose the process is in statistical control and normally distributed with mean 80 ohms and standard deviation $\sigma = 3.6$ ohms. What percentage of output would you expect to find above the upper specification limit? Below the lower specification limit? What improvement in the total percentage outside specifications would be made if the process average could be held at 78 ohms rather than at 80 ohms?

23.3.17 Subgroups of $n = 5$ items each are taken from a manufacturing process at regular intervals. A certain quality characteristic is measured, and the sample mean and sample range are computed for each subgroup. After 25 subgroups, $\Sigma \bar{x}_i = 357.50$ and $\Sigma R_i = 8.80$.
a. Compute the control chart upper and lower control limits.
b. If the specification limits are $14.40 \pm .40$, what conclusions can you draw about the ability of the existing process to produce items within these specifications? Suggest possible ways in which the situation could be improved.

23.3.18 If the width of a manufactured part is below the lower specification limit of .8750 inch, the part can be reworked to bring it within specifications. However, if the width exceeds .8800 inch, it must be scrapped. The process standard deviation is $\sigma = .0014$. A manager decided to use .8770 as the process average.
a. If the distribution of widths is normal with $\sigma = .0014$, approximately what percentage of rework and what percentage of spoilage could be expected?
b. Suppose you could adjust the mean width. Suggest an economic basis for establishing the process average in cases of this type.

23.3.19 The following data were obtained to monitor a quality characteristic of a certain manufactured product that had required a substantial amount of rework. All the figures apply to output made on a single machine by a single operator. The subgroup size was 5.

Subgroup	$\bar{x}_i$	R_i	Subgroup	$\bar{x}_i$	R_i
1	177.5	13	11	178.8	9
2	176.8	8	12	177.4	8
3	177.4	12	13	179.3	7
4	176.6	11	14	178.4	9
5	177.3	8	15	180.7	6
6	179.2	7	16	179.2	8
7	178.4	15	17	177.8	10
8	178.9	6	18	178.4	9
9	179.0	7	19	183.6	7
10	178.3	9	20	178.6	10

a. Find the grand mean.
b. Find the average of the sample ranges.
c. Find the upper and lower control limits.
d. Find the boundaries for zones 1, 2, 3, and 4.
e. Plot the 20 sample means on a control chart.
f. Use the rules for pattern analysis to determine whether the process was in control when the data were obtained. Does the chart indicate that the process is or is not under control?

23.3.20 Subgroups of four items each are taken from a manufacturing process at regular intervals. A certain quality characteristic is measured, and $\bar{X}$ and R values are computed for each subgroup. After 25 subgroups, $\Sigma \bar{x}_i = 15{,}350$ and $\Sigma R_i = 411.4$.

a. Compute the upper and lower control limits for the $\bar{x}$-chart.
b. The specification requirements for this particular quality characteristic are 610 ± 14. If the quality characteristic is normally distributed with the distribution centered at $\bar{X}$, what percentage of the product, if any, would you expect to find outside the specification limits?
c. Any product that falls below LSL = 596 must be scrapped, whereas any above USL = 624 may be brought within specifications by certain work operations. It is suggested that the process ought to be centered at a level so that not more than .1% of the product will be scrapped. If the quality characteristic is normally distributed with the dispersion indicated by the control chart data, and if it is believed that statistical control can be maintained, what should the process mean be to make this scrap exactly .1%? What percentage of rework would be expected with this centering?

23.4 *R*-Chart to Monitor Process Variation

In the previous section, we explained how the $\bar{x}$-chart is used to monitor changes in the process mean. In this section, we discuss the R-chart, which is used to monitor potential changes in process variation over time. Whereas the $\bar{x}$-chart contains a plot of the individual sample means $(\bar{x}_i)$, the R-chart contains a plot of the individual sample ranges (R_i) for successive subgroups of size n taken from the process.

The R-chart is used in conjunction with the $\bar{x}$-chart because the control limits and zone boundaries on the $\bar{x}$-chart depend on the value of $\bar{R}$, the average of the k sample ranges.

All control charts employ essentially the same logic. Like an $\bar{x}$-chart, an R-chart contains a centerline, control limits located 3 standard deviations above and below the centerline, and other zone boundaries located 1 and 2 standard deviations above and below the centerline.

The theory is based on the assumption that the sequence of sample ranges R_i are random variables selected from probability distributions that are approximately normal and that have constant mean μ_R and constant standard deviation σ_R. To construct an R-chart, we need estimates of μ_R and σ_R.

Choice of Sample Size *n* and Number of Samples *k*

As is the case with the $\bar{x}$-chart, when constructing an R-chart, the sample size n is usually kept very small. The typical sample size is $n = 4$ or $n = 5$ and is almost always below 10. The number of subgroups k should be at least 20 to 25.

Location of the Centerline

The process mean range μ_R is unknown and has to be estimated. If the process is in control, the average of the sample ranges $\bar{R}$ provides an unbiased estimate of the unknown mean range μ_R.

DEFINITION   **Average Range**

The **average range** $\bar{R}$ provides an unbiased estimate of the process mean range μ_R. The value of $\bar{R}$ is determined as follows:

$$\bar{R} = \frac{R_1 + R_2 + \cdots + R_k}{k}$$

where k is the number of samples and R_i is the sample range from the ith sample.

On the R-chart, the *centerline* is drawn at the average range $\bar{R}$.

Theoretical Control Limits

If the values of μ_R and σ_R were known, the upper and lower control limits for the R-chart would be positioned 3 standard deviations above and below the mean range μ_R. When determining the control limits for the sample ranges, it is important to use the standard deviation for the sample range (σ_R) rather than the standard deviation for an individual observation, which is σ.

The theoretical upper and lower control limits would be determined as follows:

$$\text{Theoretical upper control limit:}\quad \mu_R + 3\sigma_R$$
$$\text{Theoretical lower control limit:}\quad \mu_R - 3\sigma_R$$

Estimation of the Upper and Lower Control Limits

To determine the location of the upper and lower control limits, we must estimate the process mean range μ_R and the standard deviation σ_R. The process mean range μ_R is estimated by the average range $\bar{R}$. Quality control engineers have found that a good estimate of σ_R is provided by the estimator $\hat{\sigma}_R = d_3(\bar{R}/d_2)$, where the values of d_2 and d_3 are constants that depend on the sample size n.

The estimated upper and lower control limits are obtained as follows:

$$\text{Estimated upper control limit:}\quad \text{UCL} = \bar{R} + 3d_3\left(\frac{\bar{R}}{d_2}\right)$$

$$\text{Estimated lower control limit:}\quad \text{LCL} = \bar{R} - 3d_3\left(\frac{\bar{R}}{d_2}\right)$$

The values of d_2 and d_3 are given in Table 23.6 and in Table A.15 of the Appendix. The formulas provide unbiased estimates of the population upper and lower control limits based on the assumption that we are sampling from a normal population.

TABLE 23.6 *Factors Used When Constructing an R-Chart*

Sample size n_i	COEFFICIENT		Sample size n_i	COEFFICIENT	
	d_2	d_3		d_2	d_3
2	1.128	.853	14	3.407	.762
3	1.693	.888	15	3.472	.755
4	2.059	.880	16	3.532	.749
5	2.326	.864	17	3.588	.743
6	2.534	.848	18	3.640	.738
7	2.704	.833	19	3.689.	.733
8	2.847	.820	20	3.735	.729
9	2.970	.808	21	3.778	.724
10	3.078	.797	22	3.819	.720
11	3.173	.787	23	3.858	.716
12	3.258	.778	24	3.895	.712
13	3.336	.770	25	3.931	.709

Source: *ASTM Manual on the Presentation of Data and Control Chart Analysis,* Philadelphia, Penn.: American Society for Testing Materials, 1976.

$d_3(\bar{R}/d_2)$ is used to estimate the quantity σ_R when constructing upper and lower control limits on an R-chart.

Elements of an *R*-chart

Centerline: $$\bar{R} = \frac{1}{k} \sum_{i=1}^{k} R_i$$

Upper control limit: $$\text{UCL} = \bar{R} + 3d_3\left(\frac{\bar{R}}{d_2}\right)$$

Lower control limit: $$\text{LCL} = \bar{R} - 3d_3\left(\frac{\bar{R}}{d_2}\right)$$

Note: The values of d_2 and d_3 depend on the sample size n and are given in Table 23.6 and in Table A.15 of the Appendix.

Construction of the Zone Boundaries for an *R*-Chart

The same rules for pattern analysis that were used to analyze an $\bar{x}$-chart are to be used to analyze an R-chart. As on the $\bar{x}$-chart, the zone boundaries on an R-chart are located 1, 2, and 3 standard deviations above and below the centerline. The boundaries between the various zones are as follows:

Boundary between zones 1A and 2A: $$\bar{R} + d_3\left(\frac{\bar{R}}{d_2}\right)$$

Boundary between zones 1B and 2B: $$\bar{R} - d_3\left(\frac{\bar{R}}{d_2}\right)$$

Boundary between zones 2A and 3A: $\bar{R} + 2d_3\left(\dfrac{\bar{R}}{d_2}\right)$

Boundary between zones 2B and 3B: $\bar{R} - 2d_3\left(\dfrac{\bar{R}}{d_2}\right)$

Boundary between zones 3A and 4A (UCL): $\bar{R} + 3d_3\left(\dfrac{\bar{R}}{d_2}\right)$

Boundary between zones 3B and 4B (LCL): $\bar{R} - 3d_3\left(\dfrac{\bar{R}}{d_2}\right)$

The information in the following box summarizes how to construct and interpret an *R*-chart.

Constructing an *R*-chart: Summary

1. Collect at least $k = 20$ samples (subgroups) of data, each containing n observations. Usually $n = 4$ or $n = 5$.
2. When selecting the samples, use a method of rational subgrouping.
3. Calculate the sample range R_i for each sample.
4. Calculate the average of the sample ranges, $\bar{R}$, as follows:

$$\bar{R} = \frac{R_1 + R_2 + \cdots + R_k}{k}$$

where

$k = $ Number of samples (i.e., subgroups)

$R_i = $ Range of the *i*th sample

5. Plot the centerline, the control limits, and the zone boundaries:

Centerline: $\bar{R}$

Upper control limit: $\bar{R} + 3d_3\left(\dfrac{\bar{R}}{d_2}\right)$

Lower control limit: $\bar{R} - 3d_3\left(\dfrac{\bar{R}}{d_2}\right)$

where d_2 and d_3 are constants that depend on the same size n. Values for d_2 and d_3 are given in Table 23.6 and in the Appendix in Table A.15 for samples of size $n = 2$ to $n = 25$.
6. Plot the k sample ranges R_i on the control chart in chronological order.
7. Determine whether any of the rules for pattern analysis indicate that the process may be out of control.

EXAMPLE 23.4   **Construction of an *R*-Chart** ───────────────────────────

The data in Table 23.7 show the drained weights of fruit in 25 samples where each sample contains five observations. These data were used to construct the $\bar{x}$-chart in Example 23.2. Construct an *R*-chart to monitor process variation.

TABLE 23.7   *Fill Weights of Fruit for Example 23.4*

Sample number (hour)	Sample observations (weight in ounces)					Sample mean $\bar{x}_i$	Sample range R_i
1	14	12	11	11	14	12.4	3
2	12	13	11	11	12	11.8	2
3	16	12	11	11	11	12.2	5
4	11	12	12	12	13	12.0	2
5	14	11	11	11	12	11.8	3
6	13	10	12	11	12	11.6	3
7	12	14	12	10	11	11.8	4
8	13	13	13	12	12	12.6	1
9	11	11	15	11	13	12.2	4
10	10	14	12	14	11	12.2	4
11	13	13	11	11	12	12.0	2
12	16	13	12	11	11	12.6	5
13	12	12	11	12	14	12.2	3
14	12	12	13	12	14	12.6	2
15	11	13	12	11	14	12.2	3
16	10	13	11	13	13	12.0	3
17	13	12	11	12	12	12.0	2
18	12	14	10	10	10	11.2	4
19	12	12	13	15	10	12.4	5
20	15	11	12	14	11	12.6	4
21	13	12	14	12	11	12.4	3
22	13	13	15	13	12	13.2	3
23	17	14	11	12	12	13.2	6
24	11	10	10	11	11	10.6	1
25	14	13	12	11	11	12.2	3
						$\bar{x} = 12.16$	$\bar{R} = 3.20$

SOLUTION   First, we calculate the sample range R_i for each sample. Thus, for the first sample, the sample range is

$$R_1 = 14 - 11 = 3$$

All 25 sample ranges are shown in Table 23.7. Next, we calculate the average of the 25 sample ranges,

$$\bar{R} = \frac{3 + 2 + \cdots + 3}{25} = 3.20$$

On the R-chart, the centerline is located at $\bar{R} = 3.20$. Next, we calculate the upper and lower control limits. To do this, we need the values d_2 and d_3 from Table 23.6. When the sample size is $n = 5$, we obtain $d_2 = 2.326$ and $d_3 = .864$.

The lower control limit is

$$LCL = \bar{R} - 3d_3\left(\frac{\bar{R}}{d_2}\right) = 3.20 - 3(.864)\left(\frac{3.20}{2.326}\right) = -.366$$

Since the lower control limit is negative, we replace it with 0 (a range cannot be negative). There is no need to draw this control limit on the control chart.

The upper control limit is

$$\text{UCL} = \bar{R} + 3d_3\left(\frac{\bar{R}}{d_2}\right) = 3.20 + 3(.864)\left(\frac{3.20}{2.326}\right) = 6.766$$

On the R-chart, a horizontal line is drawn at the upper control limit. Next, the 25 sample ranges R_i are plotted in the order of sampling. This is a *trial* control chart because we have not yet established that the process is in control.

Next we calculate the other zone boundaries to check for any unusual patterns. We obtain the following:

Boundary between zones 1A and 2A: $\bar{R} + d_3\left(\frac{\bar{R}}{d_2}\right) = 3.20 + (.864)\left(\frac{3.20}{2.326}\right) = 4.389$

Boundary between zones 1B and 2B: $\bar{R} - d_3\left(\frac{\bar{R}}{d_2}\right) = 3.20 - (.864)\left(\frac{3.20}{2.326}\right) = 2.011$

Boundary between zones 2A and 3A: $\bar{R} + 2d_3\left(\frac{\bar{R}}{d_2}\right) = 3.20 + 2(.864)\left(\frac{3.20}{2.326}\right) = 5.577$

Boundary between zones 2B and 3B: $\bar{R} - 2d_3\left(\frac{\bar{R}}{d_2}\right) = 3.20 - 2(.864)\left(\frac{3.20}{2.326}\right) = .823$

Figure 23.27 shows the control chart for the data in Table 23.7. All the zone boundaries are shown, and the 25 sample ranges are plotted.

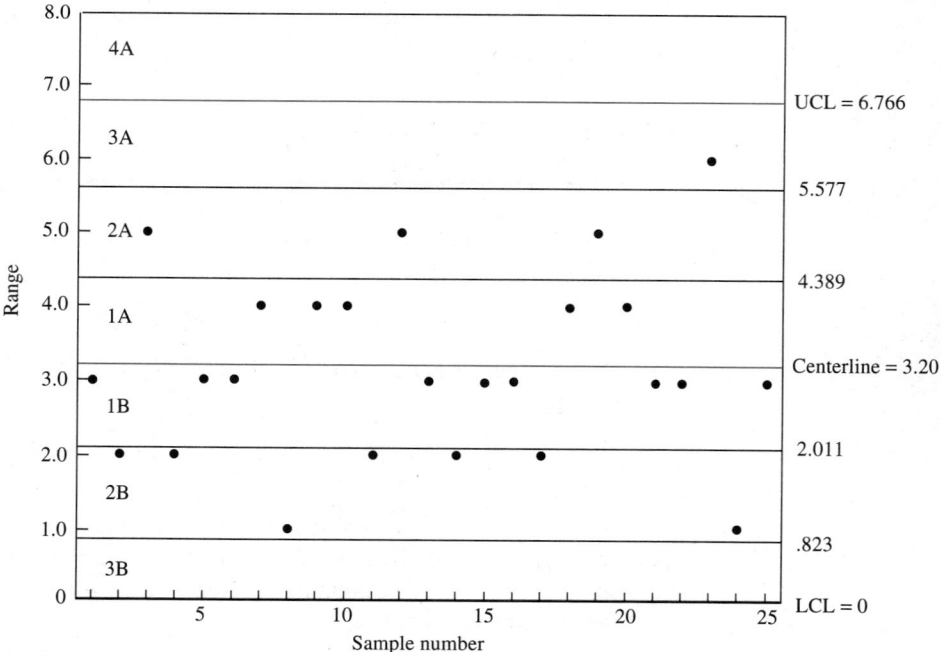

FIGURE 23.27   *R-chart for fill weights of fruit for Example 23.4; chart includes all zone boundaries*

Now we examine the rules for pattern analysis.

Pattern 1: No point falls more than 3 standard deviations from the centerline.

Pattern 2: We never observe nine consecutive sample ranges on one side of the centerline.

Pattern 3: We never observe six consecutive sample ranges steadily increasing or decreasing.

Pattern 4: We never observe 14 consecutive sample ranges alternating up and down.

Pattern 5: We never observe two out of three consecutive sample ranges more than 2 standard deviations above (below) the centerline. (To fall more than 2 standard deviations from the centerline, the sample range must be larger than 5.577 or less than .823.)

Pattern 6: We never observe four out of five consecutive sample ranges more than 1 standard deviation above (below) the centerline. (To fall more than 1 standard deviation from the centerline, the sample range must be greater than 4.389 or less than 2.011.)

Pattern 7: We never observe 15 consecutive sample ranges within 1 standard deviation of the centerline. (To fall within 1 standard deviation of the centerline, the sample range must be between 2.011 and 4.389.)

Pattern 8: We never observe eight consecutive sample ranges more than 1 standard deviation from the centerline. (To fall more than 1 standard deviation from the centerline, the sample range must be larger than 4.389 or less than 2.011.)

None of the rules for pattern analysis indicates that the process variation is out of control. Thus, we treat the process variation as though it were in control during the first 25 sample periods.

Exercises

Statistical Concepts

23.4.1 Summarize the process of constructing an R-chart and using it to monitor process quality.

23.4.2 *True or false:* Suppose all the sample ranges on an R-chart are within the upper and lower control limits. The process must be in control. Explain.

23.4.3 Explain why it is necessary to monitor the variation in the process before using the $\bar{x}$-chart to monitor the mean of the process.

Statistical Drills

23.4.4 Suppose we are constructing an R-chart. Find the values of d_2 and d_3 for each of the following sample sizes:
   a. $n = 4$ **b.** $n = 5$ **c.** $n = 6$

23.4.5 Suppose a manufacturing process is observed for $k = 20$ hours by taking a sample of size $n = 5$ items each hour. The grand mean is $\bar{x} = 18.0$ and the sample mean range is $\bar{R} = 3.4$.
   a. Use this information to construct upper and lower control limits for an R-chart.
   b. Find the boundaries for zones 1, 2, 3, and 4.
   c. Suppose a new sample of the following $n = 5$ observations is obtained:

   12.3   17.5   28.4   19.7   18.8

Does the process appear to be in control or out of control?

Statistical Applications

23.4.6   Twenty samples of size $n = 5$ were collected to construct an R-chart. The following sample means and ranges were calculated for these data:

Sample number i	Sample mean $\bar{x}_i$	Sample range R_i	Sample number i	Sample mean $\bar{x}_i$	Sample range R_i
1	58.3	6.2	11	58.5	9.3
2	59.2	8.3	12	61.7	8.9
3	62.1	5.8	13	58.3	8.4
4	60.2	5.9	14	62.0	8.9
5	59.6	8.7	15	58.8	6.7
6	61.7	8.5	16	61.2	6.1
7	61.3	6.1	17	64.7	7.8
8	57.8	6.3	18	56.8	8.7
9	60.0	7.5	19	57.2	7.2
10	64.8	6.9	20	62.7	7.7

a. Find the average of the sample ranges.
b. Find the upper and lower control limits on the R-chart.
c. Find the boundaries for zones 1, 2, 3, and 4.
d. Plot the 20 sample ranges on a control chart.
e. Use the rules for pattern analysis to determine whether the process variation was in control when the data were obtained. Does the chart indicate that the process variation is or is not under control?

23.4.7   A control chart is constructed to monitor the tensile strength in pounds of a certain yarn. Each subgroup size is 5. After 25 subgroups, we obtain the grand mean $\bar{x} = 21.7$ and $\bar{R} = 3.8$.
a. Compute the values of the upper and lower control limits for the R-chart.
b. Find the boundaries for zones 1, 2, 3, and 4.

23.4.8   A control chart is constructed to monitor the weight in ounces of the contents of a certain container. Each subgroup size is 10. The sample mean and sample range are calculated for each subgroup. After 20 subgroups, the sum of the sample means is 590.5 and the sum of the sample ranges is 6.4.
a. Compute the values of upper and lower control limits for the R-chart.
b. Find the boundaries for zones 1, 2, 3, and 4.

23.4.9   A control chart is constructed to monitor the shear strength in pounds of test spot welds. Each subgroup size is 3. The values of the sample mean and sample range are calculated for each subgroup. After 30 subgroups, the sample means sum to 12,900 and the sample ranges sum to 1,140.
a. Compute the values of upper and lower control limits for the R-chart.
b. Find the boundaries for zones 1, 2, 3, and 4.

23.4.10   A control chart is constructed to monitor the breaking strength in pounds in a certain destructive test of a type of ceramic insulator used in vacuum tubes. Each subgroup size is 10. The values of the sample mean and sample range are calculated for each subgroup. After 40 subgroups, the sample means sum to 5,600 and the sample ranges sum to 380.
a. Compute the values of upper and lower control limits for the R-chart.
b. Find the boundaries for zones 1, 2, 3, and 4.

23.4.11   The following data were obtained to monitor a quality characteristic of a certain manufactured product that had required a substantial amount of rework. All the figures apply to product made on a single machine by a single operator. The subgroup size was 5.

Subgroup	$\bar{x}_i$	R_i	Subgroup	$\bar{x}_i$	R_i
1	177.6	23	11	179.8	9
2	176.6	8	12	176.4	8
3	178.4	22	13	178.4	7
4	176.6	12	14	178.2	4
5	177.0	7	15	180.6	6
6	179.4	8	16	179.6	6
7	178.6	15	17	177.8	10
8	179.6	6	18	178.4	9
9	178.8	7	19	181.6	7
10	178.2	12	20	177.6	10

a. Find the average of the sample ranges.
b. Find the upper and lower control limits for an *R*-chart.
c. Find the boundaries for zones 1, 2, 3, and 4.
d. Plot the 20 sample ranges on a control chart.
e. Use the rules for pattern analysis to determine whether the process variation was in control when the data were obtained.

23.5 *p*-Chart to Monitor Proportion of Defectives

In this section, we discuss the construction of *p*-charts, which are used to monitor the proportion of defective items being produced by a process.

Like an $\bar{x}$-chart and an *R*-chart, the *p*-chart contains a centerline and upper and lower control limits that are calculated using the sample data. The *p*-chart is constructed by taking *k* samples from the process, where the *i*th sample size is n_i. In many cases, all the sample sizes are equal; this will simplify some of the formulas that are presented subsequently.

In each sample, each unit of output is classified as defective or nondefective. Let d_i denote the number of defective items found in the *i*th sample. Let $\hat{p}_i$ denote the proportion of defectives found in the *i*th sample. The sample proportion $\hat{p}_i$ is calculated as follows:

$$\hat{p}_i = \frac{d_i}{n_i}$$

Examples of proportions that can be monitored using a *p*-chart are the proportion of defective computer chips made by a producer, the proportion of bills that contain errors issued by a department store, the proportion of electronic circuit boards that contain soldering errors, and so forth.

DEFINITION **Process Proportion**
The proportion of defectives being produced by the process is called the **process proportion** and is denoted by *p*.

If the process proportion *p* remains constant over time, the process is said to be in control. If the process proportion is changing over time, the process is out of control.

The application of the *p*-chart is based on the sampling distribution of the sample proportion, which was discussed in Chapter 9.

When $n_i p \geqslant 5$ and $n_i(1 - p) \geqslant 5$, the sampling distribution of the sample proportion $\hat{p}_i$ approximately follows the normal distribution with mean

$$\mu_{\hat{p}_i} = p$$

and standard deviation

$$\sigma_{\hat{p}_i} = \sqrt{\frac{p(1 - p)}{n_i}}$$

Location of the Centerline

To construct the *p*-chart, we need estimates of the population proportion p and $\sigma_{\hat{p}_i}$. To estimate p, we calculate $\bar{p}$, the proportion of defective items in all k samples. The formula for calculating the estimated process proportion $\bar{p}$ is as follows.

DEFINITION   **Estimated Process Proportion** ——————————————————————

The **estimated process proportion** $\bar{p}$ provides an unbiased estimate of the process proportion p. The estimated process proportion is denoted $\bar{p}$ and is calculated as follows:

$$\bar{p} = \frac{\text{Total number of defectives}}{\text{Total number of units sampled}} = \frac{\sum_{i=1}^{k} d_i}{\sum_{i=1}^{k} n_i}$$

where k is the number of samples, n_i is the number of observations in the ith sample, and d_i is the number of defectives in the ith sample.

On the *p*-chart, the *centerline* is drawn at the estimated process proportion $\bar{p}$.

Theoretical Control Limits

On the *p*-chart, the upper and lower control limits are positioned at 3 standard deviations above and below the process proportion p. The theoretical upper and lower control limits would be determined as follows:

$$\text{Theoretical upper control limit:} \quad p + 3\sqrt{\frac{p(1 - p)}{n_i}}$$

$$\text{Theoretical lower control limit:} \quad p - 3\sqrt{\frac{p(1 - p)}{n_i}}$$

Estimation of the Upper and Lower Control Limits

The theoretical upper and lower control limits depend on the true value of the process proportion p. In general, however, p will be unknown and must be estimated using the sample data. To estimate p, we use $\bar{p}$, the proportion of defective items in the entire sample of items. The estimated upper and lower control limits are obtained by substituting the estimated sample proportion $\bar{p}$ for the process proportion p. The estimated upper and lower control limits are obtained as follows:

$$\text{Estimated upper control limit:} \quad \bar{p} + 3\sqrt{\frac{\bar{p}(1 - \bar{p})}{n_i}}$$

$$\text{Estimated lower control limit:} \quad \bar{p} - 3\sqrt{\frac{\bar{p}(1 - \bar{p})}{n_i}}$$

These formulas provide unbiased estimates of the population upper and lower control limits based on the assumption that we are sampling from a process that is in control.

When the value of $\bar{p}$ is close to 0, the value of LCL can be negative. In such cases, it is customary to replace LCL by 0.

Determination of the centerline and control limits on a *p*-chart:

Centerline:

$$\bar{p} = \frac{\text{Total number of defectives for all } k \text{ samples}}{\text{Total number of units sampled}}$$

$$\bar{p} = \frac{\displaystyle\sum_{i=1}^{k} d_i}{\displaystyle\sum_{i=1}^{k} n_i}$$

Upper control limit: $\text{UCL} = \bar{p} + 3\sqrt{\dfrac{\bar{p}(1 - \bar{p})}{n_i}}$

Lower control limit: $\text{LCL} = \bar{p} - 3\sqrt{\dfrac{\bar{p}(1 - \bar{p})}{n_i}}$

Note: If LCL is negative, replace it by 0. The number of samples is k and should be at least 20.

Selection of the Sample Sizes

For $\bar{x}$-charts and R-charts, the sample sizes usually are very small. For p-charts, the sample sizes are usually much larger, often being 1,000 or more. If p is small (say, $p \leq .01$) and a small sample size was used, it would be unlikely that we would observe any defective items. Consequently, almost all the sample proportions would be 0. In such a case, it would be virtually impossible to detect when a process has gone out of control.

As a general rule, many quality control engineers recommend that each sample size should be large enough to virtually guarantee that the lower control limit is nonnegative. The appropriate choice of sample size depends on the process proportion, which is unknown. This requires us to make an initial guess p_0 concerning the true value of p when determining the appropriate sample size.

The minimum recommended value of n_i is found by setting the value for the theoretical LCL equal to 0 and solving for n_i. We obtain

$$p - 3\sqrt{\frac{p(1 - p)}{n_i}} = 0$$

After squaring both sides of the equation, we obtain

$$p^2 = 9\frac{p(1 - p)}{n_i}$$

We can solve this equation to determine n_i as a function of p. We obtain the solution

$$n_i = \frac{9(1 - p)}{p}$$

After substituting an initial guess p_0 for p, we obtain the following rule for determining the minimum recommended sample size.

Determination of sample size

> Choose each sample size n_i such that
> $$n_i > \frac{9(1 - p_0)}{p_0}$$
> where n_i = Size of the ith sample
> p_0 = Initial estimate of the process proportion p

For example, suppose we guess that the process proportion defective is about 1% (i.e., $p = .01$). The sample size rule indicates that each sample should contain at least 891 items:

$$n_i > \frac{9(1 - .01)}{.01} = 891$$

Similarly, if the process proportion is thought to be about .05, then each sample size should be at least 171:

$$n_i > \frac{9(1 - .05)}{.05} = 171$$

In a *p*-Chart, Sample Sizes Often Vary from Sample to Sample

With $\bar{x}$-charts and R-charts, it is customary to keep the sample sizes constant from sample to sample. With p-charts, it is frequently the case that a sample will include all the items produced during an hour, a work shift, or a production run. Consequently, the sample sizes will not necessarily be equal. If you are free to choose the sample sizes, it is recommended that all the sample sizes be equal. When all the sample sizes are equal, the symbol for the sample size (n_i) can be replaced by the symbol n, which denotes the constant sample size.

Control Limits Depend on Sample Sizes

Because the formulas for the upper and lower control limits depend on the value n_i, the appropriate control limits will vary from sample to sample. Quality control engineers have adopted a compromise that greatly reduces the number of computations that must be done without sacrificing much accuracy.

When the sample sizes n_i vary from sample to sample, there are two different possibilities for calculating control limits:

1. If the sample sizes do not vary too much, use the average value of the sample sizes, $\bar{n}$, when calculating the upper and lower control limits. In the formulas for the upper and lower control limits, replace n_i by $\bar{n}$, where

$$\bar{n} = \frac{1}{k} \sum_{i=1}^{k} n_i$$

2. If the sample sizes vary substantially, then different upper and lower control limits should be constructed for each sample. The control limits will vary from sample to sample, depending on the particular sample size n_i.

Pattern Analysis

The rules for pattern analysis that were used to interpret an $\bar{x}$-chart are also used to interpret a p-chart. To apply the rules, we must calculate the control limits and the boundaries for the various zones. The control limits and other zone boundaries are located 1, 2, and 3 standard deviations above and below the centerline.

The boundaries are as follows:

$$\text{Boundary between zones 1A and 2A:} \quad \bar{p} + 1\sqrt{\frac{\bar{p}(1 - \bar{p})}{\bar{n}}}$$

$$\text{Boundary between zones 1B and 2B:} \quad \bar{p} - 1\sqrt{\frac{\bar{p}(1 - \bar{p})}{\bar{n}}}$$

$$\text{Boundary between zones 2A and 3A:} \quad \bar{p} + 2\sqrt{\frac{\bar{p}(1 - \bar{p})}{\bar{n}}}$$

$$\text{Boundary between zones 2B and 3B:} \quad \bar{p} - 2\sqrt{\frac{\bar{p}(1 - \bar{p})}{\bar{n}}}$$

$$\text{Boundary between zones 3A and 4A:} \quad \bar{p} + 3\sqrt{\frac{\bar{p}(1 - \bar{p})}{\bar{n}}}$$

$$\text{Boundary between zones 3B and 4B:} \quad \bar{p} - 3\sqrt{\frac{\bar{p}(1 - \bar{p})}{\bar{n}}}$$

The information in the following box summarizes how to construct and interpret a *p*-chart.

Constructing a *p*-chart: Summary

1. Collect at least 20 samples (subgroups) of data. Each subgroup should contain at least n_i observations, where

$$n_i > \frac{9(1 - p_0)}{p_0}$$

where

$$n_i = \text{Size of the } i\text{th sample}$$

$$p_0 = \text{Initial estimate of the process proportion } p$$

2. When selecting the samples, use a method of rational subgrouping.
3. Let d_i denote the number of defective items in the ith sample. Calculate the proportion of defective items in each sample. Denote the ith sample proportion as $\hat{p}_i$, where

$$\hat{p}_i = \frac{d_i}{n_i}$$

4. Plot the centerline, the upper and lower control limits, and the zone boundaries:

$$\text{Centerline:} \quad \bar{p} = \frac{\text{Total number of defectives}}{\text{Total number of units sampled}} = \frac{\sum_{i=1}^{k} d_i}{\sum_{i=1}^{k} n_i}$$

$$\text{Upper control limit:} \quad \text{UCL} = \bar{p} + 3\sqrt{\frac{\bar{p}(1 - \bar{p})}{\bar{n}}}$$

$$\text{Lower control limit:} \quad \text{LCL} = \bar{p} - 3\sqrt{\frac{\bar{p}(1 - \bar{p})}{\bar{n}}}$$

where $\bar{n}$ is the average sample size. (If the sample sizes differ substantially from one another, substitute n_i for $\bar{n}$ and calculate different control limits for each sample proportion. Also, if LCL is negative, replace it by 0.)
5. Plot the k sample proportions $\hat{p}_i$ on the control chart in chronological order.
6. Determine whether any of the rules for pattern analysis indicate that the process may be out of control.

EXAMPLE 23.5   **Construction of a *p*-Chart Using the Average Sample Size $\bar{n}$** ————————

The Sayette Supply Company produces magnets that are used in electrical relays. To monitor the proportion of defective magnets being produced, a sample of magnets was inspected on each of 25 consecutive days. Table 23.8 shows the number of items inspected each day (n_i) and the number of defective items obtained each day (d_i) for the 25-day period.

TABLE 23.8 *Data on Number of Defective Magnets for Example 23.5*

Day i	Sample size n_i	Number defective d_i	Proportion defective $\hat{p}_i$
1	724	48	.066
2	763	70	.092
3	748	70	.094
4	748	72	.096
5	724	45	.062
6	727	56	.077
7	726	48	.066
8	719	67	.093
9	759	37	.049
10	745	52	.070
11	736	47	.064
12	739	50	.068
13	723	47	.065
14	748	57	.076
15	770	51	.066
16	756	71	.094
17	719	53	.074
18	757	34	.045
19	760	55	.072
20	740	51	.069
21	751	39	.052
22	742	45	.061
23	725	52	.072
24	736	47	.064
25	731	38	.052
Sum	18,516	1,302	

$\bar{p}$ = Average proportion defective = 1,302/18,516 = .070
$\bar{n}$ = Average sample size = 18,516/25 = 740.64

Construct a *p*-chart for these data. Since the daily production levels are approximately equal to one another, use the average sample size method to construct the *p*-chart.

SOLUTION In Table 23.8, the average sample proportion is calculated as follows:

$$\bar{p} = \text{Average proportion defective} = \frac{1,302}{18,516} = .070$$

The average sample size is

$$\bar{n} = \text{Average sample size} = \frac{18,516}{25} = 740.64$$

Next we calculate the zone boundaries. Because the sample sizes are similar to one another, we use the average sample size $\bar{n}$ when calculating the boundaries. We obtain:

Boundary between zones 1A and 2A:

$$\bar{p} + \sqrt{\frac{\bar{p}(1 - \bar{p})}{\bar{n}}} = .07 + \sqrt{\frac{.07(1 - .07)}{740.64}} = .0794$$

Boundary between zones 1B and 2B:

$$\bar{p} - \sqrt{\frac{\bar{p}(1 - \bar{p})}{\bar{n}}} = .07 - \sqrt{\frac{.07(1 - .07)}{740.64}} = .0606$$

Boundary between zones 2A and 3A:

$$\bar{p} + 2 \sqrt{\frac{\bar{p}(1 - \bar{p})}{\bar{n}}} = .07 + 2 \sqrt{\frac{.07(1 - .07)}{740.64}} = .0888$$

Boundary between zones 2B and 3B:

$$\bar{p} - 2 \sqrt{\frac{\bar{p}(1 - \bar{p})}{\bar{n}}} = .07 - 2 \sqrt{\frac{.07(1 - .07)}{740.64}} = .0512$$

Boundary between zones 3A and 4A (UCL):

$$\bar{p} + 3 \sqrt{\frac{\bar{p}(1 - \bar{p})}{\bar{n}}} = .07 + 3 \sqrt{\frac{.07(1 - .07)}{740.64}} = .0981$$

Boundary between zones 3B and 4B (LCL):

$$\bar{p} - 3 \sqrt{\frac{\bar{p}(1 - \bar{p})}{\bar{n}}} = .07 - 3 \sqrt{\frac{.07(1 - .07)}{740.64}} = .0419$$

Figure 23.28 shows the *p*-chart for the data in Table 23.8. The 25 sample proportions are plotted in chronological order. Horizontal lines are drawn at the locations of the centerline and the zone boundaries.

Now we apply the rules for pattern analysis.

Pattern 1:   No point falls more than 3 standard deviations from the centerline.

Pattern 2:   We never observe nine consecutive sample proportions on one side of the centerline.

Pattern 3:   We never observe six consecutive sample proportions steadily increasing or decreasing.

Pattern 4:   We observe 14 consecutive sample proportions alternating up and down (observations 3–17).

Pattern 5:   We observe two out of three consecutive sample proportions more than 2 standard deviations above (below) the centerline (observations 2–4). (To fall more than 2 standard deviations from the centerline, the sample proportion must be larger than .0888 or less than .0512.)

Pattern 6:   We never observe four out of five consecutive sample proportions more than 1 standard deviation above (below) the centerline. (To fall more than 1 standard deviation from the centerline, the sample proportion must be greater than .0794 or less than .0606.)

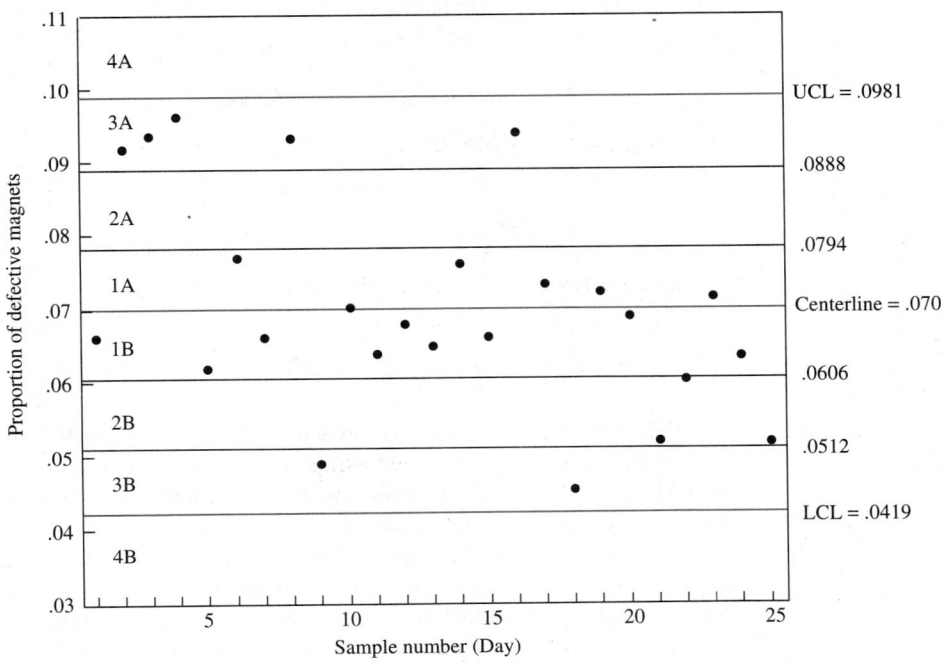

FIGURE 23.28 *p-chart for Example 23.5 showing sample proportions of defective magnets*

Pattern 7: We never observe 15 consecutive sample proportions within 1 standard deviation of the centerline. (To fall within 1 standard deviation of the centerline, the sample proportion must be between .0794 and .0606.)

Pattern 8: We never observe eight consecutive sample proportions more than 1 standard deviation from the centerline. (To fall more than 1 standard deviation from the centerline, the sample proportion must be larger than .0794 or less than .0606.)

Based on patterns 4 and 5, we conclude that it is unreasonable to treat the process as though it were in control during the first 25 sample periods.

In Example 23.5, the sample sizes for the different samples were relatively close to one another, so we used the average sample size $\bar{n} = 740.64$ in the formulas to construct the zone boundaries. The next example illustrates how to calculate the zone boundaries by using the individual sample size n_i in the formulas.

EXAMPLE 23.6 **Finding the Control Limits of a *p*-Chart Using the Exact Sample Size n_i**

Refer again to the data in Table 23.8 showing the proportions of defective magnets produced at the Sayette Supply Company. Calculate the exact control limits by substituting the exact sample size n_i into the formulas for the upper and lower control limits.

SOLUTION   To illustrate how to calculate individual control limits for each sample, let us calculate the control limits for the first sample proportion $\hat{p}_i$. For sample 1, we substitute the sample size $n_i = 724$ in the formulas. If we use the exact sample size, the control limits for sample 1 would be as follows:

Boundary between zones 3A and 4A (UCL):

$$\bar{p} + 3\sqrt{\frac{\bar{p}(1 - \bar{p})}{n_i}} = .07 + 3\sqrt{\frac{.07(1 - .07)}{724}} = .0984$$

Boundary between zones 3B and 4B (LCL):

$$\bar{p} - 3\sqrt{\frac{\bar{p}(1 - \bar{p})}{n_i}} = .07 - 3\sqrt{\frac{.07(1 - .07)}{724}} = .0416$$

For sample 1, we obtain the control limits LCL = .0416 and UCL = .0984. The control limits that were obtained using the average sample size $\bar{n} = 740.64$ were LCL = .0419 and UCL = .0981. Thus, the upper and lower control limits obtained by using the two methods are almost identical.

The same procedure can be used to find the control limits for each of the remaining samples. It is quite tedious to calculate different control limits for each sample; control limits will not change much unless there is a large variation in sample sizes.

In this example, regardless of which method is used to calculate the control limits, none of the sample proportions falls outside the control limits.

In this example, the control chart based on using the average sample size $\bar{n}$ yields approximately the same results as the control chart based on the exact sample sizes n_i because the sample sizes n_i do not vary much around their average value of 740.64. If the sample sizes differed substantially, then the individual control limits would vary from sample to sample, and the p-chart with variable control limits would be preferred over the p-chart based on the average sample size.

Exercises

Statistical Drills

23.5.1   Random samples of various sizes were selected hourly from a process judged to be in control. The average proportion of defectives was 6.2% and the average sample size was $\bar{n} = 400$.

a. Assume the sample sizes do not vary too much and find the boundaries that separate zones 1, 2, 3, and 4.

b. If the sample sizes selected from the process vary considerably, how would you calculate the control limits for a particular sample of size $n = 280$?

23.5.2   The proportion of defective items generated by a heating process is thought to be about 6%. To construct a p-chart for the process, how large must the sample size be to avoid having a negative lower control limit?

23.5.3   To construct a p-chart for a particular manufacturing process, 20 samples of size 200 were drawn from the process. The number of defectives in each sample is as follows:

Sample number	Sample size	Number defective	Sample number	Sample size	Number defective
1	200	12	11	200	15
2	200	12	12	200	10
3	200	9	13	200	7
4	200	14	14	200	12
5	200	15	15	200	13
6	200	9	16	200	11
7	200	10	17	200	8
8	200	11	18	200	6
9	200	12	19	200	12
10	200	13	20	200	14

a. Calculate the proportion defective in each sample.
b. Calculate and plot the average proportion defective $\bar{p}$.
c. Calculate and plot the upper and lower control limits for the *p*-chart.
d. Calculate and plot the boundaries that separate zones 1, 2, 3, and 4.
e. Plot the sample proportions on the *p*-chart.
f. Apply the rules for pattern analysis to determine whether the process is out of control.

23.5.4 To construct a *p*-chart at a manufacturing plant, 20 samples of size 200 were drawn from a process. The proportion of defective items found in each of the samples is as follows:

Sample	Proportion defective	Sample	Proportion defective
1	.02	11	.07
2	.04	12	.05
3	.08	13	.02
4	.02	14	.01
5	.06	15	.07
6	.09	16	.06
7	.05	17	.07
8	.06	18	.03
9	.06	19	.01
10	.08	20	.03

a. Calculate and plot the centerline and the upper and lower control limits for the *p*-chart.
b. Calculate and plot the boundaries that separate zones 1, 2, 3, and 4.
c. Plot the sample proportions on the *p*-chart.
d. Apply the rules for pattern analysis to determine whether the process is out of control.

23.5.5 In each of the following cases, determine the sample size that would be needed to avoid the construction of a *p*-chart with a negative lower control limit.
a. $p = .01$ **b.** $p = .02$ **c.** $p = .05$ **d.** $p = .08$

Statistical Applications

23.5.6 An assembly line at an automobile manufacturing plant produces car doors. A door is considered to be defective if it contains any scratches, cracks, or other obvious visual imperfections. A quality control engineer samples the output of the production line once every day by randomly selecting

100 items and noting the number of defectives. The results of the past 20 days are shown here. Construct a *p*-chart, and check to see whether the process is in control.

Sample	1	2	3	4	5	6	7	8	9	10
Number of defectives	6	5	4	5	8	2	5	6	7	4
Sample	11	12	13	14	15	16	17	18	19	20
Number of defectives	6	3	0	2	1	6	5	8	3	4

23.5.7   A tire company that manufactures racing tires for Formula I race cars is interested in monitoring the proportion of defective tires produced at its manufacturing plant. The company's quality control engineer believes that the proportion is about 7%. The quality control engineer tested a sample of 100 tires from each day's production. A total of 20 samples were obtained and the following results were obtained:

Sample	1	2	3	4	5	6	7	8	9	10
Number of Defective Tires	5	8	7	9	5	6	8	4	3	9
Sample	11	12	13	14	15	16	17	18	19	20
Number of Defective Tires	14	9	8	10	7	8	11	9	13	9

a. Construct a *p*-chart for the tire-production process.
b. Do you conclude that the process is in control?

$$ 23.5.8   Large national banks that issue credit cards must constantly upgrade their customer service to retain past customers and recruit new ones. Most banks have indicated that reducing customer billing errors is a major goal. The quality control manager at the bank checked a random sample of 500 customer bills each day for accuracy. To date, 30 samples have been evaluated. The data are shown here.

Sample	1	2	3	4	5	6	7	8	9	10
Number of Billing Errors	7	6	14	9	15	9	18	3	13	9
Sample	11	12	13	14	15	16	17	18	19	20
Number of Billing Errors	4	12	11	10	7	8	5	10	13	9
Sample	21	22	23	24	25	26	27	28	29	30
Number of Billing Errors	14	17	11	6	12	8	15	3	3	6

a. Construct a *p*-chart for the customer billing errors.
b. Do you conclude that the process is in control? Explain.

23.5.9   Compute the correct value of $\bar{p}$ from the following data:

Lot	Number inspected	Number of defectives	$\hat{p}$
1	500	27	.054
2	50	12	.240
3	800	12	.015
4	100	14	.140
5	150	15	.100
Total	1,600	80	

Compare your correct $\bar{p}$ with the unweighted average value of $\hat{p}$. Why do the two differ? Why is the unweighted average value of $\hat{p}$ unsatisfactory as a measure of the process average fraction defective?

23.5.10 The following table gives the results of daily inspection of a vacuum tube. The value of fraction defective established at the start of the inspection period was .04. The estimated daily average production was 1,600 tubes. Establish a single set of upper and lower control limits based on these figures and plot a control chart. Compute separate control limits for any points that seem to you to require them. Based on this set of data, does it appear that production was out of control during any periods?

Day	Number inspected	Number of defectives	Fraction defective
1	531	25	.0471
2	1,393	62	.0445
3	1,422	61	.0429
4	1,500	73	.0487
5	1,250	46	.0368
6	2,000	58	.0290
7	685	28	.0409
8	2,385	89	.0373
9	2,150	89	.0414
10	2,150	58	.0270
11	2,417	115	.0476
12	2,549	115	.0451
13	2,331	75	.0322
14	2,009	81	.0403
15	2,198	86	.0391
16	2,150	67	.0312
17	1,948	41	.0210
18	2,150	77	.0358
19	1,700	49	.0288
20	2,214	68	.0307
21	2,394	82	.0343
22	1,197	56	.0468
23	850	27	.0318
24	848	30	.0354
25	850	33	.0388
Total	43,271	1,591	

23.5.11 A *p*-chart is to be used to analyze production of a certain computer component. The total number of components inspected during a certain month was 2,200, and the total number of defectives was 160. Compute $\bar{p}$. Compute individual upper and lower control limits for each of the following 3 days, and state whether the fraction defective fell within the control limits for each day.

Day	Number inspected	Number of defectives
1	56	7
2	164	25
3	220	4

Chapter 23 Supplementary Exercises

23.S.1   Suppose a process is studied for $k = 40$ days by randomly sampling $n = 5$ items daily. From the sample information, we obtain $\bar{x} = 24.7$ and $\bar{R} = 5.8$.

a. Find the upper and lower control limits for the R-chart.

b. Find the upper and lower control limits for the $\bar{x}$-chart.

c. Determine the boundaries that separate zones 1, 2, 3, and 4 for both the $\bar{x}$-chart and the R-chart.

d. A few days after these control charts are constructed, the following sample of five items is obtained:

29.4	30.1	28.7	35.8	34.1

Does the process still appear to be in control?

23.S.2   The quality control engineer at an automobile assembly plant is concerned about the operating efficiency of a certain assembly line in Flint, Michigan. The engineer would like to monitor the time it takes for a certain procedure to be completed. The engineer randomly sampled four assembly operations during each hour of operation during 24 consecutive hours. The time in minutes for each assembly operation to be completed is as follows:

Sample	Time for assembly to be completed			
1	21.9	22.4	27.8	25.2
2	19.1	24.2	22.2	25.7
3	20.2	21.1	25.2	24.1
4	29.5	29.4	21.4	27.7
5	17.4	29.7	25.5	22.2
6	22.7	22.9	40.1	29.7
7	20.7	25.9	25.8	24.0
8	28.4	24.1	29.5	20.9
9	20.5	25.5	25.1	27.4
10	27.8	29.5	29.0	24.1
11	24.0	20.1	25.9	28.8
12	25.5	25.2	24.8	20.0
13	24.5	29.9	21.8	27.9
14	20.5	25.0	30.2	20.8
15	29.7	22.2	24.9	27.5
16	24.1	25.8	22.7	29.0
17	29.4	21.5	25.2	27.5
18	21.1	22.0	29.5	25.2
19	27.0	29.0	25.1	25.1
20	25.5	22.4	28.7	27.9
21	22.0	27.1	25.2	25.1
22	22.2	33.2	20.7	21.5
23	25.7	25.2	29.7	21.5
24	20.5	25.8	27.9	28.5

a. Construct an R-chart from these data.

b. Apply the rules for pattern analysis. Do you conclude that the process variation is in control?

c. Construct an $\bar{x}$-chart from these data.

d. Apply the rules for pattern analysis. Do you conclude that the process mean is in control?

23.S.3   Twenty-five samples of size $n = 5$ were collected to construct an $\bar{x}$-chart. The following sample means and sample ranges were obtained:

Sample	$\bar{x}_i$	R_i	Sample	$\bar{x}_i$	R_i
1	60.3	5.3	14	63.1	9.1
2	59.8	9.6	15	59.6	5.5
3	63.5	4.9	16	60.0	6.9
4	61.9	5.2	17	63.2	6.3
5	55.3	6.5	18	55.9	9.5
6	61.4	6.8	19	56.1	6.7
7	60.8	4.7	20	61.4	5.6
8	55.9	6.4	21	61.5	9.4
9	59.3	5.1	22	60.9	9.6
10	65.4	5.0	23	56.4	5.1
11	55.2	9.6	24	59.6	6.2
12	62.1	6.5	25	61.6	5.8
13	59.1	9.4			

a. Construct an R-chart from these data.
b. What does the R-chart suggest about the stability of the process variation?
c. Construct an $\bar{x}$-chart from these data.
d. What does the $\bar{x}$-chart suggest about the stability of the process mean?

23.S.4   A manufacturer produces precision parts for use in civilian aircraft. Specifications indicate that each part should be 47 centimeters in length. The manufacturer sampled four consecutively produced units each hour for 25 consecutive hours and recorded the lengths in centimeters to two-decimal accuracy. The data are as follows:

Sample	Length				Sample	Length			
1	47.14	47.18	46.91	46.88	14	47.18	47.17	47.31	47.16
2	46.96	47.16	46.85	46.98	15	47.14	47.16	46.89	47.13
3	47.36	47.33	46.99	47.13	16	46.95	46.98	46.91	46.99
4	47.21	47.16	47.12	46.98	17	46.97	46.96	47.36	47.31
5	46.83	46.97	46.93	47.31	18	47.33	47.16	46.98	46.93
6	47.34	46.96	47.13	46.89	19	46.88	46.99	47.13	46.96
7	47.17	46.96	46.99	47.11	20	46.91	47.35	47.19	47.11
8	47.13	46.93	46.98	47.32	21	47.13	46.96	47.15	46.96
9	47.37	47.14	46.91	47.13	22	47.19	46.95	46.94	47.32
10	46.93	46.99	46.87	47.33	23	47.11	47.12	46.95	47.16
11	46.88	47.31	47.17	47.14	24	46.99	47.17	46.91	47.12
12	47.16	46.98	46.91	46.99	25	47.31	47.14	47.13	46.91
13	46.93	47.22	47.32	47.14					

a. Construct an R-chart from these data.
b. What does the R-chart suggest about the stability of the process variation?
c. Construct an $\bar{x}$-chart from these data.
d. What does the $\bar{x}$-chart suggest about the stability of the process mean?
e. Does either chart suggest that special causes of variation are present? Justify your answer.
f. Provide an example of an assignable cause of variation that could potentially affect this manufacturing process. Do the same for a common cause of variation.

23.S.5   The thermostat in a food processing plant where perishable goods are packaged in plastic bags is set at 8 degrees Fahrenheit. Slight variations in temperature can affect the taste and longevity of the food product. The plant manager monitors the temperature inside the plant by taking sample temperature readings at five randomly chosen times per day for 20 days. The data (in degrees Fahrenheit) are as follows:

Day	Temperature					Day	Temperature				
1	8.32	8.29	8.00	7.98	7.78	11	8.01	8.07	7.68	8.21	7.93
2	7.51	7.70	7.63	6.13	7.83	12	7.86	7.97	8.38	7.80	7.72
3	8.82	8.73	7.91	7.88	8.23	13	8.13	6.90	8.06	7.81	7.72
4	8.58	7.27	8.19	7.82	8.81	14	8.06	7.81	8.60	7.81	7.98
5	7.98	8.92	7.70	7.67	7.78	15	7.92	7.33	8.28	7.93	7.80
6	8.30	8.26	6.01	8.21	8.28	16	8.20	8.07	7.79	7.81	8.31
7	8.41	7.99	8.08	8.96	8.38	17	3.82	8.09	8.27	7.61	7.78
8	7.31	7.90	8.13	7.97	7.81	18	7.33	7.88	8.20	8.08	8.17
9	8.78	8.62	7.50	8.03	7.91	19	8.06	8.83	8.29	7.78	7.77
10	8.16	8.03	7.98	8.89	8.81	20	8.26	7.60	8.37	8.71	7.70

a. Construct an R-chart from these data.
b. What does the R-chart suggest about the stability of the process variation?
c. Construct an $\bar{x}$-chart from these data.
d. What does the $\bar{x}$-chart suggest about the stability of the process mean?
e. Does either chart suggest that special causes of variation are present? Justify your answer.

23.S.6   A company that applies paint to aluminum siding has determined that, when its painting process is in control, the average thickness of the finished product is 6.28 millimeters (mm). The observed data were obtained based on 50 samples, each of which contained $n = 5$ items. The average of the sample ranges was .28 mm. During each of the next 10 days, samples of size $n = 5$ were selected from the manufacturing process. The following data were obtained:

Day	Sample mean	Range
1	6.82	.23
2	6.51	.28
3	6.23	.10
4	5.88	.16
5	5.99	.31
6	6.74	.45
7	6.51	.47
8	6.48	.42
9	6.71	.36
10	6.75	.53

a. Construct an R-chart, using the data collected during the period for which the process was assumed to be in control. Plot the data for the last 10 days. Does the production appear to be out of control during any of these 10 days?
b. Construct an $\bar{x}$-chart, using the data collected during the period for which the process was assumed to be in control. Plot the data for the last 10 days. Does the production appear to be out of control during any of these 10 days?

c. A particular customer specifies that the aluminum siding it purchases from the manufacturer must have a layer of paint between 6.20 and 6.40 mm thick. Assume the true process mean is 6.28 mm and the true process standard deviation is .05 mm. Assume output follows a normal distribution. In a shipment of 1,000 units of aluminum siding, how many can be expected not to meet the customer's specification limits?

23.S.7 A manufacturing process that makes brake pads for heavy-duty trucks is sampled every hour to monitor the variation in the thickness of the brake pads being produced. Based on contract specifications, the desired pad thickness is .300 inch and the pads must have a diameter between .290 and .310 inch. The accompanying data show the thickness measurements from 25 hourly samples of four brake pads. Use these data, and construct an *R*-chart to determine whether the variation in the pad thickness appears to be in control.

Sample	Diameters				Sample	Diameters			
1	.304	.308	.291	.298	14	.308	.307	.301	.306
2	.296	.306	.295	.298	15	.304	.306	.299	.303
3	.306	.303	.299	.303	16	.295	.298	.291	.299
4	.301	.306	.302	.298	17	.297	.296	.306	.301
5	.293	.297	.293	.301	18	.303	.306	.298	.293
6	.304	.296	.303	.299	19	.298	.299	.303	.296
7	.307	.296	.299	.301	20	.291	.305	.309	.301
8	.303	.293	.298	.302	21	.303	.296	.305	.296
9	.307	.304	.291	.303	22	.309	.295	.294	.302
10	.293	.299	.297	.303	23	.301	.302	.295	.306
11	.298	.301	.307	.304	24	.299	.307	.291	.302
12	.306	.298	.291	.299	25	.301	.304	.303	.291
13	.293	.302	.302	.304					

a. Construct an *R*-chart from these data.
b. What does the *R*-chart suggest about the stability of the process variation?
c. Construct an $\bar{x}$-chart from these data.
d. What does the $\bar{x}$-chart suggest about the stability of the process mean?
e. Does either chart suggest that special causes of variation are present? Justify your answer.
f. Provide an example of an assignable cause of variation that could potentially affect this manufacturing process. Do the same for a common cause of variation.

References

Deming, William Edwards. *Some Theory of Sampling.* New York: Dover, 1984.

Deming, William Edwards. *Sample Design in Business Research.* New York: Wiley, 1960.

Dodge, H. F., and H. G. Romig. *Sampling Inspection Tables.* 2nd ed. New York: Wiley, 1959.

Grant, E. L., and R. S. Leavenworth. *Statistical Quality Control.* 5th ed. New York: McGraw-Hill, 1979.

Juran, J. M. *Quality-Control Handbook.* New York: McGraw-Hill, 1951.

Mendenhall, William. *A Course in Business Statistics.* 2nd ed. Boston: PWS-Kent, 1988.

Shewhart, Walter A. *Statistical Method from the Viewpoint of Quality Control.* Washington, D.C.: U.S. Department of Agriculture, 1939.

Tables

A.1   Random Numbers
A.2   Binomial Distribution
A.3   Values of $e^{-\mu}$
A.4   Poisson Distribution
A.5   Areas Under the Standard Normal Distribution
A.6   Critical Values of the t Distribution
A.7   Chi-square Distribution
A.8 F Distribution
A.9   The Mann-Whitney U Distribution
A.10 Critical Values of the Wilcoxon Test Statistic
A.11 Distribution of Number of Runs
A.12 Critical Values of Spearman's Rank Correlation Coefficient
A.13 Critical Values of the Durbin–Watson Statistic
A.14 Values of A_2 Used When Constructing an $\bar{x}$-Chart
A.15 Factors Used When Constructing an R-Chart

TABLE A.1   *Random numbers*

12651	61646	11769	75109	86996	97669	25757	32535	07122	76763
81769	74436	02630	72310	45049	18029	07469	42341	98173	79260
36737	98863	77240	76251	00654	64688	09343	70278	67331	98729
82861	54371	76610	94934	72748	44124	05610	53750	95938	01485
21325	15732	24127	37431	09723	63529	73977	95218	96074	42138
74146	47887	62463	23045	41490	07954	22597	60012	98866	90959
90759	64410	54179	66075	61051	75385	51378	08360	95946	95547
55683	98078	02238	91540	21219	17720	87817	41705	95785	12563
79686	17969	76061	83748	55920	83612	41540	86492	06447	60568
70333	00201	86201	69716	78185	62154	77930	67663	29529	75116
14042	53536	07779	04157	41172	36473	42123	43929	50533	33437
59911	08256	06596	48416	69770	68797	56080	14223	59199	30162
62368	62623	62742	14891	39247	52242	98832	69533	91174	57979
57529	97751	54976	48957	74599	08759	78494	52785	68526	64618
15469	90574	78033	66885	13936	42117	71831	22961	94225	31816
18625	23674	53850	32827	81647	80820	00420	63555	74489	80141
74626	68394	88562	70745	23701	45630	65891	58220	35442	60414
11119	16519	27384	90199	79210	76965	99546	30323	31664	22845
41101	17336	48951	53674	17880	45260	08575	49321	36191	17095
32123	91576	84221	78902	82010	30847	62329	63898	23268	74283
26091	68409	69704	82267	14751	13151	93115	01437	56945	89661
67680	79790	48462	59278	44185	29616	76531	19589	83139	28454
15184	19260	14073	07026	25264	08388	27182	22557	61501	67481
58010	45039	57181	10238	36874	28546	37444	80824	63981	39942
56425	53996	86245	32623	78858	08143	60377	42925	42815	11159
82630	84066	13592	60642	17904	99718	63432	88642	37858	25431
14927	40909	23900	48761	44860	92467	31742	87142	03607	32059
23740	22505	07489	85986	74420	21744	97711	36648	35620	97949
32990	97446	03711	63824	07953	85965	87089	11687	92414	67257
05310	24058	91946	78437	34365	82469	12430	84754	19354	72745
21839	39937	27534	88913	49055	19218	47712	67677	51889	70926
08833	42549	93981	94051	28382	83725	72643	64233	97252	17133
58336	11139	47479	00931	91560	95372	97642	33856	54825	55680
62032	91144	75478	47431	52726	30289	42411	91886	51818	78292
45171	30557	53116	04118	58301	24375	65609	85810	18620	49198
91611	62656	60128	35609	63698	78356	50682	22505	01692	36291
55472	63819	86314	49174	93582	73604	78614	78849	23096	72825
18573	09729	74091	53994	10970	86557	65661	41854	26037	53296
60866	02955	90288	82136	83644	94455	06560	78029	98768	71296
45043	55608	82767	60890	74646	79485	13619	98868	40857	19415
17831	09737	79473	75945	28394	79334	70577	38048	03607	06932
40137	03981	07585	18128	11178	32601	27994	05641	22600	86064
77776	31343	14576	97706	16039	47517	43300	59080	80392	63189
69605	44104	40103	95635	05635	81673	68657	09559	23510	95875
19916	52934	26499	09821	97331	80993	61299	36979	73599	35055
02606	58552	07678	56619	65325	30705	99582	53390	46357	13244
65183	73160	87131	35530	47946	09854	18080	02321	05809	04893
10740	98914	44916	11322	89717	88189	30143	52687	19420	60061
98642	89822	71691	51573	83666	61642	46683	33761	47542	23551
60139	25601	93663	25547	02654	94829	48672	28736	84994	13071

From *A Million Random Digits with 100,000 Normal Deviates*. N.Y.: The Free Press, 1955. Reprinted with permission of the Rand Corp.

TABLE A.2 *Binomial distribution*

The following table contains selected values of the binomial cumulative distribution function

$$P(X \leq c) = \sum_{x=0}^{c} \binom{n}{x} p^x (1 - p)^{n-x}$$

n	c	.01	.05	.10	.20	.30	.40	.50	.60	.70	.80	.90	.95	.99
5	0	.951	.774	.590	.328	.168	.078	.031	.010	.002	.000	.000	.000	.000
	1	.999	.977	.919	.737	.528	.337	.188	.087	.031	.007	.000	.000	.000
	2	1.000	.999	.991	.942	.837	.683	.500	.317	.163	.058	.009	.001	.000
	3	1.000	1.000	1.000	.993	.969	.913	.812	.663	.472	.263	.081	.023	.001
	4	1.000	1.000	1.000	1.000	.998	.990	.969	.922	.832	.672	.410	.226	.049
6	0	.941	.735	.531	.262	.118	.047	.016	.004	.001	.000	.000	.000	.000
	1	.999	.967	.886	.655	.420	.233	.109	.041	.011	.002	.000	.000	.000
	2	1.000	.998	.984	.901	.744	.544	.344	.179	.070	.017	.001	.000	.000
	3	1.000	1.000	.999	.983	.930	.821	.656	.456	.256	.099	.016	.002	.000
	4	1.000	1.000	1.000	.998	.989	.959	.891	.767	.580	.345	.114	.033	.001
	5	1.000	1.000	1.000	1.000	.999	.996	.984	.953	.882	.738	.469	.265	.059
7	0	.932	.698	.478	.210	.082	.028	.008	.002	.000	.000	.000	.000	.000
	1	.998	.956	.850	.577	.329	.159	.063	.019	.004	.000	.000	.000	.000
	2	1.000	.996	.974	.852	.647	.420	.227	.096	.029	.005	.000	.000	.000
	3	1.000	1.000	.997	.967	.874	.710	.500	.290	.126	.033	.003	.000	.000
	4	1.000	1.000	1.000	.995	.971	.904	.773	.580	.353	.148	.026	.004	.000
	5	1.000	1.000	1.000	1.000	.996	.981	.937	.841	.671	.423	.150	.044	.002
	6	1.000	1.000	1.000	1.000	1.000	.998	.992	.972	.918	.790	.522	.302	.068
8	0	.923	.663	.430	.168	.058	.017	.004	.001	.000	.000	.000	.000	.000
	1	.997	.943	.813	.503	.255	.106	.035	.009	.001	.000	.000	.000	.000
	2	1.000	.994	.962	.797	.552	.315	.145	.050	.011	.001	.000	.000	.000
	3	1.000	1.000	.995	.944	.806	.594	.363	.174	.058	.010	.000	.000	.000
	4	1.000	1.000	1.000	.990	.942	.826	.637	.406	.194	.056	.005	.000	.000
	5	1.000	1.000	1.000	.999	.989	.950	.855	.685	.448	.203	.038	.006	.000
	6	1.000	1.000	1.000	1.000	.999	.991	.965	.894	.745	.497	.187	.057	.003
	7	1.000	1.000	1.000	1.000	1.000	.999	.996	.983	.942	.832	.570	.337	.077
9	0	.914	.630	.387	.134	.040	.010	.002	.000	.000	.000	.000	.000	.000
	1	.997	.929	.775	.436	.196	.071	.020	.004	.000	.000	.000	.000	.000
	2	1.000	.992	.947	.738	.463	.232	.090	.025	.004	.000	.000	.000	.000
	3	1.000	.999	.992	.914	.730	.483	.254	.099	.025	.003	.000	.000	.000
	4	1.000	1.000	.999	.980	.901	.733	.500	.267	.099	.020	.001	.000	.000
	5	1.000	1.000	1.000	.997	.975	.901	.746	.517	.270	.086	.008	.001	.000
	6	1.000	1.000	1.000	1.000	.996	.975	.910	.768	.537	.262	.053	.008	.000
	7	1.000	1.000	1.000	1.000	1.000	.996	.980	.929	.804	.564	.225	.071	.003
	8	1.000	1.000	1.000	1.000	1.000	1.000	.998	.990	.960	.866	.613	.370	.086

Example: If $p = .20$, $n = 7$, and $c = 2$, then $P(X \leq 2) = .852$.

(*continued*)

TABLE A.2   *Binomial distribution (continued)*

								p						
n	*c*	.01	.05	.10	.20	.30	.40	.50	.60	.70	.80	.90	.95	.99
10	0	.904	.599	.349	.107	.028	.006	.001	.000	.000	.000	.000	.000	.000
	1	.996	.914	.736	.376	.149	.046	.011	.002	.000	.000	.000	.000	.000
	2	1.000	.988	.930	.678	.383	.167	.055	.012	.002	.000	.000	.000	.000
	3	1.000	.999	.987	.879	.650	.382	.172	.055	.011	.001	.000	.000	.000
	4	1.000	1.000	.998	.967	.850	.633	.377	.166	.047	.006	.000	.000	.000
	5	1.000	1.000	1.000	.994	.953	.834	.623	.367	.150	.033	.002	.000	.000
	6	1.000	1.000	1.000	.999	.989	.945	.828	.618	.350	.121	.013	.001	.000
	7	1.000	1.000	1.000	1.000	.998	.988	.945	.833	.617	.322	.070	.012	.000
	8	1.000	1.000	1.000	1.000	1.000	.998	.989	.954	.851	.624	.264	.086	.004
	9	1.000	1.000	1.000	1.000	1.000	1.000	.999	.994	.972	.893	.651	.401	.096
15	0	.860	.463	.206	.035	.005	.000	.000	.000	.000	.000	.000	.000	.000
	1	.990	.829	.549	.167	.035	.005	.000	.000	.000	.000	.000	.000	.000
	2	1.000	.964	.816	.398	.127	.027	.004	.000	.000	.000	.000	.000	.000
	3	1.000	.995	.944	.648	.297	.091	.018	.002	.000	.000	.000	.000	.000
	4	1.000	.999	.987	.836	.515	.217	.059	.009	.001	.000	.000	.000	.000
	5	1.000	1.000	.998	.939	.722	.403	.151	.034	.004	.000	.000	.000	.000
	6	1.000	1.000	1.000	.982	.869	.610	.304	.095	.015	.001	.000	.000	.000
	7	1.000	1.000	1.000	.996	.950	.787	.500	.213	.050	.004	.000	.000	.000
	8	1.000	1.000	1.000	.999	.985	.905	.696	.390	.131	.018	.000	.000	.000
	9	1.000	1.000	1.000	1.000	.996	.966	.849	.597	.278	.061	.002	.000	.000
	10	1.000	1.000	1.000	1.000	.999	.991	.941	.783	.485	.164	.013	.001	.000
	11	1.000	1.000	1.000	1.000	1.000	.998	.982	.909	.703	.352	.056	.005	.000
	12	1.000	1.000	1.000	1.000	1.000	1.000	.996	.973	.873	.602	.184	.036	.000
	13	1.000	1.000	1.000	1.000	1.000	1.000	1.000	.995	.965	.833	.451	.171	.010
	14	1.000	1.000	1.000	1.000	1.000	1.000	1.000	1.000	.995	.965	.794	.537	.140
20	0	.818	.358	.122	.012	.001	.000	.000	.000	.000	.000	.000	.000	.000
	1	.983	.736	.392	.069	.008	.001	.000	.000	.000	.000	.000	.000	.000
	2	.999	.925	.677	.206	.035	.004	.000	.000	.000	.000	.000	.000	.000
	3	1.000	.984	.867	.411	.107	.016	.001	.000	.000	.000	.000	.000	.000
	4	1.000	.997	.957	.630	.238	.051	.006	.000	.000	.000	.000	.000	.000
	5	1.000	1.000	.989	.804	.416	.126	.021	.002	.000	.000	.000	.000	.000
	6	1.000	1.000	.998	.913	.608	.250	.058	.006	.000	.000	.000	.000	.000
	7	1.000	1.000	1.000	.968	.772	.416	.132	.021	.001	.000	.000	.000	.000
	8	1.000	1.000	1.000	.990	.887	.596	.252	.057	.005	.000	.000	.000	.000
	9	1.000	1.000	1.000	.997	.952	.755	.412	.128	.017	.001	.000	.000	.000
	10	1.000	1.000	1.000	.999	.983	.872	.588	.245	.048	.003	.000	.000	.000
	11	1.000	1.000	1.000	1.000	.995	.943	.748	.404	.113	.010	.000	.000	.000
	12	1.000	1.000	1.000	1.000	.999	.979	.868	.584	.228	.032	.000	.000	.000
	13	1.000	1.000	1.000	1.000	1.000	.994	.942	.750	.392	.087	.002	.000	.000
	14	1.000	1.000	1.000	1.000	1.000	.998	.979	.874	.584	.196	.011	.000	.000
	15	1.000	1.000	1.000	1.000	1.000	1.000	.994	.949	.762	.370	.043	.003	.000
	16	1.000	1.000	1.000	1.000	1.000	1.000	.999	.984	.893	.589	.133	.016	.000
	17	1.000	1.000	1.000	1.000	1.000	1.000	1.000	.996	.965	.794	.323	.075	.001
	18	1.000	1.000	1.000	1.000	1.000	1.000	1.000	.999	.992	.931	.608	.264	.017
	19	1.000	1.000	1.000	1.000	1.000	1.000	1.000	1.000	.999	.988	.878	.642	.182

(continued)

TABLE A.2   *Binomial distribution (continued)*

n	c	.01	.05	.10	.20	.30	.40	.50	.60	.70	.80	.90	.95	.99
								p						
25	0	.778	.277	.072	.004	.000	.000	.000	.000	.000	.000	.000	.000	.000
	1	.974	.642	.271	.027	.002	.000	.000	.000	.000	.000	.000	.000	.000
	2	.998	.873	.537	.098	.009	.000	.000	.000	.000	.000	.000	.000	.000
	3	1.000	.966	.764	.234	.033	.002	.000	.000	.000	.000	.000	.000	.000
	4	1.000	.993	.902	.421	.090	.009	.000	.000	.000	.000	.000	.000	.000
	5	1.000	.999	.967	.617	.193	.029	.002	.000	.000	.000	.000	.000	.000
	6	1.000	1.000	.991	.780	.341	.074	.007	.000	.000	.000	.000	.000	.000
	7	1.000	1.000	.998	.891	.512	.154	.022	.001	.000	.000	.000	.000	.000
	8	1.000	1.000	1.000	.953	.677	.274	.054	.004	.000	.000	.000	.000	.000
	9	1.000	1.000	1.000	.983	.811	.425	.115	.013	.000	.000	.000	.000	.000
	10	1.000	1.000	1.000	.994	.902	.586	.212	.034	.002	.000	.000	.000	.000
	11	1.000	1.000	1.000	.998	.956	.732	.345	.078	.006	.000	.000	.000	.000
	12	1.000	1.000	1.000	1.000	.983	.846	.500	.154	.017	.000	.000	.000	.000
	13	1.000	1.000	1.000	1.000	.994	.922	.655	.268	.044	.002	.000	.000	.000
	14	1.000	1.000	1.000	1.000	.998	.966	.788	.414	.098	.006	.000	.000	.000
	15	1.000	1.000	1.000	1.000	1.000	.987	.885	.575	.189	.017	.000	.000	.000
	16	1.000	1.000	1.000	1.000	1.000	.996	.946	.726	.323	.047	.000	.000	.000
	17	1.000	1.000	1.000	1.000	1.000	.999	.978	.846	.488	.109	.002	.000	.000
	18	1.000	1.000	1.000	1.000	1.000	1.000	.993	.926	.659	.220	.009	.000	.000
	19	1.000	1.000	1.000	1.000	1.000	1.000	.998	.971	.807	.383	.033	.001	.000
	20	1.000	1.000	1.000	1.000	1.000	1.000	1.000	.991	.910	.579	.098	.007	.000
	21	1.000	1.000	1.000	1.000	1.000	1.000	1.000	.998	.967	.766	.236	.034	.000
	22	1.000	1.000	1.000	1.000	1.000	1.000	1.000	1.000	.991	.902	.463	.127	.002
	23	1.000	1.000	1.000	1.000	1.000	1.000	1.000	1.000	.998	.973	.729	.358	.026
	24	1.000	1.000	1.000	1.000	1.000	1.000	1.000	1.000	1.000	.996	.928	.723	.222

Appendix A

TABLE A.3 *Values of $e^{-\mu}$*

$(0 < \mu < 1)$

μ	0	1	2	3	4	5	6	7	8	9
0.0	1.0000	.9900	.9802	.9704	.9608	.9512	.9418	.9324	.9231	.9139
0.1	.9048	.8958	.8869	.8781	.8694	.8607	.8521	.8437	.8353	.8270
0.2	.8187	.8106	.8025	.7945	.7866	.7788	.7711	.7634	.7558	.7483
0.3	.7408	.7334	.7261	.7189	.7118	.7047	.6977	.6907	.6839	.6771
0.4	.6703	.6636	.6570	.6505	.6440	.6376	.6313	.6250	.6188	.6126
0.5	.6065	.6005	.5945	.5886	.5827	.5770	.5712	.5655	.5599	.5543
0.6	.5488	.5434	.5379	.5326	.5273	.5220	.5169	.5117	.5066	.5016
0.7	.4966	.4916	.4868	.4819	.4771	.4724	.4677	.4630	.4584	.4538
0.8	.4493	.4449	.4404	.4360	.4317	.4274	.4232	.4190	.4148	.4107
0.9	.4066	.4025	.3985	.3946	.3906	.3867	.3829	.3791	.3753	.3716

Example: $e^{-.48} = .6188$

$(\mu = 1, 2, 3, \ldots, 10)$

μ	1	2	3	4	5	6	7	8	9	10
$e^{-\mu}$	.36788	.13534	.04979	.01832	.006738	.002479	.000912	.000335	.000123	.000045

Example: $e^{-3} = .04979$

Note: To obtain values of $e^{-\mu}$ for other values of μ, use the laws of exponents.

Example: $e^{-3.48} = (e^{-3.00})(e^{-.48}) = (.04979)(.6188) = .03081$

TABLE A.4 *Poisson distribution*

The following table gives the probability of exactly x occurrences for various values of μ, as defined by the Poisson mass function:

$$P(X = x) = \frac{e^{-\mu}\mu^x}{x!}$$

						μ				
x	0.1	0.2	0.3	0.4	0.5	0.6	0.7	0.8	0.9	1.0
0	.9048	.8187	.7408	.6703	.6065	.5488	.4966	.4493	.4066	.3679
1	.0905	.1637	.2222	.2681	.3033	.3293	.3476	.3595	.3659	.3679
2	.0045	.0164	.0333	.0536	.0758	.0988	.1217	.1438	.1647	.1839
3	.0002	.0011	.0033	.0072	.0126	.0198	.0284	.0383	.0494	.0613
4	.0000	.0001	.0002	.0007	.0016	.0030	.0050	.0077	.0111	.0153
5	.0000	.0000	.0000	.0001	.0002	.0004	.0007	.0012	.0020	.0031
6	.0000	.0000	.0000	.0000	.0000	.0000	.0001	.0002	.0003	.0005
7	.0000	.0000	.0000	.0000	.0000	.0000	.0000	.0000	.0000	.0001

						μ				
x	1.1	1.2	1.3	1.4	1.5	1.6	1.7	1.8	1.9	2.0
0	.3329	.3012	.2725	.2466	.2231	.2019	.1827	.1653	.1496	.1353
1	.3662	.3614	.3543	.3452	.3347	.3230	.3106	.2975	.2842	.2707
2	.2014	.2169	.2303	.2417	.2510	.2584	.2640	.2678	.2700	.2707
3	.0738	.0867	.0998	.1128	.1255	.1378	.1496	.1607	.1710	.1804
4	.0203	.0260	.0324	.0395	.0471	.0551	.0636	.0723	.0812	.0902
5	.0045	.0062	.0084	.0111	.0141	.0176	.0216	.0260	.0309	.0361
6	.0008	.0012	.0018	.0026	.0035	.0047	.0061	.0078	.0098	.0120
7	.0001	.0002	.0003	.0005	.0008	.0011	.0045	.0020	.0027	.0034
8	.0000	.0000	.0001	.0001	.0001	.0002	.0003	.0005	.0006	.0009
9	.0000	.0000	.0000	.0000	.0000	.0000	.0001	.0001	.0001	.0002

						μ				
x	2.1	2.2	2.3	2.4	2.5	2.6	2.7	2.8	2.9	3.0
0	.1225	.1108	.1003	.0907	.0821	.0743	.0672	.0608	.0550	.0498
1	.2572	.2438	.2306	.2177	.2052	.1931	.1815	.1703	.1596	.1494
2	.2700	.2681	.2652	.2613	.2565	.2510	.2450	.2384	.2314	.2240
3	.1890	.1966	.2033	.2090	.2138	.2176	.2205	.2225	.2237	.2240
4	.0992	.1082	.1169	.1254	.1336	.1414	.1488	.1557	.1622	.1680
5	.0417	.0476	.0538	.0602	.0668	.0735	.0804	.0872	.0940	.1008
6	.0146	.0174	.0206	.0241	.0278	.0319	.0362	.0407	.0455	.0504
7	.0044	.0055	.0068	.0083	.0099	.0118	.0139	.0163	.0188	.0216
8	.0011	.0015	.0019	.0025	.0031	.0038	.0047	.0057	.0068	.0081
9	.0003	.0004	.0005	.0007	.0009	.0011	.0014	.0018	.0022	.0027
10	.0001	.0001	.0001	.0002	.0002	.0003	.0004	.0005	.0006	.0008
11	.0000	.0000	.0000	.0000	.0000	.0001	.0001	.0001	.0002	.0002
12	.0000	.0000	.0000	.0000	.0000	.0000	.0000	.0000	.0000	.0001

						μ				
x	3.1	3.2	3.3	3.4	3.5	3.6	3.7	3.8	3.9	4.0
0	.0450	.0408	.0369	.0334	.0302	.0273	.0247	.0224	.0202	.0183
1	.1397	.1304	.1217	.1135	.1057	.0984	.0915	.0850	.0789	.0733
2	.2165	.2087	.2008	.1929	.1850	.1771	.1692	.1615	.1539	.1465
3	.2237	.2226	.2209	.2186	.2158	.2125	.2087	.2046	.2001	.1954
4	.1734	.1781	.1823	.1858	.1888	.1912	.1931	.1944	.1951	.1954
5	.1075	.1140	.1203	.1264	.1322	.1377	.1429	.1477	.1522	.1563
6	.0555	.0608	.0662	.0716	.0771	.0826	.0881	.0936	.0989	.1042
7	.0246	.0278	.0312	.0348	.0385	.0425	.0466	.0508	.0551	.0595
8	.0095	.0111	.0129	.0148	.0169	.0191	.0215	.0241	.0269	.0298
9	.0033	.0040	.0047	.0056	.0066	.0076	.0089	.0102	.0116	.0132
10	.0010	.0013	.0016	.0019	.0023	.0028	.0033	.0039	.0045	.0053
11	.0003	.0004	.0005	.0006	.0007	.0009	.0011	.0013	.0016	.0019
12	.0001	.0001	.0001	.0002	.0002	.0003	.0003	.0004	.0005	.0006
13	.0000	.0000	.0000	.0000	.0001	.0001	.0001	.0001	.0002	.0002
14	.0000	.0000	.0000	.0000	.0000	.0000	.0000	.0000	.0000	.0001

						μ				
x	4.1	4.2	4.3	4.4	4.5	4.6	4.7	4.8	4.9	5.0
0	.0166	.0150	.0136	.0123	.0111	.0101	.0091	.0082	.0074	.0067
1	.0679	.0630	.0583	.0540	.0500	.0462	.0427	.0395	.0365	.0337
2	.1393	.1323	.1254	.1188	.1125	.1063	.1005	.0948	.0894	.0842
3	.1904	.1852	.1798	.1743	.1687	.1634	.1574	.1517	.1460	.1404
4	.1951	.1944	.1933	.1917	.1898	.1875	.1849	.1820	.1789	.1755
5	.1600	.1633	.1662	.1687	.1708	.1725	.1738	.1747	.1753	.1755
6	.1093	.1143	.1191	.1237	.1281	.1323	.1362	.1398	.1432	.1462
7	.0640	.0686	.0732	.0778	.0824	.0869	.0914	.0959	.1002	.1044
8	.0328	.0360	.0393	.0428	.0463	.0500	.0537	.0575	.0614	.0653
9	.0150	.0168	.0188	.0209	.0232	.0255	.0280	.0307	.0334	.0363
10	.0061	.0071	.0081	.0092	.0104	.0118	.0132	.0147	.0164	.0181
11	.0023	.0027	.0032	.0037	.0043	.0049	.0056	.0064	.0073	.0082
12	.0008	.0009	.0011	.0014	.0016	.0019	.0022	.0026	.0030	.0034
13	.0002	.0003	.0004	.0005	.0006	.0007	.0008	.0009	.0011	.0013
14	.0001	.0001	.0001	.0001	.0002	.0002	.0003	.0003	.0004	.0005
15	.0000	.0000	.0000	.0000	.0001	.0001	.0001	.0001	.0001	.0002

TABLE A.4   *Poisson distribution (continued)*

μ

x	5.1	5.2	5.3	5.4	5.5	5.6	5.7	5.8	5.9	6.0
0	.0061	.0055	.0050	.0045	.0041	.0037	.0033	.0030	.0027	.0025
1	.0311	.0287	.0265	.0244	.0225	.0207	.0191	.0176	.0162	.0149
2	.0793	.0746	.0701	.0659	.0618	.0580	.0544	.0509	.0477	.0446
3	.1348	.1293	.1239	.1185	.1133	.1082	.1033	.0985	.0938	.0892
4	.1719	.1681	.1641	.1600	.1558	.1515	.1472	.1428	.1383	.1339
5	.1753	.1748	.1740	.1728	.1714	.1697	.1678	.1620	.1632	.1606
6	.1490	.1515	.1537	.1555	.1571	.1584	.1594	.1656	.1605	.1606
7	.1086	.1125	.1163	.1200	.1234	.1267	.1298	.1301	.1353	.1377
8	.0692	.0731	.0771	.0810	.0849	.0887	.0925	.0926	.0998	.1033
9	.0392	.0423	.0454	.0486	.0519	.0552	.0586	.0662	.0654	.0688
10	.0200	.0220	.0241	.0262	.0285	.0309	.0334	.0359	.0386	.0413
11	.0093	.0104	.0116	.0129	.0143	.0157	.0173	.0190	.0207	.0225
12	.0039	.0045	.0051	.0058	.0065	.0073	.0082	.0092	.0102	.0113
13	.0015	.0018	.0021	.0024	.0028	.0032	.0036	.0041	.0046	.0052
14	.0006	.0007	.0008	.0009	.0011	.0013	.0015	.0017	.0019	.0022
15	.0002	.0002	.0003	.0003	.0004	.0005	.0006	.0007	.0008	.0009
16	.0001	.0001	.0001	.0001	.0001	.0002	.0002	.0002	.0003	.0003
17	.0000	.0000	.0000	.0000	.0000	.0001	.0001	.0001	.0001	.0001

μ

x	7.1	7.2	7.3	7.4	7.5	7.6	7.7	7.8	7.9	8.0
0	.0008	.0007	.0007	.0006	.0006	.0005	.0005	.0004	.0004	.0003
1	.0059	.0054	.0049	.0045	.0041	.0038	.0035	.0032	.0029	.0027
2	.0208	.0194	.0180	.0167	.0156	.0145	.0134	.0125	.0116	.0107
3	.0492	.0464	.0438	.0413	.0389	.0366	.0345	.0324	.0305	.0286
4	.0874	.0836	.0799	.0764	.0729	.0696	.0663	.0632	.0602	.0573
5	.1241	.1204	.1167	.1130	.1094	.1057	.1021	.0986	.0951	.0916
6	.1468	.1445	.1420	.1394	.1367	.1339	.1311	.1282	.1252	.1221
7	.1489	.1486	.1481	.1474	.1465	.1454	.1442	.1428	.1413	.1396
8	.1321	.1337	.1351	.1363	.1373	.1382	.1388	.1392	.1395	.1396
9	.1042	.1070	.1096	.1121	.1144	.1167	.1187	.1207	.1224	.1241
10	.0740	.0770	.0800	.0829	.0858	.0887	.0914	.0941	.0967	.0993
11	.0478	.0504	.0531	.0558	.0585	.0613	.0640	.0667	.0695	.0722
12	.0283	.0303	.0323	.0344	.0366	.0380	.0411	.0434	.0457	.0481
13	.0154	.0168	.0181	.0196	.0211	.0227	.0243	.0260	.0278	.0296
14	.0078	.0086	.0095	.0104	.0113	.0123	.0134	.0145	.0157	.0169
15	.0037	.0041	.0046	.0051	.0057	.0062	.0069	.0075	.0083	.0090
16	.0016	.0019	.0021	.0024	.0026	.0030	.0033	.0037	.0041	.0045
17	.0007	.0008	.0009	.0010	.0012	.0013	.0015	.0017	.0019	.0021
18	.0003	.0003	.0004	.0004	.0005	.0006	.0006	.0007	.0008	.0009
19	.0001	.0001	.0001	.0002	.0002	.0002	.0003	.0003	.0003	.0004
20	.0000	.0000	.0001	.0001	.0001	.0001	.0001	.0001	.0001	.0002
21	.0000	.0000	.0000	.0000	.0000	.0000	.0000	.0000	.0001	.0001

μ

x	6.1	6.2	6.3	6.4	6.5	6.6	6.7	6.8	6.9	7.0
0	.0022	.0020	.0018	.0017	.0015	.0014	.0012	.0011	.0010	.0009
1	.0137	.0126	.0116	.0106	.0098	.0090	.0082	.0076	.0070	.0064
2	.0417	.0390	.0364	.0340	.0318	.0296	.0276	.0258	.0240	.0223
3	.0848	.0806	.0765	.0726	.0688	.0652	.0617	.0584	.0552	.0521
4	.1294	.1249	.1205	.1162	.1118	.1076	.1034	.0992	.0952	.0912
5	.1579	.1549	.1519	.1487	.1454	.1420	.1385	.1349	.1314	.1277
6	.1605	.1601	.1595	.1586	.1575	.1562	.1546	.1529	.1511	.1490
7	.1399	.1418	.1435	.1450	.1462	.1472	.1480	.1486	.1489	.1490
8	.1066	.1099	.1130	.1160	.1188	.1215	.1240	.1263	.1284	.1304
9	.0723	.0757	.0791	.0825	.0858	.0891	.0923	.0954	.0985	.1014
10	.0441	.0469	.0498	.0528	.0558	.0588	.0618	.0649	.0679	.0710
11	.0245	.0265	.0285	.0307	.0330	.0353	.0377	.0401	.0426	.0452
12	.0124	.0137	.0150	.0164	.0179	.0194	.0210	.0227	.0245	.0264
13	.0058	.0065	.0073	.0081	.0089	.0098	.0108	.0119	.0130	.0142
14	.0025	.0029	.0033	.0037	.0041	.0046	.0052	.0058	.0064	.0071
15	.0010	.0012	.0014	.0016	.0018	.0020	.0023	.0026	.0029	.0033
16	.0004	.0005	.0005	.0006	.0007	.0008	.0010	.0011	.0013	.0014
17	.0001	.0002	.0002	.0002	.0003	.0003	.0004	.0004	.0005	.0006
18	.0000	.0001	.0001	.0001	.0001	.0001	.0001	.0002	.0002	.0002
19	.0000	.0000	.0000	.0000	.0000	.0000	.0000	.0001	.0001	.0001

μ

x	8.1	8.2	8.3	8.4	8.5	8.6	8.7	8.8	8.9	9.0
0	.0003	.0003	.0002	.0002	.0002	.0002	.0002	.0002	.0001	.0001
1	.0025	.0023	.0021	.0019	.0017	.0016	.0014	.0013	.0012	.0011
2	.0100	.0092	.0086	.0079	.0074	.0068	.0063	.0058	.0054	.0050
3	.0269	.0252	.0237	.0222	.0208	.0195	.0183	.0171	.0160	.0150
4	.0544	.0517	.0491	.0466	.0443	.0420	.0398	.0377	.0357	.0337
5	.0882	.0849	.0816	.0784	.0752	.0722	.0692	.0663	.0635	.0607
6	.1191	.1160	.1128	.1097	.1066	.1034	.1003	.0972	.0941	.0911
7	.1378	.1358	.1338	.1317	.1294	.1271	.1247	.1222	.1197	.1171
8	.1395	.1392	.1388	.1382	.1375	.1366	.1356	.1344	.1332	.1318
9	.1256	.1269	.1280	.1290	.1299	.1306	.1311	.1315	.1317	.1318
10	.1017	.1040	.1063	.1084	.1104	.1123	.1140	.1157	.1172	.1186
11	.0749	.0776	.0802	.0828	.0853	.0878	.0902	.0925	.0948	.0970
12	.0505	.0530	.0555	.0579	.0604	.0629	.0654	.0679	.0703	.0728
13	.0315	.0334	.0354	.0374	.0395	.0416	.0438	.0459	.0481	.0504
14	.0182	.0196	.0210	.0225	.0240	.0256	.0272	.0289	.0306	.0324
15	.0098	.0107	.0116	.0126	.0136	.0147	.0158	.0169	.0182	.0194
16	.0050	.0055	.0060	.0066	.0072	.0079	.0086	.0093	.0101	.0109
17	.0024	.0026	.0029	.0033	.0036	.0040	.0044	.0048	.0053	.0058
18	.0011	.0012	.0014	.0015	.0017	.0019	.0021	.0024	.0026	.0029
19	.0005	.0005	.0006	.0007	.0008	.0009	.0010	.0011	.0012	.0014
20	.0002	.0002	.0002	.0003	.0003	.0004	.0004	.0005	.0005	.0006
21	.0001	.0001	.0001	.0001	.0001	.0002	.0002	.0002	.0002	.0003
22	.0000	.0000	.0000	.0000	.0001	.0001	.0001	.0001	.0001	.0001

TABLE A.5 *Areas under the standard normal distribution*

The following table gives the areas
under the standard normal curve
from 0 to z.

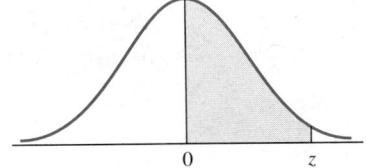

z	0	1	2	3	4	5	6	7	8	9
0.0	.0000	.0040	.0080	.0120	.0160	.0199	.0239	.0279	.0319	.0359
0.1	.0398	.0438	.0478	.0517	.0557	.0596	.0636	.0675	.0714	.0754
0.2	.0793	.0832	.0871	.0910	.0948	.0987	.1026	.1064	.1103	.1141
0.3	.1179	.1217	.1255	.1293	.1331	.1368	.1406	.1443	.1480	.1517
0.4	.1554	.1591	.1628	.1664	.1700	.1736	.1772	.1808	.1844	.1879
0.5	.1915	.1950	.1985	.2019	.2054	.2088	.2123	.2157	.2190	.2224
0.6	.2258	.2291	.2324	.2357	.2389	.2422	.2454	.2486	.2518	.2549
0.7	.2580	.2612	.2642	.2673	.2704	.2734	.2764	.2794	.2823	.2852
0.8	.2881	.2910	.2939	.2967	.2996	.3023	.3051	.3078	.3106	.3133
0.9	.3159	.3186	.3212	.3238	.3264	.3289	.3315	.3340	.3365	.3389
1.0	.3413	.3438	.3461	.3485	.3508	.3531	.3554	.3577	.3599	.3621
1.1	.3643	.3665	.3686	.3708	.3729	.3749	.3770	.3790	.3810	.3830
1.2	.3849	.3869	.3888	.3907	.3925	.3944	.3962	.3980	.3997	.4015
1.3	.4032	.4049	.4066	.4082	.4099	.4115	.4131	.4147	.4162	.4177
1.4	.4192	.4207	.4222	.4236	.4251	.4265	.4279	.4292	.4306	.4319
1.5	.4332	.4345	.4357	.4370	.4382	.4394	.4406	.4418	.4429	.4441
1.6	.4452	.4463	.4474	.4484	.4495	.4505	.4515	.4525	.4535	.4545
1.7	.4554	.4564	.4573	.4582	.4591	.4599	.4608	.4616	.4625	.4633
1.8	.4641	.4649	.4656	.4664	.4671	.4678	.4686	.4693	.4699	.4706
1.9	.4713	.4719	.4726	.4732	.4738	.4744	.4750	.4756	.4761	.4767
2.0	.4772	.4778	.4783	.4788	.4793	.4798	.4803	.4808	.4812	.4817
2.1	.4821	.4826	.4830	.4834	.4838	.4842	.4846	.4850	.4854	.4857
2.2	.4861	.4864	.4868	.4871	.4875	.4878	.4881	.4884	.4887	.4890
2.3	.4893	.4896	.4898	.4901	.4904	.4906	.4909	.4911	.4913	.4916
2.4	.4918	.4920	.4922	.4925	.4927	.4929	.4931	.4932	.4934	.4936
2.5	.4938	.4940	.4941	.4943	.4945	.4946	.4948	.4949	.4951	.4952
2.6	.4953	.4955	.4956	.4957	.4959	.4960	.4961	.4962	.4963	.4964
2.7	.4965	.4966	.4967	.4968	.4969	.4970	.4971	.4972	.4973	.4974
2.8	.4974	.4975	.4976	.4977	.4977	.4978	.4979	.4979	.4980	.4981
2.9	.4981	.4982	.4982	.4983	.4984	.4984	.4985	.4985	.4986	.4986
3.0	.4987	.4987	.4987	.4988	.4988	.4989	.4989	.4989	.4990	.4990
3.1	.4990	.4991	.4991	.4991	.4992	.4992	.4992	.4992	.4993	.4993
3.2	.4993	.4993	.4994	.4994	.4994	.4994	.4994	.4995	.4995	.4995
3.3	.4995	.4995	.4995	.4996	.4996	.4996	.4996	.4996	.4996	.4997
3.4	.4997	.4997	.4997	.4997	.4997	.4997	.4997	.4997	.4997	.4998
3.5	.4998	.4998	.4998	.4998	.4998	.4998	.4998	.4998	.4998	.4998
3.6	.4998	.4998	.4999	.4999	.4999	.4999	.4999	.4999	.4999	.4999
3.7	.4999	.4999	.4999	.4999	.4999	.4999	.4999	.4999	.4999	.4999
3.8	.4999	.4999	.4999	.4999	.4999	.4999	.4999	.4999	.4999	.4999
3.9	.5000	.5000	.5000	.5000	.5000	.5000	.5000	.5000	.5000	.5000

Example: The area between $z = 0$ and $z = 1.24$ is .3925.

TABLE A.6   *Critical values of the t distribution*

The following table contains critical values of t
for given probability levels.

Degrees of Freedom, v	CRITICAL VALUES $t\alpha$				
	$t_{.10}$	$t_{.05}$	$t_{.025}$	$t_{.01}$	$t_{.005}$
1	3.078	6.314	12.706	31.821	63.657
2	1.886	2.920	4.303	6.965	9.925
3	1.638	2.353	3.182	4.541	5.841
4	1.533	2.132	2.776	3.747	4.604
5	1.476	2.015	2.571	3.365	4.032
6	1.440	1.943	2.447	3.143	3.707
7	1.415	1.895	2.365	2.998	3.499
8	1.397	1.860	2.306	2.896	3.355
9	1.383	1.833	2.262	2.821	3.250
10	1.372	1.812	2.228	2.764	3.169
11	1.363	1.796	2.201	2.718	3.106
12	1.356	1.782	2.179	2.681	3.055
13	1.350	1.771	2.160	2.650	3.012
14	1.345	1.761	2.145	2.624	2.977
15	1.341	1.753	2.131	2.602	2.947
16	1.337	1.746	2.120	2.583	2.921
17	1.333	1.740	2.110	2.567	2.898
18	1.330	1.734	2.101	2.552	2.878
19	1.328	1.729	2.093	2.539	2.861
20	1.325	1.725	2.086	2.528	2.845
21	1.323	1.721	2.080	2.518	2.831
22	1.321	1.717	2.074	2.508	2.819
23	1.319	1.714	2.069	2.500	2.807
24	1.318	1.711	2.064	2.492	2.797
25	1.316	1.708	2.060	2.485	2.787
26	1.315	1.706	2.056	2.479	2.779
27	1.314	1.703	2.052	2.473	2.771
28	1.313	1.701	2.048	2.467	2.763
29	1.311	1.699	2.045	2.462	2.756
30	1.310	1.697	2.042	2.457	2.750
40	1.303	1.684	2.021	2.423	2.704
60	1.296	1.671	2.000	2.390	2.660
120	1.290	1.661	1.984	2.358	2.626
∞	1.282	1.645	1.960	2.326	2.576

From Merrington, Maxine. "Table of Percentage Points of the *t*-Distribution."
Biometrika, vol. 32, 1941, p. 300.

Example: The value of t with 10 degrees of freedom for which 1%
of the area is in the right-hand tail is 2.764; $t_{.01} = 2.764$ for $v = 10$.

TABLE A.7 *Chi-square distribution*

Entries in the table give χ_α^2 values, where α is the area or probability in the upper tail of the chi-square distribution.

Example: With 10 degrees of freedom and a .01 area in the upper tail, $\chi_{.01}^2 = 23.2093$.

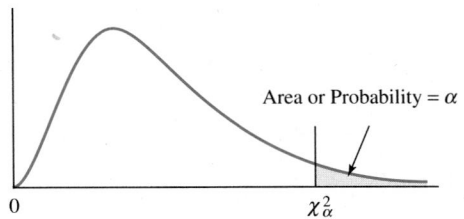

Degrees of freedom	AREA IN UPPER TAIL									
	.995	.99	.975	.95	.90	.10	.05	.025	.01	.005
1	392704×10^{-10}	157088×10^{-9}	982069×10^{-9}	393214×10^{-8}	.0157908	2.70554	3.84146	5.02389	6.63490	7.87944
2	.0100251	.0201007	.0506356	.102587	.210720	4.60517	5.99147	7.37776	9.21034	10.5966
3	.0717212	.114832	.215795	.351846	.584375	6.25139	7.81473	9.34840	11.3449	12.8381
4	.206990	.297110	.484419	.710721	1.063623	7.77944	9.48773	11.1433	13.2767	14.8602
5	.411740	.554300	.831211	1.145476	1.61031	9.23635	11.0705	12.8325	15.0863	16.7496
6	.675727	.872085	1.237347	1.63539	2.20413	10.6446	12.5916	14.4494	16.8119	18.5476
7	.989265	1.239043	1.68987	2.16735	2.83311	12.0170	14.0671	16.0128	18.4753	20.2777
8	1.344419	1.646482	2.17973	2.73264	3.48954	13.3616	15.5073	17.5346	20.0902	21.9550
9	1.734926	2.087912	2.70039	3.32511	4.16816	14.6837	16.9190	19.0228	21.6660	23.5893
10	2.15585	2.55821	3.24697	3.94030	4.86518	15.9871	18.3070	20.4831	23.2093	25.1882
11	2.60321	3.05347	3.81575	4.57481	5.57779	17.2750	19.6751	21.9200	24.7250	26.7569
12	3.07382	3.57056	4.40379	5.22603	6.30380	18.5494	21.0261	23.3367	26.2170	28.2995
13	3.56503	4.10691	5.00874	5.89186	7.04150	19.8119	22.3621	24.7356	27.6883	29.8194
14	4.07468	4.66043	5.62872	6.57063	7.78953	21.0642	23.6848	26.1190	29.1413	31.3193
15	4.60094	5.22935	6.26214	7.26094	8.54675	22.3072	24.9958	27.4884	30.5779	32.8013
16	5.14224	5.81221	6.90766	7.96164	9.31223	23.5418	26.2962	28.8454	31.9999	34.2672
17	5.69724	6.40776	7.56418	8.67176	10.0852	24.7690	27.5871	30.1910	33.4087	35.7185
18	6.26481	7.01491	8.23075	9.39046	10.8649	25.9894	28.8693	31.5264	34.8053	37.1564
19	6.84398	7.63273	8.90655	10.1170	11.6509	27.2036	30.1435	32.8523	36.1908	38.5822
20	7.43386	8.26040	9.59083	10.8508	12.4426	28.4120	31.4104	34.1696	37.5662	39.9968
21	8.03366	8.89720	10.28293	11.5913	13.2396	29.6151	32.6705	35.4789	38.9321	41.4010
22	8.64272	9.54249	10.9823	12.3380	14.0415	30.8133	33.9244	36.7807	40.2894	42.7958
23	9.26042	10.19567	11.6885	13.0905	14.8479	32.0069	35.1725	38.0757	41.6384	44.1813
24	9.88623	10.8564	12.4011	13.8484	15.6587	33.1963	36.4151	39.3641	42.9798	45.5585
25	10.5197	11.5240	13.1197	14.6114	16.4734	34.3816	37.6525	40.6465	44.3141	46.9278
26	11.1603	12.1981	13.8439	15.3791	17.2919	35.5631	38.8852	41.9232	45.6417	48.2899
27	11.8076	12.8786	14.5733	16.1513	18.1138	36.7412	40.1133	43.1944	46.9630	49.6449
28	12.4613	13.5648	15.3079	16.9279	18.9392	37.9159	41.3372	44.4607	48.2782	50.9933
29	13.1211	14.2565	16.0471	17.7083	19.7677	39.0875	42.5569	45.7222	49.5879	52.3356
30	13.7867	14.9535	16.7908	18.4926	20.5992	40.2560	43.7729	46.9792	50.8922	53.6720
40	20.7065	22.1643	24.4331	26.5093	29.0505	51.8050	55.7585	59.3417	63.6907	66.7659
50	27.9907	29.7067	32.3574	34.7642	37.6886	63.1671	67.5048	71.4202	76.1539	79.4900
60	35.5346	37.4848	40.4817	43.1879	46.4589	74.3970	79.0819	83.2976	88.3794	91.9517
70	43.2752	45.4418	48.7576	51.7393	55.3290	85.5271	90.5312	95.0231	100.425	104.215
80	51.1720	53.5400	57.1532	60.3915	64.2778	96.5782	101.879	106.629	112.329	116.321
90	59.1963	61.7541	65.6466	69.1260	73.2912	107.565	113.145	118.136	124.116	128.299
100	67.3276	70.0648	74.2219	77.9295	82.3581	118.498	124.342	129.561	135.807	140.169

From Pearson, E. S., and H. O. Hartley. Table 8, "Percentage Points of the χ^2 Distribution," *Biometrika Tables for Statisticians,* Vol. 1. Reprinted by permission of Biometrika Trustees.

TABLE A.8   *F distribution*

Entries in the following table give F_α values, where α is the area or probability in the upper tail of the F distribution.

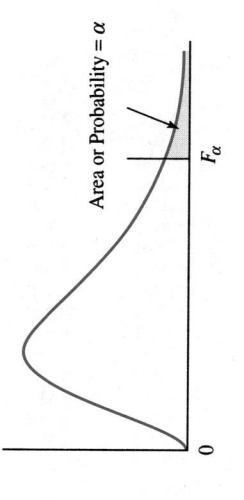

Area or Probability $= \alpha$

TABLE OF $F_{.05}$ VALUES

Denominator degrees of freedom	NUMERATOR DEGREES OF FREEDOM																		
	1	2	3	4	5	6	7	8	9	10	12	15	20	24	30	40	60	120	∞
1	161.4	199.5	215.7	224.6	230.2	234.0	236.8	238.9	240.5	241.9	243.9	245.9	248.0	249.1	250.1	251.1	252.2	253.3	254.3
2	18.51	19.00	19.16	19.25	19.30	19.33	19.35	19.37	19.38	19.40	19.41	19.43	19.45	19.45	19.46	19.47	19.48	19.49	19.50
3	10.13	9.55	9.28	9.12	9.01	8.94	8.89	8.85	8.81	8.79	8.74	8.70	8.66	8.64	8.62	8.59	8.57	8.55	8.53
4	7.71	6.94	6.59	6.39	6.26	6.16	6.09	6.04	6.00	5.96	5.91	5.86	5.80	5.77	5.75	5.72	5.69	5.66	5.63
5	6.61	5.79	5.41	5.19	5.05	4.95	4.88	4.82	4.77	4.74	4.68	4.62	4.56	4.53	4.50	4.46	4.43	4.40	4.36
6	5.99	5.14	4.76	4.53	4.39	4.28	4.21	4.15	4.10	4.06	4.00	3.94	3.87	3.84	3.81	3.77	3.74	3.70	3.67
7	5.59	4.74	4.35	4.12	3.97	3.87	3.79	3.73	3.68	3.64	3.57	3.51	3.44	3.41	3.38	3.34	3.30	3.27	3.23
8	5.32	4.46	4.07	3.84	3.69	3.58	3.50	3.44	3.39	3.35	3.28	3.22	3.15	3.12	3.08	3.04	3.01	2.97	2.93
9	5.12	4.26	3.86	3.63	3.48	3.37	3.29	3.23	3.18	3.14	3.07	3.01	2.94	2.90	2.86	2.83	2.79	2.75	2.71

10	4.96	3.71	3.48	3.33	3.22	3.14	3.07	3.02	2.98	2.91	2.85	2.77	2.74	2.70	2.66	2.62	2.58	2.54
11	4.84	3.59	3.36	3.20	3.09	3.01	2.95	2.90	2.85	2.79	2.72	2.65	2.61	2.57	2.53	2.49	2.45	2.40
12	4.75	3.49	3.26	3.11	3.00	2.91	2.85	2.80	2.75	2.69	2.62	2.54	2.51	2.47	2.43	2.38	2.34	2.30
13	4.67	3.41	3.18	3.03	2.92	2.83	2.77	2.71	2.67	2.60	2.53	2.46	2.42	2.38	2.34	2.30	2.25	2.21
14	4.60	3.34	3.11	2.96	2.85	2.76	2.70	2.65	2.60	2.53	2.46	2.39	2.35	2.31	2.27	2.22	2.18	2.13
15	4.54	3.29	3.06	2.90	2.79	2.71	2.64	2.59	2.54	2.48	2.40	2.33	2.29	2.25	2.20	2.16	2.11	2.07
16	4.49	3.24	3.01	2.85	2.74	2.66	2.59	2.54	2.49	2.42	2.35	2.28	2.24	2.19	2.15	2.11	2.06	2.01
17	4.45	3.20	2.96	2.81	2.70	2.61	2.55	2.49	2.45	2.38	2.31	2.23	2.19	2.15	2.10	2.06	2.01	1.96
18	4.41	3.16	2.93	2.77	2.66	2.58	2.51	2.46	2.41	2.34	2.27	2.19	2.15	2.11	2.06	2.02	1.97	1.92
19	4.38	3.13	2.90	2.74	2.63	2.54	2.48	2.42	2.38	2.31	2.23	2.16	2.11	2.07	2.03	1.98	1.93	1.88
20	4.35	3.10	2.87	2.71	2.60	2.51	2.45	2.39	2.35	2.28	2.20	2.12	2.08	2.04	1.99	1.95	1.90	1.84
21	4.32	3.07	2.84	2.68	2.57	2.49	2.42	2.37	2.32	2.25	2.18	2.10	2.05	2.01	1.96	1.92	1.87	1.81
22	4.30	3.05	2.82	2.66	2.55	2.46	2.40	2.34	2.30	2.23	2.15	2.07	2.03	1.98	1.94	1.89	1.84	1.78
23	4.28	3.03	2.80	2.64	2.53	2.44	2.37	2.32	2.27	2.20	2.13	2.05	2.01	1.96	1.91	1.86	1.81	1.76
24	4.26	3.01	2.78	2.62	2.51	2.42	2.36	2.30	2.25	2.18	2.11	2.03	1.98	1.94	1.89	1.84	1.79	1.73
25	4.24	2.99	2.76	2.60	2.49	2.40	2.34	2.28	2.24	2.16	2.09	2.01	1.96	1.92	1.87	1.82	1.77	1.71
26	4.23	2.98	2.74	2.59	2.47	2.39	2.32	2.27	2.22	2.15	2.07	1.99	1.95	1.90	1.85	1.80	1.75	1.69
27	4.21	2.96	2.73	2.57	2.46	2.37	2.31	2.25	2.20	2.13	2.06	1.97	1.93	1.88	1.84	1.79	1.73	1.67
28	4.20	2.95	2.71	2.56	2.45	2.36	2.29	2.24	2.19	2.12	2.04	1.96	1.91	1.87	1.82	1.77	1.71	1.65
29	4.18	2.93	2.70	2.55	2.43	2.35	2.28	2.22	2.18	2.10	2.03	1.94	1.90	1.85	1.81	1.75	1.70	1.64
30	4.17	2.92	2.69	2.53	2.42	2.33	2.27	2.21	2.16	2.09	2.01	1.93	1.89	1.84	1.79	1.74	1.68	1.62
40	4.08	2.84	2.61	2.45	2.34	2.25	2.18	2.12	2.08	2.00	1.92	1.84	1.79	1.74	1.69	1.64	1.58	1.51
60	4.00	2.76	2.53	2.37	2.25	2.17	2.10	2.04	1.99	1.92	1.84	1.75	1.70	1.65	1.59	1.53	1.47	1.39
120	3.92	2.68	2.45	2.29	2.17	2.09	2.02	1.96	1.91	1.83	1.75	1.66	1.61	1.55	1.50	1.43	1.35	1.25
∞	3.84	2.60	2.37	2.21	2.10	2.01	1.94	1.88	1.83	1.75	1.67	1.57	1.52	1.46	1.39	1.32	1.22	1.00

(continued)

Example: With 7 numerator degrees of freedom, 9 denominator degrees of freedom, and a .05 area in the upper tail, $F_{.05} = 3.29$.

TABLE A.8   *F distribution (continued)*

TABLE OF $F_{.01}$ VALUES

Denominator degrees of freedom	\multicolumn{19}{c}{NUMERATOR DEGREES OF FREEDOM}																	
	1	2	3	4	5	6	7	8	9	10	12	15	20	24	30	40	60	120	∞
1	4052	4999.5	5403	5625	5764	5859	5928	5982	6022	6056	6106	6157	6209	6235	6261	6287	6313	6339	6366
2	98.50	99.00	99.17	99.25	99.30	99.33	99.36	99.37	99.39	99.40	99.42	99.43	99.45	99.46	99.47	99.47	99.48	99.49	99.50
3	34.12	30.82	29.46	28.71	28.24	27.91	27.67	27.49	27.35	27.23	27.05	26.87	26.69	26.60	26.50	26.41	26.32	26.22	26.13
4	21.20	18.00	16.69	15.98	15.52	15.21	14.98	14.80	14.66	14.55	14.37	14.20	14.02	13.93	13.84	13.75	13.65	13.56	13.46
5	16.26	13.27	12.06	11.39	10.97	10.67	10.46	10.29	10.16	10.05	9.89	9.72	9.55	9.47	9.38	9.29	9.20	9.11	9.06
6	13.75	10.92	9.78	9.15	8.75	8.47	8.26	8.10	7.98	7.87	7.72	7.56	7.40	7.31	7.23	7.14	7.06	6.97	6.88
7	12.25	9.55	8.45	7.85	7.46	7.19	6.99	6.84	6.72	6.62	6.47	6.31	6.16	6.07	5.99	5.91	5.82	5.74	5.65
8	11.26	8.65	7.59	7.01	6.63	6.37	6.18	6.03	5.91	5.81	5.67	5.52	5.36	5.28	5.20	5.12	5.03	4.95	4.86
9	10.56	8.02	6.99	6.42	6.06	5.80	5.61	5.47	5.35	5.26	5.11	4.96	4.81	4.73	4.65	4.57	4.48	4.40	4.31
10	10.04	7.56	6.55	5.99	5.64	5.39	5.20	5.06	4.94	4.85	4.71	4.56	4.41	4.33	4.25	4.17	4.08	4.00	3.91
11	9.65	7.21	6.22	5.67	5.32	5.07	4.89	4.74	4.63	4.54	4.40	4.25	4.10	4.02	3.94	3.86	3.78	3.69	3.60
12	9.33	6.93	5.95	5.41	5.06	4.82	4.64	4.50	4.39	4.30	4.16	4.01	3.86	3.78	3.70	3.62	3.54	3.45	3.36
13	9.07	6.70	5.74	5.21	4.86	4.62	4.44	4.30	4.19	4.10	3.96	3.82	3.66	3.59	3.51	3.43	3.34	3.25	3.17
14	8.86	6.51	5.56	5.04	4.69	4.46	4.28	4.14	4.03	3.94	3.80	3.66	3.51	3.43	3.35	3.27	3.18	3.09	3.00
15	8.68	6.36	5.42	4.89	4.56	4.32	4.14	4.00	3.89	3.80	3.67	3.52	3.37	3.29	3.21	3.13	3.05	2.96	2.87
16	8.53	6.23	5.29	4.77	4.44	4.20	4.03	3.89	3.78	3.69	3.55	3.41	3.26	3.18	3.10	3.02	2.93	2.84	2.75
17	8.40	6.11	5.18	4.67	4.34	4.10	3.93	3.79	3.68	3.59	3.46	3.31	3.16	3.08	3.00	2.92	2.83	2.75	2.65
18	8.29	6.01	5.09	4.58	4.25	4.01	3.84	3.71	3.60	3.51	3.37	3.23	3.08	3.00	2.92	2.84	2.75	2.66	2.57
19	8.18	5.93	5.01	4.50	4.17	3.94	3.77	3.63	3.52	3.43	3.30	3.15	3.00	2.92	2.84	2.76	2.67	2.58	2.49
20	8.10	5.85	4.94	4.43	4.10	3.87	3.70	3.56	3.46	3.37	3.23	3.09	2.94	2.86	2.78	2.69	2.61	2.52	2.42
21	8.02	5.78	4.87	4.37	4.04	3.81	3.64	3.51	3.40	3.31	3.17	3.03	2.88	2.80	2.72	2.64	2.55	2.46	2.36
22	7.95	5.72	4.82	4.31	3.99	3.76	3.59	3.45	3.35	3.26	3.12	2.98	2.83	2.75	2.67	2.58	2.50	2.40	2.31
23	7.88	5.66	4.76	4.26	3.94	3.71	3.54	3.41	3.30	3.21	3.07	2.93	2.78	2.70	2.62	2.54	2.45	2.35	2.26
24	7.82	5.61	4.72	4.22	3.90	3.67	3.50	3.36	3.26	3.17	3.03	2.89	2.74	2.66	2.58	2.49	2.40	2.31	2.21
25	7.77	5.57	4.68	4.18	3.85	3.63	3.46	3.32	3.22	3.13	2.99	2.85	2.70	2.62	2.54	2.45	2.36	2.27	2.17
26	7.72	5.53	4.64	4.14	3.82	3.59	3.42	3.29	3.18	3.09	2.96	2.81	2.66	2.58	2.50	2.42	2.33	2.23	2.13
27	7.68	5.49	4.60	4.11	3.78	3.56	3.39	3.26	3.15	3.06	2.93	2.78	2.63	2.55	2.47	2.38	2.29	2.20	2.10
28	7.64	5.45	4.57	4.07	3.75	3.53	3.36	3.23	3.12	3.03	2.90	2.75	2.60	2.52	2.44	2.35	2.26	2.17	2.06
29	7.60	5.42	4.54	4.04	3.73	3.50	3.33	3.20	3.09	3.00	2.87	2.73	2.57	2.49	2.41	2.33	2.23	2.14	2.03
30	7.56	5.39	4.51	4.02	3.70	3.47	3.30	3.17	3.07	2.98	2.84	2.70	2.55	2.47	2.39	2.30	2.21	2.11	2.01
40	7.31	5.18	4.31	3.83	3.51	3.29	3.12	2.99	2.89	2.80	2.66	2.52	2.37	2.29	2.20	2.11	2.02	1.92	1.80
60	7.08	4.98	4.13	3.65	3.34	3.12	2.95	2.82	2.72	2.63	2.50	2.35	2.20	2.12	2.03	1.94	1.84	1.73	1.60
120	6.85	4.79	3.95	3.48	3.17	2.96	2.79	2.66	2.56	2.47	2.34	2.19	2.03	1.95	1.86	1.76	1.66	1.53	1.38
∞	6.63	4.61	3.78	3.32	3.02	2.80	2.64	2.51	2.41	2.32	2.18	2.04	1.88	1.79	1.70	1.59	1.47	1.32	1.00

From Pearson, E. S., and H. O. Hartley. Table 18, "Percentage Points of the *F*-distribution." *Biometrika Tables for Statisticians*, Vol. I. Reprinted by permission of the Biometrika Trustees.

TABLE A.9 *Mann–Whitney U Distribution*

$\Pr(U \le U_0)$: $n_1 \le n_2$; $3 \le n_2 \le 10$

		$n_2 = 3$		
n_1	1	2	3
0	.25	.10	.05
1	.50	.20	.10
U_0 2		.40	.20
3		.60	.35
4			.50

			$n_2 = 4$	
n_1	1	2	3	4
0	.2000	.0667	.0286	.0143
1	.4000	.1333	.0571	.0286
2	.6000	.2667	.1143	.0571
3		.4000	.2000	.1000
U_0 4		.6000	.3143	.1714
5			.4286	.2429
6			.5714	.3429
7				.4429
8				.5571

			$n_2 = 5$		
n_1	1	2	3	4	5
0	.1667	.0476	.0179	.0079	.0040
1	.3333	.0952	.0357	.0159	.0079
2	.5000	.1905	.0714	.0317	.0159
3		.2857	.1250	.0556	.0278
4		.4286	.1964	.0952	.0476
5		.5714	.2857	.1429	.0754
U_0 6			.3929	.2063	.1111
7			.5000	.2778	.1548
8				.3651	.2103
9				.4524	.2738
10				.5476	.3452
11					.4206
12					.5000

Example: For $n_1 = 2$, $n_2 = 4$, and $U_0 = 2$, $P(U \le 2) = .2667$.

(continued)

TABLE A.9   *Mann–Whitney U Distribution (continued)*

			$n_2 = 6$			
n_1	1	2	3	4	5	6
0	.1429	.0357	.0119	.0048	.0022	.0011
1	.2857	.0714	.0238	.0095	.0043	.0022
2	.4286	.1429	.0476	.0190	.0087	.0043
3	.5714	.2143	.0833	.0333	.0152	.0076
4		.3214	.1310	.0571	.0260	.0130
5		.4286	.1905	.0857	.0411	.0206
6		.5714	.2738	.1286	.0628	.0325
7			.3571	.1762	.0887	.0465
8			.4524	.2381	.1234	.0660
U_0 9			.5476	.3048	.1645	.0898
10				.3810	.2143	.1201
11				.4571	.2684	.1548
12				.5429	.3312	.1970
13					.3961	.2424
14					.4654	.2944
15					.5346	.3496
16						.4091
17						.4686
18						.5314

				$n_2 = 7$			
n_1	1	2	3	4	5	6	7
0	.1250	.0278	.0083	.0030	.0013	.0006	.0003
1	.2500	.0556	.0167	.0061	.0025	.0012	.0006
2	.3750	.1111	.0333	.0121	.0051	.0023	.0012
3	.5000	.1667	.0583	.0212	.0088	.0041	.0020
4		.2500	.0917	.0364	.0152	.0070	.0035
5		.3333	.1333	.0545	.0240	.0111	.0055
6		.4444	.1917	.0818	.0366	.0175	.0087
7		.5556	.2583	.1152	.0530	.0256	.0131
8			.3333	.1576	.0745	.0367	.0189
9			.4167	.2061	.1010	.0507	.0265
10			.5000	.2636	.1338	.0688	.0364
11				.3242	.1717	.0903	.0487
U_0 12				.3939	.2159	.1171	.0641
13				.4636	.2652	.1474	.0825
14				.5364	.3194	.1830	.1043
15					.3775	.2226	.1297
16					.4381	.2669	.1588
17					.5000	.3141	.1914
18						.3654	.2279
19						.4178	.2675
20						.4726	.3100
21						.5274	.3552
22							.4024
23							.4508
24							.5000

TABLE A.9 *Mann–Whitney U Distribution (continued)*

				$n_2 = 8$				
n_1	1	2	3	4	5	6	7	8
0	.1111	.0222	.0061	.0020	.0008	.0003	.0002	.0001
1	.2222	.0444	.0121	.0040	.0016	.0007	.0003	.0002
2	.3333	.0889	.0242	.0081	.0031	.0013	.0006	.0003
3	.4444	.1333	.0424	.0141	.0054	.0023	.0011	.0005
4	.5556	.2000	.0667	.0242	.0093	.0040	.0019	.0009
5		.2667	.0970	.0364	.0148	.0063	.0030	.0015
6		.3556	.1394	.0545	.0225	.0100	.0047	.0023
7		.4444	.1879	.0768	.0326	.0147	.0070	.0035
8		.5556	.2485	.1071	.0466	.0213	.0103	.0052
9			.3152	.1414	.0637	.0296	.0145	.0074
10			.3879	.1838	.0855	.0406	.0200	.0103
11			.4606	.2303	.1111	.0539	.0270	.0141
12			.5394	.2848	.1422	.0709	.0361	.0190
13				.3414	.1772	.0906	.0469	.0249
14				.4040	.2176	.1142	.0603	.0325
15				.4667	.2618	.1412	.0760	.0415
U_0 16				.5333	.3108	.1725	.0946	.0524
17					.3621	.2068	.1159	.0652
18					.4165	.2454	.1405	.0803
19					.4716	.2864	.1678	.0974
20					.5284	.3310	.1984	.1172
21						.3773	.2317	.1393
22						.4259	.2679	.1641
23						.4749	.3063	.1911
24						.5251	.3472	.2209
25							.3894	.2527
26							.4333	.2869
27							.4775	.3227
28							.5225	.3605
29								.3992
30								.4392
31								.4796
32								.5204

(continued)

TABLE A.9   *Mann–Whitney U Distribution (continued)*

					$n_2 = 9$				
n_1	1	2	3	4	5	6	7	8	9
0	.1000	.0182	.0045	.0014	.0005	.0002	.0001	.0000	.0000
1	.2000	.0364	.0091	.0028	.0010	.0004	.0002	.0001	.0000
2	.3000	.0727	.0182	.0056	.0020	.0008	.0003	.0002	.0001
3	.4000	.1091	.0318	.0098	.0035	.0014	.0006	.0003	.0001
4	.5000	.1636	.0500	.0168	.0060	.0024	.0010	.0005	.0002
5		.2182	.0727	.0252	.0095	.0038	.0017	.0008	.0004
6		.2909	.1045	.0378	.0145	.0060	.0026	.0012	.0006
7		.3636	.1409	.0531	.0210	.0088	.0039	.0019	.0009
8		.4545	.1864	.0741	.0300	.0128	.0058	.0028	.0014
9		.5455	.2409	.0993	.0415	.0180	.0082	.0039	.0020
10			.3000	.1301	.0559	.0248	.0115	.0056	.0028
11			.3636	.1650	.0734	.0332	.0156	.0076	.0039
12			.4318	.2070	.0949	.0440	.0209	.0103	.0053
13			.5000	.2517	.1199	.0567	.0274	.0137	.0071
14				.3021	.1489	.0723	.0356	.0180	.0094
15				.3552	.1818	.0905	.0454	.0232	.0122
16				.4126	.2188	.1119	.0571	.0296	.0157
17				.4699	.2592	.1361	.0708	.0372	.0200
18				.5301	.3032	.1638	.0869	.0464	.0252
19					.3497	.1942	.1052	.0570	.0313
20					.3986	.2280	.1261	.0694	.0385
21					.4491	.2643	.1496	.0836	.0470
22					.5000	.3035	.1755	.0998	.0567
23						.3445	.2039	.1179	.0680
24						.3878	.2349	.1383	.0807
25						.4320	.2680	.1606	.0951
26						.4773	.3032	.1852	.1112
27						.5227	.3403	.2117	.1290
28							.3788	.2404	.1487
29							.4185	.2707	.1701
30							.4591	.3029	.1933
31							.5000	.3365	.2181
32								.3715	.2447
33								.4074	.2729
34								.4442	.3024
35								.4813	.3332
36								.5187	.3652
37									.3981
38									.4317
39									.4657
40									.5000

U_0 labels the left column (n_1 values from 19 to 40 region).

(*continued*)

TABLE A.9 *Mann–Whitney U Distribution (continued)*

$n_2 = 10$

n_1	1	2	3	4	5	6	7	8	9	10
0	.0909	.0152	.0035	.0010	.0003	.0001	.0001	.0000	.0000	.0000
1	.1818	.0303	.0070	.0020	.0007	.0002	.0001	.0000	.0000	.0000
2	.2727	.0606	.0140	.0040	.0013	.0005	.0002	.0001	.0000	.0000
3	.3636	.0909	.0245	.0070	.0023	.0009	.0004	.0002	.0001	.0000
4	.4545	.1364	.0385	.0120	.0040	.0015	.0006	.0003	.0001	.0001
5	.5455	.1818	.0559	.0180	.0063	.0024	.0010	.0004	.0002	.0001
6		.2424	.0804	.0270	.0097	.0037	.0015	.0007	.0003	.0002
7		.3030	.1084	.0380	.0140	.0055	.0023	.0010	.0005	.0002
8		.3788	.1434	.0529	.0200	.0080	.0034	.0015	.0007	.0004
9		.4545	.1853	.0709	.0276	.0112	.0048	.0022	.0011	.0005
10		.5455	.2343	.0939	.0376	.0156	.0068	.0031	.0015	.0008
11			.2867	.1199	.0496	.0210	.0093	.0043	.0021	.0010
12			.3462	.1518	.0646	.0280	.0125	.0058	.0028	.0014
13			.4056	.1868	.0823	.0363	.0165	.0078	.0038	.0019
14			.4685	.2268	.1032	.0467	.0215	.0103	.0051	.0026
15			.5315	.2697	.1272	.0589	.0277	.0133	.0066	.0034
16				.3177	.1548	.0736	.0351	.0171	.0086	.0045
17				.3666	.1855	.0903	.0439	.0217	.0110	.0057
18				.4196	.2198	.1099	.0544	.0273	.0140	.0073
19				.4725	.2567	.1317	.0665	.0338	.0175	.0093
20				.5275	.2970	.1566	.0806	.0416	.0217	.0116
21					.3393	.1838	.0966	.0506	.0267	.0144
22					.3839	.2139	.1148	.0610	.0326	.0177
23					.4296	.2461	.1349	.0729	.0394	.0216
24					.4765	.2811	.1574	.0864	.0474	.0262
25					.5235	.3177	.1819	.1015	.0564	.0315
26						.3564	.2087	.1185	.0667	.0376
27						.3962	.2374	.1371	.0782	.0446
28						.4374	.2681	.1577	.0912	.0526
29						.4789	.3004	.1800	.1055	.0615
30						.5211	.3345	.2041	.1214	.0716
31							.3698	.2299	.1388	.0827
32							.4063	.2574	.1577	.0952
33							.4434	.2863	.1781	.1088
34							.4811	.3167	.2001	.1237
35							.5189	.3482	.2235	.1399
36								.3809	.2483	.1575
37								.4143	.2745	.1763
38								.4484	.3019	.1965
39								.4827	.3304	.2179
40								.5173	.3598	.2406
41									.3901	.2644
42									.4211	.2894
43									.4524	.3153
44									.4841	.3421
45									.5159	.3697
46										.3980
47										.4267
48										.4559
49										.4853
50										.5147

U_0

Appendix A

TABLE A.10 *Critical values of the Wilcoxon test statistic*

The following table gives critical values of T in the Wilcoxon matched-pair signed-rank test.

$n = 5, 6, 7, \ldots, 50$

1-sided	2-sided	$n = 5$	$n = 6$	$n = 7$	$n = 8$	$n = 9$	$n = 10$
$\alpha = .05$	$\alpha = .10$	1	2	4	6	8	11
$\alpha = .025$	$\alpha = .05$		1	2	4	6	8
$\alpha = .01$	$\alpha = .02$			0	2	3	5
$\alpha = .005$	$\alpha = .01$				0	2	3

1-sided	2-sided	$n = 11$	$n = 12$	$n = 13$	$n = 14$	$n = 15$	$n = 16$
$\alpha = .05$	$\alpha = .10$	14	17	21	26	30	36
$\alpha = .025$	$\alpha = .05$	11	14	17	21	25	30
$\alpha = .01$	$\alpha = .02$	7	10	13	16	20	24
$\alpha = .005$	$\alpha = .01$	5	7	10	13	16	19

1-sided	2-sided	$n = 17$	$n = 18$	$n = 19$	$n = 20$	$n = 21$	$n = 22$
$\alpha = .05$	$\alpha = .10$	41	47	54	60	68	75
$\alpha = .025$	$\alpha = .05$	35	40	46	52	59	66
$\alpha = .01$	$\alpha = .02$	28	33	38	43	49	56
$\alpha = .005$	$\alpha = .01$	23	28	32	37	43	49

1-sided	2-sided	$n = 23$	$n = 24$	$n = 25$	$n = 26$	$n = 27$	$n = 28$
$\alpha = .05$	$\alpha = .10$	83	92	101	110	120	130
$\alpha = .025$	$\alpha = .05$	73	81	90	98	107	117
$\alpha = .01$	$\alpha = .02$	62	69	77	85	93	102
$\alpha = .005$	$\alpha = .01$	55	61	68	76	84	92

1-sided	2-sided	$n = 29$	$n = 30$	$n = 31$	$n = 32$	$n = 33$	$n = 34$
$\alpha = .05$	$\alpha = .10$	141	152	163	175	199	201
$\alpha = .025$	$\alpha = .05$	127	137	148	159	171	183
$\alpha = .01$	$\alpha = .02$	111	120	130	141	151	162
$\alpha = .005$	$\alpha = .01$	100	109	118	128	138	149

1-sided	2-sided	$n = 35$	$n = 36$	$n = 37$	$n = 38$	$n = 39$	
$\alpha = .05$	$\alpha = .10$	214	228	242	256	271	
$\alpha = .025$	$\alpha = .05$	195	208	222	235	250	
$\alpha = .01$	$\alpha = .02$	174	186	198	211	224	
$\alpha = .005$	$\alpha = .01$	160	171	183	195	208	

Example: If $n = 30$, then $P(T \geq 120) = .01$ and $P(T \geq 109) = .005$.

(*continued*)

TABLE A.10 *Critical values of the Wilcoxon test statistic (continued)*

$n = 5, 6, 7, \ldots, 50$

1-sided	2-sided	$n = 40$	$n = 41$	$n = 42$	$n = 43$	$n = 44$	$n = 45$
$\alpha = .05$	$\alpha = .10$	287	303	319	336	353	371
$\alpha = .025$	$\alpha = .05$	264	279	295	311	327	344
$\alpha = .01$	$\alpha = .02$	238	252	267	281	297	313
$\alpha = .005$	$\alpha = .01$	221	234	248	262	277	292
1-sided	2-sided	$n = 46$	$n = 47$	$n = 48$	$n = 49$	$n = 50$	
$\alpha = .05$	$\alpha = .10$	389	408	427	446	466	
$\alpha = .025$	$\alpha = .05$	361	379	397	415	434	
$\alpha = .01$	$\alpha = .02$	329	345	362	380	398	
$\alpha = .005$	$\alpha = .01$	307	323	339	356	373	

From Wilcoxon, F., and R. A. Wilcox. "Some Rapid Approximate Statistical Procedures," 1964. Reprinted by permission of Lederle Labs, a division of the American Cyanamid Co.

TABLE A.11 *Distribution of number of runs*

The following table gives probabilities in the lower tail of the distribution of the number of runs r in samples of size (n_1, n_2): $P(r \le r_0)$.

(n_1, n_2)	r_0 2	3	4	5	6	7	8	9	10
(2, 3)	.200	.500	.900	1.000					
(2, 4)	.133	.400	.800	1.000					
(2, 5)	.095	.333	.714	1.000					
(2, 6)	.071	.286	.643	1.000					
(2, 7)	.056	.250	.583	1.000					
(2, 8)	.044	.222	.533	1.000					
(2, 9)	.036	.200	.491	1.000					
(2, 10)	.030	.182	.455	1.000					
(3, 3)	.100	.300	.700	.900	1.000				
(3, 4)	.057	.200	.543	.800	.971	1.000			
(3, 5)	.036	.143	.429	.714	.929	1.000			
(3, 6)	.024	.107	.345	.643	.881	1.000			
(3, 7)	.017	.083	.283	.583	.833	1.000			
(3, 8)	.012	.067	.236	.533	.788	1.000			
(3, 9)	.009	.055	.200	.491	.745	1.000			
(3, 10)	.007	.045	.171	.455	.706	1.000			
(4, 4)	.029	.114	.371	.629	.886	.971	1.000		
(4, 5)	.016	.071	.262	.500	.786	.929	.992	1.000	
(4, 6)	.010	.048	.190	.405	.690	.881	.976	1.000	
(4, 7)	.006	.033	.142	.333	.606	.883	.954	1.000	
(4, 8)	.004	.024	.109	.279	.533	.788	.929	1.000	
(4, 9)	.003	.018	.085	.236	.471	.745	.902	1.000	
(4, 10)	.002	.014	.068	.203	.419	.706	.874	1.000	
(5, 5)	.008	.040	.167	.357	.643	.833	.960	.992	1.000
(5, 6)	.004	.024	.110	.262	.522	.738	.911	.976	.998
(5, 7)	.003	.015	.076	.197	.424	.652	.854	.955	.992
(5, 8)	.002	.010	.054	.152	.347	.576	.793	.929	.984
(5, 9)	.001	.007	.039	.119	.287	.510	.734	.902	.972
(5, 10)	.001	.005	.029	.095	.239	.455	.678	.874	.958
(6, 6)	.002	.013	.067	.175	.392	.608	.825	.933	.987
(6, 7)	.001	.008	.043	.121	.296	.500	.733	.879	.966
(6, 8)	.001	.005	.028	.086	.226	.413	.646	.821	.937
(6, 9)	.000	.003	.019	.063	.175	.343	.566	.762	.902
(6, 10)	.000	.002	.013	.047	.137	.288	.497	.706	.864
(7, 7)	.001	.004	.025	.078	.209	.383	.617	.791	.922
(7, 8)	.000	.002	.015	.051	.149	.296	.514	.704	.867
(7, 9)	.000	.001	.010	.035	.108	.231	.427	.622	.806
(7, 10)	.000	.001	.006	.024	.080	.182	.355	.549	.743
(8, 8)	.000	.001	.009	.032	.100	.214	.405	.595	.786
(8, 9)	.000	.001	.005	.020	.069	.157	.319	.500	.702
(8, 10)	.000	.000	.003	.013	.048	.117	.251	.419	.621
(9, 9)	.000	.000	.003	.012	.044	.109	.238	.399	.601
(9, 10)	.000	.000	.002	.008	.029	.077	.179	.319	.510
(10, 10)	.000	.000	.001	.004	.019	.051	.128	.242	.414

TABLE A.11   *Distribution of number of runs (continued)*

(n_1, n_2)	11	12	13	14	15	16	17	18	19	20
(2, 3)										
(2, 4)										
(2, 5)										
(2, 6)										
(2, 7)										
(2, 8)										
(2, 9)										
(2, 10)										
(3, 3)										
(3, 4)										
(3, 5)										
(3, 6)										
(3, 7)										
(3, 8)										
(3, 9)										
(3, 10)										
(4, 4)										
(4, 5)										
(4, 6)										
(4, 7)										
(4, 8)										
(4, 9)										
(4, 10)										
(5, 5)										
(5, 6)	1.000									
(5, 7)	1.000									
(5, 8)	1.000									
(5, 9)	1.000									
(5, 10)	1.000									
(6, 6)	.998	1.000								
(6, 7)	.992	.999	1.000							
(6, 8)	.984	.998	1.000							
(6, 9)	.972	.994	1.000							
(6, 10)	.958	.990	1.000							
(7, 7)	.975	.996	.999	1.000						
(7, 8)	.949	.988	.998	1.000	1.000					
(7, 9)	.916	.975	.994	.999	1.000					
(7, 10)	.879	.957	.990	.998	1.000					
(8, 8)	.900	.968	.991	.999	1.000	1.000				
(8, 9)	.843	.939	.980	.996	.999	1.000	1.000			
(8, 10)	.782	.903	.964	.990	.998	1.000	1.000			
(9, 9)	.762	.891	.956	.988	.997	1.000	1.000	1.000		
(9, 10)	.681	.834	.923	.974	.992	.999	1.000	1.000	1.000	
(10, 10)	.586	.758	.872	.949	.981	.996	.999	1.000	1.000	1.000

From Swed, F., and C. Eisenhart. "Tables for Testing Randomness of Grouping in a Sequence of Alternatives." *Annals of Mathematical Statistics,* vol. 14, 1943.

TABLE A.12   *Critical values of Spearman's rank correlation coefficient*

n	$\alpha = .05$	$\alpha = .025$	$\alpha = .01$	$\alpha = .005$
5	.900	—	—	—
6	.829	.886	.943	—
7	.714	.786	.893	—
8	.643	.738	.833	.881
9	.600	.683	.783	.833
10	.564	.648	.745	.794
11	.523	.623	.736	.818
12	.497	.591	.703	.780
13	.475	.566	.673	.745
14	.457	.545	.646	.716
15	.441	.525	.623	.689
16	.425	.507	.601	.666
17	.412	.490	.582	.645
18	.399	.476	.564	.625
19	.388	.462	.549	.608
20	.377	.450	.534	.591
21	.368	.438	.521	.576
22	.359	.428	.508	.562
23	.351	.418	.496	.549
24	.343	.409	.485	.537
25	.336	.400	.475	.526
26	.329	.392	.465	.515
27	.323	.385	.456	.505
28	.317	.377	.448	.496
29	.311	.370	.440	.487
30	.305	.364	.432	.478

From Olds, E. G. "Distribution of Sums of Squares of Rank Differences for Small Numbers of Individuals." *Annals of Mathematical Statistics,* vol. 9, 1938.

Example: If $n = 20$, $P(R_s \geq .377) = .05$.

TABLE A.13 *Critical values of the Durbin–Watson statistic*

The letter k represents the number of explanatory variables in the regression equation *not* counting the constant term. The critical values in the table are for a 1-tailed test against positive serial correlation.

SIGNIFICANCE POINTS OF d_L AND d_U: 5%

n	$k=1$ d_L	d_U	$k=2$ d_L	d_U	$k=3$ d_L	d_U	$k=4$ d_L	d_U	$k=5$ d_L	d_U
15	1.08	1.36	0.95	1.54	0.82	1.75	0.69	1.97	0.56	2.21
16	1.10	1.37	0.98	1.54	0.86	1.73	0.74	1.93	0.62	2.15
17	1.13	1.38	1.02	1.54	0.90	1.71	0.78	1.90	0.67	2.10
18	1.16	1.39	1.05	1.53	0.93	1.69	0.82	1.87	0.71	2.06
19	1.18	1.40	1.08	1.53	0.97	1.68	0.86	1.85	0.75	2.02
20	1.20	1.41	1.10	1.54	1.00	1.68	0.90	1.83	0.79	1.99
21	1.22	1.42	1.13	1.54	1.03	1.67	0.93	1.81	0.83	1.96
22	1.24	1.43	1.15	1.54	1.05	1.66	0.96	1.80	0.86	1.94
23	1.26	1.44	1.17	1.54	1.08	1.66	0.99	1.79	0.90	1.92
24	1.27	1.45	1.19	1.55	1.10	1.66	1.01	1.78	0.93	1.90
25	1.29	1.45	1.21	1.55	1.12	1.66	1.04	1.77	0.95	1.89
26	1.30	1.46	1.22	1.55	1.14	1.65	1.06	1.76	0.98	1.88
27	1.32	1.47	1.24	1.56	1.16	1.65	1.08	1.76	1.01	1.86
28	1.33	1.48	1.26	1.56	1.18	1.65	1.10	1.75	1.03	1.85
29	1.34	1.48	1.27	1.56	1.20	1.65	1.12	1.74	1.05	1.84
30	1.35	1.49	1.28	1.57	1.21	1.65	1.14	1.74	1.07	1.83
31	1.36	1.50	1.30	1.57	1.23	1.65	1.16	1.74	1.09	1.83
32	1.37	1.50	1.31	1.57	1.24	1.65	1.18	1.73	1.11	1.82
33	1.38	1.51	1.32	1.58	1.26	1.65	1.19	1.73	1.13	1.81
34	1.39	1.51	1.33	1.58	1.27	1.65	1.21	1.73	1.15	1.81
35	1.40	1.52	1.34	1.58	1.28	1.65	1.22	1.73	1.16	1.80
36	1.41	1.52	1.35	1.59	1.29	1.65	1.24	1.73	1.18	1.80
37	1.42	1.53	1.36	1.59	1.31	1.66	1.25	1.72	1.19	1.80
38	1.43	1.54	1.37	1.59	1.32	1.66	1.26	1.72	1.21	1.79
39	1.43	1.54	1.38	1.60	1.33	1.66	1.27	1.72	1.22	1.79
40	1.44	1.54	1.39	1.60	1.34	1.66	1.29	1.72	1.23	1.79
45	1.48	1.57	1.43	1.62	1.38	1.67	1.34	1.72	1.29	1.78
50	1.50	1.59	1.46	1.63	1.42	1.67	1.38	1.72	1.34	1.77
55	1.53	1.60	1.49	1.64	1.45	1.68	1.41	1.72	1.38	1.77
60	1.55	1.62	1.51	1.65	1.48	1.69	1.44	1.73	1.41	1.77
65	1.57	1.63	1.54	1.66	1.50	1.70	1.47	1.73	1.44	1.77
70	1.58	1.64	1.55	1.67	1.52	1.70	1.49	1.74	1.46	1.77
75	1.60	1.65	1.57	1.68	1.54	1.71	1.51	1.74	1.49	1.77
80	1.61	1.66	1.59	1.69	1.56	1.72	1.53	1.74	1.51	1.77
85	1.62	1.67	1.60	1.70	1.57	1.72	1.55	1.75	1.52	1.77
90	1.63	1.68	1.61	1.70	1.59	1.73	1.57	1.75	1.54	1.78
95	1.64	1.69	1.62	1.71	1.60	1.73	1.58	1.75	1.56	1.78
100	1.65	1.69	1.63	1.72	1.61	1.74	1.59	1.76	1.57	1.78

TABLE A.14 *Factors Used When Constructing an $\bar{x}$-Chart*

Sample size n_i	Coefficient A_2	Sample size n_i	Coefficient A_2
2	1.880	14	.235
3	1.023	15	.223
4	.729	16	.212
5	.577	17	.203
6	.483	18	.194
7	.419	19	.187
8	.373	20	.180
9	.337	21	.173
10	.308	22	.167
11	.285	23	.162
12	.266	24	.157
13	.249	25	.153

$A_2\bar{R}$ is used to estimate the quantity $3\sigma/\sqrt{n}$ when constructing upper and lower control limits on an $\bar{x}$-chart.

TABLE A.15 *Factors Used When Constructing an R-Chart*

Sample size n_i	COEFFICIENT		Sample size n_i	COEFFICIENT	
	d_2	d_3		d_2	d_3
2	1.128	.853	14	3.407	.762
3	1.693	.888	15	3.472	.755
4	2.059	.880	16	3.532	.749
5	2.326	.864	17	3.588	.743
6	2.534	.848	18	3.640	.738
7	2.704	.833	19	3.689	.733
8	2.847	.820	20	3.735	.729
9	2.970	.808	21	3.778	.724
10	3.078	.797	22	3.819	.720
11	3.173	.787	23	3.858	.716
12	3.258	.778	24	3.895	.712
13	3.336	.770	25	3.931	.709

Source: *ASTM Manual on the Presentation of Data and Control Chart Analysis,* Philadelphia, PA: American Society for Testing Materials, 1976.

d_3 $(\bar{R}/d_2)$ is used to estimate the quantity σ_R when constructing upper and lower control limits on an R-chart.

Answers to Selected Exercises

Chapter 2

Section 2.1

2.1.3 True

2.1.5 **a.** Nominal **b.** Ordinal **c.** Ratio

2.1.7 **a.** Discrete **b.** Discrete **c.** Discrete
d. Discrete **e.** Discrete

2.1.9 **a.** A list containing the names of the 1,000 bank customers
b. Each customer
c. The account balances of the 1,000 customers
d. The account balances of the selected 50 customers

2.1.13 **a.** Continuous **b.** Continuous **c.** Discrete
d. Discrete

Section 2.2

2.2.9 "Have you cheated on your income taxes?"

2.2.11 Product marketability, television ratings, political polls

2.2.13 Determining average monthly rainfall, proportion of defective products coming off an assembly line, average birth weight of male babies

Section 2.3

2.3.3 **a.** No; they are self-selected.
b. Typically, only those with a strong opinion will reply.

Section 2.4

2.4.1 **a.** 52% male; 48% female
b. 10%; 24%; 34%; 32% **c.** 44%; 56%
d. 22%; 24%

Supplementary Exercises

2.S.3 **a.** Quantitative, discrete
b. Qualitative, dichotomous
c. Qualitative, multinomial
d. Qualitative, multinomial
e. Quantitative, continuous

2.S.5 **a.** Convenience sample
b. Judgment sample
c. Systematic random sample
d. Cluster sample with self-enumeration
e. Systematic random sample
f. Stratified random sample

2.S.15 This poll is subject to the bias of self-selection.

Chapter 3

Section 3.1

3.1.11 True

3.1.13 True; the area could also represent the relative frequency of the class.

3.1.15 **b.** 1.0125 **c.** 20

3.1.17 **a.** 36° **b.** 120°
c. Central angle $= 360° \times$ proportion

3.1.19 **a.** 24.204 million, 7.056 million, 4.704 million, 3.136 million
d. Nominal

3.1.21 **a.** Frequencies: 4, 7, 9, 3, 2.
b. Relative frequencies: .16, .28, .36, .12, .08.

3.1.23

Miles driven	Freq.	Rel. freq.	Cum. freq.
0 to under 10	5	.10	5
10 to under 20	10	.20	15
20 to under 30	19	.38	34
30 to under 40	11	.22	45
40 to under 50	5	.10	50

3.1.25

# of cars	Freq.	Rel. freq.	Cum. freq.	CRF
0	4	.08	4	.08
1	24	.48	28	.56
2	16	.32	44	.88
3	4	.08	48	.96
4	2	.04	50	1.00

Section 3.2

3.2.1 True
3.2.3 False
3.2.5 True
3.2.7 False
3.2.9 **a.** .50   **b.** True   **c.** True

Section 3.4

3.4.3 False

3.4.5
```
1 | 8 9 9
2 | 0 2 3 6 6 6 6 8
3 | 0 2 2 4 4 5 6 7 7 7 8 8
4 | 0 0 1 2 2 3 3 3 4 5 5 7 9
5 | 0 0 1 2
```

Supplementary Exercises

3.S.1 b.

Days	Freq.	Rel. freq.	CRF
0 to under 5	2	.050	.050
5 to under 10	11	.275	.325
10 to under 15	9	.225	.550
15 to under 20	12	.300	.850
20 to under 25	3	.075	.925
25 to under 30	1	.025	.950
30 to under 35	1	.025	.975
35 to under 40	1	.025	1.000

3.S.5
```
2 | 6 7 8
3 | 0 1 4 5 5 6 8 9 9
4 | 0 1 1 2 4 5 6 6 7 8 9
5 | 0 2 3 4 5
6 | 0 1
```

3.S.7
```
1 | 5.4 6.8 6.8 6.8
1 | 7.6 7.7 8.2 8.7 8.9 9.0 9.2 9.3
2 | 0.0 0.2 1.9 2.1 2.1 2.3
2 | 3.2 3.4 3.6 4.0 4.3 4.3 4.5 4.6 4.7 4.7
2 | 5.1 5.3 5.4 5.4 6.5 6.5 6.8 6.9 7.4
2 | 8.2 8.6 9.3
```

3.S.11 a. A population if we wanted information only on the NY Yankees

b. A sample if we wanted to make inferences about all major league teams

3.S.13 a. Probably skewed to the right

b. Approximately normal
c. Skewed to the right
d. Approximately uniform

Chapter 4

Section 4.1

4.1.1 100; 100
4.1.3 151; 625
4.1.5 True; 5; 5
4.1.7 **a.** 100; 121   **b.** 3,402   **c.** False
4.1.9 **a.** $2,120   **b.** $1,620
4.1.11 **a.** 7   **b.** -2 **c.** 6   **d.** 49

Section 4.2

4.2.1 False
4.2.3 False; the distribution could be bimodal.
4.2.5 False; the mean is the center of gravity.
4.2.7 True
4.2.9 Mode
4.2.13 **a.** False; the median could remain constant.
   b. True   **c.** True
4.2.15 **a.** 23,620; 24,600
   b. The deviations sum to zero.
   c. Deviations: -400; 900; $-7,500$; 2,100; 0; the deviations sum to $-4,900$.
   d. True
   e. False; the deviations from the median will sum to zero for a symmetric distribution.
4.2.17 **a.** 48.8; 56   **b.** 74.4; 56   **c.** 434.4; 56
   d. Always causes an increase in the mean; the median can increase or remain unchanged.
4.2.19 **a.** The 10 grades are: 70, 70, 80, 80, 80, 90, 90, 90, 90, 90. The mean is 83.
   b. The mean is 83.
   c. The mean is $(.2)(70) + (.3)(80) + (.5)(90) = 83$.
4.2.21 **a.** $37,360   **b.** $32,000
4.2.25 33,006.6667; 33,250; 30,500

Section 4.3

4.3.3 False
4.3.5 **a.** 2   **b.** 3   **c.** 4
4.3.7 **a.** 59.125   **b.** 58   **c.** 46.5; 71.5   **d.** 47.5

Section 4.4

4.4.3 False
4.4.5 False
4.4.9 True; the standard deviation of population 1 will be twice as large as the standard deviation of population 2.

4.4.11 a. True **b.** False **c.** True **d.** True
e. True **f.** True
4.4.13 a. 37,640 **b.** 41,100 **c.** 13,288
d. 292,743,000 **e.** 17,109.734
f. 37.640; 41.100; 13.288; 292.743; 17.109734
4.4.15 a. \$.01 **b.** 1,000 **c.** 12 lb **d.** 2 in.
4.4.17 a. 1,583.33333; 1,550; 1,500
b. 800; 69,666.667; 263.944
d. 183.333
4.4.19 a. 8.083; 7.5 **b.** 27.076 **c.** 5.203
d. 3.9444 **e.** 19

Section 4.5

4.5.1 At least 75%
4.5.3 Approximately 95%
4.5.5 True
4.5.7 True
4.5.9 a. 13.48% **b.** 2 **c.** Approximately 2.5%
d. -1 **e.** Approximately 16% **f.** 3
g. Approximately .15%
h. Approximately 68%
i. Approximately 95%
4.5.11 a. 82 **b.** 10.231 **c.** $z = .78$
d. 61.54 to 102.46; 100%
e. 12.476%
4.5.13 a. 82; $z = 1$ **b.** 46; $z = -2$ **c.** A
4.5.15 a. Approximately 97.5%
b. Approximately 16%
4.5.17 a. At least $\dfrac{8}{9}$, or 89%
b. Approximately 99.7%
4.5.19 a. 25% **b.** 5% **c.** Stock A

Section 4.6

4.6.3 True

Supplementary Exercises

4.S.1 b. 1.483 **c.** .778 **d.** 1.5
4.S.3 b.

Class	f	Rel. freq.	Cum. rel. freq.
10 to under 20	2	.04	.04
20 to under 30	4	.08	.12
30 to under 40	9	.18	.30
40 to under 50	11	.22	.52
50 to under 60	12	.24	.76
60 to under 70	6	.12	.88
70 to under 80	4	.08	.96
80 to under 90	2	.04	1.00

f. 49.18; 290.9261; 17.05656
g. 48; 35; 58
4.S.5 $k\overline{x}$; $k^2 S^2$
4.S.7 If all observations are the same, the sample variance will be 0. The sample variance cannot be negative.
4.S.13 a. Code 1 **b.** Not meaningful
c. Not meaningful
4.S.15 The mean of the standardized scores should be 0; thus, an error has been made.
4.S.17 The median or the mode; the mean
4.S.19 a. 1 car **b.** 1 car **c.** 1.417
4.S.21 a. The mean indicates the proportion of defectives.
b. 20% of the products are defective.
4.S.23 At most 16%
4.S.25 b.

Class	Freq.
131.5 to 134.5	5
134.5 to 137.5	27
137.5 to 140.5	26
140.5 to 143.5	43
143.5 to 146.5	18
146.5 to 149.5	6

c. 140.576; 3.5247
d.

Class	Cum. freq.	Cum. rel. freq.
Under 134.5	5	.040
Under 137.5	32	.256
Under 140.5	58	.464
Under 143.5	101	.808
Under 146.5	119	.952
Under 149.5	125	1.000

e. 96.8%
4.S.27 b.

Class	Freq.
15.3 to 17.3	2
17.3 to 19.3	10
19.3 to 21.3	61
21.3 to 23.3	64
23.3 to 25.3	13

c. 21.28; 1.516
d.

Class	Cum. freq.	Cum. rel. freq.
Under 17.3	2	.013
Under 19.3	12	.080
Under 21.3	73	.487
Under 23.3	137	.913
Under 25.3	150	1.000

e. 96.7%

4.S.29 a.

Class	Freq.
4.5 to 14.5	4
14.5 to 24.5	3
24.5 to 34.5	2
34.5 to 44.5	2
44.5 to 54.5	3
54.5 to 64.5	5
64.5 to 74.5	12
74.5 to 84.5	17
84.5 to 94.5	2

    b. Skewed to the left   **c.** -1.071

4.S.31 a. Not capable   **b.** Not capable
    c. Not capable   **d.** Not capable

Chapter 5

Section 5.1

5.1.1 A car buyer has five choices for the color of a new car.

5.1.3 8

5.1.5 (1, 1), (1, 2), (1, 3), (1, 4), (1, 5), (1, 6),
(2, 1), (2, 2), (2, 3), (2, 4), (2, 5), (2, 6),
(3, 1), (3, 2), (3, 3), (3, 4), (3, 5), (3, 6),
(4, 1), (4, 2), (4, 3), (4, 4), (4, 5), (4, 6),
(5, 1), (5, 2), (5, 3), (5, 4), (5, 5), (5, 6),
(6, 1), (6, 2), (6, 3), (6, 4), (6, 5), (6, 6)
Associated with each outcome is a sum, which varies from 2 to 12. Thus an alternative way of viewing the sample space would be as follows: $S = \{2, 3, \ldots, 12\}$.

Section 5.2

5.2.3 Select a person randomly from a class of 20 students; select a card randomly from a deck of 52 cards; select a numbered ball randomly from a jar containing 10 numbered balls.

5.2.5 Find the probability that the Dallas Cowboys will win the Super Bowl next year; find the probability that two companies will merge next year; find the probability that the inflation rate will exceed 4.5% next year.

5.2.7 Subjective approach

5.2.9 No

5.2.11 Yes

5.2.13 $\frac{1}{12}$

5.2.15 The answer varies from case to case.

Section 5.3

5.3.5 a. $\frac{2}{12}$ **b.** $\frac{10}{12}$

5.3.7 .8571

5.3.9 .03

5.3.11 a. .36   **b.** .81   **c.** .13

Section 5.4

5.4.3 If A and B are mutually exclusive, $P(A \text{ or } B) = P(A) + P(B)$.

5.4.5 False

5.4.7 $P(A \text{ and } B)$ cannot exceed $P(A)$; $P(A \text{ and } B)$ cannot exceed $P(B)$.

5.4.9 .8

5.4.11 .80

5.4.13 a. .48, .52, .16, .26, .35, .23
    b. 1.00   **c.** .05   **d.** .05   **e.** 0   **f.** 0

5.4.15 a. .35   **b.** .65

5.4.17 a. .04255
    b. .5106

5.4.19 a. Inflation will be 4% or more next year.
    b. A family with two children has at least one girl.
    c. The player will miss both shots.

Section 5.5

5.5.3 False

5.5.5 False

5.5.7 True

5.5.9 Assignment is invalid.

5.5.11 Yes

5.5.13 a. .06   **b.** .95   **c.** .931   **d.** .009   **e.** .180
    f. .02

5.5.15 a. .100   **b.** .3   **c.** 550

5.5.17 a. .40   **b.** .60   **c.** .50   **d.** .25   **e.** .222
    f. .125

5.5.19 a. .545   **b.** .4954

Section 5.6

5.6.3 False

5.6.5 24

5.6.7 .20

5.6.9 $\frac{1}{4}$

5.6.11 .125

5.6.13 .70248

5.6.15 .0625

5.6.17 a. .0003   **b.** .9603

5.6.19 .69

5.6.21 a. .343   **b.** .027   **c.** .973

5.6.23 a. .0081
    b. .7599

Section 5.7

5.7.3 .14
5.7.5 .09
5.7.7 **a.** .08 **b.** .8696
5.7.9 **a.** .60 **b.** .1385

Section 5.8

5.8.7 120; 360; 40,320; 42; $N!$
5.8.9 10; 10; 36; 36
5.8.11 120
5.8.13 240
5.8.15 **a.** 210 **b.** 35 **c.** .1667 **d.** .8333

Supplementary Exercises

5.S.1 **a.** .20 **b.** .80 **c.** .35 **d.** .65 **e.** .30
f. .25 **g.** .571
5.S.3 **a.** .08 **b.** .50 **c.** .167
5.S.5 .49
5.S.7 **a.** .042 **b.** .012/.042 **c.** .595
5.S.9 **a.** .55 **b.** .846
5.S.11 **a.** 0 **b.** 0 **c.** Statement is false.
d. Mutually exclusive events are not independent.
5.S.13 **a.** .10 **b.** .333 **c.** .1143
d. $P(H) \neq P(H|S)$; thus, H and S are not independent.
5.S.15 **a.** .60, .40, .85, .10
b. $P(A_1|B) = .9272$; $P(A_2|B) = .0728$
5.S.17 .2401
5.S.19 15,625
5.S.21 12
5.S.23 **a.** .31 **b.** .38 **c.** .443 **d.** No
5.S.25 **a.** .27 **b.** No **c.** .50
5.S.27 **a.** .245 **b.** .918
5.S.29 **a.** .0081 **b.** .2401 **c.** .7599
5.S.31 .67232
5.S.33 **a.** .147 **b.** .2401
5.S.35 **a.** .28 **b.** .40 **c.** .112 **d.** .432
5.S.37 **a.** .018 **b.** .50
5.S.39 .0039
5.S.41 For pigeons: $P(\text{reward}) = .58$; for rats: $P(\text{reward}) = .7$; rats are more intelligent.
5.S.43 **a.** .0092 **b.** .2608
5.S.45 .36675
5.S.47 .89989
5.S.49 .5250
5.S.51 .0039
5.S.53 .58537
5.S.55 **a.** .68 **b.** .62 **c.** .44 **d.** .70968
e. Not independent **f.** Predict Up today.

5.S.57 .000000083, or approximately 1 in 12,000,000
5.S.59 **a.** .86006 **b.** .13994

Chapter 6

Section 6.1

6.1.5 **a.** Discrete **b.** Continuous **c.** Continuous
d. Discrete **e.** Discrete **f.** Discrete
g. Continuous **h.** Discrete
6.1.7 $S = \{-30{,}000, -5{,}000, 20{,}000, 45{,}000\}$.

Section 6.2

6.2.3 The subjective probability distribution is valid.
6.2.5 **a.** $.10 + .20 + .15 + .45 + .10 = 1.0$; all probabilities are nonnegative.
b. .70, .30, .10 **c.** .55, .70, .45
6.2.7 **b.** .35 **c.** .65 **d.** .50 **e.** .35
6.2.9 **b.** .66
6.2.11 **c.** .50

d.

Y	P(Y)
0	.25
10	.25
30	.25
75	.25

e. .50

Section 6.3

6.3.5 True

6.3.7

x	P(X = x)	F(x)
2	.2	.2
4	.3	.5
6	.4	.9
8	.1	1.0

x	P(X = x)	F(x)
3	.2	.2
5	.5	.7
6	.2	.9
9	.1	1.0

x	P(X = x)	F(x)
−2	.3	.3
2	.4	.7
4	.2	.9
5	.1	1.0

6.3.9 .5; .2; .9
6.3.11 **a.** .60 **c.** .40
6.3.13 **a.** .20 **d.** .75

Section 6.4

6.4.1 False
6.4.3 True
6.4.5 **a.** 4.8 **b.** 5.2 **c.** 1.5
6.4.7 **a.** 58 **b.** 62 **c.** 25
6.4.9 **a.**

X	P(X)
5.00	$18/38 = .470$
-2.50	$2/38 = .053$
-5.00	$18/38 = .470$

    b. $-\$.1325$
    c. For each \$5 played, the casino will gain \$.1325; for 100 bets, the casino will gain \$13.25.
6.4.11 \$7,200
6.4.13 \$155
6.4.15 $E(X) = 4.4$

Section 6.5

6.5.1 False
6.5.7 True
6.5.9 **a.** $Var(X) = 3.360$; $Var(Y) = 10^2 \times Var(X) = 336.0$
    b. $Var(X) = 2.560$; $Var(Y) = 10^2 \times Var(X) = 256.0$
    c. $Var(X) = 6.250$; $Var(Y) = 10^2 \times Var(X) = 625.0$
6.5.11 3.360; 2.560; 6.250
6.5.13 **a.** 1.4; .980 **c.** 0 **d.** .1
6.5.15 Mathematical text: $E(X) = 1.60$, $\sigma = 1.88$; Non-mathematical text: $E(X) = .40$, $\sigma = .7874$
6.5.17 **a.** 38 **b.** 436

Section 6.6

6.6.1 **a.**

Class	Value	Freq.	Rel. freq.
Fresh.	1	5	.10
Soph.	2	12	.24
Junior	3	17	.34
Senior	4	16	.32
		50	

    b. 2.88

Supplementary Exercises

6.S.1

X	P(X)
74,000	.18
11,000	.42
50,000	.12
$-13,000$	.28

    b. \$20,300

6.S.3 \$60
6.S.5 \$300,000
6.S.7 **a.** 32.0; 146 **c.** $P(7.84 \leq X \leq 56.16) = .95$
6.S.9 \$5,500
6.S.11 **b.** \$15,000, \$12,000
    c. A person who cannot afford to lose \$10,000 would be wise to avoid project A.
6.S.13 **b.** \$33,470
6.S.15 \$.70
6.S.17 **a.** \$.85 **b.** \$85.00
6.S.19 **a.** \$42.86, \$0 **b.** \$71.43, $-\$20$
    c. You should bet all your money with the second person.
6.S.21 **c.** $E(X) = 15.4$; $E(Y) = 17.0$
    d. $Var(X) = 860.8400$; $Var(Y) = 2,195.0$
    f. .29 **g.** .40 **h.** .95 **i.** 1.00

Chapter 7

Section 7.1

7.1.1 True
7.1.5 True
7.1.7 .0625; .2500; .3750; .2500; .0625
7.1.9 1.00
7.1.11 2.100
7.1.13 .990
7.1.15 .194
7.1.17 **a.** .411 **b.** .794 **c.** .589 **d.** .205 **e.** .630
7.1.19 **a.** .2005 **b.** .2415
7.1.21 .5904
7.1.23 **a.** .7443 **b.** 4.2; 1.26
7.1.25 **a.** .0362
    b. If $p = .05$, then observing $X \geq 3$ is very unlikely.
7.1.27 **a.** .0005 **b.** .5220 **c.** .1095
7.1.29 **a.** .6296 **b.** .7939 **c.** .2053
7.1.31 **a.** False **b.** True **c.** False **d.** True
    e. True
7.1.33 **a.** 25,000 **b.** \$137,500

Section 7.2

7.2.5 True
7.2.7 Poisson: $P(X = 1) = .2681$, $P(X = 2) = .0536$. Binomial: $P(X = 1) = .2725$, $P(X = 2) = .0528$
7.2.9 1
7.2.11 $P(X = 0) = .2231$, $P(X = 1) = .3347$, $P(X = 2) = .2510$, $P(X = 3) = .1255$, $P(X = 4) = .0471$, $P(X = 5) = .0141$, $P(X = 6) = .0035$, $P(X = 7) = .0008$, $P(X = 8) = .0001$

7.2.13 1.5
7.2.15 a. .6767   **b.** .3233
7.2.17 a. .0758   **b.** .0758
     d. $P(3) = .0126$, $P(4) = .0016$
7.2.19 a. .2240   **b.** .5768
7.2.21 5 mufflers
7.2.23 a. .1353   **b.** .2707   **c.** .1429
7.2.25 .4335

Supplementary Exercises

7.S.1 a. .2344   **b.** .8906
7.S.3 a. .7599   **b.** .2401   **c.** .0081
7.S.5 b. 4,000, .8
7.S.7 .2618
7.S.9 .6867
7.S.11 a. .8497   **b.** .6172
7.S.13 a. .3487   **b.** .6513   **c.** .1937
7.S.15 a. .9936   **b.** .3222
7.S.17 .0498
7.S.19 .0156
7.S.21

X	$P(X)$
0	.3679
1	.3679
2	.1839
3	.0613
4	.0153
5	.0031
6	.0005
7	.0001

7.S.23 .0002
7.S.25 .6065
7.S.27 a. .0527   **b.** .6767   **c.** .1353
7.S.29 .3208
7.S.31 .0613
7.S.33 a. .0067   **b.** .0337   **c.** .7350
7.S.35 b. .6517
7.S.37 .3672
7.S.39 a. .000977   **b.** $-\$1.02$
7.S.41 a. .0313
     b. Yes, the chances of this are slim.
     c. No. Some people would be certain to receive five correct predictions.
7.S.43 .1515
7.S.45 a. .9995   **b.** .40951
7.S.47 a. Binomial   **b.** Not binomial   **c.** Binomial
7.S.49 a. .084   **b.** .0000 (approximately)
7.S.51 a. .006
     b. I would reject $p = .5$ because observing $X \geq 17$ is very unlikely.

c. I would not reject $p = .5$ because observing $X \geq 11$ is fairly unlikely.
d. $P(X \leq 4) = .006$; I would reject $p = .5$ because observing $X \leq 4$ is very unlikely.
7.S.53 a. .016   **b.** .006
7.S.55 a. .036   **b.** .816
7.S.57 a. .0183   **b.** .7852   **c.** .9084
7.S.59 .00153

Chapter 8

Section 8.1

8.1.5 True
8.1.7 .3
8.1.9 .5
8.1.11 a. Discrete   **b.** Continuous   **c.** Continuous
     d. Discrete   **e.** Continuous
8.1.13 Yes
8.1.15 a. .3   **b.** 15 minutes
8.1.17 a. 1   **b.** .5   **c.** .375

Section 8.2

8.2.3 5
8.2.5 False
8.2.7 True
8.2.9 b. 1   **c.** 15, 8.33   **d.** .40   **e.** .60   **f.** .40
     g. 0
8.2.11 a. .167   **b.** .4   **c.** .333   **d.** .167   **e.** 1.5
8.2.13 b. .667   **c.** .667   **d.** .167
8.2.15 a. $f(X) = (1/40)$ for X in [100, 140]; $f(X) = 0$ otherwise
     b. .625   **c.** .375   **d.** 120; 11.547   **e.** 0

Section 8.3

8.3.5 False
8.3.7 False
8.3.9 False
8.3.13 True
8.3.15 .9656
8.3.17 .9367
8.3.19 .9544
8.3.21 .0967
8.2.23 .74
8.3.25 2.00
8.3.27 a. .0228   **b.** .0228   **c.** .0456   **d.** .9544

Section 8.4

8.4.1 False
8.4.3 True
8.4.5 True

8.4.7 True
8.4.9 **a.** .1587 **b.** .0228 **c.** .9772 **d.** .8413
e. .8185
8.4.11 .3779
8.4.13 **a.** .0397 **b.** .4206
8.4.15 **a.** 58.225 **b.** 41.775 **c.** 61.63 **d.** 50
e. $a = 40.2$, $b = 59.8$
8.4.17 .0062
8.4.19 **a.** .9452 **b.** .3446 **c.** .9192 **d.** .3812
8.4.21 For $\sigma = 10$: .6826; for $\sigma = 16$: .4714;
for $\sigma = 25$: .3108
8.4.23 **a.** .0062 **b.** .1210
8.4.25 .0062

Section 8.5

8.5.5 Yes
8.5.7 True
8.5.9 False
8.5.11 0
8.5.13 Normal approximation: .5470; binomial: .545
8.5.15 Normal approximation: .2380; binomial: .245
8.5.17 **a.** .0838 **b.** .0166
c. Yes; .0166 is a very small probability.
8.5.19 .0019
8.5.21 .0015
8.5.23 **a.** .0861 **b.** .5987 **c.** .3151
8.5.25 **a.** .4286 **b.** 0
8.5.27 **a.** .8212 **b.** .2544
8.5.29 .0002

Supplementary Exercises

8.S.1 **a.** .4370 **b.** .4525 **c.** .9628 **d.** .0073
e. .9949 **f.** .0033 **g.** .9713 **h.** 1 **i.** 0
j. 1 **k.** 0 **l.** .0873 **m.** .0704
8.S.3 **a.** .5 **b.** .5 **c.** .6823 **d.** .9544 **e.** .9974
f. .9500
8.S.5 .0013
8.S.7 .9957
8.S.9 A = 82.8; B = 75.22; C = 64.78; D = 57.2
8.S.11 .62%
8.S.13 .24%
8.S.15 **a.** .1587 **b.** .1056 **c.** .9772 **d.** .9938
e. .0440
8.S.17 **a.** .0162 **b.** .1314
8.S.19 **a.** .0179 **b.** .0044 **c.** .1481
8.S.21 **a.** .0808 **b.** .2233
8.S.23 **a.** .0004459 **b.** .005837
8.S.25 **a.** .0314 **b.** .0029 **c.** 84
8.S.27 **a.** .0228 **b.** .8413

8.S.29 **a.** .0001 **b.** Yes, it would be an extremely rare
occurrence if the claim is true.
8.S.31 844 pounds
8.S.33 **a.** .0228 **b.** .8164 **c.** 4.852
d. 4.1065 to 5.0935
8.S.35 Normal approximation
8.S.37 **b.** .7
8.S.39 **a.** 1.28 **b.** 1.645 **c.** 1.96 **d.** 2.33
e. 2.58
8.S.41 1,320
8.S.43 **a.** .9332 **b.** .0668 **c.** .0228
8.S.45 **a.** .1517 **b.** .5403 **c.** .8849
8.S.47 $P(X \leq 390.5) \approx .0000$. The data suggest that p is
less than .3.
8.S.49 733 dolls
8.S.51 282 seconds
8.S.53 221.125 pounds
8.S.55 **a.** 2,274.6; 310.996 **b.** .0367
8.S.57 **a.** .0188 **b.** .0122
8.S.59 **a.** .1922 **b.** .9772

Chapter 9

Section 9.1

9.1.5 True; for example, when sampling from a normal
or uniform distribution, the sample median is an
unbiased estimator of the population mean.
9.1.7 False. The sample mean and sample median are
both unbiased estimators of the mean of a normal
distribution. The sample mean is preferred because
the standard deviation of the sampling distribution
of the sample mean is smaller than the standard deviation of the sampling distribution of the sample
median.
9.1.13 Procedure described in the problem is acceptable.
9.1.19 **a.** 5.75 **b.** 6 samples **f.** 16 samples
9.1.21 1.64; 1.64
9.1.23 Sample size $n = 50$, because it has the smaller
variance.
9.1.25 .9000

Section 9.2

9.2.1 True
9.2.3 False
9.2.5 False
9.2.9 $E(\bar{X}) = 220$; $\sigma_{\bar{X}} = 4.4$
9.2.11 .8081
9.2.13 **a.** 25; 2 **b.** 25; 4 **c.** 50; 1 **d.** 50; 2
9.2.15 500; 8.1; 2.8460
9.2.17 $n_2 = 1.570796n_1$

9.2.19 a. $n = 5$: 11.180; $n = 10$: 7.906; $n = 25$: 5.000; $n = 50$: 3.536; $n = 100$: 2.500

9.2.21 Smaller

Section 9.3

9.3.3 True

9.3.5 True

9.3.9 We should prefer the estimator that has the smaller variance.

9.3.11 a. .3594 **b.** .3594 **c.** .2514 **d.** .0918

9.3.13 a. .9938 **b.** .9974 **c.** .0002

9.3.15 a. $E(\overline{X}) = 274$; $\sigma_{\overline{x}} = 3.8$
   b. The sampling distribution of $\overline{X}$ is approximately normal, by the Central Limit Theorem.
   c. .8004

9.3.17 a. The sampling distribution of $\overline{X}$ is approximately normal; $E(\overline{X}) = 50$; $\sigma_{\overline{x}}^2 = 1.7778$
   b. Normal; $E(\overline{X}) = 50$; $\sigma_{\overline{x}}^2 = 14.4$

9.3.19 .6826; yes, it appears that $\overline{X}$ provides a good estimate of μ.

9.3.21 .2206

9.3.23 .0026

9.3.25 a. .2033 **b.** .4522 **c.** .0192 **d.** .3821
   e. .0000 **f.** .2709

9.3.27 a. UCL = 6.40; LCL = 6.28 **b.** .0026
   c. (Use the binomial distribution) .0257

Section 9.4

9.4.3 True

9.4.5 False

9.4.7 True

9.4.9 a. .9936 **b.** .0000 **c.** .2033

9.4.11 a. .1893 **b.** .7224 **c.** .0485

9.4.13 .4776

Section 9.5

9.5.3 Not appropriate

9.5.5 True

9.5.7 True (if we ignore the continuity correction)

9.5.9 a. False **b.** True

9.5.11 a. Yes **b.** .4; .0006 **c.** .8968 **d.** .0516

9.5.13 a. Yes **b.** .8; .0004 **c.** .6826 **d.** .8392
   e. $p_1 = .7608$, $p_2 = .8392$

9.5.15 a. .3108 **b.** .4516 **c.** .6826 **d.** .7698
   e. When we apply the continuity correction, the increase in probability will be greater when the z-score is close to 0 than when the z-score falls far from 0.
   f. True

9.5.17 .9282

9.5.19 .0918

9.5.21 a. .9708 **b.** .1233 **c.** 0

9.5.23 a. 0
   b. Yes. If $p = .9$, the probability of observing a 77% survival rate is 0.

9.5.25 .1038

Section 9.6

9.6.3 Not appropriate

9.6.5 True

9.6.7 a. .8472 **b.** .2226 **c.** .1950

9.6.9 .6730

9.6.11 .0749

Supplementary Exercises

9.S.1 .9876

9.S.3 .9992

9.S.5 .9974

9.S.7 0

9.S.9 .8098

9.S.11 .9992

9.S.13 .9974

9.S.15 .0001

9.S.17 $z = 1.645$

9.S.19 .0784

9.S.21 a. .4404 **b.** .1192 **c.** .3446

9.S.23 a. 0 **b.** 0

9.S.25 0

9.S.27 .8413

9.S.29 a. .5987 **b.** .8944 **c.** .9938

9.S.31 a. .0228 **b.** .0228 **c.** .0456 **d.** .9544

9.S.33 a. .9736 **b.** .9924

9.S.35 a. .0004 **b.** .0000 **c.** .0004

9.S.37 .9692

9.S.39 a. 74.325; 44.83325 **b.** .0721

9.S.41 True

9.S.43 False

Chapter 10

Section 10.1

10.1.1 False

10.1.3 True

10.1.5 2.58; 2.33; 1.96; 1.645; 1.28

10.1.7 1.28; 1.645; 1.96; 2.58

Section 10.2

10.2.1 False

10.2.5 False

10.2.7 Narrow interval, high level of confidence
10.2.9 Sample size must be multiplied by 4.
10.2.13 a. (118.4692, 153.5308)
 b. (123.6039, 148.3961)
 c. (127.2346, 144.7654)
 d. (129.8019, 142.1981)
 e. (131.6173, 140.3827)
10.2.15 a. (115.4, 156.6) **b.** (120.32, 151.68)
 c. (122.84, 149.16) **d.** (125.76, 146.24)
 e. (130.64, 141.36)
10.2.17 a. $(-\infty, 80.8151)$ **b.** $(-\infty, 78.5175)$
 c. $(-\infty, 77.2933)$
10.2.19 a. Yes **b.** (229.8772, 233.5228)
 c. (229.3053, 234.0948)
 d. (230.7886, 232.6114); (230.5026, 232.8974).
 As n increases, the width of the interval decreases. **e.** No
10.2.21 a. (1,174.44, 1,315.56)
 b. (1,152.12, 1,337.88)
10.2.23 a. (318.0209, 331.9791)
 b. (316.6844, 333.3156)

Section 10.3

10.3.3 True
10.3.5 False
10.3.7 False
10.3.9 True
10.3.11 b. 2.086
10.3.13 a. ± 4.303 **b.** ± 2.571 **c.** ± 2.228
 d. ± 2.086 **e.** ± 2.060
10.3.17 a. 2.764 **b.** 1.812 **c.** 2.228

Section 10.4

10.4.1 True
10.4.3 We assume the standard deviation has been estimated.
10.4.7 a. (199.2560, 224.7440)
 b. (204.0672, 219.9328)
 c. (205.7606. 218.2394)
10.4.9 a. (130.5179, 141.4821)
 b. (131.9546, 140.0454)
 c. (132.6464, 139.3536)
 d. (133.4167, 138.5833)
10.4.11 a. 30.004; 9.673 **b.** (28.024, 32.064)
 c. (28.057, 32.031)
10.4.13 (226.0413, 253.9587)
10.4.15 (25.1462, 27.9738)
10.4.17 (12.8691, 18.9623)

10.4.19 a. (8.40175, 11.59825)
 b. (8.7616, 11.2384)
10.4.21 (125.2503, 153.7497)

Section 10.5

10.5.3 True
10.5.7 False
10.5.9 False
10.5.11 a. Yes **b.** (.2936, .3864)
10.5.13 a. (.5164, .7636) **b.** (.5459, .7341)
 c. (.5610, .7190) **d.** (.5786, .7014)
 e. (.6078, .6722)
10.5.15 a. $(-\infty, .1843)$ **b.** $(-\infty, .2154)$
 c. $(-\infty, .2319)$
10.5.17 a. (.4525, .6475) **b.** (.5012, .5988)
 c. (.5256, .5744)
 d. Width of the confidence interval is cut in half.
10.5.19 a. (.3363, .4637) **b.** (.3001, .4999)
10.5.21 (.5040, .6960). Because the interval does not contain .5, it appears that the majority favor pass–fail classes (at a 95% confidence level).
10.5.23 97.08%

Section 10.6

10.6.5 97
10.6.7 a. $\sigma \approx 2$ **b.** $n = 2.90$, or 3
 c. $\sigma \approx 3$; $n = 6.54$, or 7
10.6.9 a. 554 **b.** 601
10.6.11 a. 50 **b.** 312
10.6.13 a. 16 **b.** 62 **c.** 107
10.6.15 2,436
10.6.17 a. 1,554 **b.** 1,849
10.6.19 a. 1,305 **b.** 753; this is a 43% reduction.

Section 10.7

10.7.1 True
10.7.7 a. $(-12.97, 8.77)$ **b.** $(-16.65, 12.45)$
10.7.9 (2.2853, 4.1147)
10.7.11 $(-1.8384, .6384)$

Section 10.8

10.8.3 True
10.8.5 a. $(-.175, .295)$ **b.** $(-.253, .373)$
10.8.7 (.00321, .13179)
10.8.9 a. $(-.07832, .21172)$
 b. Since this confidence interval contains 0, we should argue that the two groups are not significantly different.

Supplementary Exercises

10.S.1 (19.4984, 21.3016)

10.S.3 97; 228

10.S.5 **a.** (.5804, .6196) **b.** Yes

10.S.7 **a.** (.06319, .09681) **b.** 2,828

10.S.9 **a.** (.1792, .2208) **b.** (.1674, .2326)

10.S.11 **a.** (.05968, .16532)

    b. Yes, there is strong evidence that there are more left-handed athletes because this interval does not contain 0.

10.S.13 **a.** $(-.2960, -.1040)$

    b. Because this interval does not contain 0, we can be very confident that the proportions who favor the bill in the two groups are not equal.

10.S.15 28

10.S.17 $(-.0315, .2315)$

10.S.19 666

10.S.21 .05

10.S.23 Wider

10.S.25 **a.** (28.1603, 43.8397) **b.** (26.4699, 45.5301)

10.S.27 **b.** (8.6152, 10.9848) **c.** (8.3379, 11.2621)

10.S.29 **a.** (85.4454, 86.5546) **b.** (84.9547, 87.0453)

10.S.31 **a.** (13.8979, 15.5021) **b.** (13.4232, 15.9768)

10.S.33 **a.** (114.8377, 122.4123)

    b. (113.4868, 123.7632)

10.S.35 554

10.S.37 **a.** 17.2; 6.0882 **b.** (12.8451, 21.5549)

    c. (10.9429, 23.4571)

    d. (13.2263, 21.1737); (11.8151, 22.5849)

10.S.39 (.039, .086)

10.S.41 707

10.S.43 **a.** 1,057 **b.** 35

Chapter 11

Section 11.1

11.1.11 True

11.1.13 True

11.1.15 False

11.1.17 False

11.1.19 False

11.1.21 True

11.1.23 True

11.1.25 True

11.1.27 **a.** A Type I error occurs if we decide that an individual has the disease when he does not.

    b. A Type II error occurs if we decide that an individual does not have the disease when he actually does have the disease.

    c. Both errors are quite serious. When a Type I error occurs, the person receives medical treatment that is not needed. When a Type II error occurs, an infected individual does not receive medical treatment.

11.1.29 **a.** Level of significance should be high to reduce Type II errors. Since there are many loan applicants, we want to avoid accepting a poor risk.

    b. Level of significance should be low to reduce Type I errors. Since it is difficult to get loan applicants, we do not want to reject a good credit risk.

11.1.31 **a.** 1. Correct 2. Type II 3. Correct 4. Type I

    b. 1. Type I 2. Correct 3. Correct 4. Type II

11.1.33 **a.** $H_0: \mu \le \$11.65$ or $H_0: \mu = \$11.65$; $H_1: \mu > \$11.65$.

    Type I: Adopt new PR program when it is not better.

    Type II: Do not adopt a new PR program that is better.

    b. $H_0: \mu = 4.7$; $H_1: \mu \ne 4.7$.

    Type I: Legislation has not changed the mean, but sociologists claim the mean has changed.

    Type II: Legislation has changed the mean, but sociologists claim the mean has not changed.

    c. $H_0: \mu$(old sulfur) $\le \mu$(new sulfur); $H_1: \mu$(old sulfur) $> \mu$(new sulfur)

    Type I: Adopt new process when it does not reduce the mean sulfur content.

    Type II: Do not adopt the new process even though it reduces the mean sulfur content.

11.1.35 $H_0: p \ge .6$; $H_1: p < .6$.

11.1.37 **a.** H_0: medicine is unsafe; H_1: medicine is safe

    b. If we commit a Type I error, we assume that an unsafe medicine is safe; thus, we allow the production of an unsafe drug.

    c. If we commit a Type II error, we assume that a safe medicine is unsafe; thus, we do not allow the production of a safe drug.

    d. To ensure that unsafe medicines do not reach the market, we want to avoid a Type I error. That is, we do not want to reject H_0 when H_0 is true.

Section 11.2

11.2.5   True

11.2.9 $z = 6.24$

11.2.11   True

11.2.13   **a.** $z = 2.2$; reject H_0
  b. $z = 2.2$; do not reject H_0
  c. The results of the two tests differ because we have reduced the level of significance of the test in part **b.** When we reduce the level of significance we require stronger evidence to reject H_0.

11.2.15   **a.** $z = .75$; do not reject H_0
  b. $z = .95$; do not reject H_0
  c. $z = 1.50$; do not reject H_0
  d. $z = 3.00$; reject H_0
  e. As the sample size increases, the observed value of z increases. For very large values of n, the observed value of z exceeds the critical value, so we reject H_0.
  g. $n = 25$: 94.32, 125.68; $n = 40$: 97.604, 122.396; $n = 100$: 102.16, 117.84; $n = 400$: 106.08, 113.92
  h. $n = 25$: (100.32, 131.68)
    $n = 40$: (103.60, 128.40)
    $n = 100$: (108.16, 123.84)
    $n = 400$: (112.08, 119.92)
    We reject H_0 if the CI does not contain the value 110.

11.2.17   **a.** .1292   **b.** .0548   **c.** 3.2051; reject H_0

11.2.19   **b.** $1.95 \leq \bar{X} \leq 2.05$
  c. $\bar{X} < 1.95$ or $\bar{X} > 2.05$
  d. 1.95, 2.05   **e.** $\alpha = .4066$
  f. α increased because the acceptance region was narrowed.

11.2.21   **a.** Two-tailed   **b.** $\bar{X}$ **c.** H_0: $\mu = 21{,}000$
  d. H_1: $\mu \neq 21{,}000$
  e. Critical region: $\bar{X} < 20{,}500$ or $\bar{X} > 21{,}500$; acceptance region: $20{,}500 \leq \bar{X} \leq 21{,}500$; critical values: 20,500, 21,500
  f. $\bar{X}$ is approximately normal; mean is 21,000; std. dev. is $2{,}500/10 = 250$
  g. $\alpha = .0456$

11.2.23 $z = 4.697$; reject H_0

Section 11.3

11.3.3   False

11.3.5   True

11.3.7   False

11.3.9   True

11.3.11   A p-value of .0000 means that the probability is less than .00005 of observing a test statistic more extreme than that which was actually observed, given that H_0 is true. Such a situation provides strong evidence against H_0.

11.3.13   False

11.3.15   True

11.3.17   **a.** .012   **b.** .166
  c. The p-value exceeds α, so we do not reject H_0.

11.3.21   .0162

11.3.23 -1.645

11.3.25   .0475

11.3.27   **a.** (955.053, 1,004.947); do not reject H_0
  b. $z = -1.571$; do not reject H_0
  c. $\approx .1164$

11.3.29   **a.** $z = 6.667$; reject H_0 **b.** 0

11.3.31 p-value $= .0000$; for any value of α greater than .00005, we would reject H_0.

Section 11.4

11.4.1   True

11.4.3 $z = 2.26$; reject H_0

11.4.5 $z = -2.24$; do not reject H_0

11.4.7   **a.** $z = -2.13$; reject H_0
  b. $z = 2.67$; reject H_0
  c. $z = 21.74$; reject H_0

11.4.9 $z = -6.67$; reject H_0

11.4.11   **a.** $\bar{x} = 16$; $s = 8$
  b. (14.432, 17.568)
  c. $z = -3.75$; reject H_0
  d. In part **b**, the confidence interval does not contain the value μ_0, so we would reject H_0.

Section 11.5

11.5.3   **a.** .025   **b.** .05   **c.** .05

11.5.5   **a.** $\bar{X} = 30.4$; $s = 14.7663$
  b. $t = .086$; do not reject H_0
  c. .9282

11.5.7   **a.** $\bar{X} = 40.0$; $s = 10$
  b. $t = -3.05$; reject the null hypothesis.
  c. $P(t < -3.05) = .0011$ (approximated by area under standard normal curve)

11.5.9   **a.** $\approx .0316$ **b.** $t = 2.25$; reject H_0

11.5.11 $t = -7.74$; reject H_0

11.5.13   **a.** $\bar{X} = 315.833$; $s = 30.5133$
  b. $t = -.47$; do not reject H_0
  c. .6384

Section 11.6

11.6.3 **a.** Do not use the normal approximation
b. Use the normal approximation
c. Use the normal approximation
d. Use the normal approximation
11.6.5 **a.** .006, .006 **b.** $\alpha = .012$ **c.** .762
11.6.7 $z = .53$; do not reject H_0
11.6.9 $z = 1.86$; reject H_0
11.6.11 $z = -2.17$; reject H_0

Section 11.7

11.7.3 Increase the level of significance α.
11.7.7 True
11.7.9 True
11.7.11 False
11.7.13 **b.** 769.8 **c.** .4840 **d.** .5160 **e.** .8315
f. .1685
11.7.15 **b.** 86.58 **c.** .6554 **d.** .3446 **e.** .7764
f. .2236
11.7.17 **b.** 1,660.8 **d.** .9279 **e.** .0721
f. .9564; .9279; .8315; .6772; .4840; .2946;
.1492
i. When $\mu = 1,690$, $\beta = .9564$ and $1 - \beta = .0436$
When $\mu = 1,695$, $\beta = .9671$; $\mu = 1,690$: $\beta = .9564$; $\mu = 1,680$: $\beta = .9279$; $\mu = 1,670$: $\beta = .8869$; $\mu = 1,660$: $\beta = .8315$; $\mu = 1,650$: $\beta = .7612$; $\mu = 1,640$: $\beta = .6772$
11.7.19 **a.** 17.02, 18.98 **b.** 16.712, 19.288
c. 16.367, 19.633 **d.** .6700 **e.** .9793
f. The test becomes more powerful as the sample size increases.
11.7.23 **a.** Reject H_0 if $Z < -2.326$; alternatively, reject H_0 if $\bar{X} < 6.9534$.
b. Reject H_0
c. .9962; ≈ 1; 1; 1

Section 11.8

11.8.3 False
11.8.5 True
11.8.7 **a.** 15.5073 **b.** 20.0902 **c.** 13.3616
11.8.9 **a.** .872085 **b.** 1.63539 **c.** 2.20413
11.8.11 67.5048

Section 11.9

11.9.1 True
11.9.3 $(n - 1)$
11.9.5 **a.** $\bar{X} = 29.47$; $s^2 = 10.884556$

b. (5.14966, 36.27661)
c. $\chi^2 = 6.53$; do not reject H_0
11.9.7 **a.** (62.4208, 405.140)
b. $\chi^2 = 8.7808$; do not reject H_0
11.9.9 $\chi^2 = 70$; reject H_0
11.9.11' Let us use $\alpha = .05$. $\chi^2 = 140.875$; reject H_0

Supplementary Exercises

11.S.1 $z = -5.5$; reject H_0
11.S.3 $z = 5$; reject H_0
11.S.5 .3446
11.S.7 .0000
11.S.9 $z = 2.6$; reject H_0
11.S.11 $z = 2.83$; reject H_0
11.S.13 **a.** .5206
b. .9906; .9394; .7734; .4801; .1977
d. Critical value: .5165; powers: .9996; .9906; .9115; .6368; .2578
11.S.15 **a.** 93.6 **b.** 8.5010
c. $t = -2.38$; reject H_0
11.S.17 **a.** .4417 **b.** $z = -1.81$; reject H_0
11.S.19 **a.** $\approx .0073$ **b.** .5
11.S.21 $t = -4.47$; reject H_0
11.S.23 $z = -1.065$; do not reject H_0
11.S.25 115
11.S.27 .3897; .5; .6103; .7123; .7967; .8665; .9177; .9525
11.S.29 The statement is correct
11.S.31 **a.** $\alpha = .2$; $\beta = .4$ **b.** $\alpha = .4$; $\beta = .3$
c. Use the rule in **b.**
11.S.33 $z = 7.15$; reject H_0
11.S.35 $z = -5.31$; reject H_0
11.S.37 **a.** $t = 3.42$; reject H_0
b. .0003

Chapter 12

Section 12.1

12.1.3 True
12.1.5 $z = 3.58$; reject H_0
12.1.7 $z = -.89$; do not reject H_0
12.1.9 $z = 5$; reject H_0
12.1.11 $z = -4.71$; reject H_0
12.1.13 $z = -4.71$; reject H_0

Section 12.2

12.2.3 False. If both populations are normal and the sample sizes are small, the use of the t distribution is appropriate.

12.2.5 True

12.2.7 **a.** $s_p^2 = 21$ **b.** 18
c. $t = 2.44$; reject H_0

12.2.9 **a.** $s_p^2 = .71318$ **b.** 22
c. $t = -.41$; do not reject H_0

12.2.11 $t = -3.16$; reject H_0

Section 12.3

12.3.1 False

12.3.5 **a.** 14 **b.** $t = 5.99$; reject H_0

12.3.7 $t = 2.79$; reject H_0

12.3.9 $t = 8.57$; reject H_0

Section 12.4

12.4.3 False

12.4.5 **a.** .16667 **b.** $z = 1.2$; do not reject H_0

12.4.7 $z = .59$; do not reject H_0

12.4.9 $z = .88$; do not reject H_0

12.4.11 $z = -2.25$; reject H_0

Section 12.5

12.5.1 False

12.5.3 False

12.5.5 True

12.5.7 **a.** .0878 **b.** .0644 **c.** .0912

Section 12.6

12.6.1 False

12.6.3 **a.** 15; 20 **b.** $F = .819$
c. $F < .429$ and $F > 2.20$
d. Do not reject H_0
e. (.37227, 1.90827)

12.6.5 **a.** 24; 29 **b.** $F = .76006$
c. $F < .51546$ (approximately) and $F > 1.90$
d. Do not reject H_0
e. (.40003, 1.4745)

12.6.7 **a.** $F = 6.25$; reject H_0
b. (2.89, 13.5)

Supplementary Exercises

12.S.1 z (or t) = 4.44; reject H_0

12.S.3 $t = -.74$; do not reject H_0

12.S.5 $z = 1.13$; do not reject H_0

12.S.7 $z = 2.05$; reject H_0

12.S.9 $z = 3.33$; reject H_0

12.S.11 $t = -1.114$; do not reject H_0

12.S.13 $z = -5.83$; reject the hypothesis that the population proportions are equal

12.S.15 **a.** $H_0: \mu_1 - \mu_2 \le 0$; $H_1: \mu_1 - \mu_2 > 0$
b. $t = 3.999$; reject H_0

12.S.17 **c.** Machine A: $\bar{X} = 21.32143$; Machine B: $\bar{X} = 21.24375$
d. Machine A: $s^2 = 1.753881733$, $s = 1.324342$; Machine B: $s^2 = 2.797430193$, $s = 1.672552$
e. $t = .3122$; do not reject H_0
f. $F = .42017$; reject H_0
g. Sample A: $\hat{p} = .12857$; Sample B: $\hat{p} = .15$
h. $z = -.3773$; do not reject H_0

12.S.19 **a.** $\hat{p}_1 = .00071$; $\hat{p}_2 = .000285$
b. $\hat{p} = .0004975$
c. $z = 6.027$; reject the hypothesis that the population proportions are equal.
d. Yes
e. .000425; although this difference may appear to be small, it indicates that in Group 2 there will be approximately 425 fewer cases of polio for each 1,000,000 people. This is a significant (large) decrease in cases of polio.

12.S.21 **a.** $z = .48200$; do not reject the hypothesis that the population proportions are equal.
b. $z = -9.288$; reject H_0

Chapter 13

Section 13.1

13.1.1 False

13.1.3 True

13.1.5 True

13.1.7 **a.** 15.0863 **b.** 11.0705 **c.** 9.23635
d. 20.0902 **e.** 15.5073 **f.** 13.3616

13.1.9 **a.** 3 **b.** 7.81473 **c.** 40; 80; 200; 80
d. $\chi^2 = 3.2875$
e. Do not reject the null hypothesis.

13.1.11 **a.** 40; 40; 50; 30; 40 **b.** $\chi^2 = 24.8383$
c. 13.2767
d. Reject the null hypothesis.

13.1.13 $\chi^2 = 3.00$; do not reject the null hypothesis.

13.1.15 $\chi^2 = 14.167$; reject the null hypothesis.

13.1.17 $\chi^2 = 4.1$; do not reject the null hypothesis.

Section 13.2

13.2.5 False

13.2.9 The critical value of the chi-square test does not depend on the number of observations.

13.2.11 **a.** 2
b. 38.184; 47.816; 29.748; 37.252; 43.068; 53.932
c. 5.99147
d. $\chi^2 = 11.87929$; reject the null hypothesis.
e. .005

13.2.13 The number of degrees of freedom is 20. Critical value is 31.4104. Do not reject the null hypothesis.

13.2.15 The number of degrees of freedom is 4. Critical value is 13.2767. Reject the null hypothesis.

13.2.17 $\chi^2 = 11.034$; reject H_0

13.2.19 $\chi^2 = 14.4379$; reject H_0

13.2.21 $\chi^2 = 1.5215$; do not reject H_0

Supplementary Exercises

13.S.1 $\chi^2 = 14.54$; reject H_0

13.S.3 $\chi^2 = 15.2067$; reject H_0

13.S.5 $\chi^2 = 9.0198$; reject H_0

13.S.7 $\chi^2 = 45.0$; reject H_0

13.S.9 9

13.S.11 $\chi^2 = 4.6074$; do not reject H_0

13.S.13 $\chi^2 = 34.1911$; reject H_0

13.S.15 a. .0133 **b.** .1110 **c.** .0408 **d.** .0742
 e. .2193 **f.** .0779
 g. $\chi^2 = 186.5395$; reject H_0

13.S.17 a. $\chi^2 = .2789$; do not reject H_0

13.S.19 $\chi^2 = 77.6915$; 6 degrees of freedom; reject H_0

13.S.21 a. 3
 b. 471.817; 37.183; 295.697; 23.303; 137.189; 10.811; 34.297; 2.703
 c. $\chi^2 = 37.249$
 d. Reject H_0

13.S.23 a. Frequency distribution:

Under 370.5	8
370.5–380.5	11
380.5–390.5	20
390.5–400.5	38
400.5 or more	23

 c. $\overline{X} = 392.36$
 d. $s^2 = 201.687$; $s = 14.20166$
 e. 2
 f. Expected frequencies:

Under 370.5	6.18
370.5–380.5	13.86
380.5–390.5	24.79
390.5–400.5	26.74
400.5 or more	28.43

 g. 5.99147
 h. $\chi^2 = 7.830$; reject H_0
 i. The p-value of the test is less than .025.

13.S.25 a. 2
 b. 39.375, 65.521, 29.103, 213.625, 355.479, 157.897
 c. 5.99147

 d. $\chi^2 = 17.66$; reject H_0
 e. The approximate p-value is less than .005.

13.S.27 $\chi^2 = 16.6210$; reject H_0

13.S.29 $\chi^2 = 1.0012$; do not reject H_0

13.S.31 $\chi^2 = 618.12$; reject H_0

Chapter 14

Section 14.1

14.1.3 True

14.1.5 True

14.1.7 False

14.1.11 False

14.1.13 False

14.1.15 a. The values in Sample 1 appear to be smaller than the values in Sample 2.
 b. 6.26, 7.12, 7.88, 6.98
 c. SSW = 19.616; SSB = 6.612; SST = 26.228
 d. MSW = 1.226; MSB = 2.204
 e. $F = 1.798$; numerator degrees of freedom is 3; denominator degrees of freedom is 16
 f. Critical value is 3.24; do not reject H_0
 g. The approximate p-value exceeds .05.

14.1.17 a. SSB has degrees of freedom 4; SSW has degrees of freedom 75; SST has degrees of freedom 79.
 b. SSB = 389.2; SST = 538.3
 c. MSW = 1.988
 d. $F = 48.9437$
 f. Critical value is approximately 2.53; reject H_0

14.1.19 a. 77.6, 79.2, 80.2, 73.0
 b. SST = 279.00; SSB = 152.2; SSW = 126.8
 c. MSB = 50.7333; MSW = 7.9250
 d. $F = 6.4017$
 e. SSB has degrees of freedom 3; SSW has degrees of freedom 16
 f. Critical value is 3.24; reject H_0

14.1.21 $F = 5.25$; reject H_0

Section 14.2

14.2.3 True

14.2.5 a. The values in Sample 2 appear to be slightly smaller than the values in Sample 1.
 b. 19.87; 18.57
 c. SST = 743.612035; SSB = 8.45; SSW = 735.162035
 d. MSW = 40.8423353; MSB = 8.45
 e. $F = .2068$; degrees of freedom are 1 and 18
 f. Critical value is 4.41; do not reject H_0

g. 40.8423353

h. $t = .454855$; $t^2 = .2068 = F$

i. The t statistic has 18 degrees of freedom; critical values are $\pm t_{.025,\,18} = 2.101$; do not reject H_0

14.2.7   **a.** $F = 5.685737$

b. $t^2 = (-2.38448)^2 = 5.6857 = F$

c. 4.20

d. $t^2_{.025,\,28} = (2.048)^2 = 4.194 = F$

e. Reject H_0

Supplementary Exercises

14.S.1 $F = 1.77$; do not reject H_0

14.S.3 $F = 1.59$; do not reject H_0

14.S.5   **a.**

Source of variation	Sum of squares	Deg. of freedom	Mean square
Between samples	423.33	2	211.665
Within samples	5,600.00	12	466.67
Total	6,023.33	14	
$F = .45$			

b. Critical value is 3.89; do not reject H_0

14.S.7 $F = .29399$; do not reject H_0

14.S.9 $F = 2.43$; do not reject H_0

14.S.11   **a.** 4.52, 5.76, 5.88, 5.48

b. 5.41   **c.** $F = 6.42$

d. Critical value is 3.24; reject H_0

14.S.13 $F = 5.4032$; reject H_0

14.S.15 $F = 1.04$; do not reject H_0

14.S.17 $F = 3.37$; reject H_0

Chapter 15

Section 15.1

15.1.3   *Deterministic:* Let X denote the price of a shirt. Suppose I purchase 4 shirts. Let Y denote total expenditure. Then $Y = 0 + 4X$.

Stochastic: Let Y denote an individual's expenditures on recreation. Let X denote the person's income. Then $Y = \beta_0 + \beta_1 X + e$ where e is a random error term.

15.1.7   False

15.1.9   **b.** 3   **c.** 7

15.1.11   **b.** 3   **c.** 0

15.1.13   **a.** There is a very weak positive relationship.

b. Stochastic

c. Neither linear nor nonlinear

d. Weak

e. Slightly positive

15.1.15   **a.** As X increases, Y increases.

b. Stochastic

c. Approximately linear

d. Strong

e. Positive

Section 15.2

15.2.3   True

15.2.7   False

Section 15.3

15.3.3   True

15.3.5   True

15.3.7   True

15.3.9   **a.** There appears to be a stochastic linear relationship between Y and X.

b. $\Sigma x = 20.0$; $\Sigma y = 91.9$; $\Sigma xy = 404$; $\Sigma x^2 = 90.0$; $\Sigma y^2 = 1,829.03$

c. $\bar{X} = 4.00$; $\bar{Y} = 18.38$

d. $b_1 = 3.64$

e. $b_0 = 3.82$

f. 11.10; 14.74; 18.38; 22.02; 25.66

h. $-.1$; 1.56; -1.68; $-.92$; 1.14

i. 7.412

j. 1.571834661

15.3.11   .85465

15.3.13   **a.** $b_1 = .5827465$; $b_0 = 1.742077$

b. 167.94982   **c.** 3.05459

15.3.15   **b.** $b_1 = 9.5221$; $b_0 = 69.7548$

c. 126.89; 98.32

15.3.17   **b.** $b_1 = -16.6134$; $b_0 = 1,268.2769$

c. 603.7395

Section 15.4

15.4.1   False

15.4.3   True

15.4.5   False

15.4.7   If we have a perfect fit, then every residual will be 0 and SSE = 0.

15.4.9   False

15.4.11   True

15.4.13   True

15.4.15   **b.** $\hat{y}_i = 3.82 + 3.64 x_i$

d. SST = 139.908; SSR = 132.496; SSE = 7.412

e. $R^2 = .9470$; this value is relatively high and indicates that the data lie close to a straight line.

15.4.17 b. $\hat{y}_i = .5 + 0x_i$

c. $\hat{y}_i = .5 + 0x_i$

d. The four residuals are: $-.5, -.5, .5, .5$; their sum is 0.

d. SST = 1.0; SSR = 0.0; SSE = 1.0

e. $R^2 = 0$. This value indicates that the regression equation is a horizontal line; as a result, the values of X do not help explain the values of Y.

15.4.19 No

Section 15.5

15.5.1 True

15.5.3 a. $b_1 = .823997$; $b_0 = .2636901$

b. SST = 10,338.519; SSE = 1,914.219; SSR = 8,424.3

c. 15.468593 **d.** .1388933 **e.** 4.914743

15.5.5 a. $b_1 = .1965$; $b_0 = .4803$; $s_e^2 = .4905$

b. SST = 56; SSE = 2.9435; SSR = 53.0568

c. $R^2 = .9474$; $s_e = .7004$

d. $s^2(b_0) = .2502$; $s^2(b_1) = .0003570$

15.5.7 a. $b_1 = -113.7931$; $b_0 = 1,037.9310$

b. .5764

c. $s^2(b_0) = 63,628.0958$; $s^2(b_1) = 2,378.5463$

Section 15.6

15.6.3 False

15.6.5 a. 19 **b.** ± 2.093

c. $t = 1.411$; do not reject H_0

15.6.7 a. 8 **b.** ± 2.306

c. $t = 1.647$; do not reject H_0

d. 8 **e.** ± 2.306

f. $t = 10.33$; reject H_0

15.6.9 a. $t = 10.3999$; reject H_0

b. $t = .9602$; accept H_0

c. $t = -.1852$; accept H_0

d. $t = -2.8307$; reject H_0

15.6.11 a. $b_1 = -58.2857$; $b_0 = 829.1429$

b. $t = -2.579$; reject H_0

Section 15.7

15.7.1 True

15.7.3 If we tested $H_0: \beta_1 = 0$ against $H_1: \beta_1 \neq 0$ using a 1% level of significance, we would not be able to reject H_0.

15.7.5 If $R^2 = 1$, we have a perfect fit. Thus, the sample standard deviation is 0 and the confidence interval has 0 width.

15.7.11 a. $(-3.278, 12.918)$ **b.** $(-6.815, 16.455)$

c. $t = 1.35$; do not reject H_0

15.7.13 a. $(-2.2404, 13.4404)$ **b.** $(-5.807, 17.007)$

c. $t = 1.647$; do not reject H_0

15.7.15 b. $b_1 = -1.2645$; $b_0 = 20.6860$

c. $(-1.6740, -.8550)$

15.7.17 a. $b_1 = -.0131$; $b_0 = 3.6892$

b. $(-.1049, .0787)$

c. $t = -.657$; do not reject H_0

Section 15.8

15.8.1 True

15.8.3 False

15.8.5 a. $(450.202, 489.798)$

b. $(454.477, 485.523)$

c. $(907.509, 932.491)$

d. $(917.757, 922.243)$

15.8.7 a. $(627.079, 648.921)$

b. $(630.733, 645.267)$

c. $(887.079, 908.921)$

d. $(890.733, 905.267)$

15.8.9 a. $\hat{y}_i = -5.8603 + .008344x_i$

b. $(7.9913, 8.65922)$

Section 15.9

15.9.1 False

15.9.3 False

15.9.5 False

15.9.7 False

15.9.11 True

15.9.13 Correlation and regression results do not tell us anything about causality.

15.9.15 b. $\hat{y}_i = 7 + 0x_i$

c. $R = 0$

15.9.17 b. $\hat{y}_i = 66.308 + 14.615x_i$. Each value of Y has been multiplied by 10; thus, the slope and intercept have been multiplied by 10.

c. $R = .8482$; R does not measure the steepness of the line.

15.9.19 a. 6 **b.** 1.943

c. $t = 1.41$; do not reject H_0

d. True **e.** $t = 2.77$; reject H_0

15.9.21 a. 10 **b.** -1.812

c. $t = -2.372$; reject H_0

15.9.23 a. $R = .9864$

b. Neither; both variables are positively related to population.

15.9.25 $t = 2.3094$; reject H_0

15.9.27 We would expect a positive correlation coefficient. Theoretically, the variables appear to be causally related.

Supplementary Exercises

15.S.3 **b.** $b_1 = .1406$; $b_0 = 1.1597$
  c. .9016 **d.** .9495
  e. $t = 8.60$; reject H_0

15.S.5 **b.** $-.9006$ **c.** $t = -5.07$; reject H_0
  d. Losing teams may regard the pass as a method of catching up.

15.S.7 **b.** $R = .6970$; $t = 2.38$; reject H_0
  c. $\hat{y}_i = 32.2425 + .6695x_i$
  d. $\hat{y} = 107.2268$

15.S.9 **a.** False
  b. True
  c. True
  d. False
  e. False
  f. True
  g. True

15.S.11 **a.** $\hat{y}_i = -325.1724 + 15.2138x_i$
  b. $R^2 = .9734$ **c.** $(12.323, 18.104)$
  d. $1,196.2069$ **e.** $(1,081.6249; 1,310.7889)$

15.S.13 **a.** $\hat{y}_i = 18.1413 + 2.0372x_i$
  b. $(9.9886, 26.2940)$; $(.6424, 3.4319)$
  c. $t = 4.05$; reject H_0
  d. 30.3643 **e.** $(26.3784, 34.3502)$

15.S.15 **a.** $\hat{y}_i = 8.5536 - .1067x_i$
  b. $(-.1137, -.09961)$ **c.** 2.1545
  d. .9957; yes, the fit is extremely good.
  e. $(.8181, 1.3579)$

15.S.17 **c.** $\Delta\, G\hat{N}P = 72.4283 + .5837\Delta G$
  e. .009618; no
  g. $\Delta\, G\hat{N}P = 50.1141 + .2109\Delta M$
  i. .02369; no

15.S.21 **a.** $\hat{y}_i = -24.35494 + 1.422716x_i$
  c. .8714
  d. $t = 7.364$; reject H_0

Chapter 16

Section 16.1

16.1.3 False
16.1.5 **a.** $b_0 = 220.8235294$; $b_1 = -45.88235294$; $b_2 = -68.705882$

b. $\hat{y}_i = 220.8235294 - 45.88235294x_{i1} - 68.705882x_{i2}$
  c. -170.9882353 **d.** $b_0 = 220.8235294$
  e. $b_1 = -45.88235294$ **f.** $b_2 = -68.705882$
  g. $b_1 + b_2 = -114.5882353$

16.1.7 **a.** $\hat{y}_i = -10 + 4.5x_{i1} - 6.0x_{i2}$
  b. 2 **c.** 6.5 **d.** -4

16.1.9 **b.** A parabola would fit better than a straight line.
  c. $\hat{y}_i = 5.045455 + 1.465909x_i + 2.011364x_i^2$
  d. 86.250013

16.1.11 $\hat{y}_i = 34.971 - .6261x_i + 11.63z_i$
16.1.13 $\hat{y}_i = -150.6111 + 4.683018r_i + 1.737684t_i$

Section 16.2

16.2.1 True
16.2.5 False
16.2.7 True
16.2.9 **a.** $\hat{y}_i = -10 + 4.5x_{i1} - 6.0x_{i2}$
  b. 22 **c.** $s_e = 4.67099$

16.2.13 A \$10,000 increase in food sales leads to a \$31 increase in profits. Similarly, a \$10,000 increase in nonfood sales leads to an \$89 increase in profits. A 1,000-square-foot increase in the size of the store boosts profits by \$487.

Section 16.3

16.3.1 True
16.3.3 False
16.3.7 **a.** $R^2 = .31998$; $\bar{R}^2 = .04797$
  b. $\hat{y}_i = 43.89 + 1.11R_i$
  c. $R^2 = .03175$; $\bar{R}^2 = -.1296$; yes; R^2 decreases when T_i is omitted.
  d. $\hat{y}_i = 26.7454 + .5249T_i$
  e. $R^2 = .056243$, $\bar{R}^2 = -.10105$

Section 16.4

16.4.1 False
16.4.3 False
16.4.7 False
16.4.9 **b.** True
16.4.11 **a.** 16
  b. ± 2.120
  c. $t = 2.37$; reject H_0
  d. $t = -1.50$; do not reject H_0
  e. $F = 61.89$; reject H_0

16.4.13 **a.** $t = 2.77$; reject H_0
  b. 22 **c.** $t = -1.23$; do not reject H_0

16.4.15 **a.** $(1.3331, 3.0669)$
  b. $t_2 = 9.24$; reject H_0
  c. $(33.57, 62.71)$

d. $t = 9.24$; reject H_0. Alternatively, since the confidence interval obtained in part **c** does not contain 0, we can reject the null hypothesis that $\beta_2 = 0$.

Section 16.5

16.5.3 False
16.5.5 False
16.5.7 True
16.5.9 Yes
16.5.11 Yes
16.5.13 **a.** $R = -.7296$
  b. The coefficient of temperature should decrease because some of the negative effect from rainfall will be shifted to the temperature variable.
  c. Without the variable R: $\hat{y}_i = 26.7454 + .5249 t_i$
  With variable R: $\hat{y}_i = -150.61 + 4.683 r_i + 1.738 t_i$
16.5.15 **a.** Yes; L and K are perfectly related according to the equation $K + L = 80,000$.   **b.** No

Supplementary Exercises

16.S.1 **a.** 2,654.40   **b.** $-3,373.90$ **c.** Yes
16.S.3 **a.** $t = 4$; reject H_0 **b.** $t = 1.69$; reject H_0
  c. $(-.4, 4.8)$ **d.** $(3.3, 4.9)$
16.S.5 **b.** 36   **c.** No; we need to test the null hypothesis that the coefficient is 0.
  d. Not necessarily

Chapter 17

Section 17.1

17.1.3 **c.** When $\beta_2 > 0$ the curve bends upward as x approaches infinity. When $\beta_2 < 0$ the curve bends downward as x approaches infinity.
17.1.5 **a.** A linear relationship appears to be satisfactory.
  b. $\hat{y}_i = -30.8286 + 1.598 x_i$
  c. $\hat{y}_i = -18.7434 + .9307 x_i + .0084 x_i^2$
  d. $t = 1.515$; do not reject H_0

Section 17.2

17.2.1 True
17.2.3 True
17.2.5 True
17.2.7 **a.** $E_i = 800 + .05 W_i$
  b. $E_i = 625 + .05 W_i$
  c. 2,050
  d. 1,925

17.2.9 **a.** \$35,000
  b. Eastern
  c. Western
  d. South
  e. East, North, South, West
  f. \$25,400
17.2.11 **a.** \$14,300   **b.** \$56,500   **c.** \$40,700
17.2.13 **c.** \$30,000   **d.** \$32,000   **e.** \$27,000
  f. $t = 8.203$; reject H_0; conclude that employees with a B.A. degree have higher salaries than those with a high schoool degree, holding seniority constant.
  g. $\hat{y}_i = 20,000 + 1,000 s_i - 3,000 x_{i3} + 2,000 x_{i2}$
  h. $\hat{y}_i = 22,000 + 1,000 s_i - 2,000 x_{i1} - 5,000 x_{i3}$

Section 17.3

17.3.1 **c.** $\ln S_i = 3.2229 + .105298 x_i$
17.3.3 **b.** $\hat{y}_i = 2.575333 - .324 x_i$
  c. $R^2 = .8116$; $s_e = .47431$
  d. $t = -7.483$; reject H_0
  e. Try a logarithmic transformation or estimate a quadratic equation.

Section 17.5

17.5.3 True
17.5.5 **a.**

	PCT	SAL	REV	EXP	INC	WAGE
PCTUNION	1.00	.613	.438	.326	.460	.706
SALARY	.613	1.00	.840	.747	.767	.562
REVENUE	.438	.840	1.00	.797	.755	.425
EXPEND	.326	.747	.797	1.00	.474	.413
INCOME	.460	.767	.755	.474	1.00	.373
WAGE	.706	.562	.425	.413	.373	1.00

  b. Revenue   **c.** Salary   **d.** Salary
  e. .840; .767   **f.** .706

Supplementary Exercises

17.S.1 **b.** \$930
17.S.3 **a.** $(-25.532, -9.068)$
  b. $(.06412, .07588)$
17.S.5 **a.** $\hat{y}_i = -2.37 + 2.923 x_{i1} + 7.815 x_{i2}$
  b. $t = 7.086$; reject H_0

Chapter 18

Section 18.1

18.1.7 False
18.1.9 **a.** There may be heteroscedasticity present. The standardized residuals associated with small

fitted values tend to be more variable than the standardized residuals associated with the larger values.

b. The standardized residual -3.78 is an outlier.

Section 18.3

18.3.1 False
18.3.3 False
18.3.7 False
18.3.9 **b.** The observed data follow a parabola. The sample regression line will lie below the data for small values of X, above the data for middle values of X, and below the data for large values of X. Thus, the residuals will have a positive–negative–positive pattern.
c. Forecasts will be too high for medium values of X and too low for very small and very large values of X.

Section 18.4

18.4.3 False
18.4.7 The residuals appear to be heteroscedastic. As X increases, the absolute values of the residuals tend to increase.

Section 18.5

18.5.3 True
18.5.5 True
18.5.13 False
18.5.15 **a.** $d_L = 1.21$; $d_U = 1.55$
b. $d = 1.85$; do not reject H_0
18.5.17 **a.** $z = 1.5$; do not reject H_0
18.5.19 **a.** .632635 **c.** $b_0 = -1.2955$; $b_1 = .175171$
d. $\text{Var}(b_0) = .117766$
$\text{Var}(b_1) = .000005322$
In the previous problem, the estimated variances were $\text{Var}(b_0) = .04586$; $\text{Var}(b_1) = .000002088$. The estimated variances were larger when the Prais–Winsten estimation technique was used.
18.5.21 $d_L = 1.21$; $d_U = 1.55$

Supplementary Exercises

18.S.1. **a.** $\hat{y}_i = -1.715 + .842x_i$ **c.** 1.653
d. $s_e = .43877$; $R^2 = .64783$
e. $s(b_0) = .739$; $s(b_1) = .146$
f. (.422, 1.262); no
g. $t = 5.754$; reject H_0
h. (1.2815, 2.0245)

18.S.3 **b.** $\hat{y} = -.360 + .903x_1 - .014x_2$
The coefficient standard deviations are: (1.7), (.163), (.016).
The t statistics are: $(-.21)$, (5.55), $(-.88)$.
$R^2 = .66327$; $s_e = .441$.
c. $\hat{y} = -1.831 + .8368x_1 + .0480x_3$
The coefficient standard deviations are: (1.095), (.155), (.325).
The t statistics are: (-1.67), (5.41), (.148).
$R^2 = .64828$; $s_e = .45120$.
d. $\hat{y} = -.413 + .901x_1 - .014x_2 + .018x_3$
The coefficient standard deviations are: (2.01), (.173), (.017), (.329).
The t statistics are: $(-.21)$, (5.21), $(-.85)$, (.05).
$R^2 = .66334$; $s_e = .45503$.
18.S.5 **a.** .25363 **b.** .42074
c. $b_0 = .8826586$; $b_1 = .5038027$
d. $s(b_0) = .023605$;
$s(b_1) = .004532$
18.S.7 $F = 192.197/4.1246 = 46.598$;
reject H_0
18.S.9 **a.** Yes
b. If the errors are serially correlated, the coefficient estimates are unbiased, but they are not best.
c. If the errors are serially correlated, the variance estimates are biased.
18.S.11 **b.** The coefficient $b_4 = -.522$ has the wrong sign.
c. b_1, b_3, and b_5
d. $d = 1.046$; falls in the inconclusive range
18.S.13 **a.**

CORRELATION MATRIX

	Y	X_1	X_2	X_3
Y	1.00	.434	.310	.246
X_1	.434	1.00	.243	.015
X_2	.301	.243	1.00	.430
X_3	.246	.015	.430	1.00

b. $\hat{y} = .0197 + .0011x_1 + .1153x_2 + 8.47x_3$
e. Based on the correlation matrix, the largest sample correlation is .430 between X_2 and X_3. Thus, there are no obvious multicollinearity problems. It is possible that one variable is highly correlated with a set of 2 or more other variables, however.
f. Serial correlation is not relevant because we are using cross-section data.

Chapter 19

Section 19.1

19.1.3 True

19.1.5 There is an upward trend.

19.1.7 **a.** All of the series except for the temperature series probably contain some trend.

b. Each of the time series 1, 2, 3, 4, and 11 probably contains a seasonal component. The time series 5 and 6 contain annual data, so no seasonal component is present.

c. Each of the time series 1, 2, 4, 5, and 6 contains a cyclical component, due to interdependence with business cycles. Series 12 depends on sociological factors which may contain a cyclical component.

Section 19.2

19.2.3. **a.** $\hat{y}_t = 8.000 + 3.833t$ **b.** .97115
c. 42.50; 46.33

19.2.5 **a.** $\hat{y}_t = 7.857 + 58.893t$ **b.** .80813
c. 537.89; 596.79

19.2.7 **a.** $\hat{y}_t = .6428571 + 10.77381t$ **b.** .9879
c. 97.61; 108.38

19.2.9 **a.** $\hat{y}_t = 356.4846 + 54.23253t$ **b.** .9712
c. 1,441.1352

19.2.11 **a.** $\hat{y}_t = 2,726.318 + 78.6167t$
b. 4,298.6516; 4,377.2684

19.2.13 **a.** $\hat{y}_t = 14.1714 - .1297619t$
b. 11.83571
e. $\hat{y}_t = 11.27 + .4298t$
f. 19.01
h. R^2 for County 1 is .6617; R^2 for County 2 is .9808.

19.2.15 **b.** $\hat{y}_t = 591.8681 + 8.474725t$ **d.** .5351

Section 19.3

19.3.1 True

19.3.3 False

19.3.5 **a.** $\hat{y}_t = 24.58929 - 14.3869t + 5.589286t^2$
b. .9955
c. 347.839356; 439.64889

19.3.7 **b.** 25.7; 44.6
c. 981.2

19.3.9 $\hat{S}_t = 12.30357 - 4.363095t + 2.684524t^2$

19.3.11 **b.** $\hat{y}_{16} = 439.2$
c. $\hat{y}_{17} = 458.2$

Section 19.4

19.4.1 True

19.4.3 **a.** $\ln(\hat{y}_t) = .9652164 + .4609464t$
c. .9694
d. 166.290; 263.665

19.4.5 **a.** $\hat{y}_t = 113.0231 + 14.10835t$;
$R^2 = .9518$
b. 381.08175; 395.1901
c. $\ln(\hat{y}_t) = 4.814553 + .0709551t$;
$R^2 = .9033$
d. 357.416; 383.677

19.4.7 **c.** $\ln(\hat{y}_t) = 7.490183 + .1465278t$
d. 5,781.425

Section 19.5

19.5.1 True

19.5.5 **b.** $\hat{y}_t = 4.238637 + .909432y_{t-1}$
c. .964 **d.** 30.612165
e. 32.078319
f. $\hat{y}_t = 10.46786 + 2.365476t$;
$R^2 = .9833$
g. 31.757144

19.5.7 **a.** $\hat{y}_t = 35.63035 + .8965351y_{t-1}$;
$R^2 = .9848$

19.5.9 **a.** $\hat{y}_t = 277.699 + 1.067784y_{t-1}$;
$R^2 = .992$

Section 19.6

19.6.1 False

19.6.3 True

19.6.5 **b.** $\hat{y}_t = 1.142873 + 1.079935y_{t-1}$
d. .0226812; .4766853; .60061943

19.6.7 **a.** $\hat{y}_t = 35.63035 + .8965351y_{t-1}$
d. .0258238; 5.652206; 6.392802

19.6.9 **a.** $\hat{y}_t = 277.699 + 1.067784y_{t-1}$
d. .0202867; 65.54459; 72.93446

Supplementary Exercises

19.S.1 **a.** $\hat{y}_t = 2.50 + 10t$
c. 92.5; 102.5
d. $\hat{y}_t = 3.1431 + 1.1641y_{t-1}$
e. 105.61; 126.08
f. $\ln(\hat{y}_t) = 2.7042 + .2281t$
g. 116.437; 146.273

19.S.3 \$149,420; \$159,410; \$169,990

19.S.5 **a.** $\hat{y}_t = 2.9533 + .4085t$
c. 8.2636 (in billions)
d. $\ln(\hat{y}_t) = 1.1899 + .0783t$; 9.0957

19.S.7   **b.** $\hat{y}_t = 310.80 + 14.653t$
     c. 457.33
     d. $\ln(\hat{y}_t) = 5.756 + .037975t$
     e. 462.121182
19.S.9   **b.** $\hat{y}_t = 79.38333 + 3.616667t$
     c. $\hat{y}_t = 68.99047 - 9.285499t + .5668832t^2$
19.S.11   **a.** 60.9   **b.** 76.64   **c.** 15.74   **d.** 116.18
     e. 139.36   **f.** 23.18

Chapter 20

Section 20.1

20.1.1   False. The value is not centered correctly.
20.1.5   9-point moving average
20.1.7   **a.** No   **b.** Yes
20.1.9   **a.** 16   **b.** 21   **c.** 26

Section 20.2

20.2.1   False
20.2.3   Coefficients of the 3 dummy variables will be negative.
20.2.5   **b.** $\hat{y}_t = 1.494375 + .200625t + .379375Q_2 + .2787501Q_3 + 6.338125Q_4$
     c. $\hat{y}_t = 7.832500 + .200625t - 6.338125Q_1 - 5.95875Q_2 - 6.059375Q_3$
     e. $b_1 = .200625$ in both equations
     g. Quarter 1: $\hat{y}_t = 1.494375 + .200625t$; Quarter 2: $\hat{y}_t = 1.87375 + .200625t$; Quarter 3: $\hat{y}_t = 1.773125 + .200625t$; Quarter 4: $\hat{y}_t = 7.8325 + .200625t$
     h. Quarter 1: 5.7075; Quarter 2: 6.2875; Quarter 3: 6.3875; Quarter 4: 12.6475

Section 20.3

20.3.1   False
20.3.3   False
20.3.5   The trend equation is $\hat{y}_t = 3.363 + .059t$

Section 20.4

20.4.1   True
20.4.3 $c = .3$
20.4.5   **b.** Smoothed values for times 1–8: 9; 8; 10; 9; 9.5; 11.75; 13.875; 13.4375
     c. Each value is 13.4375.
     d. Smoothed values for times 1–8: 9; 7.2; 11.52; 8.352; 9.8352; 13.58352; 15.758352; 13.2758352
     e. Each value is 13.2758352.

20.4.7   **b.** Smoothed values for times 2–8: 6; 8.2; 8.56; 10.008; 13.4944; 13.59392; 18.17626. Trend values for times 2–8: 3; 2.6; 1.48; 1.464; 2.4752; 1.28736; 2.934848
     c. Forecasts for times 3–8: 9; 10.8; 10.04; 11.472; 15.9696; 14.88128
     d. 2.24272; 2.614535
     e. 21.1111; 24.04595; 26.9808
     f. Smoothed values for times 2–8: 6; 8.4; 9.24; 10.404; 13.1484; 13.90164; 17.53244. Trend values for times 2–8: 3; 2.7; 1.77; 1.467; 2.1057; 1.42947; 2.530137. Forecasts for times 3–8: 9; 11.1; 11.01; 11.871; 15.2541; 15.33111. MAFE = 2.193665; RMSE = 2.405011. Forecasts for periods 9–11: 20.06258; 22.59272; 25.12286
20.4.11   **b.** Smoothed values for months 2–5: 6; 12.6; 20.99; 23.954. Trend values for months 2–5: 1; 4.36; 6.78; 4.490.
     c. Forecasts for months 3–5: 7; 16.96; 27.77
     f. Smoothed values for months 2–5: 6; 11.2; 18.92; 23.532. Trend values for months 2–5: 1; 3.1; 5.41; 5.011. Forecasts for months 3–5: 7; 14.3; 24.33

Supplementary Exercises

20.S.1   **a.** $\hat{y}_t = 16.28 + .085t$
     b. .81, .85, .93, .87, 1.01, .94, .84, .95, .91, 1.21, 1.29, 1.39
     c.

	Year 1	Year 2	Year 3
Jan.	—	18.03	17.71
Feb.	—	18.05	17.78
Mar.	—	18.03	17.83
Apr.	—	17.95	17.88
May	—	17.81	17.88
June	—	17.70	17.90
July	17.98	17.68	—
Aug.	18.08	17.63	—
Sep.	18.11	17.56	—
Oct.	18.04	17.56	—
Nov.	18.01	17.60	—
Dec.	18.03	17.63	—

     g. $\hat{y}_t = 18.14 - .016t$ ($t = 1$ in January of Year 1)
     h. Jan.: 15.7343; Feb.: 16.5835; Mar.: 18.2234; Apr.: 17.1216; May: 19.9627; June: 18.6590; July: 16.7454; Aug.: 19.0190; Sep.: 18.2956; Oct.: 24.4299; Nov.: 26.1548; Dec.: 28.3004

20.S.3 **a.**

Quarter	Year 1	Year 2	Year 3	Year 4
1	—	4.3875	5.8000	7.7000
2	—	4.7000	6.2375	8.1625
3	3.8000	5.0125	6.7625	—
4	4.0750	5.3875	7.2625	—

    b. ln MA $= 1.016335 + .0793063t$ ($t = 1$ in Quarter 1 of Year 1)
    c. .87; .90; .94; 1.30
    d. 10.64; 11.52; 12.47; 13.50
    e. 9.26; 10.37; 11.72; 17.55

Chapter 21

Section 21.1

21.1.5 False
21.1.7 $P(X = 12) = .12012$; this value alone exceeds $\alpha = .05$, so do not reject H_0.
21.1.9 $z = 2.5$; reject H_0
21.1.11 $z = 2.1$; reject H_0

Section 21.2

21.2.1 True
21.2.5 **a.** $U = 380$ **b.** $z = 5.37$; do not reject H_0
21.2.7 **b.** $P(U \leq 7) = .0465$ **c.** $U = 13.5$
    d. $P(U \leq 13) = .2424$; do not reject H_0
21.2.9 $U = 73$; $z = -.27$; do not reject H_0

Section 21.3

21.3.1 True
21.3.3 True
21.3.7 True
21.3.9 **d.** $t_0 = 17$ **e.** Do not reject H_0
21.3.11 $z = -.40$; do not reject H_0
21.3.13 $T = 23$; reject H_0.
21.3.15 $T = 5$; do not reject H_0

Section 21.4

21.4.1 False
21.4.3 True
21.4.7 False
21.4.9 **b.** $K = 3.6437$ **c.** 3   **d.** Do not reject H_0.
21.4.11 $K = 6.8429$; do not reject H_0
21.4.13 $K = 14.9947$; reject H_0

Section 21.5

21.5.1 True
21.5.7 **d.** $z = -2.38$ **e.** Reject H_0

21.5.9 **b.** $r_0 = 11$; do not reject H_0
21.5.11 $z = -.37$; do not reject H_0
21.5.13 **c.** $z = .48$ **d.** Do not reject H_0

Section 21.6

21.6.3 **a.** .377   **b.** .745
21.6.5 **b.** $R_s = .885714$ **c.** Reject H_0
21.6.7 **b.** $R_s = -.3117$ **c.** Do not reject H_0
    d. $R = -.2945$
21.6.9 **a.** $R_s = .9527$ **b.** Reject H_0

Supplementary Exercises

21.S.1 **c.** $z = -1.41$; do not reject H_0
21.S.3 **a.** $z = 2.04$; reject H_0
    b. $z = -1.85$; do not reject H_0
21.S.5 $z = -2.00$; reject H_0
21.S.7 $z = -1.13$; do not reject H_0
21.S.9 $r = 5$; do not reject H_0
21.S.11 **a.** $z = -.63$; do not reject H_0
    b. $z = -1.73$; do not reject H_0
21.S.13 **a.** $R_s = .1189$ **b.** Do not reject H_0
21.S.15 **b.** $P(X \leq 7) = .617$; do not reject H_0
    c. Type II error   **d.** No

Chapter 22

Section 22.1

22.1.5 With good luck, a person might achieve a good outcome despite making an unwise decision.
22.1.9 **a.** No
    b. The opportunity losses are as follows:

0	17	18
4	9	7
13	11	0
11	0	10

22.1.11 **a.** $a_1 =$ rent now, $a_2 =$ wait and possibly rent later; $s_1 =$ win contract, $s_2 =$ do not win contract.
    Payoff table:

Actions	s_1	s_2
a_1	115,000	$-45,000$
a_2	88,000	0

    b. Opportunity loss table:

Actions	s_1	s_2
a_1	0	45,000
a_2	27,000	0

    c. No action is a dominated action.

Section 22.2

22.2.1 True
22.2.5 True
22.2.9 **a.** Action 1   **b.** Action 2   **c.** Action 2
          d. Action 2   **e.** Action 2
          f. EPPI = 5.2   **g.** EVPI = 1.0
22.2.11 **a.** a_4 **b.** a_2 or a_3 **c.** a_3 **d.** a_4 **e.** a_4
          f. EVPI = 52
22.2.13 **b.** Action a_2
          c. EPPI = 1320   **d.** EVPI = 300
22.2.15 **a.** Software   **b.** Mining   **c.** Optimist
22.2.17 **a.** Purchase 8 dozen suits.
          b. Purchase 4 dozen suits.

Section 22.3

22.3.1 False
22.3.5 Risk seeker
22.3.7 **a.** $EU_1 = 48.424$; $EU_2 = 75.064$; $EU_3 = 65.492$; $EU_4 = 66.172$
          b. Action 2
          c. $EMV_1 = .2$; $EMV_2 = 4.4$; $EMV_3 = 3.2$; $EMV_4 = 2.2$
          d. Action 2
22.3.9 **a.** $EMV(a_1) = 40$; $EMV(a_2) = 45.5$
          b. Action a_2
          c. Action a_2

Section 22.4

22.4.1 **b.** Buy Stock 4.
22.4.3 **b.** Optimal action is to wait and sell later.

Supplementary Exercises

22.S.1 Mike's optimal action is to decline the job. The present value is $66,000 more by going to graduate school. Yes, he should go to graduate school.
22.S.3 **c.** Action a_2, that is, to drill the well
          d. EVPI = 80,000
22.S.5 **b.** Optimal action is a_3, buy real estate; $E(\text{payoff}) = \$1,230$
22.S.7 **a.** a_2 **b.** a_4 **c.** a_4
22.S.9 The optimal action is a_2, invest.

Chapter 23

Section 23.1

23.1.5 Changes in machines, changes in the quality of employees, changes in temperature and humidity, and changes in the experience and education of employees are examples of assignable causes of variation.

Section 23.2

23.2.3 **b.** There is an upward trend.
          c. Perhaps the employee became tired, or bored, or stopped concentrating.

Section 23.3

23.3.3 False
23.3.7 **a.** UCL = 30.2119; LCL = 24.7881
23.3.9 **a.** $\bar{X} = 60.345$
          b. $\bar{R} = 7.495$
          c. UCL = 64.669615; LCL = 56.020385
          f. None of the sample means falls outside the control limits. Two consecutive observations, observation 18 and observation 19, both fall more than two standard deviations below the centerline.
23.3.11 **a.** UCL = 29.625; LCL = 29.425
23.3.13 .0274
23.3.15 Approximately 1.79% of the output would fall below the value of 370 and be unacceptable output.
23.3.17 **a.** UCL = 14.503104; LCL = 14.096896
          b. The specification limits fall outside the control limits; thus, almost all the output will fall between USL and LSL.
23.3.19 **c.** UCL = 183.74415; LCL = 173.41585

Section 23.4

23.4.5 **a.** UCL = 6.7888; LCL = 0
23.4.7 **a.** UCL = 8.034566; LCL = 0
23.4.9 **a.** UCL = 97.7944; LCL = 0
23.4.11 **b.** UCL = 20.72072; LCL = 0
          e. Two of the sample ranges exceed the UCL.

Section 23.5

23.5.1 **a.** UCL = .09817; LCL = .02583
          b. UCL = .10523; LCL = .01876
23.5.3 **b.** $\bar{p} = .05625$
          c. UCL = .10513; LCL = .00737
23.5.5 **a.** $n_i > 891$
          b. $n_i > 441$
          c. $n_i > 171$
          d. $n_i > 103.5$, or 104
23.5.7 **a.** $\bar{p} = .081$; UCL = .16285; LCL = 0
23.5.9 **a.** $\bar{p} = .05$; unweighted value is $\bar{p} = .1098$
23.5.11 For sample 1: UCL = .17683; LCL = $-.0313$;

the proportion of defectives is $\bar{p}_1 = .125$, which falls within the control limits.

For sample 2: UCL $= .13356$; LCL $= .01189$; the proportion of defectives is $\bar{p}_2 = .152$, which falls outside the control limits.

For sample 3: UCL $= .12525$; LCL $= .02020$; the proportion of defectives is $\bar{p}_3 = .0182$, which falls outside the control limits.

Supplementary Exercises

23.S.1 **a.** UCL $= 12.263$; LCL $= 0$
       b. UCL $= 28.0466$; LCL $= 21.3534$
       d. $\bar{X} = 31.62$. This value falls outside the control limits; it appears that the process may be out of control.

23.S.3 **a.** $\bar{R} = 6.828$; UCL $= 14.4368$; LCL $= 0$
       b. None of the sample ranges falls outside the control limits.

c. $\bar{X} = 59.972$; UCL $= 63.9118$; LCL $= 56.0322$
d. Five of the sample means fall below the LCL.

23.S.5 **a.** $\bar{R} = 1.2205$; UCL $= 2.5806$; LCL $= 0$
       b. One of the sample ranges falls outside the control limits.
       c. $\bar{X} = 7.9672$; UCL $= 8.67143$; LCL $= 7.26297$
       d. One of the sample means falls outside the control limits.
       e. Yes

23.S.7 **a.** $\bar{R} = .011$; UCL $= .0251$; LCL $= 0$
       b. None of the sample ranges falls outside the control limits.
       c. $\bar{X} = .30022$; UCL $= .308239$; LCL $= .292201$
       d. None of the sample means falls outside the control limits.
       e. No

Index

Acceptability versus uniformity, 981
Acceptance region, 391, 399
Accuracy in forecasting, 851, 855
Actions in decision theory, 945
Actual process spread, 132
Actual value in regression, 634
Additive law of probability, 147, 148–149
Additive time-series model, 823
Adjusted coefficient of determination, $\bar{R}^2$, 684
Age discrimination, 534
Age Discrimination in Employment Act of 1967, 534
Air conditioners, shipments, 818, 871
Allowable process spread, 132
Alpha, level of significance, 400, 411
Alpha, α, probability of Type I error, 400, 438
Alternative hypothesis, 393
Americans with Disabilities Act of 1990, 535
Analysis of variance, 546
  F-statistic, 557
  mean square between, 555–556
  mean square within, 555
  one-factor model, 546
  one-way table, 560
  statistical model, 565
  sum of squares between, 553
  sum of squares identity, 554
  sum of squares total, 552
  sum of squares within, 553
Arithmetic mean, 81
Assignable causes of variation in quality control, 979, 984
Assumptions of regression models, 596, 667–668
Asymptotic test for serial correlation, 790
Attitudes toward risk, 966–967
Autocorrelation, 784; *see also* Serial correlation, 784
Autoregressive forecasting model, 846
Autoregressive models, 803, 846

Average, 81, 83
  arithmetic mean, 81
  median, 83, 291, 293–294
  mode, 84
Average range in quality control, 1012

Bar charts, 54
Base case for dummy variables, 715
Basic assumptions of the multiple regression model, 667–668, 676, 761–762, 772, 784, 801
Basic assumptions of the simple linear regression model, 596
Basic outcome, 134
  assigning probabilities, 140
Bayes, Thomas, 167, 955
Bayes' theorem, 168–170
Beer production, 877
BEGIN DATA command in SPSSX, 27
Bernoulli, Daniel, 967
Best linear unbiased estimators in regression, 679
Beta, β, probability of Type II error, 400, 438
Between-groups mean square in ANOVA, 556
Between-groups variation in one-way ANOVA, 553
Bias, 14, 293
  of nonresponse, 16
  of self-selection, 17
Biased sample, 14
Bimodal distribution, 62
Binomial distribution, 220
  characteristics, 220
  continuity correction, 273, 323
  cumulative probability, 225
  formula, 222
  graphs, 224
  mean, 224
  normal approximation, 272, 324
  Poisson approximation, 235
  shape, 223
  standard deviation, 224

  table of probabilities, 1039
  use of sign test, 901–902
  use of table, 225
  variance, 224
Bivariate analysis, 70
Bivariate data set, 22, 70
Bivariate frequency distribution, 70
Box plots, 115
BREAKDOWN command in SPSSX, 330, 501
Business cycle, 821

Card selection, 138
CARDS command in SAS, 30
Careers in Statistics, 6
CASEWISE command in SPSSX, 805–806
Causes of variation in quality control, 979
  assignable, 979
  random, 979
Cell, 520
Cell frequency, 521
Census, 7
Centered 4-quarter moving average, 868
Centered 12-month moving average, 869
Centerline in quality control, 983, 1012, 1020
Central Limit Theorem, 310
Chebyshëv, Pafnuti Lvovich, 108
Chebyshëv's theorem, 108, 110
Chi-square distribution, 461
  characteristics, 461
  critical value, 462–463
  degrees of freedom, 461
  mean, 461
  table of probabilities, 1047
  test statistic, 511
  use of table, 462
  variance, 461
Chi-square goodness-of-fit test, 509
  with estimated parameters, 517
  expected frequencies, 511
  normal distribution, 515

Chi-square goodness-of-fit test (*continued*)
  observed frequencies, 511
  Poisson distribution, 513
  proportions, 509
  test statistic, 511
Chi-square test of independence, 520
  contingency table, 520
  degrees of freedom, 522
  expected frequencies, 521
  observed frequencies, 521
  test statistic, 522
Chi-square test statistic, 511
Cigarette smoking, effects, 630
Class frequency, 43
Class limits, 44–45
Class mark, 44
Class midpoint, 44
Class width, 44
Classes, 43
  number, 45
  open-ended, 44
  width, 44
Classical normal linear regression model, 596, 667–668
Cluster sampling, 18
Coal, value of, 704
Coefficient of multiple determination, R^2, 610, 683
Coefficient of variation, 111–112
Coin tossing, 138
Combinations, 175–177
Complement of an event, 145
  probability of, 146
Completely randomized design in ANOVA, 550, 565
Composite null hypothesis, 393, 410
Compound events, 145
COMPUTE command in SPSSX, 751–752, 859
Conditional mean in regression, 634
Conditional probability, 153, 156
Confidence coefficient, 339
Confidence intervals, 339
  conditional mean in regression, 635
  desirable properties, 345
  difference between two means, 377
  difference between two proportions, 381
  hypothesis test interpretation, 415
  large samples, 347, 361
  left-sided, 348, 363, 369
  lower class limit, 348, 369
  means, 342, 357
  one-sided, 348, 363, 369

Confidence intervals (*continued*)
  predictions in regression, 635
  proportions, 367
  ratio of two variances, 499
  regression coefficient, 621, 688
  regression intercept, 629
  regression slope, 629
  right-sided, 348, 363, 369
  sample size required, 373, 374–375
  small samples, 357
  unknown population variance, 357
  upper class limit, 348, 369
  variance, 464
  width, 345
Confidence limits, 348, 363, 369
  lower, 348, 363, 369
  upper, 348, 363, 369
Contemporaneously uncorrelated variable, 802
Contingency table, 520–521
Continuity correction, 272, 323
Continuous probability distributions, 248
  general characteristics, 248
  mean, 251–252
  variance, 251–252
Continuous random variable, 193, 248
Continuous variable, 9
Control, 923
  in control, 923
  out of control, 923, 994
Control chart, 995–999
  p-chart, 1019
  R-chart, 1011
  $\bar{x}$-chart, 1001
Control limits in quality control, 991
  estimation, 993
  lower control limit, LCL, 991–993
  upper control limit, UCL, 991–993
  versus specification limits, 1000, 1008
Convenience sample, 19
CORR command in SPSSX, 746, 750, 752
Correlation, 641
  causality, 650
  formula, 642, 645
  inappropriate use, 649–650
  matrix, 699, 745
  perfect, 643
  population, 641–642
  rank, 928
  sample, 641, 645

Correlation (*continued*)
  serial, 784, 878
  spurious, 649
  and statistical significance, 651
  testing hypothesis, 648
Cost analysis, 401
Counting techniques, 172
Covariance, 642
  population, 642
  sample, 642
  standardized, 642
Critical region, 399; *see also* Rejection region, 399
Critical values, 399, 406, 494
Cross-section data, 785
CROSSTABS command in SPSSX, 70–71, 180, 530
Cubic polynomial, 709
Cumulative distribution function, 199–200
Cumulative frequency, 50
Cumulative frequency distribution, 50
Cumulative relative frequency, 50
Cumulative relative frequency distribution, 50
Current Population Reports, 43, 509
Current Population Survey, 710, 719
Cyclical variation, 813, 821, 987

Data acquisition, 15
  personal interview, 15
  self-enumeration, 16
DATA command in SAS, 30
DATA LIST command in SPSSX, 26
Data set, 22
Data value, 22
Decision criteria, 952
  expected monetary value, 953, 955–956, 959
  expected opportunity loss, 953, 957, 959
  Laplace criterion, 953–954
  maximax, 953
  maximin, 953
  minimax, 953–954
  strategy of insufficient reason, 953–954
Decision making, 946, 952
  under certainty, 946
  under uncertainty, 946–948
Decision nodes, 972
Decision rule, 398–399
Decision tree analysis, 971
Deductive statistics, 5
Defective items, 980

Degrees of freedom, 101
  chi-square distribution, 461
  contingency table, 522
  F distribution, 493
  goodness-of-fit test, 509
  in multiple regression model, 680
  in one-way ANOVA, 557
  t-distribution, 353
  test of independence, 522
Deming, W. Edwards, 978
DeMoivre, Antoine, 258
Density functions, 249–250
  characteristics, 250
DEPENDENT command in SPSSX,
  702, 741, 859
Dependent events, 161–162
Dependent variable, 549, 584
  in ANOVA, 549
  heteroscedasticity, 783
  in regression, 584
Descriptive statistics, 1, 42
DESCRIPTIVES command in
  SPSSX, 119, 330, 383, 469–
  470, 746, 750, 752
Deseasonalizing of time series, 875,
  877
Desirable properties of estimators,
  290
Destructive sampling, 13
Deterministic relationship, 586
Deviations from the mean, 97, 211
Dichotomous variable, 8
Die tossing, 138
Difference between two means, 316,
  377, 476, 480
  paired samples, 486
Difference between two proportions,
  327, 381, 490
Direct relationship, 585
Discrete probability distribution, 195
Discrete probability function, 192
Discrete random variable, 193
  expected value, 202–203
  mean, 203
  standard deviation, 208
  variance, 208
Discrete variable, 9, 195
Dispersion, measures of, 95
Distribution-free statistics, 901
Dominated actions, 948
Dot diagram, 548–549
Double logarithmic transformation,
  735
Draft lottery, 138

Dummy variable trap, 731
Dummy variables, 714, 877
  base case, 715
  broad groupings of quantitative
    variables, 715
  forecasting, 877
  qualitative variables, 715
  qualitatiave variables with more
    than two cases, 725
  seasonal adjustment, 877
  spatial effects, 715
  temporal effects, 715
DURBIN command in SPSSX, 806
Durbin, James, 790
Durbin–Watson statistic, 790
Durbin–Watson table, 1061
Durbin–Watson test, 790–791

Econometric Methods, 787
Economic Review, 803
Economic Report of the President, 35,
  741, 815, 822, 833–836, 839,
  845, 852, 864
Efficient estimator, 295
Electric power, 818
Elementary units, 8
Emotional presentation of facts, 67
Empirical Rule, 109–110, 259, 269
END DATA command in SPSSX, 27
ENTER command in SPSSX, 702, 859
Equal Employment Act of 1972, 534
Equally likely approach to assigning
  probability, 137
Error of estimation, 16, 290
  sampling error, 16, 290, 339
Error sum of squares in ANOVA, 553
Estimate, 78, 289
Estimator, 78, 289
  criteria for choosing, 290
  efficient, 295
  median, 83, 293–294
  minimum variance unbiased, 296
  population mean, 82
  population proportion, 321
  population standard deviation, 100
  population variance, 99
  sample standard deviation, 100
  sample variance, 100
  unbiased, 101, 292–293, 339
Events, 134
  probability of, 142
  simple, 142
Expected frequency in chi-square test,
  510–511

Expected monetary value, EMV, 955–
  956, 959
Expected monetary value criterion,
  953, 955–956, 959
Expected opportunity loss, EOL, 953,
  957, 959
Expected payoff from perfect information, EPPI, 957–958
Expected utility criterion, 962, 968
Expected value, 283, 293
  continuous random variable, 252
  discrete random variable, 202–203
  function of a random variable, 205
  median, 293–294
  sample mean, 293
  sample proportion, 321
Expected value of perfect informa-
  tion, EVPI, 958
Expenditure on recreation, 589
Experiment, 133
Experimental design in ANOVA, 549
Explained variation in regression
  analysis, 610–611
Explanatory power of a regression
  model, 609
Exponential distribution, 61
Exponential smoothing, 888
  forecasting, 890
  Holt–Winters, 891
  two-parameter, 891
Exponential trend model, 841
Extrapolation, 639

F distribution, 493
  critical values, 494
  degrees of freedom, 493–494, 557
  shape, 494
  table of probability, 1048
F-test in multiple regression, 691
Factor in ANOVA, 549
Factor level in ANOVA, 550
Factorial notation, 172
Farm population, 815
Feasible Prais–Winsten estimator,
  798
Feasible weighted least squares, 781
Fences in box plots, 115
File, 22
FINISH command in SPSSX, 27
Finite population correction factor,
  306
First-order autoregressive scheme, 786
Fisher, R. A., 75, 390, 493
Fitted value in regression, 678

Forecast accuracy, 851, 855
Forecasting, 588
　　with autoregressive trend model, 849
　　with dummy variables, 877, 881
　　with the exponential smoothing model, 890
　　with the ratio-to-trend model, 883, 887
　　with the two-parameter exponential model, 891
Forecasting in regression, 588
Four-quarter moving average, 868
Frame, 8
FREQUENCIES command in SPSSX, 27, 48, 68–69, 80, 179, 213
Frequency
　　cumulative, 50, 264
　　cumulative relative, 50, 264
　　relative, 50
Frequency distribution, 43
　　bimodal, 62
　　common forms, 59–60
　　cumulative, 50
　　cumulative relative, 50
　　integer data, 51
　　open-ended classes, 44
　　qualitative data, 54
　　relative, 45
　　skewed, 59, 301
　　symmetrical, 59
Fundamental rule of counting, 172

Gallup poll, 17
Gauss, Karl F., 258
Gauss–Markov theorem, 679
Geis, I., 67
Goldfeld–Quandt test, 775
Goodness-of-fit in regression, 683
Goodness-of-fit tests, 509
Gosset, William, 355
Grand mean in quality control, 991
Graphical comparison of mean, median, and mode, 86–88
Graphical explanation of one-way ANOVA, 568–569
Graphing frequency distributions, 47
Griffiths, William E., 669
Gross business savings, 847
Gross Domestic Product, 815
Gross investment, 815

Handbook of Statistical Tables, 6
Heteroscedasticity, 772–774, 776
　　dependent variable, 783

Hill, R. Carter, 669
Histogram, 47
　　frequency, 43
　　graphing, 47
　　relative frequency, 45, 48
HISTOGRAM command in SPSSX, 48, 69
Holt–Winters exponential smoothing model, 891
Homoscedasticity, 596, 677, 772–773, 775
Housing starts, 822
How to Lie with Statistics, 64
Huff, D., 64, 191
Hurwicz criterion in decision making, 953
Hypothesis testing, 391
　　all regression slope coefficients jointly, 691
　　difference of two means, 476–477, 480
　　difference of two proportions, 490
　　homogeneity of variances, 775
　　paired differences, 486
　　population correlation coefficient, 648
　　population mean, 405, 425, 428
　　population proportion, 433
　　population variance, 464
　　ratio of two variances, 497
　　regression coefficient, b_i, 689
　　regression intercept, b_0, 625
　　regression slope, b_1, 623
　　subset of regression coefficients, 691

IF command in SPSSX, 751
Incorrect functional form, 765
Independence, 161
Independent events, 161
　　probability, 161
Independent variable
　　in ANOVA, 549
　　in regression, 584
Inductive statistics, 5
Inferential statistics, 2
Infinite sample spaces, probability, 143
Inner fences, 115
INPUT command in SAS, 29
Integer data, 51
Intercept of the regression line, 586
　　confidence interval, 629
　　formula, 599, 604
　　testing hypotheses, 625
Intercept of a straight line, 586

Internal Revenue Service, 38
Interquartile range, 97, 116–117
Intersection of events, 147
Interval data, 11
Interval estimation, 339
Interviewer bias, 17
Inverse relationship, 585
Irregular variation in time series, 813, 823

Johnston, John, 787
Joint frequency distribution, 153–154
Joint probabilities, 153–154
Joint probability table, 154, 169
Joint relative frequency distribution, 154
Joint test, 691
Judge, George G., 669
Judgment sample, 20

Kruskal, W. H., 918
Kruskal–Wallis test, 901, 918
　　computer output, 937

LAG command in SPSSX, 859
Lagged variables, 802
Laplace, Pierre, 953–954
Laplace criterion, 953–954
Leaf, 66
Learning and Practicing Econometrics, 669
Least squares method, 597
Left-sided confidence interval, 348, 363, 369
Level of confidence, 339, 342
Level of significance, 392, 400, 411
Life Tables, 235
Linear autoregressive model, 846
Linear regression model, 588
Linear trend model, 825
Literary Digest, 14, 16
Logarithmic models, 734
Lottery gambling, 138
Lower confidence limit, 348, 363
Lower control limit, 991–993
Lower specification limit, 131, 980

Mann, H. B., 906
Mann–Whitney test, 901, 906
　　computer output, 934–935
　　large samples, 908
　　small samples, 910
　　table, 1051
Marginal probability, 155
Marginal utility, 963

Index

Markov, A., 679
Markowitz, Harry M., 209
Matched pairs, 486
Matrix, correlation, 699, 745
Maximax criterion, 953
Maximin criterion, 953
McClesky v. Kemp, 539
Mean, 81
  continuous random variable, 252
  discrete random variable, 203
  geometric interpretation, 86
  population, 82
  sample, 82
  of the sample proportion, 321
Mean absolute deviation, 98, 118
Mean absolute forecast error, MAFE, 855, 894–895
Mean absolute percentage error, MAPE, 855–856
Mean square between in ANOVA, 555–556
Mean square due to treatments, 556
Mean square error in ANOVA, 555
Mean square within, MSW, 555
Measures of central tendency, 42
Measures of dispersion, 42, 95
Measures of location, 77
Measuring goodness of fit in regression, 683
Median, 83
  sampling distribution, 295–296
Mellon Bank, 37, 122, 502, 658
Method of least squares, 597
Method of moving averages, 867
  centered, 867
  unweighted, 868
  weighted, 867–868
Midpoint of class, 44
A Million Random Digits, 21
Minimax criterion, 953–954
Minimum variance unbiased estimator, 296
MINITAB computer program, 6, 532–533
Mode, 62, 84
Model, 133
Model inadequacies, 757
Morgenstern, Oskar, 977
Most efficient unbiased estimator, 296
Moving averages, 867
  centered, 867
  four-quarter, 868
  twelve-month, 868
  unweighted, 868
  weighted, 868

Multicollinearity, 668, 677, 691, 696, 764
  in the correlation matrix, 699, 749
  in macroeconomic models, 698
Multinomial variable, 8
Multiple regression models, 588, 667, 676
Multiplicative law of probability, 157–158, 164
Multiplicative time-series model, 823
Multivariate analysis, 22
Multivariate data set, 22
Multivariate file, 22
Mutually exclusive classes, 44
Mutually exclusive events, 149

National Football League, 67
Negative income tax system, 507
Node of decision tree, 972
Nominal data, 9
Nonlinear relationship, 766
Nonnormal errors, 761
Nonparametric statistics, 901
  Kruskal–Wallis test, 901
  Mann–Whitney test, 901, 906
  rank correlation coefficient, 902, 928
  runs test, 901, 922
  sign test, 901–902
  Wilcoxon signed-rank test, 901, 912
Nonprobability sample, 19
Nonresponse bias, 16
Normal distribution, 60, 258
  approximation to binomial, 272–275
  characteristics, 258
  continuity correction, 273, 323
  cumulative probabilities, 264
  formula, 258
  mean, 258
  median, 258
  shape, 258
  standard normal distribution, 259, 354
  table of areas, 1045
  variance, 258
NPAR TESTS command in SPSSX, 934, 936–938
Null hypothesis, 391, 393

Objective probability, 136, 139
Observation, 22
Observed frequencies in chi-square test, 510
Observed level of significance, *See P*-value, 420, 559, 576

Odds, 140
Oil drilling, 4
Omitted variables in regression, 763
One-sided confidence intervals, 348, 363, 369
One-sided test, 394
One-way ANOVA model, 565–566
  equivalence to *t* test, 569
ONEWAY command in SPSSX, 575
Open-ended classes, 44
Opportunity loss, 945, 949–950, 959
  table, 945, 959
OPTIONS command in SPSSX, 180
Ordinal data, 10
Oscillating time series data, 986
Outcomes, 134
Outer fences, 116
Outlier residuals, 759
Outliers, 111, 117, 759

p-chart in quality control, 1019
P-value, 419–420, 559, 576
Paired difference test, 486
Panel data, 785
Parabola, 709
Parameters, 14, 77, 288
Pattern analysis, 994, 1005, 1017, 1023, 1026
Payoffs, 945–946
  table, 945–947
Pearson, Karl, 509
Percentiles, 91
Perfect multicollinearity, 668, 677, 696
Permutations, 174–175
Personal interviews, 15
Pie charts, 54
Pittsburgh Press Club, 38
Plot of residuals, 788–789, 795
Point estimate, 339
Poisson distribution, 231
  approximation of binomial, 235
  characteristics, 231–232
  formula, 232
  mean, 233
  shape, 234
  table of probabilities, 1043
  test of distribution, 513
  use of table, 234
Poisson process, 231
Poisson, Simeon, 231
Polynomial models, 709–710, 836
  cubic, 709, 836
  quadratic, 710, 836
  second-degree, 709–710, 836
Polynomial trend model, 836

Pooled estimate of *p,* 490
Pooled estimate of variance, 481
Pooled time series and cross-sectional data, 785
Population, 7
Population correlation coefficient, 641–642
Population covariance, 642
Population cumulative distribution function, 50, 199, 264
Population mean, 82
Population median, 83, 295–296
Population parameters, 78, 288
Population proportion, 321
Population rank correlation coefficient, 928
Population regression equation, 593, 677
Population regression line, 593
Population regression plane, 668, 677
Population standard deviation, 99–100
Population of the United States, 815, 829–830
Population variance, 99
Portfolio Selection, 209
Portfolios, 209
Posterior probability, 167
Power curve, 446, 448, 450, 453, 456
Power of a test, 438–439
Practical significance, 401, 416, 449
Prais, S. J., 796
Prais-Winsten estimator, 796
Prais-Winsten transformation, 797
Predicted value in regression, 634, 678
Prediction using the regression model, 634
Pre-test bias, 758
Prior probabilities, 167, 955
Private housing starts, 822
Probability, 133
    of a basic outcome, 140, 142
    basic rules, 142
    of the complement of an event, 146
    of compound events, 145
    conditional, 153, 156
    of equally likely events, 137
    of an event, 142
    independent event, 161
    infinite sample spaces, 143
    joint, 154
    marginal, 155
    multiplicative law, 157–158
    objective, 136, 139
    posterior, 167
    prior, 167

Probability (*continued*)
    of the sample space, 143
    subjective, 136, 139
    of the union of events, 147–148
Probability density function, 250
Probability sample, 19
Probability of Type I error, 400, 438
Probability of Type II error, 400, 438
PROC CHART command in SAS, 29
PROC FREQ command in SAS, 531
PROC MEANS command in SAS, 120
Process, 979
Process capability, 132
Process proportion in quality control, 1019
Proportion, 321
    confidence interval, 367
    estimation of, 321

Quadratic polynomial, 710, 836
Quadratic regression model, 710
Qualitative population, 8
Qualitative variable, 8
Quality control, 978
Quality control exercises, 245, 581–582
Quantitative population, 8
Quantitative variable, 8
Quartiles, 91
    lower, 93
    middle, 93
    upper, 93
Quota sampling, 17

R^2, coefficient of determination, 610, 684
    alternative formula, 611
$\bar{R}^2$, adjusted coefficient of determination, 684–685
R-chart in quality control, 1011
Ragan, James, F., 803
Random causes of variation in quality control, 979
Random disturbance term, 594, 678
Random error term in regression, 594, 678
Random experiment, 133
Random numbers, 20
    table of, 1038
Random sample, 290
    cluster, 18
    sequential, 18
    simple, 20, 290
Random sampling, 20

Random sampling (*continued*)
    with replacement, 163, 165, 226, 306
    without replacement, 163–164, 226, 306
Random variable, 192
    continuous, 193, 248
    discrete, 193, 195
    sample space, 193
Random variation in time series, 813, 823
Range, 96, 118
    average range in quality control 1012
    interquartile, 97
    semi-interquartile, 97
    trimmed, 97
Rank correlation, 928
    alternative formula, 928
    formula, 928
    table of critical values, 1060
    testing with large samples, 929
    testing with small samples, 930
Ratio data, 11
Ratio-to-moving average method, 870
Ratio-to-trend method, 883
Rational subgrouping in quality control, 991
Raw data, 5
Reasons for sampling, 13
    accuracy, 13
    destructive sampling, 13
    expense, 13
    infinite number of observations, 13
    speed, 13
Recidivism, 507
Regression analysis, 584
    goals, 587
REGRESSION command in SPSSX, 655, 702, 741, 746, 806, 859
Regression line, 593
    population, 593
    sample, 594
Regression model with two explanatory variables, 667
Regression sum of squares, SSR, 610
Rejection region, 399
Relationships, 584
    deterministic, 584, 586
    direct, 585
    stochastic, 584, 586
    inverse, 585
    negative, 585
    positive, 585
Relative efficiency of a test, 295

Relative frequency, 45
  approach for assigning probability, 136
  distribution, 45
Relative standard deviation, 112
Residual analysis, 595, 757–758
Residual plot, 788–789, 795
Residual sum of squares, 598
Residuals, 595, 598
RESIDUALS command in SPSSX, 805–806
Response variable in ANOVA, 549
Retail sales, 818, 826, 883
Right-sided confidence interval, 348, 363, 369
Risk, 209, 966
  attitudes toward, 966
  aversion, 966
  neutral, 966
  seeking, 967
Root mean square error, RMSE, 855–856, 894–895
Roulette, 152, 204, 206
Rounding inaccuracies, 604
Rules for pattern analysis, 994
Run, 922
Runs test, 922
  computer output, 938
  large samples, 923
  small samples, 925

St. Petersburg paradox, 967
Salk polio vaccine, 507
Sample, 7
  cluster, 18
  convenience, 19
  judgment, 20
  random, 20
  sequential, 18
  simple random, 20
Sample correlation coefficient, 645
Sample correlation matrix, 699
Sample covariance, 642
Sample mean, 82, 118
  sampling distribution, 299
Sample median, 83, 118, 295–296
Sample proportion, 226, 321
  mean, 227, 321
  sampling distribution, 321
  variance, 227, 322
Sample rank correlation coefficient, 928
Sample regression equation, 594, 678
Sample regression line, 594
Sample regression plane, 668, 678

Sample size required for confidence intervals, 372
  determination of, 372
  to estimate the population mean, 372–373
  to estimate the population proportion, 374
Sample size required in a quality control p-chart, 1022
Sample space, 134, 143
  infinite, 143
Sample standard deviation, 100, 118
Sample statistic, 288
Sample trend line, 826
Sample variance, 100, 118
Sampling, 13
  reasons for, 13
  with replacement, 163, 165, 226, 306
  without replacement, 163–164, 226, 306
Sampling bias, 14
Sampling design, 7
Sampling distribution, 288–290
  difference between two means, 316, 377, 477, 480
  difference between two proportions, 327, 381
  sample mean, 295–296, 299
  sample median, 295–296
  sample proportion, 321
Sampling error, 16, 290, 339
Sampling procedures, 15, 19
SAS computer program, 6, 28–29
Savage, L. J., 953–954
Savage criterion in decision making, 953–954
Scales of measurement, 9
Scatter diagram, 584, 759, 768–769, 774–775, 777–778, 794
SCATTERGRAM command in SPSSX, 35–36, 655
Seasonal adjustment, 867, 875
  ratio-to-moving average, 870, 875
  ratio-to-trend method, 883
  using dummy variables, 877
  using moving averages, 867, 875
Seasonal index, 870
  air conditioners, 871
  beer production, 877
  retail sales, 883
  using dummy variables, 877
  using moving averages, 870
  using ratio-to-moving average, 870
  using ratio-to-trend, 883

Seasonal index numbers, 870
Seasonal variation, 813–814, 867
Second-degree polynomial, 709–710, 836
Selection bias, 17
Self-enumeration, 16
Self-selection bias, 17
Semi-interquartile range, 97
Semilogarithmic transformation, 735
Sensitive questions, 16
Sequential sampling, 18
  cluster sampling, 18
Serial correlation, 596, 784
  detection, 788
Sex discrimination, 154
Shapes of distribution, 60–61
  bimodal, 62
  exponential distribution, 61
  normal distribution, 60, 258–259
  skewness, 59, 130–131
  symmetry, 59
  uniform distribution, 61, 254
  unimodal, 63
Shewhart, Walter, A, 978
Shipments of air conditioners, 871
Sign test, 902
  large samples, 904
  small samples, 903
Simple centered moving average, 867
Simple hypothesis, 393
Simple linear regression model, 588, 593
Simple random sample, 20
Simultaneous equation model, 802
Skewness, 59, 130–131
Slope of regression line, 586
  confidence interval, 629
  formula, 598, 604
  testing hypotheses, 623
Slope of a straight line, 586
Smoothing constant, 888
Smoothing methods, 888, 890
Spatial effects and dummy variables, 715
Spearman, Charles E., 928
Spearman rank correlation coefficient, 928
  alternative formula, 928
  formula, 928
  sample, 928
  table of critical values, 1060
  testing, 929–931
Specification error in regression, 763

Specification limits, 131, 980
  lower, 980
  upper, 980
  versus control limits, 1000, 1008
Speed of response, 13
SPSSX computer program, 6, 26–27
Spurious correlation, 649
SSB in ANOVA, 553
SSE in regression, 598, 604, 678
SSR in regression, 610
SST in ANOVA, 552
Standard deviation, 100
  discrete random variable, 208
  population, 100
  sample, 100
Standard error, 309
Standard error of the regression, 603, 680
  estimate, 603
  of regression intercept, 616
  of regression slope, 616
Standard normal curve table, 1045
Standard normal distribution, 259, 354
Standardized residuals, 760, 768–770
Standardized score, 110, 118
Standardizing transformation, 266
State nodes, 972
States of nature, 946
Statistical Abstract of the United States, 32, 57–58, 200, 830
Statistical process control, 978
Statistical quality control, 978
Statistical significance, 402, 416, 449
Statistics, 1
  deductive, 5
  definition, 5
  descriptive, 1
  inductive, 5
  inferential, 2
STATISTICS command in SPSSX, 213, 530
Stem, 66
Stem-and-leaf diagram, 65, 116
STEPWISE command in SPSSX, 741, 752
Stepwise regression, 738
Stochastic relationship, 584, 586
Stock of currency, 841
Strategy of insufficient reason, 953–954
Student's *t*-distribution, 352, 430
Subgroups in quality control, 991
  rational subgroups, 991

Subjective probability, 136, 139–140
Summary statistics, 42, 118
Summation notation, 78
Sum of squares decomposition in regression, 609
Sum of squares identity in one-way ANOVA, 554, 566
Sum of squares in one-way ANOVA
  between, 553
  total, 552
  within, 553
Sums of squares in regression, 609–610
SSB in ANOVA, 553, 567
SSE, 598, 604, 610, 678
SSR, 610
SST in ANOVA, 552, 567
SST in regression, 609
SSW in ANOVA, 554, 567
Survey of Current Business, 818, 827, 848, 884
Symmetric distribution, 59

t-distribution, 352, 430
  characteristics, 353
  degrees of freedom, 353
  shape, 353–354
  table of probabilities, 1046
  use of table of probabilities, 355
TABLES MEANS command in SAS, 531
Tally sheet, 47
Temporal effects and dummy variables, 715, 877
Test, *see* Hypothesis testing
Test statistic, 391, 399
Tests of independence, 520
Third-degree polynomial, 709, 836
Time deposits, 852
Time plots in quality control, 985
Time series analysis, 813
  cyclical variation, 813, 821
  random variation, 813, 823
  seasonal variation, 813–814
  trend, 813–814
Time-series data, 785, 983
TITLE command in SPSSX, 26
Total quality management, 979
Total sum of squares
  in ANOVA, 552
  in regression, 609
Transistivity in utility theory, 963
Treatment sum of squares, 553, 567
Treatments in ANOVA, 550

Tree diagram, 163–164, 170, 173, 972–973
Trend, 813–814
Trend line, 826
Trend model
  autoregressive, 846
  exponential, 841
  linear, 825
  polynomial, 836
Trimmed range, 97
Tukey, J. W., 66
Twelve-month moving average, 869
Two-sided alternative hypothesis, 394, 412
Two-sided test, 394, 412
Two-tailed test, 394, 412
Type I error, 392, 399–400, 438
Type II error, 392, 399–400, 438

U. S. Supreme Court, 539
Unbiased estimator, 101, 292–293, 339
Uncertainty, decision making, 946–948
Uniform distribution, 61, 254
  formula, 254–255
  characteristics, 254
  mean, 255
  median, 255
  shape, 254
  standard distribution, 255
  variance, 255
Uniformity versus acceptability, 981
Unimodal distribution, 63
Union of events, 147
  probability, 148
Univariate analysis, 22
Unweighted moving average, 867
Upper confidence limit, 348, 363
Upper control limit, 991–993
Upper specification limit, 131, 980
USA Today, 400, 630
Utility, 963
  expected utility criterion, 963, 968
Utility index, 963–964
  assumptions, 963
Utility of money, 967
  marginal, 968
Utility theory, 963

Value of coal, 704
VALUE LABELS command in SPSSX, 27
Variable, 8

VARIABLES command in SPSSX, 746, 750
Variance, 99
    confidence interval, 464, 467
    of a discrete random variable, 208
    hypothesis test, 464–465
    pooled estimate, 481
    population, 99
    sample, 100
    of the sample proportion, 322
Venn, John, 134
Venn diagrams, 134–135, 146, 148–149
Vital Statistics of the United States, 235
Von Neumann, John, 977

Wainer, Howard, 76
Wald, Abraham, 953–954
Wald criterion, 953–954
Wall Street Journal, 58
Watson, Geoffrey, 790
Weighted least squares, 780
    feasible, 781
Weighted moving average, 868
Whitney, R., 906
Width of a confidence interval, 345
Wilcoxon, F., 912
Wilcoxon signed-rank test, 901, 912
    computer output, 936
    large samples, 914

Wilcoxon signed-rank test (*continued*)
    small samples, 916
    table, 1056
Winsten, C. B., 796
Winters, Holt–Winters exponential
    smoothing model, 891
Within-group mean square in ANOVA, 555

$\bar{x}$-chart in quality control, 990, 1001

Y-intercept, 586

z-score, 110–111, 259–260

Areas under the standard normal distribution

The following table gives the areas
under the standard normal curve
from 0 to z.

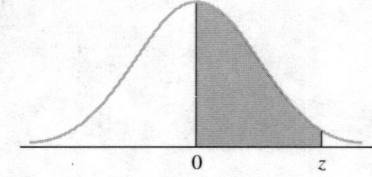

z	0	1	2	3	4	5	6	7	8	9
0.0	.0000	.0040	.0080	.0120	.0160	.0199	.0239	.0279	.0319	.0359
0.1	.0398	.0438	.0478	.0517	.0557	.0596	.0636	.0675	.0714	.0754
0.2	.0793	.0832	.0871	.0910	.0948	.0987	.1026	.1064	.1103	.1141
0.3	.1179	.1217	.1255	.1293	.1331	.1368	.1406	.1443	.1480	.1517
0.4	.1554	.1591	.1628	.1664	.1700	.1736	.1772	.1808	.1844	.1879
0.5	.1915	.1950	.1985	.2019	.2054	.2088	.2123	.2157	.2190	.2224
0.6	.2258	.2291	.2324	.2357	.2389	.2422	.2454	.2486	.2518	.2549
0.7	.2580	.2612	.2642	.2673	.2704	.2734	.2764	.2794	.2823	.2852
0.8	.2881	.2910	.2939	.2967	.2996	.3023	.3051	.3078	.3106	.3133
0.9	.3159	.3186	.3212	.3238	.3264	.3289	.3315	.3340	.3365	.3389
1.0	.3413	.3438	.3461	.3485	.3508	.3531	.3554	.3577	.3599	.3621
1.1	.3643	.3665	.3686	.3708	.3729	.3749	.3770	.3790	.3810	.3830
1.2	.3849	.3869	.3888	.3907	.3925	.3944	.3962	.3980	.3997	.4015
1.3	.4032	.4049	.4066	.4082	.4099	.4115	.4131	.4147	.4162	.4177
1.4	.4192	.4207	.4222	.4236	.4251	.4265	.4279	.4292	.4306	.4319
1.5	.4332	.4345	.4357	.4370	.4382	.4394	.4406	.4418	.4429	.4441
1.6	.4452	.4463	.4474	.4484	.4495	.4505	.4515	.4525	.4535	.4545
1.7	.4554	.4564	.4573	.4582	.4591	.4599	.4608	.4616	.4625	.4633
1.8	.4641	.4649	.4656	.4664	.4671	.4678	.4686	.4693	.4699	.4706
1.9	.4713	.4719	.4726	.4732	.4738	.4744	.4750	.4756	.4761	.4767
2.0	.4772	.4778	.4783	.4788	.4793	.4798	.4803	.4808	.4812	.4817
2.1	.4821	.4826	.4830	.4834	.4838	.4842	.4846	.4850	.4854	.4857
2.2	.4861	.4864	.4868	.4871	.4875	.4878	.4881	.4884	.4887	.4890
2.3	.4893	.4896	.4898	.4901	.4904	.4906	.4909	.4911	.4913	.4916
2.4	.4918	.4920	.4922	.4925	.4927	.4929	.4931	.4932	.4934	.4936
2.5	.4938	.4940	.4941	.4943	.4945	.4946	.4948	.4949	.4951	.4952
2.6	.4953	.4955	.4956	.4957	.4959	.4960	.4961	.4962	.4963	.4964
2.7	.4965	.4966	.4967	.4968	.4969	.4970	.4971	.4972	.4973	.4974
2.8	.4974	.4975	.4976	.4977	.4977	.4978	.4979	.4979	.4980	.4981
2.9	.4981	.4982	.4982	.4983	.4984	.4984	.4985	.4985	.4986	.4986
3.0	.4987	.4987	.4987	.4988	.4988	.4989	.4989	.4989	.4990	.4990
3.1	.4990	.4991	.4991	.4991	.4992	.4992	.4992	.4992	.4993	.4993
3.2	.4993	.4993	.4994	.4994	.4994	.4994	.4994	.4995	.4995	.4995
3.3	.4995	.4995	.4995	.4996	.4996	.4996	.4996	.4996	.4996	.4997
3.4	.4997	.4997	.4997	.4997	.4997	.4997	.4997	.4997	.4997	.4998
3.5	.4998	.4998	.4998	.4998	.4998	.4998	.4998	.4998	.4998	.4998
3.6	.4998	.4998	.4999	.4999	.4999	.4999	.4999	.4999	.4999	.4999
3.7	.4999	.4999	.4999	.4999	.4999	.4999	.4999	.4999	.4999	.4999
3.8	.4999	.4999	.4999	.4999	.4999	.4999	.4999	.4999	.4999	.4999
3.9	.5000	.5000	.5000	.5000	.5000	.5000	.5000	.5000	.5000	.5000

Example: The area between $z = 0$ and $z = 1.24$ is .3925.